EVOLUTIONARY BEHAVIORAL ECOLOGY

EVOLUTIONARY BEHAVIORAL ECOLOGY

Edited by

David F. Westneat
Charles W. Fox

2010

OXFORD
UNIVERSITY PRESS

Oxford University Press, Inc., publishes works that further
Oxford University's objective of excellence
in research, scholarship, and education.

Oxford New York
Auckland Cape Town Dar es Salaam Hong Kong Karachi
Kuala Lumpur Madrid Melbourne Mexico City Nairobi
New Delhi Shanghai Taipei Toronto

With offices in
Argentina Austria Brazil Chile Czech Republic France Greece
Guatemala Hungary Italy Japan Poland Portugal Singapore
South Korea Switzerland Thailand Turkey Ukraine Vietnam

Published by Oxford University Press, Inc.
198 Madison Avenue, New York, New York 10016

www.oup.com

Library of Congress Cataloging-in-Publication Data
Evolutionary behavioral ecology / edited by David F. Westneat, Charles W. Fox.
p. cm.
Includes bibliographical references and index.
ISBN 978-0-19-533193-6; 978-0-19-533192-9 (pbk)
1. Animal behavior—Evolution. 2. Animal ecology.
I. Westneat, David F. II. Fox, Charles W.
QL751.E87 2010
591.5—dc22 2009031325

Printed in the United States of America
on acid-free paper

Preface

An understanding of the natural world and what's in it is a source of not only a great curiosity but great fulfillment.
—Sir David Attenborough, September 2004, during a break from filming bolas spiders at the University of Kentucky's Spindletop Farm

The diversity of ways in which organisms interact with their environment, including members of their own species, other species, and the abiotic world, is astounding. The field of behavioral ecology explores the role of behavior in these interactions. Behavioral ecologists are typically interested in a functional understanding of behavior: why do animals behave the way they do, and make the decisions that they make? How are animal decisions and behaviors shaped by evolution in response to ecological and/or social conditions that they experience? Such questions have turned out to be surprisingly complicated, requiring a diversity of approaches, from mathematical modeling to biogeographic analyses to field manipulative studies, and integration among fields of study, including psychology, evolutionary biology, physiology, molecular biology, and even economics, among others. The insights gained by behavioral ecologists have generated considerable controversy over the years, spurring the field to move in new directions.

Two of the most deservedly famous individuals in behavioral ecology are John Krebs and Nick Davies. Though the study of behavioral ecology did not start with them, its definition as a field of study was in large part due to their immensely influential book, *Behavioural Ecology: An Evolutionary Approach*, published in 1978. That single book pulled together the variety of behaviors and ideas that had been the focus of researchers interested in functional explanations. Through four editions, the latest in 1997, the Krebs and Davies volumes brought both conceptual clarity and empirical excitement to the field. Probably every practicing behavioral ecologist has gained insights from these volumes. Although these editions still make for compelling reading and provide excellent overviews of the logic and methods of the field, they no longer reflect the full scope of the field: some of the material is dated, and each edition intentionally included an eclectic set of topics not broad enough in scope to form the foundation for a comprehensive, semester-long graduate class.

In *Evolutionary Behavioral Ecology* we have attempted to present the core concepts and recent elaborations upon them that define the field today—a daunting task because the field has grown substantially since Krebs and Davies' first publication. Their volumes developed in detail select topics that highlighted major conceptual developments in the field, introducing these to new behavioral ecologists and energizing the veterans. However, the field is too rich and deep to simultaneously introduce beginning students to the diversity of ideas that make up modern behavioral ecology, and present each of those fields with the thoroughness of a Krebs and Davies chapter. To accomplish our goal of making a volume accessible to students, we asked authors to focus on core concepts, presented rigorously, and not write encyclopedic reviews. We wanted authors to target their chapters directly at first- or second-year graduate students and lead them through the main ideas that have shaped the field over the past 40 (or more) years, highlighting some of the major controversies that excite us today. Our authors were thus challenged to sacrifice detail and instead focus on major ideas that have driven the field forward. We hope students find these brief introductions stimulating and that they pursue additional readings, such as those listed in "Suggestions for Further Reading" sections, to explore ideas in more detail. Most important, we hope that readers of this book—students of behavioral ecology—will take away a sense of the vitality of a field full of brilliant insights but also brimming with unanswered questions that students themselves can contribute to answering.

Unlike many books targeted to students and the classroom, we chose to edit a collection of chapters written by others rather than to write a book ourselves. A multiauthored book has strengths and weaknesses. One strength is that these chapters are written by experts in their respective areas, and so each represents a mature perspective on the important topics. A potential weakness of multiauthored books is an inconsistency of style, level of material, and breadth of focus among chapters. We encouraged authors to follow a similar outline for chapters, and we aggressively edited those chapters to provide some consistency and to highlight connections among chapters, but variation in style and level of presentation still exists. Some of these differences, such as how much mathematical modeling is included, we think are a strength of an edited volume because this diversity reflects real variation within the field, both among subjects and among researchers within subjects: some researchers do quite well without using mathematical modeling and others depend on it, yet everyone gains understanding from it. Chapters will not all be equally accessible to the intended audience, and some authors have perspectives on their topic that differ from even the majority of their colleagues (though we've strived for balanced perspectives as much as possible). Nevertheless, we think both of these problems are potentially benefits. Students grow as professionals by being confronted with topics in which there is confusion and disagreement among professionals. Often such areas of disarray are fertile ground from which new careers can sprout.

Behavioral ecology arose from scientists mixing questions and approaches among disciplines, and to understand the state of the field today it is instructive to students to learn a bit about the historical themes as well as the personalities involved in its development. We thus start with a chapter on the history of behavioral ecology, writ large (chapter 1), before delving into more specific subjects. The rest of sections I and II provide students with an introduction to the tools and broad conceptual ideas that underpin the entire field, regardless of specific subject. Together, the two sections are deliberately linked, albeit implicitly and somewhat loosely, to Tinbergen's (1963) four questions: mechanism (chapters 9 and 10), development (chapters 5 and 6), evolution (chapter 7), and function (chapters 2 and 8).

In sections III, IV, and V we march through the traditional topics in behavioral ecology: finding food and avoiding predators (section III), social behavior (section IV), and reproductive behavior (section V). We have sequenced chapters to provide broad, overarching concepts early and elaborations on those themes later in each section. Section VI is called "Extensions," but it could also be called "Cross-Cutting" or "Emerging Frontiers." Here authors take key ideas from behavioral ecology and apply them to subjects such as speciation (chapter 27), conservation biology (chapter 29), and human behavior (chapter 31), or they present new conceptual views (e.g., chapter 30 on personality) or techniques (chapter 28 on genomics) that seem likely to provide behavioral ecologists with new ways of elaborating on functional approaches.

Of necessity, we could not cover all subfields of behavioral ecology and, for most areas, we have touched only briefly on the diversity of exciting questions being asked. Moreover, we have asked authors to be sparing in their use of citations. We did this to encourage a focus on concepts and key developments and to limit the degree to which chapters became encyclopedic reviews. We regret if some subjects and key papers thus appear to be ignored, and we apologize to those whose favorite example, case study, or specific citation was left out of the book.

There are a few unique elements of this book that reaffirm to us the value of having a new summary of our field every 10–12 years. Just as each edition of Krebs and Davies highlighted new ideas and approaches that were changing the future of behavioral ecology, the emergent picture of behavioral ecology from our book is one of increasing sophistication of both questions and techniques. By far the most compelling advances in the field have been due to the increased use of genetic tools used to assess relatedness both within populations (e.g., kinship) and among taxa (as a basis for constructing phylogenies). These approaches have become so well ensconced in the field they are hardly mentioned explicitly, but their impact is broad. More compelling is the increased use of molecular tools, as described in chapters 5 and 28, for understanding the genetic basis of behaviors of ecological relevance, and the insights these provide regarding function and evolution of behavior. Finally, new advances in quantitative genetics have great potential to alter conceptual approaches and empirical research on a wide array of behaviors. General theory on plasticity (chapter 6) and on indirect genetic effects (chapter 14) enhances many traditional approaches to behavioral ecology and is likely to influence many future studies.

Preparing this book involved the assistance of many individuals. First and foremost, we thank the authors, who have produced stimulating chapters and have responded well to the many demands we placed on them. All chapters were reviewed both internally (by authors of other chapters) and externally; in addition to the authors who worked as reviewers, and a few anonymous reviewers, we thank the following people for their helpful comments: Sigal Balshine, Alex Basolo, Guy Beauchamp, Anders Berglund, Andrew Bourke, Tim Caro, Will Creswell, Jonathon Crystal, Reuven Dukas, Daniel Funk, Carl Gerhardt, Greg Grether, Paul Harvey, Ken Haynes, Natalie Hempel de Ibarra, Geoff Hill, Jerry Husak, Adam Jones, Laurent Keller, Darrell Kemp, Dov Lank, Kate Lessells, Wayne Linklater, Jeffrey Lucas, Joel McGlothlin, Molly Morris, Bryan Neff, Geoff Parker, Tomasso Pizzari, Denis Reale, David Reznick, Graeme Ruxton, Mike Ryan, William Searcy, Maria Servedio, Mike Sharkey, Ulrich Steiner, William Wagner, Jonathan Wright, and Marlene Zuk. We also thank Dov Lank and his graduate class in behavioral ecology at Simon Fraser University for providing extensive feedback on many of the chapters. Several members of our labs, notably Amanda Ensminger, Damon Orsetti, Sarah Martin, Bridget Sousa, Patricia Hartman, Ian Stewart, and Dan Wetzel, read many of the chapters at several stages and provided numerous helpful comments. The Department of Entomology at the University of Kentucky provided financial support. Both our research programs have been supported by the National Science Foundation, and that support helped us to plan and carry out this project.

David F. Westneat
Charles W. Fox

Contents

Contributors

Angeloni, Lisa. Department of Biology, Colorado State University, Fort Collins, Colorado, USA

Bell, Alison. School of Integrative Biology, University of Illinois, Urbana-Champaign, Illinois, USA

Birkhead, Tim. Department of Animal and Plant Sciences, University of Sheffield, Sheffield, UK

Blumstein, Daniel. Department of Ecology and Evolutionary Biology, University of California-Los Angeles, Los Angeles, California, USA

Boughman, Janette. Department of Zoology, Michigan State University, East Lansing, Michigan, USA

Bretman, Amanda. Centre for Ecology and Conservation, School of Biosciences, University of Exeter, Cornwall Campus, Penryn, Cornwall, UK

Briffa, Mark. Marine Biology & Ecology Research Centre, University of Plymouth, Plymouth, UK

Brooks, Robert. Evolution and Ecology Research Centre and School of Biological, Earth, and Environmental Sciences, University of New South Wales, Sydney, Australia

Calsbeek, Ryan. Department of Biological Sciences, Dartmouth University, Hanover, New Hampshire, USA

Carroll, Scott. Department of Entomology, University of California, Davis, California, USA

Chapman, Tracey. School of Biological Sciences, University of East Anglia, Norwich, UK

Dall, Sasha. Centre for Ecology and Conservation, School of Biosciences, University of Exeter, Cornwall Campus, Penryn, Cornwall, UK

Dugatkin, Lee. Department of Biology, University of Louisville, Louisville, Kentucky, USA

Earley, Ryan. Department of Biological Sciences, University of Alabama, Tuscaloosa, Alabama, USA

Enquist, Magnus. Department of Zoology and the Centre for the Study of Cultural Evolution, Stockholm University, Stockholm, Sweden

Fox, Charles W. Department of Entomology, University of Kentucky, Lexington, Kentucky, USA

Fricke, Claudia. School of Biological Sciences, University of East Anglia, Norwich, UK

Fuller, Becky. Department of Animal Biology, University of Illinois, Champaign-Urbana, Illinois, USA

Gangestad, Steven. Department of Psychology, University of New Mexico, Albuquerque, New Mexico, USA

Gardner, Andy. Department of Zoology, University of Oxford, Oxford, UK

Ghalambor, Cameron. Department of Biology, Colorado State University, Fort Collins, Colorado, USA

Ghirlanda, Stefano. Department of Psychology, University of Bologna, Bologna, Italy and the Centre for the Study of Cultural Evolution, Stockholm, Sweden

Griffin, Ashleigh. Department of Zoology, University of Oxford, Oxford, UK

Griffith, Simon. Department of Brain, Behaviour and Evolution, Macquarie University, Sydney, Australia

Grozinger, Christina. Department of Entomology, Pennsylvania State University, University Park, Pennsylvania, USA

Hamilton, Ian. Department of Evolution, Ecology, and Organismal Biology, and Department of Mathematics, Ohio State University, Columbus, Ohio, USA

Healy, Susan. Schools of Biology and Psychology, University of St. Andrews, St. Andrews, UK

Hodgson, David. Centre for Ecology and Conservation, School of Biosciences, University of Exeter, Cornwall Campus, Penryn, Cornwall, UK

Hosken, David. Centre for Ecology and Conservation, School of Biosciences, University of Exeter, Cornwall Campus, Penryn, Cornwall, UK

Hunt, John. Centre for Ecology and Conservation, School of Biosciences, University of Exeter, Cornwall Campus, Penryn, Cornwall, UK

Hurd, Peter. Department of Psychology, University of Alberta, Edmonton, Alberta, Canada

Janzen, Fred. Department of Ecology, Evolution, and Organismal Biology, Iowa State University, Ames, Iowa, USA

Jennions, Michael. Evolution, Ecology, and Genetics, Research School of Biology, The Australian National University, Canberra, Australia

Johnson, J. Chadwick. Integrated Natural Science, Arizona State University West, Phoenix, Arizona, USA

Kokko, Hanna. Department of Biological and Environmental Science, University of Helsinki, Helsinki, Finland

Kvarnemo, Charlotta. Department of Zoology, University of Gothenburg, Göteborg, Sweden

Levitan, Don. Department of Biological Science, Florida State University, Tallahassee, Florida, USA

Lieberman, Debra. Department of Psychology, University of Miami, Coral Gables, Florida, USA

Mappes, Johanna. Centre of Excellence in Evolutionary Biology, and Department of Biological and Environmental Science, University of Jyväskylä, Finland

Martins, Emília. Department of Biology, Indiana University, Bloomington, Indiana, USA

Messina, Frank. Department of Biology, Utah State University, Logan, Utah, USA

Monaghan, Pat. Department of Ecology and Evolutionary Biology, University of Glasgow, Glasgow, UK

Moore, Allen. Centre for Ecology and Conservation, School of Biosciences, University of Exeter, Cornwall Campus, Penryn, Cornwall, UK

Nonacs, Peter. Department of Ecology and Evolutionary Biology, University of California-Los Angeles, Los Angeles, California, USA

Nosil, Patrik. Department of Ecology and Evolutionary Biology, University of Colorado Boulder, USA

Ord, Terry. School of Biological, Earth, and Environmental Sciences, University of New South Wales, Sydney, Australia

Orzack, Steven Hecht. Fresh Pond Research Institute, Cambridge, Massachusetts, USA

Pitnick, Scott. Department of Biology, Syracuse University, Syracuse, New York, USA

Queller, David. Department of Ecology and Evolutionary Biology, Rice University, Houston, Texas, USA

Rowe, Candy. Centre for Behaviour and Evolution, Newcastle University, Newcastle upon Tyne, UK

Rundle, Howard. Department of Biology, University of Ottawa, Ottawa, Ontario, Canada

Runge, Michael. United States Geological Survey, and Patuxent Wildlife Research Center, Laurel, Maryland, USA

Schlaepfer, Martin A. College of Environmental Science and Forestry, State University of New York, Syracuse, New York, USA

Shaw, Kerry. Department of Neurobiology and Behavior, Cornell University, Ithaca, New York, USA

Sherman, Paul. Department of Neurobiology and Behavior, Cornell University, Ithaca, New York, USA

Shuster, Stephen. Department of Biological Sciences, Northern Arizona University, Flagstaff, Arizona, USA.

Sih, Andy. Department of Environmental Science and Policy, University of California-Davis, Davis, California, USA

Sinervo, Barry. Department of Ecology and Evolutionary Biology, University of California-Santa Cruz, Santa Cruz, California, USA

Sneddon, Lynne. School of Biological Sciences, University of Liverpool, Liverpool, UK

Stevens, Martin. Department of Zoology, University of Cambridge, Cambridge, UK

Strassmann, Joan. Department of Ecology and Evolutionary Biology, Rice University, Houston, Texas, USA

Warner, Daniel. Department of Ecology, Evolution, and Organismal Biology, Iowa State University, Ames, Iowa, USA

Webster, Michael. Cornell Lab of Ornithology and Department of Neurobiology and Behavior, Cornell University, Ithaca, New York, USA

West, Stuart. Department of Zoology, University of Oxford, Oxford, UK

Westneat, David. Department of Biology, University of Kentucky, Lexington, Kentucky, USA

Wiley, Chris. Department of Neurobiology and Behavior, Cornell University, Ithaca, New York, USA

Wolf, Jason. Department of Biology and Biochemistry, University of Bath, Bath, UK

Ydenberg, Ronald. Department of Biological Sciences, Simon Fraser University, Burnaby, British Columbia, Canada

EVOLUTIONARY BEHAVIORAL ECOLOGY

SECTION I

FOUNDATIONS

1

Ingenious Ideas: The History of Behavioral Ecology

TIM R. BIRKHEAD AND PAT MONAGHAN

The European cuckoo's (*Cuculus canorous*'s) extraordinary habit of laying its eggs in the nests of other species was known as long ago as the fourth century BC. Even more remarkably, Aristotle apparently knew that the nestling cuckoo selfishly ejects the host's eggs or young, casting "out of the nest those with whom it has so far lived" (*Historia animalium*, book IX). The discovery of this particular bit of cuckoo biology is usually credited to Edward Jenner (later of inoculation fame), who, like others before him seems to have been unaware of Aristotle's report. Jenner's paper, published as a letter to Britain's Royal Society in 1788, caused a sensation among naturalists, who were quick to check his observations and to offer explanations for the cuckoo's extraordinary habits. Gilbert White (1789) simply called it "a monstrous outrage on maternal affection" (letter IV to Daines Barrington). Alfred Brehm (1861) assumed that the foster parents reared the young cuckoo out of pity. Others, unable to make sense of brood parasitism, said that "the all-wise Creator... never intended we should penetrate into the reasons for all His actions" (Montagu 1802: 118). Jenner himself prudently avoided any explanation (Schulze-Hagen et al. 2009).

Darwin's theory of natural selection provided an alternative to divine favoritism for the existence of complex behaviors like brood parasitism and the nestling cuckoo's ejection of the host eggs and young. Indeed, using the logic of natural selection, Darwin provided a remarkably accurate interpretation of the cuckoo's brood parasitic habits, recognizing that the behaviors were advantageous for the cuckoo, that hosts are tricked into rearing a cuckoo chick—a "mistaken instinct" he called it—and that brood parasitism evolved from a nonparasitic ancestor (Darwin 1859).

Following in Darwin's footsteps and continuing to use natural selection as a guide, behavioral ecologist Nick Davies and colleagues have, since the 1980s, tested, confirmed, and extended Darwin's ideas. Their work has revealed not only an intricate web of adaptations and counteradaptations by cuckoos and their hosts, but also adaptations unanticipated by Darwin. These include the extraordinary fact that, to secure the same parental provisioning rate as the brood of host young would have obtained, the begging call of the single nestling cuckoo mimics that of an entire host brood (see also chapter 9). This enhanced auditory stimulus compensates for the fact that the cuckoo chick presents a visual stimulus of just one gaping mouth (Davies et al. 1998). Summing up his monograph on cuckoos and other brood parasites, Davies (2000: 256) says, "Cuckoos... inspire our wonder for Nature's ingenuity.... [and] allow us to unravel the forces of natural selection by combining the pleasures of bird watching with simple field experiments"—a statement that captures the spirit of behavioral ecology, combining theory, observation, and experiment to understand the fitness benefits of particular behaviors.

ORIGINS AND ANTECEDENTS

Behavioral ecology emerged as a distinct discipline in the late 1970s, but its origins go back much further. Charles Darwin seems the most obvious starting point when thinking about adaptation, but of course Darwin's ideas also had antecedents, most notably John Ray's notion of physico-theology. Ray's *Wisdom of God* (1691) changed the way people thought about the natural world. Prior to this time most people were "god-fearing," but Ray's God was benign and had created nature for mankind's enjoyment, benefit, and erudition, ensuring moreover the perfect fit between an organism and its environment. The fact that physico-theology has since been used as the basis for so-called intelligent design should not detract from Ray's contribution; at the time, it was simply the best explanation for the way the world was. By encouraging the study of natural history, Ray launched the tradition of field biology (Haffer 2007; Birkhead 2008). He also inspired William Paley, who hijacked his idea, calling it natural theology (Paley 1802), which in turn inspired Darwin. As R. Dawkins (1986) has pointed out, physico-theology, natural theology, or intelligent design—whatever we call it—was about adaptation, which is precisely why Darwin found it useful. We get some indication of Ray's prescience from the fact that virtually all of the topics in evolutionary ecology that David Lack studied during the 1940 to the 1960s, and which became part of behavioral ecology, had already been identified by Ray as important biological questions (Birkhead 2008).

Charles Darwin's revolutionary concept of evolution by natural selection (Darwin 1859) placed intraspecific variation, which until then had been viewed as little more than a biological nuisance, center stage as the raw material on which natural selection could operate (see chapter 2). However, after an initial flurry of excitement, the ideas of natural selection and sexual selection in particular lost some of their impetus (Bowler 1983). The discovery of genetics in the early 1900s ought to have revitalized evolutionary thinking, but instead generated a futile 30-year debate between the so-called Mendelians and Darwinians. The Mendelians discounted natural selection and assumed that evolution occurred through rapid, major steps, a process disparagingly referred to by the Darwinians—proponents of natural selection—as "evolution by jerks" (Provine 1971). While the Mendelian-Darwinian controversy rumbled on, a handful of field pioneers, including the British ornithologists Edmund Selous and Eliot Howard, were exploring Darwin's idea of sexual selection, and their remarkable observations of wild birds paved the way for the field study of animal behavior (Burkhardt 2005; Birkhead 2008).

In the 1920s and '30s, R. A. Fisher, J. B. S. Haldane, and Sewall Wright took the new understanding of genetics produced by laboratory studies, such as those of T. H. Morgan, and began applying it to natural populations. In doing so, they set the scene for Julian Huxley, Ernst Mayr, Theodosius Dobzhansky, and George Gaylord Simpson, who, working on different organisms and questions, combined Darwin's idea of natural selection with a clearer understanding of heredity to provide the "modern synthesis." It is ironic that Huxley, a pioneer in field studies of avian courtship, did little to foster an evolutionary way of thinking about behavior. He downplayed the role of sexual selection, which he saw as leading to the evolution of frivolous or useless traits and therefore deleterious rather than adaptive (Huxley 1938). As a result, Huxley (1942) rarely mentions animal behavior in his *Modern Synthesis*. On the other hand, by recognizing the significance of the field-based approach adopted by Selous and Howard, and by later befriending and encouraging Niko Tinbergen and Konrad Lorenz, Huxley played a pivotal role during the 1940s and '50s in the quest to develop the scientific study of animal behavior in a natural setting. This came to be known as ethology and was the stem from which behavioral ecology eventually bore fruit.

Lorenz and Tinbergen, generally considered the founding fathers of ethology, could hardly have been more different; both were charismatic, but Tinbergen was shy, retiring, and prone to depression, whereas Lorenz was bold, brash, and a self-promoter. They were deeply divided philosophically, but nonetheless able to put these divisions behind them after World War II and together create the field of ethology. With Karl von Frisch, they were rewarded for their efforts in 1973 with a Nobel Prize. Of the three, Tinbergen's contribution has been the most enduring (Kruuk 2003; Burkhardt 2005).

The early ethologists' focus was very much on *instinctive* behaviors. They wanted behaviors to be viewed as characters in the same way as morphological or physiological traits. Accordingly, ethology focused in particular on behaviors associated with communication between potential partners or rivals, or between parents and offspring, in which clear *units* of display behavior can be recognized. Tinbergen was a field naturalist. He believed strongly in the importance of detailed descriptions of behavior, if possible in a natural setting, and that an essential prerequisite to any experimentation was a complete inventory of a species' behavior patterns (Tinbergen 1951: 7). As a result, much early ethology was pre-occupied with *ethograms*, quantitative descriptions of species-specific repertoires of innate behaviors, such as displays (Radick 2007). Variation among and within individuals received little attention (Hinde 1982).

As well as conducting detailed studies on a variety of species, Tinbergen was also able to stand back and take a broader view. Building on ideas previously developed by Huxley, Tinbergen (1963) spelled out the four approaches to the study of behavior still recognized today as Tinbergen's four questions: (1) What is the physiological causation of the behavior? (2) How does behavior develop in the individual? (3) What is the function or survival value (adaptive significance) of a particular behavior? (4) How did the behavior evolve? These four questions embrace the whole biology of animal behavior, mechanistic, developmental, functional, and evolutionary approaches. No single approach is any better than the others. However, the popularity of Tinbergen's four questions as research foci has changed over time.

Tinbergen conducted studies in all four areas, but his own emphasis was very much on question 1—the causality of behavior, epitomized by *The Herring Gull's World*. His aim in this beautifully written and justifiably famous book was to show that "the behavior of herring gulls can tell us highly interesting things about the functioning of their nervous system. Their innate behavior shows wonderful adaptations, but also astonishing limitations" (Tinbergen 1953a: 3). Causation became the major focus of ethology, and the body of theory and terminology that developed was centered on representing putative physiological mechanisms. Lorenz, for example, developed a *psycho-hydraulic model* of motivation, and Tinbergen promoted hierarchical models of behavior, with drives, sign stimuli, fixed action patterns, and releasing factors, all of which were taken to be species-specific and invariant. This foundation was unsound, however, because the mechanistic framework bore little resemblance to how the nervous system actually worked, as Daniel Lehrman appositely pointed out in the early 1950s (H. Kruuk 2003: 178). As a result, Peter Klopfer (in Burkhardt 2005: 460) stated that "ethology as a coherent body of theory ceased to exist in 1950." More critically, Hinde (1970) said that the ethologists didn't have a body of theory at all, simply a "shared attitude."

Lorenz probably never recovered from Lehrman's devastating critique of ethology, whereas Tinbergen, being more open-minded, picked himself up and turned his attention in other directions. One of these was to his third question, the adaptive significance of behavior. Tinbergen's appreciation that behavior has "survival value" provides clear recognition that he understood how behavior can be shaped by natural selection, but he and Lorenz differed markedly in their view of evolution. Lorenz paid only lip service to Darwin, and Burkhardt (2005) castigates Lorenz for not reading Darwin (or indeed much else); later, R. Dawkins (1976) also took Lorenz to task for his muddled evolutionary thinking and for promoting group selection. Tinbergen, in contrast, had a much clearer idea of natural selection. His study on the adaptive significance of eggshell removal by gulls (Tinbergen et al. 1962) is pure behavioral ecology (see figure 1.1). Tinbergen must have been aware of the idea of natural selection as *individual* selection since he was on good terms with David Lack (Hinde 1982), who was a strong proponent of this way of thinking (Dawkins 1976: 115).

What is perhaps surprising is that Tinbergen and Lack did not themselves establish the discipline of behavioral ecology in the 1960s. They were both in the zoology department in Oxford University, Lack in the Edward Grey Institute of Field Ornithology (EGI) and Tinbergen in the Animal Behaviour Research Group (ABRG). The ingredients were all there, so why did it not happen? The answer seems to be partly because of differences in personality and philosophy. Lack was a prima donna and conducted science with an evangelical zeal that may have stemmed from a deep religious conviction. At coffee time in the EGI, for example, Lack held forth, preaching to submissive disciples. In the ABRG, the atmosphere was completely different. Just as passionate about his science, Tinbergen was less autocratic, welcomed criticism, and encouraged discussion that involved everyone. The two men also differed markedly in their approach to science. Tinbergen focused

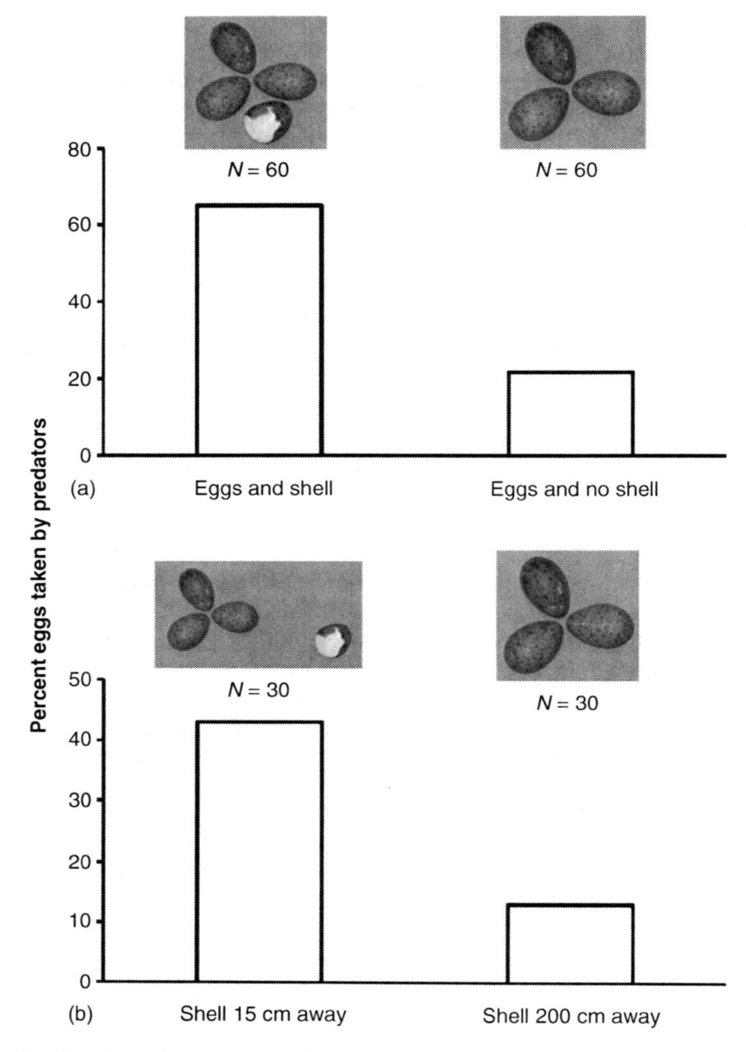

FIGURE 1.1 Soon after hatching (but not immediately), parent black-headed gulls *Larus ridibundus* remove the eggshells from their nest, flying several tens of meters away and dropping them. Tinbergen's experiments were designed to establish the function or adaptive significance of eggshell removal by parent gulls. The study examined the predation of artificial black-headed gull nests (chicken eggs painted to resemble black-headed gull eggs) placed in a black-headed gull colony. In (a) the presence of an eggshell in the nest resulted in a higher level of egg predation compared with nests with no eggshell associated with them ($N = 60$ in each case). In (b) the effect of the proximity of the eggshell was compared and showed that nests with the broken eggshell nearby (15 cm away) suffered more predation than those where the broken eggshell was 200 cm away ($N = 30$ in each case). Drawn from data in Tinbergen et al. (1962).

predominantly on problems, general issues, and ethological questions of broad significance. Lack, on the other hand, despite his uncompromising obsession with individual selection, focused more on species rather than problems. His students were allowed to choose or were allocated a species in the hope that something interesting would emerge, as indeed it invariably did because he chose his students wisely. John Krebs, who spent time in both research groups and witnessed the different research styles, recounts that when he was interviewed for a postgraduate research position in the EGI, he told Lack that he wanted to study the problem of territory, then a very topical subject. However, he was not entirely sure which species to focus on (he thought the mute swan would be good because one could see what territorial and nonterritorial swans were doing). When, after the interview, Krebs was taken to meet members of the EGI, Lack introduced him as someone who didn't know what he wanted to do (J. R. Krebs, pers. comm.).

There were further differences in the way Tinbergen and Lack approached their science. Lack was an evolutionary ecologist and rather disliked experiments. He felt very strongly that animals must be studied in the set of circumstances to which they are adapted and that this made it hard to interpret the results of even field experiments (Lack 1965). He felt that ecological adaptations vary little within species (Lack 1965) and relied instead on cross-species comparisons, exemplified by his book *Ecological Adaptations for Breeding in Birds* (Lack 1968). For Tinbergen, on the other hand, the study of function was essentially a study of outcomes: the short-term consequences of behavior. He was a great field experimenter (see figure 1.1). However, notwithstanding the brilliant simplicity of Tinbergen's eggshell removal study, in general most ethologists in the 1960s did not fully appreciate the importance of natural selection operating at the individual level. As the more explicitly evolutionary view of behavior began to take shape in the late 1960s, Tinbergen was reaching the end of his academic career and had become more interested in human behavior, including autism (Tinbergen & Tinbergen 1972). The final factor that prevented the towering intellects of Tinbergen and Lack from consolidating their science into a single discipline was their physical isolation. Even though they were in the same department, they were located in different buildings, so the opportunities to interact, to attend each other's seminar series, and to think about a joint future were surprisingly limited.

The evolutionary perspective was pushed to the fore in 1962 by the publication of Wynne-Edwards's book *Animal Dispersion in Relation to Social Behaviour*, with its group selection ideas about population regulation and behavior (Wynne-Edwards 1962):

> It is part of our Darwinian heritage to accept the view that natural selection operates largely or entirely at two levels, discriminating on the one hand in favor of *individuals* that are better adapted and consequently leave more surviving progeny than their fellows; and on the other hand between one *species* and another where their interests overlap and conflict, and where one proves more efficient in making a living than another.... Neither of these two categories of selection would be at all effective in eliciting the kind of social adaptations that concern us here. We have met already with a number of situations... in which the interests of the individual are actually submerged or subordinated to those of the community as a whole.... Survival is the supreme prize in evolution; and there is consequently great scope for selection between local groups... in the same way as there is between allied race or species.... Evolution at this level can be ascribed, therefore, to what here is termed group selection. (Wynne-Edwards 1962: 18–20)

The fallacy of this book's group selection ideas drew sharp criticism from Maynard Smith (1964), David Lack (1966), and George Williams (1966) in his book *Adaptation and Natural Selection.* Williams had apparently completed a draft of his book before even seeing Wynne-Edwards's volume, but revised it to include a critique of Wynne-Edwards (Anon. cited in Parker 2006a: 28). The appeal of group selection to Wynne-Edwards (and many others) was that it neatly (if erroneously) accounted for the puzzling phenomena of altruistic acts, including the evolution of sterile workers among social insects. Darwin (1859: 236) had anguished over altruism and the existence of sterile castes and considered them "a special difficulty, which at first appeared insuperable, and actually fatal to the whole theory [of natural selection]." Darwin's ingenious solution was to assume that kinship was the answer: if individuals carrying the trait for altruism helped others that do not express the trait, altruism might evolve.

A century later, the formal demonstration of Darwin's kinship idea came from a shy, young academic, Bill Hamilton, then based in London. Hamilton solved the paradox of altruism by thinking about the *genetic* consequences of behavior. He demonstrated mathematically that the value to individual A of the offspring of individual B varies in proportion to A's relatedness to those offspring. This is the concept of inclusive fitness. Maynard Smith (1964) coined the term *kin selection* for the same phenomenon. Interestingly, W. D. Hamilton's pair of papers (1964) was not immediately successful—his writing style and the mathematics both contain do not make for an easy read. Consequently, Hamilton's work was initially overlooked in the United Kingdom, and neither Tinbergen nor Lack appears to have appreciated its significance.

But the significance of Hamilton's papers *was* recognized by G. C. Williams (1966), E. O. Wilson (1975), and R. Dawkins (1976) in their respective books, and, because they provided the theoretical solution to one of the major issues in behavioral ecology, Hamilton's two papers have become among the most frequently cited in the discipline (Parker 2006a).

The mid-1960s also saw some other exciting developments. John Hurrell Crook conducted comparative studies of weaver birds and primates, exploring what was then the novel idea that ecology had at least as much influence as phylogeny on a species' behavior (such as nesting dispersion and mating system; Crook 1964, 1965; Crook & Gartlan 1966). This idea inspired Lack and was instrumental in his producing *Ecological Adaptations for Breeding in Birds* (1968). In the United States, Gordon Orians was also forging a union between ecology, behavior, and natural selection in his innovative studies of blackbirds (e.g., Orians 1962, 1969), as was Jerram Brown in his more theoretical studies of territoriality and later cooperative breeding; Brown's book, *The Evolution of Behaviour* (1975), which synthesizes his work and approach, has been ranked in the United States among the 10 most influential books on animal behavior by the Animal Behavior Society. In another significant development, the ecologists Robert MacArthur and Eric Pianka adopted an explicitly quantitative approach to the adaptive significance of behavior. Combining economics and population biology, they developed elegant mathematical theories to understand the choices animals make when moving between food patches (MacArthur & Pianka 1966). This was innovative work and although MacArthur and Pianka retained a rather *species*-centered approach—"We undertake to determine in which patches a species would feed and which items would form its diet if the species acted in the most economic fashion" (MacArthur & Pianka 1966)—the broad utility of optimality theory was nonetheless quickly recognized (e.g., Krebs 1974).

Using individual or gene-based selection as their guide, other researchers in the late 1960s began to revisit Darwin's idea of sexual selection, which for decades had lain virtually untouched. The key players here were Bob Trivers and Geoff Parker. Trivers (2002) recounts the wonderful story of how, on talking to his mentor Ernst Mayr about his ideas on relative parental investment, Mayr told him to go away and read Bateman's (1948) study on mate competition in *Drosophila*—one of the first experimental tests of sexual selection and the first to measure the difference in the variance in reproductive success of males and females. Trivers forgot, and on a return visit to Mayr was summarily sent away again. On reading Bateman's study that evening, Trivers says, "The proverbial scales fell from my eyes." The result was a paper linking parental investment and mate choice (Trivers 1972) that was to be extraordinarily influential: it was beautifully written, and the wealth of ideas it contained were both exciting and testable. Also clear from this story is Mayr's undisputed genius: he was well aware of the significance of Bateman's results (see also Snyder & Gowaty 2007) even though they had been published over 20 years previously.

The importance of sexual selection was further developed by Parker, who became a key player in behavioral ecology through his study of yellow dung flies *Scatophaga stercoraria* using a mix of observation of mating behavior, experiment, and mathematical modeling, all underpinned by individual-based selection thinking (Parker 2001; see chapter 2 and section V of this volume). Parker's theoretical models provided new insights, as well as clear, testable predictions, and inspired many would-be behavioral ecologists to collect data to test his ideas. His ongoing contributions to empirical studies, and in particular theory, are extraordinary in their insight and diversity, spanning animal contests, evolutionarily stable strategies, and sperm competition (Parker 2001).

Recognition of Parker's contribution was far from instantaneous: his overview of insect sperm competition (Parker 1970a) lay dormant for almost a decade before being taken up by Bob Smith and Jon Waage, whose exciting empirical studies, together with a symposium in 1980 (Smith 1984), launched the field of postcopulatory sexual selection. In 2006, writing his own account of the history of behavioral ecology, Parker conducted a straw poll among mainly 50- to 60-year-old behavioral ecologists to identify the individual researchers and papers that had most influenced the field. In the analysis Parker was at pains to avoid having himself nominated, but both he and his 1970 paper were among the most highly rated (Parker 2006a: 33–34).

Thus, by the end of the 1960s, the seeds of a new approach to behavior were sprouting. Many researchers were starting to explore patterns of variation in behavior, employing the individual-based view of selection. That ecology mattered was well established, and a more quantitative and experimental approach fuelled by optimality modeling had developed. The influence of this new approach began to be felt by students studying a broad array of behavioral phenomena. The idea of alternative *strategies*, whose fitness outcome could be frequency dependent, also took hold. Game theory (Maynard Smith & Price 1973) played a crucial role here because it allowed behavioral ecologists to consider behaviors such as fighting or courtship that optimality thinking did not adequately deal with. Game theory also provided an explanation for individual variation in adaptive terms, either because different strategies have equal fitness due to frequency dependent selection, or because of the fitness benefits of particular strategies depend on other attributes of the phenotype (i.e., adaptive phenotypic plasticity). By the early 1970s, then, the foundations of behavioral ecology were in place.

However, behavioral ecology did not immediately rise up in its finished form from these foundations. The next step came in 1975 with the publication of Edward O. Wilson's masterful tome *Sociobiology: The New Synthesis* (1975), which linked population biology, evolutionary biology, and behavior to address the evolution of social organization. As the book's title implies, this was an overt attempt to create a new discipline, and at one level it was immensely successful. It made natural history scientifically respectable by providing it with an evolutionary foundation and a body of testable theory. The next crucial step was the publication in 1976 of Richard Dawkins's *The Selfish Gene*. This was soon recognized as one of the most influential, controversial, and talked-about biology books since Darwin's *Origin of Species*. In a lucid and entertaining style, Dawkins set out the compelling case that evolutionary change is based on representation of genes in gene pools, with animal behavior being the way in which genes maximize their presence in the next generation. This clear elucidation of evolutionary processes was enthusiastically embraced by those trying to make sense of the diverse ways in which animals behave.

Behavioral ecology as a coherent discipline really came into being in 1978 with the publication of *Behavioural Ecology: An Evolutionary Approach*, devised and edited by John Krebs and Nick Davies. The attractive format, the authors' engaging and enthusiastic style, and the commonsense, realistic, quantitative cost-benefit approach proved immensely attractive to a new generation of biologists. The term that Krebs and Davies chose to define the field, *behavioral ecology*, had been used previously both by Peter Klopfer (1974) and Tinbergen (see Burkhardt 2005: 438–439), and its adoption distanced the new field from the distractions of the sociobiology debacle in the United States (discussed below). It was also, as Nick Davies (pers. comm.) told us, a conscious effort to find a title that accurately encompassed the field, which after all was not just about social behavior. Interestingly, Jerram Brown's innovative book on the evolution of behavior (1975), which covered much of the same ground as *Behavioural Ecology*, seems to have been eclipsed by Wilson's *Sociobiology* and the furor that developed around it (see below), because both appeared in the same year. Otherwise, behavioral ecology might have been born 3 years earlier. Behavioral ecology flourished in the United States and in Europe, particularly in the United Kingdom and Scandinavia, the latter being influenced by the foresight of Staffan Ulfstrand in Uppsala and later by Malte Andersson. Now, of course, the discipline has spread across the scientific world.

A NEW PARADIGM

Behavioral ecology, as we now see it, is the study of the fitness consequences of behavior. Research in this field poses the basic question: what does an animal gain, in fitness terms, by doing *this* rather than *that*? It combines the study of animal behavior with evolutionary biology and population ecology and, more recently, physiology and molecular biology. Adaptation is the central, unifying concept (see chapter 2). Tinbergen's question about *function* has been recast such that the focus is not on immediate outcomes, but on the genetic contribution to future generations—inclusive fitness. Behavioral ecology is largely theory driven, often with mathematical or graphical models being used to make testable, quantitative predictions about what animals should do in particular circumstances to maximize the fitness benefits. In their introductory chapter in the first edition of *Behavioural Ecology*, Krebs and Davies (1978) emphasize the two basic theoretical concepts: (1) kin selection and inclusive fitness, and (2) optimality modeling. These ideas pervade the entire book and continue to pervade the discipline of behavioral ecology. In practice, of course, fitness benefits are difficult to measure directly, so in most cases, some presumed correlate of fitness, such as energy gain, is used instead (see chapter 4). Furthermore, the fitness benefits may depend on what other individuals in the population are doing, a concept embodied in the idea of evolutionarily stable strategies (ESS). The main research areas in behavioral ecology combine ethology and evolutionary ecology—how animals obtain food, avoid being eaten, obtain mates, and interact with conspecifics in particular ways. Variation within and among species is central; variation in behavior, sexual ornaments, investment patterns, and so forth, as well as in evolutionary payoffs and potential conflicts of interest. Why do some male salmon mature at a small size and others much later when they have reached a large size? Why are some male starlings monogamous and others polygamous? Why be a helper rather than reproductively active? Why does species *x* live in groups whereas species *y* is solitary and territorial? One of the great strengths of behavioral ecology is that the theory crosses taxonomic boundaries; regardless of which taxon is being considered, similar problems are likely to be solved in similar ways. Topics such as the development of behavior in the sense of elucidating whether a behavior is instinctive or learned are less important in behavioral ecology; it is acknowledged that this will depend on how much flexibility is important in ensuring the response that gives the best fitness outcome. Causal mechanisms too are less important in this context, though the recognition that there is a need to understand mechanisms in order to identify capabilities, costs, and constraints is now gaining ground (e.g., see chapters 8–10 and below).

The differences between the approaches of behavioral ecology and traditional ethology are best illustrated by considering two topics, food acquisition and communication, studied from both perspectives.

1. *From feeding to foraging:* Ethologists generally study feeding behavior in terms of its motivation, releasing factors, and appetite control. In fact, for early ethologists like Tinbergen and Lorenz, feeding behavior per se was of little interest. In *The Herring Gull's World* for example, Tinbergen hardly mentions adult feeding behavior at all. When he does, it is to marvel at the gulls' inflexibility when dropping hard-shelled prey from a height to break them open. This contrasts with the behavioral ecology approach to studying the same behavior in crows, in which Reto Zach used a cost-benefit analysis of the energy expended in flying and carrying the prey in relation to the energy gained (Zach 1979). For the behavioral ecologist, the focus is on identifying the decision rules used to maximize foraging efficiency (a presumed fitness correlate) in terms of choosing the right prey, deciding when to move to a new feeding patch, and maximizing efficiency (see chapters 8 and 11). For the ethologist, the focus is on stimuli that elicit feeding. Relatively simple optimal foraging models have proved surprisingly successful, at least in stable conditions, despite the obvious difficulties in identifying the various factors likely to be involved. Moreover, the models developed in the context of foraging have proved to be widely applicable in other contexts (see Stephens & Krebs 1986).
2. *From cooperation to conflict.* During the 1950s and 1960s, ethologists viewed courtship and mating behavior as a cooperative venture between males and females, whose main function was to coordinate and synchronize the behavior and internal state of each sex to ensure that pairing and copulation occur between conspecifics (Tinbergen 1953b). In contrast, behavioral ecologists saw courtship, pair formation, and copulation more in terms of conflict, each member of the pair attempting to obtain the highest fitness return from the interaction, a situation in which partners may be informed, evaluated, deceived, and manipulated (Krebs & Dawkins 1984). In terms of courtship displays, behavioral ecologists were less interested than ethologists in the postures and fine detail of the motor patterns involved in sexual display, and more concerned with the resources required to produce the display, the consequences for other traits requiring the same resources, the information contained in the display, and the fitness benefits it confers (see chapters 20–24).

THE SHOCK OF THE NEW: CRITICS OF BEHAVIORAL ECOLOGY

As we have seen, the new discipline of behavioral ecology had a long—and largely uncontroversial—

gestation. However, the birth was traumatic. The publication of E. O. Wilson's *Sociobiology* in the early summer of 1975 released a torrent of controversy in the United States. The cause of this controversy was Wilson's final chapter, in which he advocated an evolutionary approach to the study of human behavior. To many, Wilson's human sociobiology was outrageous and a flagrant transgression of the agreement following World War II that researchers would not explore the biological bases of human behavior (Segerstråle 2000). In the early 1900s Ernst Haeckel, one of Germany's top zoologists, had been immensely successful in popularizing and extending Darwinism. He encouraged a genetic determinism that paved the way for the horrors of National Socialism in the 1930s. Hence many felt that this was not a path that should be trodden again. Wilson's final chapter was erroneously interpreted as promulgating the idea that if social behaviors have a genetic basis, they are genetically fixed. The political outrage at sociobiology thus had its roots in the fear of a return to previous right-wing regimes (Segerstråle 2000: 2, 393), including the United States' own sterilization and immigration policies of the early 1900s. Little wonder that many deemed Wilson morally irresponsible and sociobiology politically dangerous.

Just months after the publication of Wilson's book, a group calling itself the Sociobiology Study Group began a concerted attack on *Sociobiology*. They started with a letter to the *New York Review of Books* signed by several academics, including some of Wilson's colleagues at Harvard, evolutionary biologists Richard Lewontin and Stephen J. Gould. Their main accusation was that Wilson's view of human behavior had "political intent," fostering both racism and genocide. Sociobiology was also denounced as being "bad science," by which the critics meant, not only that it was morally inappropriate, but also, by ignoring mind and culture, that it was an inadequate way of explaining human behavior.

The entire sociobiology debate has been skillfully reviewed by Ullica Segerstråle (2000). She points out that after the initial criticisms, few other academics bothered to read *Sociobiology* for themselves, but instead relied on the views of its opponents, in part, she says, because these pandered to their own preconceived ideas. The debate expanded rapidly and, in simplistic terms, became a battle between nature and nurture, right and left, experimentalist and naturalist. Segerstråle also states there were few objective scientific reviews, because scientists were afraid that by criticizing the scientific content of *Sociobiology*, they would provide political support for those who opposed it. However, although there may not have been much scientific critique in the United States, this certainly was not true in Britain. In August 1976, the journal *Animal Behaviour* published no fewer than 14 separate reviews of *Sociobiology* by academics from the United States and Europe, followed by a response from Wilson. With one (American) exception, the *Animal Behaviour* reviewers focused on the science rather than the politics. Their criticisms centered on whether *Sociobiology* really represented a synthesis and why Wilson had overlooked topics such as optimization or had failed to clearly explain concepts like group selection and kin selection. There was also concern about what Wilson called the "cannibalization" of ethology by behavioral ecology and population ecology on the one hand, and neurobiology on the other (see also Hinde 1982).

In Europe, most animal behavior researchers, or at least the younger ones, looked on with disbelief as the sociobiology debate unraveled in the United States. What, they asked, was all the fuss about? Why was there no similar debate in Europe, and why hadn't Dawkins's *Selfish Gene*, published just a year later, not elicited the same furor that *Sociobiology* had in the United States? Segerstråle (2000) attributes the difference to history and a fundamental dichotomy in the way biology, including animal behavior, was studied on the two sides of the Atlantic. In the United States, and particularly among the scientific opponents of sociobiology, animal behavior had its roots in experimental psychology and B. F. Skinner's *behaviorism*—the notion that behavior was environmentally determined and infinitely flexible. In contrast, most European, and particularly British, researchers came to behavioral ecology through a more tolerant, naturalistic tradition that embraced a long-standing interest in evolutionary ideas. It was this fundamental difference in approach that resulted in many academics in the United States reading sociobiology in the "wrong" spirit, whereas those in Europe read *Sociobiology* and the *Selfish Gene* in the "right" spirit, and as a consequence could see no political intent (Segerstråle 2000: 255; see also Ruse 1979). In Britain, this "shared naturalist spirit," which had its roots in John Ray's physico-theology, moderated what little scientific skepticism there was (for example, from more senior Cambridge ethologists) and largely confined serious and informed skeptics' discussions to issues of biology rather than politics (Segerstråle 2000: 262).

The main ideological opposition to the *Selfish Gene* in Europe came from Steven Rose, a neurobiologist at the Open University in Britain, whom Segerstråle (2000: 71) charged with importing the debate from the United States. She also labeled Rose "Britain's Lewontin" (Segerstråle 2000: 71), but Rose's opposition to behavioral ecology was trivial by comparison. Far worse, however, was the fact that Rose tarred Wilson and Dawkins with the same brush, confounding their scientific views in the public's mind (Segerstråle 2000: 71). Wilson and Dawkins obviously had many ideas in common, but the aims of their respective books were different. Embracing the anti-group selection ideas of G. C. Williams and W. D. Hamilton's concept of kin selection, Dawkins's objective was to explain the logic of natural selection from a gene's perspective. Wilson's goal was broader, and he considered kin selection to be just one of several possible evolutionary processes (including group selection). Wilson placed his extrapolation from nonhumans to humans as the finale to *Sociobiology*. Interestingly, in the final chapter of *The Selfish Gene,* Dawkins also explicitly turned his attention to humans but was careful to treat the evolution of culture as a separate issue; for this he felt that we "must begin by throwing out the gene as the sole basis of our ideas on evolution." (205). This kept him out of the boiling waters that almost, but thankfully not quite, engulfed Wilson.

Like any new discipline, behavioral ecology itself has been the subject of justified and unjustified criticism. Although the immediate criticism of sociobiology was political and centered almost entirely on Wilson's book, it later broadened to include the emerging discipline of behavioral ecology. Beginning in 1978, Gould and Lewontin launched a more *scientific* attack on what they called the "adaptationist programme." Gould (1978) accused behavioral ecologists of being so convinced that all traits are adaptive that they failed to properly test their hypotheses and too willingly accepted speculations as being true. "Virtuosity in invention replaces testability as the criterion for acceptance" Gould (1978) said, accusing the behavioral ecologists of "story telling," generating post hoc adaptive explanations, or just-so stories, to account for their observations. More was to follow at a Royal Society of London meeting organized by John Maynard Smith in 1979. Gould and Lewontin joined forces to produce a forceful scientific attack on behavioral ecology (Gould & Lewontin 1979). Although they claimed to focus on scientific issues, as Segerstråle says (2000: 17), Gould and Lewontin "never quite abandoned their original moral/political condemnation of sociobiology." However, this is not to say that none of the criticisms were justified, or that the discipline has not improved as a consequence of the response to such criticism.

In *The Triumph of Sociobiology* (2001), John Alcock implies that Gould's accusation of just-so stories is not true, and Alcock vigorously defends the approach of behavioral ecology. It is now obvious that much of the initial political criticism was based on misconceptions (Alcock 2001: 217). As Alcock makes clear, overall, none of the claims made by the opposition was really true. Moreover, it eventually became obvious to the opposition that behavioral ecologists were not politically motivated, but instead were genuinely committed to discovering the truth about the natural world through rigorous scientific methodology (Segerstråle 2000).

However, in our experience, there is little doubt that in the early days budding behavioral ecologists treated the invention of adaptive explanations as a kind of intellectual gymnastics. This was a period of tremendous excitement and, fueled by the ingenuity of intellectual giants like W. D. Hamilton, there was great enthusiasm for generating adaptive explanations for particular behaviors. For the first time, researchers studying function and adaptation in animal behavior had a very clear theory base—natural selection—from which to create hypotheses, and it resulted in a huge number of new ideas and, initially at least, precious little information. There was also confusion, especially among the critics. For example, the concept of *fitness maximizing strategies* is important in behavioral ecology and animals are viewed as having *choices*. However, the use of everyday terms such as *choice*, *strategy*, or *tactics*, although making the subject accessible, also caused misunderstanding. This was partly because these terms have been used in both the conventional way—the animal may choose between different food items—and as shorthand for "natural selection has favored animals that respond to a given set of circumstances in this particular way" (see also chapters 8 and 25). This type of shorthand provided ammunition for those opposed to behavioral ecology, for whom terms such as *strategy* implied a conscious decision by an animal—something that behavioral ecologists never intended.

The central concepts of behavioral ecology—adaptation, optimization, and inclusive fitness (see chapters 2–4)—were often misunderstood in the early days, even by behavioral ecologists themselves, leaving the discipline vulnerable to criticism from evolutionary biologists. It is important to understand that adaptation is an assumption that underpins optimality modeling. Optimality models are a tool for studying adaptation. The models are not about finding out whether a particular behavior is the optimal strategy (but see box 2.1); the aim of the modeling is to identify the selective pressures that make it optimal (see Parker & Maynard Smith 1990). A different set of selective pressures can produce a different optimal outcome. As we have seen, this approach, and the so-called adaptationist program, came in for stringent criticism, even though its limitations were readily acknowledged by most of its proponents (e.g., Krebs 1985). The concept of inclusive fitness too was initially misunderstood by many, in the sense that it was often taken as the animal's own fitness plus that of its relatives, rather than the individual's own fitness *minus* any component attributable to help from others, plus the extent to which the individual has *increased* the fitness of its relatives over what it would have been without that individual's help (e.g., see Krebs & Davies 1981 and clarification from Grafen 1984). This distinction becomes very important when fitness gains and losses are modeled in a quantitative way, a hallmark of behavioral ecology. However, as the discipline has matured, so too has its theoretical basis.

With the benefit of hindsight, then, it is clear that like any new field of scientific endeavor, as well as generating excitement and enthusiasm, behavioral ecology also produced some weak science. There are several reasons for this, including the fact that, keen to get in on the act, or concerned about not throwing the baby out with the bathwater, journal editors and/or referees were less critical than they would be at other times. It is also possible that editors found it hard to locate appropriate referees for new areas of research, allowing some poor science to slip through the net. This was certainly true of behavioral ecology as a whole, but it is a recurring pattern as exemplified in later subdisciplines within behavioral ecology such as fluctuating asymmetry (Palmer 1999), human sperm competition (Birkhead et al. 1997), cryptic female choice (Birkhead 1998), and brain size trade-offs (Healy & Rowe 2007). This is not something peculiar to behavioral ecology; it is a normal part of science and plays a crucial role in how a discipline develops (Woodward & Goodstein 1996). It is probably also worth noting that, in looking back at the early phase of behavioral ecology, we should recognize that standards of scientific research (including quantification and the use of statistics) also change over time, making early research an easy target for snipers.

The criticisms of behavioral ecology by Gould, Lewontin, and others, however motivated, clearly played a vital role in forcing behavioral ecologists to smarten up their act. Confident that behavioral ecology had the potential to explain a great deal about the natural world, its proponents recognized that to convince the skeptics, they needed to elevate the standard of their science. And this, in our opinion, is precisely what happened. Segerstråle (2000) too is certain that criticism from outside the field improved behavioral ecology, but Alcock (2001: 219) feels instead that it was the competition among behavioral ecologists themselves that improved scientific standards. Both are probably true. It also probably true that as the efficacy of the behavioral ecology approach has been more widely recognized, it has attracted increasing numbers of more able scientists.

A further factor contributing to the general acceptance of behavioral ecology has been our improved understanding of the role of genes in influencing behavior. Much of the original political criticism centered on the behavioral ecologists' assumption that the behaviors and other traits they studied were heritable (Wilson 1975): for traits to evolve by natural selection, they must have a genetic basis. The assumption that behavior had a heritable basis was anathema to the political critics of behavioral ecology (e.g., Lewontin et al. 1984). They hated the idea both it for its ideological connotations but also for the lack of scientific evidence. Behavioral ecologists were not alone, however, in making assumptions. Many other areas of science, including astronomy, rely on similar basic assumptions in order to make progress. When behavioral ecology started, the genetics of animal behavior was a relatively small field, but since then progress in molecular genetics has been monumental (see chapters 5 and 28). There is now abundant evidence that many traits, including behavioral ones, have a genetic basis, though of course this does not mean that there is no environmentally induced plasticity in gene expression (see chapter 6). There is also a much better understanding of the way genetic processes interact with the environment. With the start of the human genome project in the late 1980s has come a much greater acceptance of genetic arguments (Segerstråle 2000: 2) and the recognition that nonhumans and humans are similar. As a result, the biological foundation for human behavior now seems more acceptable. Just as the political and moral objections to behavioral ecology have been resolved, a more general evolutionary controversy based on religious fundamentalism has emerged, and just as with the sociobiology debate, this too evinces a strong Atlantic divide (see Ruse 2005).

THE TRIUMPH OF BEHAVIORAL ECOLOGY

The great triumph of behavioral ecology has been twofold: its ability to explain phenomena—behavioral and other kinds of traits—that were previously inexplicable and, second, to uncover new phenomena. It seems hard to imagine, for example, that we would have discovered either extrapair paternity in birds, or the fact that animals can simultaneously integrate information on food distribution, competitors, and predators without behavioral ecology. One of the additional factors in the success of behavioral ecology has been its gradual expansion around the edges. Soon after the field started, there was concern that the obsession with *function* would eclipse Tinbergen's three other aspects of animal behavior (Dawkins 1986; Hinde 1982). The tight focus on function was to be expected; prior to behavioral ecology the adaptive significance of behavior had received only superficial treatment. The new body of theory provided an unprecedented array of research opportunities. As the field matured, and moved into what Thomas Kuhn would have called a phase of "normal science," researchers increasingly found that, on its own, function is not sufficient to fully understand biological traits, and there has since been a move back toward mechanisms. This is not to say there is a return to old concepts like fixed action patterns. No, they are dead and gone, but there has been increasing effort among some behavioral ecologists to look in more detail at anatomy, development, endocrinology, physiology, cellular processes, and neural and cognitive mechanisms. A need to incorporate realistic mechanisms at various biological levels is emerging in many areas, for example in postcopulatory sexual selection (Birkhead & Møller 1998; Eberhard 1996; Simmons 2001), maternal effects and phenotypic plasticity (Monaghan 2008; Lessells 2008), signaling, especially color (Hill & McGraw 2006), disease resistance and immune defense (Kennedy & Nager 2006), behavioral development (Bateson 2005), decision making (Fernandez-Juricic et al. 2004a; Biegler et al. 2001) and life history trade-offs (Monaghan & Haussmann 2006).

There is little doubt that behavioral ecology, even with its original, narrow focus on function, has been and remains, immensely successful (Krebs 1985; Owens 2006). There is also no doubt that the subsequent integration of function and mechanisms will be even more successful, pushing the boundaries of our understanding, raising new questions, and generating new areas of research. As this book indicates, in its new, broad guise, behavioral ecology will continue to inspire and inform. Behavioral ecology is evolving and is now much broader than before. Such expansion, of course, also carries risks. There has been criticism of the naïve ways in which behavioral ecologists "borrow" techniques, for example, in measuring immune function and presuming tradeoffs with other traits, without really understanding what they are actually measuring and the interrelationships involved (e.g., Hartley & Kennedy 2004; Kennedy & Nager 2006). However, appropriate interdisciplinary collaborations will ensure that behavioral ecology retains depth while gaining breadth. It is also important to remember that behavioral ecology is focused on diversity, explaining variation within and among species. We need to continue to study this variation. The recent surge in genomic technology and the focus on a handful of model species makes it tempting to focus on the behavioral ecology of those few species. This would be a mistake, not least because we can now obtain genetic sequences for most taxa in a matter of days or weeks. In contrast, securing time-sequence data on population biology and behavior can take decades. We should therefore concentrate more on those species for which we have long-term data in the wild, like red deer or great tits, rather than *Drosophila* and *C. elegans*. Narrowing the range of organisms studied simply because particular measurements are more easily made on the species commonly used by biomedical researchers really would be throwing out the baby with the bathwater.

Finally, it is important to stress the need for good ideas. All aspects of science have seen an increase in the use of technology, increasing quantification and reliance on more and more sophisticated statistical methods. Although new technologies and analytical techniques can certainly open up new avenues of research (Mayr 1982), they are no substitute for ideas. Theory is essential. This is what behavioral ecology brought to the study of function, and it is why the subject has stood the test of time so well. And just in case you are tempted by technology for its own sake, remember too that one of great pioneers of behavioral ecology, Nick Davies, has continued to produce exemplary research on the function and mechanism of brood parasitism, testing ingenious ideas with little more than binoculars, a notebook, chi-squared tests, and a clear understanding of the difference between good and bad science.

Acknowledgments We thank J. H. Crook, N. B. Davies, J. R. Krebs, C. M. Lessells, and G. A. Parker, as well as the editors, for their helpful comments on the manuscript.

SUGGESTIONS FOR FURTHER READING

Alcock (2001), the author of a leading animal behavior textbook, provides an enthusiastic account of the development of behavioral ecology. Readers will find Parker (2006a), one of the great architects of individual selection thinking and the "father of sperm competition," to be an engaging and personal account of the development of behavioral ecology. Finally, Uta Segerstråle's (2000) book is a monumental account of the controversies surrounding the emergence of behavioral ecology. The detail is extraordinary: it is almost as though the author recorded the conversations of the protagonists.

Alcock J (2001) The Triumph of Sociobiology. Oxford Univ Press, Oxford.

Parker GA (2006) Behavioral ecology: the science of natural history. Pp 23–56 in Lucas JR & Simmons LW (eds) Essays on Animal Behavior: Celebrating 50 Years of Animal Behavior. Elsevier, Burlington, MA.

Segerstråle U (2000) Defenders of the Truth. Oxford Univ Press, Oxford.

2

Adaptation

CHARLES W. FOX AND DAVID F. WESTNEAT

The phrase "mucking around outside" is no longer used much, but on a farm with livestock it still has literal meaning. For most, it conjures up the unpleasant task of cleaning one's shoes after trying in vain to avoid stepping in manure. People usually steer clear of such unpleasantness, but those who spend time watching insects colonizing dung, like Geoff Parker, may find out something remarkable. Fresh manure plops are where male dung flies (*Scathophaga* spp) find their mates and, as Parker discovered, male dung flies do some very interesting things (Parker 1970b). They arrive at a plop soon after it has been deposited, and wait until a female arrives. Copulation occurs on the dung itself or in the grass nearby and lasts a variable amount of time, usually between 10 and 100 minutes. Most females with whom a male may mate have already mated at least once, but the last male to copulate can, given enough time, displace most of the previous male's sperm. Curiously, though, most males do not copulate long enough to fertilize all of the eggs the female is ready to lay (Parker 1970b). That seems odd, and one might reasonably wonder why males do not stay longer to finish the job.

The specific answer to that question is less important for this chapter than the context in which it is asked, the logic used to address it, and the conclusions one can draw from the answers. This book is about the field of behavioral ecology, which tries to understand behavior, such as why male dung flies copulate for the length of time that they do. Behavioral ecologists are interested in adaptation, or how Darwin's (1859) process of natural selection shapes behavior. Surprisingly, the lowly dung fly in its stinky world reveals issues that have generated heated debate and have influenced not only biologists but philosophers as well (e.g., Holcomb 1993). The ideas involved in these debates affect what behavioral ecologists do, how they do it, and what it means. Our aim here is to better define behavioral ecology's goal of studying adaptation and to provide some perspective on the controversies that have impacted the field.

In the case of the copulation duration of male dung flies, Parker reasoned that the shorter than expected copulation duration may be an adaptation to the high frequency of encounters with females, and thus the ready availability of alternate mates. In this chapter, we examine what it means when we describe a trait as being adaptive or conclude that it is an adaptation. We also examine how behavioral ecologists study adaptation, and discuss the advantages and limitations of the ways they go about understanding the evolution of behavior.

CONCEPTS

Adaptation is a troublesome word for a variety of reasons. The word *adaptation* in common usage can describe a process per se or the outcome of a process. For example, adaptation can refer to the process of recasting of a book, play, or piece of music into a new form, or it can mean the recast version of that book, play, or piece of music (an outcome of the process). Adaptation can refer to a change in strategy by a sporting team in response to the opposing team or the processes of changing strategy. Although the common theme is about changing, usually in response to prevailing conditions, this diversity of use can create confusion.

In biology, *adaptation* has fewer uses than in the broader English language, but nonetheless is applied to some unrelated phenomena. A common usage in physiology is an adjustment of a physiological process in response to experience. For example, the response of a nerve cell to a stimulus often varies depending on the history of stimuli experienced by that cell (e.g., attenuation). Likewise, many physiological processes change in efficiency depending on experiences of the individual (acclimation), frequently due to changes in gene expression in response to environmental conditions. These uses of the word *adaptation* are generally synonymous with one of a variety of other concepts, including habituation, attenuation, accommodation, acclimation, phenotypic plasticity, and even learning (see chapter 6), all of which are processes of change that occur within an individual in a single generation.

In evolutionary biology, however, adaptation has one very important element not present in any other usage—a role of natural selection, either historically (producing the distribution of traits we observe in organisms) or currently (maintaining the distribution of traits we currently observe in organisms). Selection will be discussed in more detail in chapter 3, but there is little controversy over the basic mechanism as proposed by Darwin (1859). Variation in phenotypes exists within a population. If some variants handle difficulties in the environment better than others, then they will leave more offspring (on average) than variants that do not handle those conditions as well. Differential reproduction related to trait values is natural selection (see chapter 3). If the phenotype is inherited by offspring, then the variants that handle difficulties best will be more frequent in the next generation (figure 2.1); in other words, the distribution of phenotypes will change (evolve) due to the observed natural selection. The variant of the trait that has the highest fitness can be said to be *adaptive*, the population that evolved can be said to have *adapted* to the environmental conditions (via the process of *adaptation*), and the trait that evolved in the population (due to selection) is said to be an *adaptation* (an outcome).

It is the concept of adaptation, a consequence of natural selection, that makes dung flies so curious. Behavioral ecologists, and probably most evolutionary biologists, would find a trait that appears to lower a male's reproductive success to be rather odd, because such males should have had fewer descendants than males that stayed long enough to fertilize all of a female's eggs. Therefore, selection should have eliminated males with short copulations from the population. In considering this problem, Parker (1970b) made an implicit decision to focus on adaptive explanations despite an array of other possible explanations (e.g., they run out of sperm or the females kick them off). In particular, he reasoned that males might face a trade-off between continuing to copulate with one female versus searching for additional females with whom to copulate. The solution to this trade-off could depend rather precisely on the rates of additional eggs fertilized per unit time if a male stayed with his current female as opposed to the number of eggs fertilized per time if he left to find a new female (figure 2.2). This was an early application of the marginal value theorem (see chapter 8) that has been applied to many other types of behavior. In the case of the dung flies, once he had measured these rates, Parker (1970b, 1974a) and Parker and Stuart (1976) found that the average male dung fly copulated nearly exactly the amount of time that maximized the overall rate of fertilizations (i.e., the number of eggs fertilized per unit of time; figure 2.2). Male dung flies thus appear well adapted to the ecological situation in which a current mate decreases in value over time and other mates are available nearby. In choosing to focus on dung flies in this way, Parker joined other early behavioral ecologists in emphasizing natural selection, and thus adaptation, as an explanation for behavioral traits. Unfortunately, this emphasis became the focus of considerable criticism and debate that continues in part today (see below).

Chapter 1 describes how an explosion of interest in the study of behavior as an adaptation arose in the late '60s and early '70s. Wilson's (1975) *Sociobiology* stands as an icon of this approach, and it extended the idea of adaptation to the full array of social behavior, including that of humans. Controversy over the application of these ideas to humans soon spread to the entire approach. Gould and Lewontin (1979) took issue with what

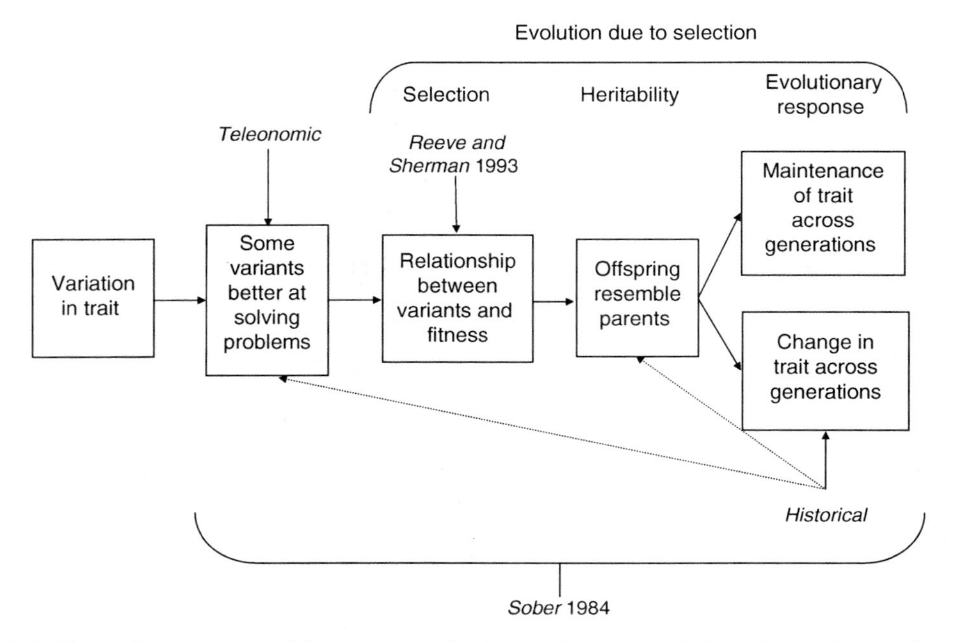

FIGURE **2.1** Flow of steps required for natural selection to lead to evolution that make up the process leading to fit between a trait and the environment in which its bearer lives (adaptation). The box describing a relationship between trait and fitness indicates selection (either directional or stabilizing). The box describing the resemblance between offspring and parents indicates heritability. The box for maintenance of a trait is the evolutionary response to stabilizing selection (which will generally reduce phenotypic variance without changing the mean), and that for a change in trait indicates the evolutionary response to directional selection. Different definitions of adaptation stress different aspects of this process. Sober's (1984) definition encompasses much of this process, although it does not include explanations for trait maintenance in populations. Teleonomic definitions focus on the problem-solving ability of particular trait values (Williams 1966). Historical definitions blend elements of change across generations with the problem-solving ability of a trait but do typically assume relationships between variation and fitness. Reeve and Sherman (1993) focus on the variant with highest fitness as their criteria for adaptation.

they perceived as adaptationism gone amok. They argued that natural selection was too often uncritically accepted as an explanation for phenomena without critical experiments distinguishing selection from other evolutionary processes. In particular, they argued that evolution is often constrained by a variety of factors and that it is these constraints on natural selection, and thus on the adaptive process, that are more interesting than whether or not a trait is adaptive. Many traits could be byproducts of other traits (e.g., the human chin is a consequence of the fit of the lower jaw to the upper), and so cannot be the direct object of selection. In addition, the reasons for a trait's origin—the history of selection versus constraints—could be very different from the selective consequences of the trait in the present. Gould and Lewontin argued that historical origin is central to the concept of adaptation. Their paper (Gould & Lewontin 1979) had a major impact, although it is as famous now for its use of metaphor and literary argument as for the biological points it actually makes (e.g., Selzer 1993). In their wake ensued substantial debate, leading to an array of new definitions of adaptation, refined techniques for how to study it, and a general increase in rigor in the study of behavioral ecology. But the field of behavioral ecology has sailed along, apparently quite successfully despite the furor (Queller 1995; Cuthill 2005). One might be tempted to conclude, as have many behavioral ecologists, that Gould and Lewontin (1979) and others who have criticized the field's approach (e.g., Orzack & Sober 1994a; Pierce & Ollason 1987) were wrong.

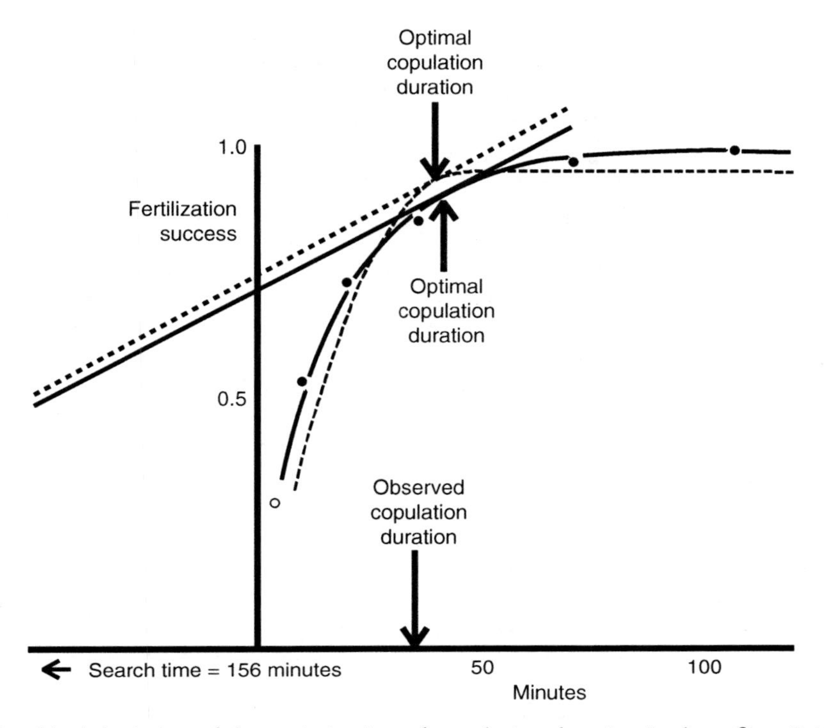

FIGURE **2.2** Graphical depiction of the optimization of copulation duration in dung flies. Points represent results from experiments assessing the fertilization success of copulations varying in duration. The tangent lines reflect the observed time needed to find a new mate (156.6 min). Two forms of the model (different equations for diminishing returns) are depicted (solid versus dashed lines) but produce a similar predicted copulation duration (41 min. versus 39 min.). The observed value of 35.5 min. is close to but slightly less than the predicted values. Redrawn from Parker and Stuart (1976).

Although we think Gould and Lewontin were mistaken in many of their specific criticisms, they raised numerous issues that needed to be addressed by behavioral ecologists (and evolutionary biologists in general). Some of these issues have not yet been resolved, and behavioral ecologists cannot ignore them if they are to articulate how their approach to studying behavior relates to other approaches and contributes to a more robust understanding of organisms. Here, we briefly review some of the diversity of perspective on adaptation (see Reeve & Sherman 1993 and Rose & Lauder 1996 for thorough discussions of alternative views). Our goal is to focus on several key issues in this debate. In doing so, we hope to better articulate what behavioral ecology seeks to address and how its approach fits within the broader domain of evolutionary biology.

Inferring Adaptation from Design

A number of authors have defined adaptation with a focus on teleonomy, the characterization of functional design (e.g., Williams 1966; Williams & Nesse 1991; West-Eberhard 1992) in which the match between the form of the trait and the task it performs is sufficiently good to rule out anything but natural selection. A focus on function is quite relevant to behavioral ecology, as we will elaborate below. However, teleonomic definitions of adaptation have been criticized for several reasons (Reeve & Sherman 1993, Reznick & Travis 1996). There is some subjectivity in the criteria for what constitutes evidence of design. For example, engineering principles commonly underlie models of optimal morphology, but many engineering principles are themselves derived from observations of design in nature, such that nature and human ideas of optimal design are not independent (Lauder 1996). Teleonomic definitions are also conservative in that fairly simple traits may not show sufficiently complex design to qualify as adaptations despite their spread due to selection. Likewise, traits may deviate substantially from optimal design and yet still be favored by natural selection relative to alternatives that are biologically available. Most importantly, the idea of *design* misrepresents how the process of evolution by selection actually works. Darwin (1859), for example, was explicit in viewing selection not as a designer but a tinkerer. A trait need not be well designed to spread—it merely must be better then the previous version to evolve through natural selection and therefore qualify as an adaptation (Reznick & Travis 1996).

Today, quantitative techniques for measuring selection (chapter 3 of this volume; Arnold & Wade 1984) and comparative analyses using phylogenies (chapter 7; Felsenstein 1985) provide alternative tools for assessing adaptation. Teleonomy is thus rarely used as a criterion. Nevertheless, as we describe in more detail below, much of what behavioral ecologists do is focused on elaborating functional problems that behaviors appear designed to solve. The teleonomic approach is beneficial because it focuses attention on the ecological or social problem giving rise to selection. As this book will illustrate, one of the great advances of behavioral ecology has been in exploring the complexity of problems faced by organisms.

Historical Definitions of Adaptation

Gould and Lewontin (1979) insisted that adaptation is a historical process and that current function (utility) is irrelevant to understanding adaptation. A division between those studying historical processes (e.g., Larson & Losos 1996) and those studying contemporary natural selection (e.g., Reeve & Sherman 1993) remains today and continues to be the subject of substantial debate and misunderstanding.

Definitions of adaptation that focus on historical process abound, but we have chosen that of Sober (1984: 208) as a starting point because it incorporates the major features of many other definitions and is more explicit than most: "A is an adaptation for task T in population P if and only if A became prevalent in P because there was selection for A, where the selective advantage of A was due to the fact that A helped perform task T." This definition illustrates three essential features of most historical definitions:

1. The trait, A, has evolved (exhibited evolutionary change) in the population (the historical component of the definition).
2. The evolutionary change in trait A was due to natural selection.
3. The natural selection on trait A was due to effects of the trait on the performance of a specific task, T; in other words, an adaptation is defined with respect to a specific trait and its specific function.

Sober's definition corresponds quite closely with the process of evolutionary change by natural selection (figure 2.1) and provides a very specific set of criteria for a trait to be labeled an *adaptation*. According to Sober, we would need to know that there has been change in A, that this was due to A having higher fitness than other variants, and that the reason A had higher fitness was because it performed T better than variants that were not A.

Establishing that any specific trait meets these criteria is difficult. We can imagine how copulation duration of dung flies fits these three criteria; copulation duration in the ancestor of present-day dung flies was likely either longer or shorter, so we could establish that a change in copulation duration occurred (criteria 1) because a different duration allowed males to either fertilize more of the current mate's eggs or find new mates sooner (criteria 3), and so the change spread because of that advantage (criteria 2). But to test each of these, several difficulties must be overcome. Here, we focus on three problems—defining the trait, defining function, and determining evolutionary history—that are particularly important.

Defining Traits

One difficulty in testing Sober's concept is in adequately defining the trait in question. Indeed, a clear definition of a trait is a critical component faced in any behavioral or evolutionary study, regardless of the definition of adaptation chosen. For example, copulation duration is not a single trait of an individual, because it may depend on traits of both the individuals that are interacting (Parker et al. 1999; chapter 14, this volume). It also may vary depending on other traits, such as male body size, ejaculate size, rate of sperm transfer, morphology of the intromittent organ, and the movement rate of a male, all of which may be under selection and evolving. Evolution of copulation duration may thus be the result of selection on these other traits rather than on copulation duration per se. Copulation duration may even be considered an emergent property of other traits; at a minimum, it arose as an emergent property of reproductive biology. Any organism that copulates must copulate for a nonzero duration, such that we must necessarily have a trait we call *copulation duration* regardless of whether copulation duration per se is under selection. Many behavioral traits have this problem—they are produced by underlying physiological and morphological attributes, and they may not be independent from the attributes of other individuals. These problems are exacerbated by the necessity in Sober's definition of maintaining the same trait definition through time. As the components influencing production of a trait change (such as due to selection on those underlying traits), the emergent trait (such as copulation duration) becomes a different trait. For example, the same copulation duration could be obtained through the action of female attributes or through male attributes, and through a variety of combinations of male attributes, such that what we perceive as one trait could reflect different traits at different times or in different populations.

There is no simple solution to these problems with defining traits and the study of adaptation. Often the appropriate approach depends on the question being asked, with problems arising when different questions about the same trait are not carefully clarified. A major debate over the function of the human clitoris illustrates some of the issues involved with defining traits, as well as the problems that arise when the goals of different researchers diverge (Gould 1987; Sherman 1988).

Defining Function

Most definitions of adaptation, including Sober's (1984) historical definition, include an importance of trait function (but see Reeve & Sherman 1993, discussed below); in other words, adaptation is defined with respect to how a trait increases fitness, rather than just whether a trait increases fitness. In Sober's (1984) historical definition, a trait must be demonstrated to have a particular function (performing task T) at its origin and that selection via this particular function is the reason for evolution of the trait. Many traits in organisms have been co-opted from their original function and now serve a new function. The courtship displays of the western grebe (*Aechmophorus occidentalis*), in which partners hold nest material in their bills but do not use it to build a nest (Nuechterlein & Storer 1982), is one example of an action that surely evolved for one reason (building a nest) but is now used for another (courtship). In the case of dung flies, one could presume that copulation arose to combine male with female gametes and that the duration of copulation functioned merely to ensure that there were enough sperm to fertilize the eggs. Subsequent evolutionary modification of copulation duration may have occurred via selection on a male's ability to fertilize additional mates. Is this a novel function (fertilizing multiple females rather than the current female)? If so, both the sources of selection on, and the function of, copulation duration have changed through evolutionary history, complicating the study of adaptation. Similar complications occur for many traits, especially those that are emergent properties of other traits.

Gould and Vrba (1982) recognized this possibility and proposed that a true adaptation be distinguished from co-opted traits by giving the co-opted case the name *exaptation*. They distinguished two versions: (1) traits that were originally adapted to a particular function but co-opted for another use (e.g., the nest material display of grebes), and (2) traits that arose through processes other than natural selection but then came to have a function. However, the concept of exaptation is complicated by whether or not selection modifies the trait after it has been co-opted. In the case of display behavior by grebes, it is clear the behavior has been modified for functioning as a display, because nest material is not placed on a nest but dangled in front of a partner and waved about. Holding nesting material in the bill is an exaptation in the context of the display, whereas the modified behavior (waving nest material about when no nest is being built) is likely an adaptation to the task of consolidating a pair bond. To most behavioral ecologists, who are interested in the current display and its effectiveness, this distinction may seem semantic. However, to those interested in the historical pathways taken by evolution, such as how particular display elements arose and in what sequences, the distinctions are important.

Determining Evolutionary History

Sober's definition of adaptation contains two major elements of history. First, the trait of interest has to have changed from one state to another. A major constraint on the application of historical definitions of adaptation to behavior is the ability to know what state a trait was in before it changed. One can infer such states from reconstructing the phylogeny of extant species (or populations within species) and comparing the trait of interest among taxa (among populations or among higher level taxa) with this reconstruction in mind. There are many more ways to do this now than when behavioral ecology first became a field, and some fascinating insights can be gained by using phylogenies (chapter 7). Indeed, a comparative study is essential for testing historical hypotheses about transitions in most behavioral traits. However, the phylogenetic approach is less useful for understanding why a character has persisted in the same state. Many behavioral ecologists find that traits are currently experiencing stabilizing selection, as appears to be the case for copulation duration in dung flies. Phylogenetic approaches have been more successful in understanding evolution as a result of directional selection than for explaining the stability of a trait due to stabilizing selection. Sober's definition, which requires evolutionary change in a trait, is ill equipped to deal with maintenance of a trait in its current state.

A second historical element to Sober's definition of adaptation occurs in criteria 3: that the change in state is due to natural selection because the new trait performed a specific function and that the change in state was due to selection via this specific function. Except for relatively short-term evolutionary events, it is very difficult to meet criteria 3 because we cannot go back and measure either selection or how a trait performed particular tasks in the past. Phylogenetic reconstruction may allow some simple hypotheses about this to be tested if, for example, it is used to compare environments among groups with and without the trait of interest (chapter 7). However, for many behaviors there is insufficient information available to accomplish this task. Returning once again to copulation in dung flies, assessing the history of the function of copulation duration in this way would require measuring copulation duration, the diminishing returns of copulation length, and the density of females in a variety of groups with known phylogenetic relationships. This is impractical. We thus generally infer historical function and historical selection from design principles or, as Parker did, from current function and patterns of selection, though this has its limitations for reasons discussed below.

Current Utility Definitions of Adaptation

Behavioral ecologists study animals behaving in their natural environments, and thus the field has developed a strong focus on the current utility of behaviors. This certainly was Parker's focus as he struggled to understand dung flies—to answer the questions that arose about their behavior, he measured the actions of existing male dung flies where they could be found, on piles of dung.

This focus on current utility of traits has led to a set of definitions of adaptation that ignore evolutionary history (at least in the specifics of the definition) and instead focus on current selection on traits. Though the diversity of definitions is large, the key points are highlighted in the definition proposed by Reeve and Sherman (1993), in which an adaptation is "a phenotypic variant that results in the highest fitness among a specified set of variants in a given environment" (9). This definition is distinct from historical definitions in that the evolutionary events that produced the trait are not essential to our understanding the current fitness consequences of the trait. Indeed, by this definition an adaptation is simply the variant that is most favored by natural selection in the environment of interest at the current time. Reeve and Sherman's definition is one of the broadest of the current utility definitions because it makes no mention of trait function, whereas most other current utility definitions make explicit the importance of trait function. A trait that evolved for one function and has been co-opted for another function, with an increase in fitness, is not a problem for current utility definitions of adaptation because the task a trait currently achieves is its contemporary function. Because behavioral ecologists use contemporary systems for study, a trait's effect on fitness can be measured and its ability to perform a particular task assessed. Thus, Reeve and Sherman's (1993) definition, or some variant of it, is implicitly accepted among most behavioral ecologists, and may underlie most usage of the term *adaptation* in behavioral ecology.

Reeve and Sherman suggest that their definition helps clarify the distinction between hypotheses about history (origin) and current utility (maintenance) and between directional selection (spread of a trait) from stabilizing selection (maintenance of a trait). They also suggest that history is accommodated by the context of their definition—that is, the evolutionary past will influence both the specified set of alternative variants and the environmental context. Finally, they argue that their definition produces testable predictions. The adaptive phenotype is predicted to predominate in the population, and if it does not, then a process other than natural selection must be at work.

Reeve and Sherman's definition does, however, sit precariously on a conceptual knife-edge with difficulties arising not over the definition itself but how it is applied. Often current utility is presented as addressing the question of why a particular phenotype predominates in a population today. This is inherently a historical question because selection cannot cause phenotypic preponderance immediately. Contemporary phenotypes exist only because of processes (be they directional or stabilizing selection, or an evolutionary process other than selection) that occurred in previous generations. We think Reeve and Sherman's definition works better as a prospective view of adaptation as opposed to a retrospective definition like that of Sober (1984). That is, a trait that exists in the population today and has highest fitness compared to alternatives would be predicted to persist in that population through stabilizing selection in the future. This is not how most behavioral ecologists use the definition, and we explore in more detail the implications of using current utility definitions in other ways. If adopting a current utility definition, behavioral ecologists must understand the limitations of such definitions.

Extrapolating from Current Utility to History

Although quantifying the current relationship between phenotype → function → fitness does not explicitly require an understanding of evolutionary history, researchers often assume that demonstrating how a trait currently affects fitness also tells one how the trait arose or has persisted in its current state in the past (origin or maintenance). A match between the mean trait value and the optimal value of that trait certainly suggests that the mean trait was shaped by similar selective pressures in the past. In many cases, this inference might be correct (Cuthill 2005), but it also could be wrong (Orzack & Sober 1994a). For example, female tungara frogs (*Physalaemus pustulosus*) prefer males with calls having a particular component (called a *chuck*). Although it is not clear whether preference for the chuck per se is adaptive, larger males produce a chuck with lower pitch and mating with larger males is advantageous to females (Ryan 2005). So it might be reasonable to suppose that female preferences have coevolved with male calls—in other words, female preference for male calls evolved as a means of selecting the largest males. However, phylogenetic reconstruction indicates that females of related species, in which the males do not produce a chuck, actually prefer calls with chucks. It is possible that this occurs because female frogs have a general preference for lower frequency calls, in which case we again have a debate over what is the trait of relevance. Regardless, the historical sequence of events appears different from that suggested by extension of current utility (a preference for chucks leads to higher fitness) into the past.

Use of current utility definitions can also be misleading about historical hypotheses of stabilizing selection, for similar reasons. A match between predicted optimal trait value and the population mean suggests a history of stabilizing selection, but it is conceivable that the two correspond by happenstance and that there has been no genetic variation for the trait in the past. This possibility is one of the justifications for employing phylogenetic contrasts (Felsenstein 1985) in comparative studies (chapter 7); if no genetic variation has been seen by selection, then multiple cases of match between conditions and trait are not independent. Although in practice there is often considerable genetic variation present for traits, most traits are constrained at some level by their genetic relationships to other parts of the phenotype (morphology, physiology, development, or other behaviors) and thus not free to evolve to any possible state. The extent of genetic variation exposed to selection is an empirical issue that, until examined for the specific traits of interest, constrains how far one can interpret a match between mean phenotype and the phenotype with the highest fitness. We thus caution that demonstrating current utility should not be extended casually to imply a history of selection or evolution of the traits under study.

The converse is also true—just as extrapolating from a modern-day match between a trait and predictions from an optimality model may mislead us about the history of selection on a trait, an absence of modern-day selection can mislead us to conclude that selection was unimportant in producing a trait. For example, the red and yellow epaulets of red-winged blackbirds (*Agelaius phoeniceus*) seem unlikely to have evolved by any mechanism other than by selection. That they are bright and flashy suggests that epaulets have a function in interactions with females (e.g., as a sexual signal) or with other males (e.g., mediating male-male competition). However, a large number of studies in several different populations have revealed little or no evidence that existing variation in epaulet size is under any current selection, either sexual or natural (reviewed by Westneat 2006). Experimental manipulation of epaulets reveal that quantitative changes in epaulet size have no apparent effect on selection but eliminating epaulets entirely has large effects on fitness (reviewed by Westneat 2006). What can we conclude about epaulets as adaptations? One problem with this example is whether field methods can measure small differences in fitness that are sufficient to maintain this trait. It is conceivable, however, that none of the observed variants in natural populations differ in fitness. A second problem is with how to interpret selection acting on entirely artificial phenotypes (no epaulets at all). Such selection is not relevant to maintenance of the trait today, because males without any epaulets apparently do not occur naturally. Under Reeve and Sherman's definition, none of the existing variants in epaulet size are adaptations. Under a historical definition, epaulets could be adaptations because individuals have higher fitness with them than without. Thus the results of experimental manipulations of trait values outside existing ranges might be relevant to how the trait arose, except that in this case we cannot know for sure that the present-day fitness (loss of territory) of a rare individual with no epaulets in a population in which all other individuals have epaulets is related to the fitness of the presumably rare ancestral individual with epaulets that occurred in a population that lacked them.

Phylogenetic analyses of the blackbird family suggest that the relatively recent transition to having epaulets is associated with nearly simultaneous transitions in breeding habitat and mating system (Searcy et al. 1999; Johnson & Lanyon 2000). A plausible hypothesis is that the function of the epaulet and the selection that caused it to spread were ephemeral events that occurred during the shift in habitat. Present-day maintenance of epaulets in red-winged blackbirds thus may depend either on weak and unmeasurable stabilizing selection or the slow action of nonadaptive processes. Regardless, this is an example in which measuring current utility is not very helpful in understanding the processes that either produced the trait or have been maintaining it. In fact, it is the combination of both approaches that produces the most interesting insights in this system.

Defining Trait Sets

Under the Reeve and Sherman definition, to demonstrate that an observed phenotype, A_1, is an adaptation, it is enough to demonstrate that A_1 results in the highest fitness among a specified set of variants of A (e.g., $A_1, A_2, \ldots A_N$), which indicates that A_1 affects fitness through some task whether or not that task is known. Defining the set of variants is critical to conclusions about the action of natural selection (Sherman & Reeve 1997) and yet doing so can be problematic. Suppose that A_2 is the most common phenotype in the population and is second in terms of fitness to A_1. Under current utility definitions, A_2 cannot be an adaptation, and yet if A_1 is a recent mutant that is currently spreading through the population, A_2 could have spread originally and been maintained for many generations by selection. As the red-winged blackbird may illustrate, the converse is also true—current utility might reveal no clear difference between variants, but the variant that predominates may have arisen through selection in the past.

For many traits we study, all individuals in a population or species express the trait and it is evolution of the presence of the trait in which we are interested (e.g., the elongated caudal fin in swordtail fish [box 24.1] or the presence of an epaulet on a male red-winged blackbird). Alternatively, individuals within populations may vary in the phenotypes they express (e.g., size of epaulets in red-winged blackbirds, duration of copulation in dung flies). For traits that are variable within populations, all phenotypes that confer higher than average fitness will increase in frequency due to selection. Thus, defining only the phenotype with the highest fitness as *the* adaptation will lead us to conclude that few individuals are adapted. The solution here is usually to compare population mean values to predictions and test whether an individual with the mean phenotype in a population is adapted (box 2.1). For example, in dung flies we can predict the exact duration of copulation that will maximize a male's fitness for a given set of criteria (e.g., female densities). We would find, however, that few male flies actually mate for the predicted time, though the population average is remarkably close to predictions. This is consistent with what we would predict when a trait is under balancing selection around a mean trait value. (Simple population genetic models instruct us that selection usually leads to maximization of mean population fitness and not maximization of the fitness of every individual. However, frequency dependent selection can lead to very different outcomes, including a reduction in population mean fitness. See chapter 3 for a discussion of frequency dependent selection.)

Understanding Function

One element that is missing from the Reeve and Sherman definition of adaptation is a reference to function. A trait is considered an adaptation if it has highest fitness. To return to dung flies, the copulation duration of 41 minutes (figure 2.2) had highest fitness and so is the most adaptive given the conditions in a typical cow pasture. No understanding of function is required for the prediction that most individuals should copulate for about 41 minutes. But the obvious next question is why 41 minutes is so good. Reeve and Sherman explicitly discuss the fact that the environmental context is critical to understanding which variant has highest fitness, and that the context may be quite complex. In the case of dung flies, Parker's key contribution was to identify that it was the specific combination of declining fertilizations with time plus the relatively short search time for new mates that produced an intermediate optimum, at 41 minutes, in the duration of copulation. Defining function, the ecological and social factors that influence the fitness consequences of a particular variant, is the major success of behavioral ecology. In our view, the greatest benefit of using the current utility concept is its utility in elaborating on function, specifically, linking behavior with ecology in more sophisticated ways.

Testing Current Utility Hypotheses

Behavioral ecologists have used two methods of applying the current utility approach to questions about adaptation. Sherman and Reeve (1997) have called these the *forward* and *backward* methods in reference to their links to the logical progression for how phenotype abundance is linked with its fitness. In the forward method, an array of alternative phenotypes is specified and their fitness predicted from specified aspects of the environment. Then data are collected to assess whether the predominant phenotype is the one with the highest fitness, as would be predicted if natural selection had occurred in the hypothesized manner. Put another way, given some information about the environment and how fitness is affected, we can identify what trait value would be best in that circumstance, and then test whether that trait value is most common. Parker's study of dung flies followed the forward method by calculating the optimal copulation duration and then measuring duration in the field to assess what most males do. The backward method also specifies an array of traits, but then asks whether the most common one has the highest fitness. Westneat's (2006) study of red-winged blackbirds employed this approach using both a correlative approach (measuring the fitness of males with different sized epaulets to test for an association) and an experimental approach (manipulating epaulet size and testing the fitness consequences). Either method is suitable for exploring aspects of adaptation, particularly with the goal of refining our understanding of trait function.

However, both approaches have weaknesses in the broader context of the study of adaptation. As discussed for red-winged blackbirds earlier, the backward approach cannot be used to test rigorously any historical element of the adaptive process, although the results may be used to infer recent processes involved in stabilizing selection. The forward approach has a different set of benefits as well as dangers. Parker's study of dung flies was so revolutionary because he envisioned the processes affecting the evolution of flies as continuous relationships (e.g., the diminishing returns in fertilization success as copulation proceeded). These produced quantitative predictions. *Optimality models* such as this revolutionized many aspects of evolutionary ecology beginning with the research of David Lack and colleagues in the 1950s (e.g., Lack 1954). Parker (and others) brought these models to the realm of behavior. Optimality models enhanced the study of behavior by adding a much-needed means of producing quantitative predictions that could be tested against real behaviors (box 2.1). Models also frequently identified nonintuitive, or even counterintuitive, relationships between traits and fitness,

BOX **2.1** Optimality Models

Steven Hecht Orzack

A trait that maximizes some measure of evolutionary performance (such as the number of descendents) is said to be optimal. Behavioral ecologists often assess the optimality of an observed behavior by comparing it to the behavior predicted to be best by a mathematical model that incorporates some set of constraints on the organism. Usually these constraints are ecological (such as a limit on the number of offspring in a given breeding attempt). Most such models omit details of the development and genetic determination of the trait. Many examples of optimality analyses can be found in Charnov (1982), Stephens and Krebs (1986), and in this book.

The use of optimality models has generated considerable controversy (e.g., Gould & Lewontin 1979; Maynard Smith 1978; Cheverton et al. 1985; Emlen 1988; Brown 1993; Orzack & Sober 1994a, 1994b, 2001a; Brandon & Rausher 1996; Grafen 2007). Many proponents acknowledge the simplicity of the approach but justify it as the only pragmatic way of understanding the tremendous diversity of adaptations observed in nature and of uncovering relationships between behaviors and ecological variables. Critics acknowledge that the approach can sometimes correctly identify an optimal trait but claim that optimality models are often misused. What constitutes a sufficiently good fit to a model is rarely specified beforehand, and a lack of fit to predictions can always be attributed to the model being incorrect (as opposed to being attributed to the trait not being optimal). Without clear grounds for rejection of a scientific claim, acceptance of the claim means little (see Orzack 1990 for examples of apparently arbitrary claims about optimality). Some critics even claim that the use of optimality models is the scientific equivalent of "theft" as opposed to "honest toil" as a means for acquiring knowledge (Brandon & Rausher 1996).

Is there a way to resolve this controversy? It is essential in this context to distinguish between three claims about the influence of natural selection on trait evolution (Orzack & Sober 1994a). The first is that natural selection is the *only* important influence on a trait's evolution. The second is that natural selection has had an important influence on a trait's evolution but other nonselective influences could also have such a role. The third is that natural selection has had some influence on a trait's evolution. If the first claim is true, the second and third claims are true, but the latter claims can be true without the first claim being true.

An optimality model is best understood as the embodiment of the first claim. It follows that this claim can be assessed in a particular instance by determining whether the optimality model under investigation provides an accurate quantitative prediction of the trait and by determining whether natural selection has eliminated all but one trait from the local population. The quantitative test is implemented using standard statistical approaches to compare numerical model predictions with observations; for example, to test an optimal sex ratio, a goodness-of-fit test or logistic regression analysis is used to compare the optimal and observed proportions of, say, males. Trait homogeneity is necessary because the premise of the model is that there is a superior trait that an individual can possess and that natural selection operates among individuals in the standard sense described by Charles Darwin. Homogeneity need not mean that a snapshot of the population reveals no variation; after all, an optimal trait of an individual could consist of a set of subtraits, each of which is manifested at different times (e.g., the production of male offspring at one time

(continued)

and of female offspring at another time). What it does mean is that all individuals are alike when assessed over their lifetime of trait expression. This is accomplished with standard statistical approaches, for example, a test of homogeneity is used to compare the sex ratios produced by a representative set of individuals in the population. If the model predictions are quantitatively accurate and individuals are statistically identical with one another, one can accept the claim that the trait is *locally* optimal, in other words, it is optimal given the constraints incorporated into the model. If one or both of these conditions is not satisfied, one should accept a weaker claim about the influence of natural selection on the trait; the choice depends upon the strength of the evidence (see Orzack & Sober 1994a for further details.)

These two elements of a test of optimality are present in the work of Brockmann and Dawkins (1979) and Brockmann et al. (1979). They showed that the proportions of two reproductive behaviors in a digger wasp matched the proportions predicted by an optimality model. They also collected data on the lifetime mixtures expressed by individuals; these mixtures are statistically identical to one another (Orzack & Sober 1994a). These two elements are also present in the work of Orzack et al. (1991), who showed that an optimal sex ratio model was quantitatively inaccurate and that isofemale strains of wasps (proxies for individuals) differed in their fit to model predictions.

This two-part test is rarely done (Orzack & Sober 1994a, 1994b), even though the structure of this test embodies the conception about the power of natural selection that most advocates of the approach have in mind. The nearly universal omission from tests of optimality of the assessment of whether individuals manifest an optimal trait is startling. Instead, sometimes there is a quantitative test of the optimality model but no assessment of whether individuals differ. For example, Milinski (1979) studied the foraging behaviors of a small fish, the stickleback; he found that the average number of fish in each of two types of patches differing in food quality matched quantitatively the numbers predicted by optimal foraging theory. However, he did not assess whether individuals each manifested the optimal distribution over time. His "snapshot" assessment of the population does not allow one to determine whether, for example, some individuals stay in one kind of patch all the time. If this were true, it would mean that natural selection had not, in fact, resulted in the evolution of an optimal set of behaviors for an individual. At present, no conclusion can be reached about optimality; additional data are needed. In this case, it is reasonable to conclude that the second claim above that natural selection has had an important influence on a trait's evolution is true.

It is more common that a qualitative criterion is used to judge whether an observed trait is optimal. For example, in the case of Milinski's study of foraging behaviors, such a test would entail assessing whether one observes the predicted inequality between the patches with respect to number of individuals present, in other words, whether there were more individuals in the patch predicted to have more individuals. Tests of this kind provide evidence for the action of natural selection but it is not a test of optimality. At best, such qualitative testing allows one to accept the second claim about the power of natural selection; this is no small accomplishment.

An understanding of what local optimality motivates in terms of the structure of a test of an optimality model is central to a resolution of the controversy over optimality models. The two-part test establishes a clear protocol that subsequent investigators can follow in order to generate results that can be compared with others based upon the same protocol. As analyses in which this protocol has been used accumulate over time, we will gain a much more meaningful and comprehensive understanding of where and when the claim of optimality is correct than we will if the protocol for testing an optimality model is left ambiguous and if claims about the power of natural selection remain unclarified. This increased understanding is the hallmark of great science.

especially when components of fitness trade-off with one another or are frequency dependent.

The incorporation of optimality models into behavioral ecology would thus seem to be a good thing. However, quantitative models were frequently used sloppily. That is, often the criteria for establishing a fit to a predicted optimum were not presented, and so whether or not a model was rejected or accepted was a qualitative and often subjective process, a disappointing contrast to the rigor intended by the quantitative model. Possibly more significant was the common practice of ignoring some sets of alternative hypotheses; continual tweaking of a model can produce a prediction that fallaciously matches our observations and misleads us to believing we have explained that which we still fail to understand. These and other criticisms of optimality modeling, as well as the arguments in its defense, are too complex to cover here; we refer you to box 2.1 and Orzack and Sober (2001b) for discussions of the role of optimality in studies of evolution.

Reconciling Disparate Uses of the Term *Adaptation*

Most of the disagreements about adaptation appear to contain, at their core, differences in the question that is most interesting to the researcher (Reeve & Sherman 1993). The process in figure 2.1 is sufficiently complex that different fields have taken different approaches and, in doing so, have emphasized different aspects of the process. For those most interested in the evolutionary process, historical definitions have more appeal because it is the nuances of history that intrigue them the most. Gould and Lewontin (1979) appeared most interested in factors that constrain the operation of natural selection. Behavioral ecologists have been much more interested in how well traits function currently, and so have been most intrigued by the power of the adaptive process as revealed by matches between form and function as well as measures of contemporary selection.

Thus, much of the confusion and debate in the literature regarding the concept of adaptation reflect use of the term *adaptation* for different aspects of the evolutionary process. Fortunately, current widely used terminology provides us the ability to clarify our language without the introduction of new terms. We propose that traits can be described as being *adaptive* when they improve current fitness relative to appropriate alternatives—the focus of most behavioral studies—and that the term *adaptation* be reserved for studies examining evolutionary history in addition to current utility. This usage of *adaptive* is already commonplace in the literature without loss of concepts or understanding in behavioral ecology.

Alternative Hypotheses and Research Paradigms

Behavioral ecologists prefer adaptive explanations, and this preference is one of the main reasons Gould and Lewontin (1979) referred to them as "Panglossian" and were critical of the approach. Behavioral ecologists have largely ignored this criticism (Cuthill 2005), but an understanding of it could influence how we collect and interpret data. The criticism raised by Gould and Lewontin (1979), and much of the debate that ensued, was about how to formulate alternative hypotheses. The debate involved elements of good scientific practice as well as aspects of broader sociological trends in science as a human endeavor.

Good scientific practice requires that one pose several alternative hypotheses and devise predictions that distinguish among them. A classic example concerns explanations for infanticide in male langur monkeys. One can collect data that support the hypothesis that males committing infanticide benefit by causing the female to enter estrous sooner and thus become available to produce an offspring sired by the infanticidal male. However, having data consistent with a hypothesis is not adequate evidence to conclude the hypothesis is correct. A result may be consistent with multiple hypotheses, all of which need to be tested and excluded before the favored adaptive hypothesis can be accepted as most likely correct (reviewed by Alcock 2005). Yet behavioral ecologists are sometimes guilty of overinterpreting experiments that test only a single hypothesis. Parker's test of the marginal value theorem and copulation duration in dung flies was not presented as a test of alternatives, although implicitly his data eliminate two. The first is that search time has no effect and the only factor causing selection on copulation duration is the fertilization rate with the present female. It could be eliminated because it predicts that males should copulate until all eggs are fertilized. The second is that copulation does not depend on diminishing returns. This is also rejected because alternative curves relating time of copulation to proportion of eggs fertilized produce very different predictions (e.g., either stay and fertilize all eggs or leave almost immediately). However, Parker did not test alternative hypotheses for an intermediate duration of copulation. Later studies of the marginal value theorem (e.g., foraging behavior of bees; Schmid-Hempel et al. 1985), on which Parker's model was based, did examine alternative adaptive hypotheses and so provided much stronger tests of the factors encapsulated in the model. Similar consideration of alternative hypotheses is absolutely necessary for behavioral ecology to be considered a rigorous science. Good scientific practice existed in behavioral ecology before Gould and Lewontin (e.g., Hoogland & Sherman 1976); since then it has become more common (though not universal) for behavioral ecologists to consider both alternative adaptive and nonadaptive hypotheses.

Yet Gould and Lewontin's (1979) objections ran deeper than concern about good scientific practice. Their focus was on whether adaptive explanations are more likely to be true than alternative nonadaptive explanations, and so addressed the use of scientific paradigms. A paradigm is a pattern of thought and set of practices that are widely adopted by scientists in a discipline (Kuhn 1962). Paradigms bias the types of hypotheses that are deemed most relevant, and so address more deeply ingrained assumptions about what questions are interesting and what approaches are likely to have more explanatory value (e.g., Holcomb 1993). They affect what we consider to be a suitable null hypothesis and how we choose among a vast array of potential alternative hypotheses. Paradigms cannot be tested by applying predictions to data—rather, they are influential as long as they continue to provide insight and so are assessed through multiple iterations of their use.

Gould and Lewontin objected to the adaptationist paradigm adhered to by most behavioral ecologists. However, the adaptationist paradigm remains dominant in behavioral ecology because, in case after case, the focus on adaptationist explanations has led to new insights. A perfect example of this was David Lack's hypothesis (e.g., Lack 1947) that clutch size in birds would be optimized to balance the number of offspring produced with the parent's ability to feed those offspring well enough to survive. In other words, parents should produce a clutch of a size that maximized the number of offspring fledged from that clutch. Experimental studies on multiple species of birds revealed that clutch sizes were close to, but did not match exactly, what Lack predicted. Most birds produced slightly smaller clutches than predicted. Lack was invested in the adaptationist paradigm so, despite the possibility that many nonadaptive hypotheses could be proposed to explain the disparity between data and theory, Lack (and many others) chose instead to hypothesize that other factors affected selection on clutch size. This search for adaptive explanations led to a diversity of new adaptive hypotheses. In particular, many studies show that parental work load is indeed important to lifetime reproductive success; current reproductive effort affects future reproductive success and, when incorporated into Lack's clutch size model, this trade-off explains the original deviation between data and theory that Lack observed (Lessells 1993). No doubt some nonadaptive processes also affect clutch size in birds, but Lack's focus on adaptive processes nonetheless led to substantial new insights. A similar record of success for the adaptationist approach can be seen across a wide range of behavioral phenomena, many of which are discussed throughout this book.

Nevertheless, behavioral ecologists may well benefit from some self-inspection about how long the adaptationist paradigm alone can sustain the field. Although we probably will never fully understand all the nuances of how selection acts on traits, we may be reaching a point at which other approaches are necessary to obtain new insights. In some areas, major adaptive explanations are fairly well understood, and what remains puzzling about traits may be better explained by the kinds of constraints that Gould and Lewontin (1979) described, or by new constraints that have yet to be imagined, than by further subtleties of selection.

Ornaments used in sexual displays provide an example. Female preferences are clearly an important selective force on these traits, leading to a widespread assumption that ornaments are an adaptation to the problem of attracting mates. However, the diversity of sexual ornaments is astounding. Why are male ornaments so diverse? Elaboration of a specific sexual ornament in a particular species may often be adaptive, but across lineages, which trait gets elaborated may be due the vagaries of starting conditions rather than any systematic advantage of particular classes of ornaments (see Prum 1997; chapter 7). Nevertheless, most research still focuses on the adaptive processes that lead to ornament elaboration within single species. This focuses attention on a subset of hypotheses for the evolution of ornaments to the exclusion of more complex hypotheses, for example, that female preferences could act ephemerally, as may be the case in red-winged blackbirds, or that starting conditions might explain why a particular trait as opposed to another was the one elaborated. A shift toward asking questions about lack of function need not require abandonment of the adaptationist paradigm in behavioral ecology; investigating constraints on adaptation, or the role of nonadaptive evolutionary processes, does not debunk the importance of selection in shaping traits. A challenge for the study of evolution, particularly of behavior, is thus integrating approaches to exploring the action of selection with those investigating other processes. Such integration is evident later on in this book, such as in chapters on phylogenetic (chapter 7) and genetic (chapters 5, 14, and 28) approaches and new perspectives on behavioral correlations (chapter 30).

FUTURE DIRECTIONS

Much of the rest of this book will introduce you to the richness of adaptationist thinking, primarily using the current utility approach. Behavioral ecology has been extremely successful in advancing our understanding not only of behavior, but of how ecology shapes selective forces. The insights we have gained have influenced thinking about evolutionary processes and have stimulated new approaches to topics as diverse as how the immune system works to the structure of the brain. However, the enthusiasm of the adaptationist program has also led to some mistakes, a primary one being a tendency to focus too strongly on one, albeit often clever, adaptive hypothesis. Sometimes simpler adaptive hypotheses exist but are not considered, and sometimes nonadaptive processes can produce apparently adaptive outcomes. The critics of the adaptationist approach have not always gotten the critique correct, but they have reminded us that a mature science must be rigorous as well as inventive.

We end here by noting that behavioral ecology has been relentlessly integrative. From Parker's use of economic model building early on, to the use of molecular tools in the '90s, to the present-day search for insights through genetics, immunology, physiology, and neurobiology, behavioral ecologists have been willing to plunge into other fields to better understand how their organism works. This integrative spirit hopefully will also erode the divisions between those studying current utility and those interested in history. Natural selection and the evolutionary process is a wonderfully simple idea with fascinatingly complex effects. It is both a contemporary and historical process, one that will require careful integration of approaches to fully understand. But we think that a thorough understanding of the ecological problems faced by animals gained through the current utility approach blended with detailed knowledge of genetic and phylogenetic context of the organism is an exciting combination. The rest of this book hopefully will illustrate this and provide the tools to do both.

Acknowledgments We thank Geoff Parker, David Reznick, Mike Ryan, Mike Sharkey, and Paul Sherman for helpful comments on this manuscript; though they did not all agree with our perspective and conclusions, their comments helped us refine our ideas. We also thank the graduate students who participated in our adaptation seminar at the University of Kentucky for an exciting semester of discussion.

SUGGESTIONS FOR FURTHER READING

For a historical overview of the concept of adaptation, from pre-Darwinian ideas through to the modern controversies, see Amundson (1996). Williams (1966) provides a primer to the role of natural selection at the level of the individual in shaping the diversity of behaviors and life histories. The book is especially significant for its role in countering the group selectionist ideas prevalent at the time of its publication, but also provides an early review and critique of adaptation in evolutionary biology. Gould and Lewontin (1979) presented the most influential challenge to the adaptationist perspective; their paper largely presents ideas initially developed by Lewontin (1978). This critique generated a large diversity of responses, most notably by Reeve and Sherman (1993), who take an extreme opposing position. Also, two recent volumes are dedicated to the study of adaptation, including *Adaptation*, edited by Rose and Lauder (1996), which presents a diversity of perspectives and approaches, and *Adaptationism and Optimality*, edited Orzack and Sober (2001b), which focuses on the use of optimality modeling in the study of adaptation. For an entertaining discussion of the controversies and some interesting insights, see Queller (1995).

Amundson R (1996) Historical development of the concept of adaptation. Pp 11–53 in Rose MR & Lauder GV (eds) Adaptation. Academic Press, San Diego.

Gould SJ & Lewontin RC (1979) The spandrels of San Marco and the Panglossian paradigm: a critique of the adaptationist program. Proc R Soc Lond B 205: 581–598.

Lewontin RC (1978) Adaptation. Sci Am 239: 212–228.

Orzack SH & Sober E (2001b) Adaptation, phylogenetic inertia, and the method of controlled comparisons. Pp 45–63 in Orzack SH & Sober E (eds) Adaptationism and Optimality. Cambridge Univ Press, Cambridge, UK.

Queller DC (1995) The spaniels of St. Marx and the Panglossian paradox: a critique of a rhetorical programme. Q Rev Biol 70:485–489.

Reeve HK & Sherman PW (1993) Adaptation and the goals of evolutionary research. Q Rev Biol 68: 1–32.

Rose MR & Lauder GV (eds) (1996) Adaptation. Academic Press, San Diego.

Williams GC (1966) Adaptation and Natural Selection: A Critique of Some Current Evolutionary Thought. Princeton Univ Press, Princeton, NJ.

3

Behavioral Concepts of Selection

BARRY SINERVO AND RYAN CALSBEEK

> *It may metaphorically be said that natural selection is daily and hourly scrutinizing, throughout the world, the slightest variations; rejecting those that are bad, preserving and adding up all that are good; silently and insensibly working, whenever and wherever opportunity offers, at the improvement of each organic being in relation to its organic and inorganic conditions of life. We see nothing of these slow changes in progress, until the hand of time has marked the lapse of ages.*
>
> —Darwin 1859

Organisms change over time, and those alive now are different from those that existed in past epochs. Much of the life that once lived is now extinct. Change is occasionally documented in the fossil record. Although all organisms share common ancestry, they have diverged into separate species over time. For Darwin, evolution by natural selection was a gradual process, taking eons to effectuate significant change. This idea was propagated by the fact that nobody had witnessed evolutionary change (or at least they hadn't made note of it). Recent advances in our ability to measure selection (molecular tools and deep field pedigrees) have challenged historical perceptions of gradual change. It is now understood that selection can drive evolution quite rapidly or, alternatively, selection can hold species in stasis, at the fulcrum of opposing forces for long periods of time. We begin with a syllogism, which is a logical outcome that follows from exposition of premise(s). Darwin's syllogism, with its three premises on selection, is succinct:

1. Populations are composed of individuals with variable phenotypes.
2. Variation among individuals causes some to "do better" than others (survive, reproduce, etc.; this is referred to as variation in fitness related traits [chapter 4]).
3. If some of the variation among individuals has a genetic basis (i.e., heritability), then heritable traits that are in individuals with higher fitness will be overrepresented in the next generation and organisms will evolve by the process of natural selection (the outcome).

Phenotypic selection provides a useful starting point for understanding the process of natural selection as described by Darwin. To measure selection, it is important to understand the concepts of fitness (see chapter 4). It is also important to distinguish between phenotypic selection in one generation and the evolutionary response across generations. We can use simple equations that predict the response to selection (R) from a selection differential (s) and heritability (h^2):

$$R = h^2 s \quad (3.1)$$

Heritability (necessarily < 1; see chapter 5) places an upper limit on how much the selection in the current generation (s) will be translated into an evolutionary response to selection (R). The term s is one measure of selection, which is referred to as the univariate selection differential, and, in the case of simple survival selection, is derived by subtracting the mean of the phenotypic distribution after selection (i.e., the mean of those that survived and/or reproduced) from the mean before selection (i.e., the mean of all individuals in the population). In the general case s = Covariance {fitness, trait} (Lande & Arnold 1983). It is important to note that although this equation will work well to describe the evolution of the vast majority of traits, key processes of selection on behavior can critically influence the amount of selection (s) we can observe. There are many potential biases in the study of selection, some of which we discuss below. Nevertheless, a perspective of phenotypic selection is quite useful in elucidating the form of selection that shapes phenotypes per se, and the ecological causes of selection, prior to the heritable transmission into the next generation of the genes that are shaped by selection.

THREE FORMS OF UNIVARIATE SELECTION

Directional selection favors one tail of a phenotype distribution and acts to change the mean phenotype of a population. Directional selection could favor either end of the distribution (e.g., largest or smallest trait value). For example, competition for resources might favor the largest progeny that secure the most resources by virtue of their large size (figure 3.1a). After one round of selection, assuming that the largest progeny have high fitness relative to the rest of the population (and that progeny size is heritable), the offspring size distribution in the next generation will have a larger mean. Alternatively, we might observe no evolutionary change in body size if, for example, the trait does not have a genetic basis, or if selection on another trait opposes selection on body size. For example, an opposing force of directional selection during another phase of the life cycle may balance the force of directional selection for large size due to offspring survival. Selection favoring large offspring is in fact often balanced by the fecundity advantage of producing small offspring, which reflects a separate episode of directional selection (m_e, figure 3.1a). Moreover, selection need not arise from simple directional forces. For example, survival selection on size may include both directional and quadratic components (l_e, figure 3.1b). This quadratic component, most often referred to as *stabilizing selection*, optimizes a phenotypic distribution and acts to reduce the variance in phenotypes of a population. (i.e., selects against both large and small values of a trait). It is also referred to as *optimizing selection* because a unique optimum is created by selection that removes individuals from the tails of the frequency distribution.

Thus, although we expect directional selection to be common in nature, it may also act in tandem with stabilizing selection, or with directional selection in the opposite direction, to create more complex surfaces. Two opposing sources or periods of directional selection acting on a trait, or opposing selection on correlated traits (see below), can create stabilizing selection; this is often referred to as *balancing selection* because the quadratic shape of the selection surface is produced by a balance between different components or periods of selection. Opposing sources of selection on single traits, or on correlated traits, are probably the norm for behavioral and life history traits, and are the conceptual basis of optimality models commonly used in behavioral ecology. In our example, fecundity selection is directional but opposes survival selection, and we can simply multiply the functions that describe selection in these different episodes (Arnold & Wade 1984) to obtain the total selection on a given trait ($m_e \times l_e$, figure 3.1a, b). Thus, stabilizing or optimizing selection might actually arise from the forces of two directional surfaces that act in opposition, or at different times in the life cycle of an organism. This form of selection is common for behavioral traits, and consideration of opposing selection is at the foundation of optimality models and many other models of behavioral evolution.

This simple view of selection across the entire life cycle is made more complex by changes in the patterns of selection in the evolutionary long run (e.g., measured across many generations). In many natural systems, selection can oscillate over time as a function of something quite simple, like changes in the prevailing weather conditions (Grant & Grant 2002). The sign of directional selection might also change over time owing to fluctuating social conditions such as changes in local population density. For example, a flip-flop in the form of selection that is tied to density regulation is observed in the side-blotched lizard, *Uta stansburiana*, in which selection favors large progeny in high-density environments, but small progeny in low-density environments (figure 3.1c, d).

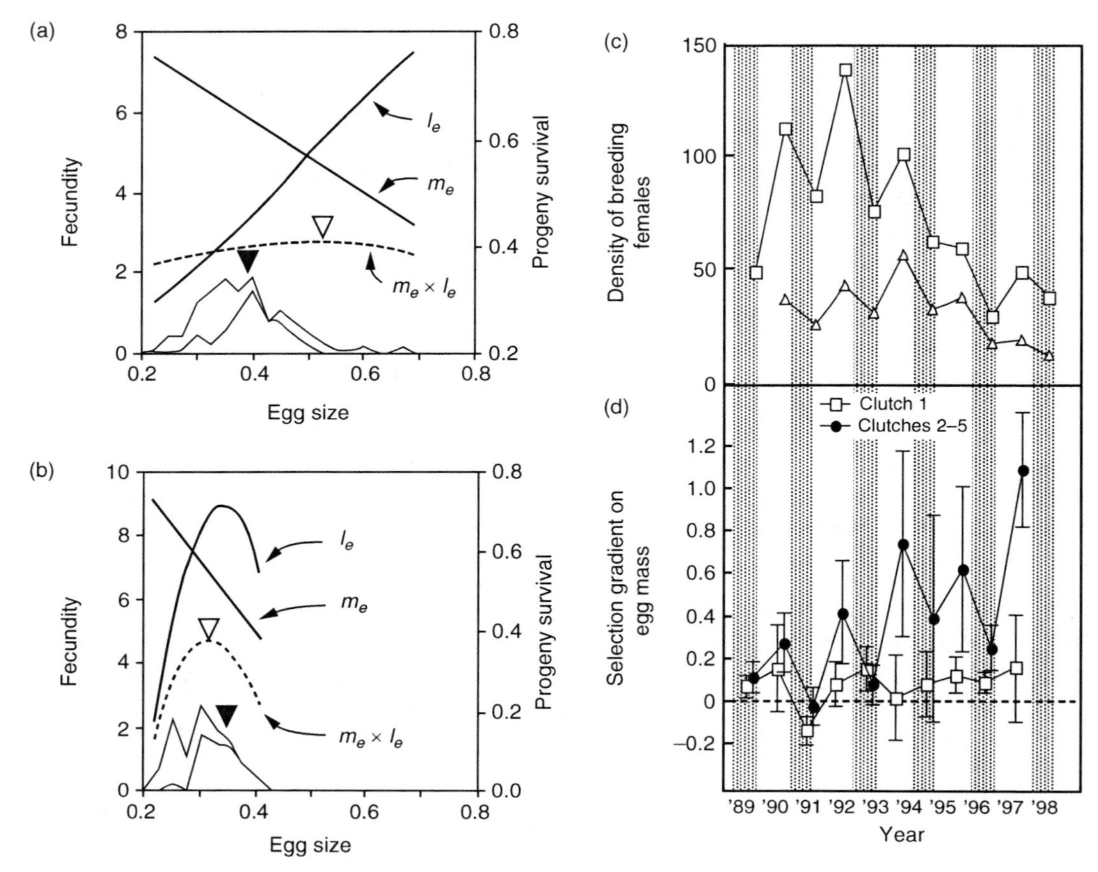

FIGURE 3.1 Experiments are always required to determine which traits are under selection. Sinervo et al. (1992) resolved selection on progeny survival by comparing survival of miniature progeny produced by removing yolk from eggs, giant progeny produced by ablating follicles on the ovary (which results in fewer giant eggs), and control progeny. (a) Selection on egg mass results from differences in strength of directional selection (l_e) on first and later clutch progeny (*a* versus *b*), balanced against directional selection for higher fecundity in the female parent (m_e) as a function of the trade-off between offspring size and number. Opposing selection surfaces generate an optimal egg size ($m_e \times l_e$) that varies with clutch. Stronger selection on later-clutch progeny causes females to lay larger eggs later in the season, a plastic adaptive response. (c, d) The strength of selection on progeny survival (l_e) expressed as selection gradients oscillates according to a density cycle. Redrawn from Sinervo et al. (1992, 2000).

The take-home message from this example is that directional selection for one component of fitness (e.g., offspring size) may be common in nature, but sustained selection in one direction is unlikely (Grant & Grant 2002). Directional selection is likely to change over time as the social context changes (Sinervo et al. 2000). Moreover, one episode of selection (offspring survival) may generate antagonistic selection on another component (fecundity) that is linked to the first by genetic mechanisms that are shared in common. To better predict the evolutionary response to all of these variable selection pressures, it is desirable to assess genetic correlations among traits. Traits covary because they are often affected by the same genes (i.e., due to pleiotropy). Again, using our example, egg size is heritable ($h^2 = 0.68$), but selection on egg size is made somewhat more complicated by its genetic correlation with another trait, clutch size ($r_G = -0.95$; Sinervo 2000). Producing large

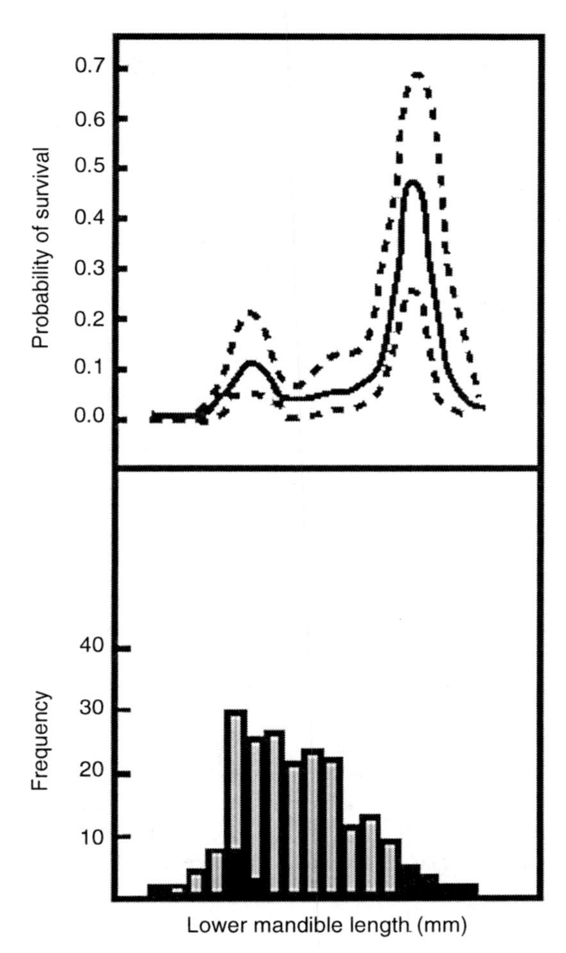

FIGURE 3.2 Disruptive selection on bill size of African seedcrackers, *Pyrenestes ostrinus*, generates two optima that are each centered on small and large bill morphs, respectively. Two fitness optima correspond to large morphs feeding on large sedge seeds and small morphs feeding on small sedge seeds. Birds with intermediate bill size and performance have poor survival. Redrawn from Smith (1993).

offspring limits the total number of offspring that can be produced (line m_e, figure 3.1b), but large offspring generally have better survival (line l_e).

In contrast to stabilizing selection, *disruptive selection* favors both tails of the phenotype distribution and selects against intermediates. Disruptive selection therefore increases the variance in phenotypes. In seedcrackers (*Pyrenestes ostrinus*), a small estrilid finch in equatorial Africa, variation in seed size leads to disruptive selection on bill dimensions (figure 3.2). This disruptive selection on bill dimensions arises from a difference in the optimal bill size in a foraging context. Tom Smith (1993) has shown that large seeds are most easily handled by large billed birds and vice versa for small seeds. In this example, we expect to see disruptive selection on both foraging behavior and morphology, but overall stabilizing selection surrounding each of the two bill types. Birds with intermediate-sized bills consume far fewer seeds than either large- or small-billed birds and die as a result (figure 3.2). As in all cases of phenotypic selection surfaces, it is useful to consider the genetic basis of the traits under selection. In this case, at least some of the variation in beak size arises from a simple Mendelian locus (Smith 1993), and is likely to be controlled, at least in part, by the gene Bmp4 (Abzhanov et al. 2004; Sinervo 2005). An understanding of the genes that underlie the trait(s) is once again very useful for understanding the outcome of selection.

EXPERIMENTAL MANIPULATION OF THE TRAIT(S) UNDER SELECTION

In the above example, one could manipulate beak shape. For example, experimental manipulations of beak morphology in crossbills, involving gradual reduction in beak curvature, have revealed the adaptive nature of beak size and shape for opening pine cones (Benkman & Lindholm 1991). Such experiments can help to resolve phenotypic traits that are the targets of selection (Sinervo & Basolo 1996). One of the first experiments to manipulate a phenotype and measure directional selection on a sexual ornament was Andersson's (1982) manipulation of tail length in widowbirds. Andersson elegantly measured the reproductive benefits of long tails by simply gluing additional feathers onto males in one experimental group. Widowbirds with experimentally lengthened tails obtained far more mates compared to control males that received a simple cut and reattachment of their own tail feathers. Experiments such as this not only help us identify causality, they can also inflate the variation on the ends of the frequency distribution, greatly enhancing our power to detect selection (Sinervo & Basolo 1996). This is important because if variation is the raw material for selection, in the absence of sufficient phenotypic variation, selection would be otherwise impossible to detect.

MULTIVARIATE SELECTION AND THE CORRELATIONAL SELECTION SURFACE

Each mode of selection discussed so far alters the mean and/or variance of the phenotypic trait in a population or species (Brodie et al. 1995). In the long term, directional and disruptive selection can have a dramatic impact on the evolution of a species, leading to large changes in the distribution of traits in a population, and even the evolution of new phenotypes. This contrasts with the action of stabilizing selection, which maintains and refines existing types without changing the mean over long periods of time (stasis). The examples of selection on foraging efficiency and selection on bill morphology in finches (figure 3.2) illustrate that selection will often operate on more than one trait at the same time. Although we discussed bill size in that example, selection can simultaneously affect different features of the bill, such as depth, width, and thickness, as well as traits correlated with bill size, such as skull musculature and body size. If these traits are truly independent and natural selection on one trait does not affect selection on another trait, we can use the breeders' equation (equation 3.1), described above, to quantify the effect of directional selection on the mean phenotype, and we can use similar univariate models to explain how variance in the phenotypic distribution will change due to selection.

However, when selection on one trait simultaneously affects other traits, univariate models of selection are inadequate. We can describe the simultaneous effects of selection on multiple traits using parametric selection coefficients analogous to the selection differential discussed above, but in a multivariate model. The *selection gradient* is the multivariate analogue of the univariate selection differential discussed above. The selection gradient is derived from the multiple regression of fitness (e.g., survival or reproduction; chapter 4) on linear (e.g., directional) and quadratic terms (e.g., stabilizing or disruptive) for two or more traits, as well as the potential interactions between traits that are reflected in a cross-product term (Lande & Arnold 1983). For example, correlational selection is a multivariate form of selection that can be visualized as a mixture of directional and quadratic forms of selection acting simultaneously on pairs of traits. Correlational selection favors successful combinations of phenotypic traits and builds genetic correlations over evolutionary time (Cheverud 1984a).

Natural selection on a single trait can simultaneously influence other traits because multiple traits are often genetically correlated. For example, if multiple traits are affected by the same genes (pleiotropy), selection will be unable to affect single traits without affecting the other traits affected by those genes (Lande 1980). Similarly, genes that are physically linked, such as those adjacent to one another on a chromosome, can behave as if they are one gene. Both of these genetic mechanisms cause traits to be genetically correlated. The generation of linkage disequilibrium via correlational selection is of particular significance in some areas of behavioral ecology (e.g., sexual selection). *Linkage equilibrium* is the multilocus version of Hardy-Weinberg equilibrium. If two loci are in linkage equilibrium, the multilocus genotypes are distributed among individuals in proportions that are expected by chance. If two loci are in linkage disequilibrium, then some multilocus genotypes are overrepresented whereas other combinations are underrepresented (Lynch & Walsh 1998). For example, the process of sexual selection is thought to generate a nonrandom association between female preference alleles and the alleles that code for male display traits (chapter 24). Other forms of social selection that act on a signaler and receiver can also generate stable linkage disequilibrium, even among unlinked genes (Sinervo & Clobert 2003, chapter 14). Correlational selection acting on two or more traits (and unlinked loci that affect those traits) that are functionally related can also produce suites of integrated traits (see below; Brodie 1992). However, although pleiotropy forms a stable coupling between traits, a genetic correlation that is due to linkage between loci will eventually decay through the normal process of segregation (if those genes reside on different chromosomes) or recombination (if they are physically linked; Lynch & Walsh 1998). Thus, continuing selection is necessary to maintain linkage disequilibrium.

Below, we discuss correlational selection in terms of its significance to behavior. The actual estimation of the selection gradients is well reviewed by Blows et al. (2003) and Lande and Arnold (1983). These gradients can be estimated using traditional regression techniques (i.e., multiple regression). However, selection gradients are now commonly estimated using path analysis, a related regression technique. Path analysis, where interactions can arise from a chain of causation, is an alternative to simple multiple linear regression. Selection acts directly on a given trait (e.g., the trait itself influences fitness), or selection can act indirectly through an intermediate mechanism (e.g., the trait influences another trait, which in turn influences fitness). Path analysis is a useful tool whose full exposition is beyond the scope of this chapter. However, Shipley (2000) provides a good summary of the technique of path analysis and we refer interested readers there for further reading.

SELECTION ON HIGHER ORDER TRAITS AND FOR WHOLE ORGANISM INTEGRATION

Traditional studies of natural selection have focused on the ways that simple phenotypic traits influence fitness. However, it is important to realize that on some level, fitness is always determined by interactions among several traits. In other words, it is not always adequate to simply consider how selection on one trait has correlated effects on other traits. We also need to consider how trait combinations interact to affect overall fitness. We thus need to look at traits that represent the culmination of many interacting physiological and behavioral processes. Arnold (1983) recognized this early on and called for studies that linked differences in morphology to changes in performance and fitness; we cannot consider behavioral traits in isolation from the other morphological and physiological traits that make up what is referred to as the *whole organism*. Only recently have studies begun to estimate the forms of natural selection that act on higher order traits like running performance (endurance and sprinting speed), immune function (immunocompetence), and social and sexual behaviors (territoriality, mate choice preferences, and social organization). We predict that estimates of selection on these higher order composite traits will reveal how sexually or socially selected signals (behaviors) are functionally integrated with performance related traits.

One recent example of selection for suites of traits and whole-organism integration comes from work on *Anolis* lizards on islands in the Bahamas. More than 140 different species of anole have arisen in an adaptive radiation on Caribbean islands (Williams 1983). *Anolis* lizards have diversified primarily through variation in limb length and body size, traits that are correlated with habitat use (Losos 1990, 1998). Morphologically similar species, which have been termed *ecomorphs* (Williams 1983), have evolved repeatedly and independently on different Caribbean Islands (Losos et al. 1998). Anole ecomorphs with relatively long limbs are usually found perching on broad diameter surfaces like tree trunks, whereas ecomorphs with relatively shorter limbs are found on narrower diameter substrates like

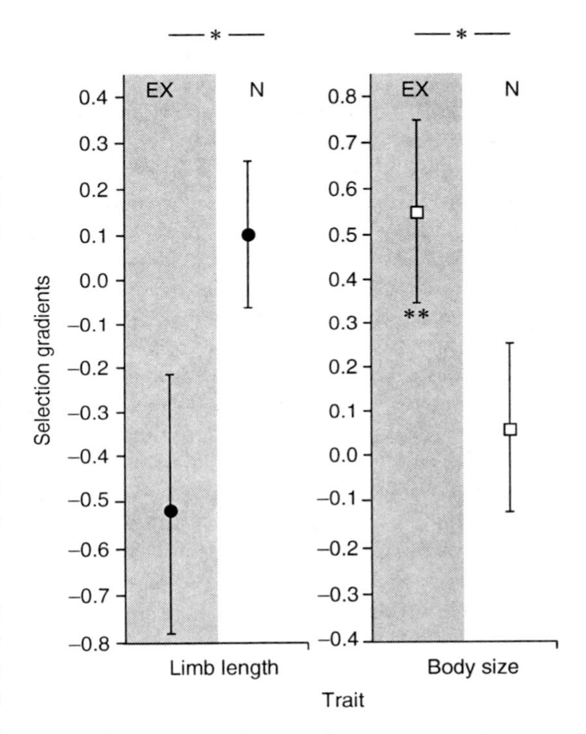

FIGURE **3.3** Drought has dramatic impacts on vegetation use by lizards, and the magnitude of both correlational selection gradients and linear selection gradients on body size. During a naturally wet year, we simulated drought effects (gray columns) on the habitat by trimming vegetation on small offshore islands. Changes in selection acting on body size were similar to those observed in naturally wet and dry years (*EX* and *N* refer to experimental and natural years, respectively). Following experimental removal of vegetation, natural selection became strong and directional, favoring larger male body size (**significant selection gradient). No significant selection on limb length was detected. However, comparison of slopes revealed significant differences between natural and experimental sites (indicated by *). Redrawn from Calsbeek et al. (2009).

twigs and bushes. Correlations between morphology and behavior (habitat use) are thought to have arisen through correlational selection favoring fast moving, long-limbed lizards in open habitats, and slow-moving short-limbed lizards in scrubby habitats. Recent work (Calsbeek & Irschick 2007, 2008) suggests that habitat preference is under strong selection along with running performance and limb morphology. Though it remains unclear whether habitat use is the direct results of preference for one perch type or the other, correlational selection can act on both habitat use and morphology, favoring lizards whose morphology "fits" the habitat type in which they reside (Calsbeek & Irschick 2007). Experimental manipulations of vegetation, the ecological variable that seems to underlie much of the natural selection on anoline performance traits, elucidate the dynamic nature of habitat use on selection (Calsbeek et al. 2009; figure 3.3). For example, narrow diameter vegetation like grasses and small shrubs are much more heavily impacted by drought conditions than are large, mature trees. Thus, during dry years, directional selection seems to favor lizards that perch on broad diameter substrates, whereas in wet years, disruptive selection reveals fitness optima on both broad and narrow surfaces. In a system in which frequent environmental disturbance (e.g., hurricane activity) can alter the landscape overnight, it is not surprising that the shape and magnitude of selection varies from year to year depending on weather conditions.

FREQUENCY-DEPENDENT SELECTION AND SOCIAL INTERACTIONS

Before examining some case studies, it is important to make a final point regarding selection acting in behavioral and social contexts. In these special cases, the traits that are under selection are behaviors, which govern interactions between individuals. We begin with two additional syllogisms (Sinervo et al. 2008). The first defines how selection on traits arises from correlational selection on behaviors expressed in two or more individuals. The second syllogism defines how selection on traits actually depends on the frequency with which a behavior is expressed in a population. To understand these perspectives, we must develop a view of how behavioral interactions arise. The theory of communication gives us this perspective. Sender and receiver coevolution is the basis of communication.

A Syllogism on Correlational Selection and the Sender-Receiver Traits of Communication

Communication is defined as a behavior (e.g., a signal) by a *sender* that impacts the behavior of a *receiver* (which could also include the self). At least some traits (and their underlying genetic loci) used in signaling will differ from the traits (loci) used in signal reception (i.e., the underlying genes are not pleiotropic in effect). Moreover, the fitness effects of behaviors that are elicited in receivers through communication are likely to arise through more than just the traits (loci) involved in communication. Social interactions generate three levels of correlational selection (Sinervo et al. 2008): (1) within sender correlational selection couples signal traits to other traits (and loci) that then either enhance or diminish the activity of signals, (2) within receiver correlational selection between traits for signal reception and the traits or loci that are invoked as a consequence of signal reception, and (3) correlational selection between sender and receiver involves their different traits/loci and generates a coevolutionary outcome (i.e., social selection; see box 14.1). The signature of these interactions is the nonrandom association of alleles (linkage disequilibrium) within individuals, or linkage disequilibrium that is measured between individuals (always relative to the frequency of alleles in the total population) during the very behavioral interactions that generate selection. For example, as noted above, the genes underlying a sexually selected male trait (male = sender) may find themselves in linkage disequilibrium with the genes governing expression of a female preference for that trait (female = receiver).

The critical components of selection on behavioral traits are multivariate and therefore generate correlational selection. We often assume that female preferences generate directional selection on male ornaments. This is true for many species; however, there are examples of more complex forms of correlational selection. For example, female preference for male colors in the guppy, *Poecilia reticulata*, generates correlational selection for three distinct signal optima (figure 3.4; Blows et al. 2003), rather than simple directional selection. Male attractiveness appears to depend on combinations of size and color traits. In other cases of selection on social interactions, male competition and/or mate preferences generate correlational selection on signals together with other morphological traits. For example, correlational selection acting on male *Junco hyemalis* generates functional integration between plumage color and male size (McGlothlin et al. 2005; figure 3.5). Functional integration is the process by which allele combinations (at many loci) are favored by selection, relative to other maladaptive combinations, owing to correlational selection. Chapter 24 explores ideas regarding female preference for good genes that are generally related to this concept of functional integration. Females may often choose functionally integrated males on the basis of multivariate combinations of traits. Likewise, male-male competition may generate functional integration, particularly in the case of alternative strategies (Sinervo et al. 2008; Miles et al. 2007). Alternative strategies may result from correlational selection, but their fitness gains often depend on the social context in which they are found. We next explore how this social context can lead to frequency dependent selection.

Syllogism on Why Correlational Selection (of Biotic Origin) Is Frequency Dependent

Selection on many behavioral traits is frequency dependent; in other words, the fitness consequence of the behavior depends on the frequency of its expression in the population. Traits experiencing positive frequency dependent selection (positive FDS) are those for which fitness increases with frequency, whereas traits experiencing negative frequency dependent selection (negative FDS) are those for which fitness declines with frequency. Selection involving signalers and receivers will necessarily be frequency dependent. This is because of the two fundamental types of interactions that are possible in signaler-receiver communication, honest and dishonest signaling (Sinervo & Clobert 2008; Sinervo et al. 2008), and because adaptive responses by intended receivers and unintended receivers are themselves frequency dependent (Maynard Smith 1982; Rowell et al. 2006). In the case of the honest signaling interactions of cooperation (chapter 18), behaviors promote positive FDS, in which the fitness of a genotype increases with its frequency. A greater frequency of the honest and cooperative signaling types fuels a positive feedback process that enhances the underlying genetic correlations between signaler and receiver loci. This in turn leads to higher frequencies of cooperation (Sinervo & Clobert 2003; Sinervo et al. 2006). Cooperators benefit from interactions with other cooperators, and any signal that enhances the frequency of this interaction is favored. By contrast, dishonest signaling relationships promote negative FDS, in which rare forms have an advantage, but become self-limiting when common. This is because dishonest strategies feed off of the success of an honest signaling system, and once dishonesty becomes too common it has a self-poisoning effect on its own strategy. An honest signaling strategy can also be governed by negative FDS. An example of self-limitation is easy to observe in mating systems in which despotic male types, which signal with an honest badge of status, have high fitness when rare. When common, despots are self-limiting owing to the negative effects on survival of despots on other despots. In social interactions, positive frequency-dependent selection generally leads to fixation of a single allele and or genotype (in the multilocus case). In contrast, negative FDS can either yield a single evolutionary equilibrium or may generate chronic cycles, in which many types are present within the same population (Sinervo & Calsbeek 2006). In such evolutionarily stable social systems (Sinervo et al. 2007) correlational selection is chronic even among unlinked loci and correlational selection generates stable linkage disequilibrium between signaler-receiver loci (Sinervo & Clobert 2008).

Social systems with cooperation should generate positive FDS, but cooperation remains vulnerable to invasion by cheaters (Trivers 1971). For example, mating reflects evolutionary cooperation in which each sex divides its genome and passes on half to their progeny. However, sex is invadable by selfish genes, such as occurs in hybridogenetic mating systems in which one genome ejects the partner's genome from zygotes: female genome ejects male (fishes, insects, amphibians; Normark 2003; Simon et al. 2003), or male genome ejects female (ants; Fournier et al. 2005). Mating cooperation is also invadable by selfish genes through genomic imprinting in which resources are secured for some progeny at the expense of others. Behavioral situations that generate positive FDS and correlational selection such as biparental care (care-related traits of two parents) are extremely vulnerable to invasion by noncare strategies.

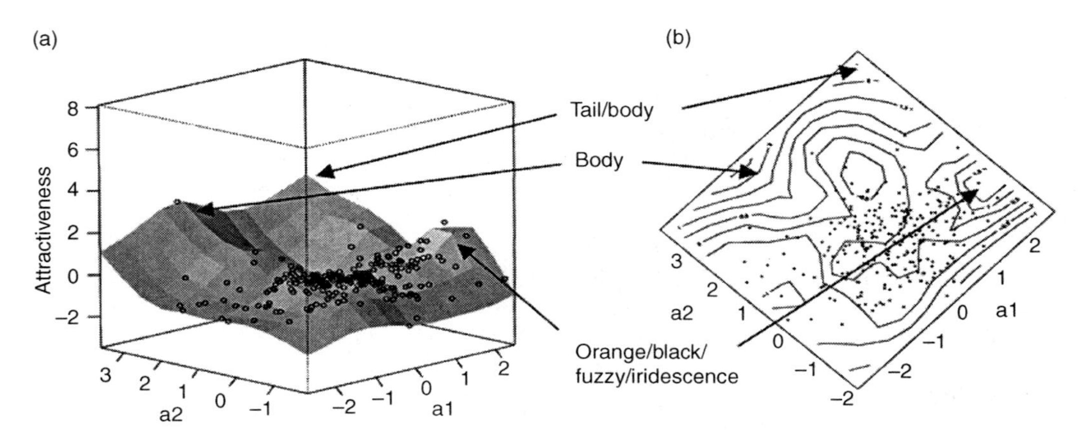

FIGURE 3.4 Correlational selection driven by female preference for male guppy, *Poecilia reticulata*; color favors three distinct peaks of highly preferred males. Reprinted from Blows et al. (2003) with permission.

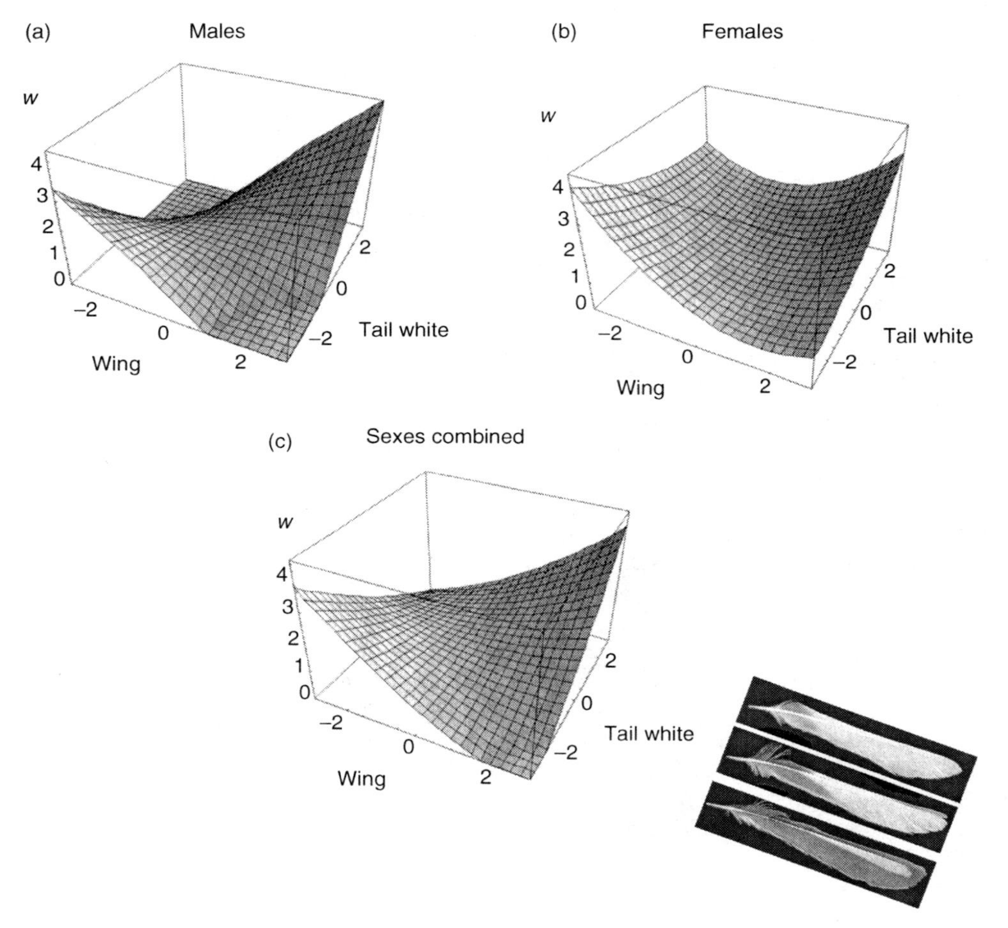

FIGURE 3.5 Correlational selection on size (i.e., wing size) and tail white variation (photos) in dark-eyed juncos, *Junco hyemalis*. Two prominent peaks of high fitness are found in two opposite corners, whereas low fitness troughs are found in the other two corners. Reprinted from McGlothlin et al. (2005) with permission.

Multivariate Analysis of Cooperation

Although the number of estimates of correlational selection within individuals is accumulating (Kingsolver et al. 2001; Sinervo & Svensson 2002), to date only a single estimate of a between-individual correlational selection gradient has been made. This is for the propensity of *Uta* males to settle next to genetically similar partners and cooperate (Sinervo & Clobert 2003; figure 3.6). Current theory typically ignores the types of correlational selection that we have described above, and their necessary coupling through FDS. However, most animals are social at least to some degree, suggesting that this pattern may be more general than currently understood. For example, many higher plants (angiosperms) are social via coevolutionary relations with pollinators. Here, we define such correlational selection in terms of traits (in this case, alleles) favored in social actors. Social interactions can generate symmetry selection (e.g., similar traits among actors such as cooperators, or between a mimic and its model), asymmetry selection (e.g., dissimilarity in traits such as resource holding potential

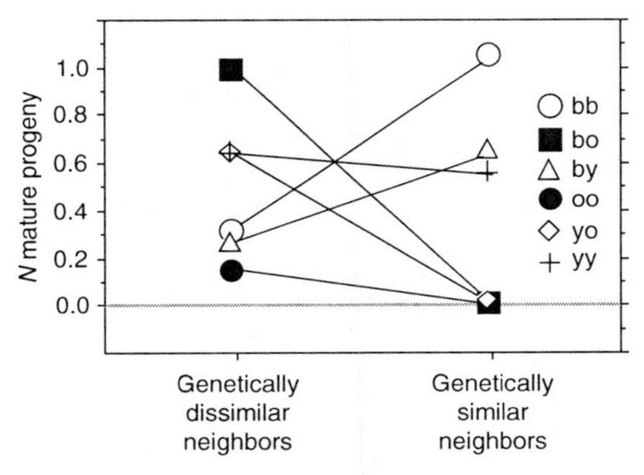

FIGURE **3.6** Correlational selection on the OBY locus with settlement beside genetically similar *Uta stansburiana* males, an example of a between-individual correlational selection gradient. Symbols indicate different color strategies (*O, B and Y*). Cooperation among genetically similar (but unrelated) *b* males (e.g., *bb* genotype) is favored: cooperation increases number of progeny recruits at maturity. The reverse is true of orange males (*oo*, *bo*, *yo*; from Sinervo 2003). Genetic similarity had no impact on fitness of yellow males (*by*, *yy*). Redrawn from Sinervo and Clobert (2003).

and badges of status) or complementarity selection (e.g., between MHC loci in mates that are favored by balancing selection [Wedekind & Füri 1997], or between pollinators and their preferred floral traits). A fruitful next phase of research in this area would be to identify ways in which correlational selection shapes traits between social actors (Sinervo & Calsbeek 2006) using experiments on FDS.

Much theory assumes that selection resides at the individual level, but in the case of communication, selection is actually a property of both inter- and intra-individual selection. Only in the case of sexual selection (chapter 24), a special form of sender-receiver communication between mating partners, has this coupling been explored in any detail. However, mate choice is really just a special form of correlational selection (Sinervo & Calsbeek 2006). In the example of Fisherian runaway described earlier in this chapter, sexual selection generates positive FDS, which fuels the conditions necessary for genetic correlations to build between the signal in one sex (often male) and the loci for preference in the other (often female).

Negative Frequency Dependent Selection and Apostatic Selection

Relations between the sexes are not invariably governed by positive FDS, and, in many cases, they involve strong negative FDS. This situation often arises in the context of antagonistic sexual selection (Rowe et al. 1994; Arnqvist & Rowe 2002a; chapter 23). For example, some damselflies have evolved a novel androchrome female morph to resemble males, and thereby elude sexual harassment. Two other female forms avoid detection via apostatic (i.e., crypsis) selection, a form of selection that gives the rare type an advantage over common types that are encountered by harassing males at higher rate. Thus, trimorphisms comprising three female types are common in damselflies of Europe and North America (Svensson et al. 2005). Female trimorphism is thought to arise from search image formation in males for common female morphs (Fincke 2004). In this case, selection imposed by male harassment behavior has generated diversifying FDS for three alternative female types.

Apostatic selection may also be common in predator-prey behaviors and their interactions. Social contexts with negative FDS generate correlational selection among alternative dominant male strategies and cryptic male strategies, as well

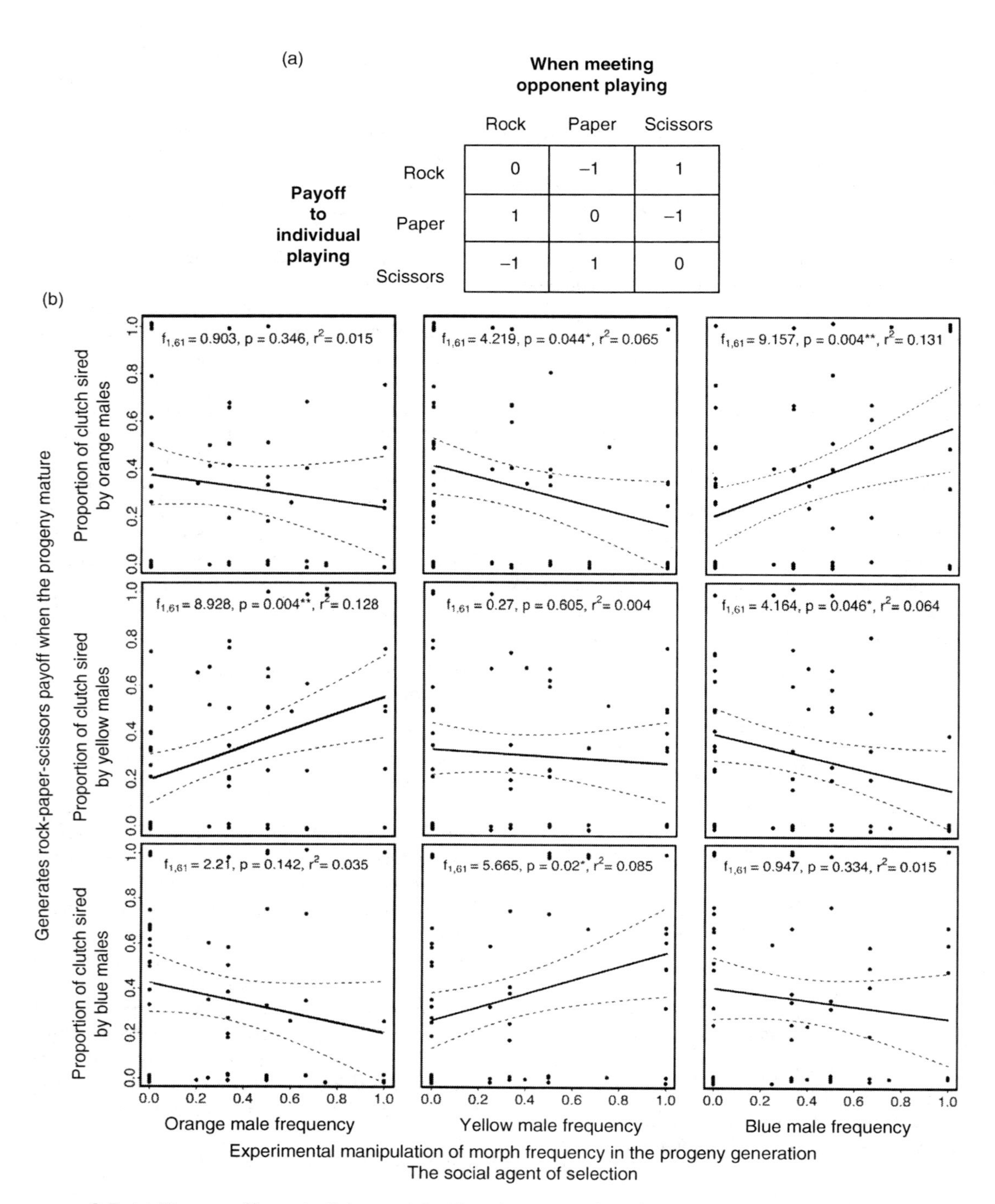

FIGURE **3.7** (a) The payoff matrix (Maynard Smith 1982) is a way to describe success or payoffs of a given strategy. The payoffs for lizards in competition with other morphs (e.g., the slope of the line) resemble the theoretical payoffs of a rock-paper-scissors game. (b) Regression plots of siring success of a male versus the frequency of each morph in a given male's local territorial neighborhood is used to characterize frequency dependent selection.

as between behaviors of a predator and its alternatively patterned and/or cryptic prey (Kokko et al. 2003a; Sinervo & Calsbeek 2006).

Negative Frequency Dependent Selection and the Rock-Paper-Scissors Game

An example of negative FDS is found in the mating system of side-blotched lizards, an evolutionary version of the familiar child's game rock-paper-scissors (RPS). The RPS game (figure 3.7a) has a simple dynamic that fits evolutionary game theory. A payoff matrix (Maynard Smith 1982; chapters 8 and 15 in this volume) expresses the fitness of a given strategy when it encounters either others playing the same strategy or another strategy. In the RPS, two opponents playing the same strategy (rock, paper, or scissors) tie, but rock loses to paper and beats scissors, paper beats rock and loses to scissors, and scissors loses to rock and beats paper. The solution to this game requires estimating total fitness, which depends entirely on the frequencies of the strategies (figure 3.7b). When rock is rare and scissors is common, then rock has higher fitness because it usually meets scissors. When paper is rare and rock is common, then paper has highest fitness, and when scissors is rare and paper is common, scissors has highest fitness. Male strategies in side-blotched lizards follow this pattern, in that polygynous males (orange throat color), mate guarders (blue color), and sneakers (yellow color) coexist in the same population. Frequency manipulations of alternative strategies at birth have been used to experimentally demonstrate the rare type advantage (Bleay et al. 2007; figure 3.7b). The slope of the relationships between paternity success and morph frequency in a given male's neighborhood varies dramatically depending on the type of competitors present in the focal male's vicinity. The slope of the line indicates whether a male type accrues higher payoffs (positive slope) as the frequency of a competitor increases, or conversely whether the male actually suffers lower payoffs (negative slope). The form of these frequency dependent payoffs closely matches the structure of the payoff matrix for the rock-paper-scissors (RPS) game and the strong effect of frequency on fitness that occurs in that game. In color morphs of side-blotched lizards, the slope of male siring success varies as a function of morph frequency on manipulated plots (points; figure 3.7b, from Bleay et al. 2007). This variation in slope reflects the fact that male siring success is frequency dependent, and each male achieves its highest fitness when rare, but only against a single strategy. Notice that the lines along the diagonal are not significant (e.g., close to zero slope, figure 3.7b), which matches the pattern in the diagonal elements of the payoff matrix (figure 3.7a). The off-diagonal elements tend to be positive and negative slopes. For example, blue males gain fitness as yellow frequency increases (central column, bottom row), whereas orange males lose fitness to yellow males (central column, top row), and yellow males do not gain paternity nor do they lose much paternity to yellow males (center diagonal element).

THE ANALYSIS OF GENOTYPIC SELECTION

Many situations arise in which selection may not be able to act on traits themselves, because relevant traits are purged from the population before they can ever be expressed. This may arise, for example, if trait expression is delayed until sexual maturity. For example, hormones affect a wide diversity of traits besides just the behaviors that may be of interest in a particular study. Thus, individuals may be selectively eliminated from a population at the onset of sexual maturity, when the reproductive endocrine system is activated for the first time, due to selection on other consequences of hormone production, but before a measurable phenotype (e.g., the behavior of interest) is even expressed. Ontogenetic shifts in the action of selection make it difficult to measure such effects with phenotypic selection. Keep in mind that genetically based traits that elevate mortality risk to progeny during maturation will likely be inherited from both the father and the mother. In addition, many traits have a sex-limited expression, as in the case of clutch/brood size produced by reproductive females. For example, although male progeny do not produce a clutch of eggs, they nevertheless produce the same hormones that control clutch size in females. Follicle-stimulating hormone and a related gonadotropin, luteinizing hormone (Sinervo 1999), are critical in males to initiate sperm production at maturity (Mills et al. 2008). Moreover, these hormones govern the expression of many sexually selected traits like color and stamina, and they even impact the immune system. Though males carry genes for the hormone, it may be impossible to measure selection on hormone levels per se. Drawing blood may reduce survival or alter behavior. Males may also die as their endocrine system is ramped up for the first time, as they begin to express secondary sexual traits at maturity.

To circumvent these problems with phenotypic selection analysis, animal models have been developed to estimate genotypic selection. The first step in an analysis of genotypic selection is to develop a pedigree with traits and fitness measured in both sexes. For example, Sinervo and McAdam (2008) estimated the genetic bases of clutch size variation in side-blotched lizards using maximum likelihood *animal models* (Kruuk et al. 2000) applied to 13 generations of data. This allowed *breeding values* (box 5.1) of clutch size to be generated for progeny regardless of whether or not the progeny survived long enough to lay a clutch and express that trait (i.e., female progeny), or if they expressed the trait clutch size at all (i.e., male progeny). Breeding values reflect trait values an individual would have expressed (such as if a male had been a female) and thus reflect the additive genetic effect of genes (figure 3.8). Selection on the clutch-size breeding values of males opposed the selection pressures acting on females (figure 3.8). This reflects antagonistic selection on genes that regulate clutch size. To survive well during this maturation period, females must carry genes for large clutches. Conversely, males that express genes for large clutch size (and thus elevated profiles in sexually selected traits; Mills et al. 2008) survive poorly. Genetic trade-offs related to the functional design of the sexes, or

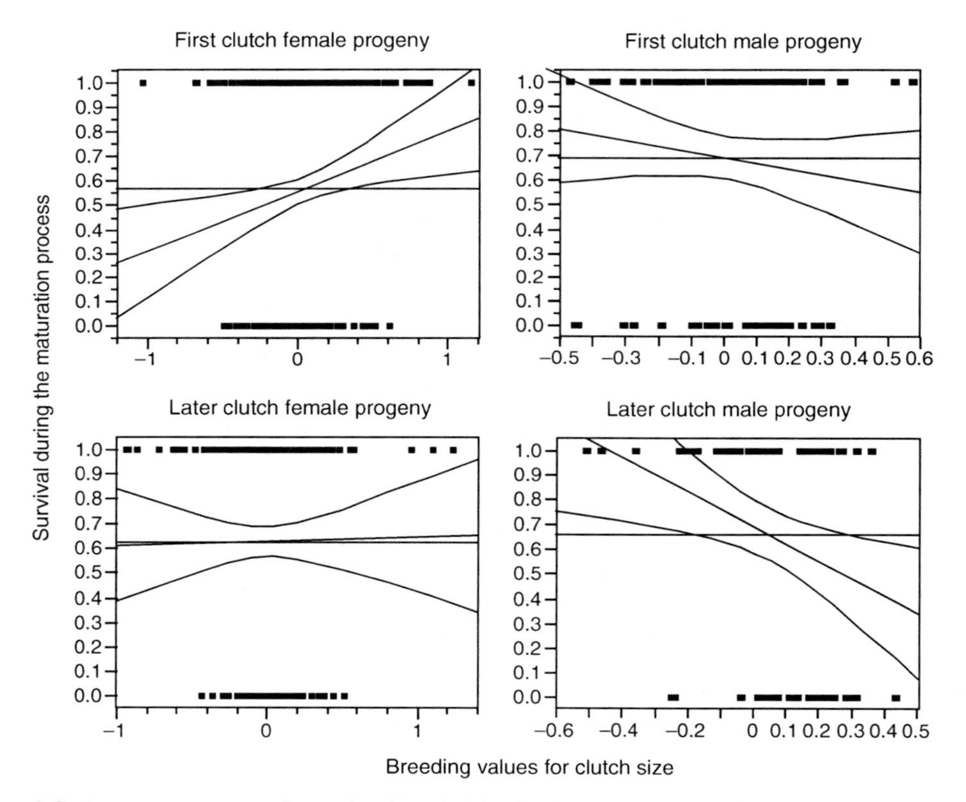

FIGURE 3.8 A 13-generation pedigree for the side-blotched lizard was used to resolve selection on the genes for clutch size in both sexes, as animals matured and expressed these genes for the first time. The pedigree consisted of a total of 7,247 individuals born between 1989 and 2002. Pedigrees in nature must be large to estimate breeding values and selection on breeding values. The pedigree resolved heritable variation in clutch size ($h^2 = 0.25 \pm 0.04$), which affected male and female progeny differently ($P = 0.01$, ANCOVA for selection differentials by sex, the effect of clutch was marginally significant, $P = 0.06$). Selection favored survival of females expressing genes for large clutches, but favored males that expressed genes for small clutches, even though males do not produce a clutch. Redrawn from Sinervo and McAdam (2008).

between juvenile and adult stages, are referred to as intralocus sexual conflict (chapter 23). In the example of selection on clutch size genes (figure 3.8), there are not only differences between females (left column) and males (right column) in the patterns of selection, but the progeny hatching on the first (top row) versus later clutches (bottom row) experience different patterns of selection. Progeny that hatched from later clutches must be carrying genes for small clutches to survive during the maturation period, relative to progeny that hatched from the first clutch. The example of clutch size genes has conflict between the sexes and conflict between different phases of the juvenile life history (born late versus early in the reproductive season).

Female progeny had not yet laid their clutch before they were selectively eliminated from the population. Moreover, male progeny appear to use the same genes involved in determining a female's clutch size, for other functions that elevate their risk of mortality during maturation such as the production of sexually selected traits (Mills et al. 2008). Such cryptic natural selection may be common if physiology or morphology changes with time and these ontogenetic changes also elevate the selective risk of mortality. Current models in which phenotypic selection is partitioned from genetic variance and covariance (Lande & Arnold 1983; Arnold & Wade 1984) may thus be inadequate for detecting selection (Endler 1986). Genotypic selection analysis (Schluter et al. 1991; Rausher 1992; Kruuk et al. 2000; Kokko 2001) circumvents many pitfalls inherent in phenotypic selection analysis and can reveal ontogenetic conflict. Studies on ontogenetic conflict can also reveal specific loci or traits under selection: clutch size (Sinervo & McAdam 2008), male body size (Calsbeek & Sinervo 2004), and dorsal pattern (Forsman & Appelqvist 1995; Lancaster et al. 2007).

SUGGESTIONS FOR FURTHER READING

For students interested in a deeper understanding of the processes that generate selection we recommend reading Lande and Arnold (1983) along with Phillips and Arnold (1989) for a more in-depth look at correlational selection. Readers interested in path analysis would benefit greatly from reading Shipley (2000). We recommend John Maynard Smith's (1982) book for those interested in game theory and the nature of frequency dependent selection. Finally, anyone interested in the evolution of cooperative behavior should investigate the important body of work by W. D. Hamilton (e.g., Hamilton 1964).

Hamilton WD (1964) The genetical evolution of social behavior. I. J Theor Biol 7: 1–16.
Lande R & Arnold SJ (1983) The measurement of selection on correlated characters. Evolution 37: 1210–1226.
Maynard Smith J (1982) Evolution and the Theory of Games. Cambridge Univ Press, Cambridge, UK.
Phillips PC & Arnold SJ (1989) Visualizing multivariate selection. Evolution 43: 1209–1222
Shipley B (2000) Cause and Correlation in Biology: A User's Guide to Path Analysis, Structural Equations, and Causal Inference. Cambridge Univ Press, Cambridge, UK.

4

What Is Fitness, and How Do We Measure It?

JOHN HUNT AND DAVE HODGSON

Fitness is a bugger.

—J. B. S. Haldane

Fitness: something everyone understands but no one can define precisely.

—S. C. Stearns 1976

Fitness is perhaps one of the most difficult to understand and frequently misinterpreted concepts in evolution. Yet no single concept is more fundamental to understanding the evolutionary process. Although most researchers are able to agree that fitness is some measure of success in contributing descendants to future generations, providing an operational definition has proved far more challenging. The problem is a simple one: no single measure is likely to reliably predict fitness in all biological contexts. Thus it is not surprising that researchers studying different organisms have difficulty finding common ground. This problem is further magnified in empirical studies because it is virtually impossible to accurately estimate fitness for most organisms under biologically relevant conditions. As a result there is still much confusion (among students) and disagreement (among scholars) over what is meant by fitness and why it matters to the evolutionary process.

The central aim of this chapter is to clarify some of the existing confusions over fitness and provide suggestions of how to measure fitness in systems asking different research questions with a variety of study organisms. To address this aim, we have structured the chapter into five major sections. We first discuss the historical confusions of fitness and build on these to formulate a general working definition of fitness for use in evolutionary studies. We then discuss some key conceptual issues regarding the definition of fitness, and then how to estimate fitness, placing particular emphasis on determining the most informative estimate of fitness to use in empirical studies. We then present three empirical case studies (black field cricket, red deer, and bacteria) measuring fitness in different study organisms, illustrating how the conceptual issues discussed in the preceding section can be put into practice in empirical studies. We conclude by outlining the key issues that researchers should consider when attempting to measure fitness and present a simple guide that may help inform researchers as to the best fitness measure to use in their particular study organism. Our hope is that students who read this chapter will give more thought to what is meant by fitness and why it matters and therefore be better informed when embarking on empirical research.

THE CONFUSIONS OF FITNESS

Contrary to popular belief, Darwin did not use the word *fitness* in early editions of *Origin of Species* (Darwin 1859). Instead he used the verb *to fit* to describe how well the phenotype of an organism matched the environment in which it lives. Using a lock-and-key analogy, the environment is a lock and the organism (the key) must fit this lock if it is to survive, grow, and reproduce (Ariew & Lewontin 2004). Fitness crept into later editions thanks to Herbert Spencer's coining of the phrase "survival of the fittest" (Spencer 1866). Darwin and Wallace both enjoyed Spencer's metaphor because it avoided the supposed anthropomorphism associated with *selection*. However, modern evolutionary biologists shy away from this metaphor of natural selection, considering it to be a gross oversimplification of evolutionary dynamics. Moreover "survival of the fittest" is often considered tautological because fitness is a function of survival; in other words, it is defined by some as "the ability to survive and reproduce" (Ariew & Lewontin 2004; Bouchard & Rosenberg 2004; Krimbas 2004).

Fitness, in its modern usage by evolutionary biologists, would ideally be a feature of an organism that determines the spread of that organism's genes in a population, through time. If this numerical spread were an entirely deterministic process, then decades of debate about how to define or measure fitness would have been unnecessary. However, even though the process of natural selection is deterministic, it occurs simultaneously with stochastic processes. Mutations are random events, and even organisms that are most able to exploit resources and cope with environmental conditions may suffer the random effects of genetic drift and natural disasters, as well as environmental and demographic stochasticity. Hence the probability of future numerical spread of traits, driven by natural selection, is a statistical property that must be linked in some way to the propensity of organisms carrying those alleles or traits to survive and reproduce in the environments in which they and their descendants live (Bouchard & Rosenberg 2004).

With this abstract model in mind, a diversity of fitness concepts has been proposed (Dobzhansky 1968a, 1968b; Endler 1986). For example, *fitness* has been defined as the propensity of an individual, or of all individuals carrying a trait or allele of interest, to survive and to produce viable offspring; the rate at which an allele or trait spreads numerically; the ability of individuals carrying alleles or traits to exploit resources and cope with environmental conditions to survive and reproduce; the ability of individuals carrying an allele or trait to adapt to all possible future environments (which we consider *adaptability* instead of fitness); or the long-term future dynamics, or persistence, of the allele or trait in a population. Each of these definitions captures some or all of the features required by the theory of natural selection, but they vary in timescale of measurement (one, few, or many generations) and in whether fitness should be ascribed to the current state of the entity or to its future possible states. However, none of them is satisfactory as written. For example, the final definition of this list confounds evolutionary dynamics (which are a consequence of fitness variation) and fitness. It attempts to predict biological dynamics into a distant uncertain future, which is fraught with all the problems of statistical extrapolation. Also, taking a pessimistic view of evolutionary dynamics, in which all traits or alleles either survive or go extinct, this definition of fitness has a binary (one or zero) outcome. Taking an even more extreme view, in which all alleles eventually go extinct, if fitness is defined as long-term persistence, it will always be zero. Defining fitness as the rate at which an allele or trait spreads numerically likewise confounds fitness with evolutionary responses to selection. This definition may be appropriate when fitness is defined with respect to specific alleles (in haploid systems), for fitness and evolutionary responses are interchangeable, but is not acceptable when we consider selection on diploid (or higher order ploidy) genotypes or phenotypes, for which evolutionary responses to selection are more complicated functions of fitness.

Despite their limitations, the first three fitness concepts above share some similarities that capture the core concept of what fitness means in evolutionary biology. Natural selection acts on the propensity of an organism to survive, reproduce, and transmit its alleles to future generations in the environment that it occupies. However, no one measures "propensities" of organisms to do things in empirical work. To be useful to empiricists, we define fitness as follows:

> a measurable feature of alleles, genotypes or traits of individuals that predicts their numerical representation in future generations.

For a population geneticist, fitness of an allele is best defined as a parameter that directly affects the change in frequency of that allele between one generation and the next. This is consistent with most of the earlier concepts of fitness because carriers and noncarriers of that allele will tend to survive differentially and produce more or fewer offspring than each other, leading to a change in allelic frequencies over the chosen timescale of one generation.

To a behavioral ecologist, fitness is usually defined for individual organisms (or the traits they express) rather than for alleles, and is defined by the relative numerical change in abundance of those individuals or traits through time. This also fits with our definition because individuals must survive and reproduce if they are to transmit heritable traits to future generations. There is actually little difference in what fitness *means* to geneticists and behavioral ecologists. However, there may be important differences in how fitness is *estimated* in these two branches of evolutionary biology.

CONCEPTUAL ISSUES

Much of the difficulty associated with defining fitness and understanding how to best measure it in different study organisms stems from the vast number of conceptual issues that underlie this concept. In the following section we outline and discuss some of the issues we view as most pertinent. It should be noted, however, that these are by no means the only ones, and we direct the readers to other important papers on the topic to gain additional perspective (Christiansen 1985; Endler 1986; de Jong 1994; Fairbairn & Reeve 2001).

Absolute or Relative Fitness: Does It Matter?

In the evolutionary literature, fitness is expressed in both absolute and relative terms. Table 4.1 presents a hypothetical example to help illustrate this point. For simplicity, we focus first on the fitness of two alleles (*A* and *B*) in a haploid organism. Although few of the organisms that are of interest to behavioral ecologists are haploid, the principles illustrated by these simple haploid models apply also to diploid species. Also, although we are focusing on alleles, these fitness measures apply equally to phenotypic traits.

TABLE 4.1 Hypothetical data illustrating the calculation of absolute fitness (W), relative fitness (w), and the selection coefficient (s) for alleles from their average fecundity and survival schedules

Alleles	Fecundity	Survival	W	w	s
A	16	0.75	12	1.00	0.00
B	18	0.50	9	0.75	0.25

Fitness is most easily conceptualized for an asexual organism that is semelparous (i.e., reproduces only once, then dies) and has nonoverlapping generations (i.e., everyone reproduces at the same time). In such an organism, the absolute fitness (W) of an allele is the average number of offspring produced by an individual with this allele and can be calculated as the product of the average fecundity times the proportion surviving. For example, the absolute fitness of allele A (W_A) in table 4.1 is simply $16 \times 0.75 = 12$. Strictly speaking, absolute fitness is almost impossible to measure for most organisms because it requires measurement of the absolute number of each genotype present immediately after reproduction (i.e., the genotypes of embryos before birth or inside a developing egg). Most studies therefore use lifetime reproductive success as a measure of absolute fitness, which typically only accounts for the average number of offspring that survive from birth to reproductive age. The relative fitness (w) of an allele is its absolute fitness relative to that of some reference allele. By convention, the reference allele is typically the one with the highest absolute fitness and is therefore assigned a relative fitness of 1.0. For example, allele *A* has the highest absolute fitness and therefore w_A is 1.0 (table 4.1). The relative fitness of allele *B* (w_B) is therefore $W_B/W_A = 9/12 = 0.75$ (table 4.1). Another common way of expressing relative fitness is relative to the mean absolute fitness of the population or as the ratio of *A*'s frequency in the next generation (p') to that in the current generation (p; $w_A = p'/p$). The selection coefficient (s) is $1 - w$ and therefore measures the strength of selection against a particular allele. The selection coefficient of 0.25 for allele *B* (table 4.1) is very strong and thus would lead to a rapid decrease in the frequency of this allele in the population.

It is generally accepted that selection operates on relative fitness and not absolute fitness *per se*. This is because evolution is a change in the frequency, not number, of each allele or phenotype in a population. Selection thus operates through fitness differences among alleles (or individuals) in the population, so that it is not an allele's absolute fitness that is important but rather how well it does relative to other alleles in the population. This can be demonstrated by calculating the expected change in allele frequencies per generation. Let the frequencies of alleles *A* and *B* in the population be $p = N_A/N$ and $q = N_B/N$, respectively, where N is the total population size at the beginning of a generation ($N_A + N_B$) and

$p + q = 1.0$. After one generation of selection, the numbers of A and B alleles in the population are $N_A W_A$ and $N_B W_B$, respectively. The new frequency of A (p') is then:

$$p' = \frac{N_A W_A}{N_A W_A + N_B W_B} = \frac{p W_A}{p W_A + q W_A} \quad (4.1)$$

and its change across this generation is:

$$\Delta p = p' - p = \frac{p W_A}{p W_A + q W_B} - p = \frac{pq(W_A - W_B)}{p W_A + q W_B} \quad (4.2)$$

Thus, if $p = 0.57$ and $q = 0.43$ and W_A and W_B are 12 and 9, respectively, as in our hypothetical example, $\Delta p = 0.069$. The same result is found if W_A and W_B are 24 and 18, 36, and 27 or 120 and 90 (i.e., as long as relative fitness remains constant at 1.0:0.75). However, if W_A and W_B are changed to 12 and 6, respectively (i.e., relative fitness changes to 1.0:0.5), Δp increases to 0.156. Consequently, it is the relative fitnesses (w_A and w_B) that determine the rate of genetic change rather than the magnitude of W_A and W_B.

The above demonstrates that absolute and relative measures of fitness tell us very different things about the evolutionary process. However, under certain conditions, absolute and relative fitness can be equally informative. In the simplest case, when the population is neither expanding nor contracting, there is a direct linear relationship between absolute and relative fitness so that maximizing arithmetic absolute fitness will similarly maximize relative fitness. This is commonly assumed to be the case in many lab-based studies that do not account for population dynamics. However, when the fitness of alleles fluctuates through time, the allele that ultimately dominates in the population is the one with the greatest geometric mean absolute fitness (Gillespie 1977).

The effect of variation in absolute fitness of A on variation in relative fitness of A (assuming the fitness of B is constant) is not linear but instead shows a curve of diminishing returns (figure 4.1; Gillespie 1977; Frank & Slatkin 1990; Grafen 1999, 2002; Orr 2007). The extent of this concave relationship depends critically on the allele frequency p (Orr 2007): when p is very small, the relationship between absolute and relative fitness approaches linear (Frank & Slatkin 1990). This is because A contributes very little to population mean fitness when it is rare, and thus small changes in absolute fitness of A affect the numerator much more than the denominator in equation 4.2. In contrast, as A becomes more common, changes in the absolute fitness of A have large effects on both the numerator and denominator and so the relationship becomes more concave as p increases in frequency. The nature of this relationship reveals an interesting, yet somewhat counterintuitive, fact about how natural selection operates: the benefit to relative fitness of producing one extra offspring is smaller than the cost of producing one fewer offspring (Orr 2007).

In summary, one of the most basic facts in the study of evolution is that fitness is a relative concept: it does not matter how many offspring a genotype (or individual) produces, only that the genotype produces more than others in the population. However, under some conditions, there is a well-defined relationship between absolute and relative fitness, meaning that the former still has some heuristic value. Under other conditions, this relationship will not be so clear and interpreting either fitness measure in isolation may provide an incorrect view of how evolution will proceed. As we will discuss later in the chapter, a classic example of this occurs when an altruistic behavior can decrease the absolute fitness of the actor but increase its relative fitness in the population. Therefore, caution should always be taken when interpreting absolute and relative fitness in evolutionary studies. In some cases, maximizing absolute fitness will be the best (or only) way to maximize relative fitness, but in other instances, this strategy may not hold.

Who Owns Fitness?

It is common practice in behavioral ecology studies to treat the number, viability, and/or reproductive success of offspring as components of maternal (or parental) fitness (e.g., Krebs & Davies 1993; figure 4.2a). For example, researchers studying the evolution of polyandry frequently assign measures of offspring fitness, such as hatching success and survival, to the mother when assessing if there are any indirect (genetic) benefits to polyandrous behavior of females (Jennions & Petrie 2000). The problem with this approach, however, is that offspring have a unique genotype that differs from each parent. Thus, when offspring fitness is assigned to the mother it does not account for the direct effect that the offspring's own genes will have on its fitness (Cheverud & Moore 1994; Wolf & Wade 2001). For this reason, evolutionary geneticists have recommended that fitness

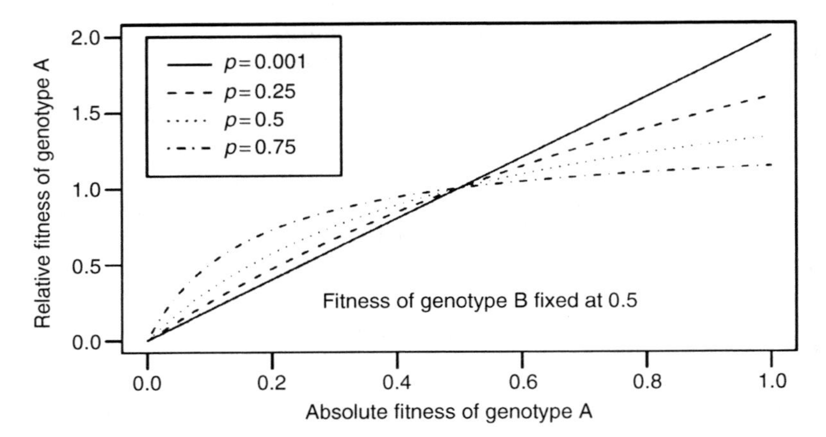

FIGURE **4.1** When the fitness of alleles fluctuates through time, relative fitness is a concave function of absolute fitness. The example shown is taken from the haploid selection model of Orr (2007) and shows the absolute and relative fitness of genotype A when $w_A = W_A/\bar{W}$, where $\bar{W}$ is the mean absolute fitness of the population calculated as $pW_A + qW_B$. For all curves, W_B is held constant at 0.5, but p and q are varied: solid curve: $p = 0.001$, dashed curve: $p = 0.25$, dotted curve: $p = 0.50$, and dot-dash curve: $p = 0.75$. However, the relationship between absolute fitness onto relative fitness will be concave whenever p, q, W_A, and W_B have nonzero values. This can be shown formally by deriving partial differentials for this curvilinear relationship:

$$\frac{\partial w_A}{\partial W_A} = \frac{qW_B}{(pW_A + qW_B)} = \frac{qW_B}{\bar{W}^2}$$

$$\frac{\partial w_A}{\partial W_A^2} = -\frac{2pqW_B}{(pW_A + qW_B)^3} = -\frac{2pqW_B}{\bar{W}^3}$$

The first derivation shows that the relationship between W_A and w_A is a positive function, and the second shows that it is concave in form. It is important to note that this concave relationship is not dependent on the fitness scheme that is used: that is, whether w_A is calculated by dividing W_A by $\bar{W}$ or by the maximum absolute fitness in the population does not alter the concave relationship between absolute and relative fitness (see appendix in Orr 2007).

be measured only from conception to death in the individual of interest (i.e., parent or offspring) so that it does not cross the generational boundary, and that the cross-generational effect be considered in evolutionary models rather than confounding estimates of fitness (Lande & Arnold 1983; Cheverud & Moore 1994; chapter 14 of this volume; figure 4.2b). In this way, fitness only ever belongs to one individual, and although it can be influenced by others in the population, it should never be reassigned to them (Cheverud & Moore 1994; chapter 14).

By comparing two simple quantitative genetic models, one in which offspring fitness is assigned to the mother and the other in which it is assigned to the offspring, Wolf and Wade (2001) showed that incorrectly assigning offspring fitness to the mother has a number of important evolutionary implications. First, they showed that the importance of maternal selection is overestimated, but only when there is environmental variance (Wolf & Wade 2001). Maternal selection occurs whenever a maternal attribute (e.g., a display that distracts a predator) directly influences the fitness of her offspring (i.e., survival; Kirkpatrick & Lande 1989), and it is overestimated when fitness is assigned across generations because the influence of environmental variation on maternal selection is not transmitted to the offspring generation (Wolf & Wade 2001). Second, their models show that the contribution of direct correlated selection on offspring traits is underestimated by a factor of one-fourth when offspring fitness is assigned to the mother. This occurs because correlated selection acting on the offspring trait

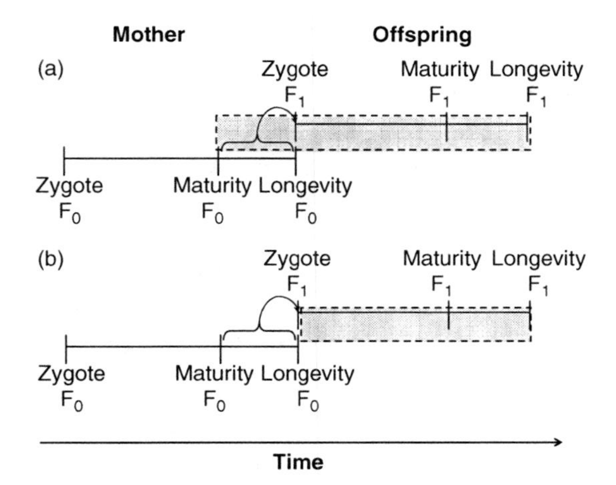

FIGURE 4.2 A schematic diagram illustrating the alternative views on the assignment of fitness. The area contained in the gray rectangle represents the time frame over which fitness is measured. The arrow represents the period of reproduction by the mother (F_0) that generated the offspring generation (F_1). (a) The predominant view used in behavioral ecology in which components of offspring fitness are assigned to the parent (in this case the mother). In this example, components of offspring fitness (namely development time and longevity) are assigned to the mother, thereby crossing the generational boundary. (b) The predominant view used in evolutionary genetics in which offspring fitness is assigned directly to the offspring. In this example, fitness is measured from conception to the death of the offspring on traits (i.e., development time, longevity, reproductive success) that belong solely to the individual. Thus, the assignment of fitness does not cross the generational boundary.

covaries with the expression of the mother's maternal character by only one-half the additive genetic variation. Also, when offspring fitness is assigned to the mother, maternal selection only influences the fitness of daughters (i.e., sex limited) and so is half that when offspring fitness is directly assigned to offspring. Collectively, these sources of over- and underestimation will not necessarily cancel out, and if maternal selection and direct selection on offspring traits are of opposite sign and different in relative strength, then incorrect dynamical equations will result that may provide incorrect conclusions about the strength and/or direction of evolution.

However, there are conditions under which assigning offspring fitness to the mother or the offspring is equivalent. In particular, the strength of selection operating on maternal trait(s) will be identical whenever all of the variance in the maternal trait is additive genetic (i.e., no environmental variance) and if there is no genetic covariance between the maternal and offspring traits (Cheverud & Moore 1994; Wolf & Wade 2001). These conditions, however, are unlikely to be true in most natural systems. Although most (if not all) phenotypic traits have an additive genetic basis, median heritability estimates range between 0.26 and 0.53 depending on the type of trait being examined (e.g., 0.32 for behavioral traits; Mousseau & Roff 1987), suggesting that the environment is likely to play an important role in the expression of most traits. Moreover, in systems in which genetic correlations between maternal and offspring traits have been measured (either for the same or different traits), nonindependence appears to be the rule rather than the exception (Roff 1997).

To illustrate the implications of this model, consider the hypothetical example of a maternal trait that influences offspring survival, such as feeding at the nest in birds (for simplicity, we assume no paternal care in this bird species). For traits like maternal feeding, it is common practice to assign offspring survival to the mother when addressing evolutionary questions (e.g., Merilä & Sheldon 2000, chapter 26). In this species, the offspring trait of interest is the growth of early downy feathers that influences thermoregulation and therefore early offspring survival. We assume that direct selection on both maternal feeding (b_d) and offspring feather growth (b_o) is positive but the latter is under stronger selection because it is linked to early offspring survival (let $b_d = 1$ and $b_o = 2$). We also assume that offspring reared by mothers that provide superior care (perhaps due to more efficient foraging) have better early survival (i.e., maternal selection, let $b_m = 1$). A researcher using a pedigree analysis estimated that the additive genetic variance (G_{mm}) and environmental variance (E_{mm}) in the maternal trait were both equal to 1 and the additive genetic covariance between maternal and offspring traits (G_{om}) was -2. For this example, we assume that there are no phenotypic maternal effects (i.e., there is Mendelian rather than maternal inheritance) so that the phenotypic variance in the maternal traits (P_{mm}) is the sum of the additive genetic and environmental component (i.e., $P_{mm} = G_{mm} + E_{mm} = 2$). We make this assumption because the maternal trait is only expressed in adulthood, in which the likelihood that the expression of the trait would be influenced by the individual's mother is greatly diminished.

We can now examine whether selection on the maternal trait differs when offspring fitness is assigned to the offspring or to the mother in this hypothetical bird species. This can be rephrased as an optimality problem to ask "Does selection favor increased or decreased maternal feeding in this species?" To answer this question, we substitute the above information into the equations derived by Wolf and Wade (2001) for the selection differential acting on the maternal trait (S_m) when offspring fitness is assigned to the offspring (equation 4.3) and to the mother (equation 4.4), calculated as follows:

$$S_m = 1/2b_m G_{mm} + b_o G_{om} + 1/2b_d P_{mm} \quad (4.3)$$

$$S_m = 1/2[b_m P_{mm} + 1/2b_o G_{om} + b_d P_{mm}] \quad (4.4)$$

We get a selection differential of −2.5 when assigning offspring fitness to the offspring and a selection differential of 1 when assigning offspring fitness to the mother. Clearly, we get very different pictures on how selection favors maternal feeding in this species depending on how fitness is assigned, and therefore we can get very different answers to our question. Although our example is hypothetical, it highlights the problems that can occur when assigning offspring fitness to the mother (or parents). Again, it is worth noting that it is *only* when all of the variance in the maternal trait is additive genetic and when there is no genetic covariance between maternal and offspring traits that these alternate ways of assigning fitness are equivalent (i.e., if E_{mm} and G_{om} are set to 0 in the above example, S_m for both fitness assignments is 1).

The models presented by Wolf and Wade (2001) clearly demonstrate that the success of either approach to assigning fitness will depend heavily on the particular components that affect fitness, the genetics of the traits involved, and the ability of the researcher to empirically separate these components or to be explicit about the assumptions they make in the analysis and interpretation of their data. Because each approach has shortcomings, one must decide on an assignment of fitness carefully and justify the assignment of fitness chosen. Researchers should always consider the possibility of genetic correlations between maternal and offspring traits. If these are high, researchers should not cross the generational boundary when assigning fitness. However, if these are low or absent, then certain components of offspring fitness may cautiously be assigned to mothers.

The Problems with Measuring Fitness in Multiple Generations

We have defined fitness to be a measurable feature of alleles, genotypes, or traits of individuals that predicts their numerical representation in future generations. This raises two important questions. First, how long into the future do we wish to predict? Second, how long should we measure fitness before using it as a predictor of future representation? Should we measure fitness only once, and then extrapolate to future periods of time, or should we measure fitness in each of multiple time periods (e.g., sequential years) to understand how fitness (and thus selection) fluctuates through time? The former question requires the researcher to define the time frame over which natural selection is expected to act (figure 4.3). This time frame should be long enough so that competition among individuals resolves itself into numerical dominance of one allele (or trait) over others, but not so long as to allow previously unidentified alleles or traits to evolve and dominate. Nor should it be so long as to allow changing environments to mask the predictive signal of fitness measurements. The latter question requires consideration of the ecological context in which fitness is expressed. The fitness of organisms in the wild will be influenced by temporal variation in climate, resource availability, social structure, and so forth; hence using an estimate of fitness from one generation to predict future fitness takes an enormous risk: if fitness is heavily influenced by the environment and it changes across generations, a snapshot estimation of fitness will be biased, if not completely uninformative, in predicting future numerical representation.

A solution to this problem is to measure fitness of several individuals, in each of several different generations of the study organism, and (if relevant) in each of several different environments. This approach satisfies two criticisms of short-term fitness measures. First, replication across many individuals helps to dissociate the fitness of an allele or trait from the schedule of survival and reproduction of each individual carrying that allele or trait. This prevents the false allocation of high fitness to a trait when it is actually caused by the overall quality of the trait carrier, which may be linked to

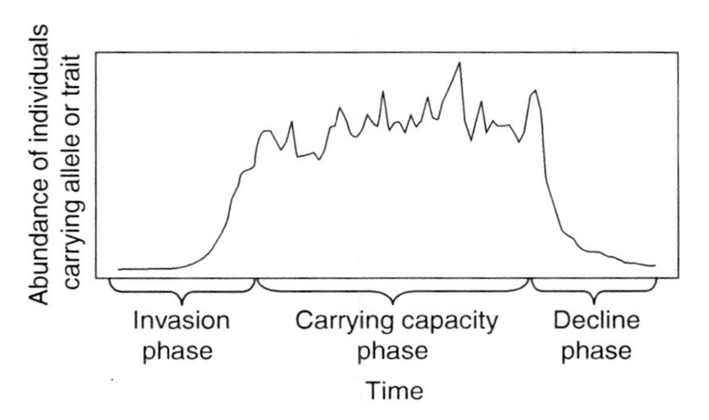

FIGURE 4.3 Lifetime dynamics of a new allele or trait. Allele initially invades a population or environment (i.e., is *fit*), then suffers stochastic fluctuations at carrying capacity, before eventually going extinct due to either replacement by another type (competitive process) or a changing environment (inability to adapt). The timescale over which fitness is measured matters because this allele is fit in the short and medium term (the allele invades and persists), but unfit in the long term (it goes extinct). During the invasion phase, exponential growth favors early reproduction; therefore fitness measures should consider rates of multiplication. Fitness during this phase may not depend on density. During the carrying capacity phase, fitness measures will depend on the magnitude of stochastic fluctuations. If the population is stable, lifetime reproductive success represents fitness. If fluctuations are large, consideration must be given to geometric mean fitness. Fitness during this phase is density dependent: populations above the carrying capacity will tend to decline, whereas those below will tend to increase. During the decline phase, late reproduction will be favored; therefore the rate of multiplication becomes important again. Fitness during this phase may be density dependent if the decline is caused by a declining carrying capacity, or density independent if the population is suffering exponential decline.

its environment. Second, replication across generations and environments helps to account for changes in the fitness value of traits across a series of temporal or physical changes in environment.

A hypothetical example serves to illustrate our point. Suppose we wish to measure the fitness of a novel phenotype of a bird, yellow legs, relative to the wild-type phenotype that has pink legs. If we measure the relative numerical increase of yellow-legged individuals over 50 generations, we risk confusing the fitness of yellow-legged individuals (a *predictor* of future numerical spread) with the process of natural selection for yellow legs (an *explanation* for the observed numerical spread). Also, the selective environment is likely to have changed over the course of 50 generations (perhaps due to environmental change, the change in frequency of yellow legs itself, or due to the appearance of further novel leg phenotypes). On the other hand, if we study only the relative fitness of yellow legs in one generation, this relative fitness will be confounded with any environmental biases that also happened during that generation. For example, perhaps we studied leg color during a generation exposed to very low rates of predation. If the survival or fecundity of yellow-legged birds is relatively low when predators are common (perhaps because yellow legs are more obvious to visual predators), we will have measured a biased estimate of relative fitness. A less biased estimate requires us to study the fitness of yellow-legged and wild-type birds in at least two different generations: one suffering high rates of predation, the other low. Even better would be to study fitness in several different generations that compose a representative sample of the natural frequency of high and low predation rates.

This is an example of a trait-by-environment interaction: the relative fitness of different leg colors depends on predation rates. It should be clear that many different environmental stressors (e.g., climate, food availability, predation, competition) can influence the relative fitness of a trait. To achieve an unbiased estimate of relative fitness, which serves as a predictor of future numerical spread through a natural population, the trait should be studied in several generations and in several relevant environments. However, combining measures of fitness from multiple different generations can be extremely problematic for organisms that engage in sexual reproduction (Wolf & Wade 2001). The problem is one of pedigrees and bookkeeping: multigenerational estimates of fitness must keep track of who mated with whom, and must dissociate the adaptive value of the allele (or trait) itself from the ancestral compounding of maternal, paternal, grandparental, and so forth effects. In such situations, the genetic context of the trait being studied becomes extremely important. To summarize, measuring fitness across multiple generations is possible, but tends to be attempted only in asexual organisms or in organisms in which pedigrees and breeding programs can be tightly controlled. Given that there are problems associated with assigning fitness to parents or offspring (Wolf & Wade 2001), that sexual reproduction dilutes patterns of inheritance between parents and offspring, and that ancestral effects can cause fitness differences that extend for many generations into the future (Beckerman et al. 2006), combining measurements of fitness from multiple generations should always be done with caution.

ESTIMATING FITNESS

There are many different ways to estimate fitness in empirical studies, each with their own pros and cons that makes them suited to different questions and study organisms. In this section, we discuss these various estimates of fitness, placing particular emphasis on determining the most informative estimate of fitness to use in empirical studies.

Fitness Is Context Dependent

In the following subsections we will show how the measurement of fitness must consider the social and ecological context in which individuals are evolving. Researchers may choose to measure fitness *in context*, either by measuring it in the wild or by attempting to recreate the social or ecological context in the laboratory. Alternatively, fitness could be measured *out of context*, for example under ideal conditions in the laboratory with plentiful food and low stress. With both approaches, the relevant context should be recognized and the appropriate estimator of fitness used. We have already described the fact that fitness will vary depending on environmental conditions and the density of competing genotypes. However, we adopt the approach of developing estimates of fitness for simple, then progressively more complicated, ecological and evolutionary scenarios. It should become clear that estimates of fitness are well defined for simple situations, but that further research is required to develop estimates for more *real*, in other words, complicated, natural systems.

Fitness in a Constant Environment with Nonoverlapping Generations

The simplest theoretical context for measuring fitness is that of a species with nonoverlapping generations, living in an environment that is stable across generations. This environmental context is hard to imagine, but is perhaps epitomized by annual plants with no seed bank or by univoltine (i.e., one generation per year) insects. It does not matter when offspring are produced by such organisms; the measurement of fitness is simply the number of viable zygotes produced over the course of one generation per viable zygote produced in the previous generation (Roff 2002). This measurement is often called lifetime reproductive success (LRS) or the net reproductive rate, R_0. R_0 is easily estimated from a life table as $\sum l_x b_x$, where l_x is the probability of surviving to age x, and b_x is the per-age-unit production of zygotes or viable offspring by individuals of age x. R_0 can be calculated for each individual (for whom l_x will be 1 for all ages until death) or averaged across individuals carrying the trait of interest. In species with nonoverlapping generations, there is no benefit to be gained by reproducing early or late, unless this decision correlates with total offspring production, because the clock will be reset at the start of the next generation. A useful analogy is that of the 100-meter sprint: the fastest sprinter may be considered most fit, but the benefit gained in one race is not transferred to the next race, when all racers commence again at the starting line.

It must be noted that LRS measured in context will be a useful measure of fitness, but LRS measured out of context (e.g., in low-stress or noncompetitive environments) will often radically overestimate fitness of organisms in their natural environment (Reznick et al. 2000). Hence LRS measured out of context will be a poor predictor of absolute fitness. Even when relative LRS is measured for two or more traits or alleles, there is a risk that ignorance of the natural context of survival and reproduction could provide false predictions of which trait is fitter than others.

When generations do not overlap, fitness can also be described by the multiplicative rate of increase of an individual, or groups of individuals that share the trait of interest. We call the multiplicative rate of increase λ, and define it as follows:

$$\lambda = \frac{N_t}{N_0} \tag{4.5}$$

where N_t is number of individuals at time t and N_0 is the initial number of individuals. In the context of nonoverlapping generations, we set t to be the length of the generation. If we are measuring fitness of an individual, $N_0 = 1$ and $\lambda = \text{LRS}$. If we are measuring fitness of groups of trait-carriers, N_0 is the parental sample size and $\lambda = \text{LRS}$.

Fitness in a Constant Environment When Generations Overlap

Many organisms have overlapping generations such that the offspring produced by an early reproducer could start their own schedule of survival, growth, and reproduction before the offspring of a late reproducer have even been born. Consider an environment without seasons, in which an organism has an allele that promotes early reproduction. Even if this trait carries a cost of reduced survival and/or fecundity, this allele may spread through the population because individuals with this allele replicate faster than those without. We can extend our previous analogy of 100-meter sprinters to the 4-×-100-meter relay: racers overlap in timing, so the winner of the first leg gains an important head-start advantage in the second, third, and final legs. Our estimate of fitness must now consider the *rate* at which offspring are produced. Several approaches to the measurement of rate-dependent fitness are possible: each has strengths and weaknesses determined by the biology and ecological context of the study species.

The Malthusian Parameter (m) *Calculated from Counts* The exponential rate of population increase is commonly used as an estimate of fitness when generations overlap and the number of individuals carrying the trait or allele of interest is growing without check. This parameter is often called m, the Malthusian parameter, named after Thomas Malthus (1798), who wrote classic essays on the exponential multiplication of unchecked population growth. This parameter is often called r in simple population dynamic models, but here we distinguish it as m when it is calculated from counts of population size. This parameter has been used by evolutionary researchers to describe the exponential rate of increase of biological entities (absolute fitness), or their rate of spread through a population (relative fitness). The simplest way to measure m is to count the density of our chosen entity at two fixed times, N_0 and N_t. The multiplicative increase in density will be $\lambda = N_t/N_0$ (also called the finite rate of increase; analogous to λ calculated above for organisms with discrete generations), and when logged to make this geometric process additive, $m = \ln(N_t/N_0)$. As we will illustrate later in the chapter, the Malthusian parameter calculated from counts has found favor with evolutionary ecologists working with microscopic, asexual organisms with large population sizes (Lenski et al. 1991).

The Intrinsic Rate of Increase and the Euler-Lotka Equation Euler (1760) and Lotka (1907) independently derived the characteristic equation that links the rate-dependent schedule of survival and reproduction of individuals to the Malthusian parameter (Stearns 1992). This equation promotes further use of life table techniques introduced above, and is also used by ecologists to link demographic measurements to predictions of growth rate of exponentially growing or declining populations. In words, r (the intrinsic rate of increase) is calculated by equating its natural exponent to the summed (if we are using discrete age intervals) or integrated (if we are studying demography in continuous time) schedule of survival and reproduction from birth to death:

$$\sum_{x=1}^{\infty} e^{-rx} l_x b_x = 1 \text{ (discrete time)} \qquad (4.6)$$

$$\int_0^{\infty} e^{-rx} l_x b_x dx = 1 \text{ (continuous time)} \qquad (4.7)$$

In practice, when measuring age-specific demography, empiricists will usually work with the discrete-time version of this equation. Whichever version is used, the equation is daunting because it can be solved only numerically: it cannot be rearranged to make the mathematical statement "r is a function of l and b." However, under certain assumptions r can be estimated as $r \sim \ln(R_0) / T$, where T is the generation time. This approximation points out the need to consider the timing of demographic events, because fitness will be higher in individuals that produce the same number of offspring but have shorter generation times. However, the approximation loses accuracy as the amount of generation overlap increases, because new generations of offspring do not completely replace previous generations. Also, T tends to be estimated as the age at which individuals first reproduce, but it is not clear that this measure of generation time is the best one: other possibilities include maximum life span, median reproductive age, and median age of first reproduction.

The Intrinsic Rate of Increase and the Eigenvalue of a Demographic Projection Matrix Perhaps more flexible and easy to calculate than the Euler-Lotka equation is the conversion of life table data into a projection matrix (Caswell 2001). This is a square matrix with rows and columns defined by the age classes of the study organism. The body of the matrix is filled with probabilities or rates at which individuals' progress through their life cycle. Columns define the age class of individuals at time t, and rows define age classes at time $t + 1$. We draw an example of a life table for an individual blue tit, from McGraw and Caswell (1996). This tit lived for 5 years and produced 48 young during its life (hence, R_0 = LRS = 48).

Age	l_x (survival from start to end of age class)	b_x (fecundity during age class)
0–1	1	3.5
1–2	1	5
2–3	1	5
3–4	1	6
4–5	0	4.5

This life history can be described as a projection matrix **A**:

$$\mathbf{A} = \begin{bmatrix} 3.5 & 5 & 5 & 6 & 4.5 \\ 1 & 0 & 0 & 0 & 0 \\ 0 & 1 & 0 & 0 & 0 \\ 0 & 0 & 1 & 0 & 0 \\ 0 & 0 & 0 & 1 & 0 \end{bmatrix}$$

The ones on the subdiagonal of **A** represent the fact that the individual survived to progress from age i to age $i + 1$. The numbers on the top row represent the contribution to age class i of that individual (i.e., the production of offspring). More complicated matrices can be created for whole samples of organisms that carry a trait or allele of interest (in which case the subdiagonal will be composed of probabilities of survival, and the top row will be average rates of reproduction, adjusted by the probability of survival of the parent). If we define a vector **x** that contains the number or density of members of a sample of individuals in each of the age classes at time t = 0, vv into the future using matrix multiplication:

$$\mathbf{x}(t = 1) = \mathbf{A}\mathbf{x}(t = 0) \qquad (4.8)$$
$$\mathbf{x}(t) = \mathbf{A}^t\mathbf{x}(0)$$

These projected dynamics will eventually settle to a stable geometric rate of increase or decline that can be described by the dominant (i.e., largest) eigenvalue of the matrix (Caswell 2001). Matrix eigenvalues are easily calculated using freely available mathematical or statistical software. We call this eigenvalue λ and note that it describes the future rate of numerical spread of the sample of individuals described by the matrix (i.e., fitness; McGraw & Caswell 1996). The link to the intrinsic rate of increase parameter is a simple one:

$$r = \ln(\lambda) \qquad (4.9)$$

For our blue tit example, we use standard mathematical software to calculate the dominant eigenvalue of the life history projection matrix. In the freely available statistical software *R* (R Development Core Team 2008), we use the command eigen(A)[1] to show that λ, for this individual, is 4.82. We can convert this to $r = \ln(4.82) = 1.57$, and if we had information on the maximum or mean fitness of individuals in the tit population, we could easily convert this to a measure of relative fitness.

This definition of r inherits all the benefits of measuring fitness via the Euler-Lotka equation, with further advantages. First, it can be written and calculated as "r is a direct function of demography." Second, it can be estimated for both populations and individuals. Third, many organisms have rates of survival and reproduction that are better predicted by stage rather than age, and demographic projection matrices are easily extended to describe transitions between stage rather than age classes. Fourth, it is not essential that age or stage classes are of equal length. Finally, it is possible to study the functional relationship between the estimate of fitness, in other words, $\ln(\lambda)$, and the magnitude of vital rates that contribute to the matrix. Hence one can use sensitivity (Caswell 2001) or transfer function analyses (Hodgson & Townley 2004) to predict the effect of different age- or stage-specific rates of survival, growth, and reproduction on estimates of fitness.

Fitness in a Competitive Environment

Up to this point, we have used r or λ to estimate the fitness of alleles, traits, or individuals with no reference to the fitness of other individuals in a population. This challenges our argument that fitness needs to be considered *relative* to the fitness of other individuals. In reality, the ability of an individual to survive and reproduce is likely to depend on the density of individuals that compete for limited resources, and/or on the frequency of each type of entity in the population. Wallace (1968, 1975) clarified the difference between density-independent and density- or frequency-dependent fitness by defining two categories of natural selection. Hard selection is density independent, acting in such a way that populations will grow if their members are fit and decline if their members are unfit. Deaths or failed reproductive attempts result in fewer individuals. Soft selection, on the other hand, is density or frequency dependent. Individuals lost from a population through selective death or low reproductive output are replaced by offspring from the same or other parents. This compensation implies that the population is at carrying capacity and that other individuals are "waiting in the wings" to exploit any newly available resources.

The difficulty that arises when measuring fitness in a context of soft selection is that the fitness of any entity is no longer a constant: instead, it is a function of the density and/or frequency of the allele or trait in the population. Directly measuring, or even describing, the complex relationship between fitness and density will often be logistically impossible. However, three useful approaches exist for estimating fitness in a competitive environment: (1) fitness at carrying capacity, (2) fitness as a measure of persistence, and (3) invasion rate.

Fitness at Carrying Capacity When population dynamics have reached equilibrium, the fitness of members of that population is defined by their ability to replace themselves. The production of a single offspring implies the loss of one other individual from the population (otherwise the population would not be at equilibrium). Crucially, it doesn't matter when this offspring is produced: the relevant estimate of fitness returns to being LRS or R_0. The mean fitness of members of a population at carrying capacity must be 1; therefore any members of the population with LRS > 1 will spread at the expense of other members of the population (Roff 2002).

Fitness as a Measure of Persistence Simple models of competitive population dynamics show that in constant but competitive environments, the trait that enables its carriers to maintain the highest density at carrying capacity (or the trait able to persist on the lowest concentration of the most limiting resource) will dominate the future population. This is an ecological concept that has been translated into a carrying-capacity measure of fitness (MacArthur 1962) and used by many researchers (Benton et al. 2002). However, this measurement carries important caveats. First, it is a multigeneration measurement of fitness; hence it carries the difficulties associated with the ownership of fitness described above (Wolf & Wade 2001). Second, it does not describe situations in which fitness is frequency dependent: entities that have a fitness advantage when rare, but a disadvantage when common, will never dominate populations in the way predicted by the persistence measure.

Invasion Rate A useful way to measure fitness in competitive situations is to ask whether the entity of interest can invade the resident population from an initially rare density (Benton & Grant 1996, 2000). This measure of fitness has several advantages. First, it can be a short-term measure (i.e., measured over a single generation of invading entity). Second, it accounts for density and frequency dependence by studying the fitness of the invader in the context of the invaded population. Third, it mimics the process of microevolution, during which novel mutants appear within populations at very low initial density and will spread numerically only if favored by natural selection (or by chance if subject to drift). Directly measuring invasion rate, however, can be difficult. Many invasions from low density will fail due to stochastic events; it can be difficult to recreate a relevant competitive context in which to test invasion and even at a theoretical level the relevant indices of invasion can be complicated. When working with models of invasion, we currently rely on an index called the *invasion exponent*. In simple terms, the invasion exponent is the density- or frequency-dependent version of λ. When individuals compete for limited resources, then values of l_x and b_x are not constants; instead they are functions of the density of competing individuals. To calculate the invasion exponent, one needs a detailed, density-dependent demographic model of the resident phenotype and of the potential invader. Many such models will have an equilibrium carrying capacity for the resident phenotype (a density at which the resident population neither increases nor declines). If so, then one plugs this density into the life history projection matrix for the invader phenotype and calculates its dominant eigenvalue, λ_{inv}. If $\lambda_{inv} > 1$, the invader phenotype should invade, and the bigger the λ_{inv} the faster this invasion will be.

The invasion exponent measures relative fitness perfectly when the population is at a stable equilibrium and the invading entity is at an infinitesimally small density. Any significant deviation away from this will defy the assumptions of invasion exponent analysis. However, the invasion exponent can be useful in revealing negative frequency-dependent selection.

Fitness in a Stochastic Environment

The environment in which organisms live is unlikely to remain constant through time. How do these environmental fluctuations impact on fitness, and how we measure it? The most important concept here is that fitness is multiplicative through time. If we accept that the fitness of a given phenotype or entity is likely to vary between environmental conditions, and assume that future environments cannot be predicted, then expected fitness must be some average of the fitnesses experienced across all possible future environments. But what kind of average should be used? Use of the arithmetic mean fitness across environments hides an important behavior of multiplicative dynamics in stochastic environments (Roff 2002). As an extreme example, consider an organism with nonoverlapping generations of 1-year and a heritable behavior that yields high fitness in "good" years but kills the bearer in "bad" years. If the arithmetic average of these fitnesses were used to describe mean fitness, it would be possible to predict long-term spread ($\bar{\lambda} > 1$) of this behavior. However, as soon as the first bad year occurs, all bearers of the behavior would be killed and the behavior would go extinct. In the short term, fitness can be high if there is a string of good years. In the medium term, fitness could easily be zero. In general, below-average performance in some generations decreases fitness more than above-average fitness in other generations. This observation is well known to professional gamblers and financiers, who will often "hedge their bets" and "spread the risk" rather than put "all eggs in one basket," trading off the benefit of potentially large income against the risk of more harmful losses (Orr 2007).

Generalizations of this argument have lead to an acceptance that the correct way to estimate fitness in stochastic environments is to use the *long-term geometric mean*. In practice, this is hard to calculate exactly, but two useful approximations are available. If we use an overbar to represent the arithmetic mean value of r or λ across all possible future environments, and σ^2 to represent variance,

$$\begin{aligned} \bar{r}_g &= E(\ln \lambda) \approx \bar{r} - \frac{\sigma_r^2}{2} \\ \bar{\lambda}_g &= e^{E(\ln \lambda)} \approx \bar{\lambda} - \frac{\sigma_\lambda^2}{2\bar{\lambda}} \end{aligned} \tag{4.10}$$

(Lacey et al. 1983; Lewontin & Cohen 1969). The striking conclusion to be drawn from these equations is that fitness in a stochastic environment can be increased in two ways: by increasing the mean arithmetic fitness across all environments, and/or by reducing the variance in fitness between environments. This offers the chance for generalists to be favored by natural selection: in the long term, behaviors that confer low average fitness but also low variance in fitness might outperform behaviors that are specialized to confer high fitness in some environments but suffer severe costs in others.

The difficulty of estimating fitness in a stochastic environment is made even more severe when one attempts to measure it. The approximation of geometric mean fitness requires knowledge of the mean and variance in r or λ across all possible future environments. Given the difficulty of measuring r or λ in any one given environment, the researcher is then faced with the enormous task of expanding these assays to a range of environments. In practice this is rarely attempted, and even then it is usually limited to pairs, or small numbers, of environmental categories.

Fitness in a Stochastic, Competitive Environment

In the real world, organisms are faced with competition from others, as well as an uncertain environment. Therefore, the fitness of most organisms will be affected by a complicated mixture of time-varying densities and frequencies of competitors and time-varying resource levels and stresses. There is no simple way to describe fitness in such an ecological context (Benton & Grant 2000). Using theoretical models, it is possible to study numerically the compounding effects of geometric mean fitness and the need to study invasion exponents rather than density-independent growth. Current wisdom states that changes to rates of survival or reproduction of individuals in stochastic, competitive environments, have effects on fitness that can differ radically from the effects of similar changes in simpler environments (Benton & Grant 1999). Clearly, much more work is needed on this topic.

Fitness in a Spatially Structured Population

So far we have considered the effects of stochastic environments and competition on fitness. In these contexts we have treated biological entities as members of a single population in which all members are assumed to interact with each other equally. It is well known in population genetics and ecology that such assumptions of *panmixis* (i.e., random mating) and *mass action* (i.e., competition mediated by total population density) will rarely hold true. Instead, populations, resources, and conditions are spatially structured (Wilson 1977). Local groups of individuals are more likely to interact within their group: interactions between groups will be determined by rates of dispersal of individuals between groups. An important consequence of spatial structure for the measurement of fitness is that numerical spread within a group need not translate directly to numerical spread between groups (West et al. 2007a). Useful examples come from species such as aphids (Hodgson 2002), which have specialized disperser morphs. Investment in the physiological and morphological traits that promote dispersal (e.g., wings) can often trade off negatively against rates of survival and reproduction (Dixon 1998). Hence dispersers may suffer low fitness within a local group, but gain increased fitness via their ability to colonize new patches of resources. Nondispersers hold the advantage within groups but are unable to colonize new patches and therefore risk extinction if local resources are exhausted or disturbance events destroy entire patches of resources.

These concepts have been synthesized theoretically by distinguishing the relative importance of *local* and *global* selection (Frank 1998). When selection is entirely local, the fitness of alleles or traits will be determined by their numerical spread through the local group. When selection is entirely global, fitness will be determined by numerical spread through the whole population. In between, when both global and local selection is important, numerical spread must be measured both within groups and between them. Crucially, if local groups contribute to the global process of population growth as a function of their own size, then individuals that contribute to *successful* groups may spread globally even if they have local fitness costs (Griffin et al. 2004). Spatial structuring and a mixture of local and global competition can, in certain contexts, promote the evolution of unselfish (altruistic) or even self-harming (spiteful) behaviors (West et al. 2007a).

Fitness in a Kin-Structured Population

The idea that population structuring can promote the evolution of locally costly but globally beneficial traits has been developed most powerfully in the context of kin selection (Hamilton 1996), in which local groups are formed by related individuals who share *inclusive* fitness (through genetic relatedness). Shared ancestry is not required for altruistic or spiteful behaviors to evolve: all that is required is that spatial structuring of the population means that individuals tend to interact with other individuals that are more, or less, likely to share the trait of interest than the global population average probability of sharing the trait. In nature, this tends to occur when family groups stick together (leading to strong positive spatial autocorrelation in relatedness, and hence to potential for altruism to evolve), or when individuals deliberately disperse away from relatives (leading to negative spatial autocorrelation in relatedness and hence to potential for spite to evolve).

Consider, for example, a behavior that reduces the survival or offspring production of an individual but benefits the survival or fecundity of other members of a social group. It is possible that this behavior is entirely altruistic. However, if the fitness context has been incorrectly identified, this conclusion may be false. One alternative scenario is that the behavior increases the between-patch spread of group members, including the actor. If this between-patch benefit outweighs the within-patch cost, then the behavior actually increases the direct fitness of the actor and the behavior is actually mutually beneficial to actor and recipient. To define a behavior as costly to the actor, the measurement of fitness must be correct. Genuinely altruistic or spiteful behaviors can still evolve via kin selection (Hamilton 1996; West et al. 2007a); selection acting on genes via behaviors of actors that benefit nonactor carriers of that gene, or harm nonactor noncarriers. The former scenario defines an altruistic act: the actor is harmed, but kin selection favors the behavior because relatives (or carriers of the same genes) benefit. The latter defines a spiteful act: the actor is harmed, but kin selection favors the behavior because nonkin (or noncarriers of the same gene) are also harmed. Therefore, kin selection acts on inclusive fitness, which is the fitness of the individual plus the fitness of all beneficiaries of the actor's behavior, weighted by their relatedness to the actor.

As with all other ecological contexts, spatial and social structure, the degree of relatedness, and the relative intensity of local and global selection will all influence which fitness estimate is most appropriate. If natural populations are patchy, then simple measures of short-term, local survival and reproduction may not predict the spread of a behavior through the global population (Grafen 1984). If fitness is measured in context, in other words, with a full recognition of the importance of relatedness, local spread, and between-patch dispersal, then predictions of the spread of traits or behaviors should be correct. If any aspect of the proper context is neglected, such as local competition, transmission processes, global resource supply, or kin structure of groups, then estimates of fitness may be very different and even misleading. Crucially, to measure the fitness benefits of behavior in a social setting requires an understanding of the consequences of that behavior for the lifetime reproductive output of all members of the social group and knowledge of the genetic relatedness of the social group. Measured in context, it could be possible to link the survival and reproduction of a social group to the behavioral traits of a focal actor. Measured out of context, ignorance of kin selected benefits might lead to the conclusion that a costly behavior will be selected against.

Simple Fitness Proxies: How Effective Are They?

Measuring the spread of alleles (or traits) across generations is often impossible or prohibitively time consuming in both wild and laboratory populations for most organisms. Given these logistic constraints, empiricists have long sought convenient and informative short-term proxies of fitness. Such fitness *proxies* are distinct from fitness *components* that have been discussed earlier in this chapter. Fitness components are those variables that occur at different stages through the life history of an organism and combine (following some mathematical function) to produce an estimate of total fitness. For example, survival and fecundity can be considered fitness components because the estimation of R_0 is a function of these two variables. In contrast, fitness proxies are those variables that are presumed to be correlated with (i.e., have major effects on) fitness components, such as energy intake, body size, or attractiveness, but do not appear in mathematical functions for estimating fitness. Both fitness proxies and fitness components can be considered *surrogates* for total fitness, but the major difference is that fitness components are logically (and necessarily) related to total fitness, whereas fitness proxies may or may not be related to total fitness.

The use of fitness proxies is particularly common in behavioral ecology research, spanning the range from morphological through to life history traits (table 4.2). Although these traits are easy to measure, they should always be interpreted with a degree of caution. The reason for this is twofold. First, each of these fitness proxies is based on its own set of assumptions on how it relates to total fitness. Therefore, proxies will inherently differ in their value as a predictor of total fitness. For example, the use of male attractiveness as a fitness proxy (e.g., Bentsen et al. 2006) assumes that an attractive male is able to successfully mate with each female he attracts and that each mating results in the production of viable offspring. Such assumptions are rarely tested and may often be incorrect. Second, individual fitness proxies, particularly when used in isolation, may ignore important trade-offs with other fitness components. The magnitude of these trade-offs is known to vary with the environment (Roff 2002). Therefore, there is no *a priori* reason to expect that a given fitness surrogate will be informative across all environments (Hunt et al. 2004). Ultimately, the predictive value of any individual fitness surrogate will depend on the strength of its relationship to total fitness. Given the importance of fitness to the study of evolution, and the strong reliance on fitness proxies in most empirical studies, it is surprising that more information does not exist on this relationship.

In the few studies that have directly addressed this issue, there appears to be little general consensus. This is, in part, due to the fact that these studies have used different measures of total fitness. In studies examining the relationship between fitness proxies and lifetime reproductive success (LRS), most fitness proxies were poor predictors

TABLE **4.2** Some common examples of fitness proxies used in behavioral ecology studies

Fitness Proxy	Description	Example
Body size	The physical magnitude of an organism.	Hunt and Simmons (2004)
Development time	The time period between birth and adulthood.	Semlitch et al. (1988)
Growth rate	The rate of increase in body size per unit time.	Lampert and Trubetskova (1996)
Sexual trait expression	A trait, either morphological or behavioral, that is the target of sexual selection.	Andersson (1982),
Attractiveness	The possession of a sexual trait(s) that is preferred by a member of the opposite sex. A male's attractiveness is typically measured as the number of females he attracts.	Bentson et al. (2006)
Dominance	The behavioral state of having high social status relative to others. Dominance is attained through either direct battles or indirect displays.	Perry et al. (2004)
Territory quality	Quality, in terms of food availability or protection from predators, of the territory maintained by an individual.	Candolin and Voigt (2001)
Energy intake	The total number of calories ingested per unit time.	Ritchie (1989)
Mating success	An individual's success in obtaining a mating.	Thusius et al. (2001)
Number of copulations	The number of successful copulations attained over a user defined period of time (i.e., per day or over the lifetime).	Arnqvist et al. (2005)
Timing of breeding	The time that an individual reproduces. Generally measured in seasonal organisms as the number of days into the reproductive season that reproduction commences.	Hau (2001)
Birth asynchrony	The temporal spread in the birth (or hatching) of offspring.	Hauber (2003)
Number of offspring reaching adulthood	The number of offspring that successfully reach adulthood. This number therefore incorporates the number of offspring produced plus their survival to adulthood.	Sheldon and Ellegren (1999)

of total fitness under both controlled laboratory conditions (Reed & Bryant 2004) and in natural populations (Fincke & Hadrys 2001). Importantly, though, both studies showed that the fitness proxies performing worst were those based on only part of the organism's life history. For example, in the housefly (*Musca domestica*) the size of the first clutch explained only 1.5% of the variance in total fitness, whereas fecundity explained 64% of this variance (Reed & Bryant 2004). Studies using the asymptotic, geometric rate of increase (λ) as a measure of total fitness also suggest that fitness proxies may often be poor predictors of total fitness (McGraw & Caswell 1996; Ehrlén 2003). For example, Ehrlén (2003) showed that although there was a large deleterious effect of mollusc herbivory on the total fitness of the perennial herb, *Lathyrus vernus*, this was largely unrelated to changes in individual fitness proxies.

One consistent pattern, however, is that survivorship appears to be a better fitness surrogate than fecundity, at least in ecological studies (Crone 2001). In a review of the published literature, Crone (2001) showed that adult survivorship consistently explained, on average, more of the variation in λ than did fecundity (i.e., adult survivorship had a higher elasticity—the predicted proportional change in fitness caused by a proportional change in vital rate—than did fecundity; Benton & Grant 1996). This finding should be interpreted with caution, however, because Crone's (2001) review is heavily biased toward longer lived species. Indeed, the only notable exception to this general pattern was for semelparous plants, in which growth typically had the highest elasticities, followed by fecundity and then survivorship (Silvertown et al. 1993). Thus, considerably more work encompassing a more diverse range of life history strategies is needed before the generality of this pattern can be determined.

The above examples paint a dire picture of the use of fitness proxies in evolutionary studies. Yet animals frequently behave as if these fitness proxies are important (chapter 11). Furthermore, in perhaps one of the most comprehensive tests of the relationship between fitness proxies and total fitness, Brommer et al. (2004) suggest that the situation may not be so bad. Using long-term genetic data sets on the short-lived collared flycatcher (*Ficedula albicolis*) and the long-lived Ural owl (*Strix uralensis*), they measured total fitness as the number of genes transmitted by a focal individual *y* years after hatching (Brommer et al. 2004). This measure of total fitness was then related to two different single generation fitness surrogates, LRS and λ, for each species. Brommer et al. (2004) showed that despite different life histories, measures of LRS and λ were good predictors of an individual's long-term genetic contribution to the population for both bird species. However, the relative success of these surrogates differed depending on whether they were based on the lifetime production of fledglings or recruits. When based on the number of recruits, both LRS and λ performed equally well. However, when these proxies were based on the number of fledglings, LRS correlated better with total fitness than did λ.

The general lack of studies explicitly testing the relationship between fitness proxies and total fitness makes it difficult to say exactly which fitness proxies should be preferentially used in empirical studies. There are, however, a number of broad recommendations that can be made when selecting fitness proxies. First, whenever possible, fitness proxies that cover more of the organism's life cycle should be used in preference to those covering less of the life cycle (e.g., body size versus number of copulations). A greater coverage of the organism's life cycle means that more potentially important fitness components will contribute to the fitness proxy being used. These proxies also have the added benefit that they are based on fewer assumptions about how they relate to total fitness. Second, it is always better to measure more fitness proxies than fewer. The possibility of trade-offs between important fitness components always exist particularly in life history data (Roff 2002; chapter 3). Thus, by measuring multiple fitness proxies, trade-offs can be tested for and taken into account when interpreting the data. Third, if a single fitness proxy is being used, it is worthwhile measuring it numerous times across the lifetime of the organism. This is because a given proxy may only be important when summed across the lifetime of the organisms (i.e., its effects are cumulative). For example, the energetic consequences of a given foraging decision (e.g., should I eat this food item or wait for another?) seem trivial, but a long sequence of such decisions may be very important for fitness (chapter 11). Finally, it is important to recognize the limitations associated with the particular fitness proxy being used. Ultimately, what fitness proxy is used will be determined by the logistical constraints of the system being studied and the particular question being asked. Therefore, some fitness proxies will have more limitations than others. However, this is not to say that these fitness proxies are without value. After all, it is far better to have an understanding of fitness from a surrogate with limitations than it is to be totally uninformed, as long as these limitations are recognized and acknowledged when drawing evolutionary conclusions.

EMPIRICAL CHALLENGES

Conceptual issues aside, perhaps the biggest challenge to evolutionary researchers is actually measuring fitness in their organism of interest. The fitness measures used in empirical studies vary greatly among the different fields of evolutionary biology. In the following section we use a series of empirical case studies, spanning the major fields of evolutionary biology (behavioral ecology, quantitative genetics, and population genetics/community ecology), to highlight the pros and cons of measuring fitness in different ways.

Condition-Dependent Sexual Advertisement in the Black Field Cricket, *Teleogryllus commodus*

How and why female mate choice evolves remains one of the most vigorously debated topics in evolutionary biology (Kokko et al. 2006a). Although many different mechanisms have been proposed to explain the evolution of female mate choice (reviewed in Kokko et al. 2003b; chapter 24), the most commonly invoked is the good-genes hypothesis (Zahavi 1975). This hypothesis proposes that if elaborate sexual traits are costly to maintain, so that only males of the highest quality are able to afford these traits, then these traits will serve as a reliable indicator of male genetic quality. Consequently, females choosing to mate with such males should secure "good genes" for their offspring (Zahavi 1975). A more specific form of this hypothesis, known as the age-based indicator mechanism, posits that females who mate with the oldest male(s) in the population will receive "good genes" because these males have already proven their viability (Hansen & Price 1995). Consequently, survival (or longevity) is often viewed as a signal of genetic quality and used as a surrogate for fitness in behavioral ecology studies. However, life history modeling coupled with empirical work on the Australian field cricket, *Teleogryllus commodus*, shows how variation in the acquisition and allocation of resources may obscure the relationship between survival and genetic quality and devalue its use as a fitness surrogate.

Using a life history model, Kokko (2001) examined how the relationship between male survival and quality varies with the allocation of resources to sexual advertisement. In both instances, the effect of allocation to sexual advertisement on survival is assumed to be constant and show diminishing returns, but the relationship between the allocation to sexual advertisement and mating success varies from strong to very strong (figure 4.4a). The model clearly illustrates that when the gains in mating success from investing in sexual advertisement increases (becomes more exponential), it is theoretically possible for survival to decrease with male quality rather than increase as predicted by "good genes" theory. In fact, the reduction in survival is more than compensated by the higher mating success associated with greater allocation to sexual advertisement.

To empirically test the predictions of this model, Hunt et al. (2004) used dietary manipulation to experimentally vary resource acquisition in *T. commodus* and examined the effects on how these resources were allocated to sexual advertisement and longevity. Using an automated electronic system to accurately record the time a given male spends producing a sexual advertisement call across his lifetime, they showed that males reared on a high-protein diet commenced calling earlier in life and called more in total compared to males reared on a low-protein diet, despite having a significantly shorter life span. Thus, when males are able to acquire more resources, they allocate more to sexual advertisement at the expense of survival (Hunt et al. 2004). Divergent artificial selection on male longevity demonstrated that this trade-off has a genetic basis, with males from short-lived genetic lines calling sooner and more overall than males from long-lived lines (Hunt et al. 2006). In this population, there is good reason for males to invest in calling at the expense of longevity (Bentsen et al. 2006). In a field experiment in which artificial calls were broadcast for various durations each night (to simulate calling effort), there was an exponential increase in the number of females attracted to a call the more it was advertised (figure 4.4b). Thus, although males that invest more resources in calling die sooner, they are likely to have a higher mating success.

Collectively, this work highlights how caution should always be taken when only a single fitness measure is being interpreted, particularly when prior knowledge of the organism being studied is lacking. Although males with more elaborate sexual traits have greater survival in a number of species (Jennions et al. 2001), and this is likely to equate with elevated fitness, work on *T. commodus* clearly shows that if male survival was interpreted without knowledge of calling effort, a very different conclusion would be reached. Thus, it is always advisable to measure more than a single fitness component in empirical studies and to consider the possibility that they may interact in different (often nonintuitive) ways across the life history of the organism being studied.

The Genetics of Fitness in a Population of Wild Red Deer, *Cervus elaphus*

Red deer (*Cervus elaphus*) on the Isle of Rum, Scotland, have served as an important model in behavioral ecology for the last 40 years (Clutton-Brock et al. 1982, 1988; figure 4.5a). This is due, in part, to the fact that the LRS of individuals can be easily monitored in the wild due to their large body size, relatively short life span (for a large mammal), and the open habitat they occupy (Clutton-Brock et al. 1982). More recently, with the advent of molecular techniques to assign parentage, red deer have also become a useful species to test a number of fundamental questions in evolutionary genetics (Kruuk et al. 2000, 2002). The value of this system for evolutionary research is that long-term data on LRS can be combined with accurate pedigrees (using the animal model) to examine evolution in the natural habitat of the red deer.

The pioneering work by Clutton-Brock and colleagues (summarized in Clutton-Brock et al. 1982, 1988) on this system resolved the major morphological and life history determinants of LRS, showed how these differed considerably between the sexes and recognized that a variety of ecological (e.g.,

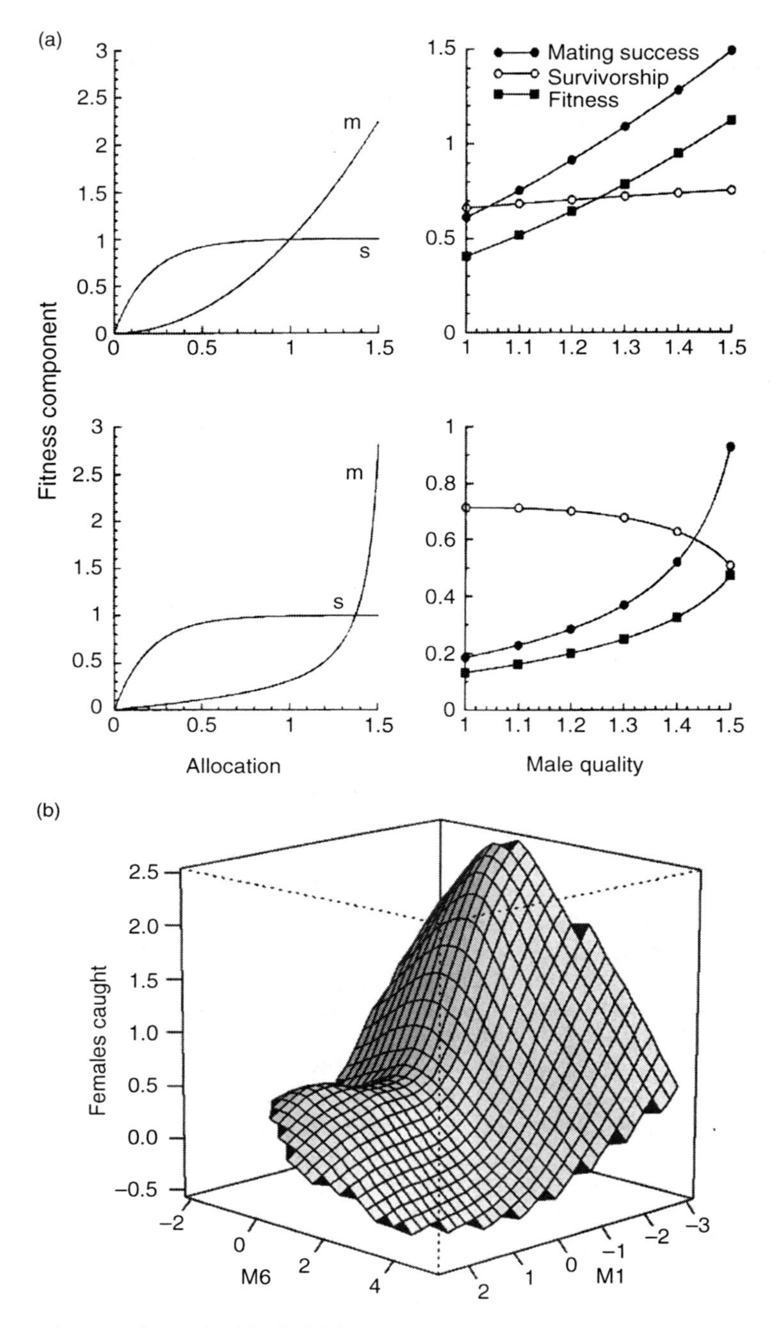

FIGURE 4.4 Empirical research on the black field cricket, *Teleogryllus commodus*. (a) Kokko's (2001) life history model illustrating how the relationship between survivorship and male quality can vary from positive (top right) to negative (bottom right) depending on the allocation of resources to mating success (top left and bottom left). (b) The individual fitness surface for artificial calls used in the field experiment. The horizontal axes (m_1 and m_6) represent the two axes of strongest nonlinear sexual selection (after canonical analysis), and the vertical axis represents the fitness surrogate measured (the number of females attracted to a call): m_6 is most heavily weighted to the structure of the call, whereas m_1 is most heavily weighted to calling effort (i.e., the amount of time per night the call was broadcast).

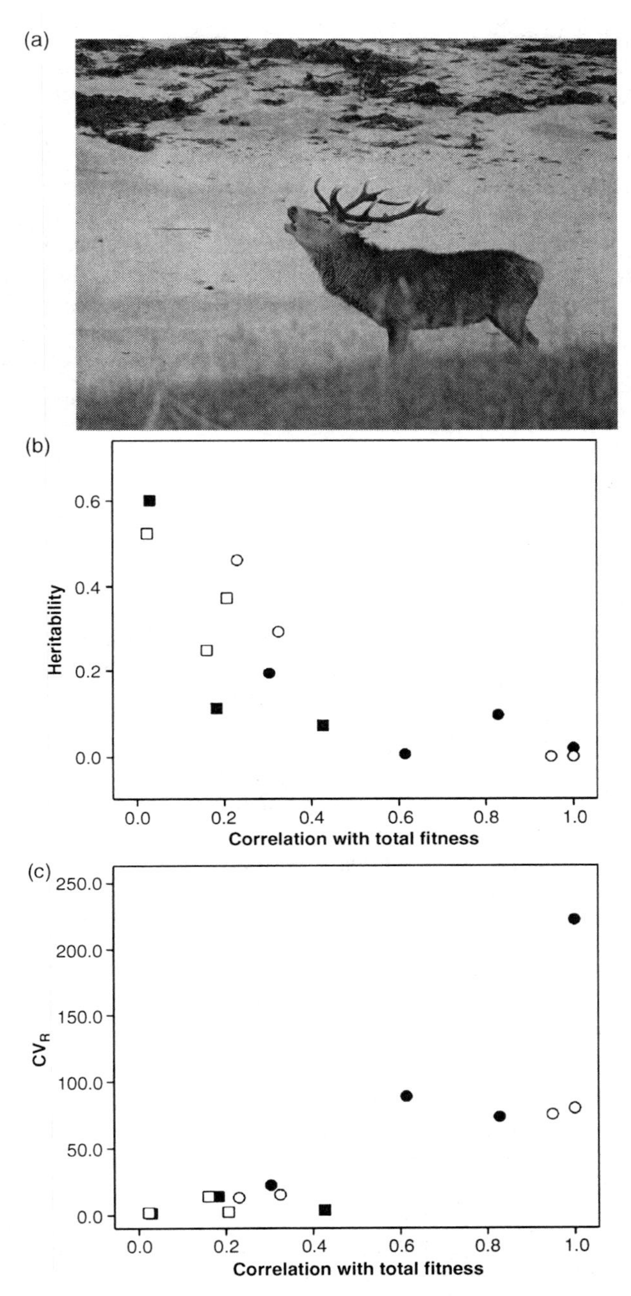

FIGURE 4.5 Red deer (*Cervus elephus*) on the Isle of Rum, Scotland. (a) A stag vocalizing to attract a mate. Photograph copyright of Loeske Kruuk. (b) A negative relationship between the heritability of a trait (defined as the ratio of additive genetic variance to total phenotypic variance in the trait) and its correlation with fitness (lifetime reproductive success). Open and closed symbols represent female and male traits, respectively. Circles and squares represent life history and morphological traits, respectively. (c) A positive relationship between the coefficient of residual variation (CV_R) of a trait (defined as the residual variance in a trait scaled to the mean size of the trait) and its correlation to fitness. CV_R should largely reflect environmental sources of variation but may also include nonadditive sources of genetic variation (i.e., dominance and epistasis). Graphs re-created from Kruuk et al. (2000).

home-range quality and food access) and social factors (e.g., group size, dominance status) also influenced individual variation in LRS. Building on this work, Kruuk et al. (2000) used genetic data to test an extension of Fisher's fundamental theorem of natural selection (Fisher 1958) that predicts that the amount of genetic variance for a trait should decrease with the traits association with fitness (Falconer 1989). Using a pedigree encompassing 2,374 individuals and spanning 30 years, Kruuk et al. (2000) estimated heritabilities for a range of morphological and life history traits that vary in the strength of their relationship to fitness. In agreement with the predictions of this theorem, they showed a significant negative relationship between the heritability of a trait and its association with fitness (figure 4.5b) and that fitness was not heritable in either sex. However, the negative relationship shown by Kruuk et al. (2000) could result from either a depletion of genetic variance in traits more closely related to fitness or that traits more closely related to fitness are more heavily influenced by the environment and therefore have lower heritability estimates. Indeed, Kruuk et al. (2000) found evidence for the latter explanation showing a negative relationship between a traits relationship to fitness and its coefficient of residual variation (CV_R; figure 4.5c). Because high CV_Rs are typical of traits with a large environmental component, this finding suggests that traits more closely related to LRS are influenced to a larger degree by the environment, although the role of dominance and epistasis cannot be fully discounted (Merilä & Sheldon 1999).

Further evidence for the importance of the environment in this system comes from a study examining the evolution of antler size in stags (Kruuk et al. 2002). Again, using a large, long-term data set (2,433 individuals over 30 years) with a known pedigree, Kruuk et al. (2002) showed that stags with larger antlers had an increased lifetime breeding success and calculated a standardized selection gradient of 0.44 ± 0.18 for antler size. Moreover, they showed that antler size was heritable (0.33 ± 0.12). Given the observed selection and heritability of antler size, quantitative genetic theory predicts that antler size should have increased per generation. However, over the 30-year period of the study Kruuk et al. (2002) actually observed a decline in antler size that closely mirrored the rise in population density. The authors suggest that this finding is consistent with the relationship between antler size and fitness being determined by environmental rather than genetic covariance. That is, an environmentally determined trait, such as nutritional status, is correlated with both antler size and fitness, and this gives the appearance of a relationship between these two traits even though a direct causal relationship does not exist.

Collectively, these studies show the important role that the environment can play in determining fitness and stress the importance of measuring fitness in the appropriate context. This is because fitness, and traits closely related to it, often has a large environmental component, meaning that the context in which it is measured is crucial for interpretation. This is particularly true when the environment influences both fitness and traits of interest in the same way as a misleading picture of how evolution should proceed may result. Wherever possible, fitness should be measured in the appropriate context for the organism being studied.

The Adaptive Radiation of Niche Specialists in the Bacterium *Pseudomonas fluorescens*

In this case study we draw examples from a growing literature on the use of bacterial microcosms to study microevolution empirically and on dramatically short timescales (Lenski et al. 1991; Rainey et al., 2000; Travisano & Rainey, 2000). Bacteria have emerged as a model system for studying microevolutionary dynamics, for several very good reasons. First, short generation times and simple laboratory growth conditions mean that microevolutionary events occur over very short timescales and are easy to study. Second, ancestral bacteria can be stored indefinitely, allowing comparisons of fitness between ancestors and derived phenotypes and genotypes. Third, as will become clear, the asexual nature of bacterial replication greatly facilitates the measurement of fitness across generations. The downside of using bacteria for evolutionary research, however, is the small size and short life span of individual bacterium that prevents the measurement of schedules of survival and reproduction for given individuals.

It would be impossible to summarize the findings of a vast literature on the evolution of bacterial phenotypes or behaviors in this chapter. Instead, we focus on a single experiment using the bacterium *Pseudomonas fluorescens* to study real-time adaptive radiations and the evolutionary coexistence of niche specialists. Rainey and Travisano

(1998) propagated bacteria, starting from a single, ancestral genotype, in each of two environments. A heterogeneous environment was created by not shaking the glass tubes in which the bacteria grew. This created three spatial niches: the body of the growth medium, the medium-air interface, and the base of the glass tube. Homogeneous environments were created by shaking the tubes, destroying the availability of distinct spatial niches. In the heterogeneous environments, the ancestral genotype diversified to form three phenotypic variants, defined by the morphology of colonies grown on agar plates: smooth morphs dominated the growth medium; wrinkly spreaders dominated the medium-air interface, and fuzzy spreaders dominated the base of the vials (figure 4.6). In the homogeneous environments, no such diversification process happened.

The ecological context in which fitness was measured was a 7-day period of microevolution and population growth. This period, which is significantly longer than other bacterial studies (e.g., Lenski et al. 1991), suggests that bacterial populations had reached their stationary phase and therefore were competing for limited resources. Indeed, this competition was probably responsible for the adaptive radiations observed: when resources become limited in the growth medium, selection for the exploitation of novel niches will be strong. Rainey and Travisano assayed the fitness of each evolved morphotype in competition with each other morphotype. Crucially, they chose not to create equal mixtures in their inocula (compare with the Lenski et al. approach). Instead, they measured the fitness of each morphotype by invading them into a larger population of the competitor, or resident, morphotype (basically, the invader and resident were inoculated at a density ratio of 1:100). Fitness of the invader was measured, after 7 days' growth, as the ratio of the Malthusian parameters of the invader versus the resident morphotype. This assay allowed Rainey and Travisano to identify a complex pattern of frequency-dependent selection in *P. fluorescens*. They showed that smooth morphs and wrinkly spreaders were mutually able to invade each other (each had Malthusian parameters > 1 when invading the other morph). Similarly, smooth

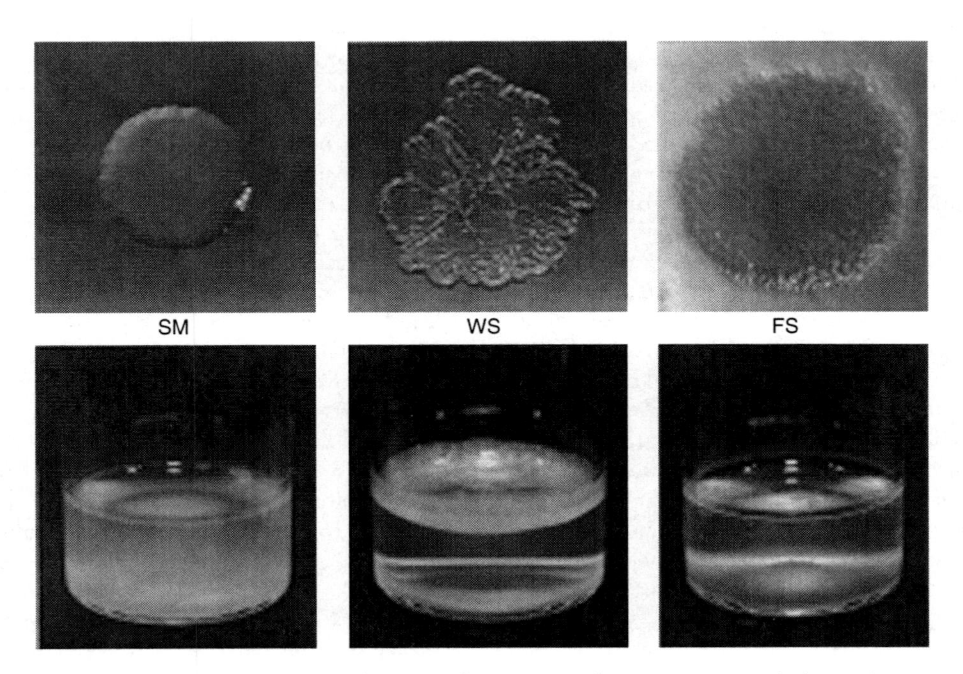

FIGURE 4.6 Evolved morphotypes of *Pseudomonas fluorescens* when grown in unshaken microcosms for 7 days. Diverse populations grown on agar plates can be categorized into smooth (SM), wrinkly spreader (WS), and fuzzy spreader (FS) morphs according to colony morphology on agar (top row) and resource/spatial partitioning within microcosm vials (bottom row). Photograph copyright of Paul Rainey and Mike Travisano, used with permission of the photographers.

morphs and fuzzy spreaders were mutually invasive. However, although wrinkly spreaders were able to invade fuzzy spreader populations, the reverse was not true. Surprisingly, fuzzy spreader morphs were selected only when smooth morphs were present. Starting from pure populations of smooth morphs, new variants of all three morphotypes were able to invade the system, resulting in a frequency-dependent coexistence.

What might have happened if Rainey and Travisano had inoculated competition assays with equal concentrations of morphotypes? The answer depends on the final ratios of morphotypes after competition: unless the final density ratio was 50:50, the ratio of Malthusian parameters would have been < 1 for one of the competitors. This would have lead to the conclusion of selection against, rather than selection for, that morphotype, even though mutual invasion was possible. In this case, the "invasion from rare" assay was the correct one because fitness was measured *in context* and *relative* to a competitor morphotype. Fitness in this example was a multigeneration summary of the growth, invasion, and persistence dynamics of each morphotype, but it sufficiently captured the numerical advantage of each morphotype in a relevant ecological background.

It should be noted that the use of the Malthusian parameter ignores the fact that populations were not increasing exponentially through the course of the assay period. Instead, the ratio of final (day 7) to initial (day 1) density ignores the lag, growth, and stationary phase dynamics of the bacterial microcosm. Had the researchers chosen a different timescale for their fitness assays (e.g., 1 day growth, or growth between days 6 and 7), entirely different answers may have emerged. Their choice of 7 days' growth was probably best, because it represented the context of the adaptive radiation experiment, but it should be noted that experiments with other organisms will rarely have the luxury of measuring multigeneration fitness over the same timescale as the microevolutionary dynamics. In such systems, more attention needs to be given to the population dynamic context of the fitness measurement.

CONCLUSIONS AND FUTURE DIRECTIONS

The aim of this chapter was to provide a rational and measurable definition of fitness for students of behavioral ecology. Fitness, as an empirical measurement, should be a predictor of the numerical spread of the allele or trait though a population and many different approaches are available to predict this each with their pros and cons (summarized in table 4.3). The emergent theme is that the best measurement of fitness will depend on the biological context that alleles, or the individuals carrying them, encounter.

Absolute fitness measures the propensity of an allele to persist or increase numerically. However, constraints on population size mean that not all alleles can be successful, and natural selection targets the difference in these alleles. We have shown that the measurement of fitness is sensitive to the timing of generations (do they overlap or not?) and to the dynamics of the population (rate of spread is important in growing and declining populations). We have shown that fitness in stochastic environments is sensitive not just to average rates of survival and reproduction but also to variance in these parameters (i.e., geometric mean is the appropriate estimate). We have shown that fitness is best measured from the start to the end of an individual's life span, and that the common practice of assigning offspring fitness to the parents can be risky in certain contexts but acceptable in others. Sexual reproduction and covariance between environment and genotype can dilute and confuse fitness measurements made across generations. Spatial structuring and social interactions biased toward kin or nonkin can also change our interpretation of fitness, since behaviors that affect the survival and reproduction of nonactors will influence the fitness of the actor both locally and globally. Fitness will depend on the resources and conditions that an individual encounters during its life span. In many or even most natural systems, these resources and conditions will themselves be influenced by the density, diversity, and relative frequencies of individuals carrying alternate alleles.

Taken together, this complexity demonstrates that there can be no single working measurement of fitness that fits all research needs. Fitness, and therefore the best estimate of fitness than can be measured empirically, is context dependent. Furthermore, fitness will usually be logistically impossible to measure, because populations are large and interactions within them are complex, and organisms are often elusive in a field setting. We must rely instead on simpler measurements that correlate

TABLE 4.3 Empirical estimates of fitness, ranked by timescale of measurement and consideration of the context-dependence of fitness

Fitness Measure	Examples	Used By	Pros	Cons
Cross-generational in competitive environment	K^* = carrying capacity on limiting resource	Community ecologists	Describes ability to invade stable system	Logistically impossible in most organisms
	Invasion exponent θ; ability to invade equilibrium system	Demographers, evolutionary geneticists.	Can include frequency dependence	Requires well-parameterized models
	Invasion assays	Microbiologists	Fitness measured in competitive context	Often requires marked reference strains
Cross-generational rate sensitive	Malthusian parameter, $\ln(N_t/N_o)/t$	Microbiologists	Measures actual numerical increase	Only for asexual organisms
	λ, dominant eigenvalue of demographic projection matrix	Demographers	Incorporates all important vital rates and life stages	Requires complete demography
Cross-generational rate insensitive	R_o, or (N_t/N_o) (number of replacements per generation)	Evolutionary demographers	Measures persistence of types in stable environments	Long timescales if long generations
Rate-sensitive LRS	Euler-Lotka equation for r	Evolutionary demographers	Uses all vital rates across life cycle	Requires detailed life table data
LRS	Survival and fecundity throughout life cycle	Demographers, evolutionary ecologists	Uses all vital rates across life cycle	Requires detailed life table data; not rate-sensitive
Reproductive success	Total offspring production per adult	Behavioral and evolutionary ecologists	Need only measure reproductive output	Ignores survival, rate of reproduction, and trade-offs
Fitness proxy	Fledging success, attractiveness, dominance, mating success	Behavioral ecologists	Easy to measure	Ignores the frequency-, density-, and time-dependent aspects of fitness; must consider that different surrogates may trade off with each other

Note: Different methods of estimating fitness are applicable in different experimental systems. The "best" estimates of fitness tend to be those that incorporate as much data on the lifetime schedule of survival and reproduction as possible. As we move down the table, estimators of fitness become simpler and less demanding of lifetime information, but can also become more divorced from fitness itself.

with this fitness parameter and therefore act as surrogates of fitness (see table 4.2).

We have identified several important gaps in our understanding of fitness. First, it is surprising how few empirical studies have even attempted to clarify the correlation between popular (and simple) fitness proxies and the observed change in frequencies of alleles or traits in natural populations. We urge researchers who have gone to the effort of describing real processes of invasion and spread of heritable traits in natural populations to ask, which (if any) proxies provide the most accurate prediction of long-term fitness? In the rare instances in which this has been attempted, the message is relatively clear: empirical information on the entire life history of the organism is required. The literature is full of examples of trade-offs among survival, longevity, and reproduction, making the measurement of any one of these proxies in isolation very restrictive and potentially misleading. Despite this, most researchers continue to use these simple proxies, often even referring to them as *fitness*, without recognizing their caveats.

Second, we have little understanding of feedbacks between population dynamics and fitness. Fundamentally, population dynamics emerge as an integration of the life histories of all members of a population. However, the population dynamics also dictate the life history schedule with highest fitness. If selection favors this optimal life history so that it spreads to dominate the population, this change will scale up to cause a change in population dynamics, and hence (possibly) a new *optimal* life history. This observation deserves both theoretical and empirical research. For example, if early reproduction is favored in growing populations, but is not important in populations at carrying capacity, and is disfavored in declining populations, there exists the possibility that selection on life history could exaggerate the naturally cyclic population dynamics displayed by many species. These observations lead to a need to consider the population dynamic processes before deciding to weight fitness according to early or late reproduction (figure 4.3). This is especially true in populations that fluctuate in size through time (Benton & Grant, 1996, 2000) and serves to highlight the fact that fitness and its population dynamic context are not independent of one another.

Third, it remains unclear how important is the need to measure variance in fitness in stochastic environments. Research on this topic is hampered by the need to measure life histories across multiple generations and in multiple environments: we believe great progress could be made using model systems and experimental evolution.

Fourth, more work is required to resolve the theoretical observation that offspring fitness should be assigned to parents only with great caution. Exactly when does it matter, and how common are the natural situations when it does?

To summarize, before embarking on research measuring fitness, a researcher must consider several key aspects of their study system and the particular question(s) being asked. First, what is the biological entity being considered: is it an allele, a trait, a behavior, or an individual? Second, how will the breeding system of the study system influence the measurement of fitness? Third, what statistical approaches should be used to measure propensity of survival and reproduction? Fourth, what are the critical components of the entity's environment, and how important is this environmental context? Fifth, is it possible to choose a relevant timescale for the study system, so that the fitness measure is a predictor, rather than a descriptor, of numerical spread? Sixth, and potentially of most interest to modern evolutionary biology, how do entity, breeding system, environment, and time interact to determine the measurement of fitness that best correlates with the spread of alleles (or traits) in biological systems?

Acknowledgments We thank Erika Newton, Iain Stott, Cheryl Mills, Mel Smee, Will Pitchers, Ruth Archer, Allen Moore, David Hosken, Clarissa House, and the editors for valuable discussions and comments on earlier versions of this chapter. JH and DH were funded by NERC, and JH was funded by a Royal Society Fellowship.

SUGGESTIONS FOR FURTHER READING

Ariew and Lewontin (2004) and Krimbas (2004) provide a well-grounded philosophical perspective on fitness (and the confusions surrounding fitness). Orr (2007) provides a strikingly clear link between absolute and relative fitness, drawing on similarities with economic theory to illustrate this connection. Wolf and Wade (2001) help settle whether offspring fitness is a property solely of the offspring or whether it can be reassigned to parents. The paper by West et al. (2007a) is bound to become a classic source because it not only discusses the evolution of social behaviors but also dispenses many of the semantic arguments that have confused research on this topic. Finally, Brommer et al. (2004) illustrate how to measure multigenerational fitness and sets a high benchmark for future empirical studies.

Ariew A & Lewontin RC (2004) The confusions of fitness. Brit J Phil Sci 55: 347–363.

Brommer JE, Gustafsson L, Pietiainen H, & Merila J (2004) Single-generation estimates of individual fitness as proxies for long-term genetic contribution. Am Nat 163: 505–517.

Krimbas CB (2004) On fitness. Biol Phil 19: 185–203.

West SA, Griffin AS, & Gardner A (2007) Social semantics: altruism, cooperation, mutualism, strong reciprocity and group selection. J Evol Biol 20: 415–432.

Wolf JB & Wade MJ (2001) On the assignment of fitness to parents and offspring: whose fitness is it and when does it matter? J Evol Biol 14: 347–356.

5

The Genetic Basis of Behavior

KERRY L. SHAW AND CHRIS WILEY

Behavioral ecologists are interested in behaviors expressed under particular ecological conditions and the evolutionary causes and consequences of such behavioral strategies. Evolution of behavior can occur only when there is a change in allele frequencies. It therefore follows that there can be no behavioral evolution without a genetic basis to behavioral variation. Thus the study of behavioral genetics is essential to our understanding of behavioral ecology and evolution because it informs us about the genetic and environmental contributions to variation in individual behavioral traits, and further may provide explanatory power for understanding major concepts in behavioral ecology. Recent advancements in molecular tools for studying genes have revolutionized the study of behavioral ecology; among a host of other uses, genetic tools have allowed us to infer paternity of offspring, indirectly assess dispersal patterns, and examine phylogenic relationships between taxa. Although these varied applications of genetic tools have had a profound impact on the study of behavioral ecology, this chapter will deal specifically with the insight that can be gained from more specific knowledge of the genetic basis of the actual behaviors being studied.

The idea that behavioral traits can have a strong genetic basis may be less intuitive than for other traits, such as morphological characters. This is perhaps illustrated most famously by the "nature-nurture" debate, in which "nurture" advocates espoused the view that human behavior is largely the effect of development, learning, and social environment (see *Philosophical Psychology*, 21[3], 2008) for a recent treatment of this debate). It is now realized that all phenotypes, including behaviors, are affected by both genes and environment. That behaviors have a genetic basis, at least partially, is vital to our expanded appreciation of the importance of natural selection in generating biodiversity (e.g., through sexual or kin selection). Many intricate and complex behaviors that serve as an endless source of fascination to the student of animal behavior defy easy comprehension as to how genes contribute to their existence (e.g., social behavior in honey bees or sexual signaling in birds). However, evidence is beginning to accumulate that even the most complex of behaviors can have genetic bases that can be dissected and understood both in terms of development and evolution.

The study of behavioral genetics has a lengthy history (Greenspan 2008) and the importance of genetic insights to theories of behavioral evolution (e.g., Hamilton 1964) cannot be overemphasized. Yet, few genetic studies of specific traits are currently integrated into our understanding of the behavioral ecology of an organism, perhaps because optimality models often omit explicit descriptions of the genetic basis of behavior, or because the study of behavioral genetics often has been undertaken in model laboratory organisms (Moore & Boake 1994; Boake et al. 2002). For many years, simple genetic assumptions (whether implicit or explicit) underlying behavioral "strategies" have been commonplace in studies of behavioral ecology, and studies of the selective forces that influence these behaviors have proceeded largely in ignorance of the underlying genetics. By and large, the approach termed the *phenotypic gambit* (Grafen 1984), in which simple genetic assumptions are made, have been successful in predicting conditions under which alternative strategies produce higher fitness payoffs. Although this simplification can be justified in some traditional approaches to behavioral ecology, we can also see a steady increase in known instances in which these simplifications are likely misleading (e.g., Hadfield et al. 2007). Furthermore, making simplified genetic assumptions greatly constrains the scope of behavioral ecology and evolution. This chapter attempts to provide examples of more complex questions of interest to behavioral ecologists that require details of, for example, the genetic origins of behavioral diversity or the evolutionary genetic processes underlying the fate of new behavioral genetic variation. Simple assumptions about the genetic inheritance of such behavioral variation would be counterproductive.

Fortunately, the approaches used to study the evolution of other phenotypes can be readily applied to the genetics of behavior, using techniques that are increasingly tractable in natural populations. Likewise, population and quantitative genetic models of behavioral trait evolution depend upon, and are improved by, knowledge of the genetic basis of behavioral variation (Moore & Boake 1994). The aims of this chapter are to (1) discuss how genetic approaches to the study of behavior, particularly when knowledge of the underlying genes is lacking, can give us deeper insight into behavioral ecology and evolution and (2) discuss what might be possible in the study of behavioral ecology and evolution if we knew the actual genes underlying behavioral variation. Throughout, the focus is on the study of behavioral trait evolution rather than the employment of molecular methods for the study of behavioral processes.

To launch a clear and meaningful discussion of the genetic basis of behavior, we must first clarify two classes of questions asked in the area of behavioral genetics. The first question asks, what is the genetic basis of the fully functional behavior (including those structures and molecular processes that combine to produce the behavioral output)? The second question, although fundamentally related to the first, represents an evolutionary perspective by asking, what is the genetic basis of behavioral variation? The scope of the latter question does not require knowledge of all genes that contribute to the structural development and expression of a behavior, but focuses on the underlying genetics currently contributing to variation in a behavior. The content of this chapter largely deals with this second question, the genetic basis of behavioral variation, because this is of primary interest to behavioral ecologists studying the fitness of organisms in contemporary or past populations. However, when considering the evolutionary origins of a behavior, the genetics of behavior and the genetics of behavioral variation are closely related because any of the genes contributing to the development and expression of a behavior may vary in a given population and thereby contribute to behavioral variation.

GENETICS OF BEHAVIOR WITHOUT THE GENES

There are many reasons that motivate investigations of the genetic basis of behavioral differences. Even in compelling behavioral systems in which little is known about the genome, we can gain important insights into questions of interest to behavioral ecologists, given some ability to perform husbandry or observe or construct pedigrees. Below, we discuss a number of interesting and important questions that may be addressed without knowledge of the specific genes underlying a behavior. Our choice of examples is not exhaustive, but rather represents an introduction to some of the important roles that behavioral genetics can play in furthering our understanding of behavioral ecology.

Causes of Behavioral Variation

Nongenetic Causes

As discussed previously, the causes of behavioral variation are profoundly important because only variation in behavior due to variation in genes can evolve. However, the presence of behavioral variation does not guarantee that there is a genetic basis to that variation. For example, hormones transferred to the egg or fetus during gestation may influence the neural development of the offspring, thereby influencing behavior through nongenetic means. In addition to such maternal effects, many behaviors have a strong propensity for cultural transmission via learning from parents, other relatives, or nonrelated individuals. Finally, behavioral variation may be induced by an underlying variability in the physical environment. Phenotypic plasticity, variation due to each of these differences in environment, is widespread and sometimes dramatic (see chapter 6). Such is the case in many social hymenoptera in which different individuals perform different tasks such as nursing, foraging, and defense of the nest (Seeley 1995), roles that are sometimes accompanied by considerable differences in morphology (e.g., as in the big-headed ant, *Pheidole megacephala*). This array of variable developmental outcomes, known as polyethism (behavioral variation) or more generally polyphenism (phenotypes in general), is due not to genetic variation but to variation in developmental or behavioral environment. Another conspicuous example of polyphenism appears in certain amphibians (Relyea 2004) in which apparently genetically identical individuals can develop into different trophic morphs with distinctive morphologies and dietary behaviors. Behavioral phenotypes, like other phenotypes, are products of genes and environment, both of which can cause behavioral variation (Mackay & Anholt 2007).

Determining a Genetic Basis to Behavioral Variation

The important goal of documenting a genetic basis to behavioral differences among populations or species, thereby ruling out environmental differences as causes of the variation, is often first achieved through a "common garden" study, in which behavioral variants are reared in an identical environment. Behavioral differences that persist, or "breed true," suggest a genetically heritable basis to the variation. For example, in a study by Simmons (2004), common garden breeding was conducted for several generations with the oceanic field cricket, *Teleogryllus oceanicus*, ultimately showing that song differences among populations are due (at least partially) to genetic differences. Common garden experiments are not restricted to species amenable to laboratory study. For example, even in wild birds, cross-fostering chicks between nests of closely related species has proven a useful tool for testing the genetic basis of behaviors ranging from mate preferences (Sæther et al. 2007) to foraging patterns (Slagsvold & Wiebe 2007).

When trait variation exists within a single population, common garden studies are of limited utility and further study must determine the degree to which variation is due to genes versus environment. This is typically carried out by comparing the phenotypes of relatives to that of nonkin, something that is achievable when pedigrees are known. In such instances, a first goal is to determine whether the behavioral variation we see is based on alternative alleles at a single locus (the realm of population genetics) or small contributions from variation at many loci (quantitative genetics). These alternatives have been studied using two very different research approaches in genetics (Greenspan 2004) that date back to the origins of the field of genetics itself. Although the research legacies of both approaches continue today, the consequences to modeling outcomes in behavioral evolution are significant (see below). If segregating variation within a family or population suggests polygenic control (i.e., variation due to the contributions of many genes), a more formal treatment to estimate heritability can be undertaken to quantitatively attribute portions of phenotypic variance to genes and environment (Freeman & Heron 2004). Examples demonstrating heritability in behavioral variation have been published for a diversity of behaviors, from dispersal behavior (e.g., ballooning in spiders: Bonte & Lens 2007; migratory behavior in birds: Pulido 2007), to male signaling (e.g., cricket song: Simmons 2004).

Genetic Architecture of Behavioral Variation

Genetic architecture refers to features characterizing the relationship between genotype and phenotypic variation, such as the number of loci involved in trait variation, the number of alleles per locus, the allelic interactions both within and between loci, the amount an allele contributes to the phenotypic variation, the degree of pleiotropy (i.e., how many phenotypes a given gene can affect), as well as the heritability of the trait. Below we discuss some of the ways in which an understanding of the genetic architecture of behavioral variation can yield insight into the evolutionary potential of the behavior.

Single Genes versus Complex Genetic Architectures

When an evolutionarily optimal endpoint lies outside standing variation in a trait, we expect a population to move more slowly toward that optimum when potential variation is controlled by a single locus than when variation is due to many genes. Under these conditions, traits with the potential for polygenic variation should evolve more quickly because there are more sources of new mutation. Conversely, traits determined by single genes may evolve more rapidly toward an adaptive optimum when the optimum is within standing variation in the trait. This is because favorable alleles are less likely to be concealed by variation at other loci, and thus can sweep to fixation more quickly. An understanding of the genetic architecture underlying phenotypic variation is therefore important for our understanding of the evolutionary potential of populations.

A phenomenal example of this was reported recently in the oceanic field cricket mentioned above. The distribution of this species includes northern coastal Australia and many of the Pacific islands, extending more recently across the Hawaiian archipelago. In Hawaii, *T. oceanicus* is now in contact with the acoustic parasitoid tachinid fly, *Ormia ochracea*, itself a recent invader from North America. Homing in on the song of male crickets, female *O. ochracea* locate and larviposit on or near the host cricket. Larvae burrow into and eventually kill the cricket. On the island of Kauai, in fewer than 20 generations, a wing mutation rendering males mute, and thus well protected from the parasitoid, swept to near fixation showing just how rapidly a population can evolve under strong selective pressures (Zuk et al. 2006).

Different methods have been developed to analyze predictions of trait evolution when phenotypic variation is due to allelic variation at single or multiple loci (Greenspan 2004). Single-locus evolution is considerably easier to predict because models focus on genotypes and allele frequency change due to particular evolutionary forces and can be studied as deviations from the Hardy-Weinberg equilibrium. With complex genetic underpinnings to behavioral variation comes more complex modeling in which phenotypic changes are predicted, often in the absence of known genotypes. Classic quantitative genetic predictions of behavioral evolution have been built around the "breeder's" equation (so-called due to its origin in the agricultural world), $R = h^2S$, where R is the response to selection, h^2 is the heritability of the trait and S is the selection differential on the trait (Freeman & Herron 2004; box 5.1). This model has been extremely useful because it offers a quantitative prediction of how much a population will respond as a function of two measurable features (heritability and selection intensity), and is most effective when phenotypic variation is due to many genes with simple, additive effects.

However, there are many different ways in which multiple genes can contribute to a phenotype that do not conform to simple, additive, quantitative genetic models, and these may further complicate an estimation of evolutionary potential. One example is a threshold trait, in which quantitative variation at multiple loci may be phenotypically visible only when a certain quantity in expression is reached (Pulido 2007). Alternatively, complex genetic architectures may arise through epistatic interactions (in which one gene modifies the expression of another) between the genes involved. Genes that operate within an epistatic network may be coadapted to function with other genes in the network, and expression in terms of timing and quantity of gene product will combine to produce a complex behavioral phenotype. Evolution of such phenotypes must proceed by mutual adjustments of multiple genetic factors. The evolution of all genes, from those contributing to additive variation in traits to those within an epistatic network, are further complicated if these genes have pleiotropic effects on other aspects of fitness. The genes *forager* and *period*, known to have pleiotropic effects in *Drosophila melanogaster*, are discussed in a later section. Such complex genetic architectures can be the source of extensive constraint on behavioral evolution. Yet despite this importance of genetic architecture to trait evolution, our current understanding of the genotype-phenotype relationship of quantitative traits is extremely limited, particularly in natural populations. What is clear is that the simplified views of both single gene control and additive quantitative genetics are frequently inappropriate in the world of behavioral genetics.

The diversity of genetic architectures of behavioral variation is illustrated nicely through the example of the genetics of animal migration. Within many species of winged insects there is variation in migratory behavior (Roff & Fairbairn 2007). In polymorphic populations, some individuals of a species undertake migratory flight, whereas others are sedentary. In some species, nonmigratory individuals are behaviorally, physiologically, and morphologically unable to fly because they lack fully developed wings and wing muscles to power flight. The genetic basis of this polymorphism differs in different insect species—both single gene

BOX 5.1 A Brief Introduction to Quantitative Genetics

Jason B. Wolf and Allen J. Moore

Quantitative genetics provides a statistical description of the various influences on measurable traits (Falconer & Mackay 1996). We typically consider *metric* or *quantitative* (continuous) traits, but there can be exceptions. In behavior, most traits, or at least most influences on traits, are continuously distributed. Moreover, researchers that study behavior typically adopt a phenotypic approach, making quantitative genetics the most common way to address how behavior is influenced by inheritance. In addition to its utility in providing a framework for understanding sources of trait variation, quantitative genetics also provides a means of understanding phenotypic evolution, including factors such as the rate of, and constraints on, adaptive evolution. Thus, while approaches such as optimality and game theory are useful for predicting expected outcomes of evolution, quantitative genetics provides insights into evolutionary processes and factors that limit optimal evolution or the attainment of an optimum.

Using quantitative genetics to understand trait evolution involves defining the various factors that can influence a trait, and then to statistically partition the effects of these influences. Influences can be very general or very specific. This is best expressed with linear equations. We begin with the most basic equation in quantitative genetics, which partitions the expression of a trait (z, which is the *phenotypic value*, or the trait value you measure on an individual) into a genetic (g, which reflects all of the genetic influences contributing to a trait) and an environmental component (e, the environmental deviation, which includes all of the environmental contributions influencing the expression of the trait):

$$z = g + e \tag{1}$$

This equation describes all traits in all organisms because all trait variation must be ultimately attributable to genetic (there are no organisms that develop or live without genes) and environmental (there are no organisms living in a vacuum) influences. Therefore, all other models for partitioning trait variation are special cases of this simple equation. For example, one may wish to partition the genetic component into those genetic influences that are independent of all other influences (i.e., an additive component, a, which is also called the additive genetic value or breeding value; see Arnold 1994a), and those influences that are dependent upon interaction between alleles at a locus (i.e., the dominance component, d, which is also called the dominance deviation) and between alleles at different loci, (the epistatic component, i, sometimes called the interaction deviation or epistatic deviation):

$$z = a + d + i + e \tag{2}$$

This partitioning is important because only additive genetic influences are passed intact from generation to generation, which is why it is the variance of a that contributes to the evolutionary response to selection (this is the all important *additive genetic variance*) and to the resemblance of all types of relatives. Because the dominance and epistatic components depend on interactions with other alleles either at the same locus (dominance) or at other loci (epistasis), they are not heritable and do not contribute to the evolutionary response to selection. Nevertheless, the dominance and epistatic components in equation 2 are sometimes of interest because of their contribution to various evolutionary processes and the covariance between certain types of relatives.

It is also possible to partition the environmental component into components attributable to various environmental factors (e.g., one might want to attribute phenotypic

(continued)

variation to a particular environmental factor that is predictably shared by individuals in the population). In addition, because it is the additive genetic (breeding) value that is the heritable component of equation 2, it is common to write equation 2 as a partitioning of heritable and nonheritable components,

$$z = a + e \tag{3}$$

where e now includes all nonheritable components contributing to trait expression, whether they be attributable to environmental or nonheritable genetic factors. We use this convention throughout.

Equations 1–3 describe the influences on trait expression. However, it is populations, not individuals, that evolve and so we need to know how the identifiable genetic and environmental factors influence variation in a population. These same equations can be used to derive expressions that partition variation in trait expression into a set of variance components. This partitioning is important for many reasons, the most significant of which is that it allows us to separate the heritable component(s) that contribute to trait evolution (the additive genetic variance) from those that do not. It also allows us to understand the contribution of components that contribute to the resemblance of relatives. For example, assuming that the terms are independent from each other, we can partition variation in the expression of the trait (i.e., the phenotypic variance, denoted P herein, also often denoted as V_p) shown in equation 3 into additive genetic (denoted G herein, also often denoted as V_a) and environmental (nonheritable) variances (denoted E herein, also often denoted as V_e):

$$P = G + E \tag{4}$$

(Falconer & Mackay 1996). Note that we have switched from examining the components that influence the trait value of an individual (z) to describing components of trait variation in a population of such individuals. The additive genetic variance (G) term in equation 4 is the only heritable component of phenotypic variation, and it is, therefore, common to express the ratio of the additive genetic variance to the phenotypic variance (G/P), which is known better as narrow-sense heritability, h^2 (see Falconer & Mackay 1996). Narrow-sense heritability expresses the proportion of variation in a trait among individuals that is attributable to the additive effects of alleles. Expressions similar to equation 4 could be derived for equations 1 and 2, and, assuming that the components are independent, one would simply separate variation in trait expression into a set of terms, each of which corresponds to the individual terms in the equation.

The terms in, for example, equation 3 are defined as being independent, but this can be violated if there is a nonrandom association between genetic and environmental effects, the so-called genotype-environment correlation. The presence of a genotype-environment association (generally measured as a covariance, rather than as a correlation) adds a term to equation 4 that accounts for the association between the additive genetic (a) and environmental values (e):

$$P = G + E + 2\mathrm{cov}(a, e) \tag{5}$$

A positive covariance between the additive genetic and environmental values inflates the phenotypic variance, whereas a negative covariance decreases the phenotypic variance. A genotype-environment covariance also alters the estimation of genetic variances

(continued)

empirically because it alters the resemblance of relatives. Genotype-environment covariances may arise when individuals "choose" or assort nonrandomly in specific environments. Such covariances may be important for behavior (Falconer & Mackay 1996) if, for example, individuals choose their social environment, but in general there is little empirical work directed at detecting these in natural populations.

It is also possible to include an interaction between the genetic terms and the environmental term, which is the so-called genotype-by-environment interaction (G × E, or sometimes GEI). The presence of G × E indicates that the effect of the genotype on the phenotype depends on the environment that the individual experiences and, likewise, that the effect of the environment on trait expression depends on the genotype of the individual (see chapter 6). It can also be thought of as a measure of the degree to which phenotypic plasticity varies with the genotype of the individual. The G × E is usually lumped into the environmental/nonheritable terms in equations 1–3 and will appear in the environmental term (E) in equation 4 because, like the dominance and epistasis terms, it is generally not heritable. The G × E can be partitioned into the interaction of various genetic values (a, d, or i) and the environment. G × E for behavior is not often investigated, but the evolution of almost all of the behaviors of interest to behavioral ecologists depends on G × E. For example, anytime a behavior is conditional on some cue, be it environmental or social, then this is a form of phenotypic plasticity (chapter 6), the evolution of which is dependent on the presence of and magnitude of G × E (Boake et al. 2002).

The methods for obtaining these partitionings of phenotypic variances depend on the statistical tools of analysis of variance (hence the name, as R. A. Fisher invented ANOVA to allow the partitioning of different variance components leading to variation in phenotypes within a population) and regression (advanced by F. Galton to understand parent-offspring resemblance). Statistical approaches to determining quantitative genetic components typically require a breeding design in which relatedness among relatives is known—that is, parent-offspring, sibling, or other relatives created by a breeding program (Falconer & Mackay 1996). More recently, with the advent of cheap computing power, pedigree-based approaches are possible in which a mixture of relatednesses in unbalanced designs can be analyzed. This opens up the possibility of examining quantitative genetic influences in natural populations as long as pedigree information is available (Kruuk 2004).

Understanding the relative importance of genetic, and especially heritable, variation and environmental or nonheritable variation is important for various problems of interest to behavioral ecologists. Many models of behavioral evolution make explicit genetic predictions, and so quantitative genetic approaches are required if one wishes to examine these problems theoretically or empirically. For example, for good genes models of indicator traits in sexual selection to work in explaining mate choice for indirect genetic benefits (see chapter 24), male genetic quality must be heritable (in the narrow sense). Other examples include the fact that measures of genetic correlations provide insights into life history trade-offs, and limits to plasticity, and that understanding the heritability of traits provides clues as to the relative importance of environments (e.g., during learning or in parent-offspring interactions).

In general, researchers interested in partitioning components of trait variation have an ultimate goal of understanding the evolutionary dynamics of trait means (either predicting them or reconstructing them). By separating the heritable from the nonheritable components of variation one can understand the relative evolutionary labiality of a trait because it is heritability that translates the change in trait mean within a generation (denoted s, the selection differential) caused by selection to an evolutionary change in the trait mean between generations (denoted $\Delta\bar{z}$, to indicate a change in the mean phenotypic value, $\bar{z}$). This is expressed in the classic *breeders' equation* that is central to quantitative genetics (see also chapter 3):

(continued)

$$\Delta\bar{z} = h^2 s \quad (6)$$

We can also view the additive variance as a measure of the (linear/additive) relationship between variation at the molecular level (allelic variation) and phenotypic variation. As such, it predicts how selection at the phenotypic level cascades down to produce changes at the molecular level, as in changes in allele frequencies that are generally regarded as the ultimate measure of evolution. This leads to an alternative view of the breeders' equation (which is mathematically equivalent to equation 6), in which the linear relationship between a trait and fitness, given as a selection gradient (β) is translated into a linear relationship between the genotype and fitness through the additive genetic variance, which then leads to the change in the mean trait value:

$$\Delta\bar{z} = G\beta \quad (7)$$

where β is equal to s/P (see chapter 3).

The breeders' equation (expressed in equations 6 and 7) can be used to understand the rate of adaptive evolution, which may be of interest to behavioral ecologists. However, because the breeders' equation makes predictions only about the rate and not the direction of evolutionary response to selection (because it always predicts that selection response is in the direction of selection at a rate defined by the heritability or additive genetic variance), the breeders' equation generally does not make predictions that conflict with the optimality models favored by behavioral ecologists. This contrasts with more complex quantitative genetic models concerned with multiple traits under selection, in which predictions of quantitative genetic theory may be at odds with predictions from optimality models. This suggests that behavioral ecologists should be concerned with multiple traits and adopt a multivariate view.

We have provided only a brief introduction to univariate quantitative genetics, which is focused on the genetics and evolution of a single trait, whereas evolutionary quantitative genetics is ultimately a multivariate problem. Organisms are not a collection of independent traits. Instead, in almost any conceivable system, individuals express multiple traits that are generally correlated genetically and phenotypically with each other and rarely, if ever, have independent effects on fitness. When considering multiple traits, one would examine components contributing to variation in each as outlined above. However, because these traits may not be independent, we need to understand the genetic and environmental components contributing to the relationship between traits. That is, we need to understand how genetic and environmental sources of variation make multiple traits correlated with each other. The relationships between traits are of interest for various reasons, most notably because they link the evolution of one trait with selection acting on other traits. It is through this linking of the evolutionary dynamics of multiple traits that predictions of quantitative genetics theory can be at odds with that of optimality theory. This is because individual traits are not free to evolve to their optimal values, but rather, one must consider the evolution of sets of correlated traits wherein conflicting selection pressures (like trade-offs) may result in constraints on adaptive evolution (see Arnold 1994a).

For example, if we consider a case of two correlated traits (which we will call trait 1 and trait 2), we could express the phenotypic value of each of the two traits as in equation 3 and define components of variation for each as in equation 4, but if we wish to understand the evolution of either trait, we would have to add a term to account for the genetic relationship between traits to either expression for the breeders' equation (equation 6 or 7). Taking the more traditional approach using equation 7, we would include a second term accounting for the genetic relationship between traits, measured by the additive

(continued)

genetic covariance (often standardized to be a genetic correlation) between the traits (G_{12}), to predict the evolutionary response to selection of trait 1 ($\Delta\bar{z}_1$):

$$\Delta\bar{z}_1 = G_{11}\beta_1 + G_{12}\beta_2 \quad (8)$$

where G_{11} is the additive genetic variance of trait 1 and the subscripts on the betas indicate independent selection gradients on traits 1 (β_1) and 2 (β_2).

The multivariate view of evolution provided by equation 8 implies that we cannot understand the evolution of a single trait in isolation and, as a result, each individual trait may not evolve in the direction predicted by selection or to the optimum predicted by an optimality model. It is ultimately the genetic relationship between traits, generally resulting from pleiotropic connections between them (owing to a shared genetic basis), that makes the viewpoint on evolution provided by quantitative genetics different from that provided by fitness/selection-based approaches such as optimality theory that are central to behavioral ecology. Because it is unarguably biological reality that individuals are not collections of independent traits, but are collections of suites of traits linked through developmental and physiological processes, one cannot lose sight of this difference, even if one does not adopt a quantitative genetic perspective when studying behavior.

Although the two-trait view of quantitative genetics is a step closer to reality, actual biological systems are likely to involve large numbers of correlated traits under complex patterns of linear and nonlinear selection. Genetic and phenotypic correlations between traits complicate selection and evolutionary dynamics in a variety of ways and, as a result, a reader wishing to eventually implement quantitative genetic techniques should ultimately become familiar with multivariate models. Arnold (1994a) provides a good introduction to multivariate quantitative genetics as well as a number of other important issues in quantitative genetics theory. Blows (2007; see also accompanying commentaries and response) also does a particularly good job of making it clear why it is critical to understand selection and genetics on sets of correlated traits and how one's understanding of trait evolution can differ significantly from that expected from a single trait perspective.

(typical for the Holometabola) and polygenic control (typical for hemimetabolous insects) have been documented (reviewed in Roff & Fairbairn 2007). Evolutionary trajectories and equilibrium frequencies of migratory and sedentary strategies have been described using single-locus and polygenic genetic models. Furthermore, the idea that a threshold architecture underlies bird migratory behavior, thereby maintaining genetic variation in nonmigratory populations until threshold expression is reached, may explain the rapid response to selection for migratory behavior (Pulido 2007). By this model, standing variation in genes underlying migratory behavior exists in the nonmigratory population below threshold values. Migratory behavior is phenotypically expressed only when recombination brings together suitable allelic variants, and under selective conditions in which migration becomes advantageous, the population can respond quickly to this change in selective pressure because genetic variation has been maintained.

Genomic Distribution of Loci

The genomes of most, but not all, animal taxa are subdivided into sex chromosomes and autosomes, and there is a growing interest in the types of genes that tend to congregate on these two types of chromosomes. Below we discuss two topics of research of particular relevance to behavioral ecologists: (1) genes underlying sex-specific traits and (2) genes involved in sexual selection.

The location of sex-determining genes impacts the evolutionary potential of sex-specific traits. For example, many traits, including behaviors, affect the fitness of the two sexes differently. When a behavior increases the fitness of one sex but reduces the fitness of the other, it leads to sexual antagonism (Arnqvist & Rowe 2005). One way organisms may avoid such sexual conflicts is through sex-specific expression of genes (Ellegren & Parsh 2007). This allows both sexes to evolve their respective optimal trait values (Lande 1980). Despite the fact that individuals of opposite sex possess almost identical genomes, sexual dimorphism (differences between the sexes in morphology and behavior) is widespread, implying that sex-specific gene expression is common. Sex-specific gene expression generally requires a physical linkage between genes determining sex and the genes regulating sex-specific expression of sexually antagonistic traits. Consequently, genes initiating the cascade of developmental processes leading to sexual dimorphism are typically located on the sex chromosomes (Fisher 1930), for example, as is found in chicken (Kaiser & Ellegren 2006). It is important to appreciate, however, that once the cascade is initiated, the regulated genes need not be on the sex chromosomes (e.g., in humans and *Drosophila* most are distributed on autosomes; reviewed by Fairbairn & Roff 2006). Even in taxa that lack sex chromosomes, the evolution of new, autosomal, sex-determining genes are facilitated if they are physically linked to genes under sexually antagonistic selection (van Doorn & Kirkpatrick 2007).

Besides being an important site for genes coding for sex-specific behaviors, the sex chromosomes are thought to be a key location for genes under sexual selection (box 5.2). Theoretical models suggest that evolution of ornaments in response to sexual selection is most likely if genes coding for the ornaments and preferences for the ornaments are located on the sex chromosomes. In particular, Z-linkage of female preference (in taxa with female heterogamety) is especially conducive to a Fisherian runaway, whereas X-linkage of female preference (in taxa with male heterogamety) coupled with autosomal inheritance of male displays favors sexual selection based on "good genes" (Reeve & Pfenning 2002; Albert & Otto 2005). Overall, the ZW sex chromosomal system in taxa with female heterogamety (e.g., birds and Lepidoptera) is thought to be especially conducive to the evolution of elaborate male ornaments and courtship behaviors (Reeve & Pfennig 2002), and this is supported by comparative analyses of several taxonomic groups (Reeve & Pfennig 2002). However, significant correlations between sex-chromosome system and ornamentation are not ubiquitous (see Mank et al. 2006), and this may reflect the facts that male ornamentation can evolve through processes other than female choice (e.g., male-male competition) and that genes underlying sexually selected traits are not always linked to sex chromosomes (Mank et al. 2006). Furthermore, male ornaments that are condition dependent are expected to be influenced by a large number of loci underlying general condition, and these are not expected to be located on sex chromosomes (Rowe & Houle 1996).

In *Drosophila* (an XY sex-determination system) most genes underlying sexually selected traits also have pleiotropic effects on other aspects of fitness; such genes do not tend to occur more often than chance on the sex chromosomes (Fitzpatrick, 2004). These results find support in crickets (Shaw et al. 2007; Shaw & Lesnick 2009) but contrast with data from birds in which genes underlying mate choice and ornamentation are located on sex chromosomes (Sæther et al. 2007). With the recent proliferation of information on the genomic distribution of loci, it is becoming increasingly apparent that genes on the sex chromosomes do much more than simply determine sex. However, what should also be clear from the studies mentioned above is that the issue of the importance of sex chromosomes for sexual selection is far from resolved. Because the genomic location of loci has important implications for sex-specific behaviors and behaviors tied to mate choice, it is a topic that will continue to be of key interest to behavioral ecologists.

Effect Sizes of Alleles

Since the modern synthesis, the neo-Darwinian view that species differences arise from small effect substitutions at multiple loci has dominated theoretical treatments of adaptive evolution (e.g., Lande 1980). The possibility that allelic substitutions of large effect cause adaptive species differences has been suggested by recent experimental work using quantitative trait locus (QTL) mapping approaches (see Broman 2001 for an accessible review of QTL methods). Behavior obviously plays a central role in the adaptive evolutionary process, but very little is known about the distribution of allelic effect sizes underlying behavioral variation or their contributions to fitness. Recent empirical results show that the dichotomy of small versus large effect sizes underlying adaptive evolution is overly simple. Furthermore, theory suggests that the size distribution of phenotypic effects of allelic substitutions that

BOX **5.2** Diversity of Sex-Determining Mechanisms

Daniel A. Warner and Fredric J. Janzen

When a given trait affects the reproductive fitness of males and females differently, sexual selection should generate phenotypic divergence between the sexes. Indeed, differences between the sexes are among the most spectacular sources of phenotypic variation within populations. How this variation between the sexes, in terms of underlying genes, arises or is maintained is of primary interest to behavioral ecologists, and is often impacted by sex-linked genes. The remarkable diversity of sex-determining mechanisms (SDMs) in animals has important consequences for primary sex ratios and inheritance of sex-linked traits, and hence has fundamental implications for the evolution of behavioral variation.

Whether an embryo develops into a male or female is the product of genes and the environment, and the degree to which these factors contribute to sex determination varies dramatically among taxa. Biologists have traditionally classified SDMs under two main categories: one in which sex is determined entirely by genotypic factors (e.g., sex chromosomes) passed from parents to offspring (i.e., genotypic sex determination, GSD), and the other in which sex is determined by environmental factors during development (i.e., environmental sex determination, ESD). Even within these categories, we see remarkable diversity in patterns of sex determination, and elements of both genotypic and environmental influences on sexual differentiation in single species or populations are common (e.g., Kozielska et al. 2006; Radder et al. 2008).

GSD typically involves sex chromosomes, which contain genes (e.g., SRY in mammals) that initiate the developmental cascade toward male or female gonadal development and further shape secondary sexual traits. In many taxa (e.g., all mammals, some reptiles, some amphibians, some fish, some invertebrates), males are the heterogametic sex containing both X and Y sex chromosomes, whereas females are homogametic (XX). The opposite pattern occurs in some other taxa (e.g., all birds, some reptiles, some amphibians, some fish, some invertebrates) in which females are heterogametic (ZW) and males are homogametic (ZZ). The diversity of GSD mechanisms does not stop here. In some invertebrates, sex determination depends on the ratio of the X chromosomes to autosomes. In other systems, multiple sex-determining genes, epigenetic factors, or sex-determining factors on autosomes may influence offspring sex (Kozielska et al. 2006). Moreover, GSD systems can involve sexes differing in overall ploidy in the absence of heteromorphic sex chromosomes. This occurs in haplodiploid insects in which females arise from fertilized eggs (diploid), but males arise from unfertilized eggs (haploid).

Under ESD, heteromorphic sex chromosomes are absent, and thus the sex of an individual cannot be predicted from its genotype. Instead, sex is determined by environmental factors during development. This mechanism evolved multiple times independently in diverse taxa, and the environmental variables involved vary broadly (e.g., temperature, photoperiod, salinity, maternal nutrition, mate availability). However, the influence of such environmental factors is not independent of genes. That is, environmental conditions interact with genes that influence molecular pathways involved in sex determination. Perhaps ESD is most widely recognized in the form of temperature-dependent sex determination (TSD), which has been well studied in reptiles, but it also occurs in a diversity of other taxa (e.g., fish, invertebrates). TSD exhibits an assortment of patterns and varies among taxonomic levels (figure I). For example, the sensitivity of sex determination to temperature, and even the direction of the temperature effect, varies among species, populations, and even individuals within populations (e.g., Ewert et al. 2005).

(continued)

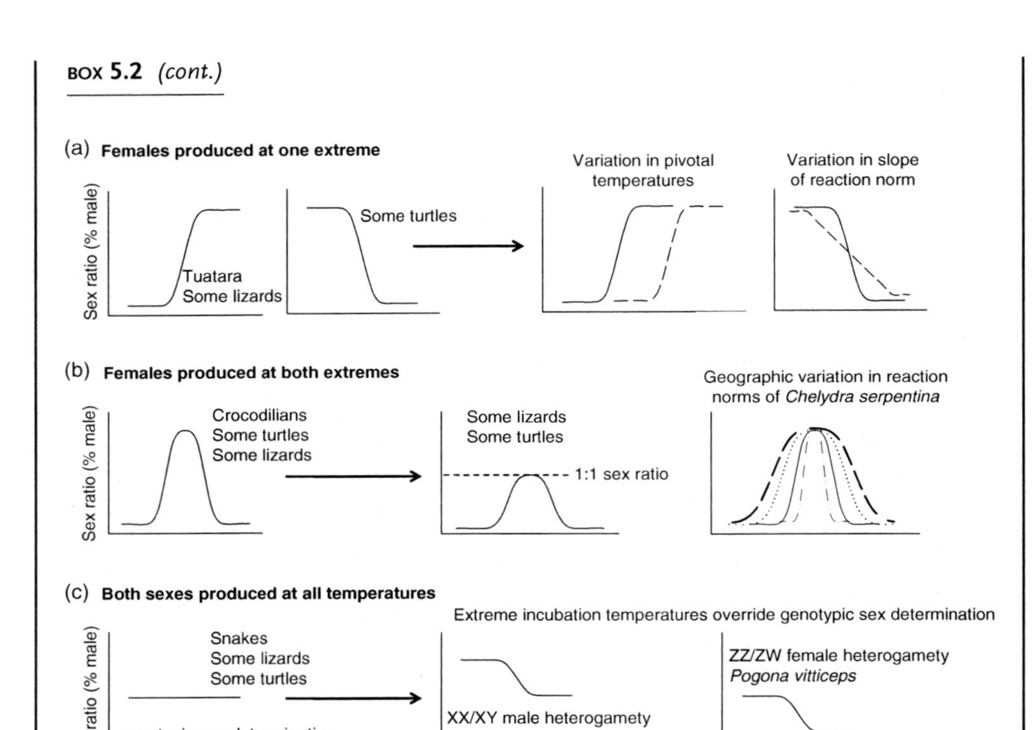

Figure I Diversity of sex-determining patterns in reptiles. All graphs show sex ratio (% male) as a function of increasing egg incubation temperature (x-axes). The three major patterns of sex determination with respect to incubation temperature are shown to the left of the arrows. Patterns to the right of the arrows are variants of those patterns. (a) Patterns of TSD in which males and females are produced at opposite temperature extremes. Pivotal temperature (i.e., temperature that produces 1:1 sex ratio) varies considerably among species, populations, and individuals. Additionally, considerable diversity occurs in the transitional range of temperatures (i.e., range of temperatures that yield mixed sex ratios) among species. (b) Pattern of TSD in which females are produced at both temperature extremes, and males at intermediate temperatures. In some species, intermediate temperatures produce mixed-sex ratios, and other species show geographic variation in shapes of reaction norms (e.g., the common snapping turtle, *Chelydra serpentina*; Ewert et al. 2005). (c) Sex ratio is not influenced by incubation temperature (genotypic sex determination). Recent studies on two distantly related Australian lizards with heteromorphic sex chromosomes demonstrate that extreme incubation temperatures can override genotypic sex determination. Such temperature effects reverse genotypic females to phenotypic males (Radder et al. 2008) and vice versa (Quinn et al. 2007).

Differences in patterns of inheritance and influences on primary sex ratios between GSD and ESD generate fundamentally different consequences for behavioral variation. For example, if genes responsible for sexually selected traits reside on sex chromosomes, then sexual dimorphism is probable (Rice 1984). Whether alleles are linked to the Y or W chromosomes influences the inheritance of traits in sons or daughters, respectively (Reeve & Pfennig 2002). Expression of these sex-linked genes generates

(continued)

sexual dimorphism, including sex-specific behaviors that are fundamental to fitness in each sex. On the other hand, pure ESD lacks sex chromosomes, so such inheritance is unlikely, yet sexual dimorphism is common in ESD species. In these systems, environmental factors trigger expression of sex-determining genes or genes controlling sex steroid hormones that elicit sex-specific phenotypes. Under these situations, mismatches sometimes occur between gonadal sex and behavioral sex (e.g., Gutzke & Crews 1988), which likely reduces individual fitness. As a consequence, selection should favor an ESD system that enables each sex to develop in its respective optimal environment (e.g., Warner & Shine 2008).

Under certain conditions, overproducing one sex enhances maternal fitness, and SDMs can influence the potential for adaptive maternal control over offspring sex ratios. The random nature of chromosome segregation during meiosis could constrain sex ratio bias in GSD systems, thereby limiting sex allocation patterns. However, a substantial literature suggests otherwise, because many species with heteromorphic sex chromosomes skew offspring sex ratios in a presumably adaptive direction (reviewed in Cockburn et al. 2002; see also box 26.3). Mechanisms that enable sex ratio biases under GSD are largely unknown, but multiple hypotheses have been proposed (Pike & Petrie 2003). On the other hand, haplodiploidy provides females with considerable control over brood sex ratios, which renders this SDM especially conducive to maternal sex ratio adjustment. Indeed, empirical work on these systems strongly conforms to predictions from sex allocation theory (reviewed in Ode & Hunter 2002). In contrast, species with ESD can influence offspring sex ratios via oviposition behaviors. For example, selection should favor genes that enable females to detect, and subsequently select, nest sites that yield desirable clutch sex ratios under the prevailing conditions. Indeed, numerous studies of TSD reptiles show that nesting behavior is nonrandom with respect to temperature, and that this behavior is repeatable and heritable (Janzen & Morjan 2001).

The impressive diversity of SDMs has critical evolutionary consequences for behavioral variation. This brief overview reflects the traditional view of SDMs by focusing on GSD versus ESD systems. However, recent evidence suggests that both GSD and ESD might co-occur within populations (Quinn et al. 2007; Radder et al. 2008). Such complexity, should it be more than transient, would have manifold ramifications for behavioral evolution. Investigating the proximate causes of SDMs and their ultimate outcomes will likely reveal greater diversity among taxa and will provide critical insights into ecological and evolutionary mechanisms driving behavioral variation.

move natural populations toward adaptive targets declines exponentially (Orr 2005). These issues are important to the question of whether population evolution is *mutation limited* (i.e., selection pressures exist but populations are unable to respond due to limited genetic variation) or *selection limited* (i.e., ample genetic variation exists but populations do not evolve substantially due to the absence of selective forces), and the potential for gradual versus punctuated change in behavior.

One area relevant to behavioral ecology that has been studied with these questions in mind is the genetic basis of song differences between closely related populations and species of insects. Many species of insects communicate acoustically in a variety of contexts. One of the most conspicuous contexts is male-female communication, in which males (and sometimes females) sing to attract prospective mates, a functional context suggesting the action of sexual selection on mate attraction. Such songs are frequently quite stereotyped in that the variability within species is usually much less than that between species. In cases in which it is possible to cross different song variants (whether within or between species), we have the opportunity to estimate the size of allelic effects that contribute to variation in acoustic behavior, affording tests of and potential insights into the past action of sexual selection (e.g., Shaw et al. 2007; Shaw & Lesnick 2009).

In a recent review, Gleason (2005) summarizes details of genetic architecture from quantitative genetic (and QTL) studies of courtship song in *Drosophila*, noting that a range of estimated effect sizes from small to large underlie the interspecific differences of various traits of the *Drosophila* courtship songs. However, large effect size estimates must be interpreted with caution for at least two reasons. First, effect sizes can be overestimated when sample sizes are low (Broman 2001). Second, without identifying the molecular cause for the behavioral difference, it is difficult to know whether one or many sequential substitutions are responsible for the difference. In a recent groundbreaking study on the evolution and development of trichome pattern, McGregor et al. (2007) found that a large difference in pattern between *D. melanogaster* and *D. simulans*, previously attributed to a single gene (*shavenbaby*), is due to three independent changes in that gene. Thus, three separate allelic substitutions have accumulated to produce the large difference, a change originally thought to be due to a single substitution of large effect.

With estimates of small effect, one is on somewhat more stable ground. Finding small effect sizes is significant because such estimates are unlikely to be biased downward; small estimates that are potentially biased upward are still small. Shaw et al. (2007) found that multiple QTL underlie differences in pulse rate of species of the Hawaiian cricket genus *Laupala*. Effect sizes were generally small and at the lower limit of the power to detect them based on the sample size of the study. Thus, in both the *Drosophila* studies and the *Laupala* study, at least some effect size estimates for the differences between species are of similar magnitude to the variation found within species, suggesting that standing variation could be the source of interspecific differences (i.e., the evolution of behavior in populations is not mutation limited). Interestingly, in the case of *Drosophila*, genes for song differences identified through mutation studies, and genes implicated through QTL studies of song in natural populations are not the same, by and large (Gleason 2005). Despite nearly 20 candidate genes identified for some involvement in courtship song through mutation studies, we are still ignorant about what genes contribute to natural variation in courtship song in this model organism.

Correlated Evolution

In animal signaling systems, such as in sexual signaling, behaviors frequently evolve in a correlated fashion. Under most models of sexual selection, the evolution of sexual signals requires a concomitant evolution of response to that signal. As a consequence, coevolving traits such as male signals and female preferences for those signals often are hypothesized to show genetic correlations within populations (i.e., a presence of allele A at one locus corresponds to a higher than random chance that the individual also contains allele B at a second locus). Genetic correlations may arise either through pleiotropy or linkage disequilibrium (due either to physical linkage or nonrandom association of alleles at unlinked loci). For example, linkage disequilibrium is expected to arise through assortative mating whenever there is genetic variation for a male trait and a female preference for that trait (Lande 1981), owing to the fact that offspring inherit both maternally derived genes for the preference and paternally derived genes for the preferred trait. This linkage disequilibrium is the basis for Fisher's (1930) runaway hypothesis of sexual selection, in which females preferring ornamented males produce sons with improved success at attracting mates that likewise carry the preference genes for those traits. Thus, enhanced mating success of these sons incidentally increases the frequency of the preference genes that they carry. In practice, assessing genetic correlations between traits is analogous to measuring heritability of a single trait, and is carried out through comparing phenotypes of relatives. To detect genetic correlations caused by assortative mating and linkage disequilibrium, it is important that individuals are allowed to exercise mate choice. Conversely, to test for physical linkage or pleiotropy, random mating should be enforced for a number of generations prior to assessment.

Whether we should expect physical linkage between loci underlying male-female processes such as sexual selection is currently unclear. In some cases, physical linkage may facilitate coevolution, when new mutations influence a trait in a favorable direction (e.g., a mutation for large ornaments in a locus tightly linked with another containing alleles for preferring large ornaments). However, equally likely is the reverse scenario, in which the new mutation reduces ornament size—a mutation that is unlikely to spread if positioned near a gene for preferring large ornaments. It is therefore unclear whether, on average, physical linkage impedes or promotes sexual selection, and explicit genetic modeling on this problem is needed.

What is clear is that there are a number of specific cases in which physical linkage between genes underlying mate choice and ornamentation has important consequences. One such case is when two populations are diverging in the face of occasional gene flow. The main force preventing the divergence of two populations that continue to exchange genes through hybridization is recombination, which breaks up coevolved genes for male traits that distinguish each population and the female preferences for those traits. Recombination is lessoned, and thus divergence facilitated, by physical linkage between loci affecting male traits and female preferences (e.g., when these loci are located near one another on a given chromosome). Given the current contention concerning the pervasiveness of sympatric speciation in nature, substantial insight may be gained by asking whether the necessary conditions (e.g., physical linkage between male traits and female preferences) are widespread.

GENETICS WITH THE GENES

There is no doubt that the study of behavioral genetics and ecology would be enriched by knowing the identity of the genes involved in natural variation in behavior. With specific knowledge of the genes contributing to the development of behavior, and the phylogenetic relationships of the organisms of study, we can begin to identify how complex behaviors may have arisen from simpler origins (Robinson et al. 2005). Knowledge of the specific genes participating in behavioral variation would enable tests to determine the sources of that variation, such as single gene versus complex genetic architectures, and the modes of gene action and interaction. Ultimately, knowledge of the genes enables insights into, and estimates of, behavioral variation that can be acted upon by selection. In a few notable examples, the genes underlying behavioral variation have been discovered, but this accomplishment is relatively rare and there remains much potential along this avenue of research. Approaches and strategies to identifying genes are discussed more fully in chapter 28. Here, we discuss a selection of topics about which we might make significant strides with specific knowledge of the genes involved, providing examples when they exist.

The Genetic Basis of Sexual Signals and Preferences

As mentioned earlier, linkage of genes underlying coevolving sexual traits may be ascertained through assessing genetic correlations between the traits, without knowledge of the actual genes involved. However, with knowledge of the genes, we can further discern whether genetic correlations arise through linkage disequilibrium, physical linkage, or pleiotropy. The last scenario asserts that the same genes underlie multiple components of mate choice. Models of sexual selection normally assume separate loci code for male traits and female preferences, although this need not be the case. It is possible that similar neural mechanisms control signal production and reception, a situation termed genetic coupling. The *desat1* gene in *Drosophila melanogaster* is one such example, in which a known inversion contributes both to the production of pheromones used in sexual signaling and to the detection of these pheromones (Marcillac et al. 2005). Genetic coupling has also been invoked to explain tight linkage between female wing coloration and male preferences for this coloration in *Heliconius* butterflies (Kronforst et al. 2006), and song and acoustic preference in *Laupala* crickets (Shaw & Lesnick 2009). However, in other systems in which genetic coupling was initially suspected, more detailed assessment revealed separate neural pathways for signal and preference (reviewed by Butlin & Ritchie 1989). Butlin and Ritchie (1989) argue that separate neural pathways may be similarly affected by common genes, and thus such a finding is not inconsistent with genetic coupling. Due to the incipient stage of such research, we currently have little idea how widespread such situations may be, or the evolutionary implications of such genetic coupling. Pleiotropy is often considered a hindrance to evolution, constraining certain traits from reaching their optimum values because fitness of the underlying gene depends on the sum of its effects across all traits (Fitzpatrick 2004). In contrast, when coevolving traits such as those involved in communication between the sexes have similar underlying genetic bases, evolution may proceed more rapidly. Knowledge about the genes affecting male traits and female preferences will allow for more informed theoretical models of evolutionary processes.

The Genetic Basis of Social Behavior

That social behavior has a biological basis and can evolve has fueled further study into the genetic basis of the constituent behavioral components. In recent years, genomic approaches have enabled much progress in understanding the genetic and molecular bases of social behavior (Robinson et al. 2005). A wide taxonomic diversity of organisms exhibit social behavior, in both simple and complex forms.

One of the most common ways that organisms engage in social behavior is in the context of sexual reproduction. It could be argued that all sexual organisms are social in this sense, because some coordination between male and female is necessary to achieve fertilization. Although many species have promiscuous mating systems, some species have evolved toward pair-bonding and monogamy, extending social interactions beyond single acts of sexual reproduction. A well-studied example of this occurs in the prairie vole (*Microtus ochrogaster*), a species in which monogamy is influenced by the neuroendocrine gene vasopressin 1a receptor (*V1aR*; Young & Hammock 2007). *V1aR* expression was significantly higher in the ventral forebrain of the monogamous *M. ochrogaster* than in its close relative, the promiscuous meadow vole, *M. pennsylvanicus* (Insel et al. 1994). In an exciting study in which *V1aR* levels were experimentally manipulated, Lim et al. (2004) showed that social behavior could be dramatically influenced by increasing the levels of this gene expressed in the ventral forebrain. These results provide an opportunity to associate naturally occurring behavioral variation with molecular causes and potential insights into the molecular mechanisms that mediate observed variation in the level of male prairie vole pair-bonding (Ophir et al. 2008).

The evolution of social behavior may well be caused, initially, by selective pressures to deal with the demands of reproduction. The communal rearing of offspring occurs in a variety of animals from arthropods to birds and mammals, illustrating one way in which the evolution of social life has become more elaborate. The social hymenoptera provide one of the best models for understanding the complexities of social life and its genetic underpinnings, largely due to the intense study of the honeybee, *Apis melifera*. In the honeybee, a conspicuous part of social life involves foraging for the hive, but this behavior shows age-related expression. Young bees typically stay in the hive performing nurse activities to developing embryos, transitioning to foraging only later in life (Seeley 1995). Noticing the similarity between the behavioral effects of the gene *forager* (*for*) in *D. melanogaster* and the behavioral transition experienced by aging honeybees, Ben-Shahar et al. (2002) examined gene expression of *Amfor* (*for*'s homolog in honeybee). A striking correlation between age (and worker task) and *Amfor* expression suggested that the onset of foraging behavior involves the expression of a cyclic GMP-dependent protein kinase, the gene product of *for* (Osborne et al. 1997). Based on this and other examples, Robinson et al. (2005) assert that an emerging theme in the genetics of social evolution is that the genes used by solitary organisms to perform vital biological functions are also used in similar contexts in social organisms, albeit with evolutionary "retooling."

Pleiotropy and Behavioral Evolution

With the identification of actual genes underlying behavioral variation comes the potential to examine the degree to which pleiotropic gene action constrains or promotes the evolution of behavior. Pleiotropic genes are genes that affect multiple phenotypes (Hall 1994) and so evolution in one trait can have additional consequences in some other vital area of the biology of the organism (Greenspan 2004). Thus, pleiotropy at a locus can both constrain and enhance evolution above and beyond a locus devoted to a single phenotype. For example, pleiotropy can constrain evolution if a new mutation that increases fitness through its effect on one trait simultaneously decreases fitness through its effect on the pleiotropic trait. Given that most new mutations are deleterious, such fitness trade-offs in pleiotropic genes are more likely than a mutation that results in fitness gains for both traits. In a recent review summarizing evidence for this idea, Fitzpatrick (2004) discussed (putatively) sexually selected genes in *D. melanogaster* and their pleiotropic effects on nonsexually selected phenotypes, building a powerful case for constrained evolution on these genes beyond the force of sexual selection. Indeed, Greenspan (2004) suggests that pleiotropic action of genes affecting behavior may be widespread. Although it is perhaps too soon to say this is demonstrated, two notable examples of the pleiotropic effects of behavioral genes support the idea. Perhaps not surprisingly, these cases come from the genetic model *D. melanogaster*, albeit through studies of natural variation. Before we discuss these in more detail, we stress that more examples illustrating functional consequences of natural allelic variation at behavioral loci are needed to critically evaluate the role of pleiotropy in behavioral evolution.

Extensive work has been conducted on the genetic basis of circadian rhythms, perhaps because daily clocks are so pervasive among life forms and because early work identified circadian mutants in *D. melanogaster* that shortened, lengthened, or abolished altogether, the circadian period (see Tauber & Kyriacou 2008 for a recent review). These circadian period mutants were subsequently cloned and characterized and were shown to be allelic variants at the same locus, named *period* (*per*). Of interest to behavioral ecologists, researchers have since found allelic variants of *per* in European, African, and Australian populations of *D. melanogaster*, stemming from variable numbers of repeats in the threonine-glycine repeat region that is characteristic of this gene (see Kyriacou et al. 2008 for a recent review). In Europe and Northern Africa, major allelic variants that differ in their ability to buffer the expression of the circadian period from fluctuating temperatures are distributed in a latitudinal cline. The allele with the most robust temperature compensation is found at highest frequency in high latitudes and greatest temperature variability, making a strong case for the action of natural selection in maintaining this variability.

In addition to having an effect on locomotor and eclosion circadian rhythms, the *per* mutants also affect a short-period rhythm in *Drosophila*, the courtship song. This short-period, behavioral effect parallels that of circadian rhythm, with short and long period mutants causing a shortening and lengthening the interpulse interval of the courtship song. The role of *per* in song variation among these mutants was solidified in an exciting transgenic study. Mutant *D. melanogaster* with arrhythmic song were transformed with the *per* homolog from *D. simulans*, restoring rhythmic song in these flies with the periodicity of *D. simulans* (Wheeler et al. 1991). Since these pioneering studies, many additional song genes have been identified, documenting the effect of additional loci on courtship song variation due to induced mutations in *Drosophila* (see Gleason 2005 for a review). Many of these are similarly pleiotropic in their effects.

The gene *forager* (*for*), discussed above in relation to social behavior, has also been shown to exhibit pleiotropic effects in *D. melanogaster*. Initial behavioral observations on flies collected from orchard populations identified two forms of larval behavior, the *sitter* and *rover* phenotypes, which exhibit relatively short and long food trails, respectively (Sokolowski 1980). Molecular investigations revealed that *for* codes for a cyclic GMP-dependent protein kinase (PKG) and that for^{R} is responsible for higher levels of PKG activity (Osborne et al. 1997). Thus the behavioral variation traces its cause to a single gene, and evidence suggests that variation in the system is maintained by frequency dependent selection (Fitzpatrick et al. 2007). Recent work has now identified a second behavioral phenotype affected by this polymorphism. Rover flies show superior short-term learning and poorer long-term memory of odors compared to sitter flies (Mery et al. 2007), and localized PKG expression to the mushroom bodies, suggesting a role for *for* in olfactory learning. These intriguing results raise the possibility of an adaptively matched, functional link between *for*'s role in feeding locomotion and a role in olfactory learning of a food source. If true, this would represent a case in which pleiotropy might facilitate adaptive evolution.

The Genetic Basis of the Mutation/Selection Balance

One of the basic results of population genetics theory is that heritable (additive) genetic variation underlying phenotypic targets of directional selection should be eliminated as advantageous alleles increase in frequency and populations evolve (Freeman & Herron 2004; chapter 3 of this volume). In general, we can predict the frequency of an allele in a given future generation (p') based on the frequency of the allele in the previous generation (p; for a 1 locus, 2 allele trait) when we know the fitness of the three genotypes in the population. An advantageous allele will increase in frequency according to its current frequency in the population and its fitness in heterozygotic and homozygotic states relative to the mean fitness of the population according to the following equation:

$$p' = (p^2 W_{AA} + qp W_{Aa}) / (p^2 W_{AA} + 2pq W_{Aa} + q^2 W_{aa})$$

in which p' is the predicted frequency of allele A in the next generation, p and q are the current frequencies of alleles A and a, respectively, and W_{AA}, W_{Aa}, and W_{aa} are the relative fitnesses of the three possible genotypes. Note that the denominator equates to the average fitness of the population (see Freeman & Herron 2004: 198, for a basic treatment). If the fitness of A is higher than average when it is matched with itself and the alternative allele, we expect its frequency to rise.

It might therefore be puzzling why lower fitness phenotypes persist in natural populations. One such example is that of behaviors associated with human mental disorders. For disorders that demonstrably affect the fitness of their bearers, such as schizophrenia, Keller and Miller (2006) argue that allelic variants causing the disorder persist due to a "mutation-selection" balance. Under the mutation-selection balance, selection removes deleterious alleles that contribute to lowering fitness, but allelic variation is nonetheless maintained by continual renewal through mutation. Thus the frequency of such deleterious alleles is expected to be a function of the mutation rate and the severity of selection removing such alleles from the population. Keller and Miller (2006) favor this evolutionary explanation over alternative hypotheses that might cause the maintenance of genetic variation, such as the persistence of ancestral polymorphisms under nonequilibrium conditions, or a past selective environment in which such traits were neutral or even favored.

Knowledge of the specific genes that lower fitness can lead to estimates of mutation rates and selection intensities against deleterious alleles, which in turn can facilitate predictions of the expected frequency of the behavioral phenotype. The mutation-selection balance is most likely to be a significant contributor to the maintenance of genetic variation when traits have a quantitative genetic basis, because the potential source of genetic variation is complex. For example, in the case of schizophrenia, many candidate genes have been suggested to be involved in the occurrence of this disorder (Sullivan 2008), in keeping with the hypothesis that the genetic underpinnings of variation in the population are quantitative.

Molecular Tests and Mechanisms of Natural Selection

Knowing the genes that affect behavioral components of fitness can enhance the power to detect selection, especially at the molecular level. A topic on which great progress has been made in the last decade is the function and evolution of reproductive proteins in *Drosophila* and other animals (see also chapter 23). For example, male accessory gland products have been shown to directly affect aspects of female reproductive behavior. In particular, the transfer of accessory gland proteins (Acp's) from male to female has been shown to cause behavioral changes between virgin and mated females. Compared to virgins, mated *Drosophila melanogaster* females show lowered receptivity to remating, have increased egg production, increased rates of ovulation and oviposition, and an increased rate of food intake (reviewed in Ram & Wolfner 2007).

In some cases, specific differences in behavior have been traced to the presence of specific accessory gland proteins. The accessory gland protein known as sex peptide (SP) contributes via oogenesis to the elevated egg-laying rate in mated females and subsequent latency to remating (reviewed in Ram & Wolfner 2007). Interestingly, a sex peptide receptor (SPR) has recently been identified (Yapici et al. 2008) that plays a role in mediating the behavioral effects of SP in females. Although SP (and SPR) is apparently highly conserved across *Drosophila,* other Acp's such as the prohormone ovulin have evolved rapidly (Haerty et al. 2007). Ovulin stimulates ovulation in mated female *D. melanogaster* (Herndon & Wolfner 1995) and shows a high rate of amino acid substitution indicative of positive selection (Tsaur et al. 1998; see Jensen et al. 2007 for a recent review of molecular tests of selection).

The Genetic Basis of Convergent versus Parallel Evolution

The study of evolution by natural selection can be approached from many other angles in addition to the molecular investigations discussed above. One of these is the study of convergent evolution, in which similar phenotypes have evolved repeatedly in the history of a group. Frequently it is claimed that convergent evolution is a signature of common selection pressures in different organisms because the same solution to a biological "problem" has been reached independently on multiple occasions. Observations of repeated outcomes in behavioral states are relatively easy to observe, but beg the question as to whether they are due to the same (*parallel* evolution) or different (*convergent* evolution) genes. Although both parallel and convergent evolution may signal the action of selection causing evolutionary outcomes, parallel evolution may also signal genetic constraint, in other words, that there are evolutionary limits to the phenotypic options available to a group of closely related organisms (Arendt & Reznick 2008). Only with knowledge of the changes in the genes involved will we be able to conclusively distinguish between parallel and convergent evolution.

It is certainly the case that many genetic pathways to phenotypic change are possible (e.g., as appears to be the case for *Drosophila* courtship song; Gleason 2005). Recently, however, extensive evidence has accumulated demonstrating parallel evolution due to single genes. Particularly intriguing is the case of melanocortin-1 receptor (*Mc1r*), which affects pigmentation in the mouse (*Mus musculus*) and apparently a variety of other animals (Fitzpatrick et al. 2005; Arendt & Reznick 2008). In both the lesser snow goose (*Anser c. caerulescens*) and arctic skuas (*Stercorarius parasiticus*), *Mc1r* correlates perfectly with the melanic polymorphisms affecting mate choice (Mundy et al. 2004), providing a rare link between genes and mating behavior in the wild.

SYNTHESIS AND CONCLUSIONS

Boake et al. (2002; see also Andersson & Simmons 2006) contrast the complementary benefits of top-down and bottom-up genetic approaches to study the evolution of behavior. A top-down approach emphasizes the genetic causes of the evolutionary process through the study of phenotypic patterns. To the extent that we can describe accurately the genetic architecture of behavioral traits, we will succeed in revealing the role that genetic architecture plays in the force of evolution. In other words, how does genetic architecture of behavior constrain or enable evolutionary change, in the context of particular processes? In a bottom-up approach, the effects of the genes underlying trait variation are under study, with the goal of elucidating the genetic mechanisms governing the expression of behavior or behavioral variation. This approach is vital to our understanding of behavioral evolution as well, in part because it is needed to identify genes contributing to the development and expression of behavior, at the genic, physiological, and behavioral levels. Not only will we gain a specific understanding of the genes underlying behavior, but we will also be able to assess the degree to which genes are evolutionarily co-opted from other roles. Although the top-down role of genetics in behavioral ecology may seem more directly relevant to behavioral ecology today, the integration of bottom-up strategies will ultimately fold back into our understanding of the process of behavioral evolution. The genetic basis of behavior is a vital component to understanding the evolutionary process of behavioral change.

SUGGESTIONS FOR FURTHER READING

For the importance of distinguishing between genetic and environmental influences on behavior, see Mackay and Anholt (2007). Grafen (1984) discusses the traditional perspective on genetics and behavior in behavioral ecology, and Moore and Boake (1994) discuss alternative contributions from evolutionary genetics. For a comparison of different genetic traditions in the study of behavioral genetics see Greenspan (2004). Recently, a synergy can be detected between the fields of behavioral genetics and behavioral ecology as techniques become more accessible in the study of naturally occurring behavior (e.g., Boake et al. 2002; Robinson et al. 2005). Although no definitive text exists, the insights to be gained from genetic approaches to behavior, behavioral ecology, and behavioral evolution are apparent from a wide range of studies (Boake et al. 2002; Fitzpatrick et al. 2005).

Boake CRB, Arnold SJ, Breden F, Meffert LM, Ritchie MG, Taylor BJ, Wolf JB, & Moore AJ (2002) Genetic tools for studying adaptation and the evolution of behavior. Am Nat 160: S143–S159.

Fitzpatrick MJ, Ben-Shahar Y, Smid, HM, Vet LEM, Robinson, GE, & Sokolowski MB (2005) Candidate genes for behavioural ecology. Trends Ecol Evol 20: 96–104.

Grafen A (1984) Natural selection, kin selection, and group selection. Pp 62–84 in Krebs JR & Davies NB (eds) Behavioral Ecology: An Evolutionary Approach. Sinauer Associates, Sunderland, MA.

Greenspan RJ (2004) Quantitative and single-gene perspectives on the study of behavior. Annu Rev Neurosci 27: 79–105.

Mackay TFC & Anholt RRH (2007) Ain't misbehavin? Genotype-environment interactions and the genetics of behavior. Trends in Genetics 23: 311–314.

Moore AJ & Boake CRB (1994) Optimality and evolutionary genetics: complementary procedures for evolutionary analysis in behavioral ecology. Trends Ecol Evol 9: 69–72.

Robinson GE, Grozinger CM, & Whitfield CW (2005) Sociogenomics: social life in molecular terms. Nat Rev Genet 6: 257–270.

6

Behavior as Phenotypic Plasticity

CAMERON K. GHALAMBOR, LISA M. ANGELONI, AND SCOTT P. CARROLL

Phenotypic plasticity is a ubiquitous attribute of organisms (West-Eberhard 2003), and adaptive plasticity in behavior provides some of the most striking examples of how animals adjust to diverse environmental conditions. Indeed, many of the major themes of behavioral ecology deal with traits that exhibit plasticity, including learning and culture, condition-dependent strategies, optimal foraging, habitat and host selection, predator avoidance, mating strategies and sex allocation, aggression, and cooperation (e.g., Wcislo 1989; Komers 1997; Carroll & Corneli 1999; Stamps 2003; West-Eberhard 2003). At the same time, and somewhat paradoxically, the remarkable flexibility of behavior, and its resulting diversity and complexity, appear to have discouraged its study under the same evolutionary framework as other phenotypically plastic traits, despite the centrality of behavior in nature (genes) versus nurture (environment) arguments (e.g., Carroll & Corneli 1999; Robinson & Dukas 1999; Pigliucci 2001).

The broadest and most commonly used definition of phenotypic plasticity is the capacity for a given genotype to produce different phenotypes in response to different environmental conditions. Thus, all phenotypic characters (e.g., morphological, behavioral, and physiological) can be studied and understood under the conceptual framework of phenotypic plasticity (West-Eberhard 2003). Indeed, there is a growing recognition that adaptive phenotypic plasticity plays an important role in ecological (e.g., Fordyce 2006) and evolutionary (e.g., Pigliucci & Murren 2003; Grether 2005; Braendle & Flatt 2006; Ghalambor et al. 2007; Crispo 2007) processes. Yet, behavioral ecologists as a whole have been slow to view adaptive or nonadaptive behavior through the lens of phenotypic plasticity (e.g., Sih 2004). In contrast, evolutionary ecologists studying morphological and life history traits have embraced and incorporated the dual role of the genotype and environment into their research programs (reviewed in Gotthard & Nylin 1995; Schlichting & Pigliucci 1998; Pigliucci 2001; West-Eberhard 2003; DeWitt & Scheiner 2004).

Why behavioral ecology has lagged in the integration of phenotypic plasticity as a foundation is open to debate (see below). However, a likely reason is that because behaviors are often reversible, and capable of rapid change, many behavioral ecologists have the impression that behaviors do not lend themselves to study under the more commonly used framework of developmental plasticity for nonreversible morphological traits (e.g., Robinson & Dukas 1999; Piersma & Drent 2003; Sih 2004). Here, we build on previous arguments (e.g., Carroll & Corneli 1999; West-Eberhard 2003; Piersma & Drent 2003; Sih 2004; Nussey et al. 2007) that phenotypic plasticity provides a powerful and appropriate framework to explore the adaptive significance and evolution of behavior in almost all contexts.

We begin with a brief conceptual introduction that defines terms and serves as a primer for the conditions that favor behavioral plasticity. Next, we discuss the behavioral reaction norm as a tool for visualizing and testing the adaptive significance of plasticity. We then discuss the potential role of behavioral plasticity as a mechanism that can facilitate adaptive evolutionary change. Finally, we discuss case studies in which the conceptual framework of behavioral plasticity has imparted key insights that might not otherwise have been appreciated.

CONCEPTUAL BACKGROUND

Phenotypic Plasticity and Behavior

All phenotypic traits, behavioral or otherwise, are the products of complex interactions between the genes individuals inherit and the environments they experience. The degree to which a genotype's phenotype is influenced by environmental variation describes a continuum of responses from traits that are *canalized* (relatively fixed or invariant expression regardless of the environment) to those that are *plastic* (labile or variable expression dependent on the environment). The reaction norm of a trait is the quantitative representation of this continuum as a line or curve that shows the particular way a genotype's phenotype varies as a function of the environment: canalized traits exhibit flat lines (figure 6.1a), whereas plastic traits exhibit lines or curves of a particular slope or shape (figures 6.1b–d). For categorical or threshold traits, such as polyphenisms or polymorphisms, the reaction norm is likely to be a *step function* in which a single phenotype is expressed across a range of environments until some critical point is reached, beyond which a different phenotype is expressed (figure 6.1b). The variation in reaction norms among the different genotypes of a population is referred to as *genotype by environment interaction* or simply G × E, and is a measure of within-population variation in plasticity (figure 6.1c, d).

Genotype by environment interactions may be striking when, for example, the phenotypic ranking among genotypes changes across environments, potentially altering fitness rankings (figure 6.1d). This last situation might imply some degree of specialization or local adaptation and the presence of nonadaptive plasticity in some environments. One explanation for such a pattern is that it is due to functional or developmental limitations that constrain individuals from performing well in different environments. This may occur when an individual encounters a new environment not previously inhabited by that population, or when the environment changes suddenly, so that its behavioral responses are not adapted to the new conditions. For example, human disturbance may contribute to *evolutionary traps*, when behavioral responses are triggered by inappropriate cues in human-modified environments (Schlaepfer et al. 2002 and chapter 29 of this volume), such as the inland (rather than seaward) migration of sea turtle hatchlings on beaches with light pollution, or oviposition on roads (rather than on the surfaces of ponds or streams) by mayflies

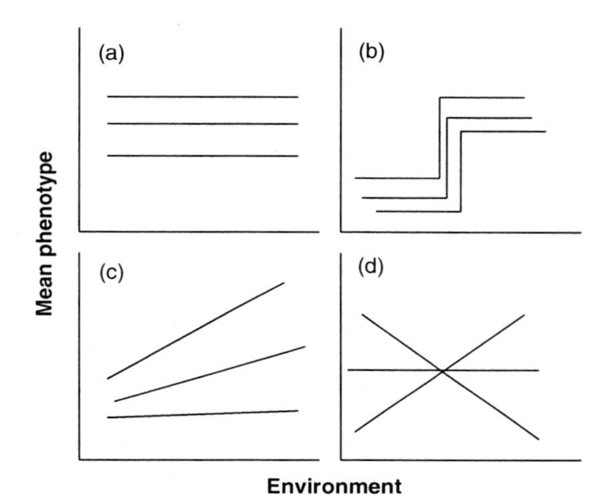

FIGURE 6.1 Shown are hypothetical reaction norms for three genotypes that exhibit variation when (a) the trait is canalized and shows no plasticity with respect to this environmental factor (i.e., the reaction norm is flat), although there are differences between genotypes; (b) there is plasticity and the trait responds to the environment in a threshold manner, such as a polyphenism (i.e., some critical level of the environmental variable triggers a change in the phenotype) and each genotype responds in a similar way, (c) there is plasticity and a genotype by environment interaction (G × E) with no change in the rank order of the genotypes (i.e., the environment causes a predictable increase in the phenotypic value of the trait, and genotypes differ in a consistent manner), and (d) there is plasticity and a G × E interaction and the rank order of the genotypes change (i.e., the environment causes an unpredictable change in the phenotypic value of different genotypes).

because of polarized light reflection from asphalt (reviewed in Robertson & Hutto 2006).

Theoretical and empirical findings suggest that adaptive plasticity evolves when several environmentally dependent conditions are met. These are when (1) populations are exposed to variable environments, (2) environments produce reliable cues, (3) different phenotypes are favored in different environments, and (4) no single phenotype exhibits high fitness across all environments (e.g., Via & Lande 1985; Via 1987; Moran 1992). Although there is general agreement that plasticity can evolve, whether adaptive plasticity evolves as a by-product of selection acting on mean trait values or directly on the coefficients describing the reaction norm remains controversial (see box 6.1). Despite the potential fitness benefits of adaptive plasticity, there is also the recognition that various costs and limits will influence how plasticity evolves (e.g., Dewitt et al. 1998; Relyea 2002a), which helps to explain why we do not find a single limitlessly plastic species dominating the planet. This general theory for phenotypic plasticity is easily extended to behavioral traits.

Behavior is no different from any other measurable component of the phenotype, except perhaps in the speed and reversibility of most behavioral change compared to some morphological and physiological characters that may only be expressed once and are nonreversible (e.g., Piersma & Drent

BOX 6.1 Contrasting Quantitative Genetic Models for the Evolution of Plasticity

There is general agreement that plasticity can evolve in response to natural selection. However, there is less agreement on how such evolution takes place (see Via et al. 1995). Two classes of quantitative genetic models have been proposed for how selection drives the evolution of plasticity, the *character state approach* and the *polynomial or reaction norm approach*. In the character state approach, the reaction norm of a particular trait is conceptualized as the *set of phenotypic values* expressed in each environment by a given genotype, and selection acts on the population mean and genetic covariances of these different character states (e.g., Via & Lande 1985; Czesak et al. 2006). In contrast, under the polynomial approach the reaction norm of a trait is described as a *polynomial function of phenotypic values* by a genotype across a range of environments and selection acts on the population means and genetic covariances of coefficients of the polynomial (e.g., Scheiner & Lyman 1989; Gavrilets & Scheiner 1993a, 1993b). Although both approaches use the generalized model for selection on multiple characters (Lande 1979) and have been shown to be mathematically equivalent under most conditions (e.g., de Jong 1995; Gavrilets & Scheiner 1993b), fundamental biological differences exist in how plasticity is envisioned to evolve in response to directional selection. Under the character state approach, plasticity evolves as a by-product of selection acting on the population mean values in different environments. Here, the character states between environments are treated as separate, but genetically correlated traits within an individual, such that the response to selection ($\Delta\bar{z}$) is determined by both the genetic variances of the character states within an environment (G_{ii}) and by the genetic covariances between the states (G_{ij}). Thus, the modeling approach developed by Via and Lande (1985) to describe the evolution of a character expressed in two environments is:

$$\begin{pmatrix} \Delta\bar{z}_1 \\ \Delta\bar{z}_2 \end{pmatrix} = \begin{pmatrix} G_{11} & G_{21} \\ G_{12} & G_{22} \end{pmatrix} \begin{pmatrix} q_1\left(\frac{\bar{w}_1}{\bar{w}}\right)\nabla_1 \ln \bar{w}_1 \\ q_2\left(\frac{\bar{w}_2}{\bar{w}}\right)\nabla_2 \ln \bar{w}_2 \end{pmatrix} \tag{1}$$

where $\Delta\bar{z}_i$ is the change in the character state expressed in the ith environment in one generation of selection, $q_i\left(\frac{\bar{w}_i}{\bar{w}}\right)\nabla_i \ln \bar{W}_i$, is the force of selection on each character (with

(continued)

a weighting function; see Via & Lande 1985), G_{ii} is the genetic variance of the character state that is expressed in the ith environment, and G_{ij} is the genetic covariance between the character states. Thus, the response to selection of the character state expressed in environment 1 is:

$$\Delta\bar{z}_1 = q_1\left(\frac{\bar{w}_1}{\bar{w}}\right) G_{11}\ \nabla_1 \ln \bar{W}_1 + q_2\left(\frac{\bar{w}_2}{\bar{w}}\right) \nabla_2 \ln \bar{W}_2 \tag{2}$$

where there is both a direct, $q_1\left(\frac{\bar{w}_1}{\bar{w}}\right) G_{11}\nabla_1 \ln \bar{W}_1$ and correlated, $q_2\left(\frac{\bar{w}_2}{\bar{w}}\right) \nabla_2 \ln \bar{W}_2$, response to selection (see also Via 1994, 1987). This same response to selection can more generally be shown as:

$$\Delta\bar{g}_{CS} = (1/\bar{w})\ G_{CS}\ b_{CS} \tag{3}$$

where $\Delta\bar{g}_{CS}$ is the vector of selection responses of character states and is determined by the genetic variance-covariance matrix $\mathbf{G}_{CS}$ of the character states across environments, as well as the selection gradient vector $\mathbf{b}_{CS}$, which contains the partial derivatives of overall mean fitness toward the mean character state per environment (de Jong 1995). The important point here is that evolutionary changes in the trait value in one environment due to the direct and indirect response to selection will necessarily alter the shape of the reaction norm without the need to treat plasticity (i.e., the slope of the reaction norm) as a trait in itself.

In contrast, under the polynomial approach the coefficients that describe the reaction norm are the targets of selection, meaning that plasticity is treated as an independent trait. A reaction norm is easily modeled as a polynomial function (Gavrilets & Scheiner 1993a, de Jong 1995), where the genotypic value in any given environment x is written as:

$$g(x) = g_0 + g_1 x = g_2 x \ldots . g_k x^k \tag{4}$$

Under the polynomial approach the coefficients g_i of the reaction norm can be treated as quantitative traits compatible with the equations (1–3) above. For example, a vector of selection responses $\Delta\bar{g}$ can contain quantitative information about the reaction norm in the form of changes in mean height, $\bar{g}_0$, mean slope $\bar{g}_1$, mean curvature $\bar{g}_2$, and so forth. Thus under the polynomial approach, selection on the reaction norm is a function of the genetic variance-covariance matrix $\mathbf{G}_{RN}$ of the coefficients in the reaction norm (see de Jong 1994, 2005). This leads to the same general equation as 3 above:

$$\Delta\bar{g}_{RN} = (1/\bar{w})\ G_{RN}\ b_{RN} \tag{5}$$

Which model to choose? We agree with De Jong (1995) that the polynomial model is most appropriate when studying a graded phenotypic response to a continuous environment, whereas the character state model is most appropriate for discrete responses to discrete environments. For example, consider how calling rate of a frog population distributed along an elevational gradient might be studied. Because temperature decreases with increasing elevation and because frogs are ectotherms, we might expect the reaction norm for calling rate to diminish with increasing elevation or temperature. Any experimental design that treated calling rate and elevation or temperature as continuous variables would lend itself well to the polynomial approach, such that the height, slope, and curvature of the reaction norm could be estimated. In contrast, if the experimental design measured calling rate of genotypes at a high and low elevation or temperature, then the character state approach would be sufficient.

2003; Sih 2004; Nussey et al. 2007). However, as Sih (2004) eloquently argues, such distinctions consider only the ends of a continuum, and actually represent interesting opportunities for studying the ecological and evolutionary consequences of different kinds of plasticity. At one end of the continuum are behaviors capable of great flexibility (i.e., exhibit reversible change) and that are frequently expressed and revised throughout the lifetime of the individual (i.e., altered after environmental assessment and experience, such as through learning). At the other end of the continuum are the strongly canalized behaviors, such as the pecking of gull chicks at the orange spot on the parent's bill, the egg-rolling behavior of geese, or other fixed action patterns (e.g., Lorenz 1965). Yet many or most traits likely show elements of both.

Consider, for example, the begging response of nestling birds, the intensity of which can be highly flexible as a function of short-term hunger (e.g., time since parents have delivered food) and long-term needs (e.g., amount of food required to fledge, which may vary due to body condition, sex, or rank in brood; Price et al. 1996). A given individual may therefore reduce or increase its begging in response to short-term hunger, but individuals of overall lower condition may always beg more than their high-condition counterparts (figure 6.2a). Intriguingly, the flexibility in this behavior can also be potentially overridden and more or less canalized by mothers who alter the androgen levels in the eggs depending on laying order (figure 6.2b). Natural and experimental results have found that increased androgens in the egg result in nestlings that will maintain high begging intensity independent of short-term hunger (figure 6.2c; see also Schwabl 1993, 1996). The degree of plasticity expressed in begging behavior is thus mediated not only by conditions in the nest environment, but also by the maternally determined embryonic environment. What are the fitness consequences and evolutionary outcomes of having a flexible versus nonflexible strategy given a predictable or unpredictable environment? If we apply such questions to the begging response of chicks, one reasonable hypothesis is that given hatching asynchrony, being the youngest offspring predictably equates to inhabiting a relatively bad environment (the last nestling hatched is often the smallest and least developed), and a nonflexible strategy in which high begging is maintained might be favored (figures 6.2b, c). At the same time, the unpredictable nature of food supply may favor a flexible strategy that allows parents to accurately assess the hunger levels of their chicks through the intensity of begging behavior (figure 6.2a), allowing food to be transferred in whatever way maximizes parental (and perhaps offspring) inclusive fitness.

Aside from the conditions that select for flexible or nonflexible behavioral strategies, the evolution of plasticity may also be constrained by various factors. For example, behavioral ecologists have recently emphasized how behavioral traits are intercorrelated within individuals, a phenomenon manifesting as *personalities* or *syndromes* (chapter 30). Personality describes the limitation of behavioral plasticity, such that behavioral expression is correlated across environments or contexts. Thus, although it might benefit animals to be less or more aggressive in specific contexts (e.g., in foraging versus predator avoidance), selective history and development may constrain individuals in how plastic they can be, via their personality. The existence of such syndromes has important implications for how selection acts on suites of behaviors and for the maintenance of behavioral variation in populations.

Finally, various authors have argued that the flexibility and reversibility of behavior has discouraged adoption by behaviorists of phenotypic plasticity theory otherwise aimed at developmental traits that are not reversible (Sih 2004; Piersma & Drent 2003). Given the evidently broad range of responses that fall under the umbrella of phenotypic plasticity, Piersma and Drent (2003) suggested a distinction between *developmental behavioral plasticity*—irreversible behavioral variation dependent on environmental cues present during development, and *behavioral flexibility*—reversible behavioral variation dependent on current environmental cues. Applying these definitions to the begging example above, the persistent maternal effect of androgen transfer resulting in consistently high begging intensity, independent of hunger, is developmental behavioral plasticity, whereas the change in begging intensity with hunger is behavioral flexibility (figure 6.2). Plasticity in sexual behavior can also be dichotomized under this framework. For example, developmental behavioral plasticity occurs when an individual's sex and all of the corresponding behavioral traits are environmentally determined very early in life, as with temperature-dependent sex determination in reptiles. In contrast, behavioral flexibility in sexual behavior can

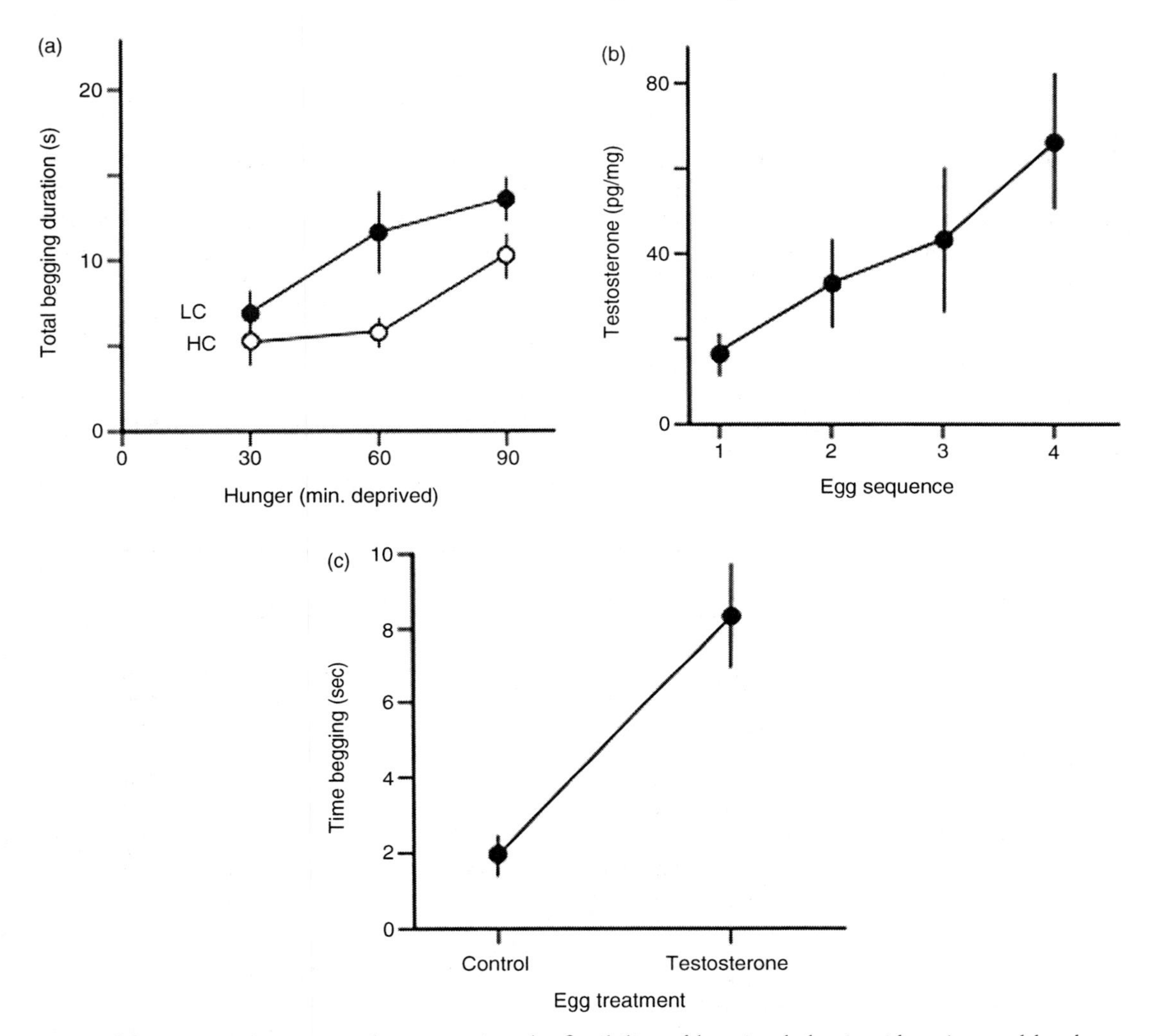

FIGURE 6.2 (a) Reaction norms demonstrating the flexibility of begging behavior (duration and loudness ± SE) of female yellow-headed blackbird nestlings in response to short-term (30, 60, or 90 minutes) food deprivation and long-term differences in condition. Redrawn from Price et al. (1996). (b) Testosterone concentration (in picograms per milligram of yolk ± SE) increases with egg sequence within clutches laid by canaries. Redrawn from Schwabl (1993). (c) Total time begging (mean ± SE) by control and testosterone-treated canary chicks within 1 hour after hatching. Redrawn from Schwabl (1996).

occur when it is continuously adjusted with changing social and environmental conditions, as in the tactical adjustment of male reproductive behavior as female availability changes (e.g., Carroll & Corneli 1999). Ultimately, we regard this segregation as overly dichotomized and focused on the ends of the continuum, as environmental cues may generate irreversible changes in behavior at all life stages, but we raise it because it helps nonetheless highlight a useful distinction.

Taking a Reaction Norm Approach

Terminology aside, the simplest means by which evolutionary behavioral ecologists may adopt the framework of phenotypic plasticity is to explicitly consider individual behaviors as reaction norms or continuous functions rather than categorical entities (see also Stamps 2003; Nussey et al. 2007). Because the reaction norm captures how ecological variation shapes behavioral variation, many of the basic questions of behavioral ecologists become quite testable, such as those probing the adaptive significance of behavioral variation within populations (e.g., between families or genotypes), and the selection pressures that drive the evolution of behavioral distinctions between populations and species. Specifically, a reaction norm approach can explicitly combine information about (1) the distribution of environments experienced by individuals (e.g., the selective influence of food, temperature, predation risk, number of mates, age), (2) the behavioral responses to these environments (i.e., the shape of the reaction norm), and (3) the extent that reaction norms reflect underlying differences among genotypes, or the amount of genetic variation in the population (i.e., the variation in slope and height of all the reaction norms measured). Thus, information regarding the selective environment, the behavioral response, and the variation among individuals in behavior can all be combined with information on fitness to understand the adaptive significance of a behavior. Such an approach serves as a powerful framework when attempting to understand the genetic basis of behavioral differences among populations or species when combined with common garden techniques, pedigree information, and/or experimental manipulations (e.g., Carroll & Corneli 1999; Ghalambor & Martin 2002; Brommer et al. 2005; Nussey et al. 2005). Furthermore, different statistical approaches are available to facilitate this type of approach (see box 6.2).

BOX 6.2 Contrasting Statistical Methods for Studying Phenotypic Plasticity

Empirical investigations of phenotypic plasticity are likely to use different statistical approaches depending on the nature of the experimental design. Here we briefly review three common approaches likely to be used by behavioral ecologists studying phenotypic plasticity. The first approach involves comparisons of genetically related groups in environments that are categorically defined (e.g., low versus high food or cold versus warm temperature) using analysis of variance (ANOVA). Here, researchers will ideally generate groups of genetically related individuals (full sibs, half sibs, or clones) that represent different families or lines. Individuals from each of these families are then split between environments where their phenotypes are measured. This experimental design lends itself well to a two-way ANOVA, in which the parameters of the model can be interpreted as follows (see also Fry 1992; Via 1994):

Source	Meaning of F-test
Genotype (V_G)	Estimates the amount of genetic variation underlying the overall mean phenotype or the covariance of genotypic values across environments (depending on how the F-tests are constructed; see Fry 1992)
Environment (V_E)	Estimates the degree to which the environment affects the phenotype (i.e., whether there is phenotypic plasticity)
G x E (V_{GxE})	Estimates the amount of variation among genotypes in response to the environment (i.e., the amount of genetic variation for phenotypic plasticity)
Error (V_e)	Estimates the variation among replicates within G x E combinations
Total (V_P)	Estimates the total phenotypic variance

A second statistical approach, related to the first, is to measure changes in the phenotype in response to a continuous environment using analysis of covariance (ANCOVA). The experimental design is as above, except that the phenotypes are measured in multiple graded environments (e.g., increasing temperature, food, predation risk). As with the ANOVA approach, the environment term estimates if there is plasticity, but does so by testing the slope of the covariate. Similarly, a significant genotype by covariate interaction would suggest there is genetic variation for plasticity. Both the ANOVA and ANCOVA approach are amenable to a diversity of field and lab experiments.

(continued)

A third statistical approach that is likely to be of interest to behavioral ecologists working with natural populations of relatively long-lived animals or intensively studied individuals is the use of random regression to describe individual reaction norms (reviewed in Nussey et al. 2007). Because many behaviors are flexible (i.e., labile or reversible) and are expressed numerous times over a lifetime, for any given individual it is possible to describe a reaction norm of how behaviors change as a function of age, condition, or other environmental factor. Variation among these individual reaction norms is obviously a product of both genetic and environmental effects, thus complicating interpretation of the ecological and evolutionary causes of population-level variation in behavioral plasticity (Nussey et al. 2007). Random regression, a mixed effect statistical model using the individual functions of continuous covariates as random effects, has recently been applied in a number of cases to describe and investigate the causes and consequences of variation in individual reaction norms (e.g., Brommer et al. 2005; Nussey et al. 2005). Following Nussey et al. (2007), consider this example: A researcher is interested in the relationship between an environmental variable E (e.g., temperature) and a behavioral trait y (e.g., timing of reproduction) both measured on occasion j. At the individual level, y_{ij} the phenotype of individual i on occasion j can be specified as follows:

$$y_{ij} = \underbrace{\mu + \beta E_j}_{\text{Fixed effect}} + \underbrace{p_i + p_{Ei} E_j}_{\text{Random effect}} + \varepsilon_{ij}$$

The fixed part of the model describes the population average response of trait y to changing environment E (here assumed to be linear), where μ is the population mean phenotype in the average environment and β is the population mean slope of y on E. Individual deviations from the population average elevation and slope of the reaction norm are modeled by the random effects structure, where p_i represents the deviation from the population mean phenotype independent of E and the elevation of the individual reaction norm, p_{Ei} represents the deviation from the population average slope of the individual's plasticity, and ε_{ij} is a residual error term. This equation is typically solved using restricted maximum likelihood (REML) to estimate the fixed effects, the residual (co)variance structure, and a variance-covariance matrix containing the variances in elevation (p_i) and slope (p_{Ei}) and the covariances between them (Nussey et al. 2007). Although interpretation of mixed models is complicated, likelihood ratio tests can be used to test the variance components in these models, where a significant V_p indicates significant differences between individuals in their average trait value, and a significant V_{pE} indicates significant variation in individual plasticity (i.e., in response to the same environment, individuals differ in the slope of their reaction norms; see Nussey et al. 2007 for more details).

Finally, if pedigree information is available, one of the major advantages of using random regression in natural populations is that it is possible to statistically separate genetic and nongenetic effects on reaction norms using the *animal model* (see Kruuk 2004; Brommer et al. 2005; Nussey et al. 2005, 2007 for more details and use of this approach). Such an approach provides the best available method to measure the evolutionary potential of plastic behaviors through the estimation of the variation in plasticity, its heritability, and the likely response to short- and long-term environmental change.

Susan Foster and colleagues have used such a reaction norm approach to understand divergence in highly plastic behaviors in three-spined stickleback fish (*Gasterosteus aculeatus*). Comparisons of oceanic (ancestral) and benthic and limnetic freshwater (derived) forms reveal that the ancestral oceanic fish exhibit behavioral plasticity in the amount of conspicuous "zigzag dancing" they perform during courtship in response to the presence of cannibalistic foraging groups. Because cannibalistic groups may detect nests based on conspicuous behavior, reducing zigzag dancing during courtship may be an adaptive means of reducing nest mortality. Shaw et al. (2007) found that among the derived freshwater ecotypes there has been divergence in the pattern of plasticity in this behavior; limnetic ecotypes (in which cannibalistic groups do not form) almost always incorporate zigzag dancing in courtship and behavioral plasticity has been lost, whereas in the benthic ecotype (in which cannibalistic groups do form) plasticity is retained and zigzag dancing decreases in the presence of cannibalistic groups (figure 6.3). Because the expression of courtship is environmentally dependent, a simplistic comparison of average behaviors would be insufficient to detect population or ecotypic differentiation. Instead, by evaluating the behavior in more than one context and looking at the slopes of the reaction norms, the pattern of divergence is revealed (figure 6.3). These results also illustrate the power of placing reaction norms in a comparative context. Focus on a single population or ecotype demonstrates that plasticity exists for a behavior, but comparisons of populations that differ in their selective environments also reveal how these plastic responses may have diverged (see also Gotthard & Nylin 1995; Carroll & Corneli 1999; Stamps 2003).

Another example in which a reaction norm approach has great potential to be informative is in the study of conditional strategies. A conditional strategy exists when an individual expresses alternative tactics depending on its status, in which status can be determined by a combination of environmental and genetic factors. Conditional strategies were initially proposed to explain cases in which alternative reproductive phenotypes were observed within a sex that did not appear to be explained by fixed genetic polymorphisms of equal fitness, as suggested by traditional game theoretic approaches. For example, in many species, males can adopt either a *bourgeois* (fighting) tactic by defending resources and attracting females, or they can adopt a *parasitic* (sneaking) tactic by stealing copulations from bourgeois males. The adoption of these alternative tactics often depends on the individual's body condition, social environment, and risk of predation. For example, male Atlantic salmon (*Salmo salar* L.) can either mature early, and at a smaller size, in order to sneak copulations, or mature later, and at a larger size, in order to compete for the opportunity to guard females. Heritability estimates have shown that juvenile growth rate, which is the main determinant of status and the subsequent tactic chosen, is shaped by genetic and environmental effects (Garant et al. 2003), and is therefore plastic.

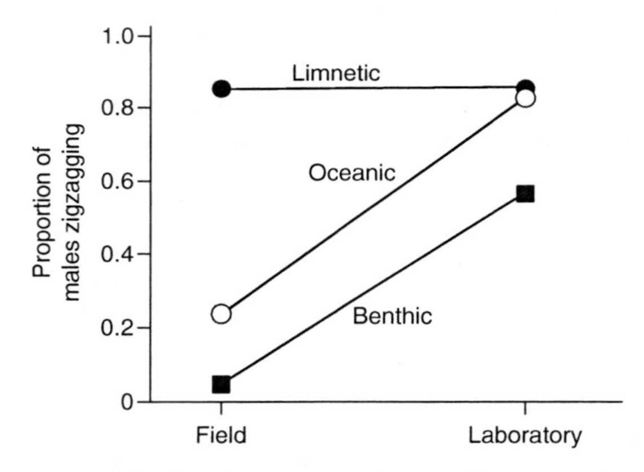

FIGURE 6.3 The proportion of males from oceanic (ancestral) and benthic and limnetic (derived) populations of three-spined stickleback that incorporated zigzag dancing into courtship displays in field and laboratory environments. Redrawn from Shaw et al. (2007). The plasticity of oceanic fish is retained in the benthic ecotype and lost in the limnetic ecotype.

Because status and tactic choice are often dependent on environmental factors, the investigation of conditional strategies is well suited to a norm of reaction approach. A particularly compelling example of this approach is work by Doug Emlen on the dung beetle, *Onthophagus acuminatus*. In these beetles large males with long horns defend burrows dug by females under dung; males mate with resident females who pull the dung into the burrows and lay eggs on them. In contrast, small males with short horns are unable to successfully defend burrows and instead adopt a sneaker strategy in which they dig side burrows or slide past territorial males to intercept and mate with females inside burrows. Which strategy is adopted by males is condition dependent and determined by the nutritional conditions during development, such that the use of standard breeding experiments that focused only on horn length revealed no heritable genetic variation between fathers and sons (Emlen 1994). Instead, Emlen (1996) took a reaction norm approach by artificially selecting on the horn length–body size relationship and was able to shift the threshold body size separating horned from hornless males. Thus, a reaction norm analysis was able to capture the standing genetic variation between individuals in the environmental switch points (i.e., the degree to which individuals differ in their response to the amount of food available during development), which is something that the traditional model of conditional strategies has lacked (see Emlen 1996 for more detail). For this reason Tomkins and Hazel (2007) have recently advocated a norm of reaction approach that incorporates quantitative genetics to understand the evolution of conditional strategies, arguing that this will provide the appropriate genetic framework for understanding selection on alternative tactics by revealing the genetic and environmental variation that underlies their expression (see also chapter 25).

As a cautionary note, there are practical constraints to a reaction norm approach, especially in an adaptive context (Stamps 2003). In particular, experimental manipulations should mimic natural conditions, and when possible multiple environmental contexts should be investigated (plasticity in response to environmental conditions outside the range an organism is likely to experience are both difficult to interpret and of less biological relevance). Second, estimating the heritability of the reaction norm is likely to be difficult for behavioral ecologists who study natural populations of animals that are not as amenable to laboratory breeding experiments. Thus, evolutionary conclusions regarding divergence in behavior should be tempered by the recognition that maternal effects, interactions among traits, experience, or other types of environmental effects can also influence the shapes of reaction norms.

ROLE OF BEHAVIORAL PLASTICITY IN ECOLOGICAL AND EVOLUTIONARY PROCESSES

That phenotypic plasticity in general (e.g., Waddington 1961; Schlichting & Pigliucci 1998; Pigliucci 2001) and behavioral plasticity specifically (Baldwin 1896; Wcislo 1989; Robinson & Dukas 1999; West-Eberhard 2003) may be advantageous ecologically when it allows a genotype to have a broader tolerance to environmental conditions, and hence higher fitness across multiple environments, has long been recognized. However, the evolutionary implications of plasticity have generated lasting controversy. Put simply, does plasticity limit evolutionary responses by shielding genotypes from selection, or does it facilitate evolution by generating novel phenotypes and altered selective regimes (West-Eberhard 2003; Pigliucci & Murren 2003; Price et al. 2003; Schlichting 2004; Ghalambor et al. 2007)? Here we explore these contrasting perspectives, focusing on the role of behavioral plasticity in determining the selective environment and its evolutionary consequences.

Behavioral Plasticity and the Selective Environment

Abiotic and biotic environments tend to vary temporally and spatially in ways both predictable (e.g., daily/seasonal changes in temperature, distribution of high versus low quality habitat) and unpredictable (e.g., a surprise attack from a predator, the sudden loss of habitat). Behavioral plasticity is typically one of the first ways in which organisms deal with such environmental variation (e.g., Wcislo 1989; Huey et al. 2003; West-Eberhard 2003). Adaptive behavioral plasticity can thus play an important role in buffering environmental heterogeneity (and thus strong directional selection) by allowing individuals to survive and persist in changed or novel environments (e.g., Baldwin 1896; Robinson & Dukas 1999; Price 2006). For example, Sol et al. (2005) argued that birds with greater behavioral flexibility have a higher probability of successful introduction beyond their native geographic range than do less flexible species. Because adaptive behavioral plasticity can broaden both the range of environments individuals occupy and the distribution of phenotypes across environments, it is likely to alter the strength and direction of natural selection when present (reviewed in Fordyce 2006; Price 2006).

Losos et al. (2004, 2006) provide an informative example of how behavioral plasticity can alter patterns of selection. Following the experimental introduction of a novel ground-dwelling predatory lizard (*Leiocephalus carinatus*) to an island inhabited by its potential prey, the brown anole (*Anolis sagrei*), the prey population experienced strong directional selection for increased running speed and longer legs (Losos et al. 2004). However, after interacting with the predator for only 6 months, a significant number of brown anoles shifted their behavior and became more arboreal. This behavioral shift resulted in a reversal of selection toward shorter limbs, which conferred agility on narrow stems in the arboreal environment (Losos et al. 2006). This result supports the classic view of behavioral plasticity initiating evolutionary shifts in morphology, physiology, or ecology of organisms by dictating how selection acts (e.g., Wcislo 1989; Huey et al. 2003).

The alternative perspective emphasizes the buffering role of behavior on selection. Rather than driving the direction of selection on morphological or physiological traits, behavior acts as a filter that shields other traits from the effects of selection. Huey et al. (2003) review this perspective and examine plasticity in behavioral thermoregulation in the lizard *Anolis cristatellus*. Body temperatures are lower than expected at low elevations and warmer than expected at high elevations because the lizards actively occupy microhabitats closer to their preferred temperature. This flexibility results in body temperatures being maintained closer to preferred values across a temperature gradient. Because body temperatures remain similar, despite different air temperatures at high and low elevation sites, it follows that the strength of selection for local adaptation in thermal physiology will be reduced. Yet, the degree to which there is reduced selection for local adaptation because of plasticity in habitat selection for preferred conditions across heterogeneous landscapes remains a largely unexplored question.

Behavioral Plasticity and Evolution

The contrasting views of behavioral plasticity as either a driver of evolutionary change or an inertial factor that inhibits evolutionary change captures the more general debate on the role of phenotypic plasticity in the evolutionary process (reviewed in Wcislo 1989; Robinson & Dukas 1999; Pigliucci & Murren 2003; West-Eberhard 2003; Schlichting 2004; Grether 2005; Braendle & Flatt 2006; Price 2006; Ghalambor et al. 2007; Crispo 2007). Baldwin (1896) was one of the first to consider how adaptive behavioral plasticity could allow for adaptive evolution in what is now often referred to as the *Baldwin effect*. Briefly, Baldwin argued that (1) many behaviors are plastic and subject to modification by the environment through learning and experience, (2) genetic variation exists in the capacity for individuals to exhibit plasticity in behavior, (3) when this plasticity is adaptive and provides a fitness advantage, there is an opportunity for selection to act on this variation, and (4) over time, behaviors that were initially plastic may become canalized, or the degree of plasticity may itself increase (see review in Crispo 2007). Thus, without invoking any Lamarckian inheritance, Baldwin argued for a mechanism by which adaptive behavioral plasticity could facilitate adaptive evolutionary change (i.e., allelic substitution) by first enabling a population to persist in response to novel environments and subsequently altering the rate and direction of evolutionary change. A similar concept to the Baldwin effect is Waddington's (1961) *genetic assimilation*, in which an environmentally induced phenotype becomes canalized by directional selection and no longer requires the environment for its induction, and West-Eberhard's (2003) less restrictive interpretation *genetic accommodation* (see more detailed discussions of this concept in Pigliucci & Murren 2003; West-Eberhard 2003; Schlichting 2004; Ghalambor et al. 2007; Crispo 2007). Both Baldwin and Waddington were contemplating mechanisms that incorporated environmentally induced plasticity as mechanisms into a neo-Darwinian framework to explain adaptive evolutionary change (West-Eberhard 2003; Crispo 2007). Yet, despite the potentially revolutionary implications of the Baldwin effect and genetic assimilation/accommodation for understanding the origins and evolution of adaptive traits, only recently have these ideas been critically discussed (Schlichting 2004).

The theoretical framework of the Baldwin effect and genetic assimilation/accommodation predicts that plasticity can drive adaptive evolutionary change, but under what set of conditions should we expect behavioral plasticity to drive evolutionary change? The general answer to this question seems to depend on how good behavioral plasticity is at tracking environmental variation. Price and colleagues (Price et al. 2003; Price 2006) provide a useful conceptual framework in the form of the *adaptive surface* for investigating how selection is facilitated by plasticity. If we consider a population in a given environment poised on an adaptive peak, and which must cross a fitness valley to access a new, higher adaptive peak, a traditional viewpoint would focus on the role of directional selection acting on existing genetic variation and causing evolutionary change via allelic substitution (Price et al. 2003). Alternatively, adaptive plasticity that matches the difference in adaptive peaks could help move the population in the direction of the new adaptive peak and effectively "smooth" the adaptive landscape by hopping over fitness valleys (Haldane 1957; Price et al. 2003; Ghalambor et al. 2007). In theory, if this plasticity is nearly perfect, it moves the population very close to the new adaptive peak, resulting in a population that would experience stabilizing selection, such that no genetic differentiation would be expected (Price et al. 2003). However, if the adaptive plasticity is incomplete (i.e., in the right direction, but still away from the peak) directional selection on extreme phenotypes would be expected to drive adaptive evolution (Price et al. 2003). Because adaptive plasticity is unlikely to produce a perfect fit to the new adaptive peak, incomplete adaptive plasticity is most likely to produce rapid evolutionary changes, in part through promoting population persistence that permits evolution (Ghalambor et al. 2007). If we apply this reasoning to the examples of predator-induced change in antipredator behavior and behavioral thermoregulation of lizards discussed above, we would draw the following inferences: In the first case, brown anoles exhibit behavioral plasticity in habitat use and shift to a more arboreal habitat in the presence of the predatory curly-tailed lizard. This behavioral shift is adaptive in the sense that it reduces predation risk, but incomplete in that many individuals continue to use the ground and experience mortality from the predatory lizard (Losos et al. 2004). Thus, variation in the plastic response provides an opportunity for directional selection and hence genetic differentiation to occur under the scenario described by Price et al. (2003). In the second case, behavioral plasticity in habitat use and thermoregulation in *Anolis cristatellus* is also adaptive in that it maintains body temperatures closer to the optimum and is invariantly applied across elevations (Huey et al. 2003). Here, adaptive plasticity appears to be so effective that it has prevented directional selection and constrained evolutionary shifts in thermal physiology along an elevational gradient (Huey et al. 2003).

In both cases above, adaptive behavioral plasticity plays a key role in either facilitating or constraining the direction of evolutionary change, yet relatively few behavioral studies have been placed in this context. One notable exception is work by Yeh, Price, and colleagues on the successful colonization of a lowland habitat by high elevation dark-eyed juncos (*Junco hyemalis*; e.g., Yeh & Price 2004). Sometime in the early 1980s, dark-eyed juncos colonized and became established in the coastal Mediterranean climate of the University of California–San Diego campus. Successful colonization appears to have been strongly facilitated by adaptive behavioral plasticity in the timing and period of reproduction in response to a breeding season that is almost twice as long as in its ancestral habitat, which in turn resulted in the production of twice as many young. Additionally, the derived population exhibits reduced aggression, lower testosterone levels, and a genetically based reduction in the amount of white in the tail feathers. Collectively these results are consistent with the Baldwin effect, whereby adaptive behavioral plasticity initially allows for expression of traits required for persistence in a novel environment and allows subsequent evolutionary change (Yeh & Price 2004). However, the degree to which behavior itself has evolved in this population remains unclear. In an analogous study of red squirrels (*Tamiasciurus hudsonicus*) in northern Canada that advanced the timing of breeding by 18 days in the 1990s, Réale et al. (2003) found that both plasticity and evolution underlie this adaptive response to climate warming. Both studies suggest that taking an explicit behavioral plasticity perspective offers key insights into understanding such important questions as how animals invade new environments and how established populations might respond to environmental change. Below, we review three case studies in more detail to illustrate how a behavioral plasticity perspective can be applied.

CASE STUDIES

The following three case studies demonstrate how a norm of reaction approach can provide additional insight into behavioral diversity at multiple levels of genetic organization: within populations, among populations, and among species. The first example examines plasticity within a population, both between individuals and between the two sexes, providing insight into the contextual origins of plasticity in female mate preferences. The second case study illustrates how plasticity can be compared between populations, as well as within each population to understand the mechanisms for adaptive variation in male mating strategy. Finally, a norm of reaction approach can be used comparatively, as seen in the third case study, to contrast the degree of plasticity between species and relate that to variation in selection pressures that favor it.

Plasticity in Female Mate Preference

Comparatively little effort has been made to apply the concepts of phenotypic plasticity to mating preferences, yet they are as likely to be flexible and developmentally plastic as are other traits. Female preferences for male traits can vary tremendously with the environment depending on such factors as the cost of sampling males, the ability to detect male signals, the density of potential mates, the sex ratio, the interactions between males, the behavior of other females, the risk of predation, the quality of resources, and the experiences and condition of the female (Jennions & Petrie 1997). A female may reap benefits for herself or her offspring by choosing an attractive male under one set of conditions, but that choice may not be optimal if the environment changes or if her offspring disperse to a different location (Greenfield & Rodriguez 2004). Plasticity in mating preference is one way to solve this problem in a highly variable environment. Here we describe a case in which reaction norms were used to characterize variation in the plasticity of mate preferences, providing new insight into the evolution of female mating strategies in the guppy, *Poecilia reticulata*.

Naturally occurring guppy populations experience a broad range of environmental conditions in the rainforest streams they inhabit in Trinidad, setting the stage for potentially high levels of adaptive behavioral plasticity. An important environmental factor that varies among streams—and among pools within streams—is forest canopy cover, which affects light level. This causes variation in the availability of the unicellular algae that serve as a primary source of carotenoid pigments in diet of guppies (Grether et al. 1999). Female guppies prefer males with greater concentrations of carotenoid pigments in the orange spots on their bodies (Grether 2000), which may indicate male quality by revealing foraging ability, immune strength, and health. However, the preference for carotenoid coloration is not fixed; there is plasticity in the strength of the preference, related to the degree of carotenoid availability in the environment. Grether et al. (2005) demonstrated that females reared without carotenoids show a stronger preference for males that consumed high levels of carotenoids (and their resulting orange body coloration) compared to females reared in a carotenoid-rich environment (figure 6.4). These findings suggest adaptive plasticity in female mate preference, because orange spots should be better indicators of male quality in low-carotenoid streams than in streams with abundant levels of carotenoids (Grether et al. 2005).

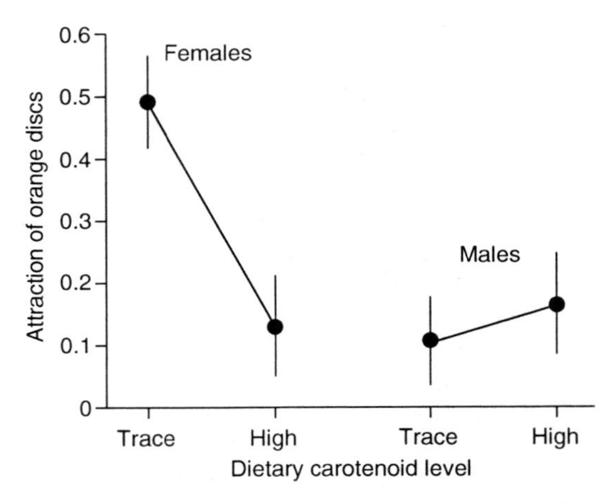

FIGURE **6.4** Reaction norms demonstrating differences in plasticity between female and male guppies in their attraction to orange coloration (mean rate of pecking at orange discs ± SE). Redrawn from Grether et al. (2005). Females reared in an environment with trace levels of carotenoids exhibited a stronger preference for orange discs than females reared in an environment with high levels of dietary carotenoids. Males, on the other hand, did not exhibit significant plasticity in response to orange coloration, suggesting that female responsiveness evolved in a mating, not a foraging, context.

In addition to characterizing adaptive plasticity in female preference, Grether et al. went one step further by using a norm of reaction approach to compare male and female responsiveness to orange coloration through their attraction to orange discs. Interestingly, they demonstrated that the plasticity of responsiveness to orange coloration was specific to females and was not present in males. This difference between males and females strengthens the argument that plasticity in the preference evolved in a mate choice context and argues against the idea that it instead evolved in a foraging context, for example, that guppies perhaps had an initial sensory bias that helped them find orange edible fruits (figure 6.4; Grether et al. 2005). This example demonstrates a novel way that analysis of plasticity—along appropriately manipulated environmental gradients that are relevant to reproduction—can advance research on mating strategies, in this case by permitting evolutionarily meaningful comparisons between males and females in responsiveness to an environmental cue.

Plasticity in Mating Tactics

Just as phenotypic plasticity in mate choice criteria used by females permits adaptation to changing environments, plasticity in male mating decisions may influence reproductive success. Mating with multiple females in a breeding season may significantly increase male reproductive success. However, in some species, seeking new females may make it physically impossible to prevent current or prior mates from pairing with new males. Defense of mates, versus seeking or controlling additional females, is the central dilemma explored in the study of male mating systems. A common approach is to compare populations in which the intensity or form of competition for mates leads to predictable differences in male mating behavior (e.g., Schuster & Wade 2003). Although such behavioral contrasts are widespread in the literature, surprisingly little work has been done to determine whether they result strictly from plastic responses to environmental disparities (the common assumption), from adaptive genetic differences in strategy among populations, or from both together. Describing norms of reaction is a good way to distinguish among these possibilities to better understand adaptive processes.

This approach has been used to investigate mating tactics of male soapberry bugs (*Jadera haematoloma*), which continue to copulate with a female after inseminating her, blocking access by other males for up to many days and uncoupling only briefly to permit oviposition. However, when mate guarding is prevented experimentally, the female usually copulates with a new male within just a few minutes, whereupon half or more of her former mate's sperm are displaced, dramatically reducing his parentage. The functional significance of guarding is clear, but it also restricts males from pursuing additional pairings.

Carroll and Corneli (1995, 1999) compared norms of reaction in mate guarding, versus searching for multiple females, in geographically isolated populations that differ in the intensity and variability of male-male competition. In the U.S. state of Florida, an even adult sex ratio on host plants results in less male-male competition than in Oklahoma, where a highly variable adult sex ratio averages 2.2 males per female. The geographic mosaic of sex ratio means that the most effective way to sire offspring likely differs between the two regions. Indeed, in nature males from Oklahoma are more likely to remain with a female across multiple daily ovipositions than are males from Florida. This could be caused by adaptive genetic differences between one population of "guarders" and another of "searchers." Alternatively, perhaps the two *ethotypes* in reality share a common, adaptively plastic reaction norm in response to female availability and the risk of sperm competition, namely, "Guard more when single females are rare, and less when they are common."

To test this possibility, greenhouse-reared virgins were arrayed in groups of 24 across a range of four sex ratios (figure 6.5). A norm of reaction approach demonstrated that Oklahoma males proved much more responsive to sex ratio, guarding less at low, and more at high, ratios, than did Florida males—a result consistent with what we know about the naturally variable sex ratio in Oklahoma. Because of reduced plasticity, Florida males were not only more likely to miss opportunities for promiscuous matings when females outnumbered them; they also did not increase their mate guarding when they outnumbered females (Carroll & Corneli 1995).

The authors further dissected this genetic difference in strategy by comparing the family reaction norms for guarding versus nonguarding within populations (figure 6.5). They found that the Oklahoma population showed significantly more additive genetic variation for the reaction norm than did the Florida population. Thus, despite the potentially buffering effects of Oklahoma plasticity on diversifying natural selection (that might otherwise lead to the evolution of guarder and searcher specialists within that population), significant genetic differences do occur between families. That variation suggests greater evolutionary potential for the behavioral strategy in Oklahoma. However, at least one Florida family ("K") increased guarding with sex ratio, suggesting that a genetic basis for adaptive plasticity is not entirely absent there.

This population comparison suggests that it is the reaction norm itself that has differentiated. Studying plastic traits with this approach distinguishes genetic from environmental contributions to phenotypic expression across a range of conditions pertinent to fitness. Because natural selection on plastic traits acts on reaction norms, evaluating the response curves and functional significance of those norms is the clearest way to model their evolution. Many natural systems offer prospects for

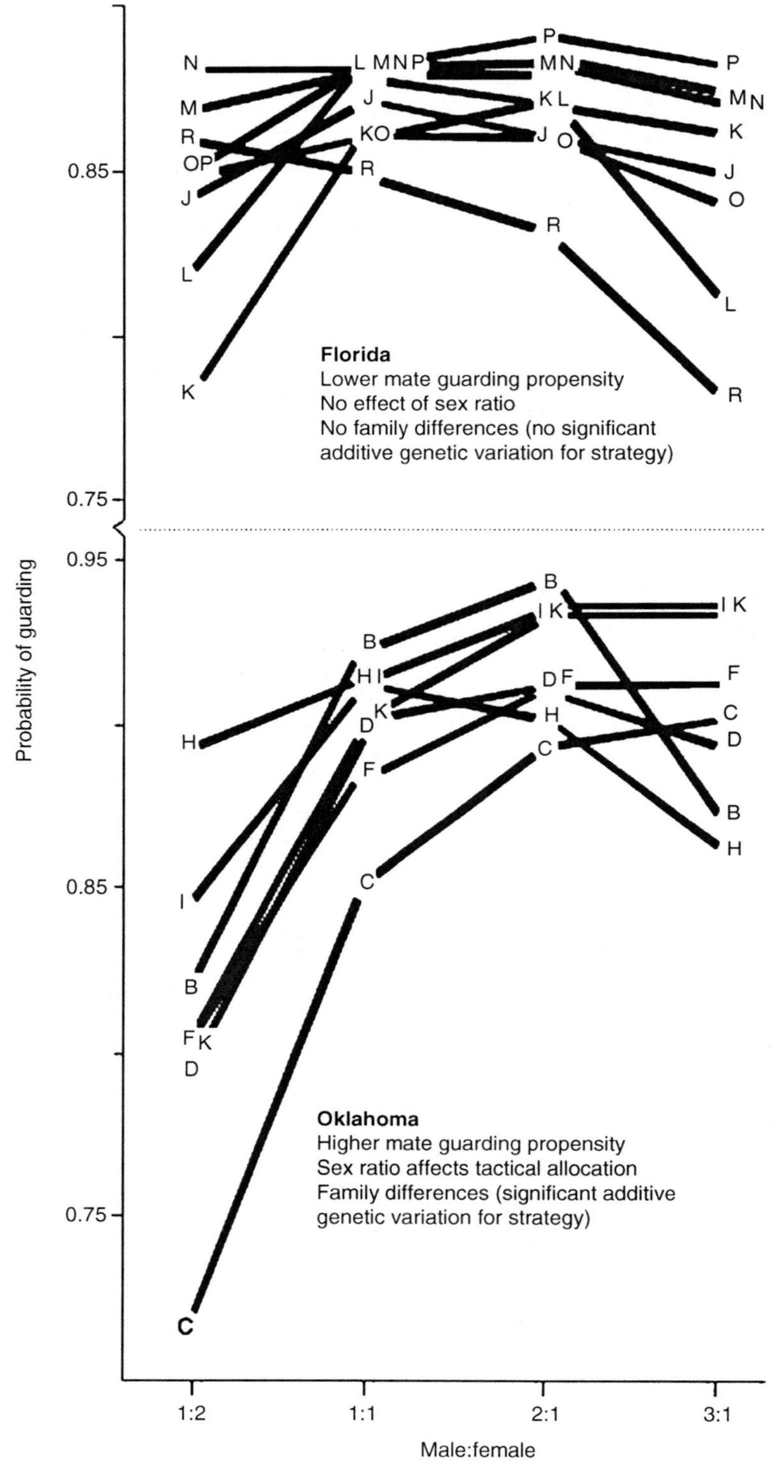

FIGURE 6.5 Influence of sex ratio on male soapberry bug allocation to serially monogamous mate guarding versus promiscuous nonguarding; shown are families (capital letters) of half-sib brothers at a series of experimental sex ratios. Letter positions show the mean probability (from four replicates) that a male continued to guard across observations made in 3-hour intervals over 8 days. Adult sex ratios in nature are naturally much more male-biased and variable in Oklahoma than in Florida. Modified from Carroll and Corneli (1999).

examining how and why individuals differ within and between populations in functional behaviors and personalities (chapter 30 of this volume; Carroll & Watters 2008).

Predator Induced Plasticity in Reproductive Strategies

It has long been recognized that in response to an increase in the perceived risk of predation, animals alter their behavior (e.g., activity level, habitat selection, time spent vigilant) to reduce that risk (e.g., Lima & Dill 1990). However, such a response often produces associated costs—for example, a reduction in foraging or provisioning of young, thus forcing a balance between competing demands (e.g., Sih 1992). Thus, the ability to exhibit behavioral plasticity under the risk of predation appears to be a common strategy for a wide range of organisms (Lima & Dill 1990). Contrasts in the decisions made by breeding birds illustrate the adaptive dynamics of plastic parenting.

Where a bird places its nest (e.g., ground, shrub, canopy, or cavity) is strongly correlated with its risk of nest predation, with ground and cavity nests being the safest and shrub and canopy nests being most vulnerable (e.g., Martin 1995). Within a substrate, successful nests tend to be more concealed than depredated nests (e.g., Martin & Roper 1988). In studies that experimentally increase the risk of nest predation when birds are selecting nest sites, birds respond plastically by selecting more concealed sites (Eggers et al. 2006; Peluc et al. 2008). Similarly, when the risk of nest predation is experimentally decreased, a diversity of bird species shift their nest sites to more conspicuous areas perceived as safe (Fontaine & Martin 2006). Although such plasticity in nest site selection appears adaptive, other fitness costs may be incurred. For example, plasticity in nest site selection has been shown to have consequences for other behavioral decisions. Eggers et al. (2006) found that Siberian jays reduced clutch size when the risk of predation was experimentally increased. Peluc et al. (2008) found that individuals that shifted to more concealed nest sites in response to experimentally elevated predation risk, in turn visited the nest to feed their young more readily that those that did not. Such plasticity suggests the cognitive ability to both monitor ongoing risks and opportunities and calculate the relative merits of each behavioral decision. Yet, how plasticity in one behavioral decision predictably alters subsequent behavioral decisions remains largely unknown for most systems.

The selective pressure imposed by nest predators on parental behavior also appears to play an important role in driving variation among species in mean and plastic responses. Comparative studies of parental activity at the nest reveal that nest visitation during the incubation (e.g., Martin & Ghalambor 1999) and nestling period (e.g., Martin et al. 2000) decrease as the vulnerability to nest predation risk increases; species that use safer nest sites appear less constrained by predators and visit their nests more frequently. What is the relationship between this continuous variation in vulnerability and behavioral plasticity? Manipulation of predation risk by presenting taxidermic mounts and vocalizations of common predators at the nests of coexisting species reveals a strong correlation between the magnitude of plasticity (i.e., the slope of the reaction norm) and the vulnerability to predation risk. For both the incubation and nestling period, the willingness of parents to visit the nest decreased as vulnerability and the risk of predation increased across species (figure 6.6; Ghalambor & Martin 2000, 2002). These results highlight how a reaction norm approach placed within a

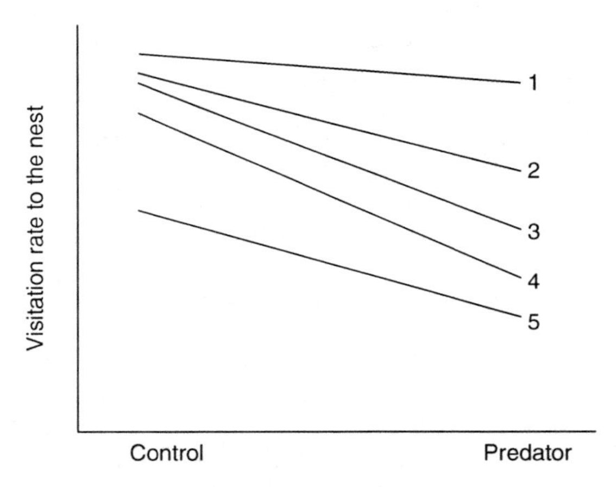

FIGURE 6.6 Reaction norms for visitation rate to the nest in the presence of a nonthreatening control and a potential nest predator. Each slope represents a different species and numbers refer to the vulnerability to nest predation (1 = lowest risk; 5 = highest risk). Species at low risk exhibit a weaker response to the predator in the form of a more shallow reaction norm, and species at higher risk exhibit a stronger response and more steep reaction norm. Modified from Ghalambor and Martin (2002).

comparative context reveals patterns not otherwise obvious. By comparing the slopes of the reaction norms, the question of behavioral response changes from "Did it significantly change?" to "How much did it change and is the variation correlated with vulnerability?" (figure 6.6). In this case, if the reaction norms are heritable, heterogeneity of slopes reflects how behavioral plasticity has evolved in response to different degrees of risk. Had the study been carried out in only a single species with either low or high vulnerability to predation, the conclusion drawn might have either supported or rejected a strong role for predation in shaping the behavioral response of parents (figure 6.5). In contrast, the comparison reveals a correlation between the magnitude of the plasticity and the vulnerability to predation risk that points to a selective mechanism underlying interspecific variation in parenting (Ghalambor & Martin 2002).

CONCLUSIONS AND FUTURE DIRECTIONS

Behavior as a Model for Studying Plasticity

To date, much of the theoretical and empirical work on phenotypic plasticity has centered on morphological, physiological, and life history traits. Yet the environmental responsiveness of many behaviors, and successful case studies of behavioral reaction norms, suggest that it is time for evolutionary behavioral ecologists to take the lead in studying behavior as plasticity. More studies are needed quantifying how behavioral expression systematically changes across environments within and between individuals, populations, and species. Mechanistically, such studies can go beyond simply looking at the phenotypic expression of behavior to incorporate hormonal influences and changes in gene expression. In animals that lend themselves to controlled breeding experiments or in which pedigrees are available, we also need to know the heritability of behavioral reaction norms if we are to predict the evolutionary response of plastic behaviors to selection. Similarly, we have relatively little data on how maternal effects and other environmental influences shape observed patterns of plasticity, and what the transgenerational consequences of this variation are. Finally, greater insight into genotype by environment interactions can be gained if behavioral ecologists report reaction norms for individuals in a population, rather than means and variances. It is often the case that behavioral ecology studies collect these types of data in the process of investigating the adaptive function of a behavior, but fail to consider their studies in the context of phenotypic plasticity.

Power of Adaptive Behavioral Plasticity in Determining Selection and Thus Adaptive Evolution

Given the importance of behavioral plasticity in shaping the selective environment, behavioral ecologists should be embracing ideas such as the Baldwin effect and genetic assimilation/accommodation and taking a leading role in empirically testing the role of behavior in evolutionary change (Price et al. 2003; West-Eberhard 2003). The explosion of phylogenetic and phylogeographic data now available for a broad range of species allows evolutionary behavioral ecologists to conduct comparative studies within a known evolutionary framework. Such frameworks allow explicit comparisons of reaction norms between ancestral and derived species or populations (see sticklebacks above), and permit hypothesis testing concerning the selection pressures that drive evolutionary changes in the reaction norm and other correlated or impacted traits. Indeed, many of the examples discussed here focus not just on how behavioral plasticity influences fitness, but also on the implications of that plasticity for evolutionary change in other traits. Thus, behavioral ecologists need to consider the integrative nature of phenotypes, and how behavior, morphology, physiology, and other traits collectively and independently respond to environmental variation.

Behavioral Plasticity, Conservation and Adaptation to Global Change

Behavioral plasticity has increasingly important implications for conservation biology (chapter 29). Behavioral plasticity can lead to conservation problems, for example, when it facilitates the introduction and establishment of exotic species in novel environments, driving native species to extinction (Hughes & Cremer 2007). It can also be problematic for the reintroduction of endangered species, because it requires that captive breeding programs provide a realistic environment for development and appropriate training, for example, with predators and prey, prior to release (e.g., Shier & Owings 2006). However, behavioral plasticity can also have important beneficial effects for the persistence of native species in disturbed and changing environments. Species with long generation times, low rates of population growth, or little genetic diversity may not be able to adapt genetically to rapid changes, such that phenotypic plasticity is the only way to mediate the negative effects of disturbance that occur over short timescales (e.g., Carroll & Watters 2008). There is a growing body of research investigating behavioral shifts of animals in response to human disturbance such as recreation and urbanization. In many cases these demonstrate behavioral avoidance of humans that effectively reduces negative human-wildlife interactions, but sometimes habituation to anthropogenic disturbances has the opposite effect by intensifying conflicts with humans and reducing survival (e.g., Harveson et al. 2007). Behavioral plasticity in response to changing climate, habitats, and landscapes will become increasingly important. Unfortunately, little is known about the extent to which behavioral plasticity will allow populations to adjust to shifts in temperature, precipitation, habitat, and resource availability. There is some evidence, however, of shifts in habitat use and flexibility in seasonal breeding and migratory behavior in response to climate change (Nussey et al. 2005; Charmantier et al. 2008). The biosphere is currently experiencing a biodiversity crisis driven by increasing human population and consumption and resultant impacts, such as habitat loss, invasive species, and overexploitation. A critical question that remains is the extent to which natural populations will be able to respond adaptively, not only through rapid evolution, but also through phenotypic plasticity (e.g., Carroll & Watters 2008; Hendry et al. 2008).

SUGGESTIONS FOR FURTHER READING

The phenotypic plasticity literature is vast, and even among researchers in the field there is disagreement on fundamental issues. Fortunately, a number of good review papers and books exist that can introduce behavioral ecologists to the major concepts and provide a gateway to more specific literatures. Pigliucci (2001) reviews a number of fundamental ecological and evolutionary concepts related to phenotypic plasticity, including a chapter dedicated to behavior. Similarly, the edited volume by DeWitt and Scheiner (2004) contains several insightful chapters, including a thoughtful treatment of behavioral plasticity by Sih (2004). Supplementing these books we recommend review papers by Wcislo (1989), Robinson and Dukas (1999), and Stamps (2003) which offer diverse yet complementary views of behavioral flexibility.

DeWitt TJ & Scheiner SM (2004) Phenotypic Plasticity: Functional and Conceptual Approaches. Oxford Univ Press, Oxford.

Pigliucci M (2001) Phenotypic Plasticity: Beyond Nature and Nurture. Johns Hopkins Univ Press, Baltimore, MD.

Robinson BW & Dukas R (1999) The influence of phenotypic modifications on evolution: the Baldwin effect and modern perspectives. Oikos 85: 582–589.

Sih A (2004) A behavioral ecological view of phenotypic plasticity. Pp 112–125 in TJ DeWitt & SM Scheiner (eds) Phenotypic Plasticity: Functional and Conceptual Approaches. Oxford Univ Press, New York.

Stamps J (2003) Behavioural processes affecting development: Tinbergen's fourth question comes of age. Anim Behav 66: 1–13.

Wcislo WT (1989) Behavioral environments and evolutionary change. Annu Rev Ecol Syst 20: 137–169.

7

Evolution of Behavior: Phylogeny and the Origin of Present-Day Diversity

TERRY J. ORD AND EMÍLIA P. MARTINS

Studying the evolution of behavior is a challenge. Behavior rarely leaves a fossil record, and selection experiments are generally feasible for only a limited number of short-lived insect and fish species. In his classic paper, Niko Tinbergen (1963) suggested two approaches to understanding behavioral evolution—studies of "survival value", or function, and evolutionary history. Many behavioral ecologists interpret evolution through the first of these and emphasize the present-day function of behavior within populations. For example, we might find that the colorful plumage and courtship dances seen in male manakin birds attracts females, and that females in turn use these traits to assess and subsequently choose among several possible mates. Or perhaps we discover that Caribbean anole lizards producing strenuous physical displays during aggressive interactions do so to deter rivals from stealing territory. In each case, we might conclude that display behavior has evolved or, more specifically, is an adaptation for attracting mates or deterring rivals. But this is essentially a reiteration of function and tells us very little about how these behavioral traits actually arose and changed over time. In particular, it is difficult to decipher why a behavioral trait takes its present form or why there is so much diversity in functionally equivalent behavior across species. If a behavioral trait serves the same function, why does it vary in different species? Are these differences adaptive, or do they reflect some other evolutionary process? To investigate evolution, we need to add Tinbergen's second emphasis on the study of closely related species and consider the differences and similarities in behavior observed across species within the context of their evolutionary history. This is the foundation of the comparative method. A similar approach was adopted by Charles Darwin trying to explain the diversity of forms he observed in the natural world, which led him to develop his theory of evolution outlined in *The Origin of Species* (1859) and other classic works.

To understand fully the origins of behavior, a behavioral ecologist needs to consider that species are the product of their evolutionary past, in addition to present-day selective pressures (see chapter 2, this volume). Two ethologists noted for first implementing a rigorous scientific approach to the study of animal behavior, Konrad Lorenz and the aforementioned Niko Tinbergen, clearly advocated the explicit consideration of evolutionary history. Indeed, one of the earliest efforts to examine the evolution of behavior was Lorenz's reconstruction of motor patterns and call types in waterfowl (figure 7.1). Without the aid of a computer, Lorenz meticulously grouped species together according to their similarities in behavior. He showed how some behavioral traits seemed to have evolved quite early in the group (i.e., traits shared by most species), whereas others were likely to be more recent innovations (traits unique to single or a subset of species). His main objective was to illustrate that some behavioral traits could be as informative as morphology for distinguishing the evolutionary relationships between species. The fact that behavior is often retained from a common ancestor is an important observation and can help explain the form that many behavior patterns take in animals today.

The difficulty that these pioneering researchers and others faced in trying to use a comparative approach to study behavioral evolution was the need for large data sets of closely related species. Unlike other aspects of an animal's phenotype that can be quantified with more ease (e.g., morphology), collecting information on behavior is often labor intensive and time consuming, taking months or even years of work to compile for a single species. Ethologists were also limited by the lack of information on the nature of evolutionary relationships between species in many groups and robust methods for integrating this information into their investigations. With the progressive culmination of empirical research over the last half-century and the onset of the information age, it is now possible to compile large inventories of behavior and ecological information. The advent of molecular techniques heralded a proliferation of "tree building" in the 1980s and *phylogenies* (evolutionary trees that illustrate the relationships between species) are now available for many taxonomic groups. There is also a daunting array of computer programs, implementing a variety of different evolutionary models and handling a variety of different types of data. In short, behavioral ecologists today have an unprecedented opportunity to investigate the evolution of behavior within an explicit and increasingly sophisticated historical framework. In this chapter, we outline how the evolutionary history of organisms can be integrated into behavioral ecology research and the unique insights that can be gained in doing so.

CONCEPTS

The study of the evolution of phenotypic traits that lack fossil records focuses on *the comparative method*, the principle of which is quite simple. We might observe that a behavioral trait in our favorite study organism seems uniquely suited or *adapted* to the environment that it currently lives in. If this is true, we might expect to see the same trait in other species living in similar environments, prompting the obvious comparison of traits and habitats across species. This is the basic concept of the comparative method in its classical form (Darwin 1859). However, drawing inferences about adaptation from species comparisons is not as straightforward as it might first seem. What can we actually conclude from an observation that species exhibiting the same trait live in similar habitats? We have no way of knowing whether this pattern reflects independent evolution of the same trait in different species or whether the trait evolved once in a common ancestor and was subsequently retained in descendent species. In the latter scenario, a trait might have little or nothing to do with the current environment a species lives in. The modern revision to the comparative method incorporates information on the evolutionary relationships among species to distinguish examples of potential adaptation from other evolutionary processes.

Many aspects of the general appearance (morphology) and lifestyle (behavior and ecology) of closely related species are often shared because of common ancestry. For example, almost all Caribbean anole lizards produce extravagant head-bobbing displays in territorial defense. In the same way Lorenz concluded that behavioral traits shared by all waterfowl were ancestral, we can say that it is highly likely that the common ancestor to the anole lizards also defended territories and used head-bob displays. If we expand our comparison even further to the entire superfamily iguana (several thousand species across the globe), we find territoriality and visual displays are generally common to all, pushing the origin of these behaviors even further back in evolutionary time (Martins 1994; Ord et al. 2001). Even so, no two species are fully alike. Species might possess the same trait through common descent, but the trait will often differ to a small or large degree in its expression. For example, each lizard species has its own unique way of performing territorial displays. One species might use a series of rapid, shallow head bobs, whereas another produces protracted, high-amplitude head bobs (e.g., compare *Anolis sagrei* and *A. carolinensis* in figure 7.4). Although similarities in behavioral traits among closely related species can reflect the age of traits, the extent of modification or variation that exists in these traits can be equally informative.

How might we explain the diversity of iguanian territorial displays? By identifying ecological, morphological, and life history factors that vary among species in conjunction with the behavioral trait in

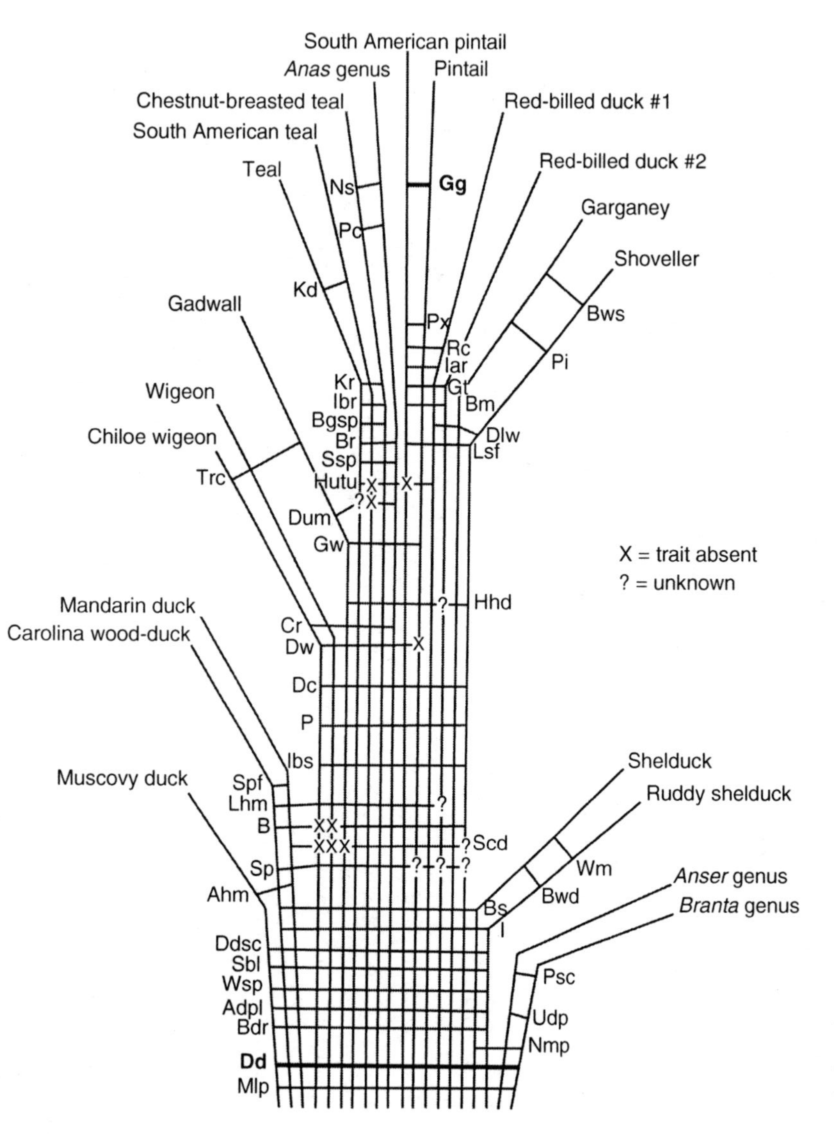

Adpl: Anatinae duckling plumage
Ahm: aiming head-movements as a mating prelude
B: "burping"
Bdr: bony drum on trachea
Bgsp: black-gold-green speculum
Bm: bill markings with spot and light-colored sides
Br: bridling
Bs: body-shaking courtship/demonstrative gesture
Bwd: black-and-white duckling plumage
Bws: blue wing secondaries
Cr: chin raising
Dc: decrescendo call of the female
Dd: **display drinking**
Ddsc: disyllabic duckling contact call
Dlw: drake lacks whistle
Dum: down-up movement
Dw: drake whistle
Gg: "**geeeeegeeeee**"
Gt: graduated tail
Gw: grunt-whistle
Hhd: hind-head display
Hutu: head-up–tail-up
I: incitement by female
Iar: incitement with anterior of body raised
Ibr: isolated bridling not coupled to head-up–tail-up
Ibs: introductory body shaking
Kd: "koo-dick"
Kr: "krick-whistle"
Lhm: lateral head movement of the inciting female
Lsf: lancet-shaped shoulder feathers
Mlp: monosyllabic "lost-piping"
Nmp: neck dipping as a mating prelude
Ns: nod swimming by female
P: pumping as prelude to mating
Pc: post-copulatory play with bridling and nod-swimming
Pi: pumping as incitement
Psc: polysyllabic gosling contact call
Px: extension of the median tail feathers
Rc: r-calls of the female in incitement and as contact call
Sbl: sieve bill with horny lamellae
Scd: social courtship
Sp: sham preening, performed behind wings
Spf: specific feather specializations serving sham-preening
Ssp: speculum same in both sexes
Trc: chin raising reminiscent of the triumph ceremony
Udp: uniform duckling plumage
Wm: black-and-white and red-brown wing marking
Wsp: wing speculum

question, it is possible to generate hypotheses on the potential cause of trait variation in the context of phylogeny. A distinction needs first to be made between the *phylogenetic* comparative method and the approach employed in classical comparative psychology. Comparative psychologists study specific and very disparate model organisms (e.g., rat, pigeon, primate) to investigate the neurological or developmental pathways of behavior or to identify universal cognitive processes, such as those involved in learning. As is already becoming evident in this chapter, phylogenetic comparative biologists typically examine phenotypic variation in a group of closely related taxa, such as a genus or clade of birds, to investigate the evolutionary causes of phenotypic diversity (or similarity). Modern phylogenetic comparative methods were born in part from a statistical problem associated with analyzing this type of data, namely, that data points for closely related species are not independent of each other (Felsenstein 1985). Statistically, analyses such as regression or ANOVA make the assumption that for a given variable, the value of one datum is not related to the value of another datum (other than by factors explicitly described by the statistical model). This assumption is violated when data are compared across species because traits are often shared through common ancestry. Ignoring this constitutes a form of pseudoreplication and inflates the degrees of freedom in statistical tests. This leads to increased Type I errors in which a conclusion is drawn that an effect exists when in fact it does not (Martins 1996a).

FIGURE **7.1** The phylogenetic reconstruction of motor patterns, calls, and morphology in waterfowl adapted from Lorenz (1941). Vertical lines correspond to different species, whereas letters and horizontal lines are traits shared by species. For example, the "geeeeegeeeee" call, **Gg**, is exhibited only by two species of pintails, *Dafila spinicauda* and *Dafila acuta*. By grouping these two species with this common call type, Lorenz effectively reconstructs the geeeeegeeeee call as a recently derived behavioral trait. Compare this with display drinking, **Dd**, which is shared by all species of waterfowl and is subsequently placed at the base of the phylogeny. The modern-day interpretation of this assignment is that display drinking evolved early in the waterfowl group and has been subsequently retained in species today.

To alleviate this problem, Joseph Felsenstein (1985) suggested a novel way of removing phylogenetic nonindependence by transforming data using a procedure that has come to be known as *phylogenetic independent contrasts* (box 7.1). A variety of free, online computer programs are available that calculate these contrasts, and the technique remains the most popular phylogenetic comparative method in use today. Commentaries by Felsenstein (1985) and others (e.g., Ridley 1983; Cheverud & Dow 1985) also heralded a reevaluation of comparative analyses of interspecific data more generally (see Harvey & Pagel 1991; Martins 1996b). This prompted the development of a series of evolutionarily explicit techniques that not only incorporated phylogeny directly into analyses to account for species relationships but now go beyond simply "controlling" for statistical nonindependence of interspecific data to examining the process of evolution itself.

Although new methods are continuously being developed, most fall into three broad categories: reconstructing the ancestor states of traits, estimating the degree of *phylogenetic signal* in traits, and correlated trait evolution. We begin first by elaborating on what a phylogeny represents, how they are made, where behavioral ecologists can obtain them, and what are the underlying assumptions of using them to study the evolution of behavior.

Phylogeny: A Primer

A phylogeny is a hypothesis summarizing the evolutionary relationships between species. Phylogenies were initially developed through comparisons of organismal traits such as morphology, but most phylogenies today are based on mitochondrial and nuclear DNA. The details of how phylogenies are put together are complex and beyond the scope of this chapter (see Felsenstein 2004). Nonetheless, a behavioral ecologist wishing to use phylogenies to inform their research needs to have a basic understanding of the terms used. It should also be recognized that phylogenies are subject to change as new data comes to hand and the techniques used to create them are refined. This is an important point to remember because changes to the phylogeny can influence the outcome of a comparative analysis (e.g., see following section).

Phylogenies are depicted as trees (figure 7.2), with the names of extant species connected by a hierarchy of bifurcating *branches* summarizing

BOX 7.1 Comparative Methods

The majority of comparative approaches that behavioral ecologists will most likely use fall into two broad categories: techniques for reconstructing ancestor states and performing correlation analyses (e.g., for testing adaptive hypotheses). We provide a brief technical introduction to these approaches below.

Ancestral Reconstruction

Parsimony is an algorithmic procedure used to reconstruct phylogenies or to map ancestor states onto a phylogeny by favoring solutions requiring the least amount of evolutionary change in traits (figure Ia). In this sense, it tends to underestimate evolutionary changes and may not be as appropriate for behavioral evolution as for other aspects of the phenotype. Parsimony methods usually also lack a statistical or probabilistic methodology; for example, assigning specific states following predefined rules without recognizing any possibility of uncertainty. Although it has come under fire for a lack of statistical rigor, and because it may not be a realistic view of how evolution occurs for many traits, parsimony approaches are often favored for their simple and straightforward computation.

Least Squares, Maximum Likelihood and Bayesian approaches are several newer methods that apply a probabilistic approach to find explicitly mathematical, evolutionary models that best fit the observed distribution of trait values across species (figure Ib). In the simplest case, these statistical methods yield results identical to those obtained using parsimony approaches. The difference being, however, that these methods also compare the probability

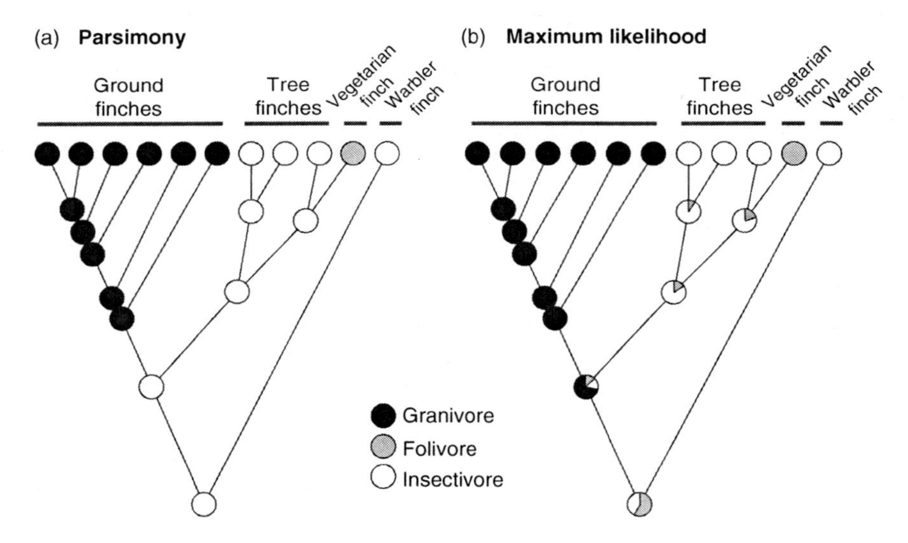

Figure I Alternative ancestor reconstructions of diet in Galapagos finches using parsimony and maximum likelihood methods. Adapted from Schluter et al. (1997). Pie charts reflect the relative support of dietary habits assigned to ancestors. The parsimony reconstruction suggests the root ancestor was an insectivore, whereas the maximum likelihood estimate implies it is more likely to have been a folivore.

(continued)

of various alternative scenarios of how evolution might occur—e.g., a model that describes trait variation as being explained by phylogenetic relationships among species (i.e., the null of Felsenstein's approach), a model assuming no relation between trait variation and phylogeny (which is the same as not including phylogenetic information at all in analyses), and various models in between these two extremes (see Martins & Lamont 1998 for further discussion and examples in a behavioral context). The probabilistic framework has the added benefit of providing confidence intervals, emphasizing, for example, that the phenotypes of ancestors deeper in the tree are unlikely to be as well estimated as more recent ancestors.

Character Correlations and "Correcting" for Phylogeny

Not incorporating phylogeny into statistical analyses of comparative (interspecific) data can create a problem of pseudoreplication and statistical nonindependence. When estimating a general pattern across a large group of organisms, data measured from several closely related taxa should not be weighed as heavily as data from distantly related taxa. Because all taxa are related in one way or another, not incorporating phylogeny can lead to inflated rates of Type I error and the finding of significant effects when in fact none exist. By "correcting for phylogeny," phylogenetic comparative methods (PCMs) correct the significance tests associated with parameter estimates for a fundamental statistical problem. In contrast, the absolute value of parameter estimates (e.g., means, correlation coefficients describing the relationship between traits) tends not to differ much, depending on the shape of the phylogeny or on whether a phylogeny is incorporated at all, unless the specific phylogeny being incorporated is extremely biased or unusual in some respect (see Martins & Housworth 2002; Rohlf 2006). The impact of incorporating phylogeny is primarily on the hypothesis test rather than on the value of the parameter being tested. Nevertheless, computer simulation studies (e.g., Martins et al. 2002) consistently confirm that PCMs nearly always perform better than not incorporating phylogeny, and even those who continue to debate exactly how and when PCMs should be applied (e.g., Rheindt et al. 2004) agree that the traits should be tested before a decision is made.

The phylogenetic independent contrasts approach is the most commonly used phylogenetic comparative method. Originally developed by Joseph Felsenstein (1985), this method corrects the basic statistical problems arising from phylogenetic relationships between species by transforming interspecific data into a set of differences or *contrasts* between immediate relatives, standardized by the amount of evolution expected along each branch of a known phylogeny (see figure II). If two sibling species express a trait value largely because of shared ancestry, then the difference score will be close to (or at) zero. Conversely, if the expression of a trait is unrelated to phylogeny, contrast values will be large. Although the computational algorithm requires estimating ancestral states along the way, these states are not intended to represent the phenotype of ancestors at phylogenetic nodes—ancestral states are better estimated using other techniques described above.

Felsenstein's (1985) method assumes that phenotypic evolution is well described by a random walk or Brownian motion process, typically used by population geneticists to describe evolution under random genetic drift or fluctuating selection, and that the rate of evolution for the particular trait of interest along each branch of a phylogeny is known. A large family of related methods have been recently developed and can be interpreted as extensions of these assumptions, for example, allowing for a broader range of evolutionary scenarios (e.g., Pagel 1994; Martins & Hansen 1997; Butler & King 2004; Housworth et al. 2004), for not knowing the complete phylogeny or the exact rate or duration of evolution along each branch (e.g., Diaz-Uriarte & Garland 1998; Housworth & Martins 2001), and for testing whether phylogeny needs to be incorporated at all (e.g., Blomberg

(continued)

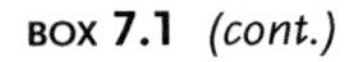

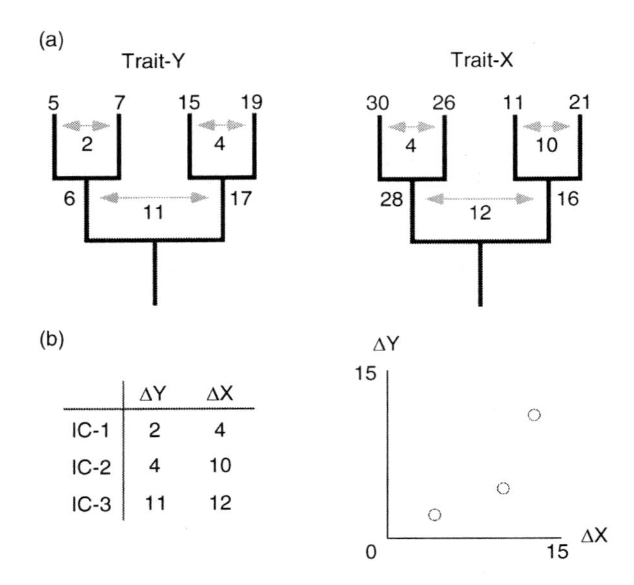

	ΔY	ΔX
IC-1	2	4
IC-2	4	10
IC-3	11	12

Figure II Stylized representation of the phylogenetic independent contrasts approach. (a) Difference scores or contrasts are calculated between species pairs and ancestor nodes on the phylogeny. (b) These contrasts are considered to be phylogenetically independent of each other and can be analyzed using standard statistical tests such as regression and ANOVA.

et al. 2003). These extensions provide additional accuracy and statistical power, clearly improving the fit and utility of Felsenstein's original approach (e.g., Martins et al. 2002). New comparative methods also address a variety of novel questions. For example, the phylogenetic mixed model (Housworth et al. 2004) can be used to estimate and compare the degree of phylogenetic heritability in two traits. Hansen's adaptation model (Hansen 1997; Butler & King 2004) estimates the degree to which a phenotype is adapted to different aspects of a complex selective regime. Others are using comparative approaches to understand how phenotypic evolution is related to species diversification (e.g., Ricklefs 2004). These new methods offer a powerful way to uncover evolutionary patterns that could not otherwise be observed.

the evolutionary relationships between species. Points at which branches split are called *nodes* and represent speciation events. A *polytomy* occurs when more than two lineages originate at a node. This can be indicative of rapid speciation (referred to as a *hard polytomy*) or when the precise relationship between lineages cannot be fully resolved (*soft polytomy*). In the latter instance, this uncertainty will reduce the accuracy of comparative methods.

The length of phylogenetic branches often reflects the extent to which lineages have diverged. For example, Clade A shown in figure 7.2 has relatively short branch lengths, suggesting that the amount of divergence among these species is relatively low. Branch lengths are typically calculated from genetic

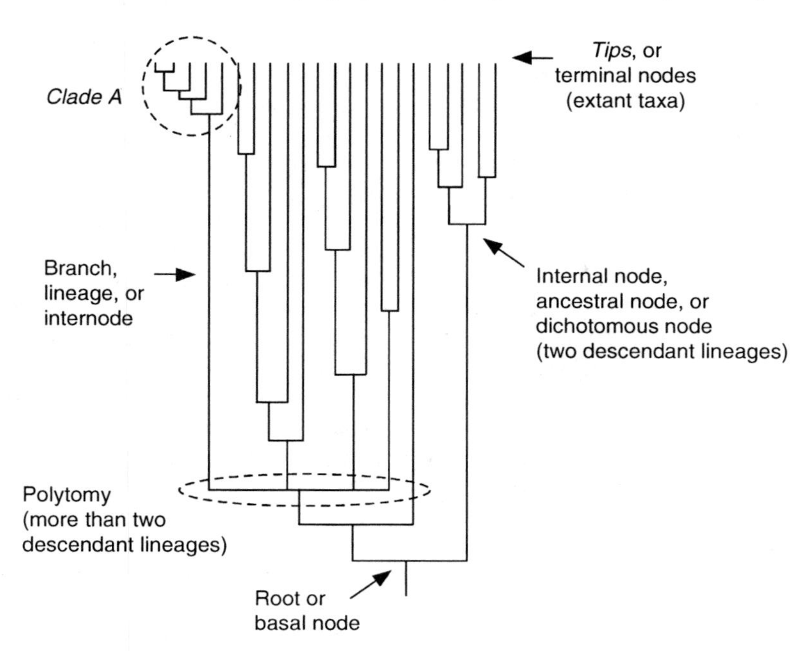

FIGURE 7.2 An example phylogeny illustrating common phylogenetic terms. Note that this and most phylogenies typically used in comparative analyses provide information only on evolutionary relationships among extant taxa and do not include extinct lineages.

sequence data, sometimes using *molecular clocks* or fossil evidence, to scale the rate of molecular divergence to units of real time (e.g., millions of years). Branch lengths are also sometimes reported in terms of number of morphological changes occurring along each branch. In comparative analyses, however, statistical methods require that branch lengths reflect the amount of evolutionary change expected for the behavioral trait being investigated. If we expect more evolutionary change along longer branches relative to shorter branches, then branch lengths in units of time may be a reasonable proxy for expected behavioral evolution. However, if we expect more change in the behavioral phenotype along particular branches of the phylogeny (perhaps because of a key innovation or adaptive radiation), change in the DNA sequences used to construct phylogenies may not correspond to rates of change occurring in the phenotypic traits under investigation, and the comparative method assumption will be violated.

There are a number of comparative methods, such as phylogenetic independent contrasts, that allow users to include information on branch lengths in analyses. In some cases, doing so assumes implicitly that trait evolution occurs via accumulated incremental changes (Brownian motion or the so-called gradualism model of evolution; Darwin 1859). This continues to be the subject of considerable debate. Paleontological evidence (Eldredge & Gould 1972) and recent analyses measuring the nature of trait evolution (e.g., Pagel et al. 2006; Atkinson et al. 2008) suggest most evolutionary change occurs when new species are formed (i.e., at phylogenetic nodes) and that certain traits may evolve more quickly along certain branches than do others. Furthermore, estimates of branch lengths are subject to error (Felsenstein 2004). The alternative is to set all branch lengths equal to 1 (a punctuated model of evolutionary change), to try several possible sets of branch lengths generated by computer simulation or using additional information (e.g., Losos 1994; Martins 1996a), or to apply a more sophisticated comparative analysis that estimates relative branch lengths from the phenotypic data themselves (e.g., spatial autocorrelation; Gittleman & Kot 1990; phylogenetic generalized least squares method; Martins & Hansen 1997; Pagel 1997; phylogenetic mixed model; Housworth et al. 2004).

In most cases, phylogenies are obtained straight from the published literature. For a large comparative dataset, a single published phylogeny will rarely cover all species of interest, requiring two or more trees to be merged together. Often this is done by extrapolating relationships between species with reference to those taxa that are common to all the sources being used. Branch length information may not be included in these composite trees because it is often difficult to know how to adequately standardize estimates between sources. With some gumption, a formal *supertree* can be constructed using genetic sequences compiled from databases such as GenBank (see Sanderson et al. 1998). Taking this approach requires some skill in phylogenetics.

Another common complication occurs when some species are not represented in any published phylogeny. One solution is to use taxonomic keys to position species. If there are large numbers in question, one remaining solution is to randomize the position of these taxa and conduct multiple analyses across all alternative phylogenies (e.g., Housworth & Martins 2001; Ord & Stuart-Fox 2006).

Ancestor Reconstructions

Fossil traces of behavior are rare, but it is still possible to infer the behavior of historical ancestors by reconstructing character states onto a phylogeny. Researchers use molecular phylogenies and a variety of computer programs to assign historical character states based on parsimony or probabilistic-based algorithms such as likelihood or Bayesian statistics (box 7.1). With these programs, it is possible to extrapolate how a particular ancestor might have lived from the behavior of descendent species that we can study today. A more powerful utility of ancestor reconstructions is the identification of broad macroevolutionary patterns. We can use phylogenetic trait reconstructions to discover whether a group of species exhibits a behavior because of common ancestry or whether it arose independently in different species through convergent evolution. The pattern of trait reconstructions on a phylogeny can in itself support or reject hypotheses on potential mechanisms or selective pressures believed to drive trait evolution (see case studies below). In particular, ancestor reconstructions are critical for testing hypotheses that assume a specific sequence of historical events.

Consider the hypothesis that preexisting biases in the sensory system of females promote the evolution and subsequent form of male traits used during courtship. This hypothesis explicitly predicts that the evolution of a trait occurs after the evolution of a sensory bias. Phylogenetic studies are the only way to investigate the timing of these events. For example, males in several closely related Central American freshwater fish species are known as swordtails because they possess an elongated caudal fin or "sword." This sword functions as an ornament for attracting females, with longer swords being preferred over shorter ones (Basolo 1990a). Surprisingly, females of several related species to the swordtails, the platyfish, also prefer males with swords even though males of their own species do not naturally develop them (Basolo 1990b). This prompted Alexandra Basolo (1990b) to test whether females had a preexisting bias toward males with long caudal fins, which leads to the clear prediction that female preference for swords should predate the evolution of the sword itself. Basolo used a phylogenetic reconstruction to pinpoint when the sword most likely arose in the group. She then showed through mate choice tests in females of several different swordtail and platyfish species that all exhibited some form of preference for swords, even in species in which conspecific males lacked swords. This information, combined with phylogenetic reconstructions, revealed that the sword evolved after the preference in females. The sword is also used by males in aggressive competition in several species, and additional phylogenetic studies confirm that both the sword (Moretz & Morris 2006) and female preference (Morris et al. 2007) evolved before males co-opted the sword as a signal for use in same-sex battles.

The swordtail example highlights several important considerations when attempting to reconstruct specific evolutionary events. Fresh scrutiny over whether the sequence of female preferences and the origin of male swords conforms to Basolo's (1990b) original findings is brought with each revision to the phylogeny and development of alternative, arguably more evolutionarily realistic methods for assigning ancestor states (e.g., Meyer et al. 1994; Schulter et al. 1997; Marcus & McCune 1999). Indeed, the question has been revisited numerous times over the last 2 decades (e.g., Basolo 1991, 1996; Wiens & Morris 1996; Morris et al. 2007), but the sensory bias hypothesis has generally stood up to the test of time. As with any phylogenetic comparative method, ancestor reconstructions rely on a phylogeny and the underlying statistical framework of the method applied (box 7.1). When either is changed, so potentially does the outcome of ancestor reconstructions (Schulter et al. 1997; Losos 1999).

Moreover, the further back in time we attempt to reconstruct ancestor traits, the greater the level of uncertainty involved in estimates (Martins 1999). Results that are consistent across alternative phylogenetic hypotheses and using different methods for assigning states to ancestors will reflect particularly robust and prominent evolutionary trends.

Phylogenetic Signal

The variability observed in behavior within species sometimes leads to the naïve perception that all behavior must be highly evolutionary labile or, at the extreme, entirely plastic in origin. In other words, changes in behavior occur so rapidly—perhaps over a matter of generations or within a single generation (e.g., behavior that is acquired through experience, such as song learning)—that little or no trace of evolutionary history or phylogenetic signal will be seen in its expression today. The term *phylogenetic signal* specifically refers to the extent trait variation across extant species reflects underlying phylogenetic relationships. When traits exhibit no phylogenetic signal, it follows that reconstructing ancestor states or applying any phylogenetic comparative method is unwarranted.

Behavior is arguably the most dynamic aspect of an animal's phenotype. Of course, when more variation exists within rather than between species, a comparative approach is probably inappropriate. With this obvious exception, we can measure how much of the behavior observed across species is predicted by phylogenetic relationships and estimate statistically the propensity for evolutionary change in traits. Different aspects of behavior can exhibit a wide range of phylogenetic signal, from traits that are retained with little modification over long periods of evolutionary time, to those that exhibit large bursts of evolutionary change. Overall, behavior does tend to be more evolutionary labile than ecological, morphological, life history, and physiological traits (Blomberg et al. 2003), but behavior is increasingly being found to exhibit more phylogenetic signal than traditionally thought (Wimberger & de Queiroz 1996; Blomberg et al. 2003; Ord & Martins 2006). Ancestor reconstructions can provide an indication of how evolutionarily conservative a trait is by revealing whether a trait is homologous (shared through common ancestry). Methods that estimate phylogenetic signal provide a more accurate picture into the potential underlying evolutionary process. Traits that are closely tied to phylogeny—have high estimates of phylogenetic signal or are homologous—could reflect a phenomenon that is sometimes referred to as *phylogenetic inertia*. This means little evolutionary change has occurred outside drift, perhaps because of low mutation rates, resulting in a trait that is retained more or less in the same form through descendent taxa. Phylogenetic inertia can also occur because of genetic, developmental, or physiological constraints limiting the amount of evolutionary change in a trait.

Alternatively, traits can appear relatively stable over long periods of evolutionary time through the entirely different phenomenon of stabilizing selection. For example, niche conservatism is the tendency for ecological traits in ancestors to carry over into descendent species, which in turn leads descendent species to settle in similar "preferred" habitat types as their ancestors. Behavioral traits that are ideally suited or adapted to a particular environment are subsequently maintained along diverging lineages (phylogenetic branches) via stabilizing selection. If the selection regime changes dramatically, either through the invasion of a novel habitat or a drastic change in the occupied habitat type (e.g., because of climatic change), either species will go extinct or adaptation will promote major evolutionary change. When two distantly related species occupy similar novel environments, this can lead to adaptive convergence. Depending on how recently and frequently these environmental changes have occurred, traits might show low or intermediate amounts of phylogenetic signal. Behavioral change that is plastic, such as behavior that is largely acquired through experience during an individual's lifetime, should intuitively have extremely little or no relationship with phylogeny. However, the extent to which a species is plastic in its behavior can be in itself an emergent trait of a species. The extent to which closely related species differ in their plasticity is an area of growing interest in evolutionary ecology and is an equally exciting avenue for future behavioral research (e.g., why are some species better learners than others?). Currently, we know little about whether behavioral plasticity as a species trait is or is not associated with phylogeny.

Correlated Trait Evolution

To establish that a specific selective pressure or constraint has led to behavioral change or stasis, we need to confirm that variance in the putative causal factor (e.g., some aspect of the environment) corresponds with changes in behavior over evolutionary time. Although phylogenetic correlation tests are typically employed to infer adaptation, they can also determine whether morphology or some developmental factor has constrained evolutionary change in behavior. For example, the beak morphology of Darwin's finches on the Galapagos Islands is a classic example of ecological adaptation. Species that feed on large, hard seeds possess large, strong beaks, whereas those feeding on insects or smaller more easily opened seeds possess small short beaks (Grant & Grant 2008). However, beak morphology also influences the mechanics of song production, such that large beaks are less capable of performing rapid movement, and in turn influences the structure of vocalizations produced. A test of correlated trait evolution by Jeffrey Podos (2001) revealed a strong historical relationship between the evolution of large beaks and the production of songs with low syllable repetition and narrow bandwidths. Although song behavior is expected to be under directional selection from female mate choice or for greater efficiency in particular acoustic environments, the evolutionary diversification of beak morphology for exploiting different food sources has apparently constrained the range of song types that can ultimately evolve (Podos 2001).

This study of beak morphology in Darwin's finches employed the most common approach in studies examining coevolution, which is Felsenstein's phylogenetic independent contrasts (box 7.1). It assumes the null hypothesis that variation across closely related species is the product of phylogeny, meaning the consequence of stochastic processes such as genetic drift and mutation that culminates in phenotypic variation in the absence of selection. The contrasts resulting from such calculations are considered to be free of the "confounding" effect of phylogeny and potentially representative of adaptive change. There are a few important points to remember with this approach and the removal of all factors associated with phylogeny. First, this interpretation of Felsenstein contrasts incorrectly attempts to separate two inextricable aspects of the underlying process generating behavioral diversity: adaptation and nonadaptive processes. In the real world, we expect traits to evolve under a complex mix of the two processes working simultaneously. Second, our ability to detect adaptive phenomena is restricted to particular types of selection. Stabilizing selection, for example, can result in adaptive evolution that tracks the phylogeny very closely (see previous section), as would constant directional selection that acts in the same direction across the phylogeny as a whole. Either of these forms of selection could be hidden if the causes of variation associated with phylogeny are not carefully considered and are subsequently removed rather than incorporated directly into a comparative analysis. Third, simplistic application of Felsenstein's method (e.g., as originally proposed) provides very little protection from possible errors in the phylogeny, branch lengths, and underlying evolutionary model (Martins et al. 2002). The calculation of individual contrasts is dependent on the position of the nodes and length of the branches being compared (box 7.1). Errors in either are subsequently magnified in the final estimate of the contrast. When branch length information is used, the scaling of contrasts assumes a Brownian motion model of evolution, which is meant to replicate gradual change in traits over evolutionary time. This is not necessarily how the traits being analyzed have in fact evolved. Most of the various extensions and additions of this method that have been developed more recently perform better statistically and are more robust to small errors (e.g., those implementing a generalized least squares approach; Martins & Hansen 1997; Pagel 1997). Fourth, Felsenstein's contrasts and many other comparative methods are explicitly designed to study the evolution of biological traits, and hence may not be appropriate for studies of species-level phenomena (e.g., degree of polygyny; Searcy et al. 1999) or environmental measures. For these, a technique that has been explicitly designed to measure the impact of environment on trait evolution may be preferred (e.g., Hansen 1997; Diniz-Filho et al. 2007). Fifth, animals evolve in a complex world and are subject to a plethora of concurrent evolutionary pressures. Behavioral phenotypes may not perfectly track a phylogeny or any single aspect of an environment because of competing selective pressures from other aspects of the physical environment (e.g., as envisioned in Hansen 1997, Butler & King 2004). For example, in lark buntings (Chaine & Lyon 2008), females prefer different aspects of male signals from year to year, choosing signals each year that best indicate a males' reproductive success in that year's particular combination of environmental features. Moreover, behavioral phenotypes often occur in *syndromes*, or groups of behavior types that are linked to each other and to other aspects of the phenotype through genetic correlations, pleiotropy, or other underlying mechanisms (e.g., chapter 30). Thus, the evolution of behavioral phenotypes may not be clearly pegged to any single environmental feature.

Treating phylogeny as a statistical problem, rather than a direct contributing factor determining how species respond to selection, skews the philosophy behind the role adaptation plays in evolutionary diversification. Instead, behavioral ecologists may learn more from recent and more sophisticated comparative methods that adopt maximum likelihood as a way to integrate phylogenetic signal, phenotypic correlations, and complex environmental forces. For example, Thomas Hansen's (1997) method offers a means for estimating the relative impact of each of several possible selective pressures (e.g., humidity, temperature, degree of vegetation) on phenotypic evolution, assuming that the organisms are living in a complex selective regime in which all of these pressures play at least an occasional role. Hansen's method provides a way to compare the variance in a trait along lineages believed to have experienced stabilizing selection in a particular type of environment, to trait variances over the rest of the phylogeny (see also Butler & King 2004). Incorporating phylogeny directly into statistical tests rather than attempting to remove it from the data beforehand gives investigators a clearer picture of how behavioral traits evolve in the context of both adaptive and nonadaptive processes.

CASE STUDIES

Identifying Modes of Selection on Behavior

There are a number of different theoretical models that explain how female mate choice might drive or constrain the evolution of male display behavior. Empirical study can identify the presence and strength of female choice, as well as the traits exhibited by males on which females base their choices. However, such investigations within populations are unable to show the precise historical mechanism through which selection has acted. This is because different evolutionary processes can produce very similar traits, making it impossible to distinguish between different models of selection with focused empirical research alone. One approach is to examine the macroevolutionary patterns of trait evolution using ancestor reconstructions. In many cases, different evolutionary processes will generate predictable patterns in how traits evolve along a phylogeny. By carefully considering the assumptions of different models, together with the natural history of a group, it is possible to use phylogenetic reconstructions to infer the predominant form of selection at work.

An early example of this approach is presented by Richard Prum's (1997) investigations of the evolution of male behavior and plumage in manakins. The group is comprised of 42 bird species and is known for its lek breeding system. Lekking males congregate at the same location and establish small display areas where they attempt to woo females through elaborate courtship displays. Although it is known that females choose among mates based on display attributes, the evolutionary mechanism behind this choosiness and how it drives differentiation of male traits across species is unclear. Prum (1997) examined two alternative sexual selection models and four hypotheses on how natural selection might contribute to display evolution in the group. By carefully considering the assumptions of each, he predicted several key macroevolutionary patterns that should be apparent once male displays are reconstructed onto a phylogeny. These predictions are summarized below and in table 7.1.

Fisherian or "Runaway" Male traits under Fisherian selection are arbitrary in the sense that they do not reflect condition or viability in males (although they may initially start off as indicator traits). Consider a population in which some males express long tail feathers and some females exhibit a preference for them. Males with long tail feathers will tend to have higher reproductive output because they are capable of acquiring matings from females both with and without the preference for long tails. On the other hand, males without the trait can mate only with females exerting no preference. Assuming that both the expression of the trait and its preference are genetically determined, they will become linked over successive generations. This mutual reinforcement will accelerate the proliferation of both the trait and preference throughout the population in a "runaway" process (Fisher 1930) and may ultimately lead to the exaggeration of the male trait beyond its initial form (chapter 24).

TABLE 7.1 Predicted macroevolutionary patterns in male behavior and plumage under different forms of selection

Mechanisms of Selection	Historically Nested Distribution	Skewed Distribution	Likelihood of Convergence	Diversity of Repertoire	Correlates of Divergence
Sexual selection					
Fisherian	Yes	Toward tips	Low	High	
Quality indicator	No	Toward base	High	Low	
Natural selection					
Mate choice efficiency			High	Potentially high	Environment (e.g., predation, search costs)
Species recognition	No		Low	Low	Sympatry
Sensory bias	Yes		High	Low High	
Narrow					
Broad					
Sensory drive	Yes		High	Potentially high	Environment (e.g., habitat structure, predation pressure)

Source: Prum (1997).

Consequently, these traits are expected to evolve freely in direction and elaboration. Prum argued that this should result in the evolution of multiple traits and diverse display repertoires across species, with few instances of convergence. The distribution of traits reconstructed on the phylogeny would also tend to be historically nested, in which traits are retained within clades and display repertoires increase as additional traits evolve (figure 7.3a). Fisherian selection also predicts rapid and continued diversification of male traits. Hence Prum argued that Fisherian selection will tend to skew trait evolution toward the tips of the phylogeny (figure 7.3d).

Quality Indicators Traits evolving under this form of selection are tied to male condition or viability. The number of possible traits that will reliably convey this type of information is limited, and Prum proposed that this should promote frequent convergence and little diversity in display repertoires across species. The potential for evolutionary change is constrained because quality indicators must provide honest information on condition and are therefore not as free to vary as Fisherian traits. Indicator traits should instead be retained with little modification in descendent taxa and, because female mate choice is believed to be ancestral in manakins, such traits should have evolved early in the group and this will tend to skew trait reconstructions toward the base of the phylogeny (figure 7.3c). Alternatively, novel traits might arise that convey more accurate information on male quality. These will replace earlier forms because indicator traits are costly for males to maintain and redundant traits will be selected against. Prum suggested that this "trait switching" will lead to a nonhistorically nested distribution and also reduce the diversity of display repertoires observed across species (figure 7.3b). A repertoire of indicator traits would evolve only if each provides unique information on different aspects of condition. However, because there is a limited pool of potential indicator traits, the evolution of unique traits will progressively decrease with increasing repertoire size and this will also skew trait evolution toward the base of the phylogeny.

Mate Choice Efficiency The longer it takes for females to find and choose among males, the greater the cost imposed by natural selection in the form of energetic expenditure and predation pressure. The strength of these selective pressures is dependent on the ecology and environment of a species. Prum thus predicted convergences in display traits among

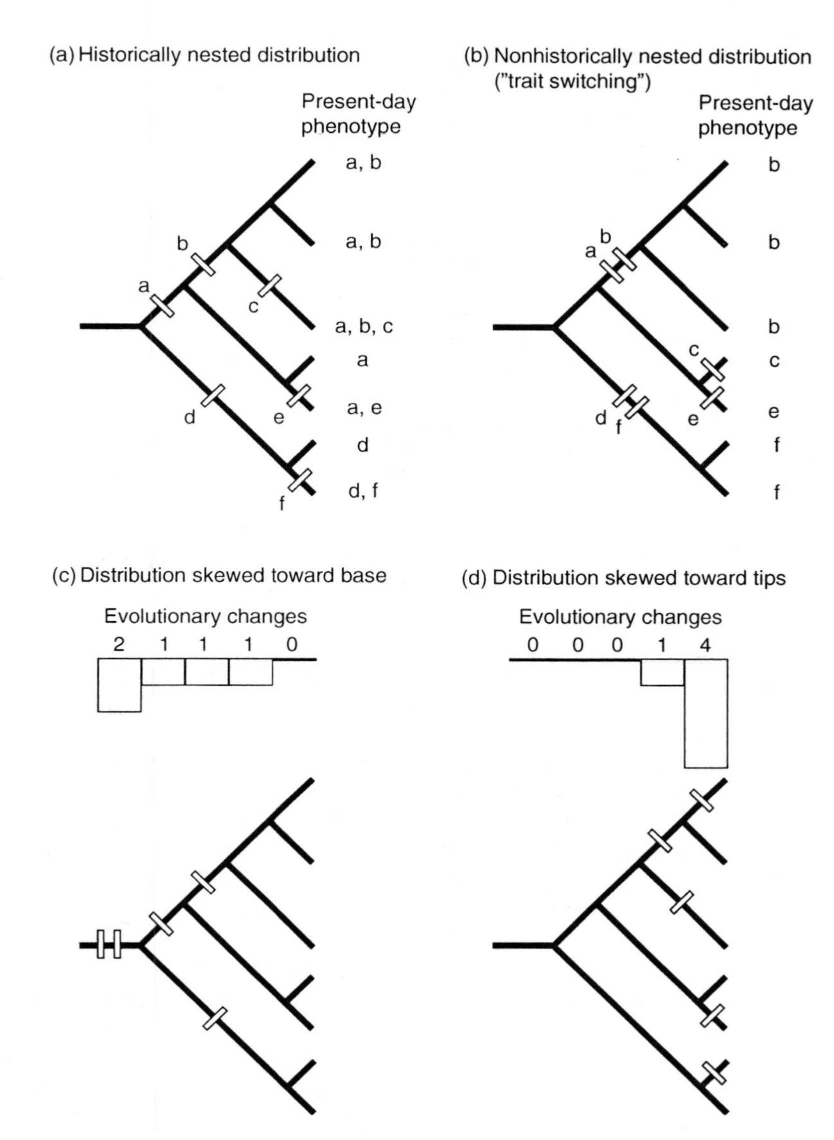

FIGURE 7.3 Hypothetical phylogenetic distributions of trait evolution modified from Prum (1997). (a) Traits exhibiting a historically nested distribution in which traits are retained within a clade in addition to the evolution of subsequent novel traits. This distribution should also lead to diverse repertoires and is predicted under Fisherian and sensory bias mechanisms of selection. (b) Traits exhibiting a nonhistorically nested distribution in which the evolution of novel traits replace those that arose earlier in a lineage's history. This distribution should lead to simple repertoires and is predicted when male traits function as quality indicators and/or in species recognition. (c) Trait evolution skewed toward the base of a phylogeny. This is predicted when traits are subject to some form of evolutionary constraint, such as traits that function as quality indicators. (d) Trait evolution skewed toward the tips of the phylogeny. This is predicted for traits that have not been constrained and have subsequently experienced rapid diversification, such as male traits subject to Fisherian or runaway selection.

species with similar ecologies and/or occupying similar habitats. Conversely, changes in the environment should promote the evolution of unique display traits. Diverse repertoires can also evolve if multiple traits contribute to mate choice efficiency or female survival.

Species Recognition Females are expected to be under considerable selection for discriminating among heterospecific and conspecific males. This would predict trait differentiation when species are sympatric and select against shared, homologous traits, which would confuse mate recognition. Prum suggests that this will constrain the evolution of diverse display repertoires, reduce the historically nested pattern in trait reconstructions (figure 7.3b), and limit display convergence.

Sensory Bias and Sensory Drive In these two related models, trait evolution is arbitrary and exploits a preexisting bias in the sensory system of females. The evolution of male traits that successfully tap into narrow biases are likely to be infrequent because the probability of such traits arising in the first place is low. Prum notes that if they do evolve, evolutionary change will be limited because trait properties are constrained by the specificity of the bias. On the other hand, a broad sensory bias could lead to the evolution of a number of different novel traits, accumulating into a diverse, historically nested display repertoire (figure 7.3a). The sensory drive model is the combined effect of sensory biases, habitat properties dictating signal efficiency and pressures from predators or parasites. Like the mate choice efficiency model, male traits are expected to diverge or converge depending on the differences or similarities in the ecology and environment of different lineages.

With these predictions in hand, Prum (1997) reconstructed male display repertoires and plumage traits onto the manakin phylogeny using parsimony. He also replicated the analysis for the same types of traits in a largely monogamous group of tyrant flycatchers, in which both sexes typically contribute to the care of offspring and in which the criteria used by females to choose between mates is potentially quite different. By doing this, Prum could compare the macrovevolutionary patterns in two very disparate groups: a polygynous lek breeding system and a monogamous, biparental care breeding system.

Phylogenetic reconstructions for the manakins revealed that male traits are historically nested and generally skewed toward the tips of the phylogeny. Convergences in behavior or plumage characteristics have been rare, and display repertoires are extremely diverse across the group as a whole. Based on this evidence, Prum (1997) argued that Fisherian selection, or potentially a broad sensory bias, has been the predominant form of selection driving the evolution of extravagant male traits in manakins. By comparison, trait evolution in male flycatchers was not historically nested and display repertoires are quite simple. This suggests that male display traits in flycatchers have been selected as quality indicators, which is consistent with the monogamous breeding system of the group (Prum 1997).

Of course, the results of this analysis are dependent on the phylogeny and reconstruction method employed. The location at which specific traits are reconstructed onto the phylogeny will vary with changes made to the phylogeny or the algorithm used to assign ancestor states (e.g., compare the analyses of Basolo 1990b, 1991, Meyer et al. 1994, and Schluter et al. 1997 for the swordtail example mentioned earlier). However, because Prum (1997) examines major macroevolutionary patterns for dozens of traits collectively (e.g., 44 display behaviors and 44 plumage traits in the case of manakins), it is unlikely that minor changes in phylogenies or methods would lead to any dramatic changes in the overall reconstructed distributions observed. Other issues arise with the way traits are initially defined, which will also affect the type of reconstructed patterns observed on a phylogeny. For example, is a feather ornament on the same part of the body that varies in color among species a homologous single trait (one of the same origin) or several different traits? There is also an implicit assumption being made that the mode of selection acting on traits is largely consistent across species. It is possible that different selective pressures have acted on the same homologous trait in different lineages, leading to different rates and types of change. This is unlikely to leave any consistent patterns across a phylogeny (which in itself would be informative).

Identifying by empirical means the selective mechanism promoting the evolution of behavior can be extremely difficult. Indeed, Prum (1997) presents a compelling argument against extrapolating evolutionary processes from the adaptive function of a behavior as observed within a population today (Gould & Vrba 1982; Baum & Larson 1991). Although empirical evidence might offer support for the hypothesis that a trait conveys honest information on male condition, it may still have originated under Fisherian selection and subsequently become linked to male quality over evolutionary time. As the manakin/flycatcher example demonstrates, selection can leave telltale signs in macroevolutionary patterns and reconstructing ancestor states is a powerful way to identify the form of these broad patterns.

Multifaceted Behavior

Individual traits do not evolve in isolation from the rest of the phenotype. Traits are often linked to each other developmentally, physiologically, or genetically. Even if traits are hypothetically free to vary independently of each other, selection acts collectively on an animal's overall phenotype and not literally on a single trait in isolation. The intrinsic factors influencing the evolution of behavior are likely to be as intricate as the specific selective pressures acting on behavior externally. An added challenge faced by researchers attempting to study a suite of traits is whether each feature being measured actually represents a separate trait or is an element of the same trait.

We can clarify the suitability of measures taken through experimental study of function. But a researcher still faces the problem of determining whether one key trait is the primary target of selection while others covary via some underlying genetic or developmental dependency, or whether each trait is under direct selection. Selection experiments in the laboratory are an excellent means to separate out intrinsic correlations and discover the focus of selection pressures (e.g., Chenoweth & Blows 2003; Mackay et al. 2005), but even in systems in which these experiments are possible (there are few in behavioral ecology), we are studying microevolutionary processes (i.e., within populations) that might not translate fully to the macroevolutionary scale (across species).

In a recent study, we investigated the evolution of a multifaceted behavior in the form of visual displays performed by West Indian anole lizards (Ord & Martins 2006). Males compete for territories using rapid up/down movements of the head, termed *head-bob displays*, and the extension of a throat fan known as a *dewlap*. The way lizards perform these displays varies across species extensively in the number, duration, and type of bobs/dewlaps incorporated (figure 7.4). This diversity suggests anole displays constitute multiple traits that are at least semi-independent of each other, with each trait potentially targeted by different forms of selection from the social and physical environment or by other factors (Ord & Martins 2006). For example, variance in the intensity of competition over territories and the need to convey species identity are both suggested to explain signal diversity across species (Jenssen 1977, 1978; Ord et al. 2001). Anoles also occupy a variety of different habitat types, and signal theory predicts animals will evolve signal forms that enhance transmission efficiency in the environment they typically communicate (Endler 1992). Each of these selective processes is unlikely to influence all signal traits equally. Genetic correlations will further complicate evolutionary outcomes, and differentiation between species is also possible in the absence of selection through mutation and genetic drift.

To explore these factors, we scored 15 display features for 53 taxa using published plots that summarize the structure of displays (e.g., figure 7.4). To highlight the presence of trait correlations, we entered display data into a principal component analysis that incorporated basic phylogenetic effects using Felsenstein's independent contrasts. This revealed several strong associations among display traits that enable us to identify key display features that summarize the diversity observed across species (and were unlikely to measure the same trait), as well as trait complexes that have potentially evolved in concert (table 7.2). To describe the relative evolutionary lability of display traits, we used two different maximum likelihood approaches to estimate how freely each trait has undergone potential adaptive change and the amount of trait variation (adaptive or otherwise) inherited directly from evolutionary ancestors (Lynch 1991; Martins & Hansen 1997; Housworth et al. 2004). We found two traits relating to how dewlap displays are performed to be tightly bound to phylogeny, indicating that variation in these traits is explained to a large degree by the evolutionary relationships between species. Change in these traits (table 7.2) has been gradual, reflecting either random genetic differentiation or some factor tracking phylogeny (e.g., stabilizing selection or physiological constraints). The majority of display traits, however, showed very little phylogenetic signal, suggesting extremely rapid

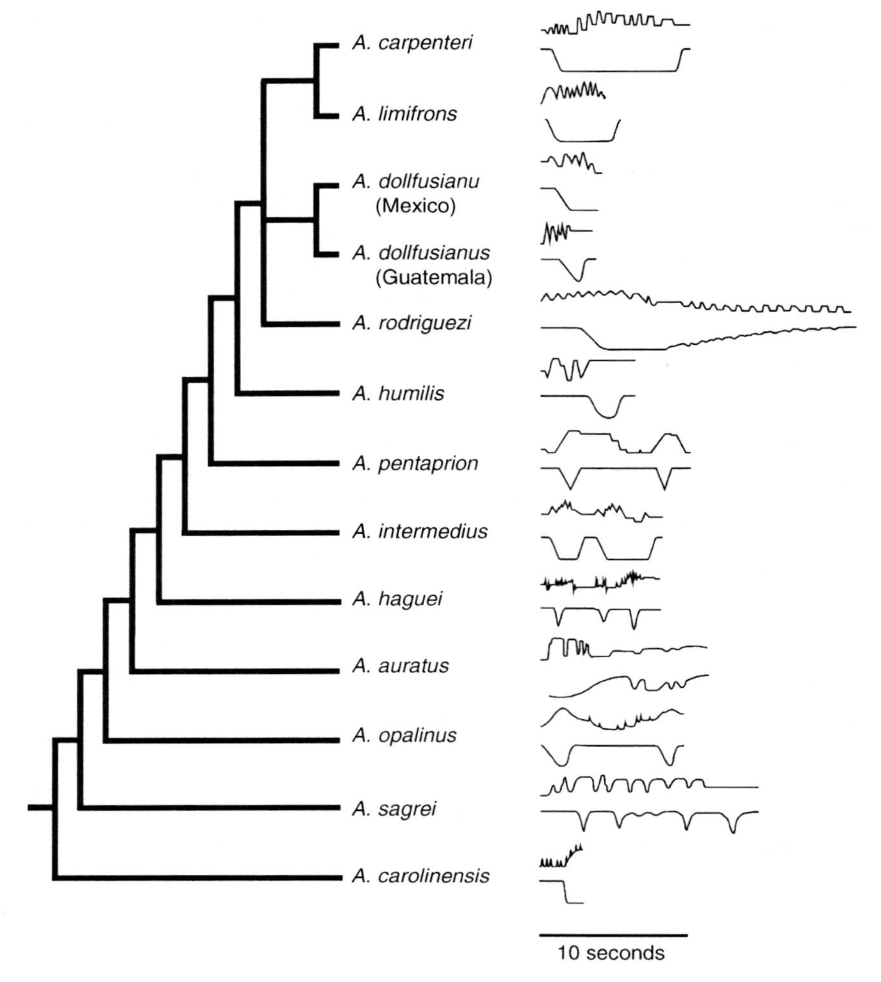

FIGURE 7.4 A subset of *Anolis* species illustrating the diversity in territorial displays. Plots represent the up/down movement of the head-bob (top line) and the extension/retraction of the dewlap (bottom line) over time. See Ord and Martins (2006) for sources used to draw display plots and phylogeny.

rates of change. We also examined several ecological traits and found that the type of habitat species occupied showed almost no phylogenetic signal (Ord & Martins 2006), confirming the absence of niche conservatism in these animals (Losos et al. 2003). Other ecological traits reflecting competitive intensity and sympatry were moderately tied to phylogeny (Ord & Martins 2006).

To explore more directly the presence of adaptive change in displays, we used several different phylogenetic regression models and found consistent evidence that competition among males, species recognition, and habitat use have all apparently promoted adaptation in different aspects of display behavior. Anoles experiencing intense competition seem to have evolved short, complex displays, which parallels what has been suggested for the evolution of bird song (e.g., Read & Weary 1992; MacDougall-Shackleton 1997). Sympatric anoles possess displays that are complex in other aspects, supporting the view that territorial displays also function in species recognition. Finally, despite some dewlap behavior being evolutionarily conservative, change in the design of this aspect of the display has nevertheless still occurred following the invasion of new habitats.

TABLE 7.2 Trait complexes and estimates of phylogenetic signal for territorial displays in anole lizards

	Trait Complexes					Phylogenetic Signal	
	1	2	3	4	5	α	h^2
Head-bob duration	+					12.38	0.25
Dewlap duration	+					15.34	0.09
Average head-bob duration	+						
Average dewlap duration	+						
Average head-bob pause duration	+	–					
Average dewlap pause duration		–					
Dewlap latency		–				6.19	0.90
% overlap		+					
Head-bob number			+			12.99	0.05
Dewlap number			+			15.33	0.02
Dewlap amplitude variation			–				
Head-bob amplitude variation				+			
Dewlap/head-bob ratio				+	–		
Head-bob uniformity				–		14.85	0.19
Dewlap uniformity					+	8.91	0.60

Note: Principal component analysis on Felsenstein's independent contrasts revealed five primary trait complexes, suggesting these traits might have evolved in concert. Signs indicate the direction of evolutionary relation between traits (positive or negative). For example, increases in the variation of head-bob amplitudes included in displays correspond to decreases in the temporal uniformity of head bobs, a trend that has been documented in other lizard groups (Martins 1993). Two estimates of phylogenetic signal are also presented for key display traits in each trait complex: a represents how free a trait is to vary along the phylogeny, with larger values indicating greater levels of change; h^2 is the extent traits are phylogenetically heritable, with high values indicating a strong correlation to phylogeny. See Ord and Martins (2006) for details on display traits scored.

We can therefore explain the evolution of diversity in anole territorial displays on two fronts. First, displays are made up of several traits, each differing in the propensity for evolutionary change. Second, different selective pressures have targeted different traits, producing changes in behavior that have either been extremely rapid or incremental over long periods of evolutionary time. By adopting several different phylogenetic approaches, we were able to show that the evolution of complex behavior is the product of an equally complex interaction between factors associated with phylogeny, correlations between traits, and an elaborate selective regime.

Thinking Phylogenetically

Even when large data sets of behavioral information are not available to a behavioral ecologist, phylogeny can still be highly informative in explaining the present-day function of behavior. Indeed, being explicitly aware that behavior carries the imprint of past selective pressures, as well as other factors associated with phylogeny, can be the only way to interpret why animals behave the way they do. For example, antipredator behavior is often assumed to be an adaptive response to predators in the current environment. However, Sih et al. (2000) observed this was not the case for the streamside salamander. Compared to other species, the high activity of larval streamside salamanders make them conspicuous targets for sunfish, their current predator, and are generally too quick to emerge from refuges when hiding from these fish predators, which increases the probability that predators are still present on emergence. Sih et al. (2000) provided an explanation for this maladaptive behavior by invoking phylogeny.

Those salamander species exhibiting effective tactics for predator avoidance (low activity, conservative emergence times from refuges) are also species known to have a long history with fish predators. The streamside salamander, however, is believed to have diverged from a recent ancestor species that was unlikely to have been exposed to fish. Sih et al. (2000) provides evidence to support the view that this predator-naive evolutionary history has subsequently limited the capacity of streamside salamanders to respond appropriately to the predatory environment they currently find themselves in.

Behavior can also be inherited from ancestors and still expressed despite serving no current adaptive function. Cowbirds and cuckoos lay eggs in the nests of other bird species to avoid the considerable investment involved in raising young. Many target species of this parasitism have evolved ways to identify and eject these foreign eggs from their nests. Rothstein (2001) discovered that the loggerhead shrike in California exhibits this same egg rejection behavior when he placed fake eggs into their nests. The surprise comes when we realize that the loggerhead shrike is not currently parasitized by other birds, meaning that this rejection behavior has no function. Rothstein was aware of the phylogeny of loggerheads and able to explain the existence of this behavior because he knew that loggerheads are nested within a large, Old World clade, all the species of which are believed to be parasitized and express egg ejection behavior. In other words, loggerheads express superfluous ejection behavior because they inherited it from evolutionary ancestors, not because it serves any adaptive function today.

Neither of these studies applies any formal phylogenetic comparative method. Nonetheless, the authors explicitly incorporate phylogeny directly into the interpretation of their experimental findings. Thinking phylogenetically, or considering carefully the phylogeny of a species being investigated, forces an investigator to acknowledge that the current environment might not necessarily explain the origin or even the precise function of the behavior being examined.

FUTURE DIRECTIONS

Behavioral phenotypes are intrinsically complex. As our ability improves to detect and study the mechanisms underlying the production and ontogeny of behavior and the environmental factors that guide behavioral evolution, it becomes increasingly difficult to make sense of an overwhelming diversity of environmental, physiological, and social factors that influence behavior. The comparative method offers a powerful tool for honing in on specific selective factors and behavioral aspects that have evolutionary consequence. For example, studies comparing behavioral syndromes in different fish populations (e.g., Bell 2005; Moretz et al. 2007) have found intriguing evidence that selection acts simultaneously on suites of behavioral traits at the population level. Phylogenetic comparative studies further exploring the habitat features that influence the evolution of such correlated suites of behavior at higher taxonomic levels promise special insight into which aspects of a complex selective regime merit special attention. Similarly, comparisons of gene expression in finch species with different levels of social behavior (Goodson et al. 2005), of endocrine response in birds with different mating systems and levels of parental care (Hirschenhauser et al. 2003), and of the genetic patterning underlying development of insect social behavior (e.g., Toth & Robinson 2007) have helped narrow down the number of genes, hormones, and receptors involved in the evolution of complex behavior.

Phylogenetic approaches are increasingly being used in sexual selection research. The study of sensory biases in mating behavior often necessitates it (e.g., Basolo 1990b). Others interested in how sexual selection might promote speciation have begun to apply phylogenetic comparative methods to estimate rates of species diversification in clades that have experienced strong sexual selection relative to those that have not (e.g., Gage et al. 2002). As Prum's (1997) study illustrates, the form that sexual selection takes is an important question and should be revisited. In particular, the recent proposal that competing models of sexual selection lie at different ends of the same continuum (Kokko et al. 2002) might be examined within the context of phylogeny and potential ecological factors that contribute to divergences in the type of selection experienced by species. Something similar could be attempted for alternative game theory models and the evolution of animal contests (chapter 15).

Confirmation through some form of phylogenetic study provides persuasive support for a behavioral trait considered to be an adaptation.

Examples might include testing the correlated evolution of mating systems, antipredator strategies, kin selection, or altruistic behavior between unrelated individuals, with historical changes in social, predatory/parasitic, or physical environments. Explaining the origins of sexual dimorphism can also be examined phylogenetically. For example, comparative studies can estimate the extent traits expressed by the sexes are historically correlated or free to vary independently, and whether the selective regimes acting on traits differ between the sexes (e.g., Stuart-Fox & Ord 2004; Ord & Stuart-Fox 2006).

We still know little about the rate and type of evolutionary change (punctuated equilibrium or gradualism) experienced by behavioral traits and whether it matches similar patterns in other phenotypic traits like morphology (Blomberg et al. 2003). Furthermore, what is the consequence of phenotypic plasticity and individual variation for macroevolutionary change? There is some debate over whether the flexibility of behavior that allows the invasion of new habitats or shifts in the use of existing habitats promotes or constrains evolution (chapter 6). With careful thought, these two alternative outcomes could be teased apart using a phylogenetic analysis estimating the extent to which traits are variable within, relative to across, species and the consequences of historical shifts in habitat use on rates of change. In addition, the extent to which species are plastic in their behavior because it is a selected trait and/or inherited from phylogenetic ancestors is unknown, yet is at the heart of understanding how plasticity influences macroevolutionary change in behavior and other phenotypic traits across species. One approach would be to quantify within species the extent to which a behavior is plastic to certain changes in the social or physical environment and repeat this for a number of closely related species to estimate the phylogenetic signal of plasticity as a species trait and what ecological/historical factors might correlate with its presence in some species and not others.

More generally, behavioral ecologists need to think more explicitly about evolutionary history and the role phylogeny plays in shaping present-day behavior. The emphasis on current utility is still prevalent in behavioral ecology. Although it is unlikely anybody would disagree with the notion that the evolutionary history of a species is important, it tends to be a post hoc discussion point rather than an intrinsic element in behavioral research. By emphasizing phylogeny during empirical investigation, researchers may become aware of better experimental designs, clearer explanations to puzzling behavioral phenomenon, or perhaps even more fundamental questions to address.

SUGGESTIONS FOR FURTHER READING

The arguments presented by Gould and Lewontin (1979) are still pertinent today. Their paper, and several responding papers that followed (see chapter 2), should be required reading for all students of behavioral ecology. The body of literature on phylogenetic comparative methodology is vast. We suggest papers by Pagel (1999), Martins (2000), Nunn and Barton (2001), and Omland et al. (2008) as starting points. There are also several books and edited volumes available. Most are somewhat dated in the specific techniques described, but they still provide a good introduction to the concepts of comparative biology and historical perspectives for studying evolution. We recommend *The Comparative Method in Evolutionary Biology* (Harvey & Pagel 1991), and *Phylogenies and the Comparative Method in Animal Behavior* (Martins, 1996b). Other useful texts include *Phylogeny, Ecology, and Behavior* (Brooks & McLennan 1991), and *New Uses for New Phylogenies* (Harvey et al. 1996). Finally, *Inferring Phylogenies* (Felsenstein 2004) provides the most recent text and comprehensive review of phylogenetic techniques. It largely focuses on approaches used to construct phylogenetic trees, but the techniques described (parsimony, likelihood, Bayesian approaches) are the same as those employed in reconstructing ancestor character states and many phylogenetic correlation tests.

Brooks DR & McLennan DA (1991) Phylogeny, Ecology, and Behaviour. Univ Chicago Press, Chicago, IL.

Felsenstein J (2004) Inferring Phylogenies. Sinauer Associates, Sunderland, MA.

Gould SJ & Lewontin RC (1979) The spandrels of San Marco and the Panglossian paradigm: a critique of the adaptationist programme. Proc R Soc Lond B 205: 581–598.

Harvey PH, Brown AJL, Smith JM, & Nee S (eds) (1996) New Uses for New Phylogenies. Oxford Univ Press, Oxford.

Harvey PH & Pagel MD (1991) The Comparative Method in Evolutionary Biology. Oxford Univ Press, New York.
Martins EP (1996b) Phylogenies and the Comparative Method in Animal Behaviour. Oxford Univ Press, New York.
Martins EP (2000) Adaptation and the comparative method. Trends Ecol Evol 15: 296–299.
Nunn CL & Barton RA (2001) Comparative methods for studying primate adaptation and allometry. Evol Anthropol 10: 81–98.
Omland KE, Cook LG, & Crisp MD (2008) Tree thinking for all biology: the problem with reading phylogenies as ladders of progress. BioEssays 30: 1–14.
Pagel M (1999) Inferring the historical patterns of biological evolution. Nature 401: 877–888.

SECTION II

DECISION MAKING

8

Decision Theory

RON YDENBERG

Behavioral ecology studies the behavior of organisms as shaped by the action of natural and sexual selection, aiming for an understanding, in fitness terms, of the behavior in the ecological system in which it evolved. Shorthand versions of this description refer to *adaptiveness*, *design*, or *function* (see also chapter 2). The subject we call *decision making* originated in this context. In this chapter I examine some of the concepts that behavioral and evolutionary ecologists have used in developing ideas about decision making. My aim is to give students who are new to behavioral ecology a starting point from which to tackle the subject of decision making, which will be addressed in much more detail throughout the rest of this book. As examples, I consider decisions animals make when foraging and interacting with other individuals.

Students learn from textbooks that the scientific method proceeds by testing the predictions of a hypothesis with experiments. The data so obtained lead either to rejection of the hypothesis or to refinements that can be tested by further empirical work. Repeating the cycle leads to development of a theory that, as it is improved, gains generality, wider application, and eventual integration with other theoretical structures in the discipline. In behavioral ecology, theory is both conceptual (logical clarification of ideas) and mathematical (in which relationships between factors are formalized in equations). Whether conceptual or mathematical, however, theory and empiricism do not always reinforce each other as immediately as students are taught—or so it sometimes seems in behavioral ecology. The literature on foraging, for example, contains repeated claims that mathematical models have led theoreticians to lose touch, or that models have raced ahead of the data (see Ydenberg et al. 2007). The histories of other disciplines such as economics and quantum physics reveal similar tensions, and most science historians would probably describe the description above as a caricature of how science really works.

Nevertheless, historical backlighting illuminates how the give and take between theory and data has driven a grand and steady increase in scientific understanding of almost all natural phenomena. This can be illustrated with an ordinary but very scientific concept that we all use every day—temperature. Before the seventeenth century, temperature could not be measured quantitatively. It took more than a century for scientists to develop a useful measuring device, two centuries to appreciate what temperature and heat really were and harness it to technological development, and three centuries to understand its fundamental laws. More than 350 years passed between the first (very inaccurate) thermometer built by Galileo in about 1600, to the (accidental) discovery of the cosmic background radiation and the hot origin of the universe by Arno Penzias and Bob Wilson in 1964 (see Lightman 2005: 412).

It would be misguided, of course, to map the development of our discipline onto that of thermodynamics or any other discipline. Each has its own peculiar history. Nevertheless, one might reflect just where behavioral ecology is along the path, from building the first measuring instruments, to formulation of fundamental laws, to rigorous and skillful application of these fundamental laws to discover, understand, and manipulate natural phenomena.

DECISION MAKING

Science is full of terms that have meanings very different from everyday usage: quarks have charm, genes are selfish, squares may be perfect, and numbers irrational. The term *decision making* is like this. When I was a graduate student in England, the thesis topic given on my foreign student visa ("Decision Making by the Great Tit") sometimes led to bemusement at the immigration counter when I entered the country—"Great tits! What the 'ell have they got to decide about?" The wry humor stemmed from the mismatch between the popular understanding of the word "*decision*" and its academic usage. The everyday meaning is that alternatives in a weighty matter are carefully considered, and a conscious choice made between them. But in behavioral ecology the term *decision* is used whenever one of two (or more) options is selected, with no implication that the choice is conscious. The choice need not be cognitive at all, and may not even use neural mechanisms. It is possible in behavioral ecology to think of a plant *deciding* about aspects of its floral display, such as investment in male versus female traits (e.g., Biernaskie & Elle 2005), and the term is commonly used in the life history literature to describe allocations of resources (e.g., *allocation decisions*).

The mechanisms that make these decisions are undoubtedly complex, involving lots of biological *wetware*—sensory, endocrine, neural, and cognitive pathways (chapters 9 and 10), built from cells, tissues, and molecules that interact electrically and chemically. However, they are very poorly known and, for the moment, are mostly hypothetical entities identified by terms such as *decision rules* or *evolved processing units* (White et al. 2007). An important development in the study of decision making is the increasing attention being paid to mechanisms. A comparison of the 1998 and 2009 editions of the edited volume *Cognitive Ecology* (Dukas 1998; Dukas & Ratcliffe 2009) will give a sense of this shift. The former devotes considerable space to some of the basic models that I will consider here, but the latter does little of this and focuses on how several model systems (the honeybee brain, bird song, spatial memory) are revealing a wealth of mechanistic detail. Most researchers would now agree that a full evaluation of decision making requires explicitly mechanistic hypotheses. Nevertheless, decision making originated with functional (i.e., survival and reproduction) questions about behavior, which stimulated model building and hence quantitative and testable predictions, with subsequent interplay between tests and model building leading to more complex views of decision making.

Clear Distinction between *How* and *Why*

Most branches of biology and almost all of medicine do not much concern themselves with the distinction between *how* and *why*. In these sciences, understanding the mechanisms involved gives mastery of the system, and there is little concern about the ecological context or evolutionary history. For example, many plant species reproduce both sexually and by apomixis; in the terminology of behavioral ecology, we could think about a decision to reproduce sexually or asexually (Sharbel & Mitchell-Olds 2001). Control of the mechanism would lead to valuable real-world applications in plant breeding—with no consideration necessary of the selection factors leading to these differing reproductive strategies.

But in behavioral ecology it is essential to distinguish carefully between the mechanistic explanations for a decision (*how*) and the selective reasons (*why*) for it. An account of the pathways whereby sensory information and endogenous factors are processed in the brain, endocrine glands, and ovaries of a great tit, to lead to the production of a clutch of 9 rather than 8 or 10 eggs, is a mechanistic explanation. An account that interprets sensory information about the time of year, abundance of food, quality of the nest cavity, or attractiveness of the mate to conclude that a clutch of 9 eggs would lead to the greatest number of surviving chicks is an explanation based on selective, or *ultimate*, factors. Both are valid and important, but they are answers to different kinds of questions, and hence are not competing explanations: the first answers *how* 9 eggs are laid, and the other *why*. Confusions between these kinds of answers are less common than they once were, but we are not yet entirely free of this error. A fuller account is given in chapter 1.

Fitness Maximizing

One can read in almost any paper, thesis, or textbook on behavioral ecology a version of the assertion that organisms "behave so as to maximize fitness." What does this mean? In the context of decision making, this phrase is shorthand for the following: the decision-making mechanisms inside the organism have been shaped by the processes of natural and sexual selection, just as have structures such as beaks, defensive morphology, and so forth. The decision-making mechanisms currently predominating are those variants that in recent evolutionary history have performed better than others.

Better performance of course refers to higher fitness, but the elephant in the room is that there is no universally applicable definition of *fitness*, and there is certainly no way to measure it easily (see chapter 4). In this regard we are not unlike seventeenth-century scientists interested in temperature. Behavioral ecologists have in practice used a wide variety of measures as currencies or substitutes for fitness, and have not worried excessively about the exact relationship of these proxies with fitness. These measures are all somehow related to fitness, some perhaps more closely than others, but none is equivalent to fitness. Even lifetime reproductive success would in many circumstances be an inadequate measure of fitness.

The foundational assumption in the study of decision making is that the mechanisms were shaped by selective processes in ecological settings, via the effects that the consequences of the decisions had on survival and reproduction. To understand decision making, we develop conceptual ideas and mathematical models that make predictions and then carry out experiments that measure the ecological consequences of the possible alternative behavioral tactics in terms of survival, reproduction, feeding rate, number of matings, or growth rate. The relationship between performance measures and fitness is usually presented in the form of an assumption such as "during winter, fitness is maximized by maximizing survival." The implications for overwinter survival of the tactics available are analyzed, and predictions made about the use of the tactics. This might be something like "forage if reserves drop below x; else rest"; or "forage at intensity $u(x)$ when reserves are x." Constraints may be added and trade-offs specified. The modus operandi is that fitness cannot be measured directly, but the *costs* and *benefits* of alternative behavioral tactics can, and we use natural history knowledge to infer the ecological task or function of the behavior of interest. It is of course preferable that the units of the costs and benefits be specified, but when they cannot, theories may resort to abstract quantities such as *net benefit* (benefit – cost). Fagerström (1987) calls these deliberate oversimplifications "productive lies," because they sometimes provide a way to evade (temporarily, anyway) the intractable parts of a problem while searching for a solution.

Strategy Sets

In any situation, the decision maker has a number (or a range) of alternative behaviors to choose from, and the first step in any *why* model of decision making is identifying the alternative actions that could be taken. In most cases this requires good natural history information on the organism. For example, a foraging flycatcher could, upon encountering a potential prey item, choose to ignore it, to catch and consume it, or to deliver it to offspring if it has some. An ovipositing parasitoid wasp could decide to make a particular offspring a son or a daughter. To elude a predator such as whimbrel, a fiddler crab could decide to remain in the safety of its burrow for a short while or a long while, or any duration in between. It would be impossible to investigate each of the situations offered by the endless variety of nature, so our goal is focus on elements common to many situations.

Two of the earliest and most basic decision-making models go by the rubric *optimal foraging*. As elaborated elsewhere (Ydenberg et al. 2007), this was an unfortunate label. Optimality theory is a set of established mathematical procedures that can be used in all branches of science to analyze which of the possible alternatives is best under a given set of assumptions. The implication drawn by many that organisms are also *optimal* generated a lot of unnecessary confusion (see chapter 2 and box 2.1). Unfortunate or not, the label has stuck, and a recent popular compendium (Trefil 2003) lists *optimal foraging theory* (as well as, under a separate entry, *marginal value theorem*) among the 200 "Laws and Principles Governing Our Universe."

The "Diet" Model

Why should a flycatcher ever ignore a passing insect? It is common to many behavioral processes that in making a choice to do one thing, the opportunity to do something else is lost. In making a decision, this lost opportunity needs to be evaluated. In his book *Swifts in a Tower*, David Lack (1956; see p 104) used this idea to explain why the meals delivered to offspring differed depending on the weather:

> It was to be expected that swifts would avoid insects that were too large for them to manage and that extremely small ones would not be taken, but a more subtle selection for size was revealed by analysis of the meals. In fine weather the meals consisted chiefly of insects 5 to 8 mm long, but in poor weather 2 to 5 mm long. Now insects 2 to 5 mm long are commoner in fine than in poor weather and are more numerous than larger insects in all types of weather. Presumably, in fine weather when larger insects are plentiful, swifts can collect a meal more quickly if they do not go out of their way to catch the smaller kinds. In bad weather, on the other hand, insects are so scarce that the swifts can not afford to be so selective.

This passage anticipates by a decade the so-called diet model presented by Emlen and by MacArthur and Pianka in back-to-back papers published in *The American Naturalist* in 1966. That they did not cite Lack's presentation of the idea was perhaps an oversight, but these models did what Lack's conceptual version did not: they formalized the situation, and made explicit that this simple process is common in one form or another to many foraging processes in the natural world.

The model is based on a very simple scenario. Upon encountering a potential prey item, a forager may decide either (a) to *attack* and handle it (pursue, open, dismember, subdue or otherwise prepare and then consume a prey item) or (b) to *reject* it and to search instead for another. Handling and searching for prey are assumed to be incompatible, and all foraging time is composed of these two activities. The hypothesis is that the mechanism inside foragers decides to attack or reject prey items in a way that maximizes the rate of energy intake while foraging.

The basic model can be derived in a variety of ways. Imagine that the density of prey is such that a forager encounters them at rate λ (meaning that during a time interval t, the forager expects to encounter a total of λt prey, with an average interval between encounters of $1/\lambda$). Each prey contains an amount of energy e, and if attacked, requires handling time h. During search time T_S the forager will encounter a total of $T_S\lambda$ prey items. If it eats all of them, it will gain total energy $E = T_S\lambda e$, and expend time $T_S\lambda h$ handling, for a total time $T = T_S + T_S\lambda h$. The rate of energy gain is

$$\frac{E}{T} = \frac{T_s\lambda e}{T_s + T_s\lambda h} = \frac{\lambda e}{1+\lambda h} \qquad (8.1)$$

One way to think about this equation is that each unit of search time (the "1" in the denominator) yields λe energy, and requires an additional λh time units of handling time. This equation is basically identical to Holling's disk equation, well known to ecologists, and to the Michaelis-Menten equation of enzyme kinetics. These equations describe the rate of a process in which *search* (of a predator for prey, or an enzyme for substrate) and *handling* (time during which the predator or enzyme is involved with the prey or substrate) are incompatible. The disk and Michaelis-Menten equations both have interesting histories (the former involves a blindfolded secretary), and both are foundations for much further work in their respective disciplines.

But these equations do not analyze *decisions*. Neither predators nor enzyme molecules *decide* to accept or reject disks or substrate: if encountered, these are attacked, and rejection is not an option. This decision is what the *diet equation* analyzes, which it does as follows. Imagine that there are two types of prey items. The first is encountered at rate λ_1, yields energy e_1, and requires handling time h_1. The second is encountered at rate λ_2, yields energy e_2, and requires handling time h_2. These prey are ranked in order of *profitability*, defined as e/h, so that $e_1/h_1 > e_2/h_2$ ($> e_3/h_{3\ldots}$, etc.). If the task is to maximize E/T as we have assumed here, it is clear that the most profitable prey (type 1, by definition) ought always to be eaten when encountered. The question is: when should the next-most profitable prey (type 2) be eaten as well?

Using the same logic as above, the rate of energy gain when eating only type 1 prey is

$$\frac{\lambda_1 e_1}{1+\lambda_1 h_1} \qquad (8.2)$$

When eating both types of prey the rate of energy gain is

$$\frac{\lambda_1 e_1 + \lambda_2 e_2}{1 + \lambda_1 h_1 + \lambda_2 h_2} \quad (8.3)$$

We now ask when eating both types of prey (called *generalizing*; expression 8.3) yields a higher rate of energy gain than eating only type 1 (*specializing*; expression 8.2) by comparing these expressions. The reader should work through the (slightly laborious) algebra and confirm that

$$\frac{e_2}{h_2} > \frac{e_1}{1/\lambda_1 + h_1} \quad (8.4)$$

This inequality tells us when prey type 2 should be "included in the diet." Inequality (8.4) compares, upon encounter with a type 2 item (i.e., the decision point), the consequences for the rate of intake of accepting the item (shown on the left-hand side of the inequality), or (on the right-hand side) of rejecting it and searching further for a type 1 item instead. Acceptance leads to energy gain of e_2 after handling time of h_2, whereas rejection leads to energy gain e_1 after handling time h_1 plus an additional time $1/\lambda_1$—the expected time until encounter with an item of type 1. The answer to the question we began with (why a flycatcher would ever ignore a passing fly) is that under the circumstances specified (8.4), the handling time required to catch and consume the fly means lost opportunity, because it requires time that is better used to search for more profitable flies. That the flycatcher searches by perching on a branch and watching the world pass by rather than by moving itself is irrelevant, because while handling a small fly it misses out when a larger one buzzes past.

One of the (initially, anyway) surprising features of this solution is that λ_2 (the rate of encounter with prey type 2) does not appear. This means that the decision to include type 2 is independent of the encounter rate with type 2; it depends on the encounter rate with type 1 (i.e., more profitable prey).

This *diet model* identifies the parameters, costs, and benefits in a clear and logical way, and analyzes their interaction in a general and therefore abstract way, but with enough clarity that empirical testing soon followed. The review by Sih and Christensen (2001) tallies 134 published studies of this model! These studies include not only carefully controlled laboratory experiments (e.g., Erichsen et al. 1980), but illuminating applications in complex field settings (e.g., Richardson & Verbeek 1986). There are two amazing aspects of these tests. First, by and large this simple set of equations does nicely predict how animals make decisions about which food items to pursue. Second, it is stunning how broadly such a simple idea applies to the endless variety of foraging processes in the natural world.

The brief development of the basic model given here does not detail the significance of each of the assumptions (full consideration and formal proofs can be found in Stephens & Krebs 1986). For example, encounters with prey are assumed to be independent and sequential. Prey encounter is assumed to be random, and predators are assumed to be fully informed. There is no variation in energy content of handling time, prey are recognized immediately, and predators are assumed to maximize the rate of energy gain. Obviously, many of these assumptions are unrealistic ("productive lies") and were made so that the basic logic could be exposed. In addition to the empirical testing, there has been a steady flow of theoretical papers that have examined consequences of changes in all of the basic assumptions (table 8.1). All in all, this perspective reveals the history of the diet model as a textbook, ongoing version of theory—data interaction. Some additional details on diet choice are provided in chapter 11.

The Patch Model

Another element common to many behavioral processes is that the rate of gain diminishes with greater investment of time, energy, or materials. In behavioral studies the paradigm-setting model is "patch" foraging: resources are concentrated in patches, among which the forager must travel. The rate of prey capture is high upon arrival, but falls as the forager spends time in the patch, so that eventually it becomes worthwhile to travel to a new patch. But when, exactly, should the forager decide to leave? This question was first addressed using a mathematical approach called the *marginal value theorem* (MVT; Charnov 1976).

Diminishing returns are ubiquitous, and the MVT has been applied (see also chapter 11) to a wide diversity of nonforaging situations. For example, Bull et al. (2004) used it to predict the timing of cell lysis by phages. Here the patch is a cell whose resources are exploited by phages in replication. The diminishing rate of new phage production after infection means that there is a point at which it is advantageous to lyse the current host cell and "travel" to new cells. In another

TABLE 8.1 A summary of subsequent modifications to the basic diet model of Emlen (1966) and MacArthur and Pianka (1966)

Issue Addressed	Modification	Source
Independent encounters	Nonindependence	McNair 1979
Sequential encounters	Encounters simultaneous	Stephens et al. 1986
Deterministic models	Stochasticity	Pulliam 1974
Immediate recognition	Recognition takes time	Erichsen et al. 1980
Energy-only nutrient considered	Other nutrients considered	Pulliam 1975
Prey consumed where taken	Prey delivered to central place	Orians & Pearson 1979
Optimization criterion	Definition of "rate"	Templeton & Lawlor 1981
Optimization criterion	Shortfall minimizing	Houston & McNamara 1985
Optimization criterion	Predation danger	Houston & McNamara 1982
Acceptance all or nothing	Partial preferences	Rechten et al. 1983
Real world too complicated	None; commentary	Zach & Smith 1981
No state dependency	Hunger included	Perry 1987

Note: In many cases, the issues were raised in empirical papers or commentaries that predate the theoretical treatments cited here.

example, Parker and Stuart (1976) used the MVT to model how long male dung flies should stay in copula before seeking a new mate (see also chapter 2).

To understand the MVT, we employ the procedure used with the diet model of creating a simple situation with clear assumptions. The habitat is assumed to be composed of *patches* (concentrations of prey) separated by distances across which the predator must travel, which requires travel time t_t. Initially we assume that all the patches are of one type. Within patches, there are diminishing returns on energy gain because the remaining prey are harder to find as more are consumed, or because they hide when they detect the predator. This is described by a function G that decelerates smoothly as a function of patch residence time t_p so that cumulative energy gain after time t_p in the patch is $G(t_p)$. An example is shown in figure 8.1. The smoothness assumption is necessary so that the function can be differentiated, which is essential in this particular analysis, but smoothness is not essential to the basic concept.

The forager chooses patch residence time t_p so as to maximize the rate of energy intake $R = E/T$. Over a single cycle of travel to and foraging in a patch, the energy gain is $G(t_p)$ and the time taken is the travel time t_t plus the patch residence time t_p. The rate of energy gain is therefore

$$R = \frac{E}{T} = \frac{G(t_p)}{t_p + t_t} \tag{8.5}$$

Because all the patches are assumed to be identical, the rate will be the same over repeated patch visits. Clearly, R depends not only on patch quality (the function G) and density (the travel time between patches) but also on the forager's decision (t_p). The MVT proves that the highest possible rate of energy gain is attained at that patch residence time with the property that the marginal rate of gain at departure is equivalent to the resultant rate of gain in the habitat as a whole.

To show this, we apply a standard optimality technique: differentiate R with respect to t_p (hint: use the *quotient rule* for differentiation):

$$\frac{dR}{dt} = \frac{G'(t_p)(t_t + t_p) - G(t_p)}{(t_t + t_p)^2}$$

and set to zero to find where R is at its maximum. This yields the following:

$$G'(t_p{}^*) = \frac{G(t_p{}^*)}{t_t + t_p{}^*} \tag{8.6}$$

The "*" indicates specifically the gain-rate maximizing value. Equation 8.6 says that when t_p is at its optimum value, the long-term rate of energy gain in the habitat (the right-hand side), it is equal to the instantaneous (i.e., marginal, hence the name of this theorem) rate of gain in the patch (the left-hand side).

We expect that the decision mechanism inside the forager somehow emulates the solution provided by the MVT. However, this does not mean

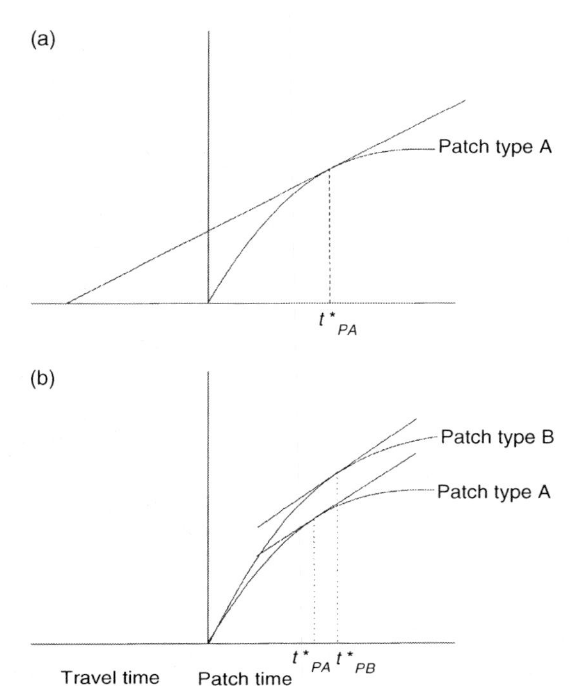

FIGURE **8.1** The marginal value theorem (MVT) applied to patch foraging. The upper panel shows the familiar presentation of a habitat with just one type of patch (type A). In the lower panel there are two patch types (A and B). The maximum rate of energy gain in each habitat is equivalent to the slope of the tangent lines. Patch-type A is the same in both habitats, and patch-type B is better than patch-type A, giving the multipatch habitat a higher attainable rate of energy gain. The MVT proves that the rate-maximizing residence time in each type of patch occurs when the marginal rate of gain is equivalent to the maximum habitat rate of gain. Note that the rate-maximizing residence time on patch-type A in the multipatch habitat (lower panel) is shorter than on patch-type A in the single-patch habitat (upper panel).

that foragers are continually calculating and updating differentials as they forage. This seems unlikely. Foragers probably measure rather different quantities to decide when to abandon patches. In an early experiment (published even before the MVT itself) Krebs et al. (1974) used the "giving-up time" (i.e., the interval between capture of the last prey and departure from the patch) as a measure of the forager's estimate of the habitat rate of intake. This led to the notion that foragers should be able to time short intervals accurately, which it turns out they do well, but not perfectly. Timing mechanisms are likely intimately involved in foraging decisions (Kacelnik & Brunner 2002).

The MVT is usually presented in its graphical version with a single type of patch and tangent line, as in figure 8.1. This simple graphical version is easy to understand, but can mislead when applied to a habitat with more than one type of patch. Then, intake-maximizing residence times in the different patches do not correspond to the tangents of lines drawn from the x-axis at the travel time. In a multipatch habitat the solution is the same (see Stephens & Krebs 1986) as in a single-patch habitat in that at the optimum value of t_p, the long-term rate of energy gain in the habitat and the instantaneous (= marginal) rate of gain in *each* type of patch are all equal, as shown in figure 8.1.

As with the diet model, there has been a history of interaction between empirical and theoretical papers. For example, Welham and Ydenberg (1988) used tractor following by gulls as a patch foraging context to study the MVT. Gulls landed immediately behind a tractor plowing or seeding a field, fed for a time, and then caught up to the tractor and repeated the cycle. Theoretical predictions and quantitative measurements of patch residence time disagreed in a very particular fashion (figure 8.2): in all cases the observed patch times were greater than energy intake rate maximizing predictions, but less than predicted by efficiency, an alternative optimality criterion under which the forager maximizes not the energy gain per unit time, but per unit energy expenditure.

Similar discrepancies had been noted by others, and a series of papers have explored the issue (summarized in Ydenberg 1998, 2007). Summarized briefly, efficiency and rate maximizing have a very particular relationship. *Efficient* tactics are indeed efficient of energy expenditure, but they are wasteful of time in that the rate of work and hence the rate of intake is low. Rate maximizing tactics are inefficient in that the amount of energy intake per unit energy expenditure is low, but the rate of work and hence the rate of gain are high. Foragers are able to gear up the rate of work from low (= efficiency maximizing) to high (= rate maximizing) depending on circumstance. When time is short, they can work hard to make best use of the available minutes; when energy is hard to obtain, they work efficiently to make best use of the available energy. Such understanding arises out of the interplay between theory and empiricism and is one of the successful features of behavioral ecology as a field.

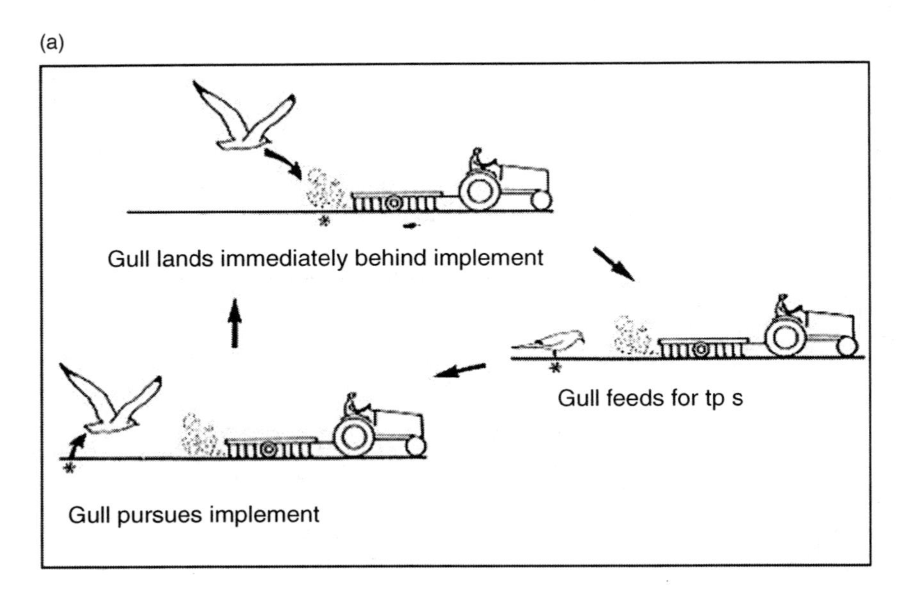

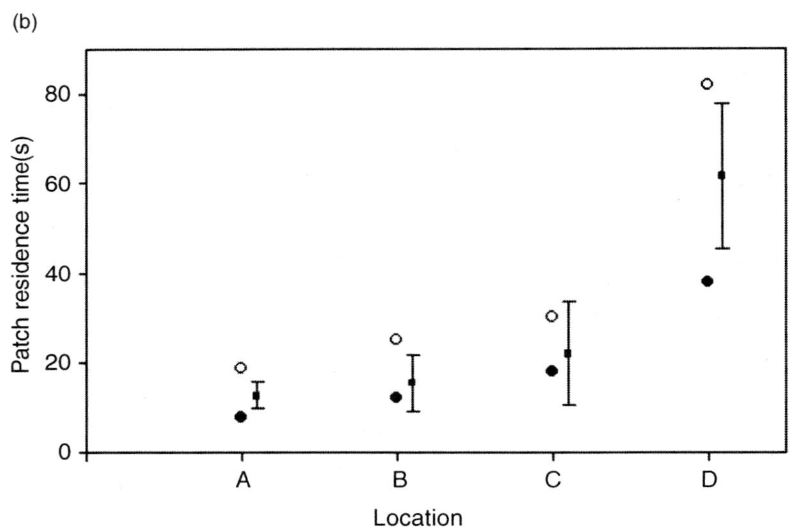

FIGURE **8.2** Tractor following as a form of patch foraging by ring-billed gulls. The foraging cycle is shown in the upper panel. The lower panel shows quantitative predictions of patch time (upper open dot: efficiency maximizing; lower solid dot: rate maximizing) in relation to patch residence times measured (mean and 95% confidence interval) at four different locations (A, B, C, and D). In each case the observations fall between predictions for the alternative maxima, a result that was explored further in other papers (see Ydenberg 1998). Reprinted from Welham and Ydenberg (1988) with permission.

Trade-Offs and State Dependence

Behavioral ecologists are very interested in trade-offs. Under a trade-off, enhanced performance in one attribute lowers performance in another (see also chapters 3 and 4). Unlike situations in which benefits (e.g., measured as energy) and costs (time) can be easily combined into the currency *rate* (energy per unit time), under many kinds of trade-offs, benefits (e.g., energy gain) and costs (e.g., mortality) are not so easily combined. Trade-offs are central to behavioral ecology, and so this is an important problem.

Early on, this problem was treated using unspecified costs and benefits (e.g., Martindale 1982), or by using a discount or *aversion* parameter to (more or less arbitrarily) devalue costs relative to the benefits (e.g., Real 1981). In the 1980s and 1990s numerical modeling techniques were developed that give more elegant ways to treat trade-off situations. Dynamic state variable (DSV; also called stochastic dynamic programming) models have roots in optimal control theory, a body of engineering concepts applied to complex systems. The central notion is that of the *state variable*. The state variable(s) may be hunger, size, or any other aspect of the organism or system modeled. In DSV models, a decision maker chooses between a number of options in each of a sequence of time intervals. Each behavioral option has consequences for the state variable, altering its value in the next time interval. For example, if a forager decides to attack and a prey item is captured, hunger is reduced or the energetic reserve increased. If no prey item is captured, hunger increases. There is usually a probabilistic element involved: for example, prey are captured with probability p, and no prey captured with probability $1 - p$. The changes in the value of the state variable are specified by so-called dynamic equations.

The key to solving a model is calculating the *decision matrix*, which specifies the optimal behavioral choice at each value of the state variable at each time interval. The decision matrix is calculated beginning in the last time interval, T. Here the fitness (= performance measure) at each value of the state variable is defined by the *terminal fitness function*. Next, the penultimate period ($T - 1$) can be considered. The consequences of each behavioral alternative at each value of the state variable can be calculated, because the consequence (the resultant value of the state variable in period T) is specified by the dynamic equations, and fitness at T is defined. Therefore, the decision that gives highest expected fitness can be specified. This procedure is repeated to fill in the entire decision matrix from the last T to the first time interval, and hence is called *backward iteration*.

To learn about DSV models, the beginning student should consult Clark and Mangel (2000), who develop and fully explain the equations and walk through the calculations. In the paradigm-setting example (also from these authors), a forager has the objective of surviving through a sequence of time intervals. The state variable is the level of energy reserves, which must not fall below a threshold. The forager has three behavioral options at each time step: the first offers the likelihood of high gain, but at high risk of mortality. The second option gives an intermediate likelihood of gain at intermediate risk, whereas the third option is safe but offers no gain. The terminal fitness function has the value of 1 (= survival) above the threshold level of reserves, whereas the forager dies (fitness = 0) if it ever falls below this threshold value.

Even though there is no extra fitness gained by having extra reserves at time T, the model shows that it is advantageous to carry some extra reserves *before* time T. The reason is that there is a probability at each time step that no food will be found, even if the decision is to feed. The forager carries some extra reserve as a hedge against this eventuality, because if starvation is imminent, large risks must be taken to obtain food. It is worth taking smaller risks well before reserves fall to zero to avoid falling into this desperate situation. But if the forager has enough reserves to reach the terminal time with no more foraging, it takes no more risks at all. In general, the best behavioral decision at each time step depends on the current size of the reserve (= value of the state variable) and the amount of time remaining. State dependency is an important and pervasive attribute of decision-making models. An empirical application of a DSV model is described in box 8.1.

Game Theory

In the flycatcher example given above, the decision of whether it is best to ignore, consume, or deliver the insect encountered is defined by attributes of the environment such as prey size, insect abundance, and distance to the nest box, and perhaps by attributes of the flycatcher itself like hunger, or the size of the brood. It does not depend on the decisions of other flycatchers. But often the best decision does depend on the actions of competing strategists. Ovipositing wasps should make sons if most other mothers are making daughters, and fiddler crabs need to take the patience of whimbrels into account in deciding how long to hide (Hugie 2003).

The study of such interactions is formalized in the branch of mathematics called *game theory*. Game theory began in economics in the 1930s, and spread to other disciplines such as political science. It made several incursions into ecology, for example, as *ideal free* and *ideal despotic* distributions, and as extraordinary sex ratios, before making a grand entrance in the form of the evolutionary stable strategy (ESS) concept (see Maynard Smith 1982). Game theory is currently widely applied in behavioral ecology, and the concepts it has generated are

BOX 8.1 A DSV Model of Clam Life History Decisions

Dynamic state variable (DSV) models have provided important new insights into decisions that have consequences over time. Here I illustrate this approach using the simple behavior of how deeply into the sand a soft-shelled clam should be buried. Soft-shell clams rely on burial for safety from crabs and other excavating predators, and experiments show that more deeply buried clams are safer. However, safety comes at a cost because deeper burial also means slower feeding. A more deeply buried clam can expect to live longer because it is safer, but it will grow more slowly. The question is, what depth trajectory should a soft-shell clam follow as it grows? Should it (a) go deep quickly and grow slowly as a consequence, (b) stay shallow for a time to grow quickly before descending to safety, or (c) something in between? The answer is not intuitively obvious because several trade-offs are involved.

A dynamic state variable model (Clark & Mangel 2000) with the above trade-off as a central assumption can be used to analyze the situation. The model also considers the allocation of resource intake between growth and reproduction. Here I outline the steps in this DSV model, which was developed with Professor Colin Clark.

Basic Time Interval

Soft-shell clams are very sessile and do not change their depth rapidly over short time intervals. We use 1 year as a basic time interval, with depth being adjusted on an annual basis. Soft-shell clams spawn in May, so we consider consecutive periods to last from June 1 in one year until the next May 31. Reproduction thus occurs at the end of the period, and the time variable t represents the clam's age in years. We chose a time horizon $T = 20$, longer than the expected life span of any soft-shell clam.

State Variable

The ultimate benefit of feeding is realized in terms of reproduction. Clams can allocate resources obtained by feeding either to immediate reproduction, or to growth, which leads to greater future reproduction (if the clam survives). The state variable was chosen to be the mass of clam (tissue excluding shell, in grams) at the start of each year, and is denoted $X(t)$, with

$$0 \leq X(t) \leq X_{max} \quad (1)$$

where $X_{max} = 30\,g$ and any clam with $X(t) = 0$ is assumed dead. It is often convenient to use the lower case x to denote the current value of $X(t)$.

Decision Variables

We use two decision variables, the first being d = burial depth (cm) with

(continued)

$$0 \leq d \leq d_{max} \tag{2}$$

where d_{max} is 20 cm.

The total food resource obtained by the clam in a year is denoted $e(x, d)$, to denote the dependence on size (x) and depth (d). The total amount also depends implicitly on location; high intertidal sites offer reduced growth potential because they are exposed by low tide for more of each day. We considered two locations (a high and a low intertidal site) and ran the model separately for each location (clams cannot move from one location to the other). The functional form of $e(x, d)$ is described below.

The second decision variable is a = proportion of annual resources allocated to growth and metabolism. We have

$$0 \leq a \leq 1 \tag{3}$$

and $(1 - a)$ denotes the proportion of annual resource intake devoted to reproductive output. Actual reproduction is given by

$$\rho (1 - a)\, e(x, d) \tag{4}$$

where ρ converts grams of reproductive tissue to number of progeny.

State Dynamics

The resource allocated to growth plus metabolism is $ae(x, d)$. We assumed that the annual metabolic cost, C, is proportional to current body size x:

$$C = c_m x \tag{5}$$

The clam is assumed to die unless enough resources are used to cover metabolic costs. Thus if $X(t) = x$

$$X(t + 1) = 0 \quad \text{if} \quad ae(x, d) < c_m x \tag{6}$$

Otherwise growth equals $ae(x, d) - c_m x$ and

$$X(t + 1) = x + ae(x, d) - c_m x \tag{7}$$

subject to $X(t + 1) < X_{max}$. Annual predation risk $\mu(d)$ is a function of depth, and also (parametrically) of intertidal location, but not of state variable x.

Fitness Characterization

The fitness function $F(x, t)$ was defined as maximum expected lifetime reproduction from time t to T, for clam with body mass $X(t) = x$ at start of period t. Assuming that any clam alive at age $t = T$ perishes before the beginning of year $T + 1$ we can express the terminal fitness condition simply as

$$F(x, T + 1) = 0 \tag{8}$$

Dynamic Programming Equation

In the model the clam "decides" on depth d and allocation a at the beginning of each period t, given its current body mass x. The reproductive consequences of those decisions

(continued)

are realized at the end of the period, if the clam survives. The dynamic programming equation is

$$F(x, T) = \max_{a,d} (1 - \mu(d)) [\rho (1 - a) e(x, d) + F(x,' t + 1)] \quad (9)$$

where

$$x' = \begin{cases} 0 & \text{if } ae(x, d) < c_m x \\ x + ae(x, d) - c_m x & \text{otherwise} \end{cases} \quad (10)$$

(subject to $x' \leq X_{max}$). This equation holds for $x > 0$ and $t \leq T$.

Measurements in the lab and field provided estimates of the model's parameter values. In combination with known relationships, we could thus estimate the rates of energy intake and mortality at different depths for each size of clam. The dynamic programming equation is used in backward iteration to produce the decision matrix, which specifies the fitness-maximizing depth and allocation for clams of each size.

The decision matrix can be also be used in a *forward iteration*. In a forward iteration, model clams are each programmed with these rules, assigned a starting state (e.g., randomly), and the consequences of following these rules computed. We could produce either the lifetime growth and burial trajectory of individual clams (figure I), or we could simulate a population of clams to ask what the body size–burial depth relationship would look like if we sampled clams on a mudflat, as Zaklan and Ydenberg (1997) did.

Many DSV models have now been published (see Clark & Mangel 2000) to help understand similarly complicated situations with interacting trade-offs. As a numerical approach to a problem, the generality of each is necessarily limited. Nonetheless, these have helped behavioral ecologists to comprehend the fundamental nature of trade-offs, and they are an indispensable tool for moving the discipline forward.

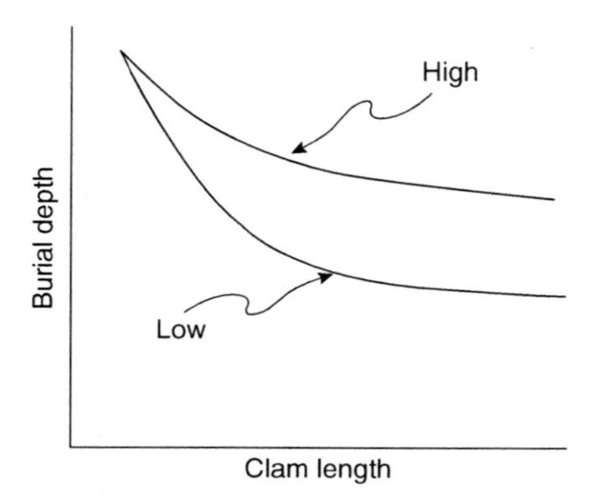

Figure I Fitness-maximizing growth–burial depth trajectories of soft-shell clams at high tidal elevation (few predators) and low tidal elevation (many predators) sites. These curves were computed using the procedure described here, and can be compared to burial depth patterns measured by Zaklan and Ydenberg (1997).

in turn creating the new discipline *evolutionary economics*. Hammerstein and Hagen (2005) discuss the relation between fundamental game theory concepts from economics and ecology, and outline the cross-fertilization between these fields. Game theory's use in behavioral ecology has progressed from simple two-player contests to population-wide "games against the field" to dynamic games, all the while led by simple and unrealistic but nevertheless powerful theoretical metaphors such as the hawk-dove game (see chapter 15).

In all the situations we have considered so far, an optimum was defined in terms of a performance measure, and we assumed that selection would drive the phenotypes toward this point. In game situations, however, the meaning of an optimum is not so clear, as illustrated by a simple "group size" game. Imagine that the fitness of lone animals is 2 fitness units, the fitness of each animal in a pair is 4 units, of each in a trio is 3 units, and in groups of four each gains 1 fitness unit. In this game we assume that individuals can decide to join groups; members of a group cannot exclude others, even if it is their self-interest to exclude a would-be joiner. (Giraldeau & Caraco 2000 discuss games of this sort, as well as those in which existing group members control membership.) We assume that singletons (e.g., recent immigrants and those whose partner has died) occur at some low rate. Their choice is between joining another loner (if one can be found) to make a duo, joining a duo and so making a trio, or remaining alone. Note that with these assumptions, groups of four never occur, as individuals would prefer to remain alone rather than join an existing trio. Also, note that although membership in a duo gives an individual highest fitness (we could call two the optimal group size), a loner would, under our assumptions, join an existing duo in preference to being alone. Depending on the rate at which singletons occur, trios form even though duos are preferred.

Game theory uses the concept of *stability* to investigate such interactions. There are several types of *stable* or *equilibrium* points. At a Nash equilibrium (NE), no individual is tempted to improve its situation by a unilateral move to another group. A population whose members were all in groups of two or all in groups of three would be at a NE, because no individual could improve its situation by moving to another group. A population some of whose members were alone would not be at NE, because these individuals improve their situation by joining another loner or duo.

Behavioral ecologists think in terms of strategies, and use another sort of stability concept, the evolutionary stable strategy (ESS). It can be shown that all ESSs are also NE, but the reverse is not necessarily so. An ESS is defined as a strategy (or mixture of strategies) that, when adopted by most members a population, cannot be invaded by (initially) rare alternative strategies (often referred to as *mutants*). A strategy is said to *invade* if when rare it does better than the prevailing strategy, and so spreads in the population. If there is negative frequency dependence (i.e., a strategy performs less well as it becomes more common; see chapter 3), it will increase to the point at which its fitness is equal to that of the other strategy, and there will be a *mixed ESS* composed of two strategies.

In the simple group size game of the above example, one strategy might be "if possible, join a lone individual to form a group of two; else join a duo to form a trio, but if the only possibility is to join a trio then go it alone" (call this strategy A). A population composed entirely of A strategists would soon be composed entirely of trios. Another possible strategy (call it B) is "if possible, join a lone individual to form a group of two; else go it alone." In a population composed entirely of B strategists, all individuals would be in pairs. An ESS analysis analyzes stability by asking whether strategy A would (when rare) be able to invade a population consisting of B strategists (and vice versa). You can see that rare A strategists would readily invade a population consisting of B strategists, because if no lone individual can be found to partner with they simply join an existing duo to form a trio, and are better off than if they had remained alone (though the original members of the duo are worse off than before!). Rare B strategists could also when rare invade a population consisting of A strategists (a singleton will eventually be found), but as they become more common, more of the B strategists will be condemned to a solitary existence because they never join pairs to form trios.

Stability is defined by two ESS conditions, first derived by John Maynard Smith (1982) as follows. Let the fitness of an individual with strategy x in a population, most of whom use strategy y, be denoted $W[x, y]$. If x is an ESS (denote this as x^*) then an x^* strategist in a population of x^* strategists must do as well or better than any other possible strategist in this population. In symbols

$$W[x^*, x^*] \geq W[x, x^*] \text{ for every possible strategy } x \quad (8.7)$$

This is the first ESS condition, and it is equivalent to a Nash equilibrium.

It is possible, however, that another strategy when rare does as well as x^* (in symbols, this means that $W[x, x^*] = W[x^*, x^*]$). For example, in the group size game outlined in the text, strategy B when rare does as well as A. When this is the case, we must use a second condition to test whether x^* is indeed an ESS. The second ESS condition specifies that for x^* to be an ESS, it must also be true that

$$W[x, x] < W[x^*, x] \quad (8.8)$$

This means that x does less well playing against itself than it does x^*. Therefore, as soon as x rises in frequency it begins to do less well than x^*. Houston et al. (2005) consider this in the context of parental care models. In the case of the group size game developed here, the reader should ask herself about the conditions under which strategy B playing against other B strategists does more poorly than does strategy A. Additional examples of game theory models are presented in chapters 15 and 16.

CASE STUDY

Conflicts between Strategists

Conflict is integral to behavioral interactions in the sense that the decisions that one individual would make if it were wholly in charge would often differ from those that another would make if it had complete control. This kind of disagreement is as integral to apparently cooperative interactions (e.g., parental care, chapter 26) as to those with life-or-death outcomes (e.g., siblicide). Here I outline a study that used game theoretic thinking as a guide.

Common eiders (a seaduck, *Somateria mollissima*) show a complex system of parental care with lone mothers as well as pairs (and occasional trios) of females cooperatively rearing young. Other females abandon their ducklings to the care of mothers who adopt them (see figure 8.1 in Öst et al. 2003). Males do not participate in care, and relatedness among females is low. Markus Öst and his colleagues were interested in the underlying strategies that give rise to this diversity, in a way parallel to the way group-joining decisions give rise to a distribution of group sizes in the example above. As in any decision-making analysis, the first step is to identify the behavioral options open to the females.

The ecological context is important. Hens nest early in the spring on small islands in the sea. They incubate for 26 days without eating, often in bad weather, in nests offering differing amounts of shelter. Consequently, they lose large and variable amounts of mass, and their body condition at hatch ranges widely. Parental care interferes with hens feeding themselves after incubation, so their body condition has a large influence on the duration and intensity of parental care they are able to give.

A period of intense socializing ensues when hens begin to arrive on the rearing area with their ducklings. Soon they sort themselves into lone-tenders, pairs, and trios and settle down to business. Each hen in a group assumes a distinct role in keeping ducklings warm, leading them to food, and defending them from predatory gulls, all of which makes central positions in the joint brood the best and safest. Though the ducklings appear identical to human observers, hens clearly recognize them, and actively manage their positions in the joint brood.

The analysis asked first about partner choice: which hens join with other hens in coalitions, and why? The important elements include state dependency (hens in better condition can take better care of ducklings), and the finding that central positions in the joint brood have higher survival value for ducklings. Two females can care better than one, and three better than two, but this relationship has strongly diminishing returns. Moreover, the good central brood positions have to be divided among the ducklings. The interaction of these costs and benefits can be simply modeled (Öst et al. 2003). To be attracted to a coalition, a female in good body condition has to obtain for her ducklings a big share of the good positions in the joint brood (called *skew*). The reason is that because her good body condition allows her to care intensely, she can expect high duckling survival from lone care.To be attractive a coalition must offer an even higher return. Consequently, two females in good condition are unlikely to be able to find a mutually agreeable division of the joint brood because they cannot both obtain high skew.

In contrast, a female in poor condition can expect little benefit from lone tending, and so is willing to accept even very unfavorable skew in a joint brood. Thus, females in very good condition can have only hens in very poor condition as partners, and the acceptable condition difference between the hens falls as their mean condition declines. These interactions set the boundaries of coalition formation, and constitute a version of the group-size game described above, with rather more complex and realistic *fitness units* attached to each group size. Due to the strongly diminishing returns, females must have even poorer prospects as lone tenders to contemplate joining trios, and these are consequently rare. Field data support all these predictions.

The intensity and duration of care vary widely among eider hens. Some are always very close to ducklings, brood them often, and defend fiercely against predators, whereas others appear far less committed. One idea to explain this is that hens are able to adjust effort strategically in coalitions. The "partner effort" game (Öst et al. 2007) asks how females should do so. It is based on parental effort games (see Houston et al. 2005), but because the partners are both females, complications arising from inherent male-female differences are sidestepped. The model assumes that the total benefit is a function of the summed effort of the partners, which rises with the now familiar diminishing returns. The benefit is split between the females by a skew parameter. Care also has costs, with each female's costs accelerating with the intensity of care she provides.

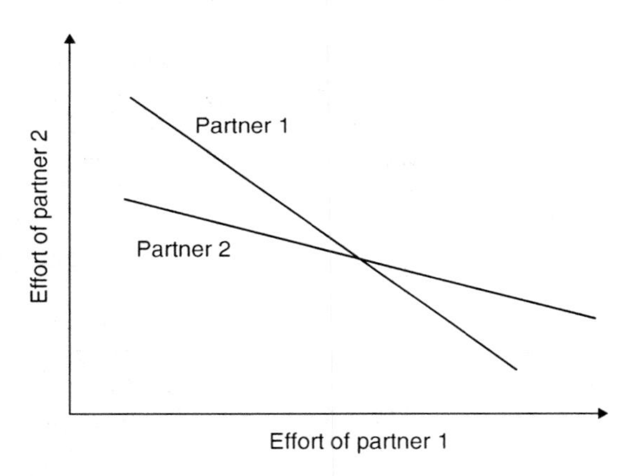

FIGURE **8.3** The Houston-Davies ESS in the eider hen partner-effort game, based on single-sealed bids. The line labeled "Partner 1" shows the best (= performance maximizing) effort of partner 1, given the effort of partner 2 on the y-axis. The line labeled "Partner 2" shows the best effort of partner 2, given the effort of partner 1 on the x-axis. The intersection of the two lines satisfies both ESS conditions.

The first step in finding the stable level of effort is to calculate the "best response" line for each partner, as shown in figure 8.3. This is done using the equations describing the costs and benefits to find the effort that would maximize an individual's fitness (i.e., performance as defined by the equations) for each level of effort that the partner could give. But an individual cannot simply choose the overall maximum, because the best choice will depend on the partner's effort, and vice versa. Therefore, this has to be a game. Examination of the intersection point of the best response lines in figure 8.3 should convince you that this is a NE: neither coalition partner can improve its position by unilaterally changing its effort level. (If either partner changes, the other follows suit and both are less well off.) To determine whether this point is also an ESS, we inquire whether when used by most members of a population this strategy for choosing effort is resistant to invasion by initially rare alternatives. We make this inquiry by applying the ESS conditions described above.

FUTURE DIRECTIONS

The intent of this chapter has been to introduce the reader to the way in which devising explicit models of behavior can lead to new empirical studies, which in turn modify the ideas at the core of the models. The rest of this book will serve to develop these ideas much further, but there are several areas that are developing exciting new insights into the decisions being made by all sorts of animals. Each is built on recent work, involves applications of developments in behavioral ecological theory, and will require ingenious empirical testing.

Negotiation

The ESS shown in figure 8.3 is called the "Houston-Davies" ESS, after the authors of the first papers that applied this reasoning to parental care (McNamara et al. 2003). There is a critical assumption here, namely, that the effort decisions are made as though each partner in advance submits a "single sealed bid," to which it must adhere. This simplifying assumption is undoubtedly unrealistic—yet another example of a "productive lie." There are many other ways that the partners could decide on the effort they will invest, and subsequent work shows that the details of this process are critical (Houston et al. 2005). For example, one partner could make an offer to the other, or the partners could exchange repeated offers (*negotiate*).

In the case of eider hens outlined above, one interpretation of the intense socializing at the onset of the brood rearing period is that hens are negotiating both the skew of the brood and the level of effort each partner is willing to put in. Öst et al. (2007) calculated the Houston-Davies efforts of each hen at every level of skew that would keep the hens in a coalition, but they had no way to determine which of these alternative combinations of effort and skew would be chosen or negotiated by the hens.

It has been suggested that it is not the level of effort itself that is adjusted over evolutionary time, but the negotiation rules that direct the interactants' behavior. Houston et al. (2005) derive intriguing testable predictions in the context of negotiation about parental care, and we might expect negotiation in any context in which there are gains to be made by some cooperation. Roulin et al. (2000), for example, proposed the "sibling negotiation" hypothesis to understand begging by nestlings. In this hypothesis, siblings signal their level of need to each other before the parent arrives with a food load. Begging-driven negotiation thus precedes and influences competition for resources, with hungrier offspring expressing their willingness to compete more fiercely at the next delivery (assuming that the value of resources exhibits a decelerating function with satiation). These and other negotiation problems represent a significant challenge for decision-making studies.

Ecological Rationality

Sih and Christensen's (2001) review of 134 published studies shows that the diet model works well in studies with immobile prey, but less well when prey are mobile. There is another very well studied situation in which the diet model does not work at all, and this has mystified investigators for several decades. When foragers face a *simultaneous* choice of options with identical rates of energy gain but differing in delay between the decision point and the food morsel becoming available, they show a very strong preference for options with short delay, even if the result is a (vastly) lowered rate of energy intake. This is called *impulsivity* or *preference for immediacy* (Stephens et al. 2004). Because this result contrasts so strongly with the predictions of a model that otherwise performs well, it has attracted much attention, and it is a sign of scientific health that several approaches are vying to explain this.

One of these is the *ecological rationality hypothesis* (see also chapter 11). A basic assumption of the approach to decision making adopted here is that the decision rules favored by natural (or sexual) selection are ones that function well in the environments in which they evolved. It follows that these rules may perform poorly in novel or changed environments. As an analogy, the rule "look left, then right when crossing the street" is rational in the ecology of pedestrians in New York, but performs poorly when the pedestrian visits London. Stephens et al. (2004) hypothesize that the preference for immediacy results from a foraging mechanism that performs well in nature in which binary, mutually exclusive choice situations are rare, but poorly in Skinner boxes that experimentally impose simultaneous choice situations. When tested in a situation in which the options are (1) to stay in the patch after finding an item, thereby reaping quickly a second small reward, or (2) to travel to a new patch and reap higher reward after a longer delay, blue jay behavior better conformed with long-term-rate maximizing than with immediacy. Stephens et al. argue that this is because their choice situation resembled the natural situation in fundamentally important ways that simultaneous choice problems in Skinner boxes do not. Diet decision mechanisms inside the organism seem to assume sequential encounter—just as the basic models do!

Behavioral Indicators

Whether one regards organisms as sophisticated decision makers or as Rube Goldberg contraptions using crude rules of thumb, behavioral ecological work of the past 40 years has made it clear that they respond to their environments. Consequently, measures of their behaviors can, with an appropriate framework, inform us of an organism's assessment of the state of the world. For example, as suggested by the first optimal foraging papers, diet choices reveal the prey richness of the environment from the forager's point of view, just as in Lack's analysis of meals delivered to offspring by parent swifts. This could be useful, because measures of prey abundance made by human researchers are generally time consuming and expensive, and are inevitably limited in their ability to capture the experiences of individual animals. It would be useful if we could learn to interpret the decisions made by animals as to what the state of the environment is.

Conservationists are also interested in using behavior as a leading indicator of population change, because that behavior should be able to reveal the nature of the stressors (Kotler et al. 2007). For example, it has often been suggested that energy acquisition rates of seabirds may be a convenient way to assess the status of their prey populations. Our understanding of foraging predator trade-offs has provided ways to assess the impact of human disturbance on animals (Frid & Dill 2002) and to ask how dangerous foragers think particular situations are (Ydenberg et al. 2004).

SUGGESTIONS FOR FURTHER READING

This brief review has touched on some of the main theoretical structures used in the study of *decision making*. I presented these in historical context, to illustrate that far from being out of step, theory and empiricism have interacted directly and intensively here and in other areas of behavioral ecology, driving steady progress in the topics considered. The range of phenomena considered by behavioral ecologists has also expanded greatly, and the subject's internal structure is becoming unified. The student may consider that she has equipped herself with the theoretical framework necessary to guide the next generation of behavioral ecologists when she has command of the material in the books by Houston and McNamara (1999) and Clark and Mangel (2000). The vast amount of work on foraging behavior that has accumulated over the past 2 decades is reviewed in Stephens et al. (2007).

Clark CW & Mangel M (2000) Dynamic State Variable Models in Ecology. Oxford Univ Press, Oxford.

Houston AI & McNamara JM (1999) Models of Adaptive Behaviour. Cambridge Univ Press, Cambridge, UK.

Stephens DW, Brown J & Ydenberg RC (eds) (2007) Foraging. Univ Chicago Press, Chicago, IL.

9

Information Use and Sensory Ecology

JOHANNA MAPPES AND MARTIN STEVENS

Animals face a range of uncertainties associated with both the abiotic and biotic environment, and have to complete a variety of tasks, such as mate location, foraging, and navigation. For example, foraging individuals must be able to locate suitable food from a variety of inedible alternatives, and make a decision about whether to consume it, search for something better, or store it for the future. In addition, or even simultaneously, they may have to find mates, avoid predation, fight competitors, and cope with a varying environment; in short, animals face a variety of tasks to which they have to respond effectively in order to maximize their fitness. Therefore, natural selection has produced a suite of *sensory systems* to utilize and process relevant information. For example, some animals are able to sense the Earth's magnetic field and use this ability in many tasks, in particular, navigation and orientation. The ability appears widespread in animals, including birds, reptiles, mammals, amphibians, and invertebrates, and can be used for long-distance navigation (*compass information*), such as in the open ocean by sea turtles, and in so-called magnetic maps for orientation on a smaller scale, such as by some lobsters (Cain et al. 2005). For other animals, gathering information through olfaction plays an important role. For example, female butterflies often choose suitable host plants for their offspring among hundreds of alternatives based on plant secondary chemicals. Those secondary compounds, such as iridoid glycosides and pyrrolizidine alkaloids can slow down the growth of larvae but also work as a "medicine" against parasitoids by reducing the growth of the parasitoid (ReudlerTalsma et al. 2007; Bernays & Singer 2005). Thus, the optimal oviposition site may depend not just on the availability of host plants in the environment, but also the density of potential enemies, which may vary among locations and years, so gathering information on those might also be important. In this chapter we use various examples to outline how sensory and information processing is linked to behavior and adaptive decision making of animals (see chapter 8). We also discuss that sensory systems have limitations, and that animal signals can be tuned to the environmental conditions across a variety of sensory modalities. However, we focus most specifically on vision because this has played a central role in many studies of evolutionary behavioral ecology.

CONCEPTS

What Information Is and Where It Comes From

Information transformation and processing are significant fields in biology, but for behavioral ecologists, how animals gain, process, and utilize information to increase fitness is of primary significance (Dall et al. 2005). Behavioral ecologists refer to *personal information* when an animal collects cues from its abiotic (e.g., the sky) or biotic (other organisms) environment. Many animals collect useful information from the behavior of other animals (from conspecifics or from other species), such as the movement of predators or a suitable foraging location. This is called *social information* (Seppänen et al. 2007; see chapters 16 and 17, this volume). For instance, migratory birds such as European flycatchers (*Ficedula* ssp.), particularly in northern areas, are limited by the availability of suitable nest sites and good quality territories. Early arrivals usually have a higher breeding success because they have greater time to gather information. Later arrivals use resident birds like tits (chickadees, *Parus* ssp.) as an information source in nest site choice to acquire direct information regarding suitable nest options (Seppänen & Forsman 2007; figure 9.1).

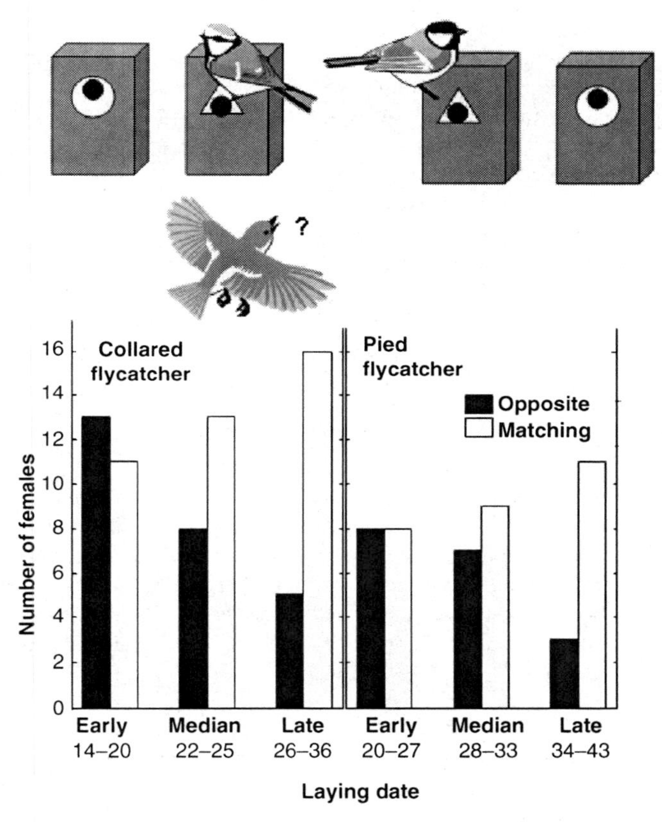

FIGURE **9.1** Social information use between species could be common in situations in which ecologically similar species have unequal access to information. This arises in boreal forests between small cavity-nesting passerines, in which breeding success is limited by the availability of suitable cavities. Seppänen and Forsman (2007) experimentally showed that migrating flycatchers (*Ficedula albicollis* and *F. hypoleuca*) can use social information from the residents (great and blue tits, *Parus major* and *Cyanistes caeruleus*). Before the arrival of flycatchers, Seppänen and Forsman created artificial, neutral nest-site preferences of tits by attaching a geometric symbol (circle or triangle) on occupied nest boxes to give the appearance that all tits within an area apparently preferred nests with a particular symbol; another empty box with the opposite symbol was placed on the nearest similar tree (2–6 m away), facing the same direction. When the first flycatcher males arrived, the researchers provided additional pairs of empty boxes with two symbols randomly assigned within the pair. Arriving flycatcher females thus had to choose between nests with the different symbols. Late-arriving females preferred the symbol that resident tits appeared to prefer. This may be because earlier arriving females tend to be older, more successful individuals with previous breeding experience and thus may possess, by individual learning or genetic quality, more and better personal knowledge so that they ignore potentially inaccurate social cues. Also, later arriving females have less time available for gathering personal information, because both flycatcher species face severe reduction in reproductive success with delayed breeding.

The information involved in making a decision or modifying a decision-making strategy (chapter 8) may stem from multiple information sources, and may involve a range of sensory modalities. Over evolutionary time, this can result in changes in the information involved in making a decision. For example, some species of birds are brood parasites (see chapter 1), laying their eggs in the nests of other species or even individuals of the same species. In many parasites, the chick often ejects (killing) all the host's own offspring, ensuring that it has the full care of the foster parents, who end up rearing a chick to which they are not related. However, although evicting the host nestlings may initially appear to enable the parasitic chick to maximize the amount of parental provisioning it receives, it can be costly because the single parasite nestling attracts a reduced level of parental care than an entire brood. The problems of reduced parental care toward a lone chick may be why some brood parasites do not evict all the host's young from the nest, but tolerate them instead. Alternatively, some cuckoos have evolved remarkable ways to manipulate the host's sensory systems into providing a greater level of care than a single chick would normally produce. For example, common cuckoo (*Cuculus canorus*) chicks that parasitize reed warblers (*Acrocephalus scirpaceus*) in the United Kingdom have begging calls that mimic the sound of an entire nest of reed warbler chicks, to compensate for the reduced visual stimulus of a single gape (Davies et al. 1998, see figure 9.2). Equally remarkably, Horsfield's hawk cuckoo (*Cuculus fugax*) chicks found in Japan have a bright yellow patch on the underside of their wings that they display to their host parents, mimicking the mouth coloration of the host chicks; coloring the wing patches black, the same color as the wing background, reduces the amount of food that the parents bring to the chicks (Tanaka & Ueda 2005)! Overall, many animals have evolved a range of tactics to transmit either honest or dishonest information, which is then processed by the receiver and used in decision making.

Receiving Information and Signal Form

Signals should be selected to maximize their effectiveness in conveying (or hiding) information. According to Guilford and Dawkins (1991) the effectiveness of a signal will depend on its *efficacy*, meaning how specific pressures stemming from the environment and the attributes of the receiver's sensory and cognitive systems have influenced how the signal has evolved to get the message across. The efficacy of a given signal relates to what makes it most effective in a specific environment and how strongly it will affect the behavior of the receiver. Efficacy is thus a measure of the signal's success (Endler 2000). At least three aspects of a signal's efficacy are particularly important: (1) how easily a signal can be distinguished from the background,

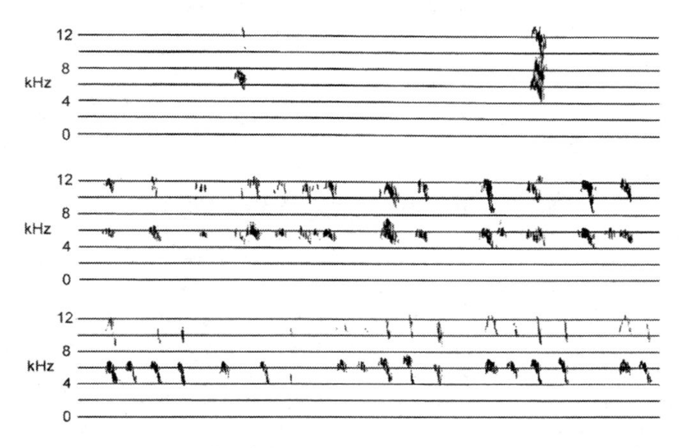

FIGURE 9.2 Sonograms lasting 2.5 seconds of the begging calls of a single reed warbler chick (top), a brood of four reed warblers (middle), and a single common cuckoo chick (bottom). The cuckoo chick calls at a rate and manner similar to a whole nest of reed warblers to induce the host parents to provision at a high rate. Adapted from Davies et al. (1998).

(2) how easily a signal is discriminated from other signals in the environment, and (3) how memorable a signal is (Guilford & Dawkins 1991). The detectability of a given signal is influenced by multiple factors, including the environment through which the signal travels and the sensory and cognitive mechanisms of the receiver. For example, ambient conditions can have a key role on the efficacy and form of a signal (see, for example, the classic works of Hailman 1977; Endler 1978; Lythgoe 1979). For instance, light conditions have experimentally been shown to affect mate choice in guppies (*Poecilia reticulate*), whereby different environmental conditions can maintain polymorphism in signal form (Gamble et al. 2003). Discriminability is also crucial if a signal is to be recognized and categorized correctly by the receiver (Guilford & Dawkins 1991). For example, for aposematism (warning signals involving color, odor, or movement combined with unprofitability) to work, the markings involved should (generally) be easily detectable and discriminable from the surrounding environment. In contrast, effective camouflage should minimize these properties (Cott 1940). Finally, memorability is also a crucial aspect of animal coloration, because warning colors should be easily learned and remembered by predators, whereas camouflaged markings should minimize memorability to prevent the rapid formation of predator search images (Ruxton et al. 2004; chapter 10, this volume). Learning is also important and a signal may be easy or difficult to learn regardless of how memorable it is.

Finally, in considering the function and efficacy of any signal, it important to avoid a human perspective because, as we shall see, many animals have sensory systems different from ours, and thus perceive the world in a different way. This is not a new realization, and scientists as far back as Alfred Russel Wallace emphasized the differences in perception among animals. For instance, signals that are conspicuous or camouflaged to humans need not necessarily be so to the intended receiver.

BOX 9.1 How Sensory Systems Work: Vision as an Example

Explaining how any sensory system works is complex, not least because such systems often involve various stages of processing and the properties of the system can differ greatly among animals. A full account of visual perception is beyond the scope of this chapter, but rather we use the example, largely based on vertebrate vision, to illustrate some general features of sensory processing. The world around us contains a range of information sources, only some of which are used by an animal. For example, vision is a way of utilizing information in the form of light, and light is merely part of the electromagnetic spectrum; forms of radiation also including gamma rays, X-rays, infrared, and radio waves. The part of the electromagnetic spectrum to which an animal is sensitive is often called *visible light*, and is something that varies among animal groups, and even within a species.

Sensory Systems Convert Information into Neural Responses via a Range of Specialist Cells

The first step in visual processing occurs in the eye, where photoreceptors transform light information into neuronal signals by trapping electromagnetic radiation by a receptor molecule, which then triggers a response in the receptor neuron (Simmons & Young 1999). The retinae of most vertebrate eyes contain several types of photoreceptors, often distinguished by their morphology and/or the visual pigment involved. These include rods and cones. Rods are generally sensitive to low levels of illumination, and usually do not give rise to color perception (although there may be exceptions); this is generally performed by the cones (Kelber et al. 2003).

(continued)

Sensory Systems Differ among Species

Color perception, an important component of many animals' vision, is essentially the ability to discriminate differences in the spectra of light, rather than simply differences in intensity. In order to be able to detect differences in the spectral composition of light, an animal may possess two or more types of receptor cells, containing pigments with different absorption spectra, or have other features to change the cells' spectral sensitivity (such as accessory structures). Most mammals have two cone types and are *dichromatic,* whereas humans, together with apes and Old World monkeys, a few New World Monkeys, and some marsupials, are *trichromatic* (Jacobs 1993). Some vertebrates—for example, birds—utilize four cone types in color vision, including sensitivity to ultraviolet (UV) light (Cuthill 2006). In invertebrates, color vision is also highly variable and can involve a diverse number of receptor types. Essentially, color vision can be explained on the basis of comparisons among the outputs of different photoreceptors in postreceptor processing stages containing color-opponent coding neurons.

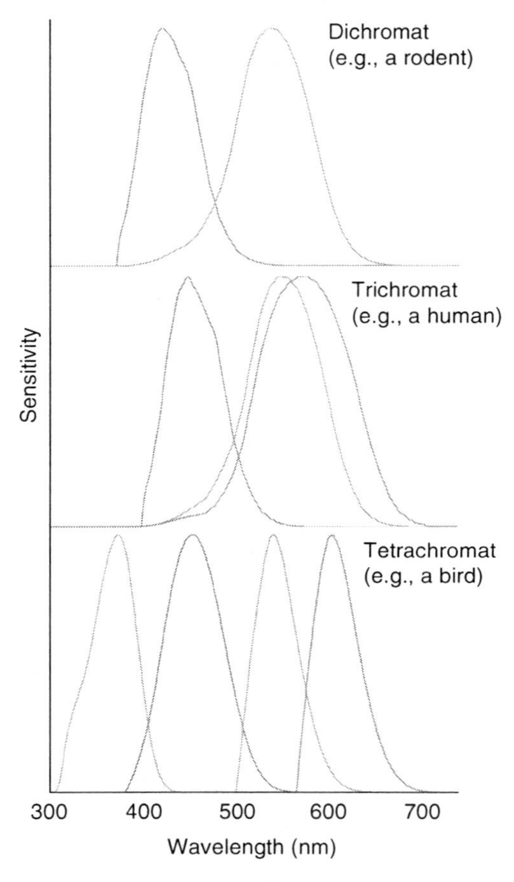

Figure I Variation in the number and spectral sensitivity of visual receptors in di-, tri-, and tetrachromatic animals.

(continued)

Diversity in Systems Lead to Differences in Perception

Although each receptor type has a point of maximum sensitivity, each is sensitive to a range of wavelengths and are "blind" to the specific wavelengths of stimulation; two different light spectra can appear the same if they produce the same receptor responses. Because color perception arises from the comparison of more than one receptor type, there are several consequences for color perception in animals with different systems. First, in principle, different light spectra can produce a sensation of the same "color" if the output of the animal's photoreceptor types (and postreceptor stages) are the same; second, the same spectra may produce different hues to animals that differ in the absorption spectra of their photoreceptors; and third, the dimensionality of color space is determined by the number of interacting receptor types (Wyszecki & Stiles 1982). This means that the same signal may look different to animals with different visual systems.

Sensory Systems Encode Multiple Forms of Information

Although color vision is undoubtedly important for many animals, it is just one component of various information types encoded by the visual system. In vertebrates, for example, the cones are also involved in achromatic or *luminance* vision (to us how "light" something appears). In humans this is based upon the combination of the longwave (LW) and mediumwave (MW) cones, but birds and fish, for example, seem to use an additional type of *double cone* (simplistically, two receptor cells linked together), whereas bees, for example, use their longer-wavelength receptors (Osorio & Vorobyev 2005). The receptors used for luminance vision also tend to be the most abundant type in the eye because they are crucial in processing pattern and texture information.

Sensory Systems Often Involve Trade-Offs

Because retinal cells have spontaneous activity in the absence of light, the brain has to somehow determine when neuronal activity corresponds simply to noise, and when it should be interpreted as a genuine signal. That residual activity or noise in the visual system can limit discrimination and has far-reaching consequences, because it results in the idea that a significant difference is demanded before neural activity is interpreted as a real signal, and represents a trade-off between reliability and sensitivity. The idea that receptor noise can limit signal detection and discrimination has been incorporated into some models of visual perception in animals (Vorobyev & Osorio 1998). Other trade-offs are also common in vision. For example, at low light levels, the eye may sacrifice some acuity for increased sensitivity. This is achieved in part by "pooling" the responses of more neurons in the retina, but comes at the cost of losing the ability to discern high detail.

Trade-offs can also result in different strategies evolving among species with similar aspects of vision, because the specific tuning of the system can represent a trade-off between balancing or optimizing opposing features, with a different outcome favored for different tasks/ecologies. For example, both Old World primates and birds apparently possess a color-processing channel sensitive to red-green contrasts, which stems from comparisons between the LW and MW cones. In birds, the separation between these two cones is large compared to Old World primates, and this likely means that the relative difference between red and green objects is probably larger to birds than primates, but because the primate cones are closer together and overlap more, the latter system is more resistant to shadows and to gradual changes in the ambient light conditions over the visual spectrum (Lovell et al. 2005).

(continued)

BOX 9.1 (cont.)

Sensory Systems May Evolve Partly in Response to Specific Tasks

Old World (and some other) primates have re-evolved the use of three cone types in color vision, compared to just two cone types in most other mammals. The evolutionary significance of trichromacy is thought to lie partially in the greater ability that it gives to detect ripe yellow or red fruit or young leaves against a dappled background of mature green leaves, perhaps in part because, as we have seen, the LW-MW system is less affected by shadows and dappled light (e.g., see Regan et al. 2001). However, sensory systems do not necessarily evolve for specific tasks; overall, they appear to have been optimized largely with respect to three main functions: the light spectra of the signal, the behavioral task, and noise that arises in the photoreceptors (Osorio & Vorobyev 2005).

There Are Many Sensory "Worlds" Humans Cannot Perceive

Sensory systems are involved in acquiring and processing information, from specific signals or the general surroundings, which serve a function for an individual in interacting with other animals and the environment. The sensory systems that have evolved in animal groups are often closely linked to the attributes of the environment in which they are found. For students of behavioral ecology, it is essential to remember that many sensory systems involve cues and information that we cannot perceive. Over time, technical innovations have allowed us to study sensory systems that had previously been a mystery to humans.

Example 1: Electric Sense

As a result of its conductive properties, many aquatic animals, especially various fish species, and some terrestrial animals that forage in water or in wet substrates, possess sensory systems involving the use of electric information. Because objects naturally emit electric cues, *electroreceptive* species can detect electric information from the environment, without necessarily using or possessing an organ which emits electricity. In addition, various species of fish have evolved the ability to actively produce electricity (*electrogenic*), of which most are electroreceptive. Electrogenic species produce electric organ discharges (EODs) from specially modified organs to *emit* electric cues to the surrounding environment (Moller 1995). This somewhat analogous to the ability of bats to produce auditory cues and monitor the changes in the intensity and structure of the returning echoes to navigate, hunt, and communicate (echolocation).

Sharks are one group of animals that are sensitive to electric cues but lack electricity-producing organs (they are electroreceptive but not electrogenic and are thus not "true" electric fish). A classic experiment by Kalmijn (1971) showed that animals can use an electric sense to find objects by detecting their associated electric field (*electrolocation*). After stimulating sharks (*Scyliorhinus canicula* and *Raja clavata*) with the odor of fish to induce active foraging, the experiment comprised a series of tests selectively discounting the use of other cues that could be potentially used by sharks to detect a hidden flatfish in the substrate. The flatfish (*Pleuronectes platessa*) was buried in the sand of an experimental setup, which discounted visual cues, and Kalmijn then placed a modified agar box around the flatfish to eliminate chemical and movement, but not electric, information. Only when the box was covered with plastic, to prevent electric cues, did the sharks fail to find the flatfish. Importantly, Kalmijn also investigated the behavior of the sharks to electrodes hidden in the sand, which had a clear effect in attracting the sharks. It thus seems that in sharks, olfactory cues are used in long-distance prey detection, whereas electric cues are useful for locating the precise location of hidden prey. Kalmijn's classic experimental protocol was recently modified to confirm that some mammals are also electroreceptive. This is best known in the platypus (*Ornithorhynchus anatinus*), but has also been discovered in the star-nosed mole (*Condylura cristata*) and the echidna and long-nosed echidna (*Tachyglossus aculeatus* and *Zaglossus bruijnii*). The platypus bill comprises a high density of mechano- and electrosensory receptors, used to forage for food in murky waters (Scheich et al. 1986; Pettigrew et al. 1998).

True electric fish have evolved a range of specific sensory organs and receptors for detecting electric fields, with different receptors often used for different tasks. In electrolocation, an electrogenic animal monitors changes in self-produced electric signals that interact with different objects in the environment. Changes in the electric field are detected by special electroreceptor cells in the skin by comparing the current flowing through the cells with that if there were no object present, and dissimilar types of objects produce different effects; those with impedance values lower than that of the surrounding water "attract" electrical current lines because more current flows through the object compared with the water, whereas objects with a higher impedance than the water cause the opposite effect (von der Emde 1999). Not only can electric fish detect the types of objects present in the environment, but they can also perceive the distance to objects, and the objects' composition, size, three-dimensional orientation, and spatial configurations!

Electric discharge is often used in other tasks, such as communication. For example, the electricity-producing organs and EODs of many species, such as various knifefish, are sexually dimorphic, and female mate selection can be undertaken on the basis of male EOD attributes, such as larger amplitude and longer duration (Curtis & Stoddard 2003). More work is needed to ascertain exactly how common female mate selection based on male EOD attributes is, and exactly what features females prefer and why, but it does appear that electric senses may have been co-opted in various species for a role in sexual selection. Given the importance of an electric sense to many fish species, it is also perhaps not surprising that some species may use their EOD ability for other forms of communication; for example, African Cornish jack (*Mormyrops anguilloides*) from Lake Malawi use EODs to maintain group cohesion in pack hunting (Arnegard & Carlson 2005).

Example 2: Ultraviolet Light

It has been known for some time that a range of animals, spanning almost every animal taxa (invertebrates and vertebrates), are sensitive to ultraviolet (UV) light (Tovée 1995). UV cues are important in many tasks, including navigation, foraging, and mate choice. Perhaps because humans cannot see UV, the function of UV signals and vision has fascinated behavioral ecologists disproportionately compared to other parts of the visible spectrum, and perhaps its role in mate choice and foraging has been overemphasized. To study whether a behavioral task involves UV vision, such as mate choice or foraging, an effective method has been to present a stimulus with or without UV light, or with the UV component of the signal modified, and to look at its effect on behavior. However, to understand the relative significance of UV to a behavioral task, it is important to investigate all parts of the spectrum to which the animal is sensitive, rather than UV alone; otherwise one cannot be sure that UV is of any more importance than other components of the signal (Stevens & Cuthill 2007). This is a common difficulty with designing behavioral experiments because it can be problematic to isolate the mechanism that underpins a behavior without creating an overly artificial situation.

The relative importance of UV, or of any other part of the spectrum, is likely to depend on the specific species in question, and there are some fascinating examples of visual signaling in animals involving UV. Perhaps some of the most exciting recent discoveries involve crab spiders, in which studies have compared the perception and detectability of spiders to both their prey and predators. Traditionally, it has been assumed that most species, which rest on flowers waiting for a prey animal to land, are highly camouflaged and thus rely on concealment to catch their prey. Recent work, however, has shown that some species, such as Australian crab spiders (*Thomisus spectabilis*), may actually use their coloration to attract prey, including honeybees, by being conspicuous in the UV and by contrasting strongly against the flower background (which does not reflect UV strongly; Heiling et al. 2003). Heiling et al. (2005) experimentally reduced the UV reflectance of the spiders (although not affecting other parts of the spectrum), which reduced the overall contrast with the flowers presumably making the spiders more camouflaged, and found that this actually decreased the likelihood that the bees would approach the flowers. The strategy of reflecting in the UV may, however, bear the cost that the spiders are also easier to detect by avian predators, as has been demonstrated in some moths (Lyytinen et al. 2004). In the future, it would be interesting to determine the relative costs and benefits of UV signaling.

Other studies have also shown that the coloration of orb web spiders may also act as lures to attract prey (Tso et al. 2006), and here, recent work has shown that this strategy may result in an increased risk of predation from wasps (Cheng & Tso 2007). Prey attraction may not always be the rule, however, because even within a species, the spiders can be either well camouflaged to both predators (birds) and prey (bees), or conspicuous in the UV, depending upon the species and exact location on the flowers on which they wait (Théry & Casas 2002).

Spiders also possess, along with some birds such as parrots and various marine organisms, another interesting signaling adaptation: fluorescence. This occurs when fluorophores absorb light of certain wavelengths (often UV) and then emit this at longer wavelengths. Various species of spiders possess fluorophores, differing in type among families, which seem to have evolved multiple times independently (Andrews et al. 2007). In the ornate jumping spider (*Cosmophasis umbratica*), males have UV-reflecting patches on the head and body, which females lack, but females have UV-induced green fluorescing palps, which males lack (figure 9.2). By using UV-blocking filters, to remove UV from the ambient light conditions, Lim et al. (2007) showed that neither non–UV-reflecting males nor nonfluorescing females were courted by the opposite sex (as under full-spectrum light conditions). By removing the UV light present at different times for males and females, such as by having males under full spectrum light and females under UV-lacking light and vice versa, Lim et al. were able to induce one sex to display normally, although eliminating the UV or fluorescent component of the signal, showing that the change in courtship of the spiders toward the UV- or fluorescent-lacking mates was due to changes in coloration as opposed to display behavior. Thus, the fluorescence of females and the UV reflectance of males would seem to be an important component of successful courtship. As discussed above, it would next be interesting to selectively remove the other parts of the spectrum in which the spiders can see to fully understand the relative importance of UV and fluorescence compared to other parts of the spectrum in this behavior.

Animals Combine Information from Multiple Sources

Animals often use information from several sources simultaneously, as can be inferred by the combination of begging and calls that enables a cuckoo chick to maximize host parental care. The way in which animals utilize information from multiple sources can have a key influence on behavior, of both individuals and groups. For example, many species of ants must often move to a new nest site when their current location becomes unusable. Such emigrations involve determining the presence, suitability, and location of a range of potential nest sites, and incorporating information about the environmental conditions, before making a collective decision about where to move to. Colonies of the ant (*Temnothorax albipennis*), which naturally live in thin fissures between rocks, weigh various sources of information about potential new nests before beginning to move. This includes information about the darkness of the cavity, the width of the nest entrance, and the area and height of the internal cavity (Franks et al. 2003a). Assessing the attributes of the nest is likely to involve several cues; for example, judging the height of the nest and the width of the entrance presumably involves simple tactile assessments, whereas ants seem to use individual-specific trail pheromones to measure the size of the internal cavity (Franks et al. 2002). The ants effectively weight the different attributes according to their relative value to choose among nests with a range of features of differing quality. Initially, individual scouts access the quality of the site, before recruiting more individuals to the location by the slow process of tandem running, physically leading others to the new site. Once recruited in this way, the new ant will then also assess the site and then begin recruiting by tandem running. On encountering a relatively poor site, scouts generally hesitate for longer before recruiting others; this delay means that colonies are less likely to move to inferior nests. *T. albipennis* uses collective "decision making" to finally accept a nest site, involving quorum sensing, in which a new nest is only accepted once a certain number of individuals are found inside the new site. The emigration then switches to a much higher rate of recruitment, involving carrying other individuals (and brood) to the nest (Pratt et al. 2002). Because of the way in which recruitment works, allowing multiple scouts to assess the nest and recruit other individuals to the site, the number of ants in the nest required to reach the threshold is usually too large to have been achieved by one ant alone (Franks et al. 2002). Such quorum sensing allows the assessments of multiple members of the colony to be combined, and for the colony to essentially reach a consensus as to whether the new nest is good enough to be accepted.

In addition to assessing the features of the nest, the ants are also able to make decisions about the emigration process based on conditions outside of the nest. Animals often face trade-offs between making an accurate decision (such as over the best nest site) and making a quick decision (for example, to reduce potential risks associated with emigrating). Such speed-accuracy trade-offs are common in nature and have also been shown in *T. albipennis*. For example, Franks et al. (2003b) forced colonies of *T. albipennis* to move to new nest sites under either benign or harsh conditions, and investigated the effect on the speed and accuracy of nest choice. Under conditions that were relatively harsh, such as with wind blowing over the nest or the presence of formic acid that may indicate the presence of predatory ants, the colonies lowered their quorum thresholds compared to the benign conditions and began rapid recruitment more quickly (figure 9.3). Essentially, they placed a greater emphasis on individual choice. Furthermore, even with knowledge of a better nest by some scouts, the colonies sometimes moved to an inferior nest when faced with a choice; as such, even when able to gather relevant information, animals can make judgment errors. Interestingly, *T. albipennis* colonies can, however, correct these errors of judgment because scouts continuously find and assess the quality of other potential nest sites in the area, even when their current site is suitable (Dornhaus et al. 2004).

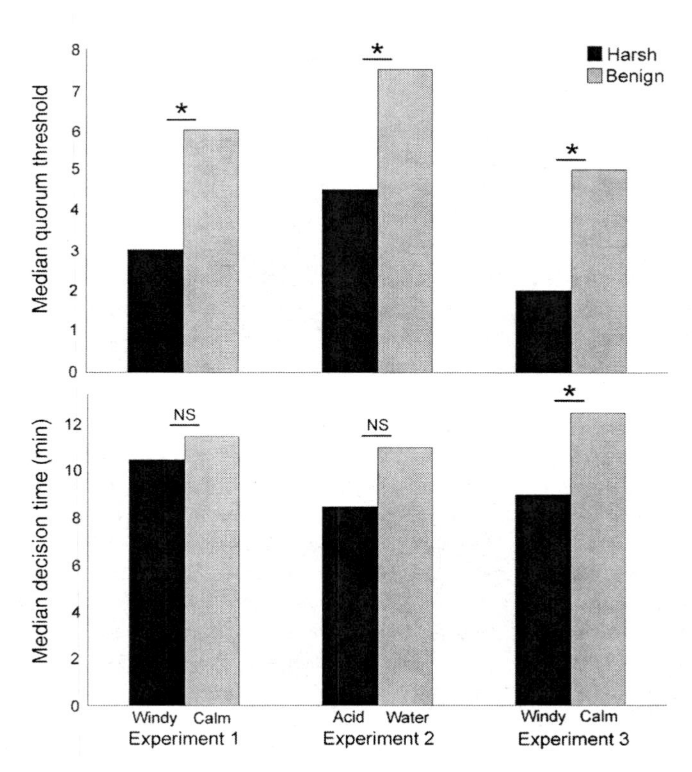

FIGURE 9.3 In decision making, speed versus accuracy trade-offs are common. When colonies of the ant *T. albipennis* have to emigrate to a new nest site, individuals will accept a new nest only once a certain number of other scouts are found at the new location (a *quorum threshold* is reached), allowing the individual assessments of nest quality by multiple scouts to be combined. By manipulating the conditions outside of artificial nests, Franks et al. (2003b) showed that under harsh conditions colonies lowered their quorum thresholds and began moving to the new nest site more quickly than under benign conditions (NS = nonsignificant, * = significant difference). However, when given a choice between two nests of differing quality, colonies would sometimes begin moving to the poorer nest under harsh conditions, indicating an error of judgment.

Thus, when a better nest site is found the colony may move, provided that a high number of scouts have agreed on the quality of the nest, resulting in a highly accurate decision, which is crucial because once the emigration process has started it is potentially dangerous. This decision is not straightforward because when a colony occupies an intact nest site, but has the option of moving to another currently unoccupied one, the colony must make a decision whether to move or not based not only on whether the new site is better than the current one, but whether it is sufficiently better to justify investing in the costs of moving. Overall, the behavioral decisions made by animals are the product of integrating multiple sources of information, often from several modalities, from both social and individual sources, and the abiotic environment. Such integration is addressed in more detail in chapter 10.

Signals Can Be Costly

Many signals used by animals can provide information for unintended receivers. For example, although conspicuous signals have evolved to attract the attention of a specific receiver(s), such as potential mates, they can also attract attention of unwelcome individuals, including predators. The evolution of conspicuous signaling is constrained by both physiological and ecological costs, although the former are sometimes surprisingly difficult to demonstrate because individuals optimize their signaling performance to the level of their physiological condition. Also, many predators and parasites are known to exploit sexual signals in prey location. An example of both physiological and ecological costs comes from drumming wolf spiders (*Hygrolycosa rubrofasciata*), in which males court females audibly by hitting their abdomen against dry leaves. One drumming comprises 30–40 separate pulses, lasting approximately for 1 second. The drumming is clearly audible to humans from up to several meters, and females choose males that drum at a high rate and produce loud and fast interval drumming bouts. Drums travel not only as substrate-borne vibrations, but also as airborne acoustic signals. Females respond sooner to drums transferred via the substrate, but the mode of signal transfer has no effect on female preference for different types of drums (Parri et al. 2002). Drumming is energetically very demanding, and the metabolic rate (the volume of CO_2: ml $CO_2 g^{-1} h^{-1}$) of drumming males is 20 times higher than their resting metabolic rate, and 4 times higher than when moving (Kotiaho et al. 1998a). Only males in good condition can maintain a high drumming rate, and therefore the drumming is a typical example of an honest sexual signal (see chapter 24); males that were induced to increase their drumming activity, by presenting them with virgin females in close proximity, died sooner than individuals kept alone (Mappes et al. 1996). In addition to physiological costs, the drumming signal can also cause an increased risk of predation, because lizards and birds can detect sexually active males by using the drumming sound as a cue for prey location (Kotiaho et al. 1998b; Lindström et al. 2006). Although female preference for drum duration exceeds the natural range of male drum duration, it would not pay males to have supernormal acoustic stimuli because these are so energetically demanding and risky in terms of predation (figure 9.4).

Animals May Exploit Hidden Channels of Communication

We have shown on several occasions that many aspects of signaling involve costs; for example, a conspicuous signal directed to a mate may attract the unwanted attention of a predator. However, in some animals natural selection has produced some clever adaptations to circumvent this problem (provided predators do not evolve counteradaptations). One such example is the idea that animals may exploit hidden channels of communication, in which they use signals to which predators are not sensitive. One remarkable case occurs in the deep ocean, where most sunlight cannot penetrate, and where a striking number of animals produce bioluminescence. Due to the way in which light is attenuated selectively, with longer wavelengths being filtered out first, residual sunlight is composed solely of shorter wavelengths (< 500 nm). Almost all deep-ocean bioluminescence is also composed of these short wavelengths. There is often an excellent match between the sensitivity of the inhabitants' visual systems and this available shortwave light. However, one group has evolved the ability to both produce and detect longwave light. At least three genera of dragonfish (*Malacosteus*, *Aristostomias*, and *Pachystomias*) are able to produce, along with shortwave bioluminescence, longer wavelength emissions ("far-red"), up to and beyond 700 nm (Partridge & Douglas 1995). In addition to this ability, the fish also possess sensitivity to longwave

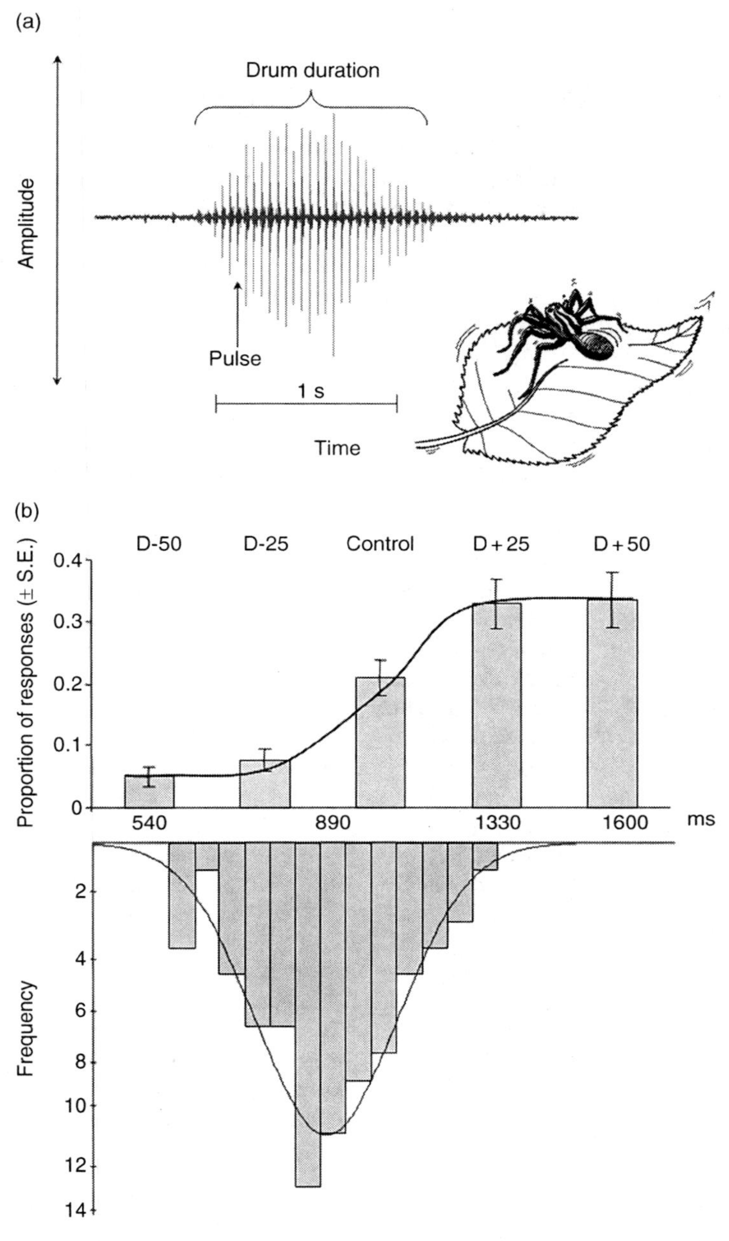

FIGURE **9.4** The unusual form of sexual signaling, the drumming, produced by the wolf spider (*Hygrolycosa rubrofasciata*) allows exceptionally detailed studies of female preference against signal characteristics, and the multiple costs that signaling conveys. (a) Because signal form is simple, it is easy to manipulate its duration and pulse rate simply by adding and removing single pulses. The female preference can be verified in playback experiments. (b) The upper part of the figure shows the proportion of female responses to the five manipulations of drum duration. Drum duration manipulations were 25% and 50% extended signal, and 25% and 50% shortened signal; in the control signal, single pulses were first removed and then inserted back without changing drum duration. The lower part of the figure shows natural range of signal duration of the males. The control level of manipulation is positioned on the population mean. The cubic regression curve of female preference, calculated from the results of the playback, is drawn over the female preference bars, and it indicates female preference over the range of natural variation. Costs due to high energy demands and signal conspicuousness are likely to constrain male drumming durations and drumming activity, and thus this example also gives an evidence of balance between natural and sexual selection. Reprinted from Parri et al. (2002, figures 1 and 4), with permission. Drawing by T. Ketola.

light, which in *Malacosteus* is based on a chlorophyll derivative but in the other two genera is based on a standard visual pigment, which seem to be ideally tuned to detect the light produced by their own bioluminescence (Douglas et al. 2000). Because the dragonfish are able to produce such far-red light, and are possibly unique among deep sea animals in being able to detect these long wavelengths, they may be able to utilize these signals to illuminate prey and communicate with one another, while being invisible to both their predators and prey. Intraspecific communication would seem more likely, however, because reflected light (such as from any prey animals) quickly becomes less detectable as distance increases.

Since the discovery of UV sensitivity in birds, and that some birds are sexually dimorphic in the UV reflecting component of their plumage, a similar suggestion has often been made that UV light represents a hidden channel of communication in songbirds. This is because most mammalian predators of songbirds are not able to see UV light, and because birds of prey have reduced sensitivity to this part of the spectrum. Blue tits (*Cyanistes caerelus*; chickadees), for example, appear the same to human eyes, but to a blue tit the males and females look very different because males reflect UV light more strongly in various regions of their plumage than do females. This idea is still controversial, however, because it is not clear that birds of prey do not retain sufficient UV sensitivity to see the songbirds effectively under many conditions, or that they do not use another channel for prey detection (such as luminance; see Stevens & Cuthill 2007).

Signals Can Combine Multiple Functions, and Result from Multiple Selection Pressures

It has been argued for some time that many animals' color patterns may combine multiple functions. For example, visual signals can have alternate effects under different conditions, such as the distance of viewing (Hailman 1977). This may be especially useful for some animals because it may allow the same color pattern to fulfill a signaling and camouflage role. For instance, some of the most striking color patterns in nature occur on coral reefs, and the fishes here frequently possess yellow and blue coloration, traditionally assumed to be invariably conspicuous. Marshall (2000) found evidence that the yellow and blue colors of many fish may be highly conspicuous when in close proximity, but that the blue color may provide effective camouflage against the background water and the yellow color against the reef, when viewed by a predator from a distance. Such an effect is possible because as an observer moves farther away from an object, the ability of the visual system to resolve pattern is accordingly reduced and the transmitting medium (e.g., water) can degrade the signal, often resulting in two or more colors of a pattern blurring into a new monochromatic appearance. If this perceived color matches the background, then a signal that is conspicuous in close proximity may become concealed from a distance. Other animals can vary their coloration over time to achieve different functions, such as either to be concealed or to attract mates. For example, chameleons are famous for being able to rapidly change their color, and a key function of their coloration is to signal to conspecifics (Stuart-Fox & Moussalli 2008). However, chameleons are also famous for modifying their coloration to match the specific background on which they are found. Remarkably, some chameleons will even fine-tune their coloration to optimize concealment against different types of predators. For example, dwarf chameleons (*Bradypodion taeniabronchum*) will change their appearance depending upon whether they are under threat from either a bird or a snake predator, which have different visual systems (Stuart-Fox et al. 2008). Such ability for rapid color change has also been well studied in various cephalopods, such as some octopus and cuttlefish. Overall, the appearance of animals can reflect a balance between opposing or multiple selection pressures, and there are a range of ways in which this is achieved in different species.

Human-Induced Changes Can Disrupt Communication

Because sensory systems have evolved in response to information from the animals' natural environment, they are also very sensitive to disturbance of many kinds. Human activity (anthropogenic disturbance) can cause both direct and indirect challenges for sensory communication. Certain pollutants can interfere with the perception of chemical stimuli either by masking the signal or by causing a physiological effect on the receptor that may influence the animal's behavior. For example, in the presence of certain chemical pollutants, some crayfish are unable to successfully locate a food source and will respond incorrectly to the alarm signal of conspecifics by directing toward, instead of away from, danger (Wolf & Moore 2002). Many marine mammals detect and use acoustic signals. During the past 50 years, human-generated ocean noise from shipping, oil and gas exploration, and sonar has doubled every decade. Such increases in ambient noise almost certainly affect, and may even disrupt, long-range acoustic signals (Richardson et al 1995). For example, male humpback whales sing longer phrases when exposed to low-frequency military sonar to increase the chance the song will be heard (Miller et al. 2000). Anthropogenic noise in cities and along major roads may also have a severe effect on birdsong, used for mate acquisition and territory defense, and may even lead to changes in song structure in species (Slabbekoorn & Ripmeester 2008).

Understanding the relationship between animal communication and sensory ecology is a complex task, and as with all areas of biology, there are many factors to consider simultaneously. Researchers have used a variety of methods, from experimental manipulations to modeling animal perception, in order to investigate the way in which animal behavior is affected by the environment, the specific sensory systems involved, and how the relevant information is processed in order to make a decision. Of course, there is more than just sensory systems underpinning most behaviors; the behavior of animals is also a result of their cognitive abilities, as the next chapter discusses.

FUTURE DIRECTION

Although direct effects of human induced disturbance (e.g., chemical and light pollution, noise, eutrophication) on animal populations is well understood, there is limited information on "fine-scale" responses of animals. For example, chemical communication that organisms use to locate food, find a mate, recognize kin, and mark territories is very vulnerable to disturbance. The same may be true when animals use visual cues in orientation, mate choice, food acquisition, or enemy avoidance under changing lighting conditions. This is, however, an understudied area, and only a few studies have addressed the problem of how light pollution affects sensory ecology, particularly in terrestrial animals. Lower levels of chemical (or light) pollution may not kill an animal but may affect information transfer at various steps in the signaling pathway, and deleteriously impact on animal behavior (Lürling & Scheffer 2007). How such fine-scale changes in behavior contribute to population level responses, and how animals can adapt their behavior in a changing world are significant challenges for future studies. Understanding such responses is especially important when predicting risks of endangered populations. This requires well-designed and -controlled laboratory experiments, as well as verifications under field conditions. Understanding population-level effects would also require substantial periods of time, which often is limited when dealing with declining populations.

SUGGESTIONS FOR FURTHER READING

Although nearly 70 years old, Cott's (1940) book is still one of the most influential and highly readable texts regarding animal coloration, in particular, antipredator strategies. Simmons and Young's (1999) work is a very good introduction to the way in which a range of sensory systems work and are linked to behavior. Bradbury and Vehrencamp (1998) present an in-depth introduction to signaling in animals. Lythgoe's (1979) text is still a very readable classic, which contains numerous excellent ideas about how visual systems and signals have evolved, and Dusenbery's (1992) book is good reading for all students working in the area of sensory ecology and signaling.

Bradbury JW & Vehrencamp SL (1998) Principles of Animal Communication. Sinauer Associates, Sutherland, MA.

Cott HB (1940) Adaptive Coloration in Animals. Methuen, London.

Dusenbery, D (1992) Sensory Ecology. Freeman, New York.

Lythgoe JN (1979) The Ecology of Vision. Clarendon Press, Oxford.

Simmons PJ & Young D (1999) Nerve Cells and Animal Behavior. Cambridge Univ Press, Cambridge, UK.

10

Information Processing: The Ecology and Evolution of Cognitive Abilities

SUSAN D. HEALY AND CANDY ROWE

In facing life's problems, such as finding food, avoiding predators, and choosing a mate, animals make decisions ranging from the simple to the complex. Many of these decisions rely on the animal's learning and memory abilities, or their cognitive abilities, which in turn depend upon the structure and function of their brains. Behavior, cognition, and neuroscience are all disciplines that converge when we attempt to understand how an animal is adapted to make decisions appropriate to its ecology and life history. This melting pot of different disciplines has already begun to provide exciting new avenues for research, by integrating different levels of understanding, as well as the interchange of techniques across researchers from different disciplines (but see box 10.1). The developing framework of what some have called *cognitive ecology* or *neuroecology* is providing some of the most exciting and novel developments in behavioral ecology today.

The aim of this chapter is to give an introduction to how cognition underpins decision making and to highlight some of the problems of studying cognition in an evolutionary and ecological context. We will first briefly review the role that cognition plays in some key behaviors: foraging, communication, and mate choice. These are not meant to be complete reviews, but to provide some examples of the current developments in the field. We will then turn to how behavior and cognition might be related to underlying neural structure, highlighting how future behavioral studies may need to incorporate methodologies in neuroscience and genetics.

WHAT EXACTLY IS COGNITION?

Cognition in its broadest sense can be defined as "information processing". It follows the acquisition of perceptual information (although the divide between the perception and cognition is often unclear; Shettleworth 1998; also see chapter 9 of this volume) and encompasses the processes that compare current sensory information with learned or unlearned knowledge to enable behavioral decisions to be made. Traditionally, behavioral ecologists have focused mainly upon cognitive processes such as learning and memory (the acquisition and retention of information), particularly in relation to foraging, communication, and mate choice. However, cognition also includes other processes commonly associated with learning, such as generalization and categorization, as well as processes assumed to be more complex, such as planning and tool making. There is increasing interest in cognitive abilities in relation to the evolution of behavior and the brain, which we will come to when we discuss the evolutionary links among behavior, cognition, and neuroscience. However, first we will review the role of cognition in studies of behavioral ecology, and illustrate how recent advances have

BOX 10.1 Testing Cognition in the Field

Most of the work on animal cognition to date comes almost entirely from laboratory experiments in which animals have been trained to perform a task before testing. For many behavioral ecologists, this testing situation may appear to miss the point because it appears highly artificial, which it is. However, the behavioral ecologist has to understand the rationale for testing in this way, which is that attempts are made to determine what an animal *can* do and not what an animal *does* do. Although these two are closely associated, there are important differences that can be missed if the subtleties are not appreciated. For example, demonstrating that a parrot can talk does not mean that parrots talk as a matter of course or that the parrot is as versatile as we are when using the words learned. A single talking parrot in a laboratory, however, is sufficient to answer the question "Can parrots talk?" When interested in ultimate or functional questions such as "Why do parrots have the ability to talk?" then we need an understanding of the natural context in which parrots vocalize and of the variation in vocalization.

Although such a natural context might ideally be the real world in which parrots operate, it is rare that cognitive abilities in vertebrates can be tested in the wild for several reasons: (1) wild, free-living animals will have had experiences prior to the testing context that may affect their performance, but in a way that cannot easily be quantified; (2) replication of a test in as similar conditions as possible is difficult; and (3) the observer/experimenter has little control over extraneous variables such as season, temperature, and humidity, which may impact on performance through their effects on the animal's motivational state (e.g., hunger, thirst, desire for sex, and so on); and (4) it can be difficult to isolate a specific cognitive ability in the field.

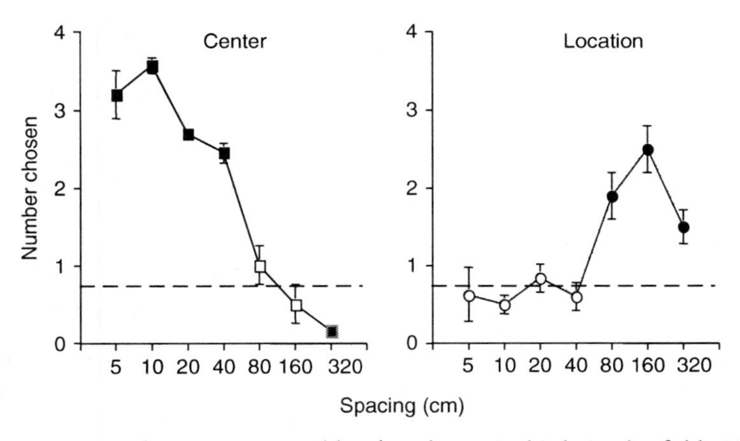

Figure I Testing spatial cognition in wild rufous hummingbirds in the field. Birds were trained to visit the central flower of an array of five flowers, which was the only flower containing a food reward (sucrose solution). Once the bird was only visiting the rewarded flower, the array was moved so that the location of the center flower was occupied by one of the outer flowers. The distance among the flowers differed on different trials from 5 cm up to 320 cm. All flowers were empty in the test trial. The data shown are the mean number of choices to the center flower (left panel) and to the flower in the correct location (right panel) for the different distances among flowers. The dashed line represents chance performance.

(continued)

BOX 10.1 *(cont.)*

One of the few tests of cognition in the field to date has been a series of experiments investigating the use of memory by foraging rufous hummingbirds (*Selasphorus rufus*). These birds can remember locations of artificial flowers they have emptied so as to avoid them, and they use different cues at different spatial scales: when rewarding flowers are within 40 cm, the birds remember the flowers' locations relative to one another; when the flowers are further apart, the birds use other more distant landmarks to return to or avoid the flower (figure I; Healy & Hurly, 2003). Rufous hummingbirds do not just learn where flowers are, but they also learn the rate at which multiple flowers refill, returning to the flowers at the time of refill or soon afterward. These experiments have also demonstrated a counterintuitive feature of hummingbird cognition: although rufous hummingbirds, like other hummingbirds, are assumed to prefer red flowers over those of any other color, they do so only when first visiting flowers. On subsequent visits, if the flower's color is dramatically changed, the birds pay little or no attention to the change.

It is not clear whether any of these results could have been obtained in the laboratory (or using any other system) because such comparisons have not been made. The laboratory provides an ideal setting in which to test what it is that animals can do but not what they do in the "real world." Perhaps more than any other aspect of animal behavior, cognition is currently understood almost entirely in a laboratory setting and it seems timely to determine whether or not that understanding is sufficient for describing variation in cognitive abilities.

contributed to our understanding of adaptive animal behavior.

THE ROLE OF COGNITION IN FORAGING, COMMUNICATION, AND MATE CHOICE

Foraging

The 1970s heralded the beginning of a long and glorious application of an optimality framework to investigating behavior, the most thoroughly developed example being that of foraging, now referred to as *optimal foraging theory* (OFT; chapters 8 and 11). Foraging decisions were examined in light of the costs and benefits to an animal of making one or more decisions, with various assumptions as to the role that information played in those decisions. Frequently, it was assumed that animals had perfect knowledge or that relatively straightforward rules could be used to describe which patches an animal chose to visit, how long they stayed in that patch, and how far they should travel to the next (Stephens & Krebs 1986). However, it became clear that foraging decisions could not all be so easily circumscribed and that in many instances animals have to learn what is profitable in their environment and what is not. Therefore, animals should be adapted to find food, to learn about its nutritional properties, and to remember where to find similar foods in the future.

Some foods are easy to find; for example, flowers and fruits advertise their presence to foragers and can be designed to be extremely conspicuous to the animals they aim to attract (Grant 1966). However, many prey try to avoid being consumed by predators and hide using various forms of crypsis. In turn, predators have evolved cognitive processes that aim to defeat this defense and break down the camouflage of their prey, for example, through the use of *search images*. This is a cognitive process by which animals selectively search for one particular cryptic prey type, which increases their probability of detecting prey of that type, while at the same time reducing the probability that the predator detects other cryptic prey types. Although the exact cognitive mechanism underlying this phenomenon is still debated (Dukas 2002), there are many studies that support the existence of search images. One successful empirical approach has been to train blue jays (*Cyanocitta cristata*) to peck at images of moths presented on different backgrounds on computer screens (or projected slides) in automated

Skinner boxes. These operant techniques consist of rewarding a jay for pecking either a key or at a touch screen when a moth is detected on the screen. In one experiment, blue jays were presented with two species of cryptic moth photographed on natural backgrounds. The birds were presented with a series of photographs that either contained only one of the species or contained a mixture of the two species. Jays were increasingly more likely to detect a moth when the successive presentations contained only one moth type, suggesting that they were using a search image that was developed when searching for one type of prey (Pietrewicz & Kamil 1979). Although this process is clearly good for the predator, it also presents an important selection pressure on prey. Not only are prey selected to become more cryptic, but also to become polymorphic, meaning that they evolve variation in their cryptic patterns to prevent predators from using a search image to find them more easily (Bond & Kamil 2002). This constant evolutionary struggle between predators and prey has led to both the evolution of cognitive mechanisms to detect prey and defense strategies to avoid detection (chapter 13).

Once food has been detected, the animal then needs to decide whether or not to eat it. Animals learn about the nutritional properties of different foods, and can select them according to their current nutritional needs. For example, locusts (*Locusta migratoria*) learn to use visual cues when choosing among foods with variable protein and carbohydrate content, which helps regulate their balance of macronutrients (Raubenheimer & Tucker 1997). Decision-making over food choices is especially important when animals are faced with toxic foods. In some cases, such as large herbivores that need to consume large amounts of food to gain sufficient nutrients, it is almost impossible for animals not to eat food that contains toxins. Grazers are good at learning about nutritional composition of plants and consume foliage that optimizes their nutrient intake relative to the amount of toxins ingested. Although most of the work on dietary cognition has been carried out in the laboratory, where it is easy to manipulate diets under controlled conditions, wild free-roaming koalas (*Phascolarctos cinereus*) also appear to spend the most time in eucalyptus trees, where the leaves are relatively high in nutrients relative to their toxin content (Moore & Foley 2005).

It is not just plants that are toxic. Many animals sequester toxins from plants or synthesize them to protect themselves from would-be predators. However, unlike plants that can often be grazed and still survive, defense is most efficient if animals can ensure that they are not damaged or eaten during a predatory attack. Defended animals have evolved warning signals, which are often conspicuous color patterns but may also be odors, secretions, or sounds. This coupling of a conspicuous label with unpalatability so as to signal defense to predators is known as aposematism (chapter 9). The conspicuous nature of the signals has been subject to some theoretical debate because it appears counterintuitive that they should make prey more detectable and thus more likely to be attacked by predators. However, experiments have shown that predators can learn to avoid defended prey more quickly if they are conspicuous as opposed to being cryptic (Gittleman & Harvey 1980; Lindström et al. 2001). In addition, when predators can detect the defenses of prey upon attack, conspicuous prey are more likely to be tasted and rejected compared to equally defended cryptic prey (Halpin et al. 2008). The value of conspicuous coloration to a prey animal, then, is dependent on the cognitive system of a would-be predator. It is not yet known whether aposematic prey have evolved to exploit a preexisting bias against conspicuous prey in predators' cognitive systems, or whether the predators' foraging strategies have evolved in response to the conspicuous aposematic coloration.

Not only can we investigate how predators' cognitive abilities affect their foraging success and prey survival strategies, we can also use foraging abilities to predict the kinds of cognitive abilities that animals might be expected to possess (if, as is usually assumed, natural selection has shaped cognitive abilities). For example, scatterhoarding of food offers an efficient method of reducing stochasticity in future food supply but only if the animal can remember where it has hidden the large number of food items. Western scrub-jays (*Aphelocoma californica*) can remember not only where the food is stored but also what they have stored and when the food was stored. These three components of a memory are consistent with the suggestion that these birds have episodic-like memory, once thought to be a uniquely human trait that was probably associated with our relatively large brain size (Clayton & Dickinson 1998). It has proved much more difficult to demonstrate episodic-like memory in other animals such as rats, which may be due to difficulties in designing appropriate experiments or because not all animals need to have episodic-like memories. In this latter case, there would have been no selection pressure leading to episodic-like memories. It is extraordinarily difficult to distinguish which is the more likely explanation.

Animals' cognitive processes are important in successful foraging, and it is tempting to think that they have evolved in response to specific environmental challenges. For example, New Caledonian crows (*Corvus moneduloides*) make tools to extract grubs from crevices (Hunt 1996), but what is it about these particular birds or their particular environment that has caused the evolution of this complex behavior? One suggestion is that they have large brains for their body size, and it is this that has enabled the kind of cognitive processing required for tool construction. The problem in testing this suggestion is that it is not possible to examine the evolution of complex behaviors in vertebrates directly (due primarily to the length of a generation relative to the life span of the experimenter). This leaves us with using comparative analyses as a way of looking at past evolutionary events. In the case of tool use, this is still not a very helpful method because tools are made by too few species for such comparative analyses to be very compelling. Frustratingly, therefore, we may never be able to fully understand the evolution of tool making. Food hoarding (scatterhoarding) has proved somewhat more useful in investigations of the evolution of cognition, but in spite of being taxonomically widespread it, too, is not sufficiently common to enable many comparisons. To really understand the evolution of cognition, we need to study behaviors that are expressed by more species so that we can carry out comparative analyses (chapter 7). Comparative analyses, of course, suffer from the problem of showing correlation and not causation. To demonstrate that variation in cognitive abilities is caused by variation in particular selective pressures, we need to turn to work on invertebrates, partly because of their relatively short generation times and also because their cognitive abilities are generally considered less complex than those of vertebrates (see work by Kawecki, Mery, and colleagues).

Communication

Cognition also plays an important role in animal communication. Tim Guilford and Marian Dawkins coined the phrase "receiver psychology" to highlight the role of cognition in the evolution and design of animal signals. They proposed that the ways in which animals detect, learn, and remember signals would be an important selection pressure on animal communication (Guilford & Dawkins 1991). Since their original publication, receiver psychology has become well-established in the field of animal communication especially with regard to how learning about signals affects signal design.

As we discussed above, warning signals are usually highly conspicuous, and only by understanding predator cognition is the adaptive nature of these signals revealed. But associative learning has also been important for understanding the evolution of mimicry, in which prey species share the same warning color pattern. Two nineteenth-century naturalists, Fritz Müller and Henry Walter Bates, were the first to realize that many species of conspicuously colored butterflies share similar warning patterns. In cases of Müllerian mimicry, both species are defended and, therefore, share the costs of predator education as predators learn to associate the color pattern with the toxicities of both comimics. However, in some cases, one species is less well defended than another and a parasitic relationship exists between a palatable mimic and a more toxic model. This is known as Batesian mimicry: predators learn that the color pattern signals variation in toxicity, with some prey (the mimics) being more palatable than others (the models). This is expected to lead to predators increasing their attacks on prey displaying the color pattern as Batesian mimics should significantly reduce the effectiveness of the model's defensive toxins. Although these two explanations for mimicry have been around for more than 100 years, it may be their very plausibility that explains why it has taken so long for researchers to conduct experiments that start to address how predator aversion learning affects the evolutionary dynamics of mimicry and warning signal evolution, and understand the complexities of signal evolution (e.g., Rowland et al. 2007).

The study of mimicry systems in natural prey species shows how complex the effects of predator learning on the evolution of warning signals and toxins can be. In the Ecuadorian Amazon, there are two species of parapatric frogs, one that lives to the south that is relatively common and more toxic (*Epipedobates parvulus*) than another species that lives to the north (*E. bilinguis*). These species are similar in appearance, having bright red spots on their backs, although *E. bilinguis* also has a yellow stripe. Both species are mimicked by a nontoxic

Batesian mimic, *Allobates zaparo*, which mimics each species within its range, including having a yellow stripe where it is sympatric with *E. bilinguis*. In the hybrid zone where both toxic models occur, we would expect that *A. zaparo* would mimic *E. parvulus*, since it is both more common and more toxic than *E. bilinguis*. Curiously, the reverse is true: *A. zaparo* mimics *E. bilinguis*, which should offer much less protection. How can predator cognition explain this? Catherine Darst and Molly Cummings suggest that the answer comes from studying how predators generalize their learned aversions to multiple prey species (Darst & Cummings 2006). *Generalization* is the use of acquired knowledge to respond to novel situations. Specifically in the case of aposematic signals, predators learn to avoid a toxic prey species with a specific warning pattern and extend this learned aversion to other prey species that have similar coloration. The degree to which predators generalize should be in part determined by the degree of toxicity—the more toxic a species is, the more the predator should avoid other prey that look even slightly similar. This appears to the case within this mimicry system, with birds showing a larger degree of generalization to prey more similar to *E. parvulus* than *E bilinguis*. Indeed, the generalization from *E. parvulus* is so strong that even *E. bilinguis* benefits from the mimicry. This leads to the conclusion that it is better for *A. zaparo* to mimic the less toxic species because it then benefits from the learned generalization from both species. Therefore, although on the face of it, mimicry systems can sometimes seem paradoxical and difficult to explain, knowledge of predator learning and generalization can provide possible evolutionary explanations.

Learning and generalization can also be important for a whole host of other communication systems. One interesting outcome of learning is the emergence of biases and preferences for particular signals that are more extreme versions of those already experienced and learned (ten Cate & Rowe 2007). A simple experimental demonstration of this was conducted using chickens pecking at colors presented on a touch-sensitive screen in the laboratory. Chickens were given a series of trials in which six colored circles were presented on a screen. Three of these circles were blue and pecks to these were not rewarded. The other three circles were shades of blue to which green had been added to varying degrees, giving three blue-green stimuli that were increasingly different from the blue stimuli. Birds were rewarded with food for pecking at any one of these three blue-green keys, regardless of which color they chose. In fact, the birds preferred to peck at the rewarding color circle that contained the most green, which was the one that was most different from the nonrewarding blue circles (Jansson & Enquist 2003). This operant experiment demonstrates the principle that learning that involves positively and negatively reinforced stimuli can elicit preferences that might drive the evolution of exaggerated signals, which could be important for the evolution of warning signals of toxic prey, but also could also be applied to a whole host of other signals, including those in mate choice (see next section).

Finally, cognitive processes have also been proposed as an important factor in the evolution of *multicomponent signals*, in which signals involve several components being sent simultaneously, often in multiple sensory modalities. Examples of multimodal signals are taxonomically widespread, from spiders and honey bees to fish, birds, and even humans (Partan & Marler 2005). There is abundant evidence, both behaviorally and from studies of sensory pathways, that animals have stronger responses to cues that are multimodal (e.g., a light and a tone presented together) compared to simple, unimodal cues (i.e., the light or tone presented alone; Calvert et al. 2004). Given the improved processing of multisensory information, we should not be surprised to find that animals have also evolved multimodal signals to communicate more effectively since such signals might be more detectable, more easy to learn, and more memorable (Rowe 1999). Why animals use multimodal signals is an area that invites further scientific investigation, although answers are likely to be embedded in the neural and cognitive pathways involved in sensory integration. Communication is, therefore, a field in which neuroscience and cognition can be integrated into behavioral studies (see section on birdsong below), and provides opportunities to study not only the mechanisms underlying behavior, but also how they might have evolved.

Sexual Selection and Mate Choice

A fundamental assumption made concerning female choice is that a female will generally attempt to choose the best male. In some cases, it appears that a female's preference is based on an unlearned bias such that a feature of a male will shape the female's decision without prior exposure to that feature. In some cases, the feature is not one even produced by conspecific males (Ryan 1998; Burley & Symanski 1998), whereas in others there are apparently general rules underpinning choices, such as preferences for symmetry (Swaddle 1999), which occur in decisions about mate choice as well as about foraging (bees prefer to feed from symmetrical flowers; Rodriguez et al. 2004).

Unlearned biases are not sufficient, however, to explain many instances of female choice and, although not yet well explored, it seems likely that cognition underpins some of the remaining decisions as to what constitutes the "best" male. One example of this is *sexual imprinting*, in which young learn about the appearance of parents or siblings and use this information when choosing a mate. Sexual imprinting is a widespread phenomenon in vertebrates, although the detail of just what the young animal cues into and how it subsequently uses this information is only just being beginning to be investigated. For example, in an experiment with white zebra finches (*Taeniopygia guttata*), the beaks of mothers and fathers were painted with different colored nail varnish: one was painted orange and the other red (the colors were reversed for half the parents). At sexual maturity, males preferred females with beaks of their own mother's color, confirming that the males had imprinted on their mothers. However, they showed their strongest preferences for those females that had even more extreme color signals than their mother, in other words, either more red or more orange (ten Cate et al. 2006). Therefore, not only do young animals learn about their parent's phenotype for use in subsequent mate choice decisions but, more importantly, this learning is translated into peak-shifted preferences for more extreme color signals when used in decision making (figure 10.1). This process might drive not only sexual signal exaggeration, but also sexual dimorphism.

Learning which attributes of a potential mate to choose may be based on information acquired when young. However, it may also be that choice will be affected by the outcomes of previous decisions (for experienced animals) or by the context in which the current choice is being made. For example, when a female is presented with a number of males simultaneously at a lek, the display of any one male may not be viewed in isolation but

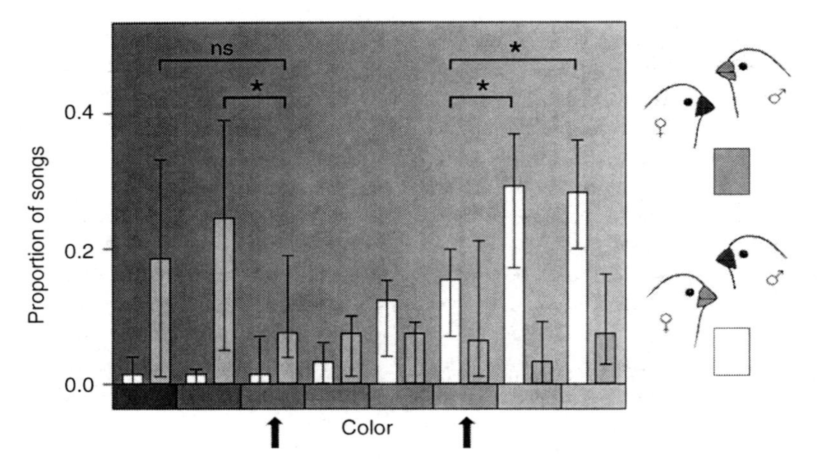

FIGURE 10.1 Peak shift in the sexual preferences of zebra finch males. Young male zebra finches were raised by white parents, whose beak colors were painted either red or orange using nail varnish. The arrows indicate the parental beak colors for two groups of males: in one the mother had an orange and the father a red beak, and for the other group, the colors were reversed. Males were the given a choice between females whose colors varied along the red (dark gray) to orange (light gray) dimension, as shown on the x-axis. White and gray bars indicate the preferences (median and interquartile ranges) for females with different beak colors. Males showed stronger preferences for females with a more extreme beak color than their mother had, demonstrating that peak shift could drive mate choice and sexual dimorphism in this and other species. Reprinted from ten Cate et al. (2006) with permission.

in relation to the displays of the other males. To appear particularly impressive, a male need only be more impressive than those males around him, in other words, a relatively good male rather than the best male in the population. This might lead to males positioning themselves on the lek where they look good relative to other males and thereby increasing a female's perception of their quality (Bateson & Healy 2005).

Although little has been done with animals to show this kind of context-dependent mate choice, studies in humans that test apparent mate preferences indicate that preferred attributes are readily modified by the range of possible mates and the range of traits. Additionally, also from the human studies, when a choice is among putative partners that vary in multiple characteristics, such as a sense of humor, physical attractiveness, or income, the decision may be reached using a rule based on the maximization of just one of those traits. Irrespective of the way in which the remaining traits vary among the possible choices, the choice of actual partner may be, for example, for the individual with the best sense of humor, rather than the individual who appears the best when averaging across all the traits. Alternatively, a choosy female may consider only the first few or last few males observed because, by virtue of their position in a sequence of possible partners, these are the more memorable males (i.e., primacy and recency effects).

There is evidence from a range of species that previous experience also affects current decision making. In zebra finches, females increased their preference for a particular male if they were exposed to males displaying at lower levels than that of the preferred male, but decreased their preference for that male if exposed to males with higher display rates (Collins 1995). Female peacock wrasse (*Symphodus tinca*) can mate with either territorial males who defend nests within those territories or with males that do not defend a nest or a territory. The benefits of paying the costs of searching for a territorial male depend on the density of nests and relative hatching success in or out of a nest. The more variable the state of the environment within and between breeding seasons, the more a female benefits from using her experience of her own success under the different conditions (Luttbeg & Warner 1999). In fruit flies, mating with a male of any size is more likely to occur after being courted by a small rather than by a large male (Dukas 2005a). Fruit flies also provide evidence that males modify sexual behavior as a result of experience; males with experience of recently mated, disinterested females speed up initiation of courtship with virgin females and males experienced with immature disinterested females approach any adult female faster (Dukas 2005b). Experience, then, is important for mate choice decisions, even in invertebrates in which it has generally been considered that, with life so short, mating should be more or less indiscriminate.

BRAIN MECHANISMS UNDERLYING COGNITION

Although it is possible to investigate cognition without understanding any of the relevant neurobiology, only by integrating the two can we gain a sensible understanding of how evolution has shaped cognition. This is illustrated by two iconic examples of the relationship between cognition and the underlying neurobiology in behavioral ecology: song learning and food hoarding.

Song Learning

Song in birds is largely used in a mating context to advertise quality either to potential mates or to rival males. The diversity across birds is stunning. Although it is usually the male who sings, there is tremendous variation in the number of songs or syllables sung by males across and sometimes within species. For example, white-crowned sparrows (*Zonotrichia leucophrys*) sing one song, whereas brown thrashers sing more than 1,000 songs, a level of difference for which we have no good explanation. Learning may occur only in the first few months of life in some species, whereas in others new songs appear throughout life (Catchpole & Slater 1995). Although we use song output as the indicator of what has been learned, it is clear that this is an underrepresentation of the full extent of the learning and memory involved. For example, females have to learn the songs they use to assess male quality, although they may never sing themselves. Migrant males remember the songs of their neighbors from the previous year when defending territory boundaries. If the old neighbor's song is sung from the appropriate boundary, the focal male will put much less effort into defense of that part of his territory (Godard 1991). However, if the same song is sung from a different part of his territory, the focal male will respond as if he has never before heard that song.

Perhaps surprisingly, given the extent of diversity that characterizes it, song has been a hugely influential behavior in our understanding of the plasticity of the brain. This is, in no small part, due to the ease of determining what is learned and of manipulating the learning environment. Deafened birds, for example, are used to show the value of auditory feedback while isolating birds in soundproof boxes enables determination of the role of a tutor. All of these behavioral manipulations can be accompanied by investigation of the neural structures involved. Fernando Nottebohm and coauthors were the first to show that the size of some of the nuclei in the brain is greatest during the early phases of song learning, then diminishing while other nuclei continue to grow. Some of these (the robust nucleus of the archistriatum [RA] and the higher vocal center [HVC], for example) grow and shrink multiple times through a bird's lifetime as a result of both neurogenesis and cell death. Although these size changes were once thought to be either directly responsible for, or in response to, changes in song learning or output, the picture is still unclear. For example, the size changes are, in at least some species, more closely correlated with hormonal changes than with any obvious changes in song. Additionally, changes in the size of song nuclei can be seen in species that do not change their songs. Nonetheless, across species, song repertoire size is correlated with the size of some song nuclei (e.g., HVC) and within species in which only males sing those males have significantly larger HVC and RA than do their conspecific females (DeVoogd et al. 1993). These size changes begin in early development with greater cell death (apoptosis) in females than in males (Kirn & DeVoogd 1989).

Zebra finches also provide evidence that song quality itself may need to be learned. Young zebra finch females need to be exposed to "normal" male song to prefer males who sing more notes and utilize greater spectral complexity (Lauay et al. 2004). Furthermore, the neural basis for this is beginning to be understood: female zebra finches that hear little or no song have significantly fewer dendritic spines in the caudomedial nidopallium (NCM), an area involved in song perception, but there is little effect of this experience on the number of dendritic spines in the HVC. Males that are not exposed to song have fewer dendritic spines in the NCM than males that are exposed to song, but there is a much greater reduction in the HVC, the area important in song production.

Plasticity is a hallmark of song because even the quality indicated by a male in his song rate, repertoire size, and so on may indicate to listeners that the male has high quality genes, that his mother provided appropriate nutrients or hormones to his egg, or that he had a well-nourished upbringing (chapter 24). By manipulating her offspring's hormone levels either in the egg or via the amount of food she provides her young, a mother has the opportunity to enhance any genetic qualities her offspring may possess. Song output is heavily affected by both the organizational (within-egg) and activational (current circulating) levels of hormones, especially testosterone and corticosterone. One effect of testosterone was strikingly demonstrated by experiments such as those in which females administered with testosterone develop a male-like brain and sing (Cunningham & Baker 1983). Experimental increases of corticosterone in zebra finches, either directly or indirectly by stressing mother or chicks, results in decreased growth rate, shorter songs, and song-reduced complexity, and in smaller HVC (Buchanan et al. 2004).

Investigating the relationship between cognition and the endocrine system in species other than laboratory rats is only just getting underway. Hormones exert a wide diversity of effects on morphology, development, cognition, and many other behaviors. Behavioral ecologists are only just beginning to understand their importance. It will also be important to determine both the organizational and the activational effects of hormones on cognition so that we can determine to what degree variation in cognitive abilities is explained by the direct effects of natural selection on behavior versus indirect effects of selection on hormones that have pleiotropic effects on behavior.

Spatial Cognition and Food Hoarding

Song learning has become an iconic system in which to study the relationship between cognition and the relevant brain regions for a combination of reasons: the learning episode can be readily identified (both what is learned and when), it is easy to determine that an animal has learned the information, and the behaviors are readily accessible to observation and manipulation. However, most contexts in which we might want to understand the role of cognition cannot be so easily circumscribed. Although we may see that animals are exposed to information of some kind, it can be extremely difficult to determine how much, what kind, and for how long the information is acquired or retained.

Spatial cognition has proved to be somewhat more amenable to investigation, not least because the context in which it is used appears obvious and because in many vertebrates there is a clear association between the processing of spatial information and a specific part of the brain, the hippocampus. For example, scatterhoarding species use memory to relocate their hidden food and have a larger hippocampus than do species that do not hoard food. Species that store more and for longer have a larger hippocampus than those species that store less. Experimental manipulations have shown that it is the experience of food storing that causes increased neurogenesis, resulting in an enlargement of the hippocampus (Clayton & Krebs 1994). Food storers remember what they have stored, when and where they did the storing (Clayton & Dickinson 1998), and can also remember food locations for longer than can nonstoring species (Biegler et al. 2001; Clayton & Dickinson 1998). The hypothesized greater selection pressure on spatial cognition in the food-storing species has also resulted in a specific enhancement in their spatial cognition (they are no better at remembering pictures of objects—or more items—than are nonstorers). Damage to the hippocampus of these birds results in difficulty with accurate retrieval although not in motivation to retrieve nor in ability to retrieve food from locations specified by learned color cues (Sherry & Vaccarino 1989). These findings support the hypothesis that the hippocampus is specifically involved in spatial cognition and not cognition more widely, nor in food storing itself.

Hippocampal damage leading to impairment of food retrieval specifically provides good evidence that the hippocampus is involved in spatial memory. But food storing is not the only instance of a hypothesized increased demand for spatial cognition driving hippocampal increase; nest-parasitic birds that search for host nests in which to lay their eggs also have a larger hippocampus. In species such as the brown-headed cowbird (*Molothrus ater*) and shiny cowbird (*M. bonariensis*), only the female searches for nests and those females have a relatively larger hippocampus than do their conspecific males (Reboreda et al. 1996). In screaming cowbirds (*M. rufoaxillaris*), however, both sexes search for nests and there is no sex difference in hippocampus size, although both sexes have larger hippocampus than either sex of the bay-winged cowbird (*Agelaioides badius*), which is not a nest parasite but is parasitized by other cowbirds. Migration, too, impacts on hippocampal volume: experienced migrant songbirds have a larger hippocampus than inexperienced migrants or residents (Healy & Hurly 2004).

It is not just in birds that there is a relationship between an increase in demand for spatial information and hippocampal volume: in several mammals, males that have larger home ranges than do their conspecific females also have a larger hippocampus and outperform them on mazes when tested in the laboratory (Galea et al. 1994). Although not so obviously explained by range size differences, human males also typically perform better than females on spatial tasks.

As occurs in song learning, the sex differences seen in spatial cognition in mammals are strongly correlated with hormone levels, especially testosterone, estrogen, and corticosterone. In polygynous voles and deer mice, the range size differences and better spatial cognition are evident only during the breeding season, and manipulations of testosterone (either in utero or later in life) result in enhanced spatial cognition in females (e.g., Galea et al. 1994; Roof 1993; Williams & Meck 1991).

To date there is more evidence supporting a relationship (albeit correlational) between an ecological demand for enhanced spatial cognition and variation in hippocampal volume than there are data demonstrating the cognitive link between the two. Compelling demonstrations of species differences in cognitive ability are, perhaps surprisingly, uncommon as yet.

THE EVOLUTION OF COGNITION

The lack of quantitative evidence for differences in cognitive ability may come as a surprise. After all, some animals can talk and some cannot, and some animals make tools and others do not. However, demonstrating that animals cannot solve certain tasks may be due to a deficit in our observational or testing skills. Euan Macphail is one who has argued rather convincingly for this lack of evidence, although there are many who would dispute this (Macphail 1982). One reason for the dispute is that we see apparently obvious variation in both complexity of abilities (e.g., talking, teaching, and so on) and in brain size. Additionally, few behavioral ecologists need convincing that natural selection via local selection pressures is likely to have resulted in cognitive adaptations. We have already pointed out that it is difficult to investigate variation (and therefore evolution) in at least some apparently complex cognitive abilities (e.g., tool construction, food storing) simply because so few animals appear to have them.

The most accessible avenue for investigating the evolution of cognition directly is via invertebrate model systems. Frédéric Mery and Tad Kawecki have used *Drosophila melanogaster* to investigate the costs and benefits of various kinds of learning. For example, flies that experienced odor conditioning resulting in long-term memory formation were significantly less able to cope with desiccation and starvation: the cost of long-term memory (requiring protein synthesis) was significantly earlier death (Mery & Kawecki 2005). Furthermore, they have determined that alleles at the foraging locus (*for*; see also chapter 5) are associated with natural variation in short- and long-term learning. Flies with the for^R allele have better short-term but poorer long-term learning than do flies with the for^s allele (Mery et al. 2007). In the presence of food, flies with the for^R allele ('rovers') move more during feeding, are more responsive to sucrose and habituate more slowly to stimulation by sucrose without reward than do "sitter" flies (those with the for^s allele). Both these variants are common in natural populations, and it is possible that their different foraging habits have selected for the variation in learning abilities.

An alternative, but indirect, method of potentially examining the evolution of cognition is associated with the long-held curiosity for understanding why humans have such large brains for their body size. Because determining the evolutionary explanation for changes in brain size in a single species is difficult, comparative analyses in which variation in brain size across multiple taxonomic levels are much more likely to yield plausible evolutionary explanations. Such correlational analyses, which were first popular in the late 1970s and early 1980s, produced a number of findings such as frugivorous primates have a larger brain size than do folivorous species and altricial bird species have larger brains as adults than those with precocial development (review in Healy & Rowe 2007). Because diet did not explain a significant degree of variation in brain size in birds, it appeared that different explanations were needed for different groups of animals. A conceptual framework for why different processes might be at work in different groups has not yet been developed. There have been a multitude of studies exploring correlations between brain size and ecological variables, with the hypothesis that social factors drive increases in brain size receiving the most enthusiasm currently.

Unfortunately, not least because some of these relationships seem very plausible, there are a number of problems with this approach that need to be considered. First, brain size itself is not readily measured and there is no single agreed method of measurement. Second, the variables typically proposed to explain brain size variation (e.g., behavioral flexibility, habitat complexity, social intelligence) are rather vague entities that are difficult to define and, therefore, to characterize numerically (discussed in Healy & Rowe 2007). Additionally, when correlations are found between brain size and explanatory variables, these are simply correlational and not convincing evidence of cause and effect. For example, the correlation between hippocampus size and food storing has been proposed to be a cause-effect relationship. This hypothesis has been supported by evidence that lesions to the hippocampus affect the memory for caches and that the hippocampus enlarges as a result of food-storing experience (Clayton & Krebs 1994; Sherry & Vaccarino 1989). This result, however, might also mean that evolution does not play a role in hippocampal expansion but rather that food storers simply have either the right kind or the sufficient amount of experience required to increase hippocampal neurogenesis and reduce apoptosis resulting in an enlarged hippocampus. Although attempts have been made to provide nonstorers with seemingly appropriate experience (not leading to hippocampal expansion), it is difficult to be sure that equivalence has been provided. More promisingly, there have been successful efforts to create animals (via artificial selection) that vary in hippocampal traits, and test these animals for variation in behavior (inbred mice strains; Sluyter et al. 1999).

Once we understand the structural and functional differences among brains, we can begin to determine the genetic changes that have led to these differences. Three significant genetic mechanisms have been implicated in changes to brain structure: addition or subtraction of whole genes, changes in levels or patterns of gene expression, and changes in gene coding sequences. Using the close similarity between the human and chimpanzee genomes, for example, it has been shown that there are regions in the human genome that show significant evolutionary acceleration in comparison to that of the chimpanzee, one of which is part of an RNA gene that is expressed in the developing human neocortex and is involved in the layering of the human cortex (Pollard et al. 2006). With the accelerating acquisition of a variety of genomes, plus the utilization of neuroscience techniques for determining function of brain regions in a wider variety of animals than rodents and primates (e.g., Smulders & DeVoogd 2000), we are rapidly approaching the situation in which we can expect to arrive at causal, rather than correlational, explanations for variation in brain size and cognitive abilities.

CONCLUSIONS AND FUTURE DIRECTIONS

Information processing is crucial for guiding adaptive behavior, and consequently we predict that it will have been shaped by natural selection. Although we can identify how cognitive processes are likely to be important in the evolution of behavior, for example, in the evolution of multimodal communication or mating strategies, we know far less about the evolution of cognition per se. This is a fundamental problem that can be addressed only by overcoming the challenges of measuring and comparing cognitive abilities across species. We have suggested a number of new avenues for research, for example, using invertebrates as model species to measure the role of selection on behavior and cognition. Their short generation times, relatively small behavioral and cognitive repertoire, and simple neural structures make them a useful group for this kind of research. We also need to understand the relationship between the different roles that evolution and ontogeny play in shaping cognitive abilities and underlying neural structure. We know that cognitive abilities and neural structures are developmentally plastic, but not yet how flexible are cognitive strategies, or how much genetic variation exists among individuals upon which evolution can act. These kinds of questions are fundamental, and in answering them we are likely to reach a deeper understanding of the evolution of cognition.

Finally, we have probably gone as far as we can trying to understand brain evolution using comparative studies alone (Healy & Rowe 2007). Instead, we need to utilize and integrate techniques from the fields of cognitive neuroscience, genetics, and the behavioral sciences if we are to fully understand the evolution of neural structures that underpin cognition and behavior (Pollen & Hoffman 2008). Understanding the trade-offs and dynamics involved among the evolution of the brain, cognitive abilities, and behavior is one of the most complex, dynamic, and exciting challenges currently facing behavioral ecologists.

SUGGESTIONS FOR FURTHER READING

Shettleworth's (1998) textbook provides an excellent and comprehensive review and comparison of the cognitive abilities of animals, and discusses some of the selection pressures that are likely to have been prominent in their evolution. Although this is the only text book on the evolution of cognition and behavior, other textbooks do have some chapters on this topic, for example, Barnard's (2004) and Pearce's (2008) books on animal behavior and psychology. As we have pointed out during this chapter, this is an emerging area for research, which naturally leads to critiques and disagreements. Perhaps the most ardent opponents of a *neuroecological* approach to the study of the evolution of cognition have been Bolhuis and Macphail (2001, 2002). Their critique stimulated an ongoing debate (see also Hampton et al. 2002, MacDougall-Shackleton and Ball 2002, and Healy et al. 2005) that provides an overview of the key issues as we attempt to integrate the fields of evolution, behavior, cognition, and neuroscience.

Barnard CJ (2004) Animal Behaviour: Function, Mechanism, Development and Evolution. Pearson/Prentice Hall, London.

Bolhuis JJ & Macphail EM (2001) A critique of the neuroecology of learning and memory. Trends Cogn Sci 4: 426–433.

Bolhuis JJ & Macphail EM (2002) Everything in neuroecology makes sense in the light of evolution. Trends Cogn Sci 6: 7–8.

Hampton RR, Healy SD, Shettleworth SJ, & Kamil AC (2002) "Neuroecologists" are not made of straw. Trends Cogn Sci 6: 6–7.
Healy SD, de Kort SR, & Clayton NS (2005) Response to Francis: puzzles are a challenge, not a frustration. Trends Ecol Evol 20: 477–477.
MacDougall-Shackleton SA & Ball GF (2002) Revising hypotheses does not indicate a flawed approach—Reply to Bolhuis and Macphail. Trends Cogn Sci 6: 68–69.
Pearce JM (2008) Animal Learning and Cognition: An Introduction, 3rd ed. Psychology Press, New York.
Shettleworth SJ (1998) Cognition, Evolution and Behavior. Oxford Univ Press, New York.

SECTION III

ECOLOGY OF BEHAVIOR

11

Foraging Theory

IAN M. HAMILTON

All living things require energy, carbon, and essential nutrients. Heterotrophic organisms acquire these from other organisms or their products. Because food sources may vary in quality and are often patchily distributed in space and time, organisms that rely on consuming other organisms or their products—foragers—face a number of decisions while searching for food. Where should a forager search for prey? Should it capture all prey that it finds, or reject some items? Should it forage alone or in a group? Should it search for prey itself, or should it rely on using information from other searchers? The field of foraging theory seeks to understand the consequences of these and many other decisions for fitness and to predict behavioral decisions about foraging that maximize relative fitness given the ecological and social setting.

In this chapter, I will summarize several of the main concepts in foraging ecology, starting with a brief introduction to the classical prey and patch models, which are explained in more detail in chapter 8. I will discuss the assumption that foragers are rational decision makers, followed by an in-depth case study of decision making by gray jays hoarding food for overwinter storage. This first part of this chapter will focus on truly solitary foragers. However, as a rule foragers are economically interdependent with one another. Even foragers that never come into close contact with others may find their foraging payoffs are influenced by the decisions of others. I discuss several social foraging models in depth: the ideal free distribution, producer-scrounger models, and models of optimal and stable group size, as well as the use of socially acquired information in foraging decisions. I will present two case studies on social foraging: one a laboratory study of producing and scrounging by nutmeg mannikins, and the other a case of social exploitation at several levels in a "gardening" reef fish in the field. I then discuss some future directions for social foraging research, focusing on incorporating the value of grouping into social foraging models and how groups come to collective foraging decisions.

All models of foraging decisions make some simplifying assumptions about fitness. Unlike, for example, avoiding predators or searching for mating opportunities, the fitness consequences of a single foraging decision are often very small. For many foragers, it is not often that a single prey item means the difference between life and death or between reproducing and not. However, a sequence of poor decisions over a long time span may well lead to starvation or inability to devote energy to reproduction. Therefore, theoretical treatments of foraging consider that each foraging decision takes place within a long sequence of similar decisions. Classically, foraging theory assumes that individuals either attempt to maximize their long-term rate of net energy intake over this sequence (although other currencies such as limited nutrients are possible) or minimize their risk of energetic shortfall over time. Both of these currencies mean that a forager will not necessarily maximize its intake rate at each and every encounter or over the short term but will do so on average over the longer term. These assumptions about fitness currencies might not always be correct (see chapter 4), but in practice have worked quite well in predicting what a forager will do in particular situations.

SOLITARY FORAGING

Concepts

The Prey Model

The classical prey model and its derivation were presented in some detail in chapter 8. To review, this simple model is based on the assumption that each prey item, i, yields some net energy reward, e_i, but gaining that reward requires some handling time, h_i. Handling time is simply the time between capture of the prey and the resumption of search. A key assumption of the model is that time spent handling is unavailable for searching and that it is not something under the control of the forager, but is a property of the prey. For example, a seed with a husk that is difficult to open would have a longer handling time for a nonspecialized seed-eating bird than would one with a softer husk.

As explained in chapter 8, the prey model makes three main predictions. First, prey types are ranked for inclusion in the diet by profitability, which is the ratio of energetic reward to handling time (e_i/h_i), with the most profitable items always consumed. Second, the decision to accept or reject a less profitable prey item does not depend on the abundance of that prey type. Rather, it depends on the abundance, via encounter rate, of more profitable items. Finally, preference for less profitable items should follow a step function (the *zero-one rule*): either a prey type is always accepted or it is never accepted.

Although the prey model predicts that optimal choice is always all-or-none, experimental tests invariably find that animals exhibit partial preferences (reviewed in McNamara & Houston 1987). Animals may prefer one item, but they typically do not eliminate less preferred items completely from their diet. Partial preferences may arise if animals base their foraging decisions on several currencies (McNamara & Houston 1987), such as both long-term rate of net energy gain and nutrient intake, or energy gain and minimizing risk of starvation, which are discussed in more detail in the risk section below and in chapter 12. Incomplete preferences may also arise if animals make errors discriminating between items or errors in estimating encounter rates. McNamara and Houston (1987) modeled prey choice decisions assuming that individuals sometimes make mistakes, and that these mistakes are more costly when the difference in fitness between options is great. In this model, less profitable items are sometimes taken and are more likely to be taken when the encounter rate with less profitable items is high, in contrast to the predictions of the rate maximizing model without errors.

Partial preferences are also predicted for foragers that consume food that is abundant but is also bulky and of poor quality, such as many herbivores (Belovsky 1978). For these foragers, foraging decisions are influenced by digestive constraints, because the gut cannot process such bulky food quickly and fills up. Foragers may also face trade-offs between energetic richness and nutrient content or the presence of toxins. Models of diet choice under digestive and nutrient constraints predict that these foragers should consume a mixture of energetically and nutrient rich foods (Belovsky 1978) or a mix of rich food and poorer quality items (Hirakawa 1997).

The Patch Model

The prey model assumes rarely encountered, discrete items of high quality that are consumed in their entirety after handling. However, some prey may be partially consumed or prey may be distributed in partially depletable patches. How long should a forager stay in the patch, and when should it leave and start searching for new patches? This is the basis for the *patch model* of optimal foraging, also known as the marginal value theorem (MVT; Charnov 1976), which readers will have already seen in chapters 2 and 8. In the simplest patch model, all patches are assumed equal, and the rate of gain of a forager on a patch declines with increasing time spent on the patch because of depletion by the forager. Maximization of the long-term rate of net energy intake can be achieved by remaining in the patch until the within-patch rate of net energy intake equals the background rate. The patch model is often presented in graphical form, as in figure 11.1. The right-hand side of the graph is the cumulative net energy gain over time from the point of arrival at the patch. Although this curve is monotonically increasing, the magnitude of this increase decreases with time. Thus the marginal increase in value of remaining on the patch declines over time. The horizontal axis on the left-hand side of figure 11.1 is the travel time to another patch. Maximization of the rate of energy intake, given a travel time *T*, is found by drawing a line that intersects the horizontal axis at *T* and is tangent to the marginal value curve. Among the key predictions of MVT is that foragers should stay longer in patches when travel times to other patches are long.

It is straightforward to combine the patch and prey models. The resulting patch-prey model retains the basic predictions of the prey model (Stephens & Krebs 1986): the zero-one rule, the independence of patch inclusion on patch abundance and the ranking of patches for inclusion based on profitability, which in this case is the rate of energy gain on the patch that would maximize long-term net energy intake if a forager used only patches of that type.

The MVT assumes that the foraging returns from staying in a patch decrease over time. However, foraging from depletable patches does not necessarily result in gradually diminishing returns, if, for example, foragers search patches systematically for randomly distributed prey. Patches that do not exhibit diminishing returns can be treated as discrete prey in the prey model (Stephens & Krebs 1986). Despite the importance of this assumption for the MVT, the shape of gain curves has rarely been measured in experimental studies. In one study, Olsson et al. (2001) found that foraging starlings did experience diminishing returns with increasing time on a patch; however, the measured gain curves were also not consistent with the assumption of random search inherent in classical prey and patch models. Rather, the data most closely fit a model in which successful foraging makes the remaining prey become more difficult to find or capture. Although the quantitative predictions of such a model may vary from those of the classical patch model, the general framework outlined above is still applicable.

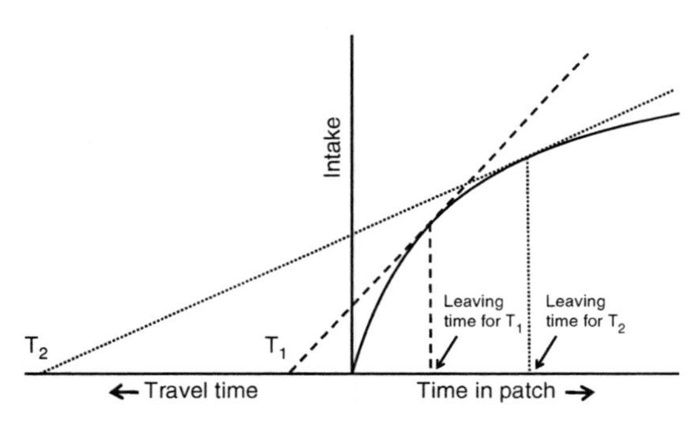

FIGURE **11.1** The marginal value theorem. On the right-hand side of the graph, intake increases with time spent on the patch. However, the marginal value of this increase decreases, so that the relationship between intake and time is concave down. On the left-hand side of the graph is time spent traveling between patches. To find the time on the patch that maximizes intake rate, including time traveling, a line is drawn tangent to the marginal value curve and intersecting the horizontal axis at the travel time, *T*. Model predictions for two different travel times, T_1 and T_2 are shown; as travel time increases, the time on the patch that maximizes intake rate also increases.

The general prediction of the marginal value theorem that patch departure time should increase with increasing travel time or costs (or decreased density of patches) has been supported in a wide variety of systems, such as nectarivores foraging on flowers (Pleasants 1989) and even in contexts other than foraging in depletable patches, such as lysis time in phages (Wang et al. 1996), dive times for air breathing animals (Walton et al. 1998), copulation duration (Parker 1992), and oviposition decisions (box 11.1). However, a review of 26 studies found that the quantitative predictions of the MVT were rarely met, with a systematic bias toward overstaying (Nonacs 2001a). Nonacs (2001a) suggests that such a bias may be a result of state-dependent predation risk on patch departure times. Thus, although the MVT has provided a useful general model of patch departure times, currencies other than long-term maximization of net energetic intake rate must also be taken into account when predicting patch use by foragers.

Variance and Risk

In a classic study, Caraco et al. (1980) allowed yellow-eyed juncos (*Junco phaenotus*), a small seed-eating bird, to choose between trays that offered either a fixed number of seeds or a variable number of seeds. The mean number of seeds presented over the course of the experiment did not differ between the fixed and variable trays. Based solely on long-term rate maximization, one would expect that birds

BOX 11.1 Allocating Eggs among Multiple Hosts by Parasitic Insects

Frank J. Messina

This chapter illustrates how foraging decisions affect an animal's intake of energy and nutrients. In some organisms, however, an individual's sequential decision making primarily affects the quality of resources available to its offspring rather than itself. This scenario applies especially to parasitic insects in which females must spread their eggs among many small, widely scattered hosts. Parasitoid wasps, seed beetles, and frugivorous flies are prominent examples. Larvae typically make no foraging decisions; they must complete development within the host chosen by their mothers. If each host provides enough resources for only one or a few larvae, it can be advantageous for a female to avoid adding eggs to occupied hosts. Ovipositing females of many species do distinguish between pristine and occupied hosts (a behavior known as *host discrimination*), and some distinguish between hosts bearing few or many eggs (Messina 2002). By reducing competition experienced by offspring, egg-laying females display a form of parental care (chapter 26).

Although the benefits to host discrimination are clear, locating unoccupied hosts in nature can be costly, and a female has a limited amount of time to distribute her eggs. Consequently, several questions arise with respect to optimal oviposition behavior. When might it be profitable to superparasitize, in other words, add eggs to an occupied host? If a female lays multiple eggs per oviposition bout, what is the optimal clutch size? How does a female's probability of accepting a host depend on her experience, age, or condition (including her current egg load)? Although a single egg-laying decision during a particular host encounter may have little consequence, females following different behavioral rules can accrue large cumulative differences in fitness.

Let us consider clutch size, the number of eggs laid on a host during a single oviposition bout. One determinant of the optimal clutch size is the shape of the *larval competition curve*, which describes the combined fitness of co-occurring larvae as a function of increasing density in a host (figure I; Smith & Lessells 1985). Assuming that a host can support more than one larva, combined fitness increases with larval density up to a *single-host*

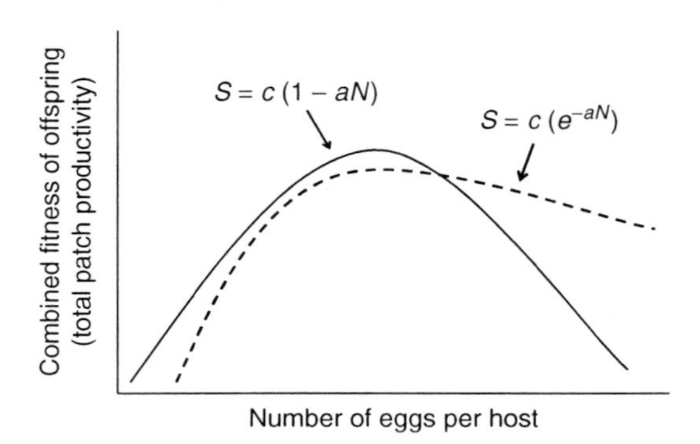

Figure I Total productivity of a host (combined offspring fitness) as a function of initial density (N). Productivity = N times per capita fitness (S) of larvae. The drop-off in patch productivity is steeper when there is a negative linear relationship between N and S (solid line) than when there is a negative exponential relationship (dashed line). The constant a determines the rate of the decline in per capita fitness; c is a proportionality constant (after Ives 1989).

(continued)

maximum, which is the number of eggs that maximizes combined fitness gain per host (analogous to the so-called Lack clutch size in birds). This value maximizes maternal fitness, but rarely maximizes the fitness of individual offspring (Nufio & Papaj 2004). Because per capita fitness decreases monotonically with each additional larva (it is highest when $N = 1$), combined fitness rises in a decelerating manner before decreasing (figure I).

A female should not add eggs to hosts so as to exceed the single-host maximum, but the penalty for doing so depends on the relationship between per capita fitness and larval density. The drop-off is steeper if competition is severe and per capita fitness declines in a linear way than if larval competition is more benign and per capita fitness declines according to a negative exponential relationship (figure I; Ives 1989). The penalty also depends on the whether current offspring in an occupied host are related to the egg-laying female, in other words, whether she runs the risk of *self*-superparasitism. Among parasitoid wasps, it is not clear whether a female can avoid self-superparasitism by distinguishing her own eggs from those laid by another female.

The single-host maximum predicts how a female should behave if her fitness depends solely on the eggs she lays on the current host. But a female must visit many hosts over her lifetime and risks dying before she exhausts her egg supply. Under these constraints, female decisions are expected to depend on search costs (Godfray 1994). When hosts are plentiful, a female maximizes her rate of fitness gain by laying one egg per host and rejecting occupied hosts. If hosts are scarce and search costs are nontrivial, females should accept some occupied hosts and lay larger clutches. Only when females are extremely time-limited (they have more eggs to lay than hosts upon which to lay them) should she deposit enough eggs to reach the single-host maximum. In general, one can apply the marginal value theorem (chapter 8); a time-limited female should leave a host when the marginal rate of fitness gain drops to a point at which she would gain fitness more rapidly by exploiting another host (figure II).

Several factors can modify expectations based on search costs alone. Female decision making may change over short time periods as a function of her physiological state or recent experience (Goubault et al. 2005). For example, recent encounters with pristine hosts (which may signal the overall quality of the local environment) can make a parasitoid

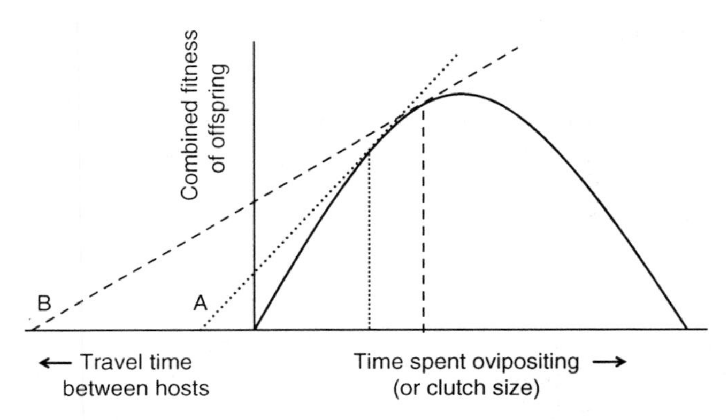

Figure II For a time-limited female, the optimal time spent ovipositing (which maximizes her *rate* of fitness gain) is represented by the tangent from the travel time between hosts to the combined fitness curve. As travel time increases (from line A to line B), so does the optimal time spent ovipositing, and hence clutch size (after Charnov & Skinner 1985; Wilson & Lessells 1994).

(continued)

BOX 11.1 *(cont.)*

wasp less likely to accept an occupied host (Hubbard et al. 1999). In some parasite-host interactions, per capita fitness of larvae does not decline monotonically with increasing density; it initially increases and is maximal at an intermediate density. For example, the presence of multiple parasitoid larvae per host may help overcome host defenses but not create substantial competition.

Two further complications need to be considered. Searching females may be *egg-limited* rather than time-limited. Here, oviposition decisions will mainly depend on a female's current supply of mature eggs. Optimal clutch sizes are determined by the point at which the fitness gain from adding another egg to a clutch is offset by the probability of egg depletion. Running out of eggs entails a cost because a female loses the opportunity to exploit better hosts encountered later. Time-limited females are expected to be less choosy than egg-limited ones, although distinguishing between time and egg limitation has been empirically difficult (Rosenheim et al. 2008).

A final point is that the larval competition curve is itself evolutionarily labile. In parasitoids and seed beetles, larvae from different species or populations can show strikingly different levels aggression within hosts. Thus, the same larval density can produce a contest outcome (winner-takes-all) or a scramble outcome (in which all larvae have roughly equal access to the resource). In the seed beetle *Callosobruchus maculatus*, the strongest host discrimination is found in populations with the most competitive larvae. Moreover, an experimentally induced shift from a small-seeded host species to a large-seeded one produced rapid, simultaneous changes in both the degree of host discrimination and competitiveness of larvae within seeds (Messina 2004). The adaptive evolution of these traits should therefore be considered jointly.

should be indifferent to variation, because the mean rewards of the two options are the same. However, Caraco et al. (1980) found that juncos that were provided more than enough food to meet their daily energy budget typically chose the fixed option, whereas birds that did not meet their daily energetic requirements typically chose the more variable option. This result has been repeated for many other organisms (reviewed in Stephens & Krebs 1986).

Why should foragers be sensitive to variation in reward? The modeling work presented so far has been based on the assumption that fitness is a linear function of the long-term rate of net energy intake. However, Caraco et al. (1980) found that the juncos' *utility* did not increase linearly with the number of seeds available. Utility is a concept from economics and refers simply to a descriptive measure of the value (often termed the *level of satisfaction*) assigned to an item; in other words, it is a measure of what consumers actually choose. In foraging theory, we assume that utility is closely linked to fitness. The utility of a resource to a foraging animal is assumed to reflect the fitness consequences of choosing a particular resource. For juncos with positive energy budgets, this utility function was decelerating (concave down); marginal utility decreased with increasing seed availability. For juncos with negative energy budgets, the utility function was accelerating (concave up). These patterns make sense if we consider that, for small birds, the risk of energetic shortfall and even starvation is high. A bird with a positive energy budget is at low risk of energetic shortfall unless it receives a very low foraging reward. Therefore, the fitness (and utility) function increases sharply with reward size when rewards are small; however, when rewards are large, there is little risk of starvation, so there is only a slight increase in fitness with increasing reward. For a bird with a negative energy budget, energetic shortfall can only be avoided by obtaining large rewards. There is little difference in fitness between a small reward and a slightly larger reward, as both are likely to lead to shortfall.

A consequence of nonlinear utility curves is that variance in prey delivery over time becomes important in foraging decisions (chapter 12). Briefly, when curves are concave down (figure 11.2a), individuals should be risk averse. That is, for the same mean reward, they should accept the option with lower risk. The reason for this is that the marginal gain from an increase in reward is small, whereas the marginal loss from a decrease in reward is great (figure 11.2a). Expected fitness, therefore, is lower when variation around the mean is high. Conversely, when curves are concave up (figure 11.2b), individuals should be risk prone. For the same mean reward, they should accept the option with higher risk. The marginal increase in fitness from an increase in reward exceeds the marginal decrease in fitness from a decrease in reward (figure 11.2b). Thus, the expected value of fitness for riskier option exceeds that of the less risky option. A variety of complications, such as exactly how the curves are shaped, will affect how risk influences the behavior of foragers (see chapter 12).

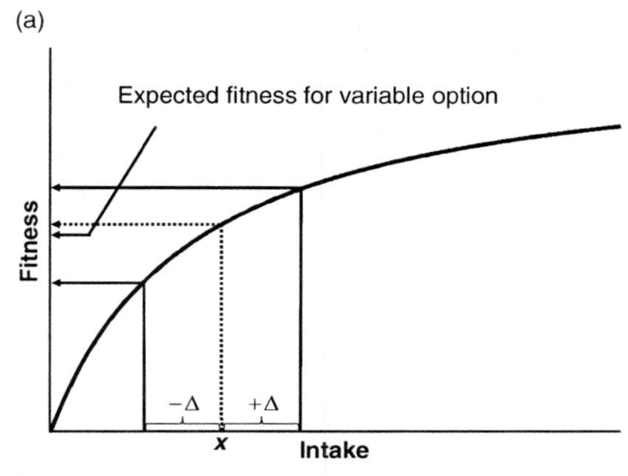

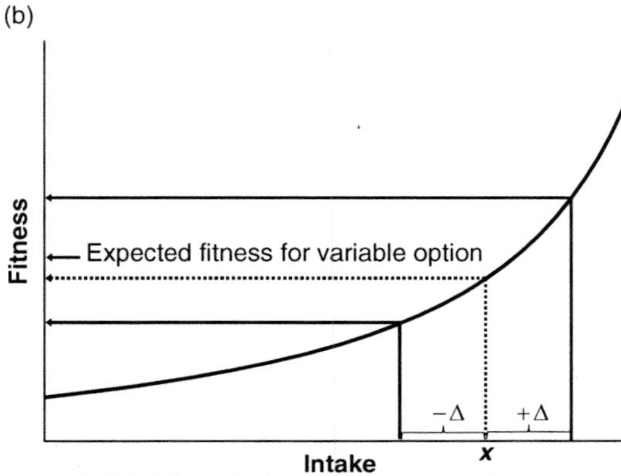

FIGURE **11.2** The relationship between fitness curves and risk-sensitive foraging. In (a), the function relating intake to fitness is concave down, whereas in (b) this function is concave up. Solid lines represent intake rates in the variable condition (intake rate varies between $x+\Delta$ and $x-\Delta$), whereas the dotted line represents intake rate from the constant option, x. In (a), the expected value of fitness for the variable option, assuming that high and low intakes are equally likely, is less than the expected value for the constant option. In (b), the expected value of fitness for the variable option is greater than that for the constant option.

FUTURE DIRECTIONS: RATIONAL CHOICE

The above analyses of foraging behavior make a critical assumption shared with classical economics. This assumption is that agents—foragers in the case of foraging ecology and consumers in the case of economics—are rational. The decisions of these agents should maximize utility or fitness. However, the consumer choice literature includes many counterexamples to this assumption of rational choice (e.g., Doyle et al. 1999). Given the parallels between economics and behavioral ecology, it is worth asking whether we observe similar apparent deviations from rationality in biological systems. A promising place to look for such deviations is among foragers.

Rational choice theory predicts that the absolute and relative preferences for two items, T and C, should be independent of the presence of a third item, D, which is less preferable than both T and C (figure 11.3a; the reason for using C, D, and T will become clear shortly). This is called *independence from irrelevant alternatives*. However, studies of consumer choice have found that the presence of the supposedly irrelevant option can have a strong effect on decisions. This is particularly intriguing when the irrelevant option is asymmetrically dominated by one of the options (Doyle et al. 1999). An option is *dominated* by another if the other option is equal to or better than it in all ways. An asymmetrically dominated item (termed the *decoy*, D) is dominated by one option (the *target*, T) but not the other (the *competitor*, C; see figure 11.3a). When this is so, the presence of the asymmetrically dominated decoy increases preference for the target over the competitor. To illustrate asymmetrically dominated choices, consider a shopper choosing between television sets. The shopper would like to purchase a large set but also one that is inexpensive. The shopper has the option of set #1: a small, inexpensive set and set #2: a large, more expensive set. Now, consider that a third option (set #3) is available; this set is just as large as set #2 (and no better than it in any other way), but is even more expensive. The third TV set is thus dominated by #2. The set is more expensive than #1, but it is better than #1 in terms of size, so it is not dominated by #1. Using the terminology above, set #3 is the decoy, set #2 the target, and set #1 the competitor.

Consumers facing a choice between a less expensive, poorer quality item and a more expensive, higher quality item are more likely to choose the more expensive item if a third, even more expensive option of similar quality is present (Doyle et al. 1999). In the above example, this effect means that choice for set #2 will increase in the presence of the third option. Several studies have reported the same effect in foraging animals (Shafir et al. 2002; Bateson et al. 2003). Proposed adaptive explanations for violation of independence from irrelevant alternatives and other apparent violations of rationality include state-dependent decision making (see

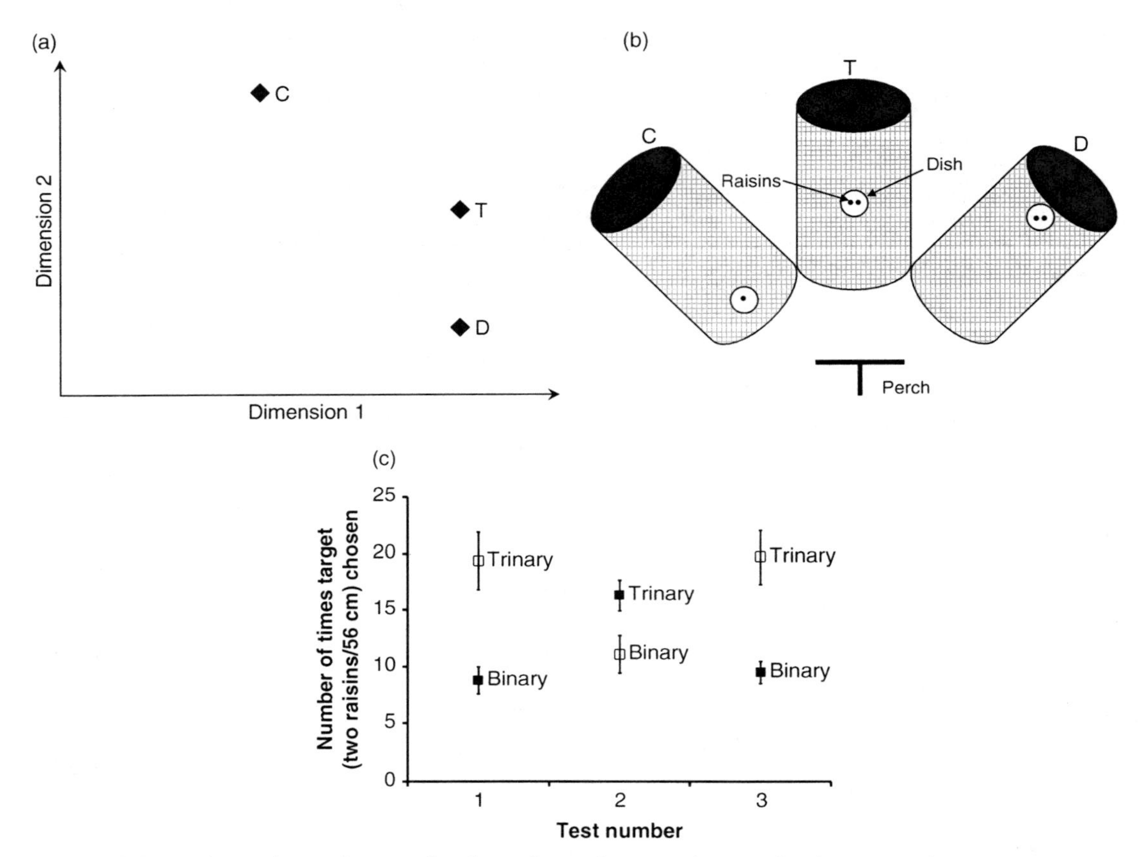

FIGURE **11.3** Evidence for violation of independence from irrelevant alternatives in foraging gray jays (redrawn from Shafir et al. 2002). (a) A schematic representation of an asymmetrically dominated option. The decoy (D) is dominated by the target (T) because the target has a greater value than the decoy on dimension 2 and the same value on dimension 1. The decoy is not dominated by the competitor (C), because the competitor has a greater value on dimension 2, but a lower value on dimension 1. (b) The trinary choice set up from Shafir et al. 2002. Gray jays had the option of choosing one raisin a short distance (28 cm) within a tube that restricted mobility (the competitor, C), two raisins further into the tube (56 cm; the target, T), and two raisins even further back in the tube (84 cm; the decoy, D). Note that the decoy is dominated by the target, which is safer (a shorter journey into the tube) and contains the same number of raisins. The decoy is not dominated by the competitor, because the competitor yields a smaller number of raisins. (c) Mean (SE) choice for the target in the presence and absence of a decoy. Test number refers to the order of testing; birds were either given the binary choice first, followed by trinary choice and then another round of binary choice (solid marker), or trinary-binary-trinary choice (open marker). In all cases, choice for the target increased in the trinary treatment.

below) and the use of rules of thumb that generally lead to fitness-maximizing decisions in nature, but not necessarily in experimental choice tests (ecological rationality; see chapter 8).

CASE STUDY

Rational Choice in Gray Jays

Gray jays (*Perisoreus canadensis*) are boreal forest birds that hoard immense quantities of food for use over winter. Many of the foraging decisions of gray jays are consistent with maximization of long-term hoarding rate (Waite & Ydenberg 1994). However, gray jays also have been implicated in a variety of apparent deviations from rational choice (see references in Shafir et al. 2002). To test whether hoarding jays' decisions were independent from irrelevant alternatives, Shafir et al. (2002) allowed birds to collect raisins positioned at various distances inside mesh tubes. To access the raisins, birds were required to hop into the tubes; these were sufficiently narrow to prevent flight, increasing the perceived riskiness of the foraging task. In the binary choice condition, birds were provided with two options: they could obtain a single raisin placed 28 cm into the tube (hereafter: 1/28) or obtain two raisins placed 56 cm into the tube (hereafter: 2/56; figure 11.3b). To test whether preferences remained consistent in the presence of an asymmetrically dominated decoy, birds were also tested in a trinary choice condition. In this treatment, a third tube was provided with two raisins available at a distance of 84 cm into the tube (hereafter: 2/84; figure 11.3b). The 2/84 option was dominated by the 2/56 option; although both options yielded the same number of raisins, the 2/56 option required a shorter foray into the mesh tube. Using the terminology presented above, the 2/56 option served as the target and the 2/84 option as the decoy in this experiment. The 2/84 option was not dominated by the 1/28 option; it was better than 1/28 on one axis of variation (number of raisins) but poorer on the other (safety). Therefore, the 1/28 options served as the competitor in this choice experiment. As expected based on experiments with human subjects, but not expected under rational choice, the birds did change their relative preferences, increasing preference for the target (2/56) in the trinary choice treatment (figure 11.3c). Incredibly, not only did they increase their relative preference for the target, but they also increased their absolute preference (Shafir et al. 2002). Even with some birds choosing the third, decoy option, birds chose the target more often in the trinary choice test than in the binary choice test. In the same paper, Shafir et al. (2002) reported similar results for honeybees (*Apis mellifera*) choosing among flowers differing in reward size and corolla length, although absolute preferences for the target did not increase in that case.

As with human consumers, the addition of an asymmetrically dominated decoy influenced choice in gray jays and honeybees. Similar results have been found in several other systems (Bateson et al. 2003; Schuck-Paim et al. 2004). However, do these results really reflect a deviation from rational choice? Schuck-Paim et al. (2004) suggested that in many tests, the addition of a third, apparently irrelevant option changes more than the background context of the decisions. As we have seen, foraging decisions can be sensitive to energetic state. Although human subjects can have the alternative options explained to them, nonhuman animals often must be trained to recognize the available options. This repeated exposure to different options means that animals trained in the presence of a decoy may differ in energetic state from animals trained without a decoy present. This argument does not explain the apparently irrational choice of the target by gray jays in the experiment by Shafir et al. (2002) because all birds were trained in the same way and each bird was exposed to both options. Birds were also retested with their original choice task after having been exposed to both conditions (e.g., those birds first exposed to the binary choice condition experienced another round of binary choice tests after having been exposed to the trinary choice condition; figure 11.3c). The increased preference for the target in the trinary choice test was found regardless of whether the bird was naïve or had previously been exposed to the binary or trinary choice test.

However, state dependence may explain apparently irrational decision making in some other experiments. Using starlings, Schuck-Paim et al. (2004) tested how state subsidies influenced responses in an asymmetrically dominated choice test. In their experiment, birds trained exclusively in an unsupplemented binary choice treatment received less food than birds trained in the trinary choice treatment. However, in one treatment, birds were trained in the binary choice treatment but also provided with supplemental food to make up for the difference in food obtained during training. The addition of this supplemental food was sufficient to eliminate differences in the relative preference for small but safe versus large but risky food rewards with and without a decoy (Schuck-Paim et al. 2004).

FORAGING WITH OTHERS

The rate-maximizing and risk-sensitive patch and prey foraging models presented so far have considered a single forager. However, in nature, many foragers are economically interdependent with one another. This is clearly so when individuals interact directly, but even foragers that never encounter one another may have an influence on one another through depletion of shared resources or changes in the behavior of prey.

The Ideal Free Distribution

The ideal free distribution (IFD) describes the expected distribution of foragers when resource patches differ in quality and foragers compete for resources (Fretwell & Lucas 1969). In this model, foragers are assumed to have perfect information about relative patch quality and the densities of foragers in each patch (they are *ideal*). They are also assumed to be able to move between patches without cost or time delay and not be excluded from entering patches by the current inhabitants (they are *free*). The simplest IFD model assumes that prey arrive randomly at different rates in different patches and are consumed immediately upon arrival (i.e., handling time is zero). If all of these assumptions are met, the long-term rate of net energy intake in patch i is the per capita encounter rate with prey, R_i. This rate is the ratio of the rate of prey input (Q_i) to the density of foragers in that patch (d_i):

$$R_i = \frac{Q_i}{d_i} \tag{11.1}$$

Because foragers are ideal and free, they can identify and move to any patch that offers a higher long-term net rate of energy intake. There is a unique internal equilibrium distribution of foragers at which the fitness returns for choosing each patch are the same. This point is a Nash equilibrium, which means that no forager could improve its fitness by moving unilaterally (see chapter 8 for more on Nash equilibria and game theory). If there are two patches that differ in quality, then this point occurs when the ratio of competitor densities in patches 1 and 2, d_1/d_2, equals the ratio of resource inputs Q_1/Q_2. This property is known as *input matching* or *habitat matching*.

Consider two ways of reaching the input matching equilibrium. The first is that all foragers spend some proportion of their time, p, in the higher quality patch (patch 1) and the remainder (i.e., $1-p$) in the lower quality patch (patch 2). To satisfy input matching, $p=d_1/(d_1+d_2)$. Assume that each forager makes this decision independently; given a sufficiently large population, following this rule will bring the population distribution of foragers very close to input matching at all times (and equal to input matching if the population is infinitely large). For the sake of argument, we will assume that p is a genetically determined strategy, although it could also represent a learned or developmental strategy. Now, suppose that through mutation or migration, a new genetically determined strategy (p')arises that spends slightly more time in the higher quality patch. By doing so, it shifts the densities in patches 1 and 2 ever so slightly, so that the density in patch 1 is now $d_1'>d_1$ and in patch 2 is $d_2'<d_2$. Inserting d_1' and d_2' into equation (3) to obtain R_1' and R_2', it is trivial to show that the intake rate of foragers using patch 1, R_1', is now slightly less than that of foragers in patch 2, R_2'. The p' mutant spends more time in patch 1 than does the average member of the population, so its fitness will be somewhat lower than the average member of the population. Thus, natural selection will not favor the p' mutant, which will not become established in the population. Using the same logic, it can be shown that a p' mutant that spends less time in the high-quality patch would also have lower relative fitness. Therefore, a population that plays p is playing an evolutionarily stable strategy (ESS; Maynard Smith 1982; see also chapter 8 of this volume).

Another way to achieve input matching is if all individuals have fixed habitat preferences, but the population contains a mix of p patch 1 occupants and $1-p$ patch 2 occupants. Again, an increase in the frequency of patch 1 occupants means that their average fitness will be low relative to the average fitness of habitat 2 occupants. The converse is true if the frequency of habitat 2 occupants increases. Thus, the system exhibits negative frequency dependent selection, in which the relative fitness of a trait (in this case, habitat preference or occupancy) decreases as the trait becomes more common. Again, p is an evolutionarily stable equilibrium.

To summarize, the *continuous-input IFD* model described above makes two testable predictions. The first is that intake rate is equal across all used patches at equilibrium. The second is the prediction of input matching: at equilibrium, consumers should be distributed so that the ratio of consumer densities across patches equals the ratio of patch resource input rates. The assumption of continuous input prey dynamics is met in some systems, such as drift-feeding stream fish (e.g., Grand 1997). However, empirical tests of the ideal free distribution have rarely supported the quantitative predictions of input matching and equal intake rates across patches (Kennedy & Gray 1993; Earn & Johnstone 1997). Typically, fewer individuals are found in high-quality patches than expected under input matching, and those individuals often have higher intake rates than do occupants of lower quality patches (Kennedy & Gray 1993).

Widespread undermatching indicates that the assumptions of this simple continuous input model are often violated. Interference models (see below) can lead to undermatching but also often predict the opposite effect, *overmatching* of resource inputs in high-quality patches. Undermatching and unequal intake rates between patches may also result from violation of the assumption of perfect information (Abrahams 1986). If animals are unable to distinguish alternatives that are very similar in payoffs, they may choose randomly when quality differences between patches are small. Because of this, perceptual constraints will always lead to overuse of low-quality patches and underuse of high-quality patches. Deviation from input matching may also result from stochastic variation in input rates (Recer et al. 1987). If one patch is variable and the other is not, ideal free consumers are expected to underuse the variable patch relative to the expectation based on the long-term average if foragers can track the state of the variable patch and respond (possibly with some delay). Additionally, in some systems, habitat use may be better described by alternative models, such as the ideal despotic distribution or ideal preemptive distribution, in which dominant or early arriving individuals, respectively, prevent others from occupying high-quality patches (Fretwell 1972; Pulliam & Danielson 1991).

In many systems, competition does not occur through instantaneous depletion of discrete prey items but through interference. Interference is a reversible decrease in intake rate with increasing density of competitors (Sutherland 1983). This is in contrast to depletion; assuming that renewal times are long, removing competitors from a depleted patch will not increase the intake rate of a forager. Often, interference is equated with aggressive defense of resources. However, there are other sources of interference, such as kleptoparasitism (the theft of resources from others), behavioral interactions that take time away from searching for food and changes in the behavior of prey so that prey become increasingly alert or hide as predator densities increase.

There are a number of ways of incorporating interference into social foraging models. One simple way of doing so is through the addition of an *interference constant*, m (Hassell & Varley 1969), so that the intake rate of a forager in the interference model is as follows:

$$R_i = \frac{Q_i}{d_i^{\,m}} \tag{11.2}$$

Here, Q_i is the foraging rate of a solitary forager. Note that when $m = 1$, this model is the same as the continuous input model. Empirically derived estimates of m in the absence of patch choice are typically much less than 1 (e.g., Sutherland & Koene 1982), although any value greater than or equal to zero is theoretically possible. The equilibrium point in the two-patch interference model is reached when the ratio of the densities of competitors in patches 1 and 2, d_1/d_2, equals $[Q_1/Q_2]^{1/m}$. If $m < 1$, the interference IFD model predicts greater use of the high-quality patch than expected under habitat matching. However, there is little empirical support for this prediction (Tregenza et al. 1996), possibly because the assumptions that all competitors are equal (see below) and that the interference constant is the same in all patches are frequently violated in natural systems. A number of other ways of modeling habitat use under interference exist (van der Meer & Ens 1997), and some of these predict undermatching rather than overmatching, consistent with the qualitative results of experimental tests (e.g., Tregenza et al. 1996).

Differences in competitive ability have a relatively minor influence on continuous input models, but greatly change the predictions of interference model (Parker & Sutherland 1986). Central to unequal competitors models is the concept of competitive weights, which are the relative competitive abilities of individual foragers. Parker and Sutherland's (1986) model of unequal competitors in a continuous input system predicts input matching of competitive weights; that is, the ratio of the sums of competitive weights per unit area in high- and low-quality patches is equal to the ratio of resource inputs into these patches. Input matching of competitive weights has been found in some systems that closely match the assumptions of the continuous input IFD (e.g., juvenile coho salmon, *Oncorhynchus kisutch*; Grand 1997).

Unequal competitors interference models predict a *truncated phenotype distribution* (Parker & Sutherland 1986), in which the best competitors exploit the best habitats. The only competitors that occupy more than one habitat are those at the *boundary phenotype*, which is the competitive phenotype for which intake rates are equal in the two habitats. Superior competitors to the boundary phenotype occupy the better habitat, whereas inferior competitors exclusively occupy the poorer quality habitat. Average intake rates are not equal across habitats at equilibrium. Rather, there is a positive correlation between habitat quality and intake rate. Truncated phenotype distributions are also predicted when competitive weights themselves are a function of habitat quality (Parker & Sutherland 1986). However, experimental evidence that unequal competitors assort themselves according to the truncated phenotype distributions is poor (Milinski et al. 1995; Tregenza & Thompson 1998). Although better competitors may tend to use those patches in which their competitive ability is best, as predicted under the truncated phenotype distribution (Milinski et al. 1995), they do not exclusively use these patches. As in other ideal free models, perceptual constraints and the need to sample multiple patches in order to choose among them may result in overuse of less profitable patches (Tregenza & Thompson 1998). There remains a need for further theoretical and empirical exploration of habitat use by competitors whose relative competitive ability changes with habitat.

As discussed in the previous sections, empirical tests of the ideal free distribution often find that the density of consumers in high-quality patches is less than that predicted by input matching and that intake rates are not equal in all patches. However, the qualitative prediction of higher densities in higher quality patches is often met. Deviation from the expectation of higher forager densities in patches with higher resource input rates suggests that foragers are considering other dimensions of habitat quality, such as safety or social factors, when choosing a habitat in which to forage. The change in forager distributions in the presence and absence of predator cues has been used to investigate how animals trade off foraging benefits and the risk of predation in the laboratory (Abrahams & Dill 1989; Grand & Dill 1997) and at the landscape level (Heithaus et al. 2002). Changes in patterns of habitat use resulting from the presence of predators, competitors, or other ecological factors provide information about how foragers perceive the landscape in fitness terms and provide insight into how animals trade off the long-term rate of net energy intake against other influences on fitness.

Social Exploitation of Resources

Foragers may benefit by obtaining prey that were discovered or otherwise produced by others. Exploitation among social foragers has been extensively explored using *producer-scrounger* models in which a population of foragers is divided into two mutually exclusive foraging tactics (Barnard & Sibly 1981). Producers search the environment for food patches. Scroungers search the environment for successful producers and join them at discovered patches. The fitness payoffs to both producers and scroungers are frequency dependent, and the payoffs to scroungers or both foraging types decline as the proportion of scroungers increases (because producers lose resources to scroungers and because scroungers compete with one another for fewer produced patches, respectively). In many cases, the fitness of scroungers is expected to be higher than that of producers when scroungers are rare, but to decline more steeply with increasing scrounger density. Thus, there is a point at which the fitness curves for the two tactics cross and fitness payoffs are equal (e.g., left panel of figure 11.4a). This point is a Nash equilibrium and is evolutionarily stable (Maynard Smith 1982; chapter 8 of this volume). In one influential model of the rate-maximizing producer-scrounger game (Vickery et al. 1991), the predicted equilibrium frequency of producers increases as the proportion of discovered resources that can be used exclusively by the producer (the *finder's share*) increases and as group size decreases. The producer-scrounger game has also been modeled in a stochastic framework, in which foragers

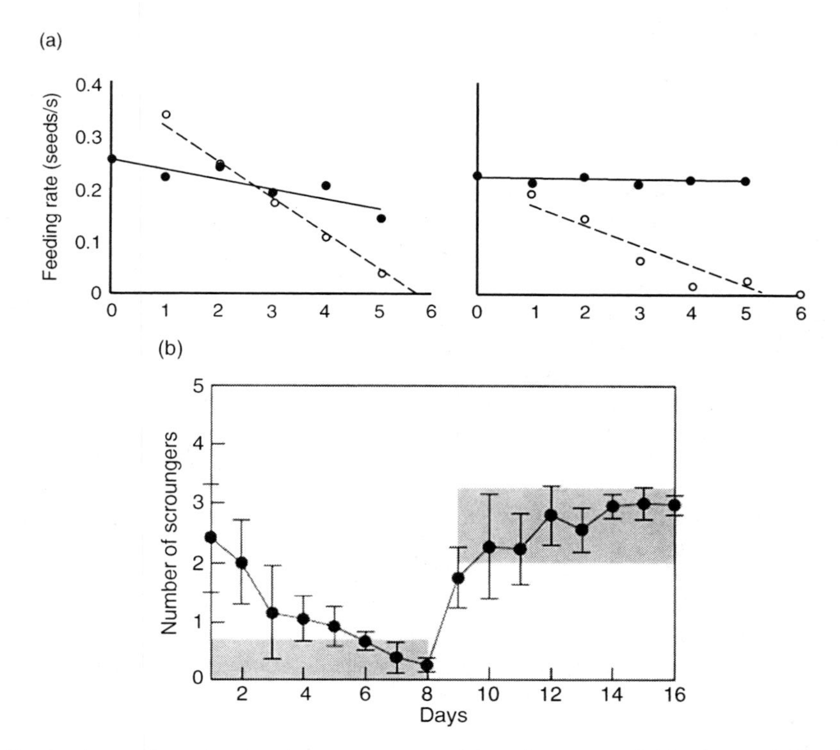

FIGURE **11.4** Producing and scrounging in nutmeg mannikins. See text and Mottley and Giraldeau (2000) for description of experimental design. (a) Mean (SE) feeding rate for birds forced to act as producers (solid circles) or scroungers (open circles) against scrounger number. On the left panel, scroungers had a higher intake rate than producers at low scrounger densities, but the decline in feeding rate with increasing scrounger density was steeper for scroungers than producers. On the right panel, scroungers' access to produced resources was restricted, resulting in lower scrounger feeding rates. Expected equilibrium frequencies of producers and scroungers occur when the intake rates of the two tactics are equal: in this example, 3 scroungers in the left panel and between 0 and 1 scrounger in the right panel. Only data from one flock are shown; for data from other flocks see figure 3 in Mottley and Giraldeau (2000). Patterns were quantitatively and qualitatively similar in other flocks. (b) The mean (SE) observed frequency of scrounging when birds were allowed to choose tactics. Results are plotted against day of observation. Shaded areas represent the predicted equilibria when scrounger access to resources was restricted (days 1–8) and not restricted (days 9–16). Birds rapidly converged to these predicted equilibria. Data from other flocks were quantitatively and qualitatively similar (see figure 4 in Mottley & Giraldeau 2000). Reprinted from Mottley and Giraldeau (2000) with permission.

seek to minimize the risk of energetic shortfall (see Giraldeau & Caraco 2000).

Case Study: Social Exploitation in Nutmeg Manikins

Nutmeg mannikins (*Lonchura punctulata*) are small, seed-eating finches from Southeast Asia that have extensively been used to study producer-scrounger games (Giraldeau et al. 1994). These birds are ideal for such studies because they are nonaggressive, so that joiners are free to recruit to discovered patches, and they appear to be unable to search simultaneously for food patches and for opportunities to join others. When searching for food patches, these birds hop with their heads down, but when searching for opportunities to join, they hop with their heads up (Coolen et al. 2001). Giraldeau and Livoreil (1998) found that the joining policy of nutmeg mannikins searching for seed patches conforms to the predictions of producer-scrounger models.

Although producing and scrounging appear to be mutually exclusive foraging tactics for these birds, all birds use a combination of the two tactics and can switch between them over very short time scales. Thus, it is difficult to show that the fitness payoffs to producers and scroungers are frequency dependent, as assumed in the producer-scrounger model (Giraldeau & Livoreil 1998; Mottley & Giraldeau 2000). Mottley and Giraldeau (2000) performed an elegant series of experiments to demonstrate frequency dependence of fitness payoffs. Two halves of a testing chamber were divided into a "producer compartment" and a "scrounger compartment" by a barrier. Seeds were held in inaccessible containers by lengths of string; birds in the producer compartment could search for seed patches by pulling the string. This released seeds into a tray accessible from both compartments. Only a single producer could access this tray because only one narrow perch per tray was available on the producer side. Because birds could be blocked from moving between compartments, the ratios of producers to scroungers could be manipulated and intake rates of all birds measured at these various ratios.

As expected, the intake rate for scroungers was strongly negatively frequency dependent (Mottley & Giraldeau 2000; figure 11.4a). The intake rate for producers did not depend on tactic frequency in most tests. Importantly, scroungers had higher intake rates than producers when scroungers were rare, but the converse was true when scroungers were common. Having measured the producer-scrounger intake curves, they then allowed birds to move freely between sides. The predicted distribution of birds would be that which equalized payoffs to producing and scrounging, in other words, the point at which the intake curves for the two tactics crossed. In both the covered and uncovered treatments, birds converged to the predicted distributions within 8 days of testing (figure 11.4b). Manipulation of the finders' share increased both the predicted and observed proportion of birds using the producer compartment. Again, there was a quantitative match between predicted and observed producer and scrounger frequencies (Mottley & Giraldeau 2000). Although the system converged on the evolutionarily stable equilibrium, adjustments to tactic frequency occurred within the same set of birds at very short time scales, rather than over a period of several generations as predicted in evolutionary models. Mottley and Giraldeau (2000) argue that the ability to reach an evolutionarily stable equilibrium in this system provides support for the existence of evolutionarily stable learning rules that allow for rapid assessment of the payoffs to and adjustment of the use of alternative behavioral tactics. What mechanisms are used, how they lead to correct decision making and the evolution of alternative decision-making tactics are key questions for future research.

Group Foraging

Foraging as part of a large group of conspecifics or heterospecifics may confer several benefits. Groups may provide protection from predators through dilution effects or increased corporate vigilance, information about the location of food patches, or access to otherwise inaccessible prey (reviewed in Krause & Ruxton 2002). Grouping may also reduce the variance in prey encounter rate, so that risk-averse foragers may minimize their risk of energetic shortfall by foraging in groups (Ekman & Rosander 1987). On the other hand, grouping is also costly. Group members are competitors for resources. Groups may also facilitate the transmission of pathogens and parasites and may be conspicuous targets for predators. Therefore, we expect that the size and structure of animal groups reflect a mix of cooperation and conflict.

Models of group foraging assume that the marginal benefits of group size decrease with increasing group size, whereas the costs of grouping increase. This leads to a dome-shaped relationship between group size and the fitness of group members (Clark & Mangel 1986). The peak of this function is the optimal group size; however, theoretical and empirical results indicate that the optimal group size is rarely expected or observed (Sibly 1983). The reason for this is straightforward. If a group exists at its optimum size, then the fitness of members of that group is greater than the fitness of solitary individuals. If a solitary forager joins the group, it will increase its fitness while decreasing the fitness of others in the group. If there are no barriers to joining the group, group members are unrelated and groups do not split, then solitary individuals should keep joining until the fitness payoffs of group and solitary foraging are the same. This is the stable group size (Sibly 1983). Group sizes that are intermediate to the stable and optimal group sizes are expected if some or all current group members have some, but incomplete, control over whether others join the group or when group members are relatives (Higashi & Yamamura 1993; Hamilton 2000).

Group members may be vulnerable to exploitation by individuals that do not contribute to obtaining group benefits. Many foraging groups appear to include a mix of cooperative and exploitative individuals. In several species of cooperative hunters (Packer & Ruttan 1988) and in at least one group-living reef fish (Hamilton & Dill 2003a; see case study below) the observed relationship between group size and intake rate best fits the predictions of models in which some individuals do not contribute to group foraging success but reap the rewards. Can apparently cooperative foraging persist in the face of such rampant cheating? A full discussion of mechanisms that can enforce cooperative behavior is beyond the scope of this chapter, but can be found in other chapters of this book, for example, chapters 14 and 18. In many foraging groups, there may be no temptation to cheat, because all group members receive direct benefits from cooperating (by-product mutualism; Connor 1995a). An example of by-product mutualism is grouping to avoid predators through numerical dilution; joining a group provides direct benefits to the joiner, whereas the benefits to other group members are incidental.

However, many foraging situations may be more prone to conflict. Although all members of the group may benefit from banding together to capture large or dangerous prey, an individual that refrains from doing so, avoiding injury and energetic costs of prey capture, but still partakes in the reward would do even better (Packer & Ruttan 1988). When the temptation to cheat exists, cooperative behavior is more difficult to explain. In such cases, cooperation can be maintained only when individuals interact nonrandomly with others in the population (see chapter 14). This could result from patterns of genotypic assortment, so that group members are genotypically more similar to one another than to the population as a whole (chapters 14 and 18). Typically, this results from grouping with kin. Nonrandom interactions could also results from behavioral interactions, including positive direct or indirect reciprocity or punishment of cheaters (chapter 18). The roles of these behavioral mechanisms in enforcing cooperation in foraging groups remain largely unexplored.

Social Information Use

The producer-scrounger and many other social foraging models assume that information about patch discoveries is transmitted to other group members. A key question, therefore, is, when should individuals use socially acquired information rather than private information? Although this is relevant to many areas of behavioral ecology (see also chapter 17), including mate choice or territorial settlement, much of the theoretical and empirical work on social information use has involved foraging studies.

One potential cost of using socially acquired information is the risk that incorrect information will be acquired. If the use of socially acquired information is prevalent, this can lead to *informational cascades*, in which possibly incorrect information is transmitted to observers who then transmit it further to others (Giraldeau et al. 2002). However, the risk of informational cascades and maladaptive traditions is not equal for all mechanisms of socially acquired information use. Informational cascades may be more likely when presence-absence cues are used than when graded information on the success of other foragers (*public information*) is available (Giraldeau et al. 2002). Socially acquired information is also expected to be most valuable in moderately unpredictable environments. In highly predictable environments, there is little advantage to using socially acquired information, whereas in highly unpredictable environments, the chance that socially acquired information will be wrong is high (Laland et al. 1996).

Social Exploitation and Group Foraging in the Field: A Case Study with Subtropical Reef Fish

The western buffalo bream (*Kyphosus cornelii*) is an herbivorous subtropical fish endemic to the eastern Indian Ocean along the coast of Western Australia. In some regions of its range, larger fish maintain algal gardens by removing less preferred algae from the patch and defending the garden from other herbivorous fish (Hamilton & Dill 2003a, 2003b). Algal gardens present rich foraging patches for their owners, but also potentially are highly profitable sites at which others can forage. Thus, in gardening systems, pilferage by so-called roving fish is common. In some cases, gardeners and rovers are of different species. For example, blue tang surgeonfish (*Acanthurus coeruleus*) band together in large roving groups to invade the territories of an herbivorous damselfish (*Stegastes dorsopunicans*; Foster 1985). In other cases, including western buffalo bream, individuals of the same species play both tactics, although the use of the tactics may differ with condition. Even most western buffalo bream that hold gardens at least occasionally invade those of others (Hamilton & Dill 2003a).

Despite the similarity to producer-scrounger systems, gardener-rover systems include some crucial differences. Algal gardens are actively defended, so that rovers cannot freely join gardeners at food patches. The fitness payoffs to investing in gardens are expected to increase as the ability to defend gardens increases. Hamilton and Dill (2003b) manipulated territory defendability by creating visual barriers across portions of defended gardens. Reducing territory defendability was expected to reduce the payoffs to engaging in gardening, prompting a shift to greater use of the roving tactic. As predicted, gardening fish were less able to defend against intruders and spent less time on their gardens when substantial visual barriers were present (Hamilton & Dill 2003b). The decision to maintain a garden versus rove has implications beyond the foraging behavior of these fish. Western buffalo bream are abundant herbivores, and the algal gardens they maintain may cover large areas of the shallow subtidal zone in their range, which are exploited by a variety of other reef herbivores (IMH unpublished data). Therefore, shifts in the frequency of gardening or the effort invested in gardens may scale up to create community-level changes.

Roving fish that do enter territories often do so in small groups; in many other reef fish, such groups appear to confer a mutual group benefit by overwhelming the territorial defenses of gardeners (Foster 1985). Data from roving buffalo bream supported the hypothesis that grouping is a response to territorial defense, as groups were larger when invading actively defended territories (Hamilton & Dill 2003a). However, the benefits of grouping were not shared by all roving fish. Per capita intake rate declined with increasing group size, although it did increase for the largest groups (figure 11.5; Hamilton & Dill 2003a). Patterns of group size and intake rate were most consistent with the predictions of a model in which some roving fish acted as scroungers, taking advantage of foraging opportunities created when other rovers distracted territory holders and in which fish had information on the likelihood that territories would be actively defended. Such information could be acquired through direct observation of the outcome of previous invaders or from signaling by garden owners. Social exploitation, rather than cooperative foraging, appears to explain group foraging in these fish, at least within the range of group sizes typically observed in this study.

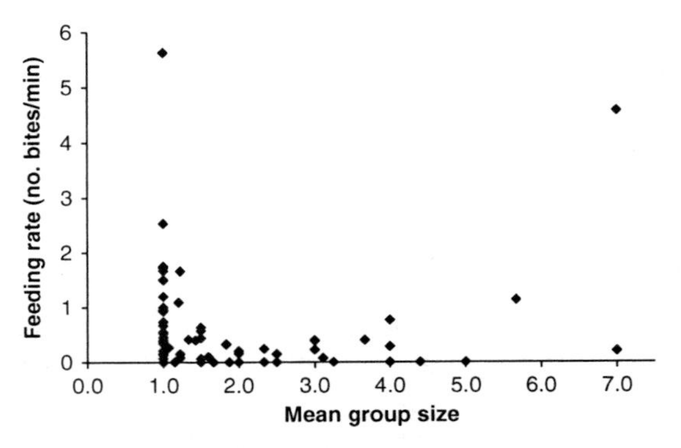

FIGURE 11.5 Observed feeding rates for roving western buffalo bream against group size. Group size is the mean size of a focal fish's foraging group over a 15-minute observation period. Intake rate declined with increasing group size except for in the largest groups. The decline in intake rate with group size was not consistent with the prediction that all group members receive a foraging benefit from group foraging. Redrawn from Hamilton and Dill (2003a).

FUTURE DIRECTIONS

Group Cohesion and Social Foraging

When group living provides fitness benefits, the need to maintain synchrony in activities must be traded off against other foraging considerations (Conradt & Roper 2000). However, few social foraging models incorporate the benefits of maintaining group cohesion. This may be particularly relevant in social exploitation, because high levels of exploitation may lead to individuals leaving the group. In the *recruiter-joiner model* (Hamilton 2000), which is based on a set of theoretical models of reproductive sharing in groups called optimal skew models (see references in Hamilton 2000 and box 19.1 of this volume), dominant individuals that could control access to resources were predicted to allow a more equitable distribution of resources to other group members when risk of predation or foraging benefits of grouping were high.

The need to maintain group cohesion may also influence foraging behavior because it influences the acquisition of social and private information. Recent modeling work suggests that local diet traditions can form even when neighbors pay no attention to the choice of others, as long as the benefits of grouping are sufficiently high (van der Post & Hogeweg 2008). In this model, foraging proficiency for particular food items, and preference for those items, increased with the frequency of exposure. Members of the same group are exposed to and become familiar with similar items when food is patchily distributed, and so develop similar diets.

Coordination among Agents with Limited Information

Following from the previous section, assuming that maintaining group cohesion is valuable, how do groups come to collective foraging decisions that are accepted by all group members? Group-level decision making has been most extensively studied in eusocial insects. In at least several systems, decision making is the result of quorum responses, in which the probability of performing an action increases with the number of conspecifics already performing the action (Sumpter 2006; chapter 9 of this volume). An example of collective decision making in a foraging context is choice of foraging patches by honeybee colonies (Seeley et al. 1991). Honeybees dance to provide information on the location and quality of food sources; however, individual foraging bees do not directly compare multiple sources (Seeley et al. 1991). Bees dancing intensely for more absolutely profitable sources attract more bees to that source, which will also dance on their return to the colony. Less profitable sources attract fewer bees, so the colony response to these decays. This leads to colony-level choice for relatively more profitable sources even when each bee has direct information about only one foraging source.

Eusocial insects are characterized by high relatedness among colony members and reproductive division of labor so that the reproductive success of group members is channeled through one or a few individuals. Thus, there is a strong fitness incentive for maximizing group-level fitness, which may facilitate efficient, low-conflict decision making. Are the insights from eusocial insects applicable to other foraging groups, which may be composed of unrelated individuals and in which group membership is more flexible? It appears that quorum decision-making rules do influence the collective movement decisions of at least some vertebrate groups. Schooling three-spined sticklebacks (*Gasterosteus aculeatus*) navigating a Y-maze showed a strongly nonlinear response to the decisions of conspecifics; as the number of fish (model fish in this case) choosing a particular direction increased, choice of that direction by others increased steeply after some threshold (Ward et al. 2008). Although much of the focus in understanding collective decision making has been at a mechanistic level (i.e., how a consensus decision is reached), this question is also pertinent at the evolutionary or ultimate level. Conradt and Roper (2007) recently presented a model predicting when shared versus unshared decision making is evolutionarily stable, and when majority, supermajority, and submajority thresholds are expected. Further development of theory, including the possibility of negotiation, threats, and promises (see chapter 8), and tests of theoretical predictions of collective decision making, would be of great value to understanding social foraging decisions.

SUGGESTIONS FOR FURTHER READING

Foraging theory has a long history in the field of behavioral ecology, and it is not possible to provide an exhaustive summary here. The classic review of foraging theory, including prey and patch models, state-dependent foraging, risk, and rules of thumb is *Foraging Theory* by Stephens and Krebs (1986) and the recent *Foraging* by Stephens et al. (2007). Social influences on foraging decisions, including optimal and stable group sizes, IFD, and producer-scrounger models are explored in depth in *Social Foraging Theory* (Giraldeau & Caraco 2000) and *Living in Groups* (Krause & Ruxton 2002). For a recent review on the use of social information in foraging and other decisions, see Dall et al. (2005).

Dall SRX, Giraldeau LA, Olsson O, McNamara JM, & Stephens DW (2005) Information and its use by animals in evolutionary ecology. Trends Ecol Evol 20: 187–193.

Giraldeau L-A & Caraco T (2000) Social Foraging Theory. Princeton Univ Press, Princeton, NJ.

Krause J & Ruxton GD (2002) Living in Groups. Oxford Univ Press, Oxford.

Stephens DW & Krebs JR (1986) Foraging Theory. Princeton Univ Press, Princeton, NJ.

Stephens DW, Brown JS, & Ydenberg RC (2007) Foraging. Univ Chicago Press, Chicago, IL.

12

Managing Risk: The Perils of Uncertainty

SASHA R. X. DALL

Risk permeates the lives of most organisms, if only for the simple reason that life is fundamentally unpredictable and so there is always the danger that things can go wrong. Indeed, almost everything that is important to an animal is uncertain. This stems from the fact that, to survive to reproduce (to maximize its fitness), an animal must solve multidimensional problems with components that can vary independently of one another over its lifetime. Furthermore, some such components are fundamentally unpredictable at the spatial and temporal scales at which organisms operate. Consider a starling foraging on the lawn in a suburban yard. While walking around, it is always on the look out for invertebrates in the grass and when probing the topsoil with its beak. Upon locating an edible morsel, the bird must assess how much nutrition it will gain from consuming it. Given this, the starling must decide whether to spend time eating it or look for more profitable fare that may or may not be located in the meantime. All the while, the starling must also consider whether to remain on the lawn or move to another yard in search of richer pickings, better company, or fewer cats, given that it could end up worse off if it does (see Stephens & Krebs 1986; Ydenberg et al. 2007; chapter 11 of this volume). Thus, wherever the bird decides to forage, it must navigate and be on the lookout for potential competitors, predators, and future mates, ever wary of errors of judgment and misfortune. And starlings are by no means unusual in the complexity of their ecological niches (Hutchinson 1957), the limits to their information processing abilities (chapters 9 and 10), or the dangers they face as a consequence.

It is easy to envisage that key aspects of almost all of a starling's day-to-day challenges will vary over its lifetime to some degree. In fact, for effective decision making and hence adaptive behavior, the current state of our starling's environment is critical, which includes the range of options open to it, the likely consequences of pursuing each option, and the probable actions of others foraging around it or of its predators. However, such features are likely to be changing continuously due to changes in weather, the behavior of other organisms, and so on. This means that any time spent in one area, or attending to a particular task, will erode the starling's ability to anticipate what is going on in the rest of its world. In this way, for instance, over its lifetime, our starling is likely to experience substantial variation in the quality of the food items it captures and so it will be unable to assess their quality exactly each time. Similarly, it faces unpredictability in whether it will find another item, or that a competitor or predator will come along, if it remains where it is. So the starling is making decisions in a world that may not be as it thought it was. Furthermore, limits to what it can detect in its environment (e.g., its sensory and cognitive capabilities; chapters 9 and 10) can generate unpredictability in important ways. For instance, even rich food patches can have some undetectably inedible parts, which will generate uncertainty in the returns gained from exploiting them. So, because it can only ever be in one place, or do a few things at once, or detect a limited range of things, a starling must resign itself to never being completely certain about anything. Key questions arise from such a perspective. To what degree does this lack of certainty matter biologically? Does it drive selection for behavioral (and other) traits directly in many species?

This chapter focuses on the fitness consequences of the uncertainty and dangers permeating the lives of animals that, like our starling, face multidimensional, varying niches equipped with limited sensory and cognitive capacities. In the process, I highlight the adaptations that can evolve in response to the dangers associated with their fundamentally unpredictable worlds.

BASIC CONCEPTS

Reducible and Irreducible Uncertainty

As far as its impact on adaptive animal behavior is concerned, *uncertainty* can be thought of as the moment-by-moment degree to which events are determined by factors that are out of an individual's control or immediate experience (i.e., *chance*) and is typically represented probabilistically (but see Carmel & Ben-Haim 2005). It is possible to distinguish two types of such instantaneous unpredictability, which I will refer to as *reducible* or *irreducible uncertainty* depending upon the degree to which current levels can be influenced by an organism's actions. On the one hand, even events with patterns of causation that can be tracked more or less deterministically by an organism (e.g., if x is experienced, then y will follow) can have uncertainty associated with them that is determined by the sheer complexity of the causal processes involved. This lies at the heart of the problems imposed by the multidimensional and variable niches that most animals exploit given that they occupy (experience) only a limited potion of their niches at any given instant. The resultant moment-by-moment stochasticity can potentially be reduced by spending time and energy sampling or observing key components of the environment when it is profitable to do so (see below for further discussion), and can therefore be thought of as *reducible* uncertainty. Alternatively, *irreducible* uncertainty is generated by causal processes underlying ecologically significant events that are fundamentally untraceable at the spatial and temporal scales at which an organism operates. In other words, some features of the world are inherently probabilistic or stochastic in nature (at least at the scale at which an organism operates, given its sensory and cognitive limitations), and hence at least some of the uncertainty associated with them cannot be altered by an organism's actions (e.g., the undetectably inedible components of food patches will always render probabilistic returns to foraging effort). This distinction is important for elucidating the types of strategy that can be deployed to maximize fitness in a risky and uncertain world (table 12.1 and below).

Risk: Uncertainty, Likelihood, and Fitness

Now, regardless of its source, uncertainty about features of the world is not necessarily important in itself. Indeed, as implied in my caricature of the problems facing a foraging starling, uncertainty is likely to be important only insofar as it impacts an organism's fitness. This interaction between uncertainty and its consequences is fundamental to the concept of risk. Moreover, in economics and statistical decision theory, risk often refers to a potential negative impact to an asset or something of value that arises from uncertain current processes, future events, or both (Knight 1921; Clark 2006). So, following a long and productive tradition in behavioral ecology of borrowing analogies and theoretical insights from economics after substituting Darwinian fitness for financial profit or utility (Stephens & Krebs 1986; chapter 8 of this volume), I am going to adopt such a definition of risk in this chapter. Specifically, as it pertains to the ecological events or outcomes associated with an animal's actions (i.e., those that result from or stimulate them):

$$\text{Risk} = \{\text{likelihood of outcome/event}\} \times \{\text{fitness cost per outcome/event}\} \quad (12.1)$$

which can be assessed on an outcome-by-outcome basis or combined over a set of outcomes to specify aggregate risk. Precisely how aggregate risk is specified depends upon whether the fitness payoffs accruing to the individuals involved add or multiply across the set of potential outcomes considered. On the one hand, for sequences of possible outcomes that include death but not reproduction (e.g., habitat selection during growth or outside the breeding season), probabilistic fitness payoffs are likely to combine multiplicatively. This is because if the organism dies, then the lost opportunity to contribute to its reproductive lineage's evolutionary persistence should dominate any valuation of risk, regardless of any other possible immediate payoffs. Alternatively, if reproduction (including that of relatives) is possible, it can make sense to assess aggregate risk additively. Such aggregate risk is particularly relevant if payoffs impact the expected reproductive output of an individual relative to the population average for its age or life history class. This mode of accounting is central to modern conceptions of reproductive value as the fundamental currency for assessing the fitness value of phenotypic traits (Houston & McNamara 1999; Grafen 2007; see chapter 4 of this volume for further discussion).

Following equation 12.1 then, there are two main components to risk, one linked to moment-by-moment (ecological) uncertainty and the other to its fitness consequences. By definition, the degree of uncertainty associated with an outcome influences its likelihood of occurring; however, the relationship between uncertainty and risk is not entirely straightforward. Uncertainty about an event often peaks when the likelihood of it occurring is moderate, whereas low levels of uncertainty can be associated with both high and low likelihoods. Indeed, generally speaking, uncertainty peaks when the likelihood of each possible outcome is equal (e.g., the frequency distribution of outcomes is at its "flattest"); for instance, for outcomes that are distributed binomially (e.g., all-or-nothing events like the presence or absence of a predator), the variance in the distribution of possible outcomes (a measure of uncertainty) peaks when the probability that a predator is present = the probability that it's absent = 0.5. Conversely, uncertainty declines with increasing bias in the likelihoods of possible outcomes; when assessed probabilistically, increasing or decreasing the likelihood of one of the potential outcomes simultaneously decreases or increases those of the others because, by definition, all probabilities of outcomes associated with each unitary event (action) must sum to one. And with less uncertainty about a potentially costly event, an animal can act more decisively to either counter the negative, or exploit the positive, potential outcomes. Without such control over its destiny, the animal is limited to acting less decisively and suffers an increased likelihood of a costly outcome. For instance, to a foraging mouse, a rustle in the bushes may indicate that it is being stalked by a cat. If it is more certain whether a particular rustle is cat-produced, the mouse can either flee to its burrow when a cat is likely or continue what it is doing if a cat is unlikely and thereby avoid actually being pounced upon or wasting valuable foraging time hiding unnecessarily. However, the less sure it is about the source of the rustle, the more likely that the mouse will experience one or other of these costly outcomes, even if it acts to maximize its fitness given its uncertainty by, say, allocating more of its attention to vigilance (Brown & Kotler 2004).

The concept of risk focuses upon the potential negative impacts of uncertain outcomes that individuals experience through their actions. For many organisms this makes risk a major selective force on behavior (and justifies the focus of this chapter) because the possible negative fitness consequences of specific actions or events are often disproportionately more severe than any potentially positive ones. This is self-evident for situations in which animals risk mortality (e.g., from predation, starvation, or aggression) and can often be true when less extreme risks are in play. For instance, short-term declines in nutrition during growth and development can force individuals into rapid, compensatory growth that often imposes substantial downstream, lifelong impairments to organismal function (Lindstrom 1999; Metcalfe & Monaghan 2001). On the other hand, gaining more nutrients than expected is unlikely to have equivalent benefits because accelerated growth rates are also not likely to be optimal (Metcalfe & Monaghan 2001). Nevertheless, it is important to remember that uncertainty can offer opportunities for substantial fitness windfalls, which can also shape adaptive behavior under risk. Key to understanding whether animals should ignore (or even encourage) uncertainty or attempt to manage the risk it engenders (see below) is the relationship between the immediate consequences of the possible outcomes in a particular context and fitness (box 12.1; chapter 11 of this volume).

From the perspective set out in equation 12.1, then, the determinants of risk can be evaluated either in terms of their impacts on either ecological uncertainty per se or the potential negative fitness consequences of such uncertainty. For instance, similar levels of risk can be ascribed to an activity that is highly likely to produce moderately negative

BOX **12.1** Fitness Consequences and Attitudes to Risk

Every biologically meaningful outcome or event (uncertain or not) will change an individual's state in some way. By *state*, I mean any quantity representing features of an organism that impact its fitness. Typical examples include stomach contents, body size, nutrient reserves, or gonadal mass, or aspects of the local environment unambiguously tied to the animal, like its territory size or how its quality is perceived by other animals (see Houston & McNamara 1999 for further discussion). However, although you can scale changes in any such measure to relate them to incremental impacts on fitness, the shapes of these scaling functions (analogous to preference-scaling or utility functions in economics) determine whether selection should favor managing (mitigating), ignoring, or encouraging uncertainty and risk in any particular context. These alternatives are illustrated in the figure I, with solid lines representing the three possibilities: labeled *averse*, *neutral*, and *promoting*, respectively. Selection should favor adaptations to mitigate risk and uncertainty (i.e., individuals should be risk averse) when the function scaling change in state (ΔState) to fitness is concave. The reason can be illustrated graphically by comparing the fitness value of a certain but moderate ΔState (= **mid**) and the expected (mean) fitness value of an uncertain alternative that yields ΔState = **high** or **low** with equal probabilities (such that E[ΔState] = **mid**). The concave *averse* curve generates fitness values of **am, al,** and **ah** for the **mid, low,** and **high** ΔState values, respectively. The expected fitness value of the uncertain E[ΔState] = **mid**, the point $\mathbf{E_a[var]}$, is the probability-weighted average of **al** and **ah** (the midpoint of the dotted line connecting them because P[**low**] = P[**high**] = 0.5) and will be lower than the fitness value of the certain ΔState = **mid**, which is the point **am.** This will always be true whenever the fitness returns to increasing changes in state are diminishing, and so selection should favor reducing the level of uncertainty and/or the fitness cost of the residual uncertainty as discussed the text. Similar reasoning can be used to show that risk

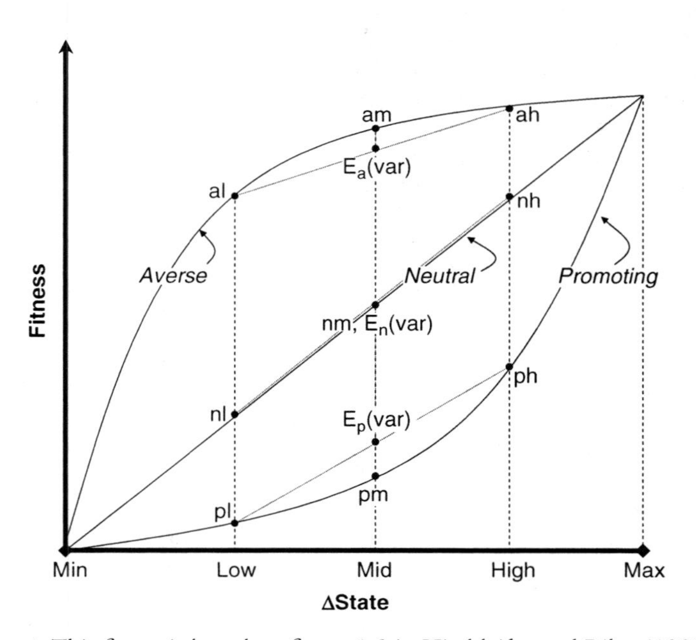

Figure I This figure is based on figure 1.3 in Hirshleifer and Riley (1992).

(*continued*)

BOX 12.1 *(cont.)*

and uncertainty are neither selected for nor against if fitness scaling functions are linear (because **nm** = E_n**[var]** for the *neutral* curve), whereas selection may even favor enhancing uncertainty (and therefore risk) when such curves are convex. The latter is true because E_p**[var]** > **pm** for the *promoting* curve, representing cases when fitness accelerates with increasing changes in state (e.g., subordinate males in mating systems characterized by high levels of reproductive skew may require dramatic improvements in body condition to improve their likelihood of ever reproducing). See chapter 11 for further discussion of how such issues influence foraging behavior (risk-sensitive foraging).

returns, as to one that is only moderately likely to generate extremely costly returns. Furthermore, this definition can encompass the more colloquial use of risk typically adopted in the literature on antipredator behavior (Lima & Dill 1990; chapter 13 of this volume) along with the more narrowly conceived, financial-style risk (probabilistic variation; Stephens & Krebs 1986) referred to in the risk-sensitive foraging literature (e.g., Bednekoff 1996; chapter 11 of this volume). The latter can be related to equation 12.1 by noting that as the probabilistic spread (e.g., standard deviation) in returns associated with an activity increases, all else being equal (keeping mean returns constant), more negative returns become more likely (as do more positive returns). Finally, thinking about risk in this way also suggests that to maximize fitness under ecological risk, organisms can act to either reduce the likelihood of experiencing a negative outcome in a particular situation or mitigate the potentially negative impact of an outcome, or both. Such insight motivates the taxonomy of adaptive risk management strategies I advocate here. Box 12.2 provides a detailed example of how such evolutionary economics of risk can elucidate animal behavior and physiology.

Adaptive Risk Management

Risk can thus be a major selective force on animal behavior (box 12.1 and box 12.2). For the remainder of this chapter I am going to explore how animals can adaptively manage the unexpected dangers they face in their day-to-day lives. As suggested by the definition of risk in equation 12.1, such outcomes select for two types of strategy to mitigate risk: animals can act to (1) minimize the likelihood of experiencing uncertain costly outcomes, or (2) insure themselves against such outcomes, or both. Gathering or providing information can be thought of as ways to minimize the likelihood of events that reduce an animal's fitness, providing its uncertainty about such events can be reduced accordingly. With increased certainty about events, animals have more control over the types of outcome they experience, which enables them to avoid dangers and sometimes even gain opportunities to grow their reproductive assets substantially under risk. Furthermore, even when information use is not possible, animals can sometimes control what they experience (and minimize the likelihood of costly events) by actively engineering key features of their environments via *niche construction* (Odling-Smee et al. 2003). On the other hand, investing in *insurance* to mitigate the potential costs of uncertainty is always a potential response even if animals have limited control over the events they experience. By *insurance*, I mean maintaining backups or buffers against the potentially negative consequences of risk. Nevertheless, although such insurance-style adaptations are always a possibility they often limit an organism's ability to exploit the potential fitness windfalls that are also associated with uncertainty. Table 12.1 summarizes this relationship between types of uncertainty and the adaptive risk management strategies available to animals.

Reducing the Likelihood of Costly Outcomes

Information Use

As one solution to the risks they face, organisms can attempt to reduce the uncertainty associated with key features of their environments by utilizing information. In this way, they can improve their control over their lives to avoid potential negative circumstances and encourage positive outcomes (see above). Adaptive information use can involve acquiring it or providing it to others, or both.

BOX 12.2 The Asset Protection Principle

While considering the fitness-based economics of animal behavior, it can be useful to think of animals working toward accumulating and utilizing assets as they go about their daily (or nightly) business. Such assets can be thought of as organismal features (e.g., fat reserves, muscle mass) contributing to an animal's propensity to increase the persistence of its reproductive lineage in the face of competition (direct or indirect) for limited resources from other such lineages. Hence, the fundamental currency used in the accounting of such assets is reproductive value (Fisher 1930; Grafen 2007; chapter 4 of this volume); for instance, muscle mass is valuable only insofar as it contributes to an individual's ability to outreproduce conspecifics in the same life history class with whom it competes for resources.

From such a perspective, Clark (1994) introduced a framework—the asset protection principle (APP)—for thinking about how animals should respond to risks to their continued survival. In a nutshell, the principle is "the larger one's current reproductive asset, the more important it is to protect it" (Clark 1994), which prescribes that selection should favor taking risks when hard up but playing it safe otherwise, given that to accumulate assets, animals must often take additional risks. For example, to increase energy reserves requires animals to leave refuges, exposing them to predators (Sih 1987). Somewhat counterintuitively, the APP also predicts that some individuals should tolerate higher levels of mortality risk during breeding than outside the breeding season. This is because during periods of reproductive activity, to have reproductive value, assets such as body condition need to be invested in reproduction. Moreover, reproductive opportunities during a breeding season are often distributed unevenly among individuals; for instance, the males of many species need to occupy dominant social positions to have the opportunity to mate, and hence the current reproductive value of the other, more subordinate males can approach zero! So most of the latter should be almost indifferent to risk during the breeding season (and as it approaches) or even seek out risky situations if they also supply opportunities to accumulate assets and make it as a breeder (Clark 1994; Bednekoff 1996). In contrast, outside (or a long time before) the breeding season, the subordinates that have (by definition) fewer assets can actually have decent odds of eventually making it as breeders without having to take too many risks. This is because current breeders might die or "runs of luck" might improve their prospects by the time females come into season. Thus, at such times, reproductive value will be distributed more evenly among individuals within demographic classes characterized by high variance in reproductive success during breeding seasons (such as males in many species). Many males will be selected to take fewer risks when breeding is not such an imperative.

Via an elegant piece of field work in the Guadarrama Mountains in central Spain, J. Martín and coworkers (2003) demonstrated that Iberian rock lizards, *Lacerta monticola*, appear to follow the APP when confronted by simulated predatory attacks during and after their mating season. Such "attacks" involved the experimenters approaching lone adults (males and virgin females) directly at either standardized fast ("high-threat") or slow ("low-threat") walks until subjects fled into cover. The latency of a lizard to emerge again was then presumed to reflect its willingness to risk that the "predator" was still around to gain opportunities to forage or mate. In this way, they documented that males during the mating season were significantly quicker to emerge after "high threat attacks" than females at any time and other males outside the breeding season (figure Ia). Furthermore, in a follow-up experiment on males during the mating season, they documented that individuals that had been presented with a tethered virgin female on their territory before facing a "high-threat attack" as before, emerged more readily than those that had been similarly presented with a tethered female-sized piece of wood (figure Ib). Thus, Iberian rock lizard males seem to take fewer risks with plenty of time to grow their assets before the next

(continued)

BOX 12.2 *(cont.)*

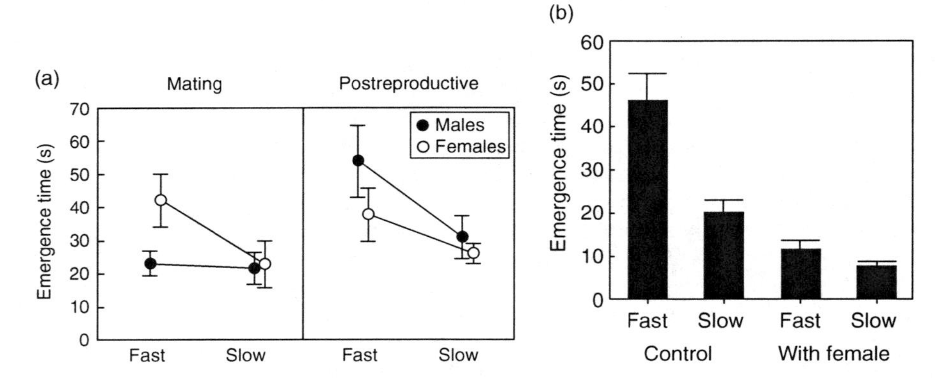

Figure I Emergence times (mean ± 1 SE) of *Lacerta monticola* lizards from refuges after being approached by an investigator. Reprinted from Martín et al. (2003) with permission. (a) Male (filled circles) and female (unfilled circles) lizards after being approached directly at one of two different approach speeds (slow versus fast) in the mating or the postreproductive seasons. (b) Male lizards in presence or absence (control) of a tethered female after being approached directly at one of two different approach speeds (slow versus fast) during the mating season.

mating opportunity, but more risks when opportunities to 'cash in' their assets were more apparent and during the breeding season, as prescribed by the APP (Martin et al. 2003).

Finally, evidence is emerging that such APP-driven attitudes to risk may also be "engineered" into animal stress physiologies. Indeed, in common fruit bats, *Artibeus jamaicansis*, physiological stress sensitivity appears to be modulated strongly by reproductive condition. Females show elevated corticosterone stress responses when they become reproductively active, which primes them to avoid risks while reproducing. Males, on the other hand, show mitigated physiological responses to stress during the breeding season, with individuals that have invested more in current reproduction (with larger testes) showing the lowest sensitivities to stress during such times (Klose et al. 2006). Thus, although the evidence is limited, it appears that, at least for the risk of death by predation, such fitness-based economics can account for variation in behavior (and physiology) under risk. Nevertheless, further evidence is required to shore up the APP, particularly as it is applied to behavior under probabilistic variation in returns while accumulating resources (e.g., risk-sensitive foraging—see chapter 11).

In many instances, animals have the option of acquiring information from their immediate surroundings. By sampling its environment regularly, an animal gains from being able to exploit options when they are productive and avoid them otherwise. In this way, collecting information can also be thought of as maximizing potential opportunities under uncertainty (Stephens 1989; Mangel 1990; Dall & Johnstone 2002). However, doing so may entail costs because valuable resources, including energy, time, and attention, must be redirected to this end at the expense of other biological demands such as immediate growth and reproduction. Formal analysis of the problem of tracking a changing environment has shown that the optimal sampling effort depends on the costs of missing productive opportunities, the costs of sampling unproductive options, and the rate at which options change states (Stephens 2007). Indeed, an optimal sampler must balance losses from sampling variable options too

TABLE 12.1 Uncertainty and the taxonomy of adaptive risk management

Uncertainty Type	Target Risk Component	Management Strategy
	Likelihood	Information use
Reducible		Niche construction
	Costs	Insurance
	Likelihood	Niche construction
Irreducible	Costs	Insurance

often (committing sample errors), and thereby overexploiting them when they are worse than other options, against the potential loss in opportunity to detect when such options offer the best returns (resulting from overrun errors). Where it strikes this balance depends on the ratio of the costs of overrun errors to those of sampling errors (figure 12.1), and how often these errors come into play, which is determined by the stability or reliability of its options (Stephens 2007).

Acquiring information by observing con- or heterospecifics can improve the efficiency of the sampling process by mitigating some of its costs; for instance, by watching the success of others

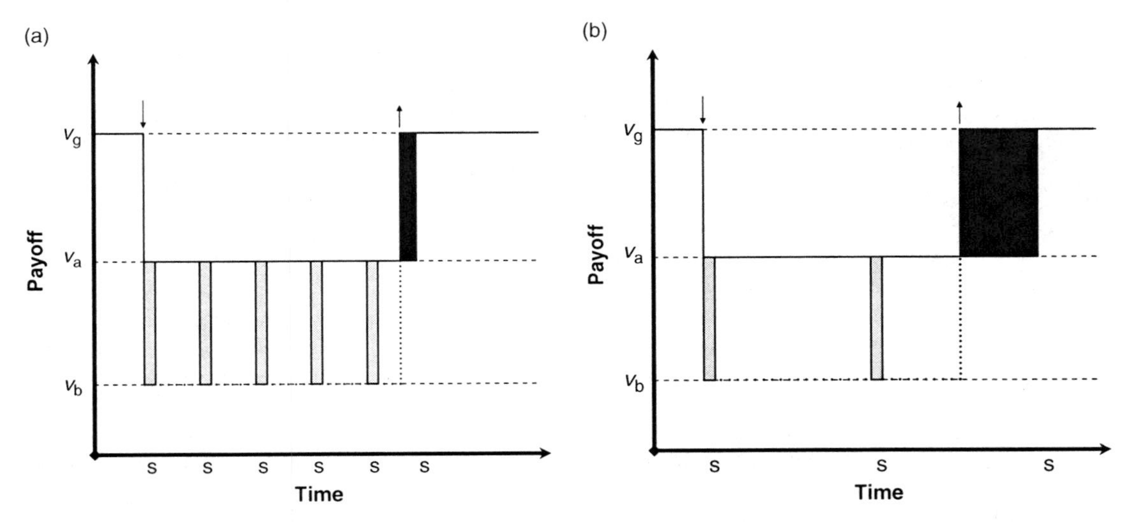

FIGURE 12.1 The hazards of tracking varying resources. The payoffs to a sequence of good (v_g) and bad (v_b) states of an unpredictably varying resource (changes in state indicated by arrows), and the average (background) payoff (v_a) offered by other options available to a forager. Because $v_g > v_a > v_b$, an optimal omniscient forager should exploit the varying resource when it is good, switch to alternative resources when it is bad, and then switch back to the varying resource when it is good again. The economic consequences of sampling in two different ways, with a (a) short or (b) long period (i.e., frequently or infrequently, respectively), are shown for a forager that is uncertain about the current state of the varying resource while foraging on alternative resources. Every sample of the varying resource (switch from the alternative resources) is indicated by "s" on the x-axis. There are two types of error a forager can make: sampling (light shaded areas) and overrun (dark shaded areas) errors. If (a) the sampling period is short relative to the length of time the varying option remains in each state, sampling errors are frequent (total area of light shading > dark area). On the other hand, (b) overrun errors are more common when foragers sample infrequently relative to the rate at which the varying resource changes states (dark area > total light shaded area). Therefore, the optimal sampling regime on a varying resource will be determined by the probability that the resource changes state each time step and the ratio of the cost of a sampling error ($v_a - v_b$) to the cost of an overrun error ($v_g - v_a$) (Stephens 2007). Based on figure 4.1 in Stephens and Krebs (1986).

exploiting variable and uncertain resources, foragers can detect when such resources are productive without paying the costs of sampling or overrun errors (figure 12.1). Such *social information use* ranges from observing the outcomes of other individuals' interactions with the local environment (e.g., their rates of food intake), through copying their decisions (e.g., following fleeing individuals), to paying attention to their deliberate signals (Danchin et al. 2004; Dall et al. 2005; Seppänen & Forsman 2007; chapters 11 and 17 of this volume). Arguably, utilizing the richest social information sources (signals and behavioral outcomes) still imposes substantial demands on an animal's time and attention, and so only barely mitigates the time and energy costs of a given level of information gain. Behavioral copying, on the other hand, can be extremely cheap, although its information content is often limited and it is not without its own potential pitfalls; when erroneous decisions are propagated through social groups individuals can get caught up in costly *information cascades* (Giraldeau et al. 2002; Danchin et al. 2004; Dall et al. 2005). For instance, in foraging flocks it can be important for birds to monitor when flock mates leave in response to perceived danger because the consequences of being left behind to face predatory attacks are particularly heinous. However, individual birds often decide to leave flocks and go to cover without having spotted any danger simply because they have eaten their fill (Lima 1995). So individuals face the dilemma of risking getting caught up in erroneous "false alarm" flights (information cascades) if they copy the leaving decisions of flock mates or becoming the lone target of a predatory attack if they ignore such flights (Giraldeau et al. 2002). Either way, in both social and asocial contexts, similar trade-offs can mean that information should not always be used when it is available (e.g., Dall et al. 1999; Eliassen et al. 2007). Indeed, sampling effort is worthwhile only when there is sufficient opportunity to utilize information after the inevitable reductions in performance associated with its acquisition; for instance, the environment must often be sampled when it is unproductive to obtain up-to-date estimates of its rate of change (figure 12.1). Therefore, as opportunities to put such hard-earned reductions in uncertainty to use become restricted, investing in information acquisition becomes less likely (Dall et al. 1999; Eliassen et al. 2007). For instance, as life expectancy drops due to heightened mortality risk or with age, immediate returns become paramount, and so strategies that ignore information come to predominate (Eliassen et al. 2007). This is because to be a resource worth investing in, information must be used to change behavior substantially and profitably at some point (Stephens 1989; Mangel 1990). As the likelihood of survival declines, the opportunity to both gather information and use it sufficiently to offset the investment costs diminishes (Dall et al. 1999; Eliassen et al. 2007).

Sometimes animals can actively inform others about some aspect of themselves (e.g., what state they are in, what they are going to do) to reduce uncertainty and thereby mitigate risk, particularly in social situations (e.g., via *signaling*; Maynard Smith & Harper 2003; chapter 16 of this volume). Indeed, social situations are often characterized by substantial levels of risk because outcomes will depend on interactions between independently motivated individuals, which often amplify the uncertainty associated with them. Furthermore, the evolutionary imperative to exploit direct competitors can often escalate the costs associated with potential negative social outcomes.

One interesting possibility in this context is that stable or consistent individual behavior (animal "personality"; chapter 30) might be selected for in response to the risks generated by social interactions. The idea is that signaling one's social intentions credibly (e.g., by behaving predictably) can bind competitors into responses that minimize the likelihood, and therefore risk, that the focal individual will be exploited. This can also provide opportunities for it to do the exploiting, which may or may not involve some form of *strategic commitment* (Nesse 2001). A clear example of how this can work comes from a variant of the classic hawk-dove game of contests over resources (chapter 15) in which contestants are allowed to choose between competing aggressively (playing "hawk") or passively (playing "dove"), depending on whether their opponent won or lost its last fight (i.e., using information from "eavesdropping" on an opponent's last fight; Johnstone 2001). Such social awareness selects for stable individual differences in aggression rather than monomorphic populations in which all individuals play hawk and dove randomly at evolutionarily stable probabilities (i.e., a mixed ESS; chapter 8). This is because with eavesdropping in the population, more consistent individual aggressiveness (high or low) is favored because by being more predictable, individuals can avoid getting into extended (costly) fights. Moreover, with increased interindividual variation in aggressiveness, increased levels of eavesdropping will be favored to minimize the likelihood of fighting with the more aggressive individuals (who are more likely to have won their last fight), and so on. This dynamic feedback will eventually result in polymorphic populations that are composed of extreme types at ESS frequencies, in which individuals are always either hawks, doves or eavesdroppers (R. A. Johnstone & S. R. X. Dall unpublished data). Thus, consistency (pure hawkishness, dovishness) can be selected for when being predictable gets competitors (eavesdroppers) to respond in the future so as to improve focal individuals' payoffs by minimizing the risk of costly escalation. Further work is needed to establish whether such outcomes are to be expected whenever social interactions can be observed by others and inform future interactions, for instance, when animals signal in communication networks (McGregor 1993) or if individual "public image" or "reputation" is constantly on the line in games involving a choice between cooperation and defection (Nowak & Sigmund 1998; Leimar & Hammerstein 2001). Nevertheless, it is perhaps telling that individual differences in aggressive tendencies are common within animal social groups across a wide range of taxa (chapter 15) and animal personalities are often more diverse and distinct in species with extensive social lives (Gosling 2001).

Niche Construction

Although the judicious use of information can often enhance an animal's control over risky situations, information is not always available. This is because an animal may not ever be able to eliminate the uncertainty associated with some environmental features no matter how much effort it devotes to sampling (or signaling): irreducible uncertainty predominates. Nevertheless, even when faced with such irreducible uncertainty, animals can still exert control over the likelihood of experiencing costly outcomes via *niche construction* (Odling-Smee et al. 2003). Thus, some organisms may actively construct features of their own ecological niches to minimize the likelihood of unfavorable conditions. Indeed, such adaptive responses to risk can lead organisms to alter their environments profoundly enough to influence the evolutionary dynamics of their own reproductive lineages (Odling-Smee et al. 2003; Laland & Sterelny 2006).

Adaptive alterations by organisms to their physical environments are the most compelling examples of risk management via niche construction (Laland & Sterelny 2006). Indeed, elaborate refuges can be constructed in response to abiotic risk, which can themselves be inherited and promote adaptation to shelter-based lifestyles. Termite nest building is a case in point. Constructing climate-controlled nests is a putative adaptive response to the vagaries of ultra-arid and thermally stressful habitats, and promotes long reproductive life spans via reduced extrinsic mortality rates (Stearns 1992). However, reliance on constructions that equilibrate temperature and humidity and resist the worst that weather or most natural enemies can throw at them limits opportunities for dispersal and the formation of independent colonies. Indeed, such opportunities are the primary route to direct fitness by the reproductive offspring of the long-lived colony reproductives: the king and queen (Roisin 1999). This binds their lineage success to colony success, which, along with vigorous intercolony competition for the inheritable fortresses, can favor individual-level traits that function to enhance group (colony) level performance (Wilson & Dugatkin 1997; Riolo et al. 2001). It is such factors, along with the high levels of inbreeding associated with termite breeding systems, that are central to evolutionary accounts of a striking feature of termite lineages: eusociality (Wilson & Hölldobler 2005).

Furthermore, some animals can actively manipulate biotic components of their niches to minimize the likelihood of adverse conditions. A diverse set of insect species actively cultivate fungus for food, including some Old World termite lineages as well as various ant and beetle species (Mueller & Gerardo 2002). As was the case with the adoption of agriculture in prehistoric human societies, such practices appear to reduce the unpredictability of food supplies, promoting ecological (and evolutionary) success. In ants, this ability has only evolved once, resulting in around 200 species of fungus-growing (attine) ants (Mueller & Gerardo 2002). By and large, such ants grow their fungi in subterranean farms fertilized with decaying vegetable matter, either gathered from the plant litter or, in the case of the famous leaf-cutter ants, cut directly from live plants. Such agricultural practices have allowed fungicultural ants to dominate their ecosystems, and leaf-cutter ants are among the most damaging pests of human agriculture in Central and South America (Mueller & Gerardo 2002). However, attine ants are totally dependent on their fungi because their helpless brood is raised on an exclusively fungal diet. This circumvents problems associated with digesting cellulose directly and minimizes the likelihood of shortfalls in nutrient supplies (compared to provisioning directly from forage), which can have profound impacts on brood success in altricial systems (Dall & Boyd 2002, 2004). Further adaptations to this fungivorous lifestyle include the production of antimicrobial secretions from specialized "organs" to ward off famine-inducing infections by the fungal parasite *Escovopsis* (Currie 2001). The evolutionary impact of this biological niche construction is driven by a fundamental constraint associated with this lifestyle—a mated female that founds a new colony must seed her new garden with a fungal inoculum from her mother's farm (Mueller et al. 2001). Thus, as a means of exerting control over the conditions that individuals experience, niche construction can have profound influences on the evolutionary dynamics of lineages that adopt such strategies, providing there is significant cross-generational *ecological inheritance* of the engineered conditions (Laland & Sterelny 2006).

Mitigating Costly Outcomes: Insuring against Adversity

There are also many circumstances in which ecological conditions are fundamentally stochastic (irreducible uncertainty predominates) and there are few opportunities for animals to engineer out the risk from their environments. Under such conditions, they must act to buffer themselves from the uncertain negative outcomes they will experience. Furthermore, because such insurance-style strategies are always an option whenever animals face risk, to assess the adaptive value of any potential risk management strategy, the fitness-based economics associated with potential insurance options must be accounted for even if there are opportunities for animals to use information or construct their niches. Accordingly, whether or not it is adaptive to adopt the latter to minimize the likelihood of costly outcomes depends on how much they will increase payoffs (e.g., reproductive value) relative to insurance-style strategies that optimize performance under unaltered levels of uncertainty (Hirshleifer & Riley 1992).

Animals can often minimize their risk by insuring themselves against any potential (uncertain) unfavorable conditions they are likely to experience. One solution is to develop and maintain a range of options as backups; the more flexible or generalist an organism is, the less likely it is to be stuck doing the wrong thing when conditions change unpredictably. However, to reap such insurance benefits from flexibility, organisms must develop and maintain the ability to exploit alternatives that are unlikely to covary positively with one another; in other words, alternatives that are unlikely to depend on the same ecological factors (Real 1980; Wilson & Yoshimura 1994). This dependence on a spread of options underlies the inefficiencies associated with developing and maintaining flexibility as a solution to the problems posed by risk (see Dall & Cuthill 1997; DeWitt et al. 1998; Bernays 2001 for further discussion). Indeed, despite retaining a range of prey types within their diets, individual generalist predators are often forced to specialize temporarily on locally abundant types to improve their foraging efficiencies under limits to the rates at which they can process sensory information ("attentional" constraints; Dukas 2004). This can lead to performance deficiencies relative to specialist predators whenever local prey abundances change (Dall & Cuthill 1997). Moreover, such "frequency-dependent predation" or "predator switching" behavior (Allen 1988; Sherratt & Harvey 1993) favors the adoption of rare antipredatory tactics by prey and may be a major selective force maintaining polymorphisms within prey populations (Bond & Kamil 1998; Dukas 2004; but see Merilaita 2006). For similar reasons, it may be that any kind of limit to the flexibility of natural enemies will select for processes that maintain substantial levels of interindividual variability in their targets, including mechanisms that generate immune individuality and egg signing in birds subject to brood parasitism (Dall 2006).

Organisms can also insure themselves against uncertain costly situations by buffering themselves against the worst-case scenarios. For instance, many organisms develop defensive morphologies, which act to mitigate the costs of encountering predators (chapter 13). In doing so, however, such organisms may limit their ability to respond adaptively in other contexts (e.g., DeWitt et al. 1998). Indeed, many species from a range of taxa show so-called inducible defenses, such that upon detecting cues that predict the proximity of predators somewhat (e.g., chemicals secreted by predatory fish), they develop protective morphologies (e.g., spines, armor plates). However, the development of such defenses often profoundly limits the subsequent lifestyles of the individuals adopting them (DeWitt et al. 1998). For instance, *Rana temporaria* tadpoles develop deeper tails in the presence of predatory dragonfly nymphs, which increases escape burst efficacy, while they also reduce their swimming activity and feeding rates to avoid detection by the visually dominated predators (Steiner 2007). These changes, however, result in reduced growth rates, longer development times (especially at high resource levels), and heightened nonpredatory mortality (especially at low resource levels) compared to tadpoles raised in the absence of potential predators (Steiner 2007). Furthermore, theoretical work suggests that individuals following slower growth trajectories should show heightened APP-style (box 12.2) attitudes to other sources of risk in their lives (Stamps 2007; Wolf et al. 2007; Biro & Stamps 2008), which will result in the adoption of substantially different lifestyles than individuals growing faster in predator-free environments (e.g., DeWitt et al. 1999).

Another form of buffer against the vagaries of chance is the development and maintenance of energy reserves (Brodin & Clark 2007). Animals can avoid starving to death if their food supplies are interrupted unpredictably by putting on fat or storing food in caches. However, caches can be pilfered, spoil, or be forgotten (Lucas et al. 2001), and being fatter makes it relatively difficult to move around, increasing the risk of injury or predation (Witter & Cuthill 1993). The trade-off between such costs and the insurance benefits of storing energy have proven useful in understanding body reserve management in general, including the influences of food supply, metabolic costs, predation risk, and social interactions (Cuthill & Houston 1997; Brodin & Clark 2007). For instance, as food encounters become more uncertain, and so energetic shortfalls become more likely, individuals should store more energy reserves (fat) to buffer against such shortfalls and thereby minimize the likelihood of starving to death. Indeed, the tendency to store fat as a buffer should increase with heightened metabolic demands, because individuals are more likely to be caught short for a given level of food predictability, but decrease with increased costs to carrying it around (Cuthill & Houston 1997). For many birds, the demands of flying impose substantial costs to carrying fat around and so they should be leaner when predators threaten because flight performance is often at a premium (Houston et al. 1993). Nevertheless, in many bird species, for which theory predicts that individuals should carry less fat as the need for antipredatory aerial maneuverability increases, individuals often respond to increased predator presence by storing more fat as if to compensate for the interruptions to feeding imposed by predator-induced hiding or reduced activity (e.g., Rands & Cuthill 2001). Intriguingly, a survey of 30 common small bird species in the United Kingdom revealed that the birds that responded to increased predation risk by losing mass were members of species with populations that have been declining over the last 30 years, whereas those in populations growing over the same period increased their masses under heightened predatory threat (MacLeod et al. 2007). This has been interpreted to suggest that when ecological conditions are favorable and food is plentiful (growing populations), individuals may minimize the risk of predation by taking time out from foraging because they will be able to catch up afterward. However, under more limited food availability (declining populations), individuals may be forced to forage in the presence of predators and therefore must reduce the level of reserves they carry to facilitate eluding predatory attacks (i.e., do the best of a bad job). Indeed, if such contrasting body mass regulation strategies can be shown to underlie these survey results, they could reinforce the influence of habitat quality on population trends, with changes in reserve levels decreasing and increasing net mortality rates, respectively (MacLeod et al. 2007).

Finally, there are important social implications that stem from considering fat storage as a buffer against foraging shortfalls. Indeed, such a perspective helps to make sense of otherwise counterintuitive findings from studies that look at individual differences in patterns of fat storage within avian social groups. From the traditional fatter-means-better-nutrition perspective, observations that the body mass and fat reserves of subordinates regularly exceed those dominants in winter flocks (e.g., Ekman & Lilliendahl 1993) make little sense: why should dominant birds that have unrestricted access to food end up lighter than birds that they often exclude from food when it is located? However, if fat levels reflect a trade-off between the risks of starvation and predation as discussed above, such observations make sense. It is precisely because dominants have more predictable access to food that they can afford to be leaner and minimize the risks of carrying weight; subordinates are caught in a bind that requires them to put themselves at heightened risk from predators to ensure an acceptable chance of avoiding starvation (Clark & Mangel 2000).

CONCLUDING REMARKS

My purpose in this chapter is to highlight the factors one needs to consider when analyzing animal behavior under risk from an adaptive perspective. Consequently, for clarity, I have treated the components of risk specified in equation 12.1 separately and outlined strategies for minimizing risk via either component independently. Nevertheless, in reality, animals can simultaneously minimize the components of the risks they face via single or multiple adaptations, and mitigating one component or source of risk can exacerbate risk elsewhere. Indeed, where the impact of risk and uncertainty on both information use and potential insurance-style adaptations have been considered together, (preliminary) analyses have revealed insights that were not apparent when the putative risk management tactics are considered separately (e.g., McNamara 1996; Dall & Johnstone 2002); such interrelationships between risk and information have long been recognized in the economics literature (Hirshleifer & Riley 1992). This suggests that to fully appreciate the influence of risk on animal behavior, we must account for the fitness-based economics of suites of traits (behavioral and otherwise) selected for by the aggregate risk faced by individuals over their entire life histories. Of course, our ability to do so in a sensible way will require that we continue to invoke Einstein's immortal axiom that analytical devices (e.g., models, hypotheses) should be "as simple as possible, but no simpler," while ensuring that they are also "explicit, quantitative and uncompromising" (Williams 1966). Nevertheless, recognizing that the dangers associated with uncertainty are pervasive for many animals and that the adaptations they promote integrate across entire life histories lends weight to calls to "de-atomize" behavioral ecology (Sih et al. 2004a; chapter 30 of this volume); under risk, animal behavior in any context must be considered functionally as a "snapshot" of an entire life history strategy.

SUGGESTIONS FOR FURTHER READING

For comprehensive, life history–based analyses of the impact of mortality risk on adaptation Clark (1994) and Bednekoff (1996) provide excellent conceptual background. Bednekoff (1996) provides a series of clearly reasoned predictions based on the APP that have so far eluded the attention of experimenters. For the more mathematically inclined, Hirshleifer and Riley (1992) provide a comprehensive introduction to the analytical economics of decision making under risk and uncertainty. Mangel and Clark (1988), Houston and McNamara (1999), and Clark and Mangel (2000) also provide excellent accounts of the theoretical approaches to such decision making from an evolutionary perspective; Mangel and Clark (1988) and Clark and Mangel (2000) are probably more accessible to beginners, whereas Houston and McNamara (1999) provide a mine of insights for the more experienced modeler. Cuthill and Houston (1997) and Brodin and Clark (2007) review work on starvation risk management and strategic body mass regulation. Furthermore, the impacts of risk and uncertainty on adaptive behavior have been investigated extensively in a foraging context, and Stephens and Krebs (1986) and Stephens et al. (2007) provide authoritative overviews of this literature. For more detail on niche construction and its evolutionary implications, Odling-Smee et al. (2003) offer a very accessible and comprehensive account.

Bednekoff PA (1996) Risk-sensitive foraging, fitness, and life histories: where does reproduction fit into the big picture? Am Zool 36: 471–483.

Brodin A & Clark CW (2007) Energy storage and expenditure. Pp 221–269 in Stephens DW, Brown JS & Ydenberg RC (eds) Foraging: Behavior and Ecology. Univ Chicago Press, Chicago, IL.

Clark CW (1994) Antipredator behavior and the asset-protection principle. Behav Ecol 5: 159–170.

Clark CW & Mangel M (2000) Dynamic State Variable Models in Ecology: Methods and Applications. Oxford Univ Press, New York.

Cuthill IC & Houston AI (1997) Managing time and energy. Pp 97–120 in Krebs JR & Davies NB (eds) Behavioural Ecology: An Evolutionary Approach. Blackwell Sci, Oxford, UK.

Hirshleifer J & Riley JG (1992) The Analytics of Uncertainty and Information. Cambridge Univ Press, Cambridge, UK.

Houston AI & McNamara JM (1999) Models of Adaptive Behaviour: An Approach Based on State. Cambridge Univ Press, Cambridge, UK.

Mangel M & Clark CW (1988) Dynamic Modeling in Behavioral Ecology. Princeton Univ Press, Princeton, NJ.

Odling-Smee FJ, Laland KN, & Feldman MW (2003) Niche Construction: The Neglected Process in Evolution. Princeton Univ Press, Princeton, NJ.

Stephens DW & Krebs JR (1986) Foraging Theory. Princeton Univ Press, Princeton, NJ.

Stephens DW, Brown JS, & Ydenberg RC (eds) (2007) Foraging: Behavior and Ecology. Univ Chicago Press, Chicago, IL.

13

Predation Risk and Behavioral Life History

PETER NONACS AND DANIEL T. BLUMSTEIN

It is generally good to be afraid because predation risk is omnipresent. If we extend our definition of *predator* to disease organisms, parasites, and herbivores (for plants), then the vast majority of organisms end their lives by falling victim to predation. Indeed, most organisms will be killed before they manage to produce even one surviving descendant offspring. The odds for a female ant leaving her nest to begin a new colony, or for an acorn navigating toward becoming an oak, are probably in excess of 10,000 to 1 against success. Even in humans, it is estimated that in an ancestral type of hunter-gatherer society the average reproductively mature woman gave birth to 8.09 children in her lifetime (Hill & Hurtado 1996). Assuming a stable population (i.e., each woman needs to produce 2 replacement adults for herself and her mate), it is obvious that the majority of children never survived to become either the "average" reproducing woman or man.

All is not lost, however. Predation may be almost inevitable, but predators can be successfully avoided, misdirected, or repelled for some time if prey act in an appropriate manner. Appropriate behavior will vary across species. For some, it will be hiding in safe refuges. For others, it will be increased vigilance and increased proclivity to flee when perceiving danger. In some cases, social behavior will create effective group defense mechanisms. Finally, for many organisms, the proper response may be a morphological one, such as making shells or other hard outer coverings, growing spines or thorns, or producing or sequestering poisonous substances.

Nevertheless, such defense mechanisms are clearly not without costs and demand trade-offs of energy or time allocated to defense or other activities. A hiding or fleeing animal is not a foraging animal. Group living can also create competitors for scarce resources or mates. Energy channeled into armaments or defensive structures is energy unavailable for attracting mates, reproducing, or parental care. Any organism that makes a poor allocation among these functions suffers on the lathe of natural selection. Indeed, given the ubiquity of predation across all life stages, one can argue that natural selection acts more strongly on antipredator behavior than anything else in the organism's repertoire. This is what makes studying predation and responses to predation risk so fascinating. To make an analogy to life history theory, those traits that are expressed earlier in life are likely under stronger selection than traits expressed late in life (Roff 1992). Thus, for all the fundamental selective importance of reproductive behavior (see chapters 20–26), many animals never have an opportunity to choose mates or produce offspring. Yet almost every animal is likely to be under some predation risk.

The effects of predation risk are evident in both flexible and fixed traits of organisms. Flexible traits (plasticity; see chapter 6) are evident in decisions about where and how long to forage, willingness to tolerate the presence of others, and the amount invested in mate choice and parental care (Lima & Dill 1990). Fixed characteristics can be general personality types (e.g., bold or shy; Sih et al. 2004b; chapter 30, this volume), developmental ontogeny of behavior, physiology, and other morphological features. The underlying assumption is that both types are shaped by natural selection. Flexible traits imply that the ability to evaluate risk and modify actions is adaptive (Nonacs & Dill 1993) and plasticity is favored (chapter 6). Fixed traits imply that predation risk is pervasive and constant in expression across multiple generations and plasticity adds little to no benefit.

Through the evolutionary perspective, we can then ask, "When does a given predation risk select for any type of antipredator defense?" The simple answer is that responding to predation risk is selectively advantageous when it produces higher fitness than ignoring predation risk. This seems like an obvious truism, but there may be many instances when ignoring predation risk is the better option. Even if being aware of or defending against predators increases survival, the functional form of that increase could have critical evolutionary implications.

Consider an organism with a fixed pool of energy, resources, or time that it can devote to either reproduction or antipredator function. If it invests a proportion (x) in antipredator defense, it will have $1 - x$ remaining for reproduction. Before we go any further, it is valuable to realize how flexible the variable x can be. The investment can be a fixed cost for morphological features such as spines, horns, shells, or toxins. It can be a behavioral energetic cost such as for fleeing from a real or perceived predator or giving loud alarm calls. Finally, the cost can be lost opportunities such as a reduced foraging rate due to vigilance behavior or foraging in safer but less rewarding food patches to avoid predators. In all cases, antipredator defenses or behavior take away energy that could be conceivably used for increasing reproductive output.

From a behavioral ecology standpoint, a most interesting question is, how can we find what the optimal level of investment, x, should be? To do so, we need to quantify the trade-off between gain in survival and loss in reproduction. Let's begin with a simple assumption: investment in antipredator function has an S-shaped payoff in survival. Initial investments do not increase survival greatly. However, as investment amount increases, survival starts to rapidly rise. Finally, at higher levels of investment, the gain in survival decreases and asymptotes to a maximum. Obviously the relationship between survival and investment need not be S-shaped, but it is a reasonable starting point. Furthermore, to describe such a relationship, we can use the general logistic curve, which has a long history in ecology and behavior (Richards 1959). Thus, we can map survival for our trade-off as follows:

$$\text{P(survival)} = 0.01 + [0.98 / (1 + e^{-b(x-m)})]^n \quad (13.1)$$

where b and m are constants. The former determines how rapidly the curve rises, and the latter determines the level of investment that produces the most rapid gain in survival. Primarily by varying m, we can produce functional relationships in which low, intermediate, or high levels of investment in antipredator defense are required to significantly increase expected survival for each predator encounter (figure 13.1). Expected overall survival will depend on the number of predator encounters (n). Notice that we have added the constants 0.01 and 0.98 to the equation. We do this in order to have survival across all predator encounters range between 0.01 and 0.99. Thus, no matter how little or great the investment in defense or how many predators are encountered, neither death nor survival are ever completely certain. Using equation 13.1, we can estimate fitness as a multiplicative relationship between overall expected survival and the proportion of resources left to allocate to offspring production, such that:

$$\text{Fitness} = (1 - x)\text{P(survival)} \quad (13.2)$$

Using equation 13.2, we can calculate the expected fitness for any level of x across the survival functions in figure 13.1 when the animal expects to encounter 1, 10, or 100 predators (figure 13.2). Figure 13.2 suggests two predictions. First, that fitness is highest when effective antipredator defense has low costs. This, in and of itself, is rather obvious and therefore unsurprising. There are some evolutionary implications. Consider that a behavioral antipredator defense mechanism may have lower maintenance costs (i.e., behavior can be turned on and off as needed, whereas a morphological response is not so flexible). Thus, one would predict that antipredator behaviors would likely be more common in nature than would antipredation physical morphologies. The latter would likely appear only when predation risk is

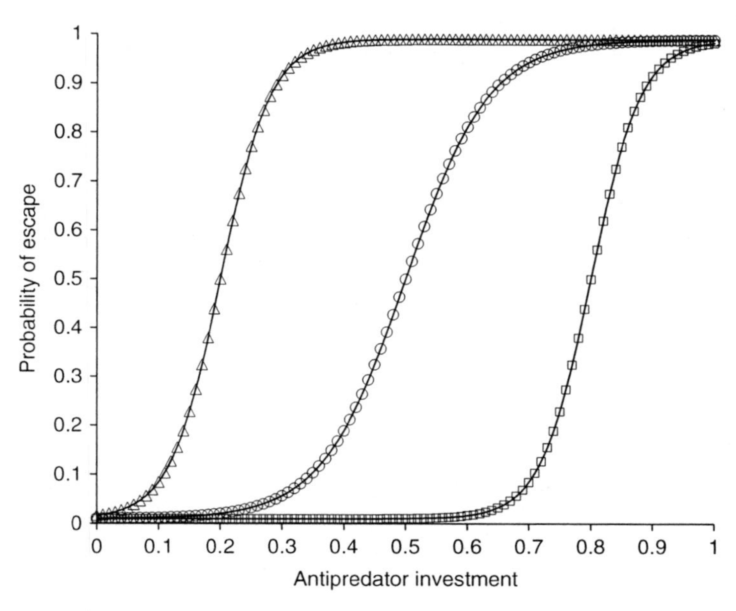

FIGURE 13.1 Probability of escaping a predator's attack per unit of investment into antipredator defenses (x). Strategies reflect where minimal investment (triangular points, $b = -25$; $m = 0.2$ in the logistic equation), moderate investment (circles, $b = -15$; $m = 0.5$), or high investment (squares, $b = -25$; $m = 0.8$) in defense assures a high probability of successful escape. For all lines, $n = 1$.

constantly significant (i.e., when antipredator defense should never be turned "off") and behavioral options produce limited success. Therefore, the world should have far more animal species that are alert, wary, and observant of their environment rather than loaded with sharp spines, poisons, and hard shells.

The second prediction is that as more predators are likely to be encountered, the optimal level of defense generally increases (and not surprisingly, expected fitness declines because more is invested in defense and survival is lower). However, looking at figure 13.2, one can see that this is not always true. For example, when the survival function requires a high level of investment to be effective, high predator encounter rates ($n = 100$) produce a radical shift in strategies: the optimal policy invests nothing in antipredator defense. Because expected survival is so low, the best evolutionary response may be to invest entirely in early reproduction and little, if any, in defense against predation.

We can better visualize the shifts in investment strategy by plotting the predicted optimal level of investment (for the three survival functions from figure 13.1) against the expected number of predator encounters (figure 13.3). This figure illustrates the central prediction from our simple model: that investment in antipredator defense is highest when individuals expect to encounter an intermediate number of predators. When predators are rare, defense investment can be lower because it is less needed. There is a point, however, when predation risk becomes so pervasive that no defense strategy is likely to be effective. Then it becomes better to simply ignore predation risk and put all resources into (rapid) reproduction.

Therefore, figure 13.3 can be considered to represent what we would call the *intermediate predation risk hypothesis* (IPRH): antipredator behaviors and morphologies are most likely to evolve and be maintained under intermediate levels of predation risk. A corollary prediction of the IPRH is that the intermediate zone is generally wider for behavioral traits than morphological traits. This follows from behavioral traits having less of a fecundity cost at low levels of investment because they may still yield high-efficacy payoffs. The results from our simple model are also very similar to a more detailed approach employed by Lima and Bednekoff (1999). They also predict that antipredator investment can be increased to be effective against any single predation attempt. If a large number of such attempts

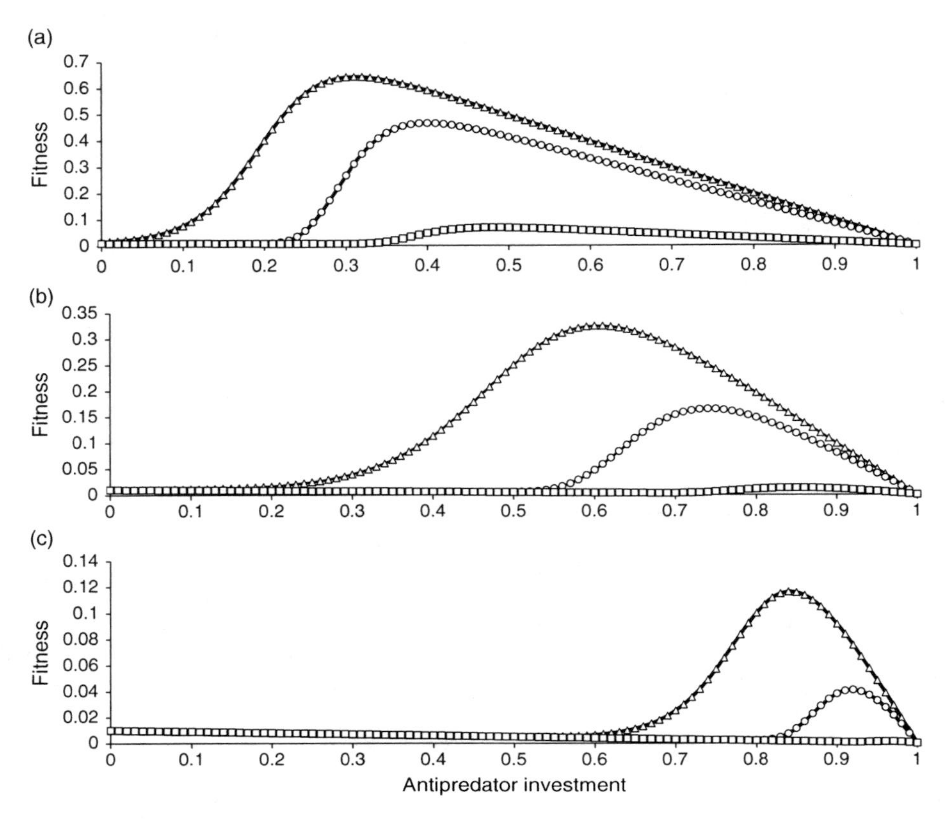

FIGURE 13.2 Fitness payoff (fecundity × survival) for various levels of investment into anti-predator defense. Panels reflect strategies that require (a) low, (b) medium, or (c) high investment to be effective (see figure 13.1). The symbols indicate where individuals expect 1 (triangles), 10 (circles) or 100 (squares) encounters with potential predators (n). The optimal investment level is found at the highest point of each curve.

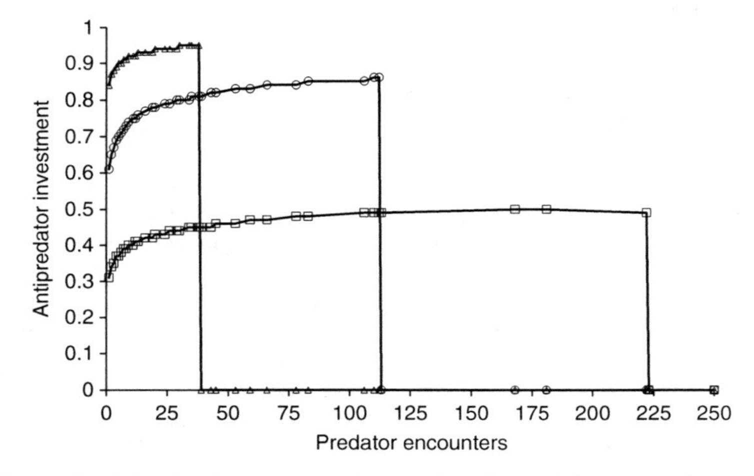

FIGURE 13.3 The optimal level of investment into antipredator defenses as the number of expected encounters with predators increases. Lines with square, circular, or triangular points reflect strategies that require low, medium, or high investment to be effective (see figure 13.1). All strategies predict an initial increase in antipredator defense as more encounters are expected. However, for all levels of defense effectiveness, there is a point at which no investment in defense becomes the optimal strategy.

are expected, however, then long-term survival will not increase enough to offset the cost of the added defense.

The concept of a trade-off between antipredator defense and other biological goals forms the basis for our review of the behavioral ecology of predation risk. First, we consider evolutionary responses to predation risks that can change on a daily or moment-to-moment basis, or remain constant across many generations. Second, we consider how both the immediate behavior and the evolutionary history of species might affect the conservation and preservation of biodiversity in a changing world.

EVOLUTIONARY BEHAVIORAL ECOLOGY OF PREDATION RISK

Vigilance

Prey cannot detect predators without some level of vigilance, but vigilance can take away from other activities. If predation risk is not constant over time, antipredator vigilance should vary dynamically and ultimately can be related to the relative time individuals are in high- and low-risk situations (Lima & Bednekoff 1999). Envision a ground squirrel or a skunk living beneath the path of a raptor migratory route. During the migration season, prey may encounter many thousands more raptors than during the rest of the year. Should animals vary vigilance and foraging during each season? Or consider the variation in risk that darkness and moon cycles create. How should animals allocate time to vigilance and foraging during risky moonlit nights?

Lima and Bednekoff's (1999) *risk allocation hypothesis* predicts that animals will be the most vigilant when high-risk situations are rare. Consider two environments, a high-risk one (H) and a low-risk one (L). The optimal times allocated to feeding in the high- and low-risk environments (f_H) and (f_L) are as follows:

$$f_H^* = \frac{R}{(\alpha_H / \alpha_L)(1-p)+p} \qquad (13.3)$$

$$f_L^* = \frac{R}{(1-p)+(\alpha_L / \alpha_H)p} \qquad (13.4)$$

where R = the average rate of foraging required to meet an energetic demand, α = the attack rates in the different environments, p = the proportion of time spent in the dangerous situation, and thus

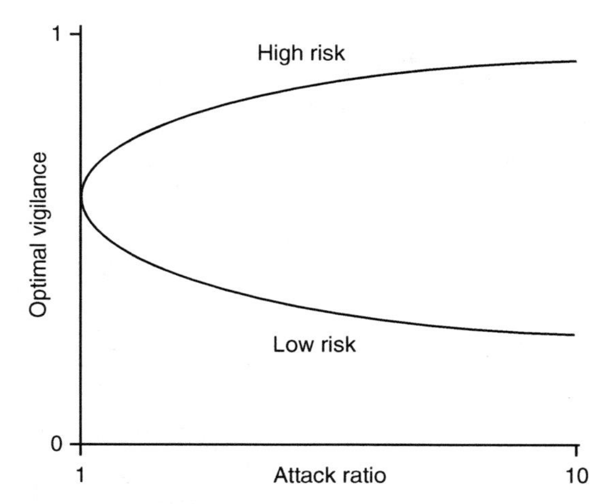

FIGURE **13.4** The optimal vigilance model. As risk in the high-risk situation relative to the low-risk situation (i.e., the attack ratio) increases, animals should become more vigilant in the high-risk situation and less vigilant in the low-risk situation. Redrawn from Lima and Bednekoff (1999).

$(1 - p)$ = the amount of time spent in the relatively safe situation. In these equations, optimal time foraging is expressed as a function of need (R) divided by a ratio of attack rates combined with the time spent in the different patches. By doing so, our units work out to time foraging.

The key to understanding risk allocation is seen by setting $\alpha_H = \alpha_L$, in which case, $f_H = f_L = R$. This shows that the average foraging rate required to meet energetic demand is not influenced by the actual level of risk, but rather the relative difference in risk between the low- and high-risk times (or environments). Thus, the *attack ratio*, (α_H / α_L), determines variation in vigilance and antipredator behavior. As the attack ratio increases, animals should feed more in the low-risk situations and the optimal level of vigilance should decrease in the low-risk situation (figure 13.4).

The situation is more complex when the relative time spent in different risk situations begins to vary. Figure 13.5 illustrates the predicted optimal levels of vigilance when the proportion of time spent in the high-risk situation is 0.3, 0.5, and 0.7. For any given attack ratio, vigilance in the low-risk situation (solid lines) is affected more by the proportion of time in the high-risk situation than is vigilance in the high-risk situation. Thus, when risks are substantial and rare, it pays to invest a lot in vigilance.

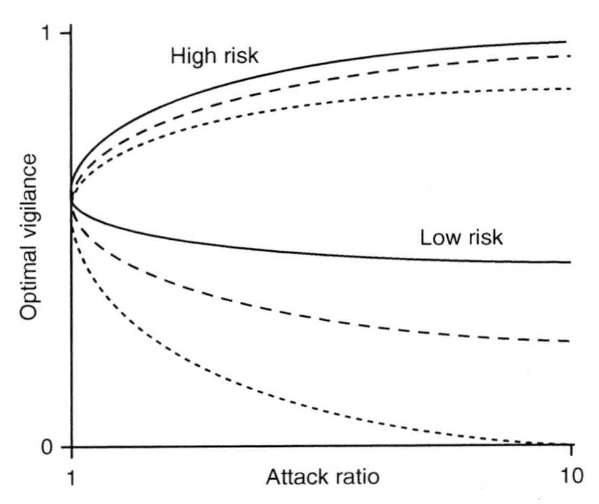

FIGURE 13.5 Optimal vigilance as a function of time spent at high risk (p). Optimal vigilance is plotted for individuals that spend 30% (solid lines), 50% (dashed lines), or 70% (dotted lines) of their time in the high-risk environment. As the proportion of time spent in the high-risk state decreases, the risk allocation hypothesis predicts more vigilance for individuals who live in low-risk situations than for those in high-risk situations. Redrawn from Lima and Bednekoff (1999).

The consequence of this is to bias foraging toward lower-risk locations or at lower-risk times. Therefore, predation risk, particularly temporally variable risk, impacts other activities and otherwise should structure time budgets. Although the risk allocation hypothesis applies generally to any antipredator behavior, behavioral correlations (i.e., syndromes, discussed below) may prevent adaptive allocation of antipredator behavior as predicted by the hypothesis (Slos & Stoks 2006).

Strong empirical support for the risk allocation hypothesis is lacking. Some of the best support comes from Sih and McCarthy (2002), who exposed snails to chemical cues associated with their predatory crayfish in different temporal scenarios. Snails living in high-risk situations were exposed to a pulse of low risk, whereas other snails typically living in low-risk situations were exposed to a pulse of high risk. Snail behavior varied based on the temporal pattern of risk. Snails typically living under high risk were relatively inactive but when risk was suddenly decreased, they foraged a lot more. Snails from low-risk situations were moderately active. Unexpectedly, these snails did not respond as predicted by the risk allocation hypothesis to pulses of high risk. In another experiment, Van Buskirk et al. (2002) kept frog tadpoles in artificial ponds with either many or few caged dragonfly larvae. Although tadpoles responded aversively to the predators, they did not vary their feeding behavior as predicted by the risk allocation hypothesis. Finally, Sundell et al. (2004) did not find strong support for the risk allocation hypothesis when they manipulated predation risk to voles by exposing them to weasels in large outdoor enclosures. Again, variation in feeding behavior was not influenced by pulses of risk, leading these authors to suggest limitations in the ability of the voles to properly assess risk. These studies clearly indicate that more work is required to understand how animals assess risk and then to manipulate risk to adequately test the risk allocation hypothesis. Failure to find support for the risk allocation hypothesis could also stem from limited phenotypic plasticity. Future comparative studies could explore the situations under which species should respond to temporally variable risk.

However, before we reject empirical tests of the hypothesis, a recent study by Creel et al. (2008) provides perhaps the best support for the risk allocation hypothesis under natural conditions. They focused on elk (*Cervus elaphus*) that had to survive snowy winters as well as the risk of predation by wolves (*Canis lupus*). In Yellowstone National Park, elk lived in areas with and without wolves and thus experienced different amounts of background predation risk. In areas with wolves, elk encountered wolves periodically. Therefore, there were pulses of predation risk. They evaluated a set of alternative statistical models that contrasted the risk allocation hypothesis against a null model (vigilance was not influenced by predation risk) and two alternative models (a risky times model and a risky places model). The risk allocation model explained the data the best: elk modified their vigilance based on both background levels of predation risk and the temporal change in predation risk.

Flight

Once detected, prey may or may not elect to flee. The decision to flee upon detecting a predator depends on both the costs and benefits of flight and thus is another antipredator behavioral trade-off (Ydenberg & Dill 1986). Eventually, all individuals will flee an approaching threat (Blumstein 2006a), and this deceptively simple observation has generated a

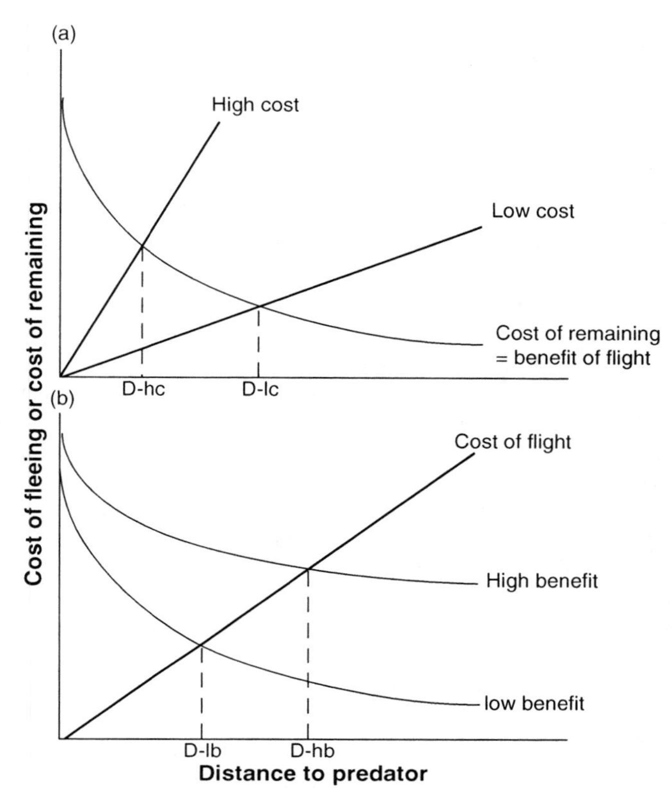

FIGURE **13.6** A summary of Ydenberg and Dill's (1986) model of flight initiation distance (FID). The intersection of cost and benefit of flight curves predicts the optimal FID. In (a), FID is found for fixed benefits of flight and variable costs of remaining. *D-hc* and *D-lc* are the optimal distances to flee when there are high or low costs of remaining, respectively. In this case, animals with high costs (e.g., those less able to move, or who will lose valuable resources by fleeing) will tolerate closer approaches. In (b), FID is found for fixed costs of flight and variable benefits from fleeing that could reflect the different risks from different predators. *D-hb* and *D-lb* are the optimal distances to flee when there are high or low benefits of flight, respectively. Thus, for a fast-moving, high-risk predator, for any distance, there would be a higher benefit of flight compared to a slower-moving, lower risk predator. Therefore, animals should flee at greater distances from higher-risk predators.

rich literature that identifies a variety of factors that influence flight (Stankowich & Blumstein 2005).

Following Ydenberg and Dill (1986), we present a simple economic model of flight (figure 13.6). This model assumes that animals should minimize overall fitness costs. The costs of fleeing or remaining, *c* and *b*, are both affected by distance to predator. The cost of remaining equals the benefit of flight. Therefore, as this cost goes up, so does the benefit of flight. In other words, the benefits of fleeing are increased fitness relative to not fleeing, whereas the costs of fleeing equals energy, opportunity, and any increased conspicuousness to predators.

Realistically, the cost of remaining (i.e., the risk of capture) is highest when the predator is very close and lower the farther away the predator is. Therefore, the cost of flight is 0 if the predator is on top of the prey, and cost increases with increasing distance. The cost of flight might be lost foraging opportunities. Like many such graphical optimality models, the crossover point is where flight distance is optimized (figure 13.6). Thus, prey should let predators get closer as the cost of flight decreases. Similarly, the relative benefit will also influence the optimal distance to flee the predator. For any given cost of flight, flight initiation distance (FID) will vary based on the cost of remaining. If some predators are more effective than others at a given distance, then this should also influence FID.

What might influence the relative cost of remaining versus leaving? Let us first focus on what might cause flight distance to increase with risk. Some species are sensitive to the speed with which a predator approaches. For species that use refugia, distance to protective cover should influence FID. Interestingly, some studies have found an effect of distance to refuge, whereas others have not (Stankowich & Blumstein 2005). The positive results are consistent with the hypothesis that prey reduce risk of predation by modifying their FID. However, not all animals seem to do this, and a more complex consideration of the costs and benefits of flight may be needed. For instance, as animals are farther from refuge, the cost to escaping increases because they have to expend more energy both moving between the cover and the presumably good foraging area.

The Ydenberg and Dill (1986) model also assumes that prey are cost minimizers. Thus, they move when the benefits of fleeing exceed the costs. However, animals that maximize the difference between the benefits and costs of fleeing will do better. The point is that both costs and benefits increase as the distance to burrow increases. If costs differ across species, this might account for some of the variation in the response of distance to safety and FID. A positive relationship between FID and distance to safety would be consistent with benefits of flight exceeding the costs.

What about costs of leaving? For an animal to leave a good patch could be costly. If you lived in a desert, it may take a lot to get you to leave an oasis! It turns out that birds on productive wetlands in generally desert-like Southern California typically are very tolerant of human approaches (Blumstein unpublished data). An individual's relative competitive ability can also influence the cost of leaving. Subordinate animals might not be able to forage at equally high rates before a disturbance as dominants and may be more likely to tolerate closer approaches. All animals are neither created equal nor remain equal throughout their lives.

Blumstein (2003) modified Ydenberg and Dill's model in a manner consistent with the intermediate predation risk hypothesis. Specifically, if the risk is too great, all animals should immediately flee. And if the risk is too low, animals should never flee. It is only in a zone of intermediate risk that the dynamics of flight become relevant and animals make trade-offs. Cooper and Frederick (2007b) further argued that the Ydenberg and Dill model was not an optimal solution because prey were assumed to break even (i.e., costs would equal benefits), rather than maximize their fitness (i.e., the location where benefits most greatly exceeded costs). They developed two optimality models for maximizing fitness. One had prey losing all residual reproductive value upon death, and the other retained it after death. Overall, they concluded that optimality models are better than break-even models in explaining the existing data on flight.

Although quite variable, flight initiation distance can be viewed as a species-specific trait (Blumstein et al. 2003). For instance, in a comparative study of 150 species of birds, body size explains a substantial portion of the variation in FID. Other variation was explained by diet (species eating living prey flee at greater distances) and sociality (cooperative breeders flee at greater distances). These results suggest some carry-over effects from foraging and sociality on FID, such that species eating living prey have better motion detection and social species are vigilant for reasons other than predator detection (Blumstein 2006a). Additional studies have explored whether birds with larger eyes are more likely to detect a threat, but in a phylogenetically corrected analysis, no significant relationship was found (Blumstein et al. 2004a). Some variation in FID was also explained by the age at first breeding (species that need to live longer to first reproduction flee at greater distances). Thus, FID is a trait influenced by many aspects of life history.

A number of recent studies also focus on dynamic "hiding games" between prey and predators (Cooper & Frederick 2007a). For species that use refuges (e.g., burrows) to reduce predation risk, the benefits of hiding eventually are outweighed by the risk of starvation and prey must eventually emerge. Like FID, the dynamics of hiding should be sensitive to the risk of predation and the benefits and costs of hiding. Anything reducing cost for remaining in a refuge (e.g., lowering metabolic rate while in a burrow) would presumably enable prey to reduce the risks of both starvation and predation.

Alarm Calling

Alarm calls are signals emitted when prey detect a predator and are remarkably plastic. Not all individuals produce alarm calls and calling appears sensitive to the trade-off between increased detectability to the predator and escape benefits (Blumstein 2007). What is particularly attractive to studying alarm calls is that we can study their trade-offs at either an ultimate or proximate level.

At the ultimate level, how does giving an alarm call increase fitness? Fitness can be increased through two nonmutually exclusive pathways. First, calling may increase the caller's likelihood of survival. For example, calls can signal predator detection and therefore discourage active pursuit. Or calling can create pandemonium among other potential prey that then allows the caller to escape. The second fitness-enhancing pathway is through warning kin, either one's own offspring or other nondescendent kin. Because alarming calling is often assumed to be altruistic (i.e., callers increase exposure to predators while warning others), many studies have focused on the conspecific warning function of alarm calls. Although the importance of signaling to predators is often downplayed, in rodents alarm calling seems to have initially evolved as detection signaling toward predators (Shelley & Blumstein 2005). Thus, in rodents, the conspecific warning functions may be exaptations (chapter 2).

Independent of its ultimate function, the proposition that alarm calls increase predation risk is often assumed but rarely tested. This involves a proximate question of call detectability. Because researchers often use alarm calls to locate callers, they assume that if it helps them, it should also help the predators (Blumstein 2007). Tests of raptors' ability to localize alarm calls, however, demonstrate that some calls may be difficult to localize (e.g., Klump et al. 1986; Wood et al. 2000). In contrast, hungry snakes are attracted to the foot thumps of banner-tailed kangaroo rats (Randall & Matocq 1997). Other evidence for a cost of calling comes from Sherman's (1985) study of Columbian ground squirrels (*Spermophilus columbianus*), which scurry for cover before or while calling when pursued by a rapidly moving aerial predator, but call in place when responding to a terrestrial predator.

Interestingly, the structure of calls shows that some are cryptic, whereas others are easy to localize. For instance, mobbing calls (not, strictly speaking, an alarm call) of many birds are rapidly paced with a wide frequency band. Their bandwidth makes them easy to localize. High frequencies predictably attenuate to allow distance estimation, and broadband sounds have more "sound" present to stand out against background noise—increasing their detectability. Thus, these calls function, in part, to recruit others to mob the predator. By contrast, many birds emit "seet" calls when they encounter an aerial predator (Marler 1955). These high-frequency, narrow bandwidth calls are difficult to localize because of their frequency and because they also fade in and out. Therefore, by examining the acoustic structure of alarm signals, we can infer something about the risk associated with producing them.

A second line of proximate investigation involves the energetic and opportunity costs of calling. At one level, calls are usually brief and there is no convincing evidence that they have an energetic cost (Blumstein 2007). Opportunity costs, however, may be quite real and must be evaluated from the signaler and recipient's perspective. If the signaler has already increased its vigilance, calling has a limited opportunity cost. From the recipient's perspective, this is an information problem. We can examine two aspects of information: (1) is a predator truly present, and (2) does the call identify a specific type of predation risk?

Suppose there is a situation in which Nervous Nelly calls at the drop of a leaf and Cool Hand Lucy calls only when certain of a predator (Blumstein 2008). We should expect receivers to use this difference in caller reliability (Blumstein et al. 2004b). In some species of primates and sciurid rodents, calls from unreliable individuals are discounted and individuals reduce their vigilance after hearing repeated calls from unreliable individuals (following the fable of "the boy who cried wolf"). In contrast, yellow-bellied marmots (*Marmota flaviventris*) respond to unreliable calls by increasing vigilance (Blumstein et al. 2004b). It might be that unreliable marmots provide limited, but possibly true, information about predation. Thus, those hearing them might need to independently validate the information.

More controversial in terms of communication is whether alarm callers specifically identify the predator. Calls that function like basic words and communicate information about external objects or events are *functionally referential* alarm calls (Evans 1997). Marler et al. (1992) defined functional reference so as to avoid implications about higher-level cognitive abilities. Evidence of functional reference requires that the calls be produced only to a specific set of stimuli and that playbacks in the absence of a predator elicit the appropriate response. This may not immediately seem that interesting, but referential ability is a key feature of human language (Hockett 1960) and once thought to be a uniquely human attribute. Subsequent research has shown some degree of referential abilities in a number of primate, rodent, and bird species.

Macedonia and Evans (1993) provided perhaps the best explanation for the evolution of functionally referential alarm calls. Species that have unique and mutually incompatible escape strategies seem especially likely to have referential alarm calls. For instance, vervet monkeys (*Cercopithicus aethiops*) must deal with snakes, which elicit investigation and avoidance; raptors, which elicit taking cover in the safety of a dense tree crown; and leopards, which elicit taking cover on peripheral tree branches (Cheney & Seyfarth 1990). Such mutually incompatible escape strategies create an opportunity to communicate specifically about them. Thus, vervet monkeys have functionally referential calls. Playback of snake-elicited calls in the absence of snakes causes monkeys to stand on their toes and look around. Playback of raptor-elicited calls causes monkeys to run to trees and hide near the trunk. And playback of leopard-elicited calls produces hiding on peripheral branches of trees. Once such referential abilities evolve, more complex cognitive processes may follow. Therefore, selection to avoid predation may be an integral pathway to complex cognitive abilities. How such abilities might be costly is unknown, although some interesting possibilities are being tested (chapter 10).

Multiple Goals and Common Currencies

The effect of predators on prey populations has been part of theoretical ecology for a long time (e.g., Volterra 1926; Lotka 1932). The incorporation of behavior into predictive models, however, took several decades longer until the rise of optimal foraging theory (OFT; Schoener 1971; Charnov 1976; Pyke et al. 1977). This body of work was the first attempt to predict behavior in an economic context based generally on the maximization of the net rate of energy intake. OFT was relatively quickly expanded by the seminal work of Sih (1980, 1982) showing that animals could balance between two simultaneous goals: the collection of food and the avoidance of predators. A multitude of studies followed that demonstrated trade-offs between avoiding predators and gathering food, habitat selection, prey choice, sociality, and group functioning (see reviews by Lima & Dill 1990; Brown & Kotler 2004).

Although it is demonstrably obvious that animals often trade off food, mating opportunities, and group status to reduce predation risk, it is more difficult to show that such trade-offs maximize fitness. This is known as the common currency problem (McNamara & Houston 1986; Gilliam & Fraser 1987; Nonacs & Dill 1990). The currency for predation is being alive or dead. A currency for foraging is energy collected over time. So how much energy should an animal give up to reduce predation risk by a given amount? One answer is to empirically ask animals themselves how much energy they are willing to sacrifice for increased safety. These are known as *behavioral titrations*, in which two factors such as food availability and predation risk are allowed to covary (Abrahams & Dill 1989). When animals are trading off across the factors, it allows experimenters to measure the value of one factor in the currency of the other factor.

Abrahams and Dill (1989) did a behavioral titration for the value of predation risk in terms of food for guppies (*Poecilia reticulata*). The fish were placed in an aquarium with two separated feeders. Initially, without a predator present, the fish arranged across the two feeders as predicted by the ideal free distribution, such that all fish fed at about equal rates (figure 13.7). A predation risk was associated with one feeder and this caused a shift in preference, measured as shift in fish distribution, toward the safer feeder. This shift created a difference in feeding rates across the two feeders that could be used to estimate the energetic value guppies assigned to predation risk. This also allowed Abrahams and Dill to estimate in a second experiment how much fish would have to receive in food to return to the risky side in numbers equivalent to the initial ideal free distribution (figure 13.7). Interestingly, only females equated food with risks. For males, even when one feeder was 17 times more rewarding, they would not go to the risky feeder. This sex difference in response to predation risk is probably due to reproductive success in female guppies being closely tied to their feeding rate. Access to females, rather than feeding rate, matters more to males.

Researchers can also use patch *giving-up density* (GUD; the amount of food left behind in a patch at the point the forager leaves the patch) to estimate the food costs of predation risk (Brown & Kotler 2004). Thus, in the presence of owls, two species of gerbils, *Gerbillus a. allenbyi* and *G. pyramidum*, have far higher GUDs in seed trays that are placed in open rather than protected microhabitats (Kotler & Blaustein 1995). Furthermore, species-level differences are evident as *G. a. allenbyi* requires twice

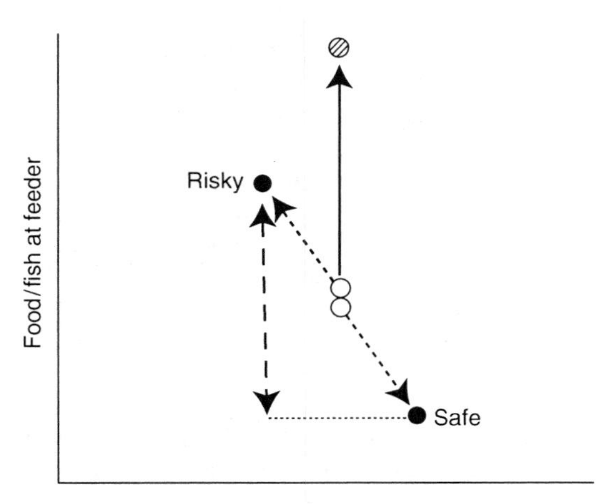

FIGURE 13.7 Measuring the energy equivalent of risk. In the experiments of Abrahams and Dill (1989), guppies were allowed to distribute themselves across two feeders. The fish arranged into an ideal free distribution such that all the fish at both feeders fed at similar rates (the two open circles). In one experiment, a predator was added to the one patch (which became the risky patch). Thereupon more guppies started using the safe feeder, and thus intake rates were no longer equivalent across the two feeders (dotted arrows pointing to black points). The difference between intake rates (the dashed arrow) is a measure of the energetic equivalent of predation risk for guppies. This measure was then used in a second experiment as the amount of food/fish (solid black arrow and hatched point) that had to be added to the riskier patch in order to have the same number of fish at the risky patch as were present in the original ideal free distribution.

as much food in risky patches as *G. pyramidum* to harvest the same number of seeds (i.e., it abandons patches with much more food left in them). Another solution with modular organisms, such as ant colonies, is to measure whether behavioral choices maximize growth and reproduction. Therefore, Nonacs and Dill (1990, 1991) presented *Lasius pallitarsis* colonies with a dichotomous choice in foraging patches that varied inversely in food quality and mortality risk. The colonies preferentially foraged in the patches that maximized colony growth rate as measured by the net of the increase in biomass due to food collected minus the loss in biomass due to predation. Whether or not foraging ants brave mortality risk is therefore a function of the expected gain in colony biomass across patches.

In conjunction with strictly empirical measures, theoretical models have been developed to predict multifaceted behavior. Particularly relevant are those that predict dynamic behavior. These models take into account that organisms have changing states, and optimal behavior should be influenced by those changes (e.g., hungry animals should take more risks for food than satiated animals). Numerical solutions, using the techniques of stochastic dynamic programming (McNamara & Houston 1986; Mangel & Clark 1988; Clark & Mangel 2000; chapter 8, this volume), have come to dominate the trade-off literature. This is true for predicting specific behaviors such as female wasp parasitoids exploiting patches of host eggs (Wajnberg et al. 2006) and broad patterns of behavior across numerous species. For example, optimal foraging models that do not simultaneously consider predation risk will consistently underestimate the time animals stay in patches (Nonacs 2001a).

Dynamic trade-off models have also been particularly useful in understanding mass regulation by animals. For example, for many species of birds, individuals seem to be on a permanent diet: they are skinnier than local food availability would predict. This becomes obvious when the environment becomes worse or more variable. Under such worsening conditions, birds get fatter (chapter 12). Mass regulation appears to be influenced by a trade-off between the risk of predation (a heavier bird is slower) and the risk of starvation. In a good, predictable environment, a bird can afford to be skinny because a meal is always readily available. Thus, starvation risk is low and birds can regulate their mass to maximize maneuverability. In poorer or unpredictable environments, birds need to carry a fat buffer because the next meal is uncertain (chapter 12). Therefore, birds will pay to reduce starvation risk by increasing predation risk (Houston et al. 1997).

Nonacs (2001a) showed, however, that predation risk could produce the same phenomenon in another way. Rather than assuming that weight correlates with predation vulnerability, Nonacs asked whether spatial location mattered. For example, if a predator employs a sit-and-wait ambush strategy, the time prey spent in a patch without being attacked increases the likelihood that this patch is predator-free. Therefore, animals may remain far longer in patches than predicted by net-energy maximizing criteria alone. A byproduct of staying where it is safe rather than always moving to find more rewarding patches is that animals stay skinny. Independent of whether location or weight itself affects predation risk, the tendency to stay skinny in the presence of plentiful food may be a quite common life history strategy in nature (although our own species appears to be a very notable exception to this rule!).

Behavioral Syndromes

Animals may behave in predictable ways that indicate behaviors are not always independent of one another (Sih et al. 2004b; chapter 30, this volume). Behavioral syndromes are seen when there is a correlation across situations or contexts. For instance, if individuals that are bold around predators are also bold when courting, a boldness syndrome has been identified. Such syndromes may inhibit the ability of traits to track environmental variation if they create a trade-off. Assume, for example, that females prefer to mate with bold males, but bold males are more likely to be killed by predators. Thus, the expression of boldness in a population will be highly dependent upon the predation risk. If this syndrome did not exist, then the traits might vary independently according to variation in the costs and benefits of their expression. Antipredator behavioral syndromes have been identified in a number of species, and in the future we will have a better understanding of their importance. Describing and understanding them can illustrate how trade-offs may explain seemingly maladaptive behavior (Sih et al. 2004b).

Social Behavior and Group Living

Groups can form for selfish reasons, in which individuals cluster in an attempt to reduce their own exposure to predators (Hamilton 1971a). Group living, however, also has tremendous potential to create effective antipredatory mechanisms (chapter 17). Regardless of whether animals aggregate to increase foraging efficiency or to reduce the probability of predation, once aggregated, novel antipredator defenses can emerge. Consider the group defenses of muskox (*Ovibos moschatus*) that form a defensive line against wolves or the group defenses of schooling fish, such as anchovies (family Engraulidae), that form a constantly moving three-dimensional mass.

Group living also allows the creation of a group phenotype that is unattainable by solitary individuals. The mutualistic benefit gained through the interactions of genetically nonidentical individuals is called social heterosis (Nonacs & Kapheim 2007, 2008). If genetic diversity within groups increases the variety of vigilance tactics or other antipredator techniques, this can reduce all group members' per capita predation risk relative to living solitarily or living in a genetically homogenous group. A possible example of the advantages of behavioral diversity could be a situation in which individuals adopt defined roles such as "sentry" (Wang et al. 2009). Indeed, a recent example with babblers (*Turdoides bicolor*) showed that sentry behavior was especially effective at reducing predation-related costs and increasing foraging intake (Hollén et al. 2008).

Sex and Alternative Reproductive Tactics

Lima and Dill (1990) argued that the degree to which predation risk might shape variation in reproductive tactics across species' life histories was greatly underappreciated. We believe this still to be the case and that predation risk may play a significant role in exaggerating behavioral and morphological differences across the sexes, alternative strategies within a sex, and species-level differences. First, consider a spider species that catches prey in a fixed web. For mating, males have to search for females. Relative to the number of expected predator encounters (x-axis in figure 13.3), males would inescapably experience considerably more risk. Hence, males and females could differ dramatically when females exhibit high investment into antipredator defense and males invest next to nothing. These differing optima could result in exaggerating sexual dimorphism such that males, when compared to females, are (1) physically unimposing and (2) behaviorally insensitive to risk to the point that they readily approach highly cannibalistic females (Andrade 2003).

Second, in many species, males (and less commonly females) exhibit alternative reproductive tactics that can differ both behaviorally and morphologically (Shuster & Wade 1991; Gross 1996; Sinervo 2001; Oliveira et al. 2008; chapter 25, this volume). A key element across these tactics is that they impose significantly different survival costs. For example, a male salmon that follows the small jack strategy will forgo an extra year of foraging in the ocean. Such males expect to survive at a higher average rate to breeding maturity, if for no other reason than that they encounter fewer predators (Gross 1996). Again, such differences in cumulative predation risk would suggest that small jack and large, nonjack males would reside at different points on the curves in figure 13.3. A (currently untested) prediction would then follow that jacks should be more likely to show antipredatory behavior and awareness of predation risk than their larger male brethren.

Species Differences

Some animals appear bolder in the face of predation risk than others. For example, you might notice that you (or your dog) can get closer to the average chipmunk than to either a jay or a deer. This certainly cannot be because the chipmunk is more able defend itself or escape your dog's jaws. However, the intermediate predation risk hypothesis may yield part of the answer from where one would plot these species on figures 13.1–13.3. Jays have a much more effective escape mechanism than chipmunks (flying versus running). Hence, in figures 13.1 and 13.2 jays may lie on the triangles, whereas chipmunks lie on the squares. Thus even though both may have many potential predators, over a large range of predation risk one would expect the jay to invest more into vigilance and predator avoidance (i.e., compare the squares and triangles in figure 13.3). In contrast, because a deer is a large animal, its size would reduce its number of potential predators relative to the chipmunk (hence the n for deer in equation 1 would be smaller than for chipmunks). Therefore, even if both hypothetically lie on the same line for "running" animals in figure 13.3, one might find the deer on the hump of the curve and the chipmunk at the bottom.

It turns out that size is a reasonably good predictor of an animal's potential life span (Roff 1992). To the degree to which predation causes this relationship, the IPRH predicts that animals of an intermediate size would invest the most in antipredatory behavior. Very small or large animals may experience so many or few predators, respectively, that high levels of vigilance would either be of little effect or not worth the cost. Blumstein (2006a) demonstrated this empirically with species-specific flight behavior in birds. Large birds generally flee a human at a greater distance. However, the largest birds (emus, *Dromaius novaehollandiae*) actually approached humans. Unarmed humans are of little risk to an emu. The smallest birds (e.g., hummingbirds) also tolerated close approaches. The relationship between body size and vulnerability may explain a substantial amount of variation in risk. Alternatively, body size may be associated with the ability of animals to detect an approaching threat (e.g., Blumstein et al. 2004c).

In all of the above examples, predation risk could create evolutionary feedbacks that increase the effects of inter- and intrasexual selection and species competition. Thus, it would be interesting to examine situations in which species have experienced prolonged evolutionary periods with reduced predation risk, such as isolated island faunas. Here all individuals may find themselves at or near zero in figure 13.3, with low investment in antipredator defenses having the highest fitness. Therefore, it would be predicted that (1) sexual dimorphism would be reduced (e.g., Blondel et al. 2002), (2) alternative reproductive strategies within a sex would be rarer, and (3) species-level characters, such as size, would not predict the remaining levels of antipredator awareness or flightiness.

PREDATION RISK AND CONSERVATION BIOLOGY

Predicting Successful Invasive Species

In 1890, Eugene Schiffelin felt that Americans suffered in not being able to experience firsthand all the birds mentioned in Shakespeare's writings. Therefore, he released bevies of larks and other nonnative birds into the New York area. Thanks to Mr. Schiffelin's efforts, we now have over 200 million starlings in North America, but despite his efforts, no nightingales. This disparity of success highlights a major question in conservation biology. Why do some introduced species survive and become pests, whereas others do not (see chapter 29)?

Following from the intermediate predation risk hypothesis, successful invasive species may be those that have evolved life histories that invest relatively less in antipredator defenses because they are on the right-hand side of the curves in figure 13.3. Effective invasiveness could occur for two reasons. First, such species are released from their natural predators. This is an obvious advantage for any introduced species. Second, if the introduced species is evolutionarily in an intermediate zone of predation risk, it still may considerably invest in antipredatory defense against absent predators. This cost would not be borne by species in which low levels of antipredatory defense have been evolutionarily favored. This saved investment could tip the competitive balance against natives. Simply not investing in antipredatory defenses, however, cannot by itself determine a successful invasive (otherwise continents would overrun by pests from distant oceanic islands rather than vice versa!). Instead, successful invasives may come from the right-hand side of figure 13.3 curves. That is, these species have adapted to strong predation pressures by evolving life histories that may maximize reproduction at the expense of long-term survival. Thus, even if such species are susceptible to novel predators in an introduced range, it may not offset their intrinsic fecundity. Therefore, in predicting which species are likely to become future economically damaging invasives, it may be helpful to gauge their boldness and their response to predation risk.

The Loss of Antipredatory Behavior

If you are lucky enough to spend time on small oceanic islands, you may find that ground-dwelling species are very tolerant of your presence. The intermediate predation risk hypothesis may provide one clue in that there is little predation here and therefore there will be little fear. But another clue may come from the observation that although most species have more than one predator, some species live in virtually predator free environments.

In general, if we assume that predators have selected for a variety of antipredator behaviors, optimality theory would predict antipredator behavior to disappear when the predators are no longer present. The loss of predators occurs naturally via colonization and extinction, and unnaturally when prey species are moved to predator-free locations. Guppies for which predation pressure was eliminated illustrate this nicely; they rapidly become sexier (or at least more colorful), and we infer that this is because the expression of this sexually selected trait is no longer traded off against predation risk (Endler 1995). We note that this seems at odds with our above suggestion that loss of predators should disfavor sexual dimorphism. The discrepancy may be that the observed changes in guppies are an evolutionarily short-term response over relatively few generations. If predators and parasites were absent for many generations, the good genes function within mate choice for male coloration would become less relevant. Whether a larger, drabber, longer-lived, more female-like male would eventually evolve in this context is an interesting, open question.

When antipredator behavior is lost, species may become particularly vulnerable to new predators and such species may be particularly vulnerable to exploitation and accidental extinction. However, in some cases, we see remarkable persistence of antipredator behavior. Antisnake adaptations persist in ground squirrels, wallabies retain group size effects and predator discrimination abilities, and pronghorn antelope (*Antilocapra americana*) retain their remarkable athletic abilities long after the extinction of important predators (Blumstein 2006b). Thus, understanding the specific conditions under which antipredator behavior is lost has important conservation implications.

It is important to realize that prey species seldom have only one species of predator (Lima 1992; Sih et al. 1998). The multipredator hypothesis capitalizes on this truism and therefore expects the evolution of linked, pleiotropic, or potentially fixed antipredator traits when the costs of expressing them in the absence of a predator are not extreme (Blumstein 2006b). Specifically, the multipredator hypothesis predicts that for species with multiple predators, the loss of a single predator may have no effect on its antipredator behavior for that predator. Why? Imagine a young ungulate that relies on both camouflage and immobility to hide from predators. Individuals not possessing both of these traits would be at a selective disadvantage. Now let us elaborate on this theme. Imagine a population of prey that had to avoid both foxes and eagles. Some could be super fox avoiders, whereas others could be super eagle avoiders. However, in an environment with both types of predators, we would expect that only those individuals that were good at avoiding both predators would persist. Thus, we expect the presence of multiple predators to select for suites (or syndromes) of antipredator behavior. Some of these suites are likely to result from linkage or pleiotropy. If so, we would expect a limited evolutionary response if suddenly one predator went extinct. In support of the multipredator hypothesis, tammar wallabies (*Macropus eugenii*) were found to retain group size effects and some degree of predator discrimination in populations in which there were some predators but not in a population in which there were no predators (Blumstein & Daniel 2002; Blumstein et al. 2004a).

The multipredator hypothesis has two important implications for future research. First, all predators that a species encounters may have effects. An individual can have the best antipredator response to a terrestrial predator, but if it never looks up in the sky it will be particularly vulnerable to aerial predators. We might expect selection to create suites of antipredator behavior because being the best responder to coyotes (*Canis latrans*) doesn't count for much in an area with eagles. Second, are antipredator traits retained following the relaxation of selection pressures located on the same chromosomes? Have these traits and genes been fixed? Our growing ability to identify and map significant genes will ultimately answer this question and provide valuable integration between phenotypic responses and genotypic architecture.

FUTURE DIRECTIONS

We believe that major advances in our understanding of how prey respond to predation risk will occur in three areas. First, both the multipredator hypothesis and the intermediate predation risk hypothesis have yet to be tested in a variety of systems. If these predictive models work, then we have made a significant advance in understanding the evolution and maintenance of antipredator behavior. Species like sticklebacks (family Gasterosteidae), in which predator loss is replicated many times, and for which there is great genomic knowledge, make an ideal system to test the multipredator hypothesis. Second, as suggested by Lima and Dill (1990), predation risk interacts with other processes that affect life history evolution. The exact nature of these interactions remains a fertile ground for new research. Third, predation risk has largely been ignored in both theory and practice of conservation biology. Many conservation actions fail, and predation is often implicated in their failure. Thus, the field of conservation behavior (Blumstein & Fernández-Juricic 2004) will profitably benefit from the cross-pollination of antipredator behavioral theory. For instance, knowledge about the ontogeny and evolution of antipredator behavior can inform captive rearing programs when animals destined to be reintroduced in natural environments that contain predators.

SUGGESTIONS FOR FURTHER READING

Tim Caro's (2005) recent book is the authoritative go-to volume on antipredator behavior in birds and mammals. Curio (1976) provides a classical ethological review of antipredator behavior. Lima (1998) and Preisser et al. (2005) provide a more ecological perspective on the consequences of antipredator behavior. General foraging theory, including examples of how understanding predation risk is essential to understand foraging decisions, is covered in Stephens and Krebs' (1986) book. More recently, in the Stephens et al. (2007) book, chapters by Ydenberg et al., Bednekoff, and Brown and Kotler discuss numerous examples of how foraging gain and predation risk trade-offs can affect both immediate behavior and community structure.

Caro T (2005) Antipredator Defenses in Birds and Mammals. Univ Chicago Press, Chicago, IL.
Curio E (1976) The Ethology of Predation. Springer-Verlag, Berlin, Germany.
Lima SL (1998) Nonlethal effects in the ecology of predator-prey interactions. BioScience 48: 25–34.
Preisser EL, Bolnick DI, & Benard MF (2005) Scared to death? The effects of intimidation and consumption in predator-prey interactions. Ecology 86: 501–509.
Stephens DW & Krebs JR (1986) Foraging Theory. Princeton Univ Press, Princeton, NJ.
Stephens DW, Brown JS, & Ydenberg RC (eds) (2007) Foraging: Behavior and Ecology. Univ Chicago Press, Chicago, IL.

SECTION IV

SOCIAL BEHAVIOR

14

Interacting Phenotypes and Indirect Genetic Effects

JASON B. WOLF AND ALLEN J. MOORE

There is a widely held notion that the genetics and evolution of behavior is somehow different from that of other sorts of traits, perhaps because it is arguably more complex or more plastic (context dependent) than traits like morphologies (Wolf 2001 West-Eberhard 2003). However, in most respects, there is no reason that behavior, when viewed from a genetic or evolutionary perspective, should be treated as being different from other types of traits. All expressed traits reflect the influences of genes and environment—organisms do not exist without genes, and no organisms exist in a vacuum. We can use standard quantitative genetic approaches to examine the specific contribution of genetic and environment effects and their interaction (i.e., the genotype-by-environment interaction, G × E, sometimes written GEI) to variation in behavior as we do with any other trait (see box 5.1). Although one might argue that some components of variation (e.g., environmental or G × E) are more important for behavioral traits than for other sorts of traits (Boake et al. 2002), this not only does not make their evolutionary dynamics fundamentally different, but, more important, it probably does not distinguish behavior from a wide array of other traits (e.g., many physiological processes in animals and a diversity of traits in plants; West-Eberhard 2003; chapter 6, this volume). How, then, do we reconcile the perception that behavioral traits are somehow more plastic, evolve faster, and are more dynamic than most morphological traits (Mayr 1964; Wilson 1975; West-Eberhard 2003; Bateson 2004)? One resolution that we have suggested is to recognize that many of the traits falling into this category involve interactions among individuals—for example, traits involved in signaling, communication, mating, parenting, social dominance, and group living (Moore et al. 1997, 1998). For behaviors (or other types of traits) affected by interactions among individuals, which we have termed *interacting phenotypes* (Moore et al. 1997), the evolutionary dynamics and underlying genetic models are clearly different.

Interacting phenotypes differ from other sorts of traits in two main respects. First, because social interactions can alter trait expression, they therefore can alter trait genetics and, consequently, evolutionary dynamics. Second, traits involved in social interactions can be both the agent and target of selection, which can have important evolutionary consequences. Traditionally, behavioral ecologists have focused on selection or how social traits influence fitness, and have analyzed the dynamics of such traits using game theory (chapters 8, 15, and 16). Our alternative approach, based on interacting phenotypes models that are built using a quantitative genetic framework (Moore et al. 1997, 1998; Wolf et al. 1999), captures both attributes and therefore provides a unified framework for modeling trait evolution. Any selection arising from social interactions falls under the class of selection termed *social selection* (Crook 1972; West-Eberhard 1979,

1983; Wolf et al. 1999), and the interacting phenotype perspective offers a formal mathematical consideration of social selection that we present in box 14.1. In this box we incorporate fitness effects caused by the traits of social partners into a quantitative selection framework to understand how social effects on fitness result in a force of selection. Each of the subsequent chapters in this and the next section of the book present specific aspects of, or concepts related to, social selection, and as a result can presumably fit into the framework presented in box 14.1. However, a consideration of selection is only part of the equation for evolutionary change, and a complete understanding also requires

BOX 14.1 Social Selection

Social interactions (i.e., social environments) can, of course, affect fitness. When this occurs, it provides the *opportunity for social selection*, which can be defined as the variance in fitness owing to variation in the social environment. Social selection occurs whenever there is a component of selection acting on a trait that can be attributed to social effects on fitness (West-Eberhard 1979, 1983)—this is consistent with the definition of social behavior, in which the behavior of one individual influences the fitness of another (Wilson 1975).

To understand social selection, we start by decomposing individual fitness (w) using a linear equation into components attributable to the variation in traits expressed by the individual itself (what we generically call *natural selection*, as this component can arise from many sources) and variation in traits expressed by other individuals that affect the fitness of the focal individual (c.f. equation 1 in Wolf et al. 1999 and equation 5 in Queller 1992a):

$$w = c + \beta_{Ni} z_i + \beta_{Sj} z_j' + \varepsilon \quad (1)$$

where c is a constant, β_{Ni} is the linear natural selection gradient associated with individual variation in the expression of trait i (denoted z_i), β_{Sj} is the linear social selection gradient associated with the expression of trait j in the individuals' social partners (denoted z_j') and ε is the error term that accounts for variation in fitness not attributable to either the measured natural or social selection. The model in equation 1 could be used in a regression analysis to detect linear components of selection, in which case the β terms are partial regression coefficients of relative fitness on the trait values (i.e., the linear relationship between trait values and fitness) of either the focal individual (in the case of natural selection) or its social partners (in the case of social selection). These selection gradients are the slope on the surface of individual fitness and measure how steep the surface is in a particular dimension (see Brodie et al. 1995 for a review of methods to quantify selection). It is critical to keep in mind that the two traits in equation 1 are measured in different individuals, but we are looking at only fitness variation in the focal individuals. Because of this, the social selection term in equation 1 is best interpreted as a measure of the opportunity for social selection, estimating whether traits expressed in other individuals affect the fitness of the focal individuals. Unlike the standard selection equation, it does not directly address whether selection acting on traits in the focal individual can be attributed to the phenotypes of social partners.

Equation 1 could be extended to include the effects of multiple traits measured in both the focal individuals and in their social partners (c.f. equations 3 and 4 in Wolf et al. 1999). When individuals interact with many social partners, we can substitute the mean value of trait j measured in the social partners as our measure of the social environment. The measurement of social selection is analogous to the contextual analysis of selection

(continued)

(Heisler & Damuth 1987; Goodnight et al. 1992), in which one measures features of the social context experienced by an individual (e.g., mean trait values in social partners) and examines whether these affect their fitness.

To understand how social effects on fitness are translated into selection acting, we need to examine how traits in the focal individuals covary with fitness (Price 1970). For such a relationship to exist, there needs to be a covariance between traits in the focal individual and the traits in the social environment that are the source of social selection. Using the model of selection shown in equation 1, Wolf et al. (1999) show that the social selection term is translated into a component of selection acting on traits when there is a covariance between traits in the focal individual and traits expressed in their social partners. That is, if we measure one trait in our focal individuals, z_i and one trait in their social partners, z_j', as in equation 1, then for social selection to act on trait z_i there must be a covariance between z_i and z_j'. Such a covariance often occurs in social interactions for a variety of reasons and is expected to be common (Wolf et al. 1999).

We can take equation 1 and express selection as a *selection differential* (s_i), which is the change in trait i within a generation caused by the components of fitness variation in equation 1. The selection differential for trait i is derived as the total covariance of trait i with fitness (Price 1970):

$$\begin{aligned} s_i &= P_{ii}\beta_{Ni} + \mathrm{cov}(z_i, z_j')\beta_{Sj} \\ &= P_{ii}\beta_{Ni} + C^{ij\prime}\beta_{Sj} \end{aligned} \tag{2}$$

where the two β terms are again the natural (subscript N) and social (subscript S) selection gradients, respectively, as in equation 1, P_{ii} is the phenotypic variance in trait i, and $C^{ij\prime}$ is the covariance between the expression of trait i in the focal individual and j in the social partner (e.g., equation 14.12). The first half of equation 2 describes standard natural selection, in which variation in focal trait expression is associated with variation in the focal individual's fitness. The second half is social selection acting on trait i provided by trait j. This provides a way of looking at how one trait (or set of traits) in a population generates a component of selection on other traits in the population through social interactions. Because social interactions are described by how they affect fitness (Hamilton 1964), this perspective allows us to understand how traits generate components of selection acting on other traits, or even the same trait, through interactions between individuals. Thus, the social selection approach allows us to understand how traits are both the targets and the *agents* of selection. In many cases, it will be the behavior of an individual that is the agent of selection acting on conspecifics. *Equation 2 illustrates a major consequence of social effects*: social effects on fitness can generate social selection on traits.

Examining social selection allows us to relate the interacting phenotype approach of quantitative genetics to the common phenotypic analysis of evolution practiced by behavioral ecologists, such as optimality and game theory. Both view individual behavior as the agent of selection acting on traits in populations by considering fitness effects. The main difference is in considering how the process of evolution will proceed; although game theory and other optimality approaches do not address the evolutionary process and instead concentrate on evolutionary outcomes, an interacting phenotype approach does both. This is because the interacting phenotype approach considers both elements of evolution separately: the change within a generation (social and natural selection) and the change across generations (evolutionary responses). The advantage of considering social selection from the perspective we have presented here is that it places a consideration of fitness into a quantitative genetic framework, providing us with both sides of the evolutionary equation in a common framework.

There are a number of reasons why we might want to decompose selection into natural and social selection components. For example, when interactions are between

(continued)

BOX **14.1** *(cont.)*

relatives, the social selection component can be viewed as kin selection, and therefore the social selection framework offers an empirical framework for measuring and understanding the origin of kin selection (see box 14.2). We may also wish to examine how behavior such as aggression or helping acts as an agent of selection. A consideration of social selection also provides insight into multilevel selection (Wolf et al. 1999; Bijma & Wade 2008), which can provide insights into the evolution of group behavior. West-Eberhard (1979, 1983) provides a general discussion of the importance of social selection.

a consideration of inheritance. Unlike game theory, interacting phenotype models also allow for a consideration of how social interactions influence trait expression (e.g., Moore et al. 1997). In this chapter we focus on the effects of social interactions on trait expression, with a goal of understanding how social interactions modify our view of trait genetics. This framework therefore can be seen as both an alternative to and as a complement to game theoretic approaches, depending on the problem.

The interacting phenotypes framework can be applied to any trait expressed, influenced by, or otherwise involved in social interactions. This interacting phenotypes framework is built by viewing social partners as part of the social environment experienced by an individual (using the term *social environment* to refer to the environment provided by conspecifics in general, whether the species or effects are social in the strict sense, whatever that may be, for example; Costa 2006). To understand the interacting phenotypes framework, we take the reader slowly through a variety of quantitative genetic models, with an ultimate goal of having the novice reader understand where the theory comes from and why it matters. We take this approach because it can be difficult to develop an appreciation for why traits like social behavior differ from other sorts of traits without developing an understanding of the structure of the models and why they differ from the traditional quantitative genetic models. We also hope that a bonus of such a treatment will be a better appreciation for the foundations of quantitative genetics in general. Because quantitative genetics generally takes a phenotype-based approach (in which phenotypic data are combined with information on relatedness to yield estimates of genetic parameters), a quantitative genetics perspective can be complementary to the phenotype-based approaches favored in behavioral ecology (Moore & Boake 1994). However, quantitative genetics theory can also make predictions about evolutionary processes or dynamics that are at odds with those generated by purely phenotypic models (see box 5.1), such as optimality models that often have implicit assumptions about genetics that may conflict with reality in some systems. Therefore, quantitative genetic approaches provide valuable or even critical data that can be integrated with other approaches used in behavioral ecology to develop a richer understanding the evolution of behavior. All that is required to start is a familiarity with basic quantitative genetics (box 5.1).

THE CLASSIC QUANTITATIVE GENETICS VIEW: ENVIRONMENTAL *OR* GENETIC?

The basic approach in quantitative genetic models is to partition variation in trait expression into a set of effects attributable to a set of measurable factors. In box 5.1 we introduce the simplest partitioning of variation, in which trait expression (z, which is the *phenotypic value*, or the trait value you measure on an individual) is partitioned into a genetic (g, which reflects all of the genetic influences contributing to a trait) and an environmental component (e, the environmental deviation, which reflects all of the environmental contributions influencing the expression of the trait):

$$z = g + e \qquad (14.1)$$

We can further partition these influences whenever we have specific information on components of genetics or the environment we know are important. For example, to start examining the importance of social environments provided by conspecifics, we partition the environmental term in equation 14.1 into separate social and nonsocial environmental effects:

$$z = a + e_n + e_\sigma \qquad (14.2)$$

where e_σ (in which the subscript sigma denotes "social"; we reserve the subscript s for later) is the effect of the social environment (i.e., the environment provided by conspecifics) on trait expression and e_n is the sum of all other environmental effects. We have also specified particular genetic effects of interest, a, or additive genetic influences (see box 5.1). Thus, our general environment term (e_n) now includes all of the nonadditive genetic effects because it is the sum of all remaining effects after removing the specific genetic or environmental effects we are measuring. We will refer to e_σ as the *social effect on trait expression* (cf. Bergsma et al. 2008). Social effects on trait expression have also been called *associate effects* (Griffing 1967, 1981a) to reflect the fact that they are the effect of an individual's associates on the expression of its phenotype. Such social effects can arise, for example, from cases in which individuals modify their behavior in response to the behavior or morphology of the individuals with whom they interact. For example, the expression of traits related to aggression are likely to be affected by the expression of aggression-associated traits in their social partners, in which individuals modulate their own aggressiveness in response to the behaviors of other individuals in their social environment (Craig & Muir 1996; Ellen et al. 2008). Indeed, at the extreme, traits like aggressiveness or social dominance (Moore et al. 2002) can't be defined outside of their relationship to social interactions, and therefore the expression of such traits are expected to be strongly tied to the social environment experienced by individuals. Thus, it is the particular properties of the social environment that make the implications of e_σ fundamentally different from those of e_n.

Because social effects on trait expression arise as a result of interacting phenotypes, the social "environment" effect can have a genetic basis and is therefore part of the genetic architecture of traits expressed in populations (Moore et al. 1997; Wolf et al. 1999). Perhaps more important, because the social environment reflects individual behavior or traits, which are themselves heritable, the social environment is also heritable and contributes to trait evolution. Thus, social environments can have profound effects on the evolutionary dynamics of traits. For example, a familiar case of interacting phenotypes and altered evolutionary dynamics occurs is sexual selection (Lande 1981), in which the expression of display traits of males trait can depend on the phenotype of the female (e.g., Meffert 1995; Moore et al. 1998; Petfield et al. 2005), perhaps expressed through her preference. However, our interacting phenotype approach is general. Below we will build on equation 14.2 to illustrate how and why the social environment affects the genetics and evolution of traits and, in doing so, illustrate the particular challenges that arise when trying to understand such traits.

THE INTERACTING PHENOTYPE VIEW: ENVIRONMENTAL *IS* GENETIC

The main point we wish to make in this section is that the effect that the social environment has on the genetics and evolution of traits depends critically on the assumptions we make about the nature of interactions among individuals. This is a consequence of the fact that, because there is no single way to model the social environment (because it will depend on the specifics of any particular species or experiment), there is no universally appropriate way to model traits affected by social environments. This is not merely a mathematical technicality—in other words, it is not simply a problem for theoreticians; it is a fundamental property of traits affected by social interactions. This means that if we wish to investigate the genetics of social traits, we need to understand where the genetic models come from and how to interpret and use them.

The simplest place to start is with a two-trait model. We define two traits of interest: trait f, which is measured in our "focal" individuals (in which the subscript f is used to indicate the focal trait) and trait s, which is measured in the focal individuals' social partner(s). These two traits have phenotypic values z_f and z_s, respectively (recall that the phenotypic value is the actual value of the trait we measure for an individual). Because these traits are measured in different individuals, we label the phenotypic value of trait s with a prime (z'_s) to make it clear that it originates from (and is ultimately the property of) another individual (see figure 14.1).

Modeling this social effect is our first problem. In the simplest case, we can assume that our focal

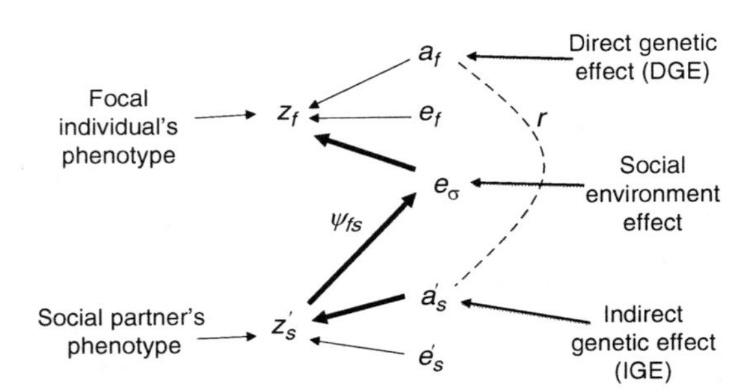

FIGURE **14.1** An illustration of direct and indirect genetic effects on trait expression in a social context. The expression of trait f by a focal individual is affected by the effect of its own genotype (a_f, the direct genetic effect), random environmental effects (e_f), and the effect of the social environment (e_σ) provided by its social partner. The social environment effect is attributed to the expression of trait s by the social partner (denoted z_s'). Trait s in the social partner is affected by genetic (a_s') and random environmental effects (e_s'). The effect of trait s expressed in the social partner on the expression of trait f by the focal individual is determined by the interaction effect coefficient, ψ_{fs}. The bold lines with arrows indicate the indirect genetic effect attributable to the mapping from the genotype of the social partner to the phenotype of the focal individual. The dashed line indicates the possibility of a correlation between the genotypes of the interacting individuals, here measured by the coefficient of relatedness (r).

individuals interact with just one partner and that some trait in the social partner has some simple additive effect on trait expression by the focal individual (cf. Moore et al. 1997). In this case, we can replace the generic term for the social environment (e_σ) in equation 14.2 with the explicit effect that the trait in a social partner (trait s) has on the expression of a trait in the focal individual (trait f; see figure 14.1):

$$z_f = a_f + e_f + \psi_{fs} z_s' \qquad (14.3)$$

The phenotypic value of the social partner (z_s'), is weighted by a coefficient, ψ_{fs} which we have called the *interaction effect coefficient* (Moore et al. 1997). Key to understanding equation 14.3 is appreciating what the interaction effect coefficient represents. This coefficient defines the extent to which the expression of a trait in one individual (trait s) affects the expression of a different trait (trait f) in their social partner (see Moore et al. 1997). Put simply, ψ_{fs} can be viewed as a measure of how important social interactions are for the expression of a trait. One could imagine that ψ_{fs} is large for some traits, such aggression (i.e., the extent that an individual is aggressive may well depend critically on traits, such as aggressiveness, expressed by their partners in the interaction) and might be very small for other traits, such as male courtship (one of the classic "stereotypic" traits, which often seems to be expressed regardless of the partner—for example, males of some species have been known to court beer bottles; Gwynne & Rentz 1983). If this coefficient is zero, then there is no effect of the social environment and we return to our standard quantitative genetic equation for a trait (box 5.1). Another way to understand this coefficient is to view equation 14.3 as a regression equation, in which ψ_{fs} would be the slope of the regression of trait f expressed in focal individuals on trait s expressed in their social partners. Generically, we can define (and measure) interaction effect coefficients for any pair of traits in a multivariate framework in which the two traits may be the same or different (Moore et al. 1997).

So far we have explicitly attributed the effect of social environment to an identifiable trait in social partners. It is common, especially in empirical studies, to examine the net influence of the social environment rather than to identify specific effects of traits in social partners (e.g., Griffing 1967; Wolf 2003; Bijma et al. 2007a, 2007b; Bijma & Wade 2008; Bergsma et al. 2008; Ellen et al. 2008). This so-called performance trait approach views the influence of the social partner only through its influence on the focal individual (in which we would refer to the effect of one individual on the phenotype of another its *social performance* phenotype, which is analogous to the concept of *maternal performance* used in models of maternal effects; Cheverud & Moore 1994; Wolf et al. 1998). In this case, the social effect on trait expression in one individual *is* the trait expressed in its social partner(s). Griffing (1967) used the performance trait approach (see recent extensions by Muir 2005; Bijma et al. 2007a, 2007b; Bijma & Wade 2008), in which social partners have an "associate effect" on the trait expressed in the focal partner. The performance trait approach has been the primary approach taken in empirical studies because researchers are often unable to measure all of the appropriate traits that make up the social environment (perhaps because they are not sure what these traits are, or because it is just not possible to measure all of the traits expressed in all of the interactions across an appropriate time span, or, as is the case in many agricultural systems, they may be uninterested in the specific traits, but rather in their net influence on trait expression and response to selection). Here we use the trait-based approach, in which we assume that the traits expressed in interactions can be measured, because most behavioral ecologists are interested in studying specific traits.

To investigate the evolutionary importance of the social environment, we decompose trait s (illustrated in figure 14.1) into heritable and nonheritable components, as in equation 14.1:

$$z_s = a_s + e_s \tag{14.4}$$

assuming for the moment that trait s is not affected by the social interaction, an assumption we relax below. Terms in equation 14.4 are not marked with a prime because we assume that we can measure trait s in any individual including the focal individual; traits are only marked with a prime when they are considered as the effect of a social partner on a focal individual. Using the definition of trait s given by equation 14.4, we can substitute the value for z_s in equation 14.3, which allows us to look at the components that affect the expression of traits in our focal individuals:

$$z_f = a_f + e_f + \psi_{fs}(a'_s + e'_s) \tag{14.5}$$

From equation 14.5 we can see the partitioning of genetic effects on the expression of trait f into so-called direct and indirect genetic effects (Moore et al. 1997; figure 14.1). The direct genetic effect, a_f, is the effect of (or mapping from) the individual's genotype onto its own phenotype. It is called *direct* (a term first used by Fisher 1918) because it corresponds to the heritable genetic influences mapping from the genotype to the phenotype within an individual independent of all other effects. The *indirect genetic effect* (a term first used by Riska et al. 1985 in the context of maternal effects), $\psi_{fs}a'_s$, corresponds to the effect of (or mapping from) the genotype of one individual on the phenotype of another individual (the prime again indicates that it is a property of the social partner of our focal individual). Thus, equation 14.9 illustrates the *first major consequence of social effects* on trait expression: traits expressed during social interactions provide the opportunity for indirect genetic effects, in which part of the genetic architecture of traits in populations can be attributed to mapping from the genotype of one individual to the phenotypes of other individuals. Additive indirect genetic effects have also been called "associative breeding values" (Bijma et al. 2007a). The last term, $\psi_{fs}e'_s$, can be considered an indirect environmental effect, and we will not consider it further because it is not part of the genetic architecture of trait expression and does not contribute to evolution.

Equation 14.5 provides a description of the factors that influence the trait of an individual. At an evolutionary level, however, we are interested in changes over time, and for this we need to consider components of variation and whether they contribute to trait evolution. Therefore, we need to partition variation in the expression of trait f into a set of genetic and environmental components (cf. box 5.1). If we assume that every individual engages in a single pairwise interaction and that there is no association (correlation or covariance) between genotype and the nonsocial environment, then variation in trait f can be expressed as (see Moore et al. 1997 for additional details):

$$\begin{aligned} P_{ff} &= G_{ff} + E_{ff} + \psi_{fs}^2 (G_{ss} + E_{ss}) \\ &\quad + 2\psi_{fs} \operatorname{cov}\left(a_f, a'_s\right) \end{aligned} \tag{14.6}$$

where P_{ff} is the phenotypic variance of trait f, G_{ff}, and E_{ff} are the direct additive genetic and environmental variances of trait f, and G_{ss} and E_{ss} are the indirect genetic and environmental variances. The last term in equation 14.6 is a special type of genotype-environment covariance—it is the covariance between the direct effect of the focal individuals' genotype and the indirect effect of their social partner's genotype on their phenotype. This genotype-social-environment covariance differs from ordinary types of genotype-environment covariances because it is an association of genetic factors in interacting individuals (as opposed to an association between the genotype and, for example, some abiotic environmental variable). This term illustrates the *second major consequence of social effects*: traits influencing social interactions can result in genetically based genotype-environment covariances, in which a genotype may experience a predictable social environment because of genetic associations between interacting individuals.

The genotype-social-environment covariance is zero when individuals interact at random, but we expect the covariance to be nonzero under a number of common circumstances. Such a covariance can arise whenever individuals choose their social partners, such as when individuals actively seek out individuals with similar (or dissimilar) phenotypes as themselves. Another possibility is that interactions occur among individuals that are related. The correlation among the genotypes of relatives can then make this covariance nonzero. When considering two different traits expressed in interacting individuals (as in equation 14.3), there must be a genetic association between the traits (i.e., a genetic correlation between the traits, perhaps due to pleiotropy) in order for relatedness to generate a genotype-social-environment covariance. That is, for trait f in one individual to predict the value of trait s in their relatives (and thereby to predict the social environment that they experience when they interact with relatives), there has to be some shared genetic basis for traits f and s (either because they are "controlled" by the same genes, or because the genes affecting the two are in linkage disequilibrium). When considering a genotype-social-environment covariance due to relatedness, we would replace the term $\text{cov}(a_f, a_s')$ in equation 14.10 with rG_{fs}, where r is the coefficient of relatedness (twice the coefficient of coancestry; Lynch & Walsh 1998) and G_{fs} is the additive genetic covariance between traits f and s. Interactions between relatives are common under many circumstances, making this genotype-social-environment covariance an important component of trait variation in many populations. For example, when mothers care for their offspring, the maternal-offspring interaction usually affects the expression of traits in the offspring. Similarly, in many systems, siblings can affect the expression of traits in one another due to sibling competition (Wolf 2003; Linkasvayer 2006). The role of relatedness is discussed further in relation to Hamilton's rule and kin selection in box 14.2 (also see box 23.1 in Frank 2006 for more on various coefficients of relatedness and similar parameters). However, it is important to remember that relatedness is not the only way to get an association between genotype and social environment (Wolf et al. 1999). Ultimately, the necessity of including this covariance is an empirical issue specific to the study organism and behavior.

Quantitative genetic models focus on the additive genetic variance component of the total phenotypic variance because it contributes to the evolutionary response to selection and to the resemblance of all types of relatives (see box 14.2). However, although there are two additive genetic variances (G_{ff} and G_{ss}) in equation 14.6, they have fundamentally different properties. To understand this, we will introduce the concept of the covariance between the additive genetic value (a; also referred to as the breeding value) of an individual and its phenotypic value (z). This is denoted C_{az} (see Kirkpatrick & Lande 1989; Arnold 1994a), and for trait f, it corresponds to the covariance between a_f and z_f. This is the heritable component of the phenotypic variance, corresponding to the association among all the additive genetic values of individuals (i.e., their breeding values, which predict the resemblance among relatives) and their own phenotypes. This can also be interpreted as a measure of the genotype-phenotype relationship (Wolf et al. 1998) as it measures the association of individual genotype and individual phenotype. For traits not affected by social interactions, the covariance between the additive genetic (breeding) value and the phenotypic value is generally assumed to be G_{ff}—in other words, the additive genetic variance (which why the additive genetic variance plays a central role in quantitative genetics). However, the covariance between breeding values and phenotypes is a general concept, and it is not so simple when social interactions are involved. Taking the covariance between a_f and z_f ($C_{a_f z_f}$) from equation 14.5:

$$C_{a_f z_f} = G_{ff} + \psi_{fs} \text{cov}(a_f, a_s') \qquad (14.7)$$

Equation 14.7 shows that the indirect genetic variance can contribute to heritable variation in trait f. Thus, associations between direct and indirect effects on trait expression caused by nonrandom social interactions can appear as part of the directly heritable variation. The heritable variance that is usually measured in standard breeding studies or from pedigrees therefore provides an inaccurate measure of the genetic influences on a trait, because they are not designed to detect indirect genetic effects. Nonrandom social interactions also result in a covariance between the additive genetic value for trait s (a_s), which is the trait having the social effect, and the phenotypic value for trait f (z_f), which is our focal trait. In standard quantitative genetics, the genetic covariance between genetic effects on one trait and the expression of a different trait is simply given by the direct additive genetic covariance (G_{fs}), which reflects either a pleiotropic relationship between direct genetic effects on traits f and s, or linkage disequilibrium between loci affecting these two traits. However, when there are social effects

BOX 14.2 An Interacting Phenotypes Perspective on Kin Selection

Kin selection analyses have traditionally focused on understanding how selection can lead to the evolution of altruism, in which expression of a particular altruistic trait reduces individual fitness while increasing the fitness of relatives. The conditions for such altruistic evolution were described by W. D. Hamilton (1964) in a form now known as Hamilton's rule, in which the negative effect of an altruistic behavior (the cost, C) has to be less than the positive effect on kin (the benefit, B) weighted by the coefficient of relatedness of the individuals (r; i.e., $rB > C$). Although most discussions of Hamilton's (1964) analysis of social evolution focus on the evolution of altruism, his original presentation was general and can be applied to any scenario in which individuals affect the fitness of each other (West et al. 2007a). The difficulty of Hamilton's rule is that defining costs and benefits is no easy task.

A variety of quantitative genetic forms of Hamilton's rule have been derived (e.g., Cheverud 1984b; Queller 1985, 1992b; Lynch 1987), providing a quantitative framework that facilitates empirical analysis of kin selection. The quantitative genetic perspective is generally a *direct fitness* perspective, in which fitness is the property of an individual, but can be affected by other individuals (see Frank 1998). This contrasts with the *inclusive fitness* approach of Hamilton (1964), in which an individual's net fitness includes its own fitness as well as a component of its kin's fitness. The two approaches yield equivalent results as long as the same assumptions are made (Frank 1998), but the direct fitness approach has a simpler accounting of fitness, in which individuals have their own fitness that is affected by various factors.

Hamilton (1964) described four scenarios that vary in the signs of the effects of a trait on the fitness of a focal individual (IND) and its social partners (SOC): (1) IND(+) and SOC(−), in which there is a positive effect on the fitness of the individual and a negative effect on the fitness of the social partner, which corresponds to selection for selfish behavior or conflict; (2) IND(−), SOC(+), which is selection for altruism; (3) IND(+) and SOC(+), which is selection for cooperation; and (4) IND(−) and SOC(−), which is selection for spite. The cases in which IND and SOC are of opposite sign have received the most interest, but again, all four scenarios can be evaluated from the kin selection perspective (West et al. 2007a). The quantitative genetic framework also allows one to examine all of these scenarios and provides a framework for including components of kin selection in evolutionary analyses (Queller 1992a; Wolf 2003).

There are alternative ways of viewing kin selection from a quantitative genetics perspective. First, we may ask whether the direct and kin components of selection favor increased trait values (e.g., Wolf et al. 1999)—in other words, whether the net force of selection acts to increase or decrease trait values. This view, focused on selection, needs to be combined with a model of trait genetics to ultimately understand the net response to selection, but it can provide a relatively simple view that facilitates empirical analysis of kin selection. Alternatively, we can look at the direct and social effects of a single trait and ask whether the expected response to selection is positive or negative. This is the approach used by Cheverud (1984b), focused on maternal effects, and Wolf (2003), focused on competition, and looks at whether, given a pattern of direct and social effects on trait expression, there should be a positive or negative response to selection on a trait.

We can illustrate the selection perspective using the social selection model (see appendix B of Wolf et al. 1999 and box 14.1, this volume, for details). To evaluate when selection will favor increased values of a trait, we can ask when the net selection on trait i (s_i; see equation 2 in box 14.1) will be > 0. We can rearrange the equation for calculating the selection differential on trait i when there is social selection (equation 2 in box 14.1) to yield the following conditions:

(continued)

$$\beta_{Ni} + \frac{C^{ij'}}{P_{ii}}\beta_{Sj} > 0 \tag{1}$$

which is analogous to Hamilton's rule, with the ratio ($C^{ij'}/P_{ii}$) as our measure of relatedness, β_{Ni} the natural selection cost (if the trait is altruistic), and β_{Sj} the benefit to social partners. β_{Ni} and β_{Sj} can have any sign, making equation 1 a general form that can be used for any of the four scenarios described above (where β_{Ni} is the IND value and β_{Sj} is the SOC value). The ratio in equation 1 is the regression of the value of trait j in the social partner(s) on the value of trait i in the focal individuals. This measure of resemblance of interacting individuals can arise from many sources (see Frank 1998, 2006), but in the simplest case, it may be due to relatedness of individuals. For example, if trait i is the same as trait j and we are looking at interactions of a parent and its offspring to understand offspring fitness (as in Cheverud 1984b), then the ratio in equation 1 is the familiar offspring-parent regression that yields $\frac{1}{2}h^2$ (i.e., one-half the heritability of trait i).

We illustrate the social effects view of kin selection using the model for trait expression given in equations 14.3–14.5. We focus on the evolution of trait s, as this is the trait that has a social effect on the expression of other traits (i.e., when there are interactions between relatives, trait s has an effect on kin, and so its evolution will be affected by kin selection). For simplicity, we can focus on the case in which selection favors an increase in trait f (i.e., $\beta_f > 0$) whereas selection favors decreased values of trait s (i.e., $\beta_s < 0$). Under this assumption, we can view β_f as a measure of the benefit of increased values of trait f whereas β_s is the cost of increased values of trait s. To examine kin selection, we first need an expression for the evolution of trait s, which is determined by the direct ($C_{a_s z_s}\beta_s$) and correlated ($C_{a_s z_f}\beta_f$) responses of trait s to selection on traits s and f respectively:

$$\begin{aligned}\Delta\bar{z}_s &= C_{a_s z_s}\beta_s + C_{a_s z_f}\beta_f \\ &= G_{ss}\beta_s + \beta_f\,(G_{fs} + r\,\psi_{fs}G_{ss})\end{aligned} \tag{2}$$

To determine whether the response to selection is positive (i.e., Δ&zbar;$_s$ > 0), we can rearrange equation 2 to yield the conditions:

$$-\beta_s < \beta_f r\,\psi_{fs} + \beta_f\frac{G_{fs}}{G_{ss}} \tag{3}$$

Because β_s is assumed to be negative ("selection against" increasing values of trait s), the left-hand side is a positive value that represents the fitness cost of increased values of trait s. If the two traits are not genetically correlated, equation 3 can be simplified to a form that is analogous to Hamilton's rule, $-\beta_s < \beta_f r\psi_{fs}$, in which the right-hand side is the fitness benefit to kin ($\beta_f\psi_{fs}$) multiplied by the coefficient of relatedness (r). When traits f and s are genetically correlated, the last term in equation 3 is nonzero and complicates the simple expression of Hamilton's rule, making the genetic covariance between traits an important determinant of the net result of kin selection (see Cheverud 1984b, 1985; Cheverud & Moore 1994).

Equation 3 depends on the causal relationship between the expression of traits f and s (measured by ψ_{fs}), and therefore this expression is valid only for the simple case in which s affects f. If the effects are reciprocal, then we get a different expression. Regardless, one can analyze the conditions for altruistic or selfish evolution by solving for the conditions in which the response to selection will be positive or negative. Of course, it is also possible

(continued)

to examine the evolution of traits that one would not typically consider in a kin selection model (Wolf 2003). For example, we could consider trait f to be some trait that is under positive selection, like body size at pupation in a fly (c.f. Wolf 2003), in which trait s in social partners has a negative effect due to social competition, and examine the conditions under which the response to selection in traits f and/or s evolve in a particular direction.

(assuming the pattern in equations 14.3–14.5), this covariance ($C_{a_s z_f}$) is again more complex:

$$\begin{aligned} C_{a_s z_f} &= G_{fs} + \psi_{fs}\,\mathrm{cov}(a_s, a_s') \\ &= G_{fs} + \psi_{fs}\, r\, G_{ss} \end{aligned} \quad (14.8)$$

The first line of this equation shows that nonrandom interactions with respect to social effects create a covariance between the genotypic values of individuals for social effects and their expression of the trait affected by social interactions. The second line shows the case in which nonrandom interactions are a result of interactions between relatives.

Equation 14.8 is important for two reasons. First, assuming that nonsocial environmental effects are random, equation 14.8 defines the phenotypic covariance (within individuals) between traits s and f (i.e., $C_{a_s z_f} = P_{fs}$, where P_{fs} is the phenotypic covariance between the traits). Second, it defines the true genetic covariance (including both direct and indirect genetic influences), which is what one would measure between traits s and f empirically. Equation 14.8 demonstrates that social effects can create a genetic and phenotypic covariance (i.e., a predictable relationship) between traits within individuals even when the two traits are not directly influenced by the same or linked loci (i.e., $G_{fs} = 0$ and there is no pleiotropic or linkage disequilibrium relationship between traits f and s). Thus, equations 14.7 and 14.8 illustrate the *third major consequence of social effects*: they can alter the genotype-phenotype relationship (Moore et al. 1997; Wolf et al. 1998). We discuss this further below when we consider other scenarios in which the change in the genotype-phenotype relationship is more pronounced.

INCREASINGLY COMPLEX SOCIAL INTERACTIONS

We have so far assumed that there is just one partner, that the social effect is independent of the direct effect, the social effect is additive, and there is no effect of trait f on the expression of trait s. All of these assumptions can be relaxed, which results in different equations for the genetics of traits and their evolution. Moore et al. (1997) present a variety of such models, allowing the numbers of traits to vary and exploring various patterns of social effects. Although these models are all based on a simple set of basic assumptions, including all interactions being pairwise and all social effects additive, they result in a diversity of genetic and evolutionary effects. We won't explore all of the possibilities here, but leave it to the interested reader to apply this introduction to interacting phenotype models to explore the additional models of Moore et al. (1997), Bijma et al. (2007a, 2007b, based on the performance trait models of Griffing 1967, 1981a, 1981b), and Bijma and Wade (2008).

Although we cannot examine every possibility, it is useful to consider what happens under some of the more common social conditions. So far, we have assumed that an individual affects its social partner, but is unaffected by the social interaction. More commonly, both individuals are affected by the interaction; that is, social interactions are reciprocal and there may be some feedback through the interaction (Moore et al. 1997). Such a scenario may occur, for example, with a trait like social dominance, in which the behavior of one individual increases as a function of the social behavior of their social partners, such as escalation of aggressive or fighting behaviors. If we assume reciprocal effects, then equation 14.3 becomes the following:

$$z_s = a_s + e_s + \psi_{sf}\, z_f' \quad (14.9)$$

where ψ_{sf} is analogous to ψ_{fs} seen above, but measures the effect of trait f in one individual on the expression of trait s in its social partner. Equation 14.9 is circular in that trait s relies on the value of trait f and vice versa (i.e., the relationship is reciprocal, and therefore the values of traits f and s are mutually interdependent). We can rewrite equation 14.9 in a form that removes circularity (for clarity, we show the expression for f as well, which simply has the subscripts reversed):

$$z_s = \frac{1}{1-\psi_{fs}\psi_{sf}}\left[a_s + e_s + \psi_{sf}(a_f' + e_f')\right]$$
$$z_f = \frac{1}{1-\psi_{fs}\psi_{sf}}\left[a_f + e_f + \psi_{fs}(a_s' + e_s')\right] \quad (14.10)$$

The ratio on the right-hand side of equation 14.10 reflects the feedback loop between the behaviors of interactants, in which changes in the behavior of one individual changes the behavior of their partner, which then feeds back onto the expression of the first individual.

The covariances between the additive genetic (a_f) and phenotypic value (z_f), which again can be viewed as a measure of the heritable variance of trait f, is defined as follows:

$$C_{a_f z_f} = \frac{1}{1-\psi_{fs}\psi_{sf}}\left[G_{ff} + \psi_{fs}\,\mathrm{cov}(a_f, a_s')\right] \quad (14.11)$$

Equation 14.11 illustrates how the presence of a feedback loop, in which the traits expressed by interacting individuals have reciprocal effects on each other, alters the genotype-phenotype relationship even when and individuals interact at random (i.e., $\mathrm{cov}(a_f, a_s') = 0$). This is a basic property of feedback between interacting individuals. For example, when traits f and s are components of social dominance, then positive feedback (in which ψ_{fs} and ψ_{sf} are both positive) would generally lead to a stronger genotype-phenotype relationship, at least as measured by the covariance. A small increase in the expression of f in the first individual increases the expression of trait s in their partner, which feeds back onto the first individual to produce an increase in the expression of trait f, and so on. This also implies that it increases the heritable variance for the trait. Thus, if one were to measure trait f in the absence of social interactions (if that is possible) then one would find a lower heritable (additive genetic) variance than in the presence of interactions between individuals. This is perhaps counterintuitive, because one might imagine that by eliminating interactions between individuals, one would actually increase the heritable variance because one would be removing environmental variation contributed by the social partners. With social traits, however, the interaction itself can increase heritable variance. It is also possible that the pair of traits could have antagonistic effects, so that the sign of ψ_{fs} and ψ_{sf} are opposite, which decreases the heritable variance. As a result, the presence of interactions between individuals in an experiment could reduce or even erase the appearance of heritable variation.

Using the example of a pair of traits with reciprocal effects on each others expression, as in equation 14.10, we will illustrate the *fourth major consequence of social effects*: they can produce a covariance (i.e., an association) between the traits expressed in interacting individuals. That is, it can make the expression of traits in one individual correlated to the expression of traits in their social partner(s). This is perhaps not surprising given that it simply results from traits of one individual altering the traits expressed by other individuals, which make the traits of the interacting individuals causally correlated. However, this covariance has a number of interesting consequences. In practical terms, by affecting the relationship between traits expressed in interactants, social effects can alter the resemblance of relatives (Wolf 2003; Bijma et al. 2007b). This can lead to an over- or underestimate of the genetic variance for a trait because the covariance of relatives is used to infer the genetic component, and if relatives are allowed to interact, their resemblance will be altered by their interaction. The covariance between interacting individuals can also have evolutionary consequences because it can lead to social selection, in which social interactions lead to components of selection acting on traits (Wolf et al. 1999). We discuss social selection in more detail below, but here we illustrate (mathematically) this fundamental consequence of social interactions.

For our example, we can return to the expressions for traits s and f shown in equation 14.10, in which the two traits have reciprocal effects. In this case, trait f in a set of focal individuals will be correlated to trait s in their social partners because f leads to changes in the expression of s and vice versa. We denote the covariance between the expression of traits f and s in social partners (i.e., $\mathrm{cov}[z_f, z_s']$, where again, we measure z_f in the focal individual and z_s' in the social partner) as $C^{fs'}$ (see Wolf et al. 1999). For our two-trait example, $C^{fs'}$ has the value:

$$C^{fs'} = \left[\frac{1}{1-\psi_{fs}\psi_{sf}}\right]^2 \begin{bmatrix} \psi_{sf}(G_{ff} + E_{ff}) \\ +\psi_{fs}(G_{ss} + E_{ss}) \\ +\mathrm{cov}\,[a_f, a_s'] \\ (1 + \psi_{sf}\psi_{fs}) \end{bmatrix} \quad (14.12)$$

The three covariance terms in the right-hand bracket account for three different phenomena. The first, $\psi_{sf}(G_{ff} + E_{ff})$, results from the effect of trait f in one individual on the expression of trait s in its social partner. The second, $\psi_{fs}(G_{ss} + E_{ss})$, accounts for the effect of trait s on the expression of trait f. The last, $\text{cov}[a_f,a_s'](1 + \psi_{sf}\psi_{fs})$, is a consequence of the covariance of traits in interacting partners that results from relatedness or nonrandom associations (although with social effects, even this term is more complex than the usual resemblance of relatives expected for "ordinary" traits). This equation shows, however, that traits cannot be assumed to be independent when they influence the outcome of social interactions. Wolf et al. (1999) discuss a number of sources for such a covariance in addition to nonrandom social interactions, such as relatedness or behavioral modification.

APPROACHING BIOLOGICAL REALITY

The examples presented above are models, a set of equations generally corresponding to a single (simplified) hypothetical scenario. This is perhaps the biggest challenge with respect to social effects on trait expression—their genetics and evolutionary dynamics (discussed below) depend critically on the structure of social interactions. We have seen that social interactions alter the phenotypic and heritable variances of traits, whereas interactions among relatives alter the resemblance of relatives. In addition, it is very likely that interactions are not pairwise. Even in cases in which single interactions are pairwise, it is of course possible that individuals engage in many interactions, in which case, one needs to consider what the trait is that they are studying. Is it the trait expressed upon the first encounter for an individual, or is it the average trait expressed across all interactions in the individual's lifetime? If an individual interacts with many individuals, we may change equation 14.3 to reflect the fact that the expression of trait f is affected by a number of social partners ($n - 1$) in a group of n individuals (Griffing 1967; Moore et al. 1997; Bijma et al. 2007a):

$$z_{fi} = a_{fi} + e_{fi} + \psi_{fs}\frac{1}{n-1}\sum_{j\neq i}^{n} z'_{sj} \qquad (14.13)$$

where the additional subscript i denotes the focal individuals and j the social partners (e.g., z_{sj}' is the phenotypic value for trait s of the jth group member) and ψ_{fs} is now the effect of the mean social environment. Moore et al. (1997) explore this in more detail, as well as consider multiple traits involved in social interactions (multivariate models).

Interactions with multiple social partners change the phenotypic variance but do not necessarily alter the heritable variance. Perhaps counterintuitively, when individuals interact with more than one social partner, the phenotypic variance is generally reduced. This is because as individuals interact with more and more partners, they experience more similar social environments. In the extreme case, individuals interact with all other individuals in the population and the average social environment experienced by all individuals is approximately the same (i.e., it is the mean social environment for the population excluding the focal individual). Thus, when considering pairwise interactions, we are looking at an extreme case in which the social environment contributes to maximal amount of phenotypic variance contributed by social effects (see Bijma et al. 2007b; Bijma & Wade 2008). Studying interactions between pairs of individuals to understand the importance of social interactions can provide a more powerful approach than looking at (potentially more) natural scenarios in which individuals interact with many social partners, but it also results in a scenario in which the variation contributed by the social environment is at its maximum.

Although we have discussed the consequences of interactions with more than one social partner and interactions with relatives, the nature of social effects themselves may be a more complex factor that will determine the genetics and evolution of traits affected by social interactions. So far, we have taken what is a typical approach of considering social effects as linear (additive) effects, in which we just add a term that accounts for the influence of the social environment. However, social effects need not follow such a simple model. For example, Moore and Wolf (unpublished) model the expression of a trait that is determined by the *relative* expression of a trait by its social partner. In this case, the effect of a trait in one individual (e.g., body size) on the expression of a trait in the social partner (e.g., aggression) is dependent on a trait size of the focal individual. We modeled this by including a term for relative body size. If we denote the size of the focal individual z_s and the size of the social partner as z_s, then we can model the expression of social aggression, z_f, as (modified from Moore and Wolf unpublished):

$$z_f = a_f + e_f + \psi_{fs}(z_s - z_s') \qquad (14.14)$$

which indicates that individuals will be more aggressive when they are larger than their social partner (where $z_s > z_s'$), and less aggressive when they are smaller than their social partner. This definition for contingent trait expression leads to very different phenotypic and heritable variances than seen in the previous scenarios and completely different evolutionary dynamics despite the fact that equation 14.14 is very similar in structure to equation 14.3. We encourage the reader to think about all of the different ways in which the expression of a trait might depend on social interactions and how those scenarios might be expressed mathematically. Such an exercise can make it clear why modeling the evolutionary dynamics of traits influenced by social interactions can be so variable and complex.

The Genotype-by-Social-Environment Interaction

There may be an infinite number of ways in which traits expressed by one individual will depend on traits expressed in their social partner(s). However, there is a specific case that may be general, in which the effect of the social environment depends on the genotype of the focal individual. This can be viewed as a special type of genotype-by-environment interaction and is often of interest to the study of behavior because it reflects how different genotypes respond differentially to various social environmental variables. Rewriting equation 14.3, assuming that there is a genotype-by-social-environment interaction such that the direct effect depends on the social environment and vice versa:

$$\begin{aligned} z_f &= e_f + \psi_{fs}\, a_f z_s' \\ &= e_f + \psi_{fs} a_f a_s' + \psi_{fs} a_f e_s' \end{aligned} \qquad (14.15)$$

This genetic interaction term, $\psi_{fs} a_f a_s'$, is interesting for a variety of reasons. First, it can be viewed as a special type of epistasis, which has been called genotype-by-genotype epistasis (Wolf 2000) and can contribute to many of the same population level processes as ordinary (within-genotype) epistasis, such as population differentiation (Wolf et al. 2004). It can also lead to the appearance of within-genotype epistasis if the interactants are related (Wolf 2000) because a_f and a_s' in the term $\psi_{fs} a_f a_s'$ can be correlated, and their interaction can be translated into the appearance of an interaction between the genetic values within individuals. For a further discussion of the evolutionary consequences of the genotype-by-social environment interaction, see Wolf et al. (2004).

EVOLUTIONARY CONSEQUENCES OF SOCIAL INTERACTIONS

Behavioral ecologists have long noted that behavior appears to evolve differently, or at least faster, than most traits (Mayr 1964, 1974; Wilson 1975; Bateson 2004). This has been suggested to reflect (among other things) behavioral scaling (Wilson 1975), unusual plasticity (West-Eberhard 2003), or a multiplier effect on behavior (because behavior is the phenotype furthest from "the gene"; Wilson 1975). We have seen that socially contingent expression of traits can influence sources of heritable variation. As a result, the rate of population or species differentiation for such traits can be much more rapid (or possibly less rapid) than that seen for other traits, which could explain, for example, why some behaviors may evolve faster than some morphological traits. Social effects can also affect the rate of coevolution of traits because of their effect on the genetic relationship between traits (e.g., their influence on C_{az} as discussed above), which could explain patterns and rates of correlated evolution of behaviors or behaviors with morphologies. It is clear that unless you wish to simply assume that genetics doesn't matter (e.g., optimality or game theory), we have to adapt our genetic models to incorporate the influence of social effects on expression of traits. We now develop equations that show how evolutionary change occurs when indirect genetic effects are involved in the expression of interacting phenotypes.

We start by looking at the expected changes in trait means across generations (the typical quantitative genetic definition of evolution) and how social effects alter the expected rate of evolution. We again first develop the simplest model of social effects shown in equation 14.3, in which a single trait is affected by one other trait and the effects are not reciprocal. We need to include two components that contribute to an evolutionary change in trait f: changes in the direct genetic contribution to f and changes in the average social environment. That is, if trait s itself evolves, the mean social environment would change, and as a result, the mean of trait f would change because trait s affects trait f through the social environment. To understand evolutionary changes in the direct genetic contribution to trait f, we use $C_{a_f z_f}$, which tells us how selection at the phenotypic level (z_f) is translated into changes at the genetic level. In addition, given that selection is acting on trait s, there can also be an evolutionary response in trait f due to a correlated response to selection on direct effects, which is accounted for by $C_{a_f z_f}$. This is the standard correlated response to selection we find in most multivariate quantitative genetic models of evolution (see chapter 3), in which selection on one trait is translated into a response in other traits through the genetic correlation between traits (here measured by a covariance parameter).

We can write the single generation change in the mean of trait f (denoted $\Delta\bar{z}_f$) in a form that separates the response due to changes in direct genetic effects from changes in social effects (the latter being the evolution of the social environment). That is, we can view the change in the mean of trait f as changes in the two genetic terms (direct and indirect) in equation 14.5, which is equivalent to:

$$\begin{aligned}\Delta\bar{z}_f &= [C_{a_f z_f}\beta_f + C_{a_f z_s}\beta_s] \\ &\quad + \psi_{fs}[C_{a_s z_s}\beta_s + C_{a_s z_f}\beta_f]\end{aligned} \qquad (14.17)$$

where β_f and β_s are the selection gradients for traits f and s, respectively (we are making no assumptions here about where such selection comes from, so these are total selection gradients). The first term in brackets is the evolutionary change in trait f due to evolution of direct genetic effects, whereas the second term in brackets is the evolution of the social environment that results in evolution of trait f. The latter highlights the *fifth major consequence of social effects*: they can lead to trait evolution due to the evolution of the (mean) social environment. This has a number of implications, including the fact that it can result in the evolution of traits without changes in (or even in the absence of) direct genetic effects and the fact that differences in trait values between populations can be at least partly due to differences in the average social environments in those populations (Moore et al. 1997).

To fully understand the evolutionary response to selection, we can expand equation 14.17 to examine the genetic components that contribute to the response to selection. To do so, we substitute the C_{az} terms in equation 14.17 with the expanded forms shown in equation 14.7, which shows $C_{a_f z_f}$, and equation 14.8, which shows $C_{a_s z_f}$. In this example, trait s is not affected by social interactions so $C_{a_f z_s} = G_{fs}$ (i.e., the direct additive genetic covariance between f and s) and $C_{a_s z_s} = G_{ss}$ (i.e., the direct additive genetic variance for trait s). Using these definitions, we can rewrite equation 14.17 to explore how various genetic effects contribute to the evolutionary response to selection:

$$\begin{aligned}\Delta\bar{z}_f &= [G_{ff}\beta_f + G_{fs}\beta_s] \\ &\quad + \psi_{fs}\,[G_{ss}\beta_s + G_{fs}\beta_f] \\ &\quad + \psi_{fs} r\,[G_{fs}\beta_f + \psi_{fs} G_{ss}\beta_f]\end{aligned} \qquad (14.18)$$

To highlight the different influences, we have separated response to selection in equation 14.18 into three components, with one on each line. The first line is the "ordinary" response to selection caused by direct genetic effects (the sum of the direct and correlated responses to selection on trait f and s respectively; see box 5.1). If there were no social influences on trait expression, this would describe the evolutionary response. The second line is the response to selection caused by the evolution of the social environment (which is again the sum of direct and correlated responses). This term corresponds to the evolutionary change in trait s (in brackets), which contributes to evolutionary change in trait f through the social effect of trait s on trait f, mediated by the strength or importance of the interaction, ψ_{fs}. The third line is an additional component that arises when interactions occur between relatives (or other phenomena that lead to an association between the genotypes of interactants). This third component has two distinct parts in the brackets: the first part ($G_{fs}\beta_f$) is the additional change in the direct genetic component, and the second part ($\psi_{fs}G_{ss}\beta_f$) is the additional change in the indirect genetic component (i.e., a change in the mean social environment) caused by interactions between relatives (these are both a consequence of the genotype-social-environment covariance that relatedness creates). This shows that when interactions are between relatives, the response to selection can be enhanced, which has been suggested as a phenomenon that could aid in the efficacy of individual selection schemes in agricultural systems (Ellen et al. 2007).

From equation 14.18 we see that the net response to selection acting only on trait f could be increased or decreased by the correlated response of the social environment (depending on the signs of ψ_{fs} and G_{sf}) and that this correlated change in the social environment to selection acting on trait f is enhanced if interactions are between relatives. This means that the trait could potentially show a response to selection in the direction opposite to that of the selection gradient. This feature is well known for maternal effects models (Kirkpatrick & Lande 1989), in which interactions are by definition between relatives, and has been demonstrated empirically (Falconer 1965).

When traits have reciprocal effects on the expression of each other, so that trait s is also influenced by the social interaction as in equations 14.14, we get additional terms in the response to selection that are caused by the feedback loop. In this case, the change in trait f would be as follows:

$$\Delta\bar{z}_f = \left[\frac{1}{1-\psi_{fs}\psi_{sf}}\right]^2 \begin{bmatrix} (G_{ff}\beta_f + G_{fs}\beta_s) + \\ \psi_{fs}(G_{ss}\beta_s + G_{fs}\beta_f) + \\ r(\psi_{fs}G_{fs}\beta_f + \psi_{sf}G_{ff}\beta_s) + \\ r\psi_{fs}(\psi_{fs}G_{ss}\beta_f + \psi_{sf}G_{fs}\beta_s) + \end{bmatrix} \quad (14.19)$$

The ratio in brackets accounts for the feedback loop and has a major effect on the evolutionary dynamics (Moore et al. 1997). If the interaction effect coefficients (ψ_{fs} and ψ_{sf}) are of opposite sign, the ratio is less than 1 (because the coefficients are bounded at +1 and −1; see Moore et al. 1997) and the feedback loop diminishes the response to selection because the effects are antagonistic, resulting in a dampened feedback loop. If they are of the same sign, then the ratio is larger than 1 and the expected response to selection is increased because the feedback loop is synergistic. The four terms in the brackets (each on its own line) correspond to the following: (1) the change in the direct genetic component, (2) the change in the indirect genetic component (i.e., the evolution of the social environment), (3) the change in the direct genetic component due to relatedness of interactants, and (4) the change in the indirect genetic component due to relatedness of interactants. Overall, equation 14.19 shows that the expected response depends on the interplay of the strength and direction of selection on the two traits, their genetic variances and covariance, and the magnitude and sign of any social effects.

HOW DO WE STUDY TRAITS AFFECTED BY SOCIAL INTERACTIONS?

The value of all models is that they identify parameters to measure and assumptions to test. For our models, there are two parameters that remain mostly unmeasured: (1) psi (ψ), the interaction effect coefficient, and (2) indirect genetic effects of social traits. Although we don't have space to provide a detailed guide to empirical analyses (see Bleakley et al. 2009), we can review some empirical studies to illustrate approaches and relevant questions.

There are at least three ways to study social effects in the context of our models. The first would be to try to characterize the interaction effect coefficients that define how traits in one individual affect the expression of traits in their social partners. Interaction effect coefficients are conceptually homologous to the maternal effect coefficient m (Kirkpatrick & Lande 1989). Thus, by analogy, the same methods used to measure m (such as multivariate model using partial regression or path analysis; Lande & Price 1989) can be used to quantify ψ_{ij} coefficients. One advantage of measuring interaction effect coefficients is that they are phenotypic effects and, consequently, do not require any genetic information to be estimated. However, it is likely to be challenging to determine a causal relationship between traits expressed in one individual and the response of the social partner.

One way to overcome the difficulty of experimentally separating cause and effect is to use artificial stimuli or to otherwise carefully control one side of the interaction to determine the response of individuals to the social environment. There are two ways this has been done. First, although few studies have explicitly set out to do this, it is possible to present artificial stimuli and measure variation to responses to artificially created social environments. In an ironic example of how this might work, Crabbe et al. (1999) inadvertently detected ψ_{ij} by testing how inbred strains of mice differed in measured behavior in different laboratories. Another common approach has been to quantify variation among individuals in response to a panel of different strains ("tester" strains). The variation in aggressiveness and its dependence on

BOX 14.3 Social Effects and the Response to Group Selection

Although the idea of group selection has been disparaged by behavioral ecologists to the point where it has become essentially a bad word, this unfavorable view of group selection theory is largely the unfortunate consequence of a number of misguided applications of group selection ideas to understanding altruism (i.e., the benefit or good of the group notion). This perspective on group selection has been called the "adaptationist school" by Goodnight and Stevens (1997), who contrast these ideas with those of the "genetic school." Although the group selection ideas championed by the adaptationist school were rightfully rejected by evolutionary biologists, the models of the genetic school are based on sound evolutionary theory and are therefore quite distinct from the adaptationist school's group selection models. Here we focus on the quantitative genetic theory that has descended from the genetic school's ideas by discussing the role of social effects in determining how traits respond to group selection (as opposed to individual selection). We avoid discussing the group selection debate (because the models we present are not controversial, in the sense that they work) and suggest that those interested in the group selectionist debate read Goodnight and Stevens (1997) and Bijma and Wade (2008).

When organisms live in groups, the mean social environment may differ among groups and influence the mean phenotype of a group. As a result, there has been an interest in the efficacy of group selection as a means of selecting on traits affected by the social environment (see Griffing 1967; Craig & Muir 1996; Muir 1996, 2005; Bijma et al. 2007a, 2007b; Bijma & Wade 2008). Group selection may be particularly important for behavioral traits and several of the best documented cases in which traits showed a strong response to group selection were for behaviors, including aggression and aggression-associated mortality (Muir 1996; Craig & Muir 1996), social dominance (Moore et al. 2002), and cannibalism (Wade 1976, 1977; Muir 1996, 2005).

When selection acts among groups, it can act directly on social effects whenever the mean social effect differs among groups. In contrast, selection on individuals results only in a response to selection of the mean social environment when there is a genetic association (covariance) between individuals' expression of the trait under selection and their own genotypic value for social effects. Because of this, there are a number of scenarios in which group selection is more efficient for producing changes in mean trait values than is ordinary individual-based selection (see Bijma et al. 2007a; Bijma & Wade 2008). Generally, whenever social (indirect) effects are not necessarily reflected in the expression of the traits of individuals (i.e., the individual phenotype does not predict their breeding value for the social effects that they have), group selection is more efficient than individual selection in producing a response through a change in the mean social environment. Indirect genetic effects can therefore contribute a component of heritable variation that can be difficult or impossible to select on at the individual level, and in these cases it may be that selection on the group mean is more efficient. Although there are a number of scenarios that can make group selection more efficient than individual selection at producing an overall change in a trait (through direct and social effects together), the most intuitive is one in which indirect effects are actually stronger than direct effects (e.g., Bergsma et al. 2008; Ellen et al. 2008). In this case, the fact that the indirect effects can be selected at the group mean level and not at the individual level makes the response to group selection larger than the response to individual selection because the indirect genetic (social effect) variation is available to the response to group selection but not to individual selection (Bijma et al. 2007a).

We illustrate the response to group selection again using the example of a single trait (f) affected by a second trait (s), as in equations 14.3–14.5, assuming that interactions are pairwise. This is a simplified scenario with evolutionary dynamics somewhat different from those expected for larger groups (the efficacy of group selection generally declines as

(continued)

group size increases), but the general conclusions are analogous (see Griffing 1967; Muir 2005; Bijma et al. 2007a, 2007b; Bijma & Wade 2008). Because the environmental effects represent random variation, they have a mean of zero, so we can write the group mean for trait f (labeling interacting group members with subscripts i and j) as:

$$\bar{z}_f = \frac{1}{2}(a_{f(i)} + a_{f(i)}) + \frac{1}{2}(a_{s(i)} + a_{s(i)}), \tag{1}$$

illustrating that the group mean reflects both direct (a_f) and indirect (a_s) genetic effects, and, consequently, that selection on the group mean will act simultaneously on the two.

We first examine the response to individual selection on trait f, assuming selection acts only on trait f (equation 14.19 showed the response of trait f to selection on both trait f and trait s):

$$\Delta\bar{z}_f = \beta_f\left[G_{ff} + \psi_{fs}G_{fs}(1+r) + r\psi_{fs}^2 G_{ss}\right] \tag{2}$$

where terms have been rearranged to facilitate comparison with the response to group selection we consider below. Equation 2 shows that, in the absence of interactions between relatives (where $r = 0$), the only response we get from a change in the social environment is the correlated response to selection, which requires traits s and f to be genetically correlated. Note that when individuals compete, it is predicted that ψ_{fs} will be negative (see Wolf 2003; Muir 2005), and as a result selection on trait f could actually result in a response to selection in the direction opposite that of the selection gradient. This scenario has been of particular interest in agricultural systems, in which competitive effects can make individual selection problematic (Muir 2005).

With group selection, the evolution of trait f is, as for individual selection, predicted by changes in the direct and indirect genetic components. However, if selection acts on group means rather than individual values, then it is not the covariance between individual additive genetic (breeding) values and trait values that predict the response to selection; rather, it is the covariance between individual additive genetic (breeding) values and the group mean (i.e., $C_{a\bar{z};}$ rather than C_{az}) that determines the response to selection. For the sake of brevity, we do not present these covariances here, but leave it to the curious reader to take the definition of the group mean given in equation 1 and to solve for the total covariance between the direct and indirect genetic effects of individuals, a_s and a_f, and the group mean (because the response to selection is ultimately governed by how selection on groups leads to changes in the genetics of individuals). Rather, we will simply examine the final equation for the response to selection acting on the group mean value of trait f (denoting the selection gradient on the group mean of trait f as β_{fG}), which is analogous to the form given for the response to individual selection in equation 14.19 (cf. equation 2 in Muir 2005 and equations 5 and 6 in Bijma 2007a, in which selection is entirely on group means, with group size of 2):

$$\Delta\bar{z}_f = \frac{1}{2}\beta_{fG}\left[G_{ff}(1+r) + 2\psi_{fs}G_{fs}(1+r) + \psi_{fs}^2 G_{ss}(1+r)\right] \tag{3}$$

Comparing the response to group selection (equation 3 above) to the response to individual selection (equation 2) shows that the genetic variance for the indirect (social) effect now shows up in the response to group selection even when individuals are not related, and that relatedness further enhances the increase response to group versus individual selection. For group selection to be more efficient than individual selection, there has to be social effects on trait expression because, in the absence of social effects, equation 2 will always be greater than equation 3 as long as the individual and group selection gradients are of the same strength. Note that equation 2 can produce a negative response to individual

(continued)

selection when there is competition or any other negative relationship between direct and indirect effects (making the relationship between direct and indirect effects antagonistic, i.e., $\psi_{fs}G_{fs} < 0$), whereas the response to group selection has to *always* be of the same sign as the selection gradient (Muir 2005). Equation 3 also includes more terms reflecting relatedness. Generally, selection on groups of relatives will lead to a greater response than selection on groups of unrelated individuals (Griffing 1981b; Bijma et al. 2007a; Ellen et al. 2007; Bijma & Wade 2008).

social context has been quantified for mice using this approach (Fuller & Hahn 1976; Hahn & Schanz 1996). Although neither of these then quantified ψ_{ij}, it should be straightforward (Bleakley et al. 2009).

The second approach is to measure *social performance*. Here, rather than measuring specific traits influencing the expression of the focal animal's phenotype, researchers simply characterize the relative importance and pattern of social effects on trait expression. Wolf (2003) used this approach to examine how social environments influenced the development of body size in *Drosophila melanogaster*. In this study, individuals were derived from a half-sib/full-sib breeding design and split into full-sib environments, environments comprised of half-sibs, or environments comprised of unrelated individuals. The results suggest that direct and indirect genetic effects may be antagonistic due to competition for limited resources, which can constrain the evolution of body size by make the evolution of direct genetic effects opposed by concerted counter evolution of the social environment.

There have been a number of recent quantitative genetic analyses of social performance in pedigreed populations that have used an *animal model* (see Kruuk 2004), in which individuals of varying relatedness are used to fit a linear model that includes both direct and indirect genetic effects on trait expression (e.g., Bijma et al. 2007b; Ellen et al. 2007; Bergsma et al. 2008). These analyses have demonstrated that indirect genetic effects (referred to as *associate effects* in animal breeding literature) contribute a larger component of heritable variance than do direct effects. For example, Bijma et al. (2007a) analyzed survival days of individual chickens living in groups and found that two-thirds of the heritable variance was due to indirect genetic effects arising from social interactions, which was almost entirely hidden to a classic quantitative genetic analysis ignoring indirect effects because there was no significant relationship between direct and indirect genetic effects. Ellen et al. (2007) found a similar pattern for mortality associated with cannibalism in a different population of chickens. Bergsma et al. (2008) found that indirect genetic effects contributed more to heritable variation than direct effects for growth rate and feed intake in domestic pigs. The animal model approach is very promising because it can be applied to natural populations (Kruuk 2004).

There have also been a number of empirical approaches that use a variety of designs to infer or estimate the importance of social effects. Petfield et al. (2005) used a half-sib breeding design and quantified direct and indirect genetic effects in a male sexual trait, the profile of a contact pheromone, in the fly *Drosophila serrata*. They examined the plastic responses of males to their mating partners and were able to estimate the indirect genetic effects that females were having on the pheromone profile of their mating partners. The pattern of direct and indirect genetic effects of females (i.e., their effect on their own pheromone profile and their effect on the profile of their mates) was almost perfectly correlated, suggesting that males may assess and alter their profile to match that of their mates. Higgins et al. (2005) found *emergent* group behavior by studying combinations of isogenic lines of *D. melanogaster*. The composition of the group, and the genetic contributions to that composition, were important for understanding the level of activity within a group, suggesting that the social environment created the persistent patterns of group behavior that they observed. Linksvayer (2006) devised a clever cross-fostering study to separate out colony, maternal, and genetic effects influencing ant mass, caste ratio, and sex ratio. Francisco García-Gonzáles and Simmons (2007) compared offspring sired by males with and without sperm competition between the father and another male and showed that paternal effects on offspring depended on indirect genetic effects of sperm competition—offspring sired by inferior males had enhanced viability if there were sperm from high viability males present, even though these sperm did not fertilize the egg. All these studies examine how the same trait is expressed in offspring experiencing different social environments.

The third approach is through experimental evolution, in which one selects on a trait and measures the selection response in terms of changes in the direct and indirect effects on trait expression. For example, Rhonda Snook developed lines of *Drosophila pseudoobscura* with over 60 generations of selection under different social environments: elevated promiscuity (E), in which each female is housed individually with six males; control promiscuity (C), in which each female is housed with three males (the average number of mates per female in nature); and monogamy (M), in which each female is housed with a single male. Clearly each treatment differs in the opportunity for sexual selection because the opportunity for interactions between the sexes differs. Line differences are not due to genetic differences, as the lines originate from the same stock, but reflect changes in ψ_{ij} (Bacigalupe et al. 2008). A related but somewhat less elegant approach is artificial selection. Moore et al. (2002) subjected lines to varying levels of social competition by competing brothers and breeding only the most dominant (high lines) or subordinate (low lines) individual from each family (within-family selection). Within each line, dominance relationships remained stable and linear, but, consistent with studies showing social behavior depends on the composition of the social environment, males from subordinate lines were very subordinate when placed with males from dominant lines, and vice versa. The group mean of the social trait had changed.

SUMMARY

Researchers have long recognized that studying the genetic basis of social behavior can be challenging because of the contingent nature of the behavior expressed. By definition, groups (albeit sometimes groups of two) are involved in social behavior, presenting difficulties for defining the phenotype of an individual (Manning 1961; Fuller & Hahn 1976; Hahn & Schanz 1996). For example, in one of the earliest evolutionary behavioral genetic studies involved artificial selection for mating speed, Manning (1961) recognized the problems presented with studying genetics of courtship (see also Meffert 1995). As Manning (1961, p. 84) wrote, "There is perhaps little reality in the heritability of a character which involves the interaction between two individuals." Mayr (1964, 1974) described behavior as "open" or "closed" to social experience. E. O. Wilson (1975) noted that there is a multiplier effect for behavior, in which its influence expands as it increases in distribution into multiple aspects of social life. Wilson also argues for behavioral scaling, in which behavior is expressed along a scale and is not genetically fixed. More recently, there has been considerable discussion regarding behavioral plasticity (West-Eberhard 2003) and alternatives (Shuster & Wade 2003), and the role of genetics in this behavioral variation. All in all, behavioral ecologists have long recognized the difficulties of separating *environment* and *genetics* in traits expressed during interactions. West-Eberhard (1979, p. 228) summarized this nicely: "Conspecific rivals are an environmental contingency that can itself evolve."

In this chapter we have attempted to provide a framework to allow researchers to clearly define, and measure, the traits of interest. By placing our analysis in a quantitative genetic framework, we can partition environmental influences into those that are nonsocial in origin, therefore not heritable, and those that are social, and therefore also heritable. Through this quantitative genetic framework, we have identified six major consequences (and there are almost certainly others) of social effects on trait expression that together summarize the primary reasons that people studying behavioral evolution need to include a role for the social environment (the sixth major consequence is presented in box 14.1): (1) social effects on trait expression provide the opportunity for indirect genetic effects, in which part of the genetic architecture of traits in populations can be attributed to mapping from the genotype of one individual to the phenotypes of other individuals, (2) social effects can result in genetically based genotype-environment covariances, in which a genotype may experience a predictable social environment because of genetic associations between interacting individuals, (3) social effects can alter the genotype-phenotype relationship, (4) social effects can produce a covariance (i.e., an association) between the traits expressed in interacting individuals, (5) social effects can contribute to trait evolution due to the evolution of the (mean) social environment, and (6) social effects on fitness can generate social selection on traits (see box 14.1). Finally, we end by reiterating the fact that there are many other scenarios to investigate, and defining the genetic effects and social selection influencing behavior can provide a number of research avenues.

SUGGESTIONS FOR FURTHER READING

Wolf et al. (1998) provide a very general (and largely verbal) introduction to the evolutionary importance of indirect genetic effects. This review takes a somewhat different approach than that we take here and therefore should be a good companion to help the reader understand how social interactions alter evolutionary processes. Cheverud and Moore (1994) focus on the effects of the environment provided by relatives to examine the importance of indirect genetic effects in behavioral evolution. This chapter should be particularly helpful for readers interested in maternal effects and/or in understanding more about the derivation of quantitative genetic models that include social effects. Kirkpatrick and Lande (1989) present the first general multivariate quantitative genetic model of maternal effects, and Moore et al. (1997) provide an analogous multivariate quantitative genetic model of indirect genetic effects resulting from interactions between nonrelatives. Together, those two papers provide a good introduction to the primary literature on indirect genetic effects. Readers interested in group selection should read Bijma and Wade (2008). Readers interested in more general questions about social evolution should read Frank (2006) and will also probably find Frank (1998) an indispensable resource. Finally, we should acknowledge and suggest the seminal work of Bruce Griffing (1981) on the evolutionary importance of interactions among individuals.

Bijma P & Wade MJ (2008) The joint effects of kin, multilevel selection and indirect genetic effects on response to genetic selection. J Evol Biol 21: 1175–1188.

Cheverud JM & Moore AJ (1994) Quantitative genetic and the role of the environment provided by relatives in behavioral evolution. Pp 67–100 in Boake CRB (ed) Quantitative Genetic Studies of Behavioral Evolution. Univ Chicago Press, Chicago, IL.

Frank SA (1998) Foundations of Social Evolution. Princeton Univ Press, Princeton, NJ.

Frank SA (2006) Social selection. Pp 350–363 in Fox CW & Wolf JB (eds) Evolutionary Genetics: Concepts and Case Studies. Oxford Univ Press, New York.

Griffing B (1981) A theory of natural selection incorporating interactions among individuals. I. The modeling process. J Theor Biol 89: 635–658.

Kirkpatrick M & Lande R (1989) The evolution of maternal characters. Evolution 43: 485–503.

Moore AJ, Brodie ED III, & Wolf JB (1997) Interacting phenotypes and the evolutionary process. I. Direct and indirect genetic effects of social interactions. Evolution 51: 1352–1362.

Wolf JB, Brodie ED III, Cheverud JM, Moore AJ, & Wade MJ (1998) Evolutionary consequences of indirect genetic effects. Trends Ecol Evol 13: 64–69.

15

Contest Behavior

MARK BRIFFA AND LYNNE U. SNEDDON

Individuals often come into conflicts of interest over access to limited resources, and when such conflicts are resolved through a direct interaction, they are termed *contests*. Therefore, the key elements of a contest are a resource, of value V, and two or more contestants or opponents, which will pay some sort of cost, C, to engage in the contest with a probability, *p*, of winning the resource. From these key elements it follows that, for an individual entering a contest, the payoff (E) is

$$E = pV - C$$

(adapted from Enquist & Leimar 1987). Contests are resolved when one of the opponents, the loser, is either forced to withdraw or makes a decision to withdraw. Although agonistic encounters sharing these key elements are very common, contests are highly variable in character. In some cases the contest is resolved only when one opponent receives significant damage. For example, male fig wasps, *Idarnes sp.*, fight by attempting to injure and in some cases decapitate their opponent. Contests over access to females in male elephant seals, *Mirounga angustirostris*, can result in fatalities or serious injuries, and male fallow deer, *Dama dama*, employ violent head-on "jump clashes" during the rut. Such examples of injurious fighting, however, are the exception rather than the rule; in most cases contests are settled through the use of ritualized noninjurious agonistic behavior. Although elephant seals can engage in dangerous fighting, most encounters between males are settled using vocalizations and displays of body size. Similarly, male red deer, *Cervus elaphus*, may engage in antler wrestling, which can cause occasional injury, but this occurs only in a small proportion of contests that become highly escalated. In most cases, the encounter is resolved though the use of intense bouts of vocalizations called roaring displays, coupled with ritualized parallel walking. Even the antler wrestling appears to serve as a trial of strength rather than an attempt to intentionally injure the opponent.

Other well-studied examples of noninjurious fighting include claw waving in male fiddler crabs during contests over territory (which influences access to females) at low tide, and shell rapping in hermit crabs during contests over the ownership of gastropod shells. Although contests frequently occur over access to mates, they may also occur over access to other resources such as food, territory, or, as in the case of hermit crabs, shelters.

The key question about contest behavior is this: how can noninjurious agonistic behaviors—especially displays—induce one opponent to give up and relinquish a valuable resource? Early explanations relied on group-selective arguments, such as "excess fatalities from dangerous fighting would be wasteful to the species as a whole," but this possibility is highly unlikely. The alternative explanation is that strategic decision rules are a result of natural selection and based on the balance between costs and benefits (C and V) that accrue to the individual as a result of engaging in a fight. This idea has been the basis of a large body of theory of contest behavior. However, optimality-based models, often used to analyze the evolution of other types of decision, are not adequate to fully explain decisions in a contest. This is because the actions of an individual's opponent will have a direct affect on the utility of an agonistic behavior that the individual might opt to employ. The best course of action will vary according to how the opponent is behaving. Models of contest behavior therefore have to incorporate this element of frequency-dependent benefits into their analysis of decision making. The approach used to achieve this, originally developed for human economics, is *game theory* (see chapter 8). This approach is now applied to a wide range of decisions in which there is a conflict of interest about the outcome of an interaction, such as group foraging, communication, and cooperation, but contests were the first example in which this technique was used in biology.

Although much effort has been devoted to understanding how the key *giving-up decision* (which determines how the encounter is resolved) is made, there are two further decisions that will be important in contests. First, there is the decision of whether to initiate a contest. Decisions about initiating or terminating a contest are said to be *strategic* because they are essentially decisions about whether to engage in a fight or not. Second, there may be a range of *tactical* decisions about how to fight. These may involve decisions regarding a choice of which agonistic behaviors to employ or may involve choices about the intensity of agonistic behavior, resulting in patterns of escalation or de-escalation as the contest progresses. Most studies of contest behavior have focused on the giving-up decision, but how this is made has consequences for fight tactics.

In this chapter we will begin by examining the theoretical logic used to explain the evolution of noninjurious agonistic behavior. Although a body of theoretical models have been developed, some as adaptations of existing models adjusted for specific situations, we focus on a core group of influential models that can be applied generally (although their suitability does vary between different examples of contest behavior). These are the hawk-dove game, the sequential assessment model, the energetic war of attrition, and the cumulative assessment model. We then discuss empirical evidence for factors that are predicted by these models to influence contest outcomes, structure, and duration. Finally, we consider the mechanistic underpinnings of decisions in contests, and how they are influenced by the ecological setting in which contests take place.

CONCEPTS

The Logic of Noninjurious Fighting: The Hawk-Dove Game

It was noted by Parker (1974) that contests are often settled by *convention*, with the use of nondangerous agonistic behaviors, including displays and trials of strength, rather than by injurious fighting. The evolution of *conventional fighting* was modeled by Maynard Smith and Parker (1976) in the hawk-dove game (box 15.1). This game theoretical model considers a hypothetical population containing two alternative strategies and asks which should be favored by natural selection. It shows that if the cost (a negative fitness consequence) of an injury (C) is greater than the value of the resource (V), then using a strategy of performing agonistic displays (dove) can provide greater payoffs than an alternative strategy of dangerous fighting (hawk). However, a population using either strategy could always be invaded by the alternative such that evolution should reach an equilibrium at which there is a mixture of hawks and doves. The ratio of hawks and doves in this *mixed evolutionarily stable strategy* (mixed ESS) would then depend on the ratio of C to V. Only in cases in which V is greater than C would we expect to see a *pure ESS* of dangerous fighting and no displays. How the different payoffs to hawks and doves can lead to a mixed ESS is explained along with the payoff matrix given in box 15.1.

The key assumption of the game theoretical approach is that the benefit of a given strategy is frequency dependent. Further, because individuals act to maximize their own fitness it is often said that the ESS is not the optimal solution for the population as a whole; this would be for all individuals to play dove, but dove cannot be a pure ESS. The basic hawk-dove game is a very simplified version of what may happen in nature. The two alternate strategies are akin to the idea of two alternate alleles that could be subject to selection. Yet such complex behaviors as attacking or performing a display are unlikely to be dependent on a single locus (see also

BOX 15.1 The Hawk-Dove Game and Evolutionarily Stable Strategies

The hawk-dove game (Maynard Smith & Parker 1976) is a development of an initial hawk-mouse model, introduced by Maynard Smith and Price (1973) and Maynard Smith (1974). Both models apply the game theory approach to the evolution of *limited wars* in animal contests, and the hawk-dove game is particularly useful as an illustration of how noninjurious fighting may be selected for over dangerous agonistic behaviors. The model considers two alternative strategies for fighting, either performing an agonistic display (dove) or attacking the opponent (hawk). The payoff matrix is as follows:

		Playing against	
		Hawk	**Dove**
Payoff to	**Hawk**	$\frac{1}{2}$ (V+C)	V
	Dove	0	$\frac{1}{2}$V

If an individual playing hawk (which we denote "hawk" for convenience) fights another hawk, then, all things considered equal, there is a 50% chance of winning and a 50% chance of picking up an injury. In this model receiving an injury triggers an immediate decision to retreat, such that the injured opponent is the loser and winners do not receive injuries. This means that the average payoff to a hawk fighting a hawk, or E(**H**,H), is half the value of the resource (V), a positive payoff, plus the cost of an injury (C), a negative payoff. In some texts, this is denoted ½(V – C) rather than ½(V + C) used here. When a dove encounters a hawk, the dove immediately retreats without performing a display. The dove avoids C but has no chance of winning V, thus E(**D**,H) = 0 and E(**H**,D) = V, as hawks always beat doves without the risk of receiving an injury. Finally, doves have a 50% chance of beating another dove. Because injuries don't figure in these encounters in which both opponents perform only displays, E(**D**,D) = ½ V. By substituting V and C in the payoff matrix illustrated above for numbers that denote the value of the resource and the cost of an injury (both in arbitrary and interchangeable "fitness units"), we can establish the expected outcome of natural selection operating on the two alternative strategies. Selection should favor the strategy that, if used by the majority of the population, cannot be replaced by the alternative strategy. This is the *evolutionarily stable strategy* (ESS) and can be thought of as the solution arrived at by analyzing the payoffs under a particular set of conditions of C and V. First, suppose V = +50 and C = –25, such that V is greater than the absolute value of C (e.g., V > |C|). In a population playing only hawk, E = +12.5. Any individual playing dove instead would encounter only hawks, so E = 0. Relative to +12.5, this is a poor payoff for doves. The dove strategy would be strongly selected against and could not replace the hawk strategy. A population playing only dove, on the other hand, could be very easily invaded by a hawk strategy. For doves, E = +25, but for a hawk that met only doves, E = +50. In this case, the hawk strategy would be strongly favored by selection and quickly spread through the population until dove was completely replaced and E = ½ (V + C) for all individuals. This is not the type of behavior usually observed in animal contests, in which displays are common and injuries are rare. Now suppose that V = +50 and C = –100 so that V < |C|. In this case, E(**H**,H) = –25 and E(**D**,H) = 0. Because 0 is better than –25, the dove strategy can invade a population of hawks. However, E(**D**,D) = +25, whereas E(**H**,D) = +50, so hawk can also invade a population of doves. When V < C, the solution is a *mixed ESS* rather than the *pure ESS* of playing hawk expected when V > |C|. If both strategies can be invaded by the alternative, evolution will reach an equilibrium in which there is

(continued)

a mixture of hawks and doves. An alternative interpretation is that individuals switch between strategies, playing hawk some of the time and dove for the remainder of the time. In either case the ratio of hawkish to dove-like behavior will depend on the difference between V and C. In the basic hawk-dove game, C refers only to the costs of injury, such that it is possible for doves to completely avoid any negative fitness consequences when engaging in a contest. This does not take into account the costs of performing an agonistic display, such as time or energy. Therefore, a refinement to the basic model has been to include display costs C_d, which accrue to doves, in addition to the injury costs C_i, which accrue to hawks. This makes the dove strategy slightly less profitable, but because $|C_i|$ will always be larger than $|C_d|$, it does not change the overall outcome of the game.

chapter 14 for a quantitative approach). Nevertheless, the model makes some very clear predictions about when we should expect to see injurious fighting, and, in a broad sense, they seem to be supported by observations of animal contests. Examples of injurious fighting in nature tend to occur when mating opportunities are dominated by a small number of males and when there is a limited window of opportunity for most males to achieve reproductive success such that V > C. But in most cases V will not be greater than C. It is difficult to imagine, for example, how ownership of a contested food item or a shelter could enhance fitness to such an extent that the benefits of winning outweigh the potential negative consequences of receiving an injury. The hawk-dove game therefore shows how noninjurious fighting is an expected result of natural selection and by considering the frequency dependence of payoffs as well as the values of C and V, explains why dangerous fighting should be rare.

Subsequent models of contest behavior have attempted to explain the functions of noninjurious agonistic behavior; specifically, they attempt to explain why such activities should induce the opponent to make a giving-up decision. Another limitation of the basic hawk-dove game is that it only considers two (extreme) strategies when it is much more likely that there would be a range of alternatives. This possibility is addressed by two extensions to the hawk-dove game, each involving the addition of a third strategy. In the hawk-dove-bourgeois game the third strategy is to respect ownership. An individual using this convention will not engage in a contest with an opponent that already owns the resource. Under this model, bourgeois is the pure ESS when V < C, but when V is high hawkish behavior becomes increasingly likely. The model is still relatively simple, but again its predictions are supported by empirical evidence. In Magellanic penguins, *Spheniscus magellanicus*, territories are highly valued in the run-up to egg laying and fights involving dangerous activities such as pecking and flipper strikes are frequent. Once egg laying is completed, the payoffs to acquiring a new territory are much reduced and fights are rarely initiated against territory owners (Renison et al. 2002). Similar results are obtained by adding an assessor strategy to the basic hawk-dove game. In the original version of the model, there is no specific role for assessment of the opponent's strength, other than the implication that doves must assess the opponent in deciding whether to perform a display or withdraw. Hawks, by contrast, do not use assessment; they simply attack or give up when injured. In the hawk-dove-assessor game, assessors can switch between hawk- and dove-like behavior according to their assessment of the opponent's fighting ability. If the opponent is weaker they play hawk, if the opponent is stronger they play dove. The additional payoffs to add into the matrix (box 15.1) are E(**A**,A) = ½ V, E(**A**,H) = ½ V, E(**A**,D) = ¾ V, E(**H**,A) = ½ V + C and E(**D**,A) = ¼ V. Using values of V < C, assessor is now the pure ESS; all individuals should assess their opponent's fighting ability or *resource holding potential* (RHP). This idea that fighting animals should attempt to assess the RHP of their opponent is central to understanding the functions of noninjurious agonistic behavior. It is also a concept that has divided the theory of fighting into two main classes: models based on the assumption that contest resolution is dependent on assessment of information about the opponent, in which agonistic behavior should advertise RHP (mutual assessment), and models without the assumption that resolution is based on assessment of information about the opponent's RHP (self-assessment).

The hawk-dove game was the first model to deal with the frequency dependence of fight strategies, and forms the basis of our understanding of noninjurious agonistic behavior. Subsequent models have focused on explaining the specific functions of agonistic behaviors; how exactly can activities that don't directly harm the opponent induce it to make a giving-up decision? Although variants on the models reviewed below have been produced, empirical studies have increasingly focused on testing the predictions of three models in particular: the sequential assessment model (Enquist & Leimar 1983), the energetic war of attrition (Payne & Pagel 1996, 1997), and the cumulative assessment model (Payne 1998).

Escalating Contests: The Sequential Assessment Model

The sequential assessment model (SAM; Enquist & Leimar 1983) is based on the assumption of *mutual assessment* (MA), in which both opponents have information about the other's RHP. It aims to explain the evolution of repeated performances of noninjurious agonistic behavior. A classic example of this is fights in the cichlid fish, *Nannacara anomala*, in which the encounter begins with repeated bouts of lateral swimming displays (Enquist & Jakobsson 1986). If even displays are costly (C_d) to perform, why should they be performed repeatedly? Enquist and Leimar (1983) suggested that during a contest, assessment of an opponent's RHP is likely to be made with a degree of error. This is reasonable to assume because contests are often characterized by frenetic activity and direct intervention, which could compromise the ability to gather information and make accurate assessments. Under such circumstances, sampling errors of information that could reveal RHP are likely to occur. However, in a manner analogous to statistical analysis of data, the error will be reduced with each additional sample. The function of repeated performances in the SAM is therefore to allow the opponent to accurately assess the performer's RHP, by attaining a rolling average of the performances with an ever decreasing degree of error. Only when sufficient repetitions for an accurate assessment of opponent RHP have been performed will the weaker individual be in a position to make the decision to withdraw from the contest and avoid the fruitless waste of time and energy and possible risk of injury that would ensue if they were to continue. This need to provide accurate information on RHP leads to the first prediction of the SAM: to maximize the accuracy of the information, agonistic behavior should be performed at a consistent level of intensity.

In many cases, however, contests involve not only repeated activities but also a transition through a series of distinct phases, characterized by the appearance of new activities. Often, the intensity of agonistic behavior escalates between these phases. *Intensity* could refer to increases in the risk of injury or in the energetic demands of performing agonistic behavior, but the key point is that C increases from phase to phase. In *N. anomala*, for example, a phase of lateral displays may escalate to a phase characterized by the appearance of tail beating, then proceed to a phase in which the opponents lock mouths and wrestle (which can cause damage to the mouth), and finally to a phase of biting and chasing in which injuries including damage to scales can occur (Enquist & Jakobsson 1986). Other examples of contests structured into distinct phases follow a typical pattern of progression from acoustic signals, through trials of strength, and finally culminating in potentially injurious fighting before the encounter is resolved. During "roaring contests" in red deer (Clutton-Brock & Albon 1979), the display phases are followed by parallel walking, antler locking and violent jump clashes in the case of fallow deer (Jennings et al 2005). In arthropods such as house crickets *Acheta domesticus* (Hack 1997) and Sierra dome spiders *Neriene litigiosa* (DeCarvalho et al. 2004), acoustic signals given by stridulatory organs or by drumming on the substrate are followed by wrestling and then biting with mandibles or chelicera. The SAM was therefore extended to account for such patterns of escalation in intensity between phases (Enquist et al. 1990). Here, the model explains the basis of the decision to swap between different tactics as well as the strategic decision to give up. In cases in which there are large differences in RHP, it may be possible for the weaker opponent to make the decision to withdraw on the basis of signals alone, and the encounter may be settled relatively easily, at little cost. In encounters in which the opponents are closely matched, however, there may be only a slight disparity in their ability to perform agonistic displays. Therefore, regardless of the number of repetitions, the difference between performances of an activity in which C is comparatively low may always be smaller than the sampling error, such that it is not possible for either opponent to discern the difference in RHP. The encounter can be resolved only if a new activity of greater intensity, which could better reveal RHP differences, is introduced and the contest is escalated to a more intense phase. Although the model predicts a pattern of escalation between phases, there is still the possibility of sampling error within phases. Therefore, within a phase we should still expect to see no changes in intensity. Rather, costs should increase in a distinct series of steps between phases as the contest progresses. In behavioral terms this might produce a consistent rate of a particular agonistic behavior during a phase or, if the phase involves multiple tactics, they should be performed at a consistent rate relative to one another.

The hawk-dove game shows that a proportion of contests should be settled by convention when $V < |C|$ and that dangerous fighting is expected only when $V > |C|$. The SAM makes the further prediction that in addition to a high resource value, opponents must be evenly matched before convention is replaced by dangerous fighting. When opponents are evenly matched, contests are more likely to be protracted and to escalate in intensity. There should therefore be a negative correlation between contest duration and RHP difference, a prediction shared with other models that assume mutual assessment. Under such models, contest duration will be determined by RHP asymmetry, the fighting ability of the loser relative to that of the winner, also called *relative RHP*. Without mutual assessment, contest duration is determined by the *absolute RHP* of the loser.

Costly Displays: Wars of Attrition

In the SAM, contests start with low intensity behaviors such as signals, escalate to noninjurious physical combat such as wrestling, and can finally involve injurious fighting. War of attrition (WOA) models focus on noninjurious agonistic behavior and are therefore applicable only to contests, or at least phases of contests, that do not involve damage. Although the SAM can deal with escalated fighting, WOAs were developed specifically to analyze the functions of agonistic displays. Somewhat counterintuitively, however, most war of attrition models make no assumption of mutual assessment. Rather, they assume that the decision to give up is based only on the level of costs that have accrued to the loser during the contest. Specifically, giving up is triggered when the costs have exceeded an individual threshold. The decision is therefore based on loser absolute RHP. Here, RHP is equivalent to a threshold of a maximum cost that an individual can potentially allocate to the contest; individuals of high RHP are capable of allocating high costs to the contest, whereas those of low RHP will be constrained in their level of allocation. Therefore, RHP influences the upper limit of the cost threshold that triggers the giving-up decision. The second factor that could influence the threshold is the value each contestant places on the resource (V). Here, we introduce C_{max} to represent this actual threshold of maximum cost used during the encounter, as opposed to the potential RHP-limited maximum. In some contests, however, C_{max} may be equal to the RHP-limited potential maximum.

Simple wars of attrition have been called waiting games, and the classic example is contests of persistence between male dung flies, *Scathophaga stercoraria,* waiting on cow pats for the arrival of females to copulate with just prior to ovipositing. As the cow pat dries out, it becomes increasingly less useful as a medium for egg laying, so males are faced with a dilemma about how long to wait. If males wait too long, females start to reject the drying cow pat and the males could have done better by moving off to a fresh one. If males don't stay long enough, they risk losing out to their rivals who persist slightly longer. The ESS here is to be unpredictable, so that rivals cannot win simply by persisting for a little longer, and observations of male dung fly wait times do indeed show a random distribution (see Maynard Smith & Parker 1976). For direct contests to be settled through differences in persistence, the opponents must match one another's level of performance of agonistic behavior. In a contest without such matching, one individual could cheat by paying a lower cost while its opponent proceeds toward C_{max}. In this case, waiting until a threshold is reached could not be an ESS. This matching implies that opponents monitor each other's behavior to some degree, but this does not necessarily mean that the display itself reveals any direct information about RHP. In the absence of such information, the contest will be settled when the first opponent to reach C_{max} makes the decision to withdraw. Therefore, in a war of attrition, RHP equates directly to endurance capacity. This type of decision rule is sometimes called *self-assessment* (SA). War of attrition models assuming both mutual assessment and self-assessment have been developed, but all are centered on the idea of matched performance costs with giving-up decisions based on either the individual's actual or predicted (in mutual assessment based wars of attrition) level of cost accrual in relation to its C_{max} cost threshold.

In the basic WOA (Bishop & Cannings 1978) it was assumed that the main cost of fighting is time allocated to the contest that would be lost to other activities. In this model each opponent decides its persistence time (= C_{max}) at the start of the encounter, and the payoffs are V - C for the winner and C_{max} for the loser. Again, the ESS here is to act in an unpredictable way, so that C_{max} varies between individuals or within individuals from contest to contest. In this way contestants avoid revealing their maximum cost threshold. Such a scenario is more applicable to the resolution of conflicts during exploitation competition rather than direct contest situations as defined above (as in the example of dung flies), but it raises an interesting point about what sort of information could be revealed during a war of attrition. Although it might pay a contestant to advertise its potential capacity for persistence—this could trigger an early giving-up decision in the opponent, minimizing the costs that would have to be expended to secure victory—it would not pay to advertise its actual C_{max}, the chosen persistence time. In a contest in which an individual advertised a willingness to persist for time t, the opponent could then win simply by persisting for a time of $t + 1$ (assuming its $C_{max} \geq t + 1$). In general, it is expected that although information on persistence or *intention* might be revealed unintentionally, fighting animals should attempt to conceal this from their opponent (see Turner & Huntingford 1986).

Extensions to this basic model involve asymmetries in costs and resource value. In a contest in which time is the major cost, the *value* of time allocated to the contest may vary between opponents. A hungry individual may value time lost to foraging more than an individual that is satiated, for example. There may also be differences in the perceived resource value (RV) of the contested resource. During a fight over a food item a hungry individual would place a higher value on the resource than a satiated individual. Thus, in the *asymmetric war of attrition* (A-WOA; Hammerstein & Parker 1982), the contestants value the resource differently and for each contestant C_{max} is chosen on the basis of perceived RV. As discussed above, they do not reveal their actual chosen C_{max} but may advertise their potential C_{max} (for example, from displays that advertise size, which is assumed to correlate with persistence capacity). Therefore, the weaker contestant may give up before C_{max} is reached such that its decision is based on mutual assessment. As in the SAM, the A-WOA therefore predicts that the duration will increase with RHP asymmetry.

Although the SAM and A-WOA incorporate information about the opponent's RHP, in many cases there may be limited opportunities for accurate information gathering. In the *war of attrition without assessment* (WOA-WA; Mesterton-Gibbons et al. 1996), each opponent has an individual threshold of time for which they are able to persist in the contest. Here, time is assumed to equate to energy expenditure, and in the absence of mutual assessment the first opponent to cross its individual energy-cost threshold will decide to withdraw. The *energetic war of attrition* (E-WOA; Payne & Pagel 1996) is based on a similar concept, but decouples the direct link between time and energy. In the WOA-WA, it is assumed that 1 unit of time in the contest requires 1 unit or energy. But in the E-WOA, there is the interesting possibility that energy expenditure per unit of time may vary between individuals, such that the rate of cost accrual may vary between opponents. Indeed, under the E-WOA, agonistic behavior is assumed to be demanding to perform; the function of noninjurious agonistic behaviors such as displays is to demonstrate endurance capacity or *stamina*. In the SAM the estimate of RHP is given by the average of the opponent's actions, whereas under the E-WOA the estimate is given by the sum of the opponent's actions—the energetically costly actions accumulate to give the overall advertisement of stamina. For this reason, the E-WOA has been referred to as a *cumulative* assessment model (e.g., Payne & Pagel 1997), but to avoid confusion with other models based on accumulated actions (see below), it has subsequently been referred to as the *energetic assessment model* or *energetic war of attrition* (e.g., in Payne's 1998 paper describing the *cumulative assessment model*).

Although models that assume mutual assessment and self-assessment make different predictions about how contest duration should be related to absolute and relative RHP (see table 15.1), another key difference between models is in their predictions about contest structure. There are three possibilities for how the intensity of agonistic behavior may change as the encounter progresses: the intensity may escalate, stay the same,

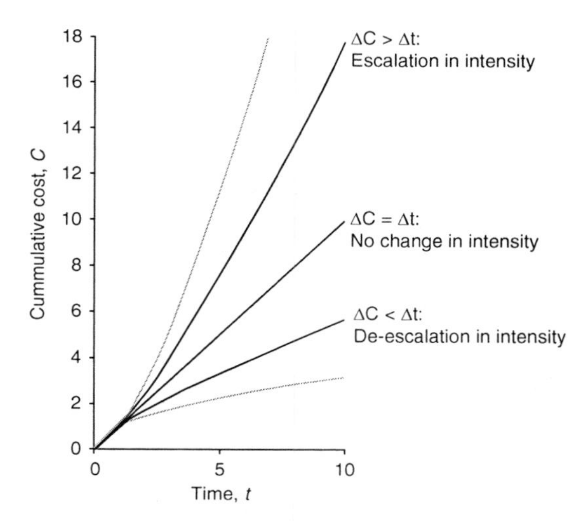

FIGURE **15.1** Relationships between accumulated costs of fighting and fight structure in the energetic war of attrition. The structure of the contest is dependent on the relationship between contest duration (change in time, Δt) and the accumulated costs (change in cost, ΔC) of performing agonistic behavior. If costs accumulate linearly with time, such that an increase from 5 to 10 time units leads to an increase from 5 to 10 cost units, then there will be no change in the intensity of agonistic behavior. If the rate of cost accrual is superlinear ($\Delta C > \Delta t$) such that a doubling of duration leads to more than a doubling of costs (e.g., between 5 and 10 time units the costs rise from 7.5 to 17.8 units), then the intensity will increase. And if the rate of cost accrual is sublinear ($\Delta C < \Delta t$), such that a doubling of time units leads to less than a doubling of costs (e.g., between 5 and 10 time units the costs rise from 3.3 to 5.6 units), then the intensity will decrease. In effect, the pattern of change in intensity is driven by the second derivative of the costs that accumulate over time (d^2C/dt^2). If $d^2C/dt^2 = 0$, there will be no change; if $d^2C/dt^2 > 0$, there will be escalation; and if $d^2C/dt^2 < 0$, there will be de-escalation. In the above example, the values of dC/dt used were 1.25, 1, and 0.75 (from upper to lower black lines). For the gray lines, $dC/dt = 1.5$ (upper) and 0.5 (lower). Here, the rates of escalation and de-escalation would be greater than for the contests represented by the black lines. Drawn from the E-WOA model described in Payne and Pagel (1996, 1997).

or de-escalate. We have already seen how the SAM predicts a pattern of escalation between phases, but no change within a phase. In the WOAs in which the key cost is time (WOA, A-WOA), or in which energetic costs are equivalent to time (WOA-WA), we also expect to see contests escalate in intensity. A limitation of these models is that they contain no specific assumptions about the structuring of contests into distinct phases, a pattern that is seen so often in nature. However, they can be applied to contests that can reasonably be considered to comprise a single phase or, alternatively, to a specific phase of a contest. In these cases, a clear difference between the wars of attrition and the SAM is that they predict escalation within phases. In the E-WOA, energetic costs drive the intensity of agonistic behavior and the relationship between time and energy can vary between contestants. Depending on how energy expenditure accrues with time, in the E-WOA the intensity of agonistic behavior can either escalate or de-escalate as the contest progresses. If energy costs increase linearly, there will be no change in intensity. If the costs increase superlinearly, the intensity of agonistic behavior will escalate, and if they increase sublinearly, the agonistic intensity will de-escalate in (figure 15.1). Note, however, that the assessment of RHP is based on accumulated actions. Therefore, even if a contestant's variation in agonistic behavior involves a reduction in intensity, the opponent's assessment of that contestant's RHP will always increase as the contest progresses (figure 15.2). Changes in intensity will, however, affect the rate at which the increase in perceived RHP occurs and should be matched between opponents.

Costly Displays and Direct Harm: The Cumulative Assessment Model

The E-WOA is a self-assessment-based model in which the costs incurred are due to an individual's performance of demanding but noninjurious agonistic activity. The SAM incorporates the possibility of injurious fighting but only in situations in which there is the possibility of mutual assessment. A final model of contest behavior, the cumulative assessment model (CAM; Payne 1998), like the E-WOA, is based on the idea that the loser should give up when the accrued costs of fighting cross an individual threshold (see above for a disambiguation of the term *cumulative assessment model*). In this case, however, there are two ways that the costs of fighting could accumulate up to the individual C_{max} threshold. First, there are the usual energetic and time costs incurred by performing agonistic

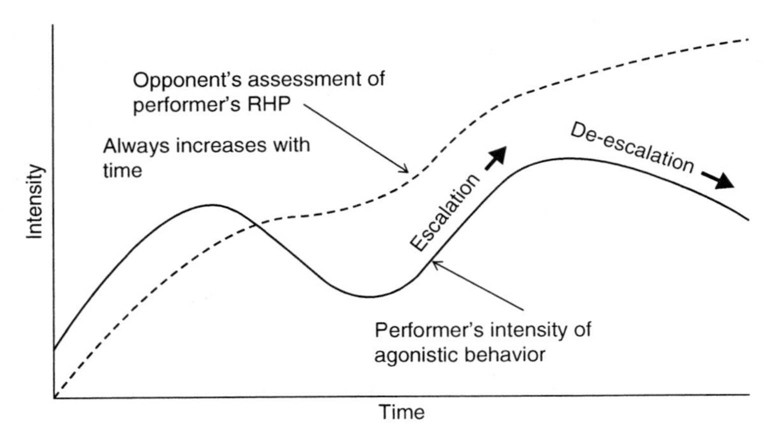

FIGURE 15.2 Relationships between advertisement of RHP and fight structure in the energetic war of attrition. The intensity of agonistic behavior may escalate or de-escalate, but the opponents demonstrate their RHP by persisting in the encounter, such that the advertised level of RHP is cumulative (the sum of all actions so far). This means that regardless of changes in the intensity, the level of RHP advertised by persistence can only increase, although the rate of increase can change. Redrawn from Payne and Pagel (1996).

behavior. Second, this model incorporates the additional possibility that costs, such as injuries, could be inflicted directly by actions of the opponent (all costs, whether derived from injuries, time, or energy, are measured in interchangeable arbitrary fitness units). Under this model, there is no assumption of mutual assessment, but it is equally clear that the actions of the opponent should directly influence the decision of the loser. Therefore, decisions in this type of contest could also be driven by relative RHP rather than by the absolute RHP of the loser, as seen in the WOA-WA and E-WOA, despite the fact that mutual assessment does not occur. This makes it difficult to distinguish between the self-assessment-based CAM and mutual assessment models on the basis of a relationship between contest duration and loser relative RHP. In the CAM, giving-up decisions are still based on an individual threshold that is independent of the opponent's activities. The difference is that the opponent's activities can influence how quickly the threshold is reached.

The CAM also makes its own set of predictions about matching between opponents and contest structure, which distinguish it from other models. This is because it does not rely simply on waiting out the opponent (in terms of who will reach C_{max} first); rather, a fighting animal can help its opponent along the way by imposing directly inflicted costs. Thus, in the CAM, opponents do not match their performances. However, in some respects the CAM closely resembles the E-WOA, another example of a cumulative assessment model, and its predictions about contest structure are rather similar. If the contest (or phase) involves only noninjurious displays, because display rate is determined by how costs accumulate with time, the intensity can escalate, de-escalate, or stay the same. The main difference between these two models, however, is that in the E-WOA an individual's costs accrue from completely internal processes (e.g., determined by elements of the contestant's physiology such as stamina or energy reserves), whereas under the CAM there is the potential for external sources of cost due to the actions of the opponent. Although the activities of the opponents are not matched, if the contest involved injurious fighting and one opponent de-escalated in intensity, a disparity in inflicted costs would open up very quickly. Rather than de-escalating, the weaker opponent should withdraw as its C_{max} threshold is reached. During dangerous fighting, then, the CAM predicts escalation within phases, but not de-escalation. Note that although the CAM can be applied to both injurious and noninjurious phases of contests, unlike the SAM it does not consider escalation between phases. As with WOA models, it can be considered to make predictions about homogeneous *single phase* contests or decisions embedded (Payne 1998) within an appropriate phase of an escalating contest.

Distinguishing between Assessment Rules

Much work on the functions of agonistic behavior aims to test the assumptions and predictions of these models (summarized in table 15.1). Factors such as contest outcome (identity of the loser), duration, and changes in intensity can be quantified during ethological observation (particularly essential for measuring duration and changes in intensity). It is also worth noting, however, that the different models are best applied to qualitatively different types of contest. It is unlikely that WOA-type models will make accurate predictions about contests that progress through a series of distinct phases if analyzed in their entirety.

Many studies have focused on testing predictions about contest duration to determine whether a mutual assessment or self-assessment decision rule is used. A negative relationship between contest duration and loser relative RHP (or *RHP asymmetry*) is assumed to indicate mutual assessment, whereas a positive relationship with loser absolute RHP is assumed to indicate self-assessment (figure 15.3). Taylor and Elwood (2003) pointed out that this approach has a number of problems. First, when absolute RHP differences are large, this can drive a weak but significant relationship between contest duration and RHP asymmetry as a mathematical artifact of a stronger positive relationship with loser absolute RHP. If both possibilities are not tested for, this could lead to a false assumption

TABLE 15.1 Assumptions and predictions of key theoretical models of contest behavior

Model	Asymmetric War of Attrition (A-WOA)	War of Attrition without Assessment (WOA-WA)	Energetic War of Attrition (E-WOA)	Sequential Assessment Model (SAM)	Cumulative Assessment Model (CAM)
Assumptions					
Decision rule based on	Own cost threshold (time) and/or actions of opponent	Own cost threshold (time ≡ energy)	Own cost threshold (energy)	Actions of opponent	Own cost threshold (energy, time, costs inflicted by opponent)
Type of assessment	MA	SA	SA	MA	SA
Function of agonistic behavior	Advertise endurance (by demonstrating persistence capacity)	Advertise endurance (by "keeping up" with opponent)	Advertise endurance (by "keeping up" with opponent)	Reduce sampling error—accurate advertisement of RHP	Inflict costs directly on opponent
Predictions					
Makes predictions concerning	Exploitation competition	Agonistic displays	Agonistic displays	Agonistic displays, trials of strength, injurious fighting	Agonistic displays, injurious fighting
Matching between opponents?	Yes	Yes	Yes	No	No
Contest structure	Escalates within phases	Escalates within phases	Constant, escalates or de-escalates within phases	Escalates between phases but not within phases	Constant, escalates or de-escalates within noninjurious phases; escalates within injurious phases
Duration correlates	– with RHP asymmetry	+ with absolute RHP of loser. But may also be artifact of a weaker—correlation with RHP asymmetry	– with RHP asymmetry	+ with absolute RHP of loser or – with RHP asymmetry; depends on costs inflicted	

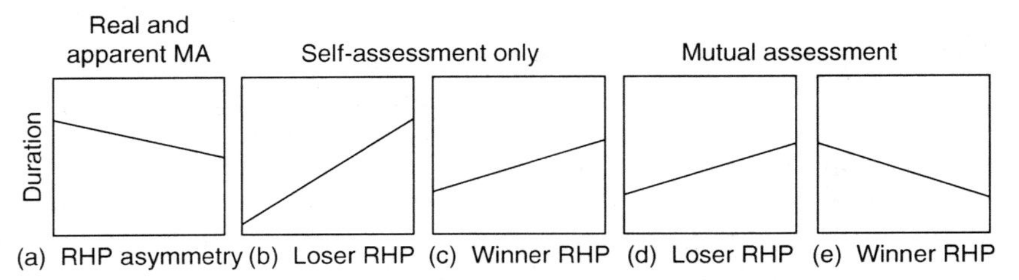

FIGURE **15.3** Distinguishing between mutual assessment (MA) and self-assessment (SA) by examining relationships between RHP and contest duration. If the contest is settled using mutual assessment, there will be (a) a negative relationship between measures of relative RHP or RHP asymmetry and the duration. However, a weaker but significant negative relationship could also occur if the contest is settled by self-assessment, in which the weaker individual decides to give up on the basis of costs accrued. This is an artifact of correlations that tend to occur between RHP difference and the RHP of the weaker individual, especially in staged encounters. To distinguish between self- and mutual assessment, it is therefore necessary to investigate the relationships between contest duration and the absolute RHP or winners and losers. If SA only is used (no assessment of opponent RHP), contest duration should increase with (b) loser RHP and to a lesser extent with (c) winner RHP. If there is mutual assessment, there should be a moderate (d) positive relationship with loser RHP and a moderate (e) negative relationship with winner RHP. Note however, that if the contestants can directly impose costs on their opponent (as assumed by the CAM), then relationships (d) and (e) will be seen in the absence of mutual assessment. See table 15.2 for a summary of models that assume MA and SA. Redrawn from Taylor and Elwood (2003) and Gammell and Hardy (2003).

of mutual assessment. Furthermore, this approach is not useful for CAM-type contests, because duration is expected to vary with relative RHP in the absence of mutual assessment. A safer approach is therefore to investigate winner and loser absolute RHP, rather than RHP asymmetry alone (see figure 15.3). Regardless of what measure of RHP is analyzed, correlations between fighting ability and contest duration can be used only to determine the mode of assessment in use (although this is made more difficult if a CAM-type rule is a possibility—see Morrell et al. 2005). To distinguish between models that assume the same mode of assessment rule, it is also necessary to determine how the contest changes in intensity as it progresses and whether there is matching between opponents (see table 15.1).

Another possibility is that the different models may be applied to different phases of contests (Payne 1998). Indeed, analysis of the relationships between absolute and relative RHP and contest duration in the killifish, *Kryptolebias marmoratus*, indicate that mutual assessment is used during early phases but self-assessment is used during escalated phases (Hsu et al. 2008). This indicates that there may be different functions for the agonistic behaviors used in different phases of the fight. Ideally, then, analysis of the suite of parameters that can be used to test the predictions (summarized in table 15.1—outcome, duration, and structure), possibly involving separate tests for each phase of the agonistic encounter, is the best way to determine how an agonistic behavior functions to induce the opponent to give up.

Contest Asymmetries

Regardless of the precise function of agonistic behavior used, the outcome is ultimately determined by several asymmetries between opponents. As we have seen above, there may be differences in capacity or ability for agonistic behavior between the opponents, which is usually referred to as *fighting ability*, *resource holding power* (Parker 1974b), or *resource holding potential* (Maynard Smith & Parker 1976), the latter two terms abbreviated to RHP. Another factor that may influence the outcome is the difference in value placed on the contested resource by each opponent, the *perceived resource value*, or RV (Maynard Smith & Parker 1976). This may cause motivational differences between opponents such that the individual with higher motivation should be prepared to persist for longer than the individual with lower motivation. RHP is related to the ability to bear and inflict costs (C in the models above), and RV is primarily determined by the absolute value of the resource (V in the models above). RHP and RV are key asymmetries, but further asymmetries may also influence contests. We have seen above that animals may adopt a bourgeois strategy in which asymmetries in ownership appear to influence fight outcomes. Opponents may also differ in age and experience, which could affect their RHP (Kemp 2002). Further, the experience of winning and losing in recent encounters can affect the chance of victory in subsequent contests (Hsu et al. 2006). Finally, opponents may differ in the quality of information about RHP and RV that they have access to. Contest asymmetries are sometimes classified as *correlated* or *uncorrelated* with respect to whether they influence the outcome or payoffs. The value of the resource and RHP are clearly correlated with payoffs. By contrast, the other asymmetries might not influence the payoffs.

In the next section we first consider empirical examples in which the effects of the asymmetries on contest outcome, structure, and duration have been investigated. We first deal with the key asymmetries of RHP and RV. Asymmetries in ownership, experience, and information are also discussed within this framework because they often covary with RHP and RV. We then discuss the proximate mechanisms involved in decision making in contests, before looking at links between these mechanisms and the ecological and social settings in which contests occur.

EMPIRICAL STUDIES: WHAT DETERMINES DURATION, STRUCTURE, AND OUTCOME OF A CONTEST?

Resource Holding Potential

All models of contest behavior assume that engaging in a contest is a costly type of activity and that the rate at which costs accrue or the ability or willingness to pay the costs is what differentiates winners and losers. Some costs that accrue due to factors such as time allocated to the contest or increased predation risk have been called *circumstantial* because they are not a direct result of performing agonistic behaviors. In contrast, the cost of receiving an injury and the costs of energy expended on the contest will accrue as a direct result of performing the specific agonistic behaviors. Direct costs are assumed to have played a major role in the evolution of contest behavior. Theory assumes that the ability to pay (wars of attrition, SAM, and CAM) and inflict (SAM, CAM) these costs will be the key factor that determines an individual's RHP. This will depend on a number of key morphological and physiological variables, the so-called correlates of RHP.

Body size has long been considered a key correlate of RHP. Indeed, it is often used synonymously with RHP, and there are several reasons that we might expect larger individuals to be better at fighting. There may be a mechanical advantage, greater strength and energy reserves, and larger individuals may be more experienced (e.g., if larger individuals are older). A vast number of studies in diverse taxa have demonstrated that overall size is important during fights, and analysis of body size continues to demonstrate its influence on contest outcome, structure, and duration. The size or strength of specific morphological structures used as weapons can be at least as important as overall body size (Sneddon et al. 1997). In addition to size and strength, war of attrition and cumulative assessment models suggest that endurance capacity or stamina should also contribute to RHP. Stamina might vary with energetic status, endocrine status, and aerobic capacity. These factors can vary temporally, and in the case of energy and hormones, during the timescale of an agonistic encounter. Changes in aerobic capacity, on the other hand, are unlikely within this timescale, as are many of the morphological features (notwithstanding injuries). It is therefore useful to distinguish between *static* and *dynamic* correlates of RHP with respect to the timescale of an agonistic encounter.

Early anecdotal evidence of high energetic costs is given by studies of contests in red deer (e.g., Clutton-Brock & Albon 1979) in which exhausted males were reported to collapse after intense bouts of roaring. Fatigue may also constrain the performance of acoustic signals during contests over territory in birds (Weary et al. 1991) and anurans (Ryan 1988). Portunid crabs, *Necora puber* and *Carcinus maenas*, show elevated ventilation and heartbeat (Rovero et al. 2000) rates during fights, indicating elevated respiration. In *C. maenas*, losers but not winners also showed increased ventilation rates during a postcontest recovery phase, indicating that costs could be greater for losers than winners. Joint respiration (by the winner and loser) has been examined using respirometry in house crickets

Acheta domesticus (Hack 1997) and sierra dome spiders *Neriene litigiosa* (DeCarvalho et al. 2004). In both cases, contests lead to a significant elevation in respiration and in both cases the metabolic rate increases as the contests escalate in intensity from displays to physical combat. This escalation in contest intensity is predicted by the SAM, but in *N. litigiosa* there is also a progressive increase in respiration rates within phases. Several studies support the predictions of the SAM, but these within-phase changes in respiration in *N. litigiosa* make it unlikely that agonistic behaviors are repeated to enhance the accuracy of assessment in this case.

The alternative explanation for costly agonistic behavior is that the contest is settled on the basis of individual cost thresholds (WOA-WA, E-WOA, CAM). Although respirometry allows costs to be tracked through the encounter, it does not allow the costs that accrue to each opponent to be measured independently. An alternative approach is to analyze postfight metabolites and by-products (although this does not allow the continuous tracking of energetic expenditures that can be achieved with respirometry). In portunid crabs there are no differences in postfight metabolites between winners and losers, and fighting does not seem to lead to exhaustion (Thorpe et al. 1995; Sneddon et al. 1999). In hermit crabs, *Pagurus bernhardus*, there are two distinct roles in fights over empty gastropod shells. *Attackers* repeatedly perform vigorous bouts of shell-rapping signals (Briffa et al. 1998), and *defenders* remain withdrawn into their shells for most of the encounter. In attackers there is no difference in postfight metabolites between outcomes, but defenders that resist being evicted from their shell have higher circulating glucose levels than evicted defenders (Briffa & Elwood 2002). Thus, in these contests in which there are two distinct roles, energetic status varies between outcomes only for defenders. Rather than being caused by a C_{max} threshold, this difference appears to be the result of an early decision about the level of resistance to offer.

Mobilization and depletion of energy reserves is one potential cost of fighting. However, this may be exacerbated by the potential need to rely on anaerobic respiration in very intense encounters. This will deplete glycogen stores more rapidly and lead to the buildup of lactic acid, which has a negative impact on muscle function. Differences in aerobic capacity, as well as in energy reserves, are therefore one way that individuals may differ in the rate at which costs accumulate as envisaged by self-assessment-based models. Indeed, the use of opercular displays during contests in fish may have evolved as a means of demonstrating aerobic capacity. Raising the gill covers disrupts ventilation, and Siamese fighting fish, *Betta splendens*, show a severely reduced display rate under hypoxic conditions (Abrahams et al. 2005). Anaerobic respiration appears to play a role in fights in other vertebrates (e.g., Schuett & Grober 2000) but appears to be particularly important in crustaceans. In *C. maenas* there is a prolonged postfight elevation in lactate (Sneddon et al. 1999), which constrains the capacity for other activities until the oxygen debt has been repaid. This indicates that if the basis of giving-up decisions is an individual threshold (C_{max}), these costs may persist after the fight has been resolved. Nevertheless, lactic acid accumulation also has the capacity to constrain agonistic behavior during the encounter, and has been investigated in relation to contest structure as outcomes. In the fiddler crab, *Uca lacteal perplexa*, claw-waving rates decline with increasing lactate as the contest progresses (Matsumasa & Maurai 2005). This is predicted by the E-WOA, which is well suited to examples such as this that are based on costly signals (see chapter 16 for a discussion of the distinction between *costly* and *conventional* signals). De-escalation is also possible under the CAM, but this model is more appropriate to situations in which the opponents directly impose costs on one another. A similar result is seen during shell fights in hermit crabs. In both attackers and defenders, lactate increases as a result of fighting but there is only a difference between outcomes in the case of attackers (Briffa & Elwood 2002). During this phase of the encounter the pattern of rapping can either escalate or de-escalate, with successful attackers increasing in intensity before evicting the defender and attackers that give up performing progressively fewer raps in each bout (Briffa et al. 1998; figure 15.4). For attackers, the accumulated costs of fighting appear to drive changes in intensity, as predicted by the E-WOA and CAM. However, these models cannot completely explain agonistic behavior in hermit crabs because they do not encompass contests with strongly asymmetric roles using different agonistic behaviors. Nevertheless, these studies on contest energetics, in a range of study systems, show that energy metabolism is a clear source of the directly incurred costs assumed by theory to be associated with noninjurious fighting.

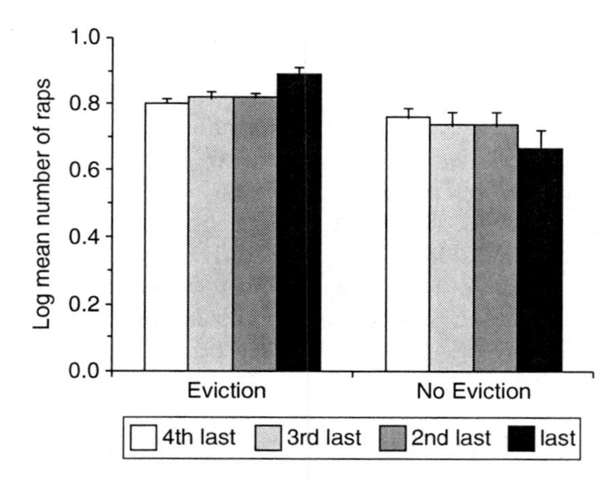

FIGURE 15.4 During shell fights in the hermit crab, *Pagurus bernhardus*, attackers that evict the defender from its shell to allow an exchange of shells perform shell rapping with increasing vigor from bout to bout. Attackers that decide to give up without effecting an eviction (no exchange) perform fewer raps in each bout, suggesting that accumulated costs increasingly constrain the vigor of their agonistic behavior. Data from Briffa et al. (1998).

Resource Value

Strategic decisions should be made on the basis of resource value as well as RHP. In other words, the willingness to commit to the contest or, in ethological terms, the motivation for fighting, should vary with the value of the resource. In general, this means that contests should be costly when V is high. Indeed, the basic hawk-dove game predicts that dangerous fighting is expected when $V > |C|$. This is borne out by examining contests over direct access to mates, especially if mating opportunities are limited. In elephant seals, for example, males can receive serious injuries in fights over access to females (Haley 1994), which are dominated by a single large harem-holding male. One of the most extreme examples of dangerous fighting is seen in fig wasps. Males are polymorphic, and although a small proportion are winged dispersers, most are wingless and do not disperse. Their only chance of mating is by copulating with females that hatch in the same fig. This places such a constraint on male reproductive success that males compete for females by a form of fatal fighting, which involves using the mandibles to decapitate the opponent (Hamilton 1967). In these cases, the value of the resource is so high that it outbalances the risk of an injury. In fact, if mating is not possible without entering a fight, the cost of being killed is equal to the cost of not entering a contest. But by engaging in a contest, the chance of mating is elevated from 0 to 0.5, assuming equal RHP between opponents.

Resource value can therefore influence the duration and structure of contests, but can it influence the outcome? For this to occur, there would need to be asymmetries in resource value between the opponents. Although contests occur over single indivisible resources (Humphries et al. 2006) such that in absolute terms the value (e.g., caloric content of a food item, size of a territory or fecundity of a mate) will be the same for each opponent, the subjective or perceived resource value (RV) may vary between contestants. This could lead to differences in motivation between opponents within contests. Such differences in subjective RV may occur when the opponents occupy distinct roles in a contest. Examples in which RV might vary between roles include old versus young individuals (especially when fighting over access to mates or territories containing mates) and owners versus intruders. In addition to the actual value of the resource for each role, opponents may differ in the availability of information about the quality of the resource and this could also influence subjective RV.

As we have discussed above, in a war of attrition it seems logical that information on RV should not be revealed intentionally, and some studies support this view. In house crickets *Acheta domesticus*, for example, components of acoustic agonistic signals vary with body size, a correlate of RHP, but not with time since mating, a correlate of RV (Brown et al. 2006). In hermit crabs, highly motivated attackers in poor quality shells do not reveal this information during a prerapping display phase of the encounter (Arnott & Elwood 2008). During the shell rapping phase, however, those with the most to gain fight more intensely (Briffa et al. 1998). Similarly, in mouthbrooders and other fishes, signals at the start of the fight correlate with persistence time in the loser (Turner & Huntingford 1986). Even if theory suggests that fighters should conceal their motivation, some information may be revealed during the contest. Irrespective of whether information on motivation is intentionally revealed by agonistic behavior, differences in subjective RV are still likely to influence the outcome and structure of contests. If the decision rule is based on a cost threshold, RHP will determine the potential maximum threshold but the actual maximum threshold used during in the encounter (C_{max}) will be set somewhere below this according to the value placed on the resource. The difference between the potential maximum cost allocated to the contest and the actual C_{max} should be proportional to RV. RV also seems to influence the chance of a contest escalating. In the bowl and doily spider, *Frontinella pyramitela*, contests between males last longer and are more likely to involve injuries when they occur over a highly valued female (Austad 1983).

RV asymmetries are most likely to be important in influencing contest outcomes when differences in RHP are low. Indeed, contests should be settled according to differences in RHP when RV is equal and by differences in RV when there is no difference in RHP (Humphries et al. 2006). Studies on a range of species (see Huntingford & Turner 1987, p. 282, for examples) show that when RHP is similar, the individual that values the resource most highly is more likely to win. Female parasitoid wasps, *Goniozus nephantidis*, fight over hosts in which to lay their eggs (butterfly larvae), and the value of the host increases with its size. Humphries et al. (2006) staged fights under two conditions: first between females that had established ownership and an intruder, then between two females that had established ownership over hosts. Host size affected contest outcomes in a different way for each scenario. For intruders, the chance of victory increased with intruder size, a correlate of RHP, and age, a correlate of RV because older females have more to gain by egg laying. The size of the host affected contest outcomes only when both opponents were owners. Although correlates of RHP such as size, weapons, or endurance can be quantified, the value of a resource is much more difficult to measure. This is because although the absolute value of a resource can be quantified (e.g., size of a female or food item), the perceived RV can vary between opponents and within contests. Our ability to determine perceived RV is limited, but techniques such as motivational probes can reveal some information about how a contestant values the resource. The study on wasps shows that RV can affect contest outcomes but that this effect is subject to subtle interactions with other asymmetries such as ownership status.

Mechanisms of Strategic Decision Making

Energetically demanding behavior requires fast physiological responses such as hormone production to be elicited both in preparation for fighting and for coping with the postcontest consequences of the energetic costs. In contests over mates, seasonal changes in circulating hormones can enhance the aggressiveness of the competing sex, particularly during the breeding season when agonistic interactions occur over mating opportunities (see Huntingford & Turner 1987). Therefore, proximate studies have sought to measure or alter sex hormones to understand their role in the enhanced aggressiveness during these key periods. In many cases the enhanced production of sex hormones has direct effects upon aggressive motivation or fighting ability, leading to success for the individual that has relatively higher hormone levels compared with its opponent. Androgens, male sex hormones such as testosterone, are strongly correlated with aggressive state in a variety of vertebrates in which male-male competition occurs over females. For example, in red jungle fowl, *Gallus gallus*, males that initiated and won fights had much higher testosterone levels relative to their opponent (Johnsen & Zuk 1995), demonstrating that their motivation to fight is higher than their opponents'. However, the effects of androgens can be less perceptible as seen in the cichlid, *Tilapia zillii*. Even though relative gonad size is related to winning, testosterone does not correlate with agonistic success (Neat & Mayer 1999).

Experimental manipulation of androgens via castration, which reduces testosterone in males and thus directly reduces agonistic behavior, or by using implants that increase hormone levels, have shown that male traits, courtship behavior, and aggression are all enhanced by androgens (Huntingford & Turner 1987). For example, sexual ornaments became larger coupled with an increase in aggressive call rate when male red grouse, *Lagopus lagopus*, were given testosterone implants, demonstrating a causal relationship between testosterone and agonistic behavior along with male trait size (Mougeot et al. 2005). In some cases, the effects of androgens can be less clear-cut, especially when competition for mating opportunities is indirect and resources such as territories have to be acquired to attract females. In *Tilapia zillii*, relative gonad size is related to winning but testosterone levels did not correlate with agonistic success (Neat & Mayer 1999). In polyandrous species in which females compete for males, testosterone also increases agonistic displays (e.g., testosterone manipulations in lizards, *Anolis carolinensis*; Woodley & Moore

1999). Female reproductive hormones have also been measured in relation to winning contests, with the winner having relatively higher levels of estradiol and progesterone (Rubenstein & Wikelski 2005).

An explanation for fluctuating androgen levels during development and/or seasonally, which correlates with increased aggression was proposed by Wingfield et al. (1990) and is called the Challenge Hypothesis. In birds with a monogamous mating system, specific changes in testosterone at puberty and during the breeding season are associated with aggressiveness. To enable individuals to compete for territories, dominance, and mates, increases in testosterone at these key stages allow reproductive behavior and aggression to be supported. When males are required to care for offspring, testosterone decreases to support paternal care behavior (Peters 2002). Polygynous birds without paternal care do not show these temporal changes in testosterone, and levels remain high throughout the breeding season (Wingfield et al. 1990). The challenge hypothesis has since been applied to many vertebrate species (Archer 2006). Much of the empirical work has demonstrated testosterone responsiveness to male-male competition and seasonality, but to fully appreciate the role of testosterone in aggression, responsiveness to receptive females, environmental variation, and physiological phenotype need to be investigated. Proximate causes of aggression are not always straightforward, and a variety of life history factors such as mating systems and parental care need to be considered in the control of agonistic behavior.

Neurohormones such as serotonin have been investigated to understand their effects on giving-up decisions during contests or why losers stop fighting. These *biogenic amines* or *catecholamines* are expected to strongly influence both motivation and RHP. In vertebrates, high serotonin is associated with low dominance status and low levels of aggression, and losers are likely to have greater serotonin levels than winners. For example, high dominance rank is correlated with low serotonin in the brain of free-ranging rhesus monkeys, *Macaca mulata* (Higley et al. 1992). However, the opposite is true in invertebrates. Sneddon et al. (2000) demonstrated that serotonin was relatively higher in winning shore crabs prior to and after fights (figure 15.5). Dopamine and octopamine were also higher

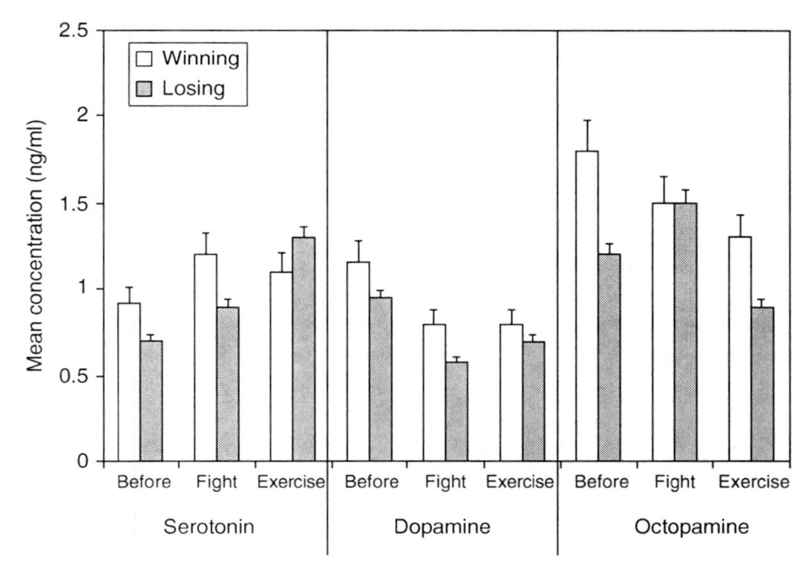

FIGURE 15.5 Mean concentrations of serotonin, dopamine, and octopamine obtained from the blood of shore crabs, *Carcinus maenas*, before fighting (Before), after engaging in pairwise fights (Fight), and after exercise (Exercise). Winning crabs have higher concentrations of all neurohormones prior to fighting. After fighting, serotonin is elevated in winners; however, dopamine and octopamine decline. In comparison, losers have lower concentrations of serotonin and dopamine after fighting, but octopamine, which elicits submissive postures, is elevated. Redrawn from Sneddon et al. (2000).

in winners before fights. Similar results were found in *P. bernhardus* (Briffa & Elwood 2007), in which defending crabs that won the shell fight had higher serotonin levels than those that gave up. Thus, biogenic amines appear to play a role in strategic decision making in contests in both vertebrates and invertebrates, such that they could be used as a predictor of fight outcome prior to a contest taking place.

These hormonal differences thus represent key proximate mechanisms that exert control over contest behavior and, along with information gathering, influence strategic decisions. In comparison to energetic status, their role is complex, with the same hormone often having opposite effects in different taxa. Nevertheless, theory suggests that contest behavior essentially involves the use of demanding activity, information gathering, and strategic decision making, and hormones appear to play a key role in the integration of these three activities. There is clear evidence to demonstrate that neurohormones are involved in giving-up decisions, and there is also evidence to suggest other endocrine mechanisms are involved in the link between information about the contest and decision making. Rapid hormonal fluctuations are seen in fish watching other males fights (Oliveira et al. 2001), thus hormones are influential in contest behavior not only during preparation but also during the interaction (e.g., Californian mice, *Peromyscus californicus*; Oyegbile & Marler 2005). Physiological or hormonal responses are vitally important in winning contests and affect motivation, persistence, and RHP through developmental effects. Therefore, investment in physiological phenotype would be advantageous to ensure competitive success particularly in the breeding season when hormone levels are so crucial in contest outcome.

Ecology of Contest Decisions

Ecological factors such as seasonal changes in temperature or resource availability, diurnal rhythms, and distribution of resources all impinge on the decisions made by contestants. Seasonal changes play an important role in hormone fluctuations, which can affect aggressiveness as previously discussed. In a seasonal context, an animal's behavior is shaped by underlying physiological variables and ecological state to enable the animal to make optimal decisions regarding how much energy to invest in fighting, which can be mediated by motivational changes (McNamara & Houston 2008). Therefore, aggressiveness may fluctuate over time and this will affect decision making during contests. Birds are known to reduce their engagement in contests during winter because they cannot afford to divert energy for this purpose and instead increase the time spent foraging (Witter & Cuthill 1993). Decreasing photoperiod leads to changes in estrogen receptors in mice augmenting their aggressiveness and frequency of fighting (Trainor et al. 2007). This enhanced aggression in winter was considered adaptive to obtain the limited amount of available food during this period. Animal contests can also be affected by daily changes in abiotic factors. For example, oxygen concentration is a limiting factor for aquatic species because low oxygen availability imposes an energetic constraint upon subsequent behavior. In intertidal rock pools, conditions can become hypoxic or even anoxic due to respiration of intertidal algae. Shore crab contests were much shorter in duration when staged under hypoxia (Sneddon et al. 1999) due to an accumulation of the anaerobic metabolite L-lactate that constrained the strategic decisions of the crabs regarding how long to fight.

Most models of game theory find that agonistic behavior will occur only if the contested resource is limited in availability. However, resources can be abundant but are spatially or temporally limited. Optimality models predict that aggression will be highest at intermediate levels of food availability because at low levels there is not enough energy to sustain conflict, and at high abundance there is no need to contest food (Grant et al. 2002). This dome-shaped aggressive response has been confirmed in juvenile convict cichlids (Grant et al. 2002). Game theory models, however, propose that aggression should increase as food availability decreases because food is more valuable when scarce (Sirot 2000). Resources are easily monopolized and defended if they are clumped in space and time (Monaghan & Metcalfe 1985). When resource availability becomes more predictable or less abundant, convict cichlids with higher RHP obtain a greater amount of food by being more aggressive and engaging in more contests (Grant et al. 2002). Female mountain gorillas were thought to be nonaggressive egalitarian foliovores; however, recent studies have shown that strong dominance hierarchies are formed when high-nutrition food is clumped and limited in availability (Scott & Lockard 2006). These studies illustrate the importance of ecological state in determining how aggressive an individual is and how motivated it is to engage in agonistic behavior.

Contests and Social Structure

Most of what we have discussed concerns two animals contesting a given resource; however, many animals live in groups or are territorial and tend to be solitary. RHP differences among individuals living in social groups lead to the formation of dominance hierarchies, and so dominance status is important in the ecology of populations. Position or rank within a dominance hierarchy influences an individual's probability of survival and overall fitness because status within the hierarchy influences an animal's access to vital resources (food, shelter, and mates). Moreover, certain individuals high in social status will be successful, survive, and reproduce and thus experience enhanced fitness. Physiological phenotype is relatively better in dominants than subordinates (e.g., standard metabolic rate, growth, stress), and so the dominant has a distinct advantage in RHP perpetuating its high social status (review in Gilmour et al. 2005). Some animals, such as fish, form simple linear hierarchies (Sneddon et al. 2005); however, complicated dominance relationships can be seen in primates in which subordinates form coalitions to outcompete the dominant and obtain access to resources such as food by allying themselves with higher ranked individuals through grooming (Watts 2000). Thus, within-group aggression in group living animals can be influenced by dominance status and by other complex social interactions.

Exerting and maintaining dominance has shown to be stressful in mammals and avian cooperative breeders with dominant individuals in hierarchies having high stress levels (Creel 2005). Relatively higher glucocorticoids have been measured in the winners of agonistic encounters and results in high stress levels in dyadic contests (Creel 2005). In dominance hierarchies of avian cooperative breeders, the dominant has the highest stress levels because it has to maintain its status while engaging in foraging and other activities (Creel 2005). Stress hormones (corticosteroids and glucocorticoids) are released during the fight-or-flight response and as such have been correlated with aggression. Infusion of corticosterone centrally into rat brain had an immediate effect and enhanced aggressive behavior (Mikics et al. 2004). However, in other nonmammalian vertebrates such as fish, the opposite is seen: winners have relatively lower corticosteroid and glucocorticoid levels compared with losing opponents, and the same is true for the dominant fish in a stable hierarchy (Gilmour et al. 2005). Stress responsiveness has been selected for in rainbow trout, *Oncorhynchus mykiss*, which is measured in terms of how much cortisol is produced in response to a stressor. These two lines show either low cortisol release (LR) or high cortisol release (HR) in response to confinement (Pottinger & Carrick 2001). LR individuals always win fights with high stress responders, and so low stress responsiveness is linked with high aggression and competitive success. Significant reductions in corticosteroids correlated with reduced aggressiveness are seen in socially isolated swordtails, *Xiphophorus helleri* (Hannes & Franck 1983). The ability to cope with stressors may be an important part of a winning phenotype in which either stress hormone production is low or these are better regulated in winning individuals. This forms part of an animal's RHP and may explain why there are no obvious differences in size or weight of individuals, why one prevails over the other. Thus individuals with low stress responsiveness may have a distinct advantage over those with relatively higher responsiveness, and this should be included into theoretical models.

Status within a dominance hierarchy or territory quality can comprise part of an animal's extended phenotype because they reflect the animal's RHP capability (McNamara & Houston 2008). The value of a territory can differ in terms of food availability as seen in kestrels (Daan et al. 1990) or proximity to feeding areas as seen in the distance of oystercatcher's territories from coastal feeding grounds (Ens et al. 1992). To be able to gain and defend a good quality territory demonstrates the high RHP of that territory holder because not only does this individual spend time and energy on defense, but it also has to forage, attract mates, reproduce, and possibly provide parental care (e.g., song sparrows, *Melospiza melodia*; Hyman et al. 2004). Animals with higher RHP acquire and successfully defend the best quality territories. Yet one of the most important observations of territory defense is the influence of prior residence in which the existing owner of a territory always wins fights with intruders (Fayed et al. 2008; Kokko et al. 2006b). Several explanations for this exist: (1) that the owner has a better RHP because it already holds a territory that it had to fight for; (2) that they value the territory more than the intruder because they know its intrinsic value; and (3) that they do not want to be displaced because this means extra costs will be incurred in obtaining another territory and disputing territory boundaries with new neighbors (Fayed et al. 2008). Therefore, the owner is more motivated to fight and escalate to keep the territory (Kokko et al. 2006b). However, empirical work in fiddler crabs has shown that physically obstructing the intruder from entering the burrow is more important than any of these other factors (Fayed et al. 2008).

Agonistic behavior that involves escalated fighting is important in social interactions to obtain resources, maintain and exert dominance, and defend territories, but it can be costly. Therefore, it would be adaptive to avoid such interactions and find an alternative solution to signaling RHP or dominance status. Badges of status are one such solution in which animals use conventional signals to assess an opponent's RHP and avoid engaging in fights they are unlikely to win (Tibbetts & Lindsay 2008). Conventional signals have no real connection to fighting ability per se, but, much like the color of medals handed out to athletes at the Olympics, one assumes that the gold medal winner is a better competitor than the winner of the silver or bronze. These badges of status can be seen in size of the black throat patch in house sparrows (Whitfield 1987) and in many other avian species or in the black facial patterns in paper wasps, *Polistes dominulus* (Tibbetts & Lindsay 2008). In the latter study, wasps preferred food patches that were guarded by wasps displaying a facial pattern indicative of poor quality and avoided those guarded by wasps with patterns suggestive of high quality (Tibbetts & Lindsay 2008). Badges of status must be used during the assessment of a rival to determine RHP and may provide a low-cost, quick means of deciding what to do in a contest.

FUTURE DIRECTIONS

A key question that arises from theory is whether noninjurious fights are settled on the basis of mutual or self-assessment. The consensus view, a result of much careful empirical work, was that mutual assessment was very pervasive. However, Taylor and Elwood's (2003) study on the "mismeasure of animal contests" shows how self-assessment could easily be mistaken for mutual assessment. It is likely that the prevalence of mutual assessment will be reassessed over the coming years. To distinguish among decision rules, beyond the question of mutual versus self-assessment, it is also necessary to investigate how contest outcomes and structure vary with measures of RHP. Analysis of contest structure may require a refocusing on ethological techniques of continuous recording of agonistic behavior during contests. Further, analysis of energetic costs, coupled with data on fight structure, could increase our understanding of decisions in war of attrition and cumulative assessment-type scenarios. Several studies have taken the approach of analyzing the predictions about outcome, duration, and structure listed in box 15.1, and this combined approach may become increasingly common. Even in cases in which all three types of prediction have been tested, it is increasingly apparent than in many situations the contest is not adequately described by any single model. Work on contests with strongly asymmetric roles (Briffa & Elwood 2002) indicates that two opponents in the same fight may use different decision rules. A recent study by Hsu et al. (2008) demonstrates that decision rules appear to change between different phases of the fight. *Variable assessment models*, incorporating different decision rules and rules for switching among them, would be a useful theoretical development. Decisions in contests are often based on repeated displays that signal an aspect of individual quality. In this respect, they may be similar to other signals of quality such as those used in courtship. Payne and Pagel (1997) pointed out that there may be similar reasons for the repetition of agonistic and sexual displays. Presumably, the same logic could be applied to other signals of quality such as food begging by offspring. Analysis of nonagonistic signals within the theoretical framework of contests may provide new insights for both areas of study.

Physiological phenotype plays a crucial role in contest structure, duration, and outcome. Current theory considers physiological costs, in terms of *energetic costs*, as a prime factor in contest resolution. As yet, however, these energetic costs have yet to be measured directly. A direct measure of energetic status, such as levels of high-energy phosphates (the immediate precursor of ATP) may provide tests of the role of energy in contest behavior that is predicted by theory. Comprehensive empirical data sets are necessary to fuel empirical modeling of physiological phenotype from energetic, hormonal, and genomic studies. Data is lacking in the area of global gene expression (see chapter 28 for some approaches that may be useful), and all physiological processes are ultimately controlled at the level of transcription. Similarly, obtaining protein expression profiles using proteomics technology will demonstrate whether the active components of cells (hormones, enzymes) were translated from the genetic code, and a full metabolomic screening can be obtained via metabolomics techniques to examine the end points of the actions of hormones and determine which metabolic pathways are correlated with aggressive decisions. Empirical data sets demonstrate that the mechanisms of contest resolution are complex, but recent work has shown it is possible to produce transcript profiles of dominance status in fish in which the brains of dominant, subdominant, and subordinate fish could clearly be distinguished upon which genes were active (Sneddon et al. 2005). By adopting these contemporary techniques (see chapter 28 for a full review of behavioral genomics), we can now direct our questions to understanding causation at the molecular level among strategic decisions made during contests, the nature of agonistic behavior including escalation, and intensity and correlate these with known physiological responses. We can also understand how interactions among different physiological responses (e.g., metabolites and hormones) affect aggressive performance, which impacts upon contest outcome. Integration of both behavioral and physiological phenotype within the theoretical framework of animal contests is vital and remains the challenge to behavioral ecologists.

SUGGESTIONS FOR FURTHER READING

Maynard Smith and Parker's (1976) paper describes the hawk-dove game, a game theoretical approach which has stimulated many models in contest behavior. A key game theoretical model, the Sequential Assessment Game is described in Enquist and Leimar (1983) and Payne and Pagel (1997) provide a comparison between this model and alternative models of contests involving repeated bouts of activity. Taylor and Elwood (2003) reappraise the question of mutual assessment, specifically considering how best to distinguish this from the possibility of decisions based on 'own-RHP.' *Animal Conflict* by Huntingford and Turner (1987) was published over 20 years ago so does not deal with the more recent theoretical developments. Nevertheless, it gives a very broad ethological overview of the subject and highlights clear links between ultimate functions and proximate mechanisms, which have stimulated much subsequent work. Past experience of winning and losing is reviewed in detail by Hsu et al. (2006) and Oliveira et al. (2001) introduce the idea of agonistic signals in communication networks and Plowes and Adams (2005) introduce contests between competing groups.

Enquist M & Leimar O (1983) Evolution of fighting behavior—Decision rules and assessment of relative strength. J Theor Biol 102: 387–410.

Hsu Y, Earley RL, & Wolf LL (2006) Modulation of aggressive behaviour by fighting experience: mechanisms and contest outcomes. Biol Rev 81: 33–74.

Huntingford FA & Turner AK (1987) Animal Conflict. Chapman & Hall, London.

Maynard Smith J & Parker G (1976) The logic of asymmetric contests. Anim Behav 24: 159–175.

Oliveira RF, Lopes M, Carneiro LA, & Canario AVM (2001) Watching fights raises hormones levels—cichlid wrestling for dominance induces an androgen surge in male spectators. Nature 409: 475.

Payne RJH & Pagel M (1997) Why do animals repeat displays? Anim Behav 54: 109–119.

Plowes NJR & Adams ES (2005) An empirical test of Lanchester's square law: mortality during battles of the fire ant *Solenopsis invicta*. Proc R Soc Lond B 272: 1809–1814.

Taylor PW & Elwood RW (2003) The mismeasure of animal contests. Anim Behav 65: 1195–1202.

16

Signaling

MAGNUS ENQUIST, PETER L. HURD, AND STEFANO GHIRLANDA

Signaling is an intriguing behavioral phenomenon observed, almost without exception, when organisms interact. Signaling occurs between a signaler using a display, often conspicuous but sometimes subtle, and a receiver that perceives the display and then responds. We see signaling between individuals with common interests, such as bees dancing to inform their sisters about distant food, and between opponents, such as paper wasps competing over a scrap of food. Signaling is at the center of courtship, begging, fighting, interactions between flower and pollinator, and between prey and predator. Sometimes more than one signaler and receiver are involved, such as in various cases of mimicry, in eavesdropping, or when several males court a female.

This chapter is a review of the theory of signal evolution. We will focus on a few main questions:

- Why is signaling so common, even when signalers and receivers have conflicting interests?
- What information do signals convey?
- What prevents signalers from displaying misleading signals, in other words, from cheating?
- Why do signals look the way they do?

We begin by adopting a working definition of what a *signal* is (see Dawkins & Krebs 1978; Wiley 1994; Searcy & Nowicki 2005 for in-depth analyses). J. B. S. Haldane noted that a signal's effect, which can be as great as the decision to mate or fight, is caused by only tiny amounts of energy conveyed to the receiver's sense organs. This contrast between minute stimulation and strong effect is so striking that we can use it as a definition of signaling (Wiley 1994). In practice, we often recognize signals just because they catch the eye: signals look like signals, even before we ask what they mean.

The study of signaling has a long history. Darwin and his contemporaries recognized that signals and responses to signals likely arose from the coevolution of signal senders and receivers, but they could not fully elucidate the selective forces driving such coevolution. In particular, they failed to realize that evolutionary conflicts of interests shape signaling within species as well as between species (one exception is Darwin's theory of sexual selection, which acknowledged competition among males for females). Classical ethologists held a similar view, that signal evolution within species is mainly driven by the need to accurately transfer information (Dawkins & Krebs 1978). Empirically, ethologists showed that the behavior of signalers can often be predicted from the signals they use (e.g., a sparrow bending its head forward is more likely to attack its opponent, if the latter does not retreat) and that signals influence receivers' behavior. They also studied how signals can evolve from nonsignaling behavior through a process they called *ritualization* (e.g., baring teeth, a necessary part of attacking, may come to signal aggressiveness; Lorenz 1970; Eibl-Eibesfeldt 1975).

In the 1970s a series of conceptual developments, including the introduction of evolutionary game theory (Maynard Smith 1982), yielded new insights into the evolution of signaling. It became clear that conflicts of interests are widespread even within species (e.g., males and females often have different interests when investing in reproduction; see chapter 23), and that natural selection does not necessarily favor organisms that reveal information about themselves. Indeed, some researchers came to question that signals can evolve to transfer information between organisms with conflicting interests (Caryl 1979; Dawkins & Krebs 1978). Subsequent theoretical work, however, has refuted this conclusion (Zahavi 1975; Enquist 1985; Grafen 1990a), and today most researchers consider signaling as exchange of information between signalers and receivers. The main conceptual framework is evolutionary game theory. It allows us to make predictions about behavior by assuming that, over evolutionary time, an equilibrium has been reached in which individuals cannot improve their fitness by adopting alternative behavior. An increasing number of studies, however, suggests that signaling evolution may not reach such evolutionary end states (evolutionarily stable strategies, ESSs), and that not all aspects of signaling can be interpreted in terms of exchange of information.

Below we summarize the modern theory of signaling evolution and some key empirical findings. Our presentation is divided into two parts. The first deals with the transfer of information between organisms and its strategic aspects, including the possibility of deceptive signals and manipulation of receiver behavior. The second part deals with the evolution of signal form. All of the theory we present rests on the following assumptions about selection pressures:

- Senders (and thus signals) are selected for their ability to elicit favorable responses from the receiver, and not for transferring information (see box 14.1 on social selection).
- Receivers are selected for their ability to make informed decisions about how to respond to a signaler. A receiver thus may respond differently to the same signal in different contexts, respond similarly to different signals, or ignore signals altogether.

Given these assumptions, it follows that signals are not necessarily selected to convey information. Cooperative situations and kinship between signalers and receivers are cases in which natural selection most strongly favors information transfer (see other chapters in section IV), but we will see below that signals can also be deceptive and manipulative.

EXCHANGE OF INFORMATION

We start our exploration of signaling evolution by considering signaling as exchange of information between signal senders and receivers. There are many examples of receivers obtaining useful information by paying attention to signals. For instance, body coloration can provide information about species and sex, displays used in fighting transmit information about motivation and ability to fight, flowers signal the availability of nectar, and prey advertise their toxicity to predators by aposematic signals. Both the evolution and persistence of signaling depend on the presence of information in the appearance and behavior of signal senders: if such information did not exist, receivers would evolve to ignore signals, and thus signaling would not persist.

It is not always obvious, however, why signalers benefit from revealing information about themselves (Dawkins & Krebs 1978). Here are two examples:

- In a scramble over food, less hungry individuals sometimes signal that they are less motivated to fight for food. Would they not benefit from using the most efficient threat display, as the most hungry individuals do, thus causing more opponents to give up the contest?
- Prey sometimes signal to predators that they have detected them, for instance, using a specific call. Consequently, the predator abandons its pursuit. Why do prey not give the "predator detected" signal all the time?

If individuals in these examples always used the "most efficient" display, the display itself would not reveal any information, because it would be unrelated to the actual state of the signaler. Consequently, receivers would evolve to ignore the display, and that would select signalers to not give the display. If a signaling system has to persist, therefore, it must include a mechanism that prevents signalers from cheating. Thus "stability of a signaling system" and "prevention of cheating" are often synonyms.

The fundamental problem in these examples, and in many other behavioral interactions, is that signalers and receivers are often in conflict about how receivers should respond to signals (Guilford & Dawkins 1995; Krebs & Dawkins 1984; Searcy & Nowicki 2005). In the above examples, the signaler always prefers that the receiver give up the scramble or the pursuit, whereas it pays the receiver to do so only under certain circumstances. When conflicts of interests are absent, or small, the evolution of signaling is easier to understand (Krebs & Dawkins 1984) and we expect organisms to evolve efficient and economic ways of exchanging information (Noble 1999). This is, for instance, expected when individuals need to coordinate actions during cooperation (see chapter 17).

At the other end of the spectrum we find interactions without any common interest, for example, between a stalking predator and its prey. We do not expect evolution to favor predators that signal their position, although such signals would benefit the prey.

Most interactions contain a mix of common and conflicting interests, sometimes in a nontrivial way. Two individuals competing for a valuable resource, for example, may seem to share no common interest. If they can injure each other in fight, however, they both have an interest in avoiding unnecessary fights and thus they may profit from signaling whether they are weak or strong, or whether they are hungry or not (see chapter 15). Below we review two theoretical approaches to the evolution of signaling and the problem of signal reliability. We consider first the costs and benefits of information transfer and then the dynamics of signaler-receiver coevolution.

Game Theory and Strategic Equilibria

Game theory is a branch of applied mathematics concerned with calculating the best strategies when one's own success depends on what others do (Gibbons 1992; Myerson 1997). Current understanding of why signaling is so common, even when interests conflict, is largely informed by game theory, complemented with insights from evolutionary biology, such as what determines the fitness of organisms (Maynard Smith 1982).

It is important to understand the scope of game theory. In biological populations, signaling is the outcome of a coevolutionary process between signalers and receivers. Organisms exchange signals and responses to signals, which contribute to their fitness (e.g., a sexual display may be more or less effective in securing a mate). Some signaling and responding strategies, which we assume to be genetically coded, confer higher fitness to the individuals that employ them, and thus increase in frequency. Additionally, genetic mutations continuously create new signals and responses. Evolutionary game theory, however, does not study this coevolutionary process in detail. Rather, it assumes that signaling systems have reached an end point, in other words, a so-called strategic equilibrium in which no individual can increase its fitness by using an alternative strategy. The rationale for such an assumption is that unless an equilibrium is reached, strategies with higher fitness can arise by mutation and replace existing strategies. Thus the only signals and responses that we expect to see in the long run are those that are part of an equilibrium.

To apply game theoretic methods, biological interactions are formalized as *games* between players that seek to maximize their gains (fitness). In many applications, games are described by *payoff matrices*, in other words, tables whose entries give the payoff of a strategy when playing against other strategies (box 15.1). To understand signaling, however, we often need a finer description of what happens during the game. For this reason, signaling games are analyzed by means of *game trees*, in other words, labeled graphs that that represent explicitly all possible sequences of actions, as well as the information that players have at each point in the game. Game trees are illustrated in box 16.1.

Game theoretic studies of signaling have identified three distinct types of signals, called performance signals, handicap signals, and conventional signals (figure 16.1). These types are distinguished by the nature of the mechanism that prevents individuals from cheating, that is, from always using the signal that would produce the response they favor (Maynard Smith & Harper 2003; Guilford & Dawkins 1995; Vehrencamp 2000; Hurd & Enquist 2005).

Performance Signals

Performance signals can be displayed only by individuals possessing a specific capability or knowledge (Hurd & Enquist 2005, Maynard Smith &

BOX 16.1 Game Trees

Analyzing a strategic model of communication requires considering all possible sequences of events, including the actions players can take (e.g., displaying a signal) and random events. This can be effectively done by drawing and analyzing *game trees* such as the one in figure I (technically, the trees are called *extensive form representations* of games—Myerson 1997; Hurd & Enquist 2005). Game trees explicitly represent the temporal structure of the game, in contrast to the more familiar payoff matrix representation (e.g., box 15.1). The tree consists of decision points, or nodes, at which players must choose among different possible actions. To each action corresponds a branch departing from the decision point, which leads to a further decision point where another action should be taken (either by the same or by a different player). A strategy specifies what action to take at each possible decision point. All signaling games include at least one player who does not have complete information at some point in the game. This means that the player will have to choose an action without knowing exactly at which decision point it is. The figure illustrates a simple signaling game in which a signaler displays a signal and a receiver chooses a response (called an *action-response* game). The signaler is in one of two states, *high* or *low*. The states can be anything that has fitness consequences (e.g., resource value or signaler quality). Whether a signaler is in *high* or *low* state is determined randomly in what is called a *move by nature*, in other words, an event that the players cannot influence. Based on its state, the signaler chooses one of two signals, *black* or *white*. The receiver witnesses the signal and chooses either a *strong* or a *weak* response. Crucially, the receiver knows which signal the signaler has chosen, but not whether the signaler's state is *high* or *low*. The dotted lines in the tree join nodes that the receiver cannot distinguish between. Such a set of nodes is referred to as an *information set*. The receiver's strategy must specify one response to use at all the nodes in the set. After the receiver responds to the signal, the interaction is over and payoffs (fitness consequences) are handed out to players. There are eight possible sequences of moves in the tree, each leading to specific payoffs for the signaler and the receiver. The example depicted in the figure indicates a mixture of conflict and common interest between signaler and receiver. The signaler always prefers the *strong* response, whereas the receiver prefers the *strong* response only when signaler state is *high*.

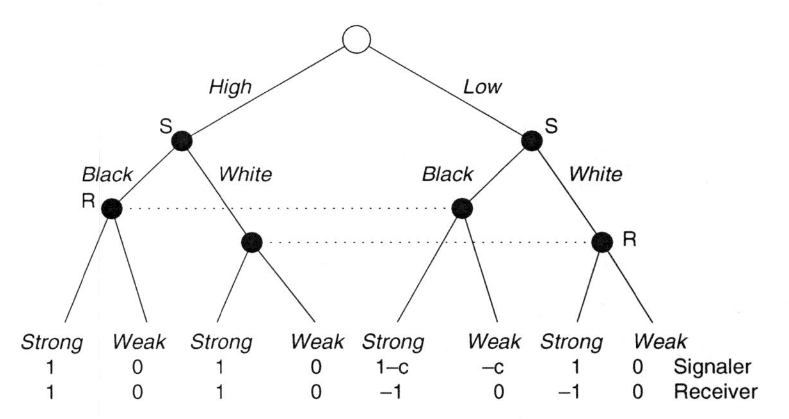

Figure I The basic action-response game. The simplest form of signaling game, in its extensive form. The signaler (S) is in one of quality two states, *high* or *low*, and chooses one of two signals, *black* or *white*. The receiver (R) witnesses the signal and chooses one of two responses, either *strong* or *weak*.

(continued)

BOX 16.1 *(cont.)*

Thus, the players have the same interest in the left part of the tree but opposing interest in the right part. The cost is a state-dependent cost that the signaler pays if it is using the *black* signal when the state is *low*.

To find possible evolutionary equilibria, or evolutionarily stable strategies (ESSs), we look at combinations of a signaler and a receiver strategy that are the best ones to use against each other. If all signaling is cost free ($c = 0$), signaling is not an ESS in this game, because the signaler would always use the signal that most often produces the response it favors (*strong*). However, if $c > 1$, a signaling ESS exists in which the signaler uses different signals in different states, and the receiver uses different responses to each signal (Hurd 1995). To see why this is, assume that the receivers respond with *strong* to the *black* signal and *weak* to *white*. The best thing the sender can do then is to use signal *black* in state *high*, which produces the preferred response. However, it does not pay to use *black* in state *low* due to the cost. Thus, the signaler will provide information about state and the receiver will continue to pay attention to the signal. Note that the cost c is a *condition-dependent handicap*—that is, a handicap that depends on the state of the signaler. This model is a simple version of Grafen's (1990a) model of state-dependent handicap signaling.

Harper 2003). When referring to a physical capability, the term *index* signal is also used. A classic example is the depth of a toad's croak. Given that body size limits how low the call can be, only the largest toads can make the deepest calls. Thus the depth of the call provides information about signaler body size (Fitch & Hauser 2003), and it influences whether receivers continue fighting or give up (Davies & Halliday 1978). Performance signals constrained by the signaler's access to information have received less attention. A potential example is a prey signaling that it has detected a stalking predator by staring directly at it. Only prey that know where the predator is can display this signal.

Unlike the types of signals described below, explaining the evolution of performance displays has never really posed a problem (Maynard Smith 1982). Cheating is simply not possible, because the existence of a physical or informational constraint makes the signal "unbluffable" or "unfakable." Performance displays can be described in a game tree by restricting the moves available to signalers,

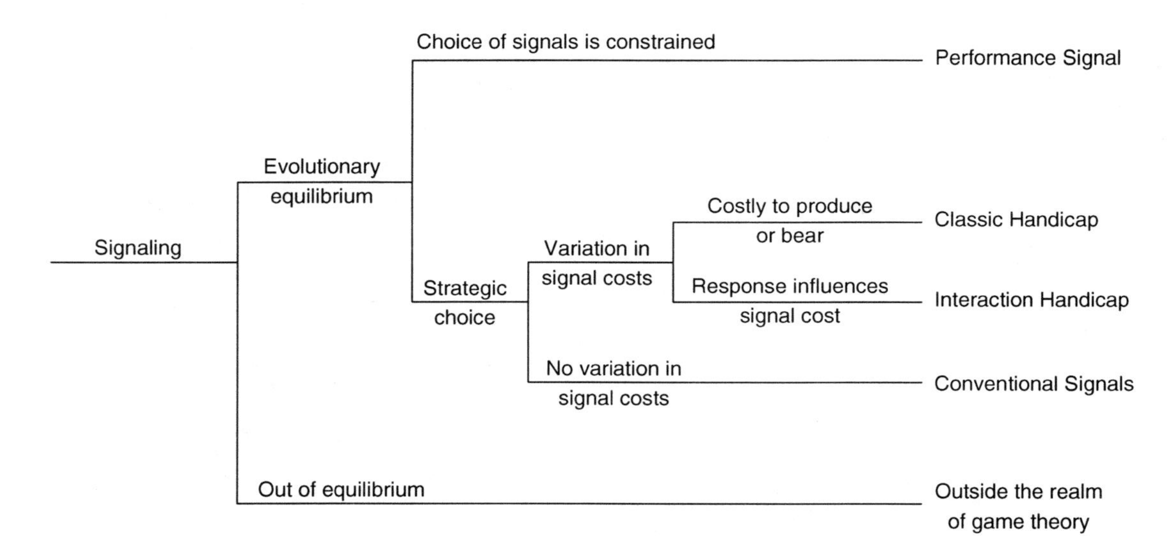

FIGURE 16.1 Proposed taxonomy of signals (after Hurd & Enquist 2005).

in other words, removing some branches from the tree. In the game in box 16.1, for instance, no cheating can occur if the *black* signal is available only to signalers in the *high* state.

Performance signals have been studied in the context of fighting in the *sequential assessment game* (chapter 15). In this model, contestants use performance displays to assess their relative strengths before deciding whether to escalate a fight. Indeed, shows of strengths, such as pushing or pulling the opponent, are very common in fights between animals. When observing actual interactions, however, it is often difficult to determine if an animal is using a less effective signal because it is unable to produce one that is more effective, or because the expected gain is not worth the cost of the signal. The former defines a performance signal, the latter a *handicap* signal (see below). For example, carotenoid-based color patterns have been suggested to be a performance display, because vertebrates are unable to produce carotenoids. Thus carotenoids must be acquired through food, and a sexual display containing a lot of carotenoid-based colors may reflect an individual's foraging ability (Johnstone 1998). Carotenoids, however, also serve an important role in the immune system. Thus the use of dietary carotenoid for display or for immune function may reflect a genetic strategy about how to ration this limited resource (Clotfelter et al. 2007). For this reason, carotenoid-based color pattern could be considered a handicap signal.

Strategic Signals 1: Handicap Signals

A strategic signal is one that any individual *could* use. Cheating is prevented because signalers that try to cheat incur a cost that overcomes the benefit gained from cheating. We distinguish between two types of strategic signals: handicap signals and conventional signals.

Handicap signals are strategic signals that are costly to make or use (Zahavi 1975; Grafen 1990a; Maynard Smith & Harper 2003). The cost is usually assumed to be an inherent part of the signal itself, and to be independent of how the receiver responds to the signals (see Adams & Mesterton-Gibbons 1995; Vehrencamp 2000; Hurd & Enquist 2005 for a *vulnerability* or *interaction* handicap that does involve the receiver's response).

The logic of handicap signaling is illustrated by the game in box 16.1. In this game, a signaler can exist either in a *high* state or a *low* state, and can display either a *black* signal or a *white* signal. All signalers, irrespective of their state, benefit from a *strong* response from a receiver, rather than from a *weak* response. Receivers, however, benefit from responding *strong* to *high* state signalers, but suffer a cost if they respond *strong* to *low* state signalers. Responding *weak* yields neither benefits nor costs. *High* state signalers and receivers, therefore, have common interests; if *high* state signalers could communicate their state to receivers, both parties would benefit. *Low* state signalers, on the other hand, would benefit from deceitfully conveying that they are in a *high* state and elicit a *strong* response, even though this would cost the receiver. Under these circumstances, we ask whether an equilibrium exists in which all signalers accurately communicate their state, that is, in which *low* state signalers would not cheat. This is possible if one signal, for example *white*, has no cost, whereas the other, *black*, has no cost for *high* state signalers but is too costly to use for *low* state signalers (with respect to the gain from a *strong* response). Given these conditions, it does not pay *low* state signalers to use *black*, and at equilibrium they use only *white*. Receivers can thus respond *strong* to *black* without incurring in any costs, and *high* state signalers can use *black* to elicit a *strong* response. Thus the signal accurately communicates signaler state to receivers.

A game with a single choice of signal and a single choice of response, such as the one just discussed, is called an *action-response* game. Variants of such a game, often with more signaler states, have been used to model as diverse phenomena as a nestling's begging (e.g., Godfray 1991), a courting male's secondary sexual traits (Hurd 1995; Számadó 1999), predator deterrent signals (Yachi 1995), and signaling between gametes (Pagel 1993). In some models, the receiver's state can also vary (*dual-state action-response games*), such as in the *Sir Philip Sidney game*, a model of begging and investment in kin (Maynard Smith & Harper 2003), and in models of stomatopod threat displays (Adams & Mesterton-Gibbons 1995). Kim's (1995) status signaling game is a model of handicap signaling with two-way communication (*mutual signaling*). In all of these games, handicap signaling can ensure stable communication. In particular, Alan Grafen's (1990a) mate choice model, in which a given level of signal costs signalers of higher phenotypic quality less than those of poorer quality, convinced most researchers that signal costs are important for the evolutionary stability of signaling.

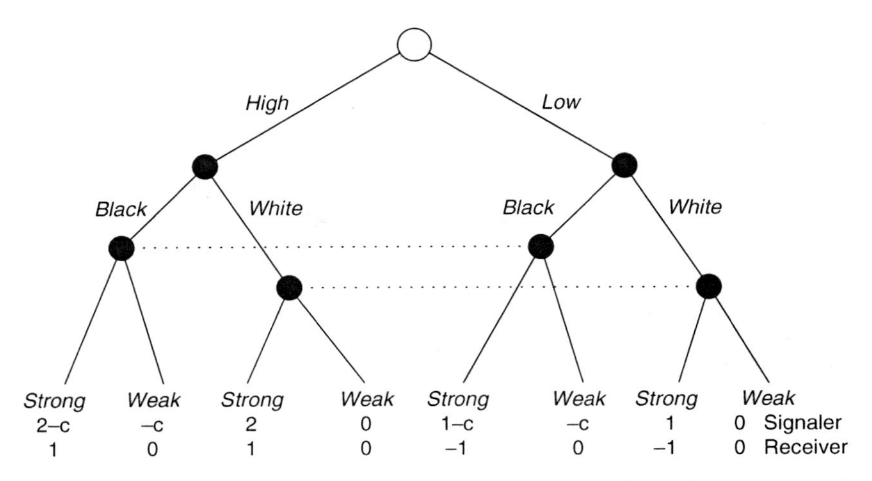

FIGURE **16.2** Handicap signaling with equal signal cost (handicap) to all signalers. Compared to the game in box 16.1, only some payoffs are changed. The cost of *black* is the same for signalers in *high* and *low* states, but the benefit of the *strong* response is higher for signalers in state *high*. If cost c is between 1 and 2, signaling is a stable equilibrium (see text).

There is still some debate over the generality of the handicap principle, partly because opinions differ about what is a handicap. Consider, for instance, the simplified version of Getty's (2006) game in figure 16.2. The cost of signal *black* is the same for all signalers (at variance with the typical formulation of handicap signaling discussed above), but signaling can still occur if the benefit of the *strong* response varies. Note that it is still necessary that the *black* signal is at least somewhat costly. Despite these variants, all action-response games share a common result: stable signaling occurs if the marginal cost of producing a signal is positive, and greater for signalers in unfavored states (Getty 2006). A perhaps counterintuitive result is that some handicap signaling games produce inefficient outcomes. That is, signal costs may result in one or both players obtaining lower fitness, compared to a situation without signaling (Rodríguez-Gironés et al. 1996; Bergstrom & Lachmann 1997).

Given that handicap signals are costly to make or use, they should be easy to recognize in nature—they ought to stand out like tails on peacocks. Many signals are indeed clearly costly. For instance, male Túngara frogs, *Physalaemus pustulosus*, may choose to make more attractive calls when singing to females, but more attractive calls also attract frog-eating bats, *Trachops cirrhosus*, and blood-sucking flies, *Corethrella spp*. (Bernal et al. 2007). Calling is also energetically costly, raising metabolic rate by about 20 times (Wells 2001). Despite many such observations, it has been difficult to produce clear-cut examples of handicap signaling that fit game theory models. One problem is that empirical studies have focused only on a particular kind of handicap (the *condition-dependent* handicap; see box 16.1). A second problem is that costly signals can also evolve for other reasons (see below).

Strategic Signals 2: Conventional Signals

A conventional signal is a strategic signal that has been chosen arbitrarily to convey a given meaning (Guilford & Dawkins 1995; Maynard Smith & Harper 2003; Hurd & Enquist 2005). Because any display can be chosen, cheap ones will be favored. In male sparrows, for example, a black patch of plumage signals aggressiveness, but a white patch could serve this purpose just as easily. In conventional signaling, cheating is prevented because a cost is eventually paid for cheating, not because the signal itself is costly (as it is in handicap signaling). Thus conventional signals can exist only in more complex interactions than considered so far, as illustrated by the following game (Enquist 1985).

Two individuals meet in a contest over a resource, such as a morsel of food. Each individual is either weak or strong, and knows its strength but not the opponent's. The purpose of the game is to show that a signaling convention can evolve to communicate strength between contestants by means of displays that do not inherently reveal strength. At the start of the contest, both individuals simultaneously display one of two signals. After observing the opponent's signal, each individual can choose one of three responses: "attack," "wait then attack," or "give up." The responses chosen determine whether the contest escalates to a physical fight according to the following rules:

- An attack from either contestant necessarily engages both in a fight (giving up is no escape).
- If one contestant responds "wait then attack" and the other gives up, the first wins without fighting. Therefore, "wait then attack" (in contrast to "attack") gives the opponent the option of not fighting.
- If both contestants respond "wait then attack," a fight necessarily follows.
- If both give up, there is no fight, and the resource goes to one contestant at random.

When a fight happens:

- A strong individual always defeats a weak one.
- Fights between individuals of equal strength are awarded at random.
- Both contestants always suffer a cost of fighting (e.g., injury or energy expenditure).

When a strong individual and a weak individual meet, they are both better off if the weak individual gives up the fight. The strong individual would get the resource whether they fight or not, but if the weak individual gives up both avoid the cost of fighting. When two individuals of equal strength meet, however, each has a 50% chance of winning a fight and thus would be better off if the other gave up. In other words, individuals of different strength have common interests when they meet in a contest, whereas individuals of equal strength have conflicting interests.

In this game, an equilibrium exists in which strong individuals display one signal, and weak individuals display the other. At this equilibrium, responses to signals are as follows. A weak individual gives up if the opponent signals being strong, and attacks if the opponent signals being weak. A strong individual attacks when the opponent signals being strong, and "waits then attacks" if the opponent signals being weak, allowing the contest to be settled without a fight. Weak individuals do not cheat (do not signal that they are strong) because, although cheating would cause other weak individuals to give up, it would also provoke attack by strong individuals.

The notion of conventional signaling is well supported by studies of threat and status displays (Hurd & Enquist 2001; Searcy & Nowicki 2005). The black patch of plumage in male sparrows, or a forward extension of the head (a threat display common to many passerines), for example, seems to carry little intrinsic cost and is not necessarily bound to a specific meaning. Observations of actual contests, moreover, corroborate the prediction that a fight is more likely to occur when individuals display the same signal. For example, male song sparrows (*Melospiza melodia cooperi*) and banded wrens (*Thryothorus pleurostictus*) have large repertoires of songs, but there is no specific song that causes the singer to be attacked by territorial neighbors. Rather, most attacks are caused by singing back the same song that the territory owner last sang (Molles & Vehrencamp 2001).

So far we have considered one-shot interactions, but conventional signaling has also been studied in repeated interactions. There is generally a higher degree of common interests between individuals who interact repeatedly, especially if cheaters can be deserted and excluded from further interactions (see other chapters in section IV). It has been suggested that the evolution of human language, the most complex system of conventional signals, has been favored by the fact that we interact repeatedly with the same individuals.

Deception and Manipulation

Although there is ample evidence that information transfer occurs in signaling, there are many examples showing that signals can also be deceitful (Bradbury & Verhencamp 1998; Searcy & Nowicki 2005): flowers that attract pollinators with a colorful display without offering any nectar, some edible prey that fool predators by mimicking poisonous prey, and anglerfish that attract prey wriggling a worm- or fish-like appendage in front of their own mouth. In these cases, the receivers' responses benefit signalers but not receivers. How can such responses evolve and persist? The answer is that deception and manipulation can exist only within a signaling system that, at least on average, benefits receivers (Johnstone & Grafen 1993; Searcy & Nowicki 2005). A simple example follows.

A female spider may consider an approaching male either as a mate or a prey, or be indifferent. A male spider, in turn, would benefit from approaching a female that is seeking a mate, but not one that is seeking food. Females could signal their own internal state to males using different signals for "hungry," "seeking a mate," and "indifferent." This would be in the males' best interest, but it is not the strategy that maximizes female fitness. The best female strategy, indeed, is to use one signal for "indifferent" and another signal for both "seeking a mate" and "hungry." Provided the risk of being eaten is small enough, a male will still benefit from approaching a signaling female, even if this sometimes entails being eaten. Thus the strategy allows females to attract males both for mating and for food.

Anglerfish, that lure prey to approach their moving bait, are another interesting example of deception. These fish exploit the fact that movement signifies a potential food source for prey and must be explored despite the risk of encountering bait and being eaten.

These examples of deceptive signaling can be studied with the same kind of cost-benefit analyses used to study informative signaling at strategic equilibria. However, we will see below that deception and manipulation can also arise in a signaling system that is not at equilibrium.

Conclusions about Game Theory

In this section we have discussed how evolutionary game theory has helped us to understand the exchange of information between individuals with common and conflicting interests. In particular, we have focused on three different mechanisms (performance, handicap, and conventional signaling) that prevent cheaters from exploiting a signaling system. Game theory has made many contributions to the study of signaling that cannot be discussed here. We have not discussed, for instance, situations with more players, for example, mimicry, eavesdropping, and the case of several senders competing for responses from a receiver (flowers competing for pollinator attention, or males competing for female attention). Although these games are more complex, they can be modeled in ways similar to the simpler ones, and the strategic considerations regarding cheating and information transfer are the same (Maynard Smith & Harper 2003; Searcy & Nowicki 2005).

Although game theory has provided tremendous insights into biological signaling, we should be aware that the analysis of evolutionary equilibria also has limitations:

- Equilibrium analysis cannot determine how receivers would respond to the introduction of novel signals. We will see below that new signals can destabilize strategic equilibria.
- The study of equilibria does not help us understand how signals and responses change during evolution.
- Signaling games can have many equilibria (e.g., signals may take many alternative forms), in which case we would like to know if some are more likely to be reached than others.
- Some games have no equilibria at all, but we would still like to know whether signaling can evolve.

These problems are not unique to signaling theory. Models that explicitly consider evolutionary change, or *dynamical models*, are being developed in many areas of biology (e.g., chapters 21 and 23; Dercole & Rinaldi 2008). We consider dynamical models of signaling in the next section.

One last shortcoming is that most models within evolutionary game theory assume that responses to signals are genetically coded. The recognition of many stimuli, however, is learned (e.g., acoustic and visual stimuli in birds and mammals; Hogan 2001). The impact of learning on signaling evolution is not well studied in general, but it is likely that the main conclusions of game theory models hold even when learning occurs. The main reason is that learning and natural selection often shape behavior in similar ways. A well-studied case is *aposematic coloration*, in which toxic prey advertise their toxicity by conspicuous signals (figure 16.5a; Leimar et al. 1986; Roper 1993; Kamo et al. 2002; Ruxton et al. 2004). If signal recognition were genetically coded, then predators would evolve to always avoid aposematic prey because of the negative fitness consequences of ingesting the toxin. In reality, signal recognition is learned, leading predators to avoid signaling prey by associating the signal with the adverse effects of the toxin. Because learning takes only a few experiences, predators will avoid the signal the vast majority of the time even if they are born naïve. Hence aposematic signals can be advantageous to prey irrespective of whether signal recognition is genetically coded or learned.

Dynamical Evolution

It has been repeatedly suggested that it is important to study signal evolution dynamically (Dawkins & Krebs 1978; Enquist et al. 2002). Dynamical models of signaling consider how signaler and receiver strategies change during the course of evolution, rather than focusing only on possible end points. These models are not as mature as game theoretical models, yet they have produced important insights:

- Signaling systems are likely to never reach equilibrium. Hence, signaler and receiver behavior may depart from the predictions of game theoretic models (Dawkins & Krebs 1978; Enquist et al. 2002).
- Signal form and responses to signals depend on the evolutionary history of a signaling system, not only on costs and benefits (Lorenz 1970; Ryan et al. 2001).
- The mechanisms behind a receiver's behavior determine the initial response to novel signals (Ryan 1990), and thus influence which novel signals are more effective. Mechanistic constraints may also prevent perfect adaptation to signals.
- Signalers can employ costly signals simply because they are more effective (Enquist et al. 2002).

We illustrate these points with a simple game, the *game of presence*. A signaler is either present or absent. A receiver, based on the presence or absence of a sender, must decide upon an *effort*, for example, an investment of energy or resources. Receivers benefit most from producing a specific effort when a signaler is present, and no effort in the absence of signalers. There is, however, a conflict of interest between signalers and receivers: signalers benefit from a larger effort than is optimal for receivers. As we have seen above, such conflicts of interests are common. For example, a begging offspring (signaler) is typically interested in getting as much food as possible from a parent (receiver), whereas the parent may benefit from distributing food among several offspring. Similarly, a courting male (signaler) benefits from the female to invest in it as much as possible, whereas the female may want to invest in other males as well.

The evolutionary equilibrium of this game is trivial: receivers produce an optimal effort when a signaler is present, and no effort in the absence of signalers. This simplicity, however, depends crucially on the assumption that signalers can use only one signal. This seems a natural assumption, because one signal suffices to communicate signaler presence. It is common in game theoretic analyses to consider only the minimum number of signals required to transfer relevant information. Nevertheless, in reality the number of possible signals is infinite. If we allow signalers to choose from many signals, the outcome of the game is strikingly different. Computer simulations show that the game has no equilibrium: signalers use a given signal for some time, then switch to a different one. By doing so, moreover, they manipulate receivers into producing a larger than optimal effort—the more signals are available, the larger the effort (figure 16.3, *left panel*). Manipulation is possible because responses to signals that are not used (i.e., that are not under selection) can become larger than the optimum by genetic drift. Such responses can thus be exploited by signalers that switch to a new signal. When this happens, receivers adapt to the new signal, but the cycle of manipulation and counteradaptation starts over whenever signalers find another signal whose response has wandered above the optimum. Figure 16.4 further illustrates this coevolutionary process.

The possibility of eliciting more favorable responses suggests that signalers may use costly signals simply because they are more effective, which is a different explanation of signal cost than given by the analysis of strategic equilibria (see above). In a version of the game of presence in which half of the signals carry a cost, signalers evolve to use costly signals even if equally informative cost-free ones are available (figure 16.3, *right panel*). Costly signals are used because they can (temporarily) elicit a large effort from receivers, to an extent that offsets their cost. This explanation of signal cost seems to agree with the empirical observation that new stimuli with costly features (e.g., larger size or intensity) are often able to elicit strong responses from animals (Ghirlanda & Enquist 2003). Note also that the information content of signals is trivial in this game (presence or absence of signalers), yet it is enough to fuel a continuous stream of new signals and counteradaptations in receivers. If signals were allowed to increase in complexity, the level of complexity reached during evolution would be unrelated to information content (see "Signal Form" section below, and Lande 1981).

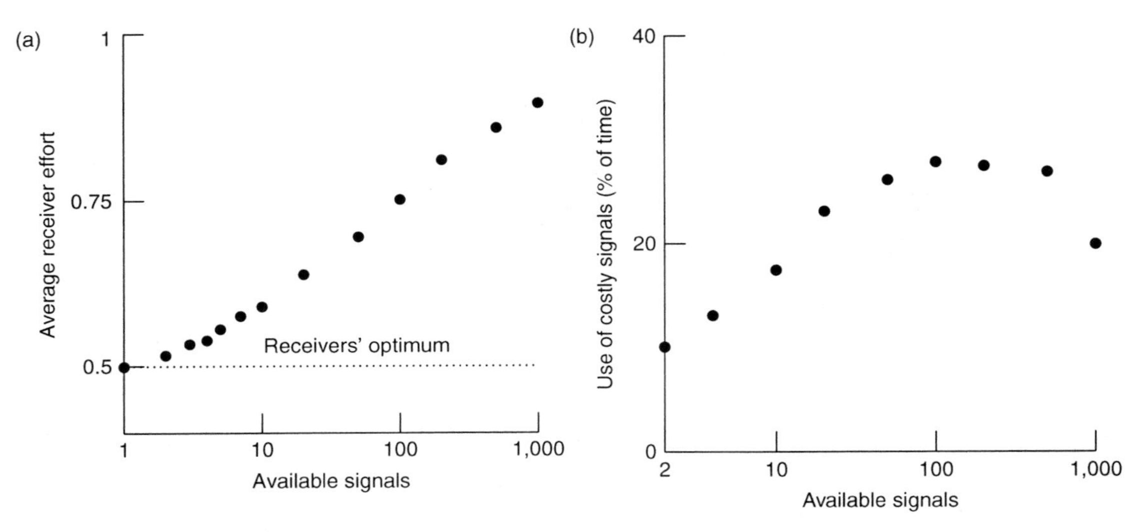

FIGURE **16.3** Simulations of the *game of presence*, using signaler and receiver populations of 1,000 individuals each. (a) Signalers could choose between a number of signals (horizontal axis). The more signals are available, the greater effort signalers elicit from receivers. Each simulation lasted 15,000 generations (we show the average of 10 simulations, with effort calculated on the last 10,000 generations). Simulations started at equilibrium: all signalers used the same signal, and all receivers responded with the optimal effort (0.5 in the simulation) to all signals. Such equilibrium proves stable only when one signal is available to signalers. With two or more signals there is no stability: receivers' effort fluctuates, and signals replace each other (the rate of change of the most common signal varied between 0.004 and 0.014 per generation). (b) The use of costly signals from a new set of simulations in which half the signals were costly to use. In all other respects the simulations were the same as above. Costly signals were used to a considerable extent. The degree of manipulation was somewhat lower compared with the previous simulations.

In the game just described, signalers can switch to any available signal. In reality, novel signals appear by mutation of existing ones, or of traits with other functions (Eibl-Eibesfeldt 1975). What novel variations can arise, therefore, depends on how signals have been shaped throughout evolution. Indeed, signs of past history are often visible in signals (see below). A similar argument holds for receivers. Responses to signals are independent of each other in the above game, but not in reality: which signal is more effective is determined by how behavior mechanisms are wired (e.g., based on similarity to existing signals), which can influence the direction of signal evolution.

Real signaling systems are more complex than the game of presence, yet share some basic elements: the possibility to always generate new signals, the fact that receivers cannot be perfectly adapted to respond to all signals, and an amount of conflict. This suggests that many signaling systems may never reach equilibrium, and that signals (at least the most effective ones) may be able to manipulate receivers. Because we seldom know exactly what is optimal for receivers (e.g., how long to wait before accepting a courting male), low levels of manipulation are hard to detect.

SIGNAL FORM

Imagine combining visual or acoustic elements at random: the result will not look or sound like the signals we see in nature (see figure 16.5). This simple thought experiment demonstrates that signal design is not arbitrary: despite their diversity, most signals follow common design principles. To tell exactly what these are, however, is not trivial. Surprisingly, no systematic study of signal form seems to exist. We list below some features that occur in many systematic groups, and that seem independent of the context in which the signal is used (see Enquist & Arak 1998).

The list, as well as our discussion below, focuses mainly on conspicuous signals:

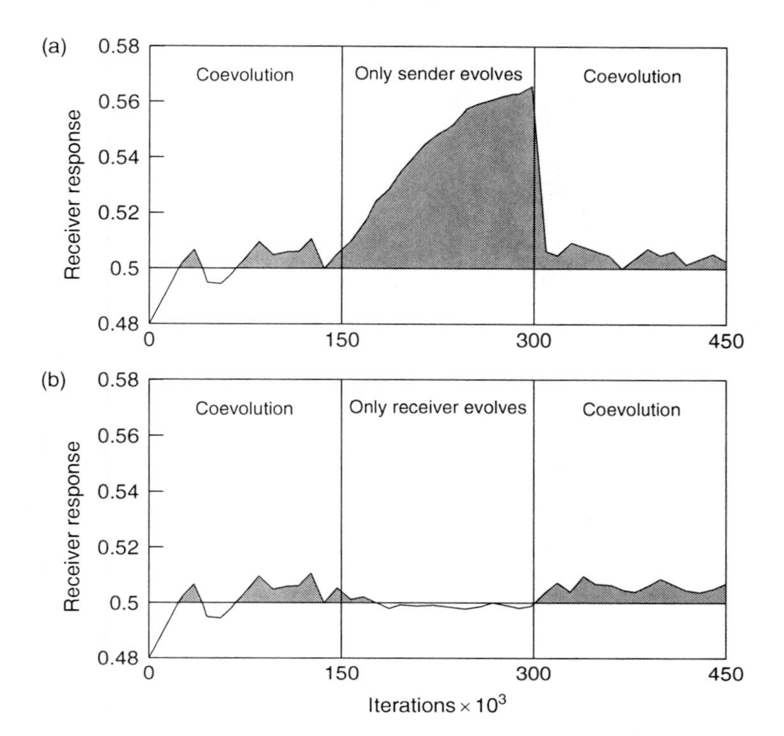

FIGURE **16.4** Simulation of the game of presence using a three-layer neural network as the receiver. Parts (a) and (b) both show how the average receiver effort changes over time. The receiver's optimal effort is 0.5. Shading shows when the receiver is manipulated (effort >0.5). First, the signaler and receiver were allowed to coevolve for 150,000 iterations of simulated evolution (both diagrams show this part). From this point the simulation was split in two. In (a), only the signalers were allowed to evolve. In (b), only the receiver evolved. After another 150,000 iterations, coevolution was again introduced. As shown, some degree of manipulation is maintained during coevolution. When the receiver's evolution was stopped, the signaler became more efficient at manipulation. When instead evolution of the signaler was stopped, the receiver evolved an optimal effort.

Distinctiveness Many signals are conspicuously different from their habitual background and from other stimuli in their environment (Cott 1940). Signals that elicit different responses appear often antithetical (figure 16.5d). Signals involving movement typically contrast with the movement of surrounding objects.

Exaggeration Many signals, the peacock's tail being a classic example, are "exaggerated" in at least two senses (Brown 1975; Andersson 1994). First, the signal has evolved from a less conspicuous ancestral state, for example, a shorter or less colored tail. Second, the signal appears excessive for the purpose of being detected, for example, because it is used at close range. Examples of exaggeration are elongated tails and plumes in birds, enlarged petals in flowers or fins in fish, saturated colors, loud calls or songs, and the production of large quantities of pheromones.

Multiple Elements and Multiple Modalities A signal is typically made up of distinct elements. For instance, ladybugs combine red (or other) color with distinct black dots. Many displays use more than one sensory modality, for example, courtship displays using both visual and acoustic stimuli.

Contrasting Elements Elements within a signal often contrast with each other. A patch of color, for instance, may neighbor patches of complementary colors or be outlined by a contrasting border.

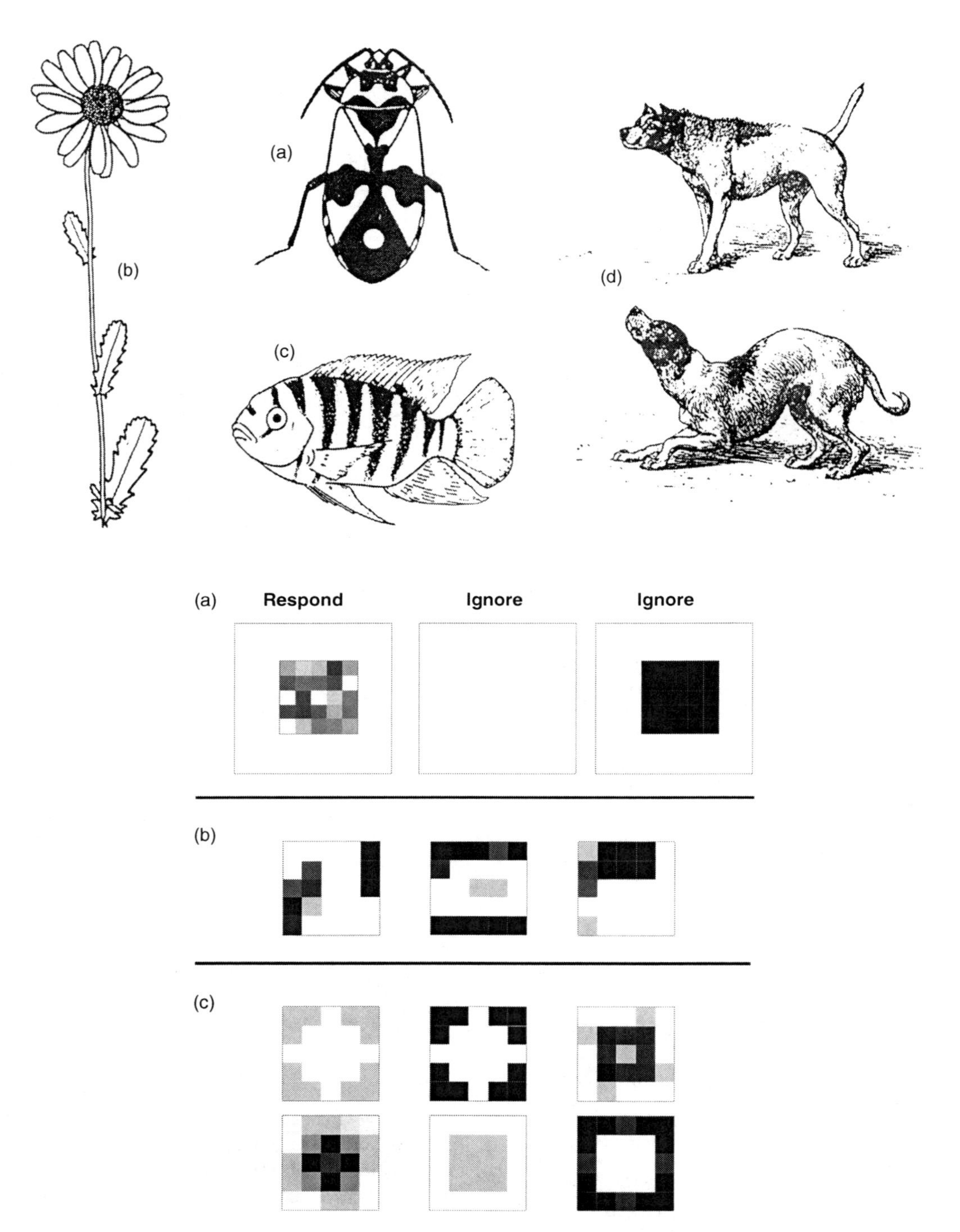

FIGURE 16.5 *Top*: Signals in nature, showing typically conspicuous features of contrast (a, c), repetition (b, c), and symmetry (a, b); (d) shows contrast in posture (e.g., ears and tails raised or lowered) between aggressive and submissive displays in dogs. Reproduced from Enquist and Arak (1998), with permission. *Bottom*: Signals evolved in computer simulations using artificial neural networks as signal receivers. Networks could change their responses by mutation of internal parameters (modeling synaptic connections), and signals could change by mutating the color of points in a 5×5 array. Signals were perceived by

Repetition A signal element may be repeated within the signal, as in the black and yellow stripes of wasps. Other examples are repeated movements within a visual display and repeated elements in bird and whale song.

Symmetry Elements of visual displays are often arranged according to bilateral, radial, or other kind of symmetry.

The three following characteristics of signals do not relate solely to form, but pose deeper puzzles regarding signal evolution:

Signal Form Provides Few Clues about Signal Function Traits like weapons and adaptations for physical endeavors (legs for locomotion) or environmental circumstances (fur for cold) are often easy to recognize. In contrast, we are unable to find systematic links between form and function in signals, as when we look at the bewildering diversity of pattern and color in coral reef organisms. Many signals, moreover, are used in several contexts (table 16.1), suggesting that they cannot be interpreted based on their form alone.

Signals Evolve Rapidly Among closely related species, signals are usually dramatically more variable than nonsignaling traits. For example, in some populations of the lizard *Urosaurus ornatus*, males show a polymorphic throat pattern (Hover 1985), and different populations show from one to five distinct patterns (Hews et al. 1997). More generally, body coloration varies to a much greater extent within genera than do other morphological features such as feet, limb, or body shape. Such variability suggests that signal form often changes rapidly.

Traces of History It is often possible to infer the history and origin of a display through observations of behavior or by comparing signals in related species. For instance, courtship rituals sometimes include elements from feeding behavior that have become stereotyped and have lost their original function (see below).

Below we explore the evolution of some these features using both equilibrium strategic and dynamical arguments. As noted above, strategic considerations cannot explain all aspect of signaling evolution. This, as we will see, is even truer for signal form than for exchange of information.

Signals Must Be Able to Carry the Information

For a signal to work, both at evolutionary equilibrium and out of equilibrium, some minimum requirements must be fulfilled. One requirement is that the signal is capable of reaching receivers and stimulating their sense organs (Bradbury & Vehrencamp 1998). For instance, signals used over long distances need greater physical intensity than signals used in short-range communication. In noisy environments such as a rain forest, signals have many adaptations that make successful reception more likely (chapter 9). A further requirement for a signaler that wants to be recognized is to use a signal that is distinct from other stimuli in its environment. Similar sympatric species, for example, tend to use different signals for species recognition, for example, frog calls or woodpecker hammering patterns. Some aspects of signals have also been related to the opposing aims of reaching desired receivers and avoiding undesired ones. For example, a speckled egg can be conspicuous to its parents at the nest and cryptic for predators farther away.

These factors can explain some, but not all, of the exaggeration, distinctiveness, and complexity of signals. For instance, great tits (*Parus major*) use exaggerated postures to threaten one another at short distance when there is little risk that the signal is not detected (Hurd & Enquist 2001). Similarly, butterflies such as the common yellow swallowtail

FIGURE **16.5** ***(continued)*** networks at different positions on a model retina (a). Networks were selected to respond to an initially random signal with a specific numerical output, and to produce no output in the absence of a signal, or when a black signal was perceived. Signals were selected to elicit maximum responding from networks (the same kind of conflict as in the game of presence; see text). During signaler-receiver coevolution, contrast-rich signals evolved (b, also featuring saturated colors in the original simulations). When networks had to recognize signals from different orientations, they evolved to respond most strongly to symmetrical signals (c). Reproduced from Hurd et al. (1995) with permission.

(*Papilio machaon*) have complex wing patterns that can be seen only at close distance. Moreover, even after fulfilling the requirement of reaching the receiver and being distinct, a signal could still take many possible forms (with the possible exception of performance signals).

Evolution of Exaggeration

In the last 15–20 years, dynamical models of signaling evolution have provided new insights into the evolution of signal form. The main result is that exaggeration (costly and noncostly), symmetry, and other properties of signals may evolve because receivers develop, during evolution, preferences for signals with such properties. These preferences are by-products of how nervous systems process stimuli, and are not necessarily related to the information content of signals.

Suppose, for instance, that a pollinator can encounter two flower species, one much richer in nectar than the other. The nectar rich species has one slightly longer petal. A flower's fitness is given by the number of visits it receives from pollinators, minus a cost that is larger for longer petals; a pollinator's fitness is proportional to the amount of nectar it gathers. Enquist and Arak (1998) simulated this scenario using neural networks as receivers. Mutations could change flowers' petal length and receivers' memory parameters that determine how the neural networks respond to signals. Nectar poor flowers did not evolve.

During the simulations, nectar-rich flowers evolved to exaggerate the initial difference with nectar-poor flowers, developing a very long petal. The reason is that when a neural network evolves (or learns; Enquist & Ghirlanda 2005) to ignore a small stimulus and to react to a large one, it will usually react even more to larger stimuli. This implies that a mutant flower displaying an elongated petal will receive more visits, and will thus be favored by selection unless petal cost is too high. Note that the exaggerated petal length conveys the same information as the initial one—a flower's species identity. Moreover, the exaggeration that evolved was not necessary for successful discrimination, as the neural networks could solve the initial discrimination equally well.

The mechanism highlighted in this study appears quite general, because discriminations among stimuli are known to cause a variety of response biases in animals (Mackintosh 1974; Ghirlanda & Enquist 2003). An animal that prefers a large stimulus to a small one, for instance, will usually prefer larger stimuli even more. Discriminations between different hues, saturations, and intensities of light all produce biases, as well as discriminations in sound frequency and intensity. Neural network models reproduce many aspects of these biases (Enquist & Ghirlanda 2005), and simulation studies using neural networks as signal receivers have successfully reproduced many aspects of signal form, such as bright, saturated, and/or contrasting colors (figure 16.4, *bottom*; Enquist & Arak 1998). Neural network simulations have also inspired animal experiments. In an experiment on the evolution of saturated colors, Jansson and Enquist (2003) allowed hens to peck color patterns on a touch screen, rewarding with food pecks to one color but not to another. Several slight variations of the rewarded color were shown on every trial, and the most pecked ones "survived" to later trials. The initial difference between rewarded and unrewarded colors was slight, but hens' pecking preferences quickly led to its exaggeration in the course of the experiment, similarly to what had been observed in network simulations. This result is also consistent with previous suggestions that learning in receivers can promote the evolution of exaggeration (Kamo et al. 2002; Guilford & Dawkins 1993; Leimar et al. 1986).

That receiver biases may influence signal evolution has been suggested several times, independent of neural network studies, and referred to as *sensory exploitation* or *sensory bias* (box 24.1 of this volume; Endler & Basolo 1998). There are now several case studies suggesting that this process has indeed contributed to the evolution of existing signals, such as male swordtails' (*Xiphophorus helleri*) elongated caudal fin (Basolo 1995b) and the mating song of the Túngara frog (Phelps & Ryan 1998). In the latter case, ancestral mating calls have been reconstructed and used to train neural networks to distinguish them from noise. Only networks trained on the correct phylogenetic sequence of calls had the same biases as real females. This shows clearly that biases that we observe today (and hence selection pressures on signals) may depend on the past history of a signaling system that has left a trace in receivers' recognition mechanisms.

In equilibrium analyses of signaling, exaggerated signals have been considered examples of handicaps, in other words, signals that, because they are costly, reveal information about signaler quality (see above). Dynamical models have shown,

TABLE **16.1** Patterns of social behavior of laughing gulls, *Larus atricilla*, and the contexts in which they occur

	Rivalry	Nest defense	Pair formation	Courtship feeding	Precopula-tion	Nest-site selection and nest-building	Nest relief	Parent-chick (filial)	Adult-chick (hostile)	Nonsocial disturbance	Flock fight	Communal feeding
Upright	X		X	X	X		X		X	X		
Oblique	X		X	X			X	X	X			
Horizontal	X		X	X								
Facing away	X		X	X			X					
Choking	X					X	X					
Stooping	X		X	X	X	X	X	X				
Head toss	X		X	X	X		X					
Long call	X		X	X			X	X	X			
Crooning	X		X	X	X	X	X	X				
Ke-hah	X		X	X			X	X	X			
Kow	X	X							X	X	X	X
Kek-kek										X	X	X
Gackering		X										
Quavering wail		X									X	
Regurgitation				X	X							

Source: Beer (1975).

however, that exaggeration may evolve without being informative. The extent to which exaggerated signals in nature are informative or manipulative remains to be determined.

Evolution of Symmetry and Repetition

Computer simulations have also been used to study the evolution of symmetry and repetition. From a strategic perspective, symmetrical or repetitive signals are a good solution to the problem of being easily recognized, because they have similar appearances under different viewing conditions. For instance, a radially symmetric flower appears similar to pollinators approaching from different directions, whereas a repetitive pattern such as stripes can be identified even when partially occluded by other objects. However, ease of recognition is not a sufficient reason for symmetrical and repetitive signals to evolve. It is necessary that a mutant signal that is slightly more symmetrical or repetitive than existing ones elicits a more favorable response from receivers. That is, receivers must have a preference for such a mutant.

Several studies show that experience with asymmetrical signals can indeed create receiver preferences that favor the evolution of more symmetrical signals. If a neural network is trained to recognize, for instance, two slightly asymmetric cross patterns, it may develop a preference for crosses with a lesser degree of asymmetry (Enquist & Arak 1998). Similar results have been obtained in animals (Jansson et al. 2002). These results imply that mutations in a signal that increase the degree of symmetry would be favored during signaler-receiver coevolution. This outcome has indeed been observed in coevolutionary simulations. In these studies, particular experiences created a preference for symmetrical patterns bearing particular relationships with experienced asymmetrical patterns, not a generalized preference for symmetry (Enquist & Arak 1998).

Kenward et al. (2004) simulated the effect of receiver preferences and mode of signal presentation on the evolution of repetition in signals. They found that a neural network evolved to recognize a particular pattern could develop a preference for patterns consisting of repetitions of the original one. Perfect or near perfect repetition evolved in coevolutionary simulations under many conditions, for instance, when signals were partially obstructed by other objects.

Evolution of Complex Signals

If all aspects of a signal carry information, a complex signal should communicate a complex message (Møller & Pomiankowski 1993). Stickleback nuptial coloration, for example, depends on at least two pigments, which have been suggested to convey different information to females (Wedekind et al. 1998). Whether or not it is possible to communicate different kinds of information through different signal elements has been studied theoretically by Johnstone (1995a) in the case of handicap signals. He concluded that multicomponent handicap signals are unlikely to convey accurate information about different qualities. At most, receivers could derive from such signals an estimate of overall signaler quality. To our knowledge, there is no similar investigation of multidimensional conventional signals, but the bee dance is an empirical example. Different aspects of the dance communicate information about the direction, distance, and quality of the food source. A multidimensional signal could inform about different qualities also in the case in which each signal component is a performance signal pertaining to a specific quality (see above).

In dynamical models of signaling, complexity can evolve without being tied to information. Rather, the complexity of a signal may derive from the accumulation of signal elements during signalers' continuous struggle to find novel, more effective signals (see "Dynamical Evolution" section above). A simulation of the game of presence, in which signals could grow in complexity (e.g., by having a physical structure that evolution can elaborate upon), would show this effect. If receivers pay attention to several modalities, signalers will evolve to effectively stimulate all available channels.

The Origin and History of Signals

The origin and history of a signal can often be inferred through comparison with behavior and signals in related species (Ryan et al. 2001; Autumn et al. 2002). This is one of the most important contributions of classical ethology to the study of animal behavior (Lorenz 1970). For instance, Lorenz reconstructed the systematic relationships among duck species comparing male courtship behavior, tracing as well the modification and accumulation of signal elements during evolution (see figure 7.1).

Sometimes the origin of a display can be traced to a behavior without signaling functions. Threat displays, for instance, sometimes contain elements of attack behavior, like advancing or baring teeth. Likewise, Schenkel (1956) traced the origin of a courtship display in peacocks to feeding behavior, by studying courtship in related species (see *ritualization* in Eibl-Eibesfeldt 1975). Traces of history can also be seen in receiver mechanisms (see "Evolution of Exaggeration" section above). Recently, historical analyses of signaling have returned with force (Ryan 1990; Ryan et al. 2001), emphasizing the importance of a dynamical approach to signal form. Equilibrium strategic theory is not informative about history and origin: all traces of history should be gone at equilibrium, when signals and response are predicted to have reached an optimum.

SUMMARY AND FUTURE DIRECTIONS

For signaling to evolve and persist, signals must contain information that allows receivers to make better responses (promoting fitness), at least on average. Deception and manipulation can occur, but cannot be so pervasive as to strip signals of all information: if signals ceased to be informative, evolution would favor receivers that ignored signals, and the signaling system would collapse.

Exchange of information between signalers and receivers can evolve in several ways. If signalers and receivers have identical interest, no special arrangement is needed: conventional (cost-free) signals are expected to evolve. When evolutionary conflicts of interests exist, signaling can evolve provided that signalers are prevented from trying to always elicit the most favorable response from the receiver. Either the access to signals must be restricted, so that some signals are available only to some signalers (performance display), or cost may be directly associated with the signal (handicap signals), or costs may be incurred from receiver responses when signalers try to exploit efficient signals (conventional signals).

The evolution of signaling cannot be completely understood in terms of information exchange and cost-benefit analyses. The reason is that signal evolution, in particular when there are evolutionary conflicts, is likely to never reach evolutionary equilibria. Signals can change in many ways, and new variants appear regularly due to mutations and other genetic changes. Some of the new variants can elicit more favorable responses from receivers, thus spreading among signalers. Receiver biases in responding are a general phenomenon, present in all systems capable of stimulus recognition. After a new effective signal has spread, receivers will be under selection to evolve an optimal response to it—with a necessary time lag. The process of influx of new signal variants and adaptation of responses has no end. Exchange of information is not necessarily prevented, and may occur to an extent similar to what is predicted from strategic models. Crucially, however, out-of-equilibrium evolution offers a new explanation for signal form and cost, in which behavior mechanisms and evolutionary history are more important than usually granted in behavioral ecology and game-theoretical approaches.

Under this explanation, the evolution of signal form is the result of a long process of signal modifications partly driven by continuously changing response biases in signal receivers. Exaggerated signals may arise in situations in which little information is exchanged, in which case little meaning can be attributed to specific signal parts and features. Our understanding of costly signals is also modified by this perspective. If signal evolution proceeds out of equilibrium, we expect signalers to use costly signals simply to exploit response biases in receivers.

In conclusion, our judgment is that a general theory of the evolution of signaling is emerging, which is applicable across all contexts. On the other hand, a number of specific theories have developed their own views and interpretations of signaling in various contexts (e.g., sexual selection, begging, aposematic coloration). An important question is whether there are general principles of signal evolution that explain, say, symmetries and repetition in all kinds of signals, or if seemingly similar features have different explanations in different contexts. For instance, sexual selection theory has been concerned with signaling that occurs between males and females in the context of reproduction. In this context, it has been argued that symmetrical displays serve to advertise to females a male's genetic quality.

We end this chapter pointing out some features of signaling that, we think, will need more consideration in the future:

Cross-Contextual Displays The observation that sometimes the same signals are used in many contexts has received little recent attention (Yabuta 2008). Table 16.1, for example, shows that laughing gulls, *Larus atricilla*, use more or less the same displays in aggressive interactions, courtship, and nest relief. What is the meaning of cross-contextual displays? Yabuta (2008), extending the theories presented in this chapter, suggests that cross-contextual displays enable animals to discover what context they are in, for example, giving them time to recognize their partner from an intruder. Cross-contextual displays might also stem from the basic design of behavioral machinery. How modularized is signal perception? Are, for instance, species and sex of potential partners assessed separately or within a single system? And how does this affect signal evolution?

Learned Recognition Most models of signal evolution assume that signal and receiver strategies are genetically coded. As noted earlier, however, the recognition of many signals is learned, at least in birds and mammals. How does learning influence signal evolution? Are the conclusions based on genetically based strategies still valid?

Comparative Analyses of Signal Form We need more detailed comparative analysis of signal systems within and among species and larger taxa, to be able to draw better empirical conclusions about the general issues discussed above. It would be particularly interesting to know whether elements of signal form are different or the same in different contexts, such as sexual, aggressive, aposematic, begging, attraction of pollinators, and so on.

Do Animals Have a Sense of Beauty? The fact that signals seem to share design features suggests that there may be general principles of animal aesthetics, whereby receivers of different species evaluate signals (Darwin 1871). Kobayashi (1999), for instance, has shown that mynahs (*Acridotheres tristis*) spontaneously choose to inspect pictures of symmetrical over asymmetrical peacock tails (see also Ghirlanda et al. 2002). Is this a general finding? And, if so, what are the underlying principles?

Acknowledgments We thank Johan Lind, Cecilia Monari Lipira and two anonymous reviewers for useful comments on previous versions of the manuscript.

SUGGESTIONS FOR FURTHER READING

The literature on signaling is vast, and it is well worth starting with classic papers. However, there are a number of books that cover the topic and are excellent guides to all the issues. We recommend Bradbury and Vehrencamp's (1998) textbook for a broad and comprehensive coverage of animal communication. A book by Searcy and Nowicki (2005) provides a more detailed look at the notion that the signaling systems contain a tension between honesty and deception. Maynard Smith and Harper (2003) provide a concise and insightful review of game theoretic approaches to signaling. Ruxton et al. (2004) present a thorough analysis of signaling by prey to their potential predators. Finally, Enquist et al. (2002) explore limits to game theoretic approaches to signaling in cases in which conflicts lead to elaborate signals to which receivers respond inappropriately.

Bradbury JW & Vehrencamp SL (1998) Principles of Animal Communication. Sinauer, Sunderland, MA.

Enquist M, Arak A, Ghirlanda S, & Wachtmeister C-A (2002) Spectacular phenomena and limits to rationality in genetic and cultural evolution. Phil Trans R Soc Lond B 357: 1585–1594.

Maynard Smith J & Harper DGC (2003) Animal Signals. Oxford Univ Press, Oxford.

Ruxton GD, Sherratt TN, & Speed MP (2004) Avoiding Attack: The Evolutionary Ecology of Crypsis, Warning Signals and Mimicry. Oxford Univ Press, Oxford.

Searcy WA & Nowicki S (2005) The Evolution of Animal Communication: Reliability and Deception in Signaling Systems. Princeton Univ Press, Princeton, NJ.

17

Behavior in Groups

RYAN L. EARLEY AND LEE ALAN DUGATKIN

If you indulge frequently in nature, be it on land or under water, you most likely will have noticed that some animals live a solitary existence whereas others appear to assemble in groups. This simple observation has inspired decades of research on the origins, maintenance, and economics of group living in diverse social contexts ranging from short-lived foraging associations and seasonal breeding aggregations to dominance hierarchies with stable membership. And, as is inevitably the case for most simple questions, the emerging story has indeed become quite complex. What sorts of environmental pressures would favor group living? What are the benefits and costs of living in a group? How are resources distributed among group members? Are social groups structured through behavioral interaction? Can group living facilitate the transfer of information between group members? The goal of this chapter is to explore these questions in some detail and to convey the richness of experimental and theoretical approaches that have been applied to understanding group living.

Throughout this chapter, we broadly define a social group as a *collection of individuals that actively cluster together, exist in close proximity in both space and time, and engage in behavioral interaction; a social group also is a discrete unit that is distinct from other such groups*. This definition encompasses both single-species and mixed-species assemblages. Although we will focus primarily on single-species groups in this chapter, we are not implying that groups consisting of two or more species are any less dynamic or worthy of investigation. Dolby and Grubb (2000), for instance, provide a compelling case in birds for reduced predation risk in satellite species (e.g., downy woodpeckers; *Picoides pubescens*) that join existing mixed-species flocks (see also Gibson et al. 2002). Mixed-species groups often adhere to the same rules of aggregation as single-species groups, and individuals within these groups likely experience a similar matrix of costs and benefits as individuals that associate exclusively with conspecifics. Our definition of a social group excludes predominantly territorial species because these species defend a defined area against conspecifics in lieu of associating with them. There is, however, plenty to learn about social behavior by studying territorial animals (Stamps & Krishnan 1999). Lastly, this chapter focuses primarily on animals other than eusocial species—those that exhibit reproductive division of labor, overlapping generations, and cooperative care of the young (e.g., termites, ants, wasps, bees, naked mole rats)—because the phenomenon of eusociality is covered in detail in chapter 19.

Social groups can consist of unrelated individuals or kin, can coalesce over various time scales (e.g., at a resource patch for a short time, seasonally, or for many years), and can exhibit a wide range of social and spatial structures. We begin the chapter with an overview of the forces that may have driven animals to live in groups, and the benefits and costs imparted by a social lifestyle. We will then explore in more detail how structure, in its many guises, is imposed upon a social group through behavioral interactions among its members. Building upon the behavioral component of social groups, we conclude the chapter by delving into some of the more complex behaviors that perhaps have evolved since grouping arose, including the evolution of sophisticated cognitive abilities that may be used to acquire information about the physical or social environments.

WHY GROUP?

Benefits and Costs to Living in Groups

Since Alexander's (1974) seminal review on the evolution of social behavior, there has been an explosion of studies seeking to understand the factors that might contribute to group living in animals. Table 17.1 provides a summary, although not exhaustive, of the benefits that animals might gain by congregating in groups. Much of this work has centered around the effects of predators on prey, and the potential benefits of grouping that prey might experience in the form of predation avoidance (Krause & Ruxton 2002). There are two ways to think about the relationship between groups and predation risk. First, one might ask whether aggregating is of higher net benefit than solitary living and, if so, how animals accomplish the task of grouping. Second, one might ask whether group living, once instituted, might favor the evolution of behavioral strategies that further reduce predation risk, facilitate emergent properties of the group that serve as added safeguards against predation, or enhance individual fitness in other ways. We address these two types of inquiry in turn.

Grouping Rules: Mechanisms of Aggregation

How are groups formed? The simple, proximate answer to this question is that animals must move toward one another. But which rules (mechanisms) do solitary animals use to decide when to aggregate and which conspecifics to move toward? And are there fitness consequences to adhering to one rule over another? The simplest mechanism is that each individual independently is attracted to a common resource. Pure aggregations are probably rare, but are typically analyzed as ideal free distributions (chapter 11). We will focus on explanations that involve attraction to other individuals.

One provocative explanation for grouping is Hamilton's (1971a) selfish herd hypothesis. He proposed that when predation risk looms, animals should move toward their nearest neighbor to reduce their *domain of danger* (DOD), the area in which the animal is most susceptible to being captured by a predator. This behavior is selfish because each animal reduces its own DOD to avoid the predator, sometimes at the expense of other group members. As a by-product of these selfish interests, a group assembles and all members could benefit from having others close by unless, for instance, predators are attracted to prey aggregations (Ale & Brown 2007). In contrast to the *dilution effect* (table 17.1), in which benefits of grouping apply equally to all members, selfish herds are characterized by uneven distribution of predation risk among group mates. Individuals at the center of the assemblage are surrounded by conspecifics who effectively buffer them from predators; individuals at the periphery, however, are still vulnerable to capture although less so than solitary individuals. In the terminology of the selfish herd, the DOD of animals at the center is small compared to those at the periphery. Such asymmetries can be established and maintained by dominance interactions within the group, with high-ranking animals often securing the safest central locations (Hemelrijk 2000).

Various aspects of the original selfish herd have been subject to intense scrutiny over the past decade. In particular, researchers have sought to establish more realistic boundaries for the domains of danger that include, for instance, limits on the distance from which a predator might assail prey (see *limited domain of danger*, LDOD; James et al. 2004). Furthermore, simulation models have pitted the *move toward the nearest neighbor rule* against other potential strategies such as the *time minimization rule*. In habitats free of movement barriers, these distance-based and time-based mechanisms of aggregation would be identical. Many species, however, inhabit complex environments with plenty of obstacles that could impede rapid affiliation with the nearest animal in space. Various models have found that (1) nearest neighbor rules are not evolutionarily stable, and (2) animals adopting time minimization rules, in which they move toward the individual that can be reached most rapidly, achieve

TABLE **17.1** Benefits of living in a social group

A. Why gravitate toward others?		
Type of Mechanism	**Description of Benefits**	**Historical Reference**
Attack abatement		Turner & Pitcher 1986
1. Avoidance effect	Prey congregate because predators can more easily locate dispersed individuals than clustered groups	
2. Dilution effect	Aggregations reduce per capita predation risk; all animals within the group have the same probability of being captured	
Selfish herd	Prey can actively seek cover (e.g., next to conspecifics) to reduce their "domain of danger"; some places within the group are safer than others, and thus predation risks are unevenly distributed among individuals in space	Hamilton 1971a

B. Emergent properties: What else can animals do once grouped?		
Behavior Type	**Description of Benefits**	**Historical Reference**
Collective detection; "many eyes"	Individual vigilance in a group (e.g., scanning for predators) significantly increases the probability of detecting predators	Pulliam 1973
Collective defense	Group members band together to "mob" or dissuade predator from attacking	Alexander 1974
Confusion effect	As prey group size increases, predators find it more difficult to identify and capture individual prey; often but not always associated with prey coordinated activities	Miller 1922
Cooperative behavior		
1. Predator inspection	Gather beneficial information about predator	Pitcher et al. 1986
2. Cooperative hunting/foraging	Capture prey items that would be unattainable for solitary foragers	Packer & Ruttan 1988
3. Care of young/cooperative breeding	Inclusive fitness benefits in kin groups; experience; safe haven when resources are limited and territories sparse	Emlen 1982
4. Grooming	Reduction in parasite load	Connor 1995b
Information transfer	Enhanced foraging efficiency; social learning about food, predators, migration routes, and so forth; cultural transmission and teaching	Ward & Zahavi 1973
Social thermoregulation	Energy conservation; maintenance of body temperature	Armitage 1999

smaller LDODs than animals adopting nearest neighbor rules (James et al. 2004; Morrell & James 2008). Thus, when it comes to avoiding predators, animals do better to implement time-based rather than distance-based rules of aggregation; the distinction between these rules is essential in relatively complex habitats. These simple rules assume that animals will move toward a single *target animal* (Morrell & James 2008). More complex rules such as those that require positional assessments of, and movement toward, two or more animals can be more successful than simple rules under certain environmental conditions (e.g., under high prey density; Morrell & James 2008).

Experimental studies support many of the predictions of selfish herd models. Viscido and Wethey (2002) monitored flocks of sand fiddler crabs (*Uca pugilator*) under natural conditions and determined the positions of individuals before and after a predator (clapper rails or humans) incited escape responses. By constructing Voronoi polygons, a method for visualizing and quantifying DODs, the authors showed that the crabs become significantly more cohesive under predation risk, which causes their DOD to shrink. A recent study on redshank (*Tringa totanus*) flocks also illustrated substantial fitness costs associated with large DOD; predators (in this case sparrowhawks, *Accipiter nisus*) preferred to target redshanks that occupied areas at a greater distance from nearest neighbors (Quinn & Cresswell 2006). The type of predator or the diversity of predator strategies also might impact the net payoff of aggregation. For instance, the benefits of grouping in Stenogastrine wasps depended critically on whether the wasps were attacked by ants, which approach nests from the periphery, or by ichneumonids (parasitoid wasps), which can dive into the center of the nest (Coster-Longman et al. 2002). It thus appears that some animals do aggregate in ways that are consistent with selfish herding, and that there can be substantial individual variation in predation risk within groups (e.g., Hilton et al. 1999; Quinn & Cresswell 2006). An influx of creative field and laboratory experiments will be essential to keep pace with theoretical advancements on this topic. Of particular importance will be systematic evaluations of which aggregation rules animals adopt and the relative benefits (and costs) associated with the diverse array of rules that have been investigated by theoretical models (e.g., nearest neighbor; see Viscido et al. 2002; Morrell & James 2008 for others).

Benefits to Living in Groups: Collective Detection

Whether individuals occupy a group or navigate their environment alone, they must partition their time among various fitness-affecting activities or else they will suffer potentially disastrous costs (e.g., starvation, predation). How this time is managed and the tasks an animal favors can depend critically on the state of the individual (e.g., hunger) and on environmental risk, as discussed in chapter 13. Furthermore, living in groups can significantly reshape individual time budgets (Caraco 1979). A classic example of this stems from the observation in bird flocks that individuals scan less for predators as group size increases (Beauchamp 2008). Interestingly, theory predicts that as groups become larger, the probability of detecting a predator increases, although with diminishing returns (figure 17.1a; Pulliam 1973). As groups become larger, more eyes and ears can scan the environment for predation threat, allowing each individual to devote less time to vigilance (and more time to other activities) without sacrificing the collective ability of the group to monitor for an attack. In order for collective detection to work, at least one individual should be scanning the environment and others must pay attention to cues emitted by their group mates that alert them about imminent attack or risk of predation.

The dynamics of collective detection have several interesting nuances. First, which cue(s) do individuals use as evidence of predator attack? In some cases, when an individual detects a predator, its best response is to seek refuge. Departure from the group may signal danger to nonvigilant animals and incite what appears to be a coordinated flushing of prey from the area. Studies on dark-eyed juncos support the view that nonvigilant animals attend to departures of individual group mates but that the departure of multiple individuals incites a greater escape response in the nonvigilant individuals (Lima 1995). This makes sense from the perspective of information reliability (chapter 12). If one group member departs, it might have done so for a number of reasons that have little to do with predation threat. If nonvigilant animals escaped each time a single member left the group, they would frequently respond when there was no predator (a false alarm). On the other hand, when several individuals depart the group simultaneously, a true threat is much more likely to be present. Under these circumstances, the cost of staying (i.e., being eaten) likely exceeds the cost of leaving (i.e., energy, time), even if it was a false alarm—it pays to depart. Although sudden departures by flock mates seem like a perfect cue, false alarms involving even several individuals are likely. If the reliability of this social information is poor, individuals might benefit more from personally acquired information (e.g., unequivocal evidence of predator attack; Giraldeau et al. 2002), in which case the predator detection benefit associated with grouping declines. The relative stock that animals place in social versus personal information has received very little attention in studies of vigilance and represents a fascinating direction for future research (for social versus personal information in a foraging context, see Templeton & Giraldeau 1996; Kendal et al. 2004; van Bergen et al. 2004; King & Cowlishaw 2007).

A second nuance of collective detection relates to how individuals decide whether and when to be vigilant. There has been a growing interest in whether individual vigilance is dependent on the behavior of other group members (Bednekoff & Lima 1998; Fernandez-Juricic et al. 2004b; Jackson & Ruxton 2006; Sirot & Touzalin 2009). If collective detection is effective only when at least one individual is scanning at any given moment, then it seems intuitive that group members should coordinate their vigilance. Collective detection benefits of grouping could dissolve if individuals engaged in entirely random bouts of vigilance. In these situations, there could be substantial periods of time when none of the individuals are being vigilant. Fernandez-Juricic and colleagues (2005) explored this idea by reducing the ability of starlings (*Sturnus vulgaris*) to view conspecifics while in a head-down posture, which is common during feeding. They found that the starlings whose view was obstructed scanned more frequently, with heads-up, particularly in the direction of conspecifics. This suggests that some birds obtain information about conspecific behavior while in a head-down position and that they will compensate for information lost due to obstruction by scanning more with their heads up. The jury is still out regarding what type of information these starlings are obtaining from conspecifics. Although it is tempting to think that starlings might be monitoring the vigilance of conspecifics—a mechanism that could potentially drive coordinated vigilance—there is very little supporting evidence (e.g., Lima & Zollner 1996; Beauchamp 2002). Notable exceptions to this include the rotational deployment in some cooperative breeding species of sentinels that watch for incoming predators and often deliver alarm calls, presumably to alert group members (e.g., Rasa 1986; Manser 1999; Hollén et al. 2008), and similar types of apparently coordinated behavior in response to conspecific intrusion in snapping shrimp (Toth & Duffy 2005). Nevertheless, recent theoretical work provides a platform for exploring the conditions under which coordinated vigilance would be favored and suggests that a comparative approach that considers, among other things, differences in the visual fields of avian species could be very fruitful (Fernandez-Juricic et al. 2004b).

A third nuance to collective detection is how the behavior of the predator influences the ways in which grouping affects vigilance (Lima 2002). Predator behavior such as movement patterns, approach speeds or trajectories, and the state (e.g., motivation to persist in an attack) of the predator could fundamentally alter the benefits and costs of vigilance, time spent scanning, and, consequently, the entire time budget of individuals in a group-living prey species. As an example, figure 17.1b shows that predator closing speed can significantly impact the return that a prey individual receives for a given level of vigilance. Prey that frequently encounter fast predators experience markedly lower probabilities of detecting attack at a given group size and for a given level of vigilance than do prey that frequently encounter slow predators. If predators are treated as uniformly perilous, we run the risk of neglecting the diversity of selection pressures exerted by predators on group living and its associated behaviors in prey. This marks an area with great promise for future theoretical and empirical study, particularly with respect to understanding the mechanisms and outcomes of collective detection.

Benefits to Group Living: Collective Defense and Confusion Effects

Once groups or at least close living quarters have been established, some animals capitalize on the power of numbers to deter potential predators. Collective defense, for example, describes seemingly paradoxical behaviors in which prey approach, attack, or mob a predator (Alexander 1974). Mobbing behavior entails active, usually close-range harassment of a predator by potential prey. There are obvious, deadly costs to approaching a predator too closely (Curio & Regelmann 1986). Nevertheless, mobbing discourages predators—a clear fitness

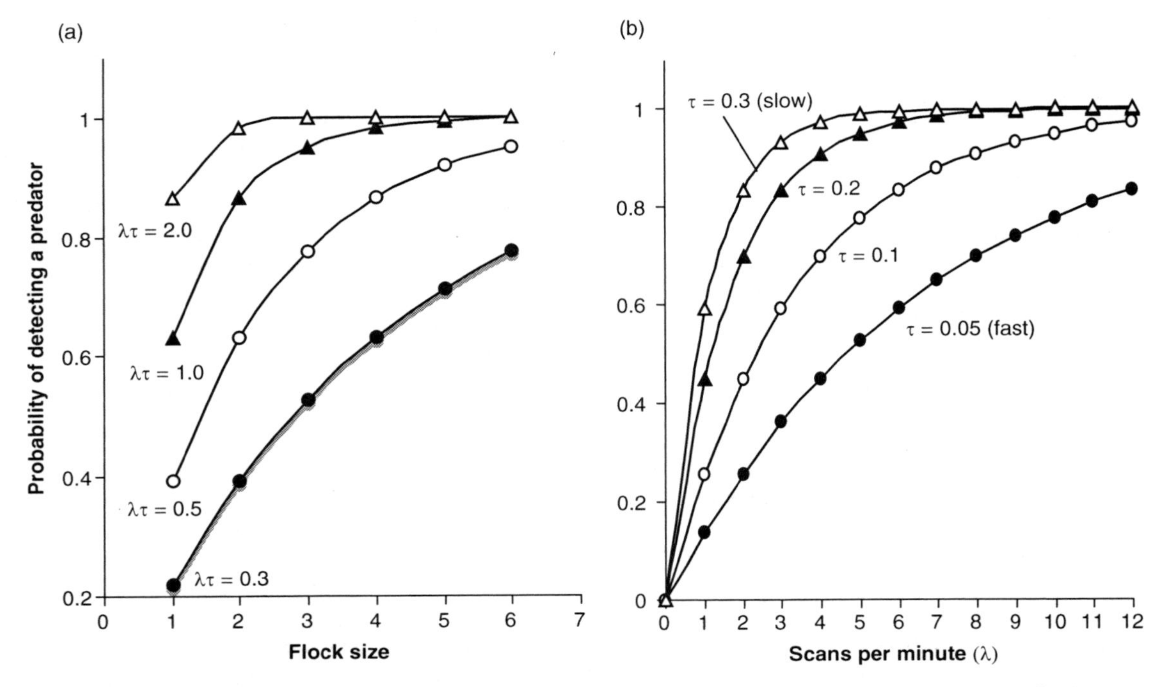

FIGURE **17.1** (a) Relationship between flock size and the probability of detecting a predator. As flock size increases, the probability of detecting the predator increases quickly but with diminishing returns after adding the second group member. $\lambda\tau$ is an index that combines the rate of scanning for a predator (λ) and how quickly the predator can maneuver into position to attack flock members (τ). As group members become more vigilant (e.g., more scans per minute) or as the predator becomes slower, $\lambda\tau$ increases. For instance, when alone, vigilant individuals that are attacked by a very slow predator ($\lambda\tau=2$) still have a high probability of successfully detecting the predator. (b) Independently varying λ and τ for a flock size of 3. As the predator becomes slower (τ increases), lower scanning rates (λ) are needed for predator detection. Alternatively, frequent scans are necessary to achieve a high probability of detection in the presence of predators that can quickly close the gap. Graphs modified from data reported in table 1 of Pulliam (1973).

benefit—and the effectiveness of mobbing increases as a function of the number of individuals recruited to the effort (Arroyo et al. 2001). Oftentimes, unrelated individuals join forces to mob a predator, which raises the question of which mechanisms favor the evolution of this apparently cooperative but not necessarily nepotistic behavior (Wheatcroft & Price 2008). What appears as cooperative mobbing could be a by-product of each individual's selfish interests to deter the predator (chapter 18). Recent evidence, however, suggests that cooperative mobbing can be maintained through reciprocity (Krams et al. 2008). In a clever experiment, Krams and colleagues established a triangle of nest boxes for three pairs of pied flycatchers (*Ficedula hypoleuca*); call the nest boxes A, B, and C. In the initial stage of the experiment, they placed a predator (tawny owl) near nest box A, and allowed only one of the other two pairs (B or C) to assist in the mobbing effort; one pair (say, B) was captured and thus could not assist. During a subsequent round, they placed a tawny owl near nest boxes B and C and determined that flycatchers from pair A exhibited an overwhelming propensity to join forces with pair C, which had assisted them previously. Pair A never joined forces to mob with pair B, which had been prevented from assisting previously (figure 17.2). In species such as the pied flycatcher that establish territories adjacent to one another, there are likely to be repeated opportunities for the

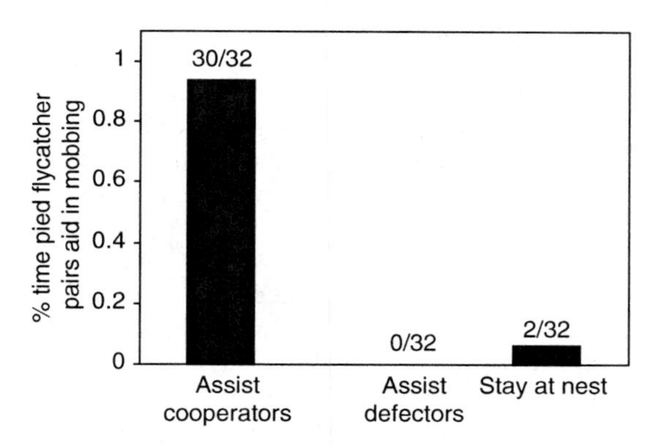

FIGURE **17.2** The probability that pied flycatcher pairs will cooperate in mobbing with pairs that had joined forces with them in the past (cooperators) or that had been prevented experimentally from cooperating with them in the past (defectors). Modified from Krams et al. (2008).

same set of nesting pairs to cooperate (or defect) in deterring the predator. In these types of potentially iterated games in which individuals keep score, as the pied flycatchers apparently do, it pays to cooperate and to reciprocate in mobbing efforts (Wheatcroft & Price 2008).

Predator approach behaviors may also serve functions other than predator deterrence (e.g., quality advertisement, information gathering; Dugatkin & Godin 1992). For instance, FitzGibbon's (1994) study on Thompson's gazelles (*Gazella thomsoni*) demonstrated that predator inspection—in which prey reduce the distance between themselves and a predator—can assist in monitoring predator movements or in gaining information about the predator. Graw and Manser (2007) observed meerkats (*Suricata suricatta*) mobbing a host of different predators (both active and opportunistic) and nonpredators and concluded that mobbing might play a role in risk assessment and information transfer among group members.

Living in groups can also reduce predation risk through the confusion effect (Krause & Ruxton 2002). This is a decreased attack efficacy by a predator when faced with choosing a particular target prey individual out of a mass of individuals. This confusion effect has been closely associated with the elaborate evasive maneuvers utilized by fish schools and other large animal aggregations. However, Ruxton et al. (2007) charged humans with preying upon virtual tadpoles on a computer screen, and showed that confusion can effectively reduce predator success even in the absence of collective, coordinated group behaviors.

Other Benefits to Living in Groups

In addition to increased predator detection, evasion, and deterrence there are a number of other benefits to group living (table 17.1). Living in groups can increase foraging efficiency in ways other than allowing animals to trade off vigilance and feeding. For instance, the group might act as an information center, in which individuals can exchange information about profitable foraging sites, or recruit companions to a patch (Ward & Zahavi 1973; Mock et al. 1988). This type of information sharing differs from situations in which attentive individuals parasitize the foraging efforts of group mates, as in the producer-scrounger game (chapter 11). Information centers, such as communal roosts, typically exist some distance from a potential food source. This requires that individuals who are knowledgeable about profitable patches return to the group and that naïve individuals follow informed group mates back to the patch. Sonerud and colleagues (2001) tested the information center hypothesis in hooded crows (*Corvus corone cornix*) by radio-tracking individuals and monitoring their whereabouts. Crows that visited an experimental food patch on day 1 roosted overnight, often with individuals that had not visited the food patch, and returned to that patch ~85% of the time on day 2. Furthermore, naïve individuals that roosted with knowledgeable crows followed them to the patch on the next day significantly more often than naïve individuals that had no knowledgeable roost mates. These data support the thesis that groups can serve as repositories for information about foraging locations, which could potentially reduce individual search times and liberate some of an animal's time budget for other fitness-related activities.

Group living can also facilitate cooperative foraging (or hunting) for prey items that would be unattainable for solitary foragers due to their size, difficulty in handling, or numbers (Packer & Ruttan 1988). Indeed, it has been argued that the fitness payoffs from cooperative foraging may have driven the evolution of sociality in some species. For this to be true, cooperative foraging must increase per capita energy gains relative to solitary foraging and these energetic profits must translate into fitness benefits in the form of, for instance, reproductive success. A recent study in colonial web-building spiders (*Anelosimus eximius*) showed that per capita prey capture decreased with web area (number of spiders) but that the size of prey increased with diminishing returns up to a colony of 10,000 spiders (Yip et al. 2008). Capture of larger prey appears to compensate for decreased prey number up to a (presumably optimal) colony size of ~500 spiders, at which point the per capita prey biomass declines precipitously. As another example, Rasmussen and colleagues (2008) conducted a large-scale study on wild dogs (*Lycaon pictus*) in Hwange National Park in western Zimbabwe and estimated both per capita net energy gains and reproductive output as a function of foraging group size. The wild dogs showed increased hunting success as foraging group size increased (figure 17.3a). Similar to the social spiders describe above, the wild dogs also showed a dome-shaped relationship between estimated per capita net energy gain and group size, with an optimal pack membership of 10 animals (figure 17.3b). Lastly, litter size increased with group size (figure 17.3c), suggesting that members of larger cooperative hunting groups might translate enhanced energy intake into fitness gains. These fitness payoffs, in turn, could have favored the evolution of sociality in the wild dogs. Rigorous documentation of the relationships between foraging group sizes, net energy gains, and reproductive investment is difficult under natural conditions as evidenced by the paucity of data like that of Rasmussen and colleagues (2008), but marks an area likely to be very fruitful in the coming years.

A last potential benefit of group living is thermoregulation. Although still lacking widespread support, social groups might provide thermal benefits, which could facilitate energy conservation and enhance survivorship (Armitage 1999). By grouping, the surface area available for heat loss can be significantly reduced. For instance, Willis and Brigham (2007) demonstrated that big brown bats (*Eptesicus fuscus*) that roost with conspecifics can conserve between 9% and 53% of their daily energy, depending on the number of roost mates, relative to solitary roosting individuals. In a peculiar twist involving notoriously asocial reptiles, Shah et al. (2003) provide evidence that Australian thick-tailed geckos (*Nephrurus milii*), and perhaps other social lizards, aggregate to combat cool, variable climates.

Costs of Living in Groups

Despite the numerous benefits to group living, there also are nontrivial costs. As we will see later in this chapter, social groups typically are structured through dominance interactions, and maintaining social structure can be physiologically taxing (e.g., increased stress hormone levels, compromised immune function) for both high- and low-ranking animals (chapter 15). Dominance interactions also can lead to an uneven distribution of resources among group members, disparate growth rates, and even complete reproductive suppression of low-ranking animals. For instance, in highly social poeciliid fishes (*Xiphophorus spp.*), maturation and secondary sexual character development is completely inhibited in subordinate males (Borowsky 1987). Thus, status-related interactions constitute a significant cost to group living (see "Dominance Hierarchies" section).

Several authors have hypothesized that disease transmission (viral, bacterial, parasitic) might also represent a potent cost to sociality, perhaps constraining group size (e.g., Ewald 1994). An empirical study on leaf-cutting ants (*Acromyrmex echinatior*), however, revealed that living in groups can reduce mortality costs, decrease transmission rates, and increase disease resistance to fungal infection (Hughes et al. 2002). The apparently conflicting positions on the cost-benefit relationship between disease transmission (or virulence) and sociality hopefully will spur some creative experimental designs or theoretical treatments in the coming years.

Both the costs and benefits of grouping may change with group size. For example, benefits such as vigilance often show diminishing returns as flock size increases (figure 17.2a). Costs could increase geometrically. There thus could be an optimal group size that is at some intermediate value. As an example, McGuire et al. (2002) showed in prairie voles (*Microtus ochrogaster*) that individuals occupying groups of intermediate size had the highest reproductive success. In addition, they showed that large groups were more susceptible to disappearing (through predation) than intermediately sized groups. Thus, although sociality may provide antipredator benefits, groups that are too large might be more conspicuous and at higher risk for predation. However, because group size is not a trait of an individual, an optimal group size might not be a stable group size (Pulliam & Caraco 1984; see also

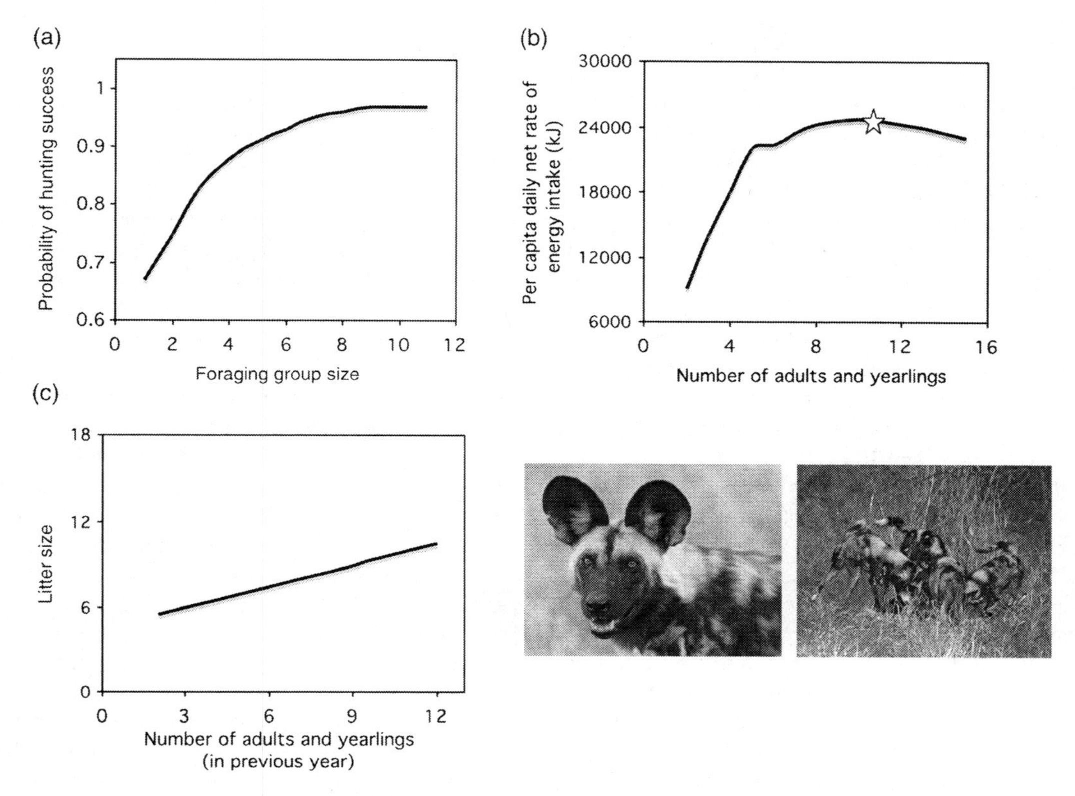

FIGURE 17.3 (a) Hunting success, (b) estimated per capita net energy gains, and (c) reproductive success in groups of wild dogs (*Lycaon pictus*) in or near the Hwange National Park in northwestern Zimbabwe. The star in (b) indicates the predicted optimal pack size for hunting wild dogs, although Rasmussen et al. (2008) documented a median pack size of six individuals. Figures redrawn from Rasmussen et al. (2008); photos courtesy of Jeff Fletcher.

figure 17.3 legend). Nevertheless, we expect the costs and benefits to influence how individuals join and remain with groups (e.g., Caraco 1979).

Origins and Maintenance of Sociality

It has often been hypothesized that sociality (or eusociality) represents an *evolutionary end point* such that once sociality evolves in a given lineage solitary living is no longer an evolutionary option (see Wcislo & Danforth 1997 for discussion). However, studies on halictine bees and communally roosting birds have revealed that secondary losses of sociality (i.e., reversion to solitary living in a lineage of primarily social species) can occur (Wcislo & Danforth 1997; Beauchamp 1999). Beauchamp (1999) entertains two hypotheses for why a secondary loss of sociality (in this case, communal roosting in birds) might occur: (1) the lack of positive selection on sociality might lead to the decay of group living over time via random drift, or (2) sociality is costly to maintain and will be selected against when conditions favor solitary living. As we discussed above, there are a number of aspects of group living that perhaps could be costly (e.g., disease transmission, competitive interactions), and there can be environmental conditions that might favor solitary states (e.g., resource availability; Emlen 1982). It is thus important to consider not only those factors that might contribute to the origins of sociality but also those responsible for its maintenance, and even its loss. Beauchamp's (1999) evaluation of communal roosting in birds provides an excellent illustration of how critical analyses of sociality will require integrative techniques, including life history assessment, behavior, and molecular phylogenetics.

Briefly, Beauchamp (1999) constructed a phylogeny based on morphological and molecular traits of a diverse set of bird species. He then mapped various characters including the degree of communality, diet, body mass, and environmental conditions onto the phylogeny to determine which factors might have contributed to the origin, maintenance, or loss of sociality (communal roosting) among birds. His analysis revealed that increased foraging efficiency provided by social lifestyles might have guided the origin of sociality among birds. Factors such as predation avoidance and thermoregulation could be responsible for the maintenance of sociality in lineages that retain communal roosting. The distinction between origin and maintenance of traits (e.g., sociality) in a population is imperative (chapter 2), because we should not assume that the factors that constitute benefits to sociality today (e.g., predation avoidance) reveal the evolutionary origins of group living. These types of questions can be approached most powerfully through the fusion of phylogenetics and behavioral ecology (chapter 7), and can be most informative in those taxa with diverse social habits, such as groups of closely related species that exhibit different types of social and/or mating system.

SOCIAL STRUCTURE

The benefits and costs of group living are not distributed equally among group members. Keen observation of social groups will reveal, for instance, that some individuals (usually dominant animals) monopolize most of the reproduction, dwell in safer areas within the group, or occupy more profitable foraging positions than others (usually subordinates). These types of fitness-related asymmetries are pervasive in animal social groups and imply that the benefits provided by group life under current environmental circumstances outweigh the costs, even for members who experience fewer perks, and that a set of rules must govern which status class or other such role (e.g., helper/breeder) a given animal acquires. In this section, we discuss some of the rules underlying the emergence of a few common types of social structure in animals.

Dominance Hierarchies

For close to a century, researchers have marveled at the fact that social groups are often rigidly organized, with individuals arranged by rank in a linear fashion. A strictly linear hierarchy is a social structure characterized by exclusively transitive relationships among group members. A transitive arrangement in its simplest form (three-member group) would show that if individual A dominates B, and B dominates C, then A also dominates C. If all possible triads in a group exhibit transitivity, then the hierarchy is said to be strictly linear; if intransitive loops exist, the hierarchy is considered less linear or nonlinear (figure 17.4; Chase 1980). The degree of linearity is thus a convenient descriptor of how individuals are arranged in a hierarchy. Assessing linearity, however, tells us very little about how resources are distributed among group members or how other factors (e.g., coalitions) might sway the distribution of power. For instance, linear hierarchies could be *top-heavy* or *despotic*, in which case a bulk of the resources (e.g., mates, food) are controlled by one or only a few high ranking animals (see "Cooperative Breeding Systems" section). On the other hand, linear hierarchies might be *egalitarian*, in which resources are spread evenly among group members, or even *reverse despotic*, in which coalitions of subordinate animals can be successful at inverting resource distribution (Boehm 1999).

If we take resource distribution out of the equation momentarily, it is still essential to understand the mechanisms underlying linear hierarchy formation in animals (see detailed review in Hsu et al. 2006). These mechanisms can be broadly classified as intrinsic, motivational, or social. Intrinsic factors

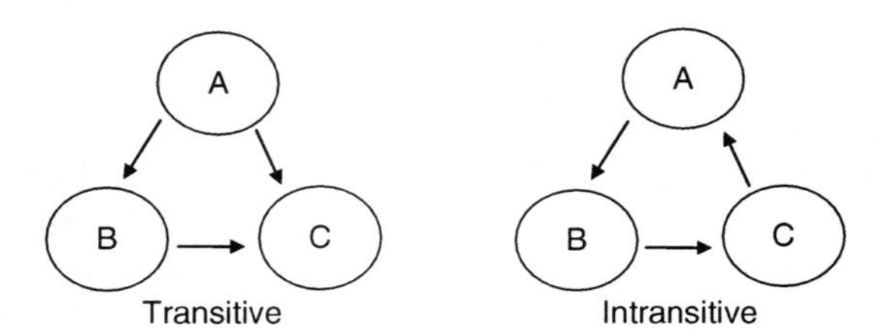

FIGURE 17.4 A diagram of (left) transitive (A beats B, B beats C, and A beats C) and (right) intransitive (A beats B, B beats C, and C beats A) relationships in the simplest dominance hierarchy, a group of three. Arrows represent the direction of dominance.

include such things as body size, weaponry, age, or genetically predisposed behavioral tendencies (e.g., aggression). Motivational factors include energy reserves, residency in the group, and condition, each of which may alter the benefit-cost structure of fighting and/or the value of contesting for status. These intrinsic and motivational factors can vary considerably among individuals in a population or within individuals over time, and asymmetries between/among contestants in these characteristics can predict with some certainty success in dyadic dominance interactions (chapter 15 of this volume; Hsu et al. 2006). Except in cases in which dramatic asymmetries exist, these factors (alone or in combination) do not appear sufficient to fully explain linear hierarchy formation. This observation spawned a new era in the analysis of social structure, one that examined the impact of winner effects, loser effects, and social eavesdropping on hierarchy establishment and maintenance (see also box 17.1).

BOX 17.1 Mechanisms of Dominance Hierarchy Formation

Imagine a group of 10 animals of equal initial fighting ability jostling for position in a dominance hierarchy. As the animals begin to fight for hierarchy position, some win and others lose, and these interactions cause the animals to update their fighting ability estimates; winners increase and losers decrease their estimates. Many theoretical models have explored whether these *winner and loser effects* can structure a dominance hierarchy. Rank order would reflect differences in the way the group members perceive their own strength, and these differences emerge through the actions of winner and loser effects. Interestingly, the manner by which winner and loser effects alter individual fighting ability estimates can have profound impacts on hierarchy structure. Some theoretical models have used additive effects (Beacham 2003), whereas other models (Dugatkin 1997) have employed multiplicative effects. The equations on the following page show simplified forms of additive and multiplicative winner-loser effects. In the additive model, wins add to fighting ability estimates at time 1, whereas losses subtract from fighting ability estimates at time 1. In the multiplicative model, fighting ability at time 1 is multiplied by some value greater than 1.0 (for wins) or less than one (for losses). To see how employing additive and multiplicative winner-loser effects can alter predictions of hierarchy structure, let's define some parameters. The magnitude of the winner and loser effects ($E_{W/L}$) will be set to 0.1 ($E_W = +0.1$; $E_L = -0.1$). All animals in a group begin with a fighting ability (FA) of 100. Animals will fight for 20 rounds. For the sake of illustration, let us assume that in the additive model, animals fight with all other group members each round. In the multiplicative model, animals fight just once per round (this will keep the numbers reasonable).

Both the additive and multiplicative models yield strictly linear hierarchies, but, as is evident from figure I below, hierarchy structure is quite different. Additive winner-loser effects promote a gradual, even differentiation of the rank order. Multiplicative effects, however, quickly drive the estimated fighting abilities of habitual winners very high and drive subordinate (those with more losses than wins) fighting abilities quickly toward zero. This, in turn leads to a hierarchy in which the upper ranks are clearly delineated but the lower ranks are cloudy at best. It is important to note that increases and decreases in fighting ability estimates are independent of an animal's starting value (e.g., 100) in the additive model; 0.1 units are simply added or subtracted from that value. On the contrary, the impact of winner-loser effects is tightly linked to fighting ability in the multiplicative models; strong animals will be most heavily impacted by wins/losses, whereas weak animals will be impacted to a significantly lesser degree.

Figure II shows asymmetries in the fighting abilities of group members occupying adjacent ranks. In the additive model, four fighting ability units separate each rank. In

(continued)

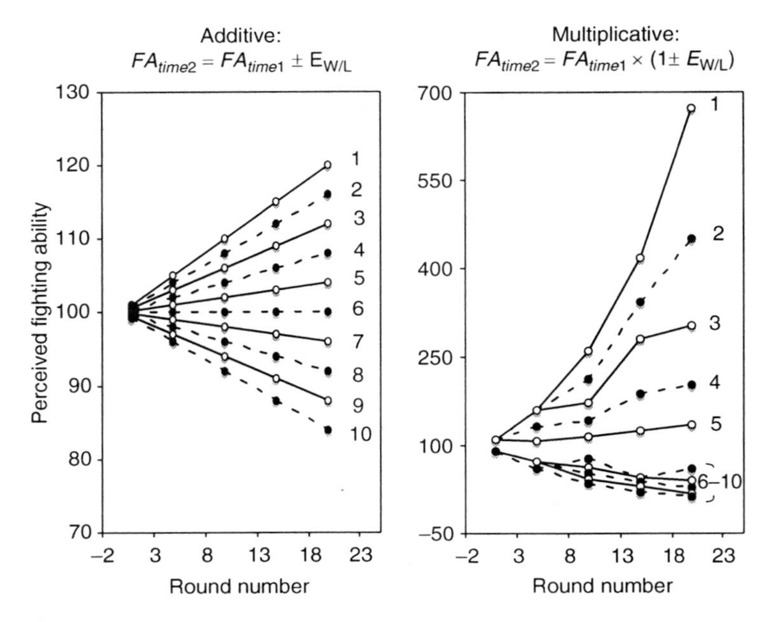

Figure I

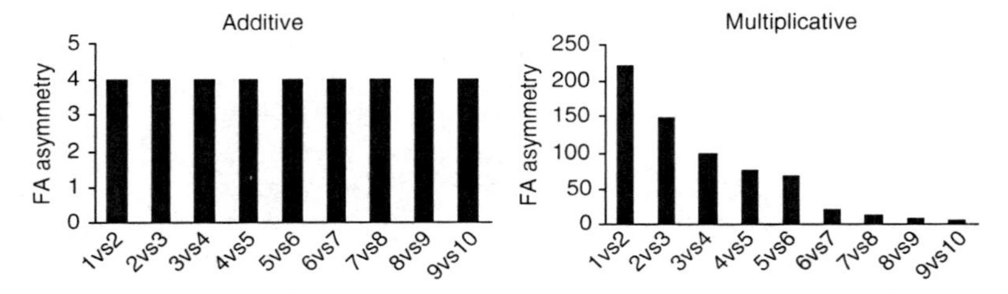

Figure II

the multiplicative model, however, large asymmetries exist near the top of the hierarchy whereas negligible asymmetries exist for the lower ranks.

Theoretical models are used as a tool to predict natural phenomena. Despite the rich set of predictions provided by these simple models of dominance hierarchy formation, empirical tests of these models are sparse. Do winner and loser effects drive hierarchy formation? To address this, we could conduct basic studies where animals are given winning or losing experiences prior to being placed in a group to examine whether previous winners occupy higher ranks than previous losers (Dugatkin & Druen 2004). Alternatively, we could monitor interactions that occur during initial stages of hierarchy formation (Chase et al. 2002). If winner and loser effects do impact hierarchy formation, do they alter perceived fighting abilities through additive or multiplicative means? An indirect test of this would be to simply establish groups and determine whether all ranks are clearly delineated (support for additive model) or whether only the top few ranks are clear (support for the

(continued)

multiplicative model). Also, recall that in the additive model, winner and loser effects will have the same influence on all animals, regardless of their strength. In the multiplicative model, however, wins and losses have a greater influence on strong animals. This could be tested, for instance, by using body size as a proxy for fighting ability. One could explore whether larger animals respond in a more pronounced way (e.g., probability of winning/losing future contests) to fighting experience than smaller animals. Lastly, it is important to recognize that different species might use different rules for establishing a dominance hierarchy; thus, it is essential to keep these and other related models (see Hsu et al. 2006) in mind when evaluating the mechanisms that govern hierarchy formation in your species.

Winner effects describe the increased probability of future contest success experienced by individuals that won in the past. Loser effects describe the decreased probability of future contest success experienced by individuals that lost in the past. There is some evidence suggesting that winner and loser effects are driven by a change in the way individuals perceive their fighting ability following victory or defeat (Hsu & Wolf 2001). In the case of dominance interactions, social eavesdropping refers to individuals updating their estimate of the fighting ability of other group members after watching them win/lose interactions (see "Social Information Use" section for more detail). Each of these effects exists across a wide range of taxonomic groups and appears to heavily impact the dynamics of hierarchy establishment (see next paragraph). But what favored the evolution of strategies that allow animals to track wins and losses and adjust their behavior accordingly? This question has gained astonishingly little attention.

Mesterton-Gibbons (1999) determined that winner and loser effects will evolve in social groups only when there are dramatic asymmetries in fitness gains between the top-ranking animal and the subordinates (e.g., when fitness benefits for the second-ranking animal are less than half that of the dominant). The costs incurred by overestimating fighting ability (e.g., engaging in a contest with a stronger opponent) and the variance of fighting abilities within the group also contributed to the evolutionary dynamics of winner and loser effects. In his model, Mesterton-Gibbons (1999) showed that winner effects cannot evolve independent of loser effects and that often loser effects alone will evolve. These predictions fit well with empirical studies: there have been no investigations to date that document the existence of only winner effects (Hsu et al. 2006). It will be essential to conduct further theoretical treatments on the evolution of winner-loser effects, and to initiate studies on the evolution of social eavesdropping. Such studies should evaluate those conditions that favor behavioral strategies, such as winner-loser effects, that ultimately impact social structure.

To date, there is no consensus as to which of these social factors might be most influential in linear hierarchy formation (Hsu et al. 2006). Simulation models, however, have determined that the presence and strength (i.e., the magnitude of post-fight increments or decrements to fighting ability; box 17.1) of winner and/or loser effects can impact both social and spatial structure in an initially unstructured group (e.g., Hemelrijk 2000). Social eavesdropping might fine-tune hierarchy structure by making perceived fighting abilities more transparent to all group members (Dugatkin 2001). Furthermore, some foundational models indicate that the relative importance of intrinsic and social factors in predicting hierarchy structure depends critically on initial group composition. Animals might be expected to simply use the most reliable indication of rank to sort out hierarchy position. For instance, social experience could be critical for hierarchy differentiation in a group whose members are all the same size, but less so in a group composed of disparately sized animals. Most of the predictions generated by simulation models (e.g., that winner effects or both winner and loser effects are required for linear hierarchy formation) remain untested in natural animal groups, in part because of the logistically daunting task of quantifying each social variable and its independent effect on social structure. Bringing our investigation of linear hierarchy formation out of the virtual world and into the real world will require crafty experimental designs and careful selection of a model animal system (e.g., Chase et al. 2002). Such efforts will be especially powerful in advancing this important niche in the study of social organization.

The relevance of these models to natural animal groups also will require that spatial structure be superimposed on social structure. In a previous section, we discussed the possibility that spatial centrality—the tendency for some animals of a group to occupy inner positions within a group—might help to reduce predation risk. Furthermore, dominant animals of a group often occupy these prime positions (Christman & Lewis 2005). Although spatial centrality of dominants has been documented for many species, this pattern may be contingent upon group size, resource distribution, or individual energetic states (Rands et al. 2004). In some circumstances, such as when groups are large and resources homogeneously distributed, the most profitable positions, which should be occupied by high-ranking animals, would be at the periphery (e.g., see Christman & Lewis 2005).

Maternal Rank Inheritance

In many cercopithecine primates (e.g., baboons, macaques, vervet monkeys) and in hyenas (*Crocuta crocuta*), there is a strong tendency for offspring to acquire similar ranks as their mother in a dominance hierarchy. Engh and colleagues (2000) tested four potential mechanisms that might drive maternal rank inheritance in hyenas, including the following: (1) genetic inheritance: some genetically encoded traits (e.g., size) predispose both mother and offspring to certain ranks, (2) directionality of unprovoked aggression: high-ranking adult females harass offspring of lower ranking females, (3) interventions: high-ranking mothers might come to the defense of their offspring more frequently than low-ranking mothers, and (4) coalition support: joining or being assisted by other group members in attacks against "appropriate" (lower ranking) individuals. In their study on hyenas, Engh and colleagues (2000) found that interventions and coalitions were the most important predictors of maternal rank inheritance. They also acknowledged that other rules (e.g., harassment) and perhaps additional factors such as observational learning of relative rank might be important in driving maternal rank inheritance in other species. Maternal rank inheritance is just one example of how social interactions can promote similar behavioral phenotypes (e.g., dominant) among related individuals. Chapter 14 provides a fascinating account, rooted in quantitative genetics, of how the social environment, and the genotypes of other individuals, can potently influence phenotypic expression and promote phenotypic covariance among individuals within the social milieu.

In addition to behavioral determinants of offspring rank (e.g., coalition support), other factors such as maternal condition might alter the competitive success of offspring. If the fitness of male offspring is impacted more heavily by maternal condition than the fitness of female offspring, we might expect mothers to somehow skew offspring sex ratios toward males when in good condition. This is the basis for Trivers and Willard's (1973) maternal sex allocation hypothesis. They hypothesized that in polygynous mating systems, male reproductive success would be highly variable relative to that of females. In these situations, only the most highly competitive males will be guaranteed to mate. Thus, natural selection would favor mothers that invest more heavily in producing male offspring when in good condition. Condition-dependent sex allocation, although still hotly debated, has gained a considerable amount of empirical support, particularly in mammals (Sheldon & West 2004). However, its underlying physiological mechanism(s) remain mysterious (see an example in Cameron et al. 2008). Integrative studies that fuse evolutionary behavioral ecology with neurobiological, endocrine, and other physiological processes should provide insights into the proximate causes for maternally driven sex ratio adjustments.

Cooperative Breeding Systems

Chapter 18 defines *cooperative behavior* as an act that "provides a benefit to individuals other than the performer, and has evolved at least partially because of this benefit." In many vertebrate and invertebrate social hierarchies, subordinates forgo (or at least curtail) reproduction and cooperate to raise the offspring of the breeding pair(s). Cooperatively breeding groups can vary in size, composition (kin versus nonkin), and in the number of breeding pairs and helpers. Research into these cooperatively breeding systems has been instrumental in shaping our understanding of the evolution of sociality and cooperation, and in identifying potential environmental constraints that might limit dispersal from a social group. The literature in this area is vast, and a complete treatment of cooperative breeding is outside the scope of this chapter, but we refer the reader to key reviews in the "Suggestions for Further Reading" section at the end of the chapter. Here, we focus on two primary questions regarding cooperative breeding systems.

Why Remain in a Group and Sacrifice Reproduction in Lieu of Seeking Independent Breeding Opportunities?

This question has been tackled from three nonmutually exclusive angles. First, because cooperatively breeding groups often consist of close relatives, helpers may receive indirect fitness benefits by either enhancing a related breeder's reproductive output or the survival of its offspring through additional provisioning or defense. Kin selection arguments garnered significant empirical attention in studies on cooperative breeding vertebrates, in part due to the success of this approach in explaining such things as worker sterility in eusocial insects (chapter 19). In some cooperatively breeding birds, however, helpers can be unrelated to the breeding pair (e.g., Cockburn 1998). In other cooperatively breeding vertebrates, relatedness does not covary consistently with helping effort (e.g., Clutton-Brock 2002), and sometimes, as is the case for white-browed scrub wrens, subordinate males prefer to help the offspring of unrelated females (Magrath & Whittingham 1997). Armed with these observations, investigators began to consider alternatives to an exclusively kin selection explanation.

The second approach to studying cooperative breeding systems evaluates the potential direct fitness benefits experienced by subordinate helpers. It is possible that the survival, antipredator, or foraging advantages to group living might trump the immediate costs of helping. This is one key aspect of the *group augmentation* hypothesis (passive augmentation; Kokko et al. 2001), the other being that actively helping to rear offspring, even if unrelated, will carry considerable future benefits. Direct benefits associated with active group augmentation come in the form of delayed reciprocity—if those individuals that were helped in turn help the helper once it ascends to dominant/breeding status (see Kokko et al. 2001 for explanation of evolutionary stability). Helpers in cooperative breeding systems also may receive other direct fitness benefits including (1) gaining parental experience, which will increase future reproductive success, (2) being tolerated by dominant animals to stay in a sheltered group environment until breeding prospects arise (*pay to stay* hypothesis), (3) avoiding costs such as punishment or eviction from the group, and (4) inheritance of the natal territory (see Clutton-Brock & Parker 1995a; Dickinson & Hatchwell 2004).

A third approach has been to explore the ecological constraints (Emlen 1982) that might dissuade subordinates from seeking independent breeding opportunities. The species' life history may be an important factor influencing the role of ecology in fostering cooperative breeding. A pair of comparative analyses (see Arnold & Owens 1999) indicate that low annual mortality (life history trait) might decrease the rate of territory turnover, reduce available nest sites (ecological constraint), and favor the evolution of cooperative breeding. Empirical investigations of the ecological constraints hypothesis have focused primarily on habitat saturation and resource distribution (habitat quality), but other environmental factors could also make dispersing from the natal territory (or living a solitary existence) quite risky. For instance, Heg et al. (2004) revealed in a cichlid fish (*Neolamprologus pulcher*) that predation risk constrains helper dispersal from the nest, perhaps promoting their cooperative breeding habits. Furthermore, in a rigorous comparative phylogenetic analysis of African starling (45 species) ecology and behavior, Rubenstein and Lovette (2007) identified temporal variability in annual precipitation as perhaps a fundamental ecological constraint driving the evolution and maintenance of cooperative breeding.

What Mechanisms Control Reproduction in Subordinate Helpers?

If individuals stay in a group until adulthood, then there may be interesting dynamics affecting whether or not they reproduce. Reproductive skew models describe how social dynamics can regulate subordinate reproduction in cooperative breeding groups in which helpers boost the reproductive output of breeders (box 17.2). There are numerous skew models (and variants thereof) that make

BOX 17.2 Reproductive Skew

Peter Nonacs

Cooperative groups are rarely egalitarian. They usually exhibit significant variance across group members in activities, resource shares, and reproductive success. Annoying the statistically inclined, this variance is called *skew*. Although many features of groups exhibit variance, cooperative breeding has attracted the most scientific curiosity, particularly in asking why less successful individuals remain in groups. It may be that cooperation increases total offspring production, but this alone does not guarantee that subordinates will do better in groups than on their own. One potential solution is that conflicts over reproductive shares can be balanced to the selfish benefit of all group members. For example, an individual that could dominate all or most of the breeding might gain more in the long run by conceding some reproduction to subordinates in order to induce staying and helping.

This dynamic can be modeled by extending Hamilton's rule to compare inclusive fitness of group living to being solitary for two interacting individuals (Nonacs 2001b, modified from Reeve & Ratnieks 1993). Mathematically, we can represent one individual (β) cooperating with another (α) by

$$k[p + r(1 - p)] \geq rx_\alpha + x_\beta$$

where k is the group's realized reproduction, p is β's proportion of group reproduction, and r is the relatedness of α to β. The right-hand side of this equation is β's fitness if both individuals live solitarily and gain x_α and x_β offspring. Rearrangement defines the proportions of group reproduction that favor β's cooperation:

$$p \geq [x_\beta - r(k - x_\alpha)] \,/\, k(1 - r)$$

A similar approach finds the proportions of reproduction (q) that α is willing to surrender to β:

$$q \leq (k - x_\alpha - r\,x_\beta)] \,/\, k(1 - r)$$

To determine the minimum total productivity required for a stable group, we solve for when $q = p$. This gives

$$k_{\min} = x_\alpha + x_\beta$$

These equations show that if $q > p$, a stable, mutually beneficial relationship between α and β is possible no matter their genetic relatedness, as long as reproduction is shared appropriately. Consider two unrelated individuals in which $x_\alpha = x_\beta > 0$ (figure Ia). Combining these minimum requirements for direct reproduction gives k_{min}. The gray outer box represents increased reproduction due to cooperation. No matter how this extra is split, p cannot be less than x_β/k between unrelated individuals. Between relatives k_{min} is the same, but $p < x_\beta/k$ is possible if cooperation produces more offspring than the minimum requirement ($k > k_{min}$). This is because losses in direct reproduction can be offset by increasing indirect fitness through relatives (figure Ib: hatched box). Thus, relatives can theoretically tolerate larger reproductive skews.

Although the Reeve and Ratnieks model defines the range of potential skews between individuals ($p <$ skew $< q$), it does not find the actual stable values. If α controls reproduction, β's share is often assumed to approximate p. If β controls its own reproduction,

(continued)

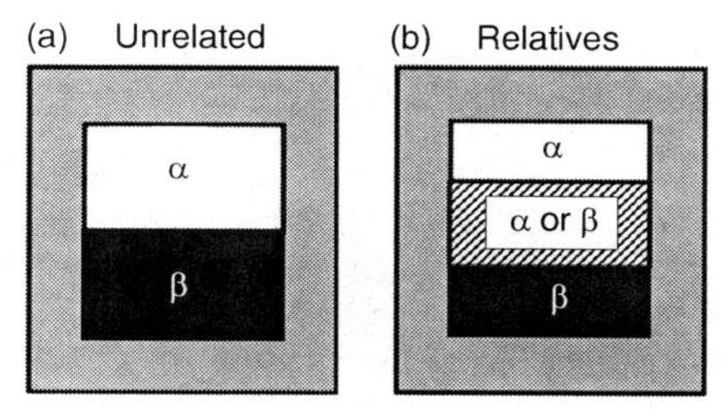

Figure I Reproductive skew in groups of two nonrelatives or relatives. The inside black and white boxes represent minimum shares of direct reproduction ($x_\alpha + x_\beta = k_{min}$), required by α and β, below which cooperation is evolutionarily unstable. The gray area represents potential extra reproduction from cooperating. This extra can be taken by one individual, shared by both, or used for competitive interactions rather than reproduction. Between relatives direct reproduction can be ceded (hatched box) to other group members, as long as enough of the gray is turned into extra offspring. Thus, direct fitness within the group can be less than direct fitness of living alone. For example, $pk < x_\beta$ is possible if $(x_\beta - pk) < r[k(1-p) - x_\alpha]$.

its share should approach q. These two contrasting predictions are sometimes presented as alternative skew models—concession versus restraint. They are, however, one model differing only in their point of view (Buston et al. 2007). To predict stable values, Reeve et al. (1998) used a game-theoretic approach. No one is assumed to completely control reproduction, and competition is costly for group productivity. Thus the gray in figure I may be used to fight for reproduction rather than actually reproducing. Under this more complex tug-of-war model (derivation in Reeve et al. 1998), α's predicted share becomes

$$p = x^* / (x^* + by^*) = 2 / [2 - r(1-b) + \sqrt{r^2(1-b)^2 + 4b}\,]$$

where b is β's competitive ability relative to α ($0 < b < 1$), and x* and y* are optimal investments in selfish effort by α and β. (Note that x^* differs in meaning from x_α or x_β.)

Since 1998, a number of variants have attempted to add to or supersede these basic skew models. Of note are modeling efforts to combine peacefully conceding some reproduction while fighting over the rest. Nonacs (2007), however, showed that such synthetic models are not evolutionarily stable except with biologically questionable assumptions or under restrictive conditions. Thus, one can distill all current reproductive skew theory to this general summary. The Reeve and Ratnieks model establishes the range of possible skews, and tug-of-war is one proximate, game-theoretic, solution for a stable skew.

How well do skew models predict the dynamics of cooperatively breeding groups? In across-species comparisons, skew theory fares well. Skew significantly increases across species as groups become composed of closer relatives (Reeve & Keller 1995). However within species, reproductive skew models have been difficult to test and interpretations of results have been contentious. For example, experimental tests often compare data across tables of contrasting predictions from skew model variants. That some model might have a post hoc match to data is almost inevitable and indicates very weak support (Taborsky 2008a: the pseudo-proof fallacy).

A problem with more specific tests is that quantitative predictions are hard to derive. This partially stems from critical variables such as b, x_α, and x_β being difficult to impossible to measure. Equally problematical is that the resolution of tug-of-war is arbitrarily

(continued)

BOX 17.2 *(cont.)*

defined to optimize a ratio of $x/(x + by)$. Although this addition to the original Reeve and Ratnieks model mathematically creates joint control over reproduction, it has not been shown to describe any known conflict resolution mechanism. Very different quantitative predictions arise if other mathematical relationships are used, or if b is a function of x or y rather than a constant. Therefore, existing skew models cannot predict some rather basic facets of group dynamics, such as the following: levels of selfish effort or aggression; which group members should be most selfish; quantitative effects of conflict on group productivity; and how strongly relatedness correlates to aggression within groups. Finally, alternative models are rarely considered, although the fitness gain of behaving optimally can be trivial relative to simpler rules of thumb for cooperation (Nonacs 2006).

Nevertheless, across all skew models there are three generalities for within-group interactions: relatedness matters, competitive ability matters, and observed skew should fall within predicted ranges. Unfortunately, a growing list of studies emphatically shows reproductive skew is unaffected by relatedness (Liebert & Starks 2006; Nonacs et al. 2006), is insensitive to competitive ability (Nonacs et al. 2004), and often falls outside predicted ranges (Nonacs et al. 2006 and, especially with regard to unrelated helpers, Queller et al. 2000). Thus, although reproductive skew theory may be occasionally relevant for some social groups, it is not a fundamental organizing principle of cooperative societies.

Do not despair. The theory's failure opens new questions as to why skew is ubiquitous. On a proximate level, frequent errors in identifying close relatives may play a role (Dani et al. 2004; Liebert & Starks 2006). Also important is to understand conflict resolution mechanisms between individuals with differing levels of power and opportunities to cheat (Cant 2006). On an ultimate level, delayed fitness benefits through inheritance may affect skew (Cant & English 2006). Finally, multilevel selection as competition between groups can favor cooperation within groups (Nonacs 2007; Reeve & Hölldobler 2007). We need to expand beyond relatedness (r) and consider how group-level benefits, such as genetic diversity, could also drive the evolution of cooperation (Nonacs & Kapheim 2007).

different predictions about the extent to which dominants control helper (subordinate) reproduction and how asymmetries in reproduction might be explained by, for instance, relatedness, physical prowess or sex of the helpers, and ecological constraints (reviewed by Nonacs 2007). Optimal skew models (or concessions models) predict that dominant breeders should provide incentives for subordinates to stay and help in the form of reproductive opportunity. Dominants might be expected to alter incentive allocations depending on such factors as relatedness (e.g., unrelated animals may require more incentive to stay), and opportunities for subordinates to disperse (e.g., if plenty of lavish territories are available). In contrast to optimal skew models in which dominants presumably have complete control over subordinate reproduction, tug-of-war models assume incomplete control, in which dominants must compete with subordinates to earn their keep. Both optimal skew and tug-of-war models thus allow for subordinate helper reproduction. A third set of models—restraint models—predict that subordinate helpers may voluntarily opt out of reproduction if dominants inflict serious punishment (e.g., eviction, injurious aggression) for doing so. From an ultimate perspective, helpers should voluntarily surrender reproduction only when the fitness benefits of a celibate but group living existence exceed the net fitness benefits of reproduction (i.e., reproduction minus punishment costs). Although some sort of decision mechanism could be involved in subordinates curtailing reproduction, punishment also could shut down the physiological machinery necessary for reproduction. Meerkats who are subordinated and aggressively evicted from the group show profound long-term increases in stress hormone levels, which is known to suppress reproductive function (Young et al. 2006).

With all of this said, there has been recent concern about the general applicability of skew models, and both their assumptions and predictions, to natural circumstances. These concerns are detailed in box 17.2 (see also Nonacs 2007), and there have been several recent attempts to streamline the conceptual and mathematical frameworks of reproductive skew theory (e.g., Buston et al. 2007).

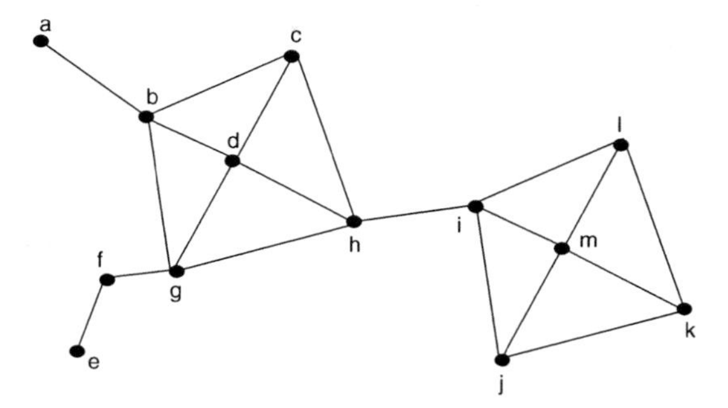

FIGURE 17.5 An example of a social network, inspired by Krause et al. (2007). In this network, the black circles (nodes) represent individuals within the network and the lines represent connections between these individuals.

Social Networks

An emerging frontier in the study of social organization is the application of social networks theory to study individual association patterns, group or population structure (Croft et al. 2007). Investigations into social organization are often simplified by assuming that individuals within a group interact randomly or that each member is equally likely to affiliate with all others. Social network theory, however, capitalizes on the fact that interactions typically are nonrandom (e.g., affiliation preferences for individuals of similar size or color), and that there is considerable variance among group members in their connectedness to others. Social network theory allows researchers to visualize connections between individuals in a network and to evaluate each individual's role in the network (figure 17.5). For instance, Krause et al. (2007) describe four parameters that can be used to characterize network members: (1) degree: the number of immediate neighbors that an individual has, (2) cluster coefficient: how connected one's immediate neighbors are, (3) path length: the number of connections on the shortest path from one individual to another, and (4) betweenness: the number of shortest paths between two individuals that pass through the focal animal. In figure 17.5, it is relatively easy to see why individuals *h* and *i* have the highest values for betweenness; to get from one cluster to the other, all roads pass through these two animals.

Social network analysis has begun to reveal some of the intricacies of group life in animals. Croft et al. (2006) documented wild shoal composition in guppies from the Arima River in Trinidad daily for 15 days (plus an additional record on days 20 and 31), constructed social network diagrams (see figure 17.5), and calculated the strength of association as the number of times individuals were found in the same shoal. They then transferred shoals of wild female guppies to an experimental arena, exposed the fish to a natural predator (pike cichlid, *Crenicichla frenata*), and determined which females joined forces for predator inspection. Females that associated more often in the shoal also were more likely to cooperate in predator inspection bouts. Repeated interactions among individuals are thought to be a requisite for the evolution and maintenance of cooperative behaviors such as predator inspection. Rarely, however, has the opportunity for repeated interaction and cooperation been evaluated in the context of overall social structure in the wild. The social networks approach employed by Croft et al. (2006) allows one to quantify natural association patterns and to make predictions about how social structure impacts the dynamics of cooperative interactions.

In an earlier section, we discussed disease transmission and information transfer as costs and benefits, respectively, of sociality. Although individual-based studies have advanced our understanding of these dynamics some, social network analysis can potentially provide much more resolution (Krause et al. 2007). For instance, we might be able to identify which animal of the group is responsible for initial inoculation of the group with the disease and determine the infectious chains that cause diseases to spread more rapidly in some groups than others. Also, we might determine which animals are information hubs of a social group, or how information travels throughout the group. For those interested in disease, or any number of other questions involving information transfer (e.g., cultural transmission, alarm calling, signaling networks, social learning) or associations within populations and communities, social networks analysis should become a staple in the empirical toolbox.

SOCIAL INFORMATION

We have focused our attention on large-scale issues in social organization such as the evolution of group living, the structure of animal groups, and the different roles (e.g., status, helper/breeder) that individuals assume within these groups as well as their fitness consequences. Now, we will hone in on some individual-level behaviors that perhaps sprouted from living in groups. As opposed to solitary living, occupying stable groups or even transient aggregations (of kin or nonkin) provides a springboard for the evolution of sophisticated types of social behavior. Some of these we discussed above, such as coalition formation, behavioral intervention, predator inspection, mobbing, and helping to raise offspring. Others include grooming, coordinated hunting activities, division of labor, adoption, and infanticide. Many of these behaviors have been studied in limited capacity, or within one or just a few lineages, and thus deserve more widespread consideration. One aspect of social life, however, has received a tremendous amount of attention across an admirably large range of social species—the transfer of information between and among individuals in a social context. In the following sections we outline some recent advances in this area.

Social Information Use

Social information is broadly defined as any information that can be extracted by a bystander from the behavior of group members (or in some cases, nearby territory holders). Social information can be partitioned into various forms, which differ essentially in the type of information gathered (Laland 2004; Bonnie & Earley 2007). By observing (or otherwise attending to) members of the social group, bystanders can gain information about their physical environment (e.g., which foods are edible, which patch is profitable), how to manage their environment (e.g., how to crack a nut), or the quality and/or behavioral decisions of animals in its vicinity (e.g., who is a good fighter, or a sought-after breeding partner). Questions regarding how this information, once mined from group mates, can impact the future behavior of the bystander have been vigorously studied in the fields of social learning, public information use, and social eavesdropping. Because we cannot realistically do justice to all of these prolific fields, we refer the reader to current reviews on these topics and the plentiful references discussed therein (Bonnie & Earley 2007; Valone 2007; chapter 11 of this volume). For the purpose of discussion, we will focus on social eavesdropping.

Social eavesdropping is the act of extracting information from signaling interactions between conspecifics (McGregor 2005). A significant majority of the work on social eavesdropping has targeted the response of male or female bystanders to witnessing male-male contests. A growing body of work clearly demonstrates that both male and female bystanders alter their behavior toward the observed contestants in a predictable manner (McGregor 2005), although the precise direction of the response (e.g., female chooses to affiliate with the winner or the loser) may depend on, for instance, social circumstance or mating system. A study by Earley and Dugatkin (2002) revealed that bystanders not only gather information about status (winner/loser) but also about contest dynamics. They showed, in green swordtail fish (*Xiphophorus helleri*), that bystanders invariably avoided observed winners but their response toward observed losers depended on how hard the loser fought. Bystanders initiated (and won) fewer encounters against observed losers that persisted longer in the witnessed interaction than losers who gave up immediately; this suggests that male swordtails keep a running tally of events that transpire during a contest.

But why should animals attend to the behavior of others if they can gather this same information on their own? The simple answer is that many of the behaviors that bystanders attend to are costly to perform. For example, as described in chapter 15, competitors assess fighting ability through a series of increasingly escalated bouts, some of which can result in energy depletion, injury, loss of time, and reduced vigilance. Bystanders pay a cost for time spent observing and perhaps risk predation, but they appear to gain accurate information about relative fighting ability without the perils of competition. This same logic has been applied to other sorts of social information gathering. However, if the information gained through observation is inaccurate or costly to obtain, we would predict that an animal should rely more heavily on personal experience (see examples in Valone 2007).

In some cases, integration of personal and observational experiences might be beneficial for a social animal. A prime example of this is the ability of some species (e.g., pinyon jays; Paz-y-Miño et al. 2004) to use transitive inference to deduce dominance relationships. Imagine this scenario—individual A interacts directly with B and loses, and then watches C beat B. If individual A retreats from C without interaction, then we might conclude that A used transitive inference to deduce that C was stronger than itself. This sort of deduction could be performed because A had a personal experience with B (loss) and gained information through eavesdropping on the interaction between B and C. Interestingly, in a species of African cichlid fish (*Astatotilapia burtoni*), individuals are able to deduce a hierarchy among others simply by watching interactions (no personal experience required), and they exhibit predictable aggressive responses to group mates that had been observed (Grosenick et al. 2007). These examples imply that social animals, or species that occupy adjacent territories, can successfully merge personal and vicarious experiences and that they make some sophisticated social decisions on the basis of this information.

Audience Effects

On the flip side of social eavesdropping, signalers are known to alter their behavior when bystanders are present versus absent. Various studies on fish show that males modulate their aggressive behavior in the presence of an audience, and in different ways when being viewed by males and females (McGregor 2005). Nonnesting male Siamese fighting fish (*Betta splendens*), for instance, perform significantly more aggressive acts in the presence of male bystanders than in the presence of females or no audience (Dzieweczynski et al. 2005). However, when both interacting males possess nests, audience effects disappear. These results suggest that individual responses to audiences can be modulated in significant ways by state-dependent (nest versus no nest) variables. In the case of fighting fish, it is possible that high levels of aggression are incompatible with tending the nest or the fry; escalated aggression could pose fitness-related risks in the form of attracting additional competitors, increasing the probability of egg cannibalism, losing mating opportunities, or inhibiting parental care of the eggs. This study and many others demonstrate that both bystanders and signalers attend to their immediate social surroundings. Audience effects also raise the question of whether bystanders might exert selection pressures on signaling characteristics, or strategies that utilize private or public signaling channels (chapter 16). Studies in birds suggest that under some circumstances, individuals might advertise their presence (or motivational state) to bystanders by increasing the loudness of their vocalizations. In other situations, especially those in which anonymity would be favored (e.g., in the presence of predators or competitors that could usurp a territory), individuals sometimes resort to quiet vocalizations (Dabelsteen 2005).

FUTURE DIRECTIONS

For over half a century, we have been wrestling with questions regarding the benefits and costs of sociality, and the evolution and maintenance of group living. We have made great strides, but these sorts of fundamental questions remain disconnected because (1) the bulk of the empirical effort has focused, until recently, on a small portion of animal groups, most notably eusocial insects and cooperative breeding vertebrates, and (2) comparative phylogenetic analyses (chapter 7) still are sparse relative to group- or population-level investigations. Molecular techniques (e.g., high-throughput sequencing) and bioinformatics (e.g., gene banks, weather databases) continue to advance at a staggering rate, and we continue to steadily accumulate knowledge of social organization in other taxa (e.g., fishes, squamate reptiles). Progress in the molecular realm is sure to outpace behavioral investigation, but their continued fusion is necessary to generate a more global view of the evolution/maintenance of sociality through comparative phylogenetics (see above descriptions of Beauchamp 1999; Arnold & Owens 1999; Rubenstein & Lovette 2007). Similarly, molecular, genetic, and neurobiological tools have proved exceptionally powerful in understanding the mechanisms that govern social behavior (Insel & Fernald 2004). Comparative behavioral neuroscience also is shedding significant light on the evolution of social behavior, and perhaps can guide us in our thinking of social structure. We should thus continue to fuel the momentum of integration by crossing disciplines in our investigations of sociality.

Interest in dominance hierarchy formation has waxed and waned for greater than 50 years. The vast majority of studies have approached this problem from mathematical or conceptual perspectives (see "Dominance Hierarchies" section and box 17.1), or have devised improved statistical measures to characterize the rank order, which are then applied to groups that have been observed *in situ*. There is a wealth of untested predictions, and very few studies that have conducted manipulations to better understand the rules governing the formation of linear dominance hierarchies (e.g., Dugatkin & Druen 2004), and perhaps their implications for spatial structure. Do animals use the same decision rules (chapter 8) in dyadic versus social settings? What is the importance of winner and loser effects in the development of social hierarchies? Could the persistence, magnitude, or decay rates of winner and loser effects be modulated by social system? Some of these questions clearly would benefit from a comparative approach, for instance, contrasting winner-loser effects among closely related species that exhibit varying degrees of sociality (Hsu et al. 2006). These are exciting and foundational areas of research in behavioral ecology—ones that we hope will garner more significant attention in the future.

Social network theory has provided new perspectives on the study of sociality. The thrust of its application to nonhuman animal groups has been relatively recent but already worthy of a comprehensive volume that describes the mathematical underpinnings of social network analysis, some of the kinks that require fine tuning, and a veritable treasure trove of questions regarding social groups that can be resolved through its use (Croft et al. 2007). As is the case with most open niches, the area of social networks is sure to be populated fast and furiously. However, we anticipate that it will be both a profitable patch to disperse to, and one that could yield some fresh, innovative insights into how group dynamics and/or composition might drive processes of information transfer and disease transmission. It will be critical to examine how the shape, distribution, and constitution of the network changes with time as individuals win or lose contests, watch interactions between others, breed, cheat in a cooperative interaction, and so forth. We suspect that there will be much promise in continuing to explore the state dependency of social behavior and its impact of group structure. For instance, the decision to eavesdrop might depend on the characteristics (e.g., size, experience, sex, status) of both the bystander and the interacting animals. Whether or not information is acquired (or used) could resonate through the group in ways that could be made more obvious by employing social network analysis.

Lastly, a great deal of work on social information use has recently been synthesized (Bonnie & Earley 2007; Valone 2007), and some interesting questions worthy of further study have emerged. If there are multiple streams of social information (e.g., about resource quality, about who occupies a patch, about their relative status), do animals integrate these streams into their behavioral decisions? Do some forms of social information trump others? When should animals use only private or only social information, and when should they use both? These are but a few of the directions that the field of social information use can take, and we encourage readers interested in this field to peruse with a critical and experimental eye the aforementioned reviews and the abundant literature cited therein.

SUGGESTIONS FOR FURTHER READING

Alexander (1974) is a classic review of the forces influencing sociality. Beauchamp (2003) takes a step beyond collective detection to explore additional mechanisms, such as competition and its impact on time budgets that could explain the observation that vigilance behavior decreases as a function of group size; Beauchamp and Ruxton (2008) devised a clever experimental approach to investigating the independent influences of collective detection and dilution on this group-size effect. A recent review by Hsu et al. (2006) dissects the influence of fighting experience (winner and loser effects) on contest dynamics and dominance hierarchy formation, and is an excellent resource for literature in each of these areas. In a short but informative review, Clutton-Brock (2002) provides some alternatives to kin selection in explaining the evolution of cooperative social groups. McGregor (2005) edits an important collection of chapters on communication networks, which focus on how group living and territorial animals extract and utilize information available in the signaling interactions of others.

Alexander RD (1974) The evolution of social behavior. Annu Rev Ecol Syst 5: 325–383.
Beauchamp G (2003) Group-size effects on vigilance: a search for mechanisms. Behav Proc 63: 111–121.
Beauchamp G & Ruxton GD (2008) Disentangling risk dilution and collective detection in the antipredator vigilance of semipalmated sandpipers in flocks. Anim Behav 75: 1837–1842.
Clutton-Brock T (2002) Breeding together: kin selection and mutualism in cooperative vertebrates. Science 296: 69–72.
Hsu Y, Earley RL, & Wolf LL (2006) Modulation of aggressive behaviour by fighting experience: mechanisms and contest outcomes. Biol Rev 81: 33–74.
McGregor PK (2005) Animal Communication Networks. Cambridge Univ Press, Cambridge, UK.

18

Altruism and Cooperation

ANDY GARDNER, ASHLEIGH S. GRIFFIN, AND STUART A. WEST

A behavior is cooperative if it provides a benefit to individuals other than the performer, and has evolved at least partially because of this benefit (West et al. 2007a). Such behaviors pose a major problem for evolutionary theory because, all else being equal, they should reduce the relative fitness of the actor (Hamilton 1964). This conflicts with the idea that natural selection favors those behaviors that improve fitness, so we might expect cooperation to be disfavored in the Darwinian struggle for existence. Yet cooperation abounds in the natural world, in spite of this difficulty. The problem, then, is to explain how cooperation can be favored by natural selection (Maynard Smith & Szathmary 1995; Sachs et al. 2004).

The problem of cooperation also extends to economics and human morality. A famous allegory, highlighting the tension between the good of the group and the selfish interests of the individual, is the "tragedy of the commons" (Hardin 1968). In this hypothetical scenario, a group of herdsmen each decide how many animals they will individually set to graze on a shared pasture, or commons. If a herdsman adds an extra animal to the pasture, this will incur both a benefit and a cost. The benefit is that the herdsman will gain from the sale of an extra animal. The cost is the potential for overgrazing, which can damage the pasture in the longer term. However, although the focal herdsman gains all of the benefit, he pays only a fraction of the cost, as this is shared among all the herdsmen. Consequently, the herdsman has more to gain than to lose from adding extra animals. The tragedy is that as a group, all the individuals would benefit from grazing fewer livestock, but such cooperation is not stable because individuals can gain by selfishly pursuing their own interests, and this leads to the ultimate destruction of the commons. There are numerous real-world examples of this problem, such as declining fish stocks due to overfishing, poor public support for vaccination programs, overuse of antibiotics resulting in the rapid evolution of multiresistant bacterial pathogens, and the current inability of countries to agree on reducing carbon emissions.

Within evolutionary biology, most attention to this problem has been focused on animals such as birds, humans, and the social insects, where cooperative behaviors can be relatively easy to identify. However, the same problem occurs at all levels of biological organization (Maynard Smith & Szathmary 1995). Animals are multicellular, and so their very existence relies upon cooperation between the eukaryotic cells that make them up. The mitochondria upon which these eukaryotic cells rely were once free-living cells, but now live cooperative symbiotic lives. Separate genes, which make up the genome, cooperate in what has been termed the "parliament of the genes" (Leigh 1971). Furthermore, the problem is not particular to animals: the tree of life is dominated by single-celled microorganisms that appear to perform a huge range of cooperative behaviors (West et al. 2007b). For example, bacteria excrete products that perform a variety of functions, such as scavenging nutrients, communication, defense, and movement. The benefits of such extracellular products can be shared by neighboring cells, and hence they represent a *public good* that is open to the problem of cooperation. Almost all of the major evolutionary transitions, from replicating molecules to complex animal societies, have relied upon the problem of cooperation being solved, and the interest of individuals becoming aligned (Maynard Smith & Szathmary 1995).

In this chapter, we show how the theory of inclusive fitness gives an insight into the evolution of cooperative behaviors. This theory highlights that there are two classes of benefits for cooperation: direct fitness benefits accruing from the individual's personal reproductive success, and indirect fitness benefits due to the reproductive success of its relatives. We review the biological mechanisms that can lead to direct and indirect fitness benefits, with application to microbes, plants, animals, and humans. Finally, we identify future research directions that will be important for further advances in this field.

CLASSIFICATION OF SOCIAL BEHAVIORS

A first step toward explaining cooperation is to provide its proper definition, and to distinguish its different forms. The classic approach to classifying social behaviors is to do so on the basis of their impact on the reproductive success (personal fitness) of the actor and recipients (table 18.1; though this has not always been carefully followed—box 18.1). Behaviors increasing the personal fitness of both the actor and the recipient (+/+) are termed *mutually beneficial*. Those behaviors increasing the personal fitness of the actor and decreasing the personal fitness of the recipient (+/–) are *selfish*. Behaviors decreasing the personal fitness of the actor and increasing the personal fitness of the recipient (–/+) are *altruistic*. And those behaviors decreasing the personal fitness of both actor and recipient (–/–) are *spiteful* (Hamilton 1964, 1970; West et al. 2007a).

In general, a behavior that improves the personal fitness of the recipient can be described as a *helping* behavior. When this helping is an evolutionary adaptation (i.e., has evolved for that purpose), the behavior is cooperative (West et al. 2007a). Thus, cooperation includes some mutually beneficial behaviors and all altruistic behaviors. It excludes helping behaviors that have not evolved for the purpose of being helpful. For instance, when an elephant produces dung, this is to its own benefit (release of waste) and also to the benefit of dung beetles, though it would be incorrect to describe this mutually beneficial behavior as cooperative (West et al. 2007a).

TABLE 18.1 A classification of social behaviors

		Fitness impact for recipient	
		+	–
Fitness impact for actor	+	Mutual benefit	Selfishness
	–	Altruism	Spite

Source: Based on Hamilton (1964, 1970), and West et al. (2007a).

HAMILTON'S RULE

The next step toward explaining the evolution of cooperation is to describe how natural selection operates upon behaviors according to the personal fitness effects felt by the actor and any recipients. Ultimately, a gene for cooperation is favored by natural selection if it successfully transmits copies of itself into future generations, no matter how this is actually achieved (Hamilton 1963). One way in which the gene can improve its own transmission is to increase the personal fitness of its bearer. This is the basis for the traditional Darwinian prediction that adaptations will normally function to promote the reproductive success of the individual bearer. Another way for the gene to improve its transmission is to increase the personal fitness of the bearer's genetic relatives, who also carry the gene (Hamilton 1963, 1964). In this way, an adaptation can promote the success of its bearer's genes, albeit indirectly. An individual can therefore gain either a direct fitness benefit from promoting its own reproductive success or an indirect fitness benefit from promoting the reproductive success of its relatives (figure 18.1; Hamilton 1964). Adding together these direct and indirect fitness components obtains a single *inclusive* fitness (Hamilton 1963, 1964, 1970), and it is this quantity that natural selection has designed organisms to maximize (Hamilton 1964, 1996; Grafen 2006).

BOX 18.1 Use and Abuse of Altruism

The scheme outlined in this chapter for classifying social behaviors (Hamilton 1964; West et al. 2007a) has not always been carefully followed, and the resulting misuse of terminology has generated much confusion and purely semantic disagreements in the social evolution literature (reviewed by West et al. 2007a). In particular, the term *altruism* is frequently misused to describe even helping behaviors that increase the direct fitness of the actor, and hence are more properly described as *mutually beneficial*. Altruism is properly defined as leading to both a direct fitness cost ($c < 0$) for the actor and a personal fitness benefit for the recipient ($b > 0$), and the only way for altruistic behavior to be favored by natural selection is if the actor and recipient are related ($r > 0$, and $rb - c > 0$; Foster et al. 2006).

It is important to emphasize that the cost and benefit terms in Hamilton's rule, upon which social behaviors are classified, describe total lifetime fitness effects, and not simply the immediate impact of the helping behavior on the actor's fecundity or survival (Lehmann & Keller 2006). Helping behaviors that involve a short-term fecundity or survival cost have often incorrectly been referred to as altruistic, even if they bring about a net increase in the total lifetime reproductive success of the actor. A recurrent example is "reciprocal altruism" (Trivers 1971), in which the helping behavior is favored because there is a sufficiently high probability of the help being reciprocated in the future, leading to a net direct fitness benefit. This type of behavior should properly be described as *reciprocal cooperation* (Axelrod & Hamilton 1981) or simply *reciprocity* (West et al. 2007a).

Another key point is that fitness effects are absolute, or expressed relative to the population as a whole, and they are not described relative to some more local, arbitrary grouping of individuals (West et al. 2007a). This is important because some researchers have incorrectly defined altruism in terms of it involving a within-group disadvantage (e.g., Wilson 1975; Gintis 2000), and thereby including even those behaviors that increase the absolute fitness of the actor when its share of the group benefit outweighs the within-group cost. Such behaviors should be described as mutually beneficial helping (West et al. 2007a). For example, female-biased sex ratios evolving due to local mate competition (Hamilton 1967) have sometimes been conceptualized as "altruistic" because individuals producing a more equal sex ratio would have greater within-group fitness (Colwell 1981). However, individuals maximize their personal fitness by producing a female-biased sex ratio, so this is actually an example of mutual benefit (Harvey et al. 1985).

Finally, we note that it will often be difficult to know what the fitness effects of a particular behavior will be without first performing careful measurements (chapter 4). Depending upon the ecological context in which the social behavior is occurring, there may be a rather complicated link between the social action and the ultimate fitness consequences for the individuals involved, particularly if localized competition means that help dispensed toward one neighbor has a negative impact upon other individuals in the vicinity (e.g., Rousset 2004). This means it is often difficult to determine whether a behavior is beneficial or deleterious for the actor or any recipients, and so, in the absence of this information, it is safest to refer to *cooperative* or *helping* behaviors, and not simply assume that they are altruistic.

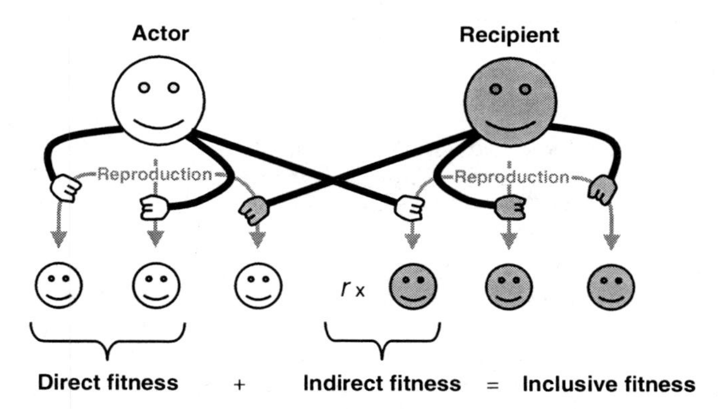

FIGURE **18.1** Inclusive fitness is the sum of direct and indirect fitness. Social behaviors affect the reproductive success of individuals beyond the actor. The impact on the actor's own reproductive success is the direct fitness effect. The impact on the reproductive success of social partners, weighted by the relatedness of the actor to the recipient, is the indirect fitness effect. In particular, inclusive fitness does not include all the reproductive success of relatives, only that which is due to the behavior of the actor. Also, inclusive fitness does not include any of the reproductive success of the actor that is due to the actions of its social partners.

The theory of inclusive fitness is encapsulated in the pleasingly simple "Hamilton's rule," named after the social evolution theorist who devised it, W. D. Hamilton. The rule states that a behavior is favored by natural selection when

$$rb - c > 0$$

(Hamilton 1963, 1964, 1970) where c represents the personal fitness cost of the behavior for the actor, b represents the personal fitness benefit for the recipient, and r represents the coefficient of genetic relatedness, which can be thought of as an exchange rate that allows the reproductive success of the recipient to be converted into the currency of the actor's inclusive fitness. So, $rb - c$ is the total inclusive fitness effect, and the behavior is favored whenever this is positive. A mathematical derivation of Hamilton's rule is given in box 18.2 (see also box 14.2 for an alternative).

The coefficient of relatedness is a statistical measure of how similar two individuals are at the genetic loci of interest, relative to other individuals in the same population (Hamilton 1963, 1964, 1970; Grafen 1985; box 18.2). If social partners are more similar than average, then the coefficient of relatedness is positive. Because r is relative to the average genetic similarity in a population, the average genetic relatedness within a population is zero, and some individuals are negatively related to each other (Hamilton 1970; Grafen 1985). Moreover, the coefficient of relatedness describes genetic similarity, irrespective of the cause of that similarity. Usually, if individuals share genes in common it is because they are close kin, but other causes for nonzero genetic relatedness are possible (e.g., *greenbeard* genes), and we discuss these in the next section. In most scenarios—for example, a large population with no inbreeding—genetic relatedness will simply reflect degree of kinship, with full siblings being related by $r = 1/2$, half siblings by $r = 1/4$, and cousins by $r = 1/8$ (Hamilton 1963, 1964). Hence J. B. S. Haldane's famous quip that he would lay down his life to save "two brothers or eight cousins," and is the reason why the theory of inclusive fitness has often been referred to as *kin selection* theory.

Hamilton's rule was devised to explain altruism, which incurs a personal cost ($c > 0$) for the actor and a personal benefit ($b > 0$) for the recipient. However, it equally well predicts when mutually beneficial cooperation ($c < 0$, $b > 0$), selfishness ($c < 0$, $b < 0$), or spite ($c > 0$, $b < 0$) will be favored. It also applies to nonsocial behaviors that do not impact the personal fitness of individuals other than actor ($b = 0$); here, it simply states that a behavior will be favored when it increases the personal fitness of the actor ($c < 0$). Thus, inclusive

BOX 18.2 Hamilton's Rule

Although we have introduced Hamilton's rule from an inclusive fitness perspective, the easiest way to derive the rule is by taking a personal (or *neighbor-modulated*) fitness approach (Hamilton 1964). The derivation presented below is based upon that of Queller (1992b; see also Frank 1998). For a similar approach, see also box 14.2.

We begin with a multiple linear regression model that predicts individual fitness (w) as a function of the individual's genetic *breeding* value (g; Falconer 1989; chapter 14, this volume) for a trait of interest, and also the breeding value of its social partner (g'; for simplicity, we assume pairwise interactions). We imagine that this model has been fitted, by the usual method of least squares, to data describing a population of social interacting individuals. The fitted model can be written as

$$w = \bar{w} + \beta_{w,g \cdot g'}(g - \bar{g}) + \beta_{w,g' \cdot g}(g' - \bar{g}) + \varepsilon$$

where $\bar{w}$ and $\bar{g}$ are the population average fitness and breeding value respectively, $\beta_{w,g \cdot g'}$ is the least-squares partial regression of fitness on one's own breeding value holding the breeding value of the social partner fixed (i.e., the average effect of the gene on one's own fitness), $\beta_{w,g' \cdot g}$ is the least-squares partial regression of fitness on the social partner's breeding value holding one's own breeding value fixed (i.e., the average effect of the social partner's gene on one's own fitness), and ε is the uncorrelated error. Note that analysis of additive (average) effects does not require any assumption of additive gene action: this derivation holds for an arbitrary number of loci with arbitrary synergistic interactions (Gardner et al. 2007).

From Price's (1970) theorem, the change in the breeding value of any character of interest due to the action of natural selection ($\Delta_S \bar{g}$) is described by

$$\bar{w}\Delta_S \bar{g} = \text{cov}(w, g)$$

where "cov" denotes a statistical covariance. This means that the direction and magnitude of change in the heritable component of any trait is given by its statistical association with personal fitness; this is true for social and nonsocial traits alike. Applying our least-squares model to Price's equation obtains

$$\bar{w}\Delta_S \bar{g} = \beta_{w,g \cdot g'} \text{cov}(g, g) + \beta_{w,g' \cdot g} \text{cov}(g', g)$$

The character of interest is favored to increase under the action of natural selection whenever this quantity is positive, and the condition for this is

$$\beta_{w,g \cdot g'} + \beta_{w,g' \cdot g} \frac{\text{cov}(g', g)}{\text{cov}(g, g)} > 0$$

This is Hamilton's rule, in its neighbor-modulated fitness guise. The cost of carrying genes for a social trait is described by $\beta_{w,g \cdot g'} = -c$; the benefit of interacting with individuals who carry genes for the social trait is described by $\beta_{w,g' \cdot g} = b$; and genetic association between social partners is described by the ratio $\text{cov}(g',g)/\text{cov}(g,g)$, which provides the proper definition for the coefficient of genetic relatedness r (Frank 1998). Substituting in the cost, benefit, and relatedness notation, and rearranging, we obtain the more familiar $rb - c > 0$. This least-squares regression approach to Hamilton's rule forms the conceptual basis

(continued)

for a powerful differential calculus approach to analyzing kin selection models, outlined in box 18.3.

We have assumed, for simplicity, that all individuals are equivalent, in other words, they are not structured into separate sex or age classes. Developments of Hamilton's rule for class-structured population are discussed by Taylor (1990). However, with the assumption of no class structure, we can most easily derive an inclusive fitness version of Hamilton's rule. Just as raising the breeding value of the social partner (g') has an impact upon the focal individual's personal fitness (w) described by $\beta_{w,g'\bullet g}$, raising the breeding value of the focal individual (g) has an impact upon the social partner's personal fitness (w'), that is described by $\beta_{w',g\bullet g'}$, and due to the symmetry of the model, these two effects are equal in value. Making this substitution, we obtain the inclusive form of Hamilton's rule:

$$\beta_{w,g\bullet g'} + \beta_{w,g'\bullet g}\frac{\mathrm{cov}(g',g)}{\mathrm{cov}(g,g)} > 0$$

This describes the impact of the focal individual's genes on its own personal fitness ($\beta_{w,g\bullet g'} = -c$), and also upon the personal fitness of its social partner ($\beta_{w',g\bullet g'} = b$), and the latter is weighted according to their genetic relatedness ($\mathrm{cov}(g',g)/\mathrm{cov}(g,g) = r$). Importantly, the inclusive fitness calculation excludes any impact upon the personal fitness of either individual that is due to genes carried by the social partner (figure 18.1). This avoids double counting of fitness increments: the components of one individual's inclusive fitness do not count toward the inclusive fitness of any other individuals (see also chapter 1). This also ensures that each actor has sole control over its own inclusive fitness. This is not true of neighbor-modulated (personal) fitness, which is potentially controlled by multiple individuals. Though inclusive fitness and neighbor-modulated fitness are equivalent, in the sense that both approaches correctly describe how natural selection operates on social traits (Taylor et al. 2007), the difference in control means that although individuals can be viewed as striving to maximize their inclusive fitness, they cannot be viewed as striving to maximize their neighbor-modulated fitness (Grafen 2006). The importance of control is also reflected in the usual assumption that the actor's phenotype is controlled only by its own genes (for an analysis of socially mediated phenotypes, see chapter 14).

fitness theory is a general theory of organismal adaptation, and it does not only apply to social evolution. See Frank (1998) for a discussion of Hamilton's rule as the central concept of social evolution theory.

BIOLOGICAL MECHANISMS

Hamilton's rule highlights the two separate ways in which behaviors, or indeed any traits, can be favored by natural selection: (1) indirect fitness benefits and (2) direct fitness benefits (Hamilton 1964). In the next two sections, we review the various biological mechanisms that can be involved in deriving direct and indirect benefits for cooperation, and hence provide an explanation for its evolutionary success. We emphasize that these explanations are not mutually exclusive, and several may be operating to favor cooperation in any given species.

Indirect Fitness Benefits

Indirect fitness benefits come from the individual enhancing the reproductive success of its genetic relatives, and these can outweigh a direct fitness cost (figure 18.2; Hamilton 1964, 1970). Thus, indirect fitness benefits are the key to altruistic behavior (Foster et al. 2006). The lives of most

BOX 18.3 How to Analyze a Kin Selection Model

Although inclusive fitness is conceptually easier, and provides the formal basis of the theory of social adaptation (Grafen 2006), the simplest way of analyzing a kin selection model is by taking a personal (or neighbor modulated) fitness approach. Here we outline the approach of Taylor and Frank (1996; see also Frank 1998; Rousset 2004; Taylor et al. 2007).

The basic optimization approach to natural selection defines an individual's personal fitness (w) as a function of its phenotype (z), and it seeks the optimum phenotype (z^*), which maximizes the fitness function within the given constraints (i.e., $\mathrm{d}w/\mathrm{d}z|_{z=z^*} = 0$ and $\mathrm{d}^2w/\mathrm{d}z^2|_{z=z^*} < 0$). However, social evolution theory is concerned with scenarios in which the phenotypes of potentially multiple individuals impacts upon fitness, and where these are potentially correlated with the phenotype of the focal individual. Fortunately, the basic optimization method can be extended to examine such cases. First, we express the individual's fitness as a function of her own phenotype z and the phenotype of her social partner z', in other words, $w(z,z')$. Then, applying the chain rule of differential calculus, we can write

$$\frac{\mathrm{d}w}{\mathrm{d}z} = \frac{\partial w}{\partial z} + \frac{\partial w}{\partial z'}\frac{dz'}{dz}$$

Here, ∂ denotes a partial derivative, or a derivative taken with respect to one variable while holding the value of any other variables constant. Thus, $\partial w/\partial z$ is the impact of the individual's phenotype on her own fitness and $\partial w/\partial z'$ is the impact of the social partner's phenotype on the focal individual's fitness, and $\mathrm{d}z'/\mathrm{d}z$ describes the correlation between the phenotypes of the two individuals. If we assume a fully heritable phenotype ($z = g$, $z' = g'$), then this chain rule becomes analogous to the neighbor-modulated fitness version of Hamilton's rule (box 18.2):

$$\beta_{w,z} = \beta_{w,z\bullet z'} + \beta_{w,z'\bullet z}\beta_{z',z}$$

In other words, here we have simply replaced each derivative with its corresponding least-squares regression coefficient, and each partial derivative with its corresponding partial regression coefficient. This becomes an exact equivalence as the genetic variation in the population tends to zero (so that the derivatives can be evaluated at $z = z' = \bar{z}$). The coefficient $\beta_{z',z} = \mathrm{cov}(z',z)/\mathrm{cov}(z,z)$ is Hamilton's coefficient of genetic relatedness (r; Taylor & Frank 1996; box 18.2 of this volume), and so the winning phenotype z^* satisfies

$$\frac{\mathrm{d}w}{\mathrm{d}z} = \frac{\partial w}{\partial z} + \frac{\partial w}{\partial z'}r = 0$$

where the derivatives are evaluated at $z = z' = \bar{z} = z^*$. Solving this equation yields an equilibrium strategy, and second derivatives can be inspected to determine its evolutionary stability (Taylor 1996).

For example, let's consider a simple model of the "tragedy of the commons" scenario (Frank 1998). If z describes the exploitation of a shared resource by a focal individual, and z' describes the average exploitation level of all individuals in her group, then a reasonable fitness function is

$$w(z,z') = \frac{z}{z'}(1 - z')$$

(continued)

The average fitness of the group is $w(z',z') = 1 - z'$, and this decreases as the average level of exploitation (z') is increased. However, each individual's share of group resources is proportional to her relative exploitation level (z/z')—in other words, individuals who exploit more than average get a larger share than do individuals who exploit less than average—and so, holding the overall exploitation by the group fixed, each individual is favored to increase her personal level of exploitation.

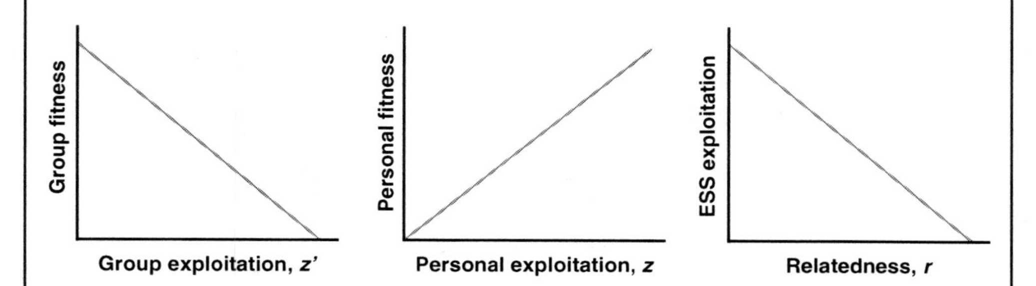

As described above, the action of natural selection is described by evaluating $\partial w/\partial z + \partial w/\partial z' \, r$ at $z = z' = \bar{z}$, which obtains

$$\frac{1-\bar{z}}{\bar{z}} - \frac{1}{\bar{z}} r$$

Setting this to zero, and solving for $\bar{z} = z^*$, the equilibrium level of exploitation is found to be $z^* = 1 - r$, and this turns out to be evolutionarily stable. Hence, in a large group of unrelated individuals ($r = 0$), selection favors total exploitation ($z^* = 1$), and the destruction of the common resource gives a group average fitness of zero ($1 - z^* = 0$). However, if an individuals is appreciably related to her group mates ($r > 0$), some self-restraint is favored ($z^* = 1 - r < 1$) and the group enjoys some success ($1 - z^* = r > 0$). Thus, the tragedy of the commons can be averted by kin selection.

organisms abound with opportunities for an individual to improve the reproductive success of its neighbors but, crucially, for this to translate into an indirect fitness benefit, the genetic relatedness of the recipients to the actor must be sufficiently greater than average. Thus, the key to cooperation being favored through indirect fitness benefits is that the individual must help its genetic relatives.

In this section, we describe the three major mechanisms that lead to indirect fitness benefits for cooperation. The first is when individuals can discriminate kin from nonkin, on the basis of either environmental or genetic cues (figure 18.2a). Because kinship will usually be correlated with genetic relatedness, this ensures that cooperation is preferentially directed toward those individuals more closely related to the actor. A second, related mechanism involves greenbeard genes, in which individuals use genetic cues to discriminate genetic relations directly rather than close kin per se (figure 18.2b). Both kin discrimination and greenbeards will be seen to have only limited explanatory power in solving the problem of altruism across the whole diversity of life. A third mechanism, limited dispersal, also ensures that individuals interact socially with their close kin (figure 18.2c). This is a straightforward way to ensure high genetic relatedness, and thus potentially could favor even indiscriminately altruistic behaviors. It promises to offer a very general explanation for altruism in the majority of taxa.

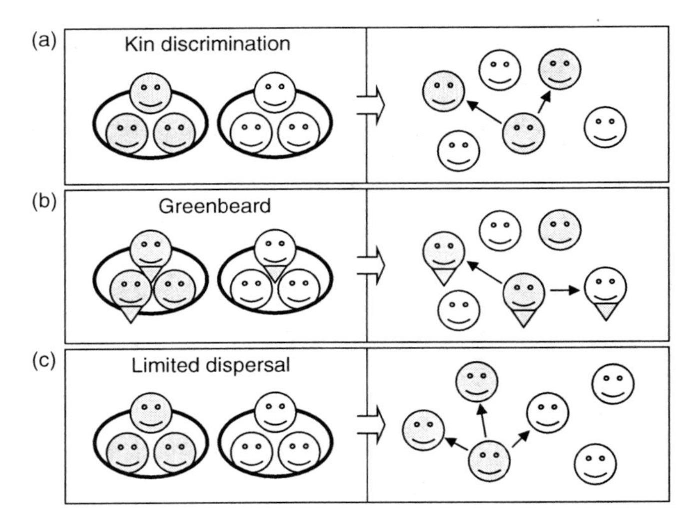

FIGURE 18.2 Indirect fitness benefits can explain the evolution of altruistic cooperation. If individuals interact socially with relatives, then even costly helping behaviors ($c > 0$, $b > 0$) can be favored by natural selection. Three main mechanisms ensure that altruistic cooperation is dispensed preferentially to one's genetic relatives. (a) Kin discrimination. For example, if the actor can remember those individuals it shared a nest with when young, and discriminate these kin (shaded) from nonkin (unshaded) after leaving the nest, then cooperation can be directed primarily toward genetic relatives. (b) Greenbeard. If the gene controlling altruism is also associated with a phenotypic marker, such as a green beard (bearded), then green-bearded individuals can identify which of their neighbors carries a copy of the gene. Altruism directed at genetic relatives can be favored by natural selection, even if these are not genealogical relatives (kin). (c) Limited dispersal. If individuals do not move far during their lifetimes, then they will tend to be surrounded by kin (shaded), and hence even indiscriminate altruism could be directed primarily toward kin rather than nonkin.

Kin Discrimination

The first mechanism that can promote the evolution of cooperation is kin discrimination, when an individual can distinguish its kin from nonkin and preferentially direct aid toward the former (figures 18.2a, 18.3, and 18.4; Hamilton 1964). Because kin tend to share genes in common ($r > 0$), nepotism can lead to an appreciable genetic relatedness between the individual and the beneficiaries of its actions. Consequently, if the benefit-to-cost ratio is high enough, then the indirect fitness benefit ($b > 0$) of cooperation will outweigh any direct fitness cost ($c > 0$), and hence increase the inclusive fitness of the individual ($rb - c > 0$). This has been demonstrated in many cooperatively breeding vertebrates such as long-tailed tits, in which individuals that fail to breed independently preferentially help at the nest of closer relatives (figure 18.3a; Russell & Hatchwell 2001). In this species, individuals distinguish between relatives and nonrelatives on the basis of vocal contact cues, which are learned from adults during the nesting period (associative learning). This leads to the situation in which individuals tend to help relatives with whom they have been associated during the nestling phase.

Inclusive fitness theory explains not only kin discrimination, but also variation in the level of kin discrimination across species (figure 18.4; Griffin & West 2003). In contrast to the long-tailed tit example given above, some cooperatively breeding vertebrates, such as meerkats (Clutton-Brock et al. 2000), have subordinates that do not show kin discrimination when helping. This is presumably because there are direct fitness benefits of helping, to be discussed in the next section. Clearly, the advantage of kin discrimination will be positively correlated with the benefit (b) provided by cooperation. In the extreme, if a supposedly "helpful" behavior provides little or no benefit to the

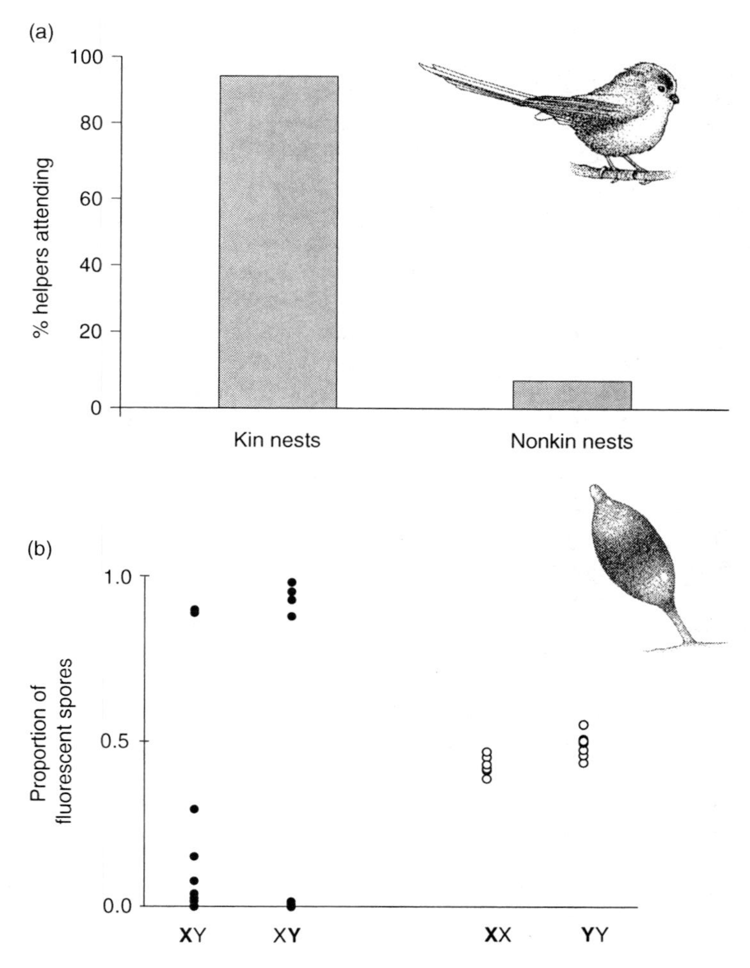

FIGURE **18.3** Kin discrimination. (a) Kin discrimination in long-tailed tits. Results of an experiment showing that ninety-six percent of helpers prefer to help at nests containing related chicks when they have the choice of where to invest their efforts (Russell & Hatchwell 2001). (b) Kin discrimination in the unicellular slime mold *Dictyostelium purpureum* (social amoeba; Mehdiabadi et al. 2006). A scatter plot shows the proportion of fluorescently labeled spores in fruiting bodies when two isolates are placed together at equal proportions and one is fluorescently labeled (*solid circles*). There is a greater variance in the experimental treatment in which the two isolates are different lineages (lineages A and B), than in the control treatment in which the isolates are the same lineage. This shows that individuals preferentially form fruiting bodies with members of their own lineage.

recipients, then there is little or no advantage in directing it preferentially toward closer relatives. This leads to the prediction that kin discrimination should be greater in those species in which cooperation provides a larger benefit for the recipient, and this pattern is indeed observed across cooperatively breeding vertebrate species (Griffin & West 2003).

Kin discrimination has also been found to occur in taxa that are not usually thought of from a social perspective. *Dictyostelium purpureum* is a unicellular slime mold (social amoeba) found in forest soils (Mehdiabadi et al. 2006). When starved of their bacterial food source, the cells of this species aggregate in tens to hundreds of thousands to form a multicellular, motile "slug." Slugs migrate to the soil surface, where they transform into a fruiting body composed of a stalk structure holding aloft a ball of spores. The nonviable stalk cells willingly sacrifice themselves to aid the dispersal of the spores. This requires explanation because cooperative cells that form stalk cells could be exploited by cheats who preferentially migrate to form spores in the fruiting body. Kin selection offers a potential solution to this problem, because stalk cells could gain an indirect fitness benefit from helping their relatives disperse. This suggests that it would be advantageous to preferentially form a slug with kin rather than nonkin, and indeed kin discrimination has recently been observed in *D. purpureum* slug formation. Specifically, when two lineages are mixed equally and allowed to form slugs on agar plates, they discriminate to the extent that the average genetic relatedness in fruiting bodies is 4/5, compared with the random expectation of 1/2 (figure 18.3b; Mehdiabadi et al. 2006).

Kin discrimination can occur through the use of environmental or genetic cues (Grafen 1990c). The most common mechanism appears to involve environmental cues in which kinship can be inferred on the basis of either a prior association or a shared environment, as in long-tailed tits and a range of other animals. For example, if two individuals were raised in the same nest, they are probably close kin. In contrast, the slime mold example is likely to involve some genetic cue of kinship (also termed *kin recognition*, *genetic similarity detection*, *matching*, or *tags*). To detect genetic similarity, an individual must have some cue that is genetically determined, such as the cuticular hydrocarbon profile of an insect (Boomsma et al. 2003), or the odor produced by scent glands in a mammal (Mateo 2002), and a *kin template* against which other individuals can be compared (Grafen 1990c). This kin template could come about through the individual's own genotype or cues (self-matching) or through learning the cues of its rearing associates (Mateo 2002).

Kin discrimination based upon genetic cues is often unlikely to be evolutionarily stable. The reason is that recognition mechanisms require genetic variability (polymorphism) at a genetic locus, or multiple loci, to provide a cue. However, individuals expressing common cues would more often be identified (correctly or incorrectly) as kin by social partners, and hence would be more likely to receive the benefits of cooperation. As long as the benefit of receiving cooperation is larger than the cost of giving cooperation, this would give already common cues an evolutionary advantage, and force less common cues to extinction (Rousset & Roze 2007). Consequently, kin discrimination is its own worst enemy, eliminating the genetic variability that it requires to work. This can explain why kin discrimination based on genetic cues is often not found where it might be expected, in both vertebrates and invertebrates (Keller 1997). In cases in which kin discrimination based upon genetic cues has been observed, it can usually be argued that

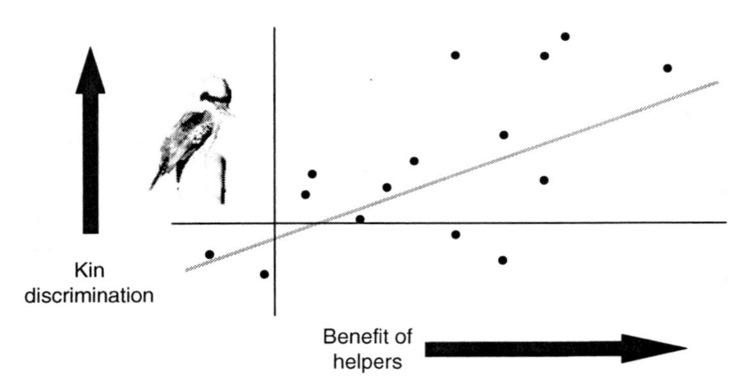

FIGURE **18.4** Kin discrimination and the benefit of helping. Helpers are more likely to discriminate in favor of relatives when the amount of help they provide increases the survival of offspring to the following year. The extent to which individuals preferentially help closer relatives (kin discrimination) is plotted against the benefit of helping. The significant positive relationship between these two variables is predicted by kin selection theory. The figure is taken from Griffin and West (2003), with two additional data points added from studies on the bell miner (*Manorina melanophrys*) and the red-cockaded woodpecker (*Picoides borealis*). The illustration shows the laughing kookaburra, a species that does not show kin discrimination in its helping behavior.

there is some other selective force maintaining variability at the recognition loci, such as host-parasite coevolution in the major histocompatibility complex (MHC) of vertebrates (Rousset & Roze 2007). Cue diversity may also be maintained if there is limited dispersal, such that interactants tend to be relatives anyway (Axelrod et al. 2004; Rousset & Roze 2007), as is likely to be the case with the slime mold example discussed above.

Greenbeards

As mentioned in the previous section, the coefficient of relatedness describes the genetic similarity between two individuals with respect to the particular locus or loci of interest, rather than over the whole genome. This means that indirect fitness benefits will be obtained if cooperation is directed toward even nonkin social partners, as long as they also carry the gene or genes for cooperation (i.e., genetic, but not necessarily genealogical, relatives; Hamilton 1964, 1975). This emphasizes that the theory of inclusive fitness applies more broadly than *kin selection*, in its literal sense. Dawkins (1976) illustrated this with a now famous hypothetical example, in which he imagined a gene that pleiotropically gives rise to a green beard and also prompts the individual to preferentially direct cooperation toward other green-bearded social partners (figure 18.2b). However, this mechanism can also occur without tags; for example, if the cooperative gene also caused some pleiotropic effect on habitat preference that led individuals carrying the same gene to settle in the same place (Hamilton 1975). Consequently, although this mechanism is usually termed a *greenbeard*, it more generally represents an assortment mechanism, requiring a single gene (or a number of tightly linked genes, forming a *supergene*) that encodes both the cooperative behavior and causes cooperators to associate (Lehmann & Keller 2006).

Greenbeards are likely to be rare, because cheats that display the green beard, or otherwise assorted with cooperators without also performing the cooperative behavior, could invade and overrun the population. Such "falsebeards" can arise owing to the appearance of a new allele at the greenbeard locus, or else as a consequence of modifier genes elsewhere in the genome (Gardner & West, in press). For this reason, greenbeard altruism is unlikely to be stable over evolutionary timescales. One of the few cases in which a cooperative greenbeard does occur is in the slime mold *Dictyostelium discoideum*, which forms fruiting bodies in a very similar way to *D. purpureum*. In *D. discoideum*, individuals with the *csa* cell-adhesion gene adhere to each other in aggregation streams (excluding mutants who lack the allele) and cooperatively form fruiting bodies (Queller et al. 2003). Here, cell adhesion commits an amoeba to the possibility of becoming a stalk cell, but is also required to be included in the fruiting body and hence to have a chance at successful spore formation. Such examples are somewhat surprising, because the concept of the greenbeard was not developed as a theory to explain altruism, but rather as a thought experiment to show that genetic relatedness, rather than genealogical relationship per se, is crucial for indirect fitness benefits.

Limited Dispersal

If individual organisms do not disperse over large distances during their lifetimes, then they will tend to stay within close proximity of their kin, who will become their social partners. In this case, altruism directed indiscriminately at all neighbors could be favored, because those neighbors tend to be relatives (figure 18.2c; Hamilton 1964, 1972). This has the potential to be important in a wide range of organisms because it does not require any mechanism of kin discrimination. Instead, all that is required is that the level of altruism evolves in response to the average genetic relatedness among individuals who tend to interact by chance.

The predicted role of limited dispersal has been supported by an experimental evolution study on cooperation in bacteria (figure 18.5; Griffin et al. 2004). Many bacteria release molecules called siderophores to scavenge for iron, because this nutrient is often limiting and sets the limit on bacterial growth. These siderophores represent a cooperative public good: they are costly to the individual to produce, but iron that is bound to siderophores can be taken up by any cell, providing a benefit to other individuals in the locality. It is unclear why cheater cells, which do not produce siderophores but benefit from the cooperation of others, do not outcompete cooperators. To address this, replicate populations of the pathogenic bacterium *Pseudomonas aeruginosa* were initiated with a mixture of a cooperative wild-type strain that produces siderophores, and a cheater mutant that does not produce siderophores, and then maintained in

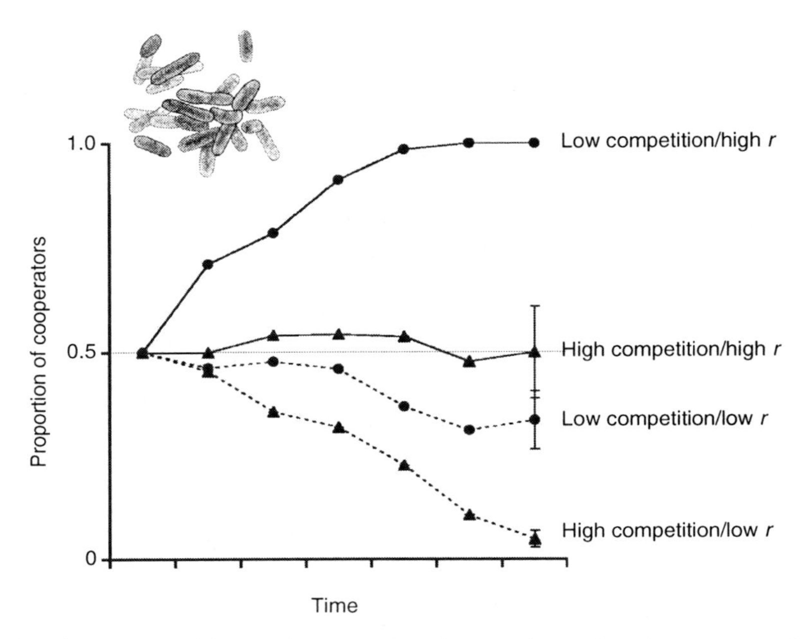

FIGURE 18.5 Relatedness, competition, and cooperation. An experimental study on cooperative siderophore production in the bacterium *Pseudomonas aeruginosa* has shown how selection for cooperation is influenced by relatedness and the extent of competition among relatives (Griffin et al. 2004). The proportion of cooperative individuals who produce siderophores is plotted against time. The different lines represent relatively high (solid lines) and low (dashed lines) relatedness. The different symbols represent relatively low (circle) and high (triangle) amounts of competition among relatives. Cooperation is favored by higher relatedness and lower competition among relatives.

conditions that led to relatively high or low genetic relatedness (Griffin et al. 2004). As predicted by inclusive fitness theory, the cooperative wild-type strain outcompeted the cheater mutant strain only when cultured under conditions of relatively high relatedness. More generally, limited dispersal is likely to be important for maintaining a range of public goods produced by bacteria to help with the gathering of resources, growth, and reproduction (West et al. 2007b).

Although limited dispersal can favor cooperation, it will not necessarily do so (Queller 1992b; Kümmerli et al. 2009). The problem is that, although limited dispersal can bring relatives together to cooperate, it can also keep them together to compete for the same resources (Hamilton 1971b). This competition among relatives can reduce or even completely remove selection for cooperation among relatives, depending upon the precise details of their ecology. One way of thinking about this is that it reduces the benefit (b) of helping relatives (Frank 1998). In the extreme case, there is no point helping one brother if his increase in fitness comes at the cost of another brother. In the simplest possible scenario, the effects of increased genetic relatedness and increased competition exactly cancel out, and so limited dispersal has no influence on the evolution of cooperation (Taylor 1992). However, a number of factors that are likely to be biologically important can reduce the relative competition among relatives, and hence allow limited dispersal to favor cooperation: for example, when cooperation allows population expansion (as with bacterial public goods discussed above; Taylor 1992), or when relatives tend to disperse together (budding viscosity; Gardner & West 2006; Kümmerli et al. 2009), or when the life cycle involves a period of local interaction with close relatives followed by dispersal before competition (alternating viscosity; Lehmann et al. 2006).

Direct Fitness Benefits

The evolution of cooperation does not require that there be indirect fitness benefits. Cooperation can also provide a direct fitness benefit to the actor (figure 18.6; Trivers 1971), corresponding to a negative cost ($c < 0$) in Hamilton's rule. In this case, cooperation is mutually beneficial ($c < 0$, $b > 0$), and not altruistic ($c > 0$, $b > 0$). We divide the direct fitness explanations for cooperation into two categories. First, cooperation may provide a benefit as a by-product (automatic consequence) of an otherwise *self-interested* act by the actor (figure 18.6a; Sachs et al. 2004). For example, cooperation could lead to an increase in group size (see chapter 17), which increases the survival of everyone, including the individual who performs the cooperative behavior, due to larger groups being better at avoiding predators or competing with other groups (Kokko et al. 2001). Second, there may be some mechanism for enforcing cooperation, by rewarding cooperation or punishing cheaters (figure 18.6a and b; Frank 2003). It can also be useful to distinguish enforcement mechanisms that are behaviorally flexible (figure 18.6b), and are adjusted conditionally in response to the level of cooperation, from those that are behaviorally inflexible (figure 18.6c). In the former case, the benefit to the actor depends upon the recipients (or a third party) adjusting their behavior toward the actor, in response to the actor's behavior.

By-product Benefits

Cooperation may provide some automatic benefit, without enforcement. One way this could occur is if social partners have some shared interest in cooperation. In many cooperatively breeding vertebrates, such as meerkats, larger group size can provide a benefit to all the members of the group through an increase in factors such as survival, foraging success, or the likelihood of winning between-group conflicts (Clutton-Brock 2002; chapter 17, this volume). In this case subordinate individuals can be selected to help rear offspring that are not their own, to increase group size (termed *group augmentation*;

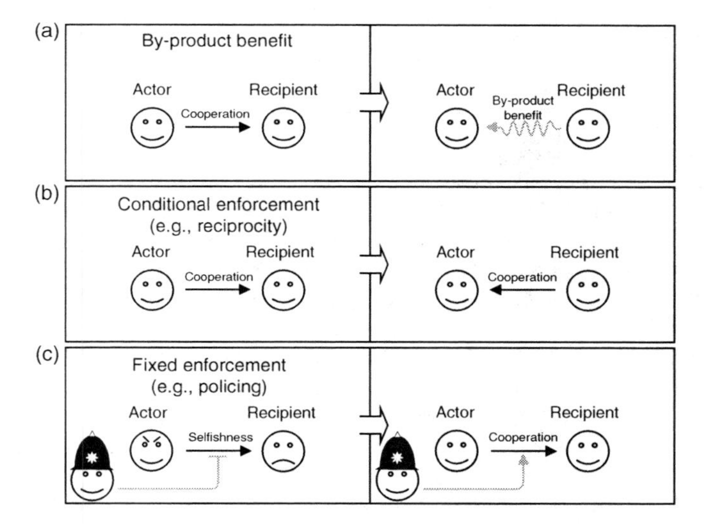

FIGURE **18.6** Direct fitness benefits can explain the evolution of mutually beneficial cooperation. Three general mechanisms can ensure that cooperation generates a direct fitness benefit for the actor. (a) By-product benefit. Helping a social partner may lead to increases in the overall fitness of all individuals in a social group; for example, in situations in which larger social groups offer better protection against predators, the actor could benefit from helping its neighbors to reproduce (group augmentation). (b) Conditional enforcement. Helping may lead to a change in the behavior of the recipient or a third party in a way that leads to an overall direct fitness benefit for the actor; for example, direct reciprocity ensures that cooperators receive more cooperation than cheats, giving a direct benefit to cooperation even if each act involves an immediate cost. (c) Fixed (unconditional) enforcement. Helping may be the only option available to an individual if the possibility for it to behave selfishly is ruled out by a successful system of policing.

Kokko et al. 2001). Selection for such helping is further increased if there is a chance that the subordinate will obtain dominance in the group at some later point, because they would then have a larger number of helpers themselves. These advantages of group augmentation would be greatest for the sex that is most likely to remain and breed in the natal group, which provides an explanation for why the level of helping is greatest in that sex for birds (males) and mammals (females; Clutton-Brock et al. 2002). Similar benefits of increased individual success in larger groups have been suggested to be important in a range of other cooperative organisms, including ants and social spiders (Avilés & Tufiño 1998; Bernasconi & Strassmann 1999). Things can get more complicated if the individuals in the group are related, because by-product benefits can then also provide indirect fitness benefits, either because the actor helps a relative (Kokko et al. 2001) or because the by-product benefits are shared with actors (Foster & Wenseleers 2006).

Direct fitness benefits also appear to be important for cooperative breeding in the wasp *Polistes dominulus* (Queller et al. 2000). In this species, colonies are initiated by one or a small number of foundress females who then form a dominance hierarchy with the dominant laying most of the eggs and the subordinates carrying out most of the more risky foraging. In one study, it was found that 35% of subordinates were unrelated to the dominant. These subordinates gain significant direct fitness benefits from staying and helping on the nest because dominants suffer an appreciable mortality rate, and so there is approximately a 10% chance that a subordinate will become the dominant in the group by the time that the workers emerge (Queller et al. 2000). It is unlikely that subordinates are merely helping nonrelatives by accident in this species, because there is sufficient between-individual variation in cuticular lipids to allow subordinates to distinguish related nest mates from unrelated nest mates (Dani et al. 2004), and so natural selection could have acted to reduce or remove cooperative behavior when subordinates are in a colony with nonrelatives. Such direct benefits of cooperation can also be important in cooperatively breeding vertebrates (Clutton-Brock 2002).

Enforced Cooperation

Cooperation can be enforced if there is a mechanism for rewarding cooperators or punishing cheats (Trivers 1971; Axelrod & Hamilton 1981; Frank 2003). This was emphasized by Trivers (1971), who showed that cooperation could be favored in reciprocal interactions, with individuals preferentially aiding those that have helped them in the past. In this case, cooperation provides a direct fitness benefit, as it is only favored if the short-term cost of being cooperative is outweighed by the long-term benefit of receiving cooperation (Lehmann & Keller 2006). Trivers (1971) termed this *reciprocal altruism*, but because it provides a direct fitness benefit it is mutually beneficial and not truly altruistic. Consequently, *reciprocity* or *reciprocal cooperation* are more appropriate terms. Here, we use the term *reciprocity* to mean the specific mechanism by which cooperation is preferentially directed at cooperative individuals, either directly (help those that help you) or indirectly (help those that help others).

Although such reciprocity has attracted a huge amount of theoretical attention, it is thought to be generally unimportant outside of humans (Hammerstein 2003). A famous possible exception is food sharing in vampire bats, in which successfully feeding individuals share their blood meals with less successful social partners who shared with them on earlier occasions (Wilkinson 1984), though even here the mechanism is somewhat obscure (e.g., Stevens & Hauser 2004). Our use of the term *reciprocity* differentiates it from other cases of enforcement that also rely on behavioral flexibility, and which has been variously termed *punishment*, *policing*, *sanctions*, *partner switching*, and *partner choice* (Frank 2003; Sachs et al. 2004). In some cases the term *reciprocity* is used more generally to cover all these cases (Lehmann & Keller 2006).

Enforcement has been suggested to be important in a number of vertebrates. A particularly ruthless example is found in meerkats, in which the dominant female suppresses reproduction in her subordinates (Young et al. 2006). If a subordinate female becomes pregnant when the dominant is also pregnant, then the dominant is likely to subject the subordinate to aggressive attack and temporarily evict her from the group; this usually leads to abortion of the subordinate's litter. In another cooperatively breeding vertebrate, the white-fronted bee-eater, harassment both encourages cooperation and reduces competition (Emlen & Wrege 1992). In this species, adults preferentially disrupt the breeding attempts of close relatives (offspring), and this harassment frequently leads to the offspring becoming a helper at the nest of the harasser. This strategy appears to work because the offspring gain approximately the same inclusive fitness return from helping at a nest as breeding independently, and so have little to lose from becoming a helper, whereas the parents have a lot to gain from obtaining a helper (Emlen & Wrege 1992).

Enforcement has also been suggested to be important in favoring cooperation in humans (figure 18.7). In one study, students were split into groups of four to play a public goods game for cash prizes, in which each person could contribute monetary units to a group project (Fehr & Gächter 2002). For every unit invested in the group project, each of the four members of the group earned 0.4 units. This leads to the problem of cooperation because although investing increased the units available to the group, the investors received less than they invested. The experiment was repeated with and without punishment; punishment was incorporated by allowing individuals to pay money to have units deducted from other players, after they were informed about each other's investments. As expected, punishment led to higher levels of cooperation, as measured by higher investment in the group project (Fehr & Gächter 2002).

Enforcement can also explain cooperation between species. An elegant example of this is how the cleaner fish *Labroides dimidiatus* removes and eats ectoparasites from its reef fish clients. Although this parasite removal and acquisition of food is clearly beneficial to the client and cleaner, respectively, there is a conflict, because the cleaners would prefer to eat the tissue or mucus of their hosts, which is costly to the host (Bshary & Grutter 2002). The clients use three mechanisms to suppress this conflict and enforce cooperative feeding on ectoparasites only: avoiding cleaners that they have observed cheating (partner choice), leaving to go to another cleaner (partner switching), and aggressively chasing the cleaner (punishment). Both observational and experimental data suggest that cleaner fish are more cooperative and less likely to feed on mucus after punishment (Bshary & Grutter 2002).

Why Enforce?

Although it is clear that behaviors such as punishment or policing favor cooperation, it is sometimes less obvious how punishment or policing can itself be favored by natural selection. To be favored, behaviors such as punishment must provide a direct or an indirect fitness benefit to the punisher. The simplest way in which such behaviors could provide a direct fitness advantage is if they lead to the termination of interactions with relatively uncooperative individuals (ostracism), and hence allow interactions to be focused on more cooperative individuals (Axelrod & Hamilton 1981; West et al. 2002; Frank 2003). This mechanism appears to be operating in cases such as the cleaner fish discussed above, and also

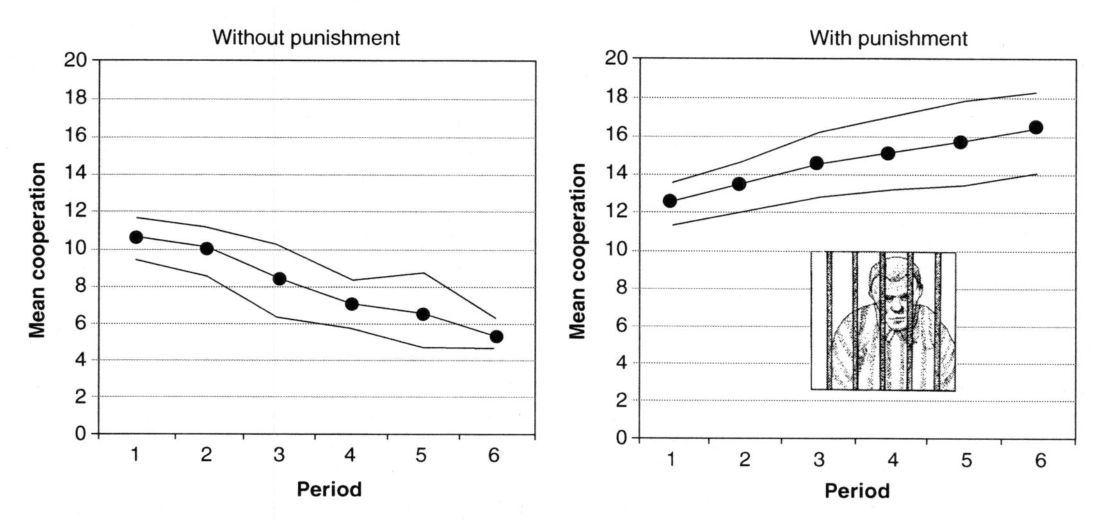

FIGURE **18.7** Humans show higher levels of cooperation in economic games when there are opportunities to punish individuals who do not cooperate (Fehr & Gächter 2002).

the interactions between legumes and rhizobia (the plant reduces investment into root nodules harboring uncooperative symbionts), between the Yucca and Yucca moth (the plant aborts fruit containing cheating moths), and between humans. In meerkats, pregnant subordinates will kill the young of other females, even those of the dominant, and so the dominant increases the survival of her offspring by harassing and evicting pregnant subordinates (Young et al. 2006). A more complicated possibility is that punished individuals change their behavior in response to punishment, and are more likely to cooperate with the punisher in future interactions (Clutton-Brock & Parker 1995a). The relative importance of such punishment remains a major problem; it is at work in cleaner fish, and could be important in species such as cooperatively breeding vertebrates or humans.

Enforcement could also be favored if it provides an indirect fitness benefit (Ratnieks 1988). The simplest way this could happen is by reducing the personal fitness of individuals who are competing with relatives, and hence freeing up resources for relatives. In some ants, bees, and wasps, a fraction of the workers lay their own eggs (Wenseleers & Ratnieks 2006a). Although the workers do not have access to mates, their sex-determination system allows unfertilized eggs to develop as males. Other workers frequently do not tolerate such selfish behavior, and selectively cannibalize or *police* worker-laid eggs. This policing behavior is favored because the workers can be more closely related to the sons of the queen (their brothers) than the sons of other workers (their nephews), so they gain an indirect fitness benefit by killing nephews and rearing brothers instead (Ratnieks 1988; Wenseleers et al. 2004). Also, worker reproduction is inefficient and reduces the total reproductive success of the colony, so policing can also be favored because it improves the overall reproductive success of the colony (Hammond & Keller 2004; Wenseleers et al. 2004). Across species, it has been shown that there are higher levels of worker cooperation in which policing is more common and effective. Specifically, the proportion of workers who lay eggs (cheats) is negatively correlated with the likelihood of worker-laid eggs being killed (figure 18.8; Wenseleers & Ratnieks 2006a). One way of conceptualizing this is that policing reduces the direct fitness gains of cheating, which is the same thing as reducing the cost (c) of cooperating in Hamilton's rule.

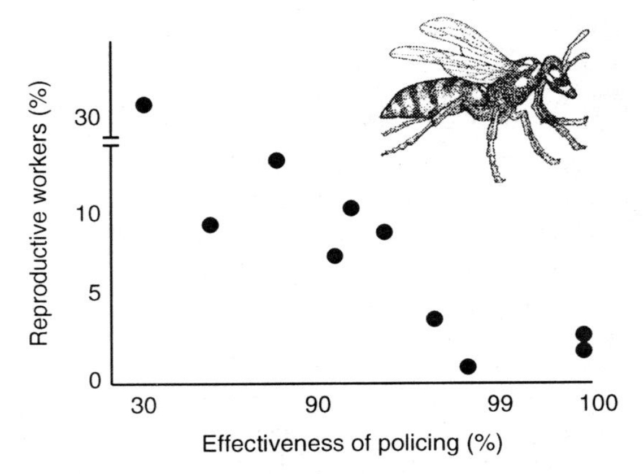

FIGURE **18.8** Lower levels of worker reproduction (cheating) are observed in wasp and bee species when worker policing is more effective. The effectiveness of policing is measured as the probability of worker-laid eggs being killed relative to queen-laid eggs. Redrawn from Wenseleers and Ratnieks (2006a).

In the examples given above, enforcement of cooperation has involved facultative, or conditional, behavior of social partners, which is adjusted in response to the behavior of the focal individual. However, cooperation can also be enforced with fixed strategies that remove the opportunity for competition or cheating (Leigh 1971; Frank 2003). If opportunities for competition or cheating are limited, then individuals can increase their own success only by increasing the success of their group (Frank 2003), and receiving their fair share of this success. Consequently, any mechanism that aligns reproductive interests or represses competition within groups will select for higher levels of cooperation. Fair meiosis may be an example of such a mechanism, and appears to have been favored because it aligns the reproductive interests of genes in a genome (Leigh 1971). Under the rules of fair Mendelian transmission, every gene in the genome has an equal probability of being passed onto the individual's offspring, so it is in the interests of all genes to maximize the reproductive success of the individual. Although selfish genes ("Mendelian outlaws") that increase their own transmission rate can arise and spread, there is often strong selection for them to be suppressed by the majority of genes elsewhere in the genome, and this policing prevents the genome dissolving into anarchy. This democratic tendency has been termed the "parliament of the genes" (Leigh 1971). An example of a Mendelian outlaw is the *t* haplotype of mice, which derives a transmission advantage by eliminating non-*t*-carrying sperm, and which has been shown to be suppressed by other genes in laboratory populations of mice (reviewed by Burt & Trivers 2006). Other examples of mechanisms that may have evolved to reduce conflict within organisms include separating symbionts into reproductive (germ line) lineages, and nonreproductive (somatic) lineages (Frank 2003), and uniparental transmission of cytoplasm genes (Hurst & Hamilton 1992).

FUTURE DIRECTIONS

In this chapter, we have seen that there is a well-developed theoretical overview for tackling the problem of cooperation in evolutionary biology. Inclusive fitness theory provides adaptive explanations of cooperation in terms of direct and indirect fitness benefits. Also, in broad outline, there is a good understanding of the types of mechanisms that can deliver such benefits. However, the relative importance of direct and indirect fitness benefits, in explaining cooperation as it occurs across the whole of the diversity of life, is far from being properly understood. We suggest four directions that will be crucial for further development of the field.

First, we need better integration between theory and empirical work. Much theoretical work is aimed at developing very general models that can be difficult to apply to real systems. Greater emphasis is needed on the development of models that can be applied to and tested in specific systems, and which make predictions for how the level of cooperation should vary, not just whether it should occur. The usefulness of this approach is clearly demonstrated by the fact that the most successful branches of social evolution theory are those in which there has been the greatest integration of theory and empirical work, such as sex allocation (Frank 1998; Boomsma et al. 2003). There is a particular need for testable predictions for many of the direct fitness explanations for cooperation, which are often only invoked when indirect fitness explanations apparently fail, and not as competing hypotheses. In addition, theory is required to make better use of real data, by providing methods for estimating parameters such as the direct and indirect fitness components of inclusive fitness (Oli 2003), and the extent to which competition between relatives reduces selection for cooperation (e.g., limited resources means that the improved fecundity of one relative may reduce survival of other relatives, which negates the advantage of cooperating; reviewed by Queller 1992b; West et al. 2002).

Second, the possible advantages of less traditional study systems need to be exploited. Previous empirical work has focused on animals, and within them, the Hymenoptera (ants, bees, and wasps) and the cooperatively breeding vertebrates. Far less attention is given to other systems such as termites, social spiders, or aphids. Beyond the animal world, the amazing opportunities offered by bacteria and other microbes have only just been realized, let alone exploited (West et al. 2007b). Different taxa offer different advantages: although bacteria are useful for genetic manipulations and multigeneration selection experiments, insects and vertebrates are more useful for studying behavior in natural populations. Furthermore, apart from work on selfish genetic elements and their suppressors, far less attention has been paid to the problems of cooperation that occupy lower levels among the major evolutionary transitions, such as the evolution of multicellularity (Michod & Roze 2001).

Third, there is a strong need for greater unification across taxa. In some cases, there is surprisingly little interaction between empirical workers and theoreticians in different areas. One could be forgiven for thinking that the human, primate, and social insect literatures on cooperation concerned completely different topics, rather than attempts to develop and test the same body of theory. This lack of unification has led to much semantic confusion and unnecessary reinvention of theory. Inclusive fitness (Hamilton 1964) provides a relatively unified body of theory on the evolution of cooperation (Frank 2003; Sachs et al. 2004; Grafen 2006; Lehmann & Keller 2006; West et al. 2007a), and the major aim for the future should be to show how this links and differentiates explanations for cooperation across the various taxa and levels of biological organization.

Fourth, it is important to emphasize both the distinction and interplay between mechanistic (proximate) and evolutionary (ultimate) approaches and explanations. It has long been appreciated in the animal behavior and evolutionary literatures that these are complementary and not competing approaches. Indeed, failing to discriminate these approaches can lead to considerable confusion, as illustrated by the recent literature on cooperation in humans (West et al. 2007a). However, this distinction has also led research on evolutionary questions to underplay the importance of mechanistic issues. This is a problem when an understanding of mechanism is crucial to explain the pattern and precision of adaptation (Shuker & West 2004). For example, the ability of ants to discriminate among kin has been shown to depend upon mechanistic constraints imposed by the cuticular hydrocarbon mechanism underlying this behavior. Ants make mistakes about how many times their queen is mated when their queen has mated with males having the same cuticular hydrocarbon profile (Boomsma et al. 2003).

Acknowledgments We thank O. Henderson for illustrations, and E. Fehr, N. Mehiabadi, and T. Wenseleers for providing data or photographs. We are all funded by Royal Society fellowships.

SUGGESTIONS FOR FURTHER READING

Maynard Smith and Szathmary (1995) provide an accessible account of cooperation at all levels of biological organization, and how it has driven the major transitions in evolution, from the origin of life, to the evolution of cells, to multicellularity, to complex human and animal societies. An excellent and comprehensive review of mechanisms promoting the evolution of cooperation is given by Sachs et al. (2004), who illustrate the theory with a range of empirical examples. A superb and general synthesis of social evolution theory is provided by Frank (1998), but this is not for the mathematically fainthearted. A recent review of recurring semantic and conceptual difficulties that frustrate scientific communication and progress in the multidisciplinary field of social evolution is given by West et al. (2007a), with particular emphasis on the application of social evolution theory to humans.

Frank SA (1998) Foundations of Social Evolution. Princeton Univ Press. Princeton, NJ.

Maynard Smith J & Szathmary E (1995) The Major Transitions in Evolution. Freeman, Oxford, UK.

Sachs JL, Mueller UG, Wilcox TP, & Bull JJ (2004) The evolution of cooperation. Q Rev Biol 79: 135–160.

West SA, Griffin AS & Gardner A (2007a) Social semantics: altruism, cooperation, mutualism, strong reciprocity and group selection. J Evol Biol 20: 415–432.

19

Evolution of Complex Societies

DAVID C. QUELLER AND JOAN E. STRASSMANN

The evolution of cooperation is sometimes viewed as something of a side topic in behavioral ecology—interesting enough, but perhaps not that widely applicable outside of some oddball social insects and cooperatively breeding vertebrates. After all, most organisms spend most of the time competing rather than cooperating. But cooperation is actually quite central and important. Cooperation among conspecifics is quite widespread (chapter 18): social insects are ecologically dominant, cooperatively breeding vertebrates are quite common, and forms of cooperation even exist among many other invertebrates and even microbes.

The focus on competition also misses a fundamental and universal kind of cooperation, that which occurs between different parts of the organism itself. It turns out that individual organisms, though they appear to be unified (*individual* means indivisible), have evolved out of groups of formerly competing entities (Szathmáry & Maynard Smith 1995). Multicellular organisms are derived from single cells. Single eukaryotic cells came from a union of several prokaryotic lineages, and prokaryotic cells arose from independent replicators. In each case, entities that worked independently and at cross-purposes have evolved interdependence, cooperation, and common purpose. But these cooperative transitions are buried deep in the past. We can study this kind of transition most directly in the relatively few taxa in which current organisms join to form superorganisms, that is, groups of individuals that interact as cooperatively and harmoniously as the cells in a metazoan body.

The best examples of superorganisms come from the more advanced social insects: those ants, bees, wasps, and termites with differentiated castes, large colonies, and complex division of labor. Many years ago, William Morton Wheeler suggested that an ant colony could be viewed as an organism because of the integration of colony members' efforts toward the colony's reproduction rather than their own (Wheeler 1911). Though this view has had its periods of favor and disfavor, there is much to be said for this analogy.

To understand the emergence of superorganismal colonies, one must also study the smaller, simpler societies. These include simpler members of the taxa named above and a scattering of others: a few thrips and aphids, at least one beetle, a parasitoid wasp, and even a social sponge-dwelling shrimp. They are similar to the advanced social insects in meeting the three requirements to be eusocial or truly social: reproductive division of labor, overlapping generations, and cooperative care of young. However, they often lack morphological castes and the selfish interests of the individuals may be more apparent.

Although we focus primarily on social insects, we use them as a springboard for shorter looks at other kinds of societies. Birds and mammals with cooperative breeding are arguably much like some of the simpler social insects in terms of complexity and degree of integration. In contrast, certain microbial groups do approach superorganismal status, but do so in ways that are significantly different from the social insects. In addition, multicellular organisms can be thought of as complex societies of cells. In this chapter we explore the processes for how individuals that cooperate can evolve into highly cooperative societies. The level of cooperation varies considerably, but can lead to colonies that function essentially as single organisms dominated by cooperative complexity.

CONCEPTS

Simple and Complex Societies

The honeybee hive is a classic complex society (Seeley 1995). The colony may have over ten thousand individuals, but only the single queen reproduces (although there are many fathers because she mates many times). Queens are raised in special cells with special food and are considerably larger than workers. The two castes also differ in specialized structures such as the pollen baskets of workers. Workers can lay eggs, but almost never do so. Instead, they do all of the foraging, brood care, and defense, including suicidal stinging attacks. They have much shorter life spans than the queen. Different workers specialize on different tasks, often in an age-related sequence, but can switch tasks according to colony needs. The colony's work is highly integrated and coordinated, and is regulated partly via queen pheromones, but largely through self-organized feedbacks in which workers respond to local cues (such as the waiting time to unload nectar) to adjust the amount of work devoted to a task. Honeybees have no solitary stage of the life cycle. New colonies are produced by swarming; the old queen leaves with the majority of the workers, leaving the old colony to a daughter queen.

Wasps of the genus *Polistes* are at the simpler end of the complexity spectrum in social insects (Reeve 1991). Colonies are smaller, typically beginning with a single female, but sometimes with several females, usually relatives. Mature colonies usually remain below 100 adults. There is typically a single queen, in the sense of a single female that lays most of the eggs. However, there is little or no morphological difference between the queen and her subordinate workers. Workers in many species can become effective queens if given the opportunity. The queen physically limits reproduction of subordinates by physical aggression, which establishes a dominance hierarchy, and also by eating subordinate-laid eggs. Workers do most of the risky foraging. Work is not as coordinated or specialized as in honeybees. For example, a worker returning with a load will not offload it to specialists; often she will use it herself to feed brood or to build the nest. The nest is not very complex, being a simple, open paper comb.

Many social insects fall between these extremes, but with varying mixtures of simple and complex traits. An example is the epiponine wasp *Parachartergus colobopterus*, a small, plump, yellow and brown wasp that lives in multiplecomb nests on trees and buildings. It lacks morphological castes, but nevertheless has rather complex societies (Strassmann et al. 2002, and references therein). As in the honeybee, each new colony is begun by a swarm from an established nest, so there is no solitary stage. Colonies are quite large, with hundreds of individuals, and they also show a fairly sophisticated division of labor. However, control of queenship is different from both honeybees (feeding) and *Polistes* (queen dominance). Here it is worker aggression on newly emerging adult females that normally keeps them from developing their ovaries and forces them to become workers. When requeening is needed, this aggression stops, and a whole cohort of individuals develops their ovaries to become queens.

Vertebrate societies tend to be at the simple end of the spectrum, with small groups, and reproductive/worker roles that remain flexible. For example, African hunting dogs (*Lycaon pictus*) are small carnivores that live in groups of 3 to 20 (Girman et al. 1997). The alpha male and the alpha female are each parents to about 90% of the young. The social organization has something in common with *Polistes* in its small group size, reproductive suppression of subordinates, and inheritance of breeding position by the oldest subordinate (Girman et al. 1997). The naked mole rat *Heterocephalus glaber* is an exception (Sherman et al. 1991). It has the most superorganism-like of vertebrate societies, with colonies averaging around 80 individuals, and a strong division of labor. Usually one female, with perhaps three mates, reproduces. Like complex social insects, the reproductives do little work, but like simple ones, they sometimes have to prod subordinates to work.

Sociality in microbes is turning out to be unexpectedly common (Foster in press). Even such mundane microbes as *Escherichia coli* and yeast

(*Saccharomyces cerevisiae*) have important cooperative interactions. Many microbial interactions involve provision of common pool resources, such as digestive enzymes. Others involve more complex specializations including suicidal killing of competitors, and specialized structures in biofilms. Some of the most advanced, for example in *Myxococcus* bacteria and the eukaryotic social amoebae *Dictyostelium* (see "Case Studies" section in this chapter), involve aggregation of cells to form multicellular fruiting bodies (see also chapter 18). Here there may be competition within the group to be among the subset of the cells that produce fertile spores, with the remainder dying.

The multicellular fruiting structures of *Dictyostelium* and *Myxococcus* are simple versions of differentiated multicellular organisms. More complex ones, such as animals and plants, can also be viewed as societies of cells (Buss 1987). These organisms are also superorganisms in the sense of having evolved from groups of organisms of a simpler stage (single cells). Here complexity can be extreme, with numerous specialized cell types rather like castes, including germ line cells specialized for reproduction.

Hamilton's Rule and Relatedness

Logically, if the allele for altruism lowers reproduction of altruists, it can increase only if it also increases reproduction of others who bear that allele. As discussed in chapter 18, Hamilton's inclusive fitness rule (Hamilton 1964) quantifies this trade-off: $-c + rb > 0$, where c is the cost to self, b is the benefit to the beneficiary, and r is their relatedness. More generally, a behavior evolves if inclusive fitness is positive: $\Sigma_i \Delta w_i r_i > 0$, where Δw_i is the change in the fitness of individual i caused by the actor and r_i is the actor's relatedness to that individual. This is a kind of optimization rule, and it can be inexact under certain conditions, but it is generally well supported by underlying population genetic and quantitative genetic models (Grafen 1985). Hamilton's inclusive fitness rule can account for all kinds of social behaviors: altruism, mutual benefit, selfishness, and spite (chapter 18). However, because our focus is on the evolution of highly cooperative societies, we concentrate on those instances in which Hamilton's rule most strongly reinforces altruism and reduces selfishness.

In social insects, much attention has been focused on the importance of relatedness. The 3/4 relatedness that applies to full sisters in haplodiploid species (see box 19.1) seemed to provide an explanation for the commonness of eusociality in the haplodiploid Hymenoptera (ants, bees, and wasps), as well as why workers are females (Hamilton 1964). Why raise your own offspring ($r = 1/2$) when you can raise sisters ($r = 3/4$)? (For diploids the two r's are both 1/2, offering no relatedness advantage to helping). But because haplodiploid females are related to their brothers by less than usual ($r = 1/4$), this explanation works only if helpers can concentrate on rearing sisters. Otherwise they rear siblings related by an average of 1/2, just like diploids. There are ways to get the explanation to work in theory (Crozier & Pamilo 1996), and it is clear that Hymenopteran workers do prefer raising full sisters over brothers (Trivers & Hare 1976), as we will see below. If queens mate with multiple males (multiple mating), the argument doesn't work as well; haplodiploid half sisters are related by only 1/4 (box 19.1). Multiple mating certainly occurs in Hymenopteran social insects, but single mating is more common and phylogenetic analysis shows that single mating was ancestral (Hughes et al. 2008).

The logic that altruism requires kinship is compelling, but it is not clear that relatedness higher than 1/2 actually provided the critical edge early in its evolution. Nor is it obvious that we need some special relatedness-based explanation for the high incidence of kin-selected altruism by Hymenopteran females. That high incidence may instead be due to unusual levels of maternal care preadaptations in the Hymenoptera (Alexander et al. 1991). If a species has already evolved sophisticated maternal care, including finding food and bringing it back to young at a protected nest, then workers can evolve by simply switching care to the offspring of others.

Some researchers have even argued that kinship is not required at all (Wilson & Hölldobler 2005). In their view, sociality might have resulted not from helping true genetic kin, but from helping anyone, kin or nonkin, who shared (but did not necessarily express) the helping allele. This would be a so-called green-beard allele, which is an allele that produces a signal (such as a green beard), recognizes the signal in others, and acts altruistically only to those with the signal (chapter 18). The problem with this hypothesis, aside from the absence of any positive evidence for it, is that green-beard altruism acts against the interests of other genes in the genome, which gain nothing

BOX 19.1 Haplodiploid Pedigree and Relatedness

The insect order Hymenoptera, and a scattering of other organisms, have haploid males and diploid females. Sex is therefore controlled by whether the mother releases stored sperm to fertilize her egg (resulting in a diploid daughter) or does not (resulting in a haploid son). This haplodiploid system leads to some unusual relatedness coefficients that are important to understand.

Figure I centers on a diploid female labeled "Self" whose pedigree connections and relatednesses are shown. Males draw their single allele from their mother. Females draw one of two maternal alleles (labeled hatched one for Self) and they also draw the only paternal allele (black for Self). All other alleles are shown in gray. Relatedness can be thought of as the identity, above random chance, to the relative, divided by identity to oneself. For example, the average identity to a half sister is 1/8, the average of 1/4 for Self's hatched maternal allele, and 0 for her black paternal allele. After dividing by 1/2, the identity of either of Self's alleles to Self, we get the relatedness of 1/4. Note that although the hatched or black allele might sometimes be identical to alleles in the gray (unrelated) zones, these would be chance identities, with expectations at the population frequencies. We should ignore these because any benefits accruing to random alleles do not affect selection.

The half-sister relatedness is the same as in diploids, but others differ, most notably relatedness to a full sister. There, Self's hatched maternal allele is still 1/4, but the black paternal allele is 1/2, because the father has only one allele and it goes to all daughters. The average of these is 3/8, divided by identity to Self (1/2), gives the famous haplodiploid 3/4 relatedness. Relatedness of Self to brothers is 1/4, essentially because, lacking a father, brothers must always be half siblings.

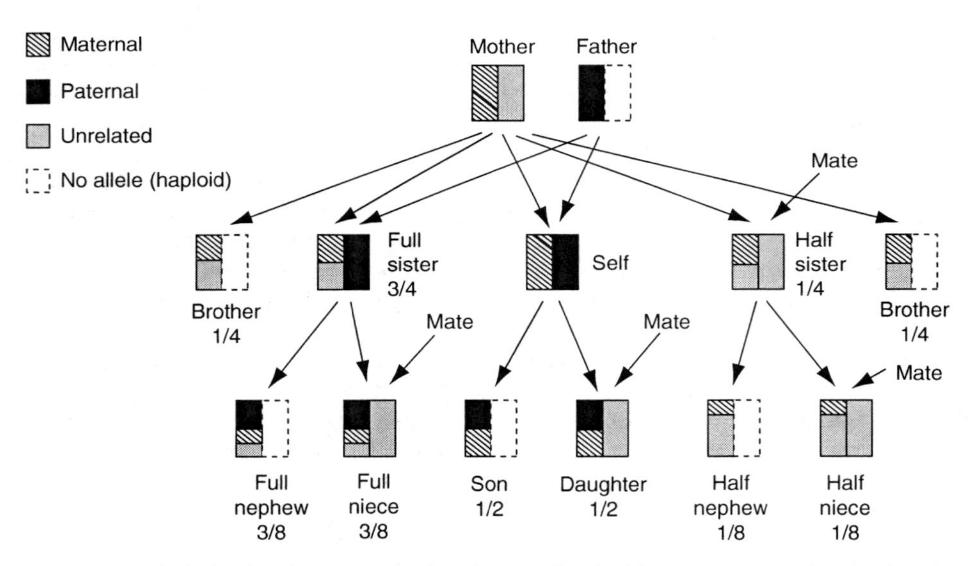

Figure I Haplodiploid pedigrees and relatedness. All relatedness values are for the female labeled as "Self," and shadings show Self's maternal and paternal alleles (after Queller & Strassmann 1998).

by helping nonkin with green-beards (Helanterä & Bargum 2007). Therefore, at any locus unlinked with the green-beard gene, selection should favor a cheater mutation that does not give help, while keeping the green-beard signal to receive help from others. Even if such cheater mutations do not arise, and the green-beard allele is selected, there is a related problem. Because unlinked loci do not gain from this altruism, there would be no selection on them to build on this foundation, so it appears that green-beards cannot generate the kinds of complex multigenic sociality that we actually observe in social insects. It is only when allele sharing is due to pedigree kinship that alleles at all loci are expected to agree on who should be helped (Grafen 1985). For every allele, the same coefficient of relationship gives the expectation that it is present in that relative above random levels.

Thus, extreme cooperation cannot be explained solely by relatedness or by processes that lack an effect of relatedness entirely. The importance of relatedness lies between these extremes. Allele sharing appears to be absolutely essential for the evolution of altruism, and kinship is the only kind of allele sharing that allows for agreement across the genome. It is surely no accident that nearly all social insect colonies are groups of relatives. Data are beginning to show this is true for altruistic microbes as well (Mehdiabadi et al. 2006; Gilbert et al. 2007), and of course, cells in most multicellular organisms are related by 1. But that does not mean that a helper has to be more closely related to its beneficiaries than to its own offspring. Instead, Hamilton's rule can also be satisfied with more modest relatedness to beneficiaries if the benefits are sufficiently large relative to costs ($b > c$). If this were not so, it would be difficult to explain altruism in diploid vertebrates and termites. Here, too, groups usually consist of relatives, often simple families of parents plus helper offspring, but there is no relatedness advantage to rearing siblings. Even in social microorganisms that are maximally related by clonal descent, there is no relatedness advantage to rearing collateral relatives ($r = 1$) over offspring ($r = 1$).

There is one kind of complex society in a scattering of ant species, *unicoloniality*, in which relatedness does seem to be too low to account for sociality (Helanterä et al. 2009). These colonies consist of many multiqueen, interlinked nests. In extreme cases, such as some introduced populations of the Argentine ant *Linepithema humile*, colonies cover hundreds of kilometers. Within the colony, there is no evident kin discrimination, and the free movement of individuals means that workers are no more closely related to those individuals they help than to random individuals in the vicinity. It appears, however, that such colonies are family groups grown abnormally large. One consequence that appears adaptive, at least in the short term, is that the lack of kin recognition also means that workers move freely among nests, making for huge cooperative units that can take over habitats very effectively. Despite the short-term advantages, unicolonial populations face a long-term problem. With relatedness near zero, there appears to be no way for worker traits to be selected (Queller & Strassmann 1998).

Hamilton's Rule: Benefits and Costs

Social insect societies clearly gain from division of labor and specialization, but it seems likely that other benefits of cooperation applied in the time before specialized behaviors evolved. There are a number of possible benefits, but we have previously suggested that social insects fall into two categories with respect to their primary initial advantages: fortress defenders and life insurers (Queller & Strassmann 1998).

Fortress defenders live and feed inside a relatively safe, defensible structure like a gall or a cavity. As long as there is space and food at the natal site, staying there may be better than risking finding a new site (Alexander et al. 1991). In other words, a helper does not have a large cost (c) if it is unlikely to succeed on its own, and the benefit (b) of staying and helping in a safe site is larger. Because there is no foraging outside the nest, helping usually takes the form of defense, which becomes increasingly important as colonies become large enough to attract more threatening predators. Although specialized defensive castes such as the attacking morphs of aphids and thrips evolve readily, fortress defenders rarely become very complex, perhaps because their nest site limits size, or because there is little to do besides defend. Termites are an exception; though some taxa feed inside their protected galleries, others forage more broadly, and some have evolved quite large and complex colonies.

Life insurers like ants, bees, and wasps, reduce the costs of mortality by pooling their risks. They forage outside the nest to feed their dependent young. Such outside foraging is dangerous, and death carries an additional cost for those who have dependent young. A single parent who dies loses all the investment she made in young that are still dependent. Groups containing caregivers do better. Individuals who die early do not necessarily waste their effort if they have completed the investments of others, or if their own investments will be completed by others (Queller 1996). The most important task here is foraging, although defense in not irrelevant. The hairy-faced hover wasp (see "Case Studies" section in this chapter) is a life insurer.

In social vertebrates, there appear to be many advantages, but these same advantages frequently appear in noncooperative species so it has been difficult to arrive at firm conclusions (Cockburn 1998). Direct benefits to the helper are common, in the form of increased experience, increased probability of inheriting a territory, increased sneaky paternity, and increased size of the group inherited. Therefore, some of these societies are not truly altruistic. However, indirect benefits to relatives are also clearly important to a varying degree (Griffin & West 2003).

Social microbes (Foster in press) appear to have entirely different kinds of benefits than social insects. Here altruism frequently appears in dire circumstances of reduced food or severe competition. Examples include social amoebae and myxobacteria, which form dispersal structures to escape patches with no food. Similarly, when competition is high, numerous bacteria release poisons, in some cases by suicidally rupturing the cell, and this reduces competition. Finally, biofilms are multicellular bacterial structures that, among other functions, provide structure at feeding sites and protection from toxins.

Asymmetries in Who Should Help

Finding a solution, or multiple solutions, to how cooperation can pay is not the end of the story. However advantageous it may be to help, it is even more advantageous to *be* helped. In *Polistes* paper wasps, two sister foundresses may compete viciously before the winner lays the eggs and the loser becomes a helper. Presumably colony efficiency suffers, especially if the loser is grudging and continues to test the dominant for weakness, or slacks off to increase the chance that she may get to become dominant.

There are two important asymmetries that can help settle the issue of who helps and who is the beneficiary. The first is in relatedness (Stubblefield & Charnov 1986). The most common case is that a mother is queen and her daughters are workers. This case is asymmetrical when the mother is singly mated. Each party is related by 1/2 to her own progeny, but the mother is related by 1/4 to her daughter's progeny, whereas the daughter is related by 1/2 to her mother's (using either diploid relatednesses or haplodiploid ones averaged over sexes). There is a bias, therefore, toward offspring helping mothers rather than vice versa. One way of looking at this bias is that the offspring, by helping her mother, is also helping her father reproduce, even though in Hymenopterans the father is present only as stored sperm. In termite and vertebrate systems, this connection is more direct, because the father is often still present as an obvious cobeneficiary with the mother.

The second kind of asymmetry biases could arise from differences in ability to give or to receive aid. In particular, less developed or less well fed individuals might not be good reproductives (West-Eberhard 1975). As long as they are not equally poor helpers, that makes their cost-benefit ratio more favorable for helping. However, studies of primitively eusocial insects have not shown this effect (Queller & Strassmann 1989; Field & Foster 1999). Still, even if this effect is not present from the beginning of sociality, it must become important later. Morphological caste is generally determined by nutrition—workers get less food than queens (Wheeler 1986)—and this is likely to be very important for limiting worker options. This benefit-cost asymmetry, like the relatedness one, can work in favor of mothers. The mother is clearly already in reproductive condition and ready to be helped, whereas her new offspring may not be fully mature. Moreover, if nutrition is important, the mother is in a position to manipulate the condition of her offspring.

It is perhaps not surprising then that social insects that have true altruism are generally structured as colonies of mothers and their offspring (and fathers in the case of termites). There are many cases in which colonies have multiple reproductives, but these are likely to be secondary. Not only does altruism not evolve out of groups of nonrelatives, it also seems not to evolve very readily out of groups of symmetrical relatives.

In a clonal society, such as a multicellular organism, the relatedness asymmetry problem does not arise. Each cell should value its clone mates as much as itself, so there is no conflict over who will be reproductive. Asymmetries of condition or nutrition might be important in microbes. In the social amoeba *Dictyostelium discoideum* cells deprived of glucose are less likely to become fertile spores, and more likely to become part of the sterile stalk that altruistically supports the spores (Leach et al. 1973).

Commitment to Being a Worker

Helping their mothers still does not fully explain why worker offspring evolve to give up more and more of their reproductive potential, ending up in some species as fully sterile. What if the mother queen weakens so that she can no longer keep the colony fully supplied with eggs? What if she dies? If either of these events is likely, it may not pay helpers to evolve specializations that would limit their chances of later becoming queen. Instead, it might pay to avoid some of the risks of helping.

For these reasons, Alexander et al. (1991) argued that specialized workers would evolve only if they could count on being able to expend their entire life effort on their mother's reproduction. Life insurers have a particularly good start on this. Adult lifetime is initially highly variable due to predation risks while foraging, and a primary effect of helping is a transfer of risk. Helpers forage and shorten their expected lifetimes, while increasing the expected lifetime of the queen to its maximum. It is less clear that fortress defenders strictly follow this rule. Even if a fortress-defender worker outlives her mother 90% of the time, it might still pay to commit as a nonreproductive soldier if in the other 10% of the time she can save the colony. The same applies to suicidal microbial altruists.

Any initial divergence of queen and worker risks can be further reinforced by senescence patterns. Workers may evolve a short maximal life span if they rarely live long anyway, whereas queens may gradually extend their maximum life span once forces of external mortality are reduced (Alexander et al. 1991). The result—short-lived workers and extremely long-lived queens—has evolved multiple times in social insects (Keller & Genoud 1997).

Various aspects of specialization, including caste differentiation and task specialization, increase with colony size (Bourke 1999). It has been suggested that workers become more specialized and committed to their roles in large colonies because each has a reduced chance of becoming queen (Alexander et al. 1991; Bourke 1999). When there is only one queen position, and many helpers, the odds are small that a given helper will become queen. This reduction in options should help select for workers that commit more fully to being workers, which in turn might increase colony size, possibly leading to a positive feedback cycle of increasing size and specialization.

We doubt this argument, however, because the reduced chance of a helper becoming queen of a large colony should be offset by an increased payoff when it does succeed. Assume that a queen can expect help from N workers that each adds b units of fitness. If such workers are initially physiologically identical to the queen, when would selection favor specialization? Assume specializing would make a worker better at helping by an amount Δb. It would also make her a worse queen, lowering her reproduction by a fraction f if she inherits the queenship. Then specialization is favored if

$$r_{sib}\Delta b > \frac{1}{2} f \frac{1}{N} bN$$

The left-hand side is the kin-selected benefit. The right-hand side is the cost, including relatedness to offspring (1/2), the probability of being the lucky worker to inherit the queenship ($1/N$), the offspring a normal queen produces (bN), and the fraction of this given up by specializing (f). The main point is that the N's cancel, making specialization independent of colony size.

How, then, do we explain the fact that large colony size is positively correlated with various dimensions of complexity in social insects? One possibility is simply that large colony size allows specialization to occur. In small colonies, random mortality would too often leave a deficit of some specialists and a surfeit of others. Unpredictable fluctuations in the tasks required at any given time would have a similar effect. This kind of effect would better explain the correlations between complexity and size in other contexts. For example, there are more cell types in larger organisms (Bell & Mooers 1997). This is similar to the greater complexity of larger insect societies, but it cannot be explained by reduced competition to reproduce, because the cells in these organisms are identical.

The distribution of complex societies among different taxa also provides clues to why complexity evolves. For example, a large number of bird and mammal species have cooperative breeding, but only two species of mole rats are generally regarded as being eusocial. In part this may be unfair—we include casteless Hymenopterans like *Polistes* as being eusocial but do not include casteless vertebrates with extensive helping like meerkats, wolves, and scrub jays. This quibble aside, it nevertheless is true that few vertebrates rise to the level of highly complex, nearly organismal societies. Why is this so?

At some risk of ecological oversimplification, one could argue that insects are commonly limited by predation, whereas vertebrates are more often limited by food. If this difference is valid, it may be important given that the two main advantages for helping in social insects are related to predation. First, if predators are not as much of a threat for vertebrates, there is little need for fortress defense and little opportunity for heroic sacrifice. Wolves, elephants, and lions are thus poor candidates for this kind of colony. Smaller vertebrates might be better candidates, but to be a fortress defender, it is also required that there be defensible and expandable nesting sites. Except for mole rats, few vertebrates appear to have both concentrated food and nest resources (Alexander et al. 1991).

It also appears that life insurance is a much weaker advantage in vertebrates, for several reasons. If predation rates are lower, a parent is less likely to die during the period of investment so the gains from life insurance are weaker. In addition, if a Hymenopteran queen survives long enough for workers to invest in her, so in effect does her mate, in the form of stored sperm. She continues to produce full siblings for her helpers. But in vertebrate groups, either the mother or the father could die, leaving the helpers with lowered relatedness. The lower relatedness might make it better to reproduce directly. Indeed, the death of one parent often leads to conflict in vertebrate societies, and even to their dissolution (Emlen 1997).

Another limitation of the life insurance benefit of eusociality is that in a bird or mammal, the mother seems locked into investing a significant portion of what is required to raise her young. The eggs of birds and the neonates of mammals are inevitably maternal investments that helpers cannot take over. This must limit the fecundity of aspiring vertebrate queens, keeping them from being able to supply a large colony with young. In contrast, an insect mother's investment in an egg is a tiny portion of what that offspring needs, and she can supply a great many eggs to helpers.

Finally, if colony size is an important prerequisite for evolving specialization, food-limited vertebrates may have more trouble reaching the size threshold. A vertebrate colony limited by food would have to add more and more territory, which may not be efficient. It seems unlikely that a pride of 100 lions or a pack of 100 hunting dogs could be sustainable. In contrast, an insect limited by predators might pack more and more individuals into one locality.

Conflicts of Interest

We have been speaking so far as if workers weighed the value of their average offspring versus their average siblings. But as we have noted above, haplodiploid females are more closely related to sisters than to brothers. So the ideal scenario might be a mixed one: raise sisters ($r = 3/4$) instead of daughters ($r = 1/2$), and raise sons ($r = 1/2$) instead of brothers ($r = 1/4$; Hamilton 1964). Curiously, full commitment to rearing their mother's offspring therefore appears to be more difficult to achieve in haplodiploids than in diploids. Diploids will give up sons for brothers if $b > c$, haplodiploids will do so only if $b > 2c$ (owing to the two-fold relatedness difference). In agreement with expectation, in many haplodiploid species workers continue to lay male eggs long after they have evolved to give up daughter production (Bourke 1988). Of course, this is also easier, as male eggs do not require prior mating.

A worker's preference to raise her own sons will put her in conflict with the queen. The queen is related only by 1/4 to the workers' sons (her grandsons), but by 1/2 to her own sons, so she will prefer her own. To the extent that this conflict or others remain unresolved and costly, the colony might not be viewed as fully organismal. Severe conflicts might even threaten cooperation itself. Completely harmonious cooperation requires that conflicts be resolved, for example in favor of maternal reproduction.

It should be pointed out, however, that a few conflicts may remain even in the most superorganismal colonies in which workers are fully committed to a life of helping. In haplodiploid societies, one such conflict is over the sex ratio. The female workers are three times more closely related to sisters than to brothers. A simple extension of sex ratio theory shows that workers prefer a 3:1 investment in sisters over brothers, as opposed to the queen's standard 1:1 preference, because her sons and daughters are equally likely to carry any of the queen's alleles (Trivers & Hare 1976; Crozier & Pamilo 1996). Numerous studies confirm that workers do bias sex ratios and perhaps usually win this conflict in the sense of achieving their preferred outcome. Evidence ranges from observations of workers killing queen-produced males (Aron et al. 1995) to comparative studies across species showing, for example, that Hymenopteran sex ratios are more female biased than those of diploid termites (Trivers & Hare 1976; Crozier & Pamilo 1996). Particularly convincing is evidence from *split sex ratios* in species in which workers vary in their preferences. For example, the relatedness asymmetry above assumes a singly mated queen, and workers in colonies with multiply-mated queens will be more evenly related to siblings (because half sisters are related by 1/4, just like brothers). The very strong evidence that colony sex ratios usually differ in the directions predicted by worker relatedness differences provides support for conflict theory, for the fact that sterile workers are still not mere appendages of the queen, and for kin selection theory in general (Queller & Strassmann 1998; Chapuisat & Keller 1999).

The other crucial conflict that remains even after worker commitment is conflict between the queen and her reproductive daughters. In most species, these new queens leave to found new colonies. But they might sometimes gain from staying at the old colony, usurping a share of the colony's work, or perhaps even killing the old queen.

The realization that there is continuing conflict and cross-purpose even in advanced social insects might suggest that their societies function less well than the superorganism should. But one should remember that even traditional organisms are not perfectly superorganismal, because their genes sometimes work at cross-purposes (Burt & Trivers 2006). For example, meiotic drive genes can battle over which ones will get into gametes, and transposable elements spread in spite of the disruptions they can cause by inserting into functional sequences. Conflicts among cells of multicellular organisms are rarer. They are certainly present in social amoebas in which cells from different clones can aggregate together (Strassmann et al. 2000) but are largely absent in multicellular organisms, like animals, that derive from division of a single cell. Buss (1987) has argued that such conflict has nevertheless been important historically. For example, early sequestration of the germ line in animals may have evolved to prevent competition. However, this seems doubtful because such competition is very weak when all the cells are related. Any mutant that gains an advantage by sneaking into the germ line gains that advantage only once; in the next generation its offspring will all have the mutation (Queller 2000). Cancer can also be viewed as a case in which a lineage of cells gains a replicative advantage within the organism, but this advantage does not persist into the next generation.

Power

When there are conflicts of interest in a group, the outcome depends on the power of the different parties. A party that gets its way in a conflict is said to be more powerful. Sometimes power depends on idiosyncratic factors, but there are some general themes (Ratnieks et al. 2006). For example, sometimes information is power, and lack of information is weakness. A worker who would readily replace a queen-laid male egg with her own might not be able to tell if the queen's egg is a male—it might be a valuable reproductive sister or a valuable worker.

Another determinant of power, physical strength, applies to many noncooperative organisms, and has been carried over into the cooperative context. Particularly in small colonies, the queen may maintain her position by physically dominating other individuals and policing their actions. This presumably becomes more difficult as colonies get larger, so how does the queen maintain her position, particularly in the face of the incentive of workers to lay male eggs? One answer is that sometimes she cannot, because in some species workers do successfully lay male eggs.

Interestingly, a shift of power from the queen to the more numerous workers does not necessarily lead to anarchy, with every worker striving to reproduce. Instead, workers sometimes collectively police each other. What should a worker do when her coworker lays an egg? It depends on the relatedness structure of the colony. If, for example a queen is multiply-mated, and workers are usually half-sisters ($r = 1/4$), the worker should be opposed to her coworker's egg laying. She will be related to the coworker's sons by 1/8, less than to her brothers (1/4). In such species, such as the honeybee, workers will often eat each other's eggs, leaving the field to the queen (Ratnieks 1988).

The balance is different under single mating because a worker is then related to her coworkers' males by 3/8, more than to her brothers. Stingless bees provide an interesting contrast in this respect. They are in the honeybee family, but in a much more speciose pantropical subfamily. As in honeybees, large colonies are headed by a single morphologically distinct queen and colony organization is quite complex. However, in stingless bees, queens mate only once and, as predicted, workers often lay eggs that are not policed and develop into adult males (Peters et al. 1999). Clearly, relatedness is not the only factor, as the percentage of worker-laid male eggs varies widely among stingless bee species. Nevertheless, this simple relatedness difference explains a good portion of the variation in worker reproduction in haplodiploid social insects (figure 19.1; Wenseleers & Ratnieks 2006b).

The power of workers to suppress other workers is probably much more broadly important. Castes in social insects are generally nutritionally determined (Wheeler 1986), and it is the workers who do most of the feeding. It may be that the evolution of morphological castes was not entirely voluntary. Instead, workers or queens provide less food to some individuals, thus limiting their options. It is important to remember that this is not a sufficient explanation of worker behavior. Suppressed individuals will not evolve to do any useful work unless that choice, however constrained it might be, benefits their inclusive fitness. For example, unrelated individuals might be forced or fooled into helping, as happens in ant species that kidnap workers or pupae from other colonies, but these workers will never evolve to further help their exploiters.

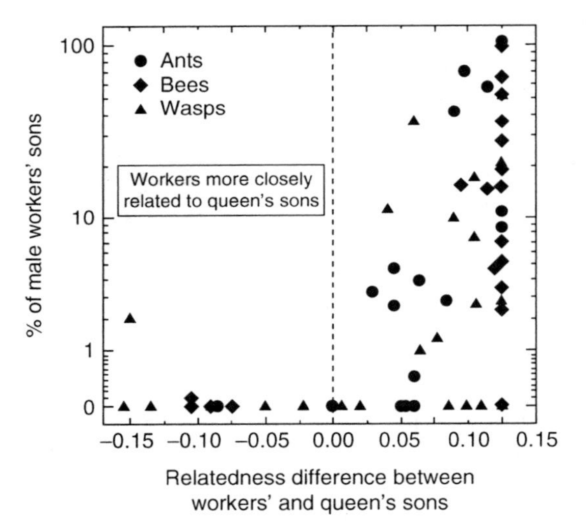

FIGURE **19.1** Relatedness and who produces the males for 90 species of social Hymenopterans. When workers are less closely related to the sons of other workers than they are to the queen's sons (left side), usually only the queen's sons are raised. When workers are more closely related to other workers' sons (right side), worker production of males is more common. Reprinted from Wenseleers and Ratnieks (2006a) with permission.

Finally, collective worker control is also important in controlling daughter reproductives who might try to reproduce in the home colony, because workers will generally pass on more copies of their genes by rearing siblings rather than the nieces and nephews that would be produced if a sister took over as queen. However, in some species, daughter queens do succeed at staying, resulting in multiqueen colonies. It is not always clear how often these queens have been able to circumvent worker control, and how often it pays workers to accept daughter queens, for example, when colonies are distributed across multiple cavities.

A clear case of circumvention of worker control is provided by one particular genus of stingless bees, *Melipona*. In all stingless bees, eggs are provisioned all at once and the cell is sealed, leaving the larva with all the food it is going to get until it becomes an adult. In most genera, caste is controlled by feeding: queens develop in larger cells with more food, and workers in smaller cells with less food. *Melipona* is the exception, with all cells being similar in size. The result is that power to control caste shifts from the feeding workers to the developing larva herself. A large fraction of female larvae (10–20%) opt to develop as queens (Wenseleers & Ratnieks 2004). Nearly all these new queens are superfluous to the colony because, like honeybees, they reproduce by swarming and need a new queen only rarely. At this point, workers reassert control and kill the superfluous queens. ESS models show that the high potential payoff of becoming a reproductive queen justifies this risk (Wenseleers & Ratnieks 2004). That is, it is justified for the potential queen, but it is a great waste of resources for the colony. The wasteful consequences of a lack of nutritional control speak to its importance.

We have been ignoring multicellular societies for the last several sections because most of them are clonal and they do not have significant conflicts of interest among their cells, and therefore no incentive to use power in the way social insects do. Their lack of conflict is likely to underlie their tremendous complexity, with far more specialized cell types than social insects have specialized castes. The exception to the rule is those microbes that form multicellular bodies after aggregation of dispersed cells, such as *Myxococcus* bacteria and *Dictyostelium* amoebas. We mentioned earlier that *Dictyostelium* cells that are better fed (more glucose) are more likely to become reproductive spores. In addition, the spore-stalk decision is mediated by secretion of a compound called DIF, which, unlike normal signaling molecules but much like many poisons, is chlorinated. It may be that cells are poisoning each other, with the stronger cells resisting best and becoming spores (Atzmony et al. 1997).

CASE STUDIES

Experimental Tests in *Liostenogaster flavolineata*, the Hairy-Faced Hover Wasp

These slender, dark wasps build mud nests under bridges and on embankments in Malaysia. As in other hover wasps (Turillazzi 1991), their colonies are tiny. With fewer than 10 adults per colony and nests open to observation, this species has proven an important experimental model system for understanding very simple societies (Field & Cant 2009 and references therein). A single dominant female lays most of the eggs, and suppresses reproduction by the others, earning her the title of queen, though she is morphologically indistinguishable from the other females, except that she has better developed ovaries.

Adult mortality drives many features of the social system. Because of queen turnover, relatedness among adult females is below 0.5, and well below the full-sister value of 0.75, though still high enough that helping behavior generates inclusive fitness benefits. Those benefits appear to be from life insurance; the combination of high adult mortality and very long juvenile developmental times (100 days), makes it very difficult for single individuals to rear young. A key assumption of the life insurance hypothesis is that an individual's investment as a member of a group continues to yield benefits even if she dies, unlike that of a solitary female, whose young depend completely on her survival. Field and coworkers were able to test this assumption by removing 1–2 helpers from 56 colonies, and comparing their subsequent production to 76 unmanipulated controls (Field et al. 2000). Figure 19.2 shows some of the results—that the investment provided by helpers continued to yield fitness after their removal, in terms of the number of large larvae reared through to the pupal stage (when they no longer need feeding). Figure 19.2a shows that a colony with, say, 4 adult females after 2 were removed, raises more large larvae than a control colony with 4 adults. In other words, productivity depends not just on the number of current females, which is the same for both colonies, but is increased by the prior work done by females who were removed. In fact, by this measure, the experimental colony does just as well as control colonies of 6 that did not have females removed (figure 19.2b). The researchers showed that because of this advantage, helping was about 1.7 times more advantageous than nesting alone, even though they would be more closely related to their own brood. For comparison, remember that Hamilton's haplodiploid hypothesis for Hymenopteran sociality generates a maximum 1.5-fold advantage.

Even though helping is better than being a solitary queen, helpers do compete to be queen in their colony. When a queen dies, the vacancy is filled by a subordinate female according to a queue based on seniority, though there is a small percentage of queue jumpers. Subordinates must therefore balance the benefits available through helping against the cost, an increased risk of not surviving to become queen. Experiments in which a high-ranked female is removed show that as subordinates move up in rank, they reduce their risky foraging, shirking on helping as their likelihood of becoming the queen increases (figure 19.3a; Field et al. 2006). That they are also sensitive to the needs of the colony is shown by experiments removing lower ranked helpers. This does not increase the rank of other females, but does increase the need for work, and the remaining females do work harder to take up the slack (figure 19.3b).

Genes for Cooperation and Conflict in a Social Microbe *Dictyostelium discoideum*

Imagine an organism that forms great rivers of life, with streams of individuals converging to a central point and then morphing into a single cohesive organism. This sounds a bit like army ants, with their foraging trails leading to a bivouac made up of their own bodies. But instead of a bivouac, 20% of these organisms construct a tall column out of their own dying bodies to support the other 80%. This is the social amoeba *Dictyostelium discoideum*,

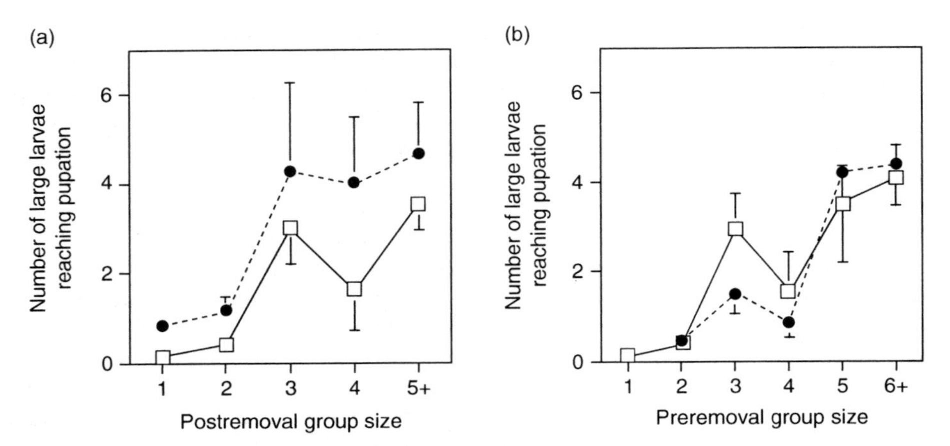

FIGURE 19.2 Life insurance in the hairy-faced hover wasp. Dark circles and dashed lines show colonies that had 1 or 2 workers removed, whereas open squares with solid lines show unmanipulated controls. (a) Removal colonies raised significantly more large larvae to pupae than controls of the same postremoval size, showing that prior work done by the removed females does add fitness benefits even after they are gone. (b) There was no difference when the experimental colonies were compared to controls of the same preremoval size. Reprinted from Field et al. (2000) with permission.

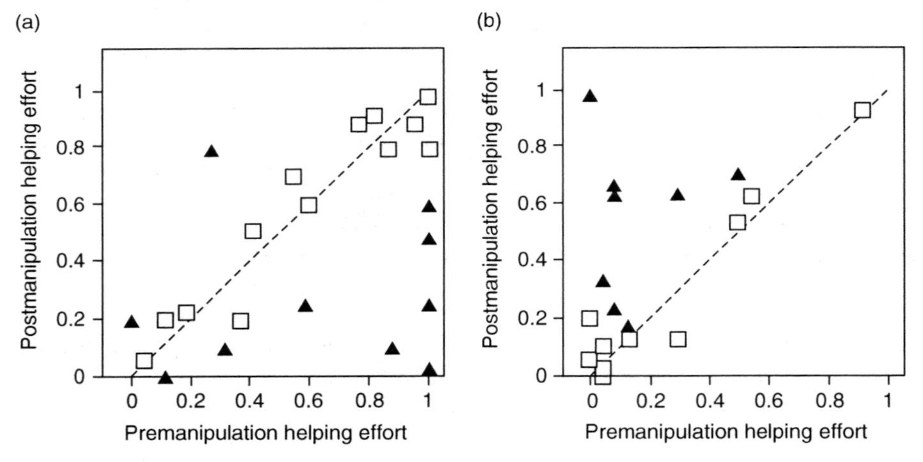

FIGURE 19.3 Change in working behavior of focal subordinates in the hairy-faced hover wasp. (a) When a subordinate moves up in rank by removal of a higher ranked wasp (triangles), she becomes a slacker, helping less than controls (squares) where a lower-ranked wasp was removed to keep group size equal. This result is consistent with her being less willing to risk her now-greater expectation of direct reproduction. (b) When a subordinate stays at the same rank but her group size is reduced by removal of lower ranked females (triangles), she takes up the slack and works harder relative to unmanipulated controls (squares), showing that helpers are sensitive to the benefits of helping. Reprinted from Field et al. (2006) with permission.

an evolutionary cousin of fungi and animals. The streams are composed of starving cells that have passed a number of generations in solitary mode, foraging for bacteria and dividing asexually. The altruistic production of a stalk by 20% allows the others, which differentiate as spores, to be dispersed to better locales by passing invertebrates.

D. discoideum thus has a multicellular stage, but differs from most multicellular organisms in that the cooperating group is formed by aggregation rather than by clonal division from a single cell. Therefore, cells from different clones may coaggregate, and this makes cheating a possible strategy; a clone can gain an advantage that makes more spores in mixtures and allows the other clones to make the stalk. Such cheating does indeed occur (Strassmann et al. 2000). Perhaps as a result, there is a cost when two or more clones aggregate together. Mixtures migrate less far during the slug stage just before fruiting bodies are formed, consistent with each clone trying to stay out of the anterior region of the slug, the area that normally produces the stalk.

D. discoideum offers an unusually good opportunity to study cooperation and conflict at the genetic level. It has a sequenced genome, techniques to manipulate genes influencing social behavior exist, and it lives happily in the laboratory where it can be subjected to many-generation evolutionary experiments. One novel result concerns the importance of pleiotropy (Foster et al. 2004). For example, the *dim*A gene is required for the response to a signal to differentiate into stalk, so knocking out this gene was predicted to produce a cheater strain that makes less stalk and more spores in mixtures. The knockout strain did contribute more cells to the prespore region of the slug, but an additional pleiotropic effect ensured that it did not actually result in more spores. Similarly, the knockout of the cell adhesion gene *csa*A creates cheats in slugs, presumably because its weaker adhesion allows it to slide back to the prespore region. However, it is not an effective cheater because its weaker adhesion prevents many of the cells from aggregating in the first place. These examples suggest that cooperative traits may often be built from elements that are essential for other reasons, which render them less susceptible to cheating.

High relatedness is also important for limiting cheating that could destroy cooperation, as shown by lab mixtures of the knockout mutant *fbxA*⁻ and its parental wildtype strain (figure 19.4; Gilbert et al. 2007; chapter 18, this volume). The first thing to notice is that *fbxA*⁻ has severe deleterious effects: the higher the percentage of *fbxA*⁻ in the mixture (the lower the percentage of wildtype), the lower spore production is. In fact, spore production is nearly zero when *fbxA*⁻ is grown by itself because it makes deformed fruiting bodies. Nevertheless, the cheater imposes an even more severe penalty on the wildtype than on itself; at every mixture frequency tested *fbxA*⁻ had higher per capita spore production than the wildtype strain. If population relatedness were zero, which means that each fruiting body would have about the same frequency of cheaters and cooperators as the population, the cheater would win at every frequency, and both cooperation and high spore production would be doomed. However this conclusion does not hold if relatedness is high enough. If relatedness were 1, then fruiting bodies would either be all cheaters with very low spore production (open circle at the origin) or all cooperators with high spore production (closed circle at upper right). Clearly cooperation would win in this case. So the crucial question is whether relatedness is high or low in nature. Individual wild fruiting bodies were collected from dung and their spores were genotyped at microsatellite loci so that relatedness could be estimated. Relatedness,

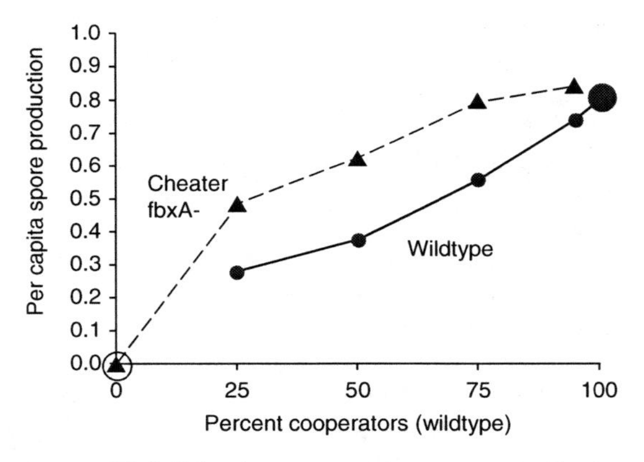

FIGURE **19.4** Selection on a cheater mutant *fbxA*– in the social amoeba, *Dictyostelium discoideum*. At every mixture with the wildtype cooperator strain, the cheater is more effective at getting into spores, so in a well-mixed (low relatedness) population, it would spread to high frequency. But this results in fruiting bodies with little or no spore production. However, if relatedness is one, all-cooperator fruiting bodies (closed circle) outproduce all-cheater fruiting bodies (open circle). Data regraphed from Gilbert et al. (2007).

at least in this environment, is very high, between 0.86 and 0.97. Cheater mutants like this one, with completely defective fruiting bodies on their own, should therefore not spread in nature. Consistent with this, screens of over 3000 clones from nature confirmed that there were none with such defective fruiting bodies (Gilbert et al. 2007).

However, this does not mean all cheating is controlled by relatedness. A genetic selection found over 100 mutants that could cheat in mixtures, but could also make normal fruiting bodies on their own (Santorelli et al. 2008). It is not yet known how important such cheating is in nature. If it has been important, it may show rapid arms race evolution as a result of conflict, so studies of molecular evolution may resolve this question.

FUTURE DIRECTIONS

In this review we have sketched out the general principles behind the evolution of increasingly organism-like colonies. Hamilton's general approach still serves us very well, even as we place less importance on relatedness and more on cost, benefits, and power. Some of our hypotheses are well supported, others less so, but all could use much more study. It is particularly useful to run the ideas up against unconventional and understudied social systems. More work is needed on what Costa (2006) has called the "other insect societies," both to understand the forces underlying their cooperation and to understand why it has not generally gone as far as in the standard social insects. Colonial invertebrates are another neglected font of cooperation and specialization. As we have stressed, cooperation in microorganisms is much more extensive than previously believed (Foster in press). The advantages of microbial systems are great because these systems can undergo experimental evolution of many generations; individual genes can be knocked out or replaced; whole genomes can be sequenced; interactions can be understood at the molecular level.

In focusing on adaptive principles, we have paid too scant attention to proximate mechanisms, but these will need to be fully explored to have a complete picture of the paths to superorganisms. Three different lines of proximate research, operating on very different levels, are worthy of note. At a very high level, there is research on how work is organized, particularly how individuals with limited information cooperate on common tasks (Bonabeau et al. 1997). Second, there is much potential to learn from historical considerations of the preadaptations or ground plans upon which social systems were built (West-Eberhard 1996). We have, for example, mentioned parental care as an important preadaptation, but much more can be said about how parental behaviors come to be differentially regulated in breeders and offspring. Phylogenetic analysis is often an important element of such studies. Finally, there is now the potential to study the mechanisms of sociality from the genes on up. Organisms with sequenced genomes like the honeybee, or the social amoeba, are particularly attractive.

SUGGESTIONS FOR FURTHER READING

Of the many fine reviews of insect sociality, the one by Alexander et al. (1991) remains most influential to our thinking. Bourke and Frank's (1995) *Social Evolution in Ants* focuses on one taxon, but it speaks very clearly to general principles. The topic of power and conflict resolution will prove increasingly important, and the best review of this topic is by Ratnieks et al. (2006). The study of microbial sociality is fast outpacing reviews, but see Foster (in press) for an introduction.

Alexander RD, Noonan KM, & Crespi BJ (1991) The evolution of eusociality. Pp 3–44 in Sherman PW, Jarvis JUM, & Alexander RD (eds) The Biology of the Naked Mole Rat. Princeton Univ Press, Princeton, NJ.
Bourke AFG & Franks NR (1995) Social Evolution in Ants. Princeton Univ Press, Princeton, NJ.
Foster K (in press) Social behavior in microorganisms. In Szekely T, Moore AJ, & Komdeur J (eds) Social Behaviour: Genes, Ecology and Evolution. Cambridge Univ Press, Cambridge, UK.
Ratnieks F, Foster KR, & Wenseleers T (2006) Conflict resolution in insect societies. Annu Rev Entomol 51: 581–608.

SECTION V

REPRODUCTIVE BEHAVIOR

20

Sexual Selection

MICHAEL D. JENNIONS AND HANNA KOKKO

In this chapter we explore the routes that link differences between the sexes to sexual selection. Why is sexual selection often stronger in one sex? Why is the choosier sex usually less sexually competitive? What is the relationship between sexual selection and parental care? We outline key theoretical models that apply to most taxa and account for the origin and evolution of sexual differences in parental care, mating effort, sexual conflict, and choosiness. This theoretical backdrop sets the scene for empirical studies that usually focus on quantifying current levels of sexual selection (Shuster & Wade 2003), identifying genetic constraints on the evolution of sexual traits (Blows 2007) and measuring the sex-specific costs and benefits of mate choice, mating rates, and alternative male mating tactics. These topics are covered in detail in chapters 21–25.

SEXUAL SELECTION: ALL PERVASIVE AND ALL POWERFUL

The theory of evolution by sexual selection has prospered from its humble origins in 1871 as a secondary refinement by Darwin to his earlier account of evolution by natural selection (Darwin 1871). Darwin's original goal was to account for elaborate male traits that could not be explained by natural selection for survival. How could the peacock's tail, the giant antlers of the extinct Irish elk, or the gaudy throat colors of male lizards have evolved through natural selection? Traits that are energetically wasteful, overly elaborate, and harmful to their bearer are difficult to reconcile with the view that natural selection continuously weeds out inferior, uneconomical variants (Cronin 1991). Darwin noted that selection will favor traits that increase male mating success and allow a male to sire more offspring, even if this is at the expense of a reduced lifespan. Today, however, sexual selection is more widely invoked to explain a broader range of pervasive and striking patterns in nature.

Sex-specific selection must be a powerful force to account for sex-limited expression of traits. Males and females often differ profoundly in size, shape, color, behavior, and life histories even though, regardless of their sex, members of the same species share the vast majority of their genome (Bonduriansky 2007). These phenotypic differences can be so extreme that taxonomists have occasionally assigned males and females of the same species to different taxa, as was the case with eclectus parrots *Eclectus roratus* (Forshaw & Cooper 1989). *Sexual divergence* is the process that leads to sexual dimorphism in traits associated with competing for fertilizations, choosing mates, and investing in offspring. Explanations for sexual divergence have expanded greatly since Darwin's initial account. We now know far more about factors determining mating success such as fighting behavior and mate choice (chapter 24), as well as postejaculation processes such as sperm competition (chapter 22), cryptic female choice, and differential allocation.

Sexual selection theory can explain not only why males (and sometimes females) bear traits that decrease their own lifespan through an increased risk of predation, parasitism, or susceptibility to disease, but also why they have evolved traits that damage the opposite sex. As sexual selection theory developed, it shifted biologists' perspectives on reproduction. Mating is no longer seen as harmonious union. Instead, it is viewed as an arena for sexual conflict in which females are under selection to evolve traits to counter males' attempts to elevate their own reproductive success (chapter 23). In the last decade, empirical studies have confirmed that sexual conflict can elevate the rate of evolution of certain genes, biochemicals, and morphological traits (e.g., Arnqvist & Rowe 2002b; Dorus et al. 2004; Andrès et al. 2006; Ramm et al. 2008). More generally, sexual selection theory explains why secondary sexual traits, such as genitalic embellishments, courtship song, and breeding coloration, show greater variation both within species and among closely related species than do other traits (Arnqvist 1998).

Recent theoretical developments have prompted reexamination of the coevolution of parental investment, sexual selection, and sex roles. The current direction and intensity of sexual selection are usually attributed to greater female than male parental care, which removes females from the mating pool (i.e., the set of individuals available to mate) and thereby forces males to compete more intensely for mates (Trivers 1972). However, it is too simple to draw the arrow of causality from parental investment to sexual selection because it can also go in the reverse direction: sexual selection is itself an important causal factor driving the initial evolution of greater postfertilization care by females (Queller 1997). Stronger sexual selection on males can therefore account for the taxonomically widespread pattern that females provide more parental care (Kokko & Jennions 2008). The details of the coevolutionary process vary greatly among species, however, and understanding unique features of the breeding biology of some taxa can also explain why they do not obey this rule, such as *sex-role-reversed* species in which females compete for males and are sometimes more ornamental in appearance (e.g., Forsgren et al. 2004; Jones et al. 2005).

The power of sexual selection to shape phenotypes is impressive when one considers that a gene for a sex-specific trait is only available to direct selection when it is expressed in the appropriate sex. For example, half the genes for a favored male trait are currently hidden from selection because they are in the body of a female. Even worse, when these genes are expressed in females they are actively selected against if they produce an inferior phenotype (Bonduriansky 2007). The reason why sexual selection remains such a powerful force is that, unlike much of natural selection, sexual selection is based on a *zero-sum game* in which one individual's win is another's loss. Sexual selection occurs only when there is competition for mates. Thus if, say, female reproduction limits population growth, then for every offspring sired by one male, another male loses out on an opportunity to reproduce (Shuster & Wade 2003; chapter 25). It is therefore impossible for every male to enjoy elevated reproductive success by competing more efficiently for females. It follows that there is no single goal or common end point to sexual selection. In contrast, natural selection sometimes selects for a single target. For example, the most aerodynamically efficient wing shape can evolve and spread to all members of a population. Sexual selection is therefore typically an open-ended process, and selection is usually directional rather than stabilizing.

SEXUAL SELECTION: PUTTING DEFINITIONS INTO PRACTICE

Darwin (1871, p. 256) stated, "That kind of selection, which I have called *sexual selection*... depends on the advantage which certain individuals have over other individuals of the same sex and species, in exclusive relation to reproduction." Darwin's definition was intentionally restricted to so-called *secondary sexual traits* that are advantageous when competing (either directly or indirectly) for mating or fertilization opportunities (Arnold 1983). *Primary sexual traits* affect fertilization success or the ability to mate irrespective of the presence of rivals. These traits are economically designed (i.e., no investment is made in offensive, defensive, or attractive components) and are solely attributable to natural selection even though they are used in a sexual context.

Darwin's definition obviously included traits that are advantageous when individual compete directly for mates, such as weapon-like tusks, horns, and spines used in fights. More controversially, he proposed that female choice selects for male traits that satisfy female aesthetic preferences for mates (Cronin 1991). In short, Darwin's definition was that sexual selection favored traits that increase mating success when individuals compete for mates. Modern definitions of sexual selection are more expansive. They accommodate the fact that mating does not guarantee parentage when females mate multiply. A more widely acceptable modern definition is therefore as follows: *sexual selection* favors investment in traits that improve the likelihood of fertilization given limited access to opposite sex gametes due to competition with members of the same sex. Traits favored by sperm competition, such as increased ejaculate size or longer sperm, are also sexually selected.

All selected traits increase the net fitness of the bearer (chapters 3 and 4). The most sensible reason to distinguish between sexual and natural selection is if these two processes generate distinctive types of traits. Succeeding in sexual competition requires investment of resources that could otherwise be used to increase naturally selected components of fitness (e.g., survival, foraging efficiency). The first problem for researchers studying sexual selection is to distinguish between primary and secondary sexual traits: is a trait more elaborate or costly than expected if its sole function is to ensure sexual union? In practice, traits such as male genitalia are a composite of features shaped by natural and sexual selection (Eberhard 1985). It is therefore a complex task to decide what features would remain if we isolated the effect of natural selection and eliminated sexual competition. Fortunately, recent experimental evolution studies offer some insights. Under experimentally imposed monogamy, natural selection favors lineages in which sexually selected traits are reduced, because this allows for greater investment into naturally selected traits that enhance the lifetime reproductive output of a pair (Holland & Rice 1999). Perfect monogamy in the absence of mate choice should eventually lead to the loss of all secondary sexual traits.

There is always a temptation to extend the scope of sexual selection. For example, it was recently suggested that there is strong sexual selection for traits ensuring female dominance in group-living species with limited breeding opportunities, such as meerkats, because only top-ranked females breed successfully (Clutton-Brock et al. 2006; Clutton-Brock 2007). However, in cases in which breeding opportunities limit female reproduction while all females have ready access to male gametes, we think the relevant female traits should be ascribed to selection due to social competition. Socially selected traits represent an anomalous investment from the perspective of natural selection acting on individuals in isolation (West-Eberhard 1983; box 14.1, this volume). The terms *social selection* and *sexual selection* are not interchangeable, however, because social competition is more inclusive and, in addition to mates/gametes, includes competition for nonsexual resources such as food items or grooming partners. Social competition for nonsexual or sexual resources selects for traits such as aggressiveness and weaponry. Because it is a zero-sum game, competing for a limited number of dominant positions in a group can select for similar traits to those favored by sexual selection.

The real problem for biologists is deciding whether a sexual trait deviates from its naturally selected optimum and has, by extension, evolved under additional sexual selection. The schoolboy error is simply to state that traits that currently increase mating or fertilization success under competition are sexually selected. This usage is far too broad. It makes the products of sexual and natural selection virtually indistinguishable because it encompasses any trait that improves reproductive success in the chain of events that begins with a nonreproducing individual and ends with the production of independent offspring. For example, foraging efficiently increases body condition, which in turn often increases male attractiveness (Tomkins et al. 2004). If all traits that increase foraging efficiency are attributed to sexual selection, we lose the meaningful distinction between economical and wasteful traits that originally drove Darwin to formulate the theory of sexual selection.

LET'S TALK ABOUT SEX

Mating Types and the Evolution of Two Sexes

Sexual selection requires sexual reproduction. Sex involves germ line cells undergoing meiosis to create sex cells (*gametes*) that contain half the parental genome. Two gametes then fuse to form a zygote that develops into another individual. Sexual reproduction is not synonymous with the existence of males and females. In some species (e.g., certain algae and fungi), all gametes look the same (*isogamy*) but are divided into self-incompatible negative and positive mating types. In a few species there are more than two mating types and whether gametes can fuse depends on their genetic compatibility, which is determined by the absence of shared alleles at each of one or more loci. Sometimes this creates only a few mating types, as in the protozoan *Tetrahymena thermophila*, in which there are seven, but in others, such as the fungi *Schizophyllum commune*, there are thousands of *mating types* (Whitfield 2004).

The most common situation in sexual species is two mating types representing two sexes: males and females. This characterization of mating types is used when individuals specialize in producing either large or small gametes (*anisogamy*). Anisogamy is widely attributed to disruptive selection when one type of gamete evolves to take advantage of the resources provided by another (box 20.1). By definition, males are members of the sex that produces smaller gametes: a seahorse whose mate inserts large gametes into its brood pouch, in which they are subsequently fertilized and protected until offspring emerge, is still a male despite its effective pregnancy. This definition works because gamete size typically has a strongly bimodal distribution in anisogamous species. This is true even when individuals are hermaphroditic and produce different types of gametes. Hermaphroditic plants, for example, still produce ova and pollen rather than a continuous size range of gametes.

Anisogamy does not always predict enormous differences in gamete size because there is sometimes secondary selection for larger sperm (chapter 22). For example, male *Drosophila bifurca* produce giant sperm that are less than an order of magnitude smaller than eggs (Bjork & Pitnick 2006). This is a tiny difference compared to that in mammals in which sperm are 4–5 orders of magnitude smaller. If *D. bifurca* sperm ever evolve to be larger than eggs, the sex currently producing sperm will be labeled females! This would probably not be remarked upon by an observer unaware of the species' history, who would follow the convention of assigning sex based on gamete size. This simple thought experiment is a reminder that, gamete size aside, there are no immutable behavioral and morphological characteristics associated with maleness and femaleness. It can sometimes appear so because relative gamete size seems to drive the evolution of other traits such as stronger sexual competitiveness or greater parental care that we closely associate with one sex. The existence of sex-role-reversed species is a stark reminder that gamete size is a fallible predictor of how sexual selection operates on

BOX 20.1 Anisogamy and the Parasitic Nature of the Origins of Sperm

Anisogamy (distinct gamete size classes that define males and females) can evolve in at least two different ways (Bulmer & Parker 2002). The first option is that the ancestral population consists of gametes that can freely fuse with each other. Disruptive selection initially creates two distinct size classes of gametes (Parker et al. 1972) before gametes evolve traits that discriminate against fusing with gametes of the same size (Parker 1978). Alternatively, an ancestral population with isogamous gametes might already have two mating types that fuse only with each other (e.g., because of chemotactic responses; Hoekstra 1982). In this case, the question is whether the two preexisting mating types evolve to differ in size. Assuming external fertilization, both cases lead to the evolution of small sperm and large eggs under relatively simple assumptions. The analysis is similar in both cases, and here we describe the latter route, which appears to hold in some taxa (e.g., evolutionary transitions of gamete size in green algae; Kirk 2006). The following is a simplified account of the model of Bulmer and Parker (2002).

Three important assumptions are required for anisogamy to evolve. First, given a fixed budget for gamete production there is a trade-off between size and number of gametes. This is usually modeled as $nm = M$, where n is the number of gametes produced, m is gamete size, and M indicates the total budget. Given a fixed budget, a clear benefit of producing many gametes is a larger potential number of offspring. Second, large gametes form well-provisioned zygotes that survive better: the size of the gamete represents parental

(continued)

investment. Third, the function that relates zygote survival to its size is nonlinear: linear relationships are, in fact, impossible because survival would eventually exceed 100%. Given suitable nonlinearity, if one parent reduces gamete size by, say, half (thus it produces what we can start to call proto-sperm), zygote survival will drop but—importantly—by less than half (figure I). Assuming that gametes form a large, well-mixed population in which other proto-sperm compete for proto-eggs, this parent has simultaneously doubled its expected offspring production (with fixed M, n will double when m is halved). Doubling of the expected number of zygotes more than compensates for their somewhat diminished survival: the parent specializing in small gametes that rely on the other gamete to provide the bulk of the necessary resources for zygotes experiences a net profit. Sperm competition is an important feature of this quantitative argument. Without it, a male would only have to produce enough proto-sperm to ensure that all the locally available proto-eggs are found with a reasonably high probability. Numerical excess beyond this would then hardly improve reproductive success. However, when many proto-males compete, a twofold increase in the quantity of sperm will double a focal male's expected fertilization success.

We have used a large change in gamete size (reduction by 50%) for illustrative purposes. Assuming suitable nonlinearity in zygote survival functions (Bulmer & Parker 2002), the argument generalizes to smaller and more realistic changes: one expects disruptive selection in which one sex becomes a "parasite" in the sense that it relies on the "host" sex to provide virtually all the resources for the zygote, and specializes in locating as many hosts as possible. The host cannot follow the same route of diminished provisioning because, if it is the only resource-provider, further diminishing zygote

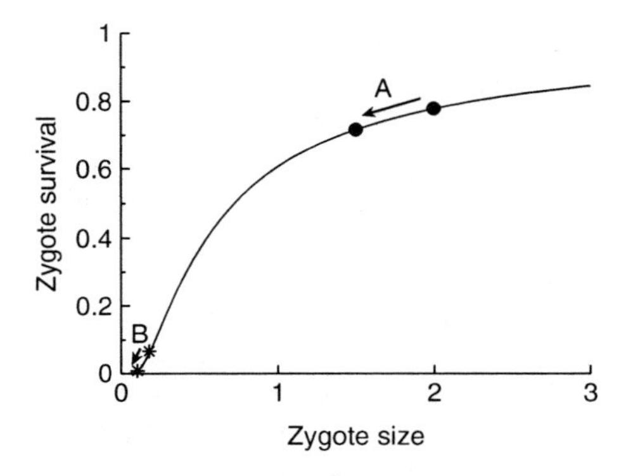

Figure I Initial steps in the evolution of anisogamy. Larger zygotes survive better, but the relationship between size and survival is nonlinear. The ancestral population is isogamous, with both parents spending their gamete budget ($M = 1$) on one gamete per time unit, thus the zygote size is 2. If one parent switches to making two gametes per time unit (arrow A), the zygote's size is $0.5 + 1 = 1.5$. Zygote survival drops from 78% to 72% in this example, but the parent's relative reproductive success has increased from 1×0.78 to $2 \times 0.72 = 1.44$. This does not assume that *all* gametes fuse with eggs, only that the expected number of eggs located doubles. It is not possible for both parents to keep diminishing gamete size in this way. If both invested very little, then further decreases eventually lead to much stronger drop in survival (arrow B). A more complete treatment of the evolution of anisogamy that takes gametic survival before fertilization into account is provided by Bulmer and Parker (2002).

(continued)

resources would harm it disproportionately: it is not possible for a host to become the parasite's parasite (figure I). If one parent does not contribute resources, the other must compensate. Sexual conflict is therefore an ancient and fundamental feature of reproduction involving two sexes (chapter 23). It will be absent only in restricted cases such as absolute monogamy, random mating, and an even adult sex ratio so that male and female reproductive interests perfectly coincide. In such cases, we might even predict a return to isogamy. Isogamy can, however, also be maintained if gamete survival requires that they remain relatively large (Bulmer & Parker 2002). The anisogamy argument can be generalized to cover internal fertilization. If females mate only with a single male, then males should evolve to provision zygotes instead of producing excess sperm. However, even if the proportion of matings involving sperm competition is very small, excess sperm production persists (Parker 1982).

Populations would grow best if all parents directed resources to zygotes, rather than trying to outcompete others of the same sex in a zero-sum game. From a population perspective (or from a female's view), resources used by males to compete are squandered. The famous *twofold cost of sex* is a consequence of the zero-sum nature of male reproduction, not a result of sex per se (see figure II). This cost halves the rate at which a sexually reproducing female's genes are transmitted to future generations, compared to that of an

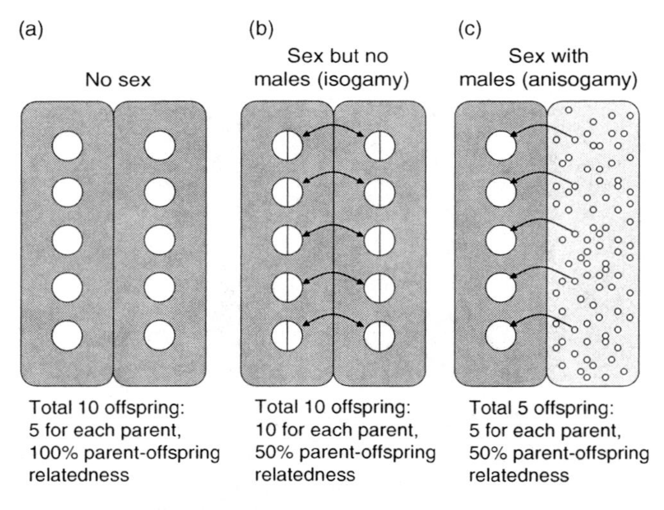

Figure II Why a twofold cost of sex arises only when males evolve. Two individuals are depicted in each population, with the shaded area indicating the resources they use. In each case the resources are sufficient to produce 5 offspring. In (a), two asexual females produce 10 offspring in total. In (b), genetic exchange occurs because of sex, but isogamy allows resources to be pooled such that total offspring production remains at 10. In (c), half the population evolve to become males and the resources they access are used to elevate success in sperm competition; these resources are not used to enhance offspring production, so the population-wide growth rate, as well as the female's fitness, is halved.

(continued)

asexual female who produces only daughters. It is the main reason that the maintenance of sex is among the most challenging questions in evolutionary biology (West et al. 1999).

The evolution of two sexes is a fascinating example of a *tragedy of the commons* in which individuals strive to secure resources for themselves (at the expense of competitors) with eventually negative consequences for the entire population. The metaphor originated with Hardin (1968) and refers to the medieval practice of allowing cattle or sheep to graze a common pasture that, without policing, leads to overgrazing (chapter 18). The tragic modern analogue is the overexploitation of marine fish stocks. Evolutionary tragedies of the commons are common (Rankin et al. 2007), and investment into winning at zero-sum games is, in general, prone to such tragedies. In the same way that a prudent fisherman who invests only in a small vessel will lose out in competition with big trawlers, a male who avoids producing excessive amounts of sperm ("excessive" because most will never fertilize an egg) will be outcompeted by males who invest more. Resources that could have provisioned offspring are "wasted" on competition that lowers the population-wide reproductive output (figure II). Of course, this does not prevent anisogamy from evolving because selection acts on individuals more strongly than it does on populations.

The evolution of males is an unusual "tragedy," though, because it contains an internal feedback that halts the spread of exploitation. In real host-parasite systems, parasites can greatly outnumber hosts, but once sperm fuse only with eggs, and vice versa (Parker 1978), Fisherian sex ratio theory (box 20.2) predicts that the rarer sex will have greater reproductive prospects so that parents will tend to invest equally in sons and daughters. Even so, anisogamy still means that up to half the individuals in a population are exploiting the rest, so there is a 50% reduction in the efficiency with which resources are converted into offspring. It should be noted that the cost of sex is twofold only if males do not interact with females or their offspring, aside from providing sperm to fertilize eggs, and all the resources males use are spent on mating effort (figure II). In the original anisogamy model the males' total budget is spent on gametes, but in reality the use of this budget itself can evolve in response to sperm competition risk (chapter 22) and trade-off with other traits, creating more prudent sperm usage in some cases (Wedell et al. 2002). Sperm limitation can cause the cost of sex to be greater than twofold (if some eggs go unfertilized). Likewise, it is reduced if there is paternal care, and increased if males actively harm females (chapter 23) or, more generally, reduce female access to food.

A reduction in the twofold cost of sex due to care is unlikely to provide a general explanation for sexual reproduction, because male parental care is a more recently derived trait than sexual reproduction, and because substantial male contributions to rearing offspring are relatively rare. It should also be noted that even though individual-level selection makes it easy to evolve from state (b) to (c) in figure II, this still does not constitute an explanation for sexual reproduction. This is because individual-level competition between asexual and sexual forms typically takes place directly between states (a) and (c), which sometimes coexist within a single species (e.g., Jokela et al. 2003).

each sex. So why are there so many species in which gamete size does predict sex roles?

Anisogamy Rather than Sex Leads to Sexual Divergence

It is conceivable that sexual selection occurs in isogamous species: for example, in *Chlamydomonas* algae positive and negative mating types are both flagellated and thus motile, presumably an outcome of selection to improve mate encounter rates. There are, however, very few phenotypic differences among mating types in most isogamous species, which is why genetic techniques are required to identify mating strains. In contrast, in anisogamous species the sexes diverge not only in the size of their gametes but in innumerable other aspects of their lives ranging from their size and shape (Fairbairn et al. 2007), to color and ornamentation (Andersson 1994), to immunocompetence and average lifespan

(Moore & Wilson 2002). This sexual divergence appears to have its origins in how sexual selection acts on each sex, so it follows that anisogamy generates additional forces of sexual selection that extend well beyond those acting on gamete size.

There are countless differences between the sexes. To explain them in a concise manner, we need to identify the most fundamental forms of sexual divergence that are likely to drive secondary differences. Theoreticians highlight three key differences. First, males are more likely to compete for a mate than are females. The most extreme weaponry in animals, such as antlers, tusks, and enlarged claws are expressed more fully, and more often, by males than females. Second, females are usually choosier than males about their mates; hence the greater occurrence of male than female sexual ornaments. Third, there is often a positive correlation between prefertilization and postfertilization parental investment. Prefertilization investment occurs at the gamete level: Trivers (1972) noted that eggs (and sperm) represent parental investment because they contain resources that improve the success of the current offspring at a cost to the parent's future reproductive success. Postfertilization parental investment occurs whenever there is costly parental care (chapter 26). In birds, mammals, reptiles, insects, and most arthropods, females provide care more often than males (Clutton-Brock 1991; Tallamy 2001; Reynolds et al. 2002). If there is biparental care, females still tend to make a greater effort than males (Schwagmeyer et al. 1999). The most notable exception to this rule occurs in fish in which male-only care is common (Reynolds et al. 2002).

FROM ANISOGAMY TO SEXUAL SELECTION

Anisogamy reflects a primordial sexual conflict whereby small male parasitic gametes exploited the rich resources provided by larger female host gametes (box 20.1). Competition among males, female choice, and sexual coercion represent the subsequent extension of this conflict to interactions among individuals rather than gametes. In this section we explain how anisogamy promotes sexual differences in the strength of sexual selection. We focus our attention on the differences between the sexes, as the evolutionary processes that lead to sexual divergence create the fundamental differences that subsequently determine how sexual selection operates within each sex. Measuring the effects on fitness of trait variation within each sex is the goal of most empirical studies that tackle issues such as how sexual competition creates niches that promote alternative mating tactics in the more competitive sex (chapter 25), or the consequences of sexual conflict over mating rates for trait evolution within each sex (chapter 23), or the benefits of mate choice for each sex (chapter 24).

Gamete Availability and Competition for Fertilizations

If the only initial difference between males and females is the type of gametes they produced, we would *a priori* predict an even *adult sex ratio* (ASR) in diploid species. This prediction arises because of the *Fisher condition* (box 20.2), which states that each offspring in a diploid species has one genetic mother and one genetic father so that the total number of offspring produced by each sex is identical. A consequence of the Fisher condition is that selection does not favor reproductive strategies that equalize the population-wide number of male and female gametes; rather it favors equal investment in individuals of either sex at conception because, on average, offspring of the rarer sex will produce more descendants (box 20.2). For the ASR to be even, male and female mortality rates must be identical. This follows from the starting premise that the only sexual difference is in the size of the gametes they produce if there is no difference in the total investment in gamete production. This last assumption will probably be violated once sexual differences other than gamete size evolve, but it remains useful to highlight the absence of any *a priori* reason for one sex to be consistently rarer in the ASR. Together these factors create the causal route from anisogamy to an excess of sperm searching for eggs: males can produce more sperm than females can produce eggs, and the number of females is not expected to greatly exceed the number of males. Many sperm will fail to locate eggs, whereas most eggs will be fertilized.

Selection can arise only if there is variance in fitness among individuals (chapters 3 and 4), and a male bias in gamete numbers sets the stage for sexual selection as variance in reproductive success is potentially far higher for males than females because fertilization success is a zero-sum game (chapter 25). The term *opportunity for sexual selection* (Shuster & Wade 2003) is often used to describe measures of variance in reproductive success, but it should be

BOX 20.2 Sex Allocation Theory and the Fisher Condition

Why create sons if most of them will produce unsuccessful gametes, whereas almost every daughter will successfully breed and have all her eggs fertilized? By analogy, to maximize their profits, chicken farmers retain no more cockerels than are needed to fertilize their hens. To date, however, animal breeders have been unable to produce strains that steadily conceive more daughters than sons (i.e., a female-biased *primary sex ratio*) despite the enormous economic incentives. Is the inability to bias offspring production toward females due to a lack of variation for selection to act upon (perhaps there is no proximate mechanism to bias sex ratio), or is there a deeper counteracting force at work?

Although a female-biased population has a higher growth rate, natural selection acts on individuals, not groups (box 20.1: the twofold cost of sex), and favors equal investment into both sexes. Consider the simple argument of R. A. Fisher, who asked what happens if the primary sex ratio deviates from 1:1 in a large, outbred population of a diploid species (Fisher 1930). Take, for example, an efficiently reproducing population in which every female produces three times as many daughters as sons. If, on average, a daughter produces n offspring ("on average" accounts for the fact that some daughters die before breeding and any variation in fecundity), how many offspring will the average son sire? No matter how many males die as juveniles or how strongly skewed mating success is, there are, on average, $3n$ offspring sired by each son. This is the only number to balance the account books when each offspring has exactly 1 genetic father and 1 genetic mother. Offspring of the rare sex, in this case, sons, are more valuable, and selection favors parents who bias production toward this sex, even though this typically reduces the population-wide growth rate.

The *Fisher condition* states that in diploid, sexually reproducing organisms, total offspring production through males cannot logically exceed that through females (or vice versa). It implies that the per capita production is higher for the rarer sex. Whenever male zygotes are rare, they grow into adults that, on average, produce more offspring than do female zygotes. Selection will therefore favor any mutation that increases the production of sons. If female zygotes are rare, the exact counterargument applies. Given certain widely applicable assumptions (e.g., that brothers do not compete for mates), negative frequency-dependent selection will result in an evolutionarily stable state in which the population invests equally into offspring of each sex. The Fisher condition holds, regardless of juvenile or adult mortality patterns, the extent of mating skew among males, or reproductive skew among females and whether either or both sex mate multiply or provide parental care. This near universal applicability in diploid species makes it an incredibly strong and fundamental force that equalizes sex allocation into males and females.

There are certain inequalities that lead to deviation from the production of equal numbers of each sex (Hardy 2002). The most important is that Fisher's argument is better framed in terms of equal parental investment of resources into each sex, rather than the production of equal numbers of each sex. This takes into account any difference in the cost of production of each sex. Given a 1:1 primary sex ratio, it is initially more profitable to produce the cheaper sex, until the lower cost of production is counterbalanced by the decline in the average number of offspring produced by this sex due to its greater abundance (hence lower average rate of offspring production). Equilibrium is reached when the primary sex ratio is the inverse of the cost ratio (i.e., when the investment per sex is identical). We do not know if animal breeders have ever attempted to manipulate the life history cost of producing males. If, for example, all offspring are used for breeding but each cow is killed after she has weaned her first son, the cost of producing sons is elevated. Fisherian sex ratio theory then predicts a shift to a female-biased primary sex ratio. This manipulation allows no cow to produce more than one valuable son, whereas those that produce daughters have a higher lifetime number of offspring (including one valuable son).

(continued)

BOX 20.2 *(cont.)*

Female-biased primary sex ratios also arise if brothers compete for mates (local mate competition; Hamilton 1967). A parent has less incentive to produce sons when the degree of competition among brothers is high. This is because parental fitness does not depend on which son mates, only on the total number of matings their sons achieve. Competition among sons is wasted effort from a parental perspective, and parents benefit if they reduce this competition by decreasing the number of sons and investing more into producing daughters. In fig wasps, females lay eggs inside figs and their offspring then mate in the fig before their inseminated daughters disperse. The average number of females laying eggs inside a fig varies among species, which affects the average likelihood that brothers will compete for mating opportunities. Local mate competition theory has been spectacularly successful in predicting the primary sex ratio in fig wasps (Herre 1985). Another factor influencing the primary sex ratio is whether parent-offspring interactions affecting the parents' fitness differ between sons and daughters. For example, in cooperative breeders, parents should bias offspring production toward the sex that is more likely to assist them in rearing offspring (Komdeur & Pen 2002). This has been shown to occur in Seychelles warblers in which a series of stylish experiments have demonstrated that the sex ratio is highly skewed toward female helpers when the territory quality and number of existing helpers makes this beneficial for the parents (Komdeur 1998). More generally, whenever there is a nonlinear relationship between total investment in each sex and the marginal returns, this can lead to modest deviations from the more simple Fisherian prediction of equal investment into both sexes (Frank 1990; Sheldon & West 2004).

These refinements to sex allocation theory are not applicable to the general argument that in the early stages of anisogamy, in which the sexes differ only in the size of the gametes they produce, there will be frequency-dependent selection favoring equal production of both sexes. So it is reasonable to conclude that there will be many more male than female gametes seeking partners. That noted, we should not forget the important caveat that a primary sex ratio of unity does not automatically lead to an even adult sex ratio if the life histories of the two sexes differ so that one sex has a higher mortality rate. Once evolution by sexual selection occurs, mortality rates might well diverge between the sexes. If such differences are limited, however, male gametes will still be more readily available than female gametes.

noted that variance is a prerequisite for sexual selection to occur. It is not a guarantee. Higher variance in male than female reproductive success is also consistent with a scenario in which male fertilization success is entirely random and no selection occurs (Sutherland 1985). For sexual selection to operate, elevated fertilization success in the face of competition must be correlated with the expression of a trait that varies among individuals (i.e., variation in mating success has a nonrandom component). Whether the trait then evolves will depend on how it is genetically correlated with other traits (Blows 2007).

If we assume that anisogamy initially evolved in an external fertilizer, the most likely immediate response to intense sexual selection on males is for increased rates of sperm production or sperm that are more efficient at finding eggs. Selection could further lead to males modifying the timing of sperm release to improve the likelihood of fertilization. These types of adaptive shifts in the timing of gamete release occur in many extant sessile marine invertebrates (chapter 21). A longer-term evolutionary response would be for males to actively locate females and ejaculate when they release eggs. This form of simultaneous spawning could eventually culminate in the evolution of copulation and internal fertilization. These selective processes could also act upon females (e.g., selection for improved ability to locate mates), but selection will be weaker if there is an excess of sperm relative to eggs.

Mate Availability and the Operational Sex Ratio

The evolution of internal fertilization raises new issues. The most obvious is how it will affect the intensity of sexual selection. The relative number of male and female gametes no longer fully captures the strength of sexual competition. Sexual selection is not tenfold stronger in a species that has 10^6 rather than 10^5 sperm per ejaculate. Once mating and internal fertilization evolve, sexual selection is usually measured as the intensity of competition for mates (and, secondarily, in terms of ongoing sperm competition if females mate multiply). Under these new conditions, does anisogamy still lead to greater competition for mates among males than among females? The data say yes, but explaining why is more difficult than it first appears. For example, sexual selection can create scenarios in which there are more adult females than males, as occurs in many mammals in which sexual selection favors increased male body size that elevates male mortality above that of females (Moore & Wilson 2002). So why do males still compete more intensely than females if gamete counts are an incorrect measure of sexual competition and there are more adult females than adult males? The short answer is that at any given time, only some adults are available as mates.

Trivers (1972) operationally defined *parental investment* (PI) as care that increases the success of the current offspring at a cost to the parent's ability to invest in future offspring. PI includes parental care as well as the cost of producing gametes as they provide the initial resources zygotes use to develop, but it excludes energy invested into competing for mates. Trivers noted that the sex with the higher PI will usually take longer to complete a breeding event and reenter the mating pool. This has been codified in terms of the duration individuals of each sex spend in *time-out* after a breeding event before they become *time-in* and return to the mating pool seeking a mate (Clutton-Brock & Parker 1992). Differences in time-out can create a marked sexual asymmetry in the availability of mates. Unless there is a counterbalancing bias in the ASR, the sex with the greater relative PI will be the limiting sex and the other, limited sex will compete for mates. This is why understanding patterns in the direction of sexual selection (which sex competes more intensely for mates) is inseparable from an understanding of why one sex, usually the female, provides relatively more parental care.

A PI asymmetry is reflected in the *operational sex ratio* (OSR), which is the instantaneous ratio of sexually active males to sexually receptive females (Emlen & Oring 1977). All else being equal between the sexes, the sex with the shorter time-out (less PI) will be more common in the mating pool and its members will compete for the rarer sex. For reasons outlined shortly, this is often the male (see below). However, it should be noted that the asymmetry in male and female time-out may decrease once there is sexual conflict over fertilization when females mate multiply (Simmons & Parker 1996). This is because polyandry is associated with the evolution of female traits that make fertilization more difficult (Birkhead et al. 1993) and sperm competition. Both factors selects for a greater number of sperm per ejaculate (chapter 22), which should increase male time-out. However, the net effect of multiple mating on male time-out is not straightforward because multiple mating also selects against male parental care, reducing this component of male PI (see below).

The OSR Is a Shortcut, So Measure the Actual Benefit of a Higher Mating Rate

A male-biased OSR together with the Fisher condition (each offspring has only two genetic parents) means that males in the mating pool have a lower mating rate than females. Even so, the assumption of a linear relationship between the OSR and sexual selection on the more common sex is rarely justified, and empirical studies have even reported negative relationships (Fitze & Le Galliard 2008). An implicit assumption is that greater competition for mates increases the strength of selection for traits that confer a mating advantage. Clearly, sexual selection relies on nonrandom variance in male mating success, not on a low average mating rate, and a male-biased OSR guarantees only the latter (Downhower et al. 1987). It is possible that changes in the OSR also increase the effect of stochastic sources of variance in mating success. The most obvious factor to take into account is the absolute density of each sex. Traits that have a causal effect on mating success and are therefore favored by sexual selection at lower densities might have no effect when the number of male-male interactions exceeds a threshold value (Kokko & Rankin 2006). For example, the ability of males to influence their own mating success by repelling rivals can break down when the OSR is highly male biased so that it becomes too costly to defend successfully a resource or a mate due to the sheer number of challengers (Mills & Reynolds 2003). A more general point is that the OSR concept was introduced as an index of the relative ease with which mates can be monopolized (Emlen & Oring 1977). However, the extent to which selection actually favors an individual who achieves a higher mating rate as a result of such short-term monopolization was not explored, and whether the OSR accurately predicts the mating skew was assumed rather than explicitly derived.

The *Bateman gradient* (BG) is a direct measure of the benefit of an elevated mating rate and therefore a more explicit predictor of the current direction of sexual selection than the OSR (see chapter 21). It is now defined as the slope of the regression of offspring production on mating rate (Arnold 1994b), and its origins lie in experiments conducted on *Drosophila melanogaster* in the 1940s by Angus Bateman (Bateman 1948). He counted how many offspring were produced by males and females that mated varying numbers of times when a small group of males and females were housed together. His best known experiments showed that offspring production by males increased linearly with mating success whereas that of females barely increased after a single mating, and that variation in mating success and offspring production was far higher for males than females (Arnold 1994b). The interpretation is straightforward. Due to the faster rate of ejaculate versus egg production, male time-out after mating is short compared to that of females. It follows that male fitness will increase more rapidly than female fitness with an elevation in mating rate. In such cases, males have a steeper BG than females and there is stronger sexual selection on males for traits that increase mating success.

In general, theory predicts a steeper BG and stronger sexual selection on the limited sex. Recent evidence that having several mates per breeding event can elevate female fitness (Jennions & Petrie 2000) means that there are species in which females also have a positive BG. Indeed, although rarely mentioned in textbooks, this was actually the case in some of Bateman's own experiments (Tang-Martinez & Ryder 2005). Even so, offspring production by females is usually more strongly dependent on the rate of egg production than the availability of mates, whereas offspring production by males is typically constrained by the availability of mates rather than a male's capacity to rear offspring. The general value of the BG in predicting the direction of sexual selection is illustrated by the switch to a steeper BG in females in sex-role-reversed species in which males provide the bulk of parental care (Jones et al. 2005). Focusing solely on sexual differences in PI or time-out to predict sexual competition is, however, a mistake (Kokko & Monaghan 2001). The ASR must also be taken into account because it, too, affects the availability of mates (e.g., Forsgren et al. 2004; Sogabe & Yanigisawa 2007).

Finally, it is worth remembering that the BG and the OSR consider different aspects of sexual competition. The BG measures how much an individual's fitness improves if it increases its mating rate. There is no consideration of how difficult this task might be. The difficulty of acquiring a mate for the average member of each sex is captured by the OSR, but this quantity, in turn, does not specify the associated fitness gain from each additional mating. In short, although they provide important clues, neither the OSR nor BG directly specify trade-offs between a current attempt to acquire a mate and other fitness-enhancing options. These trade-offs are readily apparent, however, when one investigates mate choice. It elevates offspring fitness or the number of offspring per mating, but trades off with maintaining a high mating rate. Similarly, parental care elevates offspring survival, but it again typically reduces the mating rate. We therefore now turn our attention to the evolution of mating preferences and parental care strategies.

What Is the Relationship between Sexual Competition and Mate Choice?

So far we have ignored the details of how sexual selection operates and what traits it favors (e.g., weapons or ornaments). One well-studied process is mate choice. Broadly speaking, mate choice occurs when traits create mating biases that reduce the set of potential mates (Kokko et al. 2003b). This does not necessarily involve direct rejection of a mate. For example, a preference for mating in a particular habitat can generate mate choice if only certain mates reach these locations (*indirect mate choice* sensu; Wiley & Poston 1996). There is a clear theoretical distinction between sexual selection and mate choice. The two processes are not synonymous. Mate choice generates sexual competition within the chosen sex, but it does not logically follow that competition for mates (sexual selection) must be associated with the other sex being choosy.

In many cases the evolution of mate choice means that the mating rate of the choosy sex is not maximized. If the BG is positive, fitness improves with each successive mating, which selects against the reduction of the mating rate that follows from mate choice. Choosiness will therefore more often evolve in females because they usually have a very low BG so lower gains from mate choice are sufficient to compensate for a reduction in the mating rate. If the BG is negative, then females directly benefit if they reduce their mating rate by rejecting males (a form of mate choice). It is, however, incorrect simply to assume that the sex with the lower BG will be choosy. If, for example, the female BG is lower than that of males but both are positive, both sexes still pay a mating rate cost by being choosy. Unless there are benefits to choosiness that are not captured by the BG, neither sex should be choosy. Evidence for mutual mate choice in some species (Servedio & Lande 2006) and male mate choice in other species (Wedell et al. 2002), when combined with the fact that males almost always have a positive BG, suggests that BGs fail to capture some biologically relevant factors. As stated above, the information missing from the BG (as well as the OSR) is an explicit consideration of a key trade-off: choosiness typically changes both the number and the identity of actual mates.

Consider the simple case of two rodent species whose hybrid offspring have low viability. Should males, females, or both sexes avoid mating with heterospecifics? Simply quantifying BGs does not adequately capture the situation because offspring production depends on a mate's identity. This situation requires an explicit examination of the trade-off between mating rate and the average gain per mating under different choice rules. Rejecting heterospecifics will lower the mating rate, but elevate the average number of viable offspring produced per mating. Whether a choosy individual can achieve a net benefit in the face of this trade-off will be affected by the extent to which rejecting some individuals (heterospecifics) increases the mating rate with more profitable individuals (conspecifics), as well as the relative number/fitness of offspring from each mating type, the frequency with which each type of mate is encountered, and the relative effect of hybrid and conspecific mating on mortality rates and time-out.

It is easy to see why sexual differences in mate choice might evolve given costly hybridization. If PI is low for males (e.g., sperm is cheap to produce and males do not provide parental care) so that their time-out is brief, they gain little by rejecting heterospecific females, even if hybrid matings yield fewer offspring of lower fitness. This is because a hybrid mating hardly reduces their mating rate with conspecifics. In contrast, if female PI is high (e.g., eggs are costly to produce and there is female parental care) a prolonged time-out means that there is a stronger trade-off between the gains per mating and mating rate. Female mate choice should therefore evolve more easily. It is too simplistic to claim, however, that relative PI is the sole driver of sexual differences in choosiness. The trade-off between the gains from an indiscriminate current mating and the future rate of offspring production has to be specified. If, for example, there is a hybridization asymmetry and far fewer offspring are produced when a male of species B mates with a female of species C than when a female of species B mates with a male of species C, then it is possible that species B males evolve to be choosier than species B females, even if they have a lower PI because the relative benefits from an indiscriminate mating are far smaller for males.

Hybridization is, of course, an extreme case study, but it illustrates the main principles of mate choice evolution. Choosiness does not evolve simply because individuals of the opposite sex vary in the number and/or fitness of the offspring they produce. If mating incurs no cost in terms of future reproductive success and there is a current benefit to mating, then individuals should mate with every potential partner encountered. Mate choice evolves only if mating is costly: it might elevate predation or trade off with foraging, or, perhaps most important, indiscriminate mating might reduce the mating rate with individuals that confer greater benefits per mating. Male mate choice is a weaker force than female choice because the time-out for sperm replenishment is short and, therefore, it is less likely that a male who mates will lose out on an opportunity to fertilize the eggs of a better quality female. In contrast, when females mate they often have a longer time-out for egg production. This means that females are more likely than males to lower their mating rate while waiting for better quality mates. Anisogamy is therefore a driving force of sexual divergence in mate choice.

There is one situation in which males are effectively forced to choose. Consider a case in which males who court certain females with increased vigor (elevating the likelihood of mating) cannot court other females with the same vigor (lowering the likelihood of mating): courtship feeding forms an example. Initially, selection will favor males that preferentially court the most fecund females, but there is a population-level feedback that halts the spread of such a preference. The process is analogous to that of feeding patch selection and the resultant ideal free distribution of individuals across patches that offer different rewards (chapter 11). Strong universal male preferences for fecund females are unlikely because if all males focus their mating efforts on the most fecund females, they place themselves in an increasingly competitive situation (Servedio & Lande 2006; Servedio 2007). A male with a preference for courting less fecund females (or, more generally, females who are less heavily competed over) will be favored by selection because a greater likelihood of actually mating compensates for the reduced fecundity per mating. Directional mating preferences therefore tend to be weaker in the sex that competes for mates.

The above assumes, however, that all males express the same preferences. A mating preference will evolve more readily when not mating with some individuals improves the chances of mating with other individuals. For example, in species in which males guard females prior to mating, males of low competitive ability might choose to ignore more fecund females and preferentially guard less fecund ones. They are then less likely to be displaced by more competitive males who, in turn, can afford to prefer more fecund females. If competitiveness and fecundity depend on body size, this could, under certain conditions, generate size-assortative mating (Fawcett & Johnstone 2003; Härdling & Kokko 2005).

PARENTAL CARE: LINKING ANISOGAMY AND SEXUAL SELECTION

Is Anisogamy Linked to Parental Care Because Females Invest More to Start?

Anisogamy represents a clear difference in PI between the sexes, but it is only one of several factors that determine the time-out for each sex. Trivers (1972) pointed out that the bulk of any difference in relative PI is often determined by which sex provides the most parental care. This sex will often become the limiting sex in the mating pool if an individual providing care cannot simultaneously acquire mates. (In some species, most notably nest-spawning fish, this trade-off does not occur.) The level of parental care each sex provides can readily override asymmetries in total gametic investment per offspring. In eclectus parrots, for example, the greater female cost to future reproductive output of producing eggs rather than sperm is outweighed by the subsequent provisioning of fledglings by males (Heinsohn 2008). In principle, either sex could provide the bulk of care and become the limiting sex. In most species, however, females provide more care than males (chapter 26), so the OSR is male-biased, the male BG is steeper, and sexual selection on males is stronger.

So far, we have simply stated that producing larger gametes tends to be associated with greater parental care. We have not explained why this is so. This is consistent with the traditional approach of many theoreticians studying sexual selection, whereby PI differences are taken as given. The effect of these differences for sexual selection on each sex is then explored. This practice reflects the influence of Trivers (1972), who directly ascribed greater female parental care to the initial difference in PI that defines anisogamy. Some theoreticians (e.g., Dawkins & Carlisle 1976; Queller 1997) and empiricists (e.g., Gonzalez-Voyer et al. 2008) have, however, pointed out that a plausible causal pathway was never provided. In Tanzanian cichlids, for example, changes in sexual selection have preceded evolutionary transitions between female-only, male-only, and biparental care rather than the reverse (Gonzalez-Voyer et al. 2008). More generally, recent theory shows that females are not selected to provide more parental care simply because of higher PI at the gametic stage, nor are males selected to invest more into elevating their mating/fertilization success than caring for offspring simply because an ancestrally low PI creates a male-biased OSR (Kokko & Jennions 2008).

The main argument Trivers invoked to explain why anisogamy typically leads to greater female care is summarized in a quote: "since the female already invests more than the male, breeding failure for lack of an additional investment selects more strongly against her than against the male" (Trivers 1972: 144). Unfortunately, this argument does not identify the source of selection. Past investment cannot directly determine the best decision about the future. To continue to invest simply because you have already paid a high cost is to commit the *Concorde fallacy* (named after the backers of the supersonic plane who continued to invest in its development even after being told that it would not run at a profit; Dawkins & Carlisle 1976). Of course, past investment can influence subsequent decisions if it changes residual reproductive value and thereby alters the future benefits of various courses of action (Coleman & Gross 1991). Even so, this line of reasoning has never been explicitly used to link anisogamy to greater female care. Recent attempts to do so suggest that, if anything, the greater initial investment by females into gametes makes female care less likely than male care (e.g., the effect of initial investment on body condition selects against further care; details in Kokko & Jennions 2008). The flaw in Trivers' argument is readily apparent if one considers the wider implications of the Fisher condition (box 20.2). If offspring die for lack of care, both parents lose the same number of offspring. Unless the ASR is biased (an issue we return to shortly), the proportion of an individual's lifetime breeding formed by these offspring is, on average, the same for both sexes because males and females reproduce equally often. It is incorrect to assume that males can more rapidly compensate for such a loss by remating sooner.

Does It Matter that Males Have a Higher Potential Rate of Reproduction?

Given that sexual selection is largely driven by the availability of mates, which is influenced by how long each sex spends caring for young, we need a better explanation for the general trend for female-biased parental care if we are to understand why sexual selection acts on males. Trivers (1972) had a second line of argument that invoked the trade-off between caring and mating and implied that this is of greater concern to males than females.

The *potential reproductive rate* (PRR) is the average of the maximum rate at which each individual can produce offspring, assuming unlimited mate availability (Clutton-Brock & Parker 1992; Parker & Simmons 1996). Anisogamy implies a higher PRR for males than females (assuming that producing small gametes allows males to divide the same total investment into smaller packages and invest less per mating than females). There is a widely held view that males should provide less care than females because a high PRR equates to greater mating opportunities, and thus males suffer a higher cost in terms of lost mating opportunities when they care for their current offspring instead of seeking out new mates. This sentiment is captured in the following sex-specific quote: "male-male competition will tend to operate against male parental investment in that any male investment in one female's offspring should decrease the male's chances of inseminating other females" (Trivers 1972: 144). Once we recall the Fisher condition, however, a logical flaw is apparent. The rate at which offspring are sired by males cannot differ from the rate at which they are produced by females. The PRR is a theoretical term that assumes conditions (unlimited mate availability) that are never fulfilled in nature. Selection acts on actual fertilization events. Any argument that uses the sexual difference in PRR as the sole reason to argue that males gain a greater fitness payoff by deserting offspring is simply wrong (Queller 1997; Kokko & Jennions 2003).

From an OSR perspective, male desertion of dependent offspring is a conundrum. Less male care leads to a male-biased OSR, which increases competition for mates. It is easy to see why males must then invest in competitive, sexually selected traits to succeed. It is less obvious why a male-biased OSR does not simultaneously select for a male to delay his return to the mating pool and stay to care for his current offspring. When the expected fitness gains from one activity (competition with the payoff of mating) decrease and those from another activity (caring and elevating offspring fitness) stay constant, one would expect investment to shift toward the latter activity. This is precisely what happens when an increasingly male-biased OSR makes success in the mating pool harder for males (figure 20.1). Surprisingly, general sexual selection theory rarely follows this line of argument, although it is sometimes invoked by empiricists to explain unusual breeding systems (e.g., Segoli et al. 2006).

Recent formal mathematical models have confirmed that the preceding argument is sound. When the only difference between the sexes is anisogamy, and there is a trade-off between caring and mating, the outcome is egalitarian parental care (the full model is presented in Kokko & Jennions 2008). Why? If the limiting sex provides more care, the OSR is biased toward the opposite sex, and there is now selection on members of the limited sex to care for longer instead of attempting to compete in a zero-sum game when the total benefits are smaller (fewer mates available per unit time). The model shows that whichever sex currently provides less PI is selected to provide more, until the system reaches an evolutionary stable state when the OSR is even. In short, when we derive care patterns from first principles, the Fisher condition predicts that mating systems evolve toward egalitarian parental care and an even OSR. The logic of the argument is analogous to that used by Fisher (1930) to predict an even primary sex ratio due to frequency-dependent selection (box 20.2).

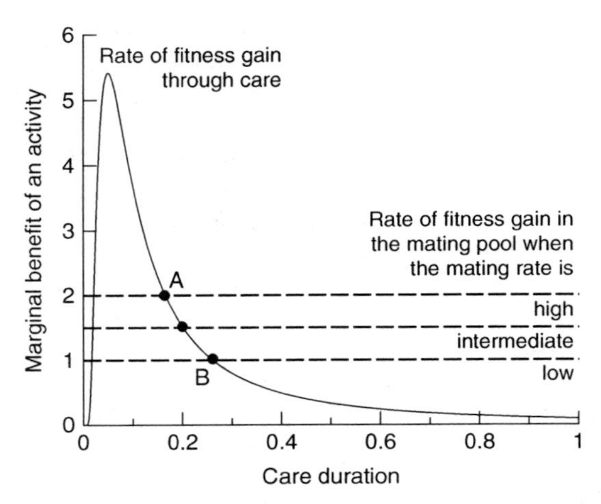

FIGURE **20.1** Frequency-dependent selection toward egalitarian care. Offspring survival depends on the care they receive: too little care hardly brings about a benefit, whereas extremely prolonged care likewise yields only a small additional (i.e., marginal) benefit. The marginal benefit therefore peaks when offspring have been given some care (here at a duration of 0.1 units), but additional care still greatly improves their survival chances (until the absolute increase in survival becomes minimal after 0.8 units of care). Parents should not desert their offspring at the peak but stay until the marginal benefit of further care falls below that which the parent can achieve after deserting its young (here determined by the time taken to remate after deserting). If two parents (a male and a female) give simultaneous care then, all else being equal, parent B, who will find it more difficult to acquire a mate (low mating rate) should desert later. Because a biased OSR lowers the mating rates for the sex that deserts earlier, there is frequency-dependent selection: whichever sex cares for a shorter time is selected to prolong its care, and vice versa. This is why additional factors must be invoked to explain sexual divergence in care. First, sexual selection can elevate the horizontal mating gain line for mated individuals. Second, low parentage within a brood can lower the marginal benefit curve for current care. Third, an ASR bias can shift the horizontal lines upward or downward.

So Why Do Females Care More Than Males? The Return of Sexual Selection

If an ancestral difference in PI due to anisogamy does not automatically become exaggerated but tends to diminish once care evolves, why is egalitarian parental care rare? A satisfactory explanation for sex roles must invoke factors other than the ancestral gametic difference in PI. In nature there are at least three additional forces that affect the outcome of the simple anisogamy scenario (Kokko & Jennions 2008) and counter frequency-dependent selection for egalitarian care. Two modifications, highlighted by Queller (1997), are consistent with patterns in nature and lead to the prediction that females are predisposed to provide more care. The third modification—relaxing the assumption that the ASR is even—produces especially interesting predictions. It is relevant because there is good evidence that sexual selection and caring create sex-specific mortality rates (Liker & Székely 2005). The effect of the ASR on sexual selection has been surprisingly poorly investigated, although Trivers (2002, p. 61) has stated that his 1972 ideas relating sexual selection to mortality rates are actually more valuable than his introduction of the concept of PI.

Sexual Selection on Males Reduces Male Parental Care

Males are usually the more common sex in the OSR, and the preceding statements about fewer mating opportunities apply to the average male. There is, however, usually nonrandom variance in male mating success so that sexual selection exists. It might appear obvious that the subset of males with a mating advantage under sexual selection should be less inclined to participate in caring if it conflicts with acquiring mates. It is important, though, to ensure that we are not making a mistake by focusing on a few successful males and ignoring the effects of selection on the rest (chapter 25). The tendency to do just this is evident in scientific jargon itself: for example, the term *polygyny* literally means "males mate with multiple females," but high variance in mating success in most polygynous systems means that most males have extremely low or even zero mating success! If most males have little chance of mating quickly once they return to the mating pool, these less attractive or competitive males might do better by caring more for their current offspring, instead of competing in a mating pool with a male-biased OSR.

Deriving predictions about parental care requires caution for at least three reasons. First, when we consider all males, it is no longer clear whether stronger sexual selection on males will favor greater sexual divergence in sex roles (i.e., males gain more by investing in sexually selected traits and females by spending more time caring for offspring). Second, it is a mistake to focus on variation in a single component of fitness within a sex without considering how it might correlate with other traits that increase fitness. If a male-biased OSR creates stronger sexual selection on males than females, this does not immediately mean that males should shy away from care: the relative benefit of caring might simultaneously become more important for those males whose prospects of succeeding in the mating pool are worse than those of females. Third, if care can be provided by either parent, when one sex provides more, the other can do less to achieve the same level of offspring survival. This is a postfertilization extension of the sexual conflict inherent in the difference in resources transferred to offspring because of anisogamy. If some males care more, then females can care less, which will feed back into the OSR. Offspring fitness depends on the parental decisions of both sexes, so we need to be precise and quantify what differences between the sexes predict female-biased care.

Having raised these complications, we note that there is some good news: models that take into account sexual selection can counter frequency-dependent selection toward egalitarian care and generate the most commonly observed sex roles. The argument is rather subtle, though. If mating success is nonrandom, then deserters (individuals who must have mated) do not have the same expected success as the average member of their sex. Every male who reenters the mating pool has already successfully induced a female to mate, despite having had to compete for mates because of a male-biased OSR. Unless obtaining a mate is a purely chance event, these males must possess sexually selected traits that confer an advantage during male-male competition or female choice. In other words, variance in mating success matters because it elevates the average expected mating success of those individuals who are most often in the position to decide whether to care for or desert young.

Males with low expected mating success have little influence on the relative amount of care provided by males (and thus females) when averaged across all mating events. For example, elephant seals live in societies with an extreme skew in mating success toward large males. A male cannot make any caring decisions when he is young and too small to sire offspring. By the time he does obtain a mate, however, this very fact indicates that he is now more competitive than the majority of other males. In effect, nonrandom variance in mating success provides a male with information about his expected future success. This is why a male who mates is unlikely to stay with the first female with whom he pairs if there is strong sexual selection on males (Queller 1997; Kokko & Jennions 2008). It is hard to succeed, but if you do, it is a sign that you will succeed again. The effective OSR for males whose decisions about parental care are exposed to selection is actually less male-biased than it appears. In a sense, males invest in competing rather than caring despite the male-biased OSR, not because of it.

The above process does not require phenotypic plasticity. It is based on the simple fact that selection only acts upon parenting traits of males who mate (and thus tend to have high mating rates). Even so, sexual selection might also lead to males showing phenotypic plasticity in how much care they provide (see Trivers 1972). If less successful males have the good fortune to acquire a mate, they should be more inclined to provide parental care than the average mated male because their future mating prospects are low. Selection for plasticity will, however, be weak if variance in mating success is high, because it is then rare for unsuccessful males to mate. Conversely, if they mate fairly often, then sexual selection must be weaker so the gains from deserting sooner are relatively larger. Plasticity also requires that males obtain reasonably accurate information about their future prospects. Behavioral details of male-female interactions might allow a male to self-assess his attractiveness when female choice is important, but, more generally, information can be gained from relatively crude measures such as the time taken to mate successfully for the first time.

Sperm Competition: More Males Than Females per Breeding Event Lowers the Benefits of Male Care

So far, we have assumed that the absolute gains from caring for the current set of offspring are the same for both sexes (e.g., in figure 20.1 the marginal fitness gains from caring are identical for males and females). This is a valid assumption given evolution from an ancestral state in which the only factor distinguishing the sexes is anisogamy. Of course, once the time devoted to care diverges between the sexes, this will select more strongly for efficient parental care by the sex that spends more time caring. This could result in sex-specific traits (e.g., female lactation in mammals; but see Kunz & Hosken 2009) that create phylogenetic inertia biasing care toward one sex in a given taxa. There are, however, numerous phylogenetic transitions in care provisioning, so shifts in relative levels of care can still occur (e.g., Reynolds et al. 2002). More important, the wider pattern despite such transitions is still a greater level of care being provided by females. This raises the possibility that the gains from current offspring care are systematically higher for female than male parents.

Why should one sex generally care more? One explanation lies with the evolution of polyandry and/or sexual selection promoting alternate male mating tactics such as sneak spawning that result in sperm competition. If there are more potential fathers than potential mothers involved in a given breeding event, then, regardless of whether fertilization is internal or external, average relatedness to young is lower for males than females. Most breeding events involve more males than females due to a male-biased OSR, sexual selection on males to evolve alternate ways of obtaining fertilization success, or multiple mating by females. Females are consequently more closely related than males to the offspring produced, and lower average relatedness must reduce the payoff from parental care (chapter 26).

This issue of how variation in paternity affects care decisions has been investigated most thoroughly in the context of male birds provisioning their young (Sheldon 2002). Simple models show that an individual does not improve his fitness by providing less care to the current brood if his paternity is, on average, always reduced by the same amount: future and current reproductive success are similarly discounted by relatedness to offspring, and the optimal solution to the life history trade-off between caring and his own survival is unchanged (Westneat & Sherman 1993). In general, there are no simple predictions about how consistent differences among males in their ability to gain paternity will be associated with how much care they provide (Sheldon 2002). It is easy to create models in which males who gain high paternity provide less care than those who gain low paternity (Houston & McNamara 2002). This is because how males respond to a lower payoff due to reduced paternity depends on what else they can do to elevate fitness aside from care for the current set of young.

When comparing average parental effort between the sexes, the Fisher condition must be taken into account. It leads to the unequivocal conclusion that, all else being equal between the sexes, the sex with lower relatedness to offspring will always care less (Queller 1997). To understand why, consider a socially monogamous bird with high levels of extra-pair paternity. So every male has low paternity with his social mate. The Fisher condition requires that males gain extra-pair young elsewhere: every offspring must have a father (Houston & McNamara 2002). Only some of the offspring a male cares for increase his fitness. So whenever there is a trade-off between investment in caring for young and increasing success at gaining extra-pair paternity with other females, we expect males to shift some resources toward the latter. In contrast, unless females dump eggs, their investment in caring returns benefits from all the young in a nest. It follows that males will care less than females if parentage per breeding event is, on average, lower because they pay the cost of care, they receive a smaller benefit, and there are fertilization opportunities available elsewhere. This example again highlights how important it is to count all the offspring of all males (see Fromhage et al. 2007).

Sexual Selection Produces Life History Differences That Bias Adult Sex Ratios

The Fisher condition precludes sexual differences in the average rate of reproduction unless the adult sex ratio differs from unity. If the ASR is biased, then "if one sex is consistently rarer [...] it will be less likely to be parental" (Queller 1997, p.1555). This is because the Fisher condition creates a higher mating rate for members of the rarer sex, who therefore suffer a greater mating opportunity cost when they care. It is worth remembering that the ASR and the OSR are not necessarily correlated. For example, in many mammals the OSR is strongly male-biased because males have a low relative PI, whereas in birds the OSR is usually only weakly male-biased because there is biparental care. The ASR patterns are very different. In mammals, there are usually many more adult females than males because of high male mortality due to male-male competition. In contrast, in birds females tend to have higher adult mortalities than males so the ASR is male-biased (Liker & Székely 2005; Donald 2007).

We highlight the role of the ASR because it has an independent effect on desertion (for details, see Kokko & Jennions 2008). The relative payoffs from caring and deserting more strongly favor desertion by individuals of the sex with fewer adults: members of this sex must reproduce more often, so the average rate of fitness return upon reentering the mating pool is higher (figure 20.1). The source of a bias in the ASR has important effects on sexual selection (figure 20.2). If the cost of breeding is the main source of adult mortality, then whichever sex provides more care will be rarer in the ASR. Members of the opposite sex then experience greater difficulty acquiring mates. But does selection then favor increased investment in sexually selected traits to succeed despite greater competition, or natural selection for greater care for offspring? An unfavorable bias in the ASR limits the absolute gains that come from competing, so the latter route to enhancing fitness is more profitable. In this case, frequency dependence arising from the Fisher condition counteracts sexual selection and sex role divergence is restrained.

The reverse situation occurs when a bias in the ASR arises due to high mortality attributable to the expression of sexually selected traits. Such traits often increase mortality during juvenile development or when competing for mates. In Kudu antelope, for example, large male body size is favored by sexual selection. This has resulted in strong selection for high male growth rates and delayed sexual maturation, with high male mortality and a heavily

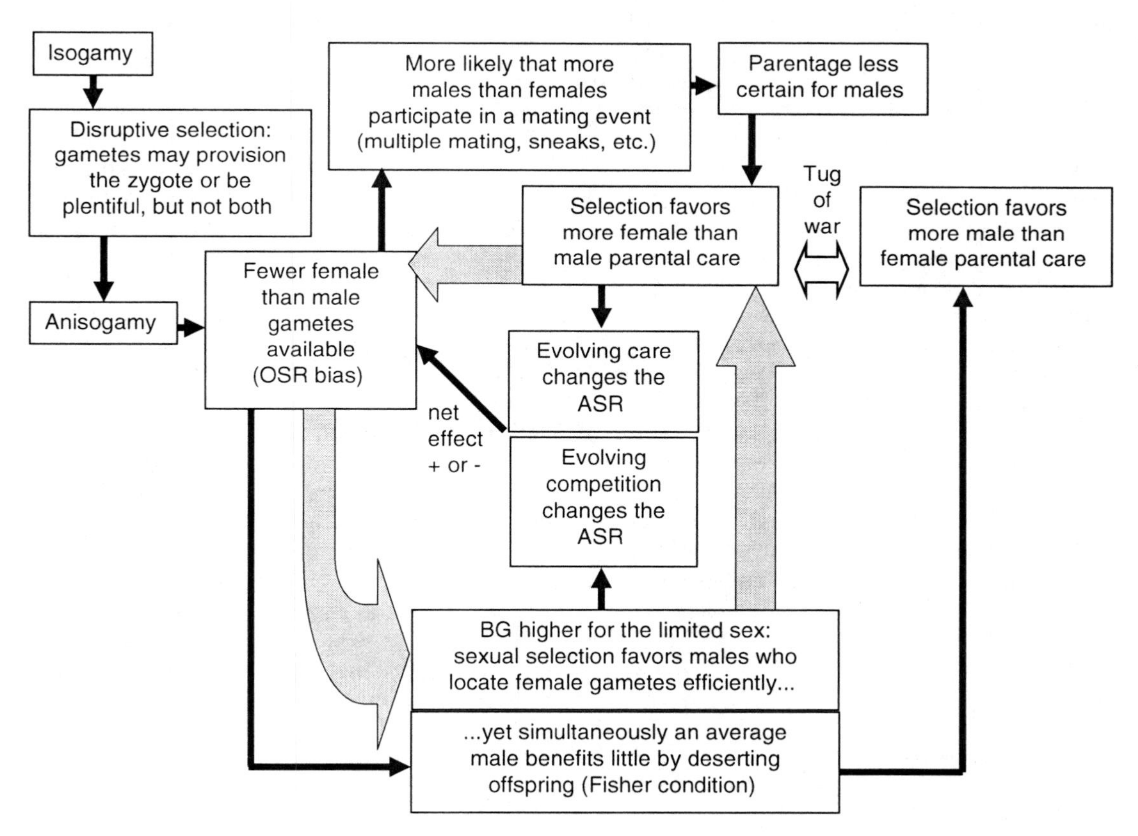

FIGURE **20.2** A graphical summary of the chapter. Broad arrows represent traditional accounts of sexual selection. They form a positive feedback loop. If they are the only forces at play, the sexes will always evolve to be maximally different. The other arrows represent forces that can be decisive in making predictions about where the positive feedback is halted. These include feedback from the ASR, the prevalence of multiple mating, and the tug-of-war between frequency-dependent selection for more care in the sex that cares less and selection for less care in favor of competitive traits. (BG = Bateman's gradient; ASR = Adult sex ratio; OSR = Operational sex ratio.)

female-biased ASR (Owen-Smith 1993). In mammals in general, sexual size dimorphism is widely attributed to strong sexual selection on males to defeat rivals during direct physical contests. Comparative studies show that as sexual dimorphism increases, males suffer greater rates of parasitism than females, which could be related to lower investment into immune defense to fuel growth. The net result is that male mammals have a shorter life span than females (Moore & Wilson 2002). When the deserting sex is the rare sex in the ASR, it must enjoy higher reproductive opportunities than the caring sex. This generates self-reinforcing selection to provide even less care and desert sooner.

To illustrate these two scenarios, compare the fate of a male in a population of finches with a male-biased ASR so that 30% of males fail to acquire a mate and that of a male antelope that faces ferocious sexual competition for mates due to a strongly male-biased OSR whereas the ASR is female-biased. Which male is more likely to evolve parental traits? Given the Fisher condition and a shortage of unpaired females, a male finch cannot be too optimistic about fertilizing offspring of many (or any) females even if he deserted his current young and increased his mating effort elsewhere. In contrast, the average male antelope in a female-biased population must, because of the Fisher condition, sire more offspring than are born to the average female. The average male therefore gains several reproductive opportunities by deserting. If there is strong sexual selection, then the future mating prospects for the average mated male are even better. This is why we do not expect male antelope to care, despite the heavily male-biased OSR that diminishes the odds that a male will succeed in the mating pool. The take-home message is that examining the OSR in isolation can mislead as to whether there is selection for the limited sex to care more. Consistent with the importance of the ASR for the evolution of sex roles, polygynous birds (i.e., birds with mammal-like breeding systems in which female compete) tend to have a more mammal-like ASR than bird species that form socially monogamous pairs (Liker & Székely 2005; Donald 2007).

Summary: The Path between Sexual Selection and Parental Care Is a Two-Way Street

The two main components of PI generate stronger sexual selection on males and make eggs less readily available than sperm: anisogamy and female-biased parental care. PI is, however, itself affected by sexual selection so that the evolutionary process involves a feedback loop (figure 20.2). Initially, anisogamy creates a numerical bias toward sperm and, even in the absence of parental care, makes the relative parental investment per breeding event greater for females. This means that females take longer to reenter the mating pool so that the OSR becomes male-biased. A male-biased OSR selects for increased male care (due to the Fisher condition), but if sexual selection creates a subset of males who are sufficiently successful at gaining matings in the face of competition, the net effect is actually reduced male parental care. If parental care is required, it is then more likely to be provided by females. This, in turn, exacerbates the difference in the speed with which females and males enter the mating pool because the relative PI of females is further increased. This makes the OSR even more male-biased, which generates still stronger sexual selection on males creating a positive feedback loop (figure 20.2). However, an increasingly male-biased OSR also means that males take longer to mate so they experience selection to become more parental instead of attempting to succeed in the increasingly difficult task of acquiring another mate. The evolving tug-of-war between these opposing forces means that the feedback loop can stop at very different evolutionary end points. These will partly depend on how strong sexual selection is as the OSR changes and on certainty of parentage (and it is worth noting that rates of polyandry or group spawning often shift with changes in the OSR). In addition, it will depend on the evolving mortality rates of the two sexes due to sex-specific investment in sexually selected traits and parental traits as well as sex-independent differences in the mortality associated with caring and competing that influence the ASR and the availability of mates (Kokko & Jennions 2008).

So Long, and Thanks for All the Fish

A key assumption in our account is that parental care trades off with mating rate. This is not always the case. Despite frequent loss of paternity by nest-building males to sneaky males in fish with external fertilization, male-only care is actually more common than female-only care (Reynolds et al. 2002). Many male fish can rapidly switch back and forth between guarding eggs and courting females. There is even evidence that parental care is sexually selected because females prefer males who are guarding eggs, due to either mate choice copying (Goulet & Goulet 2006) or a direct preference for caring males (for insect examples, see Tallamy 2000). If true, caring males pay no mating opportunity costs and might even elevate their mating rate. This effectively blurs the distinction between time-in and time-out. We therefore conclude by thanking fish for reminding us that although general theoretical models are valuable, we should never lose sight of biological contingencies. When empiricists discover that the predictions of theoretical models are not upheld, rather than rejecting modeling out of hand, a more useful approach is to look more closely at the natural history of their study organism. Working out how animals violate current modeling assumptions is often a major route to empirical and theoretical progress.

FUTURE DIRECTIONS

Most empirical work on sexual selection, with the possible exception of comparative analyses and attempts to compare Bateman gradients, has one of two goals. One is to measure the current level of sexual selection on traits in the competitive sex and understand how they function. Another goal is to quantify selection for choosiness, which requires measurement of the cost of rejecting some mates, and the magnitude and sources of variation in the benefits of mating with different individuals. Here we have taken a different path and focused on why sexual selection differs between the sexes. We chose this route because the general patterns in nature are so striking (females care, males compete, females choose, males fight) that it is easy to treat them as inevitable. In reality there are still exciting research avenues to be traveled. The unwarranted transition between a higher potential reproductive rate and a claim of greater lost mating opportunity costs when caring illustrates the tendency to jump to conclusions instead of recognizing that sexual divergence still poses theoretical and empirical questions. For this reason we have focused on the basic logic of sexual divergence. The next six chapters provide the subsequent finer-scale details of topics such as how females choose males, how male traits reliably signal offspring fitness, or what selects for alternative mating tactics in males.

In our view, the population-level consequences of the Fisher condition are still underappreciated. This has recently been emphasized for specific questions about parental care (Houston et al. 2005), but the effect of this simple fact of life on the evolution of multiple mating, adaptive sperm allocation, and other traits subject to sexual selection remains poorly explored. Too much theory, and resultant empirical work, is based on the premise that the PRR or OSR predict mating behavior. Moreover, even when models allow levels of care and competition to evolve from first principles (rather than, say, simply stating that the OSR determines which sex will compete more strongly), the evolution of the associated traits is considered in isolation (e.g., Kokko & Monaghan 2001). In reality traits coevolve due to trade-offs that should be explicitly studied. Placing numerical values on the OSR, BG, or opportunity for sexual selection does not capture the nature of their trade-offs even though the diversity of sexually selected traits is largely attributable to the vast array of potential trade-offs. For example, producing more sperm can allow a male to outcompete rivals, but a large sperm expenditure can damage a female (chapter 23) or act in the completely opposite way and provision a female's young so that the distinction between mating and parental investment becomes ambiguous.

The insightful arguments of Queller (1997) about the evolution of parental care have recently been modeled by Kokko and Jennions (2008), but the effect of the Fisher condition on sexual selection (e.g., Bateman gradients and mate choice) is still unexplored. In particular, shifts in adult sex ratios might change the net effect of the tug-of-war between sexual selection, which makes successful males care less, and the Fisher condition, which makes males care more when they are the limited sex. Quantifying relationships between the adult sex ratio, mortality rates for different activities, the amount of time each sex spends caring and competing, and variance in male mating success is therefore likely to become a fruitful area for empirical study. In sum, we still do not fully understand the sequence of events that link anisogamy to parental investment, sex roles, and different forces of sexual selection. Given the jaundiced view that we already understand sexual selection, it is exciting to realize that basic questions remain unanswered.

SUGGESTIONS FOR FURTHER READING

A highly readable account of the historic development of sexual selection theory is *The Ant and the Peacock* by Cronin (1991). A comprehensive overview of recent sexual selection theory is that of Andersson (1994), and short reviews of many topics related to sex ratios are contained in Hardy (2002). Fairbairn et al. (2007) explore the evolution of sexual size dimorphism, which offers general insights into sexual divergence. The papers of Trivers (1972) and Emlen and Oring (1977) are still classic presentations of basic ideas about the origins of sexual differences in the intensity of sexual selection (and Trivers 2002 contains amusing and rewarding introductions to his landmark papers). Clutton-Brock and Vincent (1991) introduced the concept of potential reproductive rates, and the limitations of this approach were then highlighted by Queller (1997), who drew attention to the Fisher condition (Kokko and Jennions [2008] formally model and extend Queller's ideas). Shuster and Wade (2003) provide an exhaustive overview of the importance of variance in mating success for the evolution of mating systems and male mating tactics. Finally, the question of which sex will express stronger mate choice has been tackled in a series of papers, each of which emphasize different aspects of the process (Johnstone et al. 1996; Kokko & Monaghan 2001; Fawcett & Johnstone 2003; Servedio & Lande 2006, Servedio 2007).

Andersson MB (1994) Sexual Selection. Princeton Univ Press, Princeton, NJ.
Clutton-Brock TH & Vincent AJ (1991) Sexual selection and the potential reproductive rates of males and females. Nature 351: 58–60.
Cronin H (1991). The Ant and the Peacock. Cambridge University Press, Cambridge, UK.
Emlen ST & Oring LW (1977) Ecology, sexual selection, and the evolution of mating systems. Science 197: 215–223.
Fairbairn DJ, Blanckenhorn WU, & Székely T (eds) (2007). Sex, Size and Gender Roles: Evolutionary Studies of Sexual size Dimorphism. Oxford Univ Press, Oxford.
Fawcett TW & Johnstone RA (2003) Male choice in the face of costly competition. Behav Ecol 14: 771–779.
Hardy ICW (ed) (2002) Sex Ratios: Concepts and Research Methods. Cambridge Univ Press, Cambridge, UK.
Johnstone RA, Reynolds JD, & Deutsch JC (1996) Mutual mate choice and sex differences in choosiness. Evolution 50: 1382–1391.
Kokko H & Jennions MD (2008) Parental investment, sexual selection and sex ratios. J Evol Biol 21: 919–948.
Kokko H & Monaghan P (2001) Predicting the direction of sexual selection. Ecol Lett 4: 159–165.
Queller DC (1997) Why do females care more than males? Proc R Soc Lond B 264: 1555–1557.
Servedio MR (2007). Male versus female mate choice: sexual selection and the evolution of species recognition via reinforcement. Evolution 61: 2772–2789.
Servedio MR & Lande R (2006) Population genetic models of male and mutual mate choice. Evolution 60: 674–685.
Shuster SM & Wade MJ (2003) Mating Systems and Strategies. Princeton Univ Press, Princeton, NJ.
Trivers RL (1972) Parental investment and sexual selection. Pp. 136–179 in Campbell B (ed) Sexual selection and the descent of man 1871–1971. Aldine-Atherton, Chicago, IL.
Trivers RL (2002) Natural Selection and Social Theory: Selected Papers of Robert L. Trivers. Oxford Univ Press, New York.

21

Sexual Selection in External Fertilizers

DON R. LEVITAN

Sexual selection is typically studied in organisms that have two sexes, fertilize eggs internally, and exhibit considerable dimorphism in morphology or behavior. As described in the previous chapter, the fundamental forces influencing selection on the two sexes extend beyond these characteristics. The appearance of anisogamy, copulation, and sexual dimorphism are evolutionary transitions that emerge from patterns of sexual selection. Insight into these selective forces can be found in theory, but the clearest understanding can be gained only by comparative analysis in and experiments on organisms that potentially share traits with the ancestral taxa that underwent these transitions. In this chapter, I explore patterns of sexual selection in external fertilizers. My premise is that this group of organisms is particularly informative about the forces influencing the evolutionary transition between isogamy and anisogamy, between external and internal fertilization, and between taxa that show or do not show sexual dimorphism. In particular, they can shed light on whether these evolutionary transitions set the stage for, or are a consequence of, sexual selection.

These organisms are compelling models for addressing the fundamentals of sexual selection because they have (1) gametes that show large variation in sizes, traits, and the degree of anisogamy, (2) little or no parental care, thereby avoiding many thorny issues that phenomenon creates (e.g., chapter 20), (3) patterns of mating that vary from extreme polygamy to complete monogamy, and (4) attributes that allow easy manipulation of gamete interactions and spawning behaviors for experimental investigation. So although we cannot witness the major evolutionary transitions that seem so fundamental to sexual selection, we can test hypotheses about the selective forces that may have caused them.

The Nature of Sexual Selection in External Fertilizing Species

Darwin suggested that "primitive" marine invertebrates that reproduce by external fertilization lack sexual dimorphism because they do not experience sexual selection (Darwin 1871). Sexual dimorphism is often, but not always, generated by sexual selection (Arnold 1994c). But what about the reciprocal: can sexual selection be present in taxa without the signature of sexual dimorphism? Is internal fertilization via copulation a prerequisite for sexual selection, or is the ancestral transition from external to internal fertilization perhaps a consequence of sexual selection? These questions pertain not only to studies of external fertilizers but have relevance for studies of plants, which do not copulate, and simultaneously hermaphroditic species, which cannot exhibit sexual dimorphism.

In part the answer to these questions will depend on how one defines both sexual selection and sexual dimorphism. Definitions of sexual selection involve some aspect of the fitness consequences of competition for reproductive success or by identifying the signatures of this selection such as sexual dimorphism or secondary sexual characteristics. Arnold (1994c) reviews these arguments and concludes with a simple definition: "Sexual selection is selection that arises from differences in mating success (number of mates that bear or sire progeny over some standardized time interval)" (9). This definition implicitly assumes the sexes copulate in order to combine gametes and can be problematic when mating success does not reflect success in parentage in polygamous systems. Levitan (1998a) offered an alternative definition: "selection that arises because of intra-sexual differences in the proportion of an individual's gametes that fuse to become zygotes"(179). This has a flavor similar to Jennions and Kokko's (chapter 20), "sexual selection favors traits that improve the likelihood of fertilization when there is a limited pool of opposite sex gametes." (p. 345, this volume) There are interesting nuances to all of these definitions that will be examined in this chapter, but the unifying theme is that all three are based on patterns of selection rather than the results of selection and are therefore independent of sexual dimorphism. The latter two in particular can be applied to the problem of why the typical signatures of sexual selection (anisogamy, dimorphism, behavioral differences) sometimes, but not always, appear.

Sexual dimorphism, narrowly defined to morphological variation among adults, is rare among externally fertilizing marine invertebrates (Levitan 1998a). However, considering only this form of sexual differentiation restricts the possible targets of sexual selection. Trait differences between sexes can also include dimorphism in gamete traits, physiological differences, and behavioral variation. External fertilizers show the full range of sex differences, from algal species that lack both adult and gametic differences (isogamy), to the many invertebrate and fish taxa that have gametes fully differentiated into eggs and sperm but lack adult morphological differentiation, to many fish and a few rare invertebrate species that show morphological differences among adults. Sexual differences in reproductive behavior can also be caused by sexual selection. Males and females often exhibit differences in spawning behavior even if they show no morphological differences.

The ancestral condition of plants and animals appears to be isogamy and external fertilization (Wray 1995). Jennions and Kokko (box 20.1) provide a concise account of theoretical explanations for how gamete competition and disruptive selection can result in anisogamy among external fertilizers. Empirical work using marine invertebrates has confirmed some of the critical assumptions of these models (e.g., Levitan 2000a for the nonlinearity of zygote size and fitness) and highlighted how adaptations, likely driven by sexual selection, may have further contributed to how gametes have evolved. For instance, increases in egg size may have evolved not only for zygote provisioning (chapter 20) but also to increase the likelihood of fertilization (Levitan 1993). Sperm traits can also influence the likelihood of fertilization (Levitan 2000b; Neff et al. 2003; Crean & Marshall 2008), and this not only influences gamete allocation in males but through fertilization rates can indirectly influence allocation in females (Levitan 1993). This result becomes apparent only when one considers that sperm availability can be variable and sometimes limiting. Variation in sperm availability is an important consideration often overlooked by biologists working with species with internal fertilization. The consequence of this variation is that both males and females can simultaneously be competing for fertilizations and that the nature of sexual selection and the traits under selection can be very different under conditions of sperm limitation (too few sperm), sperm saturation (just enough sperm), and sperm overabundance (sperm become destructive).

Patterns of Gamete Competition

Once gametes are released into the environment, the patterns of local sperm and egg availability can range widely and largely determine the intensity and nature of sexual selection (figure 21.1). If males and females are sparse and a female's eggs encounter only the sperm from a single male, then males compete, but the competition would be indirect. That is, males that produce a higher quantity or quality of sperm that are better at finding or fertilizing a greater number of eggs would be selected over other males (figure 21.1a) who leave some eggs unfertilized. As male density increases and eggs encounter sperm from multiple males, sperm competition becomes more direct (sperm from multiple males compete for the same egg), as late arriving or poor fertilizing sperm miss the opportunity to fertilize an egg because another male's sperm has already fertilized it (figure 21.1b). Although indirect and direct competition among males might both select for sperm better adapted to fertilizing eggs, these different conditions might select for different sperm traits. When gametes are broadcast into the environment, water or air currents largely dictate when and where gametes interact. Sperm able to fertilize over long time intervals may be favored. At higher male and sperm densities, when gamete interactions occur quickly and competition is fierce, sperm swimming speed may mediate competitive interactions (Levitan 2000b).

When sperm are at even higher densities, multiple sperm may attach and fuse with an egg before the egg can prevent them from doing so. In many invertebrate taxa, such polyspermy results in developmental failure (Gould & Stephano 2003). If all sperm surrounding an egg are from one male (figure 21.1c), this cost of developmental failure is shared by both sexes. This cost would be equally shared if mating pairs are completely monogamous. However, if males simultaneously fertilize the eggs of several females, then the cost of polyspermy might be asymmetrical because oversaturating a closer female or a female producing highly receptive eggs can be compensated by increasing the fertilization rate of a more distant female or a female producing less receptive eggs. The cost to polyspermy becomes increasingly asymmetric when sperm from multiple males approach a single egg (figure 21.1d). Here the cost is always greater in females. For females, excess sperm leading to developmental failure is a cost and eggs that can avoid excess collisions reduce the efficiency of sperm fusion or increase the speed or efficiency of the polyspermy block would have increased reproductive success. However, for males, there is never an advantage to being the second sperm to find and fuse with an egg, so sperm should be selected for rapid rates of collision and fusion. Although this might result in a reduction in the total percent of eggs fertilized, it would also result in a higher paternity share if multiple males are competing. Thus at high sperm concentrations, males are selected for a rapid fertilization rate, whereas females are selected for a slow fertilization rate. These different competing forces are a form of sexual conflict (Franke et al. 2002; Levitan & Ferrell 2006; chapter 23, this volume).

At the other end of the spectrum of sperm availability, if sperm are too sparse, many eggs may go unfertilized (figure 21.1e). This may favor eggs that are able to either attract more sperm or increase the likelihood of a fusion event given a collision (reviewed in Levitan 2006). These conditions may

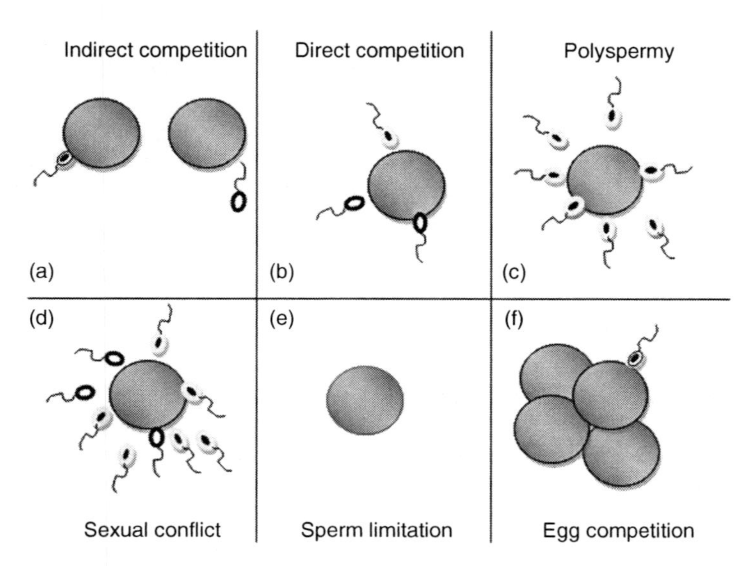

FIGURE 21.1 The gradient of gamete encounter and competition that produces different patterns of selection. All of these encounter types are possible in taxa that release gametes into the ocean for external fertilization, but the frequency distribution of these types is largely unknown. Different symbol types of sperm refer to competition within and among males. Details in text.

also select for adult spawning behaviors that favor gamete collisions such as aggregative behavior and synchronous spawning (Levitan 1998a). Under sperm-limited conditions, sexual conflict is reduced because both males and females benefit from traits that enhance fertilization success (Levitan 2004).

Not all gametes are broadcast into air or water. Eggs can often remain on or attached to the female for some time before being lifted into the water column by water flow (Yund & Meidel 2003). During this time eggs can be piled several layers thick and one egg may obstruct other eggs from sperm collisions. Eggs thus could compete for fertilizations (figure 21.1f). Evidence for egg competition has been mixed. Egg density does alter sperm-egg ratios and influences the fraction of eggs fertilized, but this effect is generally weaker than variation in sperm density (Levitan et al. 1991). Laboratory studies of within-clutch egg competition has shown that larger eggs are preferentially fertilized under sperm-limited conditions (Levitan 1996), and there is evidence from laboratory flume experiments that presence of eggs upstream can diminish fertilization of eggs released downstream (Marshall & Bolton 2007).

These different scenarios illustrated in figure 21.1 describe the spectrum of possibilities of gamete interactions in external fertilizing taxa. All of these situations can occur in nature, and likely do; the key to understanding the nature and intensity of sexual selection in a particular taxon or population is determining the frequency distribution of these different possibilities.

Variance in Fertilization Success

A critical component to understanding selective pressures on reproductive success, mating success, and the patterns of sexual selection is to estimate the variance in the proportion of eggs fertilized. Although this is likely correlated with offspring production, this measure is distinct from variation in egg production, which is influenced by energy acquisition and adult vitality, factors independent of sexual selection. Variation in fertilization success depends on gamete encounter probability and success, measures at the heart of gamete competition. For externally fertilizing anisogamous species, sperm greatly outnumber eggs, and on average competition among males will exceed competition among females for fertilizations. However, because gametes are released into the environment, each may experience different conditions. For example, some eggs may be surrounded by many sperm (from one or many males), whereas other eggs from that same spawning may drift into areas of low sperm availability (figure 21.1). This heterogeneity of gamete concentrations can also result in egg competition as small wisps of sperm pass over egg masses. Thus both males and females can simultaneously be under selection for enhanced gametic competitive ability.

Variance in fertilization success for females has been measured over the past 3 decades and provides some insight into the level of sperm limitation and polyspermy under natural conditions. More recently data on variance in male fertilization success has been collected, but only on a handful of taxa. Additional paternity studies are needed, not only to determine how different species and environments influence male success but also to examine the relative intensity of gamete competition among males and females (e.g., figure 21.2).

Estimates of female fertilization success have been gathered under three conditions: field experiments, model-based gamete kinetics and water flow, and natural observations. Field experiments at small spatial scales have been combined with models to predict larger scale events, and these can be compared to observational data from natural spawning events. Although there are many interesting nuances influencing variation in female reproductive success, the basic pattern is simple. Sperm limitation generally occurs at low population densities (less than approximately 1 male/m^2), and fertilization success becomes vanishing small when females are greater than 5 meters from the nearest spawning male (Pennington 1985). There is variation in the degree of sperm limitation experienced by different species for a given density or degree of clumping (Levitan 2002a). Some species appear able to fertilize at relatively low densities or longer mate distances (e.g., *Acanthaster*; Babcock et al. 1994). These differences may in part be caused by variation in patterns and amount of sperm release resulting from differences in body size or reproductive investment (Babcock et al. 1994), but at least some variation can be explained by differences in gamete traits (Levitan 1998b).

As the density of males increase or distance between males and females decreases, sperm become less limiting and variation in female fertilization success decreases. Nevertheless, even in dense and highly synchronized spawning events,

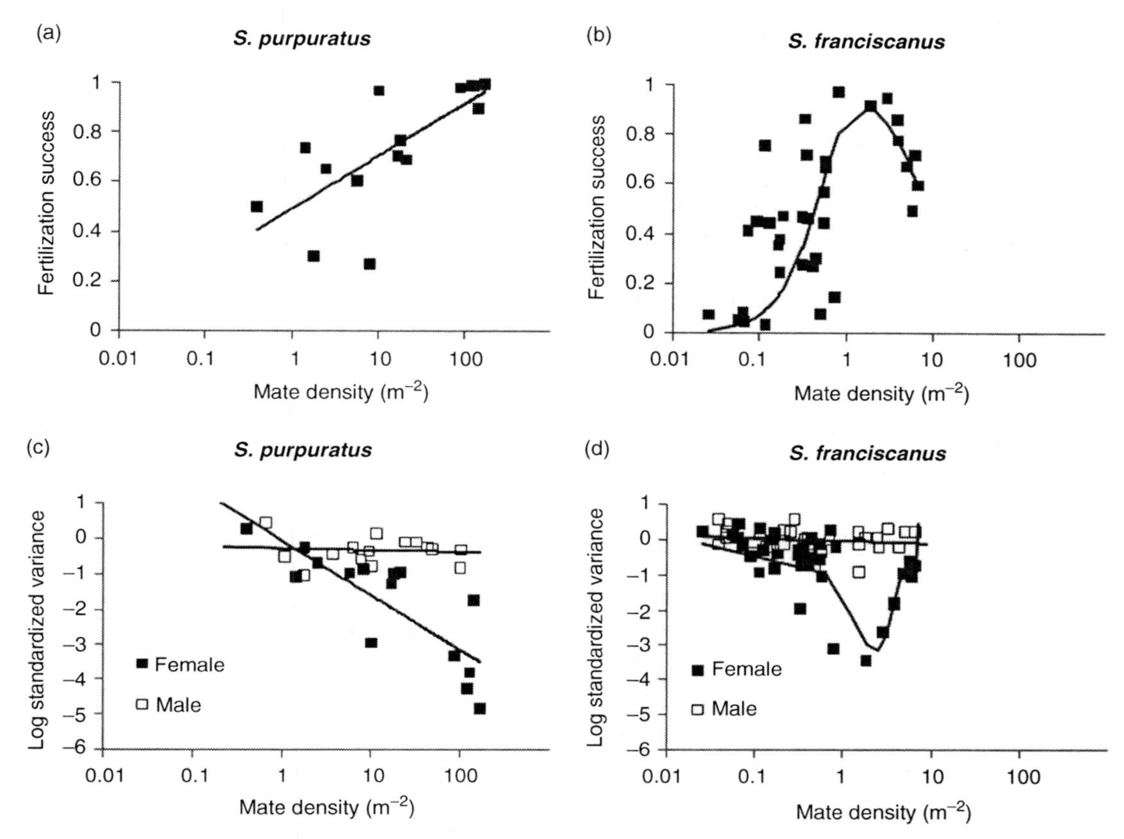

FIGURE 21.2 Average (a and b) and standardized variance (c and d) in fertilization success in the sea urchins *Strongylocentrotus purpuratus* and *S. franciscanus*. *Strongylocentrotus purpuratus* is resistant to polyspermy, and fertilization variances are similar only at low densities where sperm are limiting. As mate densities increase, fertilization variance decreases in females as they saturate with sperm. *Strongylocentrotus franciscanus* is susceptible to polyspermy. At low mate densities, male and female variance is similar because of sperm limitation, and at high mate densities, male variance remains high because of sperm competition and female variance increases because of variation in the degree of polyspermy Redrawn from Levitan (2008).

some variance in female success may occur. For instance, the sea cucumber, *Cucumaria miniata*, releases clumps of eggs in a pellet that slowly drifts to the surface. Spawning at high densities (46/m^2) resulted in an average of 97% of eggs fertilized, but ranged from 68% to 100% (Sewell & Levitan 1992). Variance has even been detected in pair and group spawning fish, in which spawning is highly synchronous and eggs and sperm can be released in very close proximity (e.g., within a centimeter). Fertilization success in bluehead wrasses (*Thalassoma bifasciatum*) averages over 95%, but varies from around near zero to 100% (Petersen et al. 1992). Variance is associated with whether spawning is a pair or group event (Marconato et al. 1997), the amount of sperm released by males in a pair spawn, and the magnitude of water flow (Petersen et al. 2001). These studies suggest that even when average sperm availability is high and males spawn in close proximity to females, variation in fertilization success can still be caused by sperm limitation. These results suggest that as sperm densities increase, selection driven by sperm limitation may diminish but not entirely vanish.

Variation in reproductive success can also be caused by too many sperm. As density increases, and sperm become highly concentrated, polyspermy becomes an increasing source of developmental failure and can lead to increased variance in reproductive success. Models of polyspermy (Styan 1998), supported by experimental spawning events (Franke et al. 2002), indicate that at intermediate densities females may experience both sperm limitation and polyspermy as some eggs released in the same clutch are surrounded by too few sperm, whereas others have too many. In such cases, there will always be some variance in female reproductive success caused by sperm availability. If such cases are the general rule, then situations in which all eggs are fertilized, but each by a single spermatozoon, might be rare. The exception would be species living at generally high densities that spawn synchronously and have highly efficient blocks to polyspermy. These species may rarely be sperm limited, and the efficient blocks to polyspermy may ameliorate the consequences of excess sperm. For example, the purple sea urchin, *Strongylocentrotus purpuratus*, is often at high densities, has high levels of fertilization, and can efficiently block polyspermy (Levitan et al. 2007; Levitan 2008; figure 21.2a). Although experimental reduction in density can decrease the average and increase the variance in fertilization success, these conditions may be uncommon in nature (Levitan 2002a, 2008).

Another possible source of variation in female success is from the interference competition that can occur among males. Female fertilization success was reduced when multiple males attempted to mate (Byrne & Roberts 1999), perhaps because as males wrestled for access to the female, they were displaced from the optimal positioning for releasing sperm over the eggs. The possibility that males might release fewer sperm under these competitive conditions was not tested. Although most theory suggests that males should release more sperm under conditions of competition (e.g., Petersen 1991), an alternative view has been suggested that males may release less sperm in competition if this allows for an increase in the number of mating events in which a male can release at least some sperm (Bode & Marshall 2007). These dynamics in variance of female success suggest again that sexual conflict over particular male or female traits may be common. However, without the complementary data on male success, these ideas cannot be tested.

Information on male fertilization success in external fertilizers is scant and restricted to patterns of success in primary and satellite males in horseshoe crabs (Brockmann et al. 1994), frogs (Roberts et al. 1999), and a few fish (Wooninck et al. 2000; Fu et al. 2001; Neff et al. 2003). For more sedentary organisms, information is available for two groups: patterns of male success are associated with distance and clone size in soft corals (Coffroth & Lasker 1998), and with distribution, abundance, and genotype in sea urchins (Levitan 2004, Levitan 2008; Levitan & Ferrell 2006).

These initial studies reveal some highly variable patterns. In systems in which there is a primary male that initiates courtship or is in a premium position during spawning and there are additional males releasing sperm, the primary male often but not always gets the majority share of fertilizations. Satellite male horseshoe crabs only occasionally have the majority of paternal success, and this is dependent on their proximity to the released eggs (Brockmann et al. 1994). The success of male bluehead wrasse is positively correlated with the amount of sperm they release during group spawns, but negatively influenced when streakers join a pair-spawn (Wooninck et al. 2000). Work on bluegill sunfish (*Lepomis macrochirus*) indicate that primary males, although having the largest absolute testes, release the smallest amount of sperm per ejaculate, compared to males that act as female mimics with intermediate ejaculate size or sneaker males with largest ejaculate size. Paternal success for a single spawning event was directly correlated with ejaculate size; sneakers generally won. However, primary males also produced the longest lived sperm and those sperm were more competitive for a given concentration of sperm (Fu et al. 2001; Neff et al. 2003). These results on mobile and socially interactive animals suggest that sperm allocation (who releases the most sperm), behavior (positional effects during gamete release), and gamete traits (e.g., longevity or compatibility; see Purchase et al. 2007) may all play a role in determining paternal fertilization success.

Reproductive and paternal success in organisms that are sessile but can reproduce asexually via fragmentation add an interesting wrinkle to mating strategies. Paternal success in the Caribbean soft coral *Plexaura kuna* was related to the size of the male clone (number of coral colonies from the same genotype) and the distance to the female colony. Clone size and distance were correlated because larger clones have the potential to cover a larger spatial area and be in closer contact with potential mates. Asexual reproduction via fragmentation not only increased genotype biomass and fecundity and reduced the likelihood of genotype mortality, but also increased the likelihood of spawning next to a potential mate (Coffroth & Lasker 1998).

In another soft coral species, individual male size and distance to females was not found to be correlated with male success (Lasker et al. 2008). This species differed from the prior study in two important ways. First, this species does not fragment clonally (at least not commonly), so there are many more male genotypes each releasing relatively few sperm compared to large clonal populations. In such cases heterogeneity in water flow might more easily swamp distance effects compared to when large clones release copious amount of sperm. Second, this species is a surface brooder and released eggs stick to the surface of females, which allows eggs to essentially filter sperm from the water column instead of drifting away. This may allow fertilization by more distant males as more dilute sperm can eventually accumulate on the surface of females.

A comparison of variance in male and female fertilization success has been performed in sea urchins (Levitan 2004, 2008; figure 21.2). Variation in male fertilization success in sea urchins appears to be moderately high across a wide range of spawning densities. Variation in female fertilization success is similar to male variance at either low densities, caused by sperm limitation, or at high densities, caused by polyspermy. Only at intermediate densities is the variance in female fertilization success lower than males. Under these intermediate densities, sperm are abundant enough to saturate females, but not too high to cause polyspermy. There are two interesting aspects to male variance in these sea urchins. The first is that male variance is not as high as one might expect, given the vast potential for single males to monopolize a spawning event (billions of sperm released per male). There are no big winners or losers: most males fertilize at least some eggs from all females within a local spawning area (e.g., 5 × 5 meter scale). The second is that male variance remains high across different densities and there is a seamless transition, in terms of reproductive variance, from sperm-limited conditions in which gamete competition is more likely to be indirect, to sexual conflict, in which sperm are more likely competing directly for the same pool of eggs.

Measuring Sexual Selection and the Relationship between Mating and Reproductive Success

The wide variation in fertilization success in both males and females across a range of sperm availabilities and population densities may create challenges for estimating the intensity of sexual selection. Measures based on mating success can fail to find evidence for sexual selection when one phenotype outcompetes another during group spawning events. For instance, if increased sperm swimming ability results in an increased chance of winning in competition with other males, but this advantage translates into larger paternity share, rather than an increase in the number of females mated, then this trait would be considered a sexually selected trait under some definitions, but not ones based on mating success.

In contrast, measures based solely on cumulative fertilization success across all mates, independent of mating success (e.g., when siring 10% of eggs from 10 females is equivalent to 100% of only 1 female), can fail to distinguish the selective forces that influence compatibility or choice. For instance, the evolution of intraspecific polymorphisms in gamete recognition proteins has been thought to be influenced by sexual selection. A male might have similar cumulative fertilization success in systems with and without variation in gametic compatibility, but in the system with variation, these fertilizations would be limited to a subset of females at a high rate of success, whereas in the other system that male might have more even success across all females. Ignoring the selection for increased mating success can fail to distinguish these differences and the cascading effects of sexual selection on compatibility, assortative mating, and speciation.

Finally, measures based on reproductive variance can fail to distinguish patterns of sexual selection when polyspermy and sexual conflict are possible. For instance figure 21.2d indicates that the classic condition of high reproductive variance in males and low reproductive variance in females occurs only at intermediate densities in which sperm are saturating, but not oversaturating. At lower and higher densities the variance among males and females is similar, which might be interpreted as no sexual selection, or that the nature of sexual selection is similar at either density extreme. However, detailed study indicates different patterns of selection at high and low density: intense selection for increase fertilization rate for both sexes at low densities and sexual conflict over fertilization rate at high densities (Levitan 2004).

These complications provide some interesting opportunities to better understand the fundamental forces at work in sexual selection. Use of a variety of measures may provide a more complete view of the nature of sexual selection. For example, a comparison of patterns of variance in reproductive success using Bateman gradients (Levitan 2008) was especially informative. Bateman gradients (Arnold & Duvall 1994; chapter 20, this volume) are plots of reproductive success (number of offspring produced) as a function of mating success (number of mates during some defined breeding period). The slope of these plots describes the intensity of sexual selection (the fitness gains associated with garnering higher number of mates). Comparison of these gradients between males and females provides an indication of which sex is under sexual selection or which sex experiences more intense sexual selection.

An illustration of the strength of this approach can be seen in the sea urchin *Strongylocentrotus franciscanus* (Levitan 2008). This broadcast spawning species occupies the shallow subtidal along the west coast of North America and lives over a wide range of population densities. Experimentally induced spawning events indicate that female fertilization success peaks at around 1–5 males/m^2 and then diminishes at higher densities because of polyspermy. Within each independent spawning event a Bateman gradient was calculated for each sex. The male gradients were always positive; males always garnered more fertilizations when they sired offspring from more females (e.g., figure 21.3a). Females had positive gradients at low population densities when sperm were limiting; increases in the number of males mated translated into increased fertilization success. But females had negative gradients at the high densities in which polyspermy was common; increases in the number of males mated reduced egg developmental success. The female Bateman gradients were zero at the same male density that resulted in peak female fertilization success and the maximum difference between male and female variance in fertilization success (compare figures 21.2b and 21.2d with 21.3b). Bateman gradients complement and provide a direction to the selective forces hinted at by the variances in male and female fertilization success. Males are always under sexual selection for increased fertilization and mating success, whereas females can have either positive or negative selection for mating

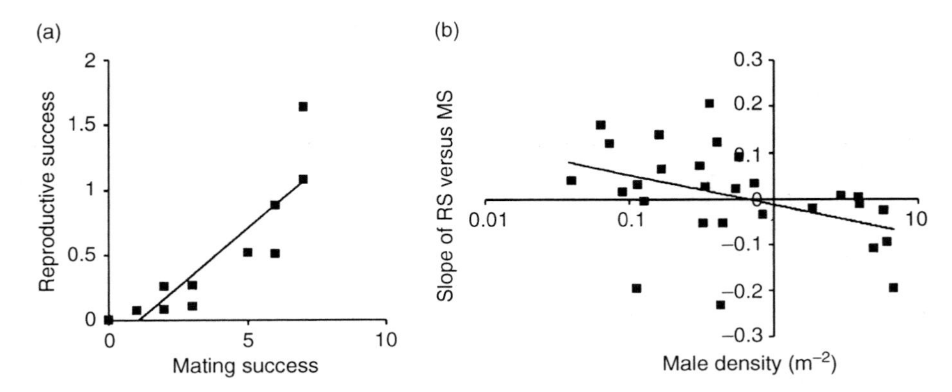

FIGURE **21.3** Bateman gradients in *Strongylocentrotus franciscanus*, a broadcast spawning sea urchin. (a) Male fertilization success as a function of the number of females with which he sired offspring. Fertilization success in males can vary between zero and the complete fertilization of all females in a spawning event (1.0* the number of females). The slope of this relationship is the Bateman gradient. (b) A series of Bateman gradients in female *Strongylocentrotus franciscanus* over a range of spawning densities. Bateman gradients are positive under sperm-limiting conditions (in which more mates equals higher fertilization success) but negative under conditions of excess sperm in which polyspermy kills embryos (more mates equals higher developmental failure). When gradients are positive in males and negative in females within the same spawning event, it is evidence of sexual conflict over mating frequency. Redrawn from Levitan (2008).

and fertilization success, and the intensity of this selection approaches zero at intermediate levels of sperm availability. These gradients also provide a means for understanding the selective forces that may generate variation in the gamete physical traits and recognition proteins that mediate compatibility (see below).

Targets of Selection

The variance in both male and female reproductive success caused by sperm limitation, sperm competition, and polyspermy provides an opportunity for selection to operate on both male and female traits. For broadcast spawners, the critical traits appear to be specific behaviors related to spawning and gamete traits. Less commonly sexual dimorphism is noted in adult morphological traits. I explore some of the forces applicable to each of these.

Spawning Behavior

During natural spawning events, not all adults release gametes and, of the fraction that do, not all spawn simultaneously (invertebrates: Babcock et al. 1992; algae: Clifton 1997). There is some data on the consequences of large-scale asynchronies (e.g., when fractions of the population spawn during different events). For instance, studies of both soft (Lasker et al. 1996) and stony (Oliver & Babcock 1992; Levitan et al. 2004) corals have shown that on the evenings leading up to and following peak spawning nights, fewer individuals spawn and fertilization success is, not surprisingly, proportionately reduced.

Data also exist on smaller scale asynchronies (variation in spawning times within a spawning event). Individuals often release gametes for seconds to minutes, but at the population level, the event can last for hours (Hamel & Mercier 1996). Even at this scale, individuals that spawn slightly out of synchrony have reduced fertilization success (Levitan et al. 2004). This can lead to intense selective pressure to spawn precisely with conspecifics. However, variance in spawning times persists, and this may be caused by additional selective pressures or constraints that prevent perfect synchrony.

One source of variation is that males often spawn before females (Levitan 1998a). Among 47 reported invertebrate taxa in which one sex initiates spawning before the other, 44 indicated male-first spawning (Levitan 1998a). Among 11 species of dioecious green algae, males always spawned first (Clifton 1997). In some cases males can initiate spawning hours before females (Hamel & Mercier 1996). Experiments investigating the costs of spawning early or late within a spawning event have been conducted with sea urchins (Levitan 2005). Manipulation of spawning times indicates that the fraction of eggs fertilized does not depend on spawning order (male or female first), but when males spawned in competition, males that spawned just before egg release had a competitive advantage over males that released sperm just after egg release. The optimal time for males to release sperm depended on the distance to females. When females were nearby, more synchronous release with eggs gave males a competitive advantage of having a high sperm concentration over males that released earlier and had a more diffuse sperm cloud. When females were more distant, males that released sperm earlier were able to permeate their sperm over a larger spatial extent and capture more fertilizations (Levitan 2005). These results indicate that the optimal male spawning strategy depends on the distribution of males and females, spawning synchrony, and the likelihood of sperm competition.

Variation among individuals in spawning times also exists within a sex (invertebrates: Babcock et al. 1992; algae: Clifton 1997) and within species that release gamete bundles containing both eggs and sperm (Levitan et al. 2004). The causes and consequences of asynchrony in spawning are not well understood and may differ for males and females. From the perspective of the egg, spawning at peak times appears to be advantageous under sperm-limited conditions (Levitan et al. 2004), but this might be tempered under high levels of sperm availability in which polyspermy is a risk. There may be sexual conflict over the optimal spawning time under conditions of high sperm availability. Females might be selected to avoid peak times and the risk of polyspermy, but any shift in their timing creates selection for males to spawn more synchronously via increased paternal success. Under the right conditions disruptive selection on spawning times driven by polyspermy are predicted to result in temporal reproductive isolation and potentially speciation (Tomaiuolo et al. 2007).

Asynchrony might also be a result of temporal differences in the reception or response to spawning cues. This might be an adaptive response, or simply that precision in spawning is difficult to achieve. Evidence for sex differences in spawning and the proposed adaptive explanations for these behaviors (see above) suggest that at least some fraction of asynchrony could have adaptive explanations. However, it has also been proposed that precision is problematic, particularly when the process of reproduction can take months to nearly a year (Fadlallah 1982). Any variance in the rates of gametogenesis, gamete packaging, cue reception, and gamete release can lead to subtle or major differences in timing. Evidence suggests that multiple cues are needed to ensure precision in synchrony (Soong et al. 2006). For instance, corals use patterns of solar irradiance (van Woosink et al. 1996), lunar irradiance (Levy et al. 2007), and sunset (Levitan et al. 2004) to, respectively, cue the month, day, and time of spawning. Rate differences in the progression to spawning are damped because individuals must wait for the next cue (removing the variance in timing) before proceeding to the next step. Thus a process that may take months will show rate differences that have accumulated only since the last cue (e.g., sunset) of the day of spawning. The use of multiple cues may also explain why different cohorts of individuals spawn on the same lunar day on different months, or different evenings on the same lunar cycle: individuals progressing too fast or slow are binned into different spawning events (like cars that miss a particular traffic light).

Constraints on synchrony induced by variance in internal clocks or by variance in rates of processes that occur after cue reception suggest that maximum synchrony would be achieved when spawning occurs immediately after cue reception. However, this may be problematic. First, time may be needed to process the cue and then prepare and proceed with spawning. Second, there may be spatial differences in the distribution of the cue (e.g., heterogeneous waterborne chemicals, gradients of light or temperature, variance in tidal flow over a topographically complex landscape). Finally, there may be adaptive reasons for spawning at a specific time, independent of precision (e.g., gamete or larval dispersal, predator avoidance, avoidance of heterospecific fertilization; Morgan 1995). These constraints may be more easily overcome in organisms with high mobility and precise sensory systems. Pair spawning fish are highly synchronized with their mates (e.g., Petersen et al. 1992), and aggregations of mobile invertebrates can use pheromones (e.g., sperm; Starr et al. 1990) that are more reliable at small spatial scales. Sessile or sedentary invertebrates spread over larger spatial scales may be more reliant on cues independent of conspecifics and subject to a higher degree of variance in synchrony with potential mates.

Gamete Traits

Variation in egg and sperm traits can also influence the probability of sperm-egg collision and fertilization after collision (Levitan 1993). Optimal egg traits may vary from those that increase collision and fertilization rates under sperm-limited conditions to those that decrease these rates and increase discrimination under sperm-competitive conditions (reviewed in Levitan 2006). Optimal sperm traits may also vary based on the likelihood of collision. Traits such as sperm velocity, which make sperm competitive in the short term, appear to trade off with traits such as sperm longevity, which are beneficial when mates are scarce (Levitan 1993, 2000b). In addition to energetic and morphological traits, intraspecific variation in gamete compatibility influences the likelihood of fertilization after collision (Palumbi 1999), and the degree to which gametes are selected to be discriminatory may be influenced by the likelihood of encounter (Levitan & Ferrell 2006).

Egg traits that have been shown to influence collision frequency involve target size. Target size can be the actual size of the egg cell but also include accessory cells, structures or chemical cues that increase the likelihood of sperm reaching the egg cell membrane. Variation in egg cell size within and often among species influences collision (Farley & Levitan 2001) and fertilization rates (Levitan 1996). Accessory structures such as jelly coats (Podolsky 2001) also increase target size and the probability of fertilization. Sperm chemotaxis has been demonstrated in a variety of, but not all, taxa (e.g., Miller 1985). Chemical attractants can initiate sperm swimming (Bolton & Havenhand 1996), alter the direction of swimming sperm (Miller 1985), and increase the likelihood of fertilization (Riffell et al. 2004). Models have predicted that egg target size and egg cell size are selected to increase when sperm are limiting (Levitan 2000a, Luttikhuizen et al. 2004; Podolsky 2004).

Sperm swimming ability can also influence collision rates. Although sperm velocity (100–200 microns/s) is almost always less than water flow velocities (cm to m/s), males with faster sperm have higher fertilization rates than slower males in laboratory experiments (Levitan 2000b; Kupriyanova &

Havenhand 2002). This suggests that if water flow mixes gametes to the point at which collisions are possible, males with faster sperm may outcompete males with slower sperm. Sperm velocity may be most important when sperm are competing with other males at high sperm concentrations. When many sperm approach an egg, the spermatozoon first able to find and fuse to the egg will have an advantage. In many cases, such as dense aggregations of spawning individuals or when streak spawning males enter into pair spawning fish, sperm velocity may be a key determinate of reproductive success.

Sperm velocity appears to trade off with sperm longevity across (Levitan 1993) and within species (Levitan 2000b). Conditions in which sperm or mates are limiting may favor males that produce sperm with increased longevity that can remain viable until collisions become likely. In many taxa sperm remain inactive while concentrated and start swimming actively when diluted in seawater. This may allow sperm to increase longevity by saving energy until the sperm becomes dispersed into the sea. In some taxa, sperm become activated by the presence of eggs. These taxa may be able to remain viable for extended periods as the energetic expense of swimming is delayed until necessary (Bolton & Havenhand 1996). Although there are scant comparative data, it is interesting that pair spawning fish, often in competition with streakers, have very short lived sperm (> 15s; Petersen et al. 1992), whereas some sedentary invertebrates have sperm that can live for hours to days (Johnson & Yund 2004).

After sperm and eggs collide, fertilization may or may not occur, and the probability of this may depend on traits present in either gamete. There is inter- and intraspecific variation in how sperm collisions translate into fertilizations. In at least some species this variation in conspecific fertilization is correlated with an egg's susceptibility to hybrid fertilization (Levitan 2002b, but see McCartney & Lessios 2002) and polyspermy (Levitan et al. 2007). This variation might be related to the distribution and abundance of sperm receptor sites on the egg surface: some sperm may just miss the receptor sites that initiate fertilization. Alternately, increasing evidence suggests that variation in fertilization rate can be attributed to variation in gametic compatibility. Gamete compatibility among related species (Zigler et al. 2005) and within species (Evans & Marshall 2005) is not black and white but gray; more compatible gametes simply require less sperm to achieve fertilization compared to less compatible gametes (Levitan 2002b).

Rapid evolution in the gamete recognition proteins that mediate gametic compatibility has been noted in several of broadcast spawning species but is not universal (reviewed in Swanson & Vacquier 2002a). Rapid rates of evolution in these proteins across species and populations might be driven by sexual selection and within populations by sexual conflict (Gavrilets & Waxman 2002; Haygood 2004; Levitan & Ferrell 2006). A protein called *sperm bindin protein* in sea urchins shows evidence of rapid evolution in about half of the species investigated (Palumbi 1999). The fitness consequences of variation in sperm bindin have been investigated in the sea urchin *Strongylocentrotus franciscanus* (figure 21.4). Individuals with common alleles do best under conditions of sperm limitation in which matches between male and female gamete recognition proteins lead to a high affinity between sperm and eggs and translate into a high probability of fertilization when gamete collisions are rare. At higher densities in which sperm concentrations can result in polyspermy, females with rare alleles are favored because they mismatch many sperm, thus having a lower affinity with most sperm. This provides eggs with the time needed to establish a block to polyspermy (by lowering the effective concentration of sperm). As the frequency of these rare females increase, males that match these rare females will also be favored. These results suggest that the degree of sperm availability can shift selection from positive to negative frequency-dependent selection (see chapter 3) and alter the rate of evolution in these proteins. This density-dependent selection provides a potential explanation for why these proteins evolve at different rates among taxa (Levitan & Ferrell 2006). This form of sexual conflict, termed *chase-away selection* (Holland & Rice 1998), in which male protein evolution follows to match novel female proteins, is predicted to lead to reproductive isolation as males chase the evolution of different female proteins in allopatry or even sympatry (Gavrilets 2000; Gavrilets & Waxman 2002; see also chapters 23, 24, and 27 of this volume).

Adult Morphological Traits

Sexual dimorphism in adult morphological traits has been noted rarely in externally fertilizing marine invertebrates (Levitan 1998a) and more commonly in externally fertilizing fish species (Thresher 1984).

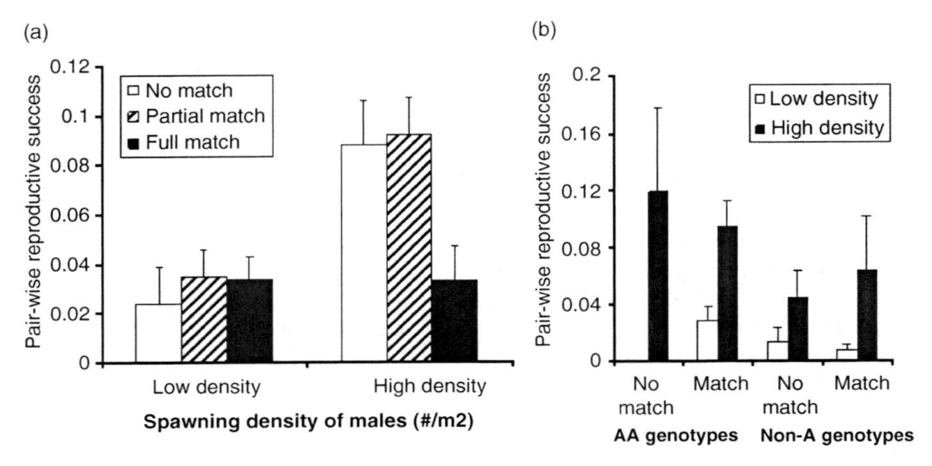

FIGURE 21.4 Fitness consequences of variation in sperm bindin genotypes in the sea urchin *Strongylocentrotus franciscanus* based on reproductive success during spawning events in the sea. Under sperm-limiting conditions, males and females that match at this locus have higher reproductive success than mismatched mates. Under polyspermic conditions (a), mismatched individuals have higher reproductive success compared to matched mates. This results in a significant interaction between spawning density and the degree of matching (Levitan & Ferrell 2006). A detailed look at this interaction (b), by comparing the most common genotype (AA) with rare genotypes, indicate that (1) males have significantly higher success at high density because females tend to be closer, (2) common males have significantly higher success than rare males because the latter rarely match with females and suffer a competitive disadvantage, and (3) males with the most common genotype do poorly with mismatched mates at low density but have the highest success with mismatched mates at high density. This occurs because mismatched females are resistant to polyspermy and they are rarely close to the rare matching male, so the common male is rarely outcompeted. Reprinted from Levitan and Ferrell (2006) with permission.

These cases appear tied to pair spawning behavior. Pair spawning sea stars, brittle stars, and horseshoe crabs show dimorphism in body size and for male horseshoe crabs specialized appendages for holding on to females (Levitan 1998b). Pair-spawning fish tend to show more pronounced sex differences in both coloration and body size compared to group spawning fish (Thresher 1984). It seems that the presence of sexual dimorphism (e.g., Darwin 1871) is tied to adult interactions. When adults compete or choose mates, adult features are under sexual selection. When gametes compete in the water column, then the targets of sexual selection are the spawning behaviors that determine the nature of gamete competition and the gamete traits that mediate this competition. This suggests that the intensity of sexual selection may be independent of whether eggs are fertilized internally or externally, but that the targets of sexual selection shift as the arena for competition shifts from eggs and sperm in the water column to adults vying for the opportunity to mate (Levitan 2005).

Evolutionary Transitions in Mating Strategies

Prior to an anisogamous condition, whether mating types had been established or not (chapter 20), gametes likely competed for successful fertilizations: some gametes fused, whereas others failed to find a match. Assuming blocks to polyspermy are a derived condition, the twin constraints of failure to find an unfertilized gamete and being killed by too many fusion events set the stage for gamete limitation, competition, and conflict over fusion rates, even before sperm evolved and sex differences were established. The issue of whether the disruptive selection that led to anisogamy was primarily driven by sperm limitation (indirect competition for fertilizations) or sperm competition (direct competition for fertilizations; figure 21.1a versus 21.1b) remains an open question. Although most models assume direct sperm competition as the mechanism (chapter 20), indirect competition under sperm limitation can also explain these patterns (Levitan 1996; Crean & Marshall 2008), and at least one model suggests that sperm limitation may be the only model to predict anisogamy, if ancestral eggs were susceptible to polyspermy (Bode & Marshall 2007). It would be interesting to determine whether isogamous algal species are more susceptible to polyspermy than related anisogamous taxa.

Most of the ideas I have discussed focus on broadcast spawners, but sometimes in external fertilizers, eggs are released from the female but remain on the surface of the female or substrate during fertilization. These *surface brooders* may represent the first step toward internal fertilization as surface brooding requires little morphological modification of adult form. Surface brooding likely evolved as means of increasing fertilization success by reducing sperm limitation. Keeping eggs in one location (as opposed to moving with water flow) allows the eggs of surface brooders to sample the water for sperm over long periods, increasing the likelihood of fertilization (Lasker 2006). A related strategy is often noted in mobile species that lay eggs on the substratum such as some amphibians and horseshoe crabs. This strategy sets the stage for adult interactions, male competition, and female choice as males via for female attachment and priority in releasing sperm as eggs are deposited (Brockmann et al. 1994).

Retaining eggs inside the body of the female represents the transition to internal fertilization and provides the female with increased control over fertilization. Arguments can be made that both sperm limitation and sexual conflict resulted in the transition to internal fertilization in some taxa. Species in which males free-spawn sperm and females retain eggs for internal fertilization (spermcasters such as many ascidians, bryozoans, and cnidarians) have the ability to maintain high levels of fertilization even when adults are sparse by filtering sperm out of the water (Bishop 1998). However, a similar case can be made that under conditions of high sperm availability, brooding allows females to collect sperm and control the rate at which eggs are exposed to sperm, possibly avoiding the risk of polyspermy. Brooding may also increase the female's ability to choose sperm from higher quality males, as has been demonstrated in a variety of copulating taxa (reviewed in Neff & Pitcher 2005). It would be interesting to determine what fraction of paternal success can be explained by sperm abundance, sperm competitive traits, sperm-egg compatibility, and female choice in spermcasting species.

For sessile species, spermcasting may be the best strategy to maximize fertilization success and control paternity. If females can discriminate among sperm from different males, it also might provide an increased ability of high-quality males, males producing high-quality sperm, and males producing more compatible sperm to be more predictably successful, because the random elements that determine which sperm arrives first may be ameliorated by female control.

Copulation is generally rare in sessile organisms, although some barnacles have penises capable of inseminating females many body lengths away. Copulation is more common among mobile species that can aggregate and pair. This is likely the most derived, and certainly the most studied, mating strategy, which allows for a high degree of mate choice and control over parentage. If spermcasting is the common transitional stage between external fertilization and copulation, then it is unlikely that sperm limitation drove this last transition, as spermcasters are rarely sperm limited (e.g., Phillippi et al 2004). Instead mate choice and sexual conflict are likely candidates. Males better at placing sperm within females and near eggs would have the advantage of priority effects. Females that can sometimes control who gets close enough to fertilize eggs, or can employ cryptic choice among sperm of different males that counters priority effects (see chapter 22) would likely be favored.

FUTURE DIRECTIONS

External fertilization is the likely the ancestral mating strategy for both animals and plants (Wray 1995) and is still the most taxonomically widespread reproductive strategy (Giese & Kanatani 1987). Yet the vast majority of research on sexual selection has been conducted on species that copulate. This skew is likely caused by two factors. First, there is a general bias to study systems with reproductive strategies closer to our own, and second, determining the fertilization success of many external fertilizers who broadcast gametes into water or air is difficult. Although the problem of sperm dilution was predicted nearly a hundred years ago (Sparck 1927), estimates of female fertilization success were not published until 1985 (Pennington 1985) and male reproductive success until a decade later (Brockmann et al. 1994). The advent of relatively inexpensive and easy to use polymorphic genetic markers (e.g., microsatellite loci) has allowed for direct estimates of male and female reproductive success. These techniques have just begun to be applied to fish and invertebrate species, and our knowledge base is just starting to grow; most of what we know about sexual selection in external fertilizing species comes from a very limited number of taxa. Several lines of research appear to be promising.

The frequency distribution of different gamete encounter probabilities and gamete competition intensities remain largely unknown (figure 21.1). As we gather additional data from more taxa with divergent life history characteristics, we will determine whether there are general principles underlying the diversity in sexual selection in external fertilizing organisms.

New advances are likely to be made both through more focused comparative studies involving a wider sampling of taxa and experimental studies of specific traits. Three examples illustrate the potential of external fertilizers for defining new approaches to sexual selection. First, the linking of gamete and adult behavioral traits with patterns of sperm and mate availability will elucidate how density may influence current reproductive success and also the patterns of selection. Second, patterns of spawning synchrony need to be addressed at a finer grain so that we can determine the costs and benefits of spawning for long or short periods at both the population and individual levels. Why do some species release gametes in a short blast, whereas others spawn for hours or more? Third, as we learn more about the molecular biology of gamete recognition proteins, experimental tests of how these proteins perform under different levels of sperm availability might provide insight into the rapid evolution in these systems, and potentially patterns of reproductive isolation and speciation. Each of these illustrates some complexities that may help flesh out a broader theory of sexual selection.

Acknowledgments I thank M. Adreani, D. Ferrell, N. Fogarty, M. Jennions, and K. Lotterhos for comments on this manuscript. Funding for the sea urchin work discussed in this chapter was provided by the National Science Foundation.

SUGGESTIONS FOR FURTHER READING

For those seeking highly detailed taxonomic coverage, I recommend the nine-volume series *Reproduction of Marine Invertebrates* edited by Giese et al. (1974–1987). For example, Giese and Kanatani, in the final volume (1987), provide a comprehensive survey of spawning in marine invertebrates. The classic paper by Thorson (1946) provides a great review of the early literature and insight into patterns of reproduction and reproductive strategies of marine invertebrates, which is brought up to date by Young (1990). The edited volume by McEdward (1995) builds upon the conceptual themes on reproductive ecology and evolution based on Thorson and later by Strathmann (1985). Strathmann (1990) also provides an insightful examination of why life histories evolve differently in the sea. A recent review on spermcasting species (Bishop & Pemberton 2006) provides details on how retaining eggs influences patterns of fertilization and life history evolution. Petersen and Warner (1998) review sperm competition in fish in much more detail than provided here.

Bishop JDD & Pemberton AJ (2006) The third way: spermcast mating in sessile marine invertebrates. Integr Comp Biol 46: 398–406.

Giese AC & Kanatani H (1987) Maturation and spawning. Pp 251–329 in AC Giese, JS Pearse, & VB Pearse (eds) Reproduction of Marine Invertebrates, Vol. IX: Seeking Unity in Diversity. Blackwell Sci/Boxwood Press, Palo Alto/Pacific Grove, CA.

McEdward L (1995) Ecology of Marine Invertebrate Larvae. CRC Press, Boca Raton, FL.

Petersen CW & Warner RE (1998) Sperm competition in fishes. Pp 435–463 in T Birkhead & A Moller (eds) Sperm Competition and Sexual Selection. Academic Press, San Diego, CA.

Strathmann RR (1985) Feeding and nonfeeding larval development and life-history in marine invertebrates. Annu Rev Ecol Syst 16: 339–361.

Strathmann RR (1990) Why life histories evolve differently in the sea. Am Zool 30: 197–207.

Thorson G (1946) Reproduction and larval development of Danish marine bottom invertebrates. Meddr Kommn Danm Fisk-og Havunders Ser. Plankton 4: 1–523.

Young CM (1990) Larval ecology of marine invertebrates: a sesquicentennial history. Ophelia 32: 1–48.

22

Postcopulatory Sexual Selection

SCOTT PITNICK AND DAVID J. HOSKEN

Darwin (1871) realized that some characters were obviously not naturally selected, leading him to synthesize his observations of sex differences in structures and behaviors into a cohesive theory of sexual selection. Darwin's theory was largely correct, but included one glaring bit of specious reasoning and one sizeable omission that relate to one another in an interesting way. First, he never discerned the fundamental cause of sex differences, or why it is typically males that compete with one another for mates and typically females that choose among potential suitors. Darwin incorrectly reasoned that males are predestined to search and compete for females because their ancestors reproduced in the ocean where sperm were the searchers for eggs (1871, pp. 273–275). As developed in chapters 20 and 21, current theory contends that anisogamy—and associated sex differences in parental investment and potential rates of reproduction—creates the conditions for sexual selection and sex differences, as numerically abundant, lower investment sperm compete to fertilize relatively rare, high investment eggs. Darwin's (1871) related omission was a failure to recognize that sexual selection could continue after copulation. As with precopulatory sexual selection, in the postcopulatory arena, male reproductive success is limited by access to females and their eggs, and the mechanisms of selection tend to be the same: female choice (here cryptic female choice) and male-male competition (here sperm competition).

The first thorough conceptualization of postcopulatory sexual selection was by Parker (1970a), who defined *sperm competition* as "the competition within a single female between the sperm from two or more males for the fertilization of the ova," later expanded to "competition between the sperm from two or more males for the fertilization of a given set of ova" so as to include externally fertilizing species (Parker 1998). In other words, sperm competition was postulated as a specific case of more generalized male-male competition. It took roughly another 20 years before serious attention was paid to the female choice component of postcopulatory sexual selection (Eberhard 1996). *Cryptic female choice* is defined as non-random paternity biases resulting from female morphology, physiology or behavior that occur after coupling. Female remating (or multiple males participating in a spawn) is an obvious prerequisite for sperm competition and cryptic female choice to occur (see box 22.1 and chapter 21). Inspired by Parker's (1970a) development of sperm competition theory, a plethora of subsequent behavioral studies, coupled with the advent of biochemical and molecular tools for estimating parentage in nature (e.g., allozymes, DNA fingerprinting), have revealed that multiple mating by females and mixed paternity are more the rule than the exception. For the majority of taxa, females store viable sperm within their

BOX 22.1 Multiple Mating by Females

Although males tend to be more promiscuous than females, it has become apparent that females of most species mate with multiple males (polyandry). Many of the costs of remating, which may include (1) expenditure of time and energy, (2) increased predation risk during mate searching, courtship, and copulating (Gwynne 1989), as well as (3) increased probability of acquiring sexually transmitted parasites and pathogens, are likely borne equally by both sexes or more so by males. The observed sex differences in remating behavior are thus best understood in terms of differential benefits accrued. The principal benefit to males of remating is obvious: because sperm are cheap relative to eggs, there is a strong positive association for males between the number of mates acquired and the number of offspring produced (Bateman 1948). In contrast, females may acquire sufficient sperm in a single copulation to fertilize all of the eggs they will produce in their lifetimes, and so the adaptive significance of female remating is more enigmatic.

A wide range of potential benefits to females of multiple mating—both direct and indirect—have been discussed and demonstrated (table I; reviewed by Arnqvist & Nilsson 2000; Jennions & Petrie 2000; Hosken & Stockley 2003; Arnqvist & Rowe 2005); which of these apply likely varies greatly among species. Direct benefits are those accrued directly to the female, such as an increase in her life span or in the number of offspring produced. Benefits such as these may occur if, for example, multiple mating increases the number of males who help rear offspring or if males provide nuptial gifts to females. Indirect benefits to polyandry are genetic benefits accrued to the female via her offspring. Several hypotheses address how benefits of this nature may underlie the evolution of polyandry by postcopulatory sexual selection (see "Ejaculate-Female Interactions, Cryptic Female Choice, and Sexually Antagonistic Coevolution" section).

Studies suggest that selection favors some moderate female remating rate determined by the trade-offs among various indirect and direct benefits and costs (Arnqvist & Nilsson 2000). Because sexual conflict over mating is widespread (Arnqvist & Rowe 2005), selection on females to maximize this ratio of benefits to costs will include any effects of male sexual coercion (Clutton-Brock & Parker 1995b). That is, an important direct "benefit" to females may in fact be minimizing the costs associated with

TABLE I Possible direct and indirect fitness benefits to females of multiple mating.

Direct Benefits	Indirect Benefits
Nuptial gift	Genetically more variable offspring
Access to male-controlled resources	Genetically more viable offspring
Additional paternal care	Genetically more resistant offspring
Sperm replenishment	Genetically more attractive and/or sexually competitive sons
Improved probability of fertilization	
Prospecting for future mates	
Reduced probability of predation through summation of antipredator defenses	
Reduced probability and/or intensity of male harassment or harm	
Reduced probability of infanticide	

(continued)

resisting the sexual advances of ardent and aggressive males (sometimes referred to as "convenience polyandry"). It is also important to consider that females may remate maladaptively, either too often (relative to a selective optimum) due either to antagonistic seduction by potential mates (Holland & Rice 1998) or to a genetic correlation between male and female mating rates (Halliday & Arnold 1987, Arnold & Halliday 1988; but see Sherman & Westneat 1988), or too infrequently due to sexual manipulation by previous mates (e.g., copulatory plugs and anti-aphrodisiacs; see "Avoiding Sperm Competition" section). The interests of an individual mated female and her most recent mate may be confluent immediately after mating, with both partners benefiting from delayed female remating. However, because the male will continue to benefit from any postponement of sperm competition, the interests of the female and this male over female remating will increasingly diverge as time progresses (figure I; Arnqvist & Rowe 2005).

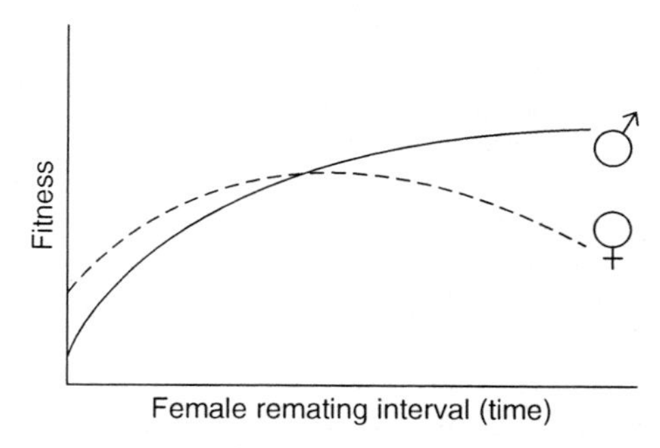

Figure I Graphical illustration of conflict over female remating rate between an individual female and her current mate in a hypothetical promiscuous species.

reproductive tracts for extended periods (days to years; reviewed by Birkhead & Møller 1993; Neubaum & Wolfner 1999), which will increase both the risk and the intensity of postcopulatory sexual selection.

THE RELATIONSHIP BETWEEN POSTCOPULATORY AND PRECOPULATORY SEXUAL SELECTION

All of sexual selection distills down to differential reproductive success covarying with the expression level of some trait(s) (i.e., ornaments, armaments, or mate preferences). There is no conceptual difference between the male-male competition occurring when two male northern elephant seals (*Mirounga angustirostris*) fight on a beach for access to receptive females and when the ejaculates of two yellow dung flies (*Scat(h)ophaga stercoraria*) compete for access to oocytes within a female's reproductive tract. Furthermore, models to explain the origin, evolution, and maintenance of female preferences (chapters 20, 23, and 24) apply equally to precopulatory mate choice and postcopulatory sire choice. Evidence suggests, for example, that the giant sperm tails observed in some *Drosophila* species are the cellular, postcopulatory equivalent of long plumage trains in peacocks, the result of "preferences" imposed by the female reproductive tract (Miller & Pitnick 2002; see "Case Studies," below).

Nevertheless, to understand the net magnitude of sexual selection, we have to know how precopulatory and postcopulatory processes interact to generate variance in reproductive success. When some males are excluded from the postcopulatory arena by precopulatory events, and males that are most attractive in the precopulatory arena also have greatest postcopulatory success, variance in reproductive success may be increased by considering both episodes of selection. For example, in the fruit fly *Drosophila simulans*, more attractive males also experience greater fertilization success during sperm competition (Hosken et al. 2008). Such a pattern could be attributable to cryptic female choice mechanisms determining sperm fate contingent upon her perception of relative male quality or due to condition dependence (see chapter 20) of male characters functioning in either selection episode. That is, males of relatively better health and vigor could invest relatively more energy into traits that influence pre- and postcopulatory success. In support of this contention, larger males of many species have been found to have higher mating and fertilization success (Birkhead & Møller 1998).

Alternatively, males that do well in one selection episode may, on average, perform poorly in the other, and hence summing both selection episodes may lead to lower variance in male reproductive success. In the water strider *Gerris lacustris*, for example, large males have an advantage in precopulatory male-male competition, but these same males have reduced fertilization success when their sperm compete with smaller males' sperm (Danielsson 2001). There are several possible (proximate and ultimate) explanations for such a pattern, one being that male adaptations function only in either competitive mating success or competitive fertilization success, and they are also energetically costly and hence trade off against each another. When such a trade-off occurs, males are predicted to conditionally or facultatively invest in one direction or the other. Indeed, such trade-offs may be optimized in different ways and lead to the evolution of alternative male reproductive strategies (e.g., precopulatory versus postcopulatory sexual selection specialists), which are expected to reduce variance in male reproductive success (see chapter 25).

The two scenarios presented above, with precopulatory and postcopulatory sexual selection either reinforcing or else trading off with one another probably represent idealized relationships. Actual associations between pre- and postcopulatory sexual selection are predicted to generally be more complex for a variety of reasons. Energetic trade-offs between an assortment of traits are possible, and patterns may be further complicated because males of different quality may face trade-offs that differ in kind, size, or shape (e.g., Getty 2002). In addition, there may be congruence of some elements of each episode but not others. For example, within the precopulatory selection episode, the male trait(s) underlying success in male-male competition for mates may differ from the trait(s) upon which female choice of mates is based, and these different traits may exhibit different patterns of covariation with the male trait(s) functioning in postcopulatory sexual selection. Finally, Simmons and Parker (1996) suggest that sperm competition will have no direct influence upon the intensity of sexual selection on male ornaments important for mating success, as the former will not impact the time costs associated with mating, which they contend ultimately determines the intensity of sexual selection.

CONCEPTS

Theoretical Foundations and Empirical Tests

In his landmark paper, Parker (1970a) convincingly argued that sperm competition was widespread among insects and frequently responsible for dramatic diversification in male behavioral, morphological, and physiological traits. To evaluate the selective consequences of sperm competition, investigators began to quantify patterns of sperm precedence. After all, the traits favored by selection will be very different if the first of two or more males to inseminate a female sires all of her offspring than if the last male sires all offspring or if paternity is shared among males. Knowledge of the sperm precedence pattern (i.e., the pattern of paternity) is of limited utility, however—a point not immediately appreciated—without accompanying knowledge of the mechanisms underlying the pattern. For example, high second male sperm precedence (referred to as "P_2") can result from the ability of copulating males to kill, remove, or displace competitor sperm residing within the female (in which case postcopulatory sexual selection on males is intense), differences in the timing of copulations, or from females depleting the supply of resident sperm to fertilize their eggs prior to remating (in which case postcopulatory sexual selection is weak).

Two clear predictions were quick to emerge from sperm competition theory (Parker 1970a). First, males should evolve adaptations serving to minimize the probability of their sperm entering into competition with the sperm of other males to fertilize eggs (often measured as sperm competition risk; see box 22.2). That is, males should evolve sperm competition avoidance traits. Second, males should evolve a level of investment in sperm production (i.e., relative testis size) that is positively associated with sperm competition risk (reviewed by Parker 1998).

Avoiding Sperm Competition

Male adaptations that result in their ejaculates avoiding competition over fertilizations with ejaculates of other males could be favored by postcopulatory sexual selection. This logic explains numerous reproductive traits, such as costly and time-consuming postinsemination associations between males and females (e.g., mate guarding by males; Alcock 1994a), that were poorly understood prior to Parker's (1970a) development of sperm competition theory. Adaptations to avoid sperm competition fall into two general categories: (1) traits that reduce the probability of a male's own sexual partner from remating (or that increase the female's remating interval), and hence acquiring the sperm of rival males, and (2) traits that eliminate or impair the competitive ability of sperm from rival males residing within the female.

In his review of postinsemination associations between males and females in insects, and their putative evolutionary maintenance by postcopulatory sexual selection, Alcock (1994a) divides known associations into four categories: (1) prolonged copulation following insemination, (2) postinsemination mating plugs, (3) prolonged contact between the male and female following copulation, and (4) monitoring of the female by the male following copulation without physically contacting her. To this list, we would expand category 2 to include the use by males of antiaphrodisiacs. This category thus includes any male-derived material

BOX 22.2 Ejaculate Expenditure and Allocation Models

Geoff Parker is responsible for most of the theoretical work on the evolution of sperm production and allocation strategies (reviewed in Parker 1998). Using game theory, these models have principally sought evolutionarily stable strategies (ESS) in (1) the level of investment in spermatogenesis and (2) the number of sperm allocated per copulation. An ESS is a strategy that cannot be beaten by an invading mutant strategy (see chapter 8 or 15). In this case, it is the strategic level of sperm production or allocation per ejaculate that maximizes male fitness and cannot therefore be beaten by any other investment/allocation strategy. In addition to the fundamental assumptions of the ESS approach (the phenotypic gambit; Grafen 1982; chapter 5, this volume), these models have assumed that investment in sperm trades off against investment in other aspects of reproductive success, such as searching for additional mates (figure Ia). Another important assumption, at least in the initial models, is that fertilization success is proportional to the relative number of sperm in competition—the "raffle principle," so named because in a lottery or raffle the number of tickets (= sperm) purchased determine the likelihood of prize winning (= gaining fertilizations). When the sperm from two males compete in some proportion (q) of females ($1 - q$ females mate with only one male), and provided that greater sperm number increases fertilization chances (either via the raffle principle or by displacement), then as q increases, the ESS is to increase sperm allocation per ejaculate. Because sperm production and sperm per ejaculate increases with testis size, one overarching conclusion from these models is that as sperm competition risk (q) increases, testis size should also increase (figure Ib). As discussed in the chapter and illustrated for bats in figures Ic and Id, this prediction has received overwhelming support from both intraspecific and interspecific studies.

(continued)

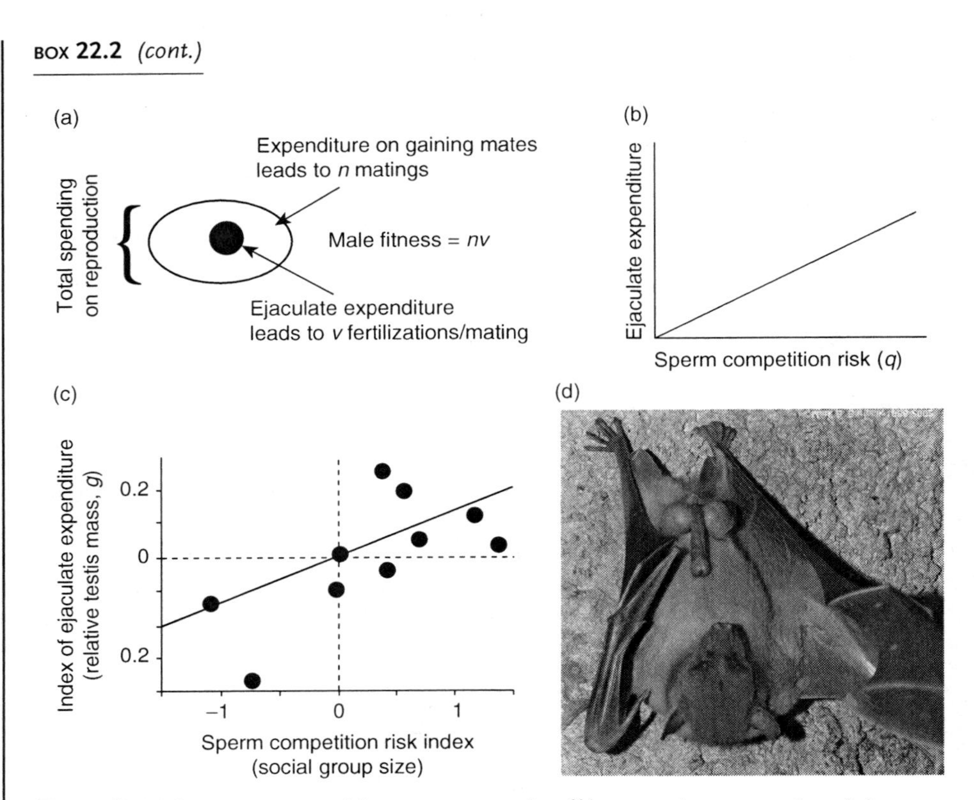

Figure II (a) In sperm competition games, a trade-off between investment in gaining mates and in gaining fertilizations (i.e., manufacturing sperm) is frequently assumed, with the product of these two fitness components determining male fitness. (b) Models predict that males should increase their expenditure on ejaculates when the risk of sperm competition increases. Specifically, ESS ejaculate expenditure (E^*) is approximately the proportion of females that mate more than once divided by 2 ($E^* = q/2$; Parker 1998). (c) The general prediction that ejaculate expenditure should increase with risk of sperm competition is upheld across many taxa, including the Megachiroptera. The graph shows $\log_{10}$ data corrected for phylogeny using independent contrasts. (d) As a consequence of intense sperm competition, the testes can weigh as much as 8.5% of total body mass on average in some bat species; shown here is a male Rousettus aegyptiacus in breeding condition, whose testes are a mere 2.15% of body mass. Illustrations (a and b) redrawn from Parker (1998) and (c) from Hosken (1998); (d) photo courtesy of Mark B. Bartosik.

In the models, each male's ejaculate enters into competition, and some mechanism determines which sperm are able to fertilize ova (= the *fertilization set*—the set of sperm that are subsequently randomly used to fertilize eggs). These games also frequently assume females have no control over male strategies (Parker 1998). The models have addressed a variety of sperm competition scenarios, such as unfair (loaded) raffles—in which each sperm does not have an equal fertilization probability because, for example, sperm entering first into the female have an advantage—and the effects of males having either fixed or random roles (e.g., first or second to mate) in a game. The games have further explored investment and allocation consequences of variation in male phenotype (e.g., *guarder* versus *sneaker* alternative mating strategies; see chapter 25), male tactics (e.g., within pair-bond mating versus extra-pair copulation), and in the information available to males (e.g., virgin versus nonvirgin female mating status; whether male is in favored or disfavored role in a loaded raffle).

or action that reduces the probability of his mate subsequently copulating with other males, despite no further association with or monitoring of the female by the male. Possible examples include seminal fluid components that render females unreceptive to male courtship (see *Drosophila melanogaster* case studies, below), the nonfertilizing ("apyrene") sperm of Lepidoptera, and physical harming of the female by the male during copulation (Johnstone & Keller 2000).

Whenever females with viable sperm in their reproductive tracts do copulate with a new sexual partner, selection may favor (depending on the relative costs/benefits) males that can avoid sperm competition by removing, destroying, or displacing the resident sperm. Relevant adaptations include penile structures that remove sperm from or relocate sperm within the female reproductive tract (see "Other Adaptations to Postcopulatory Sexual Selection: The Penis," below), frequent copulations and/or large volume ejaculates that serve to flush resident sperm from the female sperm stores, stimulatory activity that induces the female to eject her stored sperm, and "kamikaze" sperm or seminal spermicide that kills or incapacitates resident sperm. However, despite reports of evidence for the latter function of male ejaculates, all such claims remain unsupported or contentious (Moore et al. 1999; Snook & Hosken 2004).

Testis Size

Whenever sperm competition cannot be avoided, such that viable sperm from different males compete for access to ova, greater sperm numbers should enhance a male's competitive fertilization success, all other things being equal (the "raffle principle"; see box 22.2). Artificial insemination experiments that vary the ratio of the numbers of competing sperm from different males support this assumption (e.g., Martin et al. 1974). Consequently, the greater the level of sperm competition, the greater the benefits accrued by males for allocating more of their energy budget to sperm production. Because larger testes produce more sperm, male relative investment in testis mass is predicted to increase with the risk of sperm competition. This prediction has received robust empirical support through four different kinds of tests: (1) comparative/ phylogenetic analyses, (2) examination of among-male variation within populations, (3) experimental evolution studies, and (4) comparison between males employing alternative reproductive tactics.

Harcourt et al. (1981) examined primates and found significantly greater relative testis mass in species with multimale multifemale breeding systems (i.e., those expected to have a higher risk of sperm competition) than in species with monogamous or single male multifemale breeding systems (figure 22.1). This positive association between relative testis size and sperm competition risk has now been documented across numerous additional taxa (reviewed in Calhim & Birkhead 2007).

Investigations of intraspecific variation in relative testis mass or sperm production in a variety of taxa have similarly found positive relationships with both (1) success in sperm competition and (2) the risk of encountering sperm competition. First, males copulating longer, transferring larger ejaculates or greater numbers of sperm than their competitors achieve higher competitive fertilization success (Birkhead & Møller 1998). Second, studies comparing the relative testis size of males adopting alternative mating tactics within populations (see chapter 25) support the predicted relationship. In the Atlantic salmon *Salmo salar*, for example, males either leave natal rivers and head to sea, before returning as large anadromous males, or they forgo this journey and develop precociously into sexually mature but tiny parr males. Anadromous males are able to defend females and nests, whereas parr males must sneak fertilizations (Gage et al. 1995). Consequently, parr always face sperm competition, whereas anadromorphs experience it only occasionally. Theory therefore predicts that parr should invest relatively more in testis size than anadromorphs, and this is precisely what is observed (Gage et al. 1995). Similar findings have been reported in numerous other species with alternative male mating tactics (reviewed by Pitnick et al. 2009a). Even though both macroevolutionary and microevolutionary studies support the predicted evolutionary response in testis size to postcopulatory sexual selection, it is important to note such evidence is correlational, and thus does not establish cause and effect.

This concern prompted experimental tests of the relative testis mass-sperm competition risk relationship. In independent investigations using *D. melanogaster* and *S. stercoraria*, respectively, Pitnick et al. (2001) and Hosken and Ward (2001) explored how male investment in their testes evolved over many generations while experimentally eliminating sperm

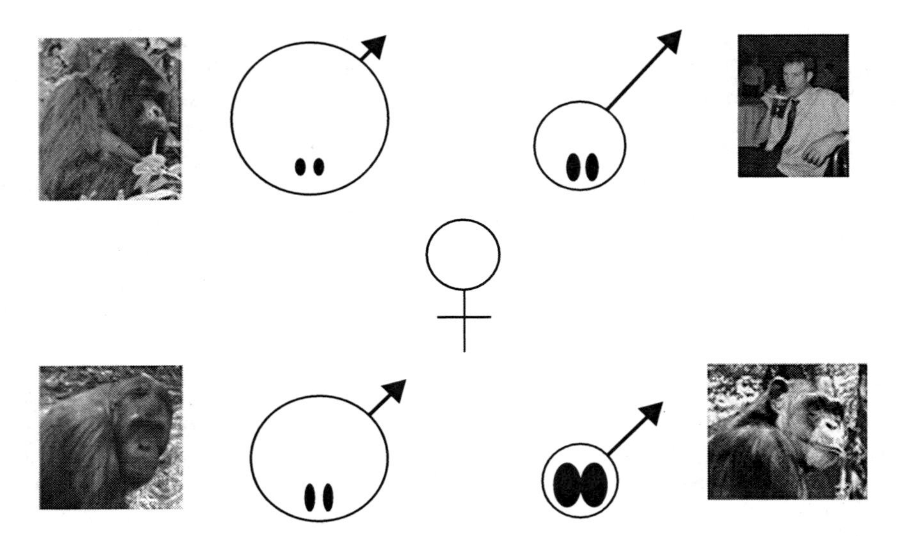

FIGURE **22.1** Relative size of the primary male sexual organs in the four largest great apes, *Gorilla gorilla*, *Pongo pygmaeus*, *Pan troglodytes*, and *Homo sapiens*. Male symbols are scaled relative to the female symbol to depict body size dimorphism; the filled ovals inside the male symbol and the length of the arrows respectively indicate relative testis and penis size. Gorillas live in single male–multifemale breeding groups, and orangutans are generally solitary, forming monogamous consortships during estrus. Hence in both species there is an extremely low risk of sperm competition. Males are much larger than females (*Gorilla*: ca. 275 kg versus 120 kg; *Pongo*: ca. 90 kg versus 50 kg) but have small penises and make a small investment in testes. Humans exhibit moderate sexual size dimorphism (male ca. 75 kg versus females 55 kg) and have breeding systems that present a mix of monogamy and varying levels of both male and female promiscuity. This breeding pattern, together with the human preoccupation with sex, suggests a moderate intensity of postcopulatory sexual selection, which is reflected in the relatively large testes (about twice the relative mass as *Gorilla*) and a large, specialized penis. Chimpanzees are highly promiscuous, with females reported to copulate up to 50 times a day with a dozen different males. They are the least sexually size dimorphic (males ca. 60 kg versus females 50 kg), with males sporting enormous testes (about 10 times the relative mass as *Gorilla*) and a moderately large specialized penis. Redrawn from Short (1977); gorilla, orangutan, and chimpanzee photos courtesy of Liz Rogers.

competition through enforced monogamy in laboratory populations. Both studies found that males from experimentally monogamous populations had relatively smaller testes than did males from polyandrous populations. On the whole, then, there is very good support for this key prediction of sperm competition theory.

Costly Ejaculates and Strategic Ejaculation

Behavioral ecologists in the 1980s and 1990s began to question another dogmatic presumption about sex differences: sperm are cheap. Traditionally, sperm have been considered energetically inexpensive and effectively limitless in supply. Dewsbury (1982) recognized, however, that whereas individual sperm may be cheap to manufacture, there can be nontrivial costs associated with producing a whole ejaculate, which in extreme cases can be as much as 30% of a male's body mass (Gwynne 1982). Moreover, there are now numerous kinds of evidence in support of the contention that spermatogenesis per se can be expensive (reviewed by Wedell et al. 2002). For example, sperm supplies can be limited, with the number of sperm per ejaculate declining with successive copulations. Dietary restriction has been shown to constrain sperm production in the Indian meal moth *Plodia interpunctella*. Increased sperm production decreases life span in the nematode *Caenorhabditis elegans*, and increases in sperm size are correlated with delayed male reproductive maturity across *Drosophila* species.

When sperm production is costly, theory suggests that males who adjust the number of sperm per ejaculate to match the risk of sperm competition they face in any particular mating would have a selective advantage (Parker 1998). There is now evidence that males from a variety of taxa strategically tailor their ejaculates based on the perceived risk of sperm competition, female mating status, or female quality (reviewed by Wedell et al. 2002). Adjustment of ejaculate size often requires only an alteration of copulation duration, because the number of sperm transferred to a female is a function how long copulation lasts. In other cases the mechanism(s) involved may be considerably more sophisticated. In some rodents, for example, innervation of the vas deferens differs among species with differing levels of sperm competition. Those that experience sperm competition have functional μ-receptors that respond to endogenous levels of circulating receptor-agonists such as ß-endorphin, which ultimately influences how much the vas deferens contracts. Because male-male interactions can affect circulating levels of these agonists, they can directly affect contractions of the vas deferens and hence stimulate the delivery of more sperm (Pound 1999).

Ejaculate-Female Interactions, Cryptic Female Choice, and Sexually Antagonistic Coevolution

In the preceding sections, mate guarding, manipulation of competitor sperm by males, and the ejaculation of a great many sperm have been given primacy as the mechanisms by which males achieve differential fertilization success via postcopulatory processes. Since publication of the influential book by Eberhard (1996), however, much more empirical attention has been paid to cryptic female choice. With internal insemination, male-female interactions during copulation and subsequent ejaculate-female interactions can influence numerous biochemical, physiological, morphological, and behavioral processes influencing insemination, sperm migration, sperm storage, the maintenance of viable sperm, and sperm modification, which must all be properly executed before fertilization can begin (reviewed by Pitnick et al. 2009b). As a consequence, there are myriad ways for females to potentially influence sperm use, although cases in which female manipulation has been unequivocally shown to generate postcopulatory variance in reproductive success (i.e., sexual selection) are limited.

Given that sexually reproducing females need viable sperm to reproduce and often possess adaptations for sustaining sperm viability, it may seem paradoxical that the female reproductive tract is often an environment that is somewhat unfavorable to, and hence potentially selective on, sperm (reviewed by Birkhead et al. 1993; Pitnick et al. 2009b). Female means of precluding some sperm reaching eggs may include (1) active sperm ejection by females, (2) physical barriers (e.g., cervix, long ducts), (3) chemical barriers (e.g., low pH and viscous mucus), and (4) leukocytic/phagocytotic responses within the female. Consequently, in many species only a small proportion of the inseminated sperm ever have the opportunity to encounter an egg. Sperm attrition within the female reproductive tract may be due to natural selection on female fecundity, for example, safeguarding against parasites, bacterial infections, and other pathogens that may enter the female reproductive tract, particularly at the time of mating. Additionally, "challenges" to sperm may be adaptations to discriminate against sperm that have abnormal morphology, weak motility, or are otherwise unfit for fertilization, or to reduce the risk of polyspermy (more than one sperm penetrating the egg, which can lower zygote viability).

Postcopulatory sexual selection is also expected to result in female reproductive tracts that select on sperm for two reasons. Indirect selection can favor multiple mating and mechanisms for biasing sperm use by females if these result in the best sperm fertilizing the female's eggs (Eberhard 1996). In fact, several experimental studies have demonstrated that postcopulatory sexual selection can enhance offspring viability (reviewed by Neff & Pitcher 2005). Three alternative models have been proposed to explain how such an adaptive process might work, with the main difference between them being how sperm vary in quality. According to the *sexually selected sperm hypothesis*, by creating a competitive fertilization environment, females increase the probability that their eggs will be fertilized by males who are good at sperm competition, and hence their sons will also be superior sperm competitors (reviewed by Keller & Reeve 1995). Alternatively, according to the *good sperm hypothesis*, females accrue indirect genetic benefits through positive covariation of sperm competitive ability and male genetic quality (e.g., Yasui 1997).

In support of this hypothesis, studies have found (1) a positive relationship between males' sperm competitive ability and the viability (i.e., development time, survival) of their offspring in the yellow dung fly (see "Case Studies," below) and the marsupial *Antechinus stuartii,* (2) positive relationships between male attractiveness or condition and sperm competitiveness in flour beetles, *D. simulans,* dung flies (see "Case Studies," below), red deer, guppies, and Atlantic cod, and (3) condition dependence of ejaculate characteristics in a dung beetle and of sperm offense ability in *D. melanogaster* (reviewed by Pitnick et al. 2009b). Finally, the *genetic compatibility hypothesis* contends the best sperm are those bearing haplotypes most compatible with the female genome (Neff & Pitcher 2005). Some of the strongest evidence in favor of this hypothesis comes from experiments demonstrating fertilization bias to minimize inbreeding (or *selfing*; reviewed by Pitnick et al. 2009b). It is important to note that this final hypothesis may not generate any selection if each male type is compatible with a female type and all occur in a population in equal frequencies.

Direct natural selection on females can also drive the evolution of female adaptations that bias paternity. To the extent that male adaptations to bias paternity are harmful to females (e.g., increasing frequency of polyspermy or physically harming the female; see below), selection could favor female adaptations that provide resistance to such male manipulation, potentially leading to arms races or perpetual coevolutionary cycles (see chapter 23). In general, aspects of cooperation and conflict are both likely to select on traits involved in ejaculate-female interactions.

It has become clear that postcopulatory variance in male reproductive success as pure male-male competition (i.e., sperm competition), in the strictest sense, is largely a fallacy, at least in species with internal fertilization. In fact, even in externally fertilizing species, sperm-female interactions may mediate the ability of sperm to reach eggs (chapter 21, e.g., salmonid sperm swim faster and live significantly longer in the presence of the ovarian fluid suffusing the eggs in a spawn; Turner & Montgomerie 2002). Thus, to the extent that there is among-male variation for traits that determine ejaculate-female compatibility, differential male competitive fertilization success will not be attributable solely to mechanisms of sperm competition per se. Simply put, a male can transfer a relatively great many sperm, but if they are not compatible (e.g., morphologically, biochemically) with the female, then they may be inferior competitors for fertilizations relative to sperm from other males that are more compatible.

We have only recently begun to appreciate the extent to which male-by-female interactions influence competitive fertilization success (see below). Moreover, because the complete set of underlying mechanisms will not be known in most cases, it is typically not possible to discriminate between sperm competition and cryptic female choice as alternative processes giving rise to differential competitive male fertilization success. As a consequence, we agree with Eberhard (1996) that the idea of sperm competition and cryptic female choice as independent phenomena largely presents a false dichotomy: they co-occur in a broad sense, and understanding their relative importance is the goal. With this philosophy in mind, we now turn our attention back to the predominant types of adaptations arising from postcopulatory sexual selection.

Other Adaptations to Postcopulatory Sexual Selection

Female Reproductive Tract Morphology

The design of the female reproductive tract suggests adaptation to control (1) the timing of fertilization and (2) paternity. First, female tracts are well suited to prevent males from ejaculating directly onto eggs and to restrict access of sperm to the ovaries. Females of most species are equipped with one or more specialized organs for storing sperm, typically referred to as a *spermatheca* (reviewed by Neubaum & Wolfner 1999). Sperm may survive within these organs for weeks, months, or years (reviewed by Birkhead & Møller 1993). Prolonged sperm storage uncouples copulation and fertilization and, combined with multiple mating by females, facilitates postcopulatory sexual selection (Keller & Reeve 1995; Eberhard 1996). The design of the female tract further prevents males from directly accessing the spermatheca(e), by virtue of the location of these organs and of the often lengthy and complex nature of their sperm ducts (for some interesting exceptions, see "The Penis," below). As a consequence, females potentially retain some influence over the fate of both a recent ejaculate and of sperm from previous mates residing within the spermatheca(e). Studies of diverse taxa have documented a variety of sperm handling options available to females, including sperm ejection from the tract, sperm digestion, discrete storage of sperm from different males in different organs, or sperm mixing, all of which could influence paternity (Eberhard 1996; Birkhead & Møller 1998). However, it is often unclear if females do any of this differentially with respect to male (or ejaculate) phenotype.

Female reproductive tract morphology, particularly the spermathecae and their associated ducts and glands, exhibits remarkable diversity and appears to evolve rapidly, as predicted of sexually selected traits (figure 22.2; reviewed by Pitnick et al. 2009b). A widespread pattern of correlated evolution between the dimensions of these female characters and sperm length has emerged from comparative/phylogenetic analyses of diverse taxa (reviewed by Pitnick et al. 2009b), and an experimental evolution study suggests that these interacting sex-specific traits coevolve (see the *Drosophila* case study, below). Although the complexity and apparent selectivity of the female reproductive tract is widely postulated to have evolved to challenge males (or more accurately, their ejaculates) or otherwise control paternity, with few exceptions, the functional relationship between variation in female reproductive morphology and sperm use patterns are unknown.

The Penis

The male intromittent organ (aedeagus or penis) is probably the fastest evolving morphological trait, and comparative studies show that male genitalic complexity increases across species with the level of sperm competition (Eberhard 1985; Hosken & Stockley 2004). With regard to male genital form and function, empirical studies suggest that there have been five generalized evolutionary responses to postcopulatory sexual selection (figure 22.3; reviewed by Eberhard 1985, 1996; Hosken & Stockley 2004). First, the penis may be adapted to positioning the ejaculate within the female in a manner that improves the probability of the sperm being stored and used for fertilization relative to competitor sperm. For example, the penis of the rove beetle *Aleochara tristis* consists of a flagellum that is twice as long as the male (figure 22.3a). It lies coiled within the male and during copula, winds up through the lengthy female spermathecal duct to guide the male's growing spermatophore (Gack & Peschke 2005). Second, in those rare cases of males having penises that can reach the spermatheca, the penis typically has adaptations to remove or reposition resident sperm, such as occurs in some odonates (figure 22.3b; Cordoba-Aguilar et al. 2003). Third, even when males cannot directly access the spermatheca(e), they may evolve structures that stimulate the female into repositioning or ejecting resident sperm (figure 22.3c; Cordoba-Aguilar et al. 2003). Fourth, males may evolve penile structures that physically harm the female, and hence presumably impact female remating and/or sperm use. For example, the end of the penis of the bean weevil *Callosobruchus maculatus* is outfitted with sclerotized spines that are everted during copulation and have been shown to punch holes in or otherwise damage the female reproductive tract (figure 22.3d). A recent analysis of variation among 13 populations of this beetle found a significant positive relationship between the length of penile spines, harm to females during mating and success in sperm competition (Hotzy & Arnqvist 2009). Fifth, males can evolve a penis designed to circumvent female control over sperm use altogether, such as the blade-like penis of bedbug *Cimex lectularius* (figure 22.3e). Males use these to pierce the abdominal wall of the female and inseminate directly into her body cavity (referred to as *traumatic* or *hypodermic insemination*), through which sperm travel to the ovaries (Siva-Jothy 2006). In addition to genitals, other structures (e.g., legs, antennae, or mouthparts), that stimulate females during copulation (= copulatory courtship) and influence sperm use can also evolve (reviewed in Eberhard 1996).

Sperm Size and "Quality"

All other things being equal, transferring relatively great numbers of sperm in a spawn or insemination should contribute to competitive fertilization success (see above, box 22.2, and chapter 21). Because sperm size and sperm number are expected to trade off, theory contends that whenever there is even weak sperm competition, males should produce the smallest size sperm possible to maximize the number of sperm produced for a given level of investment in spermatogenesis (Parker 1982; box 20.1 of this volume). However, empirical investigations of sperm function and comparative studies of variation in sperm form relative to the risk of sperm competition indicate that frequently all things are not equal when it comes to postcopulatory sexual selection.

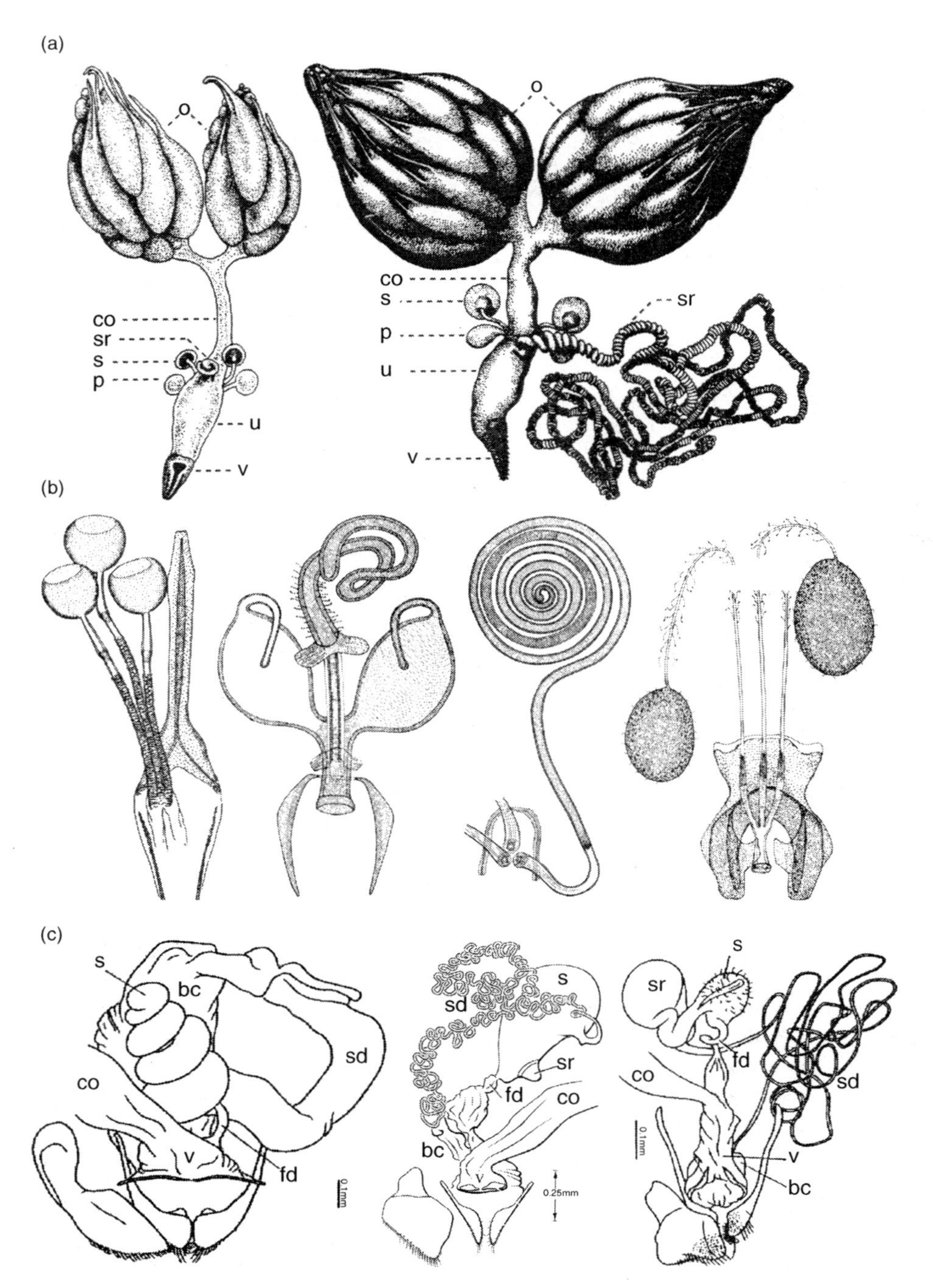

FIGURE **22.2** Interspecific variation in female sperm-storage organ morphology. (a) Female reproductive tracts of *Drosophila pseudoobscura* (left), which has a short seminal receptacle (0.41 mm) and short sperm (0.36 mm), and of *D. bifurca*, which has the longest known seminal receptacle (81.67 mm) and sperm (58.29 mm). From Patterson (1943). (b) Spermathecae of four species of robber flies (Asilidae; note that not all three spermathecae are drawn for each); the sperm of robber flies are not well studied. Reprinted from Theodor (1976) with permission from the Israel Academy of Sciences and Humanities. (c) Complex sperm-storage organ morphology of diving beetles (Dytiscidae). Sperm morphology in this beetle lineage is equally complex and rapidly divergent (see figure 22.4). *Abbreviations*: bc, bursa copulatrix; co, common oviduct; fd, fertilization duct; o, ovaries; s, spermatheca; sd, spermathecal duct; sr, seminal receptacle; p, parovarium; u, uterus; and v, vagina. Adapted from Miller (2001) and Miller et al. (2006) with permission.

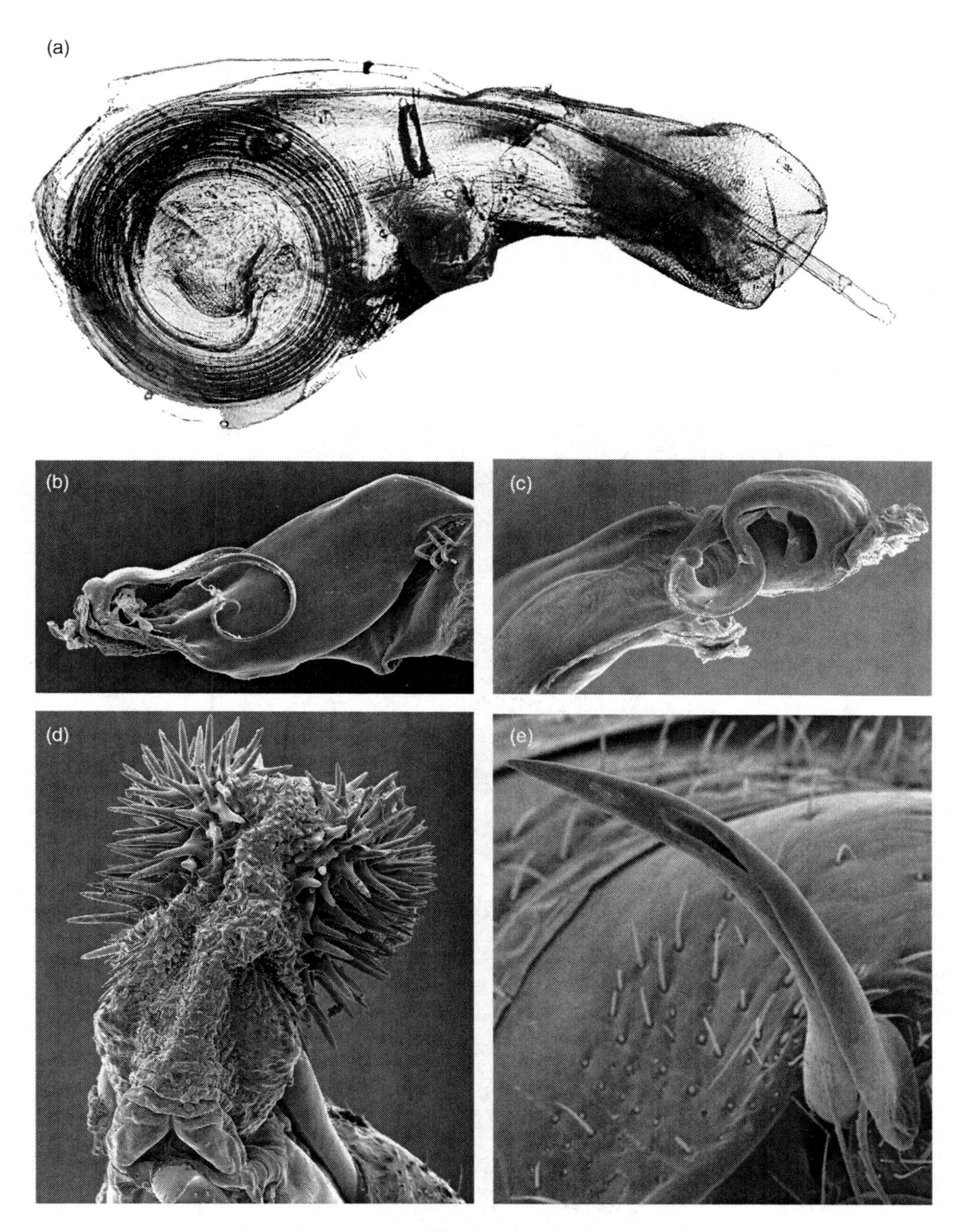

FIGURE **22.3** Generalized adaptations in penis morphology to postcopulatory sexual selection. (a) The elongate penis of a rove beetle, capable of winding up the long, narrow duct of the female's spermatheca. (b) The penis of the damselfly *Calopteryx haemorrhoidalis asturica*, bearing two lateral horn-like, spine-covered appendages, which stimulate the female to induce her ejection of resident sperm. (c) The penis of the damselfly *Ischnura elegans*, bearing two distal, thin, lateral processes that are capable of entering the female's paired spermathecae and mechanically trapping and removing resident sperm. (d) The spine-tipped penis of the bean weevil *Calosobruchus maculatus,* which has been demonstrated to damage the female reproductive tract and to correlate with male competitive fertilization success. (e) Intromittent organ (a modified paramere) of a bed bug, used to traumatically inseminate females. Reproduced with permission from (a) Gack and Peschke (2005), (b and c) Cordoba-Aguilar et al. (2003), (d) courtesy of G. Arnqvist and J. Rönn, and (e) courtesy of A. Syred, PSmicrographics.

For example, experiments controlling the number of sperm inseminated into females have found repeatable and/or heritable differences among males in ejaculate performance or the outcome of sperm competition (reviewed by Pitnick et al. 2009a). In addition, sperm viability positively covaries with the intensity of sperm competition across insect species (Hunter & Birkhead 2002). Moreover, spermatozoa are the most diverse cell type known (figure 22.4), exhibiting rapid and dramatic evolutionary divergence in form (reviewed by Pitnick et al. 2009a), which is typical of traits subject to intense sexual selection. Unfortunately, with few exceptions (e.g., sperm size and sperm conjugation; see figure 22.4 and "Case Studies," below), the adaptive significance of most of the variation in sperm form is unknown (reviewed by Pitnick et al. 2009a).

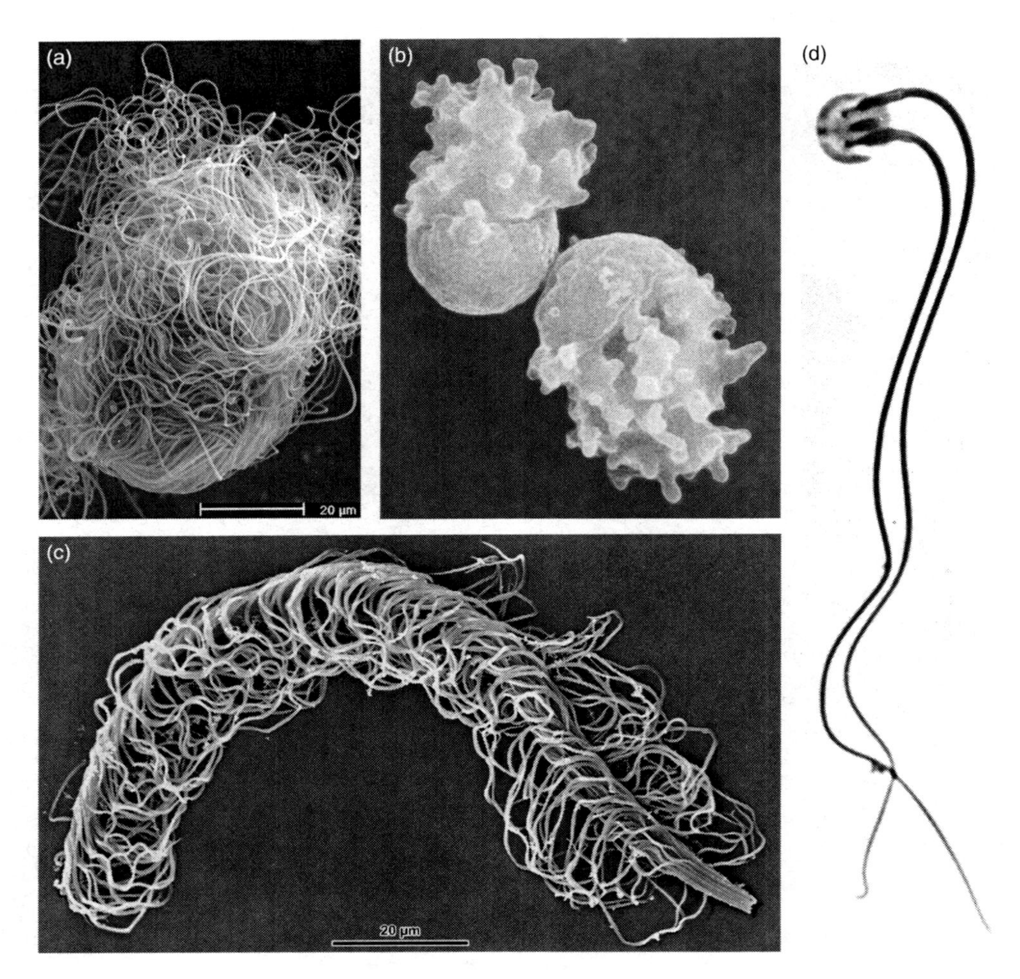

FIGURE **22.4** Sperm morphological diversity attributed to postcopulatory sexual selection. (a) SEM of single, 58-mm-long *D. bifurca* spermatozoon dissected from a male's seminal vesicle, where sperm are individually rolled into compact balls. (b) SEM of amoeboid spermatozoa of the nematode *Caenorhabditis elegans* (bar = 1 µm). (c) SEM of a spermatodesm of the insect Mantophasma zephyra. (d) Paired sperm of the gray short-tailed opossum, *Monodelphis domestica*. Sperm conjugate within the male seminal vesicles and then swim as "cooperative" bundles (c) or pairs (d) through the female reproductive tract. Reproduced with permission from (a) Bjork and Pitnick (2006; photo by R. Dallai), (b) LaMunyon and Ward (1998), (c) Dallai et al. (2003), and (d) Taggart et al. (1993).

The most thoroughly studied aspect of sperm form is their size, which appears to have evolved in response to postcopulatory sexual selection in numerous lineages (reviewed by Pitnick et al. 2009a). For example, sperm size positively correlates with fertilization success in the bulb mite *Rhizoglyphus robini*, the nematode *C. elegans* (figure 22.4b), and in the freshwater snail *Viviparus ater*. Additionally, selection lines of *C. elegans* evolved larger sperm in response to experimentally increased levels of sperm competition (for a related study, see the *Drosophila* case study, below). Further, comparative studies of nematodes, insects, fish, frogs, and some bird and some mammal lineages (but not others) have found significant positive relationships between sperm length and the risk or intensity of sperm competition (see box 22.2; reviewed by Pitnick et al. 2009a). Finally, as discussed above, sperm length appears to coevolve with dimensions of the female reproductive tract.

Seminal Fluid

Mating induces a multitude of changes in female gene expression, immunology, physiology, and behavior, essentially converting them from an *unmated* to a *mated* state (reviewed by Pitnick et al. 2009b). Many of these changes have been shown to be triggered by seminal fluid, which tends to be biochemically and functionally complex. Seminal fluid proteins influence sperm movement, storage, and survival, in some cases by inducing muscular contraction and associated changes in the conformation of the female reproductive tract or, in the case of some mammal species, by altering the penetrability of mucus within the female tract. Seminal fluid proteins also influence egg production and oviposition, as well as female sexual receptivity. Given all of these functions, it is highly likely that variation among males in the composition of their seminal fluid contributes to variation in competitive male fertilization success (reviewed by Chapman 2001).

As would be expected of targets of intense sexual selection, genes encoding seminal proteins typically evolve rapidly: at about twice the rate of nonreproductive proteins (reviewed by Pitnick et al. 2009b). For example, the SEMG2 gene of primates (encoding semenogelin II, which plays a biochemical role in the formation of a mating plug; see "Avoiding Sperm Competition," above) has been shown to evolve rapidly in some lineages, particularly in those with the highest levels of female promiscuity and hence risk of sperm competition (Dorus et al. 2004).

Male x Male and Ejaculate x Female Interactions

Thus far we have considered that competitive male fertilization success might be influenced by a multitude of male traits including sperm quantity, sperm quality, and seminal fluid composition, as well as penis size, shape, and no doubt motion, in addition to other possible forms of copulatory courtship. Numerous female traits may also be important, including reproductive tract morphology, biochemistry, and physiology, in addition to female perception of male and/or ejaculate quality. However, the efficacies of some of these sex-specific traits are expected to be influenced by the character state of the interacting traits in the opposite sex (e.g., Miller & Pitnick 2002). As a result, it may be that a given male's sperm competitiveness is difficult to predict, as it would depend upon the specific genotypes of both the male(s) he is competing with and of the female the males are competing within; this is an example of interacting phenotypes (see chapter 14). The result may be nontransitivity among males in their sperm competitive ability in a manner comparable to the rock-paper-scissors game (i.e., "a" beats "b," "b" beats "c," "c" beats "a"; see also chapter 3). Such a process should theoretically maintain genetic variation in traits that influence sperm competition (reviewed by Clark 2002).

CASE STUDIES

To illustrate some of the conceptual issues outlined above, we review studies of postcopulatory sexual selection in two models systems: (1) the yellow dung fly and (2) the fruit fly. An enormous body of relevant work has been undertaken on these species, and we touch on only a few salient aspects of this work here.

The Yellow Dung Fly, *Scathophaga stercoraria*

Since Parker's groundbreaking work on the yellow dung fly, *S. stercoraria*, this system has been a model for studies of sperm competition. These

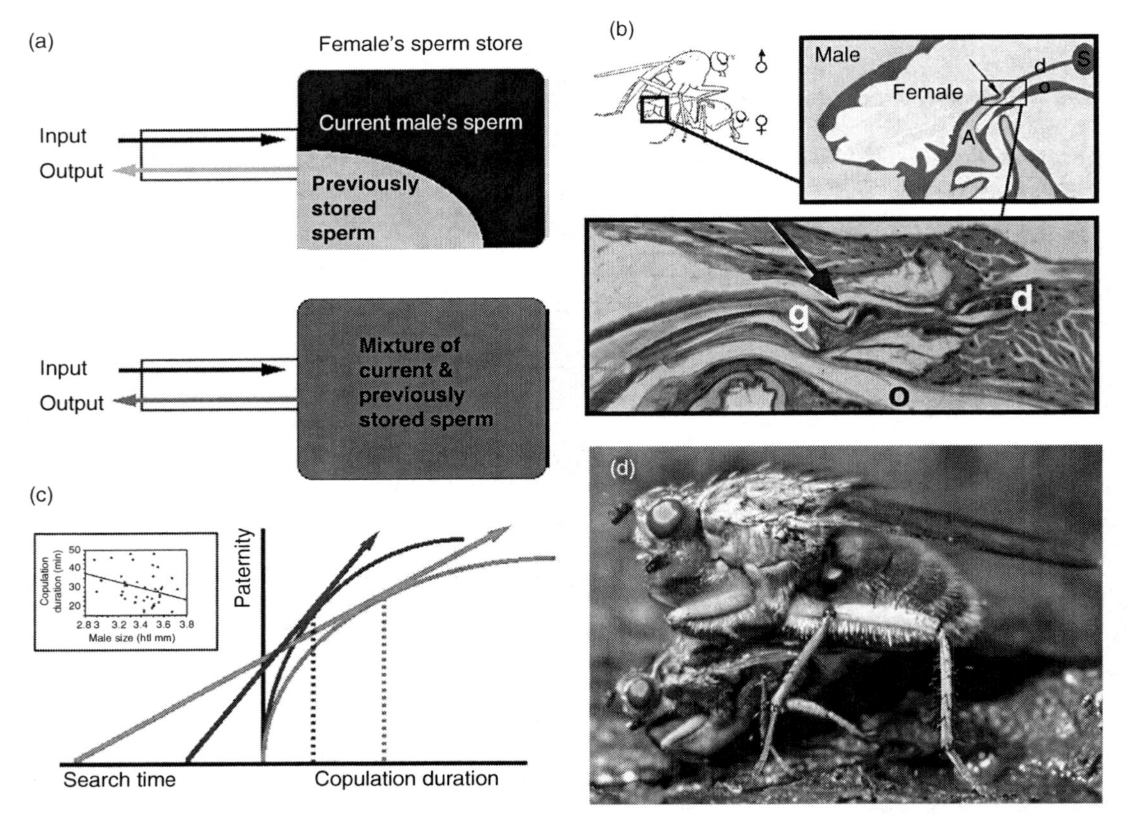

FIGURE **22.5** Investigations of postcopulatory sexual selection in the yellow dung fly, *Scathophaga stercoraria*. (a) Two models of sperm displacement in the yellow dung fly. *Top panel*: Sperm displacement is nonrandom; previously stored sperm is displaced by the incoming ejaculate before any incoming ejaculate is (self-)displaced. *Bottom panel:* Sperm displacement is random, and sperm mix instantaneously in the sperm store. As a consequence, incoming and previously stored sperm are both displaced. Additionally, displacement of previously stored sperm follows an exponentially diminishing return because as time proceeds, the current male displaces an increasing volume of self-sperm relative to previously stored sperm, which results in approximate expected paternity share of the form: $P_2 \approx [1 - \exp(-c_2 t_2)]$, where c_2 is the (constant) proportion of sperm displaced by male 2 per unit time, t_2 is the time spent copulating by male 2 and P_2 is the number of eggs fertilized by the second male. This approximation implies that the paternity share of the second male to mate is dependent only on the number of sperm transferred by him; what the previous male has done is (usually) irrelevant. The bottom panel more accurately describes empirical *S. stercoraria* data. Redrawn from Parker and Simmons (1991). (b) Yellow dung fly genitals during copulation. *Top right panel*: A line diagram close-up of the genitals of a copulating pair with internal details illustrating how the aedeagus (penis) fits into the female reproductive tract (light gray = male; white = female). *Lower panel*: A close-up, thin-section of the aedeagus tip and its proximity to the spermathecal duct (stained with hematoxylin and counterstained with eosin). Sperm can be seen exiting the male gonopore, and the female's spermathecal duct contains sperm. *Abbreviations*: A, aedeagus; S, spermathecae (one of the three); d, spermathecal duct; o, oviduct; g, male gonopore. The solid arrow marks the tip of the aedeagus at the entrance of the duct leading to a sperm store. This arrangement during copulation means sperm displacement cannot be direct as assumed by Parker and Simmons (1991). (c) Diagrammatic representation of how the marginal value approach was used to assess optimal copulation duration dung fly males of different size (black = large males; gray = small males). Curved lines represent the gain in paternity as a function of copulation duration; this relationship is steeper for large males. Lines with arrowheads are tangents from average search times for each male size class; large males have shorter search times because they have more takeovers. Dotted lines represent predicted copulation duration of small (gray) and large (black) males (drawn perpendicular from the x-axis to the intercept of the tangent flies are ubiquitous inhabitants of cow droppings throughout much of the northern hemisphere. Males aggregate on and around fresh droppings and wait for gravid females who come to the dung to lay their eggs. In the first sperm competition study of this fly, Parker (1970b) employed the sterile male technique to show that the last male to mate with a female sires around 80–85% of offspring in the subsequent clutch, and the number of offspring sired is largely independent of the number and timing of previous copulations. However, when copulations were interrupted, the percentage of offspring sired by the most recent mate increased with increasing copulation duration (figure 2.2). There are several potential explanations for these observations. Because the number of sperm moving into the female's spermathecae increased with copulation duration, and because the volume of the female spermathecae appeared fixed, the most plausible explanation is that last male's ejaculate displaces previously stored sperm. Parker and Simmons (1991) developed and tested two models of how this process could occur: sperm are displaced either randomly or nonrandomly from female sperm stores (figure 22.5a). With random displacement males will displace an ever-increasing volume of their own sperm as copulation proceeds. The models had several assumptions: sperm were transferred to females at a constant rate, 1 unit of ejaculate input resulted in 1 unit of ejaculate output, and males could directly access the spermathecae. Predictions were then tested and a model of random displacement with instantaneous mixing was best supported by the data

Although the *direct displacement with random mixing* model fits the empirical data remarkably well, it was unclear how direct displacement of rival sperm could occur, because female spermathecae are accessible only via long, narrow ducts. An extensive investigation of genital interactions during copulation unequivocally showed that males could not directly access the spermathecae (Hosken & Ward 2000; figure 22.5b). An indirect sperm displacement model was thus derived and used to predict sperm utilization patterns. Over most male body sizes (sperm transfer rate varies positively and monotonically with male size), this new model fit the empirical data as well as the direct displacement model, but additionally predicted sperm displacement more accurately for small males (Simmons et al. 1999).

In a series of optimality models using the marginal value theorem (MVT; a method to find the strategy that maximizes gains per unit time in instances where unit returns decline with time; chapter 8), Parker and colleagues further incorporated such variables as male size-dependent mate searching time, probability of takeovers (having their mate stolen during copula by another male), rate of sperm transfer, food foraging time, mating status (virgin versus mated), body size, and egg content of females to explore the relationship between male size and copulation duration (e.g., Parker 1992; Parker et al. 1999). With this approach they had previously predicted an average (across all male sizes) optimal copulation duration of about 42 minutes, which is very close to the observed average of about 36 minutes (Parker & Simmons 1991; and see chapter 2), and the inclusion of sperm production costs is likely to improve this fit. The MVT approach was subsequently applied to different sized males and predicted that larger males should copulate for less time than smaller ones, because the rate of sperm displacement, and hence the P_2 gain curve, is steeper for larger males, and becuase larger males have shorter search times due to a higher probability of successfully stealing a female from a rival male (figure 22.5c). The MVT predictions match the empirical data extraordinarily well, because larger males do have shorter copulations (figure 22.5c inset). Hence, it appears that different-sized males adjust copulation duration to achieve equal displacement and to maximize their expected fertilization gains (Parker & Simmons 1994), which explains why any male mating last fertilizes (on average) about 82% of the subsequent clutch.

In addition to optimizing paternity through adjustments in copulation duration, male *S. stercoraria* also exhibit two sperm competition avoidance

FIGURE **22.5** ***(continued)*** and the curve). The MVT model predicts that large males should have shorter duration copulations—a prediction receiving empirical support (see insert, upper left corner; D. J. Hosken, unpublished data). It therefore appears that males of different size are displacing rivals' sperm from the female sperm stores in an optimal manner. (d) Postcopulatory guarding by a male yellow dung fly of his mate while she oviposits into a dung pat. Image courtesy of P. Jann.

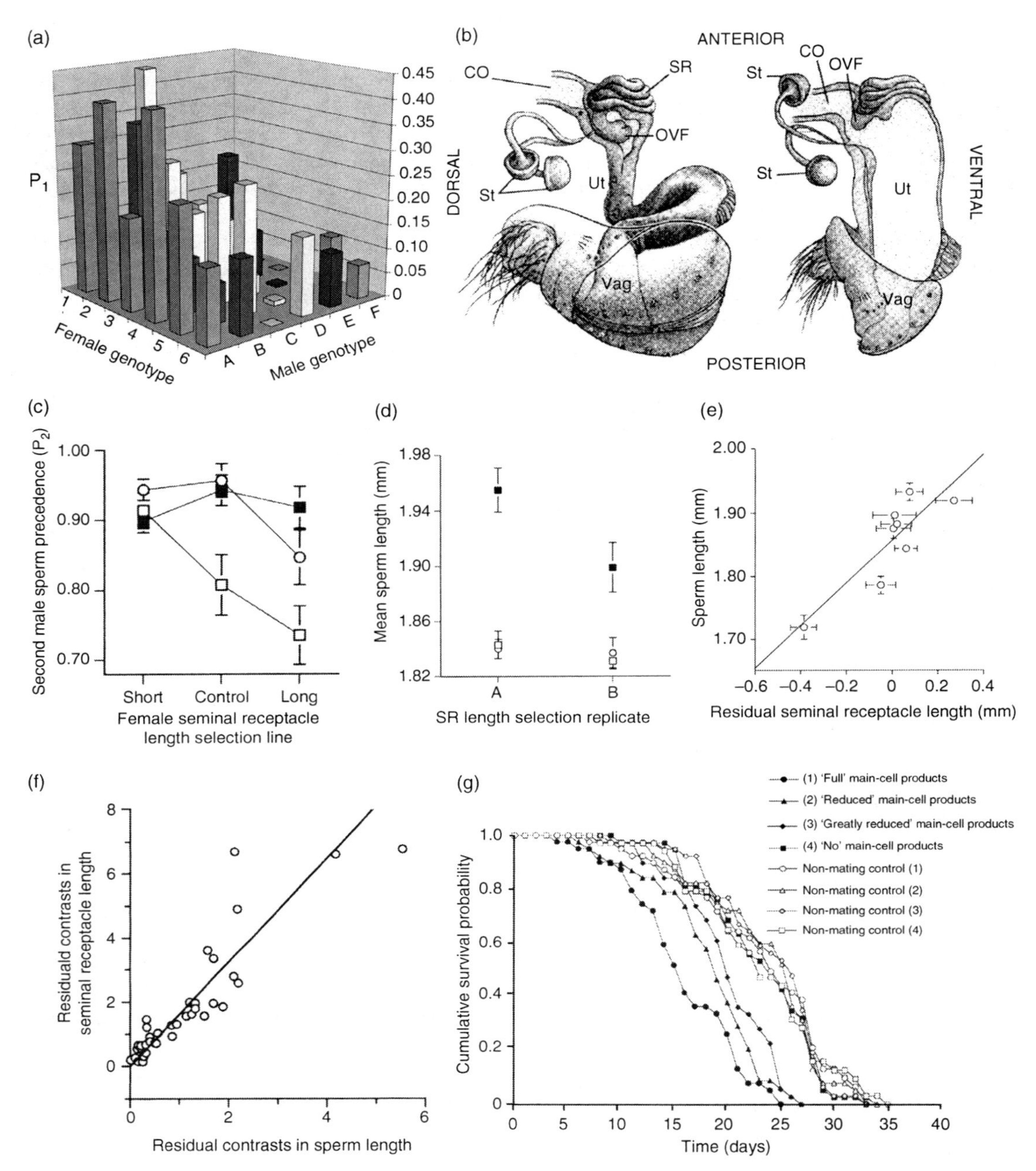

FIGURE **22.6** Investigations of postcopulatory sexual selection in the fruit fly, *Drosophila melanogaster*, and related species. (a) Male x female genotypic interactions influence competitive male fertilization success, shown here for P_1 (bar color corresponds to male genotype: S1–S6). (b) Seminal proteins trigger dramatic changes in the conformation of the female reproductive tract (left: tract prior to start of ejaculate transfer; right: postinsemination and during later stages of sperm storage). (c) Results of an experimental evolution study demonstrating that the advantage to males of producing relatively long sperm (i.e., higher P_2) increases with female seminal receptacle length (open squares: short-sperm selection line males; open circles: control-sperm selection line males; solid squares: long-sperm selection line males; bars = 1 s.e.). (d) Experimental evolution for increased SR length drove the evolution of sperm length across two experimental replicates (open squares: short-SR selection line; open circles: control-SR selection line; solid squares: long-SR selection line; bars = 1 s.e.). (e) Sperm-SR coevolution occurs rapidly in nature, mechanisms. They protect their paternity by moving away from dung pats soon after securing a mate to avoid takeovers by other males, and they continue to guard ovipositing females after copulation has ceased (Parker 1974a; figure 22.5d). More recently, males have been shown to be more effective sperm competitors in females with which they have not coevolved (Hosken et al. 2002), and males that are better sperm competitors were found to sire higher quality offspring, thus providing support for the *good sperm hypothesis* (Hosken et al. 2003).

The Fruit Fly, *Drosophila melanogaster*, and Its Relatives

Drosophila melanogaster adults interact and mate on rotting fruit, which is also the oviposition and larval substrate. They have a mating system involving scramble competition leading to polygyny, with females present on the surface of fruit often receiving courtship from several males simultaneously. Females remate on average every few days, with the most recent mate on average siring 80–85% of subsequent offspring (Gromko et al. 1984). Because they are highly amenable to both laboratory and field investigations, and the genetic and molecular tools available are unsurpassed, *D. melanogaster* and related species have been at the forefront of exploring the mechanisms underlying the pattern of sperm precedence, in addition to the role of postcopulatory sexual selection in generating species differences in reproductive characters. These studies reveal that variation in mechanical aspects of copulation, sperm numbers, sperm size, seminal fluid composition, female reproductive tract morphology, and female physiology may all contribute to the pattern of sperm precedence, and they highlight the pivotal role of male-female interactions in postcopulatory sexual selection.

Using lines of *D. melanogaster* rendered homozygous for X, second and third chromosomes, Clark and colleagues demonstrated (1) the presence of polymorphic male and female genes affecting P_2, (2) significant male-female genetic interactions influencing P_1 (figure 22.6a) and P_2, and (3) a resulting pattern of nontransitivity among males in their sperm competitive ability (reviewed by Clark 2002). In a related experiment, Bjork et al. (2007) used a population of *D. melanogaster* with natural genetic variation to show that the patterns of sperm use (P_1 and P_2) were statistically repeatable only when each male competed against the same rival male and within the same female. The outcome was less predictable when the rival male stayed the same but the female changed, and it became completely unpredictable when males competed against different rival males within different females in each trial.

Some of these male-female interactions may be mediated by mechanical interactions during copulation, which have been shown to trigger the ejection of some resident sperm from the female reproductive tract (although the contribution of sex-specific variation to the quantity of sperm ejected has not been investigated; Snook & Hosken 2004). Studies with *D. melanogaster* have further shown ejaculate-female interactions to be particularly complex and likely critical determinants of differential male fertilization success. Female *D. melanogaster* undergo dramatic physiological, morphological, and behavioral changes as a consequence of mating (figure 22.6b), and these are associated with about 13% of the female's transcriptome being altered (e.g., McGraw et al. 2004; see chapter 28). Of the 1,783 genes whose transcript level differed between mated and unmated females, 160 genes were modulated in response to seminal plasma, and another 540 genes in response to receipt of sperm (McGraw et al. 2004).

The main source of seminal plasma is the male's accessory glands, which express at least 112 of these proteins in *D. melanogaster* (known as "Acps"). These proteins have thus far been demonstrated to (1) regulate muscle contraction of the female reproductive tract during sperm storage and ovulation (figure 22.6b), (2) increase female

FIGURE 22.6 *(continued)* as indicated by covariation among eight geographic populations of *D. mojavensis* (bars = 1 s.e.). (f) This same pattern is found at the macroevolutionary level, illustrated here for 46 *Drosophila* species, after controlling for allometry and phylogeny. (g) Semen is toxic to females, experimentally demonstrated by comparing survival of females mated to males transferring "full," "reduced," "greatly reduced," and "no" proteins from the main cells of their accessory glands. Adapted with permission from (a) Clark (2002), (b) Adams and Wolfner (2007) (c and d) Miller and Pitnick (2002), (e) Pitnick et al. (2003), (f) Pitnick et al. (1999), (g) Chapman et al. (1995).

production, ovulation, and laying of eggs, (3) influence the efficiency of sperm storage and utilization, and (4) make females refractory to male courtship, all of which can influence patterns of sperm use (reviewed by Chapman 2001; Clark 2002; Pitnick et al. 2009b).

Ejaculate-female interactions influencing sperm use appear to further be mediated by sperm and female sperm-storage organ morphology. Contrary to theoretical expectation (Parker 1982), males of many *Drosophila* species produce gigantic sperm (e.g., figure 22.4a), despite having a high probability of encountering sperm competition. Comparative investigations of correlated trait evolution among numerous *Drosophila* species indicate longer sperm are relatively costly to manufacture, and trade off with the number of sperm produced and transferred to females (Pitnick et al. 2009a). By experimentally evolving populations of *D. melanogaster* with relatively long and short sperm and (separately) those with longer and shorter female sperm-storage organs, Pitnick and colleagues (Miller & Pitnick 2002; Pitnick et al. 2003; Bjork & Pitnick 2006; Pattarini et al. 2006) showed that the strength of sexual selection on males and females does not decline as sperm size increases and sperm numbers decrease (see chapters 20 and 21). In addition, greater numbers of sperm and longer sperm were shown to independently contribute to male competitive fertilization success. The advantage to males of having relatively long sperm increased with increasing length of the female sperm-storage organ (i.e., a male-female interaction on P_2; figure 22.6c). This pattern was attributable to longer sperm being better able to displace (and resist being displaced by) shorter sperm from the site of egg fertilization within the female. Finally, the evolution of longer female sperm-storage organs was experimentally demonstrated to drive the evolution of longer sperm (figure 22.6d). These results may explain the pattern of rapid coevolution between sperm and female reproductive tract morphology observed across *Drosophila* populations (figure 22.6e) and species (figure 22.6f and 22.2a), as well as in numerous other lineages (reviewed by Pitnick et al. 2009b).

The complexity of ejaculate-female interactions relevant to postcopulatory sexual selection (including sexual conflict) is well illustrated by a series of provocative studies of *D. melanogaster* that may all relate to one seminal peptide hormone known as the *sex peptide* (formally designated Acp70Aa). By comparing life span among females inseminated by males transferring normal seminal fluid and females inseminated by males carrying a transgene that prevents secretion of many seminal fluid proteins, Chapman et al. (1995) found that the ejaculate of *D. melanogaster* is toxic to females; the more seminal fluid a female receives, the sooner she dies (figure 22.6g). Subsequently, a clever experiment by Rice (1996), using a stock of *D. melanogaster* with chromosomal translocations and a breeding design that held female genomes static over time but left males free to evolve, showed that arresting the ability of females to coevolve with males results in males becoming both increasingly harmful to females and increasingly better at sperm competition. This putative link between ejaculate-induced harm to females and sperm competition (see chapter 23) has strengthened with evidence that the seminal fluid protein causing the harm is the sex peptide (Wigby & Chapman 2005). This peptide hormone crosses the female reproductive tract, enters her circulatory system, targets receptors on the female's brain, and acts as an antiaphrodisiac (see "Avoiding Sperm Competition," above). This effect can last for several days, a consequence of some sex peptide being bound to sperm flagella, from which it is cleaved off in a time-release manner (reviewed by Pitnick et al. 2009b).

FUTURE DIRECTIONS

Some of the greatest conceptual advances will come from both theoretical and empirical studies directed at discerning how adaptations to precopulatory and postcopulatory sexual selection might differ in their evolutionary dynamics. We tend to think of ornaments and armaments functioning in the competition for mates as being highly quantitative and condition-dependent traits, whereas adaptations to postcopulatory sexual selection are often single gene–single functional molecule traits (e.g., seminal proteins) and ornaments borne by single cells (e.g., long sperm flagellum). Do precopulatory and postcopulatory adaptations tend to differ in their heritability, evolvability, genetic architecture, condition dependence, ability to sweep to fixation or the extent to which their evolution is constrained by ecological (e.g., predation, foraging) selection, or antagonistic pleiotropy? Do they differ in their propensity to enter into coevolutionary cycles of sexual antagonism (see chapter 23)?

Despite detailed descriptions of penis morphology and sperm ultrastructure for thousands of species (for taxonomic and phylogenetic purposes), there has been very little examination of the adaptive significance of such variation in form. Female reproductive tract design has been studied even less, and relatively little is known about sperm flagellar motion and other sperm behavior within females. In fact, it remains to be demonstrated whether or not variation in many of the traits discussed in this chapter generates postcopulatory sexual selection. Studies that quantify how within-population variation in these traits corresponds with variation in competitive fertilization success are thus greatly needed. It would be valuable to complement such studies with experimental evolution and phenotypic engineering approaches. By "perturbing" traits known to mediate interactions between the sexes, sex-specific fitness consequences can be quantified.

Comparative investigations of interacting ejaculate and female tract traits among geographic populations and closely related species are greatly needed. In particular, a major advance would be the identification of the female receptors or female-derived proteins that target or serve as targets of specific seminal fluid proteins, coupled with evolutionary analyses to determine whether the interacting sex-specific traits coevolve. Additional comparative approaches would also be fruitful. For example, sperm and female reproductive tract form is predicted be less divergent in monogamous relative to polyandrous lineages, although this prediction has not been tested to our knowledge. It would also be interesting to consider the evolution of sperm characters in conjunction with genital elaboration: possession of an intromittent organ that places sperm closer to the site of fertilization may, for example, relax selection on sperm motility.

SUGGESTIONS FOR FURTHER READING

To learn more about postcopulatory sexual selection, we recommend reading Parker's classic paper (1970); time has not diminished its ability to interest and inspire. Next, we recommend Birkhead et al. (1993) and Keller and Reeve (1995), who helped usher in a fundamental shift toward a deeper consideration of the fitness consequences of postcopulatory sexual interactions. We further suggest a review of the costs of sperm production and the adaptive consequences of these costs (Wedell et al. 2002), as the ideas and papers reviewed similarly changed the complexion of the field. Finally, we encourage reading Birkhead's (2000) page-turning lay account of postcopulatory sexual selection.

Birkhead TR (2000) Promiscuity. Faber & Faber, London.

Birkhead TR, Møller AP, & Sutherland WJ (1993) Why do females make it so difficult for males to fertilize their eggs? J Theor Biol 161: 51–60.

Keller L & Reeve H (1995) Why do females mate with multiple males? The sexually selected sperm hypothesis. Adv Stud Behav 24: 291–315.

Parker GA (1970) Sperm competition and its evolutionary consequences in the insects. Biol Rev 45: 525–567.

Wedell N, Gage MJG, & Parker GA (2002) Sperm competition, male prudence and sperm-limited females. Trends Ecol Evol 17: 313–320.

23

Sexual Conflict

CLAUDIA FRICKE, AMANDA BRETMAN, AND TRACEY CHAPMAN

Look at a cow pat in the summer. If you watch closely, you may see that mating male dung flies sometimes drown their mates in dung—why? The study of sexual conflict provides the answer: such dramatic effects can occur because, perhaps not surprisingly, what is best for males is not always best for females. In the example of the dung flies, the competition among males for matings can sometimes be so intense that the females caught up in it pay the ultimate price. Sexual conflict makes sense of these and other seemingly counterintuitive examples of behavior. Sexual conflict, or the "conflict between the evolutionary interests of individuals of the two sexes" (Parker 1979), arises because males and females often gain different fitness benefits for any given level of a reproductive trait (figure 23.1). As a result, males and females often cannot simultaneously both achieve their potential, maximum fitness benefits because the traits over which there is sexual conflict can take only a single value (Parker 1979). The outcome is an inevitable reduction in fitness in one or both sexes. This reduction in fitness generates a novel opportunity for *sexually antagonistic selection* to reduce the fitness cost resulting from sexual conflict (figure 23.1). Provided that there is genetic variation in the traits involved and a mechanism by which fitness costs can be reduced, sexually antagonistic selection can act to reduce for each sex the fitness costs from reproduction (Parker 2006b; Chapman 2006; Lessells 2006). However, if this in turn decreases the effectiveness or impact of the trait over which there is sexual conflict, there may be subsequent selection for counteradaptations. If, for example, sexually antagonistic selection results in the elaboration of adaptations in males (e.g., fighting ability) followed by counteradaptations to reduce the cost of those adaptations in females (e.g., mating resistance), the result can be antagonistic coevolution between males and females (Parker 1979; box 23.1). Because this coevolution is focused on reproductive traits that could lead to differences in mating preferences and mating compatibilities within or between different populations, it has the potential to drive reproductive isolation and, ultimately, speciation (e.g., Parker & Partridge 1998; Gavrilets 2000).

The significance of sexual conflict was first realized by Trivers (1972), Dawkins (1976), and Parker (1979). Their pioneering studies, and particularly the groundbreaking work of Geoff Parker (1979), revealed the potential for conflicts of interest between males and females to generate evolutionary change (box 23.1). In a recent, large-scale synthesis of this subject, Arnqvist and Rowe (2005) expose the extraordinary diversity of traits that are potentially subject to selection arising from sexual conflict, and highlight the broad range of taxa in which such traits are found. Rather than try to capture this huge diversity in this short chapter, we refer the readers to Arnqvist and Rowe (2005) for the many excellent examples of sexual conflict in a wide range of different taxa (e.g., Parker 1979; Warner et al.

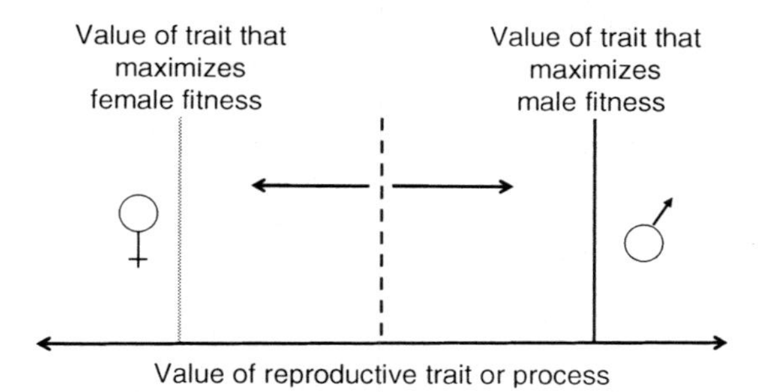

FIGURE 23.1 Schematic of fitness optima for a reproductive trait or process subject to sexual conflict. Fitness optima differ in males as compared to females (solid lines). However, the reproductive trait or process can take only one value and (unless either males or females have "won" the conflict), this is likely to lie between the male and female optima (e.g., at the dotted line). Hence there is sexual conflict because male and female optima cannot simultaneously be realized. The reduction in fitness from each sex not being at its optimum results in selection in each sex to minimize the fitness cost (in the direction of the arrows).

1995a; Rice 1996; Holland & Rice 1998, 1999; Magurran 1998; Arnqvist & Rowe 2002a; Martin & Hosken 2003; Westneat & Stewart 2003). In this chapter our aim is to focus on the concepts of sexual conflict, with a few illustrative examples. In the first section, we discuss, in turn, sexual conflict, the novel opportunity for antagonistic selection that it generates, the evolutionary potential of that selection, and finally the genetic mechanisms by which evolution resulting from sexual conflict may occur. In the second section we focus on the evolutionary potential of selection arising from sexual conflict and consider the theory and evidence that sexual conflict drives divergence both within and between species.

SEXUAL CONFLICT

Sexual conflict occurs because of differences between males and females in the optimum value of many

BOX 23.1 Key Lessons from Sexual Conflict Theory

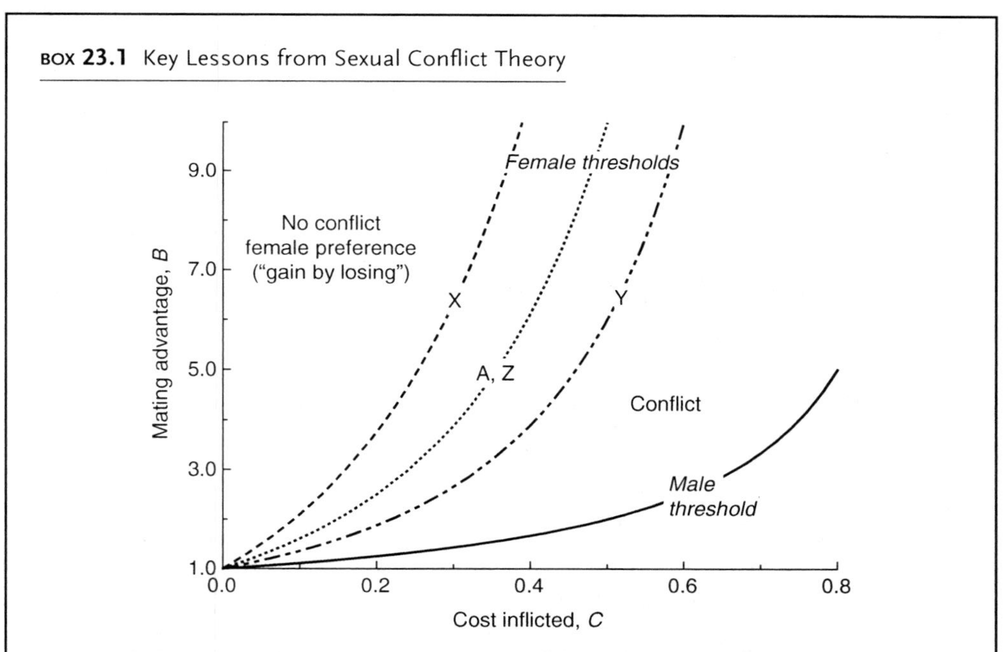

Figure I The spread of rare male mutants whose effects benefit males but incur costs to females (and which affects their joint progeny) under different genetic scenarios; A = dominant, autosomal (Parker 1979) or Y, X, or Z linked (Andres & Morrow 2003). A new mutation benefits males if B lies above the lower curve. The three upper curves are the thresholds for B above which it will pay the female to mate with males with the trait (at lower B, it pays the female to resist). Conflict occurs when B lies between the male and female thresholds. Reproduced with permission from Parker (2006b).

(continued)

Parker (1979) examined the central question of what happens when a characteristic that gives a mating advantage to males incurs a cost to the females with which they mate. Game theory models of sexual conflict were used to examine the effects of the dominance characteristic of the male trait and its frequency in the population.

The importance of the theory as illustrated by the above figure is that it identifies three zones (taken from Parker 1979, 2006b): (1) where the male trait is disadvantageous to both sexes and will not spread; (2) the sexual conflict zone—where the trait is advantageous to males but disadvantageous to females (sexually antagonistic coevolution may occur between the traits at the male locus to increase B, and those at the female locus to avoid mating with harmful males and/or to diminish harmful effects); (3) the concurrence zone—where the trait is advantageous to both sexes (selection favors both the harmful trait in males and female traits to accept or prefer males with the trait—sometimes called the "gain by losing" effect for females (Eberhard 2005).

A fundamentally important part of Parker's (1979) theory was that it showed that sexual conflict had the potential to lead to evolutionary chases between adaptations in males and counteradaptation in females. Hence sexual conflict can act as an engine for evolutionary change.

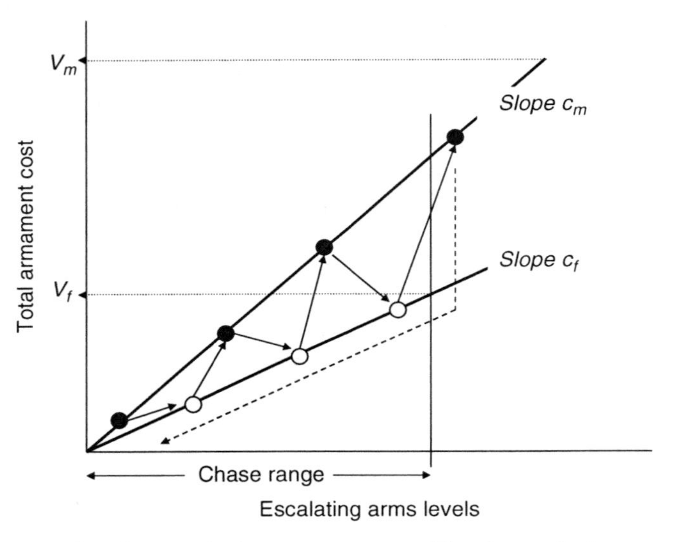

Figure II An example of an evolutionary arms race. The figure depicts an evolutionary chase in a sexual arms race model (from Parker 1979). Total arms costs are plotted against arms levels for the two sexes: at a given point on the x-axis, the total arms for each sex are exactly balanced so that the chances of winning the conflict are random; otherwise, the sex with the higher arms level wins. In this example, the value of winning for females (V_f) is lower than that for males (V_m), and the slope of the total costs with escalating arms levels is lower for females (slope c_f) than for males (slope c_m). If females start at a low arms level, males can win by a slightly greater level, which females can then outbid, and so on. As arms levels escalate, females would first reach the point where their total arms costs equal their value of winning: males can still outbid them and achieve a positive payoff. At this point females do better to reduce their arms to zero, which allows males also to reduce to a very low level. The cycle then begins again. Reproduced with permission from Parker (2006b).

aspects of reproduction (Parker 1979). Conflict occurs whenever the relationship with fitness for males and females differs for any trait (figure 23.1). Our definition is deliberately broad and includes traits having no role in interactions between the sexes as well as those that influence such interactions and result in social selection (box 23.2). Both viewpoints predict that there can be sexual conflict over virtually any reproductive trait, and indeed sexual conflict is expected to be ubiquitous among sexually reproducing organisms. The extent of sexual conflict will be exacerbated by any factors that lead the reproductive interests of the two sexes to diverge. For example, a high degree of multiple mating with different partners coupled with low relatedness between mating partners reduces the extent

BOX 23.2 Sexual Conflict as Social Selection: Insights from Selection Theory

Sexual conflict represents a difference in the fitness optima for males and females for a given reproductive process or trait (Parker 1979). However, it may also be useful to consider sexual conflict through the related and complementary view of selection theory (Arnold & Duvall 1994; Arnold & Wade 1984; Westneat 2000). Sexual conflict can lead to social selection (Wolf et al. 1999; Westneat & Stewart 2003; table 7.1 in Arnqvist & Rowe 2005; see also box 14.1 in this volume), effectively extending into the domain of the *extended phenotype* of an individual. This is because the value of many traits subject to sexual conflict has an effect not only on the fitness of the bearer, but on the fitness of the other sex (in which that trait is not expressed). Sexual conflict is therefore created when there is a positive relationship between trait value and focal individual fitness (i.e., a selection gradient), but negative relationship between trait value in the focal sex and fitness of the other sex (an *opportunity gradient*; Arnold & Wade 1984). The extent of conflict can

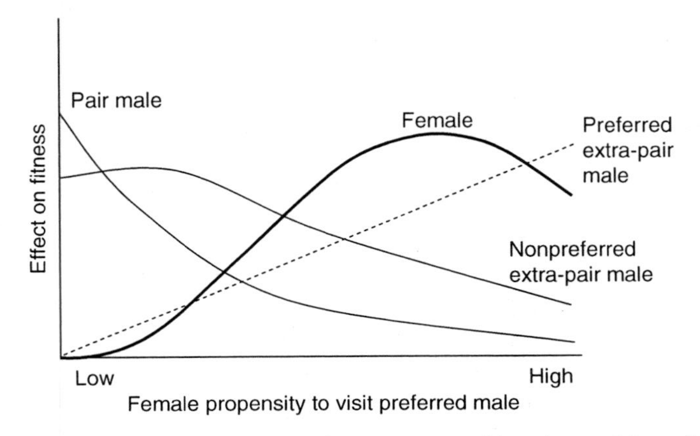

Figure I The relationship between the propensity of females to visit preferred males for EPCs and female fitness (thick line) that increases, except at high values, when the benefits diminish and costs increase. Preferred extra-pair males experience an increase in fitness as female visits increase (dashed line). However, the fitness of nonpreferred males or the pair male will decrease (thin lines) in different ways as females increase visits to preferred males. Both thin lines represent opportunity gradients describing sexual conflict on paired or nonpreferred extra-pair males due to the propensity of the female to pursue EPCs with preferred extra-pair males. Reproduced from Westneat and Stewart (2003) with permission.

(continued)

BOX 23.2 *(cont.)*

be described as the difference in slope of these two relationships. We illustrate this concept using an example of this theory as applied to sexual conflict over extra-pair copulations (EPCs) in birds (figure I, p. 403).

The social selection view emphasizes sexual conflict as a phenomenon that provides a novel opportunity for selection, rather than a special type or form of selection per se. The novel opportunity for selection could result in natural selection to increase viability or sexual selection to increase mating ability. Viewing sexual conflict as a type of social selection also provides the opportunity to more broadly encompass the types of processes that are subject to it (variously described as conflict traits, or shared traits; Lessells 2006; Rowe & Day 2006). Using the social selection framework, any adaptation whose relationship with fitness is opposing in males versus females is subject to sexual conflict. The social selection definition of sexual conflict predicts that any cost imposed indicates sexual conflict, regardless of any indirect genetic benefit that females may gain despite suffering costs. By extension, any female preference produces conflicts because some males are not preferred. For a further review of this topic, see Westneat (2000) and Westneat and Stewart (2003).

to which a mating pair have a stake in what happens beyond the current mating bout (Dawkins 1976). Hence traits that increase immediate investment in mating, at the expense of future investment by one of the current mating pair, can be selected.

That sexual conflict is widespread, however, does not imply that it automatically results in evolutionary change. Sexual conflict can create an opportunity for selection that is not realized if there is no trait variation in the affected sex. Moreover, for evolution to occur, there has to be (a) selection caused by sexual conflict and (b) genetic variation in a trait that covaries with the conflict-causing trait. In game theory models of sexual conflict, the likelihood of selection is described as *power* and *winning*, or the ratios of the benefits in males/females and costs in males/females, respectively (reviewed in Chapman 2006; Lessells 2006; Parker 2006b). What this boils down to is simply that, for sexually antagonistic selection to cause evolutionary change, the benefit-cost ratios have to be favorable. A related issue is that variation in the threshold or sensitivity of a trait subject to sexual conflict may also affect the potential for coevolution (Rowe et al. 2003, 2005). A general message is that the existence of sexual conflict cannot be assumed without knowledge of the costs and benefits of the adaptations involved (e.g., Parker 2006b).

Sexual Conflict Traits

The notion of traits that are subject to sexual conflict needs some qualification, because such traits can be of diverse origin and form. For example, traits that cause sexual conflict can be expressed in one sex (e.g., male genital claspers), both sexes (hip width), or can instead be an emergent property of both sexes (e.g., mating frequency). Hence it is difficult to define general types of traits or processes that can be subject to sexual conflict. Previous authors have referred to the subjects of sexual conflict as conflict traits or shared traits (Lessells 2006; Rowe & Day 2006) to try to capture the diversity involved. What is clear, though, is that sexual conflict can fuel selection on a very diverse range of traits and processes from pre- and postmating traits through to those that control parental investment (Lessells 2006; Parker 2006b). Interestingly, sexual conflict is predicted to be more likely over mating decisions than it is over parental investment (Lessells 2006; chapter 26 of this volume). In brief, this is because the fitness returns are higher and costs lower for males that can manipulate females into mating with them as compared to the situation in which males try to coerce their mates into increasing their parental investment (reviewed in Chapman 2006; Lessells 2006).

Selection Arising from Sexual Conflict

Sexual conflict provides an opportunity for selection because of the difference in fitness optima for males and females (or the opportunity gradient represented by the effect of the trait in one sex on the fitness of the other; box 23.2). This evolutionary tug-of-war between the sexes selects for each sex to shift the value of the trait subject to conflict to be closer to its own optimum, and hence to reduce potential costs (figure 23.1). However, this necessarily leads to increased costs in the other sex, resulting in direct selection in each sex to minimize costs and potentially to reduce the effectiveness of the original manipulative adaptation. Hence the effects of adaptations in males can select for counteradaptations in females, leading to potential cycles of adaptation followed by counteradaptation. The resulting process is sexually antagonistic coevolution, fueled by sexual conflict (Parker 1979; box 23.1).

Sexual Conflict as Fuel for Evolution

Once there is in place a cycle of sexually antagonistic coevolution, then theory shows that if the coevolution is sufficiently strong, this can lead, under certain conditions, to diversification in the traits involved within species, as well as reproductive isolation and ultimately speciation (e.g., Arak & Enquist 1995; Parker & Partridge 1998; Gavrilets 2000; Gavrilets et al. 2001; Gavrilets & Waxman 2002; Gavrilets & Hayashi 2005; box 23.3). The evolutionary consequences of sexual conflict both within and between populations are considered in more detail below.

From Sexual Conflict to Coevolution: An Example

To illustrate the different stages of the arguments above, we consider the often-used example of sexual conflict in relation to mating frequency. Mating frequency is an emergent property of males and females, and on average, in a population with equal sex ratio, the population mating

BOX 23.3 Sexual Conflict Can Fuel Evolutionary Change Leading to Reproductive Isolation

Given that sexual conflict can drive evolutionary change leading to evolutionary chases in adaptations related to mating and reproduction, the question is, to what extent is this process expected to lead to reproductive isolation and ultimately speciation? Several authors have developed theory on this (e.g., Parker & Partridge 1998; Rice 1998, Holland & Rice 1998; Gavrilets 2000; Gavrilets et al 2001; Gavrilets & Hayashi 2005). Parker and Partridge (1998) used a game theory approach to study the mating outcomes that would occur following secondary contact of populations each having undergone sexually antagonistic coevolution in allopatry. As expected, the outcomes depend on the length of allopatry. In the short term, males can gain higher fitness by mating with females from another population, as those females have no resistance to those males. However, such matings also introduce genes for female resistance into the other population, which are advantageous to females and therefore spread. In the longer term, increased levels of divergence could lead to prezygotic isolation in such matings. The theory shows that while selection on males will usually promote gene flow and hence reduce reproductive isolation, females may usually be selected to resist hybrid matings, slowing the rate of gene flow and increasing reproductive isolation. This is because it generally pays males more to search for new mates. An important result is that sexual conflict can result in higher rates of speciation in clades in which females have relatively higher armament levels (so-called female-win clades). A corollary is that when there is reinforcement, females will promote premating isolation. Lower genetic variation is expected in female-win as opposed to male-win clades. The overall conclusion is that sexual conflict can fuel reproductive isolation, but only under certain conditions.

The models developed by Sergey Gavrilets (e.g., Gavrilets & Waxman 2002) also suggest a potentially important role of sexual conflict in driving reproductive isolation. In contrast to traditional models of speciation, Gavrilets' models predict that sexual conflict can drive evolution more rapidly in large, rather than small, populations, a prediction supported by the results of a study in dung flies (Martin & Hosken 2003).

(continued)

BOX **23.3** *(cont.)*

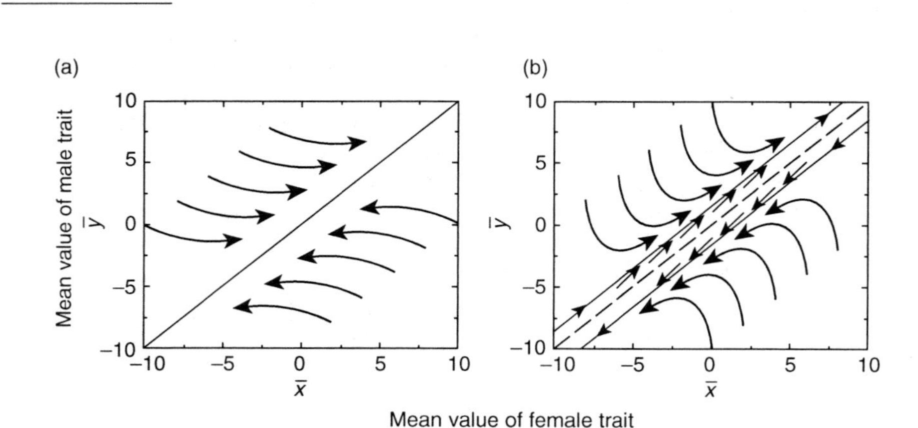

Figure I The figure shows the dynamics of mean trait values in Gavrilets' (2000) model. In (a), the costs of sexual conflict are low and the solid line depicts the line of equilibria. The trajectory of mean trait values is toward the line of equilibria, at which mean trait values then become static. In (b), the costs of sexual conflict are high. Points along the dotted line of equilibria are now unstable, and coevolutionary chase occurs along the continuous lines. In terms of the implications for speciation, the models show that where there are equilibria (a), points along the line of equilibrium are neutral, hence in populations that are isolated, allopatric genetic divergence could occur by drift along the line. Where there are coevolutionary chases (b), allopatric populations can diverge rapidly and simultaneously by selection in different directions. These models therefore support the idea that sexual conflict can drive speciation. Reproduced with permission from Gavrilets (2000) and adapted from Parker (2006b).

frequency of males must equal that of females. However, given that the variance in mating frequency is expected to be much higher for males than for females, we expect selection on male mating adaptations to be particularly strong. There will be sexual conflict if a high mating frequency is beneficial for male reproductive success but, in contrast, females show highest fitness at an intermediate mating frequency. This conflict sets up the opportunity for selection in males to increase mating frequency and in females to decrease it, because these outcomes would increase male or female fitness, respectively. Provided there is genetic variation in mating frequency and that both males and females have some control over mating frequency (i.e., have a mechanism with which to influence it), then there will be selection on males to mate with each available female, but for females sometimes to resist male mating attempts. Hence sexual conflict over mating decisions can lead to adaptation and counteradaptation and initiate antagonistic coevolution.

A good example is found among pond skaters (water striders, *Gerris* spp). Here, there is sexual conflict because males gain from higher mating frequencies and have a wide variance in mating success, whereas for females mating is costly in terms of reduced foraging time and higher predation risk (Rowe et al. 1994). Males are selected to attempt to mate with all available females, whereas females usually try to resist superfluous matings, with the result that violent premating struggles occur in which females try to shake off courting males (Rowe et al. 1994). The mechanism by which males try to increase mating frequency is through morphological changes in claspers that aid them in gaining attachment to females during mating. Males with longer abdominal claspers are likely to have higher mating success. In contrast, abdominal spines in females serve to lower mating frequency, and, in a manipulative experiment, increased female abdominal spine length led to shorter premating struggles and a lower mating rate (Arnqvist & Rowe 1995). Hence, in this example, the abdominal claspers in males and spines in females appear to be traits subject to antagonistic selection arising from sexual conflict, with longer claspers in males leading to higher male fitness and longer spines in females to higher female fitness. Experimental and comparative evidence support the idea that the armaments in males and in females across different species of pond skaters are coevolving (Arnqvist & Rowe 2002a).

Intra- or Interlocus Sexually Antagonistic Selection

The way in which traits evolve in response to antagonistic selection may be facilitated or constrained by the underlying genetic mechanisms involved. The genetic basis of sexual antagonism can therefore, in principle, have a major impact on the speed and trajectory of coevolution. The adaptations influenced by selection arising from sexual conflict can be influenced by the same (intralocus) or different (interlocus) genes in males and females, and both have the potential to drive evolutionary change that could lead to speciation (Parker 1979; Parker & Partridge 1998; box 23.3). However, the distinctions between intra- and interlocus coevolution and their relative importance in evolutionary terms, and the outcomes they generate, have not yet been explored in detail (Chapman 2006).

For example, intralocus coevolution may constrain the evolution of, or may ultimately select for, sex limitation in genes influencing traits subject to selection from sexual conflict. This is because alleles of genes that currently reside in males are prevented from reaching their male-specific optimum by counterselection whenever those genes are expressed in females. Episodes of this type of evolutionary constraint can therefore be resolved by the evolution of sex limitation in those genes. For example, if a gene that has a male beneficial function can become expressed only in males, this may prevent counterselection against the expression of that gene in females. In interlocus coevolution, on the other hand, sex-limited genes may be the starting point of conflicts. Interlocus coevolution may instead be constrained mostly by the relative costs and benefits of the adaptations that are selected.

Evidence for Intralocus Antagonistic Selection

Adult locomotory activity in the fruit fly *Drosophila melanogaster* has been used as an example of a trait whose expression is controlled by the same genes in males and females, but which may be subject to sexual conflict. Locomotion in adult flies appears to be controlled by the same sets of genes in both sexes, because there is a positive genetic correlation between movement levels in males and females. However, there is also sexual conflict over the optimum rate of locomotion for adults. High locomotory activity is beneficial to males because it increases their encounter rate with females, leading to higher courtship rate and higher reproductive success. In contrast, females who are less active have higher fitness than more active females, presumably as reduced locomotory activity is associated with increased feeding and oviposition (Long & Rice 2007). This sexual conflict offers the opportunity for antagonistic selection on the alleles involved, modified by whether they currently reside in males or females. This example provides a good demonstration of the diversity of types of traits and processes that can become subject to opposing selection pressure on males and females. Unlike mating rate, for example, the opposing selection in this case is not dependent on interactions between males and females (though that could also occur if males that are especially active impose larger reproductive costs on the females with which they interact), but is instead dependent on differences in the fitness effect of particular locomotion-affecting alleles when in one sex compared to the other (and remember these alleles are equally likely to occur in both sexes).

A number of experiments from the laboratory of Bill Rice have documented the presence of sexual conflicts that lead to selection on the same genes expressed in males and females in *D. melanogaster* (e.g., Rice 1992; Chippindale et al. 2001). For example, techniques have been used to allow alleles with sex-specific beneficial effects to accumulate in one sex by preventing counterselection against them in the other (Rice 1992). Another technique has involved testing the effects on fitness of the same genotypes expressed in males *versus* in females. This work demonstrated that fitness was positively correlated in larvae (when male and female interests are broadly similar), but negatively correlated in adults (when the sexes come into conflict over reproduction). Hence in adults, genotypes that resulted in high fitness for females resulted in low fitness for males and *vice versa*. The conclusion from such work is that the number of genes subject to intralocus sexually antagonistic selection is potentially high (Chippindale et al. 2001) and represents a considerable evolutionary constraint on each sex reaching its adaptive, optimum phenotype. For other examples, see Arnqvist and Rowe (2005).

Evidence for Interlocus Antagonistic Selection

There are many traits encoded by different loci in males and females that have the potential to be shaped by selection arising from sexual conflict. We summarize here just three of the systems that have been investigated in depth and that cover premating and mating traits through to those related to parental investment (for numerous further examples, see Arnqvist & Rowe 2005).

Pond Skaters Pond skaters have provided an excellent system in which to study sexual conflict in an ecological setting (Rowe et al. 1994). Conflicts over mating decisions in pond skaters have already been discussed above in terms of the struggles over mating. Using a combination of comparative and empirical work, Arnqvist and Rowe (2002a) demonstrated ongoing sexually antagonistic coevolution in the level of armaments in males and females, in terms of male grasping and female antigrasping behaviors. The male grasping adaptations facilitate mating, and the female antigrasping adaptations act to decrease the frequency of costly, superfluous matings. Coevolution between male adaptations and female counteradaptations occurred across 15 different species of pond skaters. Interestingly, it was not the absolute level of armaments and defenses between males and females that determined the length of premating struggles and mating rates. Mating outcomes were instead determined by the relative imbalance in armaments, in other words, if males invested more in armaments than females did in resistance, males gained greater fitness, and vice versa. The conclusion is that the absolute level of armaments and defenses is not necessarily a good indicator of the level of sexual conflict.

Fruit Flies The fruit fly *Drosophila melanogaster* has been a valuable workhorse in the study of sexual conflict. Indications that mating interactions between males and females were subject to sexual conflict originally came from studies that demonstrated significant mating costs in females that mate frequently. Experiments that used genetic manipulations showed that these costs are caused by the receipt of high levels of male seminal fluid accessory proteins (Acps) during mating (Chapman et al. 1995). Because Acps increase male reproductive success through a variety of effects on sperm competition, this suggested that a side effect of the competition among males was costly to females, leading to subsequent selection in females to reduce this cost. This is supported by the finding of a strong positive correlation between a male's ability in sperm competition and the death rate of the females with which they mated (Civetta & Clark 2000). Hence the male mating adaptations seem to be selected to increase a male's per mating share of paternity, despite the eventual cost that they may cause in females.

A number of experimental evolution studies have targeted the fruit fly mating system by altering the nature of sexually antagonistic selection. For example, Holland and Rice (1999) placed replicate lines of flies under monogamy and polyandry. In the monogamy lines, there is virtually no interlocus sexual conflict (that caused by interactions between the sexes) because the evolutionary interests of males and females become the same. After tens of generations under these selection regimes, the monogamous males became less harmful to females and monogamous females were less resistant to male-imposed mating costs. Monogamous populations also had higher net fitness. Together, these findings support the idea that when the interests of males and females become more similar, the sexes have less harmful effects on one another.

In another experimental evolution study, the adult sex ratio of males and females was altered, to investigate whether females can evolve resistance to male-imposed mating costs (Wigby & Chapman 2004). Lines of flies were set up in which the adults experienced male- or female-biased sex ratios (3 males to every female and vice versa). Females taken from male-biased populations, in which sexual conflict was predicted to be strong, were able to survive longer in the presence of males than were females from female-biased populations. These differences in survival were not found in the absence of males, which suggests that females had indeed evolved specific mechanisms to counter male mating costs. The benefits of female resistance to males have also been studied through mimicking the spread of a female resistance gene (Stewart et al. 2005). In this experiment an eye color marker was made to segregate as if it were a resistance gene that resulted in 100% reduction of mating costs in females. This trait spread rapidly through the population after only 5 generations of selection, because these females were able to avoid incurring mating costs imposed by males.

What is striking about these evolutionary studies is that adaptations and counteradaptations selected in response to underlying sexual conflicts can arise extremely rapidly, within tens of generations. This remarkable and consistent finding suggests that sexual conflict has the ability to promote rapid evolutionary change.

Mate Desertion in Birds In species in which parents look after their young, the time spent engaged in parental investment can reduce the time that males and females can spend searching for new mates (chapters 20 and 26). For this reason there is the potential for sexual conflict to arise over which sex will provide parental care, and how much of it they will give. In the Penduline tit (*Remiz pendulinus*) both males and females can perform both biparental and uniparental care. However, both sexes benefit from starting a second nest during a breeding season and there is therefore a sexual conflict over timing of nest desertion. Thirty percent of nests end up being abandoned by both parents, resulting in complete loss of reproductive investment, which suggests that the desertion mechanism is not very highly tuned. However, if either sex manages to desert before its partner, the remaining individual may stay and provide care, whereas the deserting partner can find a new mate and start a second brood, potentially gaining higher fitness. The timing of mate desertion is a balance for males between the benefits of desertion and those of staying. For example, males aim to gain a return on their investment of time and energy in nest building and holding a high-quality territory, factors that are important in attracting females. In addition, if the male deserts before the female has laid sufficient eggs in the nest, then she will desert too. Males thus benefit from assessing a female's egg-laying status. Given that females also benefit from deserting following egg laying, they apparently try to hide the number of eggs they have laid by covering them and preventing their mate from accessing the nest. If this is successful, females may sometimes desert, leaving the male to take care of the brood (Valera et al. 1997). Hence, in this example, the underlying sexual conflict gives the opportunity for selection on a male's ability to assess a female's egg-laying status and on a female's ability to disguise her egg laying.

Contrasting Selection Opportunities Arising from Sexual Selection and Sexual Conflict

Models of sexual selection and models of sexual conflict both center on the interactions between males and females during reproduction—specifically, on the ways in which males compete with each other for matings with females and the ways in which females mate with some males instead of others. There is overlap between models of sexual selection and sexual conflict because both types focus on selection for increased reproductive success. Where the models differ, however, is in the way in which the female preference is selected and specifically whether the impact of male reproductive strategies on females is beneficial, cost neutral, or costly to females (see the excellent chapter on this topic in Arnqvist & Rowe 2005). The novel opportunity for selection provided by sexual conflict occurs when the relationship with fitness for a given trait is positive for one sex and negative for the other (figure 23.1). Whenever this is not the case, then models of sexual selection, rather than conflict, are more appropriate. However, divisions between the different types of models should not be viewed as fixed. For example, the existence, sign, and magnitude of costs of mating with specific males can be environmentally dependent, changing the opportunity for selection.

Here we briefly compare the contrasting opportunity for selection on female preference under sexual selection and under sexual conflict. In models of sexual selection based on either Fisher's *runaway* or *good genes* processes, female preference genes are selected because they become associated either with genes that increase the mating success of their sons (Fisher 1930) or increase the fitness of both sexes of offspring (Zahavi 1975). Hence, female preference genes evolve under so-called indirect selection through the effects on offspring. However, Fisherian models of sexual selection cannot explain the maintenance at equilibrium of substantial costs of mating to females (e.g., Kirkpatrick & Barton 1997) and cannot explain antagonistic coevolution when there are large mating costs in one (or both) sexes. Models of sexual selection by good genes, on the other hand, assume that female mate choice will result in increased offspring fitness, which is again counter to the expectations of models of sexual conflict. Hence models of sexual selection do not easily explain what happens to female mating biases under a sexual conflict scenario.

TABLE 23.1 Summary of single species studies comparing direct costs and indirect benefits of mating for females

Study	Species	Method	Traits Measured	Fitness Measure Calculated	Son/Daughter Fitness Measured	Generation of Offspring in Which Fitness Measured	Costs Estimated?	Are Direct Costs > Indirect Benefits?
Head et al. 2005	*Acheta domesticus*	Females held with attractive or unattractive males and the number of grandchildren estimated	Sons' attractiveness, number of grandchildren produced	Rate sensitive and insensitive estimates of fitness	Sons and daughters	F1	Yes	No
Rundle et al. 2007	*Drosophila melanogaster*	Male grandchildren were bred from grandfathers that were successful or unsuccessful in gaining matings	Premating, mating success, productivity and longevity of male grandchildren	No composite fitness measure calculated	Sons and daughters	F2	No	There are indirect benefits
Orteiza et al. 2005	*Drosophila melanogaster*	To take the male offspring from the first or second mates of twice-mated females	Lifetime reproductive success of sons in a competitive environment (male postmating success)	No composite fitness measure calculated	Sons only	F1	Yes	Yes
Galliard et al. 2008	*Lacerta vivipara*	Create male and female biased sex ratio	Offspring survival and growth	Composite measure of female survival and fecundity calculated	Sons and daughters	F1	Yes	Yes
Priest et al. 2008	*Drosophila melanogaster*	Females held at three different mating frequency regimes, and fitness of mothers and daughters estimated	Female lifetime fecundity and survival and daughter fitness	Calculated fitness and inclusive fitness	Daughters only	F1	Yes	No

Note: For a gene that confers female resistance to mating costs to spread under selection arising from sexual conflict (as opposed to sexual selection) the direct costs of mating to females should be larger than the indirect benefits of mating. The table summarizes the current relevant empirical data.

The key contrast between sexual conflict and sexual selection as an explanation for female behavior is that under sexual conflict, selection on female mating decisions is direct: to avoid or reduce the costs of matings. Female mating preferences under sexual conflict are therefore best modeled by *direct benefits* theory, except that in this case the expectation is that females will exert mating biases to minimize costs rather than maximize benefits. An important issue with regard to the opportunity for selection on female preference/resistance behavior is whether indirect genetic benefits for the offspring of females that mate frequently and incur large mating costs can balance or exceed the direct cost of mating. The distinction is important because from a gene's perspective, only if direct costs to females of mating are larger than indirect genetic benefits in the offspring generation will the gene be subject to sexually antagonistic coevolution. Theory suggests that indirect genetic benefits in this situation will be small (Kirkpatrick & Barton 1997; Rowe et al. 2003, 2005), and table 23.1 provides a summary of the current empirical data from single species studies, which is mixed.

EVOLUTIONARY CONSEQUENCES OF SEXUAL CONFLICTS

In this section we consider the potential of sexual conflict to drive evolutionary change within and between species and consider the supporting theory and evidence. We focus mostly on conflicts arising from interactions between different loci (but note that conflicts arising from within loci may also have significant effects on population structure). Sexual conflict can potentially affect population fitness, the rate of adaptation, or the risk of extinction (Holland & Rice 1999; Fricke & Arnqvist 2007). However, most attention has been given to the potential of sexual conflict to generate evolutionary change resulting in population divergence. It is not yet clear, however, to what extent sexual conflict is an engine of speciation, and theory and evidence on that issue are mixed as we illustrate below.

Divergence within and between Species: Theory

The importance of sexual conflict theory lies in illuminating the evolutionary potential of sexual conflict, and showing the conditions under which it may lead to diversification, population differentiation, and potentially speciation (e.g., Parker & Partridge 1998). The key discovery of sexual conflict theory has been that it is possible for a male adaptation to spread in a population, despite the cost that this may cause in females (Parker 1979; Gavrilets et al. 2001; Holland & Rice 1998; box 23.1). In addition, theory also shows that indirect genetic benefits of mating to females are not required for the spread of adaptations in males that are harmful to females (e.g., Cameron et al. 2003), though they may also occur.

Sexual conflict can in theory generate continual evolutionary chases between the interacting parties involved (Parker 1979; Gavrilets 2000; Holland & Rice 1998). The types of dynamics resulting from these population genetic models have the potential to lead to speciation if they promote divergent evolutionary trajectories between populations. A summary of sexual conflict speciation models (Gavrilets & Hayashi 2005) shows that some dynamics that can result from sexual conflict promote speciation (e.g., endless coevolutionary chases, diversification in both sexes) but others do not (single equilibrium or line of equilibria, cycles, diversification in one sex but not the other, etc.). The differences in dynamics (diversification versus equilibrium) that result from sexual conflict are likely to depend upon the type and strength of selection that is acting on the female trait (Rowe et al. 2003), the number of loci involved, and dominance patterns.

A game theory treatment of speciation in relation to sexual conflict, by Parker and Partridge (1998; box 23.3), adds an additional and important consideration. It examines the extent to which the differential behavior of males and females may affect the extent to which sexual conflict promotes reproductive divergence. In general, females may tend to act as a force for increasing reproductive isolation and males for decreasing it. However, if females evolve insensitivity to male traits in response to sexual conflict, then gene flow due to female behavior could increase, because females would no longer discriminate between different males (Rowe et al. 2003).

If sexual conflict is driving speciation, then individuals from different allopatric populations should be divergent in the reproductive traits that are subject to sexual conflict. This might then lead to incompatibilities or interactions in crosses between individuals from different populations. Arnqvist and Rowe (2005) suggest that sexual conflict could thus be a particularly potent driver of speciation. This is because selection on both male and female adaptations and counteradaptations is direct and hence stronger than indirect selection resulting from sexual selection. Although both intra- and interlocus antagonistic coevolution have the potential to lead to reproductive isolation (Parker & Partridge 1998; Rice 1998), the mechanisms involved may differ. Intralocus coevolution can lead to sexual dimorphism or sex limitation. However, this could occur via different routes in different allopatric populations, and these mechanisms could be disrupted when previously separated populations mix, potentially leading to reproductive incompatibilities. Genetic correlations for traits related to mate choice that evolve separately in different populations could also alter the likelihood of interpopulation matings (box 23.3). Interlocus antagonistic coevolution is expected to be a powerful driver of change, particularly in internally fertilizing species in which genes involved in reproduction are predicted to perpetually coevolve in an arms race, and to diverge faster than the rest of the genome (Holland & Rice 1998; Rice 1998). This predicts an early signature of incipient speciation to be incompatibility of male and female reproductive tract proteins and physiology across different populations.

In the following two sections we discuss first the data supporting sexual conflict as a driver of divergence within species and then between species.

Within Species Divergence: Empirical Data

Evidence that sexual conflict is a major driver of diversification within species, with the potential to lead to speciation, would be exemplified by demonstrations of diversifying selection, of divergence in those traits between species, and of rapid coevolution in sexual conflict traits. We review this evidence below. A key and perhaps unresolved question is whether the reported evolution in reproductive traits is causal in population or species divergence or merely associated with it.

Diversifying Evolution in Reproductive Traits Subject to Sexual Conflict

There is much evidence that reproductive traits in general evolve rapidly (e.g., Eberhard 1985; Clark et al. 2006). Sexual conflict predicts the rapid evolution of traits that are involved in antagonistic interactions between the sexes and these could include sperm-egg recognition/binding traits, reproductive proteins, mating behavior, and reproductive morphology (Rice, 1998; chapters 21 and 22).

There is evidence for positive selection (i.e., greater variance in nucleotide sequences among taxa than expected from neutral substitutions) in reproductive proteins, especially in males, from a wide range of taxa, for example, in marine invertebrates, flies, mice, plants, birds, and mammals (reviewed in Snook et al. 2009). In addition, there is considerable evidence that reproductive morphology also evolves extremely rapidly (Eberhard 1985) and is in many cases the distinguishing feature between otherwise morphologically identical species. However, whether the selection pressure in such cases results most often from sexual selection or from sexual conflict is often unclear. For instance, rapid evolution in sperm-egg recognition molecules could result from selection to avoid sexual conflict over fertilization processes, such as the need to avoid polyspermy, or from cryptic female sperm choice or sperm competition.

It is necessary to combine studies of molecular evolution with functional information, so that the selective forces acting on the traits can be identified. For this reason, we focus here on patterns of evolution in genes that are predicted to play a role in mediating sexual conflict. The best evidence comes from the study of *Drosophila* reproductive proteins, in which it has been established that the actions of proteins made in the male accessory glands result in the expression of mating costs in females (Chapman et al. 1995). Therefore, some of these proteins are examples of adaptations that are shaped by sexual conflict.

There are over 100 Acp genes, and there is evidence that some show high levels of within-species polymorphism (e.g., Begun et al. 2000). There are also now a large number of studies that have documented positive selection on Acp genes (e.g., Begun et al. 2000; Swanson et al. 2001; Haerty et al. 2007). It is also often hard to find orthologues of Acp genes even in very close relatives (Mueller et al. 2005). Early estimates put at 11% the number of Acp genes under positive selection (Swanson et al., 2001), but recent estimates are higher (Findlay et al. 2008), reflecting the increasing statistical power that comes from the higher numbers of species comparisons that are now possible.

Is there any evidence for positive selection on any of the Acp genes likely to be subject to selection from sexual conflict? The full answer is not yet known, and there are also problems that arise from limited power to detect positive selection for short genes such as those encoding Acps. However, of six genes so far implicated in causing mating costs in females (either because of toxicity to females, increased death rate following single matings or by direct tests with mutants), four have been investigated for positive selection. Of those, two or possibly three showed evidence of nonneutrality (Begun et al. 2000; Findlay et al. 2008).

There are well-studied examples of sequence evolution in reproductive genes that mediate sperm and egg recognition in marine invertebrates such as sea urchins and abalone (for review see Swanson & Vacquier 2002b). Sperm proteins in such species show extremely rapid evolutionary change, and variation in the rates of change between different taxa. However, the evidence that these proteins are selected primarily by sexual conflict remains to be confirmed, although there is evidence that sexual conflict over polyspermy (in which too many sperm attempt to enter the egg) can drive the evolution of sperm-egg interactions (Franke et al. 2002; chapter 22, this volume).

Evidence for the strength of selection acting on female reproductive proteins is generally harder to gather than for male proteins because the female targets for male reproductive proteins do not necessarily reside in the female reproductive tract. However, evidence is now accumulating that female reproductive proteins and female reproductive tract morphology (e.g., Pitnick et al. 1999) can also evolve rapidly, although the evidence that these adaptations evolve due to selection arising from sexual conflict is in most cases still lacking. Swanson et al. (2004) detected that 6% of proteins in the female reproductive tract of *D. melanogaster* were under positive selection. A later, more detailed study (Panhuis & Swanson 2006) reported positive selection on 6 out of a set of 9 female reproductive tract genes surveyed. Similarly, Kelleher et al. (2007) conducted a survey of genes from the lower reproductive tract of *D. arizonae* and found evidence for elevated rates of evolutionary change in 31 of the 241 reproductive tract proteins detected. Across vertebrates, there is evidence for elevated evolutionary change in the female reproductive proteins of birds, humans, and other mammals (reviewed by Clark et al. 2006).

Coevolution in Reproductive Traits within Species

Abalone of the genus *Haliotis* provide one of the few examples in which the patterns of evolutionary change in both male and female interacting reproductive proteins have been studied. Male sperm contain lysin, a protein that binds to the vitelline envelope receptor for lysin (VERL) and then dissolves part of the outer layer of the egg to facilitate sperm entry. Sperm lysin is highly divergent between closely related species and data from site-specific mutagenesis shows that there are specific sites at both the N- and C-terminus of lysin that control species specificity in lysine-VERL interactions (Lyon &Vacquier 1999). Concerted evolution in VERL appears to drive positive selection in lysin. The VERL is encoded by a large and repetitive sequence, only part of which shows very strong evidence for positive selection (Galindo et al. 2003). This highlights the need to identify the functionally important parts of the interacting molecules in order not to overlook evidence for coevolution.

The best evidence for rapid coevolution between male adaptations and female counteradaptations that are subject to sexual conflict comes from studies of reproductive morphology (e.g., Arnqvist & Rowe 2002a; Rönn et al. 2007). For example, in pond skaters, there is a well-documented sexual conflict over mating decisions, which drives coevolution between male abdominal clasper morphology and female abdominal spines (Arnqvist & Rowe 2002a). Relative changes in armament levels between males and females across 15 species were associated with whether the male of any particular species is relatively better at grasping females during mating contests, as described above (Arnqvist & Rowe 2002a).

Another example of coevolution driven by sexual conflict comes from a study in the seed beetle genus *Callosobruchus* (Rönn et al. 2007). In these species there is a predicted sexual conflict over male mating frequency, with male penile spines representing adaptations that anchor males during mating but cause damage to the female reproductive tract. The female counteradaptation to that damage is a thicker lining of the female reproductive tract. The amount of spininess and amount of harm caused varies across species, and furthermore the degree of male spininess is correlated with the thickness of the connective tissue in the female reproductive tract wall. As in the pond skater example, the absolute level of armaments between males and females is independent of the degree of harm caused to females. However, the degree of harm varied instead with the relative level—in other words, harm is more evident in species in which the male genitalia was relatively more spiny and in which the female tract is relatively less robust.

A pervasive, but generally unstated, assumption underlying models of sexual conflict involving interactions between the sexes is that there will be like for like matching coevolution between male and female reproductive traits. This appears difficult to reconcile with recent findings that apparently suggest there is more genetic variation residing in male than in female reproductive or tissue specific genes. For example, Haerty et al. (2007) used data from the 12 *Drosophila* species genomes to compare rates of change in sex- or reproduction-related genes as compared to other genes. They found that genes expressed in the testis and male reproductive tract showed the most rapid patterns of gains and losses, and that genes in male reproductive tissue evolved faster than those that were female tissue specific. The fact that genes expressed only in one sex are apparently evolving faster than those expressed only in the other could mean that reproductive proteins in males are primarily subject to sexual selection among males, and that selection on females arising from sexual conflict is less strong. Alternatively, it could mean that sexual conflict is important in driving both male and female reproductive traits but there are biases that produce this result. For example, differences in expressed gene size or complexity between the sexes may make it easier to detect positive selection in male versus female reproductive genes. Alternatively, variation in the expression and function of female reproductive genes might not be encoded by nucleotide variation in the reproductive genes themselves. Finally, one sex could be more sensitive to small, subtle sequence changes, for instance, if female reproductive genes tend to be controlled at the translational level by microRNAs.

With increasing amounts of data coming from genome sequences, it would be useful to identify whether there are particular molecular signatures of sexual conflict. Although this is not yet possible, some detectable patterns are emerging; however, a fundamental problem usually remains, that functional information is needed about the selection acting on the traits involved, to distinguish the source of selection responsible. Ratios of nonsynonymous to synonymous sequence changes of between 0.5 and 1.0 may suggest evidence for evolutionary change in the recent past (Swanson et al. 2004). Gene duplication followed by positive selection may also indicate a relic of past conflicts (Kelleher et al. 2007). Rapid evolutionary change is not by itself evidence for sexual conflict. To reach that conclusion, one needs to demonstrate the relationship between the cost to females and the rate of evolutionary change in the male manipulative trait. Hence it is not currently possible to look in the sequence data for an evolutionary signature of mating rate. To do that, one would have to examine patterns in all the relevant genes that contribute to a particular phenotype.

A fruitful experimental design, which may also avoid the problem that past conflicts can be masked by current equilibria, will be to impose differing levels of sexual conflict (e.g., monogamy versus polyandry) and then, following experimental evolution, to genotype the loci subject to sexual conflict to determine whether and how they have evolved. Only then may reliable molecular signatures of sexual conflict be detected.

Between Species Divergence: Empirical Data

If sexual conflict can lead to reproductive isolation and ultimately speciation, then allopatric populations or incipient species that have been subject to it should exhibit incompatibilities or even reproductive isolation when reexposed to one another. It was proposed that such incompatibilities, as evidenced by the pattern of outcomes when crossing allopatric populations, would themselves be footprints of sexual conflict (Andrès & Arnqvist 2001; Arnqvist & Rowe 2005). However, the results of several studies in which allopatric populations have been crossed together have yielded little consistency in results (Rowe et al. 2003).

A more profitable line of inquiry has been to impose experimental evolution of differing levels of sexual selection and sexual conflict upon replicated lines and subsequently to ask whether there is any evidence for reproductive isolation when those lines are reexposed to one another. For example, Martin and Hosken (2003) conducted an artificial selection experiment with the dung fly (*Sepsis cynipsea*), with enforced monogamous and polygamous lines held under high or low population density. Females from the monogamous line showed no discrimination against males from either their own or one of the allopatric monogamous lines. In contrast, females from the polygamous lines did discriminate against males from allopatric lines, preferring their own males. This effect was more pronounced in the high density than the low density lines. This suggested that increased sexual conflict had selected for increased reproductive isolation, as predicted by theory (Gavrilets 2000).

A number of studies have conducted related experiments in *Drosophila melanogaster* (Wigby & Chapman 2006) and *Drosophila pseudoobscura* (Bacigalupe et al. 2007). However, none of these have subsequently provided support for the idea that sexual conflict leads to faster evolution of reproductive isolation. Lack of support could lie in the choice of traits examined or the amount of time that had elapsed, or arise because sexual conflict does not promote reproductive divergence in these species.

CONCLUSIONS

We have shown how sexual conflict arising from differences in the evolutionary interests of males and females can lead to antagonistic selection and coevolution between the sexes, in which male adaptations are selected despite the costs that they may cause the females with which they mate (Parker 1979). Sexual conflict can fuel evolutionary change within and between species and is more likely over traits and processes related to mating than over those related to parental investment. There is good evidence that sexual conflict can drive diversification within species, but there is currently mixed support for the idea that sexual conflict is a major engine of speciation. There is considerable evidence of rapid evolutionary change in male and female reproductive proteins, but conclusive evidence that this is driven primarily by sexual conflict is so far lacking. However, recent technical and theoretical advances will allow new experimental tests of speciation by sexual conflict, in which studies will target the relevant genes and determine evolutionary signatures of sexual conflict. There is also a need for more studies from natural populations on other species and on the influence of sexual conflict on life history traits such as life span and genomic imprinting.

SUGGESTIONS FOR FURTHER READING

Sources for further reading are Parker's (1979) original treatment of sexual conflict and Trivers' (1972) chapter, both of which illuminate fundamental concepts. Arnqvist and Rowe (2005) give an in-depth treatment of the subject and conduct a deep survey into examples of reproductive traits that are potentially subject to sexual conflict. Finally, we refer readers to three special volumes on this topic in the *American Naturalist* (2005 supplement to Vol. 165) in the *Philosophical Transaction of the Royal Society of London* (2006, Vol. 361) and in *Evolutionary Ecology* (2005, Vol. 19).

Arnqvist G & Rowe L (2005) Sexual Conflict. Princeton University Press, Princeton, NJ.

Parker GA (1979) Sexual selection and sexual conflict. Pp 123–166 in Blum MS & Blum NA (eds) Sexual Selection and Reproductive Competition in Insects. Academic Press, New York.

Trivers RL (1972) Parental investment and sexual selection. Pp 136–179 in Campbell B (ed) Sexual Selection and the Descent of Man. Heinemann, London.

24

Mate Choice

ROBERT C. BROOKS AND SIMON C. GRIFFITH

The chirping and trilling of male field crickets advertising to attract mates is a common nighttime sound in the warmer months in most parts of the world. On first inspection, mate choice in crickets may appear simple: males call as much as they can, and females use the call to localize and find the male before mating with him. Although this is true, research over the last few decades has shown that it is only part of a bigger, more fascinating story and that mate choice in field crickets is exercised at a number of junctures and in strikingly different ways. Likewise, an explosion of theoretical and empirical research on mate choice, including studies of most if not all taxa that reproduce sexually, has revealed a complex and fascinating evolutionary story. In this chapter we will introduce you to mate choice and give an overview of the rich and complex consequences that choice has via sexual selection. We will also describe some of the evolutionary forces that shape mate choice behavior, and provide a glimpse into emerging research that shows how mate choice influences population processes such as inbreeding, population growth, and speciation.

THE MANY LEVELS AT WHICH MATE CHOICE OCCURS

Halliday (1983, p. 5) defined *mate choice* operationally as "any pattern of behavior, shown by members of one sex, that leads to their being more likely to mate with certain members of the opposite sex than with others." Mate choice, according to this definition, need not be deliberate and does not even require the chooser to discriminate between two or more potential mates. In the 25 years since Halliday provided this definition, however, the frontiers of mate choice research have moved beyond mating to other factors controlling fertilization and even to the allocation of resources to the developing young. Considerable attention has since been paid to so-called cryptic choice by females in which internal fertilization is biased toward the sperm of one of two or more mates (Thornhill 1983; chapter 22 of this volume). Likewise, it has become established that the conflicting evolutionary interests of males and females (i.e., sexual conflict, chapter 23) can lead to resistance to mating by members of one sex (usually females), and that this can bias mating or fertilization toward males who are best able to overcome this resistance (Holland & Rice 1998). Despite all of this progress, the essence of Halliday's definition remains sound, although its scope needs to be expanded to include phenomena such as resistance and cryptic choice.

In figure 24.1 we illustrate when mate choice can occur in sequence of events leading to fertilization in the two organisms, Gryllid crickets and *Ficedula* flycatchers, that are the focus of most of the examples in this chapter. In crickets, for example, there are typically five stages in a successful

In the house cricket the loudness of a male's call determines how many females locate him, approach him, and assess his other qualities.

Initiation of mate choice

Locating male (pheromones, calls, display sites, color signals)

Courtship and assessment of male (through expression of ornaments, weapons, direct benefits, and mating rituals)

Female pied flycatchers pay more attention to the quality of a male's nest site than to morphological characteristics of the male himself.

Copulation

Monandry or polyandry (to mate with one or multiple males)

Even after a female has formed a pair bond with one male she has the capacity to continue to find a better father for her offspring by seeking extra-pair copulations. In the collared flycatcher extra-pair mates have larger forehead patches than cuckolded males and their offspring fledge in better condition.

Copulatory resistance (duration of spermatophore attachment, retention or rejection of ejaculate)

In Gryllid crickets the male transfers an external spermatophore that continues to transfer sperm to the female after copulation has ceased. Females exercise choice by removing the spermatophore, thus terminating the transfer of seminal fluid.

Sperm competition (sperm quality and quantity)

Cryptic female choice (physiological sperm sorting)

Fertilization

If a female mates with multiple males during one fertile period, then sperm competition will result. The outcome of fertilization will be determined by the amount of sperm inseminated by the males, the quality of that sperm, and the timings of insemination events.

Differential allocation (investing differentially in the offspring from different partners in numbers or quality of maternal care)

Female zebra finches invest more in rearing offspring sired by attractive males, thereby improving the fitness of their offspring compared to those of less attractive males.

Resulting offspring

FIGURE **24.1** Flow chart identifying the stages in the interaction between female and male at which mate choice can occur, from a female's decision about when to initiate reproduction to the production and care of her offspring. Photo credits (clockwise from top left): Felix Zajitschek, Simon Griffith, Paco Garcia-Gonzalez, Simon Griffith, Simon Griffith, and Ben Sheldon.

mating at which females may have the opportunity to exercise mate choice: locating males, assessing male courtship, influencing spermatophore attachment time, "cryptic" factors influencing fertilization success of males after insemination, and the allocation of resources to eggs. The different stages at which female crickets can exercise choice helps us to clarify what we mean by mate choice and to dispense with some confusion that choice needs to be deliberate or to involve comparison of two or more potential mates.

Male field crickets broadcast a loud advertisement via rapid stridulation of the forewings. The first and most obvious way in which female field crickets choose is in locating males on the basis of their loud advertisement call. Female field crickets locate calling males by walking up the gradient of sound toward the male. This provides two avenues of choice that result in very different patterns of sexual selection. First, the sensory receiver and neuronal processing capacities of the female tune her to particular optimum values of call properties such as the dominant frequency of the male call, the durations of the chirp pulses, and the intervals between pulses. As a result, these properties are often under strong stabilizing sexual selection (Brooks et al. 2005) and can be important in recognition of conspecific potential mates. Note that a female need not compare males for mate choice to occur; the properties of her psychosensory system have influenced the chances of her hearing and responding to the call of a male with the appropriate call properties and thus of mating with him even if he is the lone male calling in her proximity.

Second, females are only able to localize males that call for a bout that is long enough. Females are most likely to find the males that not only have the best call structure, but also those that call for the longest durations and on the most nights. This exerts strong directional selection for males that call as much as possible (i.e., that have high calling effort; Bentsen et al. 2006). Once again, this can happen without females needing to evaluate and compare males—they merely end up with the males that call the most because those males are more likely to be detected and located. The final pattern of choice and resulting sexual selection may be more complex than simply directional selection on call effort and stabilizing selection on call structure. For example, some elements of call structure, such as the intervals between calls within a bout, can influence the energetic costliness of a call to a male, and therefore trade off against bout length, adding complexity to the overall pattern of choice (Bentsen et al. 2006). The simple fact that female crickets find and locate mates on the basis of male advertisement calls results in strong mate choice, with resulting sexual selection on male call structure and effort.

Once a female has found a male he ceases his advertisement call and courts her with a second type of call: the courtship call. These calls are quieter than advertisement calls, but they have been shown, in the house cricket *Acheta domesticus*, to be twice as energetically costly per unit time as advertisement calls (Hack 1998). The courtship call is necessary before a female will mount a male for mating, and experimental evidence shows that the properties of a *G. bimaculatus* male's courtship call influence the latency to mating (Tregenza et al. 2006), and presumably females may exercise choice in the field at this juncture by breaking off the interaction on the basis of courtship properties.

Female field crickets exercise choice immediately after mating by removing (or leaving intact) the externally attached spermatophore at some point after copulation, interrupting insemination (Sakaluk 1984; Simmons 1986). This has a strong influence on fertilization success, and hence paternity that increases with spermatophore attachment time (Sakaluk 1984). This is a form of choice because females appear to favor males that are more attractive in courtship (Bussière et al. 2006) or large males (Simmons 1986) by leaving the spermatophores of these males attached for longer than those of others.

Female field crickets often mate with more than one male (Simmons 1986), and some of the seminal research on the evolution of polyandry has been done in this group (e.g., Tregenza & Wedell 2002). Although it has been suggested that multiple mating may be an incidental consequence of an adaptive tendency to mate repeatedly to obtain sperm or resources (Arnqvist & Nilsson 2000), females recognize and avoid remating with previous mates (Ivy et al. 2005). Further, experimental dissection of the number of matings from the number of mates reveals increases in fitness components in several species of insects (see list in Jennions et al. 2007, p. 1469), including at least two Gryllids (Tregenza & Wedell 1998; Ivy & Sakaluk 2005), although not in others (Jennions et al. 2007).

The excellent work on postcopulatory and especially postinsemination processes that influence paternity in field crickets illustrates the necessity of broadening Halliday's (1983) definition of choice beyond just those factors that influence the likelihood of mating with some individuals rather than others. There is a growing need to also include forces that influence the success of particular matings. Sheldon (2000) even suggests that the allocation by females of reproductive effort toward the progeny of certain males rather than others can be usefully understood as a form of mate choice. The evolution of such allocation and care strategies can best be understood in the context of models of more traditionally conceived choice such as precopulatory choice over whom to mate. For example, in house crickets, *A. domesticus*, daughters of attractive males tend to lay more eggs for large, attractive males than do daughters of unattractive males, indicating genetic variation in a form of differential allocation that mirrors the variation and evolutionary dynamics of precopulatory choice (Head et al. 2006). The integration of our understanding of pre- and postcopulatory choice and differential allocation to zygotes, embryos, and offspring remains an important challenge for the field.

Choice Is an Expression of Choosiness and Preference

As we have seen, mate choice is usually the result of a complex set of behaviors, and there are several dimensions along which these behaviors may vary. It is often useful to conceptually distinguish the tendency to choose from the traits or trait combinations that are the object of choice. We will use Jennions and Petrie's (1997) terms *choosiness* and *preference* to describe these two aspects of choice respectively. Choosiness is influenced by an individual's propensity to mate, responsiveness to potential mates, and the ability and tendency to discriminate among them. Empirically, an individual's choosiness is the time or effort that she invests in making a choice. Choosiness may, therefore, influence both how responsive an individual is to potential mates and the extent to which she discriminates among them. Preference, unlike choosiness, depends explicitly on understanding the relationship between choice and the traits that are the object of that choice. The statistical relationship between male trait values and a female's response to those males is the female's preference function.

In figure 24.2 we illustrate the concepts of choosiness and preference by highlighting how an empiricist might estimate these aspects of choice using simple descriptive statistics like mean, variance and regression coefficients. This illustration is based on work in guppies in which it is possible to quantitatively score the strength of a female's response to a given male. These methods have parallels in the population-level tools for understanding the opportunity for selection (Wade 1979) and the strength and form of linear and nonlinear selection (Lande & Arnold 1983), and there is considerable scope for extending the use of these well-established tools to dissect mate choice. Many species are not amenable to dissection in this way, however, and empiricists have to find relevant and tractable ways of dissecting choosiness and preferences that are suited to the biology of their chosen system. For example, in gray tree frogs, *Hyla versicolor*, the strength of choice for one stimulus over another can be estimated by first comparing phonotactic responses of females to the two stimuli played at equal sound pressure levels (SPLs) and then raising the SPL of the less attractive stimulus until phonotactic responses are equal. The magnitude of difference is an estimate of the strength of preference for the quieter call (Gerhardt et al. 1996).

Choosiness and the preference function interact to determine the outcome of choice (i.e., the identity of an individual's mate). Reduced choosiness, such as when an individual is under an ecological constraint to mate rapidly, often obscures preference functions as we will see when we consider mate choice in *Ficedula* flycatchers later in the chapter. It is therefore possible to see variation in the outcome of choice without any variation in preference functions. Causes of variation in choosiness have often been overlooked by both theoreticians and empiricists, despite early evidence that such variation is probably more widespread and important (Jennions & Petrie 1997) than variation in preferences. Understanding the functional building blocks of choice is crucial if we are to understand both the genetic basis of choice and how selection operates on the genes involved.

THE EVOLUTION OF MATE CHOICE

How and why choice evolves remains the most controversial subject in the study of mate choice. Much of this controversy concerns the relative importance of the various processes that may shape the

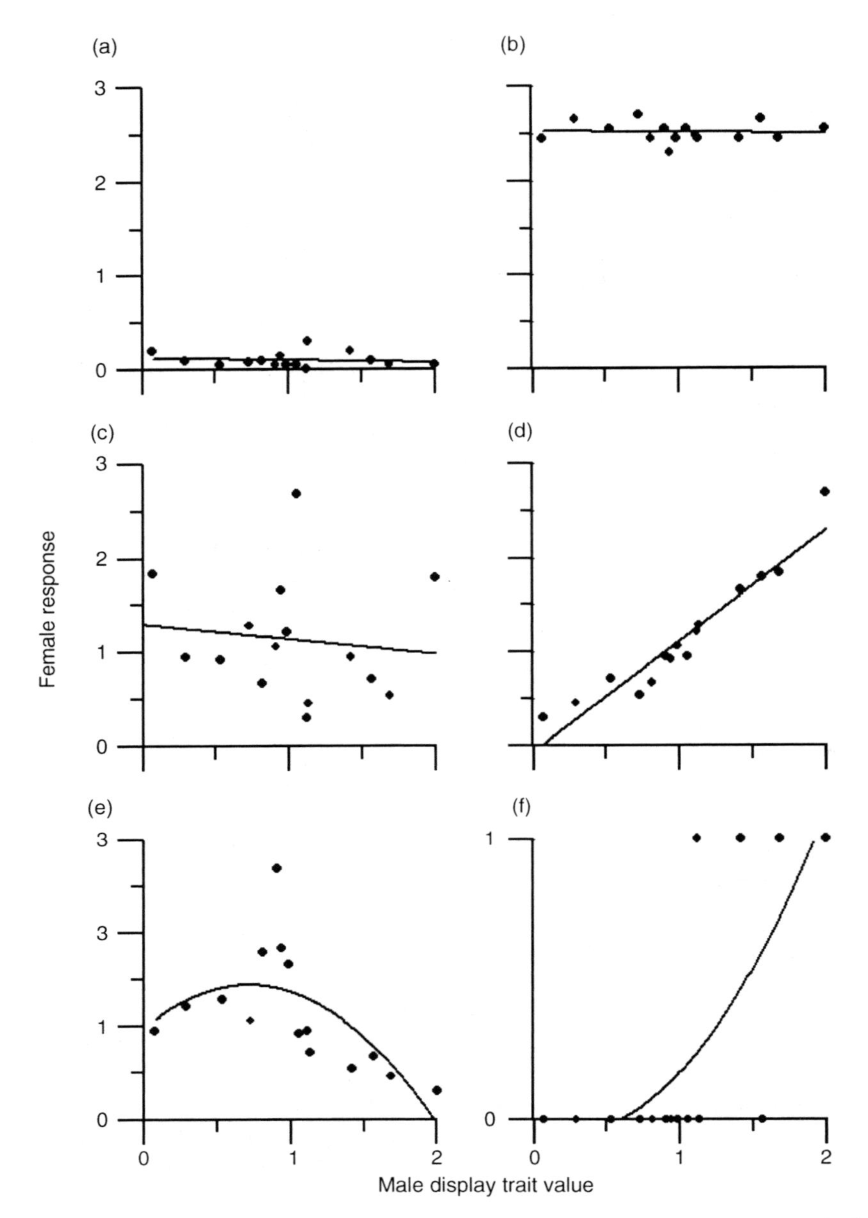

FIGURE **24.2** An illustration of female choosiness and preference functions and how they might be described using simple statistical tools. Data points are fictional observations of female response to 15 males on a continuous scale (a–e) or binary scale such as mated/unmated in (f). Two aspects of female choosiness can be extracted from this kind of data set: responsiveness and choosiness. Responsiveness is estimated by the mean response: panel (a) shows the lowest responsiveness, and panel (b) the highest compared with all remaining panels. Discrimination is estimated by variance or some other descriptor of variation about the mean response: low discrimination in panels (a) and (b) compared with the rest of the panels. Responsiveness and discrimination constrain the opportunity for preference much the same way as the variance in fitness describes the opportunity for selection (Wade 1979). When females are both responsive and discriminating, some male traits will not be the object of preferences (c), and others will (e.g., d–f). Preference functions can be linear (d), quadratic (e), or more complex. The estimation of choosiness and preference function properties can be achieved in similar ways when choice is an all-or-nothing affair (e.g., successful or unsuccessful mating attempt, f) or obeys a threshold. The estimation of preference functions using parametric regressions can be extended to the multivariate male phenotype using the methods first outlined by Lande and Arnold (1983) for estimating selection gradients.

evolution of choice. In this section we briefly introduce each of these processes, giving a sense of how each is thought to work and some of the history of the idea. Like recent reviews (Kokko et al. 2006a), we believe that the common tendency to reify the various processes involved in mate choice evolution to the status of competing alternatives is misguided. Several processes are often involved in the evolution of mate choice in any given species, and it is our aim to build a more integrated understanding of the forces acting on mate choice evolution.

The evolution of mate choice occurs in response to direct selection on genetic variation in choice and to indirect selection that acts on genetically correlated traits. The distinction between direct and indirect selection is an important one in mate choice evolution, and we will return to the importance of the fact that selection never operates on a single trait in isolation. The effects of direct selection on mate choice are intuitively easier to grasp and less controversial than the effects of indirect selection, and so we begin here.

Selection acts directly on choice when choice results in an improvement in the chooser's lifetime reproductive success (LRS), rather than in the LRS of descendants. Direct selection can operate on choice in three important ways that are worth considering separately: the *direct benefits* of choice can favor increased choosiness and alter preference functions, the *direct costs* of choice can favor reduced choosiness and altered preference functions, and direct selection on the psycho-sensory system in other ecological contexts can alter preference functions (i.e., *sensory drive*). We will discuss each of these in some detail.

Direct Selection

Direct Benefits of Choice

Direct benefits include a variety of resources provided during courtship, ranging from prey items, chemicals that confer protection from predators, and nutritious parts of the male himself. One taxonomically widespread consequence of mate choice, male parental care, is typically considered a direct benefit of mate choice. Although parental care can also have indirect effects on female fitness because good care improves offspring LRS (chapter 14), we consider here only the direct benefits to the female in terms of improved breeding success in the current and subsequent breeding events. The socially monogamous mating system typical of most human societies and very widespread in birds is driven by the direct benefits that females can receive from a prolonged social association with a male partner.

Ficedula flycatchers provide a good example of the direct effects of choice on female fitness, and, as in other monogamous birds (reviewed in Griffith & Pryke 2006), males supply females with a nest site, feeding territory, protection against harassment by other males, nest defense, and parental care of the offspring. These socially monogamous birds migrate from Africa each boreal spring to breed in the woodlands of Europe. As in many other migrant birds, the process culminating in mate choice actually starts days before the female has even arrived at the breeding ground with males typically arriving back at least a week before the females (Lundberg & Alatalo 1992). Once they arrive, each male will find and defend a nest site in a natural cavity or artificial nest box. In each woodland, some nest sites are occupied earlier and far more frequently over successive years than would be expected by chance (Askenmo 1984). Such preferred nest sites are found in areas with higher prey abundance (Lundberg & Alatalo 1992), are typically upright and dry (Slagsvold 1986), and are high off the ground and with a narrow entrance hole, providing protection from predators (Alatalo et al. 1986; Lundberg & Alatalo 1992). Favored boxes have more successful reproductive output than other, less favored boxes in the local area of the wood (Askenmo 1984). Males that arrive back early to the breeding ground will get the best choice of nest sites, and typically older males arrive earlier than yearlings (Lundberg & Alatalo 1992). Because suitable cavities are a limited resource, males arriving late are constrained in their ability to gain a nest site because all, or most, suitable cavities are already occupied. In the pied flycatcher *F. hypoleuca*, winning males are not significantly different in color from losing males (Lundberg & Alatalo 1992); however, in the collared flycatcher *F. albicollis,* the outcome of male-male competition over nest boxes is positively related to the size of the white forehead patch (Pärt & Qvarnström 1997). Before the females have even arrived on the breeding grounds, intrasexual competition among males coupled with arrival date and prior experience of the local area results in a very nonrandom distribution of males (with respect to morphological traits, age, experience, and competitive ability).

Females are also interested in settling nonrandomly throughout a wood, preferring to breed in territories and nest cavities that promise better reproductive success. Because the females arrive later, the best locations are already occupied by males. In one of the most important experimental studies in the field of mate choice, Alatalo et al. (1986) teased apart the effects of male characteristics, arrival date, and the quality of nest sites and habitat quality in the pied flycatcher, demonstrating that, on arrival, females choose territory and nest site quality and pay no heed to male characteristics, settling at the best available location with the male that happens to be occupying that space at the time. Because males are nonrandomly distributed onto territories based on sexual ornaments (through male-male competition, a week before the females even arrive), female preference for territory quality indirectly results in nonrandom pairing with respect to male phenotype. It would have been easy to interpret the observed correlation between female settlement time and male ornament size as a consequence of active female choice for ornamental traits. In fact, Alatalo et al. (1986) demonstrated that females exhibit no preference for the sexual ornament whatsoever. Most studies of mate choice in animals (before and since) have not heeded the obvious and far-reaching implications of this study.

The Costs and Constraints of Choice

The costs of choice are an important consideration in models of mate choice evolution because high costs oppose the evolution of choice. Such costs can be exacted through lost time and energy spent evaluating males (Vitousek et al. 2007), increased exposure to predation (Hedrick & Dill 1993), harassment (Magurran & Seghers 1994), sexually transmitted diseases (Boots & Knell 2002), and opportunity costs because the best males are removed from the pool of available mates as time progresses (Dale et al. 1992). Although choice often incurs substantial costs, other attempts to measure costs have suggested that choice can be cheap (Head & Brooks 2006). Despite the critical role of cost functions in determining the outcome of theoretic models (e.g., Kokko et al. 2002), there are no quantitative measures of the real costs of choice relative to the scale on which these costs are modeled, and how changes in costliness influence the outcome of mate choice evolution. These questions are among the most pressing contemporary challenges in the study of mate choice.

We illustrate some impacts of the costs of choice by taking up again the story of *Ficedula* flycatchers. When females arrive on the breeding grounds they are under strong phenological selection to breed quickly, with a marked decrease in breeding success as the spring progresses (Lundberg & Alatalo 1992). Female flycatchers make a rather restricted mate search, visiting fewer than 10 males and pairing up within 2 days of arrival. This limited mate search is driven by the physical cost of moving around and competition with other females over mating opportunities (Dale et al. 1992). Most important, in a socially monogamous species, the longer a female lingers in her choice of territory/mate, the fewer males will be available as they form socially exclusive pair-bonds. This underlines one of the major issues in socially monogamous systems such as the flycatcher. There is almost certainly a mismatch between female preference and male phenotype because even if every female in the population agrees over which is the most attractive male in the population, only one of them will be able to have him as a primary social mate.

The extent to which female preference varies across a population has yet to be determined in this or any other wild socially monogamous bird species because of this major constraint on females being able to express their preference. As the season advances toward the point when reproduction is likely to fail due to deteriorating seasonal conditions, active female mate choice becomes weaker, to the extent that many females pair with the first male they encounter. Thus where the two species occur in sympatry, some females even pair with heterospecifics (Veen et al. 2001). This final example emphasizes how difficult it can be to measure a preference function in a wild population. Females that pair with a heterospecific male may have made an adaptive mate choice decision (breed with him, or do not breed at all); however, it seems implausible that these females have a preference for males that are phenotypically like members of another species. Therefore, as a rule, in systems in which mate choice is constrained, we are unable to infer the preference function of a female by looking at the outcome of choice (Postma et al. 2006).

For those females that arrive earliest and are able to invest most time and energy in mate choice, the main challenge is to optimize all of the potential benefits that they can get from their partner, which are many and diverse in socially monogamous passerine birds (Griffith & Pryke 2006). The key problem faced by the female (with resulting challenges for the researcher) is that many of the male traits and resources are positively or negatively related to one another and even a very active and exhaustive mate choice process will be unable to optimize all traits simultaneously. For example, dominant males are best able to control resources such as territories and nest sites; however, they are also likely to invest a lot of time and effort in sexual displays and aggressive interactions, relative to subordinate males. Subordinate males, on the other hand, are more likely to invest time and energy into parental care. The way in which females resolve the trade-offs between different direct benefits can often lead to some surprising results. For example, in an island population of the house sparrow *Passer domesticus*, in which the natural variation in nest site quality was removed, females actually preferred males with the smallest sexual ornaments (Griffith et al. 1999). The complex ways in which costs and benefits are modified by context remains an important area for new and creative research, particularly but not only in socially monogamous animals.

Sensory Drive

Preferences are necessarily influenced by the sensory and neurological hardware involved in detecting signals. Much of this hardware is used in other ecological contexts, and the way that selection in these contexts influences the psychosensory system may in turn influence preference evolution. For example, guppies (*Poecilia reticulata*) use their visual system to detect predators, find prey, and choose mates. Interestingly, attraction to orange objects, a trait thought to help guppies locate nutrient-rich orange fruits, is also genetically associated with the strength of female preferences for orange males (Rodd et al. 2002). Whether the foraging-based sensory bias or the mating preference evolved first remains controversial, but it is clear that the evolutionary fate of both mate preference and sensory bias are tightly linked (see box 24.1).

Sexually Antagonistic Coevolution and "Chase Away"

A mating pair's evolutionary interests are identical only if there is true lifetime monogamy, which exists very rarely if at all. Therefore, in most animals, each sex is expected to evolve strategies that maximize their own benefit from mating, even at the partner's expense (Parker 1979). This is sexual conflict (chapter 23). One common source of sexual conflict is that females often have less to gain from mating multiply than do males, and that female resistance to male mating attempts, including courtship and coercion, may be an important feature of female mating strategies (Holland & Rice 1998; Gavrilets et al. 2001). Holland and Rice (1998) proposed that the evolution of female resistance and of male seductive strategies to overcome such resistance could lead to cycles of *chase-away coevolution*.

Clearly, female resistance to mating is well within the definition of mate choice, and like others (Gavrilets et al. 2001; Kokko et al. 2006a), we see no point in distinguishing the mating bias that resistance causes to favor seductive or coercive males from the more conventional types of choice. The extent to which sexual conflict, including the chase-away process, represents a process of mate choice evolution that is qualitatively different from the other processes we discuss in this chapter has been the subject of some debate (see Kokko et al. 2003b, 2006a for a fuller treatment). Resistance by females has probably evolved due to the costs to females of mating too many times (Gavrilets et al. 2001; Kokko et al. 2003b). However, it has been argued that if male ability to seduce or coerce females into mating is heritable (as suggested by Parker 1979), then resistant females will also benefit indirectly when they do mate because their mate will confer on her sons genes that lead to greater seductiveness (Cordero & Eberhard 2003). This claim is far more controversial than the claim that resistance is under direct selection (Cameron et al. 2003; Kokko et al. 2006a), and remains the subject of intense theoretic attention. It remains quite possible that both direct and indirect selection influence the evolution of choice in the form of female resistance, but before we get there, we must introduce you to *indirect selection*.

Indirect Selection

We have seen how selection can operate directly on mate choice in a number of ways, and we now turn our attention to the possibility that a chosen mate might have genes that give the offspring a considerable fitness advantage. If, by choosing her mate, a

BOX **24.1** Sensory Bias

Rebecca C. Fuller

The term *sensory bias* is used in two ways in the sexual selection literature. First, it is used as a hypothesis concerning which types of secondary sex traits males should evolve. Sensory bias predicts that males should evolve traits that maximally stimulate the female sensory system. Here, sensory bias is not an alternative to the other models of sexual selection (e.g., direct benefits, Fisherian model, good genes, sexual conflict) because all models of sexual selection predict the evolution of male traits to match female mating preferences to the extent possible given their costs. However, sensory bias is also used to describe a model for the evolution of female mating preference (West-Eberhard 1984; Basolo 1990b; Ryan et al. 1990; Kirkpatrick & Ryan 1991; Fuller et al. 2005). Sensory bias states that female mating preferences evolve as a by-product of natural selection on nonmating behaviors (e.g., foraging) mediated through the sensory system. Here, sensory bias is clearly an alternative model for the evolution of female mating preferences, and it is this usage of sensory bias with which we are concerned.

Historically, the sensory bias hypothesis has been tested using the comparative method. A pattern in which female mating preferences evolve prior to the evolution of male secondary sex traits has been taken as evidence for sensory bias because it cannot be explained by either the Fisherian or good genes models (Basolo 1990b; Ryan et al. 1990). Both the Fisher and good genes models rely on the buildup of gametic disequilibria either between the female preference and the male secondary sex trait (Fisher process) or between the female preference and viability as indicated through the male secondary sex trait (good genes model). Because the male secondary sex traits were absent during the establishment of the female preference, neither type of covariance could have existed, and thus neither model can explain the evolution of the female mating preferences.

Comparative data support the sensory bias model only when the male trait is absent during the establishment of the female mating preference. However, this leads us to an empirical difficulty. If the preference is defined too narrowly, then one may erroneously reject the Fisherian and/or good genes models. For example, several studies on swordtail fishes suggest that females evolved a preference for males with swords long before the swords themselves evolved (Basolo 1990b, 1995a). This pattern is consistent with sensory bias. However, female poeciliids have a general preference for large size. In at least one swordtail species, possession of the sword makes males appear larger (Rosenthal & Evans 1998). If the trait is actually size (and not swords per se), then the female mating preference may have coevolved with male size (Fisher model) or with viability as indicated through male size (good genes model). Hence, the comparative data cannot reject either the Fisherian or good genes models.

A more direct empirical approach is to test the critical assumption of the sensory bias model. Sensory bias assumes that pleiotropy creates strong genetic correlations between behaviors that share a common sensory system (Kirkpatrick & Ryan 1991; Fuller et al. 2005). Amazingly, the prediction of pleiotropy has never been tested. The best evidence for genetic correlations between behaviors sharing a common sensory system comes from studies of guppies. Rodd and colleagues (2002; Grether et al. 2005) have shown strong among-population correlations between preferences for inanimate orange objects (which are interpreted as an indicator of foraging preference) and female mating preferences for orange males (see figure I). Several possible scenarios can explain this result. If there is pleiotropy between foraging and mating preferences, then variation in selection for orange foraging preference would result in a pattern in which the strength of foraging and mating preferences for orange covary across populations. However, orange foraging preferences

(continued)

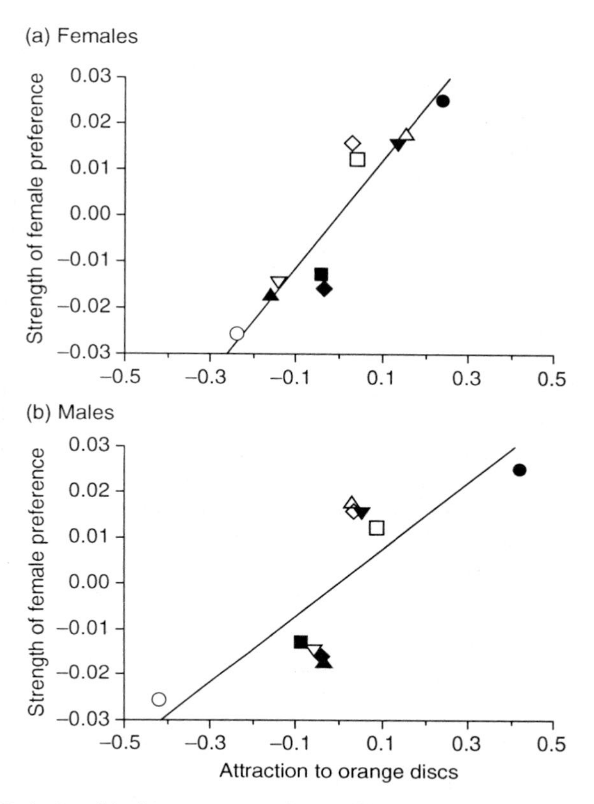

Figure I Relationship between population level strength of female mating preference for carotenoid coloration (i.e., orange) and attraction to orange discs in (a) females and (b) males across 10 guppy populations. Open symbols denote sites with open forest canopies. Closed symbols denote sites with closed forest canopies. Each symbol shape denotes a unique river drainage. Figure provided by Greg Grether and Helen Rodd, redrawn from data in Grether et al. (2005).

and orange female mating preferences may have evolved separately, in which case the among-population correlation reflects genetic disequilibrium and not pleiotropy. Many authors have suggested that female choice for males with orange spots is adaptive because males must ingest carotenoids to produce these colors. Thus, sexual selection for orange-colored males and natural selection for animals to prefer orange-colored fruits may have occurred independently. Another alternative is that selection for female mating preferences for orange males may have caused a correlated response in orange foraging preference, but this alternative seems less likely (see Rodd et al. 2002).

Again, the critical task is to determine the extent to which pleiotropy creates strong genetic correlations between behaviors that share a common sensory system. One possible approach is to randomly breed animals over a number of generations to determine the extent to which correlations between mating and nonmating behaviors remain over time. The persistence of strong genetic correlations is consistent with pleiotropy. If correlations break down over time, then this suggests that a gametic disequilibrium was present. An alterative approach is to exert selection on one behavior (e.g., foraging) over multiple

(continued)

BOX **24.1** *(cont.)*

generations and determine the extent to which mating preferences evolve as a correlated response. A correlated response across multiple generations is consistent with pleiotropy. There is good reason why such experiments have not been performed. Measuring quantitative genetic parameters on multiple behaviors is notoriously difficult. However, such experiments provide the most direct route to testing the primary assumption of the sensory bias model for the evolution of female mating preferences.

female could ensure that her offspring inherit from their father genes that are superior (on average) to the genes of other members of their cohort, then the female would experience an increase in her own fitness (long term, chapter 4). More precisely, any genes that the female has that dispose her to this kind of choice would become associated with the male's high-quality genes in the offspring and their descendents, a phenomenon known as *linkage disequilibrium*. The fitness advantage that the male's genes confer would select indirectly for the female's choice genes. There is ample evidence for the kinds of genetic consequences of choice that could lead to indirect selection. For example, female gray tree frogs (*Hyla versicolor*) prefer to mate with males with long rather than short call durations. When the eggs of a female are split into batches and experimentally fertilized by different males, the larval and juvenile offspring of males with the preferred long-duration call grow faster and survive better than their maternal half-sibs sired by males with short-duration calls (Welch et al. 1998).

Extra-Pair Paternity in Flycatchers: An Additional Mate Choice Opportunity

In the flycatcher species we introduced above, as with other socially monogamous species, the direct benefits provided by a social partner can dramatically improve reproductive success. Constraints on social mate choice mean that quite often females are not particularly choosy when finding their social mate. Once a female has paired, however, she will already be assured of most of the direct benefits to be gained through her partner. Given that mate choice also influences the sire of each fertilized ovum, the female flycatcher can still actively engage in mate choice for several weeks after pairing and up until the point that her final egg is fertilized. In fact, because her social mate will provide the important direct benefits she will need to successfully reproduce, paradoxically, her mate choice becomes less constrained as soon as she has found a social mate. Extra-pair paternity is very widespread in socially monogamous birds, accounting for about 11% of offspring across all surveyed species (Griffith et al. 2002). The collared and pied flycatchers are no exception, with about 15% and 8% of offspring being sired outside the pair-bond respectively (Gelter & Tegelström 1992; Sheldon & Ellegren 1999). Extra-pair copulations could allow a female to acquire specific genetic contributions to her offspring. An individual male flycatcher can, at most, hold two simultaneous social partners, but he could potentially have extra-pair copulations with scores of females that are socially paired with other males because such copulations are relatively quick and offer him high fitness rewards relative to the cost of raising his own offspring. As a result, all females can conceivably express the same preference and can also all potentially achieve a mating with one, or one of a few super-attractive males that best meet the shared preference. In this way extra-pair mate choice is less constrained and could be a much stronger driver of sexual selection on males than social mate choice.

Extra-pair mate paternity could result from behavior in which a female favors another male at the expense of her own partner (the social mate), and the extra-pair male may be phenotypically different from the social male. If these differences arise through female preferences, extra-pair paternity highlights the possibility that either (a) there are two different preference functions operating in the two different contexts or (b) if the female only has one preference function, then it is poorly expressed in either social pairing or during interactions with extra-pair males. In the collared flycatcher, experimental work suggests that females are expressing a different preference during precopulatory extra-pair mate choice than the one expressed during social mate choice. Collared flycatcher males have a distinctive forehead patch and small-patched social males are more likely to be cuckolded by extra-pair sires having significantly larger patches (Sheldon et al. 1997; Sheldon & Ellegren 1999). There may be indirect benefits of this; extra-pair offspring fare better in the same nests than their within-pair maternal half-siblings (Sheldon et al. 1997). Sexual selection on the forehead patch through the sperm competition route is quantitatively more important than selection through the social mate choice route (Sheldon & Ellegren 1999).

Extra-pair paternity in birds is usually studied through post hoc molecular analysis rather than detailed behavioral study. This has generated some discussion over the extent to which observed paternity represents active female choice or the result of coercion by extra-pair males (Westneat & Stewart 2003; Griffith 2007). Females clearly can exert active choice in some cases. In the collared flycatcher, an experiment that removed male social partners for 2 days during the fertile period found that extrapair paternity was negatively related to the social partner's patch size (Sheldon et al. 1997), suggesting that only females who were initially paired to unattractive, small-patched males were likely to seek extra-pair paternity. If male coercion was an important driver of extra-pair paternity in this system, then we would perhaps expect that all experimentally widowed females would have been coerced to the same extent in the absence of their partner. More direct evidence of active choice comes from the superb fairy wren *Malurus cyaneus*, a species in which over 75% of all offspring are sired by extra-pair males (Double & Cockburn 2000). Despite the extraordinary level of extra-pair paternity in this species, extra-pair copulations are conducted extremely covertly, and only four have been observed directly (Cockburn et al. 2009). Nonetheless in this territorial species, during the fertile period, a female will leave her social group before dawn and fly in a straight line to a preselected male up to several territories away. This behavior is directed to males that do sire the majority of her extra-pair offspring. The only thing that sets such males apart from others is that they molted into their nuptial plumage earlier (Dunn & Cockburn 1999), suggesting that females have identified them many months before their fertile period, indicating a high level of premeditation by the female, with respect to extra-pair mate choice (Cockburn et al. 2009).

Indirect Selection: From Simple Intuition to Theoretic Minefield

Although the idea that female behavior can bias mating and paternity to males with high-quality genes is highly intuitive, the idea of such indirect selection is very difficult to model theoretically (for example, consider the problem of modeling associations between alleles multiple generations into the future) and even more challenging to test empirically. In this section we will introduce you to a few of the indirect selection theoretic and verbal models, several of the important historic detours that this idea has taken, and some of the controversies that remain, but we intend to convince you that there is a single process of indirect selection that does not differ in essence from the original rather intuitive idea.

Although Darwin's argument that mate choice is an important and widespread agent of sexual selection remains powerful today, his failure in *The Descent* (Darwin 1871) to explain the evolution of mate choice remained an important point of criticism of the theory, one that he partially remedied only in the second and subsequent editions. It was Fisher (1930) who argued convincingly for the first time how it can be adaptive for a female to mate with males that bear a trait that is under positive selection. This is because the offspring of such a mating would bear the genes for this form of choice and for the male trait under selection. When, in the offspring generation (and in generations that follow), selection favors the chosen trait, it will also indirectly favor the choice gene. Fisher further argued that as the choice gene increased in frequency, the choices of females bearing that gene would generate additional positive sexual selection on the chosen trait, resulting in an ever-strengthening statistical association (i.e., gametic phase linkage disequilibrium) between choice and chosen trait and exaggeration of both. This has come to be known as Fisher's "runaway" process because often sexual selection would push the chosen trait well below its otherwise adaptive optimum expression.

Lande's (1981) model formalizing the relationships between mate choice, the ornaments that are the object of choice, and their underlying genetic architecture was a crucial step in the development of sexual selection theory. Lande showed that when viability selection strongly constrains the exaggeration of the male ornament, mean male ornament and female choice expression in the population

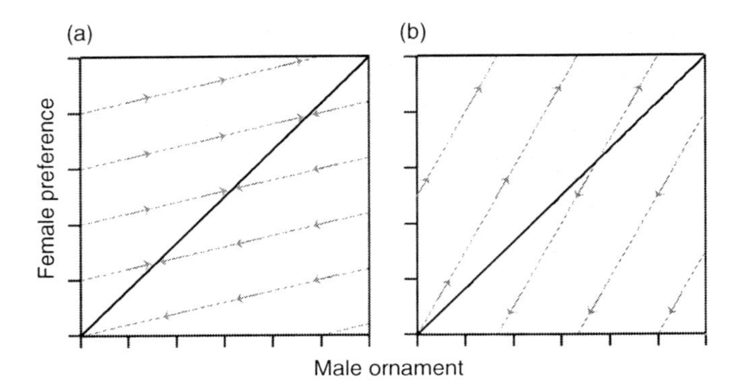

FIGURE **24.3** Lande's (1981) classic model of the coevolution of mate choice (preference) and male ornaments via indirect selection. (a) When viability selection strongly constrains the exaggeration of the male ornament, mean male ornament and female choice expression in the population evolve toward a theoretical line of equilibrium. (b) When viability selection is weaker relative to the genetic variation in and covariation between ornament and choice, the equilibrium is unstable, leading to accelerating mutual exaggeration of both choice and ornament in the way that Fisher predicted. This is the runaway process.

evolve toward a theoretic line of equilibrium (figure 24.3a). However, when viability selection is weaker relative to the genetic variation in and covariation between ornament and choice, the equilibrium is unstable (figure 24.3b), leading to accelerating mutual exaggeration of both choice and ornament in the runaway fashion that Fisher predicted.

The runaway component of Fisher's overall theory provided a compelling intellectual puzzle. The idea that choice and the object of that choice (we will now use *display* or *ornament* interchangeably to mean a chosen trait) can become locked in a mutually reinforcing cycle of exaggerated coevolution provided considerable fuel for theoretic modeling if not empirical progress. In the early 1980s the idea that held sway was that the chief adaptive benefit of choice was enhanced attractiveness of sons, and that attractiveness and choice were involved in runaway coevolution. Certainly, in many species attractive males have attractive sons, and attractiveness can be genetically correlated with female choice behavior, but as we shall see below, neither of these are exclusive predictions of the Fisherian runaway model.

More recently, the pendulum has swung back toward the first component of Fisher's idea—that choice based on traits that are under selection in other contexts can evolve via indirect selection. In many species, preferred males not only bear more extreme ornaments and display more vigorously, but they are often longer lived (Jennions et al. 2001) and appear to be in better general health. It follows from this observation that the ornamental traits may be indicators not only of a male's vigor but also of his breeding value for fitness.

Although this idea that sexual ornaments might signal genetic quality has been around for some time, it was problematic for a number of reasons. Chief among these was that such a signaling system would need to be honest (i.e., a reliable indicator of genetic quality) if it were to favor choice, yet selection on males to cheat and falsely advertise their genetic quality would be strong (Johnstone 1995b). Zahavi played a crucial role in putting the idea of sexual signals indicating fitness components other than attractiveness on a logical footing when he presented his handicap model (Zahavi 1975). The fact that ornaments and displays are often costly to express imposes a handicap in the form of reduced longevity or other fitness components on males that express these traits, and this handicap can result in the overall honesty of the signal in a range of circumstances. Zahavi's suggestion was met with skepticism in its early days until a body of theoretic models (e.g., Pomiankowski 1988; Grafen 1990a) formally demonstrated that such a handicap process can maintain honesty in signals of genetic quality. The importance of signal costs (i.e., the handicap) in maintaining signal honesty is not limited to indirect selection, but it is in this context that handicaps have received most attention from theoreticians and empiricists.

In addition to problems of the maintenance of signal honesty, the idea that preferences for indicators of genetic quality can evolve via indirect selection faced an additional hurdle: the maintenance of genetic variation in fitness components under concerted natural and sexual selection. We will turn to this problem, the so-called lek paradox, in the next section.

Since 1990, the prevailing view (with some notable exceptions) of indirect selection has pitted Fisherian selection via sons' attractiveness against indicator selection via other fitness components in sons and/or daughters. This dichotomy has flourished despite the fact that Fisher (1930) clearly took the view that both attractiveness and other fitness components were important ingredients in his model of indirect selection, and the fact that Zahavi's handicap process is mediated by the trade-offs between ornament expression and other fitness components. The polarization of the field, which can still be felt in the literature, is more the subject for a philosopher of science than for this chapter, but it may have been due to two dynamics. First is a mismatch between the way theoreticians model evolutionary processes and empiricists test theory. Much of the seminal theory in this area has involved testing particular components of the process in isolation from other components. Grafen's (1990a, 1990b) pair of papers that do this for the handicap and runaway processes, respectively, are a turning point in establishing the logical feasibility of these processes, but to do so they deal with each process in isolation. Theory also deals with one or a small number of fitness components for the sake of tractability and heuristic value. The fact that processes can be modeled in isolation from one another may have led to a misconception by empiricists that such processes might be alternative explanations for the way in which evolution occurs.

The second explanation for the way in which the dichotomy has flourished is that, as any student of rhetoric will point out, it is often convenient to erect a dichotomy when making an argument. This is especially true in science due to the appeal of the hypothetico-deductive method: by appearing to test the two models as alternatives, many papers also appeared to test, and in some cases refute, at least one hypothesis. Kokko et al. (2002) provide more comprehensive insight into this problem in a paper that shows that the "Fisher versus good genes" dichotomy is, like so many others, a false one.

Considerable research effort has been spent attempting to extract critical and mutually exclusive predictions and then using these to test Fisherian versus indicator models as competing alternatives. Although this broad approach led to many important findings and insights into mate choice evolution and sexual selection, very little evidence could be unambiguously interpreted as favoring one position over the other. Progress has lately come from an explicit integration of life history, quantitative genetic, and sexual selection theories. The theoretic prediction (Hansen & Price 1995; Kokko et al. 2002) and empirical finding (Brooks 2000; Hunt et al. 2004) that longevity (and other components of fitness) trades off with sexual signaling to such a degree that high-quality males may be shorter lived than low-quality males is particularly relevant. Once it is acknowledged that displays and ornaments exact a cost (no matter how small initially) to the bearer's residual reproductive value, and that this cost can substantially alter the relationships not only among fitness components but also between any given fitness component and total fitness, it becomes clear that any accounting for fitness effects must be integrated across all life history components (Kokko et al. 2002). This is not a new insight: the importance of total fitness was spelled out by both Fisher (1930) and Williams (1966). Indirect selection can work only if attractiveness is genetically correlated with overall fitness, and if choice therefore results in a positive genetic covariance between itself and fitness. The field needs to move beyond empirical studies that measure one, or a handful, of fitness components, and estimates of quantitative genetic covariation should replace measures of phenotypic covariation between traits.

It follows from understanding indirect selection in a life history context (Eshel et al. 2000; Kokko et al. 2002) that a single process of indirect selection can result in a range of qualitatively different outcomes, including both positive and negative genetic correlations between attractiveness and longevity (see figure 24.4). Although this position is still under debate (see, e.g., Cameron et al. 2003; Kokko et al. 2006a), we believe that it gives a more consistent and empirically useful view of the evolution of choice. We follow earlier authors (Eshel et al. 2000; Kokko et al. 2003b) who honor the contributions of both Fisher and Zahavi in reaching this unitary model by referring to it as the Fisher-Zahavi model of mate choice evolution via indirect selection.

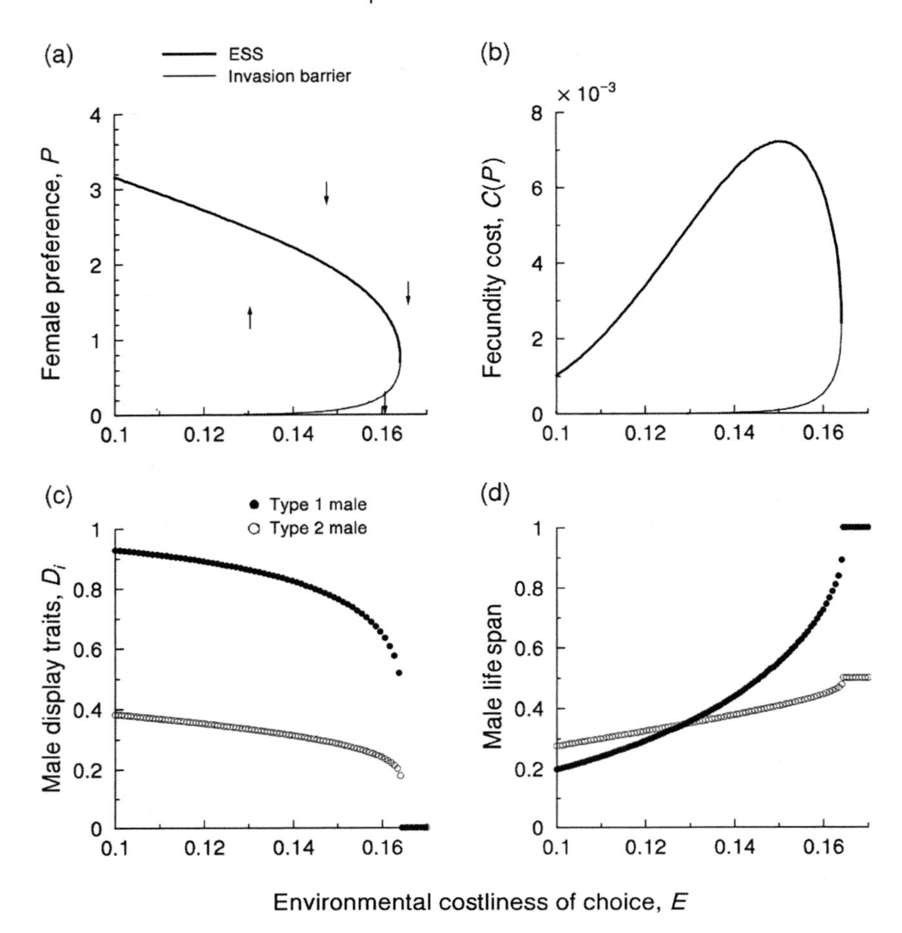

FIGURE 24.4 Evolution of female choice for high-quality (type 1) over low-quality (type 2) males, as modeled by Kokko et al. (2002, reaction norm model, figure 2) and reprinted with permission. This figure illustrates how stronger choice evolves when choice is less costly, and how under very strong (low costliness) choice, it is possible for sexual selection on male display traits to be so strong that high-quality males live shorter than low-quality males. Outcomes thought to be typical of the so-called Fisherian and good-genes alternative models are achievable from the same evolutionary process. Mate choice is measured as (a) preferences, P, and (b) fecundity cost, $C(P)$, paid by a female with a preference P. The parameter E (x-axis) is the cost function by which increasing mate preferences reduce fecundity. In (a), arrows indicate the invasion of alleles for larger or smaller female effort. (c) Display traits D_i^*, and (d) expected life spans $1/\mu_i^*$ of type 1 and 2 males at the ESS with $P^* > 0$ (or $P^* = 0$ where this is the only ESS).

The Maintenance of Genetic Variation in Fitness

The maintenance of additive genetic variation in fitness traits under selection remains a persistent problem in population genetics because even weak selection is expected to deplete additive genetic variation in most circumstances (Fisher 1930). The concerted action of selection on quality, and sexual selection on signals of quality, should, if anything, erode genetic variation more rapidly than ever. This makes the evolution of strong mate choice via apparent indirect selection paradoxical because choice and the selection that favors choice indirectly will simultaneously erode the very benefits underpinning choice. This *lek paradox* (Kirkpatrick & Ryan 1991) is typified in, but by no means confined to, highly polygynous lekking species in which there is a high skew in reproductive success among the males in a population (Widemo & Owens 1995).

Much early skepticism regarding the plausibility of indicator models of mate choice evolution was substantiated in skepticism regarding the maintenance of sufficient genetic variation in fitness (see Andersson 1994). The widespread demonstration that there is additive genetic variation in both sexual signals and direct components of fitness suggests that there may be sufficient standing variation in fitness and that all that remains to be done is to resolve the mechanistic basis of its maintenance. Emphasis has therefore turned to how such variation is maintained via processes such as condition dependence (Rowe & Houle 1996), host-pathogen coevolution (Hamilton & Zuk 1982), and genotype-environment interactions (Greenfield & Rodriguez 2004).

In recent years, evidence has emerged that the lek paradox may present, in multivariate form, as big a problem as ever for indirect selection on mate choice. Unlike laboratory selection experiments, selection in nature almost never operates on one trait at a time. When appropriate multivariate methods are used to investigate the multivariate relationships between suites of traits and genetic variation and covariation, a high proportion of studies reveal that there is little genetic variation in the direction of multivariate selection (Hall et al. 2004; Hine et al. 2004; Hunt et al. 2007; van Homrigh et al. 2007). Thus, even if there is substantial genetic variation for individual traits, there may remain little potential for indirect selection on choice because the associations between different traits, or different components of complex signals, dampen any potential for fitness gain. Two manipulative tests in different species of Australian *Drosophila* indicate that condition dependence is unlikely to resolve the lack of relevant multivariate genetic variation (Hine et al. 2004; van Homrigh et al. 2007).

Nonadditive Genetic Variation

Most research, and most of the discussion regarding indirect selection, has focused implicitly on additive genetic variation in choice, attractiveness, and other fitness components because additive variation is what predicts relationships between parental and offspring phenotypes (box 5.1). However, nonadditive genetic variation (chapter 5) can influence offspring fitness and, therefore, the indirect benefits of choice. Moreover, nonadditive genetic variation (especially overdominance) is not as prone to the same problems plaguing the maintenance of additive genetic variation. Much of the current attention on nonadditive effects of choice considers genetic incompatibility of certain mates, although historically many of the same issues have been the preserve of those studying inbreeding avoidance (Pusey & Wolf 1996).

Both genetic incompatibility and inbreeding depression result from interactions between alleles at the same locus (dominance) or at different loci (epistasis). In fact, some of the benchmark studies of genetic compatibility have used relatedness between mates to manipulate compatibility. Selection is obviously likely to favor choice if it results in a reduction in inbreeding (in which inbreeding depression affects fitness) and the incidence of other genetic incompatibilities. However, such choice is likely to be idiosyncratic at the population level and will seldom favor directional mate preferences. Instead, it is likely to favor subtler preferences and signals of compatibility such as odor-mediated cues of MHC genotype (Bennet-Clark 1989).

The extent to which genetic compatibility affects fitness is an important issue that will have far-reaching consequences, over and above its capacity to provide insight into the avoidance of inbreeding in most species, and is therefore an important subject for future research. Likewise, the ways in which selection operates on choice for genetically compatible or unrelated mates is only in its infancy, but is showing promise of being an enormously important topic. Because the genetics, and in particular the resemblance between parents and offspring, are far more complicated than the more extensively studied additive genetic case, it is crucial that progress is made in both the development of theory and in empirically understanding the quantitative genetic processes underlying nonadditive genetic variation.

Recognizing a Mate of the Correct Species

Despite a history of largely separate development, there is a strong and direct relationship between mate choice and speciation. Rundle and Boughman (chapter 27) deal with this topic in detail, and we merely note here that in many of the processes in which mate choice influences the probability of recognizing and/or mating with a heterospecific partner, there may be indirect selection on mate choice and preference functions. This is due to the commonly deleterious genetic effects of hybridization. For example, in the Swedish hybrid zone of the pied and collared flycatcher, fitness data from a long-term study demonstrate that hybrid males have approximately half the lifetime fitness of collared flycatchers whereas hybrid females are completely infertile and produce no offspring (Veen et al. 2001). These high-fitness costs of heterospecific pairing are a particularly powerful selective force underlying the evolution of both mate choice and the evolution of ornamental traits. Again, the pied and collared flycatchers provide a good example because the European distributions of these species provide areas of allopatry and sympatry. In the sympatric areas in which the risk of heterospecific pairing is high, strong selection through mate choice has reinforced the difference in ornamental plumage between the species, accentuating differences in the color of the plumage and the size of the ornamental white patches (Saetre et al. 2007). As a result, male pied flycatchers are browner and actually more female-like in their plumage in areas where they occur alongside the more strikingly black-and-white collared flycatcher than in areas in which they occur alone (Saetre et al. 2007). Even despite the displacement of the plumage pattern (called character displacement), females do occasionally form bonds with members of the wrong species. When this occurs, females significantly reduce the fitness cost of the heterospecific social pairing via conspecific extra-pair paternity, to the extent that more than half of the offspring in mixed social pairs are in fact purebred rather than hybrid offspring (Veen et al. 2001). To date, it is not clear whether the pattern of paternity observed by Veen et al. (2001) results from females knowingly accepting a heterospecific mate (perhaps because of constraints during pairing) and then increasing the level of extrapair copulations with conspecific males, or simply reflects the outcome of postcopulatory cryptic choice driven by as-yet unknown mechanisms of conspecific sperm precedence. The difference between these scenarios is interesting from a mechanistic perspective and serves to highlight the central theme of this chapter. Both processes ultimately result from an evolutionary process of mate choice but are fundamentally different with respect to female behavior, the way in which preference functions might be expressed, and the way in which selection can act upon the many different components of mate choice in the two sexes.

CONCLUSIONS AND FUTURE DIRECTIONS

We have attempted here to give a modern introduction to many of the most important topics in the evolutionary and behavioral ecology of mate choice. We suggest that those readers who are new to the field and interested in learning more about it read some of the excellent books (especially Andersson 1994) and the vast number of reviews to gain a thorough sense of the history of the ideas. We have consciously tried to provide a perspective that is different from these sources in the belief that the field of mate choice needs fresh ideas, new experimental systems, and the incorporation of tools and ways of thinking from other disciplines. In particular, new researchers should try to avoid being drawn into historic yet false dichotomies, or the testing of different processes as alternative hypotheses. Rather, it is time to build a more thoroughly integrated understanding of how multiple processes (direct benefits, costs, selection on the sensory system in other contexts, indirect selection) influence the evolution of choice. We have also steered away from explicitly addressing the causes and consequences of variation in mate choice, and the reader is advised to start with Jennions and Petrie's landmark review (1997) and then to visit Chenoweth and Blows' (2006) review of genetic variation in choice.

Several outstanding new challenges remain, and developments in genomics, quantitative genetics, bioinformatics, population biology, and life history theory look set to revolutionize our attempts to meet these challenges and forever change the way we work. For example, we have barely touched on the subject of individual variation in choice and the genetic and environmental basis of this variation. There is clearly scope to move beyond studying the evolutionary processes that shape choice in isolation and to begin resolving how they interact to influence the evolution of mate choice. This includes bringing the study of inbreeding avoidance back into the mainstream of mate choice research and reconciling it with the rapid progress in genetic compatibility research, because it seems that most cases of genetic incompatibility are likely to be due to inbreeding depression. There remains not one direct experimental test of indirect selection as an agent for the evolution of choice, but experimental tools are increasingly sophisticated and we expect that this hurdle will soon be overcome. The increasing use of demographic measures of fitness consequences (e.g., Head et al. 2005) in experiments and field studies will almost certainly lead to progress in understanding evolutionary questions. Differential allocation looks set to blur the distinction between studies of mate choice and parental care, and we believe that this will invigorate both fields as long as artificial distinctions are not erected. Last, thinking about speciation seems to have matured too and is no longer as doctrinaire or typological as it was between the modern synthesis and the 1990s, and once again there is considerable scope for our understanding of mate choice and speciation to progress beyond the question of reinforcement (see chapter 27). With such important and substantial challenges ahead, and increasingly sophisticated cross-disciplinary approaches and tools becoming available, we believe that the study of mate choice is currently entering an exciting phase.

Acknowledgments We are grateful to Tom Tregenza, Mike Jennions, Hanna Kokko, Carl Gerhardt, Matt Hall, and Andrew Cockburn for discussions and comments on the manuscript. We are both supported by grants and fellowships from the Australian Research Council.

SUGGESTIONS FOR FURTHER READING

A number of excellent reviews of mate choice and especially mate choice evolution have been published over the last 20 years. Kirkpatrick and Ryan (1991) is possibly the most influential among these, with the clearest explanation of the controversies over indirect selection and the maintenance of additive genetic variation (including the lek paradox). Andersson (1994) gives the most comprehensive treatment of sexual selection, including the evolution and consequences of mate choice. Kokko et al. (2003b, 2006a) provide comprehensive reviews of mate choice evolution, incorporating sexual conflict and recent developments, and Mead and Arnold (2004) provide a very useful review of the quantitative genetic models of mate choice evolution. Jennions and Petrie (1997) provide a comprehensive and very thought provoking review of the sources of variation in mate choice. Cotton et al. (2006) consider the consequences of condition-dependent variation in mate choice. The most useful and modern review of the genetic basis of mate choice is provided by Chenoweth and Blows (2006).

Andersson M (1994) Sexual Selection. Princeton Univ Press, Princeton, NJ.

Chenoweth SF & Blows MW (2006) Dissecting the complex genetic basis of mate choice. Nature Reviews Genetics 7: 681–692.

Cotton S, Small J, & Pomiankowski A (2006) Sexual selection and condition-dependent mate preferences. Current Biology 16: R755–R765.

Jennions MD & Petrie M (1997) Variation in mate choice and mating preferences: a review of causes and consequences. Biol Rev 72: 283–327.

Kirkpatrick M & Ryan MJ (1991) The evolution of mating preferences and the paradox of the lek. Nature 350: 33–38.

Kokko H, Brooks R, Jennions MD, & Morley J (2003b) The evolution of mate choice and mating biases. Proceedings of the Royal Society of London B 270: 653–664.

Kokko H, Jennions MD, & Brooks R (2006a) Unifying and testing models of sexual selection. Annu Rev Ecol Evol Syst 37: 43–66.

Mead LS & Arnold SJ (2004) Quantitative genetic models of sexual selection. Trends Ecol Evol 19: 264–271.

25

Alternative Mating Strategies

STEPHEN M. SHUSTER

In many animal populations, individuals exhibit discontinuous variation in their reproductive behavior and morphology (Gadgil 1972; Gross 1996; Shuster & Wade 2003). Although diversity may appear within either sex as well as within hermaphroditic populations, polymorphic mating phenotypes, also known as *alternative mating strategies*, are most commonly observed in species with separate sexes and are usually expressed among males. The staggering diversity and taxonomic breadth of mating polymorphisms have fascinated behavioral and evolutionary biologists since Darwin (1874; reviews in Gross 1996; Shuster & Wade 2003; Shuster 2007, 2008; Oliveira et al. 2008). In this chapter, I will briefly summarize this diversity and explain my view of why mating polymorphisms are biologically interesting. I will next discuss existing explanations for observed variation in mating polymorphisms, and then identify three fundamental questions that have persisted for over a quarter century in studies of alternative mating strategies. I provide answers to these questions using data from published literature as well as suggestions for new research.

DIVERSITY IN MATING STRATEGIES

Alternative male morphs appear in most major animal taxa (Shuster & Wade 2003; Oliveira et al. 2008). In certain squid, decapod crustaceans, and teleost fish, polymorphic male phenotypes are represented by small, nondescript yet fully mature individuals who lurk with bulging testes near massive harem masters, waiting for opportunities to flash into territories, steal a few matings, and then speed away unscathed (Hanlon 1998; Correa et al. 2003; Gross 1982; Taborsky 2008b). In isopod crustaceans, dung beetles, and ungulates, such "sneaker" or "satellite" males are often stealthy, insinuating themselves among females within the territories of combative males, and, again, mating when and with whom they can (Shuster 1989; Emlen 1997; Isvaran 2005). Alternative mating strategies may also include males who mimic females (isopods, Shuster 1992; sunfish, Dominey 1980; garter snakes, Shine et al. 2001; shorebirds, Delehanty et al. 1998), males who provide exceptional care to young (cichlid fish, Taborsky 1994; Awata et al. 2004), and males bearing two, three, or more distinct colors or display patterns that convey information to mates as well as to potential rivals (shorebirds, Lank et al. 1995; lizards, Sinervo & Lively 1996; fish, Taborsky 2008b). Adult male phenotypes may be determined from birth (mites, Radwan 1995; isopods, Shuster & Sassaman 1997) or they may be directed by a range of environmental influences (many species; Oliveira et al. 2008). Well-fed males in some species may delay maturation and become territorial (dung beetles, Moczek 1998), whereas rapid growth in other species leads males to mature early, live fast, and die young (salmon, Hutchings & Myers 1994).

Polymorphic body forms are easiest to notice, but males may differ only in their behavioral phenotypes (Finke 1986; Waltz & Wolf 1988). Males may attempt to woo females with gifts, food, or shelter, but, lacking these enticements, will abandon chivalry and attempt to mate by force (scorpionflies, Thornhill 1981; ducks, Gowaty & Buschhaus 1998; primates, Smuts & Smuts 1993). Males may vary in their tendencies to seek multiple mating partners or to defend individual females for extended durations (horseshoe crabs, Brockmann & Penn 1991). Male phenotypes may also be age-related, with older males attempting to signal females, whereas younger males remain silent, waiting near signalers and occasionally, only occasionally it seems, securing a chance to mate (bullfrogs, Howard 1984; tree frogs, Gerhardt et al. 1987). Even when behavior alone is polymorphic, changes in phenotype may become permanent with age, but more often behavioral polymorphism is reversible, with males changing from searchers for females, to waiters for females, and back to searching again within a single afternoon (butterflies, Alcock 1994b).

The short explanation for observed diversity in male mating phenotype, within as well as among species, is this: opportunities for acquiring multiple mates tend to be greater for individuals in species with separate sexes, and, in such species, opportunities for polygamy occur most often for males (Shuster 2007; chapter 20). Sexual selection is especially strong when each female mates only once and when the sex ratio equals 1. Under these circumstances, if some males mate more than once, other males must be excluded from mating and sexual selection is the inevitable result (Darwin 1874; Shuster & Wade 2003). Regardless of the number of times females mate, sexual selection requires that some males reproduce at the expense of others, and although sexual selection can intensify when females mate more than once, these conditions are surprisingly restrictive. For sexual selection by sperm competition to occur, females who mate more than once must all tend to mate with the same subset of males within the population, and they must also all tend to fertilize their ova using sperm produced by particular males within that subset. If such a positive covariance between male mating success and male fertilization success does not exist, multiple mating by females will ameliorate rather than intensify sexual selection (Shuster & Wade 2003). In general, sexual selection favors heritable traits that confer differential mating and fertilization success, including unconventional mating phenotypes. Stated differently, alternative mating strategies readily evolve when sexual selection is strong.

WHY STUDY ALTERNATIVE MATING STRATEGIES?

Why should we bother to study alternative mating strategies? There are at least three reasons, with the first reason described above. The diversity of alternative mating strategies is astonishing. It begs for explanation, and for this reason alone it has fueled the careers of many scientists. Yet alternative mating strategies also shed light on fundamental evolutionary processes, including a great paradox in evolutionary biology. That paradox is, how can sexual selection overcome the combined forces of natural selection on males and females that oppose it (Shuster & Wade 2003)? Highly modified male phenotypes are well known to impose great survival costs upon the males that possess them. How is it that these extreme male variants, as well as the females that mate with or produce them, are not simply eliminated by natural selection outright?

Thus a second compelling reason to study alternative mating strategies is, these traits show why sexual selection is among the most powerful evolutionary forces known (Shuster & Wade 2003). Sexual selection is well known for producing male displays, weaponry, and extreme female mate preferences (chapters 20 and 24), but it is also the primary evolutionary driver behind mating polymorphisms. As mentioned above, alternative mating strategies invade populations when relatively few conventional individuals secure mates (Gadgil 1972; Shuster 2007, 2008). As we will see, when the fraction of the population that is excluded from mating becomes large, sexual selection intensifies, increasing the likelihood that novel mating phenotypes will arise. Because alternative mating phenotypes allow their possessors to mate under these extreme circumstances and because some of these individuals mate with disproportionate success, unconventional male phenotypes experience intense sexual selection themselves. As a consequence, not only are alternative male morphs often the antithesis of conventional males in the species in which they appear (e.g., tiny, secretive sneakers coexist with huge, aggressive territory defenders) but their population frequency, as well as the frequency of conventional mating phenotypes, can rapidly and dramatically change (Gross & Charnov 1980; Shuster 1989; Sinervo & Lively 1996).

Heritable phenotypes that are not under sexual selection, but instead which confer more subtle fitness costs or advantages to their possessors may also evolve, but natural selection on these traits leaves a less easily recognized signature than when sexual selection occurs. Why is this so? The reason lies in the magnitude of fitness variance generated by each form of selection (Shuster & Wade 2003). When differences in relative fitness among individuals are absolute, variance in fitness is large and selection becomes intense. In contrast, when differences in relative fitness are subtle, variance in fitness is small and selection is relatively weak.

Under sexual selection, when some males mate more than once and/or fertilize disproportionate numbers of ova, other males must be excluded from reproducing at all. This condition makes fitness variance very large (Darwin 1874; Shuster & Wade 2003; see below). Under natural selection, differential success by some individuals within the population may occur, but success by some individuals need not obliterate the fitness of other individuals outright, as sexual selection always does. This is not to say that natural selection is never sufficiently intense to cause rapid evolutionary change. Indeed, it can be. However, when natural selection is intense, new phenotypes often sweep through populations to fixation so quickly that evolutionary change goes unnoticed until after the fact (Wootton et al. 2002). Intense natural selection may also induce phenotypic cycles, but documented cases of this process appear to take years, decades, or even millennia to complete (Kettlewell 1955; Vermeji 1987; Hori 1993).

Alternative mating strategies are different. They routinely involve multiple phenotypes that persist within populations, often with wide fluctuations in frequency. Why might this be so? The apparent answer is frequency dependent selection (Haldane & Jayakar 1963; Slatkin 1978; Crow 1986) and, in particular for mating polymorphisms, negative frequency dependent selection (see chapter 3). In this self-regulating evolutionary process, the frequency of one morph can increase, but as it becomes abundant relative to other morphs, its fitness advantage declines, progressively favoring a rarer morph whose frequency in turn increases until its own fitness decays. Fluctuating fitness differences among morphs, as we will see, are a consequence of strong sexual selection, and under such circumstances, alternative mating strategies tend to oscillate within populations instead of rushing to fixation. Moreover, because selection on polymorphic phenotypes is so strong, these oscillations tend to be large and they tend to cycle fast.

Thus, a third reason for studying alternative mating strategies is, these polymorphisms show how sexual selection can cause rapid, recurring evolutionary changes over uncommonly short periods of time. Depending on the life span of the species involved, significant population cycles of conventional and alternative mating phenotypes may be observed within a few years (Sinervo & Lively 1996); some cycles are observed annually (Lank et al. 1995; Sinervo 2001); and in some cases, oscillations may occur within months or even weeks (Shuster et al. 2001). Alternative mating strategies reveal how negative frequency dependent selection is mediated by sexual selection, and because the currencies of sexual selection are mating success and variance in offspring numbers, investigations of alternative mating strategies allow the explicit quantification of both evolutionary processes, as we will see below (Wade 1979; Shuster & Wade 2003; Shuster 2007, 2008).

FUNDAMENTAL QUESTIONS

Several schemes now exist for describing the expression of alternative mating strategies. Each has a different emphasis, but all are based on the widely recognized tendency for mating polymorphisms to differ in the degree to which genotype and the environment may influence trait expression. Since Darwin (1874) described male polymorphism at length, several authors have proposed hypotheses about their expression and persistence (Morris 1951; Gadgil 1972; Dawkins 1980; Eberhard 1982; Maynard Smith 1982). Austad (1984), after reviewing these contributions, provided a synthetic description of the observed diversity, wherein he identified issues that have remained paramount in research on alternative mating strategies ever since.

Austad's dichotomously branching diagram (figure 25.1) distinguished (1) whether polymorphisms represented genotypic versus phenotypic alternatives, (2) whether the fitnesses of each morph were expected to be equal or unequal (*isogignous* versus *allogignous* phenotypes), and (3) whether phenotypic alternatives were reversible or irreversible within individual lifetimes. Austad's (1984) diagram showed what researchers then and now have recognized: mating polymorphisms in which phenotypes show at least some flexibility in their expression are more common than those that do not. Austad (see also Dominey 1984) had begun to question whether game theory principles could be applied to the study of alternative mating strategies (or *alternative reproductive behaviors*, ARBs) because the assumptions required to find the evolutionary stable strategy (ESS; Maynard Smith 1982) did not always accord with data available for these polymorphisms. In particular, the average mating success of persistent phenotypes in many natural populations were not always found to be equivalent (Dawkins 1980; Thornhill 1981; Eberhard 1982). Then and now, equal fitnesses among morphs are a population genetic and game theory necessity if selection is to maintain genetically based polymorphism over time (Haldane & Jayakar 1963; Slatkin 1978, 1979a, 1979b; Maynard Smith 1982; Crow 1986). Nevertheless Austad (1984) explicitly identified a *genetic-nongenetic* dichotomy for alternative mating strategies consistent with Dawkins' (1980) hypothesis that fitnesses need not be equal among morphs when considering certain persistent alternative mating phenotypes (see below).

A second major development in the study of mating polymorphisms, and undeniably one of the most influential, was Gross's (1996) description of the status dependent selection (SDS) hypothesis. Gross, like Austad (1984) noticed the preponderance of mating phenotypes in nature with flexible expression, and proposed an evolutionary framework for considering them. He suggested that when the fitness consequences of alternative mating phenotypes depend on the relative competitive ability of the interactants, in other words, on their *status* relative to one another, then selection should be considered *status dependent*. Also, Gross (1996) proposed that when status did vary among individuals, this variation was primarily due to environmental influences.

Two central assumptions of the SDS hypothesis were that the distribution of status among individuals is normal within most populations, and that individuals always play the strategy that maximizes their fitness within the population given the particular local conditions of relative status. In this sense, the SDS model was conceptually linked to Dawkins' (1980) hypothesis that if sexual selection favored the most combative, showy, and vigorous males, then males unable to compete in this arena might adopt alternative sets of behaviors or morphologies that would still allow them to mate.

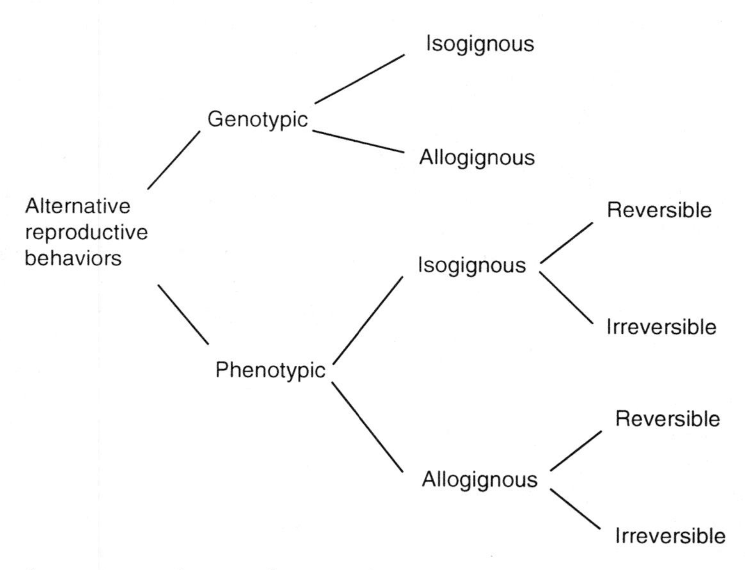

FIGURE 25.1 Austad's (1984, redrawn) diagram distinguishing whether (1) mating polymorphisms represented genotypic versus phenotypic alternatives, (2) the fitnesses of each morph were expected to be equal or unequal (*isogignous* versus *allogignous* phenotypes), and (3) phenotypic alternatives were reversible or irreversible within individual lifetimes.

These inferior males might be less successful than the dominant males in the population, but they would still do better than if they had secured no mates at all. That is, they could "make the best of a bad job," and still persist within the population. Following the form of game theory models, Dawkins (1980) proposed that if all males possessed such flexibility, that is, if this form of plasticity were fixed within the population, there would be no need for the fitness of alternative variants of male mating behavior to be equal among all male phenotypes. Gross (1996) took this notion a step further to state that if all males were genetically identical, fitness differences among them of any kind would be of no evolutionary consequence.

Gross's (1996) hypothesis elegantly captured the central elements of most observations of polymorphic phenotypes to that date; it was obvious to everyone that flexible phenotypes were more common than inflexible ones. Also, many researchers had reported that the fitnesses of the different morphs were seldom equivalent (reviews in Austad 1984; Eberhard 1982). Thus, Gross's approach reconciled, too, the study of alternative mating strategies with game theory, which had identified conditional strategies as being *nongenetic*; that is, they were *bourgeois* phenotypes in which all individuals in the population were identical in their ability to express a flexible trait (Maynard Smith 1982; Taborsky 2008b). Gross (1996) also stated explicitly what Austad (1984) and game theory implied; that individuals within a species expressing conditional phenotypes were genetically *monomorphic.*

Taborsky (1998) provided the next seminal discussion on mating polymorphisms, identifying three different levels at which alternative male phenotypes could be assessed: (1) determination, the relative degree to which genetic or environmental variation influence phenotypes, (2) plasticity, the relative degree to which phenotypes are flexible or inflexible in their expression (chapter 6), and (3) selection, the way in which fitness differences might (or might not) lead to the persistence of phenotypes, as well as how variation in male phenotype within lifetimes influences overall mating success. Regarding determination, Taborsky (1998) moderated somewhat Gross's genetic monomorphism hypothesis (as did Gross himself; see Gross & Repka 1998). However, Taborsky reiterated the disproportionate representation of condition-dependent strategies in nature, and placed environmental influences in high relief, stating, "Reproductive phenotypes may be either genetically or environmentally determined. It is highly unlikely, however, that only one of the mechanisms will be responsible for any important set of adaptive characters. Environmental modification, at least, should be ubiquitous."

Regarding plasticity, Taborsky (1998) covered similar issues as his predecessors, but he emphasized individual aspects of trait expression, particularly the tendency among individuals expressing conditional phenotypes to change or not change phenotypes within their own lifetimes. This emphasis specifically addressed Austad's (1984) third point. Taborsky (1998: 225) noted, "Reproductive phenotypes may be fixed for life, or be an expression of successive, ontogenetic stages, or be an adaptive response to momentary conditions. An important question to ask is whether behavioural plasticity exists at the population level only, or within individuals with either successive or simultaneous variation between tactics (which are not mutually exclusive)."

Regarding how selection might act on mating polymorphisms, Taborsky (1998: 225) considered both proximate and ultimate causation, and called attention to frequency dependent processes, although he allowed that inferior mating tactics might still appear within populations:

> Alternative reproductive phenotypes may be stabilized by obtaining equivalent Darwinian fitnesses, or they may reflect a disparity in the quality of individuals. The former case is based on frequency-dependent pay-offs to reproductive competitors displaying either bourgeois or parasitic tactics. The latter case is based on the common fact that the ability to monopolize access to females differs greatly between individuals because of, for example, divergent growth histories, health or reproductive experiences. Individuals of inferior competitive ability may suffer from unavoidable constraints and maximize their lifetime reproductive success by adopting parasitic rather than bourgeois tactics, even if these do not provide similar fitness rewards.

Taborsky (1998) thus encapsulated many researchers' interest in individual variation in trait expression, apart from how such variation might be represented at the population level. In this context, Taborsky suggested that "any combination is possible between the alternatives existing at the levels of determination, plasticity and selection of reproductive phenotypes" (225). He reiterated the possibility that environmental variation could lead to unequal fitnesses among phenotypes: within populations as well as within individual lifetimes. However, he also recommended that "these explanatory levels should be clearly separated from each other to avoid confusion."

Shuster and Wade (2003) made this warning more explicit, and specifically addressed the conceptual and quantitative difficulties that arise when terminology used to describe individual and population variation become confused. They attacked the concept of genetic monomorphism in favor of the view that all traits include genetic and environmental influences on their expression. They did this to echo other researchers who had made this point (Austad 1984; Hazel et al. 1990; Gross & Repka 1998; Taborsky 1998), but rather because the assumption of genetic monomorphism seemed to be made whenever conditional strategies were discussed using game theory terminology (see Tomkins & Hazel 2007; Oliveira et al. 2008, pp. 6–13 for recent examples). Maynard Smith (1982, p. 4, 21–22) had emphasized that the genetic assumptions underlying mixed evolutionarily stable strategies were made for simplicity alone. However, the simplification of genetic monomorphism seemed repeatedly to be taken at face value as an accurate representation of the genetic architecture underlying phenotypic plasticity. Most results on the inheritance of conditional polymorphisms concur that it is not (reviews in Schlicting & Pigliucci 1998; Tomkins & Brown 2004; Rowland & Emlen 2009)

Shuster and Wade (2003) argued that although descriptions of individual variation may be interesting in their own right, whether a phenotype persists in a population will depend on its relative fitness: not when examined from one life stage to the next, but rather when considered relative to all other individuals within the population. Shuster and Wade restated the fundamental evolutionary principle that phenotypic variants with relative fitnesses below the population average will inevitably be removed by selection (Maynard Smith 1982; Crow 1986), even if that selection is slowed by environmental (i.e., conditional) influences on phenotype (Haldane & Jayakar 1963). Shuster and Wade (2003) also asserted that when mating and nonmating individuals were included in fitness calculations, average fitnesses among morphs were likely to be equivalent (see below), even when phenotypic expression was condition dependent. They suggested therefore that Dawkins' (1980) "bad job" may not be so bad after all.

Shuster and Wade (2003), like Austad (1984), cast doubt on the facile use of game theory terminology and assumptions for addressing population genetic and quantitative genetic variation. However, Shuster and Wade suggested, too, that with proper considerations of fitness variance within populations, specifically by including nonmating males in estimates of relative fitness (see below), solutions obtained using game theory would indeed conform to existing population genetic theory (Haldane & Jayakar 1963; Slatkin 1978). Lastly, Shuster and Wade advocated viewing conditional mating phenotypes within the existing quantitative genetic framework for threshold traits (Roff 1996; Schlicting & Pigliucci 1998), an approach that acknowledged the ubiquity of conditional mating polymorphisms and accorded with experimental results indicating that conditional phenotypes do have a genetic basis and do oscillate in apparent response to frequency dependent selection (reviewed in Tomkins & Brown 2004; Rowland & Emlen 2009).

Most recently, Taborsky et al. (2008b) and other contributors to the volume by Oliveira et al. (2008), in keeping with the predominance of conditional polymorphisms in nature, focused their discussion on *alternative reproductive tactics*, polymorphic behavioral or developmental phenotypes whose expression depends strongly on environmental conditions. These authors presented an extensive review of theoretical and experimental results and listed 12 questions they considered central to the future study of mating polymorphisms, particularly those with conditional expression. One question (#6) concerned possible mechanisms of trait expression; another (#8) concerned whether natural selection in other contexts might shape mating polymorphisms.

However, the remaining 10 questions concerned various aspects of what appear to be the three recurrent issues in this field. After more than a quarter century of concentrated research, two of Austad's (1984) three central questions remain prominent: (1) to what degree are alternative phenotypes genetically or environmentally determined? and (2) under what circumstances must the fitnesses of alternative morphs be equal, within populations as well as within individual lifetimes? A different third question has now arisen to take the place of

Austad's, in part because the issue of reversibility seems solved (it happens, often), and also because an additional question seems more fundamental in light of recent research (Taborsky et al. 2008): (3) to what degree is frequency dependent selection responsible for maintaining mating polymorphisms within populations?

My goal for the remainder of this chapter will be to address these three issues. In my view, they are all closely linked to the same general phenomenon—sexual selection—and the controversy they incite can be reconciled by correcting a common experimental difficulty that arises when sexual selection is strong. I will first consider frequency dependent selection and explain why the operation of this evolutionary process tends to equalize fitnesses among morphs within populations (cf. Slatkin 1978). I will next provide an example of how frequency dependent selection can act on conditional mating polymorphisms such that the fitnesses of the morphs are equal over time. I will discuss, too, why the action of frequency dependent selection alone could be responsible for the preponderance of conditional mating polymorphisms. Following Shuster and Wade (2003) I will show how strong sexual selection generates a *mating niche* that can be invaded by alternative mating phenotypes, and after invasion occurs, how continued sexual selection generates frequency dependent selection that is sufficient to maintain mating polymorphism. Lastly, I will use a quantitative framework for documenting the intensity of sexual selection to show how data from existing studies claiming to verify Dawkin's (1980) "best of bad job" hypothesis are uniformly inadequate to address its central assumption.

FUNDAMENTAL CONCEPTS

Frequency Dependent Selection

Frequency dependent selection occurs when the fitness of a given phenotype depends on its frequency relative to other phenotypes within the population (see also chapter 3). The tendency for the frequencies of alternative mating phenotypes to fluctuate within populations (Shuster 1989; Lank et al. 1995) or even cycle over time (Sinervo & Lively 1996; Shuster et al. 2001) is well documented, for polymorphisms that are controlled by alleles of major effect, as well as for polymorphisms in which trait expression is contingent on environmental or social conditions (Alcock 1994a; Taborsky et al. 2008). With few exceptions (Radwan & Klimas 2001), frequency dependent selection, and in particular, negative frequency dependent selection, in which rare morphs experience the highest fitness, is widely assumed to be the fundamental process by which alternative mating strategies are maintained within natural populations (Roff 1996; Gross 1996; Shuster & Wade 2003; Oliveira et al. 2008).

For frequency dependent selection to operate, the frequency of an allele or phenotype must depend in a multiplicative way on its fitness, relative to other alleles or phenotypes within the population. In simple form, this relationship can be expressed as

$$p_i' = pi\ (w_i / W) = p_i\ \tilde{w} \qquad (25.1)$$

where p_i' is the frequency of the i-th allele (or phenotype) after selection has occurred, p_i is the frequency of the i-th allele (or phenotype) before selection, w_i is the absolute fitness of the i-th allele (or phenotype) during selection (chapter 4), and the weighted sum of all alleles (or phenotypes) and their frequencies during the selective event, is the average fitness, W, where $W = \Sigma\ p_i w_i$. The relative fitness of the i-th allele (or phenotype), $\tilde{w}$, equals the absolute fitness of each allele (or phenotype), divided by the average fitness of all alleles or phenotypes ($= w_i / W$). In a two allele system, it can be shown that when the frequencies of both alleles are equal ($p_1 = p_2 = 0.5$), and their relative fitnesses are equal ($\tilde{w}_1 = \tilde{w}_2 = 1$), both alleles will remain at equal frequency indefinitely. However, if the fitness of allele 1 decreases, say, to 0.4, the frequency of allele 2 will increase, and will eventually become fixed if its relative fitness remains greater than that of allele 1. If the relative fitness of an allele (or phenotype) decreases as it becomes common, then selection becomes negative frequency dependent and polymorphism is more likely to be maintained (Crow 1986).

The conditions necessary to maintain polymorphism have been explored extensively (reviews in Slatkin 1978, 1979a, 1979b; Fitzpatrick et al. 2007). Most researchers cite the maintenance of a 1:1 sex ratio by the equalization of fitnesses between the sexes (Fisher 1930) as the most intuitively clear example of how polymorphism can be maintained by negative frequency dependent selection (see box 20.2 and box 26.3). To visualize this process, one need only divide the total number of offspring in any generation, in turn, by the total number of individuals in each sex. Because every offspring has a mother and a father, the ratio obtained for each sex provides an estimate of its absolute fitness, and because the minority sex always generates the larger fitness ratio, the minority sex in populations with biased sex ratios will always increase in frequency. By calculating the relative fitness for each sex as described above, the expected change in sex ratio per generation can be estimated explicitly.

In his considerations of frequency dependent selection, Slatkin (1978, 1979a, 1979b) discovered a fundamental relationship. He showed that when the number of phenotypic classes within a population is greater than 2, the number of heritable factors required to produce the possible phenotypes necessarily increases. He also showed that when relative fitness fluctuates, as it does under negative frequency dependent selection, the number of parameters requiring adjustment to equalize fitnesses among the phenotypes also increases. Slatkin (1979a, 1979b) argued that when relative fitness fluctuates, modifier alleles that cause the average fitnesses of the morphs to become more similar will be favored, thereby allowing polymorphism to be more easily maintained. He therefore proposed that when multiple morphs are favored within a population (cf. Levins 1968), alleles that modify the expression of genetic systems underlying polymorphic traits will rapidly accumulate. Under such circumstances, simple genetic mechanisms controlling the expression of each phenotype will become increasingly influenced by multiple genetic factors. Several authors, e.g., Charlesworth (1971), Roughgarden (1971), Strobeck (1975), and including Haldane and Jayakar (1963) and Slatkin (1979a, 1979b), described environmental conditions in which relative fitnesses among morphs might not be equalized. In general, however, Slatkin's (1978, 1979a, 1979b) overall results were robust, and under a wide range of circumstances, frequency dependent selection appears to equalize relative fitnesses among morphs via modifier loci, that equilibrate the fitnesses of phenotypic classes within populations at a rate dependent on the complexity of the genetic system underlying trait expression.

Frequency Dependent Selection and Alternative Mating Strategies

In most descriptions of negative frequency dependent selection as it relates to alternative mating strategies, the reciprocal relationship between phenotype and fitness is represented in a bivariate plot showing two intersecting fitness functions representing alternative mating phenotypes (usually identified as α for males that sneak and β for males that fight), within the range of possible variation in a trait such as condition, age, or body size (figure 25.2). The point of intersection of the fitness functions is the switch point, at which the fitnesses of the two phenotypes are equal with respect to the character shown on the x-axis (Gross 1996). In representations of the SDS hypothesis, the distribution of a character (e.g., body size) is superimposed on the relationships between status and fitness (figure 25.3), suggesting that the population frequencies of this second trait are quantitatively related to the overlying fitness functions (Gross 1996; Tomkins & Brown 2004; Oliveira et al. 2008). However, in none of these examples is the distribution of population frequencies for the trait explicitly shown to have such a relationship to the fitness functions they accompany.

The relationships between the tactic fitness functions, the population variance in the position of the switch point, and the distribution of environmental

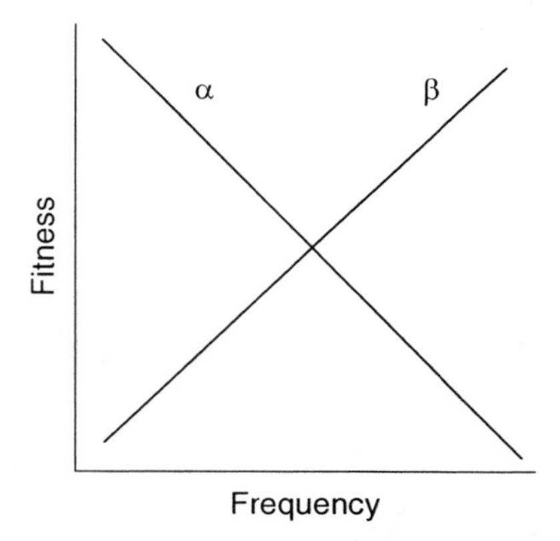

FIGURE 25.2 Bivariate plot showing reciprocal relationships between fitness and trait frequency for two mating phenotypes. α represents individuals who attempt to sneak matings, and β represents individuals who attempt to fight. The x-axis could be trait frequency (plotted), condition, age, or body size. Such descriptions are used to represent the action of negative frequency dependent selection on alternative mating strategies, but the distribution of population frequencies for traits are seldom shown to have a quantitative relationship to the fitness functions they accompany.

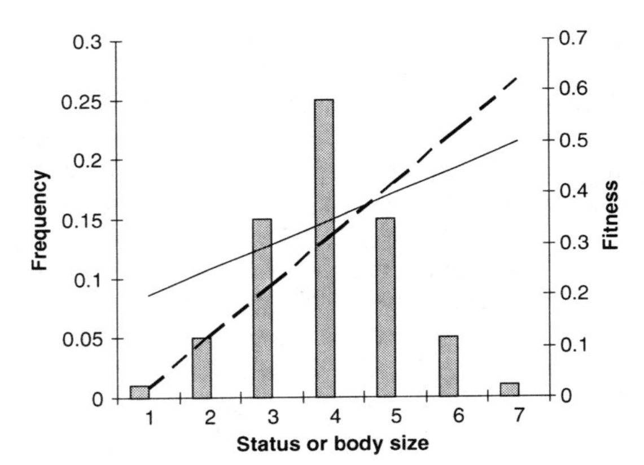

FIGURE **25.3** The status dependent selection (SDS) hypothesis (Gross 1996), in which the reciprocal relationship between status and fitness is represented in a bivariate plot showing two intersecting fitness functions within the range of possible variation in the trait; the point of intersection of the fitness functions is the switch point, at which the fitnesses of the two phenotypes are equal with respect to the character shown on the x-axis; the distribution of another trait (e.g., body size) is superimposed on the relationships between status and fitness, suggesting that the population frequencies of this second trait are somehow related to the overlying fitness functions, but such relationships are seldom explicitly shown.

cues inducing polyphenism are considered within the environmental threshold (ET) hypothesis described by Tomkins and Hazel (2007). This hypothesis explicitly combines known quantitative genetic principles and data with SDS and game theory frameworks to describe how conditional polymorphisms evolve. Tomkins and Hazel (2007) state that the combined effect of these factors on the selection differential acting on the switch point results in most cases in an equilibrium mean switch point that does not correspond to the intersection of the fitness functions. Also, Tomkins and Hazel report that the average fitness of the alternative tactics at this location will usually not be equal. Although their figures suggest that variation in status, cue reliability, and status-dependent fitness trade-offs can influence the distribution of reaction norms within a theoretical population (p. 524), the quantitative relationships among fitness, status, and traits influencing status (e.g., body size) are not clearly shown.

Tomkins and Hazel (2007) suggest that their theoretical framework identifies conditions in which conditional phenotypes can persist within populations, even when their fitnesses are unequal. They state (pp. 523–525), "For the conditional strategy to be the ESS, cues must be more reliable than random and one of the alternative phenotypes must have greater fitness in one environment and vice versa. When these conditions are met, the fitness of the conditional strategists (which is a function of the average fitnesses of the two phenotypes they produce) is greater than that of competing unconditional strategists, but the average fitness of the two phenotypes produced by the conditional strategists can be equal or unequal." It is not clear from this description whether Tomkins and Hazel envisioned fitness inequalities between conditional morphs occurring instantaneously *within* a particular environment, or if they considered fitness inequalities between morphs likely to persist over time. If frequency dependent selection indeed operates on the two phenotypes, the first condition is expected; the latter condition, however, is not.

The quantitative approach illustrated in equation 25.1 can be used to address uncertainty on this issue, as well as to verify that frequency dependent selection can indeed operate when environmental conditions have a strong influence on phenotypic expression (Taborsky et al. 2008). Following the framework of Tomkins and Hazel (2007; figure 25.4), let e_1 and e_2 represent the two possible environments in which mating contests may occur, let $w_{\alpha 1}$ and $w_{\alpha 2}$ represent the fitnesses of individuals (α_i) who attempt to sneak matings when small (α_1 in e_1) and large (α_2 in e_2), respectively, and let $w_{\beta 1}$ and $w_{\beta 2}$ be the fitnesses of individuals (β_i) who attempt to fight when small (β_1 in e_1) and large (morph β_2 in e_2), respectively. For simplicity, we will assume that each environment in which contests may occur appears with equal frequency, that each of the phenotypes in each environment occurs with equal frequency, and that only the fitness of one morph (β_i) differs in each environment.

Figure 25.4 shows that the fitness of β_2 is larger than the fitness of α_1 in environment 2; and that the fitness of α_2 is larger than the fitness of β_1 in environment 1. If the fitness of α_i is equal across the two environments, whereas the fitness of β_i is low in environment 1 and high in environment 2, then, indeed, within each environment, the average fitnesses of the two morphs are unequal (cf. Tomkins & Hazel 2007). However, because the

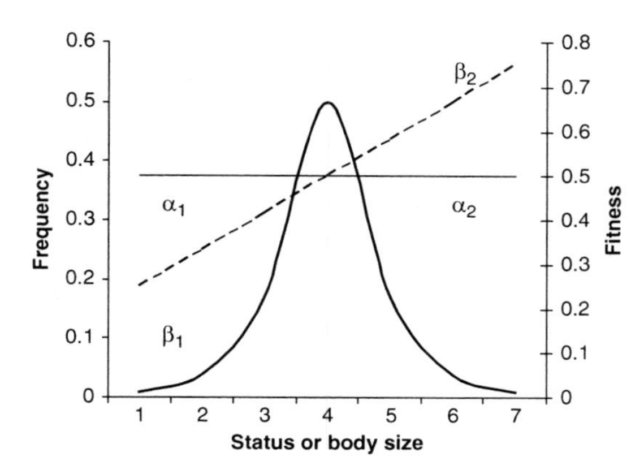

FIGURE **25.4** Frequency dependent selection on a condition-dependent mating polymorphism based on the assumptions of Tomkins and Hazel (2007); variables and relationships are described in the text.

overall frequencies of the two morphs are equal, and because the two environments appear with equal frequency, the average fitnesses of α and β across the two environments will also be equal. This can be shown as

$$[p_{\alpha1}w_{\alpha1}]+[p_{\alpha2}w_{\alpha2}]=[p_{\beta1}w_{\beta1}]+[p_{\beta2}w_{\beta2}] \quad (25.2)$$

where $p_{\alpha1}$ and $p_{\alpha2,}$ $w_{\alpha1}$ and $w_{\alpha2,}$ $p_{\beta1}$ and $p_{\beta2,}$ and $w_{\beta1}$ and $w_{\beta2}$ are the frequencies and fitnesses of the α and β phenotypes in environments 1 and 2, respectively. If the frequencies of the two morphs are indeed equal ($p_{\alpha1} = p_{\beta1} = p_{\alpha2} = p_{\beta2} = 0.25$), and if the fitnesses of α_i in each environment are also equal ($w_{\alpha1} = w_{\alpha2} = 0.5$), but the fitnesses of β_i are not equal ($w_{\beta1} = 0.25$; $w_{\beta2} = 0.75$), then by substituting these values, equation 25.2 becomes [(0.25)(0.5)] + [(0.25)(0.5)] = [(0.25)(0.25)] + [(0.25)(0.75)], and clearly the fitnesses of each morph across environments are equivalent (0.25 = 0.25).

However, if we let the environments occur with unequal frequency, then each morph's fitness must be averaged over the two environments, and because the fitnesses of the morphs differ in each environment, the fitness of each morph must be considered relative to all other morphs in all other environments. The average fitness across all morphs can now be written as

$$W = \Sigma p_{ij} w_{ij} e_j \quad (25.3)$$

where p_{ij} is the frequency of the i-th morph in the j-th environment, w_{ij} is the absolute fitness of the i-th morph in the j-th environment, and e_j is the frequency with which the j-th environment occurs. Thus, the relative fitness of the i-th morph in the j-th environment is $\tilde{w}_{ij} = w_{ij}/W$.

We can now rewrite equation 25.2 as

$$[p_{\alpha1}\tilde{w}_{\alpha1}e_1]+[p_{\alpha2}\tilde{w}_{\alpha2}e_2]=[p_{\beta1}\tilde{w}_{\beta1}e_1]+[p_{\beta2}\tilde{w}_{\beta2}e_2] \quad (25.4)$$

If the different environments appear with unequal frequency (say, $e_1 = 0.6$ and $e_2 = 0.4$), then the relative fitnesses of the different morphs will also change, as will the frequencies of the reaction norms that produce the two morphs in each environment. It is certainly true that the average fitnesses of the morphs within each environment are unequal (cf. Tomkins & Hazel 2007), but under such conditions both morphs in the same environment will no longer persist. In this example, reaction norms that produce the β morph in environment 1 will become quite rare, as will reaction norms that produce the α morph in environment 2; moreover, the relative proportion of the population that consists of each morph will change.

This result appears similar to the SDS solution because it leads to a high frequency of reaction norms that make the right choice in each environment (Gross 1996). This result is also similar to the ET solution because it assumes that each phenotype is part of a distribution of heritable reaction norms whose frequency will change under selection (Tomkins & Hazel 2007). However, unlike both the SDS and the ET hypotheses (or so it appears anyway), the results illustrated in equations 25.2–25.4 do not assume that individuals with inferior fitness somehow persist within the population indefinitely. Individuals in the above example whose fitnesses are inferior, because their reaction norms fail to respond appropriately within a given environment (i.e., they express the wrong phenotype and so have lower fitness), are either reduced to low frequency to form the tails of the normal distribution of reaction norms (figure 25.3) or they are removed from the population entirely. The rate at which inferior phenotypes are lost from this population can be slowed if mechanisms underlying trait inheritance are complex (e.g., Slatkin 1978), if individuals do not always make the correct choice in each environment (Roff 1996), or if the frequency with which each environment appears is closer to 0.5 (Haldane & Jayakar 1963). It is also possible that when a morph is common, it is less successful on average than it is when it is rare, that is, when negative frequency dependent selection occurs (Slatkin 1978, 1979a, 1979b).

What Happens When Sexual Selection Is Strong

When sexual selection occurs, two classes of males exist, males who mate and males who do not (Wade 1979; Shuster & Wade 2003, 2004). If we let p_S equal the fraction of males in the population who mate, and p_0 (= 1 − p_S) equal the fraction of males that do not mate, we can express the average fitness of the mating males as $p_S(H)$, where H is the average number of mates per mating male. The average number of mates for all males is equal to the sex ratio, R, when it is expressed as $N_{females}/N_{males}$.

We can now identify three relationships (equations 25.5a–c). The first is the average fitness of all males, which equals the sex ratio, R, rewritten as

$$R = p_0\,(0) + p_S\,(H) \qquad (25.5a)$$

Here, each term on the right side of the equation equals the fraction of males belonging to each mating class, multiplied by the average number of mates secured by members of that class. Rearranging equation 25.5a, we can see that $H = R\,/\,p_S$, indicating that the average mating success of *mating* males, H, is always greater than the average mating success of *all* males, R, except when each male mates only once.

Because $p_S = (1 - p_0)$, we can rewrite equation 25.5a to show the second relationship as

$$p_0 = 1 - (R/H) \qquad (25.5b)$$

This expression shows how p_0, the fraction of males without mates (a parameter that often goes unmeasured because males who mate are more conspicuous than males who do not) is related to parameters we can measure, specifically, R, the overall ratio of males to females, as well as H, the average number of mates per mating male.

If the ratio of females to males, R, remains at 1, equation 25.5b simplifies to

$$p_0 = 1 - (1/H) \qquad (25.5c)$$

an expression showing the fraction of males without mates, p_0, in terms of H. Again, although the fraction of nonmating males in the population, p_0, can be difficult to measure, equation 25.5c shows that p_0 is a function of H, the average number mates per mating male. The value of H usually can be measured (and is what most researchers do measure) by simply observing the males who successfully mate.

As the value of H increases, the fraction of males without mates, p_0, also must increase (figure 25.5). To place this relationship in more concrete terms, when the average number of mates per mating male, H, equals 5, as it can in African cichlids

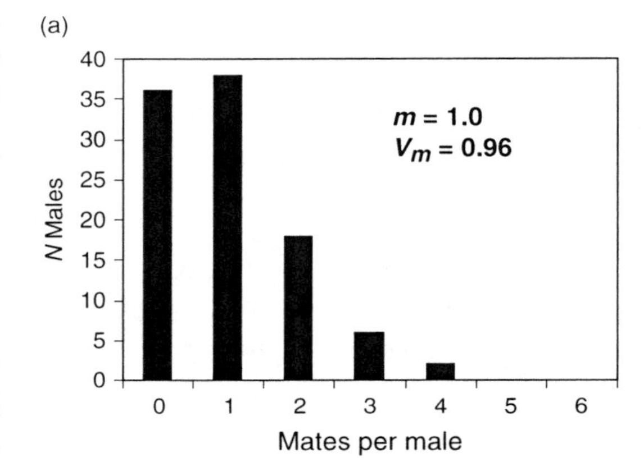

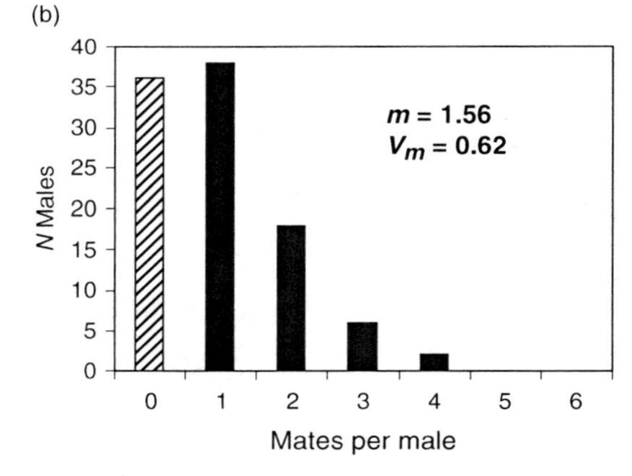

FIGURE **25.5** The mean, m, and variance, Vm, in mating success among 100 randomly mating males and females; females are assumed to mate only once; (a) when all males are included in parameter estimates; (b) when the zero class of males (hatched bars) is excluded from parameter estimates; the effect of omissions of nonmating individuals from fitness estimates tend to overestimate the average mating success and underestimate the variance in mating success for males.

(*Pelvicachromis pulcher*; E. Martin & Taborsky 1998), p_0, the proportion of males who fail to mate equals 80%. Here disproportionate mating success by 20% of males obliterates utterly the fitness of the remaining four-fifths of the male population! Certainly, if sperm competition occurs or if females mate more than once, the actual value of H will be reduced and sexual selection will be eroded. Yet such conditions can be accommodated simply by recalculating the value of H after paternity data are considered.

Despite the possibility that sexual selection can be reduced in this way, sexual dimorphism and alternative mating strategies are widespread, revealing how often sexual selection occurs in nature. As we will see, the obliteration of the fitness of some males by the disproportionate mating success of others is the source of sexual selection and of frequency dependent selection on alternative mating strategies. It is also the source of what has become a central question in the study of this phenomenon; whether mating phenotypes achieving inferior mating success can persist within natural populations.

The Variance in Mating Success

To estimate the intensity of sexual selection in terms of the variance in mate numbers among males, we must again recognize that sexual selection creates two classes of males (or two classes of females in sex role reversed species). When mating and nonmating classes of males appear, the total variance in male fitness has two components: (1) the average variance in mating success within the class of males that mate and (2) the variance in the average mating success between the classes of mating and nonmating males (Wade 1979, 1995; Shuster & Wade 2003). Because the average number of mates for nonmating males is uniformly zero (Wade 1979, 1995), following the form of equation 25.5a, the average variance in mate number within each of the two male categories, mating and nonmating, can be written as

$$V_{within} = (p_0)(0) + (p_S)V_{harem} \qquad (25.6)$$

where V_{harem} is the variance in mate numbers among the males who secure mates. By substitution from equation 25.5a, V_{within} can be rewritten as (R/H) V_{harem}, or, when the sex ratio, R, equals 1, $V_{within} = V_{harem}/H$. The median value of V_{harem} estimated from 27 studies of diverse animal taxa is 1.04, suggesting that the distribution of females with males is approximately Poisson ($V_{harem} = H = 1$). However, if females tend to cluster within the harems of particular males, V_{harem} can become disproportionately large (Wade & Shuster 2004).

The second component of the variance in male fitness exists between the classes of successful and unsuccessful males, and equals the difference in average mating success, squared, or $(H - 0)^2$, multiplied by the variance between the mating categories, $(p_0)(1 - p_0)$, or

$$V_{between} = H^2(p_0)(p_S) \qquad (25.7)$$

The total variance in male fitness in terms of male mating success, V_{mates}, is the sum of these two variance components, V_{within} and $V_{between}$, or

$$V_{mates} = (p_S)\,V_{harem} + H^2(p_0)(p_S) \qquad (25.8)$$

When data are available on the average harem size, H, the proportion of nonmating males within the population, p_0, can be estimated using equation 25.5b [i.e., $p_0 = 1 - (R\,/\,H)$]. When the variance in harem size, V_{harem} is available, equation 25.8 provides a means for estimating the total variance in male mating success, V_{mates}, even if researchers tend to focus on mating, rather than nonmating, males.

Why Alternative Mating Strategies Evolve

We can now answer more specifically why polymorphic mating strategies evolve when some males mate and other males do not. As explained above, differential mating success among males not only causes sexual selection among conventional males; it also creates a "mating niche" for males engaging in unconventional mating behaviors (Shuster & Wade 1991; 2003). By invading locations in which synchronously receptive females outnumber conventional males (i.e., *harems*), unconventional or satellite males may surreptitiously gain opportunities to mate without having to engage in combat.

To understand how such circumstances occur (cf. Shuster & Wade 2003), recall that R, the sex ratio ($= N_{females}/N_{males}$), is equal to the distribution of all females over all conventional males, mating and nonmating, and therefore equals the average fitness of conventional males, or

$$W_{\alpha} = R \qquad (25.9)$$

When satellites invade breeding territories, they usually find mates among the harems of conventional males. We can express the fitness of satellites, W_β, as

$$W_\beta = Hs \qquad (25.10)$$

where s equals the fraction of fertilizations satellite males obtain within the harems of conventional males and H equals the average harem size of conventional males. Stated differently, when satellite males mate with some of the females within the harems of conventional males, the average fitness of such males is reduced.

As Shuster and Wade (2003) showed, and as evolutionary theory (Slatkin 1978; Maynard Smith 1982) requires, for satellite males to invade a population of conventional males, the average fitness of satellite males, W_β, must exceed the average fitness of conventional males, W_α, or

$$W_\beta > W_\alpha \qquad (25.11a)$$

Note that when $R = 1$, the average and relative fitnesses are equivalent. Because persistence of a polymorphism requires equivalent fitnesses among the different phenotypes (Slatkin 1978, 1979a, 1979b), when the average fitnesses of the phenotypes differ, their relative fitnesses will also differ, and the frequency of these phenotypes will change (equations 25.1 and 25.4). In this case, when satellite males invade a population of conventional males, their numbers are small. Nevertheless, the population frequency of satellites will increase because their relative fitness exceeds that of the average conventional male. This is how negative frequency dependent selection operates on alternative mating strategies.

By substituting values from equations 25.6 and 25.7, we see that equation 25.8a can be expressed as

$$Hs > R \qquad (25.11b)$$

When the sex ratio, R, equals 1, equation 25.11b becomes $Hs > 1$, and by solving for s, we can see the minimum fertilization success that satellite males must obtain to invade a population of harem holding, conventional males is

$$s > 1/H \qquad (25.11c)$$

Thus, when the average harem size, H, equals 5 (e.g., Martin & Taborsky 1997), the fitness of satellites need only slightly exceed 20% of the fitness of the average conventional male, to satisfy the conditions required for successful invasion ($s > 1/H = 0.20$). As harem size increases, females become increasingly clustered around fewer conventional males and the invasion of alternative mating strategies becomes easier still (Shuster & Wade 2003).

By rearranging equation 25.5c, we can see that $1/H = (1 - p_0)$, and substituting from equation 25.11c, we see that the proportional success of satellite males need only exceed the proportion of mating males in the population or

$$s > 1 - p_0 \qquad (25.12)$$

We can now see why failures by conventional males to successfully defend harems provide mating opportunities for unconventional males. As p_0 increases, the mating success necessary for satellite males to successfully invade harems, s, becomes increasingly small, and negative frequency dependent selection favoring satellite males becomes increasingly intense. At evolutionary equilibrium, equation 25.12 identifies the condition necessary for both conventional males and satellite males to coexist within the same population. In effect, satellite males replace the nonmating males in the conventional male population (see below).

Why Alternative Mating Strategies Persist

After invasion occurs, how rapidly will the frequency of alternative mating strategies increase? The answer depends on how strong selection is on males of each mating phenotype. Recall that the intensity of sexual selection depends on the magnitude of fitness variance that exists among males. We have seen above how sexual selection provides a mating niche for satellite males. We can use the same framework to understand the relative intensity of selection on alternative mating strategies after invasion has occurred. Because we can partition the mating success of males into mating and nonmating classes, as well as into conventional and satellite male phenotypes, we can also use our estimates of fitness variance to determine whether the average fitnesses among male morphs are distinct. Stated differently, we can quantitatively address whether or not males "make the best of a bad job" (Dawkins 1980; Austad 1984; Gross 1996; Tomkins & Hazel 2007; Oliveira et al. 2008).

Again, when sexual selection occurs, two classes of males appear, conventional males who mate, $p_{S\alpha}$, and conventional males who do not mate, $p_{0\alpha}$, where, among conventional males $(p_{0\alpha} + p_{S\alpha}) = 1$. The fraction of conventional males in the population at any time equals $p._{\alpha}$, where the "·" indicates that all mating classes of conventional males are included. When alternative mating strategies invade conventional male populations, an additional class of males appears, $p_{S\beta}$, the successfully mating satellites, whose population frequency among all males equals $p_{S\beta} = (1 - p._{\alpha})$. As invasion proceeds and $p_{S\beta}$ increases, as it must because the average fitness of satellite males exceeds that of conventional males, satellite males will soon exhaust the supply of uninvaded harems. When this condition arises, $p_{S\beta}$ must itself divide into two classes: satellite males that mate, $p_{S\beta}$, and satellite males that fail to mate, $p_{0\beta}$, where within the satellite males $p_{0\beta} = (1 - p_{S\beta})$ and $(p_{0\beta} + p_{S\beta}) = 1$. The fraction of satellite males in the population at any time equals $p._{\beta}$, where the "·" indicates that all mating classes of satellite males are included. For all males, $[p._{\alpha} + p._{\beta}] = 1$.

With the appearance of the latter two male classes, it is now possible to partition the variance in mate numbers within each of the male phenotypes, into within- and among-male components. Then, using these variance subtotals for each morph, estimate the grand total variance in male mating success, V_{mates}, by partitioning the variance in mate numbers within and among the male morphs (Shuster & Wade 2003). This procedure is straightforward when the proportions of mating and nonmating males of each mating phenotype, as well as the average and variance number of mates per male for each of the mating classes, are available (Shuster & Wade 1991; Shuster 2007, 2008). The method becomes more difficult when field researchers focus only on mating males (table 25.1; see below), but quantification of V_{mates} is still possible provided that average number of mates per mating male, H, and the variance around that average V_{harem}, are reported. Here is how this can be done.

When multiple male phenotypes exist in the same population, we can estimate the variance in mating success, $V_{mates(i)}$, for each of the i-th mating phenotypes using equation 25.8. Note that in calculating $V_{mates(i)}$, the fractions of mating and nonmating males within each morphotype will each sum to 1 as described above. Next, we can estimate the total variance in mating success for all of the males in the population, $V_{mates(total)}$, as the sum of two components: (1) the average variance in mating success for conventional and satellite males, $V_{within(morphs)}$, and (2) the variance in the average mating success for conventional and satellite males, $V_{among(morphs)}$. The within-morph variance in mating success equals

$$V_{within(morphs)} = p._{\alpha}\,[V_{mates(\alpha)}]\; p._{\beta}\,[V_{mates(\beta)}] \quad (25.13)$$

where $p._{i}$ = the proportion of the male population comprised of each i-th male phenotype and $\Sigma\, p._{i} = 1$. The among morph variance in mating success equals

$$V_{among(morphs)} = p._{\alpha}\,(R - H_{\alpha})^2 + p._{\beta}\,(R - H_{\beta})^2 \quad (25.14)$$

where H_{j} is the average mating success of males achieved by mating males of each j-th male phenotype. Note that the proportion of males belonging to each male phenotype, $p\cdot_{j}$, the average mating success of mating males within each male phenotype, H_{j}, and the sex ratio, R, are often (but not always) available from field studies of alternative mating strategies.

CASE STUDIES

In the nearly 3 decades since Dawkins (1980) articulated his "best of a bad job" (BOBJ) hypothesis, an impressive number of researchers claim to have substantiated it, theoretically (Gross 1996; Gross & Repka 1998; Tomkins & Hazel 2007) as well as empirically (Fincke 1986; Møller & Birkhead 1993; Johnsen et al. 1998; Alcock 1996a, 1996b, 1996c; Low 2005; Beveridge et al. 2006; reviews in Oliveira et al. 2008). In contrast, Shuster and Wade (2003; Wade & Shuster 2004; Shuster 2008) have maintained that such conclusions are premature, either because the theoretical assumptions are unrealistic (see above), or because field results have inadequately accounted for the class of nonmating individuals. When clear genetic differences exist among males, there has been little dispute over whether fitnesses among morphs must be equivalent for polymorphism to be maintained (Gross 1996; Taborsky 1998; Taborsky et al. 2008). However, among conditional polymorphisms, despite considerable data suggesting that such polymorphisms represent quantitative traits with threshold inheritance (review in Rowland & Emlen 2008), uncertainty still persists (Oliveira et al. 2008).

Because the assumptions underlying theoretical analyses often require specific examination before they can be rigorously tested, there is little chance of addressing them here. However, Shuster and Wade's (2003) hypothesis regarding omission of the zero class of males in field studies generates clear predictions that are testable by combining the above framework with published results. Shuster and Wade (2003) predicted that omissions of nonmating individuals from fitness estimates would overestimate the average fitness, underestimate the variance in fitness for conventional males (figure 25.5), and cause the mating success of conventional males to appear to be significantly greater than that of satellite males. They confirmed this prediction using published data on male mating success in the marine isopod, *Paracerceis sculpta* (Shuster & Wade 1991; Shuster 2008). Unfortunately, these predictions cannot be specifically tested for other species except when information on the population frequencies, the average and variance in mating success, and the sex ratio of the species studied is available (Wade & Shuster 2004; see below).

Nevertheless, Shuster and Wade's (2003) assertion does suggest an additional, more easily testable hypothesis for any field study of a conditional mating polymorphism claiming to substantiate the BOBJ hypothesis. This prediction has two parts: (1) in studies claiming to support the BOBJ hypothesis, the zero class of males, p_0, will either be unidentified or unreported; in contrast, (2) when the zero class of males has been reported for conditional polymorphisms, equal fitnesses among male morphs will be found. The results of this analysis can be compared using Fisher's exact test.

Table 25.1 summarizes the results presented in 13 studies of conditional male polymorphisms reported since 1985. This is not an exhaustive list, but it does include prominent studies. These studies include four species of insects (four studies on the same species), one species of fish, one species of frog, and five species of birds. Eight of these 13 studies conclude that satellite males make the best of a bad job. The remaining five of the 13 studies find the fitnesses of the different male morphs to be equivalent. Consistent with the above predictions, of the 8 studies concluding that satellite males make the best of a bad job none (0/8) reported the size of the class of nonmating males in their studies. In contrast, all of the studies (5/5) in which males were found to have equal fitnesses either identified or included the class of nonmating males in their calculations of average mating success (Fisher's exact test, $P < 0.001$).

A classic study of *Hyla cinerea* by Gerhardt et al. (1987) provides another useful example for comparison. These authors recorded the mating success of calling, satellite, and noncalling males over 3 years. Of the 57 males who mated, 50 were callers and 7 males were satellites, suggesting that the average success of callers was greater than for satellites. However, because Gerhardt et al. (1987) identified mating as well as nonmating males in their analysis, they were also able to show that 416 of the 466 calling males (89%) were unsuccessful at mating, and that 50 of the 57 satellite males (88%) were also unsuccessful. Gerhard et al. (1987) concluded that the fitnesses of the two male phenotypes were equal because nearly equal proportions of each population were successful in mating (11–12%).

TABLE **25.1** Research articles addressing Dawkins' "best of a bad job" hypothesis

Author and Date	Taxon	Zero Class Quantified?	Average Fitness among Morphs
Fincke 1986	Damselflies	No	Unequal
Gerhardt et al. 1987	Tree frogs	Yes	Equal
Waltz and Wolf 1988	Dragonflies	Yes	Equal
Møller and Birkhead 1993	Birds	No	Unequal
Koprowski 1993	Birds	Yes	Equal
Alcock 1994a	Butterflies	Yes	Equal
Johnsen et al. 1998	Birds	No	Unequal (but not significant)
Alcock 1996a	Solitary bees	No	Unequal
Alcock 1996b	Solitary bees	No	Unequal
Alcock 1996c	Solitary bees	No	Unequal
Low 2005	Birds	No	Unequal
Beveridge et al. 2006	Solitary bees	No	Unequal
Rios Cardenas and Webster 2008	Pumpkinseed fish	Yes	Equal

These authors did not report the variance in mating success within the classes of mating males, so it is not possible to accurately estimate variance in mating success within and among calling and noncalling males using equation 25.8. However, it is still possible to determine whether the observed success of satellite males was sufficient for these males to persist within the population. Using equation 25.3c, we can see that if $p_{0calling}$ equalled 0.89, then the average harem size of calling males, $H_{calling}$ equalled 9.32 (not reported by Gerhardt et al.). If s represents the success satellites had to obtain by stealing mates from calling males to persist within the population, then from equation 25.12, $s > 1 - p_{0calling}$ or 0.11; indeed this value is approximately equal to the fraction of the total matings satellite males actually obtained (7/57 = 0.12).

FUTURE DIRECTIONS

There is now little doubt that both genetic and environmental factors influence the expression of alternative mating strategies (Shuster & Wade 2003; Tomkins & Hazel 2007; Oliveira et al. 2008). Thus, Austad's (1984) first question, whether behavioral differences between individuals stem from genetic differences, is for the most part solved. Slatkin's (1978, 1979a, 1979b) theoretical results indicate that the relative influences of genetic and environmental variation on trait expression mainly affect how fast, rather than whether, morph frequencies will respond to frequency dependent selection. His results also provide a simple evolutionary hypothesis for why mating polymorphisms with polygenic inheritance are overwhelmingly more common among species than mating polymorphisms with Mendelian inheritance (Austad 1984; Gross 1996; Shuster & Wade 2003; Oliveira et al. 2008). If strong negative frequency dependent selection favors the evolution of modifier alleles that equalize fitnesses among distinct phenotypes, then initially simple inheritance mechanisms underlying such traits will rapidly become polygenic. The observed rarity of simple inheritance mechanisms underlying mating polymorphisms may therefore simply reveal the lower end of the distribution of all such polymorphisms, which inevitably proceed toward greater underlying genetic complexity over time.

Whether it is possible for mating polymorphisms to persist in populations when the fitnesses of the morphs are unequal remains uncertain. However, this question, too, seems near resolution provided that researchers can agree on appropriate terminology and methods for addressing this problem. Most researchers appear to agree that frequency dependent selection operates on most if not all populations exhibiting mating polymorphism. This finding is consistent with the hypothesis that frequency dependent selection acts relentlessly to equalize fitnesses in natural populations (Slatkin 1978, 1979a, 1979b), and suggests that situations in which morph fitnesses are unequal, although plausible, may in fact be transitory. There can be little doubt that claims of unequal fitnesses among morphs are premature if based on data that excludes the zero class of males, that is, focuses only on the average success of mating males. Future research that quantifies the zero class of males and/or includes estimates of the mean and variance in mate numbers among the mating class of males will place this conclusion on firmer ground.

The central prediction that frequency-dependent selection will act on all phenotypes whose frequencies and fitnesses may vary, suggests that future research on alternative mating strategies in the following areas will be productive: (1) detailed analysis of genetic architectures underlying alternative mating phenotypes to establish whether the expression of any variable phenotype is not influenced in some way by underlying genetic variation; (2) accurate documentation of the fitnesses of mating and nonmating individuals in populations to further test the hypothesis that over time, the average fitnesses of alternative morphs will be equal; (3) investigations of how flexible phenotypes are expressed and how rapidly such traits respond to selection; (4) investigation of alternative mating phenotypes in females: if such polymorphisms are expected to appear within the sex in which fitness variance exists, females, as well as males, should express alternative strategies when fitness is variable within that sex (e.g., Berglund et al. 1989; Delehanty et al. 1998). Such variation is likely to exist in species in which males defend breeding sites and such sites become limited (e.g., pipefish, sea horses, sea spiders, shorebirds) as well as when females defend resources crucial to reproduction and these resources become limited (e.g., social insects, hyenas, primates).

Despite its appeal for devotees, the importance of alternative mating strategies in evolutionary biology is currently underappreciated and often misunderstood. Polymorphic mating phenotypes do not merely provide amusing examples of bizarre animal sex: they provide quantifiable examples of intense frequency dependent sexual selection and its rapid evolutionary consequences. Few other evolutionary phenomena are likely to be as common among species, or provide such a detailed look at how evolution proceeds. There is much exciting work to be done.

SUGGESTIONS FOR FURTHER READING

Interested readers have a wide range of possibilities for further reading on alternative mating strategies. Darwin (1874) was first to describe such variation in detail and examples abound within the second edition of *The Descent of Man and Selection in Relation to Sex*. A review of these considerations, with speculation on why Darwin found female mimicry uninteresting can be found in Shuster and Wade (2003, chapter 10). For masterful presentations of multiyear data, start either with Sinervo and Lively's (1996) account of the rock-paper-scissors polymorphism in side blotched lizards (see also chapter 3 of this volume), or Tomkins and Brown's (2004) report of earwig forcep dimorphism on Scottish islands. Lastly, the entire 2008 volume by Oliveira et al. (2008) offers unprecedented detail on how alternative mating strategies are expressed and can evolve as alternative reproductive tactics.

Darwin CR (1874) The Descent of Man and Selection in Relation to Sex, 2nd ed. Rand, McNally, New York.

Oliveira R, Taborsky M & Brockmann HJ (2008) Alternative Reproductive Tactics. Cambridge Univ Press, Cambridge, UK.

Shuster SM & Wade MJ (2003) Mating Systems and Strategies. Princeton University Press, Princeton, NJ.

Sinervo B & Lively CM (1996) The rock-paper-scissors game and the evolution of alternative male strategies. Nature 380: 40–243.

Tomkins JL & Brown GS (2004) Population density drives the local evolution of a threshold dimorphism. Nature 431: 1099–1103.

26

Parental Care

CHARLOTTA KVARNEMO

Parental care is any form of behavior by a parent that increases offspring fitness. The variation in forms of parental care among organisms is striking. Most animals provide no care other than nutrients in the egg. But among the ones that do, female care is often thought of as the rule and male care the exception. However, uni- or biparental male care is not rare. In fact, in fish, it is more common that the male alone cares for the offspring than that the female does it or that they show biparental care. In birds, biparental care is by far the most common form of care, although females generally provide more care than males. In mammals, care is always provided by the female, but in about 10% of the genera additional care is given by the male, often in the form of carrying, protecting, or feeding the young. Female care predominates in insects, female-only or biparental care is most common in reptiles, and both male-only and female-only care are relatively common in amphibians (Clutton-Brock 1991; Reynolds et al. 2002).

Variation in care behavior also occurs within a species. Some individuals provide more care than others, and levels of care differ throughout the lifetime. This diversity in parental care is intriguing and not yet well understood (Westneat & Sargent 1996; Kokko & Jennions 2008). This chapter describes attempts to explain this variation by focusing on two questions: (1) what main factors influence the evolution of major patterns of care? and (2) how is care allocated? The latter question covers allocation of care within an individual's lifetime, among multiple broods, among offspring within broods, and between parents in biparental species. What emerges is often a story of conflicts. Parental care is influenced by both conflicts between fitness enhancing activities and by different effects of care on the fitness of individuals involved.

CONCEPTS

The benefits of care behavior are usually (but not always) obvious. Parental care can be given both at the pre- and postzygotic stage of the offspring, that is, both before and after the eggs are fertilized (Clutton-Brock 1991). Prezygotic parental care may then include nutritious investment into female gametes, normally given by the female herself, but sometimes also by the male, in the form of nutritious nuptial gifts or courtship feeding, given to the female around mating. According to more narrow definitions of parental care, however, gamete investment constitutes only parental investment, but not parental care. A less controversial form of prezygotic parental care is nest building, which can be provided by the male, the female, or both. Postzygotic care includes protection and provisioning of developing young, such as gestation and lactation provided by the female in all mammals. Other examples of postzygotic care include male fish that protect and fan the developing eggs in his nest, or the more or less joint effort by male and female birds to incubate the eggs and to feed the hatchlings. Parental care can also occur after nutritional independence, for example, in the form of social assistance, as found in some primates, including humans. However, many of these behaviors have more than one function. For example, nuptial or courtship feeding can function as mating effort as well as parental investment (e.g., Simmons & Parker 1989). In fact, as explained later on in this chapter, both pre- and postzygotic care can serve multiple functions.

A likely explanation for variation in parental care is that it arises from variation in the costs of the behavior. Terminology for costs has been problematic and is often used imprecisely, leading to some confusion. Clutton-Brock (1991) attempted to clarify different measures of cost. He defined *parental effort* as the time, energy, or risk incurred by the parent when caring (Clutton-Brock 1991). The term *parental investment,* however, is more specific. It refers to the impact of parental effort on the parent's ability to invest in other offspring (Trivers 1972; Clutton-Brock 1991; box 26.1). This definition is explicitly in terms of numbers of offspring, and so it allows costs to be compared between males and females in the same currency (future offspring), regardless of the time or energy expended. A disadvantage of this concept is that future-offspring-that-could-have-been-produced-but-were-not is incredibly tricky to measure.

The study of parental care involves several difficulties with measuring the fitness consequences of behavior. As in many other areas of behavioral ecology, proxies to fitness are often used, and that can create some problems (chapter 4). One common result of this is to confuse measures of the

BOX 26.1 Parental Care and Life History

Life history theory is important for our understanding of parental care evolution. For example, why do parents not provide more care than they do, even though the young would benefit from it? Williams (1966) explains this as follows: "At the outset it must be realized that the maximization of individual reproductive success will seldom be achieved by unbridled fecundity. Even an adult tapeworm uses some of its resources for its own vegetative needs. If it were to sacrifice all of its somatic tissues for the production of gametes it might increase today's production of offspring, but it would lose all that it would have produced tomorrow and on succeeding days. Only if it had no chance of surviving until tomorrow would it be of advantage to spend all of its resources today" (161). The same applies to parental care. The more an individual spends on caring for its current offspring, the less it will have left for somatic maintenance and growth, which will have a negative influence on its own future reproduction, for example, through poorer survival or reduced fecundity. This relationship between current and future reproduction can be visualized as in the following figure:

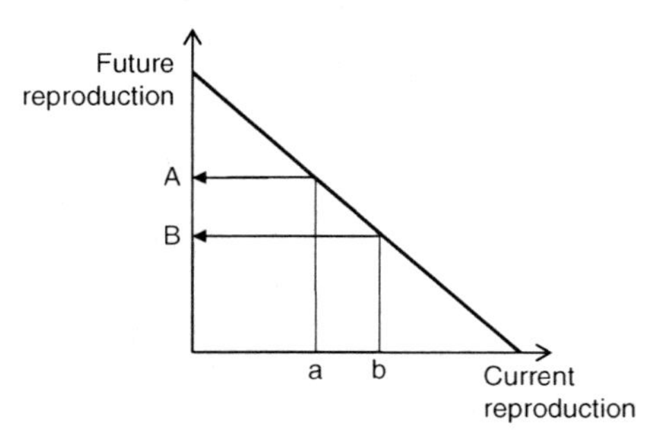

Figure I

(continued)

When current reproduction is traded off against future reproduction, an increased effort into current reproduction will result in reduced future reproduction (in terms of fecundity or survival). Figure I also serves as an illustration of parental investment, which represents a parental effort (b minus a) that comes with a cost (A minus B) in terms of reduced future reproduction.

Williams (1966) continues: "With the same expenditure of material [the tapeworm] could double its fecundity merely by halving the average size of the eggs (actually encased, dormant larvae) released into the intestine of its host. If such a curtailment of the material resources provided to each larva reduced the average survival by more than half, however, the increased fecundity would mean reduced reproductive success. I presume that egg size in tapeworms conforms to some optimum compromise between the advantages of high fecundity and of adequately providing for each of the young." (161). Thus, to maximize *lifetime reproductive success*, and produce offspring that can survive and mature into successfully reproducing adults, an individual has to balance how many offspring it produces at any one time and how much care each of these offspring receive.

To predict the optimal effort into current reproduction, one has to consider both future fecundity and the probability of survival to such future reproductive events. Data on survival and fecundity at each of several life stages can be collected in *life tables*. Expected lifetime reproductive success then depends on the anticipated reproductive success at each stage over the entire lifetime, while taking survival probability into account. Based on life tables, the average lifetime reproductive success can be calculated using $\Sigma\, l_x\, m_x$ in which l_x is the probability of surviving from birth to each particular age class x, and m_x is the average birth rate for each of the age classes (Stearns 1992; chapter 4, this volume).

We can also use life tables to calculate the *residual reproductive value* (RRV), which is the number of offspring an individual of a certain age on average can expect to have over its remaining lifetime. Young individuals with a high likelihood of surviving to a fecund adulthood have a high RRV compared to young individuals with a low probability of survival, or compared to old individuals that are reaching the end of their reproductive life span. In the latter comparison, we would expect old individuals to be more risk prone than the young one, because they have less to lose in terms of future reproductive success. Thus, reproducing now (despite, for example, high risk of predation) is better than saving it for an even more uncertain tomorrow. In iteroparous animals a declining RRV can cause old individuals to invest exceptionally much into reproduction (e.g., by producing larger clutches or by feeding the offspring more) the last season they reproduce. This is called a *terminal investment*. In fact, the high reproductive effort often shown by semelparous organisms can also be seen as terminal investment, due to their very low RRV.

One way to visualize an optimal life history at a given level of resource limitation is to place the trade-off (thick line) between two traits (e.g., parenting versus foraging for self) on a selective landscape, where the part of the trade-off that touches the highest isoline or contour (thin lines) of the nearby fitness peak gives the best fitness. The point optimizing the trade-off between trait 1 and 2 is marked with a star:

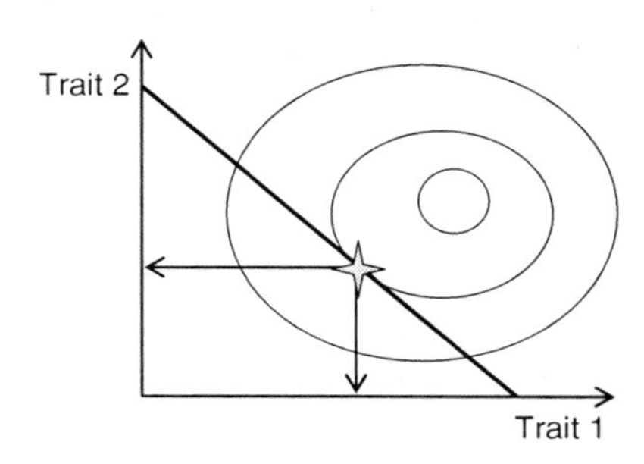

Figure II

(continued)

BOX 26.1 *(cont.)*

In this figure the trade-off is a straight line, whereas the fitness contours are curved. However, a reason we see so many different life history solutions in nature is that this relationship need not be a straight line with a slope of 1, but it can be convex, concave, sigmoid, and so on, and the slope of the fitness line or contour may also change with time. Thus the optimum of a trade-off between two life history traits may differ early and late in life:

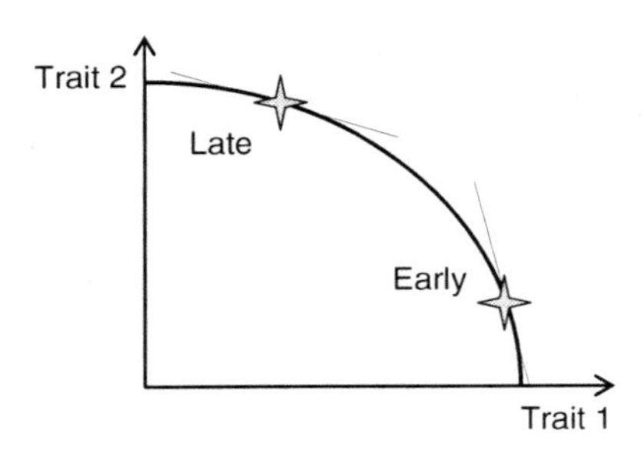

Figure III

In general, with a limited budget (which is an assumption of all life history theory), each individual has to make multiple trade-offs, both within and between reproductive events: offspring number is often traded off against offspring size, offspring provisioning against parental growth, and so on (see table 4.1 in Stearns 1992 for a more complete list of possible trade-offs among life history traits).

Although an increased level of parental care may increase the fitness of the offspring, the marginal benefit often peaks at some intermediate level of care, and then drops (figure 20.2). Therefore, even with low costs it does not pay to provide endless amounts of parental care, but in most cases a moderate level is optimal. Furthermore, in addition to life history trade-offs, factors such as parentage and opportunities for additional matings may influence the parental optimum (Kokko & Jennions 2008).

extent of the behavior with measures of the cost: for example, a parent bird that makes more trips to the nest is said to invest more, despite the fact that investment can be measured only in terms of future offspring sacrificed for present care. More fundamentally, behavioral ecologists often blur the boundaries between a parent's fitness and an offspring's when studying parental care. The majority of the parental care literature treats the fitness effects of parental care on the offspring as effects on parental fitness. This is an obvious and intuitive approach, because the essence of care is enhanced offspring fitness, which often comes at the expense of the parent's life span, for example. However, this is in contrast with views of some other evolutionary biologists who tally fitness for parents only up to fertilization—all other effects on the survival of those offspring are considered part of offspring fitness (chapter 4). Although I will tend to follow the former tradition here, it is worth exploring the alternative approach briefly.

Parental Care, Parental Effects, and Indirect Genetic Effects

Parents influence the fitness of their young through the genes they pass on to the offspring, which is called *direct genetic effects*. In addition, a parent can affect its offspring's fitness (positively or negatively) through environmental effects (Marshall & Uller 2007; chapter 14, this volume). To a developing fetus or a newborn young, the phenotypic quality of the mother can have major effects on its survival. Similarly, a nest-guarding fish by fanning and defending its offspring alters whether they can acquire sufficient oxygen for growth or survive predators. The ability of a parent bird to find carotenoid-rich food to feed the chicks can influence chick plumage coloration. Such parental behavior can obviously be under selection, but the fitness effects occur in individuals (the offspring) who are not expressing the trait. Quantitative genetics have termed these *parental* (or *maternal/paternal*) *effects*, and according to this approach, parental traits evolve because they are likely to be carried by the offspring whose fitness they influence. In addition, because variation in the parent's care behavior may have a genetic basis, any effect on the phenotype of the offspring via such care constitutes an *indirect genetic effect* (chapter 14). That is, the genes of the parent influence the phenotype of the offspring via the parent's effect on the environment that the offspring experiences. The most well studied examples of indirect genetic effects are related to maternal effects in birds and mammals (Wolf et al. 1998). Furthermore, when individuals adjust their parental allocation, for example, in relation to perceived mate quality (as discussed below), to their own age or to the age of the offspring (due to life history trade-offs; box 26.1), this can also create parental effects and (given that the reaction norm for the adjustment is heritable) indirect genetic effects on the offspring. This approach has not yet spread through the behavioral ecology literature on parental care, and so has not yet been applied to the examples I present. There is considerable potential for it to resolve some conceptual problems and generate new hypotheses, in particular when based on recent attempts to develop behavioral reaction norms, an approach that builds on both behavioral ecology and quantitative genetics (Smiseth et al. 2008; box 26.2).

Understanding the Evolution of Parental Care

When care evolves, there are multiple possible transitions of parental care patterns. Comparative analyses indicate that most often, male or female care has arisen from no care, after which biparental care has evolved if two parents did substantially better than one. In some taxa biparental care has then been lost, and uniparental care has secondarily evolved (e.g., Reynolds et al. 2002). However, for simplicity, I will focus on the reasons for transitions from no care to uniparental care.

To understand the evolution of parental care, it is instructive to focus on costs and benefits of care to the male and female parent. For care to evolve, we expect the fitness benefits of this behavior to be greater than its costs. By definition, care should increase the fitness of the young. This often happens through increased offspring survival or improved quality of the surviving young. Because all young of sexually reproducing animals have (exactly) one mother and one father, this benefit is shared equally by the two genetic parents, regardless of which sex provides the care. Thus, an individual of a sex can benefit from the care given by an individual of the other sex, and quite often manipulation occurs between the sexes of the amount of care the other sex provides, as will be dealt with later on in this chapter. Yet, to understand the evolution of care from no care, we should focus on the costs and benefits of providing care for each sex separately, assuming the other sex will not supply any care.

Five Explanations for Female Care

Over the last 40 years, a number of verbal and mathematical models and empirical reviews have tried to explain the prevalence of male and female care. Listed below are five interlinked arguments that have been used to help us to understand why female care has evolved more often than male care. In the following section I will then return to these arguments and examine how well they explain the evolution of male care.

According to the first argument anisogamy puts females into a position that selects for continued investment into the offspring they have already invested heavily into, in terms of large, expensive eggs (Trivers 1972). However, this argument does not hold, because past investment per se does not commit anyone to continued investment (Dawkins & Carlisle 1976). The mistake of basing decision arguments on past investment rather than future payoff was named the *Concorde fallacy*. The Concorde was a supersonic aircraft whose development was continued despite it being known it would never be profitable. The argument given was that

BOX 26.2 Parent-Offspring Conflict

Generally, the time (or energy) spent caring for offspring results in a large initial benefit at a small cost, but as the young grow and become more independent, the benefit levels off whereas the cost to the parent (in terms of future reproduction) increases. The optimal time from the parent's perspective to terminate care can be found at the point at which the difference between benefit and cost is the largest. However, the amount of care that is optimal for the parent (P) is usually less than optimal for the offspring (O), creating a parent-offspring conflict.

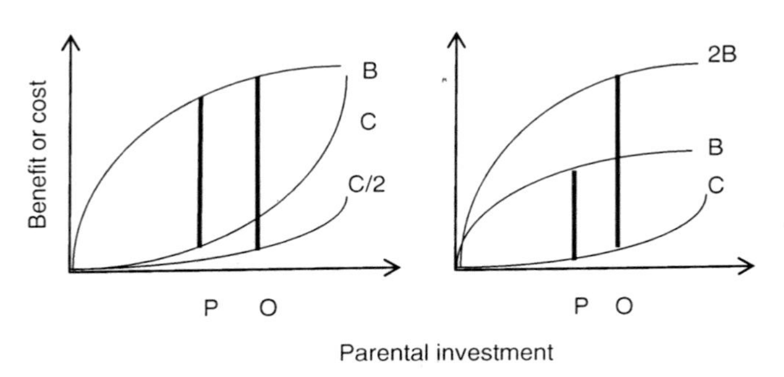

Figure I

This conclusion can be reached from two slightly different approaches (after figure 11.2 in Clutton-Brock 1991). The original idea, launched by Trivers (1974) and depicted in the left panel above, assumes that the benefit (B) of parental care shows diminishing returns and is the same for the parent and the offspring. The cost of parental care to the offspring (C/2) is only half that of the cost to the parent (C), because the offspring's relatedness to its future full sibs is 0.5, whereas its relatedness to itself is 1. However, Lazarus and Inglis (1986) presented a different argument, shown in the right panel of the figure. They suggested that the cost (C) is shared by both the parent and the offspring, because the offspring has the same relatedness to its future full siblings as the parent has to these offspring. In contrast, the benefits of care differ, being twice as high to the offspring (2B) as to the parent (B), because the offspring is twice as related to itself as it is to its parent.

A more recent attempt to understand parent-offspring conflicts has focused on behavioral response functions, that is, how parental supply increases with increased begging and how begging decreases with increased supply (Smiseth et al. 2008). The possible ways that parents may respond to offspring begging, and vice versa, can then be viewed as reaction norms with heritable variation in shape, slope, and elevation. Thus, an evolutionary change in parental supply may be caused by selection on the reaction norm for parental supply, or on the reaction norm for offspring demand, or both (Smiseth et al. 2008).

the large investment to date would be wasted if it were stopped (Dawkins & Carlisle 1976). To avoid the Concorde fallacy, the anisogamy argument was therefore rephrased, such that unequal gamete investment by males and females heavily influences the future alternatives of both sexes (whether to stay and care for the current young or to desert to seek new matings). Thus, anisogamy can influence care evolution indirectly through its effect on the intensity of sexual selection within each sex (Bateman 1948), which in turn may affect in which sex care evolves. In the sex with highest variance in mating success, some individuals will consistently mate sooner or more frequently than others. These individuals should be selected to desert rather than to care (Queller 1997; Kokko & Jennions 2008). Yet, anisogamy is far from the only determinant of which sex is under stronger sexual selection (chapter 20).

The second explanation for female care builds on the switch in emphasis, mentioned above, that follows from avoiding the Concorde fallacy. When a female is limited by her own egg production, regardless of whether she provides care or not, it should benefit her to stay and care for her offspring. Furthermore, if a female has little to gain from seeking additional matings with other males, her care behavior carries a low cost in terms of lost opportunities for additional matings.

A third explanation for widespread female care focuses on the choices available to males. Because of his small, cheap gametes, a male experiences a different consequence from caring: he might increase his fitness substantially if he can find additional females to mate instead of providing care. Thus, for such a male it would prove costly to stay and care for the offspring of his first mate if this behavior limits his ability to attract new mates. However, this logic is problematic. In fact, the relatively low gamete investment by males often results in a higher potential reproductive rate for males than for females. Because this leads to more males than females that are ready to mate at any one time, the likelihood of a male finding a new mate is less than that of a female, which reduces the benefits of desertion for males and could increase them for females (Kokko & Jennions 2008). Thus, for most males the costs of parental care may be lower than previously thought. Yet, such costs may still be large for those males that are most successful in attracting multiple females as mates (Queller 1997; Kokko & Jennions 2008).

The fourth important argument for why female care is more prevalent than male care relates to parentage (the proportion of a brood for which a male has paternity or a female maternity). Sperm competition (chapter 22), due to multiple mating by females or parasitic spawning by males, often causes a reduced level of paternity. Females thus are likely to have a higher level of maternity within a brood of young than males have paternity (Trivers 1972). This devalues the fitness benefits of providing care for males, and thus makes care less likely to evolve in the male sex (Queller 1997).

Finally, the *order of gamete release* hypothesis suggests that care should evolve such that the sex that releases its gametes first can abandon the offspring and leave the care to the other sex (Dawkins & Carlisle 1976). According to this hypothesis, female care has evolved because she is unable to desert the young, because the male has already done so. Females are in a cruel bind: desertion by the female reduces her fitness from that breeding attempt to zero, and so females do better by staying and caring. Although this hypothesis may explain the high prevalence of female care in internally fertilizing animals, it cannot explain the high prevalence of male care in externally fertilizing species, as I will return to shortly.

Five Explanations: Also for Male Care?

Male care (uniparental or biparental) is also quite common. As general explanations of care, the five hypotheses listed above should apply to these cases as well. I will therefore reexamine these five explanations using fish, in which male care is relatively common.

All fish are anisogamous, and females (by definition) produce larger gametes than do males. Therefore, based solely on anisogamy as an explanation for parental care, we would never expect male-only care to evolve. Thus, although anisogamy may indirectly contribute to the evolution of maternal care, it cannot explain the evolution of paternal care.

Like many other animals, female fish are often limited by egg production, rather than by number of mates. But, importantly, this does not mean that care comes with no cost to the female, as fish continue to grow throughout life, and care often results in reduced growth (Sargent & Gross 1986). A reduction in growth due to care often translates into a large fitness cost for females, because increases in size lead to an accelerating increase in the number of eggs produced (Sargent & Gross 1986; see figure 26.1). In contrast, in males, a similar reduction in growth due to care has only weak effects on fitness, because male fitness changes linearly or may even decelerate with body size (Sargent & Gross 1986). Thus, for females compared to males, the benefits from care need to be large to cover the accelerating costs, and this may have hampered the evolution of care in females more so than in males. Furthermore, it is worth mentioning that female fish can also be limited by mate availability, rather than by egg production, as found, for example, in species of sex-role reversed pipefish (Berglund et al. 1989).

Thus the explanation that females experience a less severe trade-off than males between current

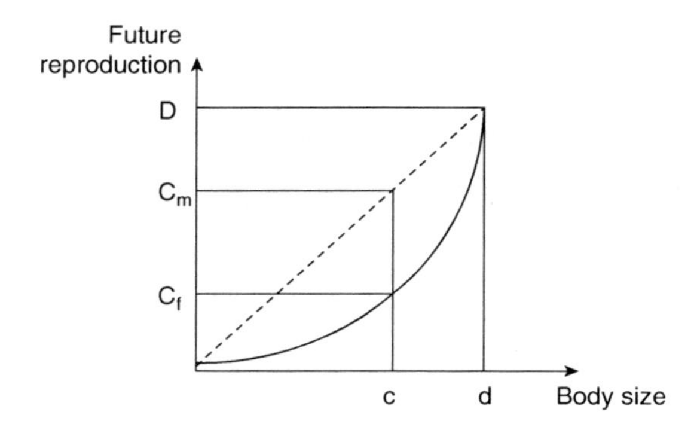

FIGURE 26.1 As most fish grow throughout their lives and time and energy invested into parental care often result in reduced growth, the cost of parental care is typically higher for females than for males (Sargent & Gross 1986, after figure 11.3): a given reduction in growth due to care (d minus c) can be expected to translate into a larger fitness cost, in terms of reduced future reproduction, for females (D minus C_f) than for males (D minus C_m). This is because larger females can produce exponentially more eggs with an increased body size (solid curve), whereas male fitness only increases linearly with body size (dashed line), or even decelerates (Sargent & Gross 1986). Males may thus more easily overcome the cost of care (through benefits accrued from it), whereas this is less likely in females (d = body size and D = future reproduction under full growth when not providing care ("deserting"); c = body size and C = future reproduction when providing care). Thus, the figure illustrates an explanation for why male care is more common than female care among fish, based on the theory about life history trade-offs between current and future reproductive success (box 26.1).

and future reproduction may not apply in fish. Another factor is that the cost of care to the male in terms of lost opportunities for additional matings is often very low. In fish, male care has evolved most often in lineages with territoriality (Ah-King et al. 2005), and generally the male can care for the clutches of several females at one time. In a growing number of species, evidence exists that females prefer to spawn with males that already care for eggs in their nests (Manica 2002). This preference may have evolved because the eggs indicate a good parental ability by the male or because the other eggs dilute the risk that the eggs of a given female are eaten if egg predation by the male himself or by others should occur. In some fish and insects with male care, males have even been shown to steal eggs from one another (Rohwer 1978; Tallamy 2000). Presumably such egg theft has evolved as a result of the female preference for egg-tending males.

Queller (1997) argued that one reason that female care is more common than male care is because females generally have higher parentage than males (increasing the fitness benefit of exhibiting care, thus making care more likely to evolve for females). Higher maternity than paternity is true also for individual clutches in fish. Each female that spawns with a male has complete maternity assurance of her eggs, whereas the male is often exposed to cuckoldry from parasitically spawning males. However, because male fish are often able to care for several female clutches at once, his total number of descendants exceeds that of the female despite partial paternity. Thus in fish, reduced paternity may have less effect on the bias in care than in other organisms. In addition, within a brood of eggs spawned by multiple females, each female's maternity is relatively low. This may explain why female-only or biparental care is quite uncommon in fish.

There is another explanation that relates to parentage. In fish and amphibians, male care is more common in species with external than internal fertilization (Gross & Shine 1981). Based on Trivers' (1972) original argument that male care should evolve only if the male has a high paternity, it has been suggested that external fertilization provides males with either higher paternity or better information about paternity than in internally fertilizing animals (e.g., Perrone & Zaret 1979). However, we now know that this is not true. Instead, external fertilization provides many opportunities for parasitic fertilizations (Petersen & Warner 1998) and males sometimes have information about this (Neff 2003), suggesting that high paternity is not the explanation for the abundance of male care in fishes.

At first, Dawkins and Carlisle's (1976) *order of gamete release* hypothesis seems plausible for fish, because male care correlates with external fertilization, which thus might allow the female to lay her eggs first and desert. However, closer scrutiny shows very little empirical support for this hypothesis in fish, because male care is still more common in species in which the sexes release gametes at the same time (Gross & Shine 1981). In fact, male care occurs even in fishes, such as in gobies, in which males deposit a mucus-embedded trail of sperm in the nest before a female attaches the eggs to the surface (Marconato et al. 1996).

An emergent lesson in trying to understand patterns of care is that it becomes almost meaningless to focus on explaining female care, and then treat male care as the exception. Similarly, by fastidiously regarding mating effort as a separate activity from parental effort, we may fail to see additional benefits to the behavior. Instead, by examining the costs and benefits of care to both sexes, we may start to understand the wider patterns of male and female care, including why male care arises not only in fish but also in other taxa.

Costs and Benefits of Care

The evolution of parental care patterns must be linked to patterns of costs and benefits. The following simple model (Kvarnemo 2006) can be used to explore such patterns. Imagine a species in which males attempt to attract females to gain matings. The males have two possible strategies, to care (C) or not to care (NC), and a male maximizes his fitness by maximizing the number of his own surviving offspring, N. Thus, for male care to be an evolutionarily stable strategy, a caring male must produce more surviving offspring than a noncaring male, in other words, $N_C > N_{NC}$.

For each strategy, N is influenced by three factors: mating success (F), paternity success (P), and offspring survival (S). Mating success F is determined by the number of females with whom a male mates and the clutch size of each female. Thus, F is calculated as the summed clutch size of all females. Paternity success, P, is the proportion of F that is fathered by the male, and offspring survival, S, is the proportion that survives to maturity, with $0 \leq P \leq 1$ and $0 \leq S \leq 1$. We may expect male care to evolve as long as it provides a higher net benefit than the no-care strategy, that is, $F_C P_C S_C > F_{NC} P_{NC} S_{NC}$. Many previous models have simply assumed that a high level of paternity is a prerequisite for male care to evolve. However, as long as $F_C P_C S_C > F_{NC} P_{NC} S_{NC}$ holds, P_C does not need to be high.

To investigate the costs or benefits incurred by the care and no-care strategies, and how each of the three factors contributes, we can compare F_C to F_{NC}, P_C to P_{NC}, and S_C to S_{NC}. A common assumption is that individual males lose opportunities for additional matings while providing care, compared to males not providing care. If this is true, then $F_C < F_{NC}$, constituting a cost of care. However, as already mentioned, this need not be the case. In the facultatively caregiving wrasse *Symphodus tinca*, $F_C >> F_{NC}$ because caregiving males can attract a huge number of spawning females to their nests compared to noncaring males (Warner et al. 1995b). Furthermore, an increased paternity as a consequence of care ($P_C > P_{NC}$) will provide an advantage to the care strategy. Such a paternity benefit of male care can arise if the nest provides better defense against sneaker males, if the size of a nuptial gift improves both offspring fitness and time for sperm transfer, or if females promote fertilizations by males that have shown that they are willing and capable of caring for offspring (Kvarnemo 2006).

If we follow the common view and say that care has to be beneficial to the offspring ($S_C > S_{NC}$), an increase in F_C or P_C is not necessary for care to evolve. However, care-like behavior can evolve even if this assumption is relaxed and we set $S_C = S_{NC}$. In the pine engraver beetle *Ips pini*, males remove frass from the tunnels where females lay eggs, and careful study has shown that this behavior increases male paternity without having a noticeable influence on offspring survival (Lissemore 1997).

The sand goby, *Pomatoschistus minutus*, provides a good illustration of costs and benefits of care, and how both natural and sexual selection can influence the level of care. A male attracts several females to spawn in his nest, dug in the sand underneath a stone or a mussel shell. He cares for the eggs until hatching by guarding and fanning, providing oxygen and removing waste products. There is a trade-off between ventilation and egg defense when it comes to nest-opening size, because males exposed to low oxygen water build larger nest openings, but if housed with an egg predator they reduce the opening (Lissåker & Kvarnemo 2006). However, nest building in sand gobies is not only a naturally selected behavior to protect eggs after spawning, but it is also sexually selected before spawning, through protection against sneaking and possibly through female choice (Svensson & Kvarnemo 2003; figure 26.2a). Fanning is another care-related behavior that males show not only to improve offspring fitness but also to impress females. Such displacement or courtship fanning can be shown even without any eggs in the nest, when in front of a female. Furthermore, when having eggs in the nest, males fan more vigorously when there is a female nearby than when they care for the eggs on their own (Pampoulie et al. 2004; figure 26.2b). This obviously suggests that females

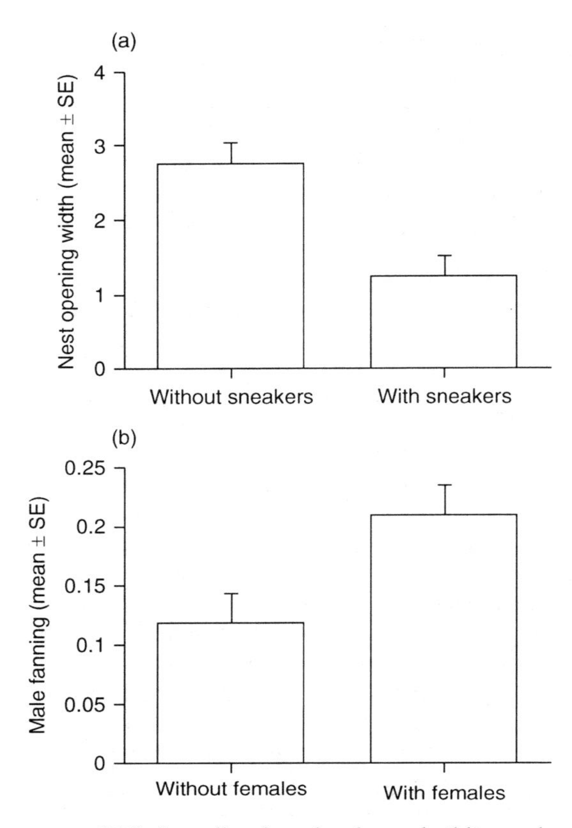

FIGURE 26.2 Sexually selected male nest building and fanning behavior in the sand goby, *Pomatoschistus minutus*. (a) Before spawning, nest-holding males built nests with wider openings (cm, mean ± SE) when housed alone, compared to when housed together with two sneaker males. Based on data from Svensson and Kvarnemo (2003). (b) Males with eggs in their nests spent more time fanning (proportion of total observation time, mean ± SE) when there were ripe females present in a compartment inside the aquarium, compared to when there were no females present. Based on data published in Pampoulie et al. (2004).

base their mate choice on male fanning behavior: indeed, females showed a strong preference for males that were experimentally manipulated to fan more by pumping low oxygen water into their nests (Lindström et al. 2006). Thus, these studies show that care, in this case nest building and fanning, are not only naturally selected behaviors, but are also clearly affected by sexual selection.

The evolution of care behavior is complex, and we are still far from fully understanding which factors really influence the evolution of care in both males and females. Models or arguments, like the ones presented above, have great heuristic values but may also prove too simple, because they generally consider just one or a few factors at a time. Kokko and Jennions (2008) have recently made a lucid attempt to investigate how a number of factors influence care evolution together. Given a few simplifying assumptions, their model suggests that for each sex (1) high marginal benefits (i.e., the additional benefit that can be gained for every small increase in care) and low mate encounter rates select for more care; (2) sexual selection and (3) low parentage select for less care; (4) a biased operational sex ratio (the ratio of males and females that are ready to mate at any one time; Emlen & Oring 1977) selects for more care by the sex in excess (due to reduced chances of finding a mate after desertion, thus reducing the cost of care for this sex; see box 26.3); and (5) the adult sex ratio can influence care evolution. A biased adult sex ratio can directly select for increased care among the sex in excess, but indirectly it can also cause selection for reduced care in this sex. Intriguingly, this is the case if the sex ratio bias is caused by a higher mortality while competing for new mates than while caring.

How Much Care to Provide, When to Provide It, and to Whom

Once care has evolved, the question about how to allocate it becomes important. Allocation of care varies at many levels, such as within an individual's lifetime, among multiple broods, and among offspring within broods. The reader should note that allocation decisions of this sort can also be made among different kinds of relatives, because siblings are as likely to carry genes for caring as are offspring in most sexually reproducing organisms. Many of the same variables that are important in decisions about parental care are also important in kin-directed altruism (chapter 18).

Distribution of Care among Broods of Offspring

When an individual gives a lot of care to the current offspring, this is generally done at the expense of future or other, simultaneous offspring (Williams 1966; Trivers 1972). Due to life history trade-offs, an individual with a long life expectancy should allocate less to reproduction early

BOX 26.3 Adaptive Offspring Sex Ratios

Michael S. Webster

One aspect of parental care that has broad implications for other areas in behavioral ecology is the extent to which parents invest in sons versus daughters. Fisher (1958) was the first to point out that under a broad range of conditions, frequency-dependent selection should lead to parents that invest equally in sons and daughters (see box 20.2). Importantly, Fisher's theory predicts that parents will produce broods/litters biased toward one sex or the other whenever certain simplifying assumptions are violated, and this forms the basis for most modern sex ratio theory. For example, as first pointed out by Hamilton (1967), if one sex does not disperse far from where it is born, then same-sex relatives will compete with one another for mates (local mate competition) and/or resources (local resource competition), and parents who overproduce the more dispersive sex will be favored by selection. Similarly, if one sex sometimes remains on the natal territory to act as "helpers" that assist in raising additional offspring, the helping sex is in effect less costly to raise (local resource enhancement), and under Fisher's framework we might expect offspring sex ratios biased toward this sex (Emlen et al. 1986).

Empirical studies have frequently shown that parents can and do produce biased offspring sex ratios, and in some cases the sex ratio is biased in the direction predicted by Fisherian theory. For example, in polygynous species, a female's physiological condition will often affect the future reproductive success of her sons more than that of her daughters (because sons compete physically for mates), and an extension of Fisher's theory (Trivers & Willard 1973) predicts that females in good condition should produce more sons, whereas females in poor condition should produce more daughters. This prediction is generally supported in studies of polygynous ungulates when appropriate measures of female condition are used, particularly in highly dimorphic species and those with long gestation periods (Sheldon & West 2004).

However, although empirical tests of sex ratio theory have met with some success (see box 20.2), in many cases offspring sex ratios have not matched theoretical expectations, particularly in studies of vertebrates (e.g., Komdeur & Pen 2002). In part, this is because the several hypotheses derived from Fisher's general theory are not mutually exclusive and several can operate simultaneously. Accordingly, recent sex ratio theory has attempted to join together various hypotheses into a single unifying framework (e.g., Wild & West 2007).

Take, for example, cooperatively breeding species in which one sex helps (e.g., sons) and the other does not. Although local resource enhancement (or "repayment"; Emlen et al. 1986) predicts that parents should overproduce the helping sex, this prediction does not hold in many studies of cooperatively breeding birds and mammals (Koenig & Walters 1999). However, in cooperative species it is typical for the nonhelping sex to disperse further than the helping sex, and under these conditions local resource and mate competition should favor overproduction of the dispersive sex, possibly canceling or even reversing the effects of local resource enhancement. The optimal offspring sex ratio therefore will depend on how dispersal and helping benefits interact (see figure I). This more general theory, which combines both local competition and enhancement, has been supported by studies showing that offspring sex ratios are biased toward the helping sex only when helpers have a strong effect on parental fitness (Griffin et al. 2005; Silk & Brown 2008).

Another factor that complicates both theory and empirical studies is interactions among siblings prior to independence: if siblings compete for parental resources, then simple predictions from models that ignore these interactions may be misleading (Uller 2006). For example, offspring competitive ability often depends on both gender and birth/hatch

(continued)

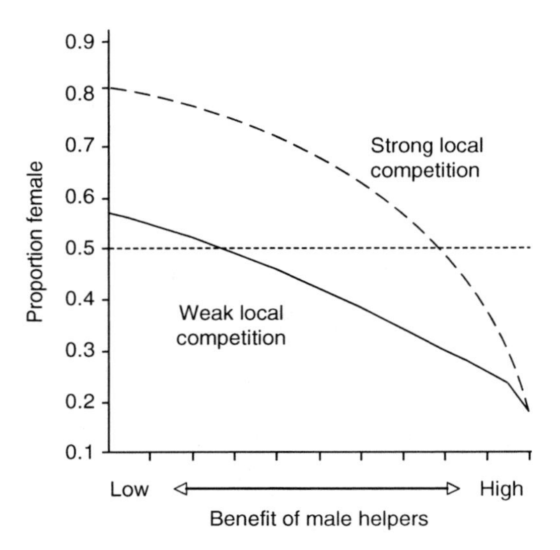

Figure I A model of adaptive offspring sex ratios that incorporates both repayment and local competition effects in a species in which sons help and daughters disperse (figure modified from Wild 2006). Under this model, optimal offspring sex ratios (y-axis) are affected by both the strength of local mate or resource competition (solid versus dashed curves), which is determined by natal dispersal distance of the less dispersive sex, and the increase in parental fitness due to presence of a helper (x-axis). When helpers have a strong effect on parental fitness, then we expect to see offspring sex ratios biased toward the helping sex (sons), but when helpers have only weak effects on parental fitness, then sex ratios should mainly be affected by local competition and we expect sex ratios biased toward the more dispersive sex (daughters). Note that, when male natal dispersal is highly limited and local mate competition is very strong (dashed line), we expect to see offspring sex ratios biased toward the helping sex only when helpers have an especially strong effect on parental fitness.

order, for example, if early-hatched nestling birds are at a competitive advantage relative to late-hatched nestlings in the same brood. Under these conditions, selection should favor females who adaptively adjust sex ratio according to laying order (e.g., by producing the less competitive sex early in the laying sequence), even if the brood sex ratio is not biased overall (Carranza 2004).

The mechanism of sex determination (see box 5.2) is yet another factor that might complicate the evolution of adaptive offspring sex ratios. Some mechanisms, such as haplodiploidy or temperature-dependent sex determination, might allow parents to control offspring sex ratios very precisely to match conditions (Warner & Shine 2008). In contrast, chromosomal sex determination and some other mechanisms may place limits on parental ability to control the primary sex ratio (West & Sheldon 2002). Accordingly, there is considerable interest in uncovering the mechanisms by which parents can bias the primary offspring sex ratio (e.g., Rutkowska & Badyaev 2008), but to date these mechanisms remain a black box for many taxa.

in life than should an individual that can expect only a short life (box 26.1). This applies when comparing reproductive decisions both within and between different species. For example, 1-year-old individuals of the short-lived sand goby, *Pomatoschistus minutus*, are more willing to reproduce under predation threat than are young adults (2–3 years old) of the black goby, *Gobius niger*, which has a life expectancy of about 5 years (Magnhagen 1990). However, 4- to 5-year-old black gobies are as willing to reproduce under predation threat as are 1-year-old sand gobies (Magnhagen 1990). For this reason, we may also expect higher level of care to be shown at any one time by individuals of short-lived species. However, because one of the major costs of parental care is a reduced life expectancy (Stearns 1989; Magnhagen 1991), a short life can be caused not only by extrinsic factors but also by parental care per se. Thus, a short life span can be both the cause and the consequence of parental care.

Parents may also adjust their level of care among broods of offspring by increasing or decreasing it according to the perceived quality of the other parent. It can be beneficial for an individual with several breeding attempts over its lifetime to invest more into an attempt when paired with a mate of superior phenotypic or genotypic quality, even though such increased investment in current reproduction comes with a cost, in terms of reduced resources for future reproduction (Sheldon 2000). Hence, females are sometimes willing to provide more care to offspring that are fathered by an attractive than an unattractive male. This behavior is called *differential allocation*, and is most well studied in biparentally caring birds (Møller & Thornhill 1998; Sheldon 2000). Classic work by Nancy Burley on the zebra finch, *Taeniopygia guttata*, showed that females prefer males ringed with red leg bands (because these colors accentuate the red colors of their beaks), and that females feed the offspring of such males more than the offspring in broods fathered by males marked with blue or green leg bands (Burley et al. 1982; Burley 1986). However, differential allocation can also occur when only one parent provides postzygotic care (Møller & Thornhill 1998). For example, in the mouth-brooding Bangaii cardinal fish, *Pterapogon kauderni*, females spawn larger eggs when they mate with large males than when they mate with smaller males (Kolm 2001).

Because differential allocation increases the fitness of attractive mates, it can contribute substantially to sexual selection (Burley 1986; Sheldon 2000). When there is a genetic component to the attractive trait, theory predicts that mate preferences should lead to a positive genetic correlation between the preference and the attractive trait, such that the two are inherited together by sons and daughters alike (Fisher 1930; Lande 1981; chapter 24). This can occur through linkage disequilibrium or through pleiotropy. This prediction applies not only to precopulatory mate preferences but also to postcopulatory actions such as differential allocation. This was elegantly shown in the house cricket, *Acheta domesticus*, in which large daughters of attractive males showed both a precopulatory preference (shorter latency to mount) and postcopulatory preference (larger number of eggs laid) for large males, a preference not shown by the daughters of unattractive males (Head et al. 2006).

To date, research on parental adjustments of care has focused almost exclusively on differential allocation, which thus involves increased investment in favorable situations (i.e., when mating with a preferred partner). However, adjustment in care can also be expected to result in increased investment in unfavorable situations. A recent model predicts increased level of care, as a way of compensating, when individuals have to mate with nonpreferred partners (Gowaty 2008). Support for such reproductive compensation has been found in a range of species, including mallards, *Anas platyrhynchos*, fruit flies, *Drosophila pseudoobscura*, pipefish, *Syngnathus typhle*, and mice, *Mus musculus* (Gowaty et al. 2007). Interestingly, an adjustment in the same direction (i.e., higher investment when in an unfavorable situation) can also be interpreted as a lower investment when in a favorable situation. In the superb fairy-wren, *Malurus cyaneus*, pairs may breed with helpers at the nest, or alone. Despite a higher feeding rate with helpers, such chicks end up with the same body mass as chicks reared by pairs alone. The mystery of this was recently solved when females with helpers at the nest were found to lay smaller eggs, containing less protein and lipids, than females without helpers (Russell et al. 2007). Hence, females with helpers clearly take advantage of a beneficial situation, because their reduced egg investment, which helpers compensate for, give them higher survival to the following breeding season (Russell et al. 2007).

Distribution of Care among Offspring within Broods

Within broods, parents may also allocate their care unevenly among offspring, according to, for example, age, size, or phenotype of the offspring, a phenomenon called *parental favoritism* or (if the two parents favor different chicks) *brood division* (Lessells 2002). In the case of a parent favoring one sex over the other, it is called *sex-biased parental care* (Lessells 1998). Sex-biased care is important also in cooperatively breeding animals, in which one sex is more likely to stay and help and the other to disperse (box 26.3). For example, in meerkats, *Suricata suricatta*, in which females are more philopatric and likely to help with care than are males, female young are fed more (Clutton-Brock et al. 2002). Similarly, in the cooperatively breeding Arabian babbler, *Turdoides squamiceps*, parents bias their feeding toward males, which is the sex that is more likely to stay and help, when the group is small. In contrast, they bias their feeding efforts toward females, the dispersing sex, when the group is large (Ridley & Huyvaert 2007).

Caring for Unrelated Young

Although generally not predicted by theory, care is sometimes given to unrelated young, a phenomenon called *alloparental care*. It is relatively common in fishes (reviewed in Wisenden 1999), but occurs also in other taxa, such as birds. In the common eider, *Somateria mollissima*, females often form a coalition with other females and rear their chicks together (chapter 8). A hypothesis that such behavior is generated through kin selection was excluded, because genetic analysis showed that females in pairs were not more related to each other than to any other female in the population. Instead, it is likely that the greatest benefit of this behavior comes from joint care allowing more time for feeding by each of the adult females, which may benefit both females but seems particularly important for females in low condition (Öst et al. 2005; chapter 8 of this volume). Similarly, in the greater rhea, *Rhea americana*, males, which herd and guard the chicks for several months after having incubated the eggs, often enlarge the brood by adopting stray chicks, that are (based on behavioral observations) unrelated (Codenotti & Alvarez 1998). The number of adopted chicks can be substantial, sometimes even outnumbering the males' own offspring. The adopted chicks seem to benefit most from the behavior, whereas it comes with low costs to the male and his own offspring (Codenotti & Alvarez 1998). Intriguingly, the male's offspring do not gain any benefit in terms of a selfish herd from the adopted young, as might have been expected (Codenotti & Alvarez 1998). However, I will return to this example below.

In species with care, individuals should be selected against expending costly care on unrelated young (Trivers 1972). This may be of greatest importance for males in species with high levels of parasitic spawning or extra-pair copulations, but it also applies to females in species with brood parasitism (Westneat & Sherman 1993). However, if the parent cannot distinguish its own offspring from the cuckolded ones, withdrawing care to an entire brood will harm its own offspring (Sheldon 2002). This might be the reason why males do not reduce care in response to reduced paternity, even in species with substantial male care, such as the house martin, *Delichon urbica* (Whittingham & Lifjeld 1995), or the common goby, *Pomatoschistus microps* (Svensson et al. 1998). Still, in other species it has been shown that when the male is given some cue that his paternity is likely to be compromised, his care is reduced (Sheldon 2002). For example, in an experimental study of the bluegill sunfish, *Lepomis macrochirus*, males were shown to adjust their level of care according to their perceived paternity (Neff 2003). Parental bluegill males, which care for the offspring both as eggs and as fry, assess their paternity indirectly at the egg stage, by observing sneaker males nearby the nest at spawning. After hatching, however, these males are able to assess their paternity more directly, possibly through chemical cues in the urine of the fry (Neff 2003). Whether or not a facultative response is expected thus depends on access to cues. If the parent is unable to assess its parentage, a nonfacultative reduction of care may still be expected to occur over evolutionary time (Westneat & Sherman 1993). The expected response should also depend on the likelihood of remating if deserting a brood (Kokko & Jennions 2008), and on the likelihood and predictability of getting a higher or lower level of parentage in subsequent broods (Westneat & Sherman 1993).

Caring for unrelated young, however, can have direct benefits to the male in terms of increased mating success. For example, in the greater rhea, the male and the chicks in his care stay together in winter flocks in which they are joined by other families and females. A larger family size has been suggested to be used by females as a signal of high paternal quality (Codenotti & Alvarez 1998), in which case adopting unrelated chicks may be directly beneficial to the male.

Conflicts over Care

Sexual reproduction involves two (usually) unrelated individuals fusing their gametes. Offspring carry some of each parent's genes but are not identical to either. Because of these patterns of relatedness, decisions about care allocation can result in conflicts of interest. For example, parents and offspring may differ in the optimal level of care to be provided. Offspring also benefit if care is allocated to them instead of a sibling. Finally, the parents may differ in the optimal level of care they provide and the amount their partner provides.

In 1974, Robert Trivers noted that the level of parental investment that benefits the offspring the most is generally greater than the parent should be selected to give to maximize its lifetime reproductive success (box 26.2). Such greediness of young can be understood if the offspring increases its direct fitness more than it reduces its indirect fitness through current or future siblings by demanding more to itself. A larger discrepancy in optimal parental investment between the offspring and the parent should lead to more conflict. Trivers (1974) predicted that offspring should have ways to manipulate the parent(s) into providing more care than is optimal for the parent. In birds, this can result in fairly large chicks harassing their parent while begging for food and in mammals, tantrums due to weaning are common.

Offspring demands for more at the expense of the fitness of its siblings also give rise to sibling competition. Such competition may be most evident in birds with asynchronous hatching of the offspring, in which the young of the first egg to hatch gets a head start, and can attract more feeding from its parents, whereas the last one to hatch often gets too little food to survive. Many different adaptive explanations have been suggested for asynchronous hatching, but the phenomenon is still poorly understood (Stoleson & Beissinger 1995). Sibling rivalry can also be substantial within synchronously hatching broods, especially if there is parentally biased favoritism, with the result that some offspring get fed more than others (Lessells 2002).

When both parents take part in offspring care, sexual conflict can occur over how much care each parent should allocate (e.g., Westneat & Sargent 1996; Lessells 1998; Houston et al. 2005; chapter 23 of this volume). However, sexual conflict over how much care to provide is not limited to species with biparental care, but can arise whenever one individual is able to manipulate its partner into providing more care to the current offspring at the expense of that partner's future reproduction. For example, in a promiscuous fruit fly, *Drosophila melanogaster*, certain toxins in the males' seminal fluid shorten the female lifetime expectancy (Chapman et al. 1995). A shorter life should affect the female life history trade-off between current and future reproduction, because it then pays the female to invest more heavily into her current brood than it would otherwise (box 26.1; Lessells 1999; Wedell et al. 2006). Consistently, females that were mated to males with toxic ejaculates produced larger broods (Chapman 2001). Although the effect in this example is larger broods, rather than larger eggs or more care, it illustrates how one sex may manipulate the other into investing more heavily into current offspring simply by reducing its life expectancy. However, this strategy is a good one for the male only if there is a low likelihood of him reproducing together with the same female again. In selection experiments with male and female fruit flies housed in monogamous pairs over 47 generations, the toxicity of the seminal fluid declined (Holland & Rice 1999). Under monogamy, toxic males share the reduction in the lifetime reproductive success of the female, resulting in reduced fitness compared to less toxic males.

FUTURE DIRECTIONS: SEXUAL SELECTION AND PARENTAL CARE

The diversity of parental care across taxa and the level of variation within populations continue to be major challenges to explain. Recent reevaluations of fundamental ideas about the link between parental care and sexual selection (Kokko & Jennions 2008; chapter 20 of this volume) could have far-reaching effects. How far reaching remains unclear, but they emphasize some unresolved issues that I will highlight here.

Until recently researchers have asked how parental care influences sexual selection, via its effects on parental investment (reflected in the potential reproductive rates of males and females, which in turn influence the operational sex ratio and the intensity and direction of mating competition in a population). However, there is growing awareness that sexual selection can also influence parental investment (e.g., Reynolds 1996; Kokko & Jennions 2003, 2008). One implication of this is that mate choice may drive the evolution of caregiving. Some attention has been given to *direct* (that is, nongenetic) benefits to female preferences. In a number of studies, females have indeed been found to prefer males that show caretaking abilities (Forsgren 1997), through courtship behavior (Östlund & Ahnesjö 1998) or nest-building ability (Soler et al. 1998). But surprisingly few studies have considered whether such mate choice may generate sexual selection on the occurrence and level of parental care. Two recent reviews of paternal care, in insects and other arthropods (Tallamy 2000) and in fish (Lindström & St. Mary 2008), provide rare exceptions. Furthermore, a phylogenetic study of cichlids has recently shown sexual selection to determine patterns of parental care, rather than the other way around (Gonzalez-Voyer et al. 2008).

Yet, some implications of male care as an object of sexual selection are intriguing. For example, what might this mean for signals of parenting ability? A meta-analysis showed relatively high direct fitness benefits to be gained from female mate choice through increased hatching success in fish, amphibians, and insects, but not through increased feeding rates in birds (Møller & Jennions 2001). As some level of honesty or reliability is required for secondary sexual traits to evolve as useful signals of future care (Westneat & Sargent 1996; Kokko 1998), does this mean that ectotherms are more (and birds less) likely to develop secondary sexual traits that honestly signal future level of paternal care? Furthermore, with the exception of a few studies (e.g., Hoelzer 1989; Price et al. 1993; Wolf et al. 1997), direct benefits have attracted relatively little attention by theoreticians. The fact that direct benefits are often seen as uncontroversial or even trivial for explaining mate choice (as pointed out by Møller & Jennions 2001; Kokko et al. 2006a) might have caused a low general interest in whether such mate choice in turn can influence the evolution and extent of parental care.

To better understand how mate choice of caring partners may evolve, we also need to better explore the *indirect* benefits that may accrue from such choosiness (chapter 24). In the green-veined white butterfly, *Pieris napi*, males make a paternal investment in the offspring through nutrients, which are provided to the female at mating. The larger the male provision, the greater number of eggs produced by the female (Wedell & Karlsson 2003). Because the male ability to provide nutrients is heritable, females that mate with high-providing males benefit both directly, through a higher fecundity, and indirectly, through the production of high-providing sons (Wedell & Karlsson 2003). Presumably, this could be the case in many other organisms and for many other kinds of parental investment.

Differential allocation is also an area that needs additional theoretical and empirical work. Differential allocation is hypothesized to evolve because it increases an individual's chances of obtaining a favorable mate or retaining that mate in a subsequent breeding attempt (Burley 1986; Sheldon 2000). The steps required for this proposition have rarely been modeled or tested. For example, how is the intention of future effort signaled to the favorable mate? What circumstances enforce reliability of the signals of higher allocation? A major assumption of the differential allocation hypothesis is that increases in allocation incur costs that must be paid later (Sheldon 2000). An individual that exhibits a large allocation thus might be expected to have less to allocate in future attempts. Differential allocation may then become similar to the handicap principle (chapters 16 and 24) except that the reliability of the signal must extend across time to influence behavior in two breeding attempts.

To conclude, many aspects of parental care (for example, nest defense and provisioning in birds) are surprisingly variable within populations, and even within individuals across attempts. Traditional ideas about parental care as a life history trait sometimes end up explaining very little of that variation. By including possible effects of sexual selection on parental care behaviors, we might account for more of this variation.

Acknowledgments I am very grateful for comments given by Colette St. Mary on an early draft of this text, by Michael Jennions and Sigal Balshine on the first full draft, and by Sami Merilaita, Niclas Kolm, Ola Svensson, and Ingrid Ahnesjö on the final draft. I have also benefited from discussions with Inês Braga Gonçalves about compensatory strategies, which provided important insights. Kai

Lindström kindly provided values for figure 26.2b, and the Swedish Research Council funded my time.

SUGGESTIONS FOR FURTHER READING

Four main works have influenced the foundation of parental care theory. Hamilton's (1964) paper on kin selection contains many of the fundamentals to understanding parental behavior. Williams' (1966) book provides a good understanding of life history theory, which is highly relevant to understanding parental behavior. Trivers established the links between parental investment and sexual selection (1972) and explored the impact of conflicts of interest on parental behavior (1974). More recently, Clutton-Brock's (1991) book is still the most thorough review of theory and empirical work on parental care. Finally, interested readers should read Kokko and Jennions (2008), who revisit the links between parental investment and sexual selection and provide a valuable new perspective.

Clutton-Brock TH (1991) The Evolution of Parental Care. Princeton Univ Press, Princeton, NJ.

Hamilton WD (1964) The genetic evolution of social behavior I and II. J Theor Biol 7: 1–52.

Kokko H & Jennions MD (2008) Parental investment, sexual selection and sex ratios. J Evol Biol 21: 919–948.

Trivers RL (1972) Parental investment and sexual selection. Pp. 136–179 in BG Campbell (ed) Sexual Selection and the Descent of Man, 1871–1971. Aldine, Chicago, IL.

Trivers RL (1974) Parent-offspring conflict. Am Zool 14: 249–264.

Williams GC (1966) Adaptation and Natural Selection. Princeton Univ Press, Princeton, NJ.

SECTION VI

EXTENSIONS

27

Behavioral Ecology and Speciation

HOWARD D. RUNDLE AND JANETTE W. BOUGHMAN

Much of the diversity of life exists among species. Understanding the mechanisms by which new species arise is therefore a fundamental goal in evolutionary biology. Central to this endeavor is a species definition. Here, as with the vast majority of speciation research, we utilize the biological species concept. According to the biological species concept, "species are groups of interbreeding natural populations, which are reproductively isolated from other such groups" (Mayr 1942, p. 120). Reproductive isolation is the result of isolating barriers, defined as those biological features of organisms that reduce or prevent gene exchange with members of other populations (Coyne & Orr 2004). Such barriers are usually based on genetic differences between populations, although not always. Increased attention of late, for example, has focused on whether learning may play a role (Irwin & Price 1999). By focusing speciation research on understanding how such barriers arise, the biological species concept has provided a highly successful framework for the investigation of Darwin's "mystery of mysteries," the origin of species.

Numerous types of reproductive barriers exist, and these are often divided into two main categories, distinguished by whether they occur before or after mating (pre- and postmating, respectively). The resulting reproductive isolation they generate can be thought of as a quantitative trait, capable of taking on intermediate values, that accumulates between populations during the speciation process. This can occur when populations are geographically separated (allopatric), when they occur together (sympatric), or may involve intermediate scenarios of partial overlap (e.g., parapatry) or secondary contact following a period of allopatry. The geographic context of speciation is of interest not only because it affects the possibility of gene flow (which tends to abolish reproductive barriers), but also because it affects the mechanisms that can contribute to the evolution of reproductive isolation (Rundle & Nosil 2005). For example, premating isolation (including behaviors such as mate preferences; see below) may be strengthened by selection in sympatry in response to reduced hybrid fitness (i.e., postmating isolation) in a process known as reinforcement (Servedio & Noor 2003).

The various types of reproductive barriers have been recently reviewed (Coyne & Orr 2004); our interest here is with those that are behavioral in nature. At least five can be distinguished, all involving differences in preferences of some sort. One of the best studied, *sexual isolation*, is caused when divergent mate preferences cause individuals to be less attracted to, and hence less likely to mate with, individuals from another population. This is termed *behavioral isolation* by Coyne and Orr (2004), but we avoid this usage here because all forms of reproductive isolation with which we are concerned involve behavior. Chapter 24 provides a detailed treatment of mate choice, including how mate preferences are defined and methods for their quantification. Demonstrating sexual isolation is straightforward, and numerous examples exist, although in many of these the specific trait(s) on which the preferences act have not been identified (Coyne & Orr 2004). A classic case comes from Darwin's finches (*Geospiza*) in the Galapagos, in which field mate choice experiments using model birds have shown that males preferentially court conspecific over heterospecific females (Ratcliffe & Grant 1983). Sexual isolation depends on body size, beak shape, and most importantly on male song (Grant & Grant 1997a). Males learn their song from their fathers, and females prefer males who sing songs similar to their father, suggesting that they imprint on their father's song. Remarkably, preference for song appears to override species discrimination in cases in which males sing heterospecific songs (i.e., because they imprinted incorrectly), indicating that hybridization depends, at least in part, on learning in both sexes (figure 27.1).

Mate preferences can also contribute to a second type of behavioral isolation if they reduce the mating success of hybrid offspring. Termed *sexual selection against hybrids* (Schluter 2000), this reproductive

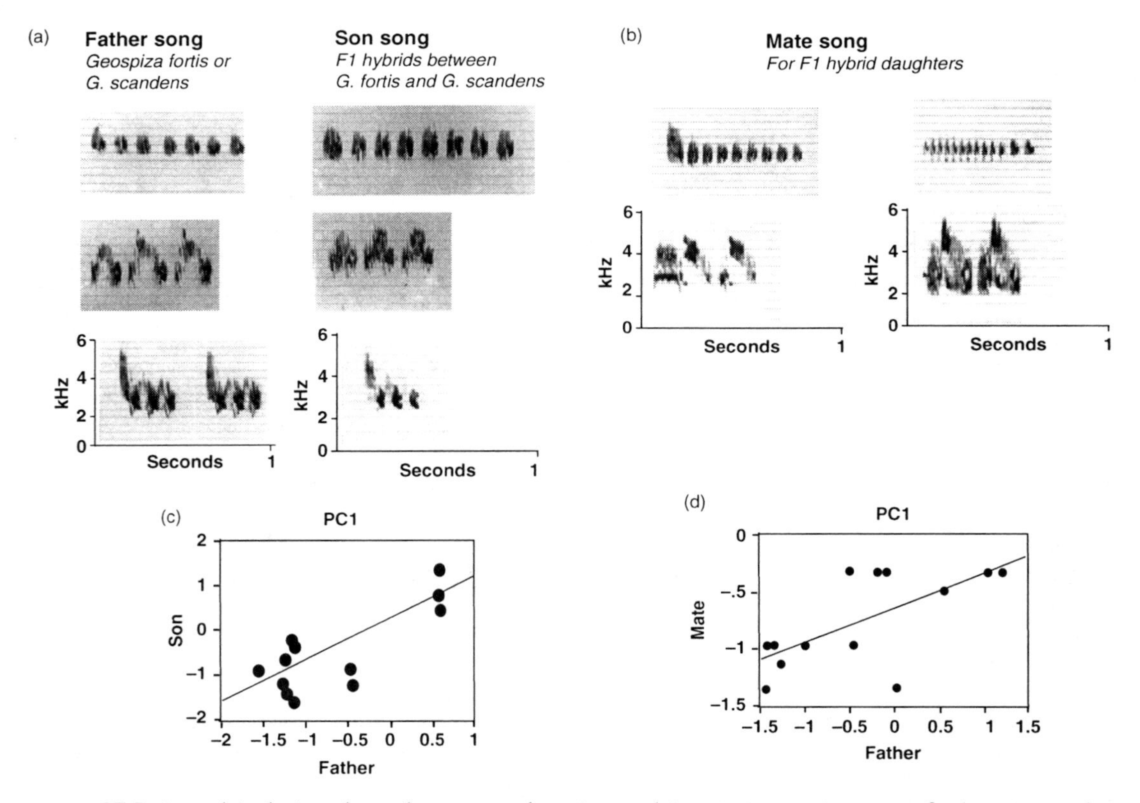

FIGURE 27.1 Sexual isolation depends on song learning and imprinting in Darwin's finches. (a) Males learn the song of their fathers. Sonograms show the song of male *Geospiza scandens* (top) and *G. fortis* (bottom) next to their F1 hybrid sons. The middle male is *G. scandens* but sings a *G. fortis* song, indicating this male is misimprinted and that song is not genetically based. (b) F1 hybrid females mate with males whose song resembles their fathers. Sonograms show songs sung by mates of two daughters of the fathers shown in the corresponding rows in panel (a). (c) Regression of song principal component 1 for sons and their fathers (slope = 1.008, $P = 0.0004$) indicating the close correspondence of song. (d) Regression of song principal component 1 for the mates of these daughters and their fathers, again indicating their close correspondence (slope = 0.29, $P = 0.014$). Panels (a–c) reprinted from Grant and Grant (1997b) with permission, and panel (d) reprinted from Grant and Grant (1997a) with permission.

barrier arises if the sexual displays of hybrids are intermediate and fail to stimulate the preferences of either parent species, or if hybrids themselves prefer rare or incompatible phenotypes (such that hybrids have reduced mating success). Reduced mating success of hybrids has been shown in a number of cases (Schluter 2000). For example, in stickleback fish inhabiting postglacial lakes of British Columbia, Canada, hybrid males are less successful at acquiring mates relative to the parent species (limnetics) alongside which they nest (Vamosi & Schluter 1999). In two species of tree frogs inhabiting the southeastern United States, first-generation hybrid males produce a sexual advertisement call that is intermediate, and hence unattractive, to females of either parent species (Hobel & Gerhardt 2003).

A third behavioral isolating barrier involves preferences not for mates, but for habitats (i.e., habitat choice). Divergent habitat preferences may create a form of premating isolation, termed *habitat isolation*, if the likelihood of between-population matings is reduced, generally because individuals mate in their preferred habitat (box 27.1). Habitat choice can produce microhabitat isolation between geographically sympatric populations and has therefore received attention as a possible route to sympatric speciation. The majority of this work has focused on phytophagous insects that feed and mate on their host plants, although rigorous demonstrations are few. Sympatric host races of the apple maggot fly (*Rhagoletis pomonella*), which utilize either apples or hawthorns, are partially isolated by differences in host preferences (Feder et al. 1994). Habitat isolation is also implicated between this species and a close relative, *R. mendax*, that is found exclusively on blueberry plants. Although hybrids are absent between these species in nature, they are readily produced when individuals are confined to the same field cage (Feder & Bush 1989).

The fourth potential form of behavioral isolation is the temporal equivalent of habitat isolation, involving preferences to mate at different times.

BOX **27.1** Habitat Preferences and the Formation of New Species

Patrik Nosil

Divergent habitat preferences can result in reproductive isolation (i.e., assortative mating) between populations in different habitats. Such habitat preference can play a unique role in speciation by overcoming some of the major theoretical criticisms of the controversial process of sympatric speciation (i.e., speciation without geographical barriers to gene flow between populations; Bush 1969; Coyne & Orr 2004). Specifically, mate preferences might often diverge between populations only because genes that affect mate preference are associated with genes conferring adaptation to different environments (i.e., divergent selection on fitness genes spills over to mate preference genes). One of the main criticisms of sympatric speciation is that gene flow between populations and subsequent recombination will destroy genetic associations between the two types of genes (preference and fitness genes), thereby preventing the evolution of preference genes. In the case of habitat preference, these two types of genes can be one and the same (i.e., habitat choice affects fitness such that habitat preference genes are also fitness genes) so that selection automatically affects habitat preference genes. When this is the case, sympatric speciation is strongly facilitated. This point aside, even when genes affecting habitat preference are different from genes affecting fitness, divergent habitat preferences can play a role in speciation similar to mating preferences, thereby contributing to speciation in organisms that do not diverge in mating preferences.

Here I describe two main ways by which divergent habitat preferences might promote speciation: (1) directly, by acting as a form of reproductive isolation, and (2) indirectly, by reducing gene flow, thereby facilitating the adaptive divergence that can incidentally drive the evolution of other forms of reproductive isolation during ecological speciation (figure I). I then discuss two distinct mechanisms by which divergent habitat preferences

(continued)

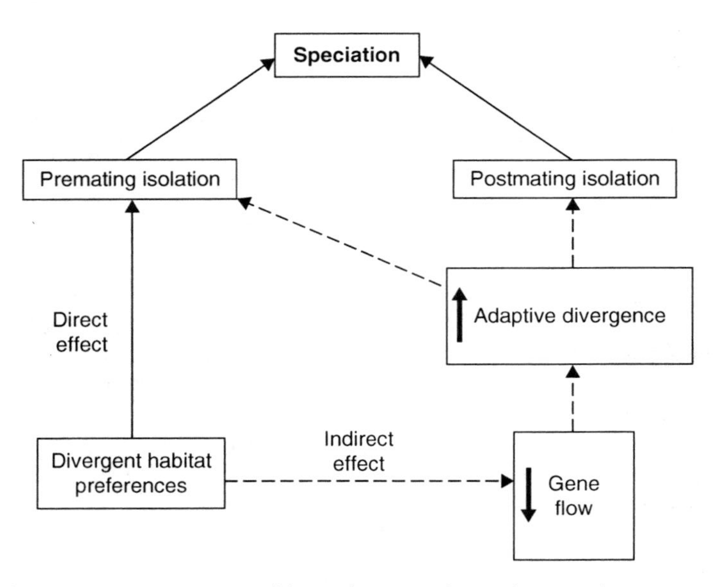

Figure I Schematic representation of how divergent host-plant preferences can promote speciation either (1) directly (solid lines), by causing premating isolation, or (2) indirectly (dashed lines), by reducing gene flow, thereby facilitating the adaptive divergence that drives the evolution of other forms of reproductive isolation during ecological speciation. This indirect role applies to any form of reproductive isolation.

themselves might evolve, and conclude with some outstanding questions concerning the role of habitat preferences in speciation. I focus on the host-plant preferences of herbivorous insects (see Funk et al. 2002 and Tilmon 2007 for thorough reviews), but the arguments apply to other forms of habitat preference.

How Do Divergent Habitat Preferences Contribute to Speciation?

When mating occurs on the host plant, divergent host-plant preferences act as a form of premating isolation between insect populations living on different plant species (Bush 1969). Specifically, premating isolation occurs because divergent host preferences reduce between-host movement, and thus reduce contact and interbreeding between individuals from different hosts (figure I). Divergent host preferences have been documented in a wide range of insect taxa, including Lepidoptera (moths and butterflies), Coleoptera (beetles), Diptera (flies) and Phasmids (walking-stick insects; Funk et al. 2002; Tilmon 2007 for reviews). An example of divergent host preferences in *Timema* walking-stick insects is depicted in figure II (Nosil 2007). Divergent host preferences can contribute to speciation in two ways, directly by causing assortative mating, and indirectly by reducing gene flow and thus facilitating overall adaptive divergence.

Directly, by Acting as a Form of Reproductive Isolation

There are only a few cases in which host preferences have been shown to actually result in assortative mating (i.e., truly result in reproductive isolation). Some examples of such studies include cage experiments showing increased assortative mating between host races of *Eurosta solidaginis* goldenrod flies when host plants are present relative to when they are absent (Craig et al. 1993), mark-recapture studies of apple and hawthorn host races of *Rhagoletis* flies, suggesting that the tendency of flies to reproduce on the same host species that they used in earlier life history stages strongly reduces gene flow between the races (Feder et al. 1994), and a combination of field and molecular data indicating that host choice reduces gene flow between alfalfa and clover races of *Acyrthosiphon pisum* pea aphids (Via 1999). Thus, in at least some cases, the degree of assortative mating caused by divergent host preferences can be adequate for subpopulations using different hosts to evolve genetic differences. However, further demonstrations that divergent host preferences cause substantial reproductive isolation are needed.

Indirectly, by Reducing the Constraining Effects of Gene Flow

Adaptive divergence is often constrained by gene flow (Nosil 2007). Thus, when divergent host preferences act as a barrier to gene flow, as described above, they can facilitate adaptive divergence: in other words, they facilitate adaptation to different habitats by different subsets of the population. In turn, adaptive divergence can cause the incidental evolution of any form of reproductive isolation (e.g., sexual isolation, hybrid sterility), via by-product models of ecological speciation: in other words, reproductive isolation evolves as a secondary (indirect) consequence of local adaptation, rather than as a direct result of selection (Funk et al. 2002; Coyne & Orr 2004). Thus, by reducing gene flow, divergent host preferences can indirectly promote speciation. However, this indirect role for host preference in promoting speciation is likely to be of importance only for cases of divergence in the face of gene flow (i.e., nonallopatric speciation). Moreover, this type of indirect role of behavior in speciation is not unique to the particular reproductive barrier of host preference, because (by definition) any form of reproductive isolation reduces gene flow, thereby potentially facilitating adaptive divergence (figure I).

Two Different Causes of Habitat Preference Evolution

For habitat preferences to lead to eventual speciation, habitat preferences must evolve to differ between populations. How can selection favor the evolution of behavioral differences within or between populations that are initially monomorphic for a single strategy (e.g., a generalist behavior, or preference for a single host plant)? Host preferences can diverge both with and without selection against switching between different, utilized hosts (the term *utilized* refers to host species that an insect species uses; other host species that the insect species cannot or does not use may exist in the environment as well). I review two hypotheses for the causes of preference evolution, but note that whenever preference evolution is driven by divergent selection, pairs of populations feeding on different host plant species will exhibit greater divergence in preference than pairs of populations feeding on the same host (Funk et al. 2002; Nosil 2007; figure II). This pattern of greater preference divergence between different-host pairs has been observed, for example, in *Timema* walking-stick insects and *Neochlamisus* leaf beetles.

Selection against Host-Switching (Fitness Trade-Offs Hypothesis)

When switching between utilized hosts is maladaptive (i.e., when local adaptation results in fitness trade-offs between hosts), host preferences can diverge via selection against individuals that switch between hosts. Under this scenario, preference for the native host is favored because individuals choosing another host suffer reduced fitness. Although selection actively favors reduced host switching, it may act indirectly on host preference loci via their genetic association with other loci conferring host-specific fitness (e.g., loci affecting color pattern or physiology). In this hypothesis, selection acts only in populations in which there is the opportunity for switching between utilized hosts (i.e., when more than one utilized host is available in the environment). One possible outcome is greater preference

(continued)

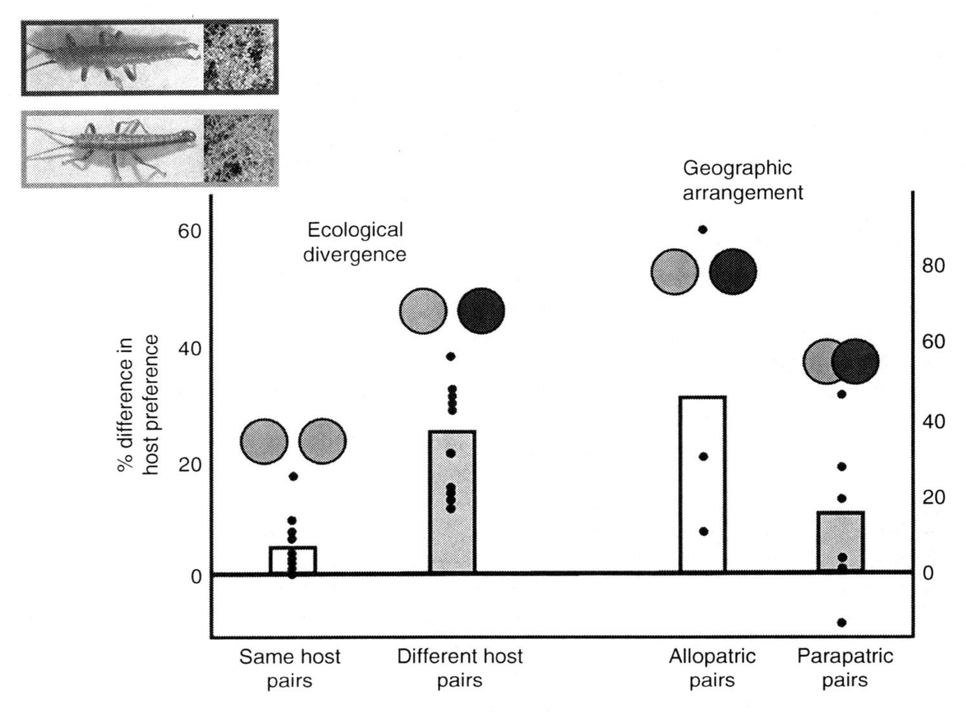

Figure II Divergent host-plant preferences between *Ceanothus* (darker circles) and *Adenostoma* (lighter circles) host ecotypes of *Timema cristinae* walking-stick insects. The y-axis refers to the mean percent difference between population pairs in individuals choosing *Ceanothus* in host choice trials (each small black dot represents the mean difference of a population pair, and the bars represent the average of the population pairs). *Ecological divergence* shows how different-host population pairs are more divergent in host preference than same-host population pairs (here, all pairs are allopatric), indicating that host preferences have differentiated due to divergent natural selection. *Geographic arrangement* shows how allopatric population pairs differ more in host preference than parapatric pairs, suggesting that gene flow constrains divergence. Thus, habitat preferences could contribute to speciation both by acting as a form of reproductive isolation and by reducing gene flow to indirectly facilitate adaptive divergence and the incidental evolution of other forms of reproductive isolation. The strong divergence in host preference observed in allopatry could arise via the *information processing* hypothesis discussed in the text. Modified from Nosil (2007) with permission.

divergence in geographic regions where multiple hosts are utilized (in sympatry or parapatry) than between geographically isolated populations that use a single, yet different, host (allopatry). This pattern can be thought of as character displacement of host preference, with preferences evolving in a reinforcement-like process.

Selection for Efficiency (Information Processing/Cognitive Constraints Hypotheses)

There is no selection against switching between hosts when only one host is utilized in the local environment. Under this scenario, search and efficiency costs can favor increased preference for the single, utilized host because individuals without strong preferences accrue lower fitness, but for reasons other than switching to an alternate host (Bernays & Wcislo 1994). For example, due to cognitive constraints associated with information processing, generalized individuals without strong preferences might take longer to locate or to decide whether to feed on the utilized host (e.g., as observed in *Neochlamisus* leaf beetles; Egan & Funk 2006), thereby wasting time and energy while increasing predation risk. Alternatively, such individuals may suffer low fitness because they attempt to use a nonutilizable host. When preference evolution is driven by such selection, populations in habitats in which only a single host is utilized still evolve preference for that host. Thus, unlike the host-switching scenario above, the cognitive constraints hypothesis can drive preference evolution within allopatric populations, and may contribute to allopatric speciation (and it does not predict character displacement). In fact, under this hypothesis, if gene flow constrains divergence, then preferences may actually be more divergent in allopatry than in parapatry/sympatry (figure II).

Summary and Outstanding Questions

Although divergent habitat preferences appear common, direct evidence that they actually cause premating isolation in nature is lacking (Funk et al. 2002). Likewise, further data on the causes of preference evolution would be useful, particularly given the implications for the geographic mode of speciation in which they are involved. In particular, the prevalence and ecological causes of fitness trade-offs between hosts remains in debate. Data on the genetic basis of habitat preferences is also lacking, and has implications for both the likelihood and rate of speciation (Coyne & Orr 2004). Finally, a major outstanding question from a behavioral perspective is whether habitat choice involves preference for the native host, avoidance of alternative hosts (e.g., Forbes et al. 2005), or both.

A number of cases of such temporal (or *allochronic*) isolation exist in the literature, including different species of corals or green algae that release their gametes in brief mass-spawning events at different times of the day (Coyne & Orr 2004). However, we are not aware of any case in which temporal isolation has been shown to involve divergent preferences for when to mate, as opposed to innate responses to external cues. An intriguing possibility is provided by an evolution experiment in melon flies (*Bactrocera cucurbitae*) in which replicate populations were selected for either fast or slow development time (Miyatake & Shimizu 1999). Time of mating diverged in response to this selection, with fast developers mating earlier in the evening than slow developers, and this generated premating isolation in mate choice trials. The degree to which divergent preferences were involved is unclear, however, and the authors suggest that it may have been a side effect (i.e., pleiotropy) of changes in genes affecting the circadian clock.

Finally, although behavioral isolation is generally restricted to animals, one form can also occur in flowering plants (angiosperms) when gene flow is reduced between species because of differential visitation by animal (usually insect) pollinators. Termed *ethological pollinator isolation*, differences in preference occur in the pollinators, usually with reference to flower characteristics such as color, odor, shape, position, and nectar content (Grant 1994). Although it is relatively straightforward to demonstrate, there are only a few rigorous examples (Coyne & Orr 2004). One of the most striking occurs in two species of North American monkeyflower. The bee-pollinated *Mimulus lewisii* has broad, pink flowers with recessed anthers and produces little nectar, whereas the hummingbird-pollinated *M. cardinalis* has red, tubular flowers with exerted anthers and high nectar volume. Despite the fact that these species can be easily crossed in the greenhouse, hybrids are extremely rare when they are sympatric in nature. Direct observation indicates that pollinator isolation between these sister species is almost complete, and genetic manipulations implicate differences in flower color and nectar volume as key traits the pollinators are targeting (Schemske & Bradshaw 1999; Ramsey et al. 2003).

FROM PATTERN TO PROCESS: THE EVOLUTION OF SEXUAL ISOLATION

A comprehensive understanding of the role of behavior in speciation goes beyond simply tallying the prevalence of the various forms of reproductive isolation involving behavior; we'd also like to know how preferences diverged to produce these barriers. Much less work has been done in this regard. In the remainder of this chapter we address this topic by focusing specifically on the evolution of sexual isolation. We chose this focus for two reasons. First, sexual isolation arises from differences in mate preferences; understanding mating behaviors and their evolution has long been the province of sexual selection, a subject that has been much studied by behavioral ecologists.

The second reason for our focus on sexual isolation is that it is often recognized as the primary source of reproductive isolation between existing species in nature (Mayr 1963). Although this may be due in part to the fact that it is easier to detect than many other barriers, and it will also have a larger effect on total isolation relative to those occurring later (e.g., postmating forms), a number of lines of evidence suggest that sexual isolation may also be key in initiating speciation (Coyne & Orr 2004). These include data from *Drosophila* and cichlids that sexual isolation is required for species to coexist in nature (Coyne & Orr 1997; Seehausen et al. 1997). In addition, fine-scale local mate preferences within species, in which individuals prefer mates from their own population over those from other populations (Ryan & Wilczynski 1988), also suggest that preference divergence may occur early during population differentiation. Finally, there is a long history of laboratory studies of speciation, largely in *Drosophila*, that have demonstrated that partial sexual isolation can evolve over relatively short timescales (Rice & Hostert 1993), although direct comparisons with the evolution of other forms of reproductive isolation have received little attention.

In the following section, we provide a conceptual overview of the various mechanisms by which mate preferences may diverge to generate sexual isolation, including the relevant models of sexual selection. We then present two case studies that have used rather different approaches to the empirical investigation of some of these mechanisms. Finally, because this is an active area of research and there are many directions for future work, we close by outlining a selection of key outstanding questions.

CONCEPTS

Sexual Selection and Mate Preference Divergence

The evolution of mate preferences is not a simple phenomenon, in part because it involves an interaction between different traits in the two sexes: generally one or more sexual display traits in one sex and a preference for these trait(s) in the other sex. Here we address the classic scenario of female choice for male display traits; although the reverse scenario of male choice also exists in nature, it has received less attention and its evolution may differ in important ways from that of female mate choice (see chapter 20). As pointed out by Fisher (1930), the existence of mate preferences in females will generate sexual selection on the preferred male display trait, and the resulting assortative mating (females with the preference mate with males with the preferred display) generates a statistical association (i.e., a positive genetic correlation) between the traits. This causes indirect selection for the preference in a self-reinforcing coevolutionary process that has come to be known as *Fisher's runaway* (see chapter 24).

This coevolutionary process can rapidly amplify small initial differences among populations in display traits and/or preferences, generating sexual isolation (Lande 1981; West-Eberhard 1983; Schluter 2000). How these differences first arise has been the topic of much theory, and models of speciation by sexual selection can be divided into two main classes depending on the mechanism by which divergence is initiated (Schluter 2000; Coyne & Orr 2004). In the first, although selection is involved, the ultimate reason that preferences diverge is chance events, including arbitrary differences in starting conditions, unique mutations, and/or the order in which they occur. It is these chance events that cause selection to follow unique evolutionary trajectories in different populations. In the second, selection differs between environments and it is this divergent selection that is the ultimate cause of divergence in mating traits.

Fisher's runaway is the original example of a chance-based mechanism. In the classic case in which preferences are neutral (i.e., females with and without the preference do not differ in survival or fecundity), the coevolutionary exaggeration of display trait and preference is eventually halted at one of many possible positions along a *line of equilibria* on which natural selection opposing the male trait precisely balances sexual selection favoring it (Kirkpatrick & Ryan 1991; see also figure 24.3). Sexual isolation can arise because chance events cause separate populations to arrive at alternate positions along this line (Lande 1981). A runaway process cannot, however, maintain a costly female preference at equilibrium. Such costs are generally considered to be inevitable (although insufficient attention has been given to quantifying them), and their presence generates direct selection on the preference that, in most cases, causes the line of equilibria to collapse to a single point at which female fitness is maximized and the preference is absent (Arnqvist & Rowe 2005). It is therefore unlikely that this process alone is responsible for mate preference divergence among populations.

Interlocus sexual conflict is another example of a chance-based mechanism by which preferences may diverge. As explained in detail in chapter 23, sexual conflict arises from differences in the evolutionary interests of males and females with respect to reproduction (Arnqvist & Rowe 2005). Interlocus sexual conflict occurs when different loci in males and females affect the value of a shared trait arising from male-female interactions. Because the trait can take on only a single value, males and females cannot simultaneously achieve their sex-specific optimum. The result is an evolutionary tug-of-war that can generate cycles of adaptation and counteradaptation in a process of sexually antagonistic coevolution. This process occurs independently in separate populations, following potentially unique evolutionary trajectories due to chance events, thereby promoting population divergence and hence reproductive isolation.

With respect to behavior, conflict can arise over such shared traits as mating rate, with selection generally favoring higher rates in males than in females. The resulting sexually antagonistic adaptations can involve behavioral traits in both sexes, including those in males that stimulate, harass, or otherwise cause females to mate at a greater rate (e.g., male persistence or exploitation of preexisting sensory biases in females), and those in females that increase their resistance to these male traits (e.g., changes in female perception). (The evolution of *preferences for* and *resistance to* male sexual traits are equivalent.) Whether sexual isolation is likely to result is not straightforward, however, and contrasting views exist in the literature (Ritchie 2007). If females evolve to resist the specific charms of conspecific males, this could conceivably cause them to prefer (i.e., be less resistant to) heterospecific males, inhibiting speciation (Coyne & Orr 2004). Likewise, males that are more persistent during courtship could increase the likelihood of heterospecific matings. On the other hand, it has been suggested that the evolutionary options available to females to decrease their response to manipulative male traits are almost infinite, including diverse routes involving changes in perception. Thus, sexually antagonistic coevolution may be particularly likely to favor a novel female perceptive trait, promoting sexual isolation (Arnqvist & Rowe 2005). Successive rounds of sexually antagonistic coevolution could also leave males in possession of a suite of ineffective traits, to which conspecific females have evolved partial resistance. If the presence of such traits is necessary to achieve matings, heterospecific males may fail to stimulate females sufficiently (Holland & Rice 1998).

Determining the contribution of interlocus sexual conflict to the evolution of sexual isolation will require empirical data, but unfortunately the few data that exist are equivocal. In the laboratory, Martin and Hosken (2003) used experimental evolution in dung flies (*Sepsis cynipsea*) to show that partial sexual isolation evolved between replicate populations in which sexual selection was permitted, but not between populations in which sexual selection was prevented (by enforcing lifelong monogamy). This is consistent with sexual conflict because in the sexual selection present treatment, behavioral isolation was stronger between larger, more dense populations in which sexual conflict was greater (as confirmed by direct measures of sexual activity), than it was between smaller, less dense populations with relaxed sexual conflict. However, it is difficult to rule out, or even determine, the potential contribution of other processes of sexual selection. In addition, similar experiments in two species of *Drosophila* have provided contrasting results, with no sexual isolation evolving between populations experiencing greater sexual conflict (Wigby & Chapman 2006; Bacigalupe et al. 2007).

In the second class of models, the divergence of mate preferences is initiated ultimately as a by-product of divergent selection adapting populations to their different niches or environments (Schluter 2000). This may occur via spatial variation in natural selection on the display trait, with preferences evolving as a correlated response to the resulting changes in the display trait (Lande 1982). Alternatively, natural selection may act on sensory or communication systems, favoring signals, sensory systems, and behaviors that function well within their specific environment, even outside of the mating context (e.g., to facilitate prey capture or predator avoidance). Adaptation of communication systems to different environments may thereby cause mate preference to diverge in a process known as sensory drive (Endler 1992; Boughman 2002). Given the great variation in habitats that exists, and the fact that subtle differences can affect sensory processes and signal transmission, divergence through sensory drive may be widespread. Several recent examples have been found of either mating trait, sensory, or preference divergence correlated with habitat. This has been well studied in *Anolis* lizards, in which the color of male dewlaps evolves to be conspicuous in different light environments. For example, background light is rich in ultraviolet in the habitat of *A. cooki*, but not *A. cristatellus*. The dewlaps of each species contrast highly against their different backgrounds, and the species differ in how well they see ultraviolet light (Leal & Fleishman 2002). Similar patterns have even been found among populations of *A. cristatellus* from different habitats (Leal & Fleishman 2004).

In all of the above models, sexual selection can amplify initial differences between populations (Kirkpatrick & Ryan 1991), causing the rapid coevolutionary diversification of display traits and preferences and thereby generating sexual isolation. Consistent with this, comparative evidence suggests that sexual selection has been central to speciation in nature (Coyne & Orr 2004). For example, closely related species, especially members of some of the most famous adaptive radiations such as the Hawaiian *Drosophila*, often exhibit spectacular diversity in traits that appear to be under sexual selection (e.g., male sexual displays) and are often characterized by striking patterns of sexual dimorphism. More compellingly, formal sister group comparisons have shown a correlation between species richness and various surrogate measures of the strength of sexual selection in a number of taxonomic groups (Coyne & Orr 2004), although not in others (Ritchie 2007). Nevertheless, although the above studies implicate sexual selection in speciation, they are unable to provide direct tests of the evolutionary mechanisms involved in preference divergence. Experimental tests of these mechanisms, however, have received limited attention. In the following section we provide an overview of the work from two systems that is beginning to address these issues.

CASE STUDIES

Species Pairs in the Threespine Stickleback

A number of ecologically and genetically differentiated, sympatric, and reproductively isolated pairs of threespine stickleback species (*Gasterosteus* species complex) occur in different regions of the globe (McKinnon & Rundle 2002; Boughman 2007a). The recent, post-Pleistocene origin of most of these species pairs allows insight into the mechanisms operating relatively early in the speciation process, minimizing the complication of trait differences that may accumulate subsequent to speciation. Moreover, the same pairs have evolved repeatedly and independently, providing evolutionary replication. These factors make them a powerful model for understanding the evolutionary mechanisms that cause new species to form. We focus here on the best studied limnetic and benthic species pairs that coexist in several lakes in coastal British Columbia, Canada. Discovered by McPhail and colleagues (McPhail 1994), the species are specialized to exploit different ecological niches within a lake (Schluter & McPhail 1992): limnetics feed primarily on zooplankton in the open water, whereas benthics feed on invertebrates in the benthic zone. They also mate in different habitats: limnetics in shallow, open areas at high density, and benthics in deeper, vegetated areas at low density, although nesting habitats overlap substantially. Although reproductive isolation arises from a number of barriers (McPhail 1994; Boughman 2007a), the fact that hybrids are viable and fertile yet extremely rare in nature implicates sexual isolation as a primary reproductive barrier.

Sexual isolation arises because limnetics and benthics strongly discriminate against heterospecific mates based on a combination of body size, male nuptial color, and odor; surprisingly, although courtship behaviors differ, they do not appear to have a strong role in causing sexual isolation (Boughman et al. 2005). Limnetics recognize conspecifics using differences in body size and color, whereas benthics use body size and odor (Boughman et al. 2005; Boughman 2007a; Rafferty & Boughman 2006). Body shape may also play a role, although this possibility requires additional work. Because there is mutual mate choice in threespine sticklebacks (Kozak et al. 2009), mate preferences in both sexes may contribute to sexual isolation. We focus largely on female preferences, however, because the majority of work has been done on them.

How Preferences Diverge: Divergent Selection between Environments

The majority of evidence implicates divergent selection between environments as the ultimate cause of preference divergence in sticklebacks (Boughman et al. 2005). Chance-based mechanisms have received less attention, largely because the biology of sticklebacks suggests that these are less likely. Sexual conflict over mating rate appears doubtful, for example, because males cannot force females to mate (females mate only when they have a clutch of eggs to spawn). Fertilization is also external, reducing the likelihood of postcopulatory manipulation via seminal fluids. However, competition for mates does occur and conflict over other aspects of mating is plausible. The role of sexual conflict in preference divergence in sticklebacks therefore warrants further study.

In contrast to the unlikely role of chance, a number of lines of evidence strongly implicate divergent selection between environments in the evolution of sexual isolation between these species (Schluter & McPhail 1992; Rundle et al. 2000). The first involves the parallel evolution of mating traits (i.e., mate preferences and the traits on which they act) among independent populations such that those adapted to different environments are sexually isolated, whereas those that are independently adapted to similar environments are reproductively compatible (Schluter 2000). Referred to as parallel speciation, such parallel evolution of mating traits strongly implicates divergent selection because chance-based mechanisms are unlikely to cause repeated trait evolution in correlation with environment. Using over 750 mating trials involving three independently evolved species pairs, Rundle et al. (2000) demonstrated that limnetics and benthics were reproductively isolated from one another no matter what combination of lakes were involved. In contrast, limnetics from the separate lakes all mated freely with one another, as did benthics, despite having independently evolved from a common ancestor. In this case, parallel speciation involved the parallel evolution of both female mate preferences and male mating traits including body size and color (Boughman et al. 2005).

The importance of selection in the divergence of mate preferences is also highlighted by two studies demonstrating reproductive character displacement of stickleback mate preferences. Reproductive character displacement is the pattern in which premating isolation between two species is stronger in areas of sympatry as compared to allopatry (Servedio & Noor 2003). This pattern has been demonstrated for both benthic females (Rundle & Schluter 1998) and limnetic males (Albert & Schluter 2004). In each case, true benthics and limnetics (i.e., the species pairs occurring sympatrically within a lake) discriminate against members of the other species, whereas morphologically benthic- or limnetic-like individuals from single-species lakes (i.e., allopatric populations) do not. Such results are consistent with a reinforcement process in which selection has caused mate preferences to divergence in response to reduced hybrid fitness, although it is difficult to rule out other potential mechanisms from correlative data alone. In the sticklebacks, however, the case appears reasonably strong (Rundle & Schluter 1998; Albert & Schluter 2004; Kozak et al. 2009).

Divergent sexual selection has also been implicated in preference divergence between limnetics and benthics and appears to arise from several sources. Differences in mating habitat reduce encounter rates between the species, and can also affect how sexual selection acts within each species, including male-male competition and female choice, and may thus contribute to divergent sexual selection between them. Studies in other stickleback populations have shown that differences in ecology can alter the intensity and targets of sexual selection (Candolin et al. 2007). Moreover, the benefits that females of other species derive from their choice of mates may also depend on ecological conditions (Welch 2003). Use of different mating habitats may therefore not only generate habitat isolation directly, but it may also contribute to divergent selection on female preferences and male traits. To date, most of the work on divergent sexual selection has focused on mechanisms of evolution in color preference and expression, focusing particularly on the hypothesis of sensory drive. More work is needed, however, concerning mechanisms of divergence in size and odor, as well as preferences for these traits. Parasite-mediated selection is a strong candidate that could easily influence both odor and color preferences.

Tests of sensory drive implicate differences in the light environment of limnetic and benthic mating habitats as the primary factor in the divergence of color preferences. In the shallow, open areas where limnetics mate, the light spectrum is broad. This contrasts with the red-shifted light characteristic of the deeper, vegetated areas where benthics mate (Boughman 2001). Sensory drive posits that differences in light environment should affect the conspicuousness of male nuptial color and the evolution of color perception (Endler 1992; Boughman 2002). These in turn are predicted to cause male color and female preference to evolve in correlation with lighting conditions in the signaling environment. Consistent with these predictions, both perception and preference for red are correlated with light environment (figure 27.2a–b; Boughman 2001). Likely because differences in habitat choice expose fish to distinct light regimes, females differ in their sensitivity to red light, with limnetic females being more sensitive than benthic females. Limnetic females have correspondingly stronger preferences for red nuptial color during mating.

The precise differences in male nuptial color are congruent with the predictions of sensory drive. Male color matches the perceptual tuning of females and the light environment (Boughman 2001) and both species appear to have conspicuous nuptial colors, although they achieve this in different ways. The large, bright red throats of limnetic males stand out by being brighter than the background in broad spectrum light, but blend into the background in red-shifted light. The reduced red or black of benthic males stands out by being darker than the background in both light environments. Consistent with this, black nuptial color and red-shifted water are correlated geographically in solitary populations of sticklebacks (Reimchen 1989), suggesting sensory drive influences color evolution broadly.

Divergence in female color preference contributes to sexual isolation in the species pairs, both because it causes sexual isolation and because it generates divergent sexual selection on male color, which has led to its evolutionary diversification (Boughman 2001). Indeed, sexual isolation increases with the extent of between species differences in both male color and female color preference (figure 27.2c; Boughman 2001) and with the strength of preference for red (Boughman 2007b). Preference evolution therefore generates sexual isolation because females with strong red preferences are more likely to mate with males expressing high red color and to reject males expressing dull or black color.

The stickleback species pairs also give us some insight into the role of learning in mate preference

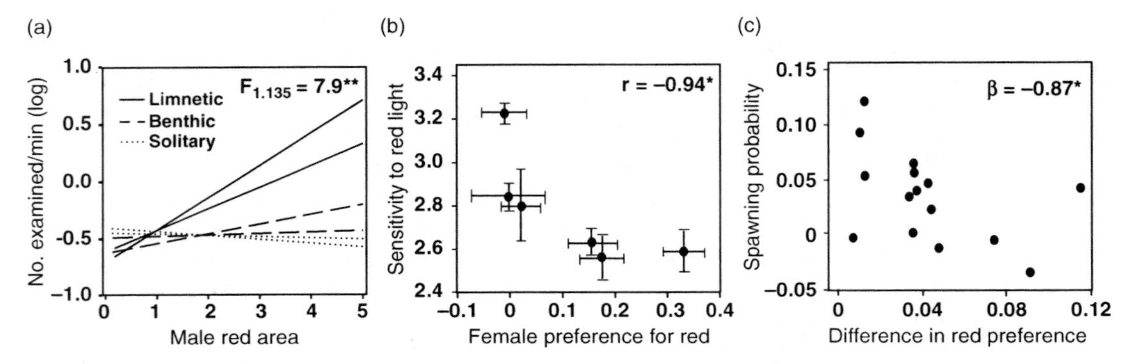

FIGURE 27.2 Preference divergence and sexual isolation in limnetic-benthic pairs of sticklebacks. (a) Color preference functions for six populations. Lines are from linear regression of preference on male color. (b) Correlation of spectral sensitivity to the strength of female preference for red in six populations. (c) Spawning probability between pairs of populations as a function of divergence in the strength of color preference. Panels (b) and (c) redrawn from Boughman (2001).

divergence. Preferences for both social partner and mates appear to depend on social environment, and the more social limnetic species shows a stronger effect. In an experiment manipulating social environment during rearing, adults of both species preferred the species they had been raised with as social partners, even if these were heterospecific individuals (Kozak & Boughman 2008). In addition, although mate preferences in benthics showed little effect of social environment, sexual isolation was undermined in limnetics raised with benthics, resulting in mate preferences for heterospecifics (Kozak & Boughman 2009). Finally, in a separate test of imprinting in first generation limnetic-benthic hybrids, males showed no effect (Albert 2005). However, the lack of experience with pure species may have contributed to the lack of preference by hybrid males. Imprinting in pure species juveniles has not yet been examined, nor have the effects of these manipulations on female species recognition and mate preference been evaluated. Both are needed for a comprehensive understanding of the role of learning in sexual isolation.

Sexual Isolation and Mate Preferences in *Drosophila serrata*

Drosophila serrata is a forest-dwelling generalist native to the eastern and northern coastal areas of Australia and extending into Papua New Guinea and surrounding northern islands (Jenkins & Hoffmann 2001). Mate preferences within *D. serrata*, which have been well characterized via a series of quantitative genetic, behavioral, and evolutionary experiments, are based, at least in part, on a suite of long-chain, nonvolatile contact pheromones composed of cuticular hydrocarbons (CHCs; figure 27.3; Blows & Allan 1998; Howard et al. 2003). Males and females express the same suite of CHCs, although they are sexually dimorphic in relative concentrations, and individuals of both sexes discriminate among potential mates using variation in the relative concentrations of these CHCs (Chenoweth & Blows 2003, 2005).

Mate preferences for CHCs are also important for sexual isolation between *D. serrata* and the closely related *D. birchii*. These species are almost identical morphologically (Schiffer & McEvey 2006), and interspecific hybrids between them are viable and fertile (Blows 1998). Nevertheless, hybridization appears to be extremely rare (Blows 1998; Higgie et al. 2000), suggesting strong sexual isolation arising from distinct mate preferences. Consistent with this, CHC profiles of the two species differ greatly and include chemically unique hydrocarbons in each (Blows & Allan 1998; Howard et al. 2003). Perfuming experiments in both sexes, in which the chemical cues of one species were physically transferred to individuals of the other, caused an increase in the frequency of hybridization from essentially zero to over 30%, strongly implicating divergent mate preferences in both sexes as a cause of sexual isolation (Blows & Allan 1998).

As a case study for the divergence of mate preferences, empirical studies with *D. serrata* have proceeded on two fronts. One uses this species as a laboratory model system to test the feasibility of the various models by which mate preferences may diverge. The other seeks to understand the specific mechanisms responsible for a naturally occurring pattern of CHC and mate preference divergence between populations that are sympatric versus allopatric with *D. birchii*.

How Preferences Diverge: Manipulative Tests of Divergent Selection between Environments

Laboratory evolution experiments have a long history in speciation research, and the majority of this work addresses the evolution of sexual isolation via various processes (Rice & Hostert 1993; Coyne & Orr 2004). These experiments generally track the evolution of assortative mating between pairs of populations and, when found, the implication is that display traits and mate preferences for them have diverged. Direct tests of the mechanisms responsible for mate preference divergence are limited, however.

Laboratory evolution with *D. serrata* has begun to address mechanisms, focusing on the second class of models by which preference divergence may be initiated: as a by-product of divergent selection between environment. Using experimental evolution, 12 replicate populations were derived from a common ancestor and allowed to independently evolve in one of three treatment environments, yielding four replicate populations in each (Rundle et al. 2005). These environments varied the food provided to the larvae, and consisted of rice, corn, or yeast, the latter being the laboratory environment to which the ancestor was already well adapted. CHCs evolved in the different treatment environments, with adaptation inferred from the parallel evolution of replicate populations within each treatment. Female mate preferences for male CHCs also diverged among populations, with a component of the divergence (at least 17%) occurring among treatments. Divergence among treatment environments indicates that preferences can evolve ultimately as a result of divergent natural selection.

The relative importance of the two classes of models (i.e., chance-based versus divergent selection) is unclear, however. Although a component of preference divergence occurred among environments, the majority was present among replicate populations within treatments, suggesting that it was not initiated by divergent selection between environments. Although this suggests a key role for the chance-based mechanisms, this within-treatment variance is confounded with measurement error because only a single preference estimate was obtained for each population; at least some of the differences among populations represent error in estimating preferences. The relative roles of these two classes of mechanisms will therefore require an experimental design that independently manipulates the opportunities for both processes within a single experiment, and that includes repeated preference measurements in each population.

How Preferences Diverge: Reinforcement in Natural Populations of D. serrata

In nature, both CHCs and mate preferences for them vary among geographic populations of *D. serrata* that span a large part of the species's range along the east coast of Australia. This variation includes a pattern of reproductive character displacement in which the CHCs of northern populations, which are sympatric with *D. birchii*, differ from those of the allopatric populations to the south (Higgie et al. 2000; Chenoweth & Blows 2008). This character displacement has been meticulously mapped in the contact zone on the Byfield Peninsula in central Queensland and is maintained over a fine spatial scale of a few kilometers (Higgie & Blows 2007). This is remarkable given the high levels of gene flow among populations estimated from microsatellite markers (Magiafoglou et al. 2002).

A similar pattern exists for female mate preferences for CHCs in males, with much of the among-population variation associated with the presence versus absence of *D. birchii* (Rundle et al. 2008). In mate choice trials in which females from four separate populations, two sympatric and two allopatric to *D. birchii*, were all presented with males from a single laboratory stock of *D. serrata*, allopatric females discriminated against males with sympatric-like CHC phenotypes, preferring allopatric-like males instead (Higgie & Blows 2007). Sympatric females, in contrast, showed no such discrimination (figure 27.3). Because the male phenotypes among which the females were choosing was held constant, this demonstrates an evolved difference in female preferences. Estimates of sexual selection generated by these different preferences correspond with the existing pattern of reproductive character displacement in male CHCs, suggesting that the different preferences may have been responsible for CHC divergence (Higgie & Blows 2007).

How did CHCs and preferences diverge between sympatry and allopatry? Manipulative tests, again exploiting the technique of experimental evolution, have focused on CHC divergence, although some insight into preference evolution can be gained. In an elegant experiment, Higgie et al. (2000) used six populations of *D. serrata* that were collected from the wild, three from locations sympatric with *D. birchii* (*field sympatric* populations) and three from allopatry (*field allopatric* populations). All six populations were brought into the laboratory and independently exposed for nine generations to experimental sympatry with *D. birchii*. CHCs evolved in all of the field allopatric populations as a consequence of experimental sympatry, and for seven of the eight CHCs measured, the direction of change in males matched the pattern of reproductive character displacement in the contact zone in nature (Higgie & Blows 2007). In contrast, experimental sympatry had little effect on CHC evolution in the field sympatric populations: changes were less pronounced, and those that did occur were inconsistent among the replicate populations, resulting in little net change overall (Higgie et al. 2000).

Heterospecific matings were extremely rare throughout this experiment, suggesting that reinforcing selection was not arising from the production of low fitness hybrids (i.e., postmating isolation). Rather, measurements of mating efficiency in *D. serrata* demonstrated that field allopatric males achieved almost 50% fewer matings with *D. serrata* females when *D. birchii* was present as compared to absent; field sympatric males were unaffected (Higgie et al. 2000). This suggests that sexual selection acted directly on males in the

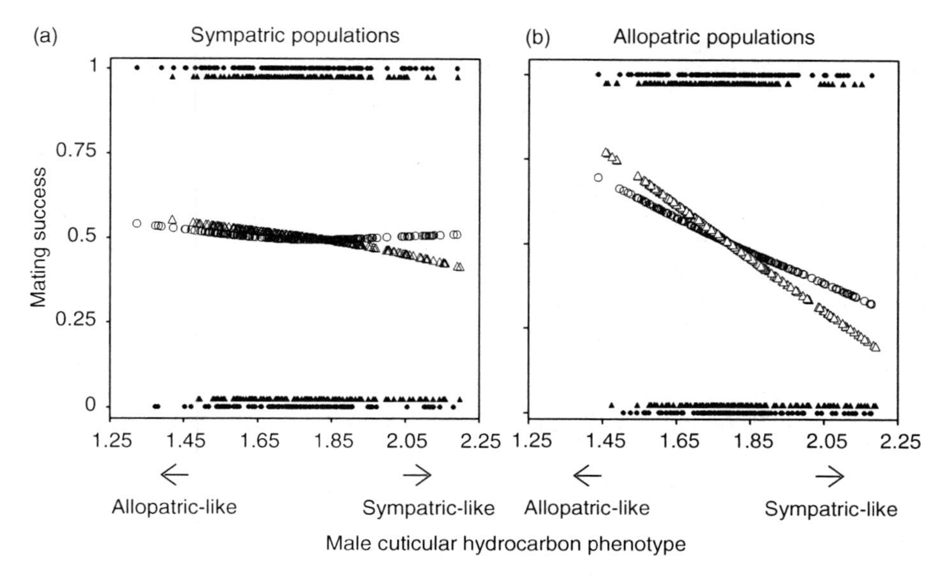

FIGURE **27.3** Mate preferences in female *D. serrata* for CHCs in males. Each panel shows females from two separate populations that are either sympatric or allopatric with *D. birchii* in nature. In each case, females chose among males from a single lab population created by hybridizing a sympatric (a) and allopatric (b) population to create males with a wide range of CHC phenotypes. Males are arrayed along an axis of reproductive character displacement for multiple CHCs in which lower scores (to the left) are characteristic of allopatric-like combinations and higher values (to the right) are characteristic of sympatric-like combinations. The actual mating success of individual males are given by filled symbols (with the two populations displaced slightly for clarity), whereas the open symbols in the middle represent the predicted mating success at the population level.

presence of *D. birchii*, although the precise mechanism is not known. Unfortunately, female mate preferences were not measured in these experimental populations, so the precise mechanism responsible is not yet known. Likely possibilities are direct selection on preferences in sympatry caused by differences in mating efficiency, or as a correlated response to altered selection on male CHCs. Plasticity of preferences in field allopatric females due to the presence of *D. birchii* is also of interest, as this could initiate divergence in a nongenetic manner.

FUTURE DIRECTIONS

As both case studies above attest, divergent selection between environments may be a powerful mechanism of mate preference divergence, driving the evolution of sexual isolation during the early stages of speciation. In addition, selection arising in sympatry from the presence of the other (incipient) species may also be an important source of divergent selection. In these systems and in many others as well, other mechanisms of preference divergence, such as sexual conflict, have received limited attention and manipulative tests are badly needed. Experimental evolution is a powerful tool is this regard that has yet to be fully exploited. By manipulating the presence of divergent natural selection, and having control over population sizes and mating systems, the full range of mechanisms of preference divergence can be explored, and interactions among them evaluated. To date, evolution experiments have provided equivocal results concerning the role of sexual conflict in the evolution of sexual isolation. Although Martin and Hosken (2003) demonstrated stronger sexual isolation among populations experiencing greater sexual conflict in dung flies, similar experiments in two species of *Drosophila* found no evidence of any reproductive isolation (Wigby & Chapman 2006; Bacigalupe et al. 2007). Additional work is clearly needed to develop a comprehensive understanding of the role of sexual conflict in the evolution of sexual isolation. As Coyne and Orr (2004) note, if chance-based mechanisms such as sexual conflict turn out to be important in preference divergence, it could fundamentally alter the common perception that divergent natural selection is the primary mechanism of speciation.

More work on natural systems that goes beyond simply asking whether sexual selection is associated with speciation is needed to address the mechanisms involved in preference divergence. There are a few examples in which this is occurring, including African cichlids (Maan et al. 2006) and Anolis lizards (Leal & Fleishman 2002), in which sensory drive (see box 24.1) has been implicated in color divergence and differences in color preference confer sexual isolation. In addition, many lacewing species (*Chrysoperla* spp) show almost no morphological or ecological divergence, yet are isolated by differences in preference and production of their vibrational songs (Wells & Henry 1998). These songs diverge in correlation with characteristics of the substrate upon which mating occurs, again implicating sensory drive. Other intriguing examples are the mimetic Heliconius butterflies that are isolated by differences in color pattern and pattern preference (Jiggins et al. 2004). Color pattern is aposematic, and many species form Mullerian complexes, implicating predation as the ultimate cause. Additional examples are needed before generalization will become feasible.

Comparative methods have also been underutilized in this regard, despite being well suited to addressing questions concerning behavior's influence on speciation. A number of studies have shown that sexual selection is associated with greater diversification rates (reviewed in Coyne & Orr 2004; Ritchie 2007). It may be possible, however, to extend such studies to address specific forms of sexual selection and other mechanisms of divergence, and to explore their association with pre- and postmating reproductive isolation.

Although we have focused here on mechanisms for the initial divergence of mate preferences, there are other important areas of speciation research in need of behavioral data. For example, whether reproductive isolation can evolve entirely in sympatry is a topic that has long occupied students of speciation. It is now generally accepted that sympatric speciation can occur, and interest has shifted instead to determining its prevalence in nature. Establishing individual cases is a difficult prospect, however, and insight may instead be gained by the empirical examination of parameters and assumptions identified by theory as having strong influences on its likelihood (Bolnick & Fitzpatrick 2007). Key ingredients in the models include a source of disruptive selection and a mechanism of assortative mating. Assortative mating can arise in a number of different ways, but behavior is often involved; habitat preferences, preferences for social partners, and mate preferences are commonly assumed mechanisms. How assortative mating arises is important because two aspects of it—its cost and whether or not it generates sexual selection—are central to the ease of speciation (Otto et al. 2008). If individuals that mate assortatively incur a fitness cost, sympatric speciation becomes much more difficult because these costs can overwhelm the indirect benefits of assortative mating that come through the reduced production of heterozygotes. Obstacles to sympatric speciation can also arise when the mechanism of assortative mating generates sexual selection on the locus under disruptive selection. This is because sexual selection can eliminate polymorphisms at key loci and can oppose disruptive selection, if the most common types of females prefer the most common types of males. Determining how conducive conditions are for sympatric speciation in nature will therefore require detailed behavioral data concerning mechanisms of assortative mating within populations, any costs associated with this, and whether or not sexual selection is generated.

Finally, behavioral ecology and speciation also intersect in the study of alternate mating tactics. There is growing interest in examining how behaviors such as sneaking may overcome sexual isolation, generating hybrids that impede the speciation process (Magurran 1998; Groning & Hochkirch 2008). Behavioral and paternity data from sympatric species for a few well-studied systems are needed to evaluate the extent of gene flow that can result from such interactions. Determining whether this is a rare or pervasive impediment to speciation will require data from a wide range of taxa. Such studies can have conservation implications, given the preponderance of hybridization between native and exotic species (Groning & Hochkirch 2008).

SUGGESTIONS FOR FURTHER READING

Comprehensive treatments of speciation are provided by Schluter (2000) and Coyne and Orr (2004);

Price (2007) tackles the topic with a focus on birds. Our understanding of how mate preferences diverge was shaped by to a large degree by a review of their evolution within populations by Kirkpatrick and Ryan (1991). Those wishing to study mate preferences should also familiarize themselves with the intricacies surrounding their estimation (e.g., Wagner 1998), statistical analysis (McGuigan et al. 2008), and genetic basis (Chenoweth & Blows 2006). A recent review of sexual selection and speciation is provided by Ritchie (2007).

Chenoweth, SF (2006) Dissecting the complex genetic basis of mate choice. Nature Reviews Genetics 7: 681–692.

Coyne J & HA Orr (2004) Speciation. Sinauer, Sunderland, MA.

Kirkpatrick M & MJ Ryan (1991) The evolution of mating preferences and the paradox of the lek. Nature 350: 33–38.

McGuigan K, Van Homrigh A, & Blows MW (2008) Genetic analysis of female preference functions as function-valued traits. Am Nat 172: 194–202.

Price T (2007) Speciation in Birds. Roberts, Greenwood Village, CO.

Ritchie MG (2007) Sexual selection and speciation. Annu Rev Ecol Evol Syst 38: 79–102.

Schluter D (2000) Ecology of Adaptive Radiation. Oxford Univ Press, Oxford.

Wagner WE Jr (1998) Measuring female mating preferences. Anim Behav 55: 1029–1042.

28

Genomic Approaches to Behavioral Ecology and Evolution

CHRISTINA M. GROZINGER

The functional approach to behavior developed by behavioral ecologists has been highly successful, as the earlier chapters in this book attest. These studies have greatly enhanced our understanding of the ecological and social factors affecting selection on behavior. The recent incorporation of physiology and neurobiology allows us to better understand the proximate mechanisms involved in behavior. Finally, conceptual and empirical integration is also providing new insights into the underlying genetics of even extremely flexible behavior (chapters 5 and 6). Here I describe additional opportunities to understand the complexity of ecologically relevant behavior through new technology. Understanding the molecular bases of behavior is critical for behavioral ecologists to better characterize the evolutionary forces and constraints operating on these traits. Recent developments in genomic technology make it possible to examine behavioral variation at the molecular level. With the advances in high-throughput sequencing, genome-wide gene expression analyses, bioinformatics, and statistics, it is now possible to associate sequence variation and expression differences on a genome-wide scale with specific behaviors or subtle differences in behavior in a wide variety of ecologically relevant species. These genomic techniques will allow researchers to begin to understand how natural genetic variation contributes to the vast amount of individual, population, and species-level variation in behavior observed in wide array of organisms, many of which can be effectively studied only in their natural ecological context.

Previously, the dominant focus in the field of behavioral genetics has been to identify the genes required to produce a particular behavior, and to understand the specific cellular functions of these individual genes; these studies have traditionally been performed in model organisms (such as *C. elegans*, *Drosophila melanogaster*, and *Mus musculus*) in which traditional quantitative genetic techniques and laboratory breeding have been feasible. With the development of new molecular tools for genome-wide studies of sequence and gene expression, it is now possible to extend these studies to nongenetic model systems (such as honeybees, cichlid fish, songbirds, etc.; Whitfield et al. 2003; Renn et al. 2008; Replogle et al. 2008) in their natural ecological contexts. As these genomic techniques are becoming more accessible, we can further expand these studies to describe large networks of genes underlying behavior. We also can understand how selective pressures modify gene function within and between species and how genes interact with each other and with the environment to produce natural variation in ecologically relevant behaviors between species, populations, and even individuals.

One advantage of genomic approaches compared to classical genetic approaches is that less manipulation is required: behaviors can be studied in a more natural context with appropriate focal organisms.

Classical genetic studies typically involve performing genetic crosses between selected lines or developing mutants in controlled genetic backgrounds, and testing behaviors under controlled environmental conditions. For many ecologically interesting species, however, such experimental control is not feasible. Genomic approaches typically rely on performing large-scale characterizations of variation in gene sequence or gene expression patterns, without the need for crosses or genetic manipulations, to find links between genes and behavior, that then can be subjected to more rigorous testing (such as manipulation of gene expression with RNA interface, RNAi) for confirmation. This methodology works best if the variations in behavior are relatively distinct.

A second advantage of genomic approaches is that multiple genes associated with a given behavior can be identified and studied simultaneously. Although there are many examples of single genes with large organizing effects on behavior (such as the vasopressin receptor in voles; Young & Hammock 2007), genes exist in the context of a genome, and most behaviors result from the interactions of genes in complex networks. Furthermore, these networks are involved in establishing physiological traits and neural pathways throughout development, and also mediate interactions with the environment. Even in cases in which a single gene or chromosomal location has been shown to have a strong effect on behavior, a change in the genetic background can negate these effects. For example, in *Drosophila melanogaster*, the introduction of a small locus from one genetic strain into a novel genetic background produced males that were significantly more aggressive than either of the parental strains or a related strain containing the same introduced region (Edwards & Mackay 2009).

Gene-by-environment interactions (chapter 6) can similarly modulate the activity of behaviorally relevant genes. For example, studies with inbred laboratory mouse strains revealed significant effects of laboratory environment on the results of standardized behavioral assays, even when care was taken to keep all conditions equivalent (Wahlsten et al. 2003). Studies on fire ants have revealed that the environmental/social context can have an even stronger effect on gene expression patterns than the genotype of the individual (Wang et al. 2008). Thus, looking at a single gene or even a handful of genes is unlikely to capture the full complexity of gene networks, or gene-by-environment interactions, or to reveal how these factors are associated with behavior. Even a seemingly clear single gene-behavior correlation undoubtedly relies on a complex genetic network for proper expression of a behavior. Although differential expression patterns of the vasopressin receptor are strongly associated with monogamous or polygamous behaviors in male voles (Young & Hammock 2007), this gene itself is not sufficient to produce these behaviors; rather, it relies on tweaking an underlying neural network, which itself is dependent on a multitude of genes for proper development and function. There are many examples in which the "organizing" gene has been identified, but the protein product of this gene is really the gatekeeper for allowing a complex behavior to be turned on or off; other genes undoubtedly are required for producing the correct neural circuits, and for producing variation in the behavior. For example, detection of pheromones by pheromone receptors in female mice is required to inhibit male-like courtship and mating behaviors; both males and females have the necessary neural circuitry to perform male mating behaviors, but in females this circuitry is inhibited by detection of pheromones (Kimchi et al. 2007). Thus, perception of pheromones and the olfactory system control the performance of these behaviors, but the full suite of genes required for generating these behaviors is unknown.

How can evolutionary behavioral ecologists benefit from the identification of the genes responsible for producing behavior? Understanding the underlying genetic pathways involved can help support or develop hypotheses of how a behavior evolved, and lend insight into the selective pressures that may be operating on this trait. For example, certain groups of genes may be under strong selection, as in the case of female and male reproductive proteins in plants and animals (reviewed in Clark et al. 2006). This supports a theory in which males and females are engaged in an evolutionary arms race in which both compete to maximize their fitness: males by increasing offspring quantity and females by increasing their offspring quality. However, in other species, such as honeybees, males mate only once, polyandry is adaptive (Oldroyd & Fewell 2007), and there is no evidence of sperm competition (Laidlaw & Page 1984). In this case, cooperation between the sexes during reproduction may be more beneficial, and thus reproductive proteins may be under different selective pressures. Other gene families, such as pheromone receptor genes, may be under negative or stabilizing selection, because variation from the norm may result in negative fitness consequences for both the males and females (Cardé & Baker 1984). Genomic studies allow the evaluation of sequence and expression differences in suites of genes or all genes in a particular functional category, such as genes involved in reproductive behavior and physiology, and allow these theories to be tested on a large scale.

Selection for behavioral variation may be constrained by the evolutionary history of the gene networks that are responsible for producing these behaviors. There are increasing numbers of examples in which apparently disparate behavioral and physiological traits are strongly linked to each other (referred to as *behavioral syndromes*; Sih et al. 2004a; chapter 30 of this volume). Whether these correlations are adaptive or simply a consequence of pleiotropic gene networks that evolved under a different context is important to distinguish. Alternatively, there may be genetic *modules* that produce specific behaviors, and differences in expression of these modules can lead to profound changes in behavior. Such a model has been exemplified by the field of "evo-devo," or evolutionary developmental biology, which has demonstrated that there are specific gene networks underlying different aspects of anatomical development (i.e., limbs or body segments). Changes in how these genetic modules are coordinated during development can lead to significant changes in body plans (i.e., different numbers of limbs or segments), but the underlying genetic modules remain essentially the same (Carroll et al. 2005; West-Eberhard 2005). Behavioral plasticity may share similar constraints, in which specific gene networks code for distinct behavioral modules, and variation in behavior is due to expression of modules at different times or in different contexts (Robinson & Ben-Shahar 2002; Hofmann 2003; Toth & Robinson 2007). Furthermore, subsequent diversification of behavioral or physiological traits may make it difficult to perform cross-species evolutionary analysis, but the traits may be regulated by similar genes or gene networks, thereby making comparisons more feasible at the molecular level.

Finally, there may be unexpected insights into the evolutionary natural history of a particular species inspired by genomic investigations. For example, the sequencing of the honeybee genome revealed a low number of immune genes relative to nonsocial insect species, suggesting that bees have developed social behaviors to deal with disease and pests (Evans et al. 2006), rather than combating these pathogens at the molecular level. To fully understand a behavior, it is necessary to be able to trace it from its genetic basis, to the neural networks, to the final expression of that behavior in an individual, in a species, and across taxa.

Behavioral genomic studies assume that variation in behavior, both within and between species, is due either to modulation of gene function, in which there are functional changes in the protein, or to modulation of gene expression, in which transcription of the gene is altered. Modulation of transcriptional regulation may be due to functional changes in a regulatory factor (altered expression or function of a transcription factor, for example), sequence changes in the promoter region of the genes, or nongenomic effects, such as methylation or modifications to chromatin structure. Regulation of gene expression can also occur posttranscription, for example, due to variation in posttranscription processing or before/during translation (the latter of which is the target of RNA interference). Although there are some striking examples in which evolution in gene families or in specific candidate genes has profound effects on behavior, many studies have found that global changes in transcription are more commonly involved in producing variation among individuals, populations, or species. Indeed, even between chimpanzees and humans, orthologous proteins are highly conserved, with 29% having identical sequences and the majority of proteins differing by only one or two amino acids (Consortium CSaA 2005). However, gene expression patterns and networks in certain brain regions are not conserved (Oldham et al. 2006). As more genomes are sequenced, it appears that there is a high degree of constraint not only on regions of DNA that code for proteins, but noncoding sequences are also subject to stabilizing selection and adaptation (Andolfatto 2005). Thus, understanding selective pressures on transcriptional regulation and gene expression patterns is likely to become increasingly important.

Genomic studies can provide information on both sequence differences and transcriptional differences, and the techniques necessary for these studies are becoming increasingly accessible. High throughput sequencing methodology is reviewed in box 28.1, whereas techniques to monitor genome-wide transcription patterns are reviewed in box 28.2. Below, I will provide examples of how different molecular techniques have been used to identify

BOX 28.1 Sequencing Technologies

There have been dramatic improvements in the speed, accuracy, and cost of DNA sequencing technologies in the last decade, due to innovations in the chemistry of the detection methods, biochemistry of the enzymes and proteins used, computational bioinformatics, and instrumentation, such as capillary electrophoresis and microfluidics. There are three main types of sequencing technologies that have been commercially developed: *Sanger sequencing, single-nucleotide addition*, and *cyclic reversible termination* (see Shendure et al. 2004; Metzker 2005; Bayley 2006; Hudson 2008).

Sanger sequencing is the most commonly used sequencing approach, and was utilized for the human genome project. In this method, complementary DNA (cDNA) is synthesized to the single-stranded DNA fragment of interest, using a DNA polymerase enzyme. The reaction includes the natural 2′-deoxynucleotides (dNTPs) and 2′,3′-dideoxynucleotides (ddNTPs), which have different fluorescent tags for each of the four nucleotide bases. Incorporation of the dNTPs results in elongation of the cDNA strand, whereas incorporation of a ddNTP terminates the strand and labels it with a corresponding fluorescent tag. Thus, at the end of the reaction, there are strands of cDNAs with a distribution of chain lengths. These strands are separated using high-resolution electrophoresis, which has a single-base resolution. The terminating bases can be determined by the fluorescent tag, thus allowing the DNA sequence to be read. This process can be automated in 384-well capillary array electrophoresis sequencing instruments.

Currently, the most commonly used approach for the single-nucleotide addition method is *pyrosequencing*. This approach takes advantage of the pyrophosphate molecule that is released as each dNTP is added to the growing cDNA strand by the DNA polymerase enzyme. An enzyme cascade is included in the polymerase reaction mix, which is activated by the pyrophosphate molecule, resulting in the production of visible light. Thus, with the addition of each nucleotide, light is produced. The reaction is controlled by adding the different dNTPs sequentially; when the correct complementary dNTP is added, the DNA polymerase extends the primer and light is emitted. The DNA polymerase then pauses at the next noncomplementary base. By matching the production of light to the introduction of different dNTPs, the sequence of the DNA can be produced. This approach produces reads of ~200 nucleotides, which is much shorter than that produced by the Sanger method (typically 1,000 nucleotides). However, the advantage of this method is that the sequencing reactions can be done in massively parallel systems and do not require electrophoresis. This approach has been used by 454 Life Sciences Corporation (acquired by Roche Applied Science; http://www.454.com/index.asp). Here, individual DNA fragments are bound to beads and amplified locally on each bead using PCR. These beads are then arrayed on a slide with hundreds of thousands of wells (1 bead/well). The pyrosequencing reaction can then be performed on each bead.

Cyclic reversible termination reactions involve modified nucleotides that contain "protecting groups" and nucleotide-specific dyes. The protecting groups prevent chain elongation, so each nucleotide is added individually. The protecting group is then removed using mild conditions, and the next nucleotide is added. Thus, the DNA sequence can be built from sequential nucleotide incorporation, imaging of the dyes, and deprotection. As in the case of pyrosequencing, this technology does not require purification of individual clones or electrophoresis, and it can be done in a parallel high-throughput system. This technology is being used by Solexa (http://www.solexa.com; recently acquired by Illumina), where the DNA fragments are bound to the surface of a microfluidic flow cell, amplified locally (resulting in millions of clusters per square centimeter), and then subjected to sequencing.

BOX 28.2 High-Throughput Gene Expression Analysis

One powerful method for identifying new genes or suites of genes that are associated with a behavior of interest is microarray technology (reviewed in Ehrenreich 2006). With this technology, expression levels of thousands of genes can be monitored simultaneously. If the microarray is constructed to contain all of the genes in the genome, a genome-wide view of the expression patterns can be obtained. Microarrays can thus be used to capture a snapshot of the gene expression patterns associated with a particular behavioral state and determine whether there are common patterns between states or species. Microarrays can identify new candidate genes whose expression levels are closely associated with a particular behavioral phenotype. Furthermore, it is possible to identify functional categories of genes that may be specifically modulated, such as olfactory receptors, genes associated with immune function, and so on. If the genome of the organism has been sequenced, it is also possible to screen the promoter regions of coregulated genes to identify common sequences that might correspond to regulatory sites (Sinha et al. 2006). For example, a particular transcription factor might be responsible for activating expression of multiple genes, and the binding site for this transcription factor should be present in the promoter region of these genes. Finally, with expression QTL, or eQTL, methods, it is possible to associate the genes on the microarray with chromosomal locations to determine whether specific chromosomal loci are regulated; these eQTL can then be related back to QTLs identified by classical genetic methods (chapter 5; reviewed in Jansen & Nap 2001).

Although there are several commercial platforms for high-throughput analysis of gene expression (Affymetrix, http://www.affymetrix.com; Agilent, http://www.agilent.com; Nimblegen, http://www.nimblegen.com; and Illumina, http://www.illumina.com/), I will focus on custom spotted DNA arrays that can be readily developed by individual research groups, which are more likely to be used for nonmodel genetic organisms. For this microarray platform, gene-specific oligonucleotides (typically 70 nucleotides long) or double-stranded PCR products can be spotted onto a glass slide with a positively charged chemical coating that immobilizes the DNA. Thus, the location of each gene-specific DNA "probe" is fixed and can be easily identified on the slide. Thousands of gene-specific probes can be spotted on the slide, and a single PCR reaction will produce enough material to be spotted on several hundred microarray slides. Furthermore, if gene sequences among species are similar enough, microarrays developed for one species can be used for another (reviewed in Bar-Or et al. 2007). Microarray developed for European honeybees (*Apis mellifera*) have been used successfully to monitor gene expression levels in other *Apis* species (Sen Sarma et al. 2007), whereas microarrays developed for cichlids have been used for other fish species (Renn et al. 2004; Cummings et al. 2008). However, sequence variation or gene duplication between species may lead to spurious results, and precautionary steps (such as hybridizing samples made from genomic DNA) should be used to assess baseline levels of hybridization expected for each gene probe.

To monitor expression levels in a given sample, the RNA must be extracted and converted into a fluorescently labeled "target" without altering the transcript abundance levels. Typically, the extracted RNA or mRNA is converted to cDNA, which is labeled with one of two fluorescent dyes, usually the cyanine dyes Cy3 or Cy5. Alternatively, the RNA can be converted to double-stranded cDNA and then linearly amplified using an *in vitro* transcription reaction to produce more RNA; this RNA can then be converted to a cDNA target or directly labeled to produce an RNA target. The labeled targets from two samples (one labeled with Cy3, the other with Cy5) are then mixed and allowed to hybridize to the array. The gene-specific targets bind to their cognate probes/spots. The excess unbound target is washed away, and the remaining bound target is visualized using a scanner that has lasers specific for the wavelengths for the two individual dyes. The amount of Cy3

(continued)

versus Cy5 bound to each spot corresponds to the amount of starting RNA for that gene in each sample. The Cy3-Cy5 dye ratios can be used to calculate relative expression levels for each gene in the two samples. Spotted DNA arrays always involve a two-sample comparison, because variation in the sample preparation, hybridization, and detection can lead to large differences in fluorescent intensity. Thus, by measuring relative expression levels, this experimental variation can be reduced. Various statistical approaches such as multivariate ANOVAs (Churchill 2004) or Bayesian analysis (Meiklejohn & Townsend 2005) are used to control for experimental variation and to identify genes with significant differences in expression levels between samples of interest.

genes associated with complex behavior and to study the evolutionary forces operating on these genes and behaviors. It is important to note that many of these studies focus simply on gene identification; characterizing the function of these genes and the specific selective pressures on these genes are undoubtedly important goals for future research. How these genes vary between populations and individuals, and how this variation is linked to fitness and adaptation, is an exciting future area of research.

COMPARATIVE GENOMICS: EVOLUTION OF GENE FAMILIES

Genomic techniques can provide new insights and allow behavioral ecologists to address novel questions in behavior. Once a genome is sequenced, a great deal of information can be obtained about the natural and evolutionary history of that species simply by comparing functional categories of genes with their counterparts in closely related species. Differences in the numbers of genes associated with certain processes can indicate that a species has developed alternative mechanisms to deal with specific environmental conditions or to produce certain behavioral responses, or that the species has been subjected to different selective pressures reflected in diverse life history traits. For example, upon sequencing the honeybee genome, it became apparent that bees have significantly lower numbers of immune genes, detoxification enzymes, and gustatory receptors (which are involved in taste) compared to solitary insects such as fruit flies (*Drosophila melanogaster*) and mosquitoes (*Anopheles gambiae*; Claudianos et al. 2006; Consortium HGS 2006; Evans et al. 2006). The decreased size of these gene families may be simply due to different evolutionary histories; honeybees are in the order Hymenoptera, whereas Drosophilidae (fruit flies) and Culicidae (mosquitoes) are in the order Diptera, and these groups diverged 300 million years ago (Grimaldi & Engel 2005). However, preliminary results of the genome sequencing project for another hymenopteran, the parasitic wasp *Nasonia vitripennis* have found that this species has similar numbers of p450 genes and gustatory receptors as fruit flies (*Drosophila melanogaster*; Johnson, Berenbaum, Robertson, & the Nasonia Sequencing Consortium, unpublished results). Thus, the reduced numbers of genes in honeybees could be a consequence of their natural history, not their evolutionary history.

The reduced number of detoxification genes and gustatory receptors may be because honeybees feed on pollen and nectar from flowering plants (reviewed in Winston 1987). This is a mutualistic relationship that benefits both the honeybees and the plants, because pollination is often necessary for plant reproduction. Thus, unlike insect species that feed on plant tissue and need to degrade toxic plant secondary compounds, honeybees have a fairly simple food source and may have little need for detoxification enzymes. Similarly, although pollen and nectar can be obtained from a variety of plant species, they are relatively uniform food sources, and it may be unnecessary for honeybees to have large repertoire of taste receptors. Honeybees do, however, possess two putative sugar receptors, which could allow them to distinguish between nectar of different sugar quantities and types (Robertson & Wanner 2006). Furthermore, bees have more olfactory receptors than *Drosophila* or *Anopheles*, which suggests that olfactory processes are more critical in honeybees. It may be necessary for bees to have a great olfactory range to identify floral sources, or it could be an adaptation to the increased amount of social chemical communication in honeybees (reviewed in Slessor et al. 2005).

In the 17 gene families implicated in immune responses, bees have approximately one-third fewer genes (71 genes) than *Drosophila* (196 genes) and *Anopheles* (209 genes; Evans et al. 2006). Because bees still possess many of the critical immune genes, it suggests that *Anopheles* and *Drosophila* greatly expanded the number of genes *via* duplication events. This is somewhat surprising, because bees do have a large number of pests and pathogens (Morse & Nowogrodski 1990), and the controlled environmental conditions of the hive, large population densities, and frequent social interactions provide ideal conditions for pest or disease transfer. However, bees exhibit a number of behavioral responses to control pathogens and pests, including *hygenic behavior*, in which diseased larvae and dead bees are physically removed from the colony (Spivak 1996); *social fever*, which is the heating of brood patches to combat chalkbrood, a bacterial pathogen in honeybees (Starks et al. 2000); and *social grooming*, which can remove ectoparasites. Thus, behavioral responses to disease may compensate for physiological immune responses.

Another example of changes in gene families correlated with major changes in life history comes from studies in primates (Gilad et al. 2004). Apes, Old World monkeys, and one New World monkey (the howler monkey) possess trichromatic vision, with three opsin genes whose protein products detect short-, medium-, and long-wavelength light. Most New World monkeys possess only two opsin genes, one of which detects short-wavelength light, and a second gene that has two alleles, which can detect either medium- or long-wavelength light. Because this second opsin gene is on the X chromosome, females can be trichromatic (by having two different alleles of this second gene) but males are dichromatic (because they have only one copy of the second gene). Trichromatic vision is thought to be adaptive by potentially increasing detection of ripe fruit from a background of foliage. A survey of the sequences of 100 olfactory receptor genes in these primate lineages revealed a dramatic increase in the number of nonfunctional pseudogenes in the apes (~33%), Old World monkeys (~29%), and howler monkeys (31%), compared to six other New World monkey species (~18%). Because howler monkeys are part of the New World monkey evolutionary lineage, this suggests that the reduction in functional olfactory receptors is linked to dramatic improvements in the visual sensory system. However, it must be noted that these studies are correlative, though the fact that the reduction in olfactory receptors occurred in two distinct evolutionary lineages is compelling.

MOLECULAR SIGNATURES OF SELECTION: EVOLUTION WITHIN A GENE FAMILY

Male and female reproductive proteins in plants, insects, and mammals have been shown to diversify rapidly, possible due to sexual selection and conflict (reviewed in Clark et al. 2006). In most animal species, male reproductive strategies would involve reducing remating by females, inseminating a larger fraction of her eggs, and commandeering a larger portion of her resources for his own offspring (chapter 23). Female reproductive strategies select for the most robust spermatozoa and increased lifetime egg-laying capacity. These differences in strategies can lead to an "arms race" between the sexes, which results in the rapid evolution of reproductive proteins, potentially differences between populations, and ultimately speciation (if there are constraints to genetic mixing; chapters 23 and 27). Of the factors that trigger postmating changes in female insects, seminal proteins have been studied most intensively, and the majority of these studies have been carried out in *Drosophila*. In *D. melanogaster,* approximately 112 accessory gland proteins (Acps) have been identified, and several have been functionally characterized (reviewed in Chapman & Davies 2004; Ravi Ram & Wolfner 2007). The transfer of specific Acps can produce specific postmating changes in females; for example, Acp70A ("sex peptide") stimulates egg-laying and reduces female receptivity to additional mating, Acp26Aa ("ovulin") stimulates egg laying, and Acp36DE is important for the storage of spermatozoa within the mated female. Many *Drosophila* Acps are involved in proteolysis, sperm binding, and immune function, which is also the case for mammalian seminal proteins (Mueller et al. 2004). Comparisons between *Drosophila* species revealed that Acps are diverging rapidly, and quantification of nucleotide changes that result in synonymous (the nucleotide changes does not result in a change in the amino acid sequence of the protein) or nonsynonymous (the nucleotide change results in an amino acid change in the protein) amino acid changes suggest that a substantial number of Acps (~20%) are under positive selection, meaning the mutations that alter protein sequence are present at higher than expected rates (i.e., Swanson et al. 2001; Mueller et al. 2005). There is also likely to be substantial variation in Acp protein sequences between populations within a single *Drosophila* species, which could lead to mating incompatibility or lowered fitness in crosses between these populations. For example, in *Drosophila mojavensis*, females produced larger eggs when mated with males from their own population versus males from a distinct population (Pitnick et al. 2003). In *Drosophila melanogaster*, there is substantial variation in sperm competitive ability related to natural genetic variation in males (Fiumera et al. 2007). Female reproductive proteins have been less extensively studied, but there is evidence that some of these are also under positive selection and diverging rapidly between *Drosophila* species (Swanson et al. 2004). Genetic variation among females also contributes to variation in sperm competition (Clark & Begun 1998). Interestingly, recent studies suggest that there are biochemical pathways to which both male and female reproductive proteins contribute together to carry out postmating changes in females (Park & Wolfner 1995; Ravi Ram et al. 2006). Thus, although reproduction may be an arms race in terms of some biological processes and genes, for others it may be more cooperative. It will be interesting to compare these different types of proteins to determine if they are under distinct selective pressures.

GENOME SCANS: EVOLUTION ACROSS THE GENOME

With the increased feasibility of whole-genome sequencing of species, populations, and individuals, it is now possible to look for signatures of selection across an entire genome. These studies can be as simple as identification of small mutations in closely related individuals or strains, or as complex as monitoring sequence differences across whole genomes of different populations or species. For example, whole-genome sequencing revealed that a single point mutation had significant impacts on the behavior and fitness of the soil proteobacterium *Myxococcus xanthus* (Fiegna et al. 2006; Velicer et al. 2006). This bacterium forms multicellular fruiting bodies under conditions of starvation, in which only a handful of cells will form long-lived spores and reproduce. Interestingly, a clone from a laboratory strain of this bacterium evolved to become an obligate "cheater" that could not form fruiting bodies in the absence of the ancestral wild-type strain, but when the wild-type strain was present, the cheater strain preferentially formed spores. During subsequent generations, this clone evolved to cooperate again, and this new clone actually had superior fitness compared to the original wild-type strain under starvation conditions. Sequencing of the parental, cheater, and evolved cooperator revealed that a single point mutation was associated with the reevolution of cooperation.

Characterizing evolution across the genomes of more genetically distant populations or species requires rigorous statistical analyses of sequence variation. A full review of these techniques is beyond the scope of this chapter, but the reader is directed to Eyre-Walker (2006) and McVean and Spencer (2006) for further information. Whole-genome association studies seek to determine whether specific nucleotide or sequence differences are associated with a particular phenotype. This approach has been used most extensively with the International HapMap Project, which seeks to identify genes associated with disease prevalence in humans. Whole-genome association studies have been applied extensively to identify genes associated with mental illness and behavioral issues. Recently, two independent whole-genome association studies identified genetic deletions associated with schizophrenia (Stefansson et al. 2008; Stone et al. 2008). However, these deletions are rare genetic variants that do not completely explain the onset of the disease. Because of the incredible complexity of these behavioral disorders and the large sample sizes required for this experimental approach, results of whole-genome association studies have often been inconclusive despite large samples sizes and meta-analyses (Burmeister et al. 2008).

Genome scans for adaptive evolution attempt to determine whether sequence variation between populations of species is neutral (either causing no change in gene function, or due to bottlenecks or drift) or caused by specific selective pressures. The majority of these studies have thus far focused on determining whether adaptive evolution can be detected and how prevalent it is (reviewed in Eyre-Walker 2006), and have not focused on genes that are specifically linked with a behavioral phenotype.

For example, comparisons of promoter regions of chimpanzee and human genes found that genes involved in neural or nutrition-associated processes display molecular signatures of positive selection (Haygood et al. 2007). Such approaches will undoubtedly be extended to identify genes or gene families associated with differences in ecologically relevant behavioral traits in a variety of species.

TRANSCRIPTIONAL PROFILING: VARIATION IN GLOBAL GENE EXPRESSION PATTERNS ASSOCIATED WITH BEHAVIOR

Changes in the timing or amount of gene expression play a central role in both the production of novel phenotypes and in phenotypic plasticity. For behavioral biologists, examining differences in the amount and timing of gene expression serves as an entry point to understanding the genetic basis of behavior and behavioral plasticity. Although transcriptional profiling with microarrays has been used frequently to identify gene expression patterns associated with behavioral differences in *Drosophila* and mice, microarray studies are now expanding into many other nongenetic model systems. From a behavioral genetics perspective, these studies allow us to identify genes associated with complex natural behaviors performed in ecologically relevant environments. From an evolutionary perspective, these studies allow us to compare similar processes (i.e., mating and reproduction) in different species that are under different selective pressures, to determine how similar or disparate these processes are at the molecular level. It also allows us to search for evidence to test evolutionary hypotheses at the molecular level. Below are summarized studies examining gene expression patterns associated with reproduction in social insects and how these relate to hypotheses of the evolution of reproductive and worker division of labor.

Comparisons of Gene Expression among Morphs/Phenotypes within Species

The shift to reproductive division of labor is considered to be one of the key transitions in the evolution of social behavior (Maynard Smith & Szathmary 1995; Keller 1999). Although in solitary species a single female performs all required tasks (egg laying, brood care, foraging), in eusocial species, such as honeybees, females differentiate into either a queen or a worker. Queens are reproductively active but do not perform other colony tasks, whereas workers specialize on brood care (nursing behavior) or foraging, and, in the case of honeybees, workers are facultatively sterile and typically do not reproduce. This caste differentiation is a dramatic example of polyphenism, in which the same genome can give rise to vastly different phenotypic outcomes, usually due to nutritional cues during development (Wheeler 1986). The *ovarian ground plan hypothesis* proposes that reproduction and maternal care (brood provisioning and foraging) became uncoupled in primitively social species, whereas in eusocial insects, ovarian development, brood care, and foraging are uncoupled, such that ovary development is maintained in queens, brood care is performed in young workers (which can also be competent to activate their ovaries), and foraging is performed by older workers (West-Eberhard 1996). Based on this model, one prediction is that gene expression patterns in young nurse bees may be similar to queens. They also predict that if workers become reproductively active, gene expression patterns will shift to become more queen-like. Other models (i.e., the *reproductive ground plan hypothesis*) suggest that reproductive traits are associated with pollen foraging behavior, assuming that reproduction in solitary species was linked with gathering pollen as a protein source for developing eggs or feeding brood (Amdam et al. 2004, 2006). This model has been supported by data in worker honeybees linking traits associated with high reproductive potential (ovariole number, levels of the egg-yolk protein vitellogenin) with increased tendencies to collect pollen rather than nectar and earlier onset of foraging (Amdam et al. 2004, 2006). However, other studies using strains of bees selected for high levels of worker reproduction (even in the presence of a queen) found no associated with pollen foraging, and these bees had a later onset of foraging (Oldroyd & Beekman 2008). These results are instead consistent with the ovarian ground plan hypothesis: these more reproductive worker bees remained in the brood care state longer.

Fundamental questions in the evolution of social behavior can best be addressed by monitoring genome-wide differences in gene expression, to capture variation in the global expression patterns and describe the relevant gene networks associated with these traits. Microarray analysis was used to monitor global gene expression patterns in the brains of same-aged virgin queens, sterile workers (with inactive ovaries), and reproductive workers (with activated ovaries containing maturing eggs; Grozinger et al. 2007). There was a large number of genes whose expression patterns were significantly different between queens and workers (~1,700 genes), whereas only ~200 genes were differentially expressed between sterile and reproductive workers. Interestingly, analysis of expression patterns of these 200 genes revealed that brain gene expression of reproductive workers was more "queen-like": reproductive workers upregulated genes that were similarly upregulated in queens (compared to sterile workers), and downregulated genes that were downregulated in queens (compared to sterile workers). This suggests that there is a core set of genes associated with reproduction in honeybees, regardless of caste. Previous microarray studies identified sets of genes that were upregulated in the brains of workers engaged in brood care (nursing-related genes) or foraging (foraging-related genes; Whitfield et al. 2003). There was a significant bias for queens (relative to workers) to upregulate nursing-related genes compared to downregulating them, and queens preferentially upregulated nursing-related genes compared to foraging-related genes. There was also a significant bias for reproductive workers (compared to sterile workers) to upregulate genes associated with nursing compared to downregulating them. Sterile workers preferentially upregulated foraging-related genes compared to nursing-related genes. These results support the ovarian ground plan model, because genes associated with brood care are also associated with reproductive potential. Of course, the number of genes in this analysis was quite small, and genes upregulated in nurse bees may not be functionally associated with the performance of nursing behavior (see below for further discussion). Furthermore, honeybees are highly derived from the ancestral solitary state, thus studies of underlying gene networks being co-opted to perform derived function would be better performed in more primitively social species, such as in paper wasps.

Comparison of Gene Expression among Species

In paper wasps (*Polistes metricus*), there is a division of labor between brood care and reproduction, but this division is much more plastic than what is observed in honeybees (reviewed in Hunt 2007). In *P. metricus*, a single reproductive female initiates a colony in the spring. The foundress is responsible for nest construction, foraging, and provisioning offspring; therefore, she behaves essentially as a solitary insect would. Once the first set of workers emerges, the foundress transitions into a queen and continues to lay eggs but ceases to forage and engage in brood care. By late summer, the workers rear gynes (reproductives) that will overwinter and become foundresses the following spring. The gynes engage in neither egg laying nor brood care.

If reproduction, foraging, and brood provisioning behaviors have become uncoupled during the life history of a *P. metricus* colony, it would be expected that genes associated with ancestral maternal care behavior are expressed similarly in foundresses and workers, because both engage in active foraging and food provisioning, whereas gynes and queens should have different expression patterns. Using a new high-throughput sequencing technology (454 Sequencing; see box 28.1), Toth, Robinson, and colleagues identified 3,000 genes from *P. metricus* (Toth et al. 2007). Thirty-two genes were selected whose corresponding honeybee orthologs (genes derived from a common ancestor) were associated with nursing versus foraging behavior in worker honeybees (Whitfield, 2003). Expression patterns of these genes in the brains of *P. metricus* foundresses, queens, gynes, and workers were monitored using quantitative real-time PCR (see box 28.3). When the overall expression patterns across all of the genes were analyzed, foundresses and workers indeed had the most similar expression patterns and were clearly distinct from the other two groups (figure 28.1). These results suggest that there are core groups of genes underlying maternal/sib care versus reproductive behaviors that have been conserved across taxa and have been selected upon to produce complex and novel behavioral phenotypes.

It is important to note that microarrays identify only gene expression patterns associated with or correlated with a specific behavioral state; further studies are necessary to determine whether these genes are functionally involved in producing a behavior. For example, in the studies outlined above, *nursing-related genes* are defined as those genes found to be expressed at higher levels in the brains of nurses bees compared to foragers, whereas *foraging-related genes* were expressed at higher levels in the brains of foragers compared

BOX **28.3** Single Gene Expression Analysis

Several processes are required for a gene to produce a functional protein product that can modify the phenotype of a cell or the behavior of an organism. The gene must first be transcribed to produce RNA. The RNA then is spliced to remove introns and produce mRNA. The mRNA is then transported from the nucleus into the cytosol, where it is translated into protein. The protein then may require posttranslation modifications (such as phosphorylation) to become fully active. Any of these steps can be regulated, and the mRNA or protein products can be degraded at different rates as well. Thus, the best measure of how gene expression is correlated to phenotype is to measure the levels of active proteins. These measurements are typically made using western blotting to quantify protein levels in cell or tissue homogenates, or using immunohistochemistry to quantify protein levels and distribution in fixed cells or tissues. Both techniques require the production of specific antibodies, which can be a laborious process. Thus, measuring protein levels is not a particularly high-throughput method for monitoring expression of multiple or novel genes.

Monitoring gene expression via RNA or mRNA levels is much more easily adapted to a high-throughput format. These methods simply require knowledge of the mRNA sequence and the production of DNA primers or probes that are complementary to this sequence. One method to monitor RNA levels is the northern blot. Here, RNA is extracted from the tissues of interest. The samples are separated by gel electrophoresis, in which the different RNA transcripts are separated by size. The RNA is then transferred to a filter, which is then incubated with a labeled probe specific for the gene of interest—the probe is labeled with radioactive nucleotides or tags that allow for subsequent chemical detection. The probe anneals to its complementary RNA transcript, and the unannealed probe is washed away. The amount of bound probe in the different tissue samples is then determined. Typically the RNA levels of the gene of interest are normalized to RNA levels of a control gene, whose expression levels are not expected to differ between the tissue samples; this controls for variation in the initial amounts of RNA between the samples.

Another method for monitoring RNA level is quantitative or semiquantitative reverse-transcriptase PCR (RT-PCR). Because this method relies on PCR amplification of the transcript of interest, much smaller quantities of RNA can be used. RNA is extracted from a set of tissue samples, and the RNA is converted to complementary DNA (cDNA) using a reverse transcriptase enzyme. The cDNA is then used in a PCR reaction with primers complementary to the gene of interest. Typically a second set of reactions is performed using primers to a control gene, whose levels should not differ between samples. The PCR reaction involves annealing of the primers to the template, extension of the primers by DNA polymerase, and then deannealing the product. The process is repeated for many cycles. Ideally, each amplification cycle doubles the amount of double-stranded gene-specific product. However, it is typically the first set of cycles that result in doubling, and eventually the reaction becomes less efficient as reagents are used up. Therefore, in later cycles the amount of product produced will plateau. Thus, to use RT-PCR to measure the starting amount of RNA in a sample, the reaction must be terminated during the exponential phase, when the amount of material is doubling during each reaction. The PCR products are then separated by gel electrophoresis, and the amount of PCR product in each sample is quantified and normalized relative to the control gene.

Real-time quantitative RT-PCR is a modification of the above method, which uses fluorescent dyes to measure the amount of double-stranded product after each PCR cycle. Thus, the cycles in the exponential amplification phase can be determined after the PCR is completed, reducing the need for preliminary testing to determine the correct number of cycles to use. Furthermore, because the amount of product is determined by monitoring the fluorescence in the reaction well, it is unnecessary to separate the PCR products by gel electrophoresis. This makes the process even more high-throughput, because it reduces the time involved, and more reactions can be run in parallel—some instruments accommodate plates containing 384 reaction wells.

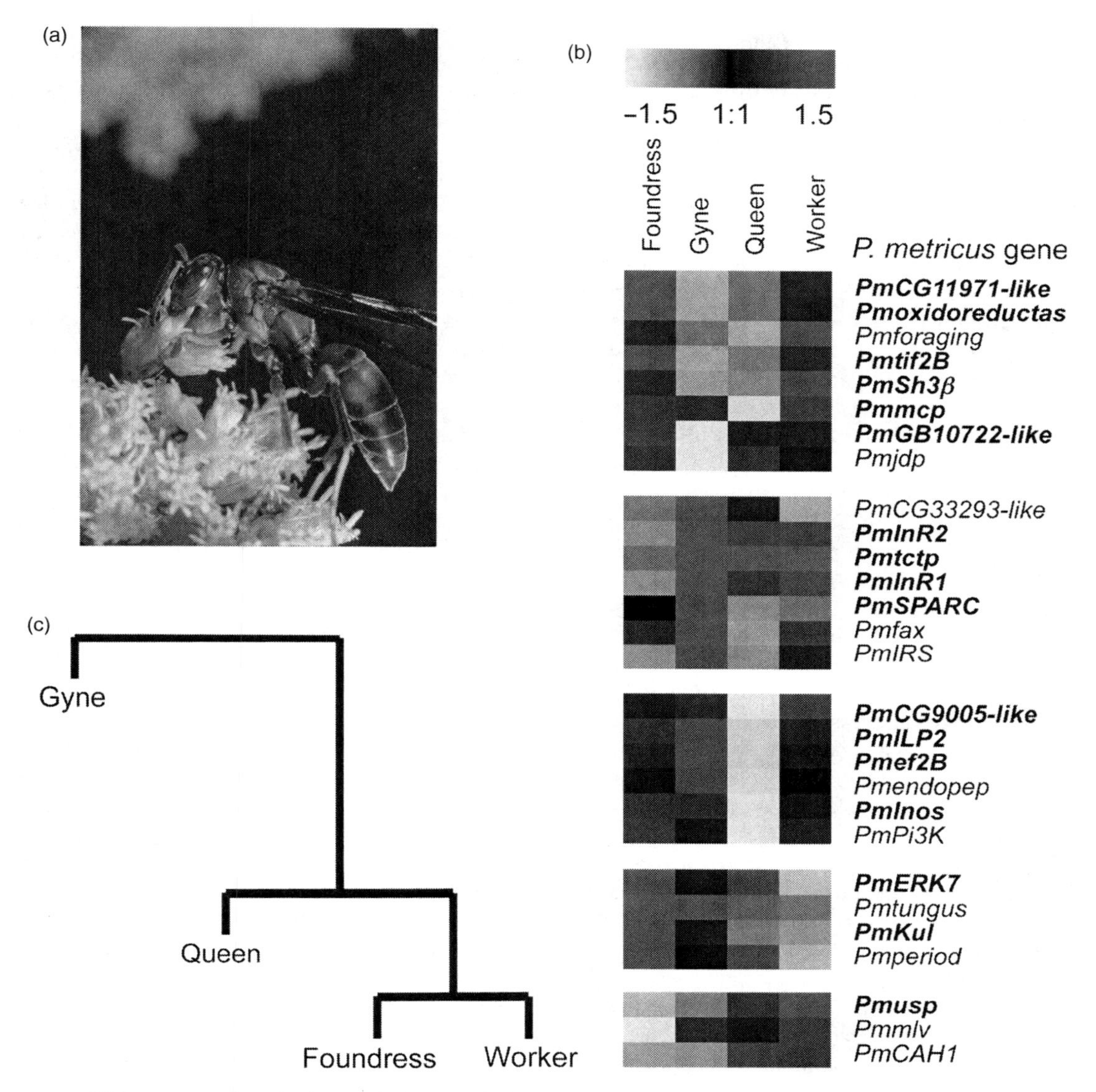

FIGURE **28.1** Maternal and sib care behavioral phenotypes share common molecular pathways in paper wasps. (a) *Polistes metricus* is a primitively social species, in which solitary foundresses initiate nests and rear workers until the first generation of workers takes over brood care, foraging, and nest maintenance. (b) Results for 28 genes selected for their known involvement in worker (honeybee) behavior. Heat map of mean expression values by group for each gene. Genes were clustered by k-means clustering; those in bold showed significant differences (based on ANOVA) between the behavioral groups. *P. metricus* gene names were assigned on the basis of orthology to honeybee genes. (c) Results of hierarchical clustering show that foundress and worker brain profiles are more similar to each other than to the other groups (based on group mean expression value for each gene). Photo of *P. metricus* courtesy of J. Hunt. Figure derived from Toth et al. (2007).

to nurses (Whitfield et al. 2003). However, a gene may be expressed at higher levels in the brains of foragers and not specifically be involved in producing foraging behavior. It may be a negative regulator of nursing behavior, and increased expression of this gene suppresses brood care. It may be an environmental response gene: although nurse bees spend the majority of their time inside the colony, which is an environmentally controlled dark environment, foragers are subjected to temperature and light extremes, and may deplete their energy stores during foraging trips. Thus, genes upregulated in foragers may be responsive to different stressors or environmental cues. Finally, a foraging-related gene may be correlated with one of the physiological traits associated with foraging and not be involved in the active production of this behavior. For example, expression levels of the transcription factor *Krüppel homolog 1* (*Kr-h1*) is significantly higher in the brains of foragers versus in-hive bees, regardless of age or foraging experience (Grozinger & Robinson 2007). However, honeybee colonies can be manipulated to cause forager bees to revert back to performing nursing behavior. *Kr-h1* levels are not reduced in these reverted nurses, suggesting that *Kr-h1* expression is linked to some factor that is likewise not altered upon reversion and therefore is not associated with the active production of foraging behavior (Fussnecker & Grozinger 2008). Several physiological traits are different between nurses and foragers and do not change upon reversion to nursing behavior, including neural branching in the mushroom bodies (the sites of learning and memory and higher level sensory processing in insects) and lipid levels in the abdominal fat bodies (Fahrbach et al. 2003; Toth et al. 2005). This example highlights the importance of detailed functional analysis of candidate genes that are identified by genomic techniques, to develop a mechanistic understanding of the behavior under investigation.

Population and Individual Variation

The majority of the microarray studies focus on identifying transcriptional profiles that are canonically associated with a particular behavioral or physiological difference. However, variation in a behavior or a physiological process between populations or individuals is really the fuel for adaptive evolution, and thus identifying the genes that produce this variation is of great interest. Below a sample of studies are highlighted in which variation in gene expression patterns between populations or individuals have been characterized.

Studies in swordtail fish have identified genes whose brain expression patterns are strongly correlated with female preference displays during mate choice assays (Cummings et al. 2008). Female swordtail fish were given choices between a small (unpreferred) and large (preferred) male, two small males, two females, or no choice. Brain microarray analysis identified ~300 genes from a 3,422 cichlid gene array (see box 28.2 for discussion of cross-species microarray experiments and Bar-Or et al. 2007) that were significantly differentially expressed between the four groups of females. The five strongest candidate genes were selected for additional analysis of brain gene expression patterns between individual female fish. These fish were again given a choice between small and large males, and the preference behavior of each fish was scored. Gene expression levels for four of these candidate genes were significantly correlated with preference behavior. Thus, making a choice leads to gene expression changes in the brain, and expression levels are affected most strongly when the choosiness of the female is higher, suggesting that variation in female preference may be linked to specific genes. These genes may be under selective pressure, and molecular evolution studies of their sequence variation could be very informative.

In honeybees, pheromonal communication is central to maintaining colony structure and function. The queen releases a pheromone (queen mandibular pheromone, QMP) that regulates a variety of behavioral and physiological responses in worker bees (Slessor et al. 2005). QMP inhibits ovary development in worker bees, and it will also elicit a *retinue response*, in which workers are attracted to the queen or to a QMP lure and lick and antennate her or the lure. There is substantial natural variation in retinue response among colonies and among supersister individuals within a colony (with a genetic relatedness of 0.75; figure 28.2; Kocher et al. submitted). This behavioral variation is strongly correlated with ovariole number, a physiological trait associated with reproductive potential in workers. Furthermore, individual variation in the retinue response and ovariole number is associated with significant global differences in brain gene expression. At the ultimate level, these data suggest that variation in retinue response may be due to reproductive conflict with the queen (workers with higher reproductive potential are trying to decrease exposure to queen pheromone to escape queen control and become reproductively active). Proximately, it is possible that retinue response may simply be linked to other behavioral and physiological traits by a network of genes involved in many aspects of worker behavioral maturation from brood care to foraging. Interestingly, although variation in the retinue response and ovariole number is strongly correlated in adult bees, regression analysis identified largely different sets of gene expression patterns in the brain associated with individual variation in each trait (figure 28.2d). Thus, the molecular pathways associated with each phenotype appear to be unrelated in adults. These traits may share a common developmental cue, which likely relies on a combination of environmental and genetic factors that canalize workers into a particular behavioral trajectory.

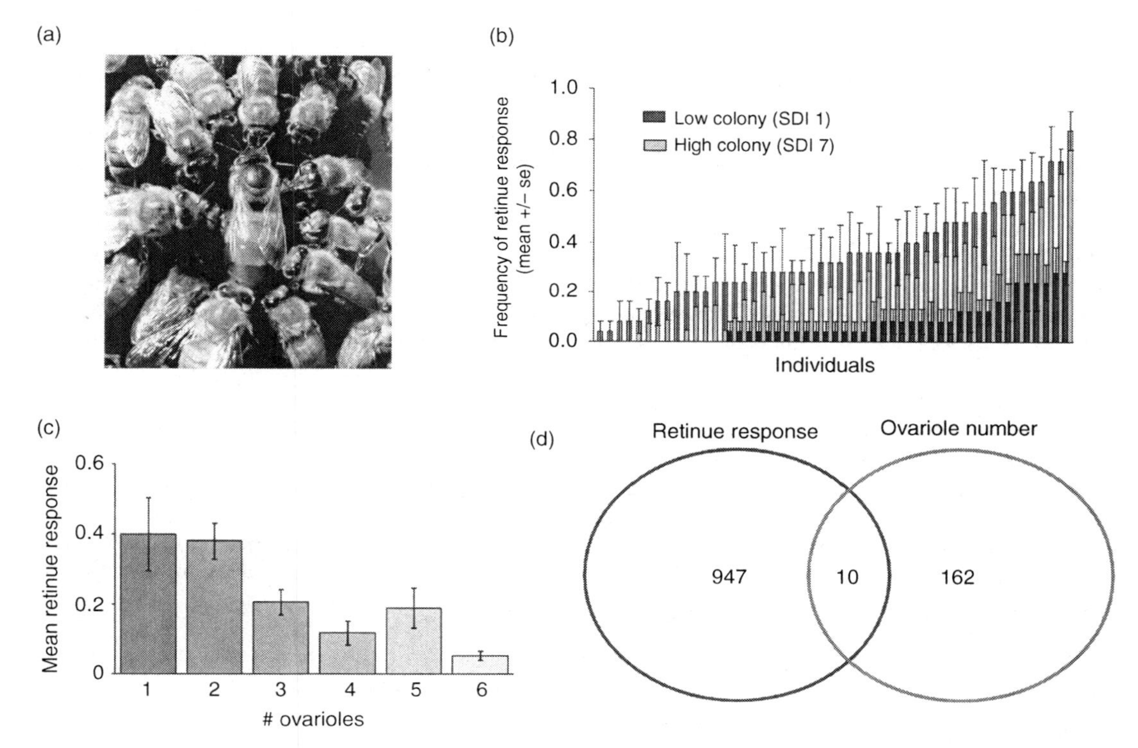

FIGURE **28.2** Natural variation in individual worker responses to queen pheromone is associated with physiological traits and brain gene expression patterns. (a) Worker bees are attracted to queens or queen pheromone and form a *retinue*, in which workers lick and antennate the queen or a pheromone lure. Photo courtesy of S. D. Kocher. (b) Individuals from two different colonies were screened for their retinue response score toward synthetic queen pheromone in a caged retinue response assay using young bees. Retinue response was scored for 4 days, five times a day (for a total of 100 scores). Significant variation in the frequency of responsiveness existed between colonies and between individuals within a colony. (c) There is a significant negative correlation between the number of ovarioles and retinue response scores of individuals (logistic regression, $P = 0.0014$). (d) Brain expression levels were measured using whole-genome microarrays. Expression patterns of the six highest and six lowest responding individuals from each colony were analyzed. Genes whose expression levels were significantly associated with differences in retinue response (960 genes) or with ovariole number (175 genes) were identified by statistical analysis. There was no significant overlap between these sets of genes (Fisher exact test, $P < 0.01$), suggesting that although pheromone response and ovariole number are associated at the organismal level, these traits are correlated with different brain gene expression patterns in adult bees. Figure derived from Kocher et al. (unpublished).

Environmental Effects on Gene Expression

The environmental context can dramatically shift the behavior, physiology, and gene expression patterns of individuals. In addition to physical environment, social context can also play a very important role. Most studies of the effects of environment on gene expression have focused on individual candidate genes. For example, in male cichlid fish, exposure to a dominant fish will cause a change in brain expression levels of the *gonadotropin-releasing hormone* neuropeptide (White et al. 2002). In songbird brains, expression of the transcription factor *zenk* is upregulated upon exposure to a novel, conspecific song, but will not be activated by exposure to a song of a different species, implying that this gene is involved in song learning (Mello et al. 1992). However, exposure to a familiar song in a novel environmental context will also stimulate expression of *zenk* (Kruse et al. 2004), suggesting that this gene is responsive to and perhaps integrating both social and physical information. Environment can also have long-term effects on gene expression. For example in rats, differences in maternal care can cause differential imprinting of the glucocorticoid receptor gene, leading to long-term effects on adult behavior (Meaney & Szyf 2005). Studies of the effects of environment on large-scale gene expression patterns have been limited so far, but undoubtedly will become more prevalent in the future as these technologies develop and become more accessible. In cichlids, whole gene expression profiles of brains of dominant and subordinate males demonstrated significant differences in brain gene expression between these two groups, as well as large individual differences within groups (Renn et al. 2008). In honeybees, exposure to social pheromones can cause changes in brain gene expression, in a behaviorally relevant manner (Grozinger et al. 2003). In fire ants, social organization can have a larger effect on gene expression than an individual's own genotype (Wang et al. 2008). Fire ant colonies can have a single queen (monogynous) or multiple queens (polygynous). These colonies differ genetically at the *Gp9* locus. Monozygous colonies have homozygous queens and workers (*BB*), whereas polygynous colonies have heterozygous queens (*Bb*) and heterozygous and homozygous workers (*Bb*, *BB*; note that the *bb* genotype is lethal). Microarray analysis of heterozygous and homozygous workers in both types of colonies revealed that social context (monogyne or polygyne) had a substantially greater effect on gene expression than genotype. It is possible that this socially mediated plasticity evolved to optimally adjust the behavior and physiology of the individual to its specific social context.

FUTURE DIRECTIONS

Natural variation in behavior among individuals in a population can lead to adaptation or, ultimately, speciation. Understanding the molecular bases of these behaviors is critical for behavioral ecologists to better to characterize the evolutionary forces and constraints operating on these traits. With the advances in high-throughput sequencing, genome-wide gene expression analyses, bioinformatics, and statistics, it is now possible to associate sequence variation and expression differences on a genome-wide scale with specific behaviors or subtle differences in behavior in a wide variety of ecologically relevant species. Although most studies have thus far focused on identifying sequence or expression differences within a population, it is becoming increasingly feasible to apply these studies across populations (see box 28.4). Furthermore, to develop a full view of the mechanisms by which genes regulate behavior, additional variables, such as developmental, environmental, and social factors, will need to be considered. All of these factors can have different effects on behavior depending on the genotype of the individual, again emphasizing the need for comparative analyses between genetic strains and populations (Mackay & Anholt 2007). Finally, gene expression patterns can be highly variable among tissues, and even genomic modifications such as DNA methylation can be tissue specific; thus, whole body atlases of gene expression patterns will be necessary to help interpret the association of gene expression and behavior. The most powerful studies of behavioral genomics will attempt to link all of these factors and correlate whole-genome sequence information with gene expression patterns, taking into account temporal factors (i.e., developmental conditions) and physical variables (such as cell and tissue type) across genetic strains or populations; such studies are already underway in *Drosophila* (Ayroles et al. 2009). However, although whole-genome studies are extremely powerful, careful analysis of candidate genes and processes identified will always be necessary to determine whether the observed

BOX 28.4 Ecological Genomic Studies

The Glanville fritillary butterfly (*Melitaea cinxia*) has long been an important model system for studying the ecological and evolutionary factors involved in population dynamics (Ehrlich & Hanski 2004). Each year there is a dramatic turnover among the 500 to 700 occupied discrete habitat patches (of a potential 4,000) in the Åland Islands in Finland, because there are roughly 100 colonization and extinction events. Interestingly, some populations are established far from other populations, suggesting that these particular foundresses have dispersed over quite large distances. Ecological factors (habitat fragmentation, seasonal variables, etc.) are obviously critical in regulating the founding and

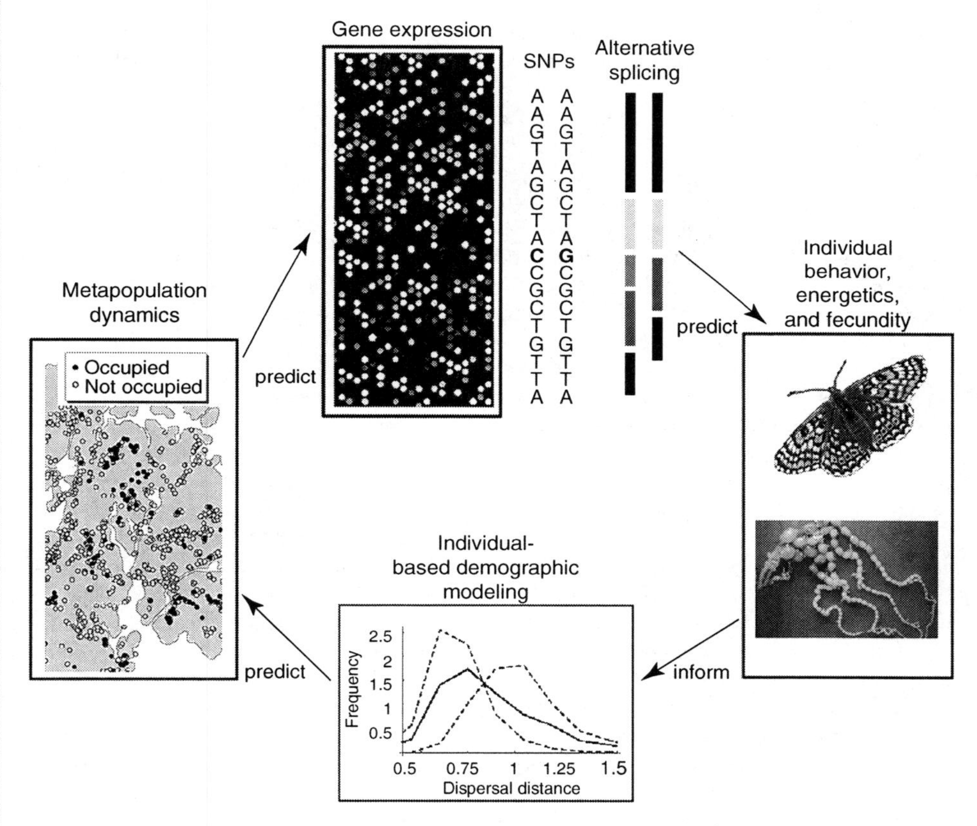

Figure 1 Genomic approaches to population dynamics. Next-generation sequencing of expressed transcripts (transcriptomics) from individuals among multiple populations allows researchers to develop information about nucleotide polymorphisms within transcripts and differences in alternative splicing of transcripts. This sequence information can subsequently be used to develop microarrays to characterize gene expression levels across tens of thousands of genes. This information will allow researchers to examine associations between specific traits and gene expression, nucleotide polymorphism, and quantitative variation in alternative splicing. In the example shown here, genomic tools allow a mechanistic examination of a butterfly species that is a model system for metapopulation dynamics but had previously lacked genomic resources. Figure derived from Vera et al. (2008) and Wheat et al. (in prep.).

(continued)

BOX 28.4 *(cont.)*

establishment of these populations, but genetic and physiological traits are important as well. It would be anticipated that traits associated with dispersal rate differ between older populations and newly established, isolated populations: individuals founding new populations are presumably more likely or better able to move long distances. Indeed, individuals within newly established, geographically isolated populations have higher flight metabolic rates than individuals from established, older populations, and these differences correlate strongly with allelic differences of a metabolic gene, *phosphoglucose isomerase* (*pgi*; Haag et al. 2005). However, it is likely that a host of physiological and genetic factors are involved in modulating dispersal rate. For example, allelic variation in the protein kinase G gene (the *foraging* gene) is linked to differences in food-associated dispersal rates in *D. melanogaster* larvae (Douglas et al. 2005).

However, it is likely that variation in brain expression levels of this gene is associated with differences in food seeking or more sedentary brood care behavior (Ben-Shahar et al. 2002). In male rhesus macaques, cerebrospinal fluid levels of a serotonin metabolite are associated with differences in aggression and age at emigration from their natal troop (Howell et al. 2007). Thus, in addition to flight metabolism or ability, there may be many different biological processes, both physiological and behavioral, that are associated with differences in dispersal rates in Glanville fritillary butterflies. Genomic approaches can help identify genetic differences, either allelic or expression, that differ between individuals in isolated, newly established populations and older populations. With the advent of new high-throughput sequencing (box 28.1) and expression analysis tools, using these genomic approaches is much more feasible. A recent 454 sequencing project (Vera et al. 2008) using 80 individuals from 8 different families, yielded ~9,000 unique annotated genes, with > 6,000 additional expressed transcripts that were not annotated based on comparisons with publicly available gene databases. A commercial microarray platform (box 28.2) was used to verify that these predicted genes were indeed expressed, and studies are underway to relate body-region-specific expression differences with variation in individual flight metabolism and population age (i.e., to determine which gene expression differences relate to dispersal ability and founding of new populations). Furthermore, because sequencing coverage of individual genes was quite high (~6.5 fold coverage for ~4,800 contigs, or contiguous sets of overlapping DNA sequences) and because individuals from many different genetic backgrounds were used, it was also possible to identify thousands of single nucleotide polymorphisms (SNPs), which can subsequently be used for studies of allelic variation.

Thus, in a species in which molecular information was previously extremely limited, it is now possible to compare expression levels and allelic differences across thousands of genes between individuals and populations. This fascinating system can now be studied at the molecular and genomic level to rapidly identify novel candidate physiological and behavioral processes that underlie population dynamics and individual fitness traits. As these genomic methodologies and techniques become more commonplace, similar approaches can be brought to bear on a variety of ecologically relevant species that can be studied in their natural context.

sequence or expression differences are functionally relevant or merely correlative. Regardless, these powerful new technologies will enable researchers to more broadly explore the molecular bases of behavior—and other phenotypes—in species that are not typically used as genetic model systems in their natural environmental context, thereby vastly expanding our understanding of the evolution of behavior.

Acknowledgments I would like to thank Mariana Wolfner and Jim Marden for helpful discussions, and members of my research group (Kevin

Donohue, Brendon Fussnecker, Sarah Kocher, Elina Niño, Freddie-Jeanne Richard, and Amy Toth), as well as Jay Evans, Jim Hunt, Harland Patch, Gene Robinson, and Kerry Shaw for critical reading of the manuscript.

SUGGESTIONS FOR FURTHER READING

For reviews that highlight examples of behaviorally relevant genes and the different mechanisms by which genes can function to regulate behavior, see Robinson et al. (2005) and Robinson et al. (2008). Hofmann (2003) describes a conceptual framework for how genes and gene networks may regulate behavioral plasticity, potentially by modifying neural pathways. Mackay and Anholt (2007) discuss the effects of genotype by environmental interactions on behavior. Feder and Mitchell-Olds (2003) describe the application of functional genomics studies to address questions in evolutionary ecology. Clayton (2004) outlines the development of genomic tools in one ecologically important, nongenetic model species, the zebra finch. Finally, Hudson (2008) describes recent breakthroughs in high throughput sequencing technologies and their applications to evolutionary biology and ecology.

Clayton DF (2004) Songbird genomics: methods, mechanisms, opportunities, and pitfalls. Ann NY Acad Sci 1016: 45–60.

Feder ME & Mitchell-Olds T (2003) Evolutionary and ecological functional genomics. Nat Rev Genet 4: 651–657.

Hofmann HA (2003) Functional genomics of neural and behavioral plasticity. J Neurobiol 54: 272–282.

Hudson ME (2008) Sequencing breakthroughs for genomic ecology and evolutionary biology. Mol Ecol Resources 8: 3–17.

Mackay TF & Anholt RR (2007) Ain't misbehavin'? Genotype-environment interactions and the genetics of behavior. Trends Genet 23: 311–314.

Robinson GE, Fernald RD, & Clayton DF (2008) Genes and social behavior. Science 322: 896–900.

Robinson GE, Grozinger CM, & Whitfield CW (2005) Sociogenomics: social life in molecular terms. Nature Genetics 6: 257–70.

29

Decision Making, Environmental Change, and Population Persistence

MARTIN A. SCHLAEPFER, PAUL W. SHERMAN, AND MICHAEL C. RUNGE

Conservation biology is concerned primarily with documenting and preserving biodiversity, whereas behavioral ecology investigates the ecological and evolutionary forces that shape behavior. It might seem, at first glance, that these fields are disparate with little common ground. Recently, however, there has been a growing realization that behavioral ecology has much to contribute to conservation biology (e.g., Clemmons & Buchholz 1997; Caro 1999; Linklater 2004; Sutherland 2006; Angeloni et al. 2008). Behavior plays an essential role in determining an organism's survival and reproductive success because it modulates interactions between individuals and their environment and, by extension, the growth and persistence of natural populations. Therefore, management and recovery plans are most likely to succeed if they are designed in light of, rather than in spite of, the behavioral ecology of the targeted populations or species.

This chapter focuses on the importance of an unseen and therefore largely unappreciated aspect of behavior, namely the decision-making rules (chapters 8–10) that organisms rely upon to process information from their biotic and abiotic environments and that result in particular behavioral outputs. Understanding the evolutionary forces that have shaped these decision-making rules, or *Darwinian algorithms* (Cosmides & Tooby 1987), is not just interesting academically, it is also important if we are to predict whether the behaviors of organisms will be adaptive—and whether their populations will persist—in a rapidly changing world.

Consider a simple decision-making rule with an important contribution to individual fitness: immediately after emerging from their nest, hatchling green sea turtles (*Chelonia mydas*) must disperse toward the safety of the ocean to avoid predation. How do hatchlings decide which direction to go? Natural selection favors reliance on environmental cues that, over time, have correlated with positive fitness outcomes. Thus, hatchling sea turtles rely on the downward slope of the beach and the star-lit horizon to orient themselves (Mrosovsky & Carr 1967). These are cues that, under normal circumstances, predictably guide the young to the ocean, where they are far less clumsy and thus safer than on land.

In environments that have experienced recent anthropogenic changes, however, evolved decision-making rules do not necessarily result in adaptive outcomes. On beaches with human housing developments, for example, electric lights can lure hatchling green sea turtles inland, where they perish (Witherington & Bjorndal 1991). This is an example of an evolutionary trap (Schlaepfer et al. 2002) because although a suitable option remains available (going toward the ocean), the young turtles' formerly adaptive decision-making rule now leads them to turn inland—in other words, to behave maladaptively in the altered environment.

Evolutionary traps can threaten populations and species if they result in widespread reductions in survival and reproduction. The unprecedented rate at which humans are altering the world's habitats, climate, nutrient cycles, and plant and animal distributions (Vitousek et al. 1997; Palumbi 2001), has the potential to disrupt organisms' behavioral decision-making rules globally. Thus, understanding the circumstances in which evolutionary traps can arise, how organisms respond to them, and what can be done to mitigate their effects are some of the most important conservation questions of our times.

We begin this chapter with a more formal introduction to the evolutionary trap concept, and explore it at four *levels of analysis* (Tinbergen 1963; Sherman 1988). We argue that it is informative to consider evolutionary traps from the proximate (mechanistic) standpoint to predict the initial behavioral outputs of decision-making rules in novel environmental circumstances. We then explore evolutionary traps from the ultimate (evolutionary) perspective, and discuss effects of decision-making rules that are relatively rigid (so-called "canalized" behaviors or "closed programs"; Mayr 1974; Arnold 1994d; Linklater 2004) versus decision-making rules that are more "open" to environmental influences and thus exhibit phenotypic plasticity (Harvell 1990; Ghalambor et al. 2007; chapter 6 of this volume). We also ask whether reaction norms of phenotypically plastic behaviors can yield adaptive outcomes when confronted with environmental cues that exceed the range of values historically encountered. Finally, we suggest that viewing conservation and management issues through an evolutionary lens will increase the range of mitigation options available to help vulnerable populations escape from traps and make successful transitions to novel selective regimes (Blumstein & Fernández-Juricic 2004).

ENVIRONMENTAL CUES AND DARWINIAN ALGORITHMS

Darwinian algorithms consist of the sensory and cognitive processes that perceive and prioritize cues within an individual's perceptual range, and then translate those inputs into motor outputs (i.e., behaviors; see chapters 9 and 10). Organisms rely upon them to process the virtually limitless quantities of information from their physical and social environments, and to guide key behavioral and life history decisions. A Darwinian algorithm may involve a stimulus threshold (e.g., "when the day length exceeds 10 h, migrate north") or depend on the occurrence and intensity of a cue that is normally associated with a fitness-enhancing outcome (e.g., "build nests in dense vegetation where chick survival is predictably high"). Darwinian algorithms are shaped through evolutionary time by the specific selective regime of each population. Which cues are relied upon will depend on three factors: the certainty with which a cue can be recognized, the reliability of the relationship between the cue and the anticipated environmental outcome, and the fitness benefits of making a correct decision versus the costs of making an incorrect decision (Reeve 1989; Neff & Sherman 2002).

In reality, of course, a Darwinian algorithm is a heuristic term to describe a *black box*—a collection of sensory organs, neural connections, cognitive processing centers (e.g., the brain), and motor connections that result in the expression of behavior. Although each component may be under multiple and potentially conflicting selective pressures (Arnold 1994d), given sufficient time we expect the components to be shaped by selection to work together harmoniously, thus creating fitness-enhancing outcomes. The operative words in the previous sentence, of course, are "sufficient time." A key challenge is to determine how rapidly and by what means organisms' phenotypes (including its Darwinian algorithms) can be altered to become adaptive again after a sudden change in selective regime (Visser 2008).

CONCEPTS

Ecological and Evolutionary Traps

The evolutionary trap concept has been applied both at the level of the individual, to describe an animal that makes a suboptimal decision (Schlaepfer et al. 2002; Robertson & Hutto 2006), and at the level of the population, to describe a preferred habitat that has been altered in a way that transforms it into a population sink (Delibes et al. 2001; Battin 2004; Runge et al. 2006). At the individual level, an evolutionary trap occurs when an organism makes a suboptimal choice because it is relying on a formerly adaptive decision-making rule. At the population level, we believe that use of the population sink criterion of a trap is too simple because sinks can also arise naturally, making it difficult to identify a bona fide evolutionary trap. Instead, we propose that populations have fallen into an evolutionary trap if (1) they exhibit lower projected population growth rates (λ) than populations that remain in predisturbance conditions and (2) it can be shown that the lower population growth rate is due to individuals behaving suboptimally. Thus, a thorough understanding of evolutionary traps will come from studies that include analyses of effects at both individual and population levels.

Of course, populations also may decline due to massive alterations in their native habitats (agricultural conversion, deforestation, desertification, urbanization, etc.). Such blatant disturbances (Sherman & Runge 2002) differ from evolutionary traps because individuals are making the best possible choice given the current circumstances (i.e., not making a mistake). Organisms experience reduced fitness due to a blatant disturbance when there are no alternative, better options available. By contrast, when organisms have choices, yet still express suboptimal preferences, they likely have fallen into an evolutionary trap.

Given the extent and speed of human-induced environmental changes, one might expect evolutionary traps to be widespread. Surprisingly, however, there are few clear examples (Robertson & Hutto 2006). This may be because evolutionary traps actually are rare, the evolutionary trap concept is relatively new, or traps are difficult to identify (Schlaepfer et al. 2002, Robertson & Hutto 2006). We believe that the latter two are the most likely explanations: traps actually are becoming more common, but they have only recently been recognized as such (Schlaepfer et al. 2002). Moreover, several critical components of an evolutionary trap, such as *preferences* and *fitness*, are conceptually easy to grasp but require considerable time and resources to document. An ideal study would track individuals for sufficiently long periods to (1) quantify the mode and range of Darwinian algorithms (preferences) in a population, (2) measure the fitness outcomes of these preferences (taking into account that different age and sex classes of individuals likely will have different optimal strategies), and (3) estimate population demographic parameters. Moreover, there are several reasons that some individuals may make what seem to be maladaptive choices (e.g., environmental stochasticity, competitive exclusion from high quality habitat; Tufto 2000; Kristan 2003; Arlt & Part 2007). Thus, preferences must be evaluated through time, at different population densities, and in light of possible social hierarchies (e.g., Marra & Holmes 1997) before deciding whether individuals are evolutionarily trapped or simply making the best of a bad situation.

Levels of Analysis

Evolutionary traps are complex phenomena at the intersection of evolution, ecology, and behavior, and it is essential to identify the level of analysis at which one is working. There are four levels of analysis: two proximate (mechanistic and ontogenetic) and two ultimate (fitness effects and evolutionary history; Tinbergen 1963; Sherman 1988; chapter 1 of this volume). Investigators who are interested in mechanisms attempt to elucidate the cues used in decision-making rules and how they are prioritized (Visser 2008). For example, what environmental cues does a female lizard rely on when deciding where to lay her eggs? To address this question, one might combine observations of the microhabitats where eggs are found in nature with laboratory choice experiments in which one measures the fitness consequences of oviposition choices by gravid females when presented with microhabitats differing in relevant environmental cues (wet soil versus dry soil, leaf-litter versus no leaf-litter, etc.; Andrews et al. 2000; Socci et al. 2005). Armed with an understanding of which environmental cues are integrated, how they are prioritized into a Darwinian algorithm, and the variations in such choices within a population, we can begin to predict how individuals and populations will respond to sudden changes in the quality or quantity of those cues.

Investigators interested in the ontogenetic level of analysis attempt to elucidate how cue priorities and decision-making rules change across individuals' lifetimes. For example, in contrast to the light and beach slope cues they relied upon to find the ocean after hatching, adult green sea turtles utilize the earth's magnetic field to navigate in the open oceans (Lohmann et al. 2004). Such juvenile-adult differences are widespread (e.g., in the many insects and anurans that undergo complete metamorphosis), and they highlight the importance of keeping the developmental stage and physiological state of an organism in mind when investigating the optimality of its decision-making algorithms.

Investigators interested in fitness effects attempt to understand the evolutionary forces that shaped Darwinian algorithms in unaltered environments and how novel environments can lead individuals into evolutionary traps and populations into decline. Most investigations of traps have focused on this level of analysis.

By contrast, the fourth level of analysis, namely, the importance of phylogeny and evolutionary history in affecting the likelihood that genera or species will be trapped, remains largely unexplored. We hypothesize that long periods of strong stabilizing selection would result in Darwinian algorithms that are relatively inflexible and difficult to modify when confronted with a novel selective regime (i.e., "closed" behavioral programs; Mayr 1974, 1976a), making species especially vulnerable to being trapped. In contrast, populations that have been exposed to highly variable selective regimes that have favored the evolution of phenotypic plasticity or maintenance of additive genetic variance for different behavioral outcomes (Mayr's "open" programs) might more readily be able to respond to novel conditions. These taxa are predicted to be less likely to fall into an evolutionary trap, and also to escape them more quickly if they do.

Individual and Population Level Responses to Novel Selective Regimes

Evolutionary traps are ephemeral situations, and there are three nonmutually exclusive ways individuals and populations can escape them: (1) individuals might disperse to places where the former selective regime (to which they are adapted) persists; (2) individuals might cope with the novel selective regime *in situ* provided that they are phenotypically plastic (e.g., capacity for learning, or inducible morphologies) and that the reaction norm underlying their response serendipitously produces an adaptive outcome under the novel circumstances; or (3) a fraction of individuals might serendipitously be adapted to the novel environment (i.e., not be trapped) and reproduce successfully there, resulting in evolution via natural selection.

Dispersal

Individuals can escape an evolutionary trap via dispersal provided that unaffected areas are reachable. For example, dragonflies that are trapped by polarized light patterns into ovipositing on polished stone surfaces (gravestones) can escape if they fly or are blown by the wind out of the cemetery and over the nearest water (Horváth et al. 2007). Likewise, natal or postbreeding dispersal (Nunes 2007) may serendipitously take small mammals and birds beyond the area affected by a trap. Alternatively, individuals might recognize suboptimal local conditions and disperse to escape from them. In instances in which a population's original selective regime persists elsewhere, dispersal—and possibly human-assisted colonization (Hoegh-Guldberg et al. 2008)—offers a rapid solution to an evolutionary trap.

Phenotypic Plasticity

Phenotypic plasticity is defined as the ability of a given genotype to express different phenotypes (including behaviors) under different environmental conditions (Harvell 1990; DeWitt & Scheiner 2004; chapter 6 of this volume). A reaction norm defines the relationship between the expressed phenotype and relevant environmental cues. Reaction norms can vary between individuals and are partially heritable. Thus, phenotypic plasticity is a trait that can evolve in response to novel environments (Visser 2008). At the population level, the ability to respond to a novel selective regime will be dictated by the variation and heritability of reaction norms, which in turn are determined by the nature of past selective pressures. One would expect reaction norms to show low intrapopulation variation and low flexibility in instances in which there was a long-standing, predictable correspondence between an environmental cue (e.g., day length) and an adaptive phenotype (e.g., timing of migration).

The shape of the reaction norm, however, can also vary over the course of an individual's lifetime, due to maturation and learning. Kokko and Sutherland (2001) showed that learning can help populations escape an evolutionary trap by following simple behavioral rules—for example, preferring the cues of habitats in which breeding was successful or selecting habitats with cues that differ from habitats in which breeding was unsuccessful. By enabling individuals to select breeding habitats that are especially suitable, learning may represent the most effective mechanism for escaping evolutionary traps, especially among long-lived species or those with little genetic variation.

Evolution

Provided that not all individuals in a population are trapped, evolution by natural selection is another mechanism by which a population can escape a trap (Carroll & Watters 2008). Our understanding of the rates at which populations can evolve has matured in the last decade. Some populations can evolve rapidly (on the order of tens of generations), both under natural conditions (Reznick et al. 1997; Thompson 1998; Hendry & Kinnison 1999) and in response to novel selective pressures caused by humans, such as size-selective harvesting regimes (Coltman et al. 2003; Baskett et al. 2005; Conover et al. 2005), competition and predation from non-native species (Strauss et al. 2006), and global climate change (Bradshaw & Holzapfel 2006). Rapid evolution offers further hope that some populations can escape from traps rather than go extinct.

Some of the best examples of evolutionary responses to traps come from instances in which the trap was caused by an introduced species (Schlaepfer et al. 2005). For example, the cane toad (*Bufo marinus*) was introduced to Australia in 1935. All life stages of cane toads contain a bufotoxin that serves as a chemical defense. There are, however, no toad species native to Australia (Tyler 1994). Thus, cane toads are both evolutionarily novel and toxic to native Australian predators (Phillips et al. 2003). Naïve predators will attack cane toads, presumably because of their superficial morphological resemblance to native frog species, and the predators often sicken or die as a result of ingesting the toxic chemicals. Declines in native snakes, lizards, and marsupials that follow the invasion of cane toads result, at least partly, from this evolutionary trap (Phillips et al. 2003). Interestingly, recent evidence suggests that several species of Australian snakes are adapting to cane toads via evolutionary reduction of their gape-width-to-body-length ratio resulting in the snakes being unable to eat toads large enough to kill them (Phillips & Shine 2004), and also the evolution of resistance to cane toad toxins (Phillips & Shine 2006).

Native anuran larvae (tadpoles) from the western United States provide another example of an evolutionary response to a trap, this one created by American bullfrogs (*Rana catesbeiana*, or *Lithobates catesbeianus*), which are native to the northeastern United States but were transported westward by humans. Native anuran larvae typically appear naïve to the predation threat that American bullfrogs represent. Kiesecker and Blaustein (1997) showed that red-legged frog (*Rana aurora*) tadpoles are vulnerable to American bullfrogs in experimental enclosures if they come from populations that have never been exposed to bullfrogs (only 64.7% survived). By contrast, red-legged frog tadpoles from populations that have been sympatric with American bullfrogs for 60 years were significantly more likely to survive in the presence of the predator (87.7% survival, nearly equivalent to control enclosures without American bullfrogs), suggesting that predator recognition and evasion mechanisms have evolved in the intervening period (an alternative, but less likely, explanation is that these populations were pre-adapted to bullfrogs).

These two examples suggest that successful transition to a novel selective regime can occur when the negative effects of the trap are not too severe, the native population is large (enabling it to persist long enough for adaptive shifts in its behavior to occur), and if there is sufficient genetic variation or behavioral plasticity within the native population in its responses to the novel cues of the introduced species (Schlaepfer et al. 2002; Strauss et al. 2006). Theoretical studies suggest that natural populations must number 10,000–100,000 individuals to contain sufficient genetic variability to adapt to even slowly changing, unidirectional, selective regimes (Nunney 2000; Reed & Frankham 2003). Furthermore, theoretical and empirical studies suggest that rapid adaptation is most likely to occur in populations that are growing quickly, possibly because it offsets the demographic cost of selection (the proportion of individuals experiencing reduced fitness due to the trap; Reznick & Ghalambor 2001).

The proximate pathways by which these evolutionary responses occur are unknown, but we speculate that selection will modify Darwinian algorithms to either refine the cue values they rely upon, or incorporate new cues that now correlate with positive outcomes in the novel environment (figure 29.1). In the example of the sea turtle hatchlings that are trapped by beach front lights, natural selection will favor individuals with a narrower acceptance threshold that distinguishes between natural and artificial light sources. Selection may also favor adding additional cues to the recognition template to ensure that the original behavior is elicited only in its adaptive context. For example, as day lengths and average temperatures become increasingly dissociated (Visser et al. 2004), passerine birds might

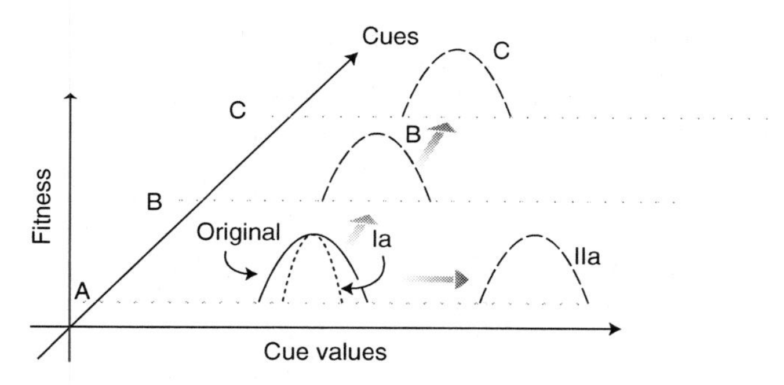

FIGURE 29.1 Responses of Darwinian algorithms to evolutionary traps. In the original, unaltered environment (solid line), a key behavior is elicited within a certain value range of cue A, and the elicited behavior is adaptive under these circumstances. In a suddenly altered environment, however, the formerly adaptive Darwinian algorithm can lead an organism into an evolutionary trap. But Darwinian algorithms can be modified (through learning or evolution) to match novel circumstances. If cue A still carries relevant information in the novel environment, the range of cue values under which the behavior is elicited can evolve (dashed line IIa). For example, with rising global temperatures, turtle species with temperature-dependent sex determination can be selected to adjust their critical temperature upward or alter their nesting behavior to maintain a balanced sex ratio (Janzen 1994). Furthermore, a narrower acceptance threshold (Sherman et al. 1997; dotted line Ia) or additional cues (B, C, etc.) can be added to the recognition template to ensure that the original behavior is elicited only in suitable circumstances. For example, mayflies might use a refined set of criteria, including the size, color, or smell of an object to discriminate among shiny polarizing surfaces and suitable ponds for ovipositioning.

evolve or learn to rely on different environmental cues (e.g., leaf flush) to optimize egg-laying dates.

Extinctions

In the absence of dispersal, phenotypic plasticity, or rapid evolutionary changes, trapped populations may go extinct. All extinctions and population extirpations can be viewed as evolutionary failures (Holt & Gomulkiewicz 2004). In other words, a declining population, by definition, is composed of individuals that are maladapted to their environment. The vast majority of extinctions are probably caused by blatant disturbances, in which a change in the mean characteristic of some environmental attribute affects all individuals negatively and there is no possibility of escape.

To our knowledge, no population extinction has been unambiguously linked to an evolutionary trap. However, such cases may have occurred but gone undocumented, probably because they were chalked up to blatant disturbances. There are, nonetheless, numerous circumstances that could hinder escape from an evolutionary trap. Trapped small populations will be especially vulnerable to extinction if the fitness costs of making a suboptimal choice are large. Small population size and limited genetic variability can also prevent populations from adapting to sudden, large, or sustained changes in selective regimes (Reed & Frankham 2003; Holt & Gomulkiewicz 2004). Indeed, natural selection exacts a demographic cost on populations (Nunney 2003), so even those that have the genetic variability to potentially adapt to a novel selective regime may drop below a critical size threshold that exposes them to inbreeding, drift, and Allee effects, which can exacerbate the declines and ultimately lead to extinction (Holt & Gomulkiewicz 2004).

Management Implications

Behaviorally Informed Management

Wildlife management practices can potentially lead organisms into evolutionary traps if they do not consider past evolutionary forces and resulting Darwinian algorithms. For example, Semel and Sherman (2001) reported that erecting nest boxes for wood ducks *Aix sponsa* in clusters over open marshes (i.e., the traditional management practice) had detrimental effects on reproduction because it did not consider the birds' Darwinian algorithm for nest-site selection. Wood ducks nest normally in cavities of living or dead trees, and their clutch size is 10–12 eggs. Because suitably large, relatively predator-proof cavities in old trees have been widely dispersed and difficult to locate over the course of the species' evolution, young females often follow established nesters to active nests. A follower will sometimes lay eggs in the cavity and then either contest ownership of it or simply leave the eggs behind. By placing boxes in groups over open water sites, managers attempted to make cavities easier to find. Unfortunately, the conspicuousness of nest locations made it too easy for females to follow others to their active nests, resulting in supernormal intraspecific parasitism (egg dumping), with some nests containing 30–50 eggs. Supernormal clutches could not be incubated properly: some nests were abandoned, and in others eggs were crushed and fungal infections occurred. As a result, individual reproductive success suffered and population productivity declined. When nest boxes were repositioned individually on tree trunks in dense woodlands, the natural difficulty of following conspecifics through occluded habitats was reestablished. Parasitism was reduced to its normal, nondeleterious levels (3–4 eggs per nest), and productivity increased (Semel et al. 1988). This example illustrates the importance of elucidating behavioral decision-making rules in developing evolutionarily informed management strategies.

Managing Trapped Species

The most straightforward way to manage species that are known or suspected to be caught in evolutionary traps is to restore the native environment and its associated selective regime, so that the species' Darwinian algorithms and their cues yield adaptive outcomes once again. This is what managers have done by repositioning wood duck nest boxes in forests. However, it is often impractical or impossible to restore native selective regimes. In such cases, we suggest another approach: subsidizing the survival of native species until they have adapted to their novel environmental circumstances and are able to persist on their own (Schlaepfer et al. 2005; Carroll & Watters 2008). We can envision three ways in which this could be accomplished: (1) deliberate exposure to the new selective agent while mitigating its negative effects, (2) introduction of adapted individuals from other populations of the trapped species, and (3) use of learning techniques.

First, conservation biologists and wildlife managers might be able to create conditions in which native species are exposed to sufficient selective pressure to drive an evolutionary change in behavior or Darwinian algorithms, but not so strong as to extirpate a local population. For example, this might be accomplished by creating temporal or spatial refugia. In the case of naïve anuran larvae, introducing a dense lattice work of aquatic roots and stems could offer spatial refugia from predation by evolutionarily novel predators such as American bullfrogs or nonnative predatory fishes. In areas with a mix of refugium and nonrefugium habitat, natural selection will favor behaviors and decision rules within prey populations that will enable them to escape novel predators, thus facilitating long-term coexistence—for example, increased predator-detection abilities, increased use of refugia, changes in foraging times or behaviors to avoid temporal overlap with the predator, or increased escape swimming speed. This type of intervention would be particularly useful for species or populations with small ranges, in which refugia can be added to most or all of the inhabited areas.

In addition to creating refugia, managers could attempt to temporarily reduce the abundance—but not eradicate—an evolutionarily novel selective agent, such as a nonnative predator. Again, the goal is to maintain sufficient selective pressure to favor the emergence and spread traits within the prey population that are likely to facilitate long-term coexistence with the novel predator, but without jeopardizing the persistence of the prey population. One way to guard against the inadvertent extinction of the prey population is to combine the novel selection with short-term population growth (Reznick & Ghalambor 2001) through, for example, supplemental food, short-term habitat enhancement, or head-starting life stages that are not normally subject to predation. The goal of the short-term population boost is to compensate for the loss of poorly adapted individuals and to favor the spread of behaviors (whether learned or genetically encoded) that are better adapted to the novel selective regime. Once this has been achieved, intensive management efforts would no longer be necessary. The important difference between this approach and traditional management is that it deemphasizes indefinite efforts to manage or eradicate the invader (which are costly and often futile) in favor of short-term environmental manipulations designed to help native populations successfully transition to the new selective environment that includes the invader (Rice & Emery 2003).

The second general approach we suggest is to manipulate the genetic composition of native populations to help guide and increase their rate of evolution. This could be done by inoculating naïve populations with conspecifics from experienced populations that contain morphological or behavioral traits that potentially are useful in escaping the trap. Provided that the desirable behavioral trait (e.g., predator avoidance behavior, disease resistance) is partially heritable, the hybridization of naïve and experienced individuals will increase the likelihood that at least some of the natives will produce better adapted offspring. For example, experienced red-legged frog tadpoles that have somehow survived the invasion of American bullfrogs (e.g., by evolving a more restrictive predator-recognition template or more effective escape behaviors) could be used to inoculate naïve populations before the front of the expanding American bullfrog invasion reaches them. The intentional hybridization of conspecifics from populations with different characteristics is a tool that has been used in crop and agricultural science, but to our knowledge has not been tried in a conservation or wildlife management context. Although there are certain risks (e.g., disease transmission, outbreeding depression, irrevocable changes in genetic composition), such evolutionary tactics could prove essential for managing populations in a rapidly changing world (Ashley et al. 2003; Carroll & Watters 2008).

The third general approach we suggest is using learning techniques to help trapped organisms survive in their novel environments (Griffin 2004). Learning has been used, for example, to facilitate the reintroduction of endemic prey from small, predator-free islands to areas where they face a suite of novel predators (Griffin et al. 2000; Griffin 2004). Thus, rufous hare-wallabies *Lagorchestes hirsutus* increased their vigilance toward models of foxes *Vulpes vulpes* and cats *Felis catus* if previous exposure to the models occurred in conjunction with alarm sounds or water squirts (McLean et al. 1996). Similarly, the frog-eating bat *Trachops cirrhosus* can learn to associate cues of novel prey with unpalatability (Page & Ryan 2005), which raises the possibility of training the bats to avoid threatened or endangered species of anurans. Many vertebrates can rapidly learn to associate chemical, visual, and auditory cues with a novel predator or prey (Crossland 2001; Griffin 2004). For example, Chivers and Smith (1995) showed that free-living fathead minnows from a pike-free lake were initially naïve to this potential predator, in that they showed no predator avoidance response. Two weeks after the release of pike into the lake, however, fathead minnows showed strong predator-avoidance responses to the cues of pike. Collectively, these examples suggest that learning may provide a useful tool in many taxa for initially coping with novel situations, particularly in a predator-prey context.

CASE STUDY: BREEDING CHRONOLOGY OF GREAT TITS

One consequence of global climate change is that the phenological response of species that rely on temperature cues to initiate important life-history events such as breeding and hibernation will be altered, potentially causing mismatches in timing. In many temperate bird species laying date is phenotypically plastic, with females laying earlier in years with warmer spring temperatures, so that chick development coincides with peak resource availability (Visser et al. 2004). But what will happen as the global climate warms? Because breeding pairs must commence nest building and egg laying well before prey abundance peaks, the environments in which key life history decisions are being made are separated temporally from the selective environment that will affect chick growth and survival.

Between 1973 and 1995 early spring temperatures increased significantly in parts of Holland. However, mean egg hatching dates of Dutch great tits (*Parus major*) did not, in part because the egg-incubation period is relatively inflexible. In contrast, the peak abundance of caterpillars, which are the primary food source for great tit chicks, advanced by an average of 9 days (Visser et al. 1998). The result is that chicks are now hatching after the peak availability of insect biomass. All great tits lay earlier during warmer springs, but recent evidence indicates that those females that show greater phenotypic plasticity relative to the population mean (i.e., advance their egg laying date the most in response to warmer spring temperatures) enjoy higher reproductive success. Thus, selection on the heritable component of the great tit's reaction norm is resulting in earlier egg-laying dates and decreasing the mismatch between the timing of insects and chicks (Nussey et al. 2005). Spring temperatures, however, continue their warming trend, and caterpillar peak dates are advancing three times faster than the rate of change in egg-laying dates, resulting in a decline in mean fitness of great tits (figure 29.2). It remains uncertain whether the great tit's reaction norms will be able to keep pace with the warming trends they are experiencing.

It should be noted that population declines have not occurred in all great tit populations. Egg laying in at least one population in the United Kingdom appears to be closely tracking temperatures (Charmantier et al. 2008). In this case, the birds start laying earlier in warmer springs and later in colder springs, indicating considerable intraspecific geographic variation in phenotypic plasticity in breeding phenology.

Will the Dutch great tit population go extinct? It is conceivable that some Dutch great tits could migrate farther north each spring, thus reestablishing the timing between hatching date and seasonal phenologies of trees and insect flushes. Alternatively, local migrations between different habitat types or shifts to novel prey items (that emerge later) may help rescue some populations. Should populations of passerines become seriously threatened by the effects of global warming, conservation biologists might consider capturing some birds, translocating them several hundred kilometers northward, and releasing them in suitable breeding habitats. If the translocated birds breed successfully, and the young imprint on their natal habitat and return there subsequently (Davis 2008; Stamps et al. 2009), the match between Darwinian algorithms and selective outcomes would be reestablished, at least temporarily.

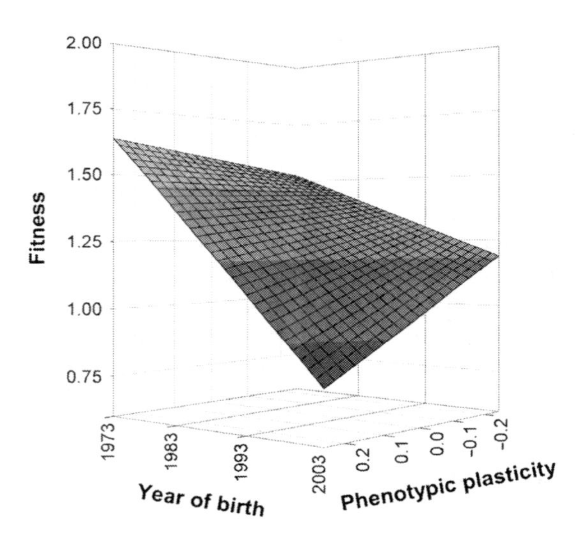

FIGURE **29.2** The average fitness of great tits (*Parus major*) has been declining since 1973 because of an increasing mismatch between timing of chick hatching and the peak abundance of their primary prey. Females that advance their laying date more strongly than average during warm springs (denoted as negative plasticity), however, experience a small reduction in fitness. Thus, selection appears to favor females with greater plasticity. Figure provided by Daniel Nussey; redrawn from data in Nussey et al. (2005).

REVIEW AND SYNTHESIS

In general, Darwinian algorithms underlying behavioral and life history decisions are only as complex as is necessary to yield adaptive outcomes under normal environmental circumstances, but not so complex as to cover all experimentally or anthropogenically induced contingencies (Schlaepfer et al. 2002). Thus, novel selective regimes are likely to elicit evolutionary traps because the efficacy of the underlying Darwinian algorithms has not yet been vetted by natural selection in the altered circumstances.

The evolutionary trap concept illustrates the importance of understanding how past evolutionary pressures shape the behavioral and life history decision rules of organisms and how sudden changes in selective circumstances can result in maladaptations. In a world that is rapidly changing under the dominion of humans, the prevailing genotype and phenotype of any population can no longer be assumed to be adaptive. Behavioral ecologists will increasingly be challenged to explain whether the behaviors they are observing are suited only to yesterday's environments, and conservation biologists will increasingly be challenged to pry open the jaws of evolutionary traps to help species survive and reproduce in novel environments.

FUTURE DIRECTIONS

To advance our understanding of evolutionary traps and how to manage them, we need both ways to predict what environmental alterations are most likely to create traps and how to help populations escape them. To address these objectives, we have identified three lines of research:

1. Are certain past selective regimes, behaviors, or life history strategies particularly vulnerable to traps (Rice & Emery 2003)? A start on this ambitious agenda was made by Ashley et al. (2003), who predicted that populations exposed to long periods of stabilizing selection will not have the underlying additive genetic variation to adapt to a novel selective regime, and Battin (2004), who suggested that small population size, low learning ability, and slow rate of evolution will make an organism vulnerable to evolutionary traps.
2. Which species can respond adaptively to invasions by non-native predators, competitors, or pathogens via learned changes in behavior? Among plants, and animals that do not readily learn, how strong must selection be and how much time must pass before adaptive responses are established?
3. Do traps affect animals that can disperse with relative ease (e.g., birds, insects) less, or in different ways, than animals and plants with low dispersal capabilities? Do traps affect animals that can readily learn less, or in different ways, than plants and animals with limited learning abilities?

SUGGESTIONS FOR FURTHER READING

Schlaepfer et al. (2002) introduced the evolutionary trap concept, and Schlaepfer et al. (2005) applied it to effects of introduced species on native species. Ghalambor et al. (2007) and Visser (2008) provide good overviews of phenotypic plasticity and the evolution of underlying reaction norms. Holt and Gomulkiewicz's (2004) chapter illustrates the challenges to a population when confronted with a novel selective regime. A recent book on conservation biology (Carroll & Fox 2008) embodies many of the novel ideas at the intersection of conservation biology, ecology, evolution, and behavior.

Carroll SP & Fox CW (2008) Conservation Biology: Evolution in Action. Oxford Univ Press, New York.

Ghalambor CK, McKay JK, Carroll SP, & Reznick DN (2007) Adaptive versus non-adaptive phenotypic plasticity and the potential for contemporary adaptation in new environments. Funct Ecol 21: 394–407.

Holt RD & Gomulkiewicz R (2004) Conservation implications of niche conservatism and evolution in heterogeneous environments. Pp 244–264 in Ferrière R, Dieckmann U, & Couvet D (eds) Evolutionary Conservation Biology. Cambridge Univ Press, Cambridge, UK.

Schlaepfer MA, Runge MC, & Sherman PW (2002) Ecological and evolutionary traps. Trends Ecol Evol 17: 474–480.

Schlaepfer MA, Sherman PW, Blossey B, & Runge MC (2005) Introduced species as evolutionary traps. Ecol Lett 8: 241–246.

Visser ME (2008) Keeping up with a warming world; assessing the rate of adaptation to climate change. Proc Roy Soc Lond B 275: 649–659.

30

Behavioral Syndromes

ANDREW SIH, ALISON BELL, AND J. CHADWICK JOHNSON

Animals often exhibit individual variation in behavioral tendencies. For example, although individuals typically alter their boldness or aggressiveness depending on the ecological or social context, often, some are consistently more aggressive or bold than others across multiple contexts. We experience this in our own lives in the sense that people have reasonably consistent personalities. Although human personalities seem to embody more complex emotions or intentions than we attribute to other animals (Gosling 2001), in statistical terms, the concept of personality involves within- and between-individual behavioral consistency that can be quantified as suites of behavioral correlations across multiple contexts and over time. Individuals that are more bold (or aggressive, or extroverted, or neurotic) than others in one context retain that tendency in other contexts. Related concepts include temperament, behavioral profiles, or psychological constructs.

In evolutionary ecology, we refer to suites of correlated traits as syndromes, for example, life history (Roff 1992), or dispersal syndromes (Dingle 2001). Similarly, suites of correlated behaviors across multiple contexts can be called *behavioral syndromes* (Sih et al. 2004a, 2004b). A behavioral syndrome is a property of a group; individual units in the group each have a *behavioral type* (e.g., a more versus less aggressive behavioral type). In most of this chapter, the group is a population, in which each individual in the population has a behavioral type; however, in principle, the concept could apply to groups of species in which each species has a behavioral type.

Personality has long been a topic of study in several primates, domesticated animals (dogs, cats, farm animals), and laboratory rodents, with much of the focus on genetic and neuroendocrine correlates of personality (Gosling 2001; Koolhaas et al. 1999). Despite previous calls for increased attention to individual differences (e.g., Clark & Ehlinger 1987; Huntingford 1976; Magurran 1993; Riechert & Hedrick 1993; Wilson et al. 1994) and implicit consideration of individual differences in game theory and alternative mating strategies, until recently, this was not a major topic of study in behavioral ecology. Although many people observed that their study subjects appeared to show consistent individual differences in behavior, there was some sense that talking about *animal personalities* was anthropomorphic, perhaps unscientific.

Recently, there has been a surge of interest in animal personality due, at least in part, to important ecological and evolutionary implications of behavioral carryovers and correlations (Sih et al. 2004a, 2004b). If behavior within any given context reflects a behavioral type that carries over to other contexts, then both the mean and variation in behavior in any given context can be shaped by events and selection in other contexts. A male's aggressiveness with a mate can be shaped not only by its benefits and costs in that context, but by selection favoring high aggressiveness in earlier male-male contests, low aggressiveness during later parental care, or, for that matter, by selection on aggressiveness during the nonmating season. If behaviors come in suites or packages, then this suggests the value of thinking about behavior in a broader, multicontext, whole-life-history framework.

Behavioral syndromes imply the possibility of limited behavioral plasticity that can result in suboptimal behavior. Consider a boldness syndrome in which some individuals are bolder than others across multiple contexts. In humans, the very fact that we can identify bold versus shy individuals is associated with the idea that bold individuals sometimes take inappropriate risks, whereas shy individuals sometimes miss out on rewarding opportunities because they are too slow to react. In animals, bold animals might not hide as much as they should when predators are present, and shy individuals might hide too much when predators are absent. The notion that behavioral carryovers happen, and that they can explain suboptimal behavior, has been invoked to explain inappropriate risk taking by prey (Sih et al. 2003), excessive sexual cannibalism (Arnqvist & Henriksson 1997; Johnson & Sih 2005), inappropriate aggressiveness (Duckworth 2006), and poor parental care (Ketterson & Nolan 1999).

In practical terms, behavioral syndromes suggest the need to both quantify individual variation in behavior, and to follow individuals across situations. Many experimental studies in behavioral ecology focus on the group's average behavior (e.g., average female preferences), or compare shifts in average behavior across different treatments (e.g., refuge use in the presence/absence of risk). Behavioral syndrome studies focus on the individual variation that is invariably observed. Studies that compare treatments typically use different individuals in the different treatments (for statistical independence). Behavioral syndrome studies note the value of tracking the same individuals to look for limited plasticity and correlations across treatments.

Another exciting aspect of behavioral syndromes lies in the fact that it provides a framework for integrating traditional behavioral ecology with several related fields. It offers an opportunity to bring together proximate mechanisms (genetics, development, neuroendocrine) that underlie not just one type of behavior, but suites of correlated behaviors, and it examines effects (e.g., on fitness) of this suite over multiple contexts, ideally over the lifetime of the organism. It draws together behavioral ecology and evolutionary theory involving phenotypic plasticity and genetic correlations (see chapters 5 and 6). And it pulls in insights from the human personality literature into the study of animal behavior. Finally, it offers an integrated view on how a broad range of behaviors might have important ecological implications. In essence, it invokes the Tinbergian, four-pronged approach of integrating current function, evolution, development, and proximate mechanisms.

In this chapter, we summarize emerging insights on the integration of proximate and evolutionary views on behavioral syndromes, describe three particularly detailed case studies on behavioral syndromes, and suggest directions for future study that both relate to classic issues, and new questions not previously emphasized in behavioral ecology.

WHAT IS A BEHAVIORAL SYNDROME?

A behavioral syndrome occurs when individual differences in behavior in one context or situation are correlated with individual differences in behavior in a different context or situation (figure 30.1). Behavioral syndromes can be visualized either via a reaction norm framework (figure 30.2a) or as a behavioral correlation (figure 30.2b; also see chapters 5 and 6). A reaction norm line quantifies a genotype's traits in two (or more) environments. Reaction norms can be plastic (have a nonzero slope) or nonplastic (i.e., show the same trait value in all environments). When genotypes differ in their response to environmental variation (i.e., if reaction norms differ in slope), this results in a significant genotype × environment (G × E) interaction (chapter 6).

Because our plots show the environment-dependent behavior of individuals and not of genotypes, technically speaking, our lines are not reaction norms. We refer to them instead as *plasticity plots* (cf. Sih et al. 2004a). Still, it is useful to use the reaction norm framework to illustrate key concepts. A behavioral syndrome exists when individuals tend to exhibit parallel reaction norms that result in significant differences among individuals in their behavioral type (Nussey et al. 2007). This produces a positive correlation between behavior in the two environments. It is important to note here that a behavioral syndrome can occur even

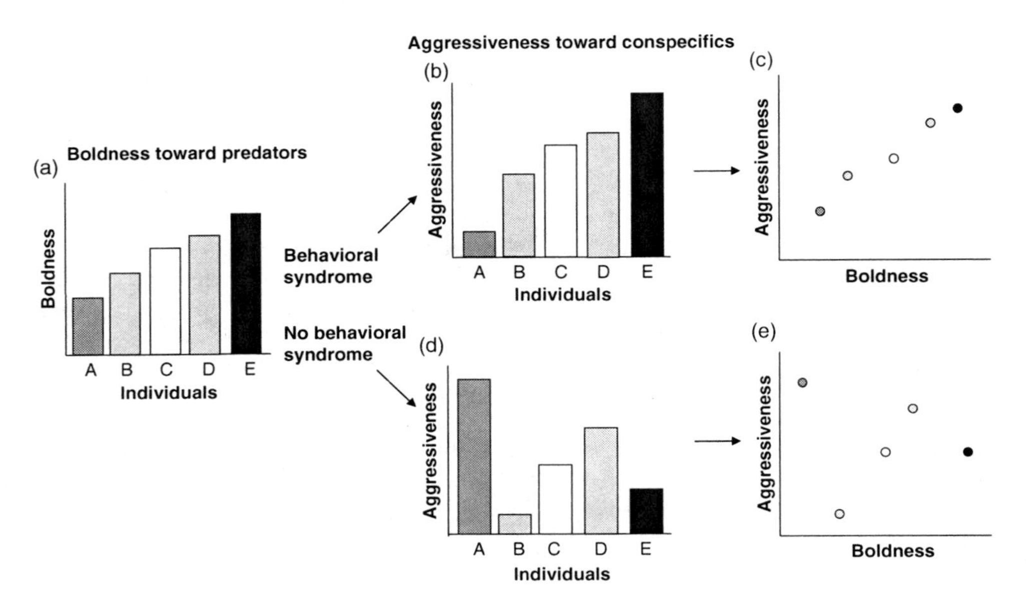

FIGURE **30.1** A behavioral syndrome occurs when individual differences in behavior are correlated across contexts or situations. For example, the individuals that show the highest level of boldness toward predators (a) are also the ones that are especially aggressive toward conspecifics (b). This produces a correlation between boldness and aggressiveness across individuals (c). When there is no relationship between behaviors across contexts or situations (d), then behaviors are not correlated and there is not a behavioral syndrome (e).

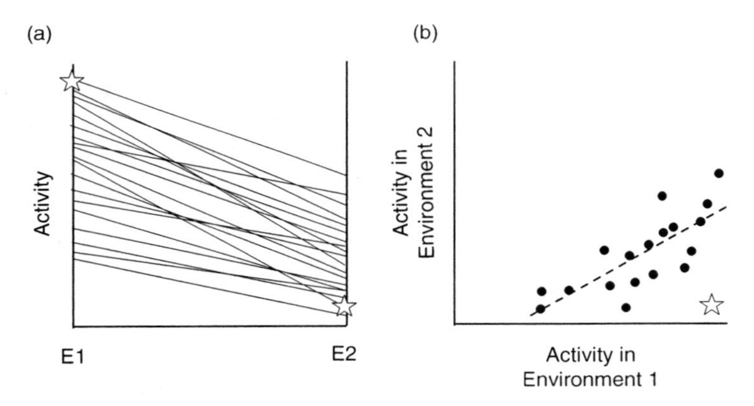

FIGURE **30.2** Graphical representations of behavioral syndromes showing activity in two environments. E1 and E2 might represent, for example, activity in the absence and presence of predators. These plots are analogous to reaction norm (a) or genetic correlation plots (b) of quantitative genetics, except that for our purposes, each line (a) or point (b) can represent either a genotype or simply an individual. The stars are the optima for each environment. In (a), the optimal reaction norm connects the two stars, but these hypothetical data show that although all the individuals reduce activity in E2 (show plasticity), no individuals hit the optimal value in both environments. This point is also illustrated in (b), in which the same data as in (a) are plotted, showing that there is a positive correlation between activity in both environments, and that no individuals show the optimal level of activity (indicated by the star) in both environments.

when individuals plastically modify their behavior in response to the environment (figure 30.3). In contrast, if reaction norm lines frequently cross, so that the rank order among individuals differs significantly in the two environments, then individuals are plastic with no behavioral syndrome.

A behavioral syndrome can involve the following: (1) behavioral consistency through time (e.g., a positive correlation between feeding voracity in juveniles and adults in which the same individuals that were more voracious than others in the juvenile stage remain more voracious than others as adults); (2) correlations involving the same basic type of behavior but in different contexts (e.g., aggressiveness in parental and competitive contexts; Ketterson & Nolan 1999); or (3) correlations involving different behaviors (e.g., a positive correlation between aggressiveness with conspecifics when predators are absent and boldness when predators are present; Riechert & Hedrick 1993; Bell 2005). For illustrative purposes, we often compare the behavior of two extreme behavioral types (e.g., bold versus shy, proactive versus reactive, or more aggressive versus less aggressive). In principle, however, variation in behavioral types could be either continuously distributed or classified into two (or a few) discrete categories.

Although a few axes of behavioral variation have dominated the literature to date on behavioral syndromes (Reale et al. 2007), for example, aggressiveness, exploratory behavior and boldness, there is often important and interesting behavioral variation along other, relatively unexplored axes, such as parental behavior, choosiness, and environmental sensitivity. Indeed, the work on coping styles (discussed below) has shown that individuals differ not only in aggressiveness, but in how they cope with changes in the environment (Benus et al. 1987, 1990). Although the relatively unaggressive and timid individuals react strongly to changes in the

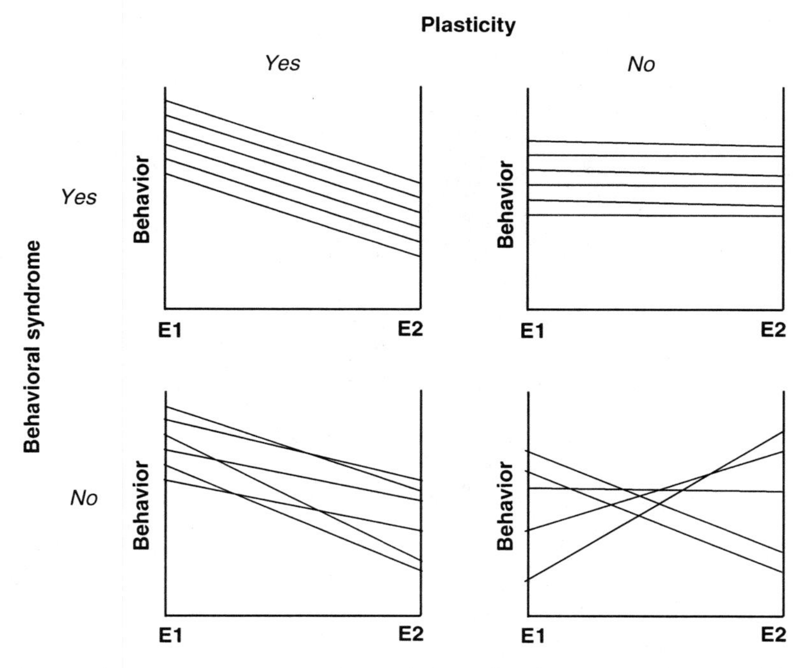

FIGURE **30.3** Behavioral syndromes can be compatible with phenotypic plasticity. This table shows different combinations of behavioral syndromes (*yes/no*) and plasticity (*yes/no*). The behavior of different individuals in different environments (E1 and E2) is represented by different lines. When there is plasticity, on average, individuals shift their behavior "down" in E2. When there is not plasticity, behavior does not differ between the two environments. When there is a behavioral syndrome, the rank order differences between individuals is maintained in both environments.

environment, the behavior of the aggressive type is more rigid and stereotyped. Variation in environmental sensitivity then forms part of the behavioral type (see also Kralj-Fiser et al. 2007; Wolf et al. 2008). In that case, an axis of behavioral variation (the y-axis in a reaction norm diagram) is not aggressiveness or boldness, but instead is behavioral plasticity per se.

Some clarification of misunderstandings about our definition of a behavioral syndrome should prove useful. First, although behavioral correlations can be particularly interesting when they result in suboptimal behavior, our definition does not require suboptimality or limited plasticity relative to the optimal. In fact, behavioral correlations can be adaptive. Second, behavioral syndromes do not require stability of behavioral types over an individual's entire lifetime. They can be particularly important when they persist over a lifetime; however, even a short-term carryover of aggressiveness can make the difference between life and death if, for example, it means that a male that is pumped up on testosterone behaves inappropriately in the presence of a predator. And third, behavioral syndromes do not, by definition, have to be genetically based. They are more likely to have evolutionary implications if there is a genetic basis to the syndrome, but even if they are shaped primarily or even exclusively by experience, behavioral carryovers can be ecologically important.

DRAWING INSIGHTS FROM OTHER FIELDS

Neuroendocrine and Genetic Basis of Behavioral Syndromes

The root of a behavioral syndrome could lie in a shared causal mechanism underlying different behaviors, in which either the same genes (pleiotropy) or the same hormones act on several targets (Ketterson & Nolan 1999). A particularly influential view is that individual differences in behaviors such as aggression, exploration, and response to novelty are manifestations of different hormonally based styles for coping with environmental challenges (Koolhaas et al. 1999). In a broad range of vertebrate species, individuals that are particularly aggressive toward conspecifics also explore the environment more superficially, respond quickly to novelty, and are less attentive to changes in the environment compared to less aggressive individuals. These different styles, variously named proactive-reactive, active-passive, and fast-slow, have been associated with differences in stress reactivity. Proactive individuals have low HPA (hypothalamic-pituitary-adrenal) axis activity and reactivity, high sympathetic reactivity, low parasympathetic reactivity, and high corticosterone reactivity (Koolhaas et al. 1999). Much of the work on coping styles has been motivated by the notion that these styles differ in their susceptibility to disease (Koolhaas et al. 1999).

Some of the most productive work on proximate mechanisms underlying coping styles has centered on artificially selected lines (Koolhaas et al. 1999; van Oers et al. 2005). Mice selected for either quickly attacking an opponent (*proactive*) versus for slower aggressive responses (*reactive*) differ in HPA regulation, serotonin neurotransmission, and hippocampal cell proliferation rate. Reactive mice show greater responsiveness of physiological and molecular stress markers (Veenema et al. 2003). Detailed studies have focused on how structural differences in the hippocampus underlie differences between lines, and on the serotonergic system and corticosterone as important regulators.

An alternative to using selected lines involves studying candidate genes that have been associated with personality variants in other species. For example, polymorphism in a dopamine receptor (DRD4) has been associated with personality differences in humans (Reif & Lesch 2003), and has recently been shown to differ between lines of great tits selected for exploratory behavior (Fidler et al. 2007). Another candidate gene of interest is the serotonin transporter because of the important role that serotonin uptake plays in mediating aggression and other risk-taking behaviors (Champoux et al. 2002). Although the candidate gene approach (Fitzpatrick et al. 2005) is appealing because it allows findings from research on model organisms to be applied to less well-studied organisms, the jury is still out on when the candidate gene approach is most likely to be useful (e.g., only for genes of large effect?), and on the severity of problems such as population substructure (reviewed in Savitz & Ramesar 2004).

Evolution of Correlated Characters

The notion that correlations among traits are common and evolutionarily significant is not new.

Holistic studies of the integrated phenotype (Pigliucci & Preston 2004) and allometry have a long history within biology. For example, it has long been appreciated that a shared developmental origin can produce covariance between two or more seemingly unrelated traits (Olson & Miller 1958). Moreover, a well-appreciated observation from the animal breeding literature is that genetic correlations between traits are important because selection on one trait can produce a correlated response to selection in the other trait (Falconer & Mackay 1996; chapter 3 and box 5.1, this volume). If genetic correlations are strong, they can act as a constraint on adaptation.

This literature has typically focused on correlations among morphological and life history traits, perhaps because behavioral traits are often more plastic compared to other types of traits, but it offers several insights for the evolution of correlated behaviors. One insight is that if a genetic correlation acts as a hard constraint, we would expect the correlation to be fixed within a species (Armbruster & Schwaegerle 1996; Lande & Arnold 1983). For example, although different populations of the same species might differ in the mean level of two behaviors, the correlation between the two behaviors should be the same within each population (Bell 2005). This was tested recently in three-spined sticklebacks (*Gasterosteus aculeatus*) and funnel web spiders (*Agelenopsis aperta*) with mixed results (see "Case Studies," below). Interestingly, in some systems, behavioral correlations appear to have evolved in response to putative selection pressures (Bell 2005; Bell & Sih 2007; Dingemanse et al. 2007).

However, studying correlated behaviors in this framework presents special challenges. Although the evolutionary genetics literature on correlated traits has focused primarily on genetic rather than phenotypic correlations, behavioral ecologists generally collect data on phenotypic correlations among behaviors, which may or may not reflect genetic correlations (Roff 1995). More interestingly, behavior has properties that differ fundamentally from morphological traits that are often studied by evolutionary geneticists (Sih 2004). Because behavioral plasticity can be rapid and reversible (e.g., prey changing activity levels as predators come and go), it is possible to quickly measure the pattern of plasticity exhibited by each individual (see figure 30.2). In contrast, because induced morphologies (and other developmentally plastic traits) are often slow and irreversible (i.e., each individual expresses only one trait in response to one environment), quantifying the pattern of plasticity requires looking at many individuals of the same genotype. Quantifying behavioral correlations as plasticity within an individual can take substantially more effort, however, than quantifying correlations among multiple morphological characters expressed by each individual. Putting a number on an individual's overall aggressiveness can be much harder than measuring the size of its body parts. This is particularly problematic because large sample sizes are required to accurately assess genetic correlations (Lynch & Walsh 1998).

CASE STUDY: PERSONALITY IN GREAT TITS

A particularly comprehensive data set on behavioral syndromes in the laboratory and the field comes from work on a songbird, the great tit (*Parus major*). These birds exhibit repeatable individual differences in exploratory behavior in a novel laboratory environment (Verbeek et al. 1994). Exploratory behavior is correlated with other behaviors: *fast* explorers are also more aggressive, less neophobic, and quicker to form routines relative to *slow* explorers (Verbeek et al. 1996).

By artificially selecting fast and slow exploration in a captive population of great tits, Drent and colleagues showed that individual differences in exploratory behavior are heritable and genetically correlated with aggressiveness (reviewed in van Oers et al. 2005). Detailed physiological analyses of the two selected lines showed that birds from lines selected for slow exploration showed significantly heightened reactions to social stress as measured by fecal corticosteroid concentrations (Carere et al. 2003). Birds from this same slow exploratory line, however, also showed reduced physiological impairment following social defeat relative to birds selected for fast exploration (Carere et al. 2001).

Studies on a wild population of great tits confirmed that exploratory behavior is heritable (Dingemanse et al. 2002) and added the insight that individual differences in exploratory behavior have fitness consequences in terms of dispersal (Dingemanse et al. 2003), survival (Dingemanse et al. 2004), and offspring recruitment/condition (Both et al. 2005). Interestingly, effects of personality on fitness depended on dominance rank, sex, and resource abundance. For example, among nonterritory-holding juveniles, fast-exploring individuals dispersed to greater distances (Dingemanse et al. 2003) and had lower dominance ranking relative to slow explorers (Dingemanse & De Goede 2004). In years of resource limitation, when competition for resources was intense, slow-exploring males and fast-exploring females had significantly higher fitness. In contrast, when competition for resources was relaxed (due to beech masting), fast-exploring males but also slow-exploring females had higher fitness (Dingemanse et al. 2004; reviewed in Dingemanse & Reale 2005).

In other studies, Both et al. (2005) demonstrated that slow-exploring females nested more successfully and reared larger offspring than fast-exploring females. In addition, mating pairs with the same personality type (slow-slow or fast-fast) produced significantly heavier offspring than mating pairs of different personality (Both et al. 2005). However, again temporal fluctuation in resource abundance affected the fitness (offspring recruitment) of these mating pairs. In years of resource abundance, pairs with matching personalities had higher fitness, whereas in years of relative food scarcity, pairs with mismatched personalities had higher fitness. This temporal fluctuation in resource abundance has been invoked as a possible explanation for the maintenance of variation in great tit personality (Dingemanse & De Goede 2004).

Another key set of studies on the fitness consequences of temperament in the wild has also suggested that temporal variation (in this case, in predation pressure) can maintain behavioral variation in bighorn sheep (Reale & Festa-Bianchet 2003; Reale et al. 2000).

CASE STUDY: PERSONALITY IN STICKLEBACKS

In one of the first published cases of an ecologically relevant behavioral syndrome, Huntingford (1976) showed that individual male sticklebacks (*Gasterosteus aculeatus*) that were particularly bold toward a predator during the nonbreeding season were also more aggressive toward rival males during the breeding season. Huntingford hypothesized that a common causal mechanism (e.g., hormonal basis) might be the source of the covariation. The stickleback system has since offered several insights into the evolution of behavioral syndromes because the correlation between boldness and aggressiveness is variable in this species.

Sticklebacks are renowned for their geographic variation in behavior and morphology (Bell & Foster 1994). In particular, predation pressure has a strong effect on stickleback behavior (Huntingford et al. 1994). Notably, sticklebacks from high versus low predation populations differ with respect to the correlation between boldness and aggressiveness. Boldness and aggressiveness are positively correlated (both phenotypically and genetically) in high predation populations, but not in low predation populations (Bell 2005; Dingemanse et al. 2007). Predation was directly identified as the cause of the correlation in an experiment by Bell and Sih (2007), who found that exposure to real predation caused the behaviors to become phenotypically correlated. Altogether, these results show that the relationship between boldness and aggressiveness is evolutionarily labile.

Further evidence that boldness and aggressiveness in sticklebacks are not tightly locked together comes from neuroendocrine analyses. Although both fighting with conspecifics and exposure to predators elicited a cortisol stress response, bold and aggressive behaviors were associated with different neuroendocrine responses. Whereas serotonin level was negatively correlated with frequency of attacking a conspecific, it was positively associated with predator inspection (Bell et al. 2007). These data suggest that *bold* and *aggressive* behaviors are not simply different manifestations of the same underlying tendency.

The development of individual differences in behavior in sticklebacks also illustrates the tension between continuity and change (Roberts et al. 2006; Suomi et al. 1996), or constraint and flexibility, underlying correlated behaviors or personality. For example, individual sticklebacks that were particularly bold as juveniles did not necessarily become exceptionally bold as adults (Bell and Stamps 2004). However, Bell and Stamps (2004) found that the relationship between boldness and aggressiveness was stable over development in one population.

Another interesting twist is that behavioral correlations generally weakened at the onset of sexual maturity (Bell and Stamps 2004). Perhaps the associated hormonal reorganization triggered a restructuring of a suite of behaviors. This study suggests the interesting possibility that behavioral correlations might be especially likely to be broken apart during major transitional periods because different suites of traits might be favored at different ages, especially if different age groups inhabit very different selective environments.

CASE STUDY: PERSONALITY IN SPIDERS

Riechert and colleagues studied behavioral syndromes in the funnel web spider, *Agelenopsis aperta*. Funnel web spiders demonstrate positive correlations among aggression toward prey, including superfluous killing, boldness toward predators, aggression toward conspecifics, and the size of territory defended. Controlled crosses revealed a genetic basis (probably pleiotropy) to these behavioral correlations (see references in Maupin & Riechert 2001).

In addition to being correlated at the individual level, these behaviors are also correlated at the population level. Riparian populations of *A. aperta* enjoy six times higher prey availability but suffer significantly higher predation risk relative to their counterparts from grassland populations. As one might predict then from a traditional adaptive paradigm, spiders from riparian populations are significantly shyer toward predation risk, less voracious toward prey, less agonistic toward conspecifics, and defend smaller territories than grassland spiders. However, what is most interesting is that within the adaptive modulation going on across populations, we continue to see correlations at the individual level. In other words, although riparian spiders are, on average, shy toward predation risk, some individuals from riparian populations continue to show high levels of boldness toward predators, and this individual variation can be explained by the fact that these bolder individuals are also more voracious toward prey. Conversely, grassland populations are, on average, more voracious toward prey, but some individuals continue to show low voracity toward prey, and this can be explained, at least in part, by the fact that these individuals are also exceedingly shy toward predators (Riechert & Hedrick 1993). This work provides a beautiful illustration of the tug-of-war occurring between adaptation to one's environment at the population level and potential constraints arising from behavioral correlations occurring at the level of individuals.

More recently, Johnson and Sih (2005, 2007), working with the semiaquatic fishing spider *Dolomedes triton*, investigated a connection between behavioral syndromes and precopulatory cannibalism by females on courting males. Precopulatory sexual cannibalism can be puzzling from an evolutionary point of view particularly when it results in females not mating enough to fertilize their eggs (Arnqvist & Henriksson 1997). The *adaptive foraging hypothesis* (Johnson & Sih 2005) explains precopulatory sexual cannibalism by positing that females that had poor foraging success on other prey (e.g., crickets) might be willing to risk the cost of lost reproduction in favor of the benefits of food. Contrary to this hypothesis, Johnson and Sih found that females that most readily attack males are the voracious females that had the highest levels of recent foraging success on other prey. That is, female spiders exhibited positive behavioral correlations between voracity toward heterospecific prey, boldness toward predation, and tendency to attack males. These data are consistent with the hypothesis that precopulatory sexual cannibalism reflects a carryover of voraciousness toward prey, in general, on to voraciousness toward potential mates. Subsequent studies showed that this carryover is not absolute: if a female grew up experiencing low availability of males, then she is less likely to engage in sexual cannibalism (Johnson & Sih 2005).

INSIGHTS AND FUTURE DIRECTIONS FOR BEHAVIORAL ECOLOGY

Suboptimal Behavior

One key insight well illustrated by the fishing spider example above is that behaviors (such as precopulatory sexual cannibalism) might often appear suboptimal when viewed in an isolated context, but can make sense when seen as part of a package. In essence, if behaviors are correlated across contexts, and if the correlations are stable over an ecologically relevant timescale, then it is the package of behaviors (not isolated behaviors) that should be shaped by natural selection to be adaptive. Depending on the correlation structure and relative selection pressures in different contexts, behaviors in some situations can even appear strikingly maladaptive. For example, salamander larvae exhibit excess boldness that results in high mortality rates when predatory sunfish are present. This can be explained as a carryover from strong selection favoring high activity in the absence of predators (Sih et al. 2003). In patchy populations in which only a low to moderate proportion of larvae encounter predatory fish, selection favoring high activity apparently overrides the cost of high activity in the presence of predators. Similarly, in male water striders, selection favoring aggressiveness can result in hyperaggressive behavior that disrupts the entire social system (Sih et al. 2002; Sih & Watters 2005), and in some birds, selection favoring aggressiveness (and high testosterone levels) might carry over to result in poor parental care (Ketterson & Nolan 1999). We are not suggesting that animals routinely exhibit behavior that appears maladaptive in one context, but when they do, behavioral syndromes might provide a plausible explanation.

Explaining the Existence and Structure of Behavioral Syndromes

If behavioral syndromes sometimes result in limited plasticity and maladaptive behavioral carryovers, why do they exist? Why do individuals show consistent boldness or aggressiveness when it can get them into trouble? Why do humans have a personality? Note that although behavioral syndromes often occur, they are not ubiquitous. In some situations, animals show consistent behavioral types, but in other cases, they do not. What explains the existence of behavioral syndromes and what factors govern variation in behavioral correlations? Next, we discuss alternative approaches to addressing these questions.

Proximate Constraints

One approach to explaining behavioral syndromes invokes constraints based on inflexible proximate mechanisms. For example, individuals might be highly aggressive because they have a particular hormonal profile, and the correlation between boldness and aggressiveness might be due to shared hormonal control.

In general, understanding the relative lability of the mechanisms underlying behavior can help explain why certain behaviors covary whereas others do not. For example, it is likely that neuroendocrine substrates that are relatively fixed, or that are the product of organizational effects of hormones early in development (for example, receptor density or structural differences in the brain), underlie behavioral syndromes that are stable over time. In contrast, short-term fluctuations in hormone levels that are related to several behaviors are less likely to cause long-term behavioral consistency. One of the advantages of this approach is that it can not only explain why behavioral syndromes occur (because they are caused by a proximate constraint), but it makes clear predictions about which behaviors are likely to covary (the ones that are affected by the same neuroendocrine pathways; Bell 2007).

New computational and genomic tools hold promise for identifying the proximate basis for behavioral syndromes. For example, we could use unbiased, whole-genome approaches such as expression microarrays to identify candidate genes and pathways that are expressed following the performance of different kinds of behaviors. Other genetic approaches such as QTL mapping can be used to determine whether individual variation in two or more behaviors is associated with the same genomic regions. Given the ubiquity of G × E interactions, pleiotropy and epistasis, comparing the structure of genetic networks in different environments could potentially help to explain variation in behavioral syndromes.

Adaptive Approaches

An alternative approach assumes that behavioral syndromes are adaptive (within constraints). Recent theoretical papers have proposed three types of adaptive hypotheses to explain why individuals have a consistent behavioral type: (1) the benefits of predictability, (2) the benefits of specialization, and (3) what we will call *individual status quo* selection.

With regard to predictability, in any game, predictable individuals run the risk of falling victim to being exploited. Prey that use predictable escape routes will be relatively easily captured, and predators that use predictable attack patterns can be more easily evaded. In social circumstances, if you are reliably cooperative, you can easily be cheated. So when is it beneficial to be predictable? Dall et al. (2004) suggest that consistent behavior can be favored if it allows one to manipulate the behavior of others via credible threats or promises. A threat to fight to the death can induce a rational opponent to back off rather than engage in a deadly fight. However, such a threat should be taken seriously only if it is reliable (credible), and that requires predictability. Similarly, a promise to reciprocate can induce a partner to be altruistic, but only if that promise is reliable.

McNamara et al. (2009) explored the joint evolution of trustworthiness (a personality axis) and social sensitivity about the trustworthiness of others. They found that if the population includes both cooperators and cheaters, then this favors the evolutionary maintenance of social sensitivity. Sensitive individuals accept sampling costs (e.g., the effort required to evaluate and remember the trustworthiness of others) to gain the benefits of extra social information. Insensitive individuals save on the costs of sampling, but know less about their social partners. Conversely, if the population includes some socially sensitive individuals, then this favors the persistence of both cheaters and reliable, trustworthy cooperators. Cheaters exploit insensitive cooperators, whereas reliable cooperators gain the trust of sensitive reciprocal altruists. In humans, there are data on individual variation in trustworthiness in economic games; developing a parallel literature on other animals should prove insightful.

With regard to behavioral specialization, Sih et al. (2004b) suggested some ideas drawing on an analogy with models of adaptive plasticity. They speculated that in a variable environment, it might be adaptive for individuals to show limited plasticity because of a cost of switching behavioral types, particularly when individuals experience uncertainty about current environmental conditions and thus about the optimal behavior at any given time. McElreath and Strimling (2006) confirmed the second part of this conjecture with a formal model (based on Sih 1992). If an animal is uncertain about whether predators are present or not, then fixed behavior might be favored over inaccurate tracking of a changing environment. McElreath and Strimling added the insight that differences between individuals in a state variable (e.g., size or vigor) that determines relative ability to be bold versus shy can explain why some individuals are bold, whereas others are shy. State variables (size, energy reserves, life history stage, information state, skill level) carry over across time. They change, but relatively slowly. Although behavior, in theory, can be infinitely plastic, if the best behavior is connected to a slower, more stable state variable, then that connection can explain stability of the behavioral type. McElreath and Strimling (2006), however, did not address how differences in state might arise or why they should be maintained.

Wolf et al. (2007) examined a specific mechanism for generating variation in state. Individuals that are more exploratory early in life should accumulate more energy reserves (assets, a state variable), whereas those that explore less should have lower reserves. Following the asset protection principle (Clark 1994; box 12.2, this volume), animals that have more to lose (more assets to protect) should be less bold and less aggressive rather than risk losing their substantial assets. In contrast, animals with little to lose should take greater risks (be more bold and more aggressive). As long as assets do not change appreciably, animals should maintain a stable behavioral type. McElreath et al. (2007) noted, however, that asset protection is a negative feedback process that should not tend to maintain differences among individuals in state. Individuals that are more bold and aggressive should garner more resources and increase in state, whereas those that play it safe (are shy and unaggressive) should gradually decline in state. That is, energy state should converge over time, and thus, so should behavior. Via this process, if behavior affects state, then over longer periods, there should be no stable behavioral types.

To maintain stable differences in state and thus behavioral type, we need positive feedback between behavioral type and state. For example, boldness generally results in higher feeding rates, which increase the individual's state. Positive feedback occurs if higher state then increases the tendency to be bold, which further increases state and so on (and vice versa for shyness). A simple mechanism generating this positive feedback loop might be state-dependent safety, in which higher state (e.g., larger size, greater vigor) reduces the actual risk associated with being bold. This allows bold, high-state animals to continue to forage boldly, garner energy, and further increase their high state. This is in contrast to standard asset protection in which high-state individuals should be less bold (to protect assets) and thus decline in state over time. Conversely, if it is too dangerous for low-state individuals to forage boldly, then they might be stuck with low state and a shy behavioral type.

The positive feedback loop in the previous paragraph can explain the existence of two behavioral types—shy versus bold. To explain a continuous range of behavioral types, we might need *individual status quo selection*, in which individuals do best if they continue to do what they have been doing. An obvious mechanism that can produce this effect is the benefits of learning a particular behavioral style, for example, very bold individuals might learn how to be good at being bold, which should favor them continuing to be bold, and so on. Alternatively, shy individuals learn how to be shy, and moderately bold learn how to be moderately bold, and so on.

Stamps (2007) proposed a variation on the status quo theme based on the well-known growth-mortality trade-off. She suggested that a range of alternative life histories might yield roughly equal fitness—some exhibiting high growth offset by high mortality, others with low growth but low mortality, and yet others somewhere in between for both. The key is that selection favors individuals maintaining a consistent growth rate (see references in Stamps 2007). Fluctuating growth rates (e.g., slow growth followed by compensatory rapid growth) can be costly (Metcalfe & Monaghan 2001). Individual differences in preferred, optimal growth rate would then explain individual differences in life history and behavioral type. If different life history-behavioral type strategies yield similar fitness, genetic variation can, in theory, be maintained by environmental variation, a mutation-selection balance, pleiotropy, or negative frequency dependent selection (see the following section, and chapters 3 and 5).

Although we now have several hypotheses for explaining the existence of behavioral consistency, other fundamental questions about the structure of behavioral syndromes remain largely unexplored. What explains variation in behavioral correlations? For example, why are boldness and aggressiveness sometimes correlated, but other times not (Bell 2005; Bell & Sih 2007; Dingemanse et al. 2007)? What explains variation in the stability of behavioral correlations? A first step toward a general explanation for variation in behavioral correlations was suggested by Stamps (2007). If individual variation in behavioral type reflects individual variation in optimal growth rate (and associated mortality risk), then all behaviors that together influence growth rate and mortality risk should be part of a behavioral syndrome. Boldness should be part of a behavioral syndrome when it involves increased foraging (which yields resources); however, if boldness is predator inspection (which yields information about risk, but takes time and energy away from foraging), this should be less likely to be part of a behavioral syndrome. Notably, Bell (2005), Dingemanse et al. (2007), and Bell and Sih (2007) found that the correlation between boldness and aggressiveness tends to be stronger when animals have been exposed to high predation risk. Theory that explains the adaptive mechanism underlying this pattern and other aspects of the structure of animal personalities remains to be developed.

Game Theory and Behavioral Syndromes

Given that behavioral syndromes exist, an obvious question is, what maintains variation in behavioral types? One way to maintain alternative behavioral types in a population is via negative frequency dependence, in which each strategy does best when it is rare. In behavioral ecology, frequency dependence is often analyzed using game theory. Game theory has long implicitly invoked the existence of behavioral types—hawks versus doves, cooperators versus defectors, producers versus scroungers, sneakers versus territorial males. In the behavioral syndrome context, we are most interested in mixed ESSs that consist of a mix of individuals with distinct behavioral types. Despite a long history of interest in the social ecology of behavioral types, except for when there are clear morphotypes (e.g., large territorial versus small satellite males), few studies have identified individual variation in behavioral types and studied game dynamics involving mixes of individuals of different behavioral type. Notably few studies have experimentally manipulated the mix of behavioral types in social groups to examine individual and group outcomes to quantify how the fitness of behavioral types depends on their relative frequency. Recent studies suggest that the social ecology of behavioral types might be much more rich and complex than the simple scenarios painted by game theory (Mottley & Giraldeau 2000; Sih & Watters 2005). Real individuals are not pure hawks or pure doves, and populations do not have just two or three discrete types, but instead a continuous range of behavioral types with varying levels of social plasticity. Although some individuals have relatively fixed aggressiveness (identifiable hawks versus doves), others alter their behavior depending on social conditions. Note that if the fitness of different behavioral types depends on the mix of behavioral types in the group, then individuals should actively choose social conditions that are better for them. That is, individuals might show behavioral type-dependent social situation choice. Individual fitness might depend heavily on social skill involving both adaptive social plasticity and adaptive social situation choice. At this time, these possibilities remain largely unexplored, and should be fertile ground for more theory and experiments.

Correlational Selection

A complementary evolutionary approach involves direct measurements of selection on behavior. The literature on phenotypic integration suggests that patterns of covariation among traits might reflect the direct operation of selection for those patterns (Pigliucci & Preston 2004). For example, Berg (1960) predicted that correlations between floral and vegetative traits should be weaker in specialized species that rely on a precise fit between their flowers and one or a few pollinator species. This coevolution should require specialists to evolve different, specialized modules for floral traits and vegetative traits (Armbruster & Schwaegerle 1996).

Perhaps something similar is happening with behavioral correlations: if particular combinations of behaviors work well together, then they could be favored by correlational selection (chapter 3). Multiple, alternative combinations of traits then share in yielding the highest fitness. Theory predicts that correlational selection will eventually produce an adaptive genetic correlation between traits because alleles influencing one will be coinherited together with alleles influencing the other trait (linkage disequilibrium). Over time, the covariance structure is expected to evolve such that the major axis is aligned with the direction of selection.

In the same way that we can measure how directional selection favors a particular trait, we can measure how correlational selection favors particular correlations between traits using phenotypic selection analysis (Arnold & Wade 1984), which measures the strength of the relationship between fitness and traits, independent of other measured correlated traits (Brodie et al. 1995). The basic procedure is to regress standardized trait values, their squared terms, and their cross products on relative fitness. The partial regression coefficient of the cross-product term measures correlational selection (Phillips & Arnold 1989).

One of the most famous examples of correlational selection is between color patterns and escape behaviors in garter snakes (Brodie 1992). Although correlational selection is probably one of the most important selective forces in nature, we know little about it, especially for behavioral traits (Kingsolver et al. 2001). This is probably due to the fact that it is difficult to detect statistically and can require large sample sizes ($n > 600$; Kingsolver et al. 2001). Therefore, it remains an open question whether it will prove to be useful for most animals and traits of interest to behavioral ecologists.

Note that if correlational selection favors a behavioral correlation, then individuals should exhibit behavioral plasticity, if necessary, to move closer to the line of alternative behavioral combinations that yield high fitness. Although we know that both selection and plasticity can interact to shape single traits (Relyea 2002b; West-Eberhard 2003), and that correlations can be produced in response to the environment (Armbruster & Schwaegerle 1996; Sgro & Hoffmann 2004), few studies have looked at the relative importance of selection and plasticity in generating correlations. In the context of behavioral syndromes, when selection favors a positive correlation between boldness and aggressiveness, individuals that are mismatched (i.e., that are bold and unaggressive or shy and aggressive) have low fitness. Accordingly, those individuals should shift one or both behaviors toward the adaptive line of high-fitness behavioral combinations.

For example, Bell and Sih (2007) found that exposure to predators generated a positive correlation between boldness and aggressiveness for sticklebacks that was due in part to selection (bold, unaggressive animals tended to get eaten) and in part to correlational plasticity in which shy, aggressive animals changed their behavior (figure 30.4). Further study on the relative roles of selection and plasticity in generating and maintaining behavioral correlations should prove insightful.

In addition, to date, few studies have quantified the mechanism underlying either correlational selection or correlational plasticity. Knowing the mechanism underlying trait correlations is important for understanding the evolutionarily stability of a correlation. Correlations due to pleiotropy will be relatively difficult to uncouple without changing the shared genetic mechanism. In contrast, a correlation that reflects linkage disequilibrium caused by correlational selection can be uncoupled relatively quickly (over evolutionary time) with random mating. A correlation that reflects a plastic response to the environment can be modified even more rapidly—within an organisms' lifetime, particularly if the response is reversible.

Mate Choice Based on Personality

Behavioral syndromes also offer new insights for classic issues in behavioral ecology such as the study of mate choice and sexual selection. In this

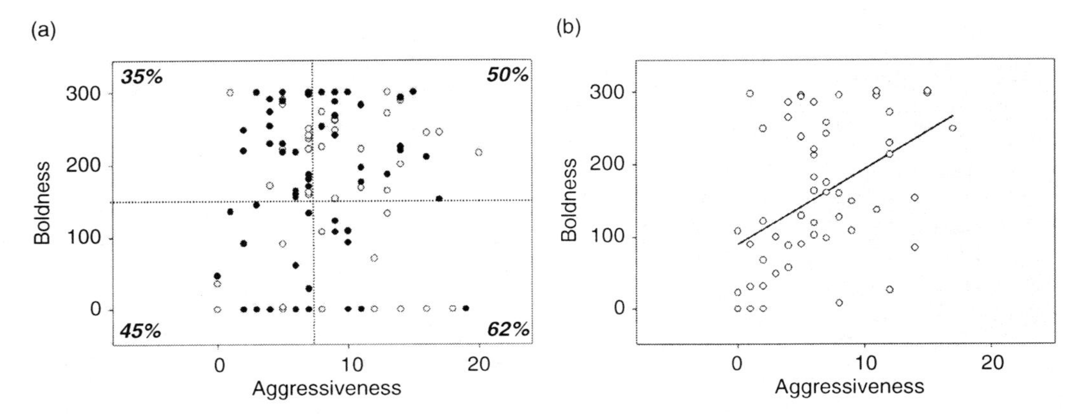

FIGURE 30.4 Exposure to predation generated the boldness-aggressiveness behavioral syndrome. (a) Prior to exposure to predation, individual differences in aggressiveness (number of orients to an unfamiliar intruder) were not correlated with boldness under predation risk (time spent eating following a simulated attack by an egret). Survivors are represented by open circles, and individuals that were consumed by the trout are represented by closed circles. Rates of survivorship in each quartile are marked. Note that survival was nonrandom, but was particularly low for bold and nonaggressive individuals. (b) After exposure to predation, boldness and aggressiveness were positively correlated with each other among the survivors. Reprinted from Bell and Sih (2007) with permission.

field, a key prediction is that females should prefer male traits that are indicators of something that will enhance female fitness—either via good genes or direct benefits. Although in humans, it seems obvious that mate choice is often based not just on resources or looks, but on personality, studies have only just begun to look at the analog in other animals. A male's behavioral type indicates his style of coping with various environmental factors, for example, boldness and aggressiveness influence foraging, antipredator, social (agonistic and cooperative), and dispersal behavior. Given that personality is typically heritable (van Oers et al. 2005), mate choice based on personality can affect offspring fitness by influencing how offspring will cope with various ecological and social factors. Females with different behavioral types might differ in their assessment of what are good male behavioral type genes (e.g., assortative or disassortative mating by behavioral type), or adaptive female choice for male behavioral type might depend on the future environment faced by offspring (which could depend on the female's nest site selection).

Alternatively, females can choose males based on direct benefits, for example, resources provided by males. Female preference for more aggressive males that win contests presumably reflects the ability of those males to monopolize resources provided for females or offspring. Interestingly, in some cases, females prefer less aggressive males because they are less likely to engage in costly sexual coercion that can injure females (Ophir et al. 2005), or perhaps because they are more likely to be good parents (Ketterson & Nolan 1999). The latter issue is particularly interesting because it involves the use of courtship displays to signal deferred direct benefits, for example, future parental care, or cooperation. A key issue here is honesty. When a male promises to deliver future benefits, what keeps him from being deceptive? One idea is that honesty is enforced by high signal costs. In the behavioral syndrome context, if there is a negative correlation between aggressiveness and parental care (Ketterson & Nolan 1999), a male's behavioral type during male-male competition or courtship displays might carry over to be an accurate indicator or index of his future affiliative behavior (e.g., future cooperation in parental care).

Another aspect of male behavioral type that has been shown to influence female choice is the male's responsiveness to her signals during courtship (Patricelli et al. 2006). Male bowerbirds that display too vigorously, regardless of the female's signals, can drive females away. Using a robot female to manipulate female signals, Patricelli et al. (2006) showed that variation among males in social responsiveness can explain a substantial amount of variation in male mating success. Overall, further studies on the possibility of mate choice based on behavioral type should prove insightful.

Going beyond the Standard Syndromes

To date, almost all studies of behavioral syndromes have focused on individual variation in boldness, aggressiveness, activity, exploratory tendency, or proactive/reactive behavior. In general, higher values for each of these are associated with higher resource intake, and higher risk. A fundamentally different, major type of behavior is cooperation. Theory on cooperation models individuals that are either cooperators or defectors; however, few studies have looked for individual variation in cooperativeness. In the syndrome context, an interesting, unexplored question is whether cooperative tendency carries over across multiple cooperative contexts (e.g., social foraging, group vigilance, biparental care, resource sharing, and cooperative breeding).

Of course, theory does not suggest that animals should be unconditionally cooperative. Instead, subtle cheating can be favored, which in turn favors social sensitivity (evaluating the cooperativeness of potential partners). The interplay between cooperation and deception suggests that behavior can be affected by an intersection of several behavioral axes. Consider, for example, predator inspection. In several species of small, schooling fish, individuals leave the group and approach predators apparently to gain information about the risk posed by the predator (Pitcher 1992). Individuals often inspect in pairs, taking turns as the lead individual. This fascinating behavior is a possible example of reciprocal altruism. A given individual's behavior during predator inspection could reflect the intersection of three syndromes: (1) boldness per se (measured as predator inspection when the individual is alone); (2) cooperative tendency (continuing to inspect in response to predator inspection by a partner), and (3) social sensitivity (responding to whether the partner cooperates or cheats). The notion that behavior reflects multiple behavioral axes is familiar in humans—our behavior in any given context is thought to reflect five personality axes (McCrae & Costa 1999). Parallel work needs to be done on how multiple personality axes influence individual variation in behavior in other animals.

In the previous paragraph, we invoked the possibility of a social sensitivity, or more generally, an environmental sensitivity syndrome. The work on bowerbirds suggests that social sensitivity to females could play a major role in governing variation in male mating success (Patricelli et al. 2006). The literature on proactive/reactive coping styles (Koolhaas et al. 1999) also touches on this—reactive individuals are thought to be more environmentally sensitive than proactive individuals. Reactive animals respond to changes in their environment, whereas proactive ones form set routines and take longer to respond to environmental changes. An interesting unexplored question is whether sensitive individuals are sensitive across a broad range of contexts, including both social contexts (interactions involving competition/contests, mating, and cooperation) and nonsocial environmental contexts (evaluating food types, predators, or habitats)?

A related idea is variation in choosiness, the tendency to evaluate more options before making a choice. Choice has been studied in many isolated contexts, for example, diet choice (Sih & Christensen 2001), mate choice (Forstmeier & Birkhead 2004), or habitat choice. Numerous studies show data on a group's average preference, but few emphasize variation in choosiness. For example, for mate choice, even when females, on average, prefer larger males, some prefer smaller ones, and others appear to be nonselective (Morris et al. 2003). To date, studies have not looked for carryovers in choosiness across contexts. Are the same individuals more versus less choosy in multiple contexts? If optimal choosiness varies across contexts, then an individual's choosiness syndrome could result in suboptimal choosiness in some situations.

Finally, note that syndromes of social sensitivity or choosiness emphasize variation in social plasticity, but not necessarily adaptive plasticity. A highly sensitive or choosy individual does not necessarily make good choices. Thus another potential behavioral axis is variation in social skill. Theory in behavioral ecology routinely assumes social skill—the ability to shift behavior adaptively depending on the social context. Few studies, however, have looked at individual variation in social skill and how it might affect fitness (but see Sih & Watters 2005). A valuable area for future study should be to quantify costs, benefits, and correlates of social skill.

ECOLOGICAL IMPLICATIONS

Although most analyses of behavioral syndromes focus on individual variation in behavioral types, as noted earlier, in principle, one could also compare the behavioral types of different species. A species' behavioral type reflects multiple behavioral components of its overall niche (e.g., foraging, competitive, antipredator, and dispersal tendencies) that, in turn, governs its distribution and abundance (e.g., Schöpf Rehage et al. 2005; Sih et al. 2003). A species' behavioral type can also affect its interactions with other species. For example, prey activity syndromes influence their ability to persist with predators (Sih et al. 2003), and forager voracity syndromes influence their competitive ability (Dame & Petren 2006) and impacts on prey (Schöpf Rehage et al. 2005). Differences among species in behavioral type can then help explain species coexistence and patterns of species diversity.

The study of behavioral syndromes can be particularly important for understanding species' responses to anthropogenic change. A population's ability to thrive in a rapidly changing environment might depend largely on its behavioral repertoire, including the extent to which critical behaviors are coupled or decoupled. Macroevolutionary studies on birds show that behavioral flexibility, measured indirectly by brain size and number of foraging innovations, predicts both invasion success and species richness (Sol et al. 2002). However, the converse, that a lack of behavioral flexibility leads to extinction risk, was not supported (Nicolakakis et al. 2003). Interestingly, Sol et al. (2002) also report that bird species that successfully live in close contact with human disturbance are more successful as invaders than birds that fare poorly in human habitations. Overall, species that tend to be non-neophobic and exploratory appear to have a behavioral type that succeeds in disturbed environments. Successful invasive species may display suites of cognitive abilities that allow them to adapt to the rapidly changing environments characteristic of urbanization.

Other studies have noted that successful invasive species are those that, in fact, show coupling of key behavioral traits. For example, one key trait for invasive species is dispersal ability (Lodge 1993), and high dispersal ability has been shown to be correlated with boldness (e.g., Fraser et al. 2001). Thus, disruptive species (e.g., native pests and nonnative invasive species) may have behavioral types that favor their spread, whereas threatened species may have behavioral types that compete poorly in their environment. Understanding the link between behavioral syndromes and response to anthropogenic change will likely aid our efforts to control the spread of pest organisms and conserve threatened species.

SUGGESTIONS FOR FURTHER READING

For classic work that highlights the evolutionary significance of individual differences in behavior, see Clark and Ehlinger (1987). David Sloan Wilson's papers in the 1990s also brought attention to the importance of individual differences (Wilson 1998). For a discussion of the fitness consequences of behavioral syndromes, see Dingemanse and Reale (2005). Sih et al. (2004) provide a conceptual foundation for studying behavioral syndromes, and shows how behavioral syndromes can create limited plasticity and explain seemingly maladaptive behavior. For a complementary framework for conceptualizing individual differences in behavior (temperament), see Reale et al. (2007). Bell (2007) outlines different approaches for studying behavioral syndromes, and Sih and Bell (2008) set an agenda for the future of studies on behavioral syndromes.

Bell AM (2007) Future directions in behavioral syndromes research. Proc R Soc Lond B 274: 755–761.

Clark AB & Ehlinger TJ (1987) Pattern and adaptation in individual behavioral differences. Pp 1–47 in Bateson PPG & Klopfer PH (eds) Perspectives in Ethology. Plenum Press, New York.

Dingemanse NJ & Reale D (2005) Natural selection and animal personality. Behaviour 142: 1159–1184.

Reale D, Reader SM, Sol D, McDougall PT, & Dingemanse NJ (2007) Integrating animal temperament within ecology and evolution. Biol Rev 82: 291–318.

Sih A, Bell AM, Johnson JC, & Ziemba R (2004b) Behavioral syndromes: an integrative overview. Q Rev Biol 79: 241–277.

Sih A & Bell AM (2008) Insights for behavioral ecology from behavioral syndromes. Adv Study Behav 38: 227–281.

Wilson DS (1998) Adaptive individual differences within single populations. Phil Trans R Soc Lond B 353: 199–205.

31

Evolution and Human Behavior

DEBRA LIEBERMAN AND STEVEN W. GANGESTAD

In the distant future I see open fields for far more important researches. Psychology will be based on a new foundation, that of the necessary acquirement of each mental power and capacity by gradation. Light will be thrown on the origin of man and his history.

—Darwin 1859

As foretold by Darwin (1859), scientists across multiple disciplines including anthropology, biology, and psychology have started to apply the same evolutionary principles (e.g., sexual selection, parental investment, parent-offspring and intragenomic conflict, kin selection, reciprocal altruism, and life history theory) used to understand the behavior of nonhuman species to explain human behavior, cognition, and culture. Currently, a handful of overlapping fields of research—evolutionary psychology, human behavioral ecology, evolutionary developmental biology, and gene-culture coevolutionary perspectives—share this common goal. Though there is great diversity in methodological approach and the particular focus of investigation, together the human evolutionary behavioral sciences have taken great strides in uncovering the elements of human nature.

Our aim in this chapter is to provide an overview of two major avenues of research: human behavioral ecology and evolutionary psychology. Both approaches share the central tenets of evolutionary biology and an appreciation of the distinction between ultimate and proximate questions of causality as outlined by Tinbergen (1963) and Mayr (1976b). They differ mainly according to the types of research questions they ask and the data viewed as appropriate to answering these questions. To illustrate the strengths of each approach and how they complement the work being done on nonhumans, we provide examples of active research in each discipline: in human behavioral ecology, patterns of food production in the Hadza; in evolutionary psychology, investigations into the architecture of human kin detection; and how women's sexuality changes across the menstrual cycle. Although advancements in research and theory on these and other topics in humans arise from principles grounded in evolutionary biology, work on humans now has the potential to contribute novel insights to behavioral ecology and evolutionary biology by providing key tests of ideas within the field and by contributing to important comparative perspectives. Throughout our discussion we point out ways this might occur. In addition, we touch on current debates and conclude with our thoughts on future directions of the field.

APPROACHES TO STUDYING HUMAN BEHAVIOR

Approaches to studying human behavior parallel that of nonhuman behavioral ecology in many respects. Both human behavioral ecology and evolutionary psychology maintain that human nature, just like wasp, finch, or mole-rat nature, is a product of historical selective forces. Accordingly, these fields apply evolutionary principles to derive empirically testable predictions regarding the selection pressures that may have shaped human cognition and behavior. For example, selection pressures relating to the differential parental investments by men and women guide research on human mate choice; selection pressures posed by deleterious recessive mutations and pathogens guide investigations of inbreeding avoidance in humans; life history theory informs the study of human maturation rates and senescence; and kin selection and reciprocal altruism yield predictions about the nature of various forms of human cooperation. Thus, human behavioral ecology and evolutionary psychology draw upon the same pool of theoretical tools to study human nature.

Despite a shared focus on the same species, human behavioral ecology and evolutionary psychology nevertheless differ in the kinds of research questions they emphasize and, as a result, the kinds of data they collect. We address each discipline next but wish to point out that, rather than existing as two entirely separate fields, human behavioral ecology and evolutionary psychology complement one another and, in our view, represent two sides of the same Darwinian coin—one focusing primarily on the structure of our evolved psychological adaptations (evolutionary psychology), and the other primarily on their behavioral outputs (human behavioral ecology)—both sharing an emphasis on function.

Human Behavioral Ecology: A Focus on Behavior

Human behavioral ecologists apply the same approach as nonhuman behavioral ecologists, investigating how ecological and social variation within and between populations account for the variation of particular behavioral strategies and resulting individual reproductive success (for a more in-depth discussion, see chapters 1 and 2). In this way, human behavioral ecology attempts to model the selection pressures that existed throughout human evolution and that played a role patterning modern human behavior. Tools used by behavioral ecologists to generate predictions about and interpret data on humans mirror those used to study nonhumans. They include quantitative modeling for identifying optimal behavioral and reproductive strategies within a certain socioecological context and the assessment of the trade-offs humans make, not only in the behavioral strategies employed, but also in the investment in various physiological processes. For instance, increased levels of testosterone might facilitate mating effort at the expense of parenting effort or immunocompetence (e.g., Ellison 2003).

Human behavioral ecologists investigate social and ecological selection pressures by collecting data on the fitness consequences of human behavior in purportedly "natural" contexts, that is, contexts similar in key ways to those experienced by human ancestors dating back at least thousands of years. This approach, however, has not been without controversy. For instance, there has been much discussion regarding the utility of measuring current reproductive success to identify ancestral selection pressures (e.g., Borgerhoff Mulder 2007; Symons 1992; see also chapter 2, of this volume). These discussions have helped clarify the strengths and weaknesses of the behavioral ecology approach in humans. In general, we point out that human behavioral ecology and the conceptual tools they employ have shed light on many patterns of human behavior (e.g., see Smith & Winterhalder 1992). For example, calories returned to camp provide insights on foraging and hunting strategies. Similarly, allocation of these calories to others in the group sheds light on patterns of cooperation, mating effort, and parenting effort (see below). Finally, indices of health such as fluctuating asymmetry and body mass index have been used to investigate mating behavior (e.g., Sherry & Marlowe 2007).

Evolutionary Psychology: A Focus on Psychological Mechanisms

Though exceptions exist, typically human behavioral ecologists are less focused on identifying the proximate mechanisms that govern decision making and regulate behavior. As was pointed out in chapters 8–10, however, natural selection shapes behavior by shaping developmental and psychological processes that, in turn, generate behavior. Evolutionary psychologists are interested in understanding the architecture of these processes (Tooby & Cosmides 1992). More specifically, the aim of evolutionary psychology is to identify and detail the structure of human cognitive adaptations based on evidence of functional design. The brain, however, is not as straightforward a structure to reverse engineer as the heart, eye, or wing; its form does not provide obvious clues to function. To uncover the nature of our psychological adaptations, evolutionary psychologists employ a set of guiding principles. These include the notion that our neural circuits were designed by natural selection to solve *adaptive problems*—conditions generating long enduring selection pressures whose successful navigation impacted survival and reproductive success, however distally. Examples of adaptive problems (with associated selection pressures in parentheses) include avoiding predators (various predators in the environment), avoiding sex with close genetic relatives (deleterious recessive mutations), finding nutritious food (nutritional requirements for growth and development and variation in available nutrients), caring for offspring (kin selection), and selecting a mate (differential parental investments by the sexes and sexual selection).

A second guiding principle is that of functional specialization. Evolutionary psychologists posit that different neural circuits likely exist to solve different adaptive problems. This is because psychological mechanisms well designed to perform one task are unlikely to perform additional diverse tasks equally well. This principle is seen throughout the human body: bodily functions are carried out by dedicated organs, not one general-purpose organ. The same logic applies to psychological adaptations. A system for avoiding the ingestion of pathogens is unlikely to process information regarding reciprocal altruism or govern kin detection. For this reason, evolutionary psychologists expect the human neuro-computational architecture to contain a constellation of integrated, functionally specialized psychological mechanisms. Much debate has occurred over the degree of functional specialization (often framed as modularity), especially in the human brain given the broad scope of human intellectual capacities (e.g., Buller 2005; Elman et al. 1996; Panksepp & Panksepp 2000). We address this topic in more depth in box 31.1.

In general, the concepts of adaptive problem and functional specialization help generate hypotheses regarding evolved behaviors as well as our brain's information-processing architecture. To investigate the information-processing systems that natural selection shaped to enable a particular ability, evolutionary psychologists start with the question of what a well-engineered system designed to perform a particular function might look like. It is not expected that natural selection shaped systems to perform perfectly, because evolutionary responses always are subject to trade-offs and historical constraints (e.g., take the existence of a blind spot in our visual system). Nevertheless, models of the cognitive procedures required to perform a task can sharpen predictions and reveal new relationships.

Information-processing models of psychological adaptations bear much similarity to those developed for animals (see chapter 10). They include a description of ancestrally available cues, both external (e.g., social information) and internal (e.g., physiological states), that would have provided information about a particular recurring state. For instance, seeing one's mother breast-feeding a newborn provides information regarding probable siblingship; the detection of one's mate in bed with someone else provides information regarding infidelity; and low blood glucose levels signal the need for additional food. Models also postulate how these cues are transformed into internal representations and they specify the decision-making and motivational systems that use these representations to regulate behavior (Tooby, Cosmides, Sell, Lieberman & Sznycer 2008). Empirical investigations in the laboratory and in the field can help refine proposed computational models—for example, by integrating additional mechanisms that compute newly discovered or hypothesized cost-benefit trade-offs—which, in turn, can generate new lines of inquiry. Importantly, information-processing models and associated findings in humans can inform investigations in nonhuman species that faced similar adaptive problems and therefore might employ similar cognitive solutions.

In summary, according to evolutionary psychologists, psychological adaptations can be viewed as a set of information-processing systems that were designed by natural selection in response to the various adaptive problems faced by our ancestors. Taking advantage of the empirical techniques employed across the psychological sciences including those found within social and cognitive psychology, neuropsychology, and neuroscience, this approach has shed light on a wide range of human behaviors and cognitive abilities such as those relating to mate choice, social exchange, aggression, human face recognition, social categorization, theory of mind, intuitive physics, and various emotions (e.g., see Buss 2005). Below, we illustrate how evolutionary psychologists approach human behavior and cognition, using two examples: human kin detection and female mating preferences across the menstrual cycle.

BOX 31.1 Modularity in Human Psychology

Much debate surrounds the topic of modularity, especially as it relates to human psychology. The notion of modularity is of particular concern for those interested in the computational theory of mind, which, based on the work of philosophers such as Turing, proposes that mental *outputs* (thoughts, perceptions, behaviors) derive from computations or algorithmic operations on information. The philosopher Jerry Fodor (1983) highlighted the idea that certain computations require operations that are *modular*. However, Fodor had a particular conception of modularity in mind and suggested that only computations possessing the following features could be considered modular:

1. Modules are domain specific, that is, they process only a select type of information (e.g., linguistic or visual information).
2. Modules operate in a mandatory fashion, that is, they automatically process information specific to their domain. For instance, when your eyes are directed toward a flower, you cannot help but see the flower—the visual information is necessarily taken as input.
3. Modules do not generate representations that are accessible to central (i.e., conscious) processes, but instead generate lower level representations of which we have no conscious awareness.
4. Modules are informationally encapsulated. This means that of the range of information that could, in principle, contribute to a particular analysis, only a small proportion is in fact accessed. Put another way, computations that modules perform are likely not affected by feedback from higher level processes (e.g., ones that generate representations regarding expectations or beliefs). For instance, visual processes that create the Muller-Lyre illusion (lines with different arrow ends that appear to be different lengths, yet are in fact identical) do not have access to outputs of higher level feedback processes. Subjects can know that the lines are identical (by measuring them), but this does not affect the way the lines are perceived. This is one example of how visual processes draw upon limited, encapsulated, lower level information to generate percepts.
5. Modules are associated with a fixed neural architecture.
6. Modules exhibit characteristic breakdown patterns.
7. Modules follow an innately specified developmental trajectory.

Importantly, Fodor proposed that modularity characterizes perceptual input systems including vision, audition, olfaction, and language, but not more *central* cognitive processes such as those governing judgment and decision making (*higher cognition*). For instance, our visual system contains various computational procedures that are modular, including those that permit us to see color or infer objects from two dimensional arrays of stimulation on our retinas. They are modular because, for example, we cannot help but see color, and our perceptions of color are not influenced by higher level decisions. Fodor's criteria, more central systems cannot qualify as modular. This is because, among other things, central systems have access to a wide range of inputs and do not necessarily operate in a mandatory fashion.

As discussed in this chapter, evolutionary psychologists also argue that psychological systems are modular. Contrary to Fodor, however, evolutionary psychologists argue that even computational procedures downstream of perceptual processes are governed by modules (e.g., those guiding judgment, decision making, and emotions). For instance, evolutionary psychologists have proposed that kin detection mechanisms are evolved modules. But the information processed by these modules (e.g., duration of coresidence) is downstream to perceptual input systems and available to more central systems.

So do evolutionary psychologists argue against Fodor that central processing can be modular? In fact, not in any straightforward sense: Fodor and evolutionary psychologists have used the terms *module* and *modularity* in different ways. Fodor's modules are defined by structural criteria (e.g., information encapsulation and limited access to more central mental representations). By contrast, evolutionary psychologists define modules in terms of *functional specialization* (Barrett & Kurzban 2006). Using kin detection as an example, kin detection mechanisms take as input information relevant to the task of categorizing according to kinship (and not just any input). These inputs, far from being shallow perceptual inputs, are likely higher level representations. For instance, receiving care from the same caregivers throughout childhood has been proposed as one cue to siblingship that is taken as input by kin detection mechanisms. Assessments of sharing the same caregivers (typically one's mother and father) likely rely on a suite of complex higher level systems. The system is specialized, because it relies on specific information from which it yields a specific decision (the reliability of an inference that a particular individual is a sibling). But the system is not modular in a strict Fodorian sense.

In general, rather than using *a priori* criteria for generating predictions regarding likely structures of cognitive systems, evolutionary psychologists maintain it is more useful to consider functionality, as a system's function will, to a large extent, dictate its information processing structure (Sperber 2005). Thus, evolutionary psychologists do not primarily concern themselves with whether functionally specialized processing of information is modular in a strict Fodorian sense. Rather, they see that many decisions must rely on functionally specialized systems. For example, incest aversion appears to be a function of specific cues such as coresidence duration and seeing one's mother caring for a newborn, and women's attraction to men's masculine features appears to depend on their ovulatory status in ways that other decisions are not. It is difficult to see how these computations could be the result of a highly general processing system (e.g., one system that took as input both kinship cues and ovulatory status and generated appropriate behavior toward kin and potential mates).

This is not to say that evolutionary psychologists do not concern themselves with issues of how cognitive systems are structured. Indeed, they are interested in coming to understand how, in general, computational systems can be functionally specialized without being limited by the structural criteria set forth by Fodor, and how their structure is instantiated in our neural networks. Barrett (2005), for instance, proposes one perspective on this topic.

Once functional specialization is taken as the key concept and not Fodor's strict criteria for modularity, other related issues become easier to dissect. Take, for instance, debates about cognitive flexibility. It would seem that the greater the number of Fodor-type modules we relied upon, the less flexible and adaptable our decision making would be. That's because many of our decisions would rely on automatic, encapsulated procedures, with no opportunity for modification based on our experiences or reflection, and no ability to deal with anything more than a very narrow range of information. Given humans' ability to learn from experiences and, through careful reflection, to solve a vast array of novel problems never before encountered in ancestral or modern environments, it may seem a natural conclusion that our minds are not characterized by massive modularity.

Again, however, evolutionary psychologists do not claim that functional specialization is modular in Fodor's sense. Rather, the adaptationist principle of functional specialization provides a starting point for understanding the flexibility generated from our evolved cognitive mechanisms. Cognitive mechanisms may be specialized to take particular subsets of information as input (consider the paralyzing alternative of taking any and all information as input continuously). Flexibility may arise because some information may be distinct from that which played a role in the mechanism's evolution, but be sufficiently similar to

(continued)

BOX 31.1 *(cont.)*

be processed. For instance, as discussed by Sperber (2005), given that cognitive mechanisms evolved to process information in a particular domain, one can distinguish between that mechanism's *proper domain*, the set of inputs the system evolved to process, and that mechanism's *actual domain*, the set of inputs the system actually processes, regardless of whether they played a causal role in the evolution of that mechanism. It is therefore possible that evolved mechanisms process a wider range of inputs than what they were originally selected to process, leading to a wider range of behavioral outputs. A similar point may be made with regard to nonpsychological adaptations. The human hand may have been modified by selection to manipulate tools using materials available in ancestral environments, but it may now be used to manipulate a much larger range of modern objects, for example, to open refrigerator doors.

Importantly, some human capacities involve abilities to learn and innovate solutions, in instances across a wide array of specific content domains (e.g., *improvisational intelligence*; Barrett et al. 2007). These capacities, too, however, possess specialized information-processing structure as reflected, for instance, in psychological theories of various kinds of learning (Gallistel 2000). Hence, in theory they can be characterized in ways similar to other psychological adaptations.

In sum, evolutionary psychologists bypass the debate on modularity by deemphasizing structural criteria as typified by Fodor's concept of modularity, and stressing evolved functional specialization. This allows them to ask questions such as, what would the information-processing structure of a system well-designed to perform function X look like? In this way, evolutionary psychologists are able to generate new hypotheses regarding a wide array of cognitive processes underlying human behavior.

EXAMPLES OF CURRENT RESEARCH IN HUMAN EVOLUTIONARY BEHAVIORAL SCIENCE

Human behavioral ecology and evolutionary psychology emphasize different sequelae of historical selection pressures, namely, the association of adaptive behavioral strategies with particular socioecologies versus the existence of functionally specialized information-processing procedures, respectively. However, these approaches complement one another. Together they provide the methods for detailing many aspects of human nature. Next, we highlight three examples of recent research within the human behavioral sciences. Though we discuss each example as originating with researchers identified as either human behavioral ecologists or evolutionary psychologists, questions pertinent to the goals of both disciplines should be applied to each, a topic we return to in our conclusions.

Patterns of Food Production in Human Foragers

Most human foraging groups rely substantially on the production of meat through hunting or fishing. In most of these groups, men hunt and fish more than do women. Accordingly, men tend to produce more calories than women. Marlowe (2001) reported that, on average, across 95 foraging groups, men produced 64% of all calories. In 9 groups on which careful food-weighing techniques were applied, men's production accounted for, on average, 66% of all calories (Kaplan et al. 2000). Notable exceptions do exist. In particular, in many groups within the matrilineal belt of Africa and in the insular Pacific, women produce as many or more calories as men. In these areas, hunting of large game is relatively uncommon, and horticulture, an activity women commonly engage in, is relatively common (e.g., Schlegel & Barry 1986).

A critical question that behavioral ecologists have asked is, why do men hunt? That is, what ancestral fitness benefits led men to generate, through hunting, calories that exceed what they consume?

Hunting in the Hadza

The Study Population The Hadza of Tanzania is one group whose foraging activities have been investigated extensively through research by Nick Blurton-Jones, Kristen Hawkes, and their colleagues and students (notably, Frank Marlowe). The Hadza number about 1,000 and live a nomadic lifestyle in a savanna-woodland habitat. They live in small groups, averaging about 30 individuals per group, with migration between groups common. Typically, groups move to a new location every month or two. Women dig tubers, gather berries, and collect other fruits. Men collect honey and fruit and hunt mammals and birds with bows and arrows. Children's diets are subsidized by the efforts of men and postreproductive women (though the Hadza are at the low end of the cross-cultural distribution of men's contributions to subsistence, with men producing only about 40% of the calories overall). At age 10, children produce about half of the calories they consume through gathering of fruit (Hawkes et al. 1995).

The Paternal Care Hypothesis One view of men's hunting is that it functions as paternal care—that is, it benefits men's fitness through increased offspring survival or quality. According to this view, men and women have evolved to cooperatively raise offspring. In so doing, they divide their labor efforts, with women in traditional societies performing most of the direct child care, and men supplying food through activities that women cannot readily do while caring for children, primarily hunting. Kaplan et al. (2000) argue that biparental care and a sexual division of labor evolved in humans as part of a larger coadapted complex of features, including extractive foraging of high-quality foods, a long period of juvenile dependency, intensive learning in childhood, and a long productive life span that renders skill-intensive means of production (which requires many years to acquire to peak levels) worth pursuing.

The Show-Off Hypothesis Hadza hunters share meat across the band instead of giving it directly to their nuclear families, particularly when the captured game is large. Hunters have greater control over the consumption of small game and can direct larger shares to their own families. In a series of papers in the 1990s, Hawkes and colleagues argued that men's hunting, especially large-game hunting, has been selected to advertise (show off) a male's mate value because successful hunting results in benefits in the currency of mating opportunities (Hawkes 2004). Put otherwise, hunting functions (largely) as mating effort, especially hunting of large game. In support, good hunters are more likely to remate following the death of a spouse and are more likely to take a second wife when a first wife reaches menopause (Marlowe 2000). Good hunters in the Ache, another widely studied foraging group, have more extra-pair mates than do less successful hunters. In addition, Hadza men purportedly allocate greater time on hunting large game than is profitable, given relative rates of return from hunting large game and smaller game, respectively (see Hawkes 2004; though see also Marlowe 2003). Second, although children's nutritional status is predicted by their mothers' food production, it is not predicted by the rate of food produced by fathers (Hawkes 2004).

Hawkes and colleagues accept that men's hunting does subsidize the diets of women and children. However, they argue that men have not been selected to hunt because hunting large game resulted in greater offspring fitness through direct nutritional benefits. The benefits of men's surplus production are a fortuitous side effect for women and children, not benefits that directly shaped men's hunting efforts. Hawkes et al. essentially suggest that female preferences for good hunters may have both direct and possibly indirect effects for females (see chapter 24). For example, Marlowe (2005) found that Hadza women rate hunting prowess as more important in potential mates than any other feature. Thus, Hawkes and colleagues argue that men's hunting is a sexually selected trait via female preference (see Hawkes 2004).

A Blended View The male-hunting-as-parental-effort and the male-hunting-as-mating-effort theories sometimes are presented in extreme forms, if not by their proponents, then by their critics. But good hunting may have both natural and sexually selected effects on men. Historically, men's hunting may have yielded reproductive benefits by enhancing parental investment as well as mate attraction. Accordingly, the modulation of hunting behavior may arise from psychological adaptations with two different functions—parental effort and mating effort—served at least partly by distinct adaptations.

Under a mixed model, different hunting endeavors may differentially benefit men through parental investment and mating effort. Hawkes et al. (2001) emphasize that men's large game hunting is not an effective or efficient means of provisioning offspring. Large-game hunting may hence benefit men substantially in the form of mating effort (though, we note, alternative explanations, such as benefits achieved through building of social alliances, are possible). By contrast, men in foraging societies have much more control of the distribution of captured small game and may preferentially direct it toward primary partners and offspring. Thus the hunting of small game may function as an effective means by which men exert parental investment.

Marlowe (2003) presents data on the Hadza that are consistent with a blended view. Overall, married Hadza women produce as many calories as do married Hadza men. Women with small offspring, however, produce far fewer calories than produced by women without young offspring. Compared to all other married women, women whose youngest child is 3 years of age or younger harvest fewer calories, and women with an infant 1 year of age or younger harvest only about half as many calories as other married women. These outcomes are consistent with the argument that women's child care interferes with effective foraging. When women have young children, however, their husbands' production of calories increases, and substantially exceeds that of their wives. Hence, whereas in couples without a child 3 years of age or younger, wives produce more calories than husbands do, in couples with an infant younger than 1 year of age, men produce almost 70% of the calories (see figure 31.1). These increases in production by Hadza men presumably function to make up for deficits in the harvesting of calories by wives with small children.

Hadza men, then, appear to facultatively adjust their work efforts (and perhaps the prey items they target) in response to the direct food production of wives, as it varies with the presence or absence of young children. This pattern is not well explained by the idea that men's work functions solely as mating effort (see Marlowe 2003 for a discussion of possible alternative explanations). Moreover, separation of men into fathers and stepfathers provides additional evidence that men's production functions

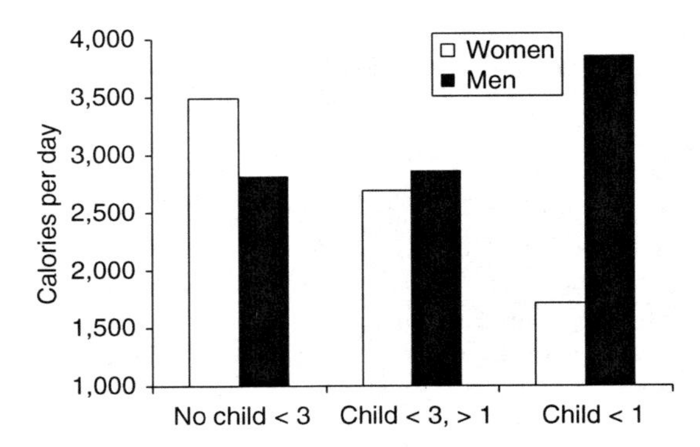

FIGURE **31.1** Hadza male and female caloric production as a function of a couple having no child under age 3, a child older than 1 but younger than 3, and a child younger than 1 year old. Adapted from Marlowe (2003).

partly as parental effort. Approximately 30% of Hadza children have stepfathers. Stepfathers do not show the same pattern of enhanced food production in response to the presence of young children in the household that children's genetic fathers demonstrate.

Summary of Human Foraging

In sum, human behavioral ecologists have studied patterns of men's and women's foraging in considerable detail. Of particular interest are the evolved functions of men's hunting. Sharp debate has led to focused investigation of when and how men decide to allocate their foraging efforts. At this time, it appears likely that men's hunting does partly function as parental effort, and hence benefits of food production for offspring have at least partly shaped the decision processes that underlie men's foraging efforts. Nonetheless, men's hunting may also function to attract mates. Additional research into the psychological systems responsible for foraging decisions—an evolutionary psychological approach—may sharpen our understanding of the forces of selection that shaped them. (For further discussion, see Smith 2004.)

Kin Detection and Kin-Directed Behavior

Another area of active research within the human evolutionary behavioral sciences is the investigation of the cognitive mechanisms governing kin detection and kin-directed behavior. As in other species that regularly encountered close genetic relatives across the life span, mechanisms for detecting kin are expected to exist in humans for at least two reasons: to avoid choosing a close genetic relative as a sexual partner, thereby avoiding the deleterious consequences associated with inbreeding, and to regulate altruistic and competitive effort according to the probability of genetic relatedness as indicated by inclusive fitness theory. But how might evolution have engineered a psychological system to discriminate according to genetic relatedness and then regulate sexual aversions and altruistic motivations accordingly?

An evolutionary psychological approach to this question involves developing models of our evolved computational architecture. At minimum, a system designed to avoid inbreeding and to allocate assistance according to genetic relatedness would need a way to assess relatedness and systems governing sexual attraction and altruism. One recently proposed model of kin detection and kin-directed behaviors suggests the following components: (1) procedures that monitor for and take as input the particular cues signaling relatedness, (2) a *kinship estimator* that takes as input the detected cues and, based on these cues, computes for each individual, *i*, an estimate of relatedness, or *kinship index* (KI), and (3) procedures that feed the KI into two separate motivational systems: one regulating sexual attraction/avoidance and one regulating altruistic/competitive behavior (see figure 31.2; adapted from Lieberman et al. 2007). Thus, according to this model, the same kin detection procedures can regulate two distinct classes of behavior.

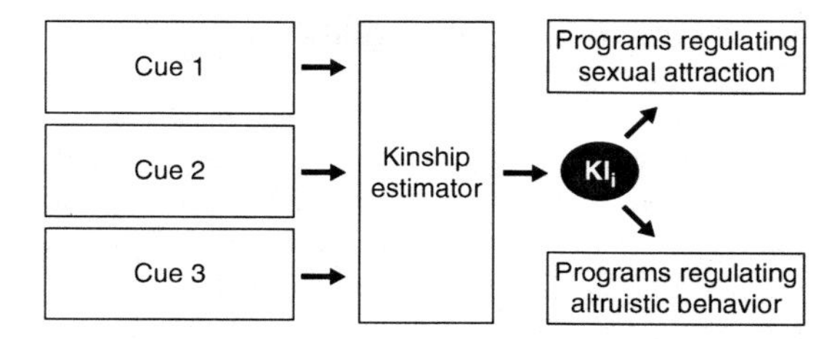

FIGURE 31.2 A model of the information-processing architecture of human kin detection. Cues that correlated with relatedness throughout our species' evolutionary history are taken as input by a *kinship estimator* that computes for each individual, *i*, an estimate of kinship or *kinship index* (KI). The KI is then taken as input by two distinct motivational systems: one regulating sexual attraction and one regulating altruistic motivations. Specifically, the computed KI is capable of having multiple effects: as KI increases (i.e., as the probability an individual is likely to be kin increases), programs guiding sexual attraction can be downregulated and those guiding sexual avoidance upregulated. With respect to helping behavior, as KI increases, programs guiding altruistic motivations can be upregulated and programs guiding competitive motivations downregulated. Thus, according to this model, the same kin detection procedures can regulate two distinct classes of behavior. Redrawn from Lieberman et al. (2007).

Cues Used to Assess Genetic Relatedness

There are a number of constraints that confine the set of cues selection might have favored to engineer kin detection systems. For instance, barring recent medical technology, we are not able to directly compare genomes to assess kinship. However, other possible kinship cues exist. One possibility includes the use of more evolutionarily novel cultural information such as linguistic kin terms. But these are unlikely to be the primary cues used to detect kin because kin terms can blur genetic boundaries (e.g., *aunt* in our culture refers to a parent's sister, a blood relative, and a parent's brother's wife, a nonblood relative). Furthermore, it is unlikely that phylogenetically prior kin detection mechanisms that functioned in the absence of linguistic information were overwritten by more variable and potentially less reliable cultural information.

Rather, it is likely we rely on ecologically valid cues that correlated with genetic relatedness in human ancestral environments. Importantly, the cues mediating kin detection might differ depending on the type of kin in question. To the extent that different cues signaled an individual was a specific type of close genetic relative (e.g., mother, father, offspring, or sibling), different detection mechanisms are likely to exist. Additionally, males and females might use distinct cues to identify the same type of kin. For example, because men can never be fully certain of their relatedness to offspring, the cues signaling that an infant is indeed one's own are likely to differ for men and women.

A range of kinship cues have been identified by evolutionary biologists investigating inbreeding avoidance and altruism in nonhuman species (for review, see Hepper 1991). For instance, early association, a spatial cue that identifies likely siblings in species in which offspring require extended maternal care, predicts patterns of social preferences and mate choice in species such as voles, mice, macaques, and chimps. In some species, chemical cues guide kin detection and associated kin-directed behaviors. Studies on house mice, for example, show that mate preferences are guided by assessments of similarity at loci controlling the major histocompatibility complex (MHC). That is, males and females prefer to mate with individuals who are MHC dissimilar from themselves, a preference thought to protect against the negative effects of pathogens. For MHC disassortative mating to occur, however, individuals require a referent, either themselves or a close relative, to determine what counts as MHC dissimilar. A series of cross-fostering experiments in which individuals were raised by MHC-dissimilar parents showed that individuals preferred to mate with others who were dissimilar from their foster parent's MHC composition. Thus MHC-guided mate preferences appear to use parental phenotypes as referents of one's own genetic composition (e.g., Penn & Potts 1999).

In the human evolutionary literature, the majority of research has focused on the detection of siblings and the associated development of sexual aversions and sibling-directed altruism. Next, we briefly discuss some recent findings from this literature.

Sibling Detection: Exposure to Mother-Infant Association and Coresidence Duration

The ancestral social environment of humans was such that a likely reliable cue to siblingship would have been seeing one's own mother caring for (e.g., breast-feeding) a newborn. Indeed, the intense mother-child association that typically occurs surrounding the natal period and continues throughout the first few years of life would have served as a stable anchor point for others to infer relatedness. Thus, if an individual observed an infant receiving care from the individual's own mother (at least the female categorized as one's own mother), then it was highly probable that that infant was the individual's sibling. Further, exposure to this cue would have signaled genetic relatedness regardless of coresidence (or association) duration. That is, regardless of whether one was 5, 10, or 15 years old, maternal-infant directed care would have cued probable genetic relatedness. However, as potent a cue as mother-infant association might be, it is available only to older siblings already present in the social environment; the arrow of time forbids a younger sibling from having seen his or her older sibling born and cared for as an infant. For younger siblings, then, what cue or cues might evolution have used to identify probable older siblings?

One solution is to track the flow of parental effort. Any child regularly receiving care from one's own mother and father had a higher probability of being kin than children receiving care from other individuals. Moreover, the longer the care, the more likely the individual would have been a sibling. This cue, operationalized as childhood coresidence duration, was first proposed by Edward Westermarck, a Finnish social scientist who noted that children reared in close physical proximity during childhood tend to develop a sexual aversion toward one another later in adulthood (Westermarck 1891/1921). This idea, known as the *Westermarck hypothesis* (WH), has received support from various anthropological and psychological investigations (see review in Lieberman et al. 2003). Perhaps most notable are the cases of the Israeli kibbutzim and Taiwanese minor marriages, two natural experiments inadvertently created by cultural institutions in which unrelated children were reared in close physical proximity throughout childhood. As the WH predicts, children reared together throughout childhood rarely marry one another (Israeli kibbutzim: Shepher 1983), and if forced to marry suffer decreased rates of fertility and increased rates of divorce and extramarital affairs (Taiwanese minor marriages: Wolf 1995). Together, these studies point to early coresidence as one cue our mind uses to assess relatedness and to dampen sexual desires. However, they raise many questions. For example, does coresidence duration predict sexual aversions differently for the younger and older sibling in a sib-pair? As suggested above, older siblings might rely on a different cue to identify probable younger siblings, one that operates independent of coresidence duration. Also, do the same kinship cues that regulate inbreeding avoidance also regulate kin-directed altruism, the other suite of behaviors relying on assessments of relatedness? These and other questions are being addressed in the psychological sciences.

Psychological Investigation of Kinship Cues

It is not ethical to subject humans to the life-altering experiments used by evolutionary biologists to study kin recognition in nonhuman animals. For this reason, scientists have either had to look for natural experiments, such as those mentioned above, or take advantage of the natural variation that exists in families composed of actual genetic relatives. To investigate whether a proposed cue serves as a signal of relatedness, it is possible to match individual variation in exposure to the specific cue (e.g., coresidence duration, maternal-infant association) to behaviors and reactions relating to sexual behaviors with family members. Converging lines of evidence that we use a particular cue to categorize individuals according to genetic relatedness can be found through investigations of altruism. If the same kin detection mechanism serves to regulate both sexual avoidance and altruism, then a cue to kinship should show parallel effects across these two distinct motivational systems.

Recently, a team of researchers set out to investigate whether our mind evolved to use coresidence duration and maternal-infant association as separate cues for detecting older siblings and younger siblings, respectively. Using surveys to collect information, Lieberman et al. (2007) found that individuals not exposed to their mother caring for their sibling as a newborn (as it is typically for the younger sibling in a sib-pair), their duration of coresidence with an opposite sex sibling predicted aversions to sibling incest, as measured by disgust at imagining sex with one's own sibling and moral sentiments relating to third-party sibling incest. By contrast, for individuals exposed to their mother caring for their sibling as an infant (the older siblings in a sib-pair), coresidence duration with an opposite sex sibling *did not* predict aversion to incest. When measures of altruistic attitudes and behavior were analyzed, the same pattern emerged. That is, coresidence duration with a sibling predicted altruism more strongly for individuals without access to the more potent cue of seeing their mother caring for their sibling as a newborn (see figure 31.3). These data provide compelling evidence that the mind uses two different cues for identifying older versus younger siblings and for regulating sexual aversions and altruistic motivations.

Upon inspecting the levels of aversions and altruistic inclinations reported, Lieberman et al. (2007) found that older siblings exposed to the cue of seeing their mother care for a younger sibling as a newborn reported intense levels of disgust toward sexual acts with that sibling as well as increased levels of altruism across all durations of coresidence. That is, regardless of whether a subject resided for 15 or only 3 years with their younger sibling, the level of sexual aversion reported in response to sibling incest and altruistic inclinations were close to the maximum. In contrast, for subjects for whom this cue was not available and who relied on coresidence duration as a cue to siblingship (the younger siblings in the dataset), disgust at sexual acts with their older sibling and sibling-directed altruism were low for shorter periods of coresidence and gradually increased with extended periods of coresidence. In fact, data suggest it takes approximately 14–15 years of coresidence for younger siblings to reach the same level of sexual aversions and altruistic effort reported by older siblings who were exposed to the cue of seeing their mother care for their sibling as a newborn.

Taken together, these data provide a first glimpse into the cognitive procedures governing kin detection and kin-directed behavior in humans. The findings indicate that the mind uses at least two cues to detect siblings and mediate inbreeding avoidance and kin-directed altruism: exposure to maternal investments in a newborn (used by older siblings to detect younger siblings) and duration of coresidence throughout periods of shared parental investment (typically used by younger siblings to detect older siblings). Because these same cues were found to regulate aversions and altruism in the same way, it suggests the existence of a single set of kinship-estimating procedures that feed motivational systems guiding mate choice and, separately, altruistic effort.

Additional Lines of Inquiry

Cues aside from coresidence duration and exposure to maternal-infant association may also play a role in sibling detection. For example, facial resemblance has been found to predict trustworthiness as well as attractiveness (DeBruine 2005). And olfactory cues such as those derived from the major histocompatibility complex (MHC) have been found to influence mate choice (e.g., Wedekind & Füri 1997; it should be noted here that MHC similarity may not function to cue kinship in this context but rather may cue compatible MHC alleles). No matter what the cues, if the model of kin detection

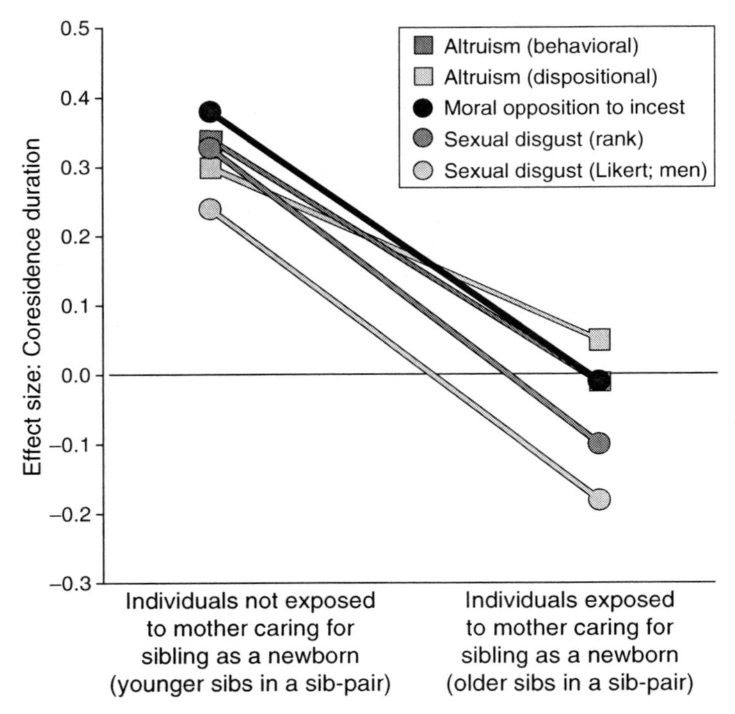

FIGURE **31.3** Two different cues to siblingship, exposure to maternal-neonate association and coresidence, regulate sexual aversions and sibling-directed altruism in the same manner. For individuals not exposed to the potent cue of seeing their mothers caring for a younger sibling as a newborn (typically the younger siblings in a sib-pair), coresidence duration is used as the cue to siblingship and, as the left-hand points on the chart show, significantly predicts disgust toward engaging in sexual acts with that sibling (two different measures), moral opposition to third-party sibling incest, and sibling-directed altruism (two different measures). By contrast, individuals exposed to their mother caring for their sibling as a newborn (typically the older sibling in a sib-pair), coresidence duration does not predict sexual aversions, moral opposition to incest, or altruism (right-hand points on the chart that hover around an effect size of zero). These data suggest that the detection of older and younger siblings relies on different types of cues. Redrawn from Lieberman et al. (2007).

and kin-directed behaviors outlined above is correct, these cues should regulate both sexual aversions and altruism.

Of course, many questions remain unanswered. For example, if coresidence duration mediates sibling detection, is a specific period of coresidence (e.g., ages <5) required, as some have suggested (Shepher 1983; Wolf 1995)? Or does each year of coresidence contribute in equal increments to a computed kinship estimate as research by Lieberman et al. (2007) suggests? Furthermore, what cues might distinguish a full biological sibling from maternal and paternal half siblings? Are there specific neural circuits associated with kin detection procedures? Do impairments in these regions explain certain psychopathologies underlying, for instance, incestuous interests? These questions can be addressed through interdisciplinary collaborations within the human evolutionary behavioral sciences but also with researchers interested in kin recognition mechanisms in animals (e.g., Holmes & Sherman 1983; Gerlach et al. 2008). In this way, we can develop a more complete picture of the psychological adaptations governing human kin detection and kin-directed behaviors.

Implications for Inquiry into Kin Detection Mechanisms in Nonhuman Species

The application of evolutionary psychological approaches to human kin recognition can inform investigations of similar processes in nonhuman species (Rendall 2004; Hepper 1991). For instance, in species that regularly encountered close kin over the life cycle, inbreeding avoidance mechanisms are expected to exist. Thus, computational procedures that estimate relatedness based on the cues that would have carved kin from nonkin over that species' evolutionary history should also exist. But kin detection doesn't buy inbreeding avoidance. Motivational systems must exist that use estimates of relatedness to regulate behavior. This view can be helpful in understanding the processes governing the behavior of nonhuman animals observed in natural and laboratory environments.

For example, there has been discussion in the animal literature regarding the extent to which dispersal of one or both sexes serves as an evolved mechanism for inbreeding avoidance. Some researchers have suggested that dispersal patterns evolved specifically to decrease the probability of close kin matings (Cockburn et al. 1985; Costello et al. 2008; Wolff 1992). From this point of view, dispersal patterns are behaviors that evolved to solve the problem of inbreeding avoidance. Alternately, others have suggested that dispersal patterns do not function as an adaptation for inbreeding avoidance per se, but instead are a result of intrasexual competition and territory choice that achieve inbreeding avoidance as a byproduct (Moore & Ali 1984). Although these latter forces may indeed influence dispersal patterns in species of birds and mammals, in those species in which fertile close genetic relatives had a high probability of encountering one another during maturity and choosing one another as mates, systems for kin detection and sexual inhibition are expected to exist. The operation of systems producing a sexual aversion, or otherwise rendering those individuals categorized as close kin as unacceptable mates, could result in the dispersal of individuals of either sex (or both sexes) to seek alternate mates. That is, dispersal behaviors may, at least in part, be a consequence of cognitive systems inhibiting the choice of close kin as sexual partners, not the evolved function per se. In this case and in other areas of joint interest, we suggest that considerations of the computational architecture underlying a particular set of behaviors can provide additional hypotheses for investigating evolved adaptations in human and nonhuman species alike.

Changes in Women's Sexuality across the Cycle

A final example of how evolutionary approaches to humans are stimulating new ideas concerns patterns of female sexuality. Similar to many other species, human females are fertile during a brief window of their cycles, from a few days prior to the day of ovulation up until the day of ovulation itself (e.g., Wilcox et al. 1995). Based on this fact, what has become known as the *ovulatory shift hypothesis* was proposed (see Gangestad & Thornhill 2008; for a first statement of the basis for this idea, see Grammer 1993). If ancestral females benefited from multiple matings to obtain genetic benefits for offspring, but at some potential cost of losing social mates, selection may have shaped female preferences for male features indicative of those benefits to vary as a function of fertility status: to be maximal at peak fertility and less pronounced outside the fertile period. The logic is that of conflicting demands (see chapter 8): if the costs of multiply mating (e.g., losing a partner) exceed benefits at that point in time (e.g., whenever fertilization is not possible), then females should avoid extra-pair partners at those times. Cycle shifts in preferences should also be more extreme when women evaluate men along dimensions relating to genetic benefits (e.g., their sexiness) rather than dimensions related to long-term mateship compatibility (Penton-Voak et al. 1999).

Over a dozen studies in the past few years show that female preferences clearly do shift. At mid-cycle, normally ovulating, nonpill-using women particularly prefer a number of male traits perceived through a variety of sensory modalities: the scents associated with male symmetry and social dominance, facial and bodily masculinity, taller height, masculine vocal qualities, and masculine behavioral displays. Symmetry, masculine facial, body, and vocal qualities, intrasexual competitiveness, and various forms of talent ancestrally may have been indicators of intrinsic good genes. (Intrinsic good genes are those that have additive effects on fitness and hence could benefit the offspring of any female; see Jennions & Petrie 2000 and chapter 24 of this volume). Not all positive traits are sexier mid-cycle, however. Traits particularly valued in long-term mates (e.g., promise of material benefits) appear to be preferred as strongly by infertile women as by fertile women. In one study, Gangestad et al. (2007) had women rate the attractiveness of men shown on videotapes. Independent samples of women rated men on a variety of qualities desirable in mates. Whereas women were particularly sexually attracted to men seen to be arrogant, intrasexually competitive, muscular, and physically attractive when they were in a fertile phase of their cycles, no cycle shifts were observed in women's attraction to men seen to be successful financially, intelligent, or kind and warm. Interestingly, men who appeared to be sexually faithful were less sexually attractive when women were fertile.

Shifts in Women's Sexual Interests

Patterns of women's sexual interests also shift across the cycle. In one study, normally ovulating women reported thoughts and feelings over the previous 2 days twice: once when fertile (as confirmed by a luteinizing hormone surge, 1–2 days before ovulation) and once when infertile. When fertile, women reported greater sexual attraction to and fantasy about men other than primary partners—but not primary partners (for reviews, see Gangestad & Thornhill 2008; Thornhill & Gangestad 2008). The study was not able to examine what kinds of men women were attracted to, but the studies on shifts of female preferences provide good reason to think that these men tended to possess masculine faces, voices, scents, and behavioral displays.

In fact, however, the ovulatory shift hypothesis expects a more finely textured pattern. On average, ancestral women could have garnered genetic benefits through extra-pair mating, but those women whose primary partners had good genes could not. Selection thus should have shaped interest in extra-pair men mid-cycle to itself depend on partner features; only women with men who, relatively speaking, lack purported indicators of genetic benefits should be particularly attracted to extra-pair men when fertile. Findings support this prediction. In one study, for instance, women with asymmetrical partners were more attracted to extra-pair men when fertile; not so of women with symmetrical partners, who were more attracted to their partner mid-cycle (see Gangestad et al. 2005).

Though most work on the ovulatory shift hypothesis has examined women's preferences for men who vary with respect to features thought to possibly have been associated with "intrinsic" good genes ancestrally, women may also particularly prefer men who possess another form of good genes when mid-cycle: compatible genes, genes that work well together with genes women possess and thereby enhance offspring fitness. It has been conjectured that men who possess alleles at major histocompatibility complex (MHC) loci that differ from women's own MHC alleles possess a form of compatible genes (e.g., Penn & Potts 1999). Indeed, women appear to be particularly attracted to the scent of men who possess MHC alleles dissimilar to their own (see review in Garver-Apgar et al. 2006). A recent study showed that women who share MHC alleles with romantic partners, and thereby have partners with incompatible genes, were less sexually responsive to their partners and more likely to have had sex with a man other than their partners while romantically involved with their partners. They furthermore reported particularly enhanced sexual attraction to men other than partners when fertile in their cycles (Garver-Apgar et al. 2006; see figure 31.4).

Male Counterstrategies across the Cycle

If women have been under selection to seek good genes mid-cycle, men should have been under selection to take additional steps to prevent them from seeking extra-pair sex at this time (see chapters 22 and 23). Multiple studies indicate that they do so by being more vigilant, proprietary, or monopolizing of mates' time (see Gangestad & Thornhill 2008; Thornhill & Gangestad 2008).

Men might use one or more of several candidate cues of fertility status. Men find the scent of ovulating women particularly attractive, judge women's faces more attractive mid-cycle, and may detect subtle behavioral changes (see Thornhill & Gangestad 2008). Whatever the cues, the view that women are not benefited by men detecting their cycle-related fertility status suggests that women are unlikely to have been designed through selection to send them (see Thornhill & Gangestad 2008; this general view, however, may be subject to debate). Men, nonetheless, should be selected to detect by-products of fertility status that women do not fully suppress, and apparently can detect at least one. Consistent with this idea, men are particularly vigilant of their partner during mid-cycle when they are paired with women who should least want them to be vigilant—those particularly attracted to extra-pair men mid-cycle (see Thornhill & Gangestad 2008).

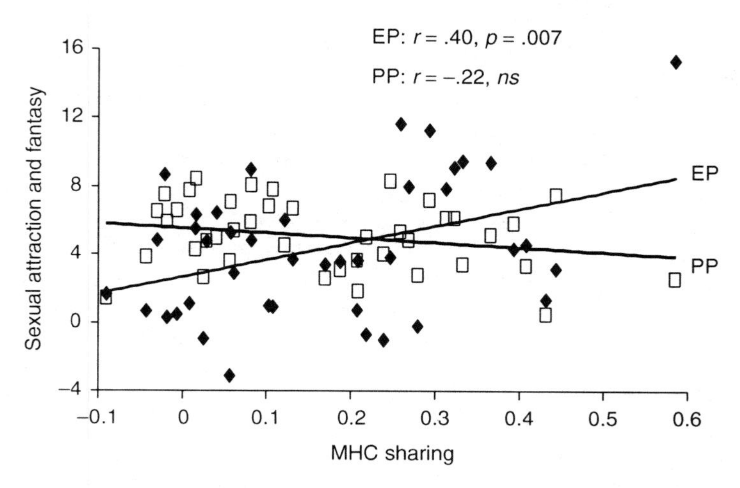

FIGURE 31.4 Women's reported sexual attraction to and fantasy about men other than primary partners and primary partners during the fertile phase as a function of proportion of MHC alleles shared across partners. MHC loci assessed were the A, B, and DRβ loci. Age and relationship status were statistically controlled, such that plotted values are residuals (with these variables controlled) plus the sample mean. *Solid line*: Attraction to extra-pair men. *Dashed line*: Attraction to primary partners. Interaction $F_{1,36} = 11.1$, $P < 0.002$. Redrawn from Garver-Apgar et al. (2006).

Do Women Possess Estrus? A Comparative Perspective

A long-standing conclusion about the evolution of human sexuality is that the lack of estrus (a distinctive fertile phase of the ovarian cycle characterized by intensified sexual receptivity and proceptivity) is a derived feature. Indeed, a key issue that scholars pursued through the 1970s and 1980s is what insights the evolutionary loss of women's estrus reveals about the evolution of human sociality more generally. For instance, some scholars proposed that loss of estrus and its replacement with continuous sexual receptivity across the cycle promoted pair-bonding, a crucial evolutionary novelty in the hominid lineage (e.g., Symons 1979).

The recent findings about changes in women's sexuality across the cycle led Thornhill and Gangestad (2008) to revisit the claim that women lost estrus. Women's fertile-phase sexuality, they argued, is estrus—that is, it shares homologies with fertile-phase sexuality in close relatives and mammals in general. Though women's estrous sexuality may have been modified in the context of pair-bonding, it was never evolutionarily lost.

At the same time, Thornhill and Gangestad (2008) proposed that reproductive biologists' understanding of the functions of female estrus must be sharpened in light of both comparative data and evolutionary theory. A common conception of estrous sexual proceptivity ("heat") and attractivity is that they function to obtain sperm from males in general, permitting conception. Thornhill and Gangestad (2008) argue that, in fact, because sexual selection operates strongly on males to find and inseminate fertile females, females typically need not pay the costs for such adaptations designed to arouse male sexual interest. Accordingly, they propose, female estrous behavior is discriminating sexuality; it at least partly functions to attract females to males who can provide genetic benefits to offspring when females may conceive.

Women's fertile-phase sexuality can be understood in this context. During estrus, women are particularly attracted to men who display purported ancestral indicators of genetic benefits to offspring (e.g., masculinity). Women are not more attracted to men in general during this phase. Outside of estrus, women retain sexual interests, but this interest is not characterized by precisely the same pattern of attraction. This extended sexuality (sexual receptivity during nonconceptive periods) functions differently from estrus; that is, it evolved because it enhanced reproductive success in other ways. Arguably, according to Rodriguez-Girones and Enquist

(2001), it functions to obtain direct benefits, largely from long-term male partners.

Findings on other species, however, can also be interpreted in light of these claims. An illustration is provided by common chimpanzees. Females are actually more sexually receptive and initiate sex with more males outside of the period of peak fertility than during the most fertile period (Stumpf & Boesch 2005). Sex during the period of infertility appears to function to reduce male aggression toward offspring by confusing paternity, which females do by having sex with most any resident male. At peak fertility, by contrast, females are actually choosier and their preferences tend to converge on the same males, ones that may offer the best genes for offspring.

More generally, Thornhill and Gangestad (2008) conjecture, women's estrus possesses homologies not only with other female mammals, but with female vertebrates in general. All vertebrates possess receptors for estrogen, named for it being the "gen" (or generator) of estrus. Estrogen's effects on female sexuality may be homologous across (nearly) all vertebrates and function, in some ways, similarly (which is not to deny that other reproductive hormones, such as progesterone and testosterone, phylogenetically almost as old as estrogen, also play important roles in modulating female sexuality in these species). For instance, females of many bird species exhibit different mating preferences during their fertile period than during the period preceding peak fertility. Do these changes across their fertile periods possess homologies with women's estrus? According to Thornhill and Gangestad's (2008) proposals, they do.

Full evaluation of these claims obviously requires much more research. Claims about the function of estrus illustrate, however, how research on human behavior may not only be inspired by evolutionary, reproductive, and comparative biology, but may also, through comparative and phylogenetic considerations, lead to broad evolutionary perspectives.

FUTURE DIRECTIONS IN HUMAN BEHAVIORAL EVOLUTIONARY SCIENCE

As we have emphasized, human behavioral ecologists rely heavily on optimality models of how selection pressures affect behavioral strategies, models rooted in the recognition that organisms possess limited budgets of time and energy. They focus on trying to understand the behavioral outcomes of implicit allocation decisions. Evolutionary psychologists, by contrast, are fundamentally concerned with identifying the proximate mechanisms that were shaped by selection and that guide decision making and behavior. They often discuss selection pressures and the adaptive problems they created and predict the form of the behaviors and cognitive mechanisms that evolved in response. And, they rely heavily on arguments relating to functional design to assess claims about adaptation.

In the coming years, we foresee increasing cross-fertilization and, ultimately, integration of approaches. Increasingly, human behavioral ecologists appear to be interested in specific physiological mechanisms through which resource allocations are made. One topic of interest is the hormonal regulation of resource allocation. Testosterone, for instance, may be conceptualized as a messenger in a distributed communication system, one that can simultaneously upregulate and downregulate specific activities (e.g., energy dedicated to muscle growth, the neural underpinnings of status competition). Increases in male testosterone and its utilization lead to increases in broadly conceived mating effort, but at the expense of somatic maintenance and parental effort. Guided by this model, behavioral and reproductive ecologists have investigated the conditions that lead to decreases (e.g., fatherhood) and increases (e.g., divorce) in men's testosterone levels (for a review, see Ellison 2003). Based on similar thinking, factors that affect women's estrogen levels across their cycles (which reflect allocations to reproductive effort) have been explored (e.g., Ellison 2003). Some behavioral ecologists have also become interested in the psychological adaptations responsible for allocation decisions (e.g., Marlowe 2003, 2005; Cashdan 1993).

At the same time, evolutionary psychologists increasingly utilize optimality models and concepts of trade-offs in their thinking about selection (e.g., see DeScioli & Kurzban 2007). For instance, evolutionary psychologists interested in understanding adaptations that regulate mating effort recognize that they should be sensitive not only to the benefits of mating effort but also to the opportunity costs of lost parental or somatic effort (e.g., Ellison 2003). Examples such as this suggest a growing overlap and synergy between human behavioral ecology and evolutionary psychology. We foresee increased attention to a variety of other phenomena by the human evolutionary sciences, many that will benefit from similar conceptual cross-fertilization.

Phenomena of Culture

Thirty years ago, discussions of evolution and culture typically pitted them as alternative influences on human behavior. Today, evolutionary scientists are more interested in understanding how various cultural phenomena reflect ancestral selection. What selection pressures shaped human abilities involved in the horizontal and vertical transmission of information? What selection pressures are responsible for traits leading to the importance of regulation of behavior through group norms? What led to human abilities to innovate, and how, precisely, can we characterize these abilities? Behavioral ecologists, as well as evolutionary psychologists, have addressed these issues (see, for instance, readings in Gangestad & Simpson 2007).

One conceptual approach emphasizes the coevolution of genetic information and culture (e.g., Richerson & Boyd 2005). For instance, selection may give rise to adaptations that solve a particular social problem. In turn, the behavior generated by these adaptations may alter, as a by-product, group-level socioecologies, which then set the stage for new selection. A complete understanding of the evolved bases of culture requires not only an understanding of adaptations that give rise to culture (see, e.g., Sperber 2005), it also requires an understanding of human adaptations *to* cultural phenomena. This enterprise demands an understanding of how individual-level "strategies" generate group-level phenomena, which may require agent-based modeling and simulation (e.g., Aktipis 2004).

Multilevel Selection. A related topic concerns levels of selection. Many evolutionary behavioral scientists have focused on understanding adaptations produced by selection on individuals. Others have suggested that selection on groups of individuals may be responsible for a variety of human social characteristics (e.g., Sober & Wilson 1998; box 14.3 of this volume). The processes through which multilevel selection can occur are now understood analytically (e.g., Price 1970). Nevertheless, questions concerning whether selection other than that on the genic and individual level explains adaptations remain (for more on group level selection, see Richerson & Boyd 2005; Sober & Wilson 1998).

Development and Phylogeny

Tinbergen famously identified four levels at which behavioral phenomena can be explained (see chapters 1 and 29). Human evolutionary behavioral scientists have primarily focused on understanding two levels of explanation: function and proximate mechanism. The processes through which adaptations develop and the phylogenetic history of the evolution of adaptations and their by-products have received much less attention to date. We foresee increased attention to these phenomena in the future.

SUMMARY

The evolution-based behavioral and cognitive science of Darwin's premonition has been born, and has fledged into a multidisciplinary effort. The field of human evolutionary behavioral science is still in its youth, but great progress has been made in the few decades scientists have applied evolutionary principles to human nature. We anticipate the coming years will see continued growth, maturity, and enhanced productivity.

SUGGESTIONS FOR FURTHER READING

Laland and Brown (2002) have written a valuable introduction to major approaches to the study of human behavior from evolutionary perspectives: sociobiology, behavioral ecology, evolutionary psychology, and gene-culture coevolutionary theory. Barkow et al.'s (1992) book contains classic statements of the perspective of evolutionary psychology—a melding of adaptationist thinking and a computational theory of mind. A number of handbooks of evolutionary psychology that provide overviews of theory and empirical findings on humans generated by the perspective have recently appeared, including those by Buss (2005), Crawford and Krebs (2007), and Dunbar and Barrett (2007). All are important summaries of the current state of the field. Gangestad and Simpson (2007) posed 10 specific questions that remain debated to key contributors to the field. The essays that resulted address major methodological and metatheoretical issues including reconstructing the evolution of the human mind, measuring reproductive success, modularity of mind, development, group selection, intelligence, and culture. Hirschfeld and Gelman's (1994) edited collection illustrates why and how evolutionary psychologists identify functional specialization.

Barkow JH, Cosmides L, & Tooby J (1992) The Adapted Mind: Evolutionary Psychology and the Generation of Culture. Oxford Univ Press, New York.
Buss DM (2005) Handbook of Evolutionary Psychology. Wiley, New York.
Crawford CB & Krebs D (2007) Handbook of Evolutionary Psychology, 2nd ed. Erlbaum, Mahwah, NJ.
Dunbar RIM & Barrett L (2007) Oxford handbook of evolutionary psychology. Oxford Univ Press, New York.
Gangestad SW & Simpson JA (2007) The Evolution of Mind: Fundamental Questions and Controversies. Guilford Press, New York.
Hirschfeld LA & Gelman SA (eds) (1994) Mapping the Mind: Domain Specificity in Cognition and Culture. Cambridge Univ Press, New York.
Laland KN & Brown G (2002) Sense and Nonsense: Evolutionary Perspectives on Human Behaviour. Oxford Univ Press, New York.

References

Abrahams MV (1986) Patch choice under perceptual constraints: a cause for departures from an ideal free distribution. Behav Ecol Sociobiol 19: 409–415 [Chap 11]

Abrahams MV & Dill LM (1989) A determination of the energetic equivalence of the risk of predation. Ecology 70: 999–1007 [Chaps 11, 13]

Abrahams MV, Robb TL, & Hare JF (2005) Effect of hypoxia on opercular displays: evidence for an honest signal? Anim Behav 70: 427–432 [Chap 15]

Abzhanov A, Protas M, Grant R, Grant PR, & Tabin CJ (2004) *Bmp4* and morphological variation of beaks in Darwin's finches. Science 305: 1462–1465 [Chap 3]

Adams EA & Wolfner MF (2007) Seminal proteins but not sperm induce morphological changes in the *Drosophila melanogaster* female reproductive tract during sperm storage. J Insect Physiol 53: 319–331 [Chap 22]

Adams ES & Mesterton-Gibbons M (1995) The cost of threat displays and the stability of deceptive communication. J Theor Biol 175: 405–421 [Chap 16]

Ah-King M, Kvarnemo C, & Tullberg BS (2005) The importance of territoriality and mating system for the evolution of male care: a phylogenetic study on fish. J Evol Biol 18: 371–382 [Chap 26]

Ahnesjö I, Kvarnemo C, & Merilaita S (2001) Using potential reproductive rates to predict mating competition among individuals qualified to mate. Behav Ecol 12: 397–401 [Chap 20]

Aktipis CA (2004) Know when to walk away: contingent movement and the evolution of cooperation. J Theor Biol 231: 249–260 [Chap 31]

Alatalo RV, Lundberg A, & Glynn C (1986) Female pied flycatchers choose territory quality and not male characteristics. Nature 323: 152–153 [Chap 24]

Albert AYK (2005) Mate choice, sexual imprinting, and speciation: a test of a one-allele isolating mechanism in sympatric sticklebacks. Evolution 59: 927–931 [Chap 27]

Albert AYK & Otto SP (2005) Sexual selection can resolve sex-linked sexual antagonism. Science 310: 119–121 [Chap 5]

Albert AYK & Schluter D (2004) Reproductive character displacement of male stickleback mate preference: reinforcement or direct selection? Evolution 58: 1099–1107 [Chap 27]

Alcock J (1994a) Postinsemination associations between male and females in insects: the mate-guarding hypothesis. Annu Rev Entomol 39: 1–21 [Chap 22]

Alcock J (1994b) Alternative mate-locating tactics in *Chlosyne californica* (Lepidoptera, Nymphalidae). Ethology 97: 103–118 [Chap 25]

Alcock J (1996a) Provisional rejection of three alternative hypotheses on the maintenance of a size dichotomy in males of Dawson's burrowing bee, *Amegilla dawsoni* (Apidae, Apinae, Anthophorini). Behav Ecol Sociobiol 39: 181–188 [Chap 25]

Alcock J (1996b) The relation between male body size, fighting, and mating success in Dawson's burrowing bee, *Amegilla dawsoni* (Apidae, Apinae, Anthphorini). J Zool 239: 663–674 [Chap 25]

Alcock J (1996c) Male size and survival: the effects of male combat and bird predation in Dawson's burrowing bees, *Amegilla dawsoni*. Ecol Entomol 21: 309–316 [Chap 25]

Alcock J (2001) The Triumph of Sociobiology. Oxford Univ Press, New York [Chap 1]

Alcock J (2005) Animal Behavior: An Evolutionary Approach, 8th ed. Sinauer, Sunderland, MA [Chap 2]

Ale SB & Brown JS (2007) The contingencies of group size and vigilance. Evol Ecol Res 9: 1263–1276 [Chap 17]

Alexander RD (1974) The evolution of social behavior. Annu Rev Ecol Syst 5: 325–383 [Chap 17]

Alexander RD, Noonan KM, & Crespi BJ (1991) The evolution of eusociality. Pp 3–44 in Sherman PW, Jarvis JUM, & Alexander RD (eds) The Biology of the Naked Mole Rat. Princeton Univ Press, Princeton, NJ [Chap 19]

Allen JA (1988) Frequency-dependent selection by predators. Phil Trans R Soc Lond B 319: 485–503 [Chap 12]

Amdam GV, Norberg K, Fondrk MK, & Page RE Jr (2004) Reproductive ground plan may mediate colony-level selection effects on individual foraging behavior in honey bees. Proc Natl Acad Sci USA 101: 11350–11355 [Chap 28]

Amdam GV, Csondes A, Fondrk MK, & Page RE Jr (2006) Complex social behaviour derived from maternal reproductive traits. Nature 439: 76–78 [Chap 28]

Amundson R (1996) Historical development of the concept of adaptation. Pp 11–53 in Rose MR & Lauder GV (eds) Adaptation. Academic Press, San Diego [Chap 2]

Andersson M (1982) Female choice selects for extreme tail length in a widowbird. Nature 299: 818–820 [Chap 4]

Andersson M (1994) Sexual Selection. Princeton Univ Press, Princeton, NJ [Chaps 16, 20, 24]

Andersson M & Simmons LW (2006) Sexual selection and mate choice. Trends Ecol Evol 21: 296–302 [Chap 5]

Andolfatto P (2005) Adaptive evolution of non-coding DNA in *Drosophila*. Nature 437: 1149–1152 [Chap 28]

Andrade MCB (2003) Risky mate search and male self-sacrifice in redback spiders. Behav Ecol 14: 531–538 [Chap 13]

Andrès JA & Arnqvist G (2001) Genetic divergence of the seminal signal-receptor system in houseflies: the footprints of sexually antagonistic coevolution? Proc R Soc Lond B 268: 399–405 [Chap 23]

Andrès JA, Maroja LS, Bogdanowicz SM, Swanson WJ, & Harrison RG (2006) Molecular evolution of seminal proteins in field crickets. Molec Biol Evol 23: 1574–1584 [Chap 20]

Andrès JA & Morrow EH (2003) The origin of interlocus conflict: is sex linkage important? J. Evol Biol 16: 219–223 [Chap 23]

Andrews K, Reed SM, & Masta SE (2007) Spiders fluoresce variably across many taxa. Biol Lett 3: 265–267 [Chap 9]

Andrews RM, Mathies T, & Warner DA (2000) Effect of incubation temperature on morphology, growth, and survival of juvenile *Sceloporus undulatus*. Herp Monogr 14: 420–431 [Chap 29]

Angeloni L, Schlaepfer MA, Lawler JJ, & Crooks KR (2008) A reassessment of the interface between conservation and behaviour. Anim Behav 75: 731–737 [Chap 29]

Arak A & Enquist M (1995) Conflict, receiver bias and the evolution of signal form. Phil Trans R Soc Lond B 349: 337–344 [Chap 23]

Archer J (2006) Testosterone and human aggression: an evaluation of the challenge hypothesis. Neurosci Biobehav Rev 30: 319–345 [Chap 15]

Arendt J & Reznick D (2008) Convergence and parallelism reconsidered: what have we learned about the genetics of adaptation? Trends Ecol Evol 23: 26–32 [Chap 5]

Ariew A & Lewontin RC (2004) The confusions of fitness. Brit J Phil Sci 55: 347–363 [Chap 4]

Arjan J, de Visser JAGM, Hermisson J, Wagner GP, Meyers LA, Bagheri-Chaichian H, Blanchard JL, Chao L, Cheverud JM, Elena SF, Fontana W, Gibson G, Hansen TF, Krakauer D, Lewontin RC, Ofriao C, Rice SH, von Dassow G, Wagner A, & Whitlock MC (2003) Perspective: evolution and detection of genetic robustness. Evolution 57: 1959–1972 [Chap 31]

Arlt D & Part T (2007) Nonideal breeding habitat selection: a mismatch between preference and fitness. Ecology 88: 792–801 [Chap 29]

Armbruster WS & Schwaegerle KE (1996) Causes of covariation of phenotypic traits among populations. J Evol Biol 9: 261–276 [Chap 30]

Armitage KB (1999) Evolution of sociality in marmots. J Mammal 80: 1–10 [Chap 17]

Arnegard ME & Carlson BA (2005) Electric organ discharge patterns during group hunting by a mormyrid fish. Proc R Soc Lond B 272: 1305–1314 [Chap 9]

Arnold KE & Owens IPF (1999) Cooperative breeding in birds: the role of ecology. Behav Ecol 10: 465–471 [Chap 17]

Arnold SJ (1983a) Morphology, performance and fitness. Am Zool 23: 347–361 [Chap 3]

Arnold SJ (1983b) Sexual selection: the interface of theory and empiricism. Pp 67–107 in Bateson PG (ed) Mate Choice. Cambridge Univ Press, Cambridge [Chap 20]

Arnold SJ (1994a) Multivariate inheritance and evolution: a review of concepts. Pp 17–48 in

Boake CRB (ed) Quantitative Genetic Studies of Behavioral Evolution. Univ Chicago Press, Chicago [Chaps 5, 14]

Arnold SJ (1994b) Bateman's principles and the measurement of sexual selection in plants and animals. Am Nat 144: S126–S149 [Chap 20]

Arnold SJ (1994c) Is there a unifying concept of sexual selection that applies to both plants and animals? Am Nat 144: S1–S12 [Chap 21]

Arnold SJ (1994d) Constraints on phenotypic evolution. Pp 258–278 in Real LA (ed) Behavioral Mechanisms in Evolutionary Ecology. Univ Chicago Press, Chicago [Chap 29]

Arnold SJ & Duvall D (1994) Animal mating systems: a synthesis based on selection theory. Am Nat 143: 317–348 [Chaps 21, 23]

Arnold SJ & Halliday T (1988) Multiple mating: natural selection is not evolution. Anim Behav 36: 1547–1548 [Chap 22]

Arnold SJ & Wade MJ (1984) On the measurement of natural and sexual selection: theory. Evolution 38: 709–719 [Chaps 2, 3, 23, 30]

Arnott G & Elwood RW (2008) Fighting for shells: how private information about resource value changes hermit crab pre-fight displays and escalated fight behaviour. Proc R Soc Lond B 274: 3011–3017 [Chap 15]

Arnqvist G (1998) Comparative evidence for the evolution of genitalia by sexual selection. Nature 393: 784–786 [Chap 20]

Arnqvist G & Henriksson S (1997) Sexual cannibalism in the fishing spider and a model for the evolution of sexual cannibalism based on genetic constraints. Evol Ecol 11: 255–273 [Chap 30]

Arnqvist G & Nilsson T (2000) The evolution of polyandry: multiple mating and female fitness in insects. Anim Behav 60: 145–164 [Chaps 22, 24]

Arnqvist G, Nilsson T, & Katvala M (2005) Mating rate and fitness in female bean weevils. Behav Ecol 16: 123–127 [Chap 4]

Arnqvist G & Rowe L (1995) Sexual conflict and arms races between the sexes: a morphological adaptation for control of mating in a female insect. Proc R Soc Lond B 261: 123–127 [Chap 23]

Arnqvist G & Rowe L (2002a) Antagonistic coevolution between the sexes in a group of insects. Nature 415: 787–789 [Chaps 3, 23]

Arnqvist G & Rowe L (2002b) Correlated evolution of male and female morphologies in water striders. Evolution 56: 936–947 [Chap 20]

Arnqvist G & Rowe L (2005) Sexual Conflict. Princeton Univ Press, Princeton, NJ [Chaps 5, 20, 22, 23, 27]

Aron S, Vargo EL, & Passera L (1995) Primary and secondary sex ratios in monogyne colonies of the fire ant *Solenopsis invicta*. Anim Behav 49: 749–757 [Chap 19]

Arroyo B, Mougeot F, & Bretagnolle V (2001) Colonial breeding and nest defence in Montagu's harrier (*Circus pygargus*). Behav Ecol Sociobiol 50: 109–115 [Chap 17]

Ashley MV, Willson MF, Pergams ORW, O'Dowd DJ, Gende SM, & Brown JS (2003) Evolutionarily enlightened management. Biol Conserv 111: 115–123 [Chap 29]

Askenmo CEH (1984) Polygyny and nest site selection in the pied flycatcher. Anim Behav 32: 972–980 [Chap 24]

Atkinson QD, Meade A, Venditti C, Greenhill SJ, & Pagel M (2008) Languages evolve in punctuational bursts. Science 319: 588 [Chap 7]

Atzmony D, Zahavi A, & Nanjundiah V (1997) Altruistic behaviour in *Dictyostelium discoideum* explained on the basis of individual selection. Curr Sci 72: 142–145 [Chap 19]

Austad SN (1983) A game theoretical interpretation of male combat in the bowl and doily spider (*Frontinella pyramitela*). Anim Behav 31: 59–73 [Chap 15]

Austad SN (1984) A classification of alternative reproductive behaviors, and methods for field testing ESS models. Am Zool 24: 309–320 [Chap 25]

Autumn K, Ryan MJ, & Wake DB (2002) Integrating historical and organismal biology enhances the study of adaptation. Q Rev Biol 77: 383–408 [Chap 16]

Avilés L & Tufiño P (1998) Colony size and individual fitness in the social spider *Anelosimus eximius*. Am Nat 152: 403–418 [Chap 18]

Awata S, Heg D, Munehara H, & Kohda M (2004) Testis size depends on social status and the presence of male helpers in the cooperatively breeding cichlid *Julidochromis ornatus*. Behav Ecol 17: 372–379 [Chap 25]

Axelrod R & Hamilton WD (1981) The evolution of cooperation. Science 211: 1390–1396 [Chap 18]

Axelrod R, Hammond RA, & Grafen A (2004) Altruism via kin-selection strategies that rely on arbitrary tags with which they coevolve. Evolution 58: 1833–1838 [Chap 18]

Ayroles JF, Carbone MA, Stone EA, Jordan KW, Lyman RF, Magwire MM, Rollmann SM, Duncan LH, Lawrence F, Anholt RR, & Mackay TF (2009) Systems genetics of complex traits in *Drosophila melanogaster*. Nat Genet 41: 299–307 [Chap 28]

Babcock RC, Mundy C, Keesing J, & Oliver J (1992) Predictable and unpredictable spawning events: in situ behavioural data from free-spawning coral reef invertebrates. Invert Reprod Dev 22: 213–228 [Chap 21]

Babcock RC, Mundy CN, & Whitehead D (1994) Sperm diffusion models and in situ confirmation of long distance fertilization in the free-spawning asteroid *Acanthaster planci*. Biol Bull 186: 17–28 [Chap 21]

Bacigalupe LD, Crudgington HS, Hunter F, Moore AJ, & Snook RR (2007) Sexual conflict does not drive reproductive isolation in experimental populations of *Drosophila pseudoobscura*. J Evol Biol 20: 1763–1771 [Chaps 23, 27]

Bacigalupe LD, Crudgington HS, Slate J, Moore AJ, & Snook RR (2008) Sexual selection and interacting phenotypes in experimental evolution: a study of *Drosophila pseudoobscrua* mating behavior. Evolution 62: 1804–1812 [Chap 14]

Baldwin JM (1896) A new factor in evolution. Am Nat 30: 441–451, 536–553 [Chap 6]

Barkow JH, Cosmides L, & Tooby J (1992) The Adapted Mind: Evolutionary Psychology and the Generation of Culture. Oxford Univ Press, New York [Chap 31]

Barnard CJ (2004) Animal Behaviour: Function, Mechanism, Development & Evolution. Pearson/Prentice Hall, London [Chap 10]

Barnard CJ & Sibly RM (1981) Producers and scroungers: a general model and its application to captive flocks of house sparrows. Anim Behav 29: 543–550 [Chap 11]

Bar-Or C, Czosnek H, & Koltai H (2007) Cross-species microarray hybridizations: a developing tool for studying species diversity. Trends Genet 23: 200–207 [Chap 28]

Barrett HC (2005) Enzymatic computation and cognitive modularity. Mind Lang 20: 259–287 [Chap 31]

Barrett HC, Cosmides L, & Tooby J (2007) The hominid entry into the cognitive niche. Pp 241–248 in Gangestad SW & Simpson JA (eds) The Evolution of Mind. Guilford Press, New York [Chap 31]

Barrett HC & Kurzban R (2006) Modularity in cognition: framing the debate. Psych Rev 113: 628–647 [Chap 31]

Barton RA & Harvey PH (2000) Mosaic evolution of brain structure in mammals. Nature 405: 1055–1058 [Chap 10]

Baskett ML, Levin SA, Gaines SD, & Dushoff J (2005) Marine reserve design and the evolution of size at maturation in harvested fish. Ecol Appl 15: 882–901 [Chap 29]

Basolo AL (1990a) Female preference for male sword length in the green swordtail, *Xiphophorus helleri* (Pisces: Poeciliidae). Anim Behav 40: 332–338 [Chap 7]

Basolo AL (1990b) Female preference predates the evolution of the sword in swordtail fish. Science 250: 808–810 [Chaps 7, 24]

Basolo AL (1991) Male swords and female preferences. Science 253: 1427 [Chap 7]

Basolo AL (1995a) A further examination of a pre-existing bias favouring a sword in the genus *Xiphophorus*. Anim Behav 50: 365–375 [Chaps 7, 24]

Basolo AL (1995b) Phylogenetic evidence for the role of a pre-existing bias in sexual selection. Proc R Soc Lond B 259: 307–311 [Chaps 7, 16]

Basolo AL (1996) The phylogenetic distribution of a female preference. Syst Biol 45: 290–307 [Chap 7]

Basolo AL (2002) Congruence between the sexes in preexisting receiver responses. Behav Ecol 13: 832–837 [Chap 7]

Bateman AJ (1948) Intra-sexual selection in *Drosophila*. Heredity 2: 349–368 [Chaps 1, 20, 26]

Bateson M & Healy SD (2005) Comparative evaluation and its implications for mate choice. Trends Ecol Evol 20: 659–664 [Chap 10]

Bateson M, Healy SD, & Hurly TA (2003) Context-dependent foraging decisions in rufous hummingbirds. Proc R Soc Lond B 270: 1271–1276 [Chap 11]

Bateson P (2004) The active role of behaviour in evolution. Biol Phil 19: 283–298 [Chap 14]

Bateson P (2005) The return of the whole organism. J Biosciences 30: 31–39 [Chap 1]

Battin J (2004) When good animals love bad habitats: ecological traps and the conservation of animal populations. Conserv Biol 18: 1482–1491 [Chap 29]

Baum DA & Larson A (1991) Adaptation reviewed: a phylogenetic methodology for studying character macroevolution. Syst Zool 40: 1–18 [Chap 7]

Bayley H (2006) Sequencing single molecules of DNA. Curr Opin Chem Biol 10(6): 628–637.

Beacham JL (2003) Models of dominance hierarchy formation: effects of prior experience and intrinsic traits. Behaviour 140: 1275–1303 [Chap 17]

Beauchamp G (1999) The evolution of communal roosting in birds: origin and secondary losses. Behav Ecol 10: 675–687 [Chap 17]

Beauchamp G (2002) Little evidence for visual monitoring of vigilance in zebra finches. Can J Zool 80: 1634–1637 [Chap 17]

Beauchamp G (2003) Group-size effects on vigilance: a search for mechanisms. Behav Proc 63: 111–121 [Chap 17]

Beauchamp G (2008) What is the magnitude of the group-size effect on vigilance? Behav Ecol 19: 1361–1368 [Chap 17]

Beauchamp G & Ruxton GD (2008) Disentangling risk dilution and collective detection in the antipredator vigilance of semipalmated sandpipers in flocks. Anim Behav 75: 1837–1842 [Chap 17]

Beckerman AP, Benton TG, Lapsley CT, & Koesters N (2006) How effective are maternal effects at having effects? Proc R Soc London B 273: 485–493 [Chap 4]

Bednekoff PA (1996) Risk-sensitive foraging, fitness, and life histories: where does reproduction fit into the big picture? Am Zool 36: 471–483 [Chap 12]

Bednekoff PA & Lima SL (1998) Randomness, chaos and confusion in the study of antipredator vigilance. Trends Ecol Evol 13: 284–287 [Chap 17]

Beer CG (1975) Multiple functions and gull display. Pp 16–54 in Baerends CG, Beer GP & Manning A (eds) Function and Evolution in Behavior: Essays in Honour of Professor Niko Tinbergen. Clarendon Press, Oxford [Chap 16]

Begun DJ, Whitley P, Todd BL, Waldrip-Dail HM, & Clark AG (2000) Molecular population genetics of male accessory gland proteins in *Drosophila*. Genetics 156: 1879–1888 [Chap 23]

Bell AM (2005) Behavioural differences between individuals and two populations of stickleback (*Gasterosteus aculeatus*). J Evol Biol 18: 464–473 [Chaps 7, 30]

Bell AM (2007) Future directions in behavioral syndromes research. Proc R Soc Lond B 274: 755–761 [Chap 30]

Bell AM & Sih A (2007) Exposure to predation generates personality in threespined sticklebacks. Ecol Lett 10: 828–834 [Chap 30]

Bell AM & Stamps JA (2004) The development of behavioural differences between individuals and populations of stickleback. Anim Behav 68: 1339–1348 [Chap 30]

Bell AM, Backstrom T, Huntingford FA, Pottinger TG, & Winberg S (2007) Variable behavioral and neuroendocrine responses to ecologically-relevant challenges in sticklebacks. Physiol Behav 91: 15–25 [Chap 30]

Bell G & Mooers AO (1997) Size and complexity among multicellular organisms. Biol J Linn Soc 60: 345–363 [Chap 19]

Bell MA & Foster SA (1994) The Evolutionary Biology of the Threespine Stickleback. Oxford Univ Press [Chap 30]

Belovsky GE (1978) Diet optimization in a generalist herbivore. Theor Pop Biol 14: 105–134 [Chap 11]

Benkman CW & Lindholm AK (1991) The advantages and evolution of a morphological novelty. Nature 349: 519–520 [Chap 3]

Bennet-Clark HC (1989) Song and the physics of sound production. Pp 227–261 in Huber F, Moore TE & Loher W (eds) Cricket Behaviour and Neurobiology. Cornell Univ Press, Ithaca, NY [Chap 24]

Ben-Shahar Y, Robichon A, Sokolowski MB, & Robinson GE (2002) Influence of gene action across different time scales on behavior. Science 296: 741–744 [Chaps 5, 28]

Benton TG & Grant A (1996) How to keep fit in the real world: elasticity analyses and selection pressures on life histories in a variable environment. Am Nat 147: 115–139 [Chap 4]

Benton TG & Grant A (1999) Elasticity analysis as an important tool in evolutionary and population ecology. Trends Ecol Evol 14: 467–471 [Chap 4]

Benton TG & Grant A (2000) Evolutionary fitness in ecology: comparing measures of fitness in stochastic, density-dependent environments. Evol Ecol Res 2: 769–789 [Chap 4]

Benton TG, Lapsley CT, & Beckerman AP (2002) The population response to environmental noise: population size, variance and correlation in an experimental system. J Anim Ecol 71: 320–332 [Chap 4]

Bentsen CL, Hunt J, Jennions MD, & Brooks R (2006) Complex multivariate sexual selection on male acoustic signaling in a wild population of *Teleogryllus commodus*. Am Nat 167: E102–E116 [Chaps 4, 24]

Benus RF, Den Daas S, Koolhaas JM, & Van Oortmerssen GA (1990) Routine formation and flexibility in social and non-social behaviour of aggressive and nonaggressive male mice. Behaviour 112: 176–193 [Chap 30]

Benus RF, Koolhaas JM, & Van Oortmerssen GA (1987) Individual differences in behavioural reaction to a changing environment in mice and rats. Behaviour 100: 105–122 [Chap 30]

Berg RL (1960) The ecological significance of correlation pleiades. Evolution 14: 171–180 [Chap 30]

Berglund A, Rosenqvist G, & Svensson I (1989) Reproductive success of females limited by males in two pipefish species. Am Nat 133: 506–516 [Chaps 25, 26]

Bergsma R, Kanis E, Knol EF, & Bijma P (2008) The contribution of social effects to heritable variation in finishing traits of domestic pigs (*Sus scrofa*). Genetics 178: 1559–1570 [Chap 14]

Bergstrom CT & Lachmann M (1997) Signaling among relatives. I. Is costly signaling too costly? Phil Trans R Soc Lond B 352: 609–617 [Chap 16]

Bernal XE, Page RA, Rand AS, & Ryan MJ (2007) Cues for eavesdroppers: do frog calls indicate prey density and quality? Am Nat 169: 409–415 [Chap 16]

Bernasconi G & Strassman JE (1999) Cooperation among unrelated individuals: the ant foundress case. Trends Ecol Evol 14: 477–482 [Chap 18]

Bernays EA (2001) Neural limitations in phytophagous insects: implications for diet breadth and evolution of host affiliation. Annu Rev Entomol 46: 703–727 [Chap 12] rnays EA & Singer MS (2005) Insect defences: taste alteration and endoparasites. Nature 436: 476 [Chap 9]

Bernays EA & Wcislo WT (1994) Sensory capabilities, information processing, and resource specialization. Q Rev Biol 69: 187–204 [Chap 27]

Beveridge M, Simmons LW, & Alcock J (2006) Genetic breeding system and investment patterns within the nests of Dawson's burrowing bee (*Amegilla dawsoni*) (Hymenoptera: Anthophorini). Mol Ecol 15: 3459–3467 [Chap 25]

Biegler R, McGregor A, Krebs JR, & Healy SD (2001) A larger hippocampus is associated with longer-lasting spatial memory. Proc Natl Acad Sci USA 98: 6941–6944 [Chaps 1, 10]

Biernaskie JM & Elle E (2005) Conditional strategies in an animal-pollinated plant: size-dependent adjustment of gender and rewards. Evol Ecol Res 7: 901–913 [Chap 8]

Bijma P, Muir WM, & Van Arendonk JAM (2007a) Multilevel selection 1: quantitative genetic of inheritance and response to selection. Genetics 175: 277–288 [Chap 14]

Bijma P, Muir WM, Ellen ED, Wolf JB, & Van Arendonk JAM (2007b) Multilevel selection 2: estimating the genetic parameters determining inheritance and response to selection. Genetics 175: 289–299 [Chap 14]

Bijma P & Wade MJ (2008) The joint effects of kin, multilevel selection and indirect genetic effects on response to genetic selection. J Evol Biol 21: 1175–1188 [Chap 14]

Birkhead TR (1998) Cryptic female choice: criteria for establishing female sperm choice. Evolution 52: 1212–1218 [Chap 1]

Birkhead TR (2008) The Wisdom of Birds. Bloomsbury, London [Chap 1]

Birkhead TR & Møller AP (1993) Sexual selection and the temporal separation of reproductive events: sperm storage data from reptiles, birds and mammals. Biol J Linn Soc 50: 295–311 [Chap 22]

Birkhead TR & Møller AP (eds) (1998) Sperm Competition and Sexual Selection. Academic Press, London [Chap 1]

Birkhead TR, Møller AP, & Sutherland WJ (1993) Why do females make it so difficult for males to fertilize their eggs? J Theor Biol 161: 51–60 [Chaps 20, 22]

Birkhead TR, Moore HDM, & Bedford JM (1997) Sperm wars: by R. R. Baker. Trends Ecol Evol 12: 121–122 [Chap 1]

Biro PA & Stamps JA (2008) Are animal personality traits linked to life-history productivity? Trends Ecol Evol 23: 361–368 [Chap 12]

Bishop DT & Cannings C (1978) Generalised war of attrition. J Theor Biol 70: 85–124 [Chap 15]

Bishop JDD (1998) Fertilization in the sea: are the hazards of broadcast spawning avoided when free-spawned sperm fertilize retained eggs? Proc R Soc Lond B 265: 725–731 [Chap 21]

Bishop JDD & Pemberton AJ (2006) The third way: spermcast mating in sessile marine invertebrates. Integr Comp Biol 46: 398–406 [Chap 21]

Bjork A & Pitnick S (2006) Intensity of sexual selection along the anisogamy-isogamy continuum. Nature 441: 742–745 [Chaps 20, 22]

Bjork A, Starmer WT, Higginson DM, Rhodes CJ, & Pitnick S (2007) Complex interactions with females and rival males limit the evolution of sperm offence and defence. Proc R Soc Lond B 274: 1779–1788 [Chap 22]

Bleakley BH, Wolf JB, & Moore AJ (2009) Evolutionary quantitative genetics of social behaviour. In Székely T, Moore AJ, & Komdeur J (eds) Social Behaviour: Genes, Ecology and Evolution. Cambridge Univ Press, Cambridge [Chap 14]

Bleay C, Comendant T, & Sinervo B (2007) An experimental test of frequency-dependent selection on male mating strategy in the field. Proc R Soc Lond B 274: 2019–2025 [Chap 3]

Blomberg SP, Garland T Jr, & Ives AR (2003) Testing for phylogenetic signal in comparative data: behavioral traits are more labile. Evolution 57: 717–745 [Chap 7]

Blondel J, Perret P, Anstett M-C, & Thébaud C (2002) Evolution of sexual size dimorphism in birds: test of hypotheses using blue tits in contrasted Mediterranean habitats. J Evol Biol 15: 440–450 [Chap 13]

Blows MW (1998) Evolution of a mate recognition system after hybridization between two *Drosophila* species. Am Nat 151: 538–544 [Chap 27]

Blows MW (2007) A tale of two matrices: multivariate approaches in evolutionary biology. J Evol Biol 20: 1–8 [Chaps 5, 20, 30]

Blows MW & Allan RA (1998) Levels of mate recognition within and between two *Drosophila* species and their hybrids. *Am Nat* 152: 826–837 [Chap 27]

Blows MW, Brooks R, & Kraft PG (2003) Exploring complex fitness surfaces: multiple ornamentation and polymorphism in male guppies. Evolution 57: 1622–1630 [Chap 3]

Blumstein DT (2003) Flight initiation distance in birds is dependent on intruder starting distance. J Wildl Manag 67: 852–857 [Chap 13]

Blumstein DT (2006a) Developing an evolutionary ecology of fear: how life history and natural history traits affect disturbance tolerance in birds. Anim Behav 71: 389–399 [Chap 13]

Blumstein DT (2006b) The multi-predator hypothesis and the evolutionary persistence of antipredator behaviour. Ethology 112: 209–217 [Chap 13]

Blumstein DT (2007) The evolution of alarm communication in rodents: structure, function, and the puzzle of apparently altruistic calling in rodents. Pp 317–327 in Wolff JO & Sherman PW (eds) Rodent Societies. Univ of Chicago Press, Chicago [Chap 13]

Blumstein DT (2008) Fourteen lessons from antipredator behavior. Pp 147–158 in Sagarin R & Taylor T (eds) Natural Security: A Darwinian Approach to a Dangerous World. Univ California Press, Berkeley [Chap 13]

Blumstein DT & Daniel JC (2002) Isolation from mammalian predators differentially affects two congeners. Behav Ecol 13: 657–663 [Chap 13]

Blumstein DT & Fernández-Juricic E (2004) The emergence of conservation behavior. Conserv Biol 18: 1175–1177 [Chaps 13, 29]

Blumstein DT, Daniel JC, & Springett BP (2004a) A test of the multi-predator hypothesis: rapid loss of antipredator behavior after 130 years of isolation. Ethology 110: 919–934 [Chap 13]

Blumstein DT, Verenyre L, & Daniel JC (2004b) Reliability and the adaptive utility of discrimination among alarm callers. Proc R Soc Lond B 271: 1851–1857 [Chap 13]

Blumstein DT, Fernandez-Juricic E, LeDee O, Larsen E, Rodriguez-Prieto I, & Zugmeyer C (2004c) Avian risk assessment: effects of perching height and detectability. Ethology 110: 273–285 [Chap 13]

Blumstein DT, Anthony LL, Harcourt RG, & Ross G (2003) Testing a key assumption of wildlife buffer zones: is flight initiation distance a species-specific trait? Biol Conserv 110: 97–100 [Chap 13]

Boake CRB, Arnold SJ, Breden F, Meffert LM, Ritchie MG, Taylor BJ, Wolf JB, & Moore AJ (2002) Genetic tools for studying adaptation and the evolution of behavior. Am Nat 160: S143-S159 [Chaps 5, 14]

Bode M & Marshall DJ (2007) The quick and the dead? Sperm competition and sexual conflict in the sea. Evolution 61: 2693–2700 [Chap 21]

Boehm C (1999) Hierarchy in the Forest: The Evolution of Egalitarian Behavior. Harvard Univ Press, Cambridge, MA [Chap 17]

Bolhuis JJ & Macphail EM (2001) A critique of the neuroecology of learning and memory. Trends Cogn Sci 5: 426–433 [Chap 10]

Bolhuis JJ & Macphail EM (2002) Everything in neuroecology makes sense in the light of evolution. Trends Cogn Sci 6: 7–8 [Chap 10]

Bolnick DI & B Fitzpatrick (2007) Sympatric speciation: theory and empirical data. Annu Rev Ecol Syst 38: 459–487 [Chap 27]

Bolton TF & Havenhand JN (1996) Chemical mediation of sperm activity and longevity in the solitary ascidians *Ciona intestinalis* and *Ascidiella aspersa*. Biol Bull 190: 329–335 [Chap 21]

Bonabeau E, Theraulaz G, Deneubourg JL, Aron S, & Camazine S (1997) Self organization in social insects. Trends Ecol Evol 12: 188–193 [Chap 19]

Bond AB & Kamil AC (1998) Apostatic selection by blue jays produces balanced polymorphism in virtual prey. Nature 395: 594–596 [Chap 12]

Bond AB & Kamil AC (2002) Visual predators select for crypticity and polymorphism in virtual prey. Nature 415: 609–613 [Chap 10]

Bond AB, Cook RG, & Lamb MR (1981) Spatial memory and the performance of rats and pigeons in the radial-arm maze. Anim Learn & Behav 9: 575–580 [Chap 10]

Bonduriansky R (2007) The genetic architecture of sexual dimorphism: the potential roles of genomic imprinting and condition dependence. Pp 176–184 in DJ Fairbairn, WU Blanckenhorn, & T Székely (eds) Sex, Size and Gender Roles: Evolutionary Studies of Sexual Size Dimorphism. Oxford Univ Press, Oxford [Chap 20]

Bonnie KE & Earley RL (2007) Expanding the scope for social information use. Anim Behav 74: 171–181 [Chap 17]

Bonte D & Lens L (2007) Heritability of spider ballooning motivation under different wind velocities. Evol Ecol Res 9: 817–827 [Chap 5]

Boomsma JJ, Nielsen J, Sundstrom L, Oldham NJ, Tentschert J, Petersen HC, & Morgan ED (2003) Informational constraints on optimal sex allocation in ants. Proc Natl Acad Sci USA 100: 8799–8804 [Chap 18]

Boots M & Knell RJ (2002) The evolution of risky behaviour in the presence of a sexually transmitted disease. Proc R Soc Lond B 269: 585–589 [Chap 24]

Borgerhoff Mulder M (2007) On the utility, not the necessity, of tracking current fitness. Pp 78–85 in Gangestad SW & Simpson JA (eds) The Evolution of Mind: Fundamental Questions and Controversies. Guilford Press, New York [Chap 31]

Borowsky RL (1987) Agonistic behavior and social inhibition of maturation in fish of the genus *Xiphophorus* (Poeciliidae). Copeia 1983: 792–796 [Chap 17]

Both C, Dingemanse NJ, Drent PJ, & Tinbergen JM (2005) Pairs of extreme avian personalities have highest reproductive success. J Anim Ecol 74: 667–674 [Chap 30]

Bouchard F & Rosenberg A (2004) Fitness, probability and the principles of natural selection. Brit J Phil Sci 55: 693–712 [Chap 4]

Boughman JW (2001) Divergent sexual selection enhances reproductive isolation in sticklebacks. Nature 411: 944–948 [Chap 27]

Boughman JW (2002) How sensory drive can promote speciation. Trends Ecol Evol 17: 571–577 [Chap 27]

Boughman JW (2007a) Speciation in sticklebacks. Pp 83–126 in Ostlund-Nilsson S, Mayer I, & Huntingford F (eds) The Biology of the Three-Spined Stickleback. CRC Press, Boca Raton, FL [Chap 27]

Boughman JW (2007b) Condition-dependent expression of red color differs between stickleback species. Journal of Evol Biol 20: 1577–1590 [Chap 27]

Bourke AFG (1988) Worker reproduction in the higher eusocial hymenoptera. Q Rev Biol 63: 291–311 [Chap 19]

Bourke AFG (1999) Colony size, social complexity, and reproductive conflict in social insects. J Evol Biol 12: 245–257 [Chap 19]

Bourke AFG & Franks NR (1995) Social Evolution in Ants. Princeton Univ Press, Princeton, NJ [Chap 19]

Bowler P (1983) Evolution: The History of an Idea. Univ California Press, Berkeley [Chap 1]

Bradbury JW & Vehrencamp SL (1998) Principles of Animal Communication. Sinauer, Sunderland, MA [Chaps 9, 16]

Bradshaw WE & Holzapfel CM (2006) Climate change: evolutionary response to rapid climate change. Science 312: 1477–1478 [Chap 29]

Braendle C & Flatt T (2006) A role for genetic accommodation in evolution? Bioessays 28: 868–873 [Chap 6]

Brandon RN & Rausher MD (1996) Testing adaptationism: a comment on Orzack and Sober. Am Nat 148: 189–201 [Chap 2]

Brehm AE (1861) Das Leben der Vögel. Glogau, C. Flemming [Chap 1]

Briffa M & Elwood RW (2002) Power of shell-rapping signals influences physiological costs and subsequent decisions during hermit crab fights. Proc R Soc Lond B 269: 2331–2336 [Chap 15]

Briffa M & Elwood RW (2007) Monoamines and decision making during contests in the hermit crab *Pagurus bernhardus*. Anim Behav 73: 605–612 [Chap 15]

Briffa M, Elwood RW, & Dick JTA (1998) Analysis of repeated signals during shell fights in the hermit crab *Pagurus bernhardus*. Proc R Soc Lond B 265: 1467–1474 [Chap 15]

Brockmann HJ & Dawkins R (1979) Joint nesting in a digger wasp as an evolutionarily stable preadaptation to social life. Behaviour 71: 203–245 [Chap 2]

Brockmann HJ & Penn D (1991) Conditional mating strategies in *Limulus polyphemus*. Abstract, 22nd Int. Ethology Conf., Otani Univ, Kyoto, Japan, p 10 [Chap 25]

Brockmann HJ, Colsen T, & Potts W (1994) Sperm competition in horseshoe crabs (*Limulus polyphemus*). Behav Ecol Sociobiol 35: 153–160 [Chap 21]

Brockmann HJ, Grafen A, & Dawkins R (1979) Evolutionarily stable nesting strategy in a digger wasp. J Theor Biol 77: 473–496 [Chap 2]

Brodie ED III (1992) Correlational selection for color pattern and antipredator behavior in the garter snake *Thamnophis ordinoides*. Evolution 46: 1284–1298 [Chaps 3, 30]

Brodie ED III, Moore AJ, & Janzen FJ (1995) Visualizing and quantifying natural selection. Trends Ecol Evol 10: 313–318 [Chaps 3, 14, 30]

Brodin A & Clark CW (2007) Energy storage and expenditure. Pp 221–269 in Stephens DW, Brown JS, & Ydenberg RC (eds) Foraging: Behavior and Ecology. Univ Chicago Press, Chicago [Chap 12]

Broman KW (2001) Review of statistical methods for QTL mapping in experimental crosses. Laboratory Animal 30: 44–52 [Chap 5]

Brommer JE, Gustafsson L, Pietiainen H, & Merilä J (2004) Single-generation estimates of individual fitness as proxies for long-term genetic contribution. Am Nat 163: 505–517 [Chap 4]

Brommer JE, Merilä J, Sheldon BC, & Gustafsson L (2005) Natural selection and genetic variation for reproductive reaction norms in a wild bird population. Evolution 59: 1362–1371 [Chap 6]

Brooks DR & McLennan DA (1991) Phylogeny, Ecology, and Behaviour. Univ Chicago Press [Chap 7]

Brooks R (2000) Negative genetic correlation between male sexual attractiveness and survival. Nature 406: 67–70 [Chap 24]

Brooks R, Hunt J, Blows MW, Smith MJ, Bussière LF, & Jennions MD (2005) Experimental evidence for multivariate stabilizing sexual selection. Evolution 59: 871–880 [Chap 24]

Brown JL (1975) The Evolution of Behaviour. Norton, New York [Chap 1, 16]

Brown JS (1993) Model validation: optimal foraging theory. Pp 360–377 in Scheiner SM & Gurevitch J (eds) Design and Analysis of Ecological Experiments. Chapman & Hall, New York [Chap 2]

Brown JS & Kotler BP (2004) Hazardous duty pay and the foraging cost of predation. Ecol Lett 7: 999–1014 [Chaps 12, 13]

Brown WD, Smith AT, Moskalik B, & Gabriel J (2006) Aggressive contests in house crickets: size, motivation and the information content of aggressive songs. Anim Behav 72: 225–233 [Chap 15]

Bshary R & Grutter AS (2002) Asymmetric cheating opportunities and partner control in a cleaner fish mutualism. Anim Behav 63: 547–555 [Chap 18]

Buchanan KL, Leitner S, Spencer KA, Goldsmith AR, & Catchpole CK (2004) Developmental stress selectively affects the song control nucleus HVC in the zebra finch. Proc R Soc Lond B 271: 2381–2386 [Chap 10]

Bull JJ, Pfennig DW, & Wang IN (2004) Genetic details, optimization and phage life histories. Trends Ecol Evol 19: 76–82 [Chap 8]

Buller DJ (2005) Adapting minds: evolutionary psychology and the persistent quest for human nature. MIT Press, Cambridge, MA [Chap 31]

Bulmer MG & Parker GA (2002) The evolution of anisogamy: a game-theoretic approach. Proc R Soc Lond B 269: 2381–2388 [Chap 20]

Burkhardt RW (2005) Patterns of behavior: Konrad Lorenz, Niko Tinbergen and the founding of ethology. Univ Chicago Press, Chicago [Chap 1]

Burley N (1986) Sexual selection for aesthetic traits in species with biparental care. Am Nat 127: 415–445 [Chap 26]

Burley N, Kratzberg G, & Radman P (1982) Influence of colour-banding on the conspecific preferences of zebra finches. Anim Behav 30: 444–455 [Chap 26]

Burley NT & Symanski R (1998) "A taste for the beautiful": latent aesthetic mate preferences for white crests in two species of Australian grassfinches. Am Nat 152: 792–802 [Chap 10]

Burmeister M, McInnis MG, & Zollner S (2008) Psychiatric genetics: progress amid controversy. Nat Rev Genet 9: 527–540 [Chap 28]

Burt A & Trivers R (2006) Genes in Conflict: The Biology of Selfish Genetic Elements. Harvard Univ Press, Cambridge, MA [Chaps 18, 19]

Bush GL (1969) Sympatric host-race formation and speciation in frugivorous flies of the genus *Rhagoletis* (Diptera, Tephritidae). Evolution 23: 237–251 [Chap 27]

Buss DM (2005) Handbook of Evolutionary Psychology. Wiley, New York [Chap 31]

Buss LW (1987) The Evolution of Individuality. Princeton Univ Press, Princeton, NJ [Chap 19]

Bussière LF, Hunt J, Jennions MD, & Brooks R (2006) Sexual conflict and cryptic female choice in the black field cricket, *Teleogryllus commodus*. Evolution 60: 792–800 [Chap 24]

Buston PM, Reeve HK, Cant MA, Vehrencamp SL, & Emlen ST (2007) Reproductive skew and the evolution of group dissolution tactics: a synthesis of concession and restraint models. Anim Behav 74: 1643–1654 [Chap 17]

Butler MA & King AA. (2004) Phylogenetic comparative analysis: a modeling approach for adaptive evolution. Am Nat 164: 683–695 [Chap 7]

Butlin RK & Ritchie MG (1989) Genetic coupling in mate recognition systems: what is the evidence? Biol J Linn Soc 37: 237–246 [Chap 5]

Byrne PG & Roberts JD (1999) Simultaneous mating with multiple males reduces fertilization success in the myobatrachid frog *Crinia Georgiana*. Proc R Soc Lond B 266: 717–721 [Chap 21]

Caceres CE & Tessier AJ (2003) How long to rest: the ecology of optimal dormancy and environmental constraint. Ecology 84: 1189–1198 [Chap 4]

Cain SD, Boles LC, Wang JH, & Lohmann KJ (2005) Magnetic orientation and navigation in marines turtles, lobsters and molluscs: concepts and conundrums. Integr Comp Biol 45: 539–546 [Chap 9]

Calhim S & Birkhead TR (2007) Testes size in birds: quality versus quantity—assumptions, errors, and estimates. Behav Ecol 18: 271–275 [Chap 22]

Calsbeek R (2008) An ecological twist on the morphology, performance, fitness axis. Evol Ecol Res 10: 197–212 [Chap 3]

Calsbeek R, Buermann W, & Smith TB (2009) Parallel shifts in ecology and natural selection in an island lizard. BMC Evol Biol 9: 3 [Chap 3]

Calsbeek R & Irschick DJ (2007) The quick and the dead: correlational selection on morphology, performance, and habitat use in island lizards. Evolution 61: 2493–2503 [Chap 3]

Calsbeek R & Sinervo B (2004) Within clutch variation in offspring sex determined by differences in sire body size: cryptic mate choice in the wild. J Evol Biol 17: 464–470 [Chap 3]

Calvert G, Spence C, & Stein BE (2004) The Handbook of Multisensory Processes. MIT Press Cambridge MA [Chap 10]

Cameron E, Day T, & Rowe L (2003) Sexual conflict and indirect benefits. J Evol Biol 16: 1055–1060 [Chaps 23, 24]

Cameron EZ, Lemons PR, Bateman PW, & Bennet NC (2008) Experimental alteration of litter sex ratio in a mammal. Proc R Soc Lond B 275: 323–327 [Chap 17]

Candolin U, Salesto T, & Evers M (2007) Changed environmental conditions weaken sexual selection in sticklebacks. J Evol Biol 20: 233–239 [Chap 27]

Candolin U & Voigt H-R (2001) Correlation between male size and territory quality: consequences of male competition or predation susceptibility. Oikos 95: 225–230 [Chap 4]

Cant MA (2006) A tale of two theories: parent-offspring conflict and reproductive skew. Anim Behav 71: 255–263 [Chap 17]

Cant MA & English S (2006) Stable group size in cooperative breeders: the role of inheritance and reproductive skew. Behav Ecol 17: 560–568 [Chap 17]

Cant MA & Johnstone RA (2008) Reproductive conflict and the separation of reproductive generations in humans. Proc Natl Acad Sci USA 105: 5332–5336 [Chap 4]
Caraco T (1979) Time budgeting and group size: a theory. Ecology 60: 611–617 [Chap 17]
Caraco T, Martindale S, & Whittam TS (1980) An empirical demonstration of risk-sensitive foraging preferences. Anim Behav 28: 820–830 [Chap 11]
Carde RT & Baker TC (1984) Sexual communication with pheromones. Pp 355–383 in Bell WJ & Carde RT (eds) Chemical Ecology of Insects. Chapman & Hall, London [Chap 28]
Carere C, Welink D, Drent PJ, Koolhaas JM, & Groothuis TGG (2001) Effect of social defeat in a territorial bird (*Parus major*) selected for different coping styles. Physiol Behav 73: 427–433 [Chap 30]
Carere C, Groothuis TGG, Moestl E, Daan S, & Koolhaas JM (2003) Fecal corticosteroids in a territorial bird selected for different personalities: daily rhythm and the response to social stress. Horm Behav 43: 540–548 [Chap 30]
Carmel Y & Ben-Haim Y (2005) Info-gap robust-satisficing model of foraging behavior: do foragers optimize or satisfice? Am Nat 166: 633–641 [Chap 12]
Caro T (1999) The behaviour-conservation interface. Trends Ecol Evol 14: 366–369 [Chap 29]
Caro T (2005) Antipredator Defenses in Birds and Mammals. Univ Chicago Press, Chicago [Chap 13]
Carranza J (2004) Sex allocation within broods: the intrabrood sharing-out hypothesis. Behav Ecol 15: 223–232 [Chap 26]
Carroll SP & Corneli PS (1995) Divergence in male mating tactics between two populations of the soapberry bug: II. Genetic change and the evolution of a plastic reaction norm in a variable social environment. Behav Ecol 6: 46–56 [Chap 6]
Carroll SP & Corneli PS (1999) The evolution of behavioral norms of reaction as a problem in ecological genetics. Pp 53–68 in Foster S & Endler J (eds) Geographic Variation in Behavior. Oxford Univ Press, New York [Chap 6]
Carroll SP & Fox CW (2008) Conservation Biology: Evolution in Action. Oxford Univ Press, New York [Chap 29]
Carroll SB, Grenier JK, & Weatherbee SD (2005) From DNA to Diversity: Molecular Genetics and the Evolution of Animal Design. Blackwell Sci, Malden, MA [Chap 28]
Carroll SP & Watters JV (2008) Managing phenotypic variability with genetic and environmental heterogeneity: adaptation as a first principle. Pp 181–198 in Caroll SP & Fox CW (eds) Conservation Biology: Evolution in Action. Oxford Univ Press, New York [Chaps 6, 29]
Caryl PG (1979) Communication by agonistic displays: what can games theory contribute to ethology? Behaviour, 68: 136–169 [Chap 16]
Cashdan E (1993) Attracting mates: effects of paternal investment on mate attraction strategies. Ethol Sociobiol 14: 1–24 [Chap 31]
Caswell H (2001) Population Matrix Models. Sinauer, Sunderland, MA [Chap 4]
Catchpole CK & Slater PJB (1995) Bird Song: Biological Themes and Variations. Cambridge Univ Press, Cambridge [Chap 10]
Chaine AS & Lyon BE (2008) Adaptive plasticity in female mate choice dampens sexual selection on male ornaments in the lark bunting. Science 319: 459–462 [Chap 7]
Champoux M, Bennett A, Shannon C, Higley JD, Lesch KP, & Suomi SJ (2002) Serotonin transporter gene polymorphism, differential early rearing, and behavior in rhesus monkey neonates. Mol Psychiat 7: 1058–1063 [Chap 30]
Chapman T (2001) Seminal fluid-mediated fitness traits in *Drosophila*. Heredity 87: 511–521 [Chaps 22, 26]
Chapman T (2006) Evolutionary conflicts of interest between males and females. Curr Biol 16: R744–R754 [Chap 23]
Chapman T & Davies SJ (2004) Functions and analysis of the seminal fluid proteins of male *Drosophila melanogaster* fruit flies. Peptides 25: 1477–1490 [Chap 28]
Chapman T, Liddle LF, Kalb JM, Wolfner MF, & Partridge L (1995) Cost of mating in *Drosophila melanogaster* females is mediated by male accessory gland products. Nature 373: 241–244 [Chaps 22, 23, 26]
Chapuisat M & Keller L (1999) Testing kin selection theory with sex allocation data in eusocial Hymenoptera. Heredity 82: 473–478 [Chap 19]
Charlesworth B (1971) Selection in density regulated populations. Ecology 52: 469–474 [Chap 25]
Charmantier A, McCleery RH, Cole LR, Perrins C, Kruuk LEB, & Sheldon BC (2008) Adaptive phenotypic plasticity in response to climate change in a wild bird population. Science 320: 800–803 [Chaps 6, 29]
Charnov EL (1976) Optimal foraging, marginal value theorem. Theor Pop Biol 9: 129–136 [Chaps 8, 11, 13]
Charnov EL (1982) The Theory of Sex Allocation. Princeton Univ Press, NJ [Chap 2]
Charnov EL & Skinner SW (1985) Complementary approaches to the understanding of parasitoid oviposition decisions. Envir Entomol. 14: 383–391 [Chap 11]
Chase ID (1980) Social process and hierarchy formation in small groups: a comparative

perspective. Am Sociol Rev 45: 905–924 [Chap 17]
Chase ID, Tovey C, Spangler-Martin D, & Manfredonia M (2002) Individual differences versus social dynamics in the formation of animal dominance hierarchies. Proc Natl Acad Sci USA 99: 5744–5749 [Chap 17]
Cheney DL & Seyfarth RM (1990) How Monkeys See the World. Univ Chicago Press, Chicago [Chap 13]
Cheng R-C & Tso I-M (2007) Signaling by decorating webs: luring prey or deterring predators? Behav Ecol 18: 1085–1091 [Chap 9]
Chenoweth SF & Blows MW (2003) Signal trait sexual dimorphism and mutual sexual selection in *Drosophila serrata*. Evolution 57: 2326–2334 [Chaps 7, 27]
Chenoweth SF & Blows MW (2005) Contrasting mutual sexual selection on homologous signal traits in *Drosophila serrata*. Am Nat 165: 281–289 [Chap 27]
Chenoweth SF & Blows MW (2006) Dissecting the complex genetic basis of mate choice. Nat Rev Genet 7: 681–692 [Chap 24]
Chenoweth SF & Blows MW (2008) Q(st) meets the G matrix: The dimensionality of adaptive divergence in multiple correlated quantitative traits. Evolution 62: 1437–1449 [Chap 27]
Cheverton J, Kacelnik A, & Krebs JR (1985) Optimal foraging: constraints and currencies. Pp 109–126 in Hölldobler B & Lindauer M (eds) Experimental Behavioral Ecology and Sociobiology. Gustav Fischer Verlag, Stuttgart [Chap 2]
Cheverud JM (1984a) Quantitative genetics and developmental constraints on evolution by selection. J Theor Biol. 110: 155–171 [Chap 3]
Cheverud JM (1984b) Evolution by kin selection: a quantitative genetic model illustrated by maternal performance in mice. Evolution 38: 766–777 [Chap 14]
Cheverud JM (1985) A quantitative genetic model of altruistic selection. Behav Ecol Sociobiol 16: 239–243 [Chap 14]
Cheverud JM & Dow MM (1985) An autocorrelation analysis of genetic variation due to lineal fission in social groups of Rhesus Macaques. Am J Phys Anthrop 67: 113–121 [Chap 7]
Cheverud JM & Moore AJ (1994) Quantitative genetic and the role of the environment provided by relatives in behavioral evolution. Pp 67–100 in Boake CRB (ed) Quantitative Genetic Studies of Behavioral Evolution. Univ Chicago Press, Chicago [Chaps 4, 14]
Chippindale AK, Gibson JR, & Rice WR (2001) Negative genetic correlation for adult fitness between sexes reveals ontogenetic conflict in *Drosophila*. Proc Natl Acad Sci USA 98: 1671–1675 [Chap 23]
Chivers DP & Smith RJF (1995) Free-living fathead minnows rapidly learn to recognize pike as predators. J Fish Biol 46: 949–954 [Chap 29]
Chown SL & Gaston KJ (2008) Macrophysiology for a changing world. Proc R Soc Lond B 275: 1469–1478 [Chap 6]
Christiansen FB (1985) The definition and measurement of fitness. Pp. 65–79 in Shorrocks B (ed) Evolutionary Ecology. Blackwell Sci, Oxford [Chap 4]
Christman MC & Lewis D (2005) Spatial distribution of dominant animals within a group: comparison of four statistical tests of location. Anim Behav 70: 73–82 [Chap 17]
Churchill GA (2004) Using ANOVA to analyze microarray data. Biotechniques 37: 173–175, 177 [Chap 28]
Civetta A & Clark A (2000) Correlated effects of sperm competition and postmating female mortality. Proc Natl Acad Sci USA 97: 13162–13165 [Chap 23]
Clark AB & Ehlinger TJ (1987) Pattern and adaptation in individual behavioral differences. Pp 1–47 in Bateson PPG & Klopfer PH (eds) Perspectives in Ethology. Plenum Press, New York [Chap 30]
Clark AG (2002) Sperm competition and the maintenance of polymorphism. Heredity 88: 148–153 [Chap 22]
Clark AG & Begun DJ (1998) Female genotypes affect sperm displacement in *Drosophila*. Genetics 149: 1487–1493 [Chap 28]
Clark CW (1994) Antipredator behavior and the asset-protection principle. Behav Ecol 5: 159–170 [Chaps 12, 30]
Clark CW (2006) The Worldwide Crisis in Fisheries: Economic Models and Human Behaviour. Cambridge Univ Press, Cambridge [Chap 12]
Clark CW & Mangel M (1986) The evolutionary advantages of group foraging. Theor Pop Biol 30: 45–75 [Chap 11]
Clark CW & Mangel M (2000) Dynamic State Variable Models in Ecology: Methods and Applications. Oxford Univ Press, New York [Chaps 8, 12, 13]
Clark NL, Aagaard JE, & Swanson WJ (2006) Evolution of reproductive proteins from animals and plants. Reproduction 131: 11–22 [Chaps 23, 28]
Claudianos C, Ranson H, Johnson RM, Biswas S, Schuler MA, et al. (2006) A deficit of detoxification enzymes: pesticide sensitivity and environmental response in the honeybee. Insect Mol Biol 15: 615–636 [Chap 28]
Clayton DF (2004) Songbird genomics: methods, mechanisms, opportunities, and pitfalls. Ann NY Acad Sci 1016: 45–60 [Chap 28]
Clayton NS & Dickinson A (1998) Episodic-like memory during cache recovery by scrub jays. Nature 395: 272–274 [Chap 10]

Clayton NS & Krebs JR (1994) Hippocampal growth and attrition in birds affected by experience. Proc Natl Acad Sci USA 91: 7410–7414 [Chap 10]

Clemmons JR & Buchholz R (eds) (1997) Behavioral Approaches to Conservation in the Wild. Cambridge Univ Press, Cambridge [Chap 29]

Clifton KE (1997) Mass spawning by green algae on coral reefs. Science 275: 1116–1118.

Clotfelter ED, Ardia DR, & McGraw KJ (2007) Red fish, blue fish: tradeoffs between pigmentation and immunity in *Betta splendens*. Behav Ecol 18: 1139–1145 [Chap 16]

Clutton-Brock T (2002) Breeding together: kin selection and mutualism in cooperative vertebrates. Science 296: 69–72 [Chap 17]

Clutton-Brock TH (1988) Reproductive Success: Studies of Individual Variation in Contrasting Breeding Systems. Univ Chicago Press, Chicago [Chap 4]

Clutton-Brock TH (1991) The Evolution of Parental Care. Princeton Univ Press, Princeton, NJ [Chaps 20, 26]

Clutton-Brock TH (2002) Breeding together: kin selection, reciprocity and mutualism in cooperative animal societies. Science 296: 69–72 [Chap 18]

Clutton-Brock TH (2007) Sexual selection in males and females. Science 318: 1882–1885 [Chap 20]

Clutton-Brock TH & Albon SD (1979) The roaring of red deer and the evolution of honest advertisement. Behaviour 69: 145–170 [Chap 15]

Clutton-Brock TH & Parker GA (1992) Potential reproductive rates and the operation of sexual selection. Q Rev Biol 67: 437–456 [Chap 20]

Clutton-Brock TH & Parker GA (1995a) Punishment in animal societies. Nature 373: 209–216 [Chaps 17, 18]

Clutton-Brock TH & Parker GA (1995b) Sexual coercion in animal societies. Anim Behav 49: 1345–1365 [Chap 22]

Clutton-Brock TH & Vincent AJ (1991) Sexual selection and the potential reproductive rates of males and females. Nature 351: 58–60 [Chap 20]

Clutton-Brock TH, Brotherton PNM, Oriain MJ, Griffin AS, Gaynor D, Sharpe L, Kansky R, Manser MB, & McIlrath GM (2000) Individual contributions to babysitting in a cooperative mongoose, *Suricata suricatta*. Proc R Soc Lond B 267: 301–305 [Chap 18]

Clutton-Brock TH, Guineness FE, & Albon SD (1982) Red Deer: Behavior and Ecology of the Two Sexes. Univ Chicago Press, Chicago [Chap 4]

Clutton-Brock TH, Guineness FE, & Albon SD (1988) Reproductive success in male and female red deer. Pp 325–343 in Clutton-Brock TH (ed) Reproductive Success: Studies of Individual Variation in Contrasting Breeding Systems. Univ Chicago Press, Chicago [Chap 4]

Clutton-Brock TH, Hodge SJ, Spong G, Russell AF, Jordan NR, Bennett NC, Sharpe LL, & Manser MB (2006) Intrasexual competition and sexual selection in cooperative mammals. Nature 444: 1065–1068 [Chap 20]

Clutton-Brock TH, Russell AF, Sharpe LL, Young AJ, Balmforth Z, & McIlrath GM (2002) Evolution and development of sex differences in cooperative behavior in meerkats. Science 297: 253–256 [Chaps 18, 26]

Cockburn A (1998) Evolution of helping behavior in cooperatively breeding birds. Annu Rev Ecol Syst 29: 141–177 [Chaps 17, 19]

Cockburn A, Dalziell AH, Blackmore CSJ, Double MC, Kokko H, Osmond HL, Beck NR, Head ML, & Wells K (2009) Superb fairy-wren males aggregate into hidden leks to solicit extra-group fertilisations before dawn. Behav Ecol 20: 501–510 [Chap 24]

Cockburn A, Legge S, & Double MC (2002) Sex ratios in birds and mammals: can the hypotheses be disentangled? Pp 266–286 in Hardy ICW (ed) Sex Ratios: Concepts and Research Methods. Cambridge Univ Press, Cambridge [Chap 5]

Cockburn A, Scott MP, & Scotts DJ (1985) Inbreeding avoidance and male-biased natal dispersal in *Antechinus* spp. (Marsupialia: Dasyuridae). Anim Behav 33: 908–915 [Chap 31]

Codenotti TL & Alvarez F (1998) Adoption of unrelated young by greater rheas. J Ornithol 69: 58–65 [Chap 26]

Coffroth MA & Lasker HR (1998) Larval paternity and male reproductive success of a broadcast-spawning gorgonian, *Plexaura kuna*. Mar Biol 131: 329–337 [Chap 21]

Coleman RM & Gross MR (1991) Parental investment theory—the role of past investment. Trends Ecol Evol 6: 404–406 [Chap 20]

Collins SA (1995) The effect of recent experience on female choice in zebra finches. Anim Behav 49: 479–486 [Chap 10]

Coltman DW, O'Donoghue P, Jorgenson JT, Hogg JT, Strobeck C, & Festa-Bianchet M (2003) Undesirable evolutionary consequences of trophy hunting. Nature 426: 655–658 [Chap 29]

Colwell RK (1981) Group selection is implicated in the evolution of female-biased sex ratios. Nature 290: 401–404 [Chap 18]

Connor RC (1995a) The benefits of mutualism: a conceptual framework. Biol Rev 70: 427–457 [Chap 11]

Connor RC (1995b) Altruism among non-relatives: alternatives to the "Prisoner's Dilemma." Trends Ecol Evol 10: 84–86 [Chap 17]

Conover DO, Arnott SA, Walsh MR, & Munch SB (2005) Darwinian fishery science: lessons from the Atlantic silverside (*Menidia menidia*). Can J Fish Aq Sci 62: 730–737 [Chap 29]

Conradt L & Roper TJ (2000) Activity synchrony and social cohesion: a fission-fusion model. Proc R Soc Lond B 267: 2213–2218 [Chap 11]

Conradt L & Roper TJ (2007) Democracy in animals: the evolution of shared group decisions. Proc R Soc Lond B 274: 2317–2326 [Chap 11]

Consortium CSaA (2005) Initial sequence of the chimpanzee genome and comparison with the human genome. Nature 437: 69–87 [Chap 28]

Consortium HGS (2006) Insights into social insects from the genome of the honeybee *Apis mellifera*. Nature 443: 931–949 [Chap 28]

Coolen I, Giraldeau LA, & Lavoie M (2001) Head position as an indicator of producer and scrounger tactics in a ground-feeding bird. Anim Behav 61: 895–903 [Chap 11]

Cooper WE Jr & Frederick WG (2007a) Optimal time to emerge from refuge. Biol J Linn Soc 91: 375–382 [Chap 13]

Cooper WE Jr & Frederick WG (2007b) Optimal flight initiation distance. J Theor Biol 244: 59–67 [Chap 13]

Cordero C & Eberhard WG (2003) Female choice of sexually antagonistic male adaptations: a critical review of some current research. J Evol Biol 16: 1–6 [Chap 24]

Cordoba-Aguilar A, Uhia E, & Rivera AC (2003) Sperm competition in odonata (Insecta): the evolution of female sperm storage and rivals' sperm displacement. J Zool 261: 381–398 [Chap 22]

Correa C, Baeza JA, Hinojosa IA, & Thiel M (2003) Male dominance hierarchy and mating tactics in the rock shrimp, *Rhynchocinetes typus* (Decapoda: Caridea). J Crust Biol 23: 33–45 [Chap 25]

Cosmides L & Tooby J (1987) From evolution to behavior: evolutionary psychology as the missing link. Pp 277–306 in Dupré J (ed) The Latest on the Best: Essays on Evolution and Optimality. MIT Press, Cambridge, MA [Chap 29]

Costa JT (2006) The Other Insect Societies. Harvard Univ Press, Cambridge, MA [Chaps 14, 19]

Costello CM, Creel SR, Kalinowski ST, Vu NV, & Quigley HB (2008) Sex-biased natal dispersal and inbreeding avoidance in American black bears as revealed by spatial genetic analyses. Mol Ecol 17: 4713–4723 [Chap 31]

Coster-Longman C, Landi M, & Turillazzi S (2002) The role of passive defense (selfish herd and dilution effect) in the gregarious nesting of *Liostenogaster* wasps (Vespidae, Hymenoptera, Stenogastrinae). J Insect Behav 15: 331–350 [Chap 17]

Cott HB (1940) Adaptive Coloration in Animals. Methuen, London [Chaps 9, 16]

Cotton S, Small J, & Pomiankowski A (2006) Sexual selection and condition-dependent mate preferences. Curr Biol 16: R755-R765 [Chap 24]

Coyne JA & Orr HA (1997) "Patterns of speciation in *Drosophila*" revisited. Evolution 51: 295–303 [Chap 27]

Coyne JA & Orr HA (2004) Speciation. Sinauer, Sunderland, MA [Chap 27]

Crabbe JC, Wahlsten D, & Dudek BC (1999) Genetics of mouse behavior: interactions with laboratory environment. Science 284: 1670–1672 [Chap 14]

Craig JV & Muir WM (1996) Group selection for adaptation to multiple-hen cages: beak-related mortality, feathering and body weight responses. Poult Sci 75: 294–302 [Chap 14]

Craig TP, Itami JK, Abrahamson WG, & Horner JD (1993) Behavioral evidence for host-race formation in *Eurosta solidaginis*. Evolution 47: 1696–1710 [Chap 27]

Crawford CB & Krebs D (2007) Handbook of Evolutionary Psychology, 2nd ed. Erlbaum, Mahwah, NJ [Chap 31]

Crean AJ & Marshall DJ (2008) Gamete plasticity in a broadcast spawning marine invertebrate. Proc Natl Acad Sci USA 105: 13508–13513 [Chap 21]

Creel SF (2005) Dominance, aggression and glucocorticoid levels in social carnivores. J Mammol 86: 255–264 [Chap 15]

Creel S, Winnie JA Jr, Christianson D, & Liley S (2008) Time and space in general models of antipredator response: tests with wolves and elk. Anim Behav 76: 1139–1146 [Chap 13]

Crispo E (2007) The Baldwin effect and genetic assimilation: revisiting two mechanisms of evolutionary change mediated by phenotypic plasticity. Evolution 61: 2469–2479 [Chap 6]

Croft DP, James R, & Krause J (2007) Exploring Animal Social Networks. Princeton Univ Press, Princeton, NJ [Chap 17]

Croft DP, James R, Thomas POR, Hathaway C, Mawdsley D, Laland KN, & Krause J (2006) Social structure and co-operative interactions in a wild population of guppies (*Poecilia reticulata*). Behav Ecol Sociobiol 59: 644–650 [Chap 17]

Crone EE (2001) Is survivorship a better fitness surrogate than fecundity? Evolution 55: 2611–2614 [Chap 4]

Cronin H (1991) The Ant and the Peacock. Cambridge Univ Press, Cambridge [Chap 20]

Crook JH (1964) The evolution of social organisation and visual communication in the weaver birds (Ploceinae). Behaviour Suppl 10: 1–178 [Chap 1]

Crook JH (1965) The adaptive significance of avian social organisations. Symp Zool Soc Lond 14: 181–218 [Chap 1]

Crook JH (1972) Sexual selection, dimorphism, and social organization in the primates. Pp 231–281 in Campbell B (ed) Sexual Selection and the Descent of Man, 1871–1971. Aldine, Chicago [Chap 14]

Crook JH & Gartlan JS (1966) Evolution of primate societies. Nature 210: 1200–1203 [Chap 1]

Crossland MR (2001) Ability of predatory native Australian fishes to learn to avoid toxic larvae of the introduced toad *Bufo marinus*. J Fish Biol 59: 319–329 [Chap 29]

Crow JF (1986) Basic Concepts in Population, Quantitative and Evolutionary Genetics. Freeman, New York [Chap 25]

Crozier RH & Pamilo P (1996) Evolution of Social Insect Colonies: Sex Allocation and Kin Selection. Oxford Univ Press [Chap 19]

Crudgington HS & Siva-Jothy MT (2000) Genital damage, kicking and early death. Nature 407: 855–856 [Chap 22]

Cummings ME, Larkins-Ford J, Reilly CR, Wong RY, Ramsey M, et al. (2008) Sexual and social stimuli elicit rapid and contrasting genomic responses. Proc R Soc Lond B 275: 393–402 [Chap 28]

Cunningham MA & Baker MC (1983) Vocal learning in white-crowned sparrows: sensitive phase and song dialects. Behav Ecol Sociobiol 13: 259–269 [Chap 10]

Curio E (1976) The Ethology of Predation. Springer-Verlag, Berlin [Chap 13]

Curio E & Regelmann K (1986) Predator harassment implies a real deadly risk: a reply to Hennessy. Ethology 72: 75–78 [Chap 17]

Currie CR (2001) A community of ants, fungi, and bacteria: a multilateral approach to studying symbiosis. Annu Rev Microbiol 55: 357–380 [Chap 12]

Curtis CC & Stoddard PK (2003) Mate preference in female electric fish, *Brachyhypopomus pinnicaudatus*. Anim Behav 66: 329–336 [Chap 9]

Cuthill IC (2005) The study of function in behavioural ecology. Anim Biol 55: 399–417 [Chap 2]

Cuthill IC (2006) Color perception. Pp 3–40 in Hill GE & McGraw KJ (eds) Bird Coloration, Vol. I: Mechanisms & Measurements. Harvard Univ Press, Cambridge, MA [Chap 9]

Cuthill IC & Houston AI (1997) Managing time and energy. Pp 97–120 in Krebs JR & Davies NB (eds) Behavioural Ecology: An Evolutionary Approach, 4th ed. Blackwell Sci, Oxford [Chap 12]

Czesak, ME, Fox CW, & Wolf JB (2006) Experimental evolution of phenotypic plasticity: how predictive are cross-environment genetic correlations? Am Nat 168: 323–335 [Chap 6]

Daan S, Dijkstra C, & Tinbergen JM (1990) Family planning in the kestrel (*Falco tinnunculus*)—the ultimate control of covariation of laying date and clutch size. Behaviour 114: 83–116 [Chap 15]

Dabelsteen T (2005) Public, private or anonymous? Facilitating and countering eavesdropping. Pp 38–62 in McGregor PK (ed) Animal Communiction Networks. Cambridge Univ Press, Cambridge [Chap 17]

Dale S, Rinden H, & Slagsvold T (1992) Competition for a mate restricts mate search of female pied flycatchers. Behav Ecol Sociobiol 30: 165–176 [Chap 24]

Dall SRX (2006) Evolution: mothers "sign" their eggs where cuckoos lurk. Curr Biol 16: R162–R165 [Chap 12]

Dall SRX & Boyd IL (2002) Provisioning under the risk of starvation. Evol Ecol Res 4: 883–896 [Chap 12]

Dall SRX & Boyd IL (2004) Evolution of mammals: lactation helps mothers to cope with unreliable food supplies. Proc R Soc Lond B 271: 2049–2057 [Chap 12]

Dall SRX & Cuthill IC (1997) The information costs of generalism. Oikos 80: 197–202 [Chap 12]

Dall SRX & Johnstone RA (2002) Managing uncertainty: information and insurance under the risk of starvation. Phil Trans R Soc Lond B 357: 1519–1526 [Chap 12]

Dall SRX, Giraldeau L-A, Olsson O, McNamara JM, & Stephens DW (2005) Information and its use by animals in evolutionary ecology. Trends Ecol Evol 20: 187–193 [Chaps 9, 11, 12]

Dall SRX, Houston AI, & McNamara JM (2004) The behavioural ecology of personality: consistent individual differences from an adaptive perspective. Ecol Lett 7: 734–739 [Chap 30]

Dall SRX, McNamara JM, & Cuthill IC (1999) Interruptions to foraging and learning in a changing environment. Anim Behav 57: 233–241 [Chap 12]

Dallai R, Frati F, Lupetti P, & Adis J (2003) Sperm ultrastructure of *Mantophasma zephyra* (Insecta, Mantophasmatodea). Zoomorphology 122: 67–76 [Chap 22]

Dame EA & Petren K (2006) Behavioural mechanisms of invasion and displacement in Pacific island geckos (Hemidactylus). Anim Behav 71: 1165–1173 [Chap 30]

Danchin E, Giraldeau LA, Valone TJ, & Wagner RH (2004) Public information: from nosy neighbors to cultural evolution. Science 305: 487–491 [Chap 12]

Dani FR, Foster KR, Zacchi F, Seppä P, Massolo A, Carelli A, Arévalo E, Queller DC, Strassmann JE, & Turillazzi S (2004) Can cuticular

lipids provide sufficient information for within-colony nepotism in wasps? Proc R Soc Lond B 271: 745–753 [Chaps 17, 18]
Danielsson I (2001) Antagonistic pre- and post-copulatory sexual selection on male body size in a water strider (*Gerris lacustris*). Proc R Soc Lond B 268: 77–81 [Chap 22]
Darst CR & Cummings ME (2006) Predator learning favours mimicry of a less-toxic model in poison frogs. Nature 440: 208–211 [Chap 10]
Darwin C (1859) On the Origin of Species by Means of Natural Selection, or the Preservation of Favoured Races in the Struggle for Life. John Murray, London [Chaps 1, 2, 3, 4, 7, 31]
Darwin C (1871) The Decent of Man and Selection in Relation to Sex. John Murray, London [Chaps 16, 20, 21, 22, 24]
Darwin CR (1874) The Descent of Man and Selection in Relation to Sex, 2nd ed. Rand, McNally, New York [Chap 25]
Davies NB (2000) Cuckoos, Cowbirds and Other Cheats. Poyser, London [Chap 1]
Davies NB & Halliday TR (1978) Deep croaks and fighting assessment in toads (*Bufo bufo*). Nature 274: 683–685 [Chap 16]
Davies NB, Kilner RM, & Noble DG (1998) Nestling cuckoos *Cuculus canorus* exploit hosts with begging calls that mimic a brood. Proc R Soc Lond B 265: 673–678 [Chaps 1, 9]
Davis JM (2008) Patterns of variation in the influence of natal experience on habitat choice. Q Rev Biol 83: 363–380 [Chap 29]
Dawkins M (1986) Unravelling Animal Behaviour. Longman, London [Chap 1]
Dawkins R (1976) The Selfish Gene. Oxford Univ Press, New York [Chaps 1, 18, 23]
Dawkins R (1980) Good strategy or evolutionarily stable strategy? Pp 331–367 in Barlow GW & Silverberg J (eds) Sociobiology: Beyond Nature/Nurture? Westview, Boulder, CO [Chap 25]
Dawkins R (1986) The Blind Watchmaker. Norton, New York [Chap 1]
Dawkins R & Carlisle TR (1976) Parental investment, mate desertion and a fallacy. Nature 262: 131–133 [Chaps 20, 26]
Dawkins R & Krebs JR (1978) Animal signals: information or manipulation? Pp 282–309 in Krebs JR & Davies NB (eds) Behavioural Ecology: An Evolutionary Approach. Blackwell, Oxford [Chap 16]
DeBruine LM (2005) Trustworthy but not lust-worthy: context-specific effects of facial resemblance. Proc R Soc Lond B 272: 919–922 [Chap 31]
DeCarvalho TN, Watson PJ, & Field SA (2004) Costs increase as ritualized fighting progresses within and between phases in the sierra dome spider, *Neriene litigiosa*. Anim Behav 68: 473–482 [Chap 15]
de Jong G (1994) The fitness of fitness concepts and the description of natural selection. Q Rev Biol 69: 3–29 [Chaps 4, 6]
de Jong G (1995) Phenotypic plasticity as a product of selection in a variable environment. Am Nat 145: 493–512 [Chap 6]
de Jong, G (2005) Evolution of phenotypic plasticity: patterns of plasticity and the emergence of ecotypes. New Phytologist 166: 101–117 [Chap 6]
Delehanty DJ, Fleischer RC, Colwell MA, & Oring LW (1998) Sex-role reversal and the absence of extra-pair fertilization in Wilson's phalaropes. Anim Behav 55: 995–1002 [Chap 25]
Delibes M, Gaona P, & Ferreras P (2001) Effects of an attractive sink leading into maladaptive habitat selection. Am Nat 158: 277–285 [Chap 29]
Dercole F & Rinaldi S (2008) Analysis of Evolutionary Processes: The Adaptive Dynamics Approach and Its Applications. Princeton Univ Press, Princeton, NJ [Chap 16]
DeScioli P & Kurzban R (2007) The games people play. Pp 130–136 in Gangestad SW & Simpson JA (eds) The Evolution of Mind. Guilford Press, New York [Chap 31]
DeVoogd TJ, Krebs JR, Healy SD, & Purvis A (1993) Relations between song repertoire size and the volume of brain nuclei related to song—comparative evolutionary analyses amongst oscine birds. Proc R Soc Lond B 254: 75–82 [Chap 10]
DeWitt TJ & Scheiner SM (2004) Phenotypic Plasticity: Functional and Conceptual Approaches. Oxford Univ Press [Chaps 6, 29]
DeWitt TJ, Sih A, & Hucko JA (1999) Trait compensation and cospecialization in a freshwater snail: size, shape and antipredator behaviour. Anim Behav 58: 397–407 [Chap 12]
DeWitt TJ, Sih A, & Wilson DS (1998) Costs and limits of phenotypic plasticity. Trends Ecol Evol 13: 77–81 [Chaps 6, 12]
Dewsbury DA (1982) Ejaculate cost and male choice. Am Nat 119: 601–610 [Chap 22]
Diaz-Uriarte R & Garland T Jr (1998) Effects of branch length errors on the performance of phylogenetically independent contrasts. Syst Biol 47: 654–672 [Chap 7]
Dickinson JL & Hatchwell BJ (2004) Fitness consequences of helping. Pp 48–66 in Koenig WD & Dickinson JL (eds) Ecology and Evolution of Cooperative Breeding in Birds. Cambridge Univ Press [Chap 17]
Dingemanse NJ & De Goede P (2004) The relation between dominance and exploratory behavior is context-dependent in wild great tits. Behav Ecol 15: 1023–1030 [Chap 30]
Dingemanse NJ & Reale D (2005) Natural selection and animal personality. Behaviour 142: 1159–1184 [Chap 30]
Dingemanse NJ, Both C, Drent PJ, & Tinbergen JM (2004) Fitness consequences of avian

personalities in a fluctuating environment. Proc R Soc Lond B 271: 847–852 [Chap 30]

Dingemanse NJ, Both C, Drent PJ, van Oers K, & van Noordwijk AJ (2002) Repeatability and heritability of exploratory behaviour in great tits from the wild. Anim Behav 64: 929–938 [Chap 30]

Dingemanse NJ, Both C, van Noordwijk AJ, Rutten AL, & Drent PJ (2003) Natal dispersal and personalities in great tits (*Parus major*). Proc R Soc Lond B 270: 741–747 [Chap 30]

Dingemanse NJ, Wright J, Kazem AJN, Thomas DK, Hickling R, & Dawnay N (2007) Behavioural syndromes differ predictably between twelve populations of three-spined stickleback. J Anim Ecol 76: 1128–1138 [Chap 30]

Dingle H (2001) The evolution of migratory syndromes in insects. Pp 159–181 in Woiwod IP, Reynolds DR, & Thomas CD (eds) Insect Movement: Mechanisms and Consequence: Proceedings of the Royal Entomological Society's 20th Symposium. CABI,New York [Chap 30]

Diniz-Filho JAF, Bini LM, Rodríguez MA, Rangel TFLVB, & Hawkins BA (2007) Seeing the forest for the trees: partitioning ecological and phylogenetic components of Bergmann's rule in European Carnivora. Ecogr 30: 598–608 [Chap 7]

Dixon AFG (1998) Aphid Ecology, 2nd ed. Chapman & Hall, London [Chap 4]

Dobzhansky T (1968a) Adaptedness and fitness. Pp 109–121 in Lewontin RC (ed) Population Biology and Evolution. Syracuse Univ Press, Syracuse, NY [Chap 4]

Dobzhansky T (1968b) On some fundamental concepts of Darwinian biology. Pp 1–24 in Dobzhansky TH, Hecht MK, & Steere WC (eds) Evolutionary Biology, vol 2. Appleton Century-Crofts, New York [Chap 4]

Dolby AS & Grubb TC Jr (2000) Social context affects risk taking by a satellite species in a mixed-species foraging group. Behav Ecol 11: 110–114 [Chap 17]

Dominey WJ (1980) Female mimicry in male bluegill sunfish-a genetic polymorphism? Nature 284: 546–548 [Chap 25]

Dominey WJ (1984) Alternative mating tactics and evolutionarily stable strategies. Am Zool 24: 385–396 [Chap 25]

Donald PF (2007) Adult sex ratios in wild bird populations. Ibis 149: 671–692 [Chap 20]

Dornhaus A, Franks NR, Hawkins RM, & Shere HNS (2004) Ants move to improve: colonies of *Leptothorax albipennis* emigrate whenever they find a superior nest site. Anim Behav 67: 959–963 [Chap 9]

Dorus S, Evans PD, Wyckoff GJ, Choi SS, & Lahn BT (2004) Rate of molecular evolution of the seminal protein gene SEMG2 correlates with levels of female promiscuity. Nature Genetics 36: 1326–1329 [Chaps 20, 22]

Double M & Cockburn A (2000) Pre-dawn infidelity: females control extra-pair mating in superb fairy-wrens. Proc R Soc Lond B 267: 465–470 [Chap 24]

Douglas RH, Mullineaux CW, & Partridge JC (2000) Long-wave sensitivity in deep-sea stomiid dragonfish with far-red bioluminescence: Evidence for a dietary origin of the chlorophyll-derived retinal photosensitizer of *Malacosteus niger*. Phil Trans R Soc Lond B 355: 1269–1272 [Chap 9]

Douglas SJ, Dawson-Scully K, & Sokolowski MB (2005) The neurogenetics and evolution of food-related behaviour. Trends Neurosci 28: 644–652 [Chap 28]

Downhower JF, Blumer LS, & Brown L (1987) Opportunity for selection: an appropriate measure for evaluating variation in the potential for selection? Evolution 41: 1395–1400 [Chap 20]

Doyle JR, O'Connor DJ, Reynolds GM, & Bottomley PA (1999) The robustness of the asymmetrically dominated effect: buying frames, phantom alternatives, and in-store purchases. Psychology and Marketing 16: 225–243 [Chap 11]

Duckworth RA (2006) Behavioral correlations across breeding contexts provide a mechanism for a cost of aggression. Behav Ecol 17: 1011–1019 [Chap 30]

Dugatkin LA (1997) Winner and loser effects and the structure of dominance hierarchies. Behav Ecol 8: 583–587 [Chap 17]

Dugatkin LA (2001) Bystander effects and the structure of dominance hierarchies. Behav Ecol 12: 348–352 [Chap 17]

Dugatkin LA & Druen M (2004) The social implications of winner and loser effects. Proc R Soc Lond B 271: S488–S489 [Chap 17]

Dugatkin LA & Godin JGJ (1992) Prey approaching predators—a cost-benefit perspective. Ann Zool Fenn 29: 233–252 [Chap 17]

Dukas R (1998) Cognitive Ecology: The Evolutionary Ecology of Information Processing and Decision Making. Univ Chicago Press [Chap 8]

Dukas R (2002) Behavioural and ecological consequences of limited attention. Phil Trans R Soc Lond B 357: 1539–1547. [Chap 10]

Dukas R (2004) Causes and consequences of limited attention. Brain Behav Evol, 63: 197–210 [Chap 12]

Dukas R (2005a) Learning affects mate choice in female fruit flies. Behav Ecol 16: 800–804 [Chap 10]

Dukas R (2005b) Experience improves courtship in male fruit flies. Anim Behav 69: 1203–1209 [Chap 10]

Dukas R & Ratcliffe JM (eds) (2009) Cognitive Ecology II. Univ Chicago Press, Chicago [Chap 8]

Dunbar RIM & Barrett L (2007) Oxford Handbook of Evolutionary Psychology. Oxford Univ Press, New York [Chap 31]

Dunn PO & Cockburn A (1999) Extrapair mate choice and honest signaling in cooperatively breeding superb fairy-wrens. Evolution 53: 938–946 [Chap 24]

Dusenbery D (1992) Sensory Ecology. Freeman, New York [Chap 9]

Dwyer DM & Clayton NS (2002) A reply to the defenders of the faith. Trends Cogn Sci 6: 109–111 [Chap 10]

Dzieweczynski TL, Earley RL, Green TM, & Rowland WJ (2005) Audience effect is context dependent in Siamese fighting fish, *Betta splendens*. Behav Ecol 16: 1025–1030 [Chap 17]

Earley RL & Dugatkin (2002) Eavesdropping on visual cues in green swordtail (*Xiphophorus helleri*) fights: a case for networking. Proc R Soc Lond B 269: 943–952 [Chap 17]

Earn DJD & Johnstone RA (1997) A systematic error in tests of ideal free theory. Proc R Soc Lond B 264: 1671–1675 [Chap 11]

Eberhard WG (1982) Beetle horn dimorphism: making the best of a bad lot. Am Nat 119: 420–426 [Chap 25]

Eberhard WG (1985) Sexual Selection and Animal Genitalia. Harvard Univ Press, Cambridge, MA [Chaps 20, 22, 23]

Eberhard WG (1996) Female Control: Sexual Selection by Cryptic Female Choice. Princeton Univ Press, Princeton, NJ [Chaps 1, 22]

Eberhard WG (2005) Evolutionary conflicts of interest: are female sexual decisions different? Am Nat 165: S19–S25 [Chap 23]

Edwards AC & Mackay TFC (2009) Quantitative trait loci for aggressive behavior in *Drosophila melanogaster*. Genetics (Epub) [Chap 28]

Egan SP & Funk DJ (2006) Individual advantages to ecological specialization: insights on cognitive constraints from three conspecific taxa. Proc R Soc Lond B 273: 843–848 [Chap 27]

Eggers S, Griesser M, Nystrand M, & Ekman J (2006) Predation risk induces changes in nest-site selection and clutch size in the Siberian jay. Proc R Soc Lond B 273: 701–706 [Chap 6]

Ehrenreich A (2006) DNA microarray technology for the microbiologist: an overview. Appl Microbiol Biotechnol 73: 255–273 [Chap 28]

Ehrlén J (2003) Fitness components versus total demographic effects: evaluating herbivore impacts on a perennial herb. Am Nat 162: 796–810 [Chap 4]

Ehrlich PA & Hanski I (2004) On the Wings of Checkerspots: A Model System for Population Biology. Oxford Univ Press, New York [Chap 28]

Eibl-Eibesfeldt I (1975) Ethology: The Biology of Behavior. Holt, Rinehart & Winston, New York [Chap 16]

Ekman J & Lilliendahl K (1993) Using priority to food access: fattening strategies in dominance-structured willow tit (*Parus montanus*) flocks. Behav Ecol 4: 232–238 [Chap 12]

Ekman J & Rosander B (1987) Starvation risk and flock size of the social forager: when there is a flocking cost. Theor Pop Biol 31: 167–177 [Chap 11]

Eldredge N & Gould SJ (1972) Punctuated equilibria: an alternative to phyletic gradualism. Pp 82–115 in Schopf TJM (ed) Models in Paleobiology. Doubleday, New York [Chap 7]

Eliassen S, Jorgensen C, Mangel M, & Giske J (2007) Exploration or exploitation: life expectancy changes the value of learning in foraging strategies. Oikos 116: 513–523 [Chap 12]

Ellegren H & Parsch J (2007) The evolution of sex-biased genes and sex-biased gene expression. Nature Rev Genet 8: 689–698 [Chap 5]

Ellen ED, Muir WM, Teuscher F, & Bijma P (2007) Genetic improvement of traits affected by interactions among individuals: sib selection schemes. Genetics 176: 489–499 [Chap 14]

Ellen ED, Visscher J, van Arendonk JAM, & Bijma P (2008) Survival of laying hens: genetic parameters for direct and associative effects in three purebred layer lines. Poultry Science 87: 233–239 [Chap 14]

Ellison PT (2003) Energetics and reproductive effort. Am J Hum Biol 15: 342–351 [Chap 31]

Elman JL, Bates EA, Johnson MH, Karmiloff-Smith A, Parisi D, & Plunkett K (1996) Rethinking Innateness: A Connectionist Perspective on Development. MIT Press, Cambridge MA [Chap 31]

Emlen DJ (1994) Environmental control of horn length dimorphism in the beetle *Onthophagus acuminatus* (Coleoptera: Scarabaeidae). Proc Roy Soc Lond B 256: 131–136 [Chap 6]

Emlen DJ (1996) Artificial selection on horn length-body size allometry in the horned beetle *Onthophagus acuminatus* (Coleoptera: Scarabaeidae). Evolution 50: 1219–1230 [Chap 6]

Emlen DJ (2008) The roles of genes and the environment in the expression and evolution of alternative tactics. Pp 85–108 in Oliviera RF, Taborsky M, & Brockmann HJ (eds) Alternative Reproductive Tactics. Cambridge Univ Press, Cambridge [Chap 25]

Emlen JM (1966) The role of time and energy in food preference. Am Nat 100: 611–617 [Chap 8]

Emlen JM (1988) Evolutionary ecology and the optimality assumption. Pp 163–177 in Dupre J (ed) The Latest on the Best. MIT Press, Cambridge, MA [Chap 2]

Emlen ST (1982) The evolution of helping. I. an ecological constraints model. Am Nat 119: 29–39 [Chap 17]

Emlen ST (1997) Predicting family dynamics in social vertebrates. Pp 228–253 in Krebs J & Davies N (eds) Behavioral Ecology: An Evolutionary Approach, 4th ed. Blackwell Sci, Oxford [Chap 19]

Emlen ST & Oring LW (1977) Ecology, sexual selection, and the evolution of mating systems. Science 197: 215–223 [Chaps 20, 26]

Emlen ST & Wrege PH (1992) Parent-offspring conflict and the recruitment of helpers among bee-eaters. Nature 356: 331–333 [Chap 18]

Emlen ST, Emlen JM, & Levin SA (1986) Sex-ratio selection in species with helpers-at-the-nest. Am Nat 127: 1–8 [Chap 26]

Endler JA (1978) A predator's view of animal color patterns. Evol Biol 11: 319–364 [Chap 9]

Endler JA (1986) Natural selection in the wild. Princeton Univ Press, Princeton, NJ [Chaps 3, 4]

Endler JA (1992) Signals, signal conditions, and the direction of evolution. Am Nat 139: S125–S153 [Chaps 7, 27]

Endler JA (1995) Multiple-trait coevolution and environmental gradients in guppies. Trends Ecol Evol 10: 22–29 [Chap 13]

Endler JA (2000) Evolutionary implications of the interaction between animal signals and the environment. Pp 11–46 in Espmark Y, Amundsen T & Rosenqvist G (eds) Animal Signals: Signalling and Signal Design in Animal Communication. Tapir Academic Press, Trondheim, Norway [Chap 9]

Endler JA & Basolo AL (1998) Sensory ecology, receiver biases and sexual selection. Trends Ecol Evol 13: 415–420 [Chap 16]

Engh AL, Esch K, Smale L, & Holekamp KE (2000) Mechanisms of maternal rank "inheritance" in the spotted hyaena, *Crocuta crocuta*. Anim Behav 60: 323–332 [Chap 17]

Enquist M (1985) Communication during aggressive interactions with particular reference to variation in choice of behaviour. Anim Behav 33: 1152–1161 [Chap 16]

Enquist M & Arak A (1998) Neural representation and the evolution of signal form. Pp 21–87 in Dukas R (ed) Cognitive Ethology: The Evolutionary Ecology of Information Processing and Decision Making. Univ Chicago Press, Chicago [Chap 16]

Enquist M, Arak A, Ghirlanda S, & Wachtmeister C-A (2002) Spectacular phenomena and limits to rationality in genetic and cultural evolution. Phil Trans R Soc Lond B 357: 1585–1594 [Chap 16]

Enquist M & Ghirlanda S (2005) Neural Networks and Animal Behavior. Princeton Univ Press, Princeton, NJ [Chap 16]

Enquist M & Jakobsson S (1986) Decision-making and assessment in the fighting behavior of *Nannacara anomala* (Cichlidae, Pisces). Ethology 72: 143–153 [Chap 15]

Enquist M & Leimar O (1983) Evolution of fighting behavior: decision rules and assessment of relative strength. J Theor Biol 102: 387–410 [Chap 15]

Enquist M & Leimar O (1987) Evolution of fighting behavior: the effect of variation in resource value. J Theor Biol 127: 187–205 [Chap 15]

Enquist M, Leimar O, Ljungberg T, Mallner Y, & Segerdahl N (1990) A test of the sequential assessment game: fighting in the cichlid fish *Nannacara anomala*. Anim Behav 40: 1–14 [Chap 15]

Ens BJ, Kersten M, Brenninkmeijer A, & Hulscher JB (1992) Territory quality, parental effort, and reproductive success of oystercatchers (*Haematopus ostralegus*) J. Anim. Ecol 61: 703–715 [Chap 15]

Erichsen JT, Krebs JR, & Houston AI (1980) Optimal foraging and cryptic prey. J Anim Ecol 49: 271–276 [Chap 8]

Eshel I, Volovik I, & Sansone E (2000) On Fisher-Zahavi's handicapped sexy son. Evol Ecol Res 2: 509–523 [Chap 24]

Euler L (1760) Recherches generales sur la mortalite: la multiplication du genre humaine. Mem Acad Sci, Berlin 16: 144–164 [Chap 4]

Evans CS (1997) Referential communication. Perspectives in Ethology 12: 99–143 [Chap 13]

Evans JD, Aronstein K, Chen YP, Hetru C, Imler JL, et al. (2006) Immune pathways and defence mechanisms in honey bees *Apis mellifera*. Insect Mol Biol 15: 645–656 [Chap 28]

Evans JP & Marshall DJ (2005) Male-by-female interactions influence fertilization success and mediate the benefits of polyandry in the sea urchin *Heliocidaris erythrogramma*. Evolution 59: 106–112 [Chap 21]

Ewald PW (1994) Evolution of Infectious Disease. Oxford Univ Press, New York [Chap 17]

Ewert MA, Lang JW, & Nelson CE (2005) Geographic variation in the pattern of temperature-dependent sex determination in the American snapping turtle (*Chelydra serpentina*). J Zool 265: 81–95 [Chap 5]

Eyre-Walker A (2006) The genomic rate of adaptive evolution. Trends Ecol Evol 21: 569–575.

Fadlallah YH (1982) Reproductive ecology of the coral *Astrangia lajollaensis:* sexual and asexual patterns in a kelp forest habitat. Oecologia 55: 379–388 [Chap 21]

Fagerström T (1987) On theory, data and mathematics in ecology. Oikos 50: 258–261 [Chap 8]

Fahrbach SE, Farris SM, Sullivan JP, & Robinson GE (2003) Limits on volume changes in the mushroom bodies of the honey bee brain. J Neurobiol 57: 141–151 [Chap 28]

Fairbairn DJ, Blanckenhorn WU, & Székely T (eds) (2007) Sex, Size and Gender Roles: Evolutionary Studies of Sexual Size Dimorphism. Oxford Univ Press, New York [Chap 20]

Fairbairn DJ & Reeve J (2001) Natural selection. Pp. 29–43 in Fox CW, Roff DA, & Fairbairn DJ (eds.) Evolutionary Ecology: Concepts and Case Studies. Oxford Univ Press, New York [Chap 4]

Fairbairn DJ & Roff DA (2006) The quantitative genetics of sexual dimorphism: assessing the importance of sex-linkage. Heredity 97: 319–328 [Chap 5]

Falconer DS (1965) Maternal effects and selection response. Pp 763–774 in Geerts SJ (ed) Genetics Today, Proceedings of the XI International Congress on Genetics, Vol. 3. Pergamon Press, Oxford [Chap 14]

Falconer DS (1989) Introduction to Quantitative Genetics. Longman Publishing, New York [Chap 4]

Falconer DD & Mackay TFC (1996) Introduction to Quantitative Genetics, 4th ed. Longman, Essex [Chaps 5, 14, 25, 30]

Farley GS & Levitan DR (2001) The role of jelly coats in sperm-egg encounters, fertilization success, and selection on egg size in broadcast spawners. Am Nat 157: 626–236 [Chap 21]

Fawcett TW & Johnstone RA (2003) Male choice in the face of costly competition. Behav Ecol 14: 771–779 [Chap 20]

Fayed SA, Jennions MD, & Backwell PRY (2008) What factors contribute to an ownership advantage? Biol Lett 4: 143–145 [Chap 15]

Feder JL & Bush GL (1989) A field test of differential host-plant usage between two sibling species of *Rhagoletis pomonella* fruit flies (Diptera: Tephritidae) and its consequences for sympatric models of speciation. Evolution 43: 1813–1819 [Chap 27]

Feder JL, Opp SB, Wlazlo B, Reynolds K, Go W, & Spisak S (1994) Host fidelity is an effective premating barrier between sympatric races of the apple maggot fly. Proc Natl Acad Sci USA 91: 7990–7994 [Chap 27]

Feder ME & Mitchell-Olds T (2003) Evolutionary and ecological functional genomics. Nat Rev Genet 4: 651–657 [Chap 28]

Fehr E & Gächter S (2002) Altruistic punishment in humans. Nature 415: 137–140 [Chap 18]

Fellowes MDE, Kraaijeveld AR, & Godfray HCJ (1999) Association between feeding rate and parasitoid resistance in *Drosophila melanogaster*. Evolution 53: 1302–1305 [Chap 4]

Felsenstein J (1985) Phylogenies and the comparative method. Am Nat 125: 1–15 [Chaps 2, 7]

Felsenstein J (2004) Inferring Phylogenies. Sinauer, Sunderland, MA [Chap 7]

Fernandez-Juricic E, Erichsen J, & Kacelnik A (2004a) Visual perception and social foraging in birds Trends Ecol Evol 19: 25–31 [Chap 1]

Fernandez-Juricic E, Kerr B, Bednekoff PA, & Stephens DW (2004b) When are two heads better than one? Visual perception and information transfer affect vigilance coordination in foraging groups. Behav Ecol 15: 898–906 [Chap 17]

Fernandez-Juricic E, Smith R, & Kacelnik A (2005) Increasing the costs of conspecific scanning in socially foraging starlings affects vigilance and foraging behaviour. Anim Behav 69: 73–81 [Chap 17]

Fidler AE, van Oers K, Drent PJ, Kuhn S, Mueller JC, & Kempenaers B (2007) Drd4 gene polymorphisms are associated with personality variation in a passerine bird. Proc R Soc Lond B 274: 1685–1691 [Chap 30]

Fiegna F, Yu YTN, Kadam SV, & Velicer GJ (2006) Evolution of an obligate social cheater to a superior cooperator. Nature 441: 310–314 [Chap 28]

Field J & Cant M (2009) The ecology and evolution of hover wasps (Hymenoptera: Stenogastrinae). In Korb J & Heinze J (eds) Ecology of Social Evolution. Springer, Berlin [Chap 19]

Field J, Cronin A, & Bridge C (2006) Future fitness and helping in social queues. Nature 441: 214–217 [Chap 19]

Field J & Foster W (1999) Helping behavior in facultatively eusocial hover wasps: an experimental test of the subfertility hypothesis. Anim Behav 57: 633–636 [Chap 19]

Field J, Shreeves G, Sumner S, & Casiraghi M (2000) Insurance-based advantage to helpers in a tropical hover wasp. Nature 404: 869–871 [Chap 19]

Fincke OM (2004) Polymorphic signals of harassed female odonates and the males that learn them support a novel frequency-dependent model. Anim Behav 67: 833–845 [Chap 3]

Fincke OM (1986) Lifetime reproductive success and the opportunity for selection in a nonterritorial damselfly (Odonata: Coenagrionidae). Evolution 40: 791–803 [Chap 25]

Fincke OM & Hadrys H (2001) Unpredictable offspring survivorship in the damselfly, *Megaloprepus coerulatus*, shapes parental behavior, constrains sexual selection and challenges traditional fitness estimates. Evolution 55: 762–772 [Chap 4]

Findlay GD, Yi XH, MacCoss MJ, & Swanson WJ (2008) Proteomics reveals novel *Drosophila* seminal fluid proteins transferred at mating. PLoS Biol 6: 1417–1426 [Chap 23]

Finlay BL & Darlington RB (1995) Linked regularities in the development and evolution of mammalian brains. Science 268: 1578–1584 [Chap 10]

Fisher RA (1918) The correlation among relatives on the supposition of Mendelian inheritance. Trans R Society, Edinburgh 52: 399–433 [Chap 14]

Fisher RA (1930) The Genetical Theory of Natural Selection. Clarendon Press, Oxford [Chaps 4, 5, 7, 12, 20, 23, 24, 25, 26, 27]

Fisher RA (1958) The Genetical Theory of Natural Selection, 2nd ed. Dover,New York [Chap 26]
Fitch WT & Hauser MD (2003) Unpacking honesty: generating and extracting information from acoustic signals. Pp 65–137 in Simmons AM, Popper AN, & Fay RR (eds) Animal Communication. Springer-Verlag, Berlin [Chap 16]
Fitze PS & Le Galliard J-F (2008) Operational sex ratio, sexual conflict and the intensity of sexual selection. Ecol Lett 11: 432–439 [Chap 20]
FitzGibbon CD (1994) The costs and benefits of predator inspection behaviour in Thomson's gazelles. Behav Ecol Sociobiol 34: 139–148 [Chap 17]
Fitzpatrick MJ (2004) Pleiotropy and the genomic location of sexually selected genes. Am Nat 163: 800–808 [Chap 5]
Fitzpatrick MJ, Ben-Shahar Y, Smid HM, Vet LEM, Robinson GE, & Sokolowski MB (2005) Candidate genes for behavioural ecology. Trends Ecol Evol 20: 96–104 [Chaps 5, 30]
Fitzpatrick MJ, Feder E, Rowe L, & Sokolowski MB (2007) Maintaining a behaviour polymorphism by frequency-dependent selection on a single gene. Nature 447: 210–212 [Chaps 5, 25]
Fiumera AC, Dumont BL, & Clark AG (2007) Associations between sperm competition and natural variation in male reproductive genes on the third chromosome of Drosophila melanogaster. Genetics 176: 1245–1260 [Chap 28]
Flombaum JI, Santos LR, & Hauser MD (2002) Neuroecology and psychological modularity. Trends Cogn Sci 6: 106–108 [Chap 10]
Fluri P, Luscher M, Wille H, & Gerig L (1982) Changes in the weight of the pharyngeal gland and haemolymph titres of juvenile hormone, protein and vitellogenin in worker honey bees. J Insect Physiol 28: 61–68 [Chap 28]
Fodor J (1983) The modularity of mind. MIT Press, Cambridge, MA [Chap 31]
Foerster K, Coulson T, Sheldon BC, Pemberton JM, Clutton-Brock TH, & Kruuk LEB (2007) Sexually antagonistic genetic variation in the red deer. Nature 447: 1107–1111 [Chap 3]
Fontaine JJ & Martin TE (2006) Parent birds assess nest predation risk and adjust their reproductive strategies. Ecol Lett 9: 428–434 [Chap 6]
Forbes AA, Fisher J, & Feder JL (2005) Habitat avoidance: overlooking an important aspect of host specific mating and sympatric speciation? Evolution 59: 1552–1559 [Chap 27]
Fordyce JA (2006) The evolutionary consequences of ecological interactions mediated through phenotypic plasticity. J Exp Biol 209: 2377–2383 [Chap 6]
Forsgren E (1997) Female sand gobies prefer good fathers over dominant males. Proc R Soc Lond B 264: 1283–1286 [Chap 26]
Forsgren E, Amundsen T, Borg ÅA, & Bjelvenmark J (2004) Unusually dynamic sex roles in a fish. Nature 429: 551–554 [Chap 20]
Forshaw JM & Cooper WD (1989) Parrots of the World. Lansdowne Press, Australia [Chap 20]
Forsman A & Appelqvist S (1995) Experimental manipulation reveals differential effects of colour pattern on survival in male and female pygmy grasshoppers. J Evol Biol 12: 391–401 [Chap 3]
Forstmeier W & Birkhead TR (2004) Repeatability of mate choice in the zebra finch: consistency within and between females. Anim Behav 68: 1017–1028 [Chap 30]
Foster K (in press) Social behavior in microorganisms. In Szekely T, Moore AJ, & Komdeur J (eds) Social Behaviour: Genes, Ecology and Evolution. Cambridge Univ Press, Cambridge [Chap 19]
Foster KR, Shaulsky G, Strassmann JE, Queller DC, & Thompson CRL (2004) Pleiotropy as a mechanism to stabilize cooperation. Nature 431: 693–696 [Chap 19]
Foster KR & Wenseleers T (2006) A general model for the evolution of mutualisms. J Evol Biol 19: 1283–1293 [Chap 18]
Foster KR, Wenseleers T, & Ratnieks FLW (2006) Kin selection is the key to altruism. Trends Ecol Evol 21: 57–60 [Chap 18]
Foster SA (1985) Group foraging by a coral reef fish: a mechanism for gaining access to defended resources. Anim Behav 33: 782–792 [Chap 11]
Fournier D, Estoup A, Orivel J, Foucaud J, Jourdan H, Le Breton J, & Keller L (2005) Clonal reproduction by males and females in the little fire ant. Nature 435: 1230–1234 [Chap 3]
Frank SA (1990) Sex allocation theory for birds and mammals. Annu Rev Ecol Syst 21: 13–55 [Chap 20]
Frank SA (1998) Foundations of Social Evolution. Princeton Univ Press, Princeton, NJ [Chaps 4, 14, 18]
Frank SA (2003) Repression of competition and the evolution of cooperation. Evolution 57: 693–705 [Chap 18]
Frank SA (2006) Social selection. Pp 350–363 in Fox CW & Wolf JB (eds) Evolutionary Genetics: Concepts and Case Studies. Oxford Univ Press, New York [Chap 14]
Frank SA & Slatkin M (1990) Evolution in a variable environment. Am Nat 136: 244–260 [Chap 4]
Franke ES, Babcock RC, & Styan CA (2002) Sexual conflict and polyspermy under sperm-limited conditions: in situ evidence from field simulations with the free-spawning marine echinoid *Evechinus chloroticus*. Am Nat 160: 485–496 [Chap 21, 23]

Franks NR, Mallon EB, Bray HE, Hamilton MJ, & Mischler TC (2003a) Strategies for choosing between alternatives with different attributes: exemplified with house-hunting ants. Anim Behav 65: 215–223 [Chap 9]

Franks NR, Dornhaus A, Fitzsimmons JP, & Stevens M (2003b) Speed versus accuracy in collective decision making. Proc R Soc Lond B 270: 2457–2463 [Chap 9]

Franks NR, Pratt SC, Mallon EB, Britton NF, & Sumpter DJT (2002) Information flow, opinion polling and collective intelligence in house-hunting social insects. Phil Trans R Soc Lond B 357: 1567–1583 [Chap 9]

Fraser DF, Gilliam JF, Daley MJ, Le AN, & Skalski GT (2001) Explaining leptokurtic movement distributions: intrapopulation variation in boldness and exploration. Am Nat 158: 124–135 [Chap 30]

Freeman S & Herron JC (2004) Evolutionary Analysis, 3rd ed. Pearson Education, Upper Saddle River, NJ [Chap 5]

Fretwell SD (1972) Populations in a seasonal environment. Monogr Popul Biol 5: 1–217 [Chap 11]

Fretwell SD & Lucas HL (1969) On territorial behavior and other factors influencing habitat distribution in birds. Acta Biotheor 19: 16–36 [Chap 11]

Fricke C & Arnqvist G (2007) Rapid adaptation to a novel host in a seed beetle (*Callosobruchus maculatus*): the role of sexual selection. Evolution 61: 440–454 [Chap 23]

Frid A & Dill LM (2002) Human-caused disturbance stimuli as a form of predation risk. Cons Ecol 6: 11 [Chap 8]

Fromhage L, McNamara JM, & Houston AI (2007) Stability and value of male care for offspring: is it worth only half the trouble? Biol Lett 3: 234–236 [Chap 20]

Fry JD (1992) The mixed-model analysis of variance applied to quantitative genetics: biological meaning of the parameters. Evolution 46: 540–550 [Chap 6]

Fu P, Neff BD, & Gross MR (2001) Tactic-specific success in sperm competition. Proc R Soc Lond B 268: 1105–1112 [Chap 21]

Fuller JL & Hahn ME (1976) Issues in the genetics of social behavior. Behav Genet 6: 391–406 [Chap 14]

Fuller RC, Houle D, & Travis J (2005) Sensory bias as an explanation for the evolution of mate preferences. Am Nat 166: 437–446 [Chap 24]

Funk DJ, Filchak KE, & Feder JL (2002) Herbivorous insects: model systems for the comparative study of speciation ecology. Genetica 116: 251–267 [Chap 27]

Fussnecker B & Grozinger CM (2008) Dissecting the role of Kr-h1 brain gene expression in foraging behavior in honey bees (*Apis mellifera*). Insect Mol Biol 17: 515–522 [Chap 28]

Gack C & Peschke K (2005) 'Shouldering' exaggerated genitalia: a unique behavioural adaptation for the retraction of the elongate intromittant organ by the male rove beetle (*Aleochara tristis* Gravenhorst). Biol J Linn Soc 84: 307–312 [Chap 22]

Gadgil M (1972) Male dimorphism as a consequence of sexual selection. Am Nat 106: 574–558 [Chap 25]

Gage MJG, Parker GA, Nylin S, & Wiklund C (2002) Sexual selection and speciation in mammals, butterflies and spiders. Proc R Soc Lond B 269: 2309–2316 [Chap 7]

Gage MJG, Stockley P, & Parker GA (1995) Effects of alternative male mating strategies on characteristics of sperm production in the Atlantic salmon (*Salmo salar*): theoretical and empirical investigations. Phil Trans R Soc Lond B 350: 391–399 [Chap 22]

Galea LAM, Kavaliers M, Ossenkopp K-P, Innes D, & Hargreaves EL (1994) Sexually dimorphic spatial learning varies seasonally in two populations of deer mice. Brain Res 635: 18–26 [Chap 10]

Galindo BE, Vacquier VD, & Swanson WJ (2003) Positive selection in the egg receptor for abalone sperm lysin. Proc Natl Acad Sci USA 100: 4639–4643 [Chap 23]

Gallistel CR (2000) The replacement of general-purpose learning models with adaptively specialized learning modules. Pp 1179–1191 in Gazzaniga MS (ed) The Cognitive Neurosciences, 2nd ed. MIT Press, Cambridge, MA [Chap 31]

Gamble S, Lindholm AK, Endler JA, & Brooks R (2003) Environmental variation and the maintenance of polymorphism: the effect of ambient light spectrum on mating behavior and sexual selection in guppies. Ecol Lett 6: 463–472 [Chap 9]

Gammell MP & Hardy ICW (2003) Contest duration: sizing up the opposition? Trends Ecol Evol 18: 491–493 [Chap 15]

Gangestad SW, Garver-Apgar CE, Simpson JA, & Cousins AJ (2007) Changes in women's mate preferences across the ovulatory cycle. J Personality Soc Psychol 92: 151–163 [Chap 31]

Gangestad SW & Simpson JA (2007) The Evolution of Mind: Fundamental Questions and Controversies. Guilford Press, New York [Chap 31]

Gangestad SW & Thornhill R (2008) Human oestrus. Proc R Soc B 275: 991–1000 [Chap 31]

Gangestad SW, Thornhill R, & Garver-Apgar CE (2005) Women's sexual interests across the ovulatory cycle depend on primary partner

fluctuating asymmetry. Proc R Soc London B 272: 2023–2027 [Chap 31]
Garant D, Dodson JJ, & Bernatchez L (2003) Differential reproductive success and heritability of alternative reproductive tactics in wild Atlantic salmon (*Salmo salar* L.). Evolution 57: 1133–1141 [Chap 6]
García-Gonzáles F & Simmons LW (2007) Paternal indirect genetic effects on offspring viability and the benefits of polyandry. Curr Biol 17: 32–36 [Chap 14]
Gardner A & West SA (2006) Demography, altruism, and the benefits of budding. J Evol Biol 19: 1707–1716 [Chap 18]
Gardner A & West SA (in press) Greenbeards. Evolution [Chap 18]
Gardner A, West SA, & Barton NH (2007) The relation between multilocus population genetics and social evolution theory. Am Nat 169: 207–226 [Chap 18]
Garver-Apgar CE, Gangestad SW, Thornhill R, Miller RD, & Olp J (2006) Major histocompatibility complex alleles, sexually responsivity, and unfaithfulness in romantic couples. Psych Sci 17: 830–835 [Chap 31]
Gavrilets S (2000) Rapid evolution of reproductive barriers driven by sexual conflict. Nature 403: 886–889 [Chaps 21, 23]
Gavrilets S, Arnqvist G, & Friberg U (2001) The evolution of female mate choice by sexual conflict. Proc R Soc Lond B 268: 531–539 [Chaps 23, 24]
Gavrilets S & Hayashi TI (2005) Speciation and sexual conflict. Evol Ecol 19: 167–198 [Chap 23]
Gavrilets S & Scheiner SM (1993a) The genetics of phenotypic plasticity. V. Evolution of reaction norm shape. J Evol Biol 6: 31–48 [Chap 6]
Gavrilets S & Scheiner SM (1993b) The genetics of phenotypic plasticity. VI. Theoretical predictions for directional selection. J Evol Biol 6: 49–68 [Chap 6]
Gavrilets S & Waxman D (2002) Sympatric speciation by sexual conflict. Proc Natl Acad Sci USA 99: 10533–10538 [Chaps 21, 23]
Gelter HP & Tegelström H (1992) High frequency of extra-pair paternity in Swedish pied flycatchers revealed by allozyme electrophoresis and DNA fingerprinting. Behav Ecol Sociobiol 31: 1–7 [Chap 24]
Gerhardt HC, Daniel RE, Perrill SA, & Schramm S (1987) Mating behavior and male mating success in the gree treefrog. Anim Behav 35: 1490–1503 [Chap 25]
Gerhardt HC, Dyson ML, & Tanner SD (1996) Dynamic properties of the advertisement calls of gray tree frogs: patterns of variability and female choice. Behav Ecol 7: 7–18 [Chap 24]
Gerlach G, Hodgins-Davis A, Avolio C, & Schunter C (2008) Kin recognition in zebrafish: a 24-hour window for olfactory imprinting. Proc R Soc Lond B 275: 2165–2170 [Chap 31]
Getty T (2002) Signalling health versus parasites. Am Nat 159: 363–371 [Chap 22]
Getty T (2006) Sexually selected signals are not similar to sports handicaps. Trends Ecol Evol 21: 83–88 [Chap 16]
Ghalambor CK & Martin TE (2000) Parental investment strategies in two species of nuthatch vary with stage-specific predation risk and reproductive effort. Anim Behav 60: 263–267 [Chap 6]
Ghalambor CK & Martin TE (2001) Fecundity-survival trade-offs and parental risk-taking in birds. Science 292: 494–497 [Chap 6]
Ghalambor CK & Martin TE (2002) Comparative manipulation of predation risk in incubating birds reveals variability in the plasticity of responses. Behav Ecol 13: 101–108 [Chap 6]
Ghalambor CK, McKay JK, Carroll SP, & Reznick DN (2007) Adaptive versus non-adaptive phenotypic plasticity and the potential for contemporary adaptation in new environments. Funct Ecol 21: 394–407 [Chaps 6, 29]
Ghirlanda S & Enquist M (2003) A century of generalization. Anim Behav 66: 15–36 [Chap 16]
Ghirlanda S, Jansson L, & Enquist M (2002) Chickens prefer beautiful humans. Hum Nat 13: 383–389 [Chap 16]
Gibbons R (1992) A primer in game theory. Harvester Wheatsheaf, London [Chap 16]
Gibson RM, Aspbury AS, & McDaniel LL (2002) Active formation of mixed-species grouse leks: a role for predation in lek evolution? Proc R Soc Lond B 269: 2503–2507 [Chap 17]
Giese AC & Kanatani H (1987) Maturation and spawning. Pp 251–329 in AC Giese, JS Pearse, & VB Pearse (eds) Reproduction of Marine Invertebrates, Vol IX: Seeking Unity in Diversity. Blackwell Sci/Boxwood Press, Palo Alto/Pacific Grove, CA [Chap 21]
Gilad Y, Wiebe V, Przeworski M, Lancet D, & Pääbo S (2004) Loss of olfactory receptor genes coincides with the acquisition of full trichromatic vision in primates. PLoS Biol 2: 120–125 [Chap 28]
Gilbert OM, Foster KR, Mehdiabadi NJ, Strassmann JE, & Queller DC (2007) High relatedness maintains multicellular cooperation in a social amoeba by controlling cheater mutants. Proc Natl Acad Sci USA 104: 8913–8917 [Chap 19]
Gillespie JH (1977) Natural selection for variances in offspring number: a new evolutionary principle. Am Nat 111: 1010–1014 [Chap 4]
Gilliam JF & Fraser DF (1987) Habitat selection under predation hazard: test of a model with foraging minnows. Ecology 68: 1856–1862 [Chap 13]
Gilmour KM, DiBattista JD, & Thomas, JB (2005) Physiological causes and consequences of

social status in salmonid fish. Integ Comp Biol 45: 263–273 [Chap 15]
Gintis H (2000) Strong reciprocity and human sociality. J Theor Biol 206: 169–179 [Chap 18]
Giraldeau L-A & Caraco T (2000) Social Foraging Theory. Princeton Univ Press, Princeton, NJ [Chaps 8, 11]
Giraldeau L-A & Livoreil B (1998) Game theory and social foraging. Pp 16–37 in Dugatkin LA & Reeve HK (eds) Game Theory and Animal Behavior. Oxford Univ Press, New York [Chap 11]
Giraldeau LA, Soos C, & Beauchamp G (1994) A test of the producer-scrounger foraging game in captive flocks of spice finches, *Lonchura punctulata*. Behav Ecol Sociobiol 34: 251–256 [Chap 11]
Giraldeau LA, Valone TJ, & Templeton JJ (2002) Potential disadvantages of using socially acquired information. Phil Trans R Soc Lond B 357: 1559–1566 [Chaps 11, 12, 17]
Girman D, Mills M, Geffen E, & Wayne R (1997) A molecular genetic analysis of social structure, dispersal, and interpack relationships of the African wild dog (*Lycaon pictus*). Behav Ecol Sociobiol 40: 187–198 [Chap 19]
Gittleman JL & Harvey PH (1980) Why are distasteful prey not cryptic. Nature 286: 149–150 [Chap 10]
Gittleman JL & Kot M (1990) Adaptation: statistics and a null model for estimating phylogenetic effects. Syst Zool 39: 227–241 [Chap 7]
Gleason JM (2005) Mutations and natural genetic variation in the courtship song of *Drosophila*. Behav Genet 35: 265–277 [Chap 5]
Godard R (1991) Long-term memory of individual neighbors in a migratory songbird. Nature 350: 228–229 [Chap 10]
Godfray HCJ (1991) Signaling of need by offspring to their parents. Nature 352: 328–330 [Chap 16]
Godfray, HCJ (1994) Parasitoids: Behavioral and Evolutionary Ecology. Princeton Univ Press, Princeton, NJ [Chap 11]
Gonzalez-Voyer A, Fitzpatrick JL, & Kolm N (2008) Sexual selection determines parental care patterns in cichlid fishes. Evolution 62: 2015–2026 [Chaps 20, 26]
Goodnight CJ, Schwartz JM, & Stevens L (1992) Contextual analysis of models of group selection, soft selection, hard selection and the evolution of altruism. Am Nat 140: 743–761 [Chap 14]
Goodnight CJ & Stevens L (1997) Experimental studies of group selection: what do they tell us about group selection in nature? Am Nat 150: S59–S79 [Chap 14]
Goodson JL, Evans AK, Lindberg L, & Allen, CD (2005) Neuro-evolutionary patterning of sociality. Proc R Soc Lond B 272: 227–235 [Chap 7]
Gosling SD (2001) From mice to men: what can we learn about personality from animal research? Psych Bull 127: 45–86 [Chaps 12, 30]
Gotthard K & Nylin S (1995) Adaptive plasticity and plasticity as an adaptation: a selective review of plasticity in animal morphology and life history. Oikos 74: 3–17 [Chap 6]
Goubault M, Outreman Y, Poinsot D, & Cortesero AM (2005) Patch exploitation strategies of parasitic wasps under intraspecific competition. Behav Ecol 16: 693–701 [Chap 11]
Gould MC & Stephano JL (2003) Polyspermy prevention in marine invertebrates. Microsc Res Tech 61: 379–388 [Chap 21]
Gould SJ (1978) Sociobiology: The art of storytelling. New Sci 80: 530–533 [Chap 1]
Gould SJ (1987) This view of life: Freudian slip. Natural History 96: 14–21 [Chap 2]
Gould SJ & Lewontin RC (1979) The spandrels of San Marco and the Panglossian paradigm: a critique of the adaptationist programme. Proc R Soc Lond B 205: 581–598 [Chaps 1, 2, 7]
Gould SJ & Vrba ES (1982) Exaptation: a missing term in the science of form. Paleobiology 8: 4–15 [Chaps 2, 7]
Goulet D & Goulet TL (2006) Non-independent mating in a coral reef damselfish: evidence of mate choice copying in the wild. Behav Ecol 17: 998–1003 [Chap 20]
Gowaty PA (2008) Reproductive compensation. J Evol Biol 21: 1189–1200 [Chap 26]
Gowaty PA & Buschhaus N (1998) Ultimate causation of aggressive and forced copulation in birds: female resistance, the CODE hypothesis, and social monogamy. Am Zool 38: 207–225 [Chap 25]
Gowaty PA, Anderson WW, Bluhm CK, Drickamer LC, Kim YK, & Moore AJ (2007) The hypothesis of reproductive compensation and its assumptions about mate preferences and offspring viability. Proc Natl Acad Sci USA 104: 15023–15027 [Chap 26]
Grafen A (1982) How not to measure inclusive fitness. Nature 298: 425 [Chap 22]
Grafen A (1984) Natural selection, kin selection and group selection. Pp 62–84 in Krebs JR & Davies NB (eds) Behavioural Ecology: An Evolutionary Approach, 2nd ed. Blackwell Sci, Oxford [Chaps 1, 4, 5, 16]
Grafen A (1985) A geometric view of relatedness. Oxford Surv Evol Biol 2: 28–89 [Chaps 18, 19]
Grafen A (1990a) Biological signals as handicaps. J Theor Biol 144: 517–546 [Chaps 16, 24]
Grafen A (1990b) Sexual selection unhandicapped by the Fisher process. J Theor Biol 144: 473–516 [Chap 24]

Grafen A (1990c) Do animals really recognize kin? Anim Behav 39: 42–54 [Chap 18]

Grafen A (1999) Formal Darwinism, the individual-as-maximising-agent analogy and bet-hedging. Proc R Soc London B 266: 799–803 [Chap 4]

Grafen A (2002) A first formal link between the Price equation and an optimization program. J Theor Biol 217: 75–91 [Chap 4]

Grafen A (2006) Optimization of inclusive fitness. J Theor Biol 238: 541–563.

Grafen A (2007) The formal Darwinism project: a mid-term report. J Evol Biol 20: 1243–1254 [Chaps 2, 12]

Grammer K (1993) 5-alpha-androst-16en-3alpha-on: a male pheromone? A brief report. Ethol Sociobiol 14: 201–214 [Chap 31]

Grand TC (1997) Foraging site selection by juvenile coho salmon: ideal free distributions of unequal competitors. Anim Behav 53: 185–196 [Chap 11]

Grand TC & Dill LM (1997) The energetic equivalence of cover to juvenile coho salmon (*Oncorhynchus kisutch*): ideal free distribution theory applied. Behav Ecol 8: 437–447 [Chap 11]

Grant JWA, Girard IL, Breau C, & Weir LK (2002) Influence of food abundance on competitive aggression in juvenile convict cichlids. Anim Behav 63: 323–330 [Chap 15]

Grant KA (1966) A hypothesis concerning the prevalence of red coloration in California hummingbird flowers. Am Nat 100: 85–97 [Chap 10]

Grant PR & BR Grant (1997a) Hybridization, sexual imprinting, and mate choice. Am Nat 149: 1–28 [Chap 27]

Grant PR & BR Grant (1997b) Mating patterns of Darwin's finch hybrids determined by song and morphology. Biol J Linn Soc 60: 317–343 [Chap 27]

Grant PR & Grant BR (2002) Unpredictable evolution in a 30-year study of Darwin's finches. Science 296: 707–711 [Chap 3]

Grant PR Grant BR (2008) How and Why Species Multiply. Princeton Univ Press, Princeton, NJ [Chap 7]

Grant V (1994) Modes and origins of mechanical and ethological isolation in angiosperms. Proc Natl Acad Sci USA 91: 3–10 [Chap 27]

Graw B & Manser MB (2007) The function of mobbing in cooperative meerkats. Anim Behav 74: 507–517 [Chap 17]

Greenfield MD & Rodriguez RL (2004) Genotype-environment interactions and the reliability of mating signals. Anim Behav [Chaps 6, 24]

Greenspan RJ (2004) E pluripus unum, ex uno plura: Quantitative and single-gene perspectives on the study of behavior. Annu Rev Neurosci 27: 79–105 [Chap 5]

Greenspan RJ (2008) The origins of behavioral genetics. Curr Biol 18: R192–R198 [Chap 5]

Grether GF (2000) Carotenoid limitation and mate preference evolution: a test of the indicator hypothesis in guppies (*Poecilia reticulata*). Evolution 54: 1712–1724 [Chap 6]

Grether GF (2005) Environmental change, phenotypic plasticity, and genetic compensation. Am Nat 166: E115–E123 [Chap 6]

Grether GF, Hudon J, & Millie DF (1999) Carotenoid limitation of sexual coloration along an environmental gradient in guppies. Proc R Soc Lond B 266: 1317–1322 [Chap 6]

Grether GF, Kolluru GR, Rodd FH, De La Cerda J, & Shimazaki K (2005) Carotenoid availability affects the development of a colour-based mate preference and the sensory bias to which it is genetically linked. Proc R Soc Lond B 272: 2181–2188 [Chaps 6, 24]

Griffin AS (2004) Social learning about predators: A review and prospectus. Learn Behav 32: 131–140 [Chap 29]

Griffin AS, Blumstein DT, & Evans CS (2000) Training captive-bred or translocated animals to avoid predators. Conserv Biol 14: 1317–1326 [Chap 29]

Griffin AS & West SA (2003) Kin discrimination and the benefit of helping in cooperatively breeding vertebrates. Science 302: 634–636 [Chaps 18, 19]

Griffin AS, West SA, & Buckling A (2004) Cooperation and competition in pathogenic bacteria. Nature 430: 1024–1027 [Chaps 4, 18]

Griffin AS, Sheldon BC, & West SA (2005) Cooperative breeders adjust offspring sex ratios to produce helpful helpers. Am Nat 166: 628–632 [Chap 26]

Griffing B (1967) Selection in reference to biological groups. I. Individual and group selection applied to populations of unordered groups. Aust J Biol Sc. 20: 127 [Chap 14]

Griffing B (1981a) A theory of natural-selection incorporating interaction among individuals. I. The modeling process. J Theor Biol 89: 635–658 [Chap 14]

Griffing B (1981b) A theory of natural-selection incorporating interaction among individuals. II. Use of related groups. J Theor Biol 89: 659–677 [Chap 14]

Griffith SC (2007) The evolution of infidelity in socially monogamous passerines: neglected components of direct and indirect selection. Am Nat 169: 274–281 [Chap 24]

Griffith SC, Owens IPF, & Burke T (1999) Female choice and annual reproductive success favour less-ornamented male house sparrows. Proc R Soc Lond B 266: 765–770 [Chap 24]

Griffith SC, Owens IPF, & Thuman KA (2002) Extra-pair paternity in birds: a review of interspecific variation and adaptive function. Mol Ecol 11: 2195–2212 [Chap 24]

Griffith SC & Pryke SR (2006) Benefits to female birds of assessing color displays. Pp 233–279 in GE Hill & KJ McGraw (eds) Bird Coloration, Vol 2: Function and Evolution. Harvard Univ Press, Cambridge, MA [Chap 24]

Grimaldi D & Engel MS (2005) Evolution of the Insects. Cambridge Univ Press, Cambridge [Chap 28]

Gromko MH, Gilbert DG, & Richmond RC (1984) Sperm transfer and use in the multiple mating system of *Drosophila*. Pp 371–426 in Smith RL (ed) Sperm Competition and the Evolution of Animal Mating Systems. Academic Press, New York [Chap 22]

Groning J & Hochkirch A (2008) Reproductive interference between animal species. Q Rev Biol 83: 257–282 [Chap 27]

Grosenick L, Clement TS, & Fernald RD (2007) Fish can infer social rank by observation alone. Nature 445: 429–432 [Chap 17]

Gross MR (1982) Sneakers, satellites and parentals: Polymorphic mating strategies in North American sunfishes. Z Tierpsychol 60: 1–26 [Chap 25]

Gross MR (1996) Alternative reproductive strategies and tactics: diversity within sexes. Trends Ecol Evol 11: 92–98 [Chaps 13, 25]

Gross MR & Charnov EL (1980) Alternative male life histories in bluegill sunfish. Proc Natl Acad Sci USA 77: 6937–6940 [Chap 25]

Gross MR & Repka J (1998) Game theory and inheritance of the conditional strategy. Pp 168–187 in LA Dugatkin & HK Reeve (eds) Game Theory and Animal Behavior. Oxford Univ Press, New York [Chap 25]

Gross MR & Shine R (1981) Parental care and mode of fertilization in ectothermic vertebrates. Evolution 35: 775–793 [Chap 26]

Grozinger CM, Fan Y, Hoover SE, & Winston ML (2007) Genome-wide analysis reveals differences in brain gene expression patterns associated with caste and reproductive status in honey bees (*Apis mellifera*). Mol Ecol. 16: 4837–4848 [Chap 28]

Grozinger CM & Robinson GE (2007) Endocrine modulation of a pheromone-responsive gene in the honey bee brain. J Comp Physiol A 193: 461–470 [Chap 28]

Grozinger CM, Sharabash NM, Whitfield CW, & Robinson GE (2003) Pheromone-mediated gene expression in the honey bee brain. Proc Natl Acad Sci USA 100: 14519–14525 [Chap 28]

Guilford TC & Dawkins MS (1991) Receiver psychology and the evolution of animal signals. Anim Behav 42: 1–14 [Chaps 9, 10, 16]

Guilford TC & Dawkins MS (1995) What are conventional signals? Anim Behav 49: 1689–1695 [Chap 16]

Gutzke WHN & Crews D (1988) Embryonic temperature determines adult sexuality in a reptile. Nature 332: 832–834 [Chap 5]

Gwynne DT (1982) Mate selection by female katydids (Orthoptera: Tettigoniidae, *Conocephalus nigropleurum*). Anim Behav 30: 734–738 [Chap 22]

Gwynne DT (1989) Does copulation increase the risk of predation? Trends Ecol Evol 4: 54–56 [Chap 22]

Gwynne DT & Rentz CF (1983) Beetles on the bottle: male buprestids mistake stubbies for females (Coleoptera). J Aust Entomol Soc 22: 79–80 [Chap 14]

Haag CR, Saastamoinen M, Marden JH, & Hanski I (2005) A candidate locus for variation in dispersal rate in a butterfly metapopulation. Proc Biol Sci 272: 2449–2456 [Chap 28]

Hack MA (1997) Assessment strategies in the contests of male crickets, *Acheta domesticus* (L). Anim Behav 53: 733–747 [Chap 15]

Hack MA (1998) The energetics of male mating strategies in field crickets (Orthoptera: Gryllinae: Gryllidae). J Insect Behav 11: 853–867 [Chap 24]

Hadfield JD, Nutall A, Osorio D, & Owens IPF (2007) Testing the phenotypic gambit: phenotypic, genetic and environmental correlations of colour. J Evol Biol 20: 549–557 [Chap 5]

Haerty W, Jagadeesham S, Kulathinal RJ, Wong A, Ram KA, Sirot LK, Levesque L, Artieri CG, Wolfner MF, Civetta A, & Singh RS (2007) Evolution in the fast land: rapidly evolving sex-related genes in *Drosophila*. Genetics 177: 1321–1335 [Chaps 5, 23]

Haffer J (2007) The development of ornithology in central Europe. J Ornithol 148: S125–S153 [Chap 1]

Hahn ME & Schanz N (1996) Issues in the genetics of social behavior: revisited. Behav Genet 26: 463–470 [Chap 14]

Hailman JP (1977) Optical Signals: Animal Communication and Light. Indiana Univ Press, Bloomington [Chap 9]

Haldane JBS (1957) The cost of natural selection. J Genet 55: 511–524 [Chap 6]

Haldane JBS & Jayakar SD (1963) Polymorphism due to selection of varying direction. J Genet 58: 237–242 [Chap 25]

Haley MP (1994) Resource-holding power asymmetries, the prior residence effect, and reproductive payoffs in male northern elephant seal fights. Behav Ecol Sociobiol 34: 427–434 [Chap 15]

Hall JC (1994) Pleiotropy of behavioral genes. Pp 15–27 in Greenspan RJ & Kyriacou CP (eds) Flexibility and Constraints in Behavioral Systems. Wiley, New York [Chap 5]

Hall M, Lindholm AK, & Brooks R (2004) Direct selection on male attractiveness and female preference fails to produce a response. BMC Evol Biol 4: 1 [Chap 24]

Halliday T & Arnold SJ (1987) Multiple mating by females: a perspective from quantitative genetics. Anim Behav 35: 939–941 [Chap 22]

Halliday TR (1983) The study of mate choice. Pp 3–32 in P Bateson (ed) Mate Choice. Cambridge Univ Press, Cambridge [Chap 24]

Halpin CG, Skelhorn J, & Rowe C (2008) Naïve predators and selection for rare conspicuous defended prey: the initial evolution of aposematism revisited. Anim Behav 75: 771–781 [Chap 10]

Hamel J-F & Mercier A (1996) Gamete dispersion and fertilisation success of the sea cucumber *Cucumaria frondosa*. SPC Beche-de-mer Info Bull 8: 34–40 [Chap 21]

Hamilton IM (2000) Recruiters and joiners: using optimal skew theory to predict group size and the division of resources within groups of social foragers. Am Nat 155: 684–695 [Chap 11]

Hamilton IM & Dill LM (2003a) Group foraging by a kleptoparasitic fish: a strong inference test of social foraging models. Ecology 84: 3349–3359 [Chap 11]

Hamilton IM & Dill LM (2003b) The use of territorial gardening versus kleptoparasitism by a subtropical reef fish (Kyphosus cornelii) is influenced by territory defendability. Behav Ecol 14: 561–568 [Chap 11]

Hamilton WD (1963) The evolution of altruistic behaviour. Am Nat 97: 354–356 [Chap 18]

Hamilton WD (1964) The genetical evolution of social behaviour I and II. J Theor Biol 7: 1–16 and 17–52 [Chaps 1, 3, 5, 14, 18, 19, 26]

Hamilton WD (1967) Extraordinary sex ratios. Science 156: 477–488 [Chaps 15, 20, 26]

Hamilton WD (1970) Selfish and spiteful behaviour in an evolutionary model. Nature 228: 1218–1220 [Chap 18]

Hamilton WD (1971a) Geometry for the selfish herd. J Theor Biol 31: 295–311 [Chap 13, 17]

Hamilton WD (1971b) Selection of selfish and altruistic behaviour in some extreme models. Pp 57–91 in Eisenberg JF & Dillon WS (eds) Man and Beast: Comparative Social Behavior. Smithsonian Press, Washington DC [Chap 18]

Hamilton WD (1972) Altruism and related phenomena, mainly in social insects. Annu Rev Ecol Syst 3: 193–232 [Chap 18]

Hamilton WD (1975) Innate social aptitudes of man: an approach from evolutionary genetics. Pp 133–155 in Fox R (ed) Biosocial Anthropology. Wiley, New York [Chap 18]

Hamilton WD (1996) Narrow Roads of Gene Land: The Collected Papers of W. D. Hamilton Volume 1: Evolution of Social Behaviour. Freeman, Oxford [Chaps 4, 18]

Hamilton WD & Zuk M (1982) Heritable true fitness and bright birds: a role for parasites? Science 218: 384–387 [Chap 24]

Hammerstein P (2003) Genetic and Cultural Evolution of Cooperation. MIT Press, Cambridge, MA [Chap 18]

Hammerstein P & Hagen EH (2005) The second wave of evolutionary economics in biology. Trends Ecol Evol 20: 604–609 [Chap 8]

Hammerstein P & Parker GA (1982) The asymmetric war of attrition. J Theor Biol 96: 647–682 [Chap 15]

Hammond RL & Keller L (2004) Conflict over male parentage in social insects. PLoS Biol 2: 1472–1482 [Chap 18]

Hampton RR, Healy SD, Shettleworth SJ, & Kamil AC (2002) "Neuroecologists" are not made of straw. Trends Cogn Sci 6: 6–7 [Chap 10]

Hanlon RT (1998) Mating systems and sexual selection in the squid *Loligo:* How might commercial fishing on spawning squids affect them? Calcofi Rep 39: 92–100 [Chap 25]

Hannes RP & Franck D (1983) The effect of social isolation on androgen and corticosteroid levels in a cichlid fish (*Haplochromis burtoni*) and in swordtails (*Xiphophorus helleri*). Horm Behav 17: 292–301 [Chap 15]

Hansen TF (1997) Stabilizing selection and the comparative analysis of adaptation. Evolution 51: 1341–1351 [Chap 7]

Hansen TF & Price DK (1995) Good genes and old age: do old mates provide superior genes? J Evol Biol 8: 759–778 [Chaps 4, 24]

Harcourt AH, Harvey PH, Larson SG, & Short RV (1981) Testis weight, body weight and breeding system in primates. Nature 293: 55–57 [Chap 22]

Hardin G (1968) The tragedy of the commons. Science 162: 1243–1248 [Chaps 18, 20]

Härdling R & Kokko H (2005) The evolution of prudent choice. Evol Ecol Res 7: 697–715 [Chap 20]

Hardy ICW (ed) (2002) Sex Ratios: Concepts and Research Methods. Cambridge Univ Press, Cambridge [Chap 20]

Hartley RC & Kennedy MW (2004) Are carotenoids a red herring in sexual display? Trends Ecol Evol 19: 353–354 [Chap 1]

Harvell CD (1990) The ecology and evolution of inducible defenses. Q Rev Biol 65: 323–340 [Chap 29]

Harveson PM, Lopez RR, Collier BA, & Silvy NJ (2007) Impact of urbanization on Florida Key deer behavior and population dynamics. Biol Conserv 134: 321–331 [Chap 6]

Harvey PH, Brown AJL, Smith JM, & Nee S (eds) (1996) New Uses for New Phylogenies. Oxford Univ Press, New York [Chap 7]

Harvey PH & Pagel MD (1991) The Comparative Method in Evolutionary Biology. Oxford Univ Press, New York [Chap 7]

Harvey PH, Partridge L, & Nunney L (1985) Group selection and the sex ratio. Nature 313: 10–11 [Chap 18]

Hassell MP & Varley GC (1969) New inductive population model for insect parasites and its bearing on biological control. Nature 223: 1133–1137 [Chap 11]

Hastings A & Caswell H (1979) Role of environmental variability in the evolution of life history strategies. Proc Natl Acad Sci USA 76: 4700–4703 [Chap 4]

Hau M (2001) Timing of breeding in variable environments: tropical birds as a model system. Anim Behav 40: 281–290 [Chap 4]

Hauber ME (2003) Hatching asynchrony, nesting competition and the cost of interspecific brood parasitism. Behav Ecol 14: 227–235 [Chap 4]

Hawkes K (2004) Mating, parenting, and the evolution of human pairbonds. Pp 443–473 in Chapais B & Berman CM (eds) Kinship and Behavior in Primates. Oxford Univ Press, New York [Chap 31]

Hawkes K, O'Connell JF, & Blurton Jones NG (1995) Hadza children's foraging: juvenile dependency, social arrangements and mobility among hunter-gatherers. Curr Anthrop 36: 688–700 [Chap 31]

Hawkes K, O'Connell JF, & Blurton Jones NG (2001) Hunting and nuclear families. Curr Anthrop 42: 681–709 [Chap 31]

Haygood R (2004) Sexual conflict and protein polymorphism. Evolution 58: 1414–1423 [Chap 21]

Haygood R, Fedrigo O, Hanson B, Yokoyama KD, & Wray GA (2007) Promoter regions of many neural- and nutrition-related genes have experienced positive selection during human evolution. Nat Genet 39: 1140–1144 [Chap 28]

Hazel WN, Smock R, & Johnson MD (1990) A polygenic model for the evolution and maintenance of conditional strategies. Proc R Soc Lond B 242: 181–187 [Chap 25]

Head ML & Brooks R (2006) Sexual coercion and the opportunity for sexual selection in guppies. Anim Behav 71: 515–522 [Chap 24]

Head ML, Hunt J, & Brooks R (2006) Genetic association between male attractiveness and female differential allocation. Biol Lett 2: 341–344 [Chap 24, 26]

Head ML, Hunt J, Jennions MD, & Brooks R (2005) The indirect benefits of mating with attractive males outweigh the direct costs. PLoS Biol 3: 289–294 [Chaps 23, 24]

Healy SD, de Kort SR, & Clayton NS (2005) Response to Francis: Puzzles are a challenge, not a frustration. Trends Ecol Evol 20: 477–477 [Chap 10]

Healy SD & Hurly TA (2003) Cognitive ecology: foraging in hummingbirds as a model system. Adv Stud Behav 32: 325–359 [Chap 10]

Healy SD & Hurly TA (2004) Spatial learning and memory in birds. Brain Behav Evol 63: 211–220 [Chap 10]

Healy SD & Rowe C (2007) A critique of comparative studies of brain size. Proc R Soc Lond B 274: 453–464 [Chap 10]

Hedrick AV & Dill LM (1993) Mate choice by female crickets is influenced by predation risk. Anim Behav 46: 193–196 [Chap 24]

Heg D, Bachar Z, Brouwer L, & Taborsky M (2004) Predation risk is an ecological constraint for helper dispersal in a cooperatively breeding cichlid. Proc R Soc Lond B 271: 2367–2374 [Chap 17]

Heiling AM, Cheng K, Chittka L, Goeth A, & Herberstein ME (2005) The role of UV in crab spider signals: effects on perception by prey and predators. J Exp Biol 208: 3925–3931 [Chap 9]

Heiling AM, Herberstein ME, & Chittka L (2003) Pollinator attraction: Crab spiders manipulate flower signals. Nature 421: 334 [Chap 9]

Heinsohn R (2008) The ecological basis of unusual sex roles in reverse-dichromatic eclectus parrots. Anim Behav 76: 97–103 [Chap 20]

Heisler IL & Damuth JD (1987) A method for analyzing selection in hierarchically structured populations. Am Nat 130: 582–602 [Chap 14]

Heithaus MR, Dill LM, Marshall GJ, & Buhleier B (2002) Habitat use and foraging behavior of tiger sharks (Galeocerdo cavier) in a seagrass ecosystem. Mar Biol 140: 237–248 [Chap 11]

Helanterä H & Bargum K (2007) Pedigree relatedness, not greenbeard genes, explains eusociality. Oikos 116: 217–220 [Chap 19]

Helanterä H, Strassmann JE, Carillo J, & Queller DC (2009) Unicolonial ants: where do they come from, what are they, where are they going? Trends Ecol Evol 24: 341–349 [Chap 19]

Hemelrijk CK (2000) Towards the integration of social dominance and spatial structure. Anim Behav 59: 1035–1048 [Chap 17]

Hendry AP, Farrugia TJ, & Kinnison MT (2008) Human influences on rates of phenotypic change in populations of animals. Mol Ecol 17: 20–29 [Chap 6]

Hendry AP & Kinnison MT (1999) Perspective: the pace of modern life: measuring rates of contemporary microevolution. Evolution 53: 1637–1653 [Chap 29]

Hepper PG (1991) Kin Recognition. Cambridge Univ Press, Cambridge [Chap 31]

Herndon LA & Wolfner MF (1995) A *Drosophila* seminal fluid protein, Acp26aa, stimulates egg-laying in females for 1 day after mating. Proc Natl Acad Sci USA 92: 10114–10118 [Chap 5]

Herre EA (1985) Sex ratio adjustment in fig wasps. Science 228: 896–898 [Chap 20]

Hews DK, Thompson CW, Moore IT, & Moore MC (1997) Population frequencies of alternative male phenotypes in tree lizards: geographic variation and common-garden rearing studies. Behav Ecol Sociobiol 41: 371–380 [Chap 16]

Higashi M & Yamamura N (1993) What determines animal group size? Insider-outsider conflict and its resolution. Am Nat 142: 553–563 [Chap 11]

Higgie M & Blows MW (2007) Are traits that experience reinforcement also under sexual selection? Am Nat 170: 409–420 [Chap 27]

Higgie M, Chenoweth S, & Blows MW (2000) Natural selection and the reinforcement of mate recognition. Science 290: 519–521 [Chap 27]

Higgins LA, Jones KM, & Wayne ML (2005) Quantitative genetics of natural variation in *Drosophila* melanogaster: the possible role of the social environment on creating persistent patterns of group activity. Evolution 59: 1529–1539 [Chap 14]

Higley JD, Mehlman PT, Taub DM, Higley SB, Suomi J, Linnoila M, & Vickers JH (1992) Cerebrospinal fluid monoamine and adrenal correlates of aggression in free ranging rhesus monkeys. Arch Gen Psychiat 49: 436–441 [Chap 15]

Hill GE & McGraw KJ (2006) Bird Coloration. Harvard Univ Press, Cambridge, MA [Chap 1]

Hill K & Hurtado AM (1996) Ache Life History: The Ecology and Demographics of a Foraging People. Aldine de Gruyter, New York [Chap 13]

Hilton GM, Cresswell W, & Ruxton GD (1999) Intraflock variation in the speed of escape-flight response on attack by an avian predator. Behav Ecol 10: 391–395 [Chap 17]

Hinde RA (1970) Animal Behavior. McGraw Hill, New York [Chap 1]

Hinde RA (1982) Ethology. Oxford Univ Press, New York [Chap 1]

Hine E, Chenoweth SF, & Blows MW (2004) Multivariate quantitative genetics and the lek paradox: genetic variance in male sexually selected traits of *Drosophila serrata* under field conditions. Evolution 58: 2754–2762 [Chap 24]

Hirakawa H (1997) Digestion-constrained optimal foraging in generalist mammalian herbivores. Oikos 78: 37–47 [Chap 11]

Hirschenhauser K, Winkler H, & Oliveira RF (2003) Comparative analysis of male androgen responsiveness to social environment in birds: the effects of mating system and paternal incubation. Horm Behav 43: 508–519 [Chap 7]

Hirschfeld LA & Gelman SA (eds) (1994) Mapping the Mind: Domain Specificity in Cognition and Culture. Cambridge Univ Press, Cambridge [Chap 31]

Hirshleifer J & Riley JG (1992) The Analytics of Uncertainty and Information. Cambridge Univ Press, Cambridge [Chap 12]

Hobel G & Gerhardt HC (2003) Reproductive character displacement in the acoustic communication system of green tree frogs (*Hyla cinerea*). Evolution 57: 894–904 [Chap 27]

Hockett CF (1960) The origin of speech. Scientific American 203: 89–96 [Chap 13]

Hodgson DJ (2002) An experimental manipulation of the growth and dispersal strategy of a parasitic infection using monoclonal aphid colonies. Evol Ecol Res 4: 133–145 [Chap 4]

Hodgson DJ & Townley S (2004) Linking management changes to population dynamic responses: the transfer function of a projection matrix perturbation. J Appl Ecol 41: 1155–1161 [Chap 4]

Hoegh-Guldberg O, Hughes L, McIntyre S, Lindenmayer DB, Parmesan C, Possingham HP, & Thomas CD (2008) Assisted colonization and rapid climate change. Science 321: 345–346 [Chap 29]

Hoekstra RF (1982) On the asymmetry of sex: evolution of mating types in isogamous populations. J Theor Biol 98: 427–451 [Chap 20]

Hoelzer GA (1989) The good parent process of sexual selection. Anim Behav 38: 1067–1078 [Chap 26]

Hofmann HA (2003) Functional genomics of neural and behavioral plasticity. J Neurobiol 54: 272–282 [Chap 28]

Hogan J (2001) Development of behavior systems. Pp 229–279 in Blass E (ed) Developmental Psychobiology, Vol. 13 of Handbook of Behavioral Neurobiology. Kluwer Academic, New York [Chap 16]

Holcomb HR (1993) Sociobiology, Sex, and Science. State Univ New York Press, Albany, NY [Chap 2]

Holland B & Rice WR (1998) Chase-away sexual selection: antagonistic seduction versus resistance. Evolution 52: 1–7 [Chaps 21, 22, 23, 24, 27]

Holland B & Rice WR (1999) Experimental removal of sexual selection reverses intersexual antagonistic coevolution and removes a reproductive load. Proc Natl Acad Sci USA 96: 5083–5088 [Chaps 20, 23, 26]

Hollén LI, Bell MBV, & Radford AN (2008) Cooperative sentinel calling? Foragers gain increased

biomass intake. Curr Biol 18: 576–579 [Chaps 13, 17]
Holmes WG & Sherman PW (1983) Kin recognition in animals. Am Sci 71: 46–55 [Chap 31]
Holt RD & Gomulkiewicz R (2004) Conservation implications of niche conservatism and evolution in heterogeneous environments. Pp 244–264 in Ferrière R, Dieckmann U, & Couvet D (eds) Evolutionary Conservation Biology. Cambridge Univ Press, Cambridge [Chap 29]
Hoogland JL & Sherman PW (1976) Advantages and disadvantages of Bank Swallow coloniality. Ecol Monogr 46: 33–58 [chap 2]
Hori M (1993) Frequency-dependent natural selection in the handedness of scale-eating cichlid fish. Science 260: 216–219 [Chap 25]
Horváth G, Malik P, Kriska G, & Wildermuth H (2007) Ecological traps for dragonflies in a cemetery: the attraction of *Sympetrum* species (Odonata: Libellulidae) by horizontally polarizing black gravestones. Freshwater Biology 52: 1700–1709 [Chap 29]
Hosken DJ (1998) Testes mass in megachiropteran bats varies in accordance with sperm competition theory. Behav Ecol Sociobiol 44: 169–177 [Chap 22]
Hosken DJ, Blanckenhorn WU, & Garner TWJ (2002) Heteropopulation males have a fertilization advantage during sperm competition in the yellow dung fly (*Scathophaga stercoraria*). Proc R Soc Lond B 269: 1701–1707 [Chap 22]
Hosken DJ, Garner TWJ, Tregenza T, Wedell N, & Ward PI (2003) Superior sperm competitors sire higher-quality young. Proc R Soc Lond B 270: 1933–1938 [Chap 22]
Hosken DJ & Stockley P (2003) Benefits of polyandry: a life history perspective. Evol Biol 33: 173–194 [Chap 22]
Hosken DJ & Stockley P (2004) Sexual selection and genital evolution. Trends Ecol Evol 19: 87–93 [Chap 22]
Hosken DJ, Taylor ML, Hoyle K, Higgins S, & Wedell N (2008) Attractive males have greater success in sperm competition. Curr Biol 18: R553–R554 [Chap 22]
Hosken DJ & Ward PI (2000) Copula in yellow dung flies (*Scathophaga stercoraria*): investigating sperm competition models by direct observation. J Insect Physiol 46: 1355–1363 [Chap 22]
Hosken DJ & Ward PI (2001) Experimental evidence for testis size evolution via sperm competition. Ecol Lett 4: 10–13 [Chap 22]
Hotzy C & Arnqvist G (2009) Sperm competition favors harmful males in seed beetles. Curr Biol 19: 404–407 [Chap 22]
Houston AI & McNamara J (1982) A sequential approach to risk-taking. Anim Behav 30: 1260–1261 [Chap 8]
Houston AI & McNamara J (1985) The choice of two prey types that minimises the probability of starvation. Behav Ecol Sociobiol 17: 135–141 [Chap 8]
Houston AI & McNamara JM (1999) Models of Adaptive Behaviour: An Approach Based on State. Cambridge Univ Press, Cambridge [Chaps 8, 12, 16]
Houston AI & McNamara JM (2002) A self-consistent approach to paternity and parental effort. Phil Trans R Soc Lond B 357: 351–362 [Chap 20]
Houston AI, McNamara JM, & Hutchinson JMC (1993) General results concerning the trade-off between gaining energy and avoiding predation. Phil Trans R Soc Lond B 341: 375–397 [Chap 12]
Houston AI, Szekely T, & McNamara JM (2005) Conflict between parents over care. Trends Ecol Evol 20: 33–38 [Chaps 8, 20, 26]
Houston AI, Welton NJ, & McNamara JM (1997) Acquisition and maintenance costs in the long-term regulation of avian fat reserves. Oikos 78: 331–340 [Chap 13]
Housworth EA & Martins EP (2001) Random sampling of constrained phylogenies: conducting phylogenetic analyses when the phylogeny is partially known. Syst Biol 50: 628–639 [Chap 7]
Housworth EA, Martins EP, & Lynch M (2004) The phylogenetic mixed model. Am Nat 163: 84–96 [Chap 7]
Hover EL (1985) Differences in aggressive behavior between two throat color morphs in a lizard, *Urosaurus ornatus*. Copeia 1985: 933–940 [Chap 16]
Howard RD (1984) Alternative mating behaviors of young male bullfrogs. Am Zool 24: 397–406 [Chap 25]
Howard RW, Jackson LL, Banse H, & Blows MW (2003) Cuticular hydrocarbons of *Drosophila birchii* and *D. serrata:* Identification and role in mate choice in *D. serrata*. J Chem Ecol 29: 961–976 [Chap 27]
Howell S, Westergaard G, Hoos B, Chavanne TJ, Shoaf SE, et al. (2007) Serotonergic influences on life-history outcomes in free-ranging male rhesus macaques. Am J Primatol 69: 851–865 [Chap 28]
Hsu Y, Earley RL, & Wolf LL (2006) Modulation of aggressive behaviour by fighting experience: mechanisms and contest outcomes. Biol Rev 81: 33–74 [Chaps 15, 17]
Hsu Y, Lee S-P, Chen M-H, Yang S-Y, & Cheng K-C (2008) Switching assessment strategy during a contest: fighting in killifish *Kryptolebias marmoratus*. Anim Behav 75: 1641–1649 [Chap 15]
Hsu Y & Wolf LL (2001) The winner and loser effect: what fighting behaviours are influenced? Anim Behav 61: 777–786 [Chap 17]

Hubbard SF, Harvey IF, & Fletcher JP (1999) Avoidance of superparasitism: a matter of learning? Anim Behav 57: 1193–1197 [Chap 11]
Hudson ME (2008) Sequencing breakthroughs for genomic ecology and evolutionary biology. Mol Ecol Resources 8: 3–17 [Chap 28]
Huey RB, Hertz P, & Sinervo B (2003) Behavioral drive versus behavioral inertia in evolution: a null model approach. Am Nat 161: 357–366 [Chap 6]
Hughes DP & Cremer (2007) Plasticity in anti-parasite behaviours and its suggested role in invasion biology. Anim Behav 74: 1593–1599 [Chap 6]
Hughes W, Oldroyd B, Beekman M, & Ratnieks F (2008) Ancestral monogamy shows kin selection is key to the evolution of eusociality. Science 320: 1213–1216 [Chap 19]
Hughes WHO, Eilenberg J, & Boomsma JJ (2002) Trade-offs in group living: transmission and disease resistance in leaf-cutting ants. Proc R Soc Lond B 269: 1811–1819 [Chap 17]
Hugie DM (2003) The waiting game: a "battle of waits" between predator and prey. Behav Ecol 14: 807–817 [Chap 8]
Humphries EL, Hebblethwaite AJ, Batchelor, & Hardy ICW (2006) The importance of valuing resources: host weight and contender age as determinants of parasitoid wasp contest outcomes. Anim Behav 72: 891–898 [Chap 15]
Hunt GR (1996) Manufacture and use of hook-tools by New Caledonian crows. Nature 379: 249–251 [Chap 10]
Hunt J, Brooks R, Jennions MD, Smith MJ, Bentsen CL, & Bussiere LF (2004) High-quality male field crickets invest heavily in sexual display but die young. Nature 432: 1024–1027 [Chaps 4, 24]
Hunt J, Blows MW, Zajitschek F, Jennions MD, & Brooks R (2007) Reconciling strong stabilizing selection with the maintenance of genetic variation in a natural population of black field crickets (*Teleogryllus commodus*). Genetics 177: 875–880 [Chap 24]
Hunt J, Jennions MD, Spyrou N, & Brooks R (2006) Artificial selection on male longevity influences age-dependent reproductive effort in the black field cricket, *Teleogryllus commodus*. Am Nat 168: E72–E86 [Chap 4]
Hunt J & Simmons LW (2004) Optimal maternal investment in the dung beetle *Onthophagus taurus?* Behav Ecol Sociobiol 55: 302–312 [Chap 4]
Hunt JH (2007) The Evolution of Social Wasps. Oxford Univ Press, New York [Chap 28]
Hunter FM & Birkhead TR (2002) Sperm viability and sperm competition in insects. Curr Biol 12: 121–123 [Chap 22]
Huntingford FA (1976) The relationship between anti-predator behaviour and aggression among conspecifics in the three-spined stickleback. Anim Behav 24: 245–260 [Chap 30]
Huntingford FA, Lazarus J, Barrie BD, & Webb S (1994) A dynamic analysis of cooperative predator inspection in sticklebacks. Anim Behav 47: 413–423 [Chap 30]
Huntingford FA & Turner AK (1987) Animal Conflict. Chapman & Hall, London [Chap 15]
Hurd PL (1995) Communication in discrete action-response games. J Theor Biol 174: 217–222 [Chap 16]
Hurd PL & Enquist M (1998) Conventional signaling in aggressive interactions: the importance of temporal structure. J Theor Biol 192: 197–211 [Chap 16]
Hurd PL & Enquist M (2001) Threat display in birds. Can J Zool 79: 931–942 [Chap 16]
Hurd PL & Enquist M (2005) A strategic taxonomy of biological communication. Anim Behav 70: 1155–1170 [Chap 16]
Hurd PL, Wachtmeister C-A, & Enquist M (1995) Darwin's principle of antithesis revisited: a role for perceptual biases in the evolution of intraspecific signals. Proc R Soc Lond B 259: 201–205 [Chap 16]
Hurst LD & Hamilton WD (1992) Cytoplasmic fusion and the nature of the sexes. Proc R Soc Lond B 247: 189–194 [Chap 18]
Hutchings JA & Myers RA (1994) The evolution of alternative mating strategies in variable environments. Evol Ecol 8: 256–268 [Chap 25]
Hutchinson GE (1957) Population studies—animal ecology and demography—concluding remarks. Cold Spring Harbor Symposia on Quantitative Biology 22: 415–427 [Chap 12]
Huxley JS (1938) The present standing of the theory of sexual selection. Pp 11–42 in de Beer GR (ed) Evolution: Essays on Aspects of Evolutionary Biology Presented to Professor E S Goodrich on his Seventieth Birthday. Clarendon Press, Oxford [Chap 1]
Huxley JS (1942) Evolution, the Modern Synthesis. Allen & Unwin, London [Chap 1]
Hyman J, Hughes M, Searcy WA, & Nowicki S (2004) Individual variation in the strength of territory defense in male song sparrows: correlates of age, territory tenure, and neighbor aggressiveness. Behaviour 141: 15–27 [Chap 15]
Insel TR & Fernald RD (2004) How the brain processes social information: searching for the social brain. Annu Rev Neurosci 27: 697–722 [Chap 17]
Insel TR, Wang ZX, & Ferris CF (1994) Patterns of brain vasopressin receptor distribution associated with social-organization in microtine rodents. J Neuroscience 14: 5381–5392 [Chap 5]
Irwin DE & Price T (1999) Sexual imprinting, learning and speciation. Heredity 82: 347–354 [Chap 27]

Isvaran K (2005) Variation in male mating behaviour within ungulate populations: patterns and processes. Current Science 89: 1192–1199 [Chap 25]

Ives AR (1989) The optimal clutch size of insects when many females oviposit per patch. Am Nat 133: 671–687 [Chap 11]

Ivy TM & Sakaluk SK (2005) Polyandry promotes enhanced offspring survival in decorated crickets. Evolution 59: 152–159 [Chap 24]

Ivy TM, Weddle CB, & Sakaluk SK (2005) Females use self-referent cues to avoid mating with previous mates. Proc R Soc Lond B 272: 2475–2478 [Chap 24]

Jackson AL & Ruxton GD (2006) Toward an individual-level understanding of vigilance: the role of social information. Behav Ecol 17: 532–538 [Chap 17]

Jacobs GH (1993) The distribution and nature of color vision among the mammals. Biol Rev 68: 413–471 [Chap 9]

James R, Bennett PG, & Krause J (2004) Geometry for mutualistic and selfish herds: the limited domain of danger. Journal of Theoretical Biology 228: 107–113 [Chap 17]

Jansen RC & Nap JP (2001) Genetical genomics: the added value from segregation. Trends Genet 17: 388–391 [Chap 28]

Jansson L & Enquist M (2003) Receiver bias for colourful signals. Anim Behav 66: 965–971 [Chap 10, 16]

Jansson L, Forkman B, & Enquist M (2002) Experimental evidence of receiver bias for symmetry. Anim Behav 63: 617–621 [Chap 16]

Janzen FJ (1994) Climate change and temperature-dependent sex determination in reptiles. Proc Natl Acad Sci USA 91: 7487–7490 [Chap 29]

Janzen FJ & Morjan CL (2001) Repeatability of microenvironment-specific nesting behaviour in a turtle with environmental sex determination. Anim Behav 62: 73–82 [Chap 5]

Jenkins NL & Hoffmann AA (2001) Distribution of *Drosophila serrata* Malloch (Diptera: Drosophilidae) in Australia with particular reference to the southern border. Aust J Entomol 40: 41–48 [Chap 27]

Jenner E (1788) Observations of the natural history of the cuckoo. Phil Trans R Soc Lond 78: 219–237 [Chap 1]

Jennings D, Gammell MP, Payne RJH, & Hayden TJ (2005) An investigation of assessment games during fallow deer fights. Ethology 111: 511–525 [Chap 15]

Jennions MD, Drayton JM, Brooks R, & Hunt J (2007) Do female black field crickets *Teleogryllus commodus* benefit from polyandry? J Evol Biol 20: 1469–1477 [Chap 24]

Jennions MD, Møller AP, & Petrie M (2001) Sexually selected traits and adult survival: a meta-analysis. Q Rev Biol 76: 3–36 [Chaps 4, 24]

Jennions MD & Petrie M (1997) Variation in mate choice and mating preferences: a review of causes and consequences. Biol Rev 72: 283–327 [Chap 6, 24]

Jennions MD & Petrie M (2000) Why do females mate multiply? A review of the genetic benefits. Biol Rev 75: 21–64 [Chaps 4, 20, 31]

Jensen JD, Wong A, & Aquadro CF (2007) Approaches for identifying targets of positive selection. Trends in Genetics 23: 568–577 [Chap 5]

Jenssen TA (1977) Evolution of anoline lizard display behavior. Am Zool 17: 203–215 [Chap 7]

Jenssen TA (1978) Display diversity in anoline lizards and problems in interpretation. Pp 269–285 in Greenberg N & MacLean PD (eds) Behavior and Neurology of Lizards. National Institute of Mental Health, Washington, DC [Chap 7]

Jiggins CD, Estrada C, & Rodrigues A (2004) Mimicry and the evolution of premating isolation in *Heliconius melpomene* Linnaeus. J Evol Biol 17: 680–691 [Chap 27]

Johnsen A, Lifjeld JT, Rohde PA, Primmer CR, & Ellegren H (1998) Sexual conflict over fertilizations: female bluethroats escape male paternity guards. Behav Ecol Sociobiol 43: 401–408 [Chap 25]

Johnsen TS & Zuk M (1995) Testosterone and aggression in male red jungle fowl. Horm Behav 29: 593–598 [Chap 15]

Johnson JC & Sih A (2005) Pre-copulatory sexual cannibalism in fishing spiders (*Dolomedes triton*): a role for behavioral syndromes. Behav Ecol Sociobiol 58: 390–396 [Chap 30]

Johnson JC & Sih A (2007) Fear, food, sex and parental care: a syndrome of boldness in the fishing spider, *Dolomedes triton*. Anim Behav 74: 1131–1138 [Chap 30]

Johnson KP & Lanyon SM (2000) Evolutionary changes in color patches of blackbirds are associated with marsh nesting. Behav Ecol 11: 515–519 [Chap 2]

Johnson SL & Yund PO (2004) Remarkable longevity of dilute sperm in a free-spawning colonial ascidian. Biol Bull 206: 144–151 [Chap 21]

Johnstone RA (1995a) Honest advertisement of multiple qualities using multiple signals. J Theor Biol 177: 87–94 [Chap 16]

Johnstone RA (1995b) Sexual selection, honest advertisement and the handicap principle: reviewing the evidence. Biol Rev 70: 1–65 [Chap 24]

Johnstone RA (1998) Game theory and communication. Pp 94–116 in Dugatkin LA & Reeve HK (eds) Game Theory and Animal Behavior. Oxford Univ Press, New York [Chap 16]

Johnstone RA (2001) Eavesdropping and animal conflict. Proc Natl Acad Sci USA 98: 9177–9180 [Chap 12]

Johnstone RA & Grafen A (1993) Dishonesty and the handicap principle. Anim Behav 46: 759–764 [Chap 16]

Johnstone RA & Keller L (2000) How males can gain by harming their mates: sexual conflict, seminal toxins, and the cost of mating. Am Nat 156: 368–377 [Chap 22]

Johnstone RA, Reynolds JD, & Deutsch JC (1996) Mutual mate choice and sex differences in choosiness. Evolution 50: 1382–1391 [Chap 20]

Jokela J, Lively CM, Dybdahl MF, & Fox JA (2003) Genetic variation in sexual and clonal lineages of a freshwater snail. Biol J Linn Soc 79: 165–181 [Chap 20]

Jones AG, Rosenqvist G, Berglund A, & Avise JC (2005) The measurement of sexual selection using Bateman's principles: an experimental tests in the sex-role-reversed pipefish *Syngnathus typhle*. Integr Comp Biol 45: 874–884 [Chap 20]

Kacelnik A & Brunner D (2002) Timing and foraging: Gibbon's scalar expectancy theory and optimal patch exploitation. Learn Motiv 33: 177–195 [Chap 8]

Kaiser VB & Ellegren H (2006) Nonrandom distribution of genes with sex-biased expression in the chicken genome. Evolution 60: 1945–1951 [Chap 5]

Kalmijn AJ (1971) The electric sense of sharks and rays. J Exp Biol 55: 371–383 [Chap 9]

Kamo M, Ghirlanda S, & Enquist M (2002) The evolution of signal form: effects of learned vs. inherited recognition. Proc R Soc Lond B 269: 1765–1771 [Chap 16]

Kaplan H, Hill K, Lancaster J, & Hurtado AM (2000) A theory of human life history evolution: diet, intelligence, and longevity. Evol Anthrop 9: 156–185 [Chap 31]

Kelber A, Vorobyev M, & Osorio D (2003) Animal color vision—behavioral tests and physiological concepts. Biol Rev 78: 81–118 [Chap 9]

Kelleher ES, Swanson WJ, & Markow TA (2007) Gene duplication and adaptive evolution of digestive proteases in *Drosophila arizonae* female reproductive tracts. PLoS Genet 3: 1541–1549 [Chap 23]

Keller L (1997) Indiscriminate altruism: unduly nice parents and siblings. Trends Ecol Evol 12: 99–103 [Chap 18]

Keller L (1999) Levels of Selection in Evolution. Princeton Univ Press, Princeton, NJ [Chap 28]

Keller L & Genoud M (1997) Extraordinary lifespans in ants: a test of evolutionary theories of ageing. Nature 389: 958–960 [Chap 19]

Keller L & Reeve H (1995) Why do females mate with multiple males? The sexually selected sperm hypothesis. Adv Stud Behav 24: 291–315 [Chap 22]

Keller MC & Miller G (2006) Resolving the paradox of common, harmful, heritable mental disorders: which evolutionary genetic models work best? Behavioral and Brain Sciences 29: 385–452 [Chap 5]

Kemp DJ (2002) Butterfly contests and flight physiology: why do older males fight harder? Behav Ecol 13: 456–461 [Chap 15]

Kendal RL, Coolen I, & Laland KN (2004) The role of conformity in foraging when personal and social information conflict. Behav Ecol 15: 269–277 [Chap 17]

Kennedy M & Gray RD (1993) Can ecological theory predict the distribution of foraging animals? A critical analysis of experiments on the ideal free distribution. Oikos 68: 158–166 [Chap 11]

Kennedy MW & Nager R (2006) The perils and prospects of using phytohaemagglutinin in evolutionary ecology. Trends Ecol Evol 21: 653–655 [Chap 1]

Kenward B, Wachtmeister C-A, Ghirlanda S, & Enquist M (2004) Spots and stripes: the evolution of repetition in visual signal form. J Theor Biol 230: 407–419 [Chap 16]

Ketterson ED & Nolan V (1999) Adaptation, exaptation, and constraint: a hormonal perspective. Am Nat 154: S4–S25 [Chap 30]

Kettlewell HBD (1955) Selection experiments on industrial melanism in the Lepidoptera. Heredity 9: 323–342 [Chap 25]

Kidd NAC & Tozer DJ (1985) On the significance of post-reproductive life in aphids. Ecol Entomol 10: 357–359 [Chap 4]

Kiesecker JM & Blaustein AR (1997) Population differences in responses of red-legged frogs (*Rana aurora*) to introduced bullfrogs. Ecology 78: 1752–1760 [Chap 29]

Kim Y-G (1995) Status signaling games in animal contests. J Theor Biol 176: 221–231 [Chap 16]

Kimchi T, Xu J, & Dulac C (2007) A functional circuit underlying male sexual behaviour in the female mouse brain. Nature 448: 1009–1014 [Chap 28]

King AJ & Cowlishaw G (2007) When to use social information: the advantage of large group size in individual decision making. Biol Lett 3: 137–139 [Chap 17]

Kingsolver JG, Hoekstra HE, Hoekstra JM, Berrigan D, Vignieri SN, Hill CE, Hoang A, Gibert P, & Beerli P (2001) The strength of phenotypic seleciton in natural populations. Am Nat 157: 245–261 [Chaps 3, 30]

Kirk DL (2006) Oogamy: inventing the sexes. Curr Biol 16: R1028–R1030 [Chap 20]

Kirkpatrick M & Barton NH (1997) The strength of indirect selection on female mating preferences. Proc Natl Acad Sci USA 94: 1282–1286 [Chap 23]

Kirkpatrick M & Lande R (1989) The evolution of maternal characters. Evolution 43: 485–503 [Chaps 4, 14]

Kirkpatrick M & Ryan MJ (1991) The evolution of mating preferences and the paradox of the lek. Nature 350: 33–38 [Chaps 24, 27]

Kirn JR & Devoogd TJ (1989) Genesis and death of vocal control neurons during sexual-differentiation in the zebra finch. J Neurosci 9: 3176–3187 [Chap 10]

Klopfer PH (1974) An introduction to animal behavior: ethology's first century. Prentice-Hall Englewood Cliffs, NJ [Chap 1]

Klose SM, Smith CL, Denzel AJ, & Kalko EKV (2006) Reproduction elevates the corticosterone stress response in common fruit bats. J Comp Physiol A 192: 341–350 [Chap 12]

Klump GM, Kretzschmar E, & Curio E (1986) The hearing of an avian predator and its avian prey. Behav Ecol Sociobiol 18: 317–323 [Chap 13]

Knight FH (1921) Risk, Uncertainty and Profit. Houghton Mifflin Company, Chicago [Chap 12]

Kobayashi T (1999) Do Mynahs prefer peacock feathers of more regular pattern? Ornis Svecica 9: 59–64 [Chap 16]

Kocher SD, Ayroles JF, Stone EA, & Grozinger CM (unpublished) Individual variation in pheromone response correlates with reproductive traits and brain gene expression in worker honey bees [Chap 28]

Koenig WD & Walters JR (1999) Sex-ratio selection in species with helpers at the nest: the repayment model revisited. Am Nat 153: 124–130 [Chap 26]

Kokko H (1998) Should advertising parental care be honest? Proc R Soc Lond B 265: 1871–1878 [Chap 26]

Kokko H (2001) Fisherian and "good genes" benefits of mate choice: how (not) to distinguish between them. Ecol Lett 4: 322–326 [Chaps 3, 4]

Kokko H & Jennions MD (2003) It takes two to tango. Trends Ecol Evol 18: 103–104 [Chaps 20, 26]

Kokko H & Jennions MD (2008) Parental investment, sexual selection and sex ratios. J Evol Biol 21: 919–948 [Chaps 20, 26]

Kokko H & Monaghan P (2001) Predicting the direction of sexual selection. Ecol Lett 4: 159–165 [Chap 20]

Kokko H & Rankin DJ (2006) Lonely hearts or sex in the city? Density-dependent effects in mating systems. Phil Trans R Soc Lond B 361: 319–334 [Chap 20]

Kokko H & Sutherland WJ (2001) Ecological traps in changing environments: ecological and evolutionary consequences of a behaviourally mediated Allee effect. Evol Ecol Res 3: 537–551 [Chap 29]

Kokko H, Brooks R, McNamara JM, & Houston AI (2002) The sexual selection continuum. Proc R Soc Lond B 269: 1331–1340 [Chaps 7, 24]

Kokko H, Mappes J, & Lindstrom L (2003a) Alternative prey can change model-mimic dynamics between parasitism and mutualism. Ecol Lett 6: 1068–1076 [Chap 3]

Kokko H, Brooks R, Jennions MD, & Morley J (2003b) The evolution of mate choice and mating biases. Proc R Soc Lond B 270: 653–664 [Chaps 4, 20, 24]

Kokko H, Jennions MD, & Brooks R (2006a) Unifying and testing models of sexual selection. Annu Rev Ecol Evol Syst 37: 43–66 [Chaps 4, 24, 26]

Kokko H, Lopez-Sepulcre A, & Morrell LJ (2006b) From hawks and doves to self consistent games of territorial behavior. Am Nat 167: 901–912 [Chap 15]

Kokko H, Johnstone RA, & Clutton-Brock TH (2001) The evolution of cooperative breeding through group augmentation. Proc R Soc Lond B 268: 187–196 [Chaps 17, 18]

Kolm N (2001) Females produce larger eggs for large males in a paternal mouthbrooding fish. Proc R Soc Lond B 268: 2229–2234 [Chap 26]

Komdeur J (1998) Long-term fitness benefits of egg sex modification by the Seychelles warbler. Ecol Lett 1: 56–62 [Chap 20]

Komdeur J & Pen I (2002) Adaptive sex allocation in birds: the complexities of linking theory and practice. Phil Trans R Soc Lond B 357: 373–380 [Chaps 20, 26]

Komers PE (1997) Behavioral plasticity in variable environments. Can J Zool 75: 161–169 [Chap 6]

Koolhaas JM, Korte SM, De Boer SF, Van Der Vegt BJ, Van Reenen CG, Hopster H, De Jong IC, Ruis MAW, & Blokhuis HJ (1999) Coping styles in animals: current status in behavior and stress-physiology. Neurosci Biobehav Rev 23: 925–935 [Chap 30]

Koprowski JL (1993) Alternative reproductive tactics in male eastern gray squirrels: "making the best of a bad job." Behav Ecol 4: 165–171 [Chap 25]

Kotiaho JS, Alatalo RV, Mappes J, Nielsen MG, Parri S, & Rivero A (1998a) Energetic costs of size and sexual signalling in a wolf spider. Proc R Soc Lond B 265: 2203–2209 [Chap 9]

Kotiaho JS, Alatalo RV, Mappes J, Parri S, & Rivero A (1998b) Male mating success and risk of predation in a wolf spider: a balance between sexual and natural selection? J Anim Ecol 67: 287–291 [Chap 9]

Kotler BP & Blaustein L (1995) Titrating food and safety in a heterogeneous environment: when are the risky and safe patches of equal value? Oikos 74: 251–258 [Chap 13]

Kotler BP, Morris DW, & Brown JS (2007) Behavioral indicators and conservation: wielding "the biologist's tricorder." Israel J Ecol Evol 53: 237–244 [Chap 8]

Kozak GM & JW Boughman (2008) Experience influences shoal member preference in a species pair of sticklebacks. Behav Ecol 19: 667–676 [Chap 27]

Kozak GM & Boughman JW (2009) Learned conspecific mate preference in a species pair of sticklebacks. Behav Ecol (in press) [Chap 27]

Kozak GM, Reisland M, & Boughman JW (2009) Sex differences in mate recognition for species with mutual mate choice. Evolution 63: 353–365 [Chap 27]

Kozielska M, Pen I, Beukeboom LW, & Weissing FJ (2006) Sex ratio selection and multi-factorial sex determination in the housefly: a dynamic model. J Evol Biol 19: 879–888 [Chap 5]

Kralj-Fiser S, Scheiber IBR, Blejec A, Möstl E, & Kotrschal K (2007) Individualities in a flock of free-roaming greylag geese: behavioral and physiological consistency over time and across situations. Horm Behav 51: 239–248 [Chap 30]

Krams I, Krama T, Igaune K, & Mänd R (2008) Experimental evidence of reciprocal altruism in the pied flycatcher. Behav Ecol Sociobiol 62: 599–605 [Chap 17]

Krause J, Croft DP, & James R (2007) Social network theory in the behavioural sciences: potential applications. Behav Ecol Sociobiol 62: 15–27 [Chap 17]

Krause J & Ruxton GD (2002) Living in groups. Oxford Univ Press, New York [Chaps 11, 17]

Krebs JR (1985) Sociobiology ten years on: E. O. Wilson. New Scientist 108: 40–43 [Chap 1]

Krebs JR & Davies NB (1978) Behavioural Ecology: An Evolutionary Approach. Blackwell, Oxford [Chap 1]

Krebs JR & Davies NB (1981) Behavioural Ecology: An Evolutionary Approach, 2nd ed. Blackwell, Oxford [Chap 1]

Krebs JR & Davies NB (1993) Behavioural Ecology: An Evolutionary Approach, 3rd ed. Blackwell, Oxford [Chap 4]

Krebs JR & Dawkins R (1984) Animal signals: mind-reading and manipulation. Pp 380–402 in Krebs JR & Davies NB (eds) Behavioural Ecology: An Evolutionary Approach, 2nd ed. Blackwell, Oxford [Chaps 1, 16]

Krebs JR, Ryan JC, & Charnov EL (1974) Hunting by expectation or optimal foraging? A study of patch use by chickadees. Anim Behav 22: 953–964 [Chaps 1, 8]

Krimbas CB (2004) On fitness. Biol Phil 19: 185–203 [Chap 4]

Kristan WB III (2003) The role of habitat selection behavior in population dynamics: source-sink systems and ecological traps. Oikos 103: 457–468 [Chap 29]

Kronforst MR, Young, LG, Kapan, DD, McNeely C, O'Neill RJ, & Gilbert LE (2006) Linkage of butterfly mate preference and wing color preference cue at the genomic location of wingless. Proc Natl Acad Sci USA 103: 6575–6580 [Chap 5]

Kruse AA, Stripling R, Clayton DF (2004) Context-specific habituation of the zenk gene response to song in adult zebra finches. Neurobiol Learn Mem 82: 99–108 [Chap 28]

Kruuk H (2003) Niko's Nature: The life of Niko Tinbergen and his science of animal behavior. Oxford Univ Press [Chap 1]

Kruuk LEB (2004) Estimating genetic parameters in natural populations using the "animal model." Phil Trans R Soc Lond B 359: 873–890 [Chaps 5, 6, 14]

Kruuk LEB, Clutton-Brock TH, Slate J, Pemberton JM, Brotherstone S, & Guinness FE (2000) Heritability of fitness in a wild mammal population. Proc Nat Acad Sci USA 97: 698–703 [Chaps 3, 4]

Kruuk LEB, Slate J, Pemberton JM, Brotherstone S, Guiness F, & Clutton-Brock TH (2002) Antler size in red deer: heritability and selection but no evolution. Evolution 56: 1683–1695 [Chap 4]

Kuhn TS (1962) The Structure of Scientific Revolutions. Univ Chicago Press, Chicago [Chap 2]

Kümmerli R, Gardner A, West SA, & Griffin AS (2009) Limited dispersal, budding dispersal and cooperation: an experimental study. Evolution 63: 939–949 [Chap 18]

Kunz TH & Hosken DJ (2009) Male lactation: Why, why not, and is it care? Trends Ecol Evol 24: 80–85 [Chap 20]

Kupriyanova E & Havenhand JN (2002) Variation in sperm swimming behaviour and its effect on fertilization success in the serpulid polychaete *Galeolaria caespitosa*. Invert Reprod Develop 41: 21–26 [Chap 21]

Kvarnemo C (2006) Evolution and maintenance of male care: is increased paternity a neglected benefit of care? Behav Ecol 17: 144–148 [Chap 26]

Kyriacou CP, Peixoto AA, Sandrelli F, Costa R, & Tauber E (2008) Clines in clock genes: fine-tuning circadian rhythms to the environment. Trends in Genetics 24: 124–132 [Chap 5]

Lacey EP, Real L, Antonovics J, & Heckel DG (1983) Variance models in the study of life histories. Am Nat 122: 114–131 [Chap 4]

Lack D (1947) The significance of clutch size. Ibis 89: 302–352 [Chap 2]

Lack D (1954) Natural Regulation of Animal Numbers. Clarendon Press, Oxford [Chap 2]

Lack D (1956) Swifts in a Tower. Chapman & Hall, London [Chap 8]

Lack D (1965) Evolutionary ecology. J Ecol 53: 237–245 [Chap 1]

Lack D (1966) Population Studies of Birds. Clarendon Press, Oxford [Chap 1]

Lack D (1968) Ecological Adaptations for Breeding in Birds. Chapman Hall, London [Chap 1]

Laidlaw HH & Page RE (1984) Polyandry in honey hees (*Apis mellifera* L.): sperm utilization and intracolony genetic relationships. Genetics 108: 985–997 [Chap 28]

Laland KN (2004) Social learning strategies. Learn Behav 32: 4–14 [Chap 17]
Laland KN & Brown G (2002) Sense and nonsense: evolutionary perspectives on human behaviour. Oxford Univ Press [Chap 31]
Laland KN, Richerson PJ, & Boyd R (1996) Developing a theory of animal social learning. Pp 129–154 in Heyes CM & Galef BG (eds) Social Learning in Animals: The Roots of Culture. Academic Press, New York [Chap 11]
Laland KN & Sterelny K (2006) Seven reasons (not) to neglect niche construction. Evolution 60: 1751–1762 [Chap 12]
Lampert W & Trubetskova I (1996) Juvenile growth rate as a measure of fitness in *Daphnia*. Funct Ecol 10: 631–635 [Chap 4]
LaMunyon CW & Ward S (1998) Larger sperm outcompete smaller sperm in the nematode *Caenorhabditis elegans*. Proc R Soc Lond B 265: 1997–2002 [Chap 22]
Lancaster LT, McAdam AG, Wingfield JC, & Sinervo BR (2007) Adaptive social and maternal induction of antipredator dorsal patterns in a lizard with alternative social strategies. Ecol Lett 10: 798–808 [Chap 3]
Lande, R. (1979) Quantitative genetic analysis of multivariate evolution, applied to brain: body size allometry. Evolution 33: 402–416 [Chap 6]
Lande R (1980) Sexual dimorphism, sexual selection, and adaptation in polygenic characters. Evolution 34: 292–305 [Chaps 3, 5]
Lande R (1981) Models of speciation by sexual selection on polygenic traits. Proc Natl Acad Sci USA 78: 3721–3725 [Chaps 5, 14, 16, 24, 26, 27]
Lande R (1982) Rapid origin of sexual isolation and character divergence in a cline. Evolution 36: 213–223 [Chap 27]
Lande R & Arnold SJ (1983) The measurement of selection on correlated characters. Evolution 37: 1210–1226 [Chaps 3, 4, 24, 25, 30]
Lande R & Price T (1989) Genetic correlations and maternal effect coefficients obtained from offspring-parent regression. Genetics 122: 915–922 [Chap 14]
Lank DB, Smith CM, Hanotte O, Burke T, & Cooke F (1995) Genetic polymorphism for alternative mating behaviour in lekking male ruff *Philomachus pugnax*. Nature 378: 59–62 [Chap 25]
Larson A & Losos JB (1996) Phylogenetic systematics of adaptation. Pp 187–220 in Rose MR & Lauder GV (eds) Adaptation. Academic Press, San Diego [Chap 2]
Lasker HR (2006) High fertilization success in a surface-brooding Caribbean gorgonian. Biol Bull 210: 10–17 [Chap 21]
Lasker HR, Brazeau DA, Calderon J, Coffroth MA, Comia R, & Kim K (1996) In situ rates of fertilization among broadcast spawning gorgonian corals. Biol Bull 190: 45–55 [Chap 21]
Lasker HR, Gutierrez-Rodriguez C, Bala K, Hannes A, & Bilewitch JP (2008) Male reproductive success during spawning events of the octocoral *Pseudopterogorgia elisabethae*. Mar Ecol Prog Ser 367: 153–161 [Chap 21]
Lauay C, Gerlach NM, Adkins-Regan E, & Devoogd TJ (2004) Female zebra finches require early song exposure to prefer high-quality song as adults. Anim Behav 68: 1249–1255 [Chap 10]
Lauder GV (1996) The argument from design. Pp 55–91 in Rose MR & Lauder GV (eds) Adaptation. Academic Press, San Diego [Chap 2]
Lazarus J & Inglis IR (1986) Shared and unshared parental investment, parent-offspring conflict and brood size. Anim Behav 34: 1791–1804 [Chap 26]
Leach CK, Ashworth JM, & Garrod DR (1973) Cell sorting out during the differentiation of mixtures of metabolically distinct populations of *Dictyostelium discoideum*. J Embryol Exp Morphol 29: 647–661 [Chap 19]
Leal M & Fleishman LJ (2002) Evidence for habitat partitioning based on adaptation to environmental light in a pair of sympatric lizard species. Proc R Soc Lond B 269: 351–359 [Chap 27]
Leal M & Fleishman LJ (2004) Differences in visual signal design and detectability between allopatric populations of *Anolis* lizards. Am Nat 163: 26–39 [Chap 27]
LeGalliard JF, Cote J, & Fitze PS (2008) Lifetime and intergenerational fitness consequences of harmful male interactions for female lizards. Ecology 89: 56–64 [Chap 23]
Lehmann L & Keller L (2006) The evolution of cooperation and altruism: A general framework and classification of models. J Evol Biol 19: 1365–1376 [Chap 18]
Lehmann L, Perrin N, & Rousset F (2006) Population demography and the evolution of helping behaviors. Evolution 60: 1137–1151 [Chap 18]
Leigh EG (1971) Adaptation and Diversity. Freeman, Cooper, San Francisco [Chap 18]
Leimar O & Hammerstein P (2001) Evolution of cooperation through indirect reciprocity. Proc R Soc Lond B 268: 745–753 [Chap 12]
Leimar O, Enquist M, & Sillén-Tullberg B (1986) Evolutionary stability of aposematic coloration and prey unprofitability: a theoretical analysis. Am Nat 128: 469–490 [Chap 16]
Lenski RE, Rose MR, Simpson, SC, & Tadler SC (1991) Long-term experimental evolution in *Escherichia coli*. 1. Adaptation and divergence during 2,000 generations. Am Nat 138: 1315–1341 [Chap 4]
Lessells CM (1993) The evolution of life histories. Pp 32–68 in JR Krebs & Davies NB (eds) Behavioural Ecology: An Evolutionary Approach, 3rd ed. Blackwell Sci, Oxford [Chap 2]

Lessells CM (1998) A theoretical framework for sex-biased parental care. Anim Behav 56: 395–407 [Chap 26]

Lessells CM (1999) Sexual conflict in animals. Pp 75–99 in Keller L (ed) Levels of Selection in Evolution. Princeton Univ Press, Princeton, NJ [Chap 26]

Lessells CM (2002) Parentally biased favouritism: why should parents specialize in caring for different offspring? Phil Trans R Soc Lond B 357: 381–403 [Chap 26]

Lessells CM (2006) The evolutionary outcome of sexual conflict. Phil Trans R Soc Lond B 361: 301–317 [Chap 23]

Lessells CM (2008) Neuroendocrine control of life histories: what do we need to know to understand the evolution of phenotypic plasticity? Phil Trans R Soc Lond 363: 1589–1598 [Chap 1]

Levins R (1968) Evolution in Changing Environments: Some Theoretical Explorations. Princeton Univ Press, Princeton, NJ [Chap 25]

Levitan DR (1993) The importance of sperm limitation to the evolution of egg size in marine invertebrates. Am Nat 141: 517–536 [Chap 21]

Levitan DR (1995) The ecology of fertilization in free-spawning invertebrates. Pp 123–156 in McEdward LR (ed) Ecology of Marine Invertebrate Larvae. CRC Press, Boca Raton, FL [Chap 21]

Levitan DR (1996) Effects of gamete traits on fertilization in the sea and the evolution of sexual dimorphism. Nature 382: 153–155 [Chap 21]

Levitan DR (1998a) Sperm limitation, gamete competition and sexual selection in external fertilizers. Pp 175–217 in Birkhead T & Møller A (eds) Sperm Competition and Sexual Selection. Academic Press, London [Chap 21]

Levitan DR (1998b) Does Bateman's Principle apply to broadcast-spawning organisms? Egg traits influence in situ fertilization rates among congeneric sea urchins. Evolution 52: 1043–1056 [Chap 21]

Levitan DR (2000a) Optimal egg size in marine invertebrates: theory and phylogenetic analysis of the critical relationship between egg size and development time in echinoids. Am Nat 156: 175–192 [Chap 21]

Levitan DR (2000b) Sperm velocity and longevity trade-off and influence fertilization in the sea urchin *Lytechinus variegatus*. Proc R Soc Lond B 267: 531–534 [Chap 21]

Levitan DR (2002a) Density-dependent selection on gamete traits in three congeneric sea urchins. Ecology 83: 464–479 [Chap 21]

Levitan DR (2002b) The relationship between conspecific fertilization success and reproductive isolation among three congeneric sea urchins. Evolution 56: 1599–1609 [Chap 21]

Levitan DR (2004) Density-dependent sexual selection in external fertilizers: variances in male and female reproductive success along the continuum from sperm limitation to sexual conflict in the sea urchin *Strongylocentrotus franciscanus*. Am Nat 164: 298–309 [Chap 21]

Levitan DR (2005) Sex specific spawning behavior and its consequences in an external fertilizer. Am Nat 165: 682–694 [Chap 21]

Levitan DR (2006) The relationship between egg size and fertilization success in broadcast spawning marine invertebrates. Integr Comp Biol 46: 298–311 [Chap 21]

Levitan DR (2008) Gamete traits influence the variance in reproductive success, the intensity of sexual selection, and the outcome of sexual conflict among congeneric sea urchins. Evolution 62: 1305–1316 [Chap 21]

Levitan DR & Ferrell DL (2006) Selection on gamete recognition proteins depends on sex, density and genotype frequency. Science 312: 267–269 [Chap 21]

Levitan DR, Fukami H, Jara J, Kline D, McGovern TA, McGhee KM, Swanson CA, & Knowlton N (2004) Mechanisms of reproductive isolation among sympatric broadcast-spawning corals of the *Montastraea annularis* complex. Evolution 58: 308–323 [Chap 21]

Levitan DR, Sewell MA, & Chia F-S (1991) Kinetics of fertilization in the sea urchin *Strongylocentrotus franciscanus:* interaction of gamete dilution, age, and contact time. Biol Bull 181: 371–378 [Chap 21]

Levitan DR, terHorst CP, Fogarty ND (2007) The risk of polyspermy in three congeneric sea urchins and its implications for gametic incompatibility and reproductive isolation. Evolution 61: 2007–2014 [Chap 21]

Levy O, Appelbaum L, Leggat W, Gothliff Y, Hayward DC, Miller DJ, & Hoegh-Guldberg O (2007) Light-responsive crytochromes from a simple multicellular animal, the coral *Acropora millepora*. Science 318: 467–470 [Chap 21]

Lewontin RC (1978) Adaptation. Sci Am 239: 212–228 [Chap 2]

Lewontin RC & Cohen D (1969) On population growth in a randomly varying environment. Proc Natl Acad Sci USA 62: 1056–1060 [Chap 4]

Lewontin RC, Rose S, & Kamin LJ (1984) Not in Our Genes: Biology, Ideology, and Human Nature. Pantheon Books, New York [Chap 1]

Lieberman D, Tooby J, & Cosmides L (2003) Does morality have a biological basis? An empirical test of the factors governing moral sentiments relating to incest. Proc R Soc Lond B 270: 819–826 [Chap 31]

Lieberman D, Tooby J, & Cosmides L (2007) The architecture of human kin detection. Nature 445: 727–731 [Chap 31]

Liebert AE & Starks PT (2006) Taming of the skew: transactional models fail to predict reproductive partitioning in the paper wasp *Polistes dominulus*. Anim Behav 71: 913–923 [Chap 17]

Lightman A (2005) The discoveries: great breakthroughs in 20th century science. Pantheon Books, New York [Chap 8]

Ligon JD & Zwartjes PW (1995) Ornate plumage of male red junglefowl does not influence mate choice by females. Anim Behav 49: 117–125 [Chap 2]

Liker A & Székely T (2005) Mortality costs of sexual selection and parental care in natural populations of birds. Evolution 59: 890–897 [Chap 20]

Lim MLM, Land MF, & Li D (2007) Sex-specific UV and fluorescence signals in jumping spiders. Science 315: 481 [Chap 9]

Lim MM, Wang ZX, Olazabal DE, Ren XH, Terwilliger EF, & Young LJ (2004) Enhanced partner preference in a promiscuous species by manipulating the expression of a single gene. Nature 429: 754–757 [Chap 5]

Lima SL (1992) Life in a multi-predator environment: some considerations for anti-predatory vigilance. Annales Zoologici Fennici 29: 217–226 [Chap 13]

Lima SL (1995) Collective detection of predatory attack by social foragers: fraught with ambiguity? Anim Behav 50: 1097–1108 [Chaps 12, 17]

Lima SL (1998) Nonlethal effects in the ecology of predator-prey interactions: what are the ecological effects of anti-predator decision-making? BioScience 48: 25–34 [Chap 13]

Lima SL (2002) Putting predators back into behavioral predator-prey interactions. Trends Ecol Evol 17: 70–75 [Chap 17]

Lima SL & Bednekoff PA (1999) Temporal variation in danger drives antipredator behavior: the predation risk allocation hypothesis. Am Nat 153: 649–659 [Chap 13]

Lima SL & Dill LM (1990) Behavioral decisions made under the risk of predation: a review and prospectus. Can J Zool 68: 619–640 [Chaps 6, 12, 13]

Lima SL & Zollner PA (1996) Anti-predatory vigilance and the limits to collective detection: visual and spatial separation between foragers. Behav Ecol Sociobiol 38: 355–363 [Chap 17]

Lindstrom J (1999) Early development and fitness in birds and mammals. Trends Ecol Evol 14: 343–348 [Chap 12]

Lindström K & St. Mary CM (2008) Parental care and sexual selection. Pp 660 in Magnhagen C, Braithwaite VA, Forsgren E, & Kapoor BG (eds) Fish Behaviour. Science Publishers, Enfield, NH [Chap 26]

Lindström K, St. Mary CM, & Pampoulie C (2006) Sexual selection for male parental care in the sand goby, *Pomatoschistus minutus*. Behav Ecol Sociobiol 60: 46–51 [Chap 26]

Lindström L, Ahtiainen J, Alatalo RV, Kotiaho JS, Lyytinen A, & Mappes J (2005) Negatively condition dependent predation cost of a positively condition dependent sexual signalling. J Evol Biol 19: 649–656 [Chap 9]

Lindström L, Alatalo RV, Mappes J, Riipi M, & Vertainen (2001) Can aposematic signals evolve by gradual change? Nature 397: 249–251 [Chap 10]

Linklater WL (2004) Wanted for conservation research: behavioral ecologists with a broader perspective. BioScience 54: 352–360 [Chap 29]

Linksvayer TA (2006) Direct, maternal, and sibsocial genetic effects on individual and colony traits in an ant. Evolution 60: 2552–2561 [Chap 14]

Lissåker M & Kvarnemo C (2006) Ventilation or nest defence: parental trade-offs in a fish with male care. Behav Ecol Sociobiol 60: 864–873 [Chap 26]

Lissemore FM (1997) Frass clearing by male pine engraver beetles (*Ips pini;* Scolytidae): paternal care or paternity assurance? Behav Ecol 8: 318–325 [Chap 26]

Lodge DM (1993) Biological invasions: lessons for ecology. Trends Ecol Evol 8: 133–137 [Chap 30]

Lohmann KJ, Lohmann CMF, Ehrhart LM, Bagley DA, & Swing T (2004) Geomagnetic map used in sea-turtle navigation. Nature 428: 909–910 [Chap 29]

Long TAF & Rice WR (2007) Adult locomotory activity mediates intralocus sexual conflict in a laboratory-adapted population of *Drosophila melanogaster*. Proc R Soc Lond B 274: 3105–3112 [Chap 23]

Lorenz K (1941) Comparative studies of the motor patterns of Anatinae. Stud Anim Hum Behav 2: 14–18, 106–114 [Chap 7]

Lorenz K (1965) Evolution and Modification of Behavior. Univ Chicago Press, Chicago [Chap 6]

Lorenz K (1970) Studies in Animal and Human Behaviour, Vol. 1. Methuen, London [Chap 16]

Losos JB (1990) Ecomorphology, performance capability, and scaling of West Indian *Anolis* lizards: an evolutionary analysis. Ecol Monog 60: 369–388 [Chap 3]

Losos JB (1994) An approach to the analysis of comparative data when a phylogeny is unavailable or incomplete. Syst Biol 43: 117–123 [Chap 7]

Losos JB (1999) Uncertainty in the reconstruction of ancestral character states and limitations on the use of phylogenetic comparative methods. Anim Behav 58: 1319–1324 [Chap 7]

Losos JB, Jackman TR, Larson A, DeQueiroz K, & Rodriguez-Shettino L (1998) Contingency and determinism in replicated adaptive radiations of island lizards. Science 279: 2115–2118 [Chap 3]

Losos JB, Leal M, Glor RE, de Queiroz K, Hertz PE, Schettino LR, Lara AC, Jackman TR, & Larson A (2003) Niche lability in the evolution of a Caribbean lizard community. Nature 424: 542–545 [Chap 7]

Losos JB, Schoener TW, & Spiller DA (2004) Predator-induced behaviour shifts and natural selection in field-experimental lizard populations. Nature 432: 505–508 [Chap 6]

Losos JB, Schoener TW, Langerhans RB, & Spiller DA (2006) Rapid temporal reversal in predator-driven natural selection. Science 314: 1111 [Chap 6]

Lotka AJ (1907) Studies on the mode of growth of maternal aggregates. Am J Sci 24: 199–216 [Chap 4]

Lotka AJ (1932) The growth of mixed population: two species competing on a common food supply. J Wash Acad Sci 22: 461–469 [Chap 13]

Lovell PG, Tolhurst DJ, Párraga CA, Baddeley R, Leonards U, Troscianko J, & Troscianko T (2005) Stability of the color-opponent signals under changes of illuminant in natural scenes. J Optical Soc Am A 22: 2060–2071 [Chap 9]

Low M (2005) Factors influencing mate guarding and territory defense in the stitchbird (hihi) *Notiomystis cincta*. New Zeal J Ecol 29: 231–242 [Chap 25]

Lucas JR, Pravosudov VV, & Zielinski DL (2001) A reevaluation of the logic of pilferage effects, predation risk, and environmental variability on avian energy regulation: the critical role of time budgets. Behav Ecol 12: 246–260 [Chap 12]

Lundberg A & Alatalo RV (1992) The Pied Flycatcher. Poyser, London [Chap 24]

Lürling M & Scheffer M (2007) Info-disruption: pollution and the transfer of chemical information between organisms. Trends Ecol Evol 22 (7): 374–379.

Luttbeg B & Warner RR (1999) Reproductive decision-making by female peacock wrasses: flexible versus fixed behavioral rules in variable environments. Behav Ecol 10: 666–674 [Chap 10]

Luttikhuizen PC, Honkoop PJH, Drent J, & van der Meer J (2004) A general solution for optimal egg size during external fertilization, extended scope for intermediate optimal egg size and the introduction of Don Ottavio "tango." J Theor Biol 231: 333–343 [Chap 21]

Lynch M (1987) Evolution of intrafamilial interactions. Proc Natl Acad Sci USA 84: 8507–8511 [Chap 14]

Lynch M (1991) Methods for the analysis of comparative data in evolutionary biology. Evolution 45: 1065–1080 [Chap 7]

Lynch M & Walsh B (1998) Genetics and analysis of quantitative traits. Sinauer, Sunderland, MA [Chaps 3, 14, 30]

Lyon JD & Vacquier VD (1999) Interspecies chimeric sperm lysins identify regions mediating species-specific recognition of the abalone egg vitelline envelope. Developmental Biology 214: 151–159 [Chap 23]

Lythgoe JN (1979) The Ecology of Vision. Clarendon Press, Oxford [Chap 9]

Lyytinen A, Lindström L, & Mappes J (2004) Ultraviolet reflection and predation risk in diurnal and nocturnal Lepidoptera. Behav Ecol 15: 982–987 [Chap 9]

Maan ME, Hofker KD, van Alphen JJM, & Seehausen O (2006) Sensory drive in cichlid speciation. Am Nat 167: 947–954 [Chap 27]

MacArthur RH (1962) Some generalised theorems of natural selection. Proc Natl Acad Sci USA 48: 1893–1897 [Chap 4]

MacArthur RH & Pianka ER (1966) On optimal use of a patchy environment. Am Nat 100: 603–609 [Chaps 1, 8]

MacDougall-Shackleton SA (1997) Sexual selection and the evolution of song repertoires. Curr Ornith 14: 81–124 [Chap 7]

MacDougall-Shackleton SA & Ball GF (2002) Revising hypotheses does not indicate a flawed approach—Reply to Bolhuis and Macphail. Trends Cognit Sci 6: 68–69 [Chap 10]

Macedonia JM & Evans CS (1993) Variation among mammalian alarm call systems and the problem of meaning in animal signals. Ethology 93: 177–197 [Chap 13]

Mackay TFC & Anholt RRH (2007) Ain't misbehavin? Genotype-environment interactions and the genetics of behavior. Trends in Genetics 23: 311–314 [Chaps 5, 28]

Mackay TFC, Heinsohn SL, Lyman RF, Moehring AJ, Morgan TJ, & Rollmann SM (2005) Genetics and genomics of *Drosophila* mating behavior. Proc Natl Acad Sci USA 102: 6622–6629 [Chap 7]

Mackintosh NJ (1974) The Psychology of Animal Learning. Academic Press, London [Chap 16]

MacLeod R, Lind J, Clark J, & Cresswell W (2007) Mass regulation in response to predation risk can indicate population declines. Ecol Lett 10: 945–955 [Chap 12]

Macphail EM (1982) Brain and Intelligence in Vertebrates. Clarendon Press, Oxford [Chap 10]

Magiafoglou A, Carew ME, & Hoffmann AA (2002) Shifting clinal patterns and microsatellite variation in *Drosophila serrata* populations: a comparison of populations near the southern border of the species range. J Evol Biol 15: 763–774 [Chap 27]

Magnhagen C (1990) Reproduction under predation risk in the sand goby, *Pomatoschistus minutus,* and the black goby, *Gobius niger:* the effect of age and longevity. Behav Ecol Sociobiol 26: 331–335 [Chap 26]

Magnhagen C (1991) Predation risk as a cost of reproduction. Trends Ecol Evol 6: 183–185 [Chap 26]

Magrath RD & Whittingham LA (1997) Subordinate males are more likely to help if unrelated to the breeding female in cooperatively breeding white-browed scrubwrens. Behav Ecol & Sociobiology 41: 185–192 [Chap 17]

Magurran AE (1993) Individual differences and alternative behaviours. Pp 441–477 in Pitcher TJ (ed) Behaviour of Teleost Fishes, 2nd ed. Chapman & Hall, New York [Chap 30]

Magurran AE (1998) Population differentiation without speciation. Phil Trans R Soc Lond B 353: 275–286 [Chaps 23, 27]

Magurran AE & Seghers BH (1994) A cost of sexual harassment in the guppy, *Poecilia reticulata.* Proc R Soc Lond B 258: 89–92 [Chap 24]

Malthus TJ (1798) *An Essay on the Principle of Population.* J. Johnson, London [Chap 4]

Mangel M (1990) Dynamic information in uncertain and changing worlds. J Theor Biol 146: 317–332 [Chap 12]

Mangel M & Clark CW (1988) Dynamic Modeling in Behavioral Ecology. Princeton Univ Press, Princeton, NJ [Chaps 12, 13]

Manica A (2002) Filial cannibalism in teleost fish. Biol Rev 77: 261–277 [Chap 26]

Mank JE, Hall DW, Kirkpatrick M, & Avise JC (2006) Sex chromosomes and male ornaments: a comparative evaluation in ray-finned fishes. Proc R Soc Lond B 273: 233–236 [Chap 5]

Manning A (1961) The effects of artificial selection for mating speed in *Drosophila melanogaster.* Anim Behav 9: 82–92 [Chap 14]

Manser MB (1999) Response of foraging group members to sentinel calls in suricates, *Suricata suricatta.* Proc R Soc Lond B 266: 1013–1019 [Chap 17]

Mappes J, Alatalo RV, Kotiaho JS, & Parri S (1996) Viability costs of condition-dependent sexual male display in a drumming wolf spiders. Proc R Soc Lond B 263: 785–789 [Chap 9]

Marcillac F, Grosjean Y, & Ferveur JF (2005) A single mutation alters production and discrimination of *Drosophila* sex pheromones. Proc R Soc Lond B 272: 303–309 [Chap 5]

Marconato A, Rasotto MB, & Mazzoldi C (1996) On the mechanism of sperm release in three gobiid fishes (Teleostei: Gobiidae). Environ Biol Fishes 46: 321–327 [Chap 26]

Marconato A, Shapiro DY, Petersen CW, Warner RR, & Yoshikawa T (1997) Methoodlogical analysis of fertilization rate in the bluehead wrasse *Thalassoma bifasciatum:* pair versus group spawns. Mar Ecol Prog Ser 161: 61–70 [Chap 21]

Marcus JM & McCune AR (1999) Ontogeny and phylogeny in the northern swordtail clade of *Xiphophorus.* Syst Biol 48: 491–522 [Chap 7]

Marler P (1955) Characteristics of some animal calls. Nature 176: 6–8 [Chap 13]

Marler P, Evans CS, & Hauser MD (1992) Animal signals: motivational, referential, or both? Pp 66–86 in Papoušek H, Jürgens U, & Papoušek M (eds) Nonverbal vocal communication: comparative and developmental approaches. Cambridge Univ Press [Chap 13]

Marlowe FW (2000) The patriarch hypothesis: an alternative explanation for menopause. Hum Nat 11: 27–42 [Chap 31]

Marlowe FW (2001) Male contribution to diet and female reproductive success among foragers. Curr Anthrop 42: 755–760. [Chap 31]

Marlowe FW (2003) A critical period for provisioning by Hadza men: implications for pair bonding. Evol Hum Behav 24: 217–229 [Chap 31]

Marlowe FW (2005) Mate preferences among Hadza hunter-gatherers. Hum Nat 15: 365–375 [Chap 31]

Marr D (1982) Vision. H. Freeman, New York [Chap 31]

Marra PP & Holmes RT (1997) Avian removal experiments: do they test for habitat saturation or female availability? Ecology 78: 947–952 [Chap 29]

Marshall DJ & Bolton TF (2007) Sperm release strategies in marine broadcast spawners: the costs of releasing sperm quickly. J Exp Biol 210: 3720–3727 [Chap 21]

Marshall DJ & Uller T (2007) When is a maternal effect adaptive? Oikos 116: 1957–1963 [Chap 26]

Marshall NJ (2000) Communication and camouflage with the same "bright" colors in reef fishes. Phil Trans R Soc B. 355: 1243–1248 [Chap 9]

Martin E & Taborsky M (1997) Alternative male mating tactics in a cichlid, *Pelvicachromis pulcher:* A comparison of reproductive effort and success. Behav Ecol Sociobiol 41: 311–319 [Chap 25]

Martin J, Lopez P, & Cooper WE (2003) Loss of mating opportunities influences refuge use in the Iberian rock lizard, *Lacerta monticola.* Behav Ecol Sociobiol 54: 505–510 [Chap 12]

Martin OY & Hosken DJ (2003) The evolution of reproductive isolation through sexual conflict. Nature 423: 979–982 [Chaps 23, 27]

Martin PA, Reimers TJ, Lodge JR, & Dziuk PJ (1974) The effect of ratios and numbers of spermatozoa mixed from two males on proportions of offspring. J Reprod Fert 39: 251–258 [Chap 22]

Martin TE (1995) Avian life history evolution in relation to nest sites, nest predation, and food. Ecological Monographs 65: 101–127 [Chap 6]

Martin TE & Ghalambor CK (1999) Males feeding females during incubation. I. Required by microclimate or constrained by nest predation? Am Nat 153: 131–139 [Chap 6]

Martin TE & Roper JJ (1988) Nest predation and nest-site selection of a western population of hermit thrush. Condor 90: 51–57 [Chap 6]

Martin TE, Scott J, & Menge C (2000) Nest predation increases with parental activity: separating nest site and parental activity effects. Proc R Soc Lond B 267: 2287–2293 [Chap 6]

Martindale S (1982) Nest defense and central place foraging: a model and experiment. Behav Ecol Sociobiol 10: 85–89 [Chap 8]

Martins EP (1993) A comparative study of the evolution of *Sceloporus* push-up displays. Am Nat 142: 994–1018 [Chap 7]

Martins EP (1994) Phylogenetic perspectives on the evolution of lizard territoriality. Pp 117–144 in Vitt LJ & Pianka ER (eds) Lizard Ecology: Historical and Experimental Perspectives. Princeton Univ Press, Princeton, NJ [Chap 7]

Martins EP (1996a) Phylogenies, spatial autoregression, and the comparative method: a computer simulation test. Evolution 50: 1750–1765 [Chap 7]

Martins EP (1996b) Phylogenies and the Comparative Method in Animal Behaviour. Oxford Univ PressNew York [Chap 7]

Martins EP (1999) Estimation of ancestral states of continuous characters: a computer simulation study. Syst Biol 48: 642–650 [Chap 7]

Martins EP (2000) Adaptation and the comparative method. Trends Ecol Evol 15: 296–299 [Chap 7]

Martins EP, Diniz-Filho JA, & Housworth EA (2002) Adaptive constraints and the phylogenetic comparative method: a computer simulation study. Evolution 56: 1–13 [Chap 7]

Martins EP & Hansen TF (1997) Phylogenies and the comparative method: a general approach to incorporating phylogenetic information into the analysis of interspecific data. Am Nat 149: 646–667. ERRATUM Am Nat 153: 448 [Chap 7]

Martins EP & Housworth EA (2002) Phylogeny shape and the phylogenetic comparative method. Syst Biol 51: 873–880 [Chap 7]

Martins EP & Lamont J (1998) Estimating ancestral states of a communicative display: a comparative study of *Cyclura* rock iguanas. Anim Behav 55: 1685–1706 [Chap 7]

Mateo JM (2002) Kin-recognition abilities and nepotism as a function of sociality. Proc R Soc Lond B 269: 721–727 [Chap 18]

Matsumasa M & Murai M (2005) Changes in blood glucose and lactate levels of male fiddler crabs: effects of aggression and claw waving. Anim Behav 69: 569–577 [Chap 15]

Maupin JL & Riechert SE (2001) Superfluous killing in spiders: a consequence of adaptation to food-limited environments? Behav Ecol 12: 569–576 [Chap 30]

Maynard Smith J (1964) Group selection and kin selection. Nature 201: 1145–1147 [Chap 1]

Maynard Smith J (1974) The theory of games and the evolution of animal conflicts. J Theor Biol 47: 209–221 [Chap 15]

Maynard Smith J (1978) Optimization theory in evolution. Annu Rev Ecol Syst 9: 31–56 [Chap 2]

Maynard Smith J (1982) Evolution and the Theory of Games. Cambridge Univ Press [Chaps 3, 8, 11, 16, 25]

Maynard Smith J (1991) Honest signaling: the Philip Sidney game. Anim Behav 42: 1034–1035 [Chap 16]

Maynard Smith J (1994) Must reliable signals always be costly? Anim Behav 47: 1115–1120 [Chap 16]

Maynard Smith J & Harper D (2003) Animal Signals. Oxford Univ Press, New York [Chaps 12, 16]

Maynard Smith J & Parker GA (1976) The logic of asymmetric contests. Anim Behav 24: 159–175 [Chap 15]

Maynard Smith J & Price GR (1973) The logic of animal conflict. Nature 246: 15–18 [Chaps 1, 15]

Maynard Smith J & Szathmary E (1995) The Major Transitions in Evolution. Oxford Univ Press, New York [Chap 18]

Mayr E (1942) Systematics and the Origin of Species. Columbia Univ Press [Chap 27]

Mayr E (1963) Animal Species and Evolution. Belknap Press, Cambridge MA [Chap 27]

Mayr E (1964) The evolution of living systems. Proc Natl Acad Sci USA 51: 934–941 [Chap 14]

Mayr E (1974) Behavior programs and evolutionary strategies. Amer Sci 62: 650–659 [Chaps 14, 29]

Mayr E (1976a) Evolution and the diversity of life. Belknap Press, Cambridge, MA [Chap 29]

Mayr E (1976b) Cause and effect in biology. Pp 359–371 in Mayr E (ed) Evolution and the Diversity of Life. Harvard Univ Press, Cambridge, MA [Chap 31]

Mayr E (1982) The Growth of Biological Thought. Belknap Press, Cambridge, MA [Chap 1]

McAuliffe K & Whitehead H (2005) Eusociality, menopause and information in matrilineal whales. Trends Ecol Evol 20: 650–650 [Chap 4]

McCartney MA & Lessios HA (2002) Quantitative analysis of gametic incompatibility between closely related species of neotropical sea urchins. Biol Bull 202: 166–181 [Chap 21]

McCrae RR & Costa PT Jr (1999) A five-factor theory of personality. Pp 139–153 in Pervin

LA & John OP (eds) Handbook of Personality: Theory and Research, 2nd ed. Guilford Press, New York [Chap 30]

McEdward L (1995) Ecology of Marine Invertebrate Larvae. CRC Press, Boca Raton, FL [Chap 21]

McElreath R, Luttbeg B, Fogarty SP, Brodin T, & Sih A (2007) Evolution of animal personalities. Nature 450: E5 [Chap 30]

McElreath R & Strimling P (2006) How noisy information and individual asymmetries can make "personality" an adaptation: A simple model. Anim Behav 72: 1135–1139 [Chap 30]

McGlothlin JW, Parker PG, Nolan V, & Ketterson ED (2005) Correlational selection leads to genetic integration of body size and an attractive plumage trait in dark-eyed juncos. Evolution 59: 658–671 [Chap 3]

McGraw JB & Caswell H (1996) Estimation of individual fitness from life-history data. Am Nat 147: 47–64 [Chap 4]

McGraw LA, Gibson G, Clark AG, & Wolfner MF (2004) Genes regulated by mating, sperm, or seminal proteins in mated female *Drosophila melanogaster*. Curr Biol 14: 1509–1514 [Chap 22]

McGregor AP, Orgogozo V, Delon I, Zanet J, Srinivasan DG, Payre F, & Stern DL (2007) Morphological evolution through multiple cis-regulatory mutations at a single gene. Nature 448: 587–590 [Chap 5]

McGregor PK (1993). Signalling in territorial systems: a context for individual identification, ranging and eavesdropping. Phil Trans R Soc Lond B 340: 237–244 [Chap 12]

McGregor PK (2005) Animal Communication Networks. Cambridge Univ Press, Cambridge [Chap 17]

McGuigan K, Van Homrigh A, & Blows MW (2008) Genetic analysis of female preference functions as function-valued traits. Am Nat 172: 194–202 [Chap 27]

McGuire B, Getz LL, & Oli MK (2002) Fitness consequences of sociality in prairie voles, *Microtus ochrogaster*: influence of group size and composition. Anim Behav 64: 645–654 [Chap 17]

McKinnon JS & Rundle HD (2002) Speciation in nature: the threespine stickleback model systems. Trends Ecol Evol 17: 480–488 [Chap 27]

McLean IG, Lundie-Jenkins G, & Jarman PJ (1996) Teaching an endangered mammal to recognise predators. Biol Conserv 75: 51–62 [Chap 29]

McNair JN (1979) A generalized model of optimal diets. Theor Popul Biol 15: 159–170 [Chap 8]

McNair JN (1983) A class of patch-use strategies. Amer Zool 23: 303–313 [Chap 8]

McNamara JM (1996) Risk-prone behaviour under rules which have evolved in a changing environment. Am Zool 36: 484–495 [Chap 12]

McNamara JM & Houston AI (1986) The common currency for behavioral decisions. Am Nat 127: 358–378 [Chap 13]

McNamara JM & Houston AI (1987) Partial preferences and foraging. Anim Behav 35: 1084–1099 [Chap 11]

McNamara JM & Houston AI (2008) Optimal annual routines: behaviour in the context of physiology and ecology. Phil Trans R Soc B 363: 301–319 [Chap 15]

McNamara JM, Houston AI, Barta Z, & Osorno J-L (2003) Should young ever be better off with one parent than with two? Behav Ecol 14: 301–310 [Chap 8]

McNamara JM, Stephens PA, Dall SRX, & Houston A (2009) Evolution of trust and trustworthiness: social awareness favours personality differences. Proc R Soc Lond B 276: 605–613 [Chap 30]

McPhail JD (1994) Speciation and the evolution of reproductive isolation in the sticklebacks (*Gasterosteus*) of south-western British Columbia. Pp 399–437 in Bell MA & Foster SA (eds) The Evolutionary Biology of the Threespine Stickleback. Oxford Univ Press, New York [Chap 27]

McVean GA & Spencer CC (2006) Scanning the human genome for signals of selection. Curr Opin Genet Dev 16: 624–629 [Chap 28]

Mead LS & Arnold SJ (2004) Quantitative genetic models of sexual selection. Trends Ecol Evol 19: 264–271 [Chap 24]

Meaney MJ & Szyf M (2005) Maternal care as a model for experience-dependent chromatin plasticity? Trends Neurosci 28: 456–463 [Chap 28]

Meffert LM (1995) Bottleneck effects on genetic variance for courtship repertoire. Genetics 139: 365–374 [Chap 14]

Mehdiabadi NJ, Jack CN, Farnham TT, Platt TG, Kalla SE, Shaulsky G, Queller DC, & Strassmann JS (2006) Kin preference in a social microbe. Nature 442: 881–882 [Chaps 18, 19]

Meiklejohn CD & Townsend JP (2005) A Bayesian method for analysing spotted microarray data. Brief Bioinform 6: 318–330 [Chap 28]

Mello CV, Vicario DS, & Clayton DF (1992) Song presentation induces gene expression in the songbird forebrain. Proc Natl Acad Sci USA 89: 6818–6822 [Chap 28]

Merilä J & Sheldon BC (1999) Genetic architecture of fitness and nonfitness traits: empirical patterns and development of ideas. Heredity 83: 103–109 [Chap 4]

Merilä J & Sheldon BC (2000) Lifetime reproductive success and heritability in nature. Am Nat 155: 301–310 [Chap 4]

Merilaita S (2006) Frequency-dependent predation and maintenance of prey polymorphism. J Evol Biol 19: 2022–2030 [Chap 12]
Méry F, Belay AT, So AKC, Sokolowski MB, & Kawecki TJ (2007) Natural polymorphism affecting learning and memory in *Drosophila*. Proc Natl Acad Sci USA 104: 13051–13055 [Chaps 5, 10]
Méry F & Kawecki TJ (2005) A cost of long-term memory in *Drosophila*. Science 308: 1148–1148 [Chap 10]
Messina FJ (2002) Host discrimination by seed parasites. Pp 65–87 in Lewis EE, Campbell JF, & Sukhdeo MVK (eds) The Behavioural Ecology of Parasites. CAB International, New York [Chap 11]
Messina FJ (2004) Predictable modification of body size and competitive ability following a host shift by a seed beetle. Evolution 58: 2788–2797 [Chap 11]
Mesterton-Gibbons M (1999) On the evolution of pure winner and loser effects: a game-theoretic model. Bull Math Biol 61: 1151–1186 [Chap 17]
Mesterton-Gibbons M, Marden JH, & Dugatkin LA (1996) On wars of attrition without assessment. J Theor Biol 181: 65–83 [Chap 15]
Metcalfe NB & Monaghan P (2001) Compensation for a bad start: grow now, pay later? Trends Ecol Evol 16: 254–260 [Chaps 12, 30]
Metzker ML (2005) Emerging technologies in DNA sequencing. Genome Res 15: 1767–1776 [Chap 28]
Meyer A, Morrissey JM, & Schartl M (1994) Recurrent origin of a sexually selected trait in *Xiphophorus* fishes inferred from a molecular phylogeny. Nature 368: 539–542 [Chap 7]
Michod RE & Roze D (2001) Coopration and conflict in the evolution of multicellularity. Heredity 86: 1–7 [Chap 18]
Mikics E, Kruk MR, & Haller J (2004) Genomic and non-genomic effects of glucocorticoids on aggressive behavior in male rats. Psychoneuroendochronology 29: 618–635 [Chap 15]
Miles DB, Sinervo B, Hazard LC, Svensson EI, & Costa D (2007) Relating endocrinology, physiology, and behaviour using species with alternative mating strategies. Func Ecol 21: 653–665 [Chap 3]
Milinski, M (1979) An evolutionarily stable feeding strategy in sticklebacks. Zeitschrift für Tierpsychologie 51: 36–40 [Chap 2]
Milinski M, Boltshauser P, Buchi L, Buchwalder T, Frischknecht M, Hadermann T, Kunzler R, Roden C, Ruetschi A, Strahm D, & Tognola M (1995) Competition for food in swans: an experimental test of the truncated phenotype distribution. J Anim Ecol 64: 758–766 [Chap 11]
Miller GT & Pitnick S (2002) Sperm-female coevolution in *Drosophila*. Science 298: 1230–1233 [Chap 22]
Miller KB (2001) On the phylogeny of the Dytiscidae (Insecta: Coleoptera) with emphasis on the morphology of the female reproductive system. Insect Syst Evol 32: 45–92 [Chap 22]
Miller KB, Wolfe GW, & Biström O (2006) The phylogeny of the Hydroporinae and classification of the genus *Peschetius* Guignot 1942 (Coleoptera: Dytiscidae). Insect Syst Evol 37: 257–279 [Chap 22]
Miller PJO, Biassoni N, Samuels A, & Tyack P (2000) Whale songs lengthen in response to sonar. Nature 405: 903 [Chap 9]
Miller RC (1922) The significance of the gregarious habit. Ecology 3: 122–126 [Chap 17]
Miller RL (1985) Demonstration of sperm chemotaxis in Echinodermata: Asteroidea, Holothuroidea, Ophiuroidea. J Exp Zool 234: 383–414 [Chap 21]
Mills SC & Reynolds JD (2003) Operational sex ratio and alternative reproductive behaviours in the European bitterling, *Rhodeus sericeus*. Behav Ecol Sociobiol 54: 98–104 [Chap 20]
Mills SC, Hazard LC, Lancaster L, Mappes T, Miles DB, Oksanen TA, & Sinervo B (2008) Gonadotropin hormone modulation of testosterone, immune function, performance and behavioral trade-offs among male morphs of the lizard, *Uta stansburiana*. Amer Nat 171: 339–357 [Chap 3]
Miyatake T & Shimizu T (1999) Genetic correlations between life-history and behavioral traits can cause reproductive isolation. Evolution 53: 201–208 [Chap 27]
Mock DW, Lamey TC, & Thompson DBA (1988) Falsifiability and the information centre hypothesis. Ornis Scand 19: 231–248 [Chap 17]
Moczek AP (1998) Horn polyphenism in the beetle Onthophagus taurus: larval diet quality and plasticity in parental investment determine adult body size and male horn morphology. Behav Ecol 9: 636–641 [Chap 25]
Møller AP & Birkhead TR (1993) Cuckoldry and sociality: a comparative study of birds. Am Nat 142: 118–140 [Chap 25]
Møller AP & Jennions MD (2001) How important are direct benefits of sexual selection? Naturwissenschaften 88: 401–415 [Chap 26]
Møller AP & Pomiankowski A (1993) Why have birds got multiple sexual ornaments? Behav Ecol Sociobiol 32: 167–176 [Chap 16]
Møller AP & Thornhill R (1998) Male parental care, differential parental investment by females and sexual selection. Anim Behav 55: 1507–1515 [Chap 26]
Moller P (1995) Electric fishes history and behavior. Chapman & Hall, London [Chap 9]

Molles LE & Vehrencamp SL (2001) Songbird cheaters pay a retaliation cost: evidence for auditory conventional signals. Proc R Soc Lond B 268: 2013–2019 [Chap 16]

Monagahan P (2008) Early growth conditions, phenotypic development and environmental change. Phil Trans R Soc Lond 363: 1635–1645 [Chap 1]

Monaghan P & Haussmann MF (2006) Do telomere dynamics link lifestyle and lifespan? Trends Ecol Evol 21: 47–53 [Chap 1]

Monaghan P & Metcalfe NB (1985) Group foraging in wild brown hares: effects of resource distribution and social status. Anim Behav 33: 993–999 [Chap 15]

Montagu G (1802) Ornithological Dictionary. White, London [Chap 1]

Moore AJ & Boake CRB (1994) Optimality and evolutionary genetics: complementary procedures for evolutionary analysis in behavioral ecology. Trends Ecol Evol 9: 69–72 [Chaps 5, 14]

Moore AJ, Brodie ED III, & Wolf JB (1997) Interacting phenotypes and the evolutionary process: I. direct and indirect genetic effects of social interactions. Evolution 51: 1352–1362 [Chap 14]

Moore AJ, Haynes KF, Preziosi RF, & Moore PJ (2002) The evolution of interacting phenotypes: genetics and evolution of social dominance. Am Nat 160: S186–S197 [Chap 14]

Moore AJ, Wolf JB, & Brodie ED III (1998) The influence of direct and indirect genetic effects on the evolution of behavior: social and sexual selection meet maternal effects. Pp 22–41 in Mousseau TA & Fox CW (eds) Maternal Effects as Adaptations. Oxford Univ Press, New York [Chap 14]

Moore BD & Foley WJ (2005) Tree use by koalas in a chemically complex landscape. Nature 435: 488–490 [Chap 10]

Moore HDM, Martin M, & Birkhead TR (1999) No evidence for killer sperm or other selective interactions between human spermatozoa in ejaculates of different males in vitro. Proc R Soc Lond B 266: 2343–2350 [Chap 22]

Moore J & Ali R (1984) Are dispersal and inbreeding related? Anim Behav 32: 94–112 [Chap 31]

Moore SL & Wilson K (2002) Parasites as a viability cost of sexual selection in natural populations of mammals. Science 297: 2015–2018 [Chap 20]

Moran NA (1992). The evolutionary maintenance of alternative phenotypes. Amer Nat 139: 971–989 [Chap 6]

Moretz JA, Martins EP, & Robison BD (2007) Behavioral syndromes and the evolution of correlated behavior in zebrafish. Behav Ecol 18: 556–562 [Chap 7]

Moretz JA & Morris MR (2006) Phylogenetic analysis of the evolution of a signal of aggressive intent in northern swordtail fishes. Am Nat 168: 336–349 [Chap 7]

Morgan SG (1995) The timing of larval release. Pp 157–191 in McEdward L (ed) Ecology of Marine Invertebrate Larvae. CRC Press, Boca Raton, FL [Chap 21]

Morrell LJ, Backwell PRY, & Metcalfe NB (2005) Fighting in fiddler crabs *Uca mjoebergi*: what determines duration? Anim Behav 70: 653–662 [Chap 15]

Morrell LJ & James R (2008) Mechanisms for aggregation in animals: rule success depends on ecological variables. Behav Ecol 19: 193–201 [Chap 17]

Morris D (1951) Homosexuality in the ten-spined stickleback (*Pygosteus pungitius* L.). Behaviour 4: 233–261 [Chap 25]

Morris MR, Nicoletto PF, & Hesselman E (2003) A polymorphism in female preference for a polymorphic male trait in the swordtail fish *Xiphophorus cortezi*. Anim Behav 65: 45–52 [Chap 30]

Morris MR, Tudor MS, & Dubois NS (2007) Sexually selected signal attracted females before deterring aggression in rival males. Anim Behav 74: 1189–1197 [Chap 7]

Morse RA & Nowogrodski R (1990) Honey Bee Pests, Predators and Diseases. Cornell Univ Press, Ithaca, NY [Chap 28]

Mottley K & Giraldeau LA (2000) Experimental evidence that group foragers can converge on predicted producer-scrounger equilibria. Anim Behav 60: 341–350 [Chaps 11, 30]

Mougeot F, Dawson A, Redpath SM, & Leckie F (2005) Testosterone and autumn territorial behaviour in male red grouse *Lagopus lagopus scoticus*. Horm Behav 47: 576–584 [Chap 15]

Mousseau TA & Roff DA (1987) Natural selection and the heritability of fitness components. Heredity 59: 181–197 [Chap 4]

Mrosovsky N & Carr A (1967) Preference for light of short wavelengths in hatchling green sea turtles, *Chelonia mydas*, tested on their natural nesting beaches. Behaviour 28: 217–231 [Chap 29]

Mueller JL, Ram KR, McGraw LA, Qazi MCB, Siggia ED, Clark AG, Aquadro CF, & Wolfner MF (2005) Cross-species comparison of *Drosophila* male accessory gland protein genes. Genetics 171: 131–143 [Chaps 23, 28]

Mueller JL, Ripoll DR, Aquadro CF, & Wolfner MF (2004) Comparative structural modeling and inference of conserved protein classes in Drosophila seminal fluid. Proc Natl Acad Sci USA 101: 13542–13547 [Chap 28]

Mueller UG & Gerardo N (2002) Fungus-farming insects: multiple origins and diverse

evolutionary histories. Proc Natl Acad Sci USA 99: 15247–15249 [Chap 12]
Mueller UG, Schultz TR, Currie CR, Adams RMM, & Malloch D (2001) The origin of the attine ant-fungus mutualism. Q Rev Biol 76: 169–197 [Chap 12]
Muir WM (1996) Group selection for adaptation to multiple-hen cages: selection program and direct responses. Poult Sci 75: 447–458 [Chap 14]
Muir WM (2005) Incorporation of competitive effects in forest tree or animal breeding programs. Genetics 170: 1247–1259 [Chap 14]
Mundy NI, Badcock NS, Hart T, Scribner K, Janssen K, & Nadeau NJ (2004) Conserved genetic basis of a quantitative plumage trait involved in mate choice. Science 303: 1870–1873 [Chap 5]
Myerson RB (1997) Game theory: Analysis of conflict. Harvard Univ Press, Cambridge MA [Chap 16]
Neat FC & Mayer I (1999) Plasma concentrations of sex steroids and fighting in male *Tilapia zillii*. J Fish Biol 54: 695–697 [Chap 15]
Neff BD (2003) Decisions about parental care in response to perceived paternity. Nature 422: 716–719 [Chap 26]
Neff BD, Fu P, & Gross MR (2003) Sperm investment and alternative mating tactics in bluegill sunfish (*Lepomis machrochirus*). Behav Ecol 14: 634–641 [Chap 21]
Neff BD & Pitcher TE (2005) Genetic quality and sexual selection: an integrated framework for good genes and compatible genes. Mol Ecol 14: 19–38 [Chaps 21, 22]
Neff BD & Sherman PW (2002) Decision making and recognition mechanisms. Proc R Soc Lond B 269: 1435–1441 [Chap 29]
Nesse RM (2001) Evolution and the Capacity for Commitment. Russell Sage Foundation, New York [Chap 12]
Neubaum DM & Wolfner MF (1999) Wise, winsome, or weird? Mechanisms of sperm storage in female animals. Curr Top Devel Biol 41: 67–97 [Chap 22]
Nicolakakis N, Sol D, & Lefebvre L (2003) Behavioural flexibility predicts species richness in birds, but not extinction risk. Anim Behav 65: 445–452 [Chap 30]
Noble J (1999) Cooperation, conflict and the evolution of communication. Adaptive Behavior 7: 349–370 [Chap 16]
Nonacs P (2001a) State dependent behavior and the marginal value theorem. Behav Ecol 12: 71–83 [Chaps 11, 13]
Nonacs P (2001b) A life-history approach to group living and social contracts between individuals. Ann Zool Fenn 38: 239–254 [Chap 17]
Nonacs P (2006) The rise and fall of transactional skew theory in the model genus *Polistes*. Ann Zool Fenn 43: 443–455 [Chap 17]
Nonacs P (2007) Tug-of-war has no borders: it is the missing model in reproductive skew theory. Evolution 61: 1244–1250 [Chap 17]
Nonacs P & Dill LM (1990) Mortality risk vs. food quality trade-offs in a common currency: ant patch preferences. Ecology 71: 1886–1892 [Chap 13]
Nonacs P & Dill LM (1991) Mortality risk versus food quality trade-offs in ants: patch use over time. Ecol Entomol 16: 73–80 [Chap 13]
Nonacs P & Dill LM (1993) Is satisficing an alternative to optimal foraging theory? Oikos 24: 371–375 [Chap 13]
Nonacs P & Kapheim KM (2007) Social heterosis and the maintenance of genetic diversity. J Evol Biol 20: 2253–2265 [Chaps 13, 17]
Nonacs P & Kapheim KM (2008) Social heterosis and the maintenance of genetic diversity at the genome level. J Evol Biol 21: 631–635 [Chap 13]
Nonacs P, Liebert AE, & Starks PT (2006) Transactional skew and assured fitness return models fail to predict patterns of cooperation in wasps. Am Nat 167: 467–480 [Chap 17]
Nonacs P, Reeve HK, & Starks PT (2004) Optimal reproductive-skew models fail to predict aggression in wasps. Proc R Soc Lond B 271: 811–817 [Chap 17]
Normark BB (2003) The evolution of alternative genetic systems in insects. Annu Rev Entomol 48: 397–423 [Chap 3]
Nosil P (2007) Divergent host-plant adaptation and reproductive isolation between ecotypes of *Timema cristinae* walking-sticks. Am Nat 169: 151–162 [Chap 27]
Nowak MA & Sigmund K (1998) The dynamics of indirect reciprocity. J Theor Biol 194: 561–574 [Chap 12]
Nuechterlein GL & Storer RW (1982) Pair formation displays of the western grebe. Condor 84: 350–369 [Chap 2]
Nufio CR & Papaj DR (2004) Superparasitism of larval hosts by the walnut fly, *Rhagoletis juglandis*, and its implications for female and offspring performance. Oecologica 141: 460–467 [Chap 11]
Nunes S (2007) Dispersal and philopatry. Pp 150–162 in Wolff J & Sherman PW (eds) Rodent Societies. Univ Chicago Press, Chicago [Chap 29]
Nunn CL & Barton RA (2001) Comparative methods for studying primate adaptation and allometry. Evol Anthropol 10: 81–98 [Chap 7]
Nunney L (2000) The limits to knowledge in conservation genetics: the value of effective population size. Evol Biol 32: 179–194 [Chap 29]
Nunney L (2003) The cost of natural selection revisited. Ann Zool Fenn 40: 185–194 [Chap 29]
Nussey DH, Postma E, Gienapp P, & Visser ME (2005) Selection on heritable phenotypic

plasticity in a wild bird population. Science 310: 304–306 [Chaps 6, 29]
Nussey DH, Wilson AJ, & Brommer JE (2007) The evolutionary ecology of individual phenotypic plasticity in wild populations. J Evol Biol 20: 831–844 [Chaps 6, 30]
Ode PJ & Hunter MS (2002) Sex ratios of parasitic Hymenoptera with unusual life-histories. Pp 218–234 in Hardy ICW (ed) Sex Ratios: Concepts and Research Methods. Cambridge Univ Press, Cambridge [Chap 5]
Odling-Smee FJ, Laland KN, & Feldman MW (2003) Niche Construction: The Neglected Process in Evolution. Princeton Univ Press, Princeton, NJ [Chap 12]
Oldham MC, Horvath S, & Geschwind DH (2006) Conservation and evolution of gene coexpression networks in human and chimpanzee brains. Proc Natl Acad Sci USA 103: 17973–17978 [Chap 28]
Oldroyd BP & Beekman M (2008) Effects of selection for honey bee worker reproduction on foraging traits. PLoS Biol 6: 463–470 [Chap 28]
Oldroyd BP & Fewell JH (2007) Genetic diversity promotes homeostasis in insect colonies. Trends Ecol Evol 22: 408–413 [Chap 28]
Oli MK (2003) Hamilton goes empirical: estimation of inclusive fitness from life-history data Proc R Soc Lond B 270: 307–311 [Chap 18]
Oliveira RF, Lopes M, Carneiro LA, & Canario AVM (2001) Watching fights raises fish hormone levels—cichlid fish wrestling for dominance induces an androgen surge in male spectators. Nature 409: 475 [Chap 15]
Oliveira RF, Taborsky M, & Brockman HJ (eds) (2008) Alternative Reproductive Tactics: An Integrative Approach. Cambridge Univ Press, Cambridge [Chaps 13, 25]
Oliver J & Babcock R (1992) Aspects of the fertilization ecology of broadcast spawning corals—sperm dilution effects and in situ measurements of fertilization. Biol Bull 183: 409–417 [Chap 21]
Olson EC & Miller RL (1958) Morphological Integration. Univ Chicago Press, Chicago [Chap 30]
Olsson O, Brown JS, & Smith HG (2001) Gain curves in depletable food patches: a test of five models with European starlings. Evol Ecol Res 3: 285–310 [Chap 11]
Omland KE, Cook LG, & Crisp MD (2008) Tree thinking for all biology: the problem with reading phylogenies as ladders of progress. BioEssays 30: 854–867 [Chap 7]
Ophir AG, Persaud KN, & Galef BG (2005) Avoidance of relatively aggressive male Japanese quail (*Coturnix japonica*) by sexually experienced conspecific females. J Comp Psychol 119: 3–7 [Chap 30]
Ophir AG, Wolff JO, & Phelps SM (2008) Variation in neural V1aR predicts sexual fidelity and space use among male prairie voles in semi-natural settings. Proc Natl Acad Sci USA 105: 1249–1254 [Chap 5]
Ord TJ & Blumstein DT (2002) Size constraints and the evolution of display complexity: why do large lizards have simple displays? Biol J Linn Soc 76: 145–161 [Chap 7]
Ord TJ, Blumstein DT, & Evans CS (2001) Intrasexual selection predicts the evolution of signal complexity in lizards. Proc R Soc Lond B 268: 737–744 [Chap 7]
Ord TJ, Blumstein DT, & Evans CS (2002) Ecology and signal evolution in lizards. Biol J Linn Soc 77: 127–148 [Chap 7]
Ord TJ & Martins EP (2006) Tracing the origins of signal diversity in anole lizards: phylogenetic approaches to inferring the evolution of complex behaviour. Anim Behav 71: 1411–1429 [Chap 7]
Ord TJ & Stuart-Fox D (2006) Ornament evolution in dragon lizards: multiple gains and widespread losses reveal a complex history of evolutionary change. J Evol Biol 19: 797–808 [Chap 7]
Orians GH (1962) Natural selection and ecological theory. Am Nat 96: 257–263 [Chap 1]
Orians GH (1969) On the evolution of mating systems in birds and mammals. Am Nat 103: 589–603 [Chap 1]
Orians GH & Pearson NE (1979) On the theory of central place foraging. Pp 155–177 in Horn DH, Mitchell R, & Stairs GR (eds) Analysis of Ecological Systems. Ohio State Univ Press, Columbus [Chap 8]
Orr HA (2005) The genetic theory of adaptation: a brief history. Nat Rev Genet 6: 119–127 [Chap 5]
Orr HA (2007) Absolute fitness, relative fitness and utility. Evolution 61: 2997–3000 [Chap 4]
Orteiza N, Linder JE, & Rice WR (2005) Sexy sons from re-mating do not recoup the direct costs of harmful male interactions in the *Drosophila melanogaster* laboratory model system. J Evol Biol 18: 1315–1323 [Chap 23]
Orzack SH (1990) The comparative biology of second sex ratio evolution within a natural population of a parasitic wasp, *Nasonia vitripennis*. Genetics 124: 385–396 [Chap 2]
Orzack SH, Parker ED Jr & Gladstone J (1991) The comparative biology of genetic variation for conditional sex ratio behavior in a parasitic wasp, *Nasonia vitripennis*. Genetics 127: 583–599 [Chap 2]
Orzack SH & Sober E (1994a) Optimality models and the test of adaptationism. Am Nat 143: 361–380 [Chap 2]
Orzack SH & Sober E (1994b) How (not) to test an optimality model. Trends Ecol Evol 9: 265–267 [Chap 2]

Orzack SH & Sober E (2001a) Adaptationism and Optimality. Cambridge Univ Press, Cambridge [Chap 2]

Orzack SH & Sober E (2001b) Adaptation, phylogenetic inertia, and the method of controlled comparisons. Pp 45–63 in Orzack SH & Sober E (eds) Adaptationism and Optimality. Cambridge Univ Press, Cambridge [Chap 2]

Osborne KA, Robichon A, Burgess E, Butland S, Shaw RA, Coulthard A, Pereira HS, Greenspan RJ, & Sokolowski MB (1997) Natural behaviour polymorphism due to a cGMP-dependent protein kinase of *Drosophila*. Science 277: 834–836 [Chap 5]

Osorio D & Vorobyev M (2005) Photoreceptor spectral sensitivities in terrestrial animals: adaptations for luminance and color vision. Proc R Soc Lond B 272: 1745–1752 [Chap 9]

Öst M, Clark CW, Kilpi M, & Ydenberg RC (2007) Parental effort and reproductive skew in coalitions of brood rearing female common eiders. Am Nat 169: 73–86 [Chap 8]

Öst M, Vitikainen E, Waldeck P, Sundström L, Lindström K, Hollmén T, Franson JC, & Kilp M (2005) Eider females form non-kin brood-rearing coalitions. Mol Ecol 14: 3903–3908 [Chap 26]

Öst M, Ydenberg RC, Lindstrom K, & Kilpi M (2003) Body condition and the grouping behaviour of brood-caring female common eiders (*Somateria mollissima*). Behav Ecol Sociobiol 54: 451–457 [Chap 8]

Östlund S & Ahnesjö I (1998) Female fifteen-spined sticklebacks prefer better fathers. Anim Behav 56: 1177–1183 [Chap 26]

Otto SP, Servedio MR, & Nuismer SL (2008) Frequency-dependent selection and the evolution of assortative mating. Genetics 179: 2091–2112 [Chap 27]

Owens IPF (2006) Where is behavioral ecology going? Trends Ecol Evol 21: 356–361 [Chap 1]

Owen-Smith N (1993) Age, size dominance, and reproduction among male Kudus: mating enhancement by attrition of rivals. Behav Ecol Sociobiol 32: 177–184 [Chap 20]

Oyegbile TO & Marler CA (2005) Winning fights elevates testosterone levels in California mice and enhances future ability to win fights. Horm Behav 48: 259–267 [Chap 15]

Packer C (1979) Inter-troop transfer and inbreeding avoidance in *Papio anubis*. Anim Behav 27: 1–36 [Chap 31]

Packer C & Ruttan L (1988) The evolution of cooperative hunting. Am Nat 132: 159–198 [Chaps 11, 17]

Page RA & Ryan MJ (2005) Flexibility in assessment of prey cues: frog-eating bats and frog calls. Proc R Soc Lond B 272: 841–847 [Chap 29]

Pagel M (1993) Honest signaling among gametes. Nature 363: 539–541 [Chap 16]

Pagel M (1994) Detecting correlated evolution on phylogenies: a general method for the comparative analysis of discrete characters. Proc R Soc Lond B 255: 37–45 [Chap 7]

Pagel M (1997) Inferring evolutionary processes from phylogenies. Zoologica Scripta 26: 331–348 [Chap 7]

Pagel M (1999) Inferring the historical patterns of biological evolution. Nature 401: 877–884 [Chap 7]

Pagel M, Venditti C, & Meade A (2006) Large punctuational contribution of speciation to evolutionary divergence at the molecular level. Science 314: 119–121 [Chap 7]

Paley W (1802) Natural Theology, 2nd ed. Vincent, Oxford [Chap 1]

Palmer AR (1999) Detecting publication bias in meta-analyses: a case study of fluctuating asymmetry and sexual selection. Am Nat 154: 220–233 [Chap 1]

Palumbi SR (1999) All males are not created equal: fertility differences depend on gamete recognition polymorphisms in sea urchins. Proc Natl Acad Sci USA 96: 12632–12637 [Chap 21]

Palumbi SR (2001) Humans as the world's greatest evolutionary force. Science 293: 1786–1790.

Pampoulie C, Lindström K, & St Mary CM (2004) Have your cake and eat it too: male sand gobies show more parental care in the presence of female partners. Behav Ecol 15: 199–204 [Chap 26]

Panhuis TM & Swanson WJ (2006) Molecular evolution and population genetic analysis of candidate female reproductive genes in *Drosophila*. Genetics 173: 2039–2047 [Chap 23]

Panksepp J & Panksepp JB (2000) The seven sins of evolutionary psychology. Evol Cogn 6: 108–131 [Chap 31]

Park M & Wolfner MF (1995) Male and female cooperate in the prohormone-like processing of a Drosophila melanogaster seminal fluid protein. Dev Biol 171: 694–702 [Chap 28]

Parker GA (1970a) Sperm competition and its evolutionary consequences in the insects. Biol Rev 45: 525–567 [Chaps 1, 22]

Parker GA (1970b) Sperm competition and its evolutionary effect on copula duration in the fly *Scatophaga stercoraria*. J Insect Physiol 16: 1301–1328 [Chaps 2, 22]

Parker GA (1974a) Courtship persistence and female guarding as male time investment strategies. Behaviour 48: 157–184 [Chaps 2, 22]

Parker GA (1974b) Assessment strategy and the evolution of fighting behaviour. J Theor Biol 47: 223–243 [Chap 15]

Parker GA (1978) Selection on non-random fusion of gametes during the evolution of anisogamy. J Theor Biol 73: 1–28 [Chap 20]

Parker GA (1979) Sexual selection and sexual conflict. Pp 123–166 in MS Blum & NA Blum (eds) Sexual Selection and Reproductive Competition in Insects. Academic Press, New York [Chaps 23, 24]
Parker GA (1982) Why are there so many tiny sperm? Sperm competition and the maintenance of two sexes. J Theor Biol 96: 281–294 [Chaps, 20, 22]
Parker GA (1992) Marginal value theorem with exploitation time costs: diet, sperm reserves, and optimal copula duration in dung flies. Am Nat 139: 1237–1256 [Chaps 11, 22]
Parker GA (1998) Sperm competition and the evolution of ejaculates: towards a theory base. Pp 3–54 in Birkhead TR & Møller AP (eds) Sperm Competition and Sexual Selection. Academic Press, London [Chap 22]
Parker GA (2001) Golden flies, sunlit meadows: a tribute to the yellow dungfly. Pp 3–26 in Dugatkin LA (ed) Model Systems in Behavioral Ecology: Integrating Conceptual, Theoretical, and Empirical Approaches. Princeton Univ Press, Princeton, NJ [Chap 1]
Parker GA (2006a) Behavioral ecology: the science of natural history. Pp 23–56 in Lucas JR & Simmons LW (eds) Essays on Animal Behavior: Celebrating 50 Years of Animal Behavior. Elsevier, Burlington, MA [Chap 1]
Parker GA (2006b) Sexual conflict over mating and fertilisation: an overview. Phil Trans R Soc Lond B 361: 235–260 [Chap 23]
Parker GA, Baker RR, & Smith VGF (1972) The origin and evolution of gamete dimorphism and the male-female phenomenon. J Theor Biol 36: 529–553 [Chap 20]
Parker GA & Maynard Smith J (1990) Optimality theory in evolutionary biology. Nature 348: 27–33 [Chap 1]
Parker GA & Partridge L (1998) Sexual conflict and speciation. Phil Trans R Soc Lond B 353: 261–274 [Chap 23]
Parker GA & Simmons LW (1991) A model of constant random sperm displacement during mating: evidence from *Scatophaga*. Proc R Soc Lond B 246: 107–115 [Chap 22]
Parker GA & Simmons LW (1996) Parental investment and the control of sexual selection: predicting the direction of sexual competition. Proc R Soc Lond B 263: 315–321 [Chap 20]
Parker GA, Simmons LW, Stockley P, McChristie DM, & Charnov EL (1999) Optimal copula duration in yellow dung flies: effects of female size and egg content. Anim Behav 57: 795–805 [Chaps 2, 22]
Parker GA & Stuart RA (1976) Animal behavior as a strategy optimizer: evolution of resource assessment strategies and optimal emigration thresholds. Am Nat 110: 1055–1076 [Chaps 2, 8]
Parker GA & Sutherland WJ (1986) Ideal free distributions when individuals differ in competitive ability: phenotype-limited ideal free models. Anim Behav 34: 1222–1242 [Chap 11]
Parri S, Alatalo RV, Kotiaho J & Mappes J, & Rivero A (2002) Sexual selection in the wolf spider *Hygrolycosa rubrofasciata*: female preference for drum duration and pulse rate. Behav Ecol 13: 615–621 [Chap 9]
Pärt T & Qvarnström A (1997) Badge size in collared flycatchers predicts outcome of male competition over territories. Anim Behav 54: 893–899 [Chap 24]
Partan SR & Marler P (2005) Issues in the classification of multimodal communication signals. Am Nat 166: 231–245 [Chap 10]
Partridge JC & Douglas RH (1995) Far-red sensitivity of dragonfish. Nature 375: 21–22 [Chap 9]
Patricelli GL, Coleman SW, & Borgia G (2006) Male satin bowerbirds, *Ptilonorhynchus violaceus,* adjust their display intensity in response to female startling: an experiment with robotic females. Anim Behav 71: 49–59 [Chap 30]
Pattarini JM, Starmer WT, Bjork A, & Pitnick S (2006) Mechanisms underlying the sperm quality advantage in *Drosophila melanogaster*. Evolution 60: 2064–2080 [Chap 22]
Patterson JT (1943) Studies in the genetics of *Drosophila*. III. The Drosophilidae of the southwest. Univ Texas Publ 4313: 7–203 [Chap 22]
Payne RJH (1998) Gradually escalating fights and displays: the cumulative assessment model. Anim Behav 56: 651–662 [Chap 15]
Payne RJH & Pagel M (1996) Escalation and time costs in displays of endurance. J Theor Biol 183: 185–193 [Chap 15]
Payne RJH & Pagel M (1997) Why do animals repeat displays? Anim Behav 54: 109–119 [Chap 15]
Paz-y-Miño CG, Bond AB, Kamil AC, & Balda RP (2004) Pinyon jays use transitive inference to predict social dominance. Nature 430: 778–781 [Chap 17]
Pearce JM (2008) Animal Learning and Cognition: An Introduction, 3rd ed. Psychology Press, East Sussex, UK [Chap 10]
Pellmyr O & Huth CJ (1994) Evolutionary stability of mutualism between yuccas and yucca moths. Nature 372: 257–260 [Chap 18]
Peluc SI, Sillett TS, Rotenberry JT, & Ghalambor CK (2008) Adaptive phenotypic plasticity in island songbird exposed to a novel predation risk. Behav Ecol 19: 830–835 [Chap 6]
Penn DJ & Potts WK (1999) The evolution of mating preferences and major histocompatibility complex genes. Am Nat 153: 145–164 [Chap 31]
Pennington JT (1985) The ecology of fertilization of echinoid eggs: the consequence of sperm

dilution, adult aggregation, and synchronous spawning. Biol Bull 169: 417–430 [Chap 21]

Penton-Voak IS, Perrett DI, Castles DL, Kobayashi T, Burt DM, Murray LK, & Minamisawa R (1999) Menstrual cycle alters face preference. Nature 399: 741–742 [Chap 31]

Perrone M Jr & Zaret TM (1979) Parental care patterns of fishes. Am Nat 113: 351–361 [Chap 26]

Perry DM (1987) Optimal diet theory: behavior of a starved predatory snail. Oecologia 72: 360–365 [Chap 8]

Perry G, LeVering K, Girard I, & Garland T Jr (2004) Locomotory performance and social dominance in male *Anolis cristatellus*. Anim Behav 67: 37–47 [Chap 4]

Peters A (2002) Testosterone and the trade-off between mating and paternal effort in extrapair-mating superb fairy-wrens. Anim Behav 64: 103–112 [Chap 15]

Peters JM, Queller DC, Imperatriz-Fonseca VL, Roubik DW, & Strassmann JE (1999) Mate number, kin selection and social conflicts in stingless bees and honey bees. Proc R Soc Lond B 266: 379–384 [Chap 19]

Petersen CW (1991) Sex allocation in hermaphroditic sea basses. Am Nat 138: 650–667 [Chap 21]

Petersen CW & Warner RR (1998) Sperm competition in fishes. Pp 435–464 in Birkhead TR & Møller AP (ed) Sperm Competition and Sexual Selection. Academic Press, London [Chaps 21, 26]

Petersen CW, Warner RR, Shapiro DY, & Marconato A (2001) Components of fertilization success in the bluehead wrasse, *Thalassoma bifasciatum*. Behav Ecol 12: 237–245 [Chap 21]

Petersen CW, Warner RR, Cohen S, Hess HC, & Sewell AT (1992) Variable pelagic fertilization success: implications for mate choice and spatial patterns of mating. Ecology 73: 391–401 [Chap 21]

Petfield D, Chenoweth SF, Rundle HD, & Blows MW (2005) Genetic variance in female condition predicts indirect genetic variance in male sexual display traits. Proc Natl Acad Sci USA 102: 6045–6050 [Chap 14]

Pettigrew JD, Manger PR, & Fine SLB (1998) The sensory world of the platypus. Phil Trans R Soc Lond B 353: 1199–1210 [Chap 9]

Phelps SM & Ryan MJ (1998) Neural networks predict response biases of female Túngara frogs. Proc R Soc Lond B 265: 279–185 [Chap 16]

Philippi T & Seger J (1989) Hedging one's evolutionary bets, revisited. Trends Ecol Evol 4: 41–44 [Chap 4]

Phillippi A, Hamann E, & Yund PO (2004) Fertilization in an egg-brooding colonial ascidian does not vary with population density. Biol Bull 206:152–160 [Chap 21]

Phillips BL & Shine R (2004) Adapting to an invasive species: toxic cane toads induce morphological change in Australian snakes. Proc Natl Acad Sci USA 101: 17150–17155 [Chap 29]

Phillips BL & R Shine (2006) An invasive species induces rapid adative change in a native predator: cane toads and black snakes in Australia. Proc R Soc Lond B 273: 1545–1550 [Chap 29]

Phillips BL, Brown GP, & Shine R (2003) Assessing the potential impact of cane toads on Australian snakes. Conserv Biol 17: 1738–1747 [Chap 29]

Phillips PC & Arnold SJ (1989) Visualizing multivariate selection. Evolution 43: 1209–1222 [Chaps 3, 30]

Pierce GJ & Ollason JG (1987) Eight reasons why optimal foraging theory is a complete waste of time. Oikos 49: 111–118 [Chap 2]

Piersma T & Drent J (2003) Phenotypic flexibility and the evolution of organismal design. Trends Ecol Evol 18: 228–233 [Chap 6]

Pietrewicz AT & Kamil AC (1979) Search image-formation in the blue jay (*Cyanocitta cristata*). Science 204: 1332–1333 [Chap 10]

Pigliucci M (2001) Phenotypic Plasticity: Beyond Nature and Nurture. Johns Hopkins Univ Press, Baltimore, MD [Chap 6]

Pigliucci M & Murren CJ (2003) Genetic assimilation and a possible evolutionary paradox: can macroevolution sometimes be so fast as to pass us by? Evolution 57: 1455–1464 [Chap 6]

Pigliucci M & Preston K (2004) Phenotypic Integration: Studying the Ecology and Evolution of Complex Phenotypes. Oxford Univ Press, New York [Chap 30]

Pike TW & Petrie M (2003) Potential mechanisms of avian sex determination. Biol Rev 78: 553–574 [Chap 5]

Pishcedda A & Chippindale AK (2006) Intralocus conflict diminishes the benefits of sexual selection. PLoS Biol 4: 2099–2103 [Chap 3]

Pitcher T (1992) Who dares, wins: the function and evolution of predator inspection behaviour in shoaling fish. Neth J Zool 42: 371–391 [Chap 30]

Pitcher TJ, Green DA, & Magurran AE (1986) Dicing with death—predator inspection behavior in minnow shoals. J Fish Biol 28: 439–448 [Chap 17]

Pitnick S, Markow TA, & Spicer GS (1999) Evolution of multiple kinds of female sperm-storage organs in *Drosophila*. Evolution 53: 1804–1822 [Chaps 22, 23]

Pitnick S, Miller GT, Reagan J, & Holland B (2001) Males' evolutionary responses to experimental removal of sexual selection. Proc R Soc Lond B 268: 1071–1080 [Chap 22]

Pitnick S, Miller GT, Schneider K, & Markow TA (2003) Ejaculate-female coevolution in *Drosophila mojavensis*. Proc R Soc Lond B 270: 1507–1512 [Chaps 22, 28]

Pitnick S, Hosken DJ & Birkhead TR (2009a) Sperm morphological diversity. Pp 69–149 in Birkhead TR, Hosken DJ, & Pitnick S (eds) Sperm Biology: An Evolutionary Perspective. Academic Press, London [Chap 22]

Pitnick S, Wolfner MF, & Suarez SS (2009b) Ejaculate- and sperm-female interaction. Pp. 247–304 in Birkhead TR, Hosken DJ, & Pitnick S (eds) Sperm Biology: An Evolutionary Perspective Academic Press, London [Chap 22]

Pleasants JM (1989) Optimal foraging by nectarivores—a test of the marginal-value theorem. Am Nat 134: 51–71 [Chap 11]

Plowes NJR & Adams ES (2005) An empirical test of Lanchester's square law: mortality during battles of the fire ant *Solenopsis invicta*. Proc R Soc Lond B 272: 1809–1814 [Chap 15]

Podolsky RD (2001) Evolution of egg target size: an analysis of selection on correlated characters. Evolution 55: 2470–2478 [Chap 21]

Podolsky RD (2004) Life history consequences of investment in free-spawned eggs and their accessory coats. Am Nat 163: 735–753 [Chap 21]

Podos J (2001) Correlated evolution of morphology and vocal signal structure in Darwin's finches. Nature 409: 185–188 [Chap 7]

Pollard KS, Salama SR, Lambert N, Lambot MA, Coppens S, Pedersen JS, Katzman S, King B, Onodera C, Siepel A, Kern AD, Dehay C, Igel H, Ares M, Vanderhaeghen P, & Haussler D (2006) An RNA gene expressed during cortical development evolved rapidly in humans. Nature 443: 167–172 [Chap 10]

Pollen AA & Hofmann HA (2008) Beyond neuroanatomy: novel approaches to studying brain evolution. Brain, Behav Evol 72: 145–158 [Chap 10]

Pomiankowski A (1988) The evolution of female mate preferences for male genetic quality. Oxford Survey Evolutionary Biology 5: 136–184 [Chap 24]

Postma E, Griffith SC, & Brooks R (2006) Evolution of mate choice in the wild. Nature 444: E16–17 [Chap 24]

Pottinger TG & Carrick TR (2001) Stress responsiveness affects dominant-subordinate relationships in rainbow trout. Horm Behav 40: 419–427 [Chap 15]

Pound N (1999) Effects of morphine on electrically evoked contractions of the vas deferens in two congeneric rodent species differing in sperm competition intensity. Proc R Soc Lond B 266: 1755–1758 [Chap 22]

Pratt SC, Mallon EB, Sumpter DJT, & Franks NR. (2002) Quorum sensing, recruitment, and collective decision-making by the ant *Leptothorax albipennis*. Behav Ecol Sociobiol 52: 117–127 [Chap 9]

Preisser EL, Bolnick DI, & Benard MF (2005) Scared to death? The effects of intimidation and consumption in predator-prey interactions. Ecology 86: 501–509 [Chap 13]

Price GR (1970) Selection and covariance. Nature 227: 520–521 [Chaps 14, 18, 31]

Price K, Harvey H, & Ydenberg R (1996) Begging tactics of nestling yellow-headed blackbirds, *Xanthocephalus xanthocephalus*, in relation to need. Anim Behav 51: 421–435 [Chap 6]

Price T (2007) Speciation in Birds. Roberts, Greenwood Village, CO [Chap 27]

Price T, Schluter D, & Henckman NE (1993) Sexual selection when the female directly benefits. Biol J Linn Soc 48: 187–211 [Chap 26]

Price TD (2006) Phenotypic plasticity, sexual selection and the evolution of colour patterns. J Exp Biol 209: 2368–2376 [Chap 6]

Price TD, Qvarnstrom A, & Irwin DE (2003) The role of phenotypic plasticity in driving genetic evolution. Proc R Soc Lond B 270: 1433–1440 [Chap 6]

Priest NK, Galloway LF, & Roach DA (2008) Mating frequency and inclusive fitness in *Drosophila melanogaster*. Am Nat 171: 10–21 [Chap 23]

Provine WB (1971) The Origins of Theoretical Population Genetics. Univ Chicago Press, Chicago [Chap 1]

Prum RO (1997) Phylogenetic tests of alternative intersexual selection mechanisms: trait macroevolution in a polygynous clade (Aves: Pipridae). Am Nat 149: 668–692 [Chaps 2, 7]

Pulido F (2007) The genetics and evolution of avian migration. Bioscience 57: 165–174 [Chap 5]

Pulliam HR (1973) On the advantages of flocking. J Theor Biol 38: 419–422 [Chap 17]

Pulliam HR (1974) Theory of optimal diets. Am Nat 108: 59–74 [Chap 8]

Pulliam HR (1975) Diet optimization with nutrient constraints. Am Nat 109: 765–768 [Chap 8]

Pulliam HR & Caraco T (1984) Living in groups: is there an optimal group size? Pp 122–147 in Krebs JR & Davies NB (eds) Behavioural Ecology: An Evolutionary Approach, 2nd ed. Sinauer, Sunderland, MA [Chap 17]

Pulliam HR & Danielson BJ (1991) Sources, sinks, and habitat selection—a landscape perspective on population-dynamics. Am Nat 137: S50–S66 [Chap 11]

Purchase CF, Hasselman DJ, & Weir LK (2007) Relationship between fertilization success and the number of milt donors in rainbow smelt *Osmerus mordax* (Mitchell): implications for population growth rates. J Fish Biol 70: 934–946 [Chap 21]

Pusey A & Wolf M (1996) Inbreeding avoidance in animals. Trends Ecol Evol 11: 201–206 [Chap 24]

Pyke GH, Pulliam HR, & Charnov EL (1977) Optimal foraging a selective review of theory and tests. Q Rev Biol 52: 137–154 [Chap 13]

Queller DC (1985) Kinship, reciprocity and synergism in the evolution of social behavior. Nature 318: 366–367 [Chap 14]

Queller DC (1992a) A general model for kin selection. Evolution 46: 376–380 [Chaps 14, 18]

Queller DC (1992b) Does population viscosity promote kin selection? Trends Ecol Evol 7: 322–324 [Chap 18]

Queller DC (1995) The spaniels of St. Marx and the Panglossian paradox: a critique of rhetorical programme. Q Rev Biol 70: 485–489 [Chap 2]

Queller DC (1996) The origin and maintenance of eusociality: the advantage of extended parental care. Pp 218–234 in Turillazzi S & West-Eberhard MJ (eds) Natural History and Evolution of Paper Wasps. Oxford Univ Press, New York [Chap 19]

Queller DC (1997) Why do females care more than males? Proc R Soc Lond B 264: 1555–1557 [Chaps 20, 26]

Queller DC (2000) Relatedness and the fraternal major transitions. Phil Trans R Soc Lond B 355: 1647–1655 [Chap 19]

Queller DC & Strassmann JE (1989) Measuring inclusive fitness in social wasps. Pp 103–122 in Breed MD & Page RE (eds) The Genetics of Social Evolution. Westview Press, Boulder, CO [Chap 19]

Queller DC & Strassmann JE (1998) Kin selection and social insects. Bioscience 48: 165–175 [Chap 19]

Queller DC, Ponte E, Bozzaro S, & Strassmann JE (2003) Single-gene greenbeard effects in the social amoeba *Dictyostelium discoideum*. Science 299: 105–106 [Chap 18]

Queller DC, Zacchi F, Cervo R, Turillazzi S, Henshaw MT, Santorelli LA, & Strassmann JE (2000) Unrelated helpers in a social insect. Nature 405: 784–787 [Chaps 17, 18]

Quinn AE, Georges A, Sarre SD, Guarino F, Ezaz T, & Marshall Graves JA (2007) Temperature sex reversal implies sex gene dosage in a reptile. Science 316: 411 [Chap 5]

Quinn JL & Cresswell W (2006) Testing domains of danger in the selfish herd: sparrowhawks target widely spaced redshanks in flocks. Proc R Soc Lond B 273: 2521–2526 [Chap 17]

Radder RS, Quinn AE, Georges A, Sarre SD, & Shine R (2008) Genetic evidence for co-occurrence of chromosomal and thermal sex-determining systems in a lizard. Biol Lett 4: 176–178 [Chap 5]

Radick G (2007) The Simian Tongue. Univ Chicago Press, Chicago [Chap 1]

Radwan J (1995) Male morph determination in two species of acarid mites. Heredity 74:669–673 [Chap 25]

Radwan J & Klimas M (2001) Male dimorphism in the bulb mite. *Rhizoglyphus robini:* fighters survive better. Ethology, Ecology and Evolution 13: 69–79 [Chap 25]

Rafferty NE & Boughman JW (2006) Olfactory mate recognition in a sympatric species pair of three-spined sticklebacks. Behav Ecol 17: 965–970 [Chap 27]

Rainey PB, Buckling A, Kassen R, & Travisano M (2000) The emergence and maintenance of diversity: insights from experimental bacterial populations. Trends Ecol Evol 15: 243–247 [Chap 4]

Rainey PB & Travisano M (1998) Adaptive radiation in a heterogeneous environment. Nature 394: 69–72 [Chap 4]

Ram KR & Wolfner MF (2007) Seminal influences: *Drosophila* Acps and the molecular interplay between male and females during reproduction. Integr Comp Biol 47: 427–445 [Chap 5]

Ramm SA, Oliver PL, Ponting CP, Stockley P, & Emes RD (2008) Sexual selection and the adaptive evolution of mammalian ejaculate proteins. Molec Biol Evol 25: 207–219 [Chap 20]

Ramsey J, Bradshaw HD, & Schemske DW (2003) Components of reproductive isolation between the monkeyflowers *Mimulus lewisii* and *M. cardinalis* (Phrymaceae). Evolution 57: 1520–1534 [Chap 27]

Randall JA & Matocq MD (1997) Why do kangaroo rats (*Dipodomys spectabilis*) footdrum at snakes? Behav Ecol 8: 404–413 [Chap 13]

Rands SA & Cuthill IC (2001) Separating the effects of predation risk and interrupted foraging upon mass changes in the blue tit *Parus caeruleus*. Proc R Soc Lond B 268: 1783–1790 [Chap 12]

Rands SA, Pettifor RA, Rowcliffe JM, & Cowlishaw G (2004) State-dependent foraging rules for social animals in selfish herds. Proc R Soc Lond B 271: 2613–2620 [Chap 17]

Rankin DJ, Bargum K, & Kokko H (2007) The tragedy of the commons in evolutionary biology. Trends Ecol Evol 22: 643–651 [Chap 20]

Rasa OAE (1986) Coordinated vigilance in dwarf mongoose family groups: the "watchman song" hypothesis and the costs of guarding. Ethology 71: 340–344 [Chap 17]

Rasmussen GSA, Gusset M, Courchamp F, & Macdonald DW (2008) Achilles' hell of sociality revealed by energetic poverty trap in cursorial hunters. Am Nat 172: 508–518 [Chap 17]

Ratcliffe LM & Grant PR (1983) Species recognition in Darwin's finches (*Geospiza*, Gould). 1. Discrimination by morphological cues. Anim Behav 31: 1139–1153 [Chap 27]

Ratnieks FLW (1988) Reproductive harmony via mutual policing by workers in eusocial Hymenoptera. Am Nat 132: 217–236 [Chaps 18, 19]
Ratnieks F, Foster KR, & Wenseleers T (2006) Conflict resolution in insect societies. Annu Rev Entomol 51: 581–608 [Chap 19]
Raubenheimer D & Tucker D (1997) Associative learning by locusts: pairing of visual cues with consumption of protein and carbohydrate. Anim Behav 54: 1449–1459 [Chap 10]
Rausher MD (1992) The measurement of selection on quantitative traits: biases due to the environmental covariances between traits and fitness. Evolution 46: 616–626 [Chap 3]
Ravi Ram K, Sirot LK, & Wolfner MF (2006) Predicted seminal astacin-like protease is required for processing of reproductive proteins in *Drosophila melanogaster*. Proc Natl Acad Sci USA 103: 18674–18679 [Chap 28]
Ravi Ram K & Wolfner MF (2007) Seminal influences: *Drosophila* Acps and the molecular interplay between males and females during reproduction. Integr Comp Biol 47: 427–445 [Chap 28]
Ray J (1691) The Wisdom of God Manifested in the Works of Creation. Smith, London [Chap 1]
R Development Core Team (2008). R: A language and environment for statistical computing. R Foundation for Statistical Computing [Chap 4]
Read AF & Weary DM (1992) The evolution of bird song: comparative analyses. Phil Trans R Soc London B 338: 165–187 [Chap 7]
Real LA (1980) Fitness, uncertainty, and the role of diversification in evolution and behavior. Am Nat 115: 623–638 [Chap 12]
Real LA (1981) Uncertainty and pollinator-plant interactions: the foraging behavior of bees and wasps on artificial flowers. Ecology 62: 20–26 [Chap 8]
Réale D & Festa-Bianchet M (2003) Predator-induced natural selection on temperament in bighorn ewes. Anim Behav 65: 463–470 [Chap 30]
Réale D, Gallant BY, Leblanc M, & Festa-Bianchet M (2000) Consistency of temperament in bighorn ewes and correlates with behaviour and life history. Anim Behav 60: 589–597 [Chap 30]
Réale D, McAdam AG, Boutin S, & Bertreaux D (2003) Genetic and plastic responses of a northern mammal to climate change. Proc R Soc Lond B 270: 591–596 [Chap 6]
Réale D, Reader SM, Sol D, McDougall PT, & Dingemanse NJ (2007) Integrating animal temperament within ecology and evolution. Biol Rev 82: 291–318 [Chap 30]
Reboreda JC, Clayton NS, & Kacelnik A (1996) Species and sex differences in hippocampus size in parasitic and non-parasitic cowbirds. Neuroreport 7: 505–508 [Chap 10]
Recer GM, Blanckenhorn WU, Newman JA, Tuttle EM, Withiam ML, & Caraco T (1987) Temporal resource variability and the habitat matching rule. Evol Ecol 1: 363–378 [Chap 11]
Rechten C, Avery M, & Stevens A (1983) Optimal prey selection: why do great tits show partial preferences? Anim Behav 31: 576–584 [Chap 8]
Reed DH & Bryant EH (2004) Phenotypic correlations among fitness and its components in a population of the housefly. J Evol Biol 17: 919–923 [Chap 4]
Reed DH & Frankham R (2003) Correlation between fitness and genetic diversity. Conserv Biol 17: 230–237 [Chap 29]
Reeve HK (1989) The evolution of conspecific acceptance thresholds. Am Nat 133: 407–435 [Chap 29]
Reeve HK (1991) *Polistes*. Pp 99–148 in Ross KG & Matthews RW (eds) The Social Biology of Wasps. Cornell Univ Press, Ithaca, NY [Chap 19]
Reeve HK, Emlen ST, & Keller L (1998) Reproductive sharing in animal societies: reproductive incentives or incomplete control by dominant breeders? Behav Ecol 9: 267–278 [Chap 17]
Reeve HK & Hölldobler B (2007) The emergence of a superorganism through intergroup competition. Proc Natl Acad Sci USA 104: 9736–9740 [Chap 17]
Reeve HK & Keller L (1995) Partitioning of reproduction in mother-daughter versus sibling associations—a test of optimal skew theory. Am Nat 145: 119–132 [Chap 17]
Reeve HK & Pfennig DW (2002) Genetic biases for showy males: are some genetic systems especially conducive to sexual selection? Proc Natl Acad Sci 100: 1089–1094 [Chap 5]
Reeve HK & Ratnieks FLW (1993) Queen-queen conflict in polygynous societies: mutual tolerance and reproductive skew. Pp 45–85 in Keller L (ed) Queen Number and Sociality in Insects. Oxford Univ Press [Chap 17]
Reeve HK & Sherman PW (1993) Adaptation and the goals of evolutionary research. Q Rev Biol 68: 1–32 [Chap 2]
Reeve HK & Sherman PW (2001) Optimality and phylogeny: a critique of current thought. Pp 64–113 in SH Orzack & Sober E (eds) Adaptationism and Optimality. Cambridge Univ Press, Cambridge [Chap 2]
Regan BC, Julliot C, Simmen B, Viénot F, Charles-Dominique P, & Mollon JD (2001) Fruits, foliage and the evolution of primate color vision. Phil Trans R Soc Lond B 356: 229–283 [Chap 9]

Reif A & Lesch KP (2003) Toward a molecular architecture of personality. Behav Brain Res 139: 1–20 [Chap 30]

Reimchen TE (1989) Loss of nuptial color in threespine sticklebacks (*Gasterosteus aculeatus*). Evolution 43: 450–460 [Chap 27]

Relyea RA (2002a) Costs of phenotypic plasticity. Am Nat 159: 272–282 [Chap 6]

Relyea RA (2002b) The many faces of predation: how induction, selection and thinning combine to alter prey phenotypes. Ecology 83: 1953–1964 [Chap 30]

Relyea RA (2004) Fine-tuned phenotypes: tadpole plasticity under 16 combinations of predators and competitors. Ecology 85: 172–179 [Chap 5]

Rendall D (2004) "Recognizing" kin: mechanisms, media, minds, modules, and muddles. Pp 295–316 in Chapais B & Berman CM (ed) Kinship and Behavior in Primates. Oxford Univ Press, New York [Chap 31]

Renison D, Boersma D, & Martella MB (2002) Winning and losing: causes for variability in outcome of fights in male Magellanic penguins (*Spheniscus magellanicus*). Behav Ecol 13: 462–466 [Chap 15]

Renn SCP, Aubin-Horth N, & Hofmann HA (2004) Biologically meaningful expression profiling across species using heterologous hybridization to a cDNA microarray. BMC Genomics 5: 42 [Chap 28]

Renn SCP, Aubin-Horth N, & Hofmann HA (2008) Fish and chips: functional genomics of social plasticity in an African cichlid fish. J Exp Biol 211: 3041–3056 [Chap 28]

Replogle K, Arnold AP, Ball GF, Band M, Bensch S, et al. (2008) The Songbird Neurogenomics (SoNG) Initiative: community-based tools and strategies for study of brain gene function and evolution. BMC Genomics 9: 131 [Chap 28]

Reudler Talsma J, Biere A, Harvey JA, & van Nouhuys S (2008) Oviposition cues for a specialist butterfly: plant chemistry and size. J Chem Ecol 34: 1202–1212

Reynolds JD (1996) Animal breeding systems. Trends Ecol Evol 11: 68–72 [Chap 26]

Reynolds JD, Goodwin NB, & Freckleton RP (2002) Evolutionary transitions in parental care and live bearing in vertebrates. Phil Trans R Soc Lond B 357: 269–281 [Chaps 20, 26]

Reznick DN & Ghalambor CK (2001) The population ecology of contemporary adaptations: what empirical studies reveal about the conditions that promote adaptive evolution. Genetica 112: 183–198 [Chap 29]

Reznick DN, Nunney L, & Tessier A (2000). Big houses, big cars, superfleas and the costs of reproduction. Trends Ecol Evol 15: 421–425 [Chap 4]

Reznick, DN, Shaw FH, Rodd FH, & Shaw RG (1997) Evaluation of the rate of evolution in natural populations of guppies (*Poecilia reticulata*). Science 275: 1934–1937 [Chap 29]

Reznick D & Travis J (1996) The empirical study of adaptation in natural populations. Pp 243–289 in Rose M & Lauder GV (eds) Adaptation. Academic Press, San Diego, CA [Chap 2]

Rheindt FE, Grafe TU, & Abouheif E (2004) Rapidly evolving traits and the comparative method: how important is testing for phylogenetic signal? Evol Ecol Res 6: 377–396 [Chap 7]

Rice KJ & Emery NC (2003) Managing microevolution: restoration in the face of global change. Front Ecol Envir 1: 469–478 [Chap 29]

Rice WR (1984) Sex chromosomes and the evolution of sexual dimorphism. Evolution 38: 735–742 [Chap 5]

Rice WR (1992) Sexually antagonistic genes—experimental-evidence. Science 256: 1436–1439 [Chap 23]

Rice WR (1996) Sexually antagonistic male adaptation triggered by experimental arrest of female evolution. Nature 381: 232–234 [Chaps 22, 23]

Rice WR (1998) Intergenomic conflict, interlocus antagonistic coevolution and the evolution of reproductive isolation. Pp. 261–270 in DJ Howard & SH Berlocher (eds) Endless Forms: Species and Speciation. Oxford Univ Press, New York [Chap 23]

Rice WR & Hostert EE (1993) Laboratory experiments on speciation: What have we learned in 40 years. Evolution 47: 1637–1653 [Chap 27]

Richards FJ (1959) A flexible growth function for empirical use. J Exp Bot 10: 290–301 [Chap 13]

Richardson H & Verbeek NAM (1986) Diet selection and optimization by northwestern crows feeding on Japanese littleneck clams. Ecology 67: 1219–1226 [Chap 8]

Richardson WJ, Greene CR, Malme CI, & Thomson DH (1995) Marine Mammals and Noise. Academic Press, San Diego, CA [Chap 9]

Richerson PJ & Boyd R (2005) Not by genes alone: how culture transformed human evolution. Univ Chicago Press, Chicago [Chap 31]

Ricklefs RE (2004) Cladogenesis and morphological diversification in passerine birds. Nature 430: 338–341 [Chap 7]

Ridley AR & Huyvaert P (2007) Sex-biased preferential care in the cooperatively breeding Arabian babbler. J Evol Biol 20: 1271–1276 [Chap 26]

Ridley M (1983) The explanation of organic diversity: the comparative method and adaptations for mating. Oxford Univ Press, New York [Chap 7]

Riechert SE & Hedrick AV (1993) A test for correlations among fitness-linked behavioural traits in the spider *Agelenopsis aperta*. Anim Behav 46: 669–675 [Chap 30]

Riffell JA, Krug PJ, & Zimmer RK (2004) The ecological and evolutionary consequences of sperm chemoattraction. Proc Natl Acad Sci USA 101: 4501–4506 [Chap 21]

Riolo RL, Cohen MD, & Axelrod R (2001) Evolution of cooperation without reciprocity. Nature 414: 441–443 [Chap 12]

Rios-Cardenas O & Webster MS (2008) A molecular genetic examination of the mating system of pumpkinseed sunfish reveals high pay-offs for specialized sneakers. Mol Ecol 17: 2310–2320 [Chap 25]

Riska B, Rutledge JJ, & Atchley WR (1985) Covariance between direct and maternal genetic effects in mice, with a model of persistent environmental influences. Genet Res 45: 287–297 [Chap 14]

Ritchie ME (1989) Optimal foraging and fitness in Columbian ground squirrels. Oecologia 82: 56–67 [Chap 4]

Ritchie MG (2007) Sexual selection and speciation. Annu Rev Ecol Evol Syst 38: 79–102 [Chap 27]

Roberts BW, Walton KE, & Viechtbauer W (2006) Patterns of mean-level change in personality traits across the life course: a meta-analysis of longitudinal studies. Psych Bull 132: 1–25 [Chap 30]

Roberts JD, Standish RJ, Byrne PB, & Doughty P (1999) Synchronous polyandry and multiple paternity in the frog *Crinia georgiana* (Anura: Myobatrachidae). Anim Behav 57: 721–726 [Chap 21]

Robertson BA & Hutto RL (2006) A framework for understanding ecological traps and an evaluation of existing evidence. Ecology 87: 1075–1085 [Chaps 6, 29]

Robertson HM & Wanner KW (2006) The chemoreceptor superfamily in the honey bee, *Apis mellifera:* expansion of the odorant, but not gustatory, receptor family. Genome Res 16: 1395–1403 [Chap 28]

Robinson BW & Dukas R (1999) The influence of phenotypic modifications on evolution: the Baldwin effect and modern perspectives. Oikos 85: 582–589 [Chap 6]

Robinson GE & Ben-Shahar Y (2002) Social behavior and comparative genomics: new genes or new gene regulation? Genes Brain Behav 1: 197–203 [Chap 28]

Robinson GE, Fernald RD, & Clayton DF (2008) Genes and social behavior. Science 322: 896–900 [Chap 28]

Robinson GE, Grozinger CM, & Whitfield CW (2005) Sociogenomics: social life in molecular terms. Nat Rev Genet 6: 257–270 [Chaps 5, 28]

Rodd FH, Hughes KA, Grether GF, & Baril CT (2002) A possible non-sexual origin of a mate preference: are male guppies mimicking fruit? Proc R Soc Lond B 269: 475–481 [Chap 24]

Rodriguez I, Gumbert A, de Ibarra NH, Kunze J, & Giurfa M (2004) Symmetry is in the eye of the "beeholder": innate preference for bilateral symmetry in flower-naive bumblebees. Naturwissenschaften 91: 374–377 [Chap 10]

Rodriguez-Girones MA & Enquist M (2001) The evolution of female sexuality. Anim Behav 61: 695–704 [Chap 31]

Rodríguez-Gironés MA, Cotton PA, & Kacelnik A (1996) The evolution of begging: signaling and sibling competition. Proc Natl Acad Sci USA 93: 14637–14641 [Chap 16]

Roff DA (1992) The Evolution of Life Histories: Theory and Analysis. Chapman & Hall, New York [Chaps 13, 30]

Roff DA (1995) The estimation of genetic correlations from phenotypic correlations: a test of Cheverud's conjecture. Heredity 74: 481–490 [Chap 30]

Roff DA (1996) The evolution of threshold traits in animals. Q Rev Biol 71: 3–35 [Chap 25]

Roff DA (1997) Evolutionary Quantitative Genetics. Chapman and Hall, New York [Chap 4]

Roff DA (2002) Life History Evolution. Sinauer, Sunderland, MA [Chap 4]

Roff DA & Fairbairn DJ (2007) The evolution and genetics of migration in insects. Bioscience 57: 155–164 [Chap 5]

Rohlf FJ (2006) A comment on phylogenetic correction. Evolution 60: 1509–1515 [Chap 7]

Rohwer S (1978) Parent cannibalism of offspring and egg raiding as a courtship strategy. Am Nat 112: 429–440 [Chap 26]

Roisin Y (1999) Philopatric reproduction, a prime mover in the evolution of termite sociality? Insectes Sociaux 46: 297–305 [Chap 12]

Rönn J, Katvala M, & Arnqvist G (2007) Coevolution between harmful male genitalia and female resistance in seed beetles. Proc Natl Acad Sci USA 104: 10921–10925 [Chap 23]

Roof RL (1993) Neonatal exogenous testosterone modifies sex difference in radial arm and Morris water maze performance in prepubescent and adult rats. Behavioural Brain Research 53: 1–10 [Chap 10]

Roper T (1993) Effects of novelty on taste-avoidance learning in chicks. Behaviour 125: 265–281 [Chap 16]

Rose MR & Lauder GV (eds) (1996) Adaptation. Academic Press, San Diego [Chap 2]

Rosenheim JA, Jepsen SJ, Matthews CE, Smith DS, & Rosenheim MR (2008) Time limitation, egg limitation, the cost of oviposition, and lifetime reproduction by an insect in nature. Am Nat 172: 486–496 [Chap 11]

Rosenthal GG & Evans CS (1998) Female preference for swords in *Xiphophorus helleri* reflects a bias for large apparent size. Proc Natl Acad Sci USA 95: 4431–4436 [Chap 24]

Rothstein SI (2001) Relic behaviours, coevolution and the retention versus loss of host defences after episodes of avian brood parasitism. Anim Behav 61: 95–107 [Chap 7]

Roughgarden J (1971) Density-dependent natural selection. Ecology 52: 453–468 [Chap 25]

Roulin A, Kölliker M, & Richner H (2000) Barn owl siblings vocally negotiate resources. Proc R Soc Lond B 267: 459–463 [Chap 8]

Rousset F (2004) Genetic structure and selection in subdivided populations. Princeton Univ Press, Princeton, NJ [Chap 18]

Rousset F & Roze D (2007) Constraints on the origin and maintenance of genetic kin recognition. Evolution 61: 2320–2330 [Chap 18]

Rovero F, Hughes RN, Whiteley NM, & Chelazzi G (2000) Estimating the energetic cost of fighting in shore crabs by noninvasive monitoring of heartbeat rate. Anim Behav 59: 705–713 [Chap 15]

Rowe C (1999) Receiver psychology and the evolution of multicomponent signals. Anim Behav 58: 921–931 [Chap 10]

Rowe L & Day T (2006) Detecting sexual conflict and sexually antagonistic coevolution. Phil Trans R Soc Lond B 361: 277–285 [Chap 23]

Rowe L & Houle D (1996) The lek paradox and the capture of genetic variance by condition dependent traits. Proc R Soc Lond B 263: 1415–1421 [Chaps 5, 24]

Rowe L, Arnqvist G, Sih A, & Krupa JJ (1994) Sexual conflict and the evolutionary ecology of mating patterns: water striders as a model system. Trends Ecol Evol 9: 289–293 [Chaps 3, 23]

Rowe L, Cameron E, & Day T (2003) Detecting sexually antagonistic coevolution with population crosses. Proc R Soc Lond B 270: 2009–2016 [Chap 23]

Rowe L, Cameron E, & Day T (2005) Escalation, retreat, and female indifference as alternative outcomes of sexually antagonistic coevolution. Am Nat 165: S5–S18 [Chap 23]

Rowell JT, Ellner SP, & Reeve HK (2006) Why animals lie: how dishonesty and belief can coexist in a signaling system. Am Nat. 168: E180–E204 [Chap 3]

Rowland HM, Ihalainen E, Lindstrom L, Mappes J, & Speed MP (2007) Co-mimics have a mutualistic relationship despite unequal defences. Nature 448: 64–67 [Chap 10]

Rowland JM & Emlen DJ (2009) Two thresholds, three male forms result in facultative male trimorphism in beetles. Science 323: 773–776 [Chap 25]

Rubenstein DR & Lovette IJ (2007) Temporal environmental variability drives the evolution of cooperative breeding in birds. Curr Biol 17: 1414–1419 [Chap 17]

Rubenstein DR & Wikelski M (2005) Steroid hormones and aggression in female Galapagos marine iguanas. Horm Behav 48: 329–341 [Chap 15]

Rundle HD, Chenoweth SF, & Blows MW (2008) Comparing complex fitness surfaces: among-population variation in mutual sexual selection in *Drosophila serrata*. Am Nat 171: 443–454 [Chap 27]

Rundle HD, Chenoweth SF, Doughty P, & Blows MW (2005) Divergent selection and the evolution of signal traits and mating preferences. PLoS Biol 3: 1988–1995 [Chap 27]

Rundle HD, Nagel L, Boughman JW, & Schluter D (2000) Natural selection and parallel speciation in sympatric sticklebacks. Science 287: 306–308 [Chap 27]

Rundle HD & Nosil P (2005) Ecological speciation. Ecol Lett 8: 336–352 [Chap 27]

Rundle HD, Ödeen A, & Mooers AØ (2007) An experimental test for indirect benefits in *Drosophila melanogaster*. BMC Evolutionary Biology 7: 36 [Chap 23]

Rundle HD & Schluter D (1998) Reinforcement of stickleback mate preferences: sympatry breeds contempt. Evolution 52: 200–208 [Chap 27]

Runge JP, Runge MC, & Nichols JD (2006) The role of local populations within a landscape context: defining and classifying sources and sinks. Am Nat 167: 925–938 [Chap 29]

Ruse M (1979) Sociobiology: Sense or Nonsense? Reidel, Dordrecht [Chap 1]

Ruse M (2005) The Evolution/Creation Struggle. Harvard Univ Press, Cambridge, MA [Chap 1]

Russell AF & Hatchwell BJ (2001) Experimental evidence for kin-biased helping in a cooperatively breeding vertebrate. Proc R Soc Lond B 268: 2169–2174 [Chap 18]

Russell AF, Langmore NE, Cockburn A, Astheimer LB, & Kilner RM (2007) Reduced egg investment can conceal helper effects in cooperatively breeding birds. Science 317: 941–944 [Chap 26]

Rutkowska J & Badyaev AV (2008) Meiotic drive and sex determination: molecular and cytological mechanisms of sex ratio adjustment in birds. Phil Trans Roy Soc B 363: 1675–1686 [Chap 26]

Ruxton GD, Jackson AL, & Tosh CR (2007) Confusion of predators does not rely on specialist coordinated behavior. Behav Ecol 18: 590–596 [Chap 17]

Ruxton GD, Sherratt TN, & Speed MP (2004) Avoiding attack: The evolutionary ecology of crypsis, warning signals and mimicry. Oxford Univ Press, Oxford [Chaps 9, 16]

Ryan MJ (1988) Energy, calling, and selection. Am Zool 28: 885–898 [Chap 15]
Ryan MJ (1990) Sexual selection, sensory systems and sensory exploitation. Oxford Survey of Evolutionary Biology 7: 157–195 [Chap 16]
Ryan MJ (1998) Sexual selection, receiver biases, and the evolution of sex differences. Science 281: 1999–2003 [Chap 10]
Ryan MJ (2005) The evolution of behaviour, and integrating it toward a complete and correct understanding of behavioural biology. Anim Biol 55: 419–439 [Chap 2]
Ryan MJ, Fox JH, Wilczynski W, & Rand AS (1990) Sexual selection for sensory exploitation in the frog *Physalaemus pustulosus*. Nature 343: 66–67 [Chap 24]
Ryan MJ, Phelps SM, & Rand AS (2001) How evolutionary history shapes recognition mechanisms. Trends in Cognitive Science 5: 143–148 [Chap 16]
Ryan MJ & Wilczynski W (1988) Coevolution of sender and receiver: effect on local mate preference in cricket frogs. Science 240: 1786–1788 [Chap 27]
Sachs JL, Mueller UG, Wilcox TP, & Bull JJ (2004) The evolution of cooperation. Q Rev Biol 79: 135–160 [Chap 18]
Sæther SA, Sætre GP, Borge T, Wiley C, Svedin N, Andersson G, Veen T, Haavie J, Servedio MR, Bures S, Král M, Hjernquist MB, Gustafsson L, Träff J, & Qvarnström A (2007) Sex chromosome-linked species recognition and evolution of reproductive isolation in flycatchers. Science 318: 95–97 [Chap 5]
Saetre GP, Moum T, Bures S, Kral M, Adamjan M, & Moreno J (2007) A sexually selected character displacement in flycatchers reinforces premating isolation. Nature 387: 589–592 [Chap 24]
Sakaluk SK (1984) Male crickets feed females to ensure complete sperm transfer. Science 223: 609–610 [Chap 24]
Sanderson MJ, Purvis A, & Henze C (1998) Phylogenetic supertrees: assembling the trees of life. Trends Ecol Evol 13: 105–109 [Chap 7]
Santorelli LA, Thompson CRL, Villegas E, Svetz J, Dinh C, Parikh A, Sucgang R, Kuspa A, Strassmann JE, Queller DC, & Shaulsky G (2008) Facultative cheater mutants reveal the genetic complexity of cooperation in social amoebae. Nature 451: 1107–1110 [Chap 19]
Sargent RC & Gross MR (1986) William's principle: An explanation of parental care in teleost fishes. Pp 275–293 in Pitcher TJ (ed) Behaviour of Teleost Fishes. Croom Helm, London [Chap 26]
Savitz JB & Ramesar RS (2004) Genetic variants implicated in personality: a review of the more promising candidates. Am J Med Genet B 131B: 20–32 [Chap 30]
Scheich H, Langner G, Tidemann C, Coles RB, & Guppy A (1986) Electroreception and electrolocation in platypus. Nature 319: 401–402 [Chap 9]
Scheiner SM & Lyman RF (1989) The genetics of phenotypic plasticity I. Heritability. J Evol Biol 2: 95–107 [Chap 6]
Schemske DW & Bradshaw HD (1999) Pollinator preference and the evolution of floral traits in monkeyflowers (*Mimulus*). Proc Natl Acad Sci USA 96: 11910–11915 [Chap 27]
Schenkel R (1956) Zur Deutung der Phasianidenbalz. Ornitologische Beobachtung 53: 182 [Chap 16]
Schiffer M & McEvey SF (2006) *Drosophila bunnanda*—a new species from northern Australia with notes on other Australian members of the *montium* subgroup (Diptera: Drosophilidae). Zootaxa 1333: 1–23 [Chap 27]
Schlaepfer MA, Runge MC, & Sherman PW (2002) Ecological and evolutionary traps. Trends Ecol Evol 17: 474–480 [Chap 29]
Schlaepfer MA, Sherman PW, Blossey B, & Runge MC (2005) Introduced species as evolutionary traps. Ecol Lett 8: 241–246 [Chap 29]
Schlegel A & Barry H III (1986) The cultural consequences of female contribution to subsistence. Amer Anthro 88: 142–150 [Chap 31]
Schlichting CD (2004) The role of phenotypic plasticity in diversification. Pp 191–200 in DeWitt TJ & Scheiner SM (eds) Phenotypic Plasticity: Functional and Conceptual Approaches. Oxford Univ Press, New York [Chap 6]
Schlichting CD & Pigliucci M (1998) Phenotypic Evolution: A Reaction Norm Perspective. Sinauer, Sunderland, MA [Chaps 6, 25]
Schlichting CD & Smith H (2002) Phenotypic plasticity: linking molecular mechanisms with evolutionary outcomes. Evol Ecol 16: 189–211 [Chap 6]
Schluter D (1988) Estimating the form of natural selection on a quantitative trait. Evolution 42: 849–861 [Chap 3]
Schluter D (2000) Ecology of Adaptive Radiation. Oxford Univ Press, Oxford [Chap 27]
Schluter D & McPhail JD (1992) Ecological character displacement and speciation in sticklebacks. Am Nat 140: 85–108 [Chap 27]
Schluter D, Price TD, & Rowe L (1991) Conflicting selection pressures and life history trade-offs. Proc R Soc Lond B. 246: 11–17 [Chap 3]
Schluter D, Price T, Mooers AO, & Ludwig D (1997) Likelihood of ancestor states in adaptive radiation. Evolution 51: 1699–1711 [Chap 7]
Schmid-Hempel P, Kacelnik A, & Houston AI (1985) Honeybees maximize efficiency by not filling their crop. Behav Ecol Sociobiol 17: 61–66 [Chap 2]
Schoener T (1971) Theory of feeding strategies. Annu Rev Ecol Syst 2: 369–404 [Chap 13]

Schöpf Rehage J, Barnett BK, & Sih A (2005) Behavioral responses to a novel predator and competitor in invasive mosquitofish and their non-invasive relatives (*Gambusia* sp). Behav Ecol Sociobiol 57: 256–266 [Chap 30]

Schuck-Paim C, Pompilio L, & Kacelnik A (2004) State-dependent decisions cause apparent violations of rationality in animal choice. PLoS Biol 2: 2305–2315 [Chap 11]

Schuett GW & Grober MS (2000) Post-fight levels of plasma lactate and corticosterone in male copperheads, *Agkistrodon contortrix* (Serpentes, Viperidae): differences between winners and losers. Physiol Behav 71: 335–341 [Chap 15]

Schulze-Hagen K, Stokke BG, & Birkhead TR (2009) Reproductive biology of the European cuckoo *Cuculus canorus:* early insights, persistent errors and the acquisition of knowledge. J Ornithology 150: 1–16 [Chap 1]

Schuster SM & Wade MJ (2003) Mating Systems and Strategies. Princeton Univ Press, Princeton, NJ [Chap 6]

Schwabl H (1993) Yolk is a source of maternal testosterone for developing birds. Proc Nat Acad Sci USA 90: 11446–11450 [Chap 6]

Schwabl H (1996) Maternal testosterone in the avian egg enhances postnatal growth. Comp Bioch Physiol 114: 271–276 [Chap 6]

Schwagmeyer PL, St Clair RC, Moodie JD, Lamey TC, Schnell GD, & Moodie MN (1999) Species differences in male parental care in birds: A reexamination of correlates with paternity. Auk 116: 487–503 [Chap 20]

Scott J & Lockard JS (2006) Captive female gorilla agonistic relationships with clumped defendable food resources. Primates 47: 199–209 [Chap 15]

Scotti ML & Foster SA (2007) Phenotypic plasticity and the ecotypic differentiation of aggressive behavior in threespine stickleback. Ethology 113: 190–198 [Chap 6]

Searcy WA & Nowicki S (2005) The Evolution of Animal Communication: Reliability and Deception in Signaling Systems. Princeton Univ Press, Princeton, NJ [Chap 16]

Searcy WA, Yasukawa K, & Lanyon S (1999) Evolution of polygyny in the ancestors of red-winged blackbirds. Auk 116: 5–19 [Chaps 2, 7]

Seehausen O, van Alphen JJM, & Witte F (1997) Cichlid fish diversity threatened by eutrophication that curbs sexual selection. Science 277: 1808–1811 [Chap 27]

Seeley TD (1995) The Wisdom of the Hive: The Social Physiology of Honey Bee Colonies. Harvard Univ Press, Cambridge, MA [Chaps 5, 19]

Seeley TD, Camazine S, & Sneyd J (1991) Collective decision-making in honey bees: how colonies choose among nectar sources. Behav Ecol Sociobiol 28: 277–290 [Chap 11]

Segerstråle U (2000) Defenders of the Truth. Oxford Univ Press, Oxford [Chap 1]

Segoli M, Harari AR, & Lubin Y (2006) Limited mating opportunities and male monogamy: a field study of white widow spiders, *Latrodectus pallidus* (Theridiidae) Anim Behav 72: 635–642 [Chap 20]

Selzer J (ed) (1993) Understanding Scientific Prose. Univ Wisconsin Press, Madison [Chap 2]

Semel B & Sherman PW (2001) Intraspecific parasitism and nest-site competition in wood ducks. Anim Behav 61: 787–803 [Chap 29]

Semel B, Sherman PW, & Byers SM (1988) Effects of brood parasitism and nest-box placement on wood duck breeding ecology. Condor 90: 920–930 [Chap 29]

Semlitsch RD, Scott DE, & Pechmann JHK (1988) Time and size at metamorphosis related to adult fitness in *Ambystoma talpoideum*. Ecology 69: 184–192 [Chap 4]

Sen Sarma M, Whitfield CW, & Robinson GE (2007) Species differences in brain gene expression profiles associated with adult behavioral maturation in honey bees. BMC Genomics 8: 202–215 [Chap 28]

Seppänen J-T & Forsman JT (2007) Interspecific social learning: novel preference can be acquired from a competing species. Curr Biol 17: 1248–1252 [Chaps 9, 12]

Seppänen J-T, Forsman JT, Mönkkönen M, & Thomson RL (2007) Social information use is a process across time, space and ecology, reaching heterospecifics. Ecology 88: 1622–1633 [Chap 9]

Serrão EA & Havenhand J (2009) Fertilization strategies. Pp 149–164 in Wahl M (ed) Marine Hard Bottom Communities: Patterns, Scales, Dynamics, Functions, Shifts. Springer Verlag, Berlin [Chap 21]

Servedio MR (2007) Male versus female mate choice: sexual selection and the evolution of species recognition via reinforcement. Evolution 61: 2772–2789 [Chap 20]

Servedio MR & Lande R (2006) Population genetic models of male and mutual mate choice. Evolution 60: 674–685 [Chap 20]

Servedio MR & Noor MAF (2003) The role of reinforcement in speciation: theory and data. Annu Rev Ecol Evol Syst 34: 339–364 [Chap 27]

Sewell MA & Levitan DR (1992) Fertilization success in a natural spawning of the dendrochirote sea cucumber *Cucumaria miniata*. Bull Mar Sci 51: 161–166 [Chap 21]

Sgro CM & Hoffmann AA (2004) Genetic correlations, tradeoffs and environmental variation. Heredity 93: 241–248 [Chap 30]

Shafir S, Waite TA, & Smith BH (2002) Context-dependent violations of rational choice in honeybees (Apis mellifera) and gray jays (*Perisoreus canadensis*). Behav Ecol Sociobiol 51: 180–187 [Chap 11]
Shah B, Shine R, Hudson S, & Kearney M (2003) Sociality in lizards: why do thick-tailed geckos (*Nephrurus milii*) aggregate? Behaviour 140: 1039–1052 [Chap 17]
Sharbel TF & Mitchell-Olds T (2001) Recurrent polyploid origins and chloroplast phylogeography in the *Arabis holboellii* complex (Brassicaceae). Heredity 87: 59–68 [Chap 8]
Shaw KA, Scotti ML, & Foster SA (2007) Ancestral plasticity and the evolutionary diversification of courtship behaviour in threespine sticklebacks. Anim Behav 73: 415–422 [Chap 6]
Shaw KL & Lesnick SC (2009) Genomic linkage of male song and female acoustic preference QTL underlying a rapid species radiation. Proc Natl Acad Sci USA 106: 9737–9742 [Chap 5]
Shaw KL, Parsons YM, & Lesnick SC (2007) QTL analysis of a rapidly evolving speciation phenotype in the Hawaiian cricket *Laupala*. Mol Ecol 16: 2879–2892 [Chap 5]
Sheldon BC (2002) Relating paternity to paternal care. Phil Trans R Soc Lond B 357: 341–350 [Chaps 20, 26]
Sheldon BC & Ellegren H (1999) Sexual selection resulting from extrapair paternity in collared flycatchers. Anim Behav 57: 285–298 [Chaps 4, 24]
Sheldon BC (2000) Differential allocation: tests, mechanisms and implications. Trends Ecol Evol 15: 397–401 [Chaps 24, 26]
Sheldon BC, Merilä J, Qvarnstrom A, Gustafsson L, & Ellegren H (1997) Paternal genetic contribution to offspring condition predicted by size of male secondary sexual character. Proc R Soc Lond B 264: 297–302 [Chap 24]
Sheldon BC & West SA (2004) Maternal dominance, maternal condition, and offspring sex ratio in ungulate mammals. Am Nat 163: 40–54 [Chap 17, 20, 26]
Shelley EL & Blumstein DT (2005) The evolution of vocal alarm communication in rodents. Behav Ecol 16: 169–177 [Chap 13]
Shendure J, Mitra RD, Varma C, & Church GM (2004) Advanced sequencing technologies: methods and goals. Nat Rev Genet 5: 335–344 [Chap 28]
Shepher J (1983) Incest: a biosocial view. Academic PressNew York [Chap 31]
Sherman PW (1985) Alarm calls of Belding's ground squirrels to aerial predators: nepotism or self-preservation? Behav Ecol Sociobiol 17: 313–323 [Chap 13]
Sherman PW (1988) The levels of analysis. Anim Behav 36: 616–619 [Chaps 2, 29]
Sherman PW, Jarvis JUM, & Alexander RD (1991) The biology of the naked mole rat. Princeton Univ Press, Princeton, NJ [Chap 19]
Sherman PW & Reeve HK (1997) Forward and backward: Alternative approaches to studying human social evolution. Pp 147–158 in Betzig L (ed) Human Nature: A Critical Reader. Oxford Univ Press, New York [Chap 2]
Sherman PW, Reeve HK, & Pfennig DW (1997) Recognition systems. Pp 69–96 in Krebs JR & Davies NB (eds) Behavioural Ecology: An Evolutionary Approach, 4th ed. Blackwell Sci, Oxford [Chap 29]
Sherman PW & Runge MC (2002) Demography of a population collapse: the Northern Idaho ground squirrel (*Spermophilus brunneus brunneus*). Ecology 83: 2816–2831 [Chap 29]
Sherman PW & Westneat DF (1988) Multiple mating and quantitative genetics. Anim Behav 36: 1545–1547 [Chap 22]
Sherratt TN & Harvey IF (1993) Frequency-dependent food selection by arthropods: a review. Biol J Linn Soc 48: 167–186 [Chap 12]
Sherry DF & Vaccarino AL (1989) Hippocampus and memory for food caches in black-capped chickadees. Behavioral Neuroscience 103: 308–318 [Chap 10]
Sherry DS & Marlowe FW (2007) Anthropometric data indicate nutritional homogeneity in Hadza foragers of Tanzania. Am J Hum Biol 19: 107–118 [Chap 31]
Shettleworth SJ (1998) Cognition, Evolution and Behavior. Oxford Univ Press [Chap 10]
Shettleworth SJ (2003) Memory and hippocampal specialization in food-storing birds: Challenges for research on comparative cognition. Brain Behav Evol 62: 108–116 [Chap 10]
Shier DM & Owings DH (2006) Effects of predator training on behavior and post-release survival of captive prairie dogs (*Cynomys ludovicianus*). Biol Conserv 132: 126–135 [Chap 6]
Shine R, Phillips B, Waye H, Lemaster M, & Mason RT (2001) Animal behaviour: benefits of female mimicry in snakes. Nature 414: 267 [Chap 25]
Shipley B (2000) Cause and Correlation in Biology: A User's Guide to Path Analysis, Structural Equations, and Causal Inference. Cambridge Univ Press, Cambridge [Chap 3]
Short RV (1977) Sexual selection and the descent of man. Pp. 3–19 in Calaby JH & Tyndale-Biscoe CH (eds) Reproduction and Evolution. Australian Academy of Science, Sydney [Chap 22]
Shuker DM & West SA (2004) Information constraints and the precision of adaptation: sex ratio manipulation in wasps. Proc Natl Acad Sci USA 101: 10363–10367 [Chap 18]
Shuster SM (1989) Male alternative reproductive strategies in a marine isopod crustacean

(*Paracerceis sculpta*): the use of genetic markers to measure differences in fertilization success among α-, β- and γ-males. Evolution 43: 1683–1698 [Chap 25]
Shuster SM (1992) The reproductive behaviour of α, β-, and γ-males in *Paracerceis sculpta*, a marine isopod crustacean. Behaviour 121: 231–258 [Chap 25]
Shuster SM (2007) The evolution of crustacean mating systems. Pp 29–47 in Duffy EJ & Thiel M (eds) Evolutionary Ecology of Social and Sexual Systems: Crustaceans as Model Organisms. Oxford Univ Press, New York [Chap 25]
Shuster SM (2008) The expression of crustacean mating strateges. Pp 224–250 in Oliviera RF, Taborsky M, & Brockmann HJ (eds) Alternative Reproductive Tactics. Cambridge Univ Press, Cambridge [Chap 25]
Shuster SM, Ballard JOW, Zinser G, Sassaman C, & Keim P (2001) The influence of genetic and extrachromosomal factors on population sex ratio in the marine isopod, *Paracerceis sculpta*. Pp 313–326 in Kensley B & Brusca RC (eds) Isopod Systematics and Evolution, Crustacean Issues, Vol. 13. Balkema Press, Amsterdam [Chap 25]
Shuster SM & Sassaman C (1997) Genetic interaction between male mating strategy and sex ratio a marine isopod. Nature 388: 373–376 [Chap 25]
Shuster SM & Wade MJ (1991) Equal mating success among male reproductive strategies in a marine isopod. Nature 350: 608–610 [Chaps 13, 25]
Shuster SM & Wade MJ (2003) Mating Systems and Strategies. Princeton Univ Press, Princeton, NJ [Chaps 14, 20, 25]
Sibly RM (1983) Optimal group-size is unstable. Anim Behav 31: 947–948 [Chap 11]
Sih A (1980) Optimal behavior, can foragers balance two conflicting demands? Science 210: 1041–1043 [Chap 13]
Sih A (1982) Foraging strategies and the avoidance of predation by an aquatic insect, *Notonecta hoffmanni*. Ecology 63: 786–796 [Chap 13]
Sih A (1987) Predator and prey lifestyles: an evolutionary and ecological overview. Pp 203–225 in Kerfoot WC & Sih A (eds) Predation: Direct and Indirect Impacts on Aquatic Communities. Univ Press of New England, Hanover, NH [Chap 12]
Sih A (1992) Prey uncertainty and the balancing of antipredator and feeding needs. Am Nat 139: 1052–1069 [Chaps 6, 30]
Sih A (2004) A behavioral ecological view of phenotypic plasticity. Pp 112–125 in DeWitt TJ & Scheiner SM (eds) Phenotypic Plasticity: Functional and Conceptual Approaches. Oxford Univ Press, New York [Chaps 6, 30]
Sih A & Bell AM (2008) Insights for behavioral ecology from behavioral syndromes. Adv Study Behav 38: 227–281 [Chap 30]
Sih A, Bell A, & Johnson JC (2004a) Behavioral syndromes: an ecological and evolutionary overview. Trends Ecol Evol 19: 372–378 [Chaps 12, 28, 30]
Sih A, Bell AM, Johnson JC, & Ziemba R (2004b) Behavioral syndromes: an integrative overview. Q Rev Biol 79: 241–277 [Chaps 13, 30]
Sih A & Christensen B (2001) Optimal diet theory: when does it work, and when and why does it fail? Anim Behav 61: 379–390 [Chap 8, 30]
Sih A, Englund G, & Wooster D (1998) Emergent impacts of multiple predators on prey. Trends Ecol Evol 13: 350–355 [Chap 13]
Sih A, Kats LB, & Maurer EF (2000) Does phylogenetic inertia explain the evolution of ineffective antipredator behavior in a sunfish-salamander system? Behav Ecol Sociobiol 49: 48–56 [Chap 7]
Sih A, Kats LB, & Maurer EF (2003) Behavioral correlations across situations and the evolution of antipredator behaviour in a sunfish-salamander system. Anim Behav 65: 29–44 [Chap 30]
Sih A, Lauer M, & Krupa JJ (2002) Path analysis and relative importance of male-female conflict, female choice and male-male competition in water striders. Anim Behav 63: 1079–1089 [Chap 30]
Sih A & McCarthy TM (2002) Prey responses to pulses of risk and safety: testing the risk allocation hypothesis. Anim Behav 63: 437–443 [Chap 13]
Sih A & Watters JV (2005) The mix matters: behavioural types and group dynamics in water striders. Behaviour 142: 1417–1431 [Chap 30]
Silk JB & Brown GR (2008) Local resource competition and local resource enhancement shape primate birth sex ratios. Proc R Soc Lond B 275: 1761–1765 [Chap 26]
Silvertown J, Franco M, Pisanty I, & Mendoze A (1993) Comparative demography: relative importance of lifecycle components to the finite rate of increase in woody and herbaceous perennials. J Ecol 81: 465–476 [Chap 4]
Simmons LW (1986) Female choice in the field cricket *Gryllus bimaculatus* (De Geer). Anim Behav 34: 1463–1470 [Chap 24]
Simmons LW (2001) Sperm Competition and Its Evolutionary Consequences in the Insects. Princeton Univ Press, Princeton, NJ [Chap 1]
Simmons LW (2004) Genotypic variation in calling song and female preferences of the field cricket *Teleogryllus oceanicus*. Anim Behav 68: 313–322 [Chap 5]
Simmons LW & Parker GA (1989) Nuptial feeding in insects: mating effort versus paternal investment. Ethology 81: 332–343 [Chap 26]

Simmons LW & Parker GA (1996) Parental investment and the control of sexual selection: can sperm competition affect the direction of sexual competition? Proc R Soc Lond B 263: 515–519 [Chaps 20, 22]

Simmons LW, Parker GA, & Stockley P (1999) Sperm displacement in the yellow dung fly, *Scatophaga stercoraria:* an investigation of male and female processes. Am Nat 153: 302–314 [Chap 22]

Simmons PJ & Young D (1999) Nerve Cells and Animal Behavior. Cambridge Univ Press, Cambridge [Chap 9]

Simon J-C, Delmotte F, Rispe C, & Crease T (2003) Phylogenetic relationships between parthenogens and their sexual relatives. Biol J Linn Soc 79: 151–163 [Chap 3]

Sinervo B (1993) The effect of offspring size on physiology and life history: manipulation of size using allometric engineering. Bioscience 43: 210–218 [Chap 3]

Sinervo B (1999) Mechanistic analysis of natural selection and a refinement of Lack's and William's principles. Am Nat 154: S26–S42 [Chap 3]

Sinervo B (2000) Adaptation, natural selection, and optimal life history allocation in the face of genetically-based trade-offs. Pp 41–64 in Mousseau T, Sinervo B, & Endler JA (eds) Adaptive Genetic Variation in the Wild. Oxford Univ Press, New York [Chap 3]

Sinervo B (2001) Runaway social games, genetic cycles driven by alternative male and female strategies, and the origin of morphs. Genetica 112: 417–434 [Chaps 13, 25]

Sinervo B (2005) Darwin's finch beaks, *Bmp4,* and the developmental origins of novelty. Heredity 94: 141–142 [Chap 3]

Sinervo B & Basolo AL (1996) Testing adaptation using phenotypic manipulations. Pp 149–185 in Rose MR & Lauder GV (eds) Adaptation. Academic PressNew York [Chap 3]

Sinervo B & Calsbeek R (2006) The developmental, physiological, neural and genetical causes and consequences of frequency dependent selection in the wild. Annu Rev Ecol Syst 37: 581–610 [Chap 3]

Sinervo B & Clobert J (2003) Morphs, dispersal behavior, genetic similarity, and the evolution of cooperation. Science 300: 1949–1951 [Chap 3]

Sinervo B & Clobert J (2008) Life history strategies, multi-dimentional trade-offs and behavioural syndromes. Pp 135–183 in Danchin E, Giraldeau L-A, & Cézilly F (eds) Behavioural Ecology. Oxford Univ Press, Oxford [Chap 3]

Sinervo B & Lively CM (1996) The rock-paper-scissors game and the evolution of alternative male strategies. Nature 380: 240–243 [Chap 25]

Sinervo B & McAdam AG (2008) Maturational costs of reproduction due to clutch size and ontogenetic conflict as revealed in the invisible fraction. Proc R Soc Lond B 275: 629–638 [Chap 3]

Sinervo B & Svensson E (2002) Correlational selection and the evolution of genomic architecture. Heredity 89: 329–338 [Chap 3]

Sinervo B, Doughty P, Huey RB, & Zamudio K (1992) Allometric engineering: a causal analysis of natural selection on offspring size. Science 258: 1927–1930 [Chap 3]

Sinervo B, Svensson E, & Comendant T (2000) Density cycles and an offspring quantity and quality game driven by natural selection. Nature 406: 985–988 [Chap 3]

Sinervo B, Chaine A, Clobert J, Calsbeek R, Hazard L, Lanccaster L, McAdam AG, Alonzo S, Corrigan G, & Hochberg ME (2006) Self-recognition, color signals, and cycles of greenbeard mutualism and altruism. Proc Nat Acad Sci USA 103: 7372–7377 [Chap 3]

Sinervo B, Huelin B, Surget-Groba Y, Clobert J, Miles DB, Corl A, Chaine A, & Davis A (2007) Models of density-dependent genic selection and a new rock-paper-scissors social system. Am Nat 170: 663–680 [Chap 3]

Sinervo B, Clobert J, Miles DB, McAdam AG, & Lancaster LT (2008) The role of pleiotropy versus signaler-receiver gene epistasis in life history trade-offs: dissecting the genomic architecture of organismal design in social systems. Heredity 101: 197–211 [Chap 3]

Sinha S, Ling X, Whitfield CW, Zhai C, & Robinson GE (2006) Genome scan for cis-regulatory DNA motifs associated with social behavior in honey bees. Proc Natl Acad Sci USA 103: 16352–16357 [Chap 28]

Sirot E (2000) An evolutionarily stable strategy for aggressiveness in feeding groups. Behav Ecol 11: 351–356 [Chap 15]

Sirot E & Touzalin F (2009) Coordination and synchronization of vigilance in groups of prey: the role of collective detection and predators' preference for stragglers. Am Nat 173: 47–59 [Chap 17]

Siva-Jothy MT (2006) Trauma, disease and collateral damage: conflict in cimicids. Phil Trans R Soc Lond B 361: 269–275 [Chap 22]

Skelhorn J & Rowe C (2007) Predators' toxin burdens influence their strategic decisions to eat toxic prey. Curr Biol 17: 1479–1483 [Chap 10]

Slabbekoorn H & Ripmeester EAP (2008)[0]. Birdsong and anthropogenic noise: implications and applications for conservation. Mol Ecol 17: 72–83 [Chap 9]

Slagsvold T (1986) Nest site settlement by the pied flycatcher: does the female choose her mate for the quality of his house or himself? Ornis Scandinavica 17: 210–220 [Chap 24]

Slagsvold T & Wiebe KL (2007) Learning the ecological niche. Proc R Soc Lond B 274: 19–23 [Chap 5]
Slatkin M (1978) On the equilibration of fitnesses by natural selection. Am Nat 112: 845–859 [Chap 25]
Slatkin M (1979a) The evolutionary response to frequency and density dependent interactions. Am Nat 114: 384–398 [Chap 25]
Slatkin M (1979b) Frequency- and density-dependent selection on a quantitative character. Genetics 93: 755–771 [Chap 25]
Slessor KN, Winston ML, & Le Conte Y (2005) Pheromone communication in the honeybee (*Apis mellifera* L.). J Chem Ecol 31: 2731–2745 [Chap 28]
Slos S & Stoks R (2006) Behavioural correlations may cause partial support for the risk allocation hypothesis in damselfly larvae. Ethology 112: 143–151 [Chap 13]
Sluyter F, Marican CCM, & Crusio WE (1999) Further phenotypical characterisation of two substrains of C57BL/6J inbred mice differing by a spontaneous single-gene mutation. Behav Brain Res 98: 39–43 [Chap 10]
Smiseth PT, Wright J, & Kölliker M (2008) Parent-offspring conflict and co-adaptation: behavioural ecology meets quantitative genetics. Proc R Soc Lond B 275: 1823–1830 [Chap 26]
Smith EA (2004) Why do good hunters have higher reproductive success? Human Nature 15: 343–364 [Chap 31]
Smith EA & Winterhalder B (1992) Evolutionary Ecology and Human Behavior: Foundations of Human Behavior. Aldine Transaction, Chicago [Chap 31]
Smith RH & Lessells CM (1985) Oviposition, ovicide, and larval competition in granivorous insects. Pp 423–448 in Sibly RM & Smith RH (eds) Behavioural Ecology: Ecological Consequences of Adaptive Behaviour. Blackwell, Oxford [Chap 11]
Smith RL (1984) Sperm Competition and the Evolution of Animal Mating Systems. Academic Press, Orlando, FL [Chap 1]
Smith TB (1993) Disruptive selection and the genetic basis of bill size polymorphisms in the African Finch *Pyrenestes*. Nature 363: 618–620 [Chap 3]
Smulders TV & DeVoogd TJ (2000) Expression of immediate early genes in the hippocampal formation of the black-capped chickadee (*Poecile atricapillus*) during a food-hoarding task. Behavioural Brain Research 114: 39–49 [Chap 10]
Smuts BB & Smuts RW (1993) Male aggression and sexual coercion of females in nonhuman primates and other mammals: evidence and theoretical implications. Adv Stud Behav 22: 1–63 [Chap 25]
Sneddon LU, Huntingford FA, & Taylor AC (1997) Weapon size versus body size as a predictor of winning in fights between shore crabs, *Carcinus maenas* (L.). Behav Ecol Sociobiol 41: 237–242 [Chap 15]
Sneddon LU, Margareto J, & Cossins AR (2005) The use of transcriptomics to address questions in behaviour: production of a suppression subtractive hybridisation library from dominance hierarchies of rainbow trout. Physiol Biochem Zool 78: 695–705 [Chap 15]
Sneddon LU, Taylor AC, & Huntingford FA (1999) Metabolic consequences of agonistic behaviour: crab fights in declining oxygen tensions. Anim Behav 57: 353–363 [Chap 15]
Sneddon LU, Taylor AC, Huntingford FA, & Watson DG (2000) Agonistic behaviour and biogenic amines in shore crabs. J Exp Biol 203: 537–545 [Chap 15]
Snook RR & Hosken DJ (2004) Sperm death and dumping in *Drosophila*. Nature 428: 939–941 [Chap 22]
Snook RR, Chapman T, Moore PJ, Wedell N, & Crudgington HS (2009) Interactions between the sexes: new perspectives on sexual selection and reproductive isolation. Evol Ecol 23: 71–91 [Chap 23]
Snyder BF & Gowaty PA (2007) A reappraisal of Bateman's classic study of intrasexual selection. Evolution 61: 2457–2468 [Chap 1]
Sober E (1984) The Nature of Selection. MIT Press, Cambridge, MA [Chap 2]
Sober E & Wilson DS (1998) Unto others: the evolution and psychology of unselfish behavior. Harvard Univ Press, Cambridge, MA [Chap 31]
Socci AM, Schlaepfer MA, & Gavin TA (2005) The importance of soil moisture and leaf cover in a female lizard's (*Norops polylepis*) evaluation of potential oviposition sites. Herpetologica 61: 233–240 [Chap 29]
Sogabe A & Yanagisawa Y (2007) Sex-role reversal of a monogamous pipefish without higher potential reproductive rate in females. Proc R Soc Lond B 274: 2959–2963 [Chap 20]
Sokolowski MB (1980) Foraging strategies of *Drosophila melanogaster:* a chromosomal analysis. Behav Genet 10: 291–302 [Chap 5]
Sol D, Duncan, RP, Blackburn TM, Cassey P, & Lefebvre L (2005) Big brains, enhanced cognition, and response of birds to novel environments. Proc Natl Acad Sci 102: 5460–5465 [Chap 6]
Sol D, Timmermans S, & Lefebvre L (2002) Behavioural flexibility and invasion success in birds. Anim Behav 63: 495–502 [Chap 30]
Soler JJ, Cuervo JJ, Møller AP, & De Lope F (1998) Nest building is a sexually selected behaviour in the barn swallow. Anim Behav 56: 1435–1442 [Chap 26]

Sonerud GA, Smedshaug CA, & Bråthen O (2001) Ignorant hooded crows follow knowledgeable roost-mates to food: support for the information centre hypothesis. Proc R Soc Lond B 268: 827–831 [Chap 17]

Soong K, Chen J, & Tsao CJ (2006) Adaptation for accuracy or for precision? Diel emergence timing of the intertidal insect *Pontomyia oceana* (Chironomidae). Mar Biol 150: 173–181 [Chap 21]

Sparck R (1927) Studies on the biology of the oyster (*Ostrea edulis*) IV. On fluctuations in the oyster stock in the Limfjord. Rep Danish Biol Sta 33: 60–65 [Chap 21]

Spencer H (1866) *The Principles of Biology*. Appleton, London [Chap 4]

Sperber D (2005) Modularity and relevance: how can a massively modular mind be flexible and context-sensitive? Pp 53–68 in Carruthers P, Laurence S, & Stich S (eds) The innate mind: structure and contents. Oxford Univ Press, New York [Chap 31]

Spetch ML & Edwards CA (1986) Spatial memory in pigeons (*Columba livia*) in an open-field feeding environment. J Com Psych 100: 266–278 [Chap 10]

Spivak M (1996) Honey bee hygienic behavior and defense against *Varroa jacobsoni*. Apidologie 27: 245–260 [Chap 28]

Stamps J (2003) Behavioural processes affecting development: Tinbergen's fourth question comes of age. Anim Behav 66: 1–13 [Chap 6]

Stamps J, Luttbeg B, & Krishnan VV (2009) Effects of survival on the attractiveness of cues to natal dispersers. Am Nat 173: 41–46 [Chap 29]

Stamps JA (2007) Growth-mortality tradeoffs and personality traits in animals. Ecol Lett 10: 355–363 [Chaps 12, 30]

Stamps JA & Krishnan VV (1999) Learning-based model of territory establishment. Q Rev Biol 74: 291–318 [Chap 17]

Stankowich T & Blumstein DT (2005) Fear in animals: a meta-analysis and review of risk assessment. Proc R Soc Lond B 272: 2627–2634 [Chap 13]

Starks PT, Blackie CA, & Seeley TD (2000) Fever in honeybee colonies. Naturwissenschaften 87: 229–231 [Chap 28]

Starr M, Himmelman JH, & Therriault JC (1990) Direct coupling of marine invertebrate spawning with phytoplankton blooms. Science 247: 1071–1074 [Chap 21]

Stearns SC (1989) Trade-offs in life-history evolution. Funct Ecol 3: 259–268 [Chap 26]

Stearns SC (1992) The Evolution of Life Histories. Oxford Univ Press, Oxford [Chaps 4, 12, 26]

Stefansson H, Rujescu D, Cichon S, Pietilainen OP, Ingason A, et al. (2008) Large recurrent microdeletions associated with schizophrenia. Nature. 455: 232–236 [Chap 28]

Steiner UK (2007) Investment in defense and cost of predator-induced defense along a resource gradient. Oecologia 152: 201–210 [Chap 12]

Stephens DW (1989) Variance and the value of information. Am Nat 134: 128–140 [Chap 12]

Stephens DW (2007) Models of information use. Pp 31–58 in Stephens DW, Brown JS, & Ydenberg RC (eds) Foraging: Behavior and Ecology. Univ Chicago Press, Chicago [Chap 12]

Stephens DW & Krebs JR (1986) Foraging Theory. Princeton Univ Press, Princeton, NJ [Chaps 1, 2, 8, 10, 11, 12, 13]

Stephens DW, Brown J, & Ydenberg RC (eds) (2007) Foraging: Behavior and Ecology. Univ Chicago Press, Chicago [Chaps 8, 11, 12, 13]

Stephens DW, Kerr B, & Fernandez-Juricic E (2004) Impulsiveness without discounting: the ecological rationality hypothesis. Proc R Soc Lond B 271 : 2459–2465 [Chap 8]

Stephens DW, Lynch JF, Sorensen AE, & Gordon C (1986) Preference and profitability: theory and experiment. Am Nat 127: 533–553 [Chap 8]

Steppan SJ, Phillips PC, & Houle D (2002) Comparative quantitative genetics: evolution of the G matrix. Trends Ecol Evol 17: 320–327 [Chap 30]

Sterelny, K (2000) Development, evolution, adaptation. Phil Sci 67: S369-S387 [Chap 31]

Stevens JR & Hauser MD (2004) Why be nice? Psychological constraints on the evolution of cooperation. Trends Cogn Sci 8: 60–65 [Chap 18]

Stevens M & Cuthill IC (2007) Hidden messages: are ultraviolet signals a special channel in avian communication? BioScience 57: 501–507 [Chap 9]

Stewart AD, Morrow EH, & Rice WR (2005) Assessing putative interlocus sexual conflict in *Drosophila melanogaster* using experimental evolution. Proc R Soc Lond B 272: 2029–2035 [Chap 23]

Stoleson SH & Beissinger SR (1995) Hatching asynchrony and the onset of incubation in birds, revisited: when is the critical period? Pp 191–270 in Power DM (ed) Current Ornithology. Plenum PressNew York [Chap 26]

Stone JL, O'Donovan MC, Gurling H, Kirov GK, Blackwood DH, et al. (2008) Rare chromosomal deletions and duplications increase risk of schizophrenia. Nature 455: 237–241 [Chap 28]

Strassmann J, Sullender B, & Queller D (2002) Caste totipotency and conflict in a large-colony social insect. Proc R Soc Lond B 269: 263–270 [Chap 19]

Strassmann JE, Zhu Y, & Queller DC (2000) Altruism and social cheating in the social amoeba, *Dictyostelium discoideum*. Nature 408: 965–967 [Chap 19]

Strathmann RR (1985) Feeding and nonfeeding larval development and life-history in marine invertebrates. Annu Rev Ecol Syst 16: 339–361 [Chap 21]

Strathmann RR (1990) Why life histories evolve differently in the sea. Am Zool 30: 197–207 [Chap 21]

Strauss SY, Lau JA, & Carroll SP (2006) Evolutionary responses of natives to introduced species: what do introductions tell us about natural communities? Ecol Lett 9: 357–371 [Chap 29]

Strobeck C (1975) Selection in fine-grained environments. Am Nat 109: 31–55 [Chap 25]

Stuart-Fox D & Moussalli A (2008) Selection for social signalling drives the evolution of chameleon colour change. PLoS Biol 6: 22–29 [Chap 9]

Stuart-Fox D, Moussalli A, & Whiting MJ (2008) Predator-specific camouflage in chameleons. Biol Lett 4: 326–329 [Chap 9]

Stuart-Fox D & Ord TJ (2004) Sexual selection, natural selection and the evolution of dimorphic coloration and ornamentation in agamid lizards. Proc R Soc Lond B 271: 2249–2255 [Chap 7]

Stubblefield JW & Charnov EL (1986) Some conceptual issues in the origin of eusociality. Heredity 57: 181–187 [Chap 19]

Stumpf RM & Boesch C (2005) Does promiscuous mating preclude female choice? Female sexual strategies in chimpanzees (Pan troglodytes verus) of the Taï National Park, Côte d'Ivoire. Behav Ecol Sociobiol 57: 511–524 [Chap 31]

Styan CA (1998) Polyspermy, egg size, and the fertilization kinetics of free-spawning marine invertebrates. Am Nat 152: 290–297 [Chap 21]

Sullivan PF (2008) Schizophrenia genetics: the search for a hard lead. Curr Opin Psychiatr 21: 157–160 [Chap 5]

Sumpter DJT (2006) The principles of collective animal behaviour. Phil Trans R Soc Lond B 361: 5–22 [Chap 11]

Sundell J, Dudek D, Klemme I, Koivisto E, Pusenius J, & Ylönen H (2004) Variation in predation risk and vole feeding behaviour: a field test of the risk allocation hypothesis. Oecologia 139: 157–162 [Chap 13]

Suomi SJ, Novak MA, & Well A (1996) Aging in rhesus monkeys: different windows on behavioral continuity and change. Dev Psych 32: 1116–1128 [Chap 30]

Sutherland WJ (1983) Aggregation and the ideal free distribution. J Anim Ecol 52: 821–828 [Chap 11]

Sutherland WJ (1985) Chance can produce a sex difference in variance in mating success and explain Bateman's data. Anim Behav 33: 1349–1353 [Chap 20]

Sutherland WJ (2006) Predicting the ecological consequences of environmental change: a review of the methods. J Appl Ecol 43: 599–616 [Chap 29]

Sutherland WJ & Koene P (1982) Field estimates of the strength of interference between oystercatchers *Haematopus ostralegus*. Oecologia 55: 108–109 [Chap 11]

Svensson E, Abbott J, & Hardling R (2005) Female polymorphism, frequency dependence, and rapid evolutionary dynamics in natural populations. Am Nat 165: 567–576.

Svensson O & Kvarnemo C (2003) Sexually selected nest building—*Pomatoschistus minutus* males build smaller nest-openings in the presence of sneaker males. J Evol Biol 16: 896–902 [Chap 26]

Svensson O, Magnhagen C, Forsgren E, & Kvarnemo C (1998) Parental behaviour in relation to the occurrence of sneaking in the common goby. Anim Behav 56: 1285–1290 [Chap 26]

Swaddle JP (1999) Visual signalling by asymmetry: a review of perceptual processes. Phil Trans R Soc Lond 354: 1383–1393 [Chap 10]

Swanson WJ, Clark AG, Waldrip-Dail HM, Wolfner MF, & Aquadro CF (2001) Evolutionary EST analysis identifies rapidly evolving male reproductive proteins in *Drosophila*. Proc Natl Acad Sci USA 98: 7375–7379 [Chap 23, 28]

Swanson WJ & Vacquier VD (2002a) Reproductive protein evolution. Annu Rev Ecol Syst 33: 161–179 [Chap 21]

Swanson WJ & Vacquier VD (2002b) The rapid evolution of reproductive proteins. Nature Rev Genet 3: 137–144 [Chap 23]

Swanson WJ, Wong A, Wolfner MF, & Aquadro CF (2004) Evolutionary expressed sequence tag analysis of *Drosophila* female reproductive tracts identifies genes subjected to positive selection. Genetics 168: 1457–1465 [Chaps 23, 28]

Symons D (1979) The Evolution of Human Sexuality. Oxford Univ Press [Chap 31]

Symons D (1992) On the use and misuse of Darwinism in the study of human behavior. Pp 137–159 in Barkow JH, Cosmides L, & Tooby J (eds) The Adapted Mind: Evolutionary Psychology and the Generation of Culture. Oxford Univ Press, New York [Chap 31]

Számadó S (1999) The validity of the handicap principle in discrete action-response games. J Theor Biol 198: 593–602 [Chap 16]

Szathmáry E & Maynard Smith J (1995) The major evolutionary transitions. Nature 374: 227–232 [Chap 19]

Taborsky M (1994) Sneakers, satellites, and helpers: parasitic and cooperative behavior in fish reproduction. Adv Stud Behav 23: 1–100 [Chap 25]

Taborsky M (1998) Sperm competition in fish: 'bourgeois' males and parasitic spawning. Trends Ecol Evol 13: 222–227 [Chap 25]

Taborsky M (2008a) The use of theory in behavioural research. Ethology 114: 1–6 [Chap 17]

Taborsky M (2008b) Alternative reproductive tactics in fish. Pp 251–299 in Oliviera RF, Taborsky M, & Brockmann HJ (eds) Alternative Reproductive Tactics. Cambridge Univ Press, Cambridge [Chap 25]

Taborsky M, Oliviera RF, & Brockmann HJ (2008) The evolution of alternative reproductive tactics: concepts and questions. Pp 1–21 in RF Oliviera, M Taborsky, & HJ Brockmann (eds) Alternative Reproductive Tactics. Cambridge Univ Press, Cambridge [Chap 25]

Taggart DA, O'Brien HP, & Moore HDM (1993) Ultrastructural characteristics of in vivo and in vitro fertilization in the grey short-tailed opossum, *Monodelphis domestica*. Anat Rec 237: 21–37 [Chap 22]

Tallamy DW (2000) Sexual selection and the evolution of exclusive paternal care in arthropods. Anim Behav 60: 559–567 [Chaps 20, 26]

Tallamy DW (2001) Evolution of exclusive parental care in arthropods. Annu Rev Entomol 46: 139–165 [Chap 20]

Tanaka KD & Ueda K (2005) Horsfield's hawk-cuckoo nestlings simulate multiple gapes for begging. Science 308: 653 [Chap 9]

Tang-Martinez Z & Ryder TB (2005) The problem with paradigms: Bateman's worldview as a case study. Integr Comp Biol 45: 821–830 [Chap 20]

Tauber E & Kyriacou CP (2008) Genomic approaches for studying biological clocks. Funct Ecol 22: 19–29 [Chap 5]

Taylor PD (1990) Allele-frequency change in a class-structured population. Am Nat 135: 95–106 [Chap 18]

Taylor PD (1992) Inclusive fitness in a homogeneous environment. Proc R Soc Lond B 249: 299–302 [Chap 18]

Taylor PD (1996) Inclusive fitness arguments in genetic models of behaviour. J Math Biol 34: 654–674 [Chap 18]

Taylor PD, Wild G, & Gardner A (2007) Direct fitness or inclusive fitness: how shall we model kin selection? J Evol Biol 20: 301–309 [Chap 18]

Taylor PW & Elwood RW (2003) The mis-measure of animal contests. *Anim Behav* 65: 1195–1202 [Chap 15]

Templeton AR & Lawlor LR (1981) The fallacy of the averages in ecological optimization theory. Am Nat 117: 390–393 [Chap 8]

Templeton JJ & Giraldeau L-A (1996) Vicarious sampling: the use of personal and public information by starlings foraging in a simple patchy environment. Behav Ecol Sociobiol 38: 105–114 [Chap 17]

ten Cate C & Rowe C (2007) Biases in signal evolution: learning makes a difference. Trends Ecol Evol 22: 380–387 [Chap 10]

ten Cate C, Verzijden MN, & Etman E (2006) Sexual imprinting can induce sexual preferences for exaggerated parental traits. Curr Biol 16: 1128–1132 [Chap 10]

Theodor O (1976) On the Structure of the Spermathecae and Aedeagus in the Asilidae and Their Importance in the Systematics of the Family. The Israel Academy of Sciences and Humanities, Jerusalem [Chap 22]

Théry M & Casas J (2002) Predator and prey views of spider camouflage. Nature 415: 133–133 [Chap 9]

Thompson JN (1998) Rapid evolution as an ecological process. Trends Ecol Evol 13: 329–332 [Chap 29]

Thornhill R (1981) *Panorpa* (Mecoptera: Panorpidae) scorpionflies: systems for understanding resource-defense polygyny and alternative male reproductive efforts. Annu Rev Ecol Syst 12: 355–386 [Chap 25]

Thornhill R (1983) Cryptic female choice and its implications in the scorpion fly *Harpobittacus nigriceps*. Am Nat 122: 765–788 [Chap 24]

Thornhill R & Gangestad SW (2008) The Evolutionary Biology of Human Female Sexuality. Oxford Univ Press, New York [Chap 31]

Thorpe KE, Taylor AC, & Huntingford FA (1995) How costly is fighting? Physiological effects of sustained exercise and fighting in swimming crabs, *Necora puber* (L) (Brachyura, Portunidae). Anim Behav 50: 1657–1666 [Chap 15]

Thorson G (1946) Reproduction and larval development of Danish marine bottom invertebrates. Meddr Kommn Danm Fisk- og Havunders Ser. Plankton 4: 1–523 [Chap 21]

Thresher RE (1984) Reproduction in reef fishes. T.F.H. Publications, Neptune City, NJ [Chap 21]

Thusius KJ, Peterson AK, Dunn PO, & Whittingham LA (2001) Male mask size is correlated with mating success in the common yellowthroat. Anim Behav 62: 435–446 [Chap 4]

Tibbetts EA & Lindsay R (2008) Visual signals of status and rival assessment in *Polistes dominulus* paper wasps. Biol Lett 4: 237–239 [Chap 15]

Tilmon K (2007) Specialization, Speciation, and Radiation: The Evolutionary Biology of Herbivorous Insects. California Univ Press, Cambridge [Chap 27]

Tinbergen N (1951) The Study of Instinct. Oxford Univ Press [Chap 1]

Tinbergen N (1953a) The Herring Gull's World. Collins, London [Chap 1]

Tinbergen N (1953b) Social Behavior in Animals. London: Methuen [Chap 1]

Tinbergen N (1963) On aims and methods of ethology. Z Tierpsychol 20: 410–433 [Chaps 1, 7, 29, 31]
Tinbergen N, Broekhuysen GJ, Feekes F, Houghton JCW, Kruuk H, & Szulc E (1962) Egg shell removal by the black-headed gull *Larus ridibundus*. Behavior 19: 74–118 [Chap 1]
Tinbergen N & Tinbergen EA (1972) Early Childhood Autism: An Ethological Approach. Parey, Berlin [Chap 1]
Tomaiuolo M, Hansen TF, & Levitan DR (2007) A theoretical investigation of sympatric evolution of temporal reproductive isolation as illustrated by marine broadcast spawners. Evolution 61: 2584–2595 [Chap 21]
Tomkins JL & Brown GS (2004) Population density drives the local evolution of a threshold dimorphism. Nature 431: 1099–1103 [Chap 25]
Tomkins JL & Hazel W (2007) The status of the conditional evolutionarily stable strategy. Trends Ecol Evol 22: 522–528 [Chaps 6, 25]
Tomkins JL, Radwan J, Kotiaho JS, & Tregenza T (2004) Genic capture and resolving the lek paradox. Trends Ecol Evol 19: 323–328 [Chap 20]
Tooby J & Cosmides L (1992) The psychological foundations of culture. Pp 19–136 in Barkow JH, Cosmides L, & Tooby J (eds) The Adapted mind: Evolutionary Psychology and the Generation of Culture. Oxford Univ Press, New York [Chap 31]
Tooby J, Cosmides L, Sell A, Lieberman D, & Sznycer D (2008). Internal Regulatory Variables and the Design of Human Motivation: a Computational and Evolutionary Approach. Pp. 251–271 in Elliot A (ed.) Handbook of Approach and Avoidance Motivation. Lawrence Erlbaum, New York [Chap 31]
Toth AL, Kantarovich S, Meisel AF, & Robinson GE (2005) Nutritional status influences socially regulated foraging ontogeny in honey bees. J Exp Biol 208: 4641–4649 [Chap 28]
Toth AL & Robinson GE (2007) Evo-devo and the evolution of social behavior. Trends Genet 23: 334–341 [Chaps 7, 28]
Toth AL, Varala K, Newman TC, Miguez FE, Hutchison SK, et al. (2007) Wasp gene expression supports an evolutionary link between maternal behavior and eusociality. Science 318: 441–444 [Chap 28]
Toth E & Duffy JE (2005) Coordinated group response to nest intruders in social shrimp. Biol Lett 1: 49–52 [Chap 17]
Tovée MJ (1995) Ultraviolet photoreceptors in the animal kingdom: their distribution and function. Trends Ecol Evol 10: 455–460 [Chap 9]
Trainor BC, Rowland MR, & Nelson RJ (2007) Photoperiod affects estrogen receptor alpha, estrogen receptor beta and aggressive behavior. Eur J Neurosci 26: 207–218 [Chap 15]
Travisano M & Rainey PB (2000) Studies of adaptive radiation using model microbial systems. Am Nat 156: S35–S44 [Chap 4]
Trefil J (2003) The Nature of Science: An A–Z Guide to the Laws and Principles Governing Our Universe. Houghton Mifflin Company, Boston [Chap 8]
Tregenza T, Parker GA, & Thompson DJ (1996) Interference and the ideal free distribution: models and tests. Behav Ecol 7: 379–386 [Chap 11]
Tregenza T, Simmons LW, Wedell N, & Zuk M (2006) Female preference for male courtship song and its role as a signal of immune function and condition. Anim Behav 72: 809–818 [Chap 24]
Tregenza T & Thompson DJ (1998) Unequal competitor ideal free distribution in fish? Evol Ecol 12: 655–666 [Chap 11]
Tregenza T & Wedell N (1998) Benefits of multiple mates in the cricket *Gryllus bimaculatus*. Evolution 52: 1726–1730 [Chap 24]
Tregenza T & Wedell N (2002) Polyandrous females avoid costs of inbreeding. Nature 415: 71–73 [Chap 24]
Trivers RL (1971) The evolution of reciprocal altruism. Q Rev Biol 46: 35–57 [Chaps 3, 18]
Trivers RL (1972) Parental investment and sexual selection. Pp 136–179 in Campbell B (ed) Sexual Selection and the Descent of Man 1871–1971. Aldine-Atherton, Chicago [Chaps 1, 20, 23, 26]
Trivers RL (1974) Parent-offspring conflict. Am Zool 14: 249–264 [Chap 26]
Trivers RL (2002) Natural Selection and Social Theory: Selected Papers of Robert L. Trivers. Oxford Univ Press, New York [Chap 1, 20]
Trivers RL & Hare H (1976) Haplodiploidy and the evolution of the social insects. Science 191: 249–263 [Chap 19]
Trivers RL & Willard DE (1973) Natural selection of parental ability to vary the sex ratio of offspring. Science 179: 90–92 [Chap 17, 26]
Tsaur SC, Ting CT, & Wu CI (1998) Positive selection driving the evolution of a gene of male reproduction, Acp26Aa, of *Drosophila:* II. Divergence versus polymorphism. Molec Biol Evol 15: 1040–1046 [Chap 5]
Tso IM, Liao CP, Huang RP, & Yang EC (2006) Function of being colorful in web spiders: attracting prey or camouflaging oneself? Behav Ecol 17: 606–613 [Chap 9]
Tufto J (2000) The evolution of plasticity and nonplastic spatial and temporal adaptations in the presence of imperfect environmental cues. Am Nat 156: 121–130 [Chap 29]
Turillazzi S (1991) The stenogastrinae. Pp 74–98 in Ross KG & Matthews RW (eds) The Social

Biology of Wasps. Cornell Univ Press, Ithaca, NY [Chap 19]
Turner E & Montgomerie R (2002) Ovarian fluid enhances sperm movement in arctic charr, *Salvelinus alpinus*. J Fish Biol 60: 1570–1579 [Chap 22]
Turner GF & Huntingford FA (1986) A problem for game-theory analysis—assessment and intention in male mouthbrooder contests. Anim Behav 34: 961–970 [Chap 15]
Turner GF & Pitcher TJ (1986) Attack abatement: a model for group protection by combined avoidance and dilution. Am Nat 128: 228–240 [Chap 17]
Tyler MJ (1994) Australian Frogs. Reed Books Australia, Chatswood, NSW [Chap 29]
Uller T (2006) Sex-specific sibling interactions and offspring fitness in vertebrates: patterns and implications for maternal sex ratios. Biol Rev 81: 207–217 [Chap 26]
Valera F, Hoi H, & Schleicher B (1997) Egg burial in penduline tits, *Remiz pendulinus:* Its role in mate desertion and female polyandry. Behav Ecol 8: 20–27 [Chap 23]
Valone TJ (2007) From eavesdropping on performance to copying the behavior of others: a review of public information use. Behav Ecol Sociobiol 62: 1–14 [Chap 17]
Vamosi SM & Schluter D (1999) Sexual selection against hybrids between sympatric stickleback species: evidence from a field experiment. Evolution 53: 874–879 [Chap 27]
van Bergen Y, Coolen I, & Laland KN (2004) Nine-spined sticklebacks exploit the most reliable source when public and private information conflict. Proc R Soc Lond B 271: 957–962 [Chap 17]
van Buskirk J, Müller C, Portmann A, & Surbeck M (2002) A test of the risk allocation hypothesis: tadpole responses to temporal change in predation risk. Behav Ecol 13: 526–530 [Chap 13]
van der Meer J & Ens BJ (1997) Models of interference and their consequences for the spatial distribution of ideal and free predators. J Anim Ecol 66: 846–858 [Chap 11]
van der Post DJ & Hogeweg P (2008) Diet traditions and cumulative cultural processes as side-effects of grouping. Anim Behav 75: 133–144 [Chap 11]
van Doorn GS & Kirkpatrick M (2007) Turnover of sex chromosomes induced by sexual conflict. Nature 449: 909–912 [Chap 5]
van Homrigh A, Higgie M, McGuigan K, & Blows MW (2007) The depletion of genetic variance by sexual selection. Curr Biol 17: 528–532 [Chap 24]
van Oers K, de Jong G, van Noordwijk AJ, Kempenaers B, & Drent PJ (2005) Contribution of genetics to the study of animal personalities: a review of case studies. Behaviour 142: 1185–1206 [Chap 30]
van Woesik R, Lacharmoise F, & Koksal S (1996) Annual cycles of solar insolation predict spawning times of Caribbean corals. Ecol Lett 9: 390–398 [Chap 21]
Veen T, Borge T, Griffith SC, Saetre GP, Bures S, Gustafsson L, & Sheldon BC (2001) Hybridization and adaptive mate choice in flycatchers. Nature 411: 45–50 [Chap 24]
Veenema AH, Meijer OC, de Kloet ER, & Koolhaas JM (2003) Genetic selection for coping style predicts stressor susceptibility. J Neuroendocrin 15: 256–267 [Chap 30]
Vehrencamp SL (2000) Handicap, index, and conventional signal elements of bird song. Pp 277–300 In Espmark E, Amundsen T, & Rosenqvist G (eds) Animal Signals: Signaling and Signal Design in Animal Communication. Tapir, Trondheim [Chap 16]
Vehrencamp SL (2001) Is song-type matching a conventional signal of aggressive intentions? Proc R Soc Lond B 268: 1637–1642 [Chap 16]
Velicer GJ, Raddatz G, Keller H, Deiss S, Lanz C, Dinkelacker I, & Schuster SC (2006) Comprehensive mutation identification in an evolved bacterial cooperator and its cheating ancestor. Proc Natl Acad Sci USA 103: 8107–8112 [Chap 28]
Vera JC, Wheat CW, Fescemyer HW, Frilander MJ, Crawford DL, Hanski S, & Marden JH (2008) Rapid transcriptome characterization for a nonmodel organism using 454 pyrosequencing. Mol Ecol 17: 1636–1647 [Chap 28]
Verbeek MEM, Boon A, & Drent PJ (1996) Exploration, aggressive behaviour and dominance in pair-wise confrontations of juvenile male great tits. Behaviour 133: 945–963 [Chap 30]
Verbeek MEM, Drent PJ, & Wiepkema PR (1994) Consistent individual differences in early exploratory behaviour of male great tits. Anim Behav 48: 1113–1121 [Chap 30]
Vermeji GJ (1987) Evolution and Escalation: an Ecological History of Life. Princeton Univ Press, Princeton, NJ [Chap 25]
Via S (1987) Genetic constraints on the evolution of phenotypic plasticity. Pp 46–71 in Loeschcke V (ed) Genetic Constraints on Adaptive Evolution. Springer, Berlin [Chap 6]
Via S (1994) The evolution of phenotypic plasticity: what do we really know? Pp 35–57 in Real LA (ed) Ecological Genetics. Princeton Univ Press, Princeton, NJ [Chap 6]
Via S (1999) Reproductive isolation between sympatric races of pea aphids. I. Gene flow restriction and habitat choice. Evolution 53: 1446–1457 [Chap 27]
Via S, Gomulkiewicz R, de Jong G, Scheiner SM, Schlichting CD, & Van Tienderen PH (1995)

Adaptive plasticity: consensus and controversy. Trends Ecol Evol 10: 212–217 [Chap 6]
Via S & Lande R (1985) Genotype-environment interaction and the evolution of phenotypic plasticity. Evolution 39: 505–522 [Chap 6]
Vickery WL, Giraldeau L-A, Templeton JJ, Kramer DL, & Chapman CA (1991) Producers, scroungers and group foraging. Am Nat 137: 847–863 [Chap 11]
Viscido SV & Wethey DS (2002) Quantitative analysis of fiddler crab flock movement: evidence for "selfish herd" behavior. Anim Behav 63: 735–741 [Chap 17]
Viscido SV, Miller M, & Wethey DS (2002) The dilemma of the selfish herd: the search for a realistic movement rule. J Theor Biol 217: 183–194 [Chap 17]
Visser ME (2008) Keeping up with a warming world; assessing the rate of adaptation to climate change. Proc R Soc Lond B 275: 649–659 [Chap 29]
Visser ME, Both C, & Lambrechts MM (2004) Global climate change leads to mistimed avian reproduction. Adv Ecol Res 35: 89–110 [Chap 29]
Visser ME, Van Noordwijk AJ, Tinbergen JM, & Lessells CM (1998) Warmer springs lead to mistimed reproduction in great tits (*Parus major*). Proc R Soc Lond B 265: 1867–1870 [Chap 29]
Vitousek MN, Mitchell MA, Woakes AJ, Niemack MD, & Wikelski M (2007) High costs of female choice in a lekking lizard. PLoS One 2: e567 [Chap 24]
Vitousek PM, Mooney HA, Lubchenco J, & Melillo JM (1997) Human domination of Earth's ecosystems. Science 277: 494–499 [Chap 29]
Volterra V (1926) Variations and fluctuations of the number of individuals in animal species living together. Reprinted 1931 in RN Chapman, Animal Ecology, McGraw-Hill, New York [Chap 13]
von der Emde G (1999) Active electrolocation of objects in weakly electric fish. J Exp Biol 202: 1205–1215 [Chap 9]
Vorobyev M & Osorio D (1998) Receptor noise as a determinant of colour thresholds. Proc R Soc Lond B 265: 351–358 [Chap 9]
Waddington CH (1961) Genetic assimilation. Adv Genet 10: 257–293 [Chap 6]
Wade MJ (1976) Group selection among laboratory populations of *Tribolium*. Proc Natl Acad Sci USA 73: 4604–4607 [Chap 14]
Wade MJ (1977) An experimental study of group selection. Evolution 31: 134–153 [Chap 14]
Wade MJ (1979) Sexual selection and variance in reproductive success. Am Nat 114: 742–764 [Chaps 24, 25]
Wade MJ (1995) The ecology of sexual selection: mean crowding of females and resource-defence polygyny. Evol Ecol 9: 118–124 [Chap 25]
Wade MJ & Arnold SJ (1980) The intensity of sexual selection in relation to male sexual behaviour, female choice, and sperm precedence. Anim Behav 28: 446–461 [Chap 25]
Wade MJ & Shuster SM (2004) Sexual selection: harem size and the variance in male reproductive success. Am Nat 164: E83–E89 [Chap 25]
Wade MJ, Shuster SM, & Demuth JP (2003) Sexual selection favors female-biased sex ratios: the balance between the opposing forces of sex-ratio selection and sexual selection. Am Nat 162: 403–414 [Chap 25]
Wagner GP, & Altenberg L (1996) Perspective: complex adaptations and the evolution of evolvability. Evolution 50: 967–976 [Chap 31]
Wagner WE Jr (1998) Measuring female mating preferences. Anim Behav 55: 1029–1042 [Chap 27]
Wahlsten D, Metten P, Phillips TJ, Boehm SL II, Burkhart-Kasch S, et al. (2003) Different data from different labs: lessons from studies of gene-environment interaction. J Neurobiol 54: 283–311 [Chap 28]
Waite TA & Ydenberg RC (1994) What currency do scatter-hoarding gray jays maximize? Behav Ecol Sociobiol 34: 43–49 [Chap 11]
Wajnberg E, Bernhard P, Hamelin F, & Boivin G (2006) Optimal patch time allocation for time-limited foragers. Behav Ecol Sociobiol 60: 1–10 [Chap 13]
Wallace B (1968) Polymorphism, population size, and genetic load. Pp 87–108 in Lewontin RC (ed) Population Biology and Evolution. Syracuse Univ Press, Syracuse, NY [Chap 4]
Wallace B (1975) Hard and soft selection revisited. Evolution 29: 465–473 [Chap 4]
Walton P, Ruxton GD, & Monaghan P (1998) Avian diving, respiratory physiology and the marginal value theorem. Anim Behav 56: 165–174 [Chap 11]
Waltz EC & Wolf LL (1988) Alternative mating tactics in male white-faced dragonflies (*Leucorrhinia intacta*): plasticity of tactical options and consequences for reproductive success. Evol Ecol 2: 205–231 [Chap 25]
Wang IN, Dykhuizen DE, & Slobodkin LB (1996) The evolution of phage lysis timing. Evol Ecol 10: 545–558 [Chap 11]
Wang J, Ross KG, & Keller L (2008) Genome-wide expression patterns and the genetic architecture of a fundamental social trait. PLoS Genet 4: e1000127 [Chap 28]
Wang TB, Nonacs P, & Blumstein DT (2009) Social skew as a measure of the costs and benefits of group-living in marmots. Pp 114–133 in Hager R & Jones CB (eds) Reproductive Skew in Vertebrates: Proximate and Ultimate Causes. Cambridge Univ Press, Cambridge [Chap 13]

Ward AJW, Sumpter DJT, Couzin ID, Hart PJB, & Krause J (2008) Quorum decision-making facilitates information transfer in fish shoals. Proc Natl Acad Sci USA 105: 6948–6953 [Chap 11]

Ward P & Zahavi A (1973) The importance of certain assemblages of birds as "information-centres" for food finding. Ibis 115: 517–534 [Chap 17]

Warner DA & Shine R (2008) The adaptive significance of temperature-dependent sex determination in a reptile. Nature 451: 566–568 [Chaps 5, 26]

Warner RR (1975) The adaptive significance of sequential hermaphroditism in animals. Am Nat 109: 61–82 [Chap 6]

Warner RR, Shapiro DY, Marcanato A, & Petersen, CW (1995a) Sexual conflict—males with highest mating success convey the lowest fertilisation benefits to females. Proc R Soc Lond B 262: 135–139 [Chap 23]

Warner RR, Wernerus F, Lejeune P, & van den Berghe E (1995b) Dynamics of female choice for parental care in a fish where care is facultative. Behav Ecol 6: 73–81 [Chap 26]

Watts DP (2000) Grooming between male chimpanzees at Ngogo, Kibale National Park. II. Influence of male rank and possible competition for partners. Int J Prim 21: 211–238 [Chap 15]

Wcislo WT (1989) Behavioral environments and evolutionary change. Annu Rev Ecol Syst 20: 137–169 [Chap 6]

Wcislo WT & Danforth BN (1997) Secondarily solitary: the evolutionary loss of social behavior. Trends Ecol Evol 12: 468–474 [Chap 17]

Weary DM, Lambrechts MM, & Krebs JR (1991) Does singing exhaust male great tits. Anim Behav 41: 540–542 [Chap 15]

Weber KE (1992) How small are the smallest selectable domains of form? Genetics 130: 345–353 [Chap 30]

Wedekind C & Füri S (1997) Body odour preferences in men and women: do they aim for specific MHC combinations or simply heterozygosity? Proc R Soc London B 264: 1471–1479 [Chaps 3, 31]

Wedekind C, Meyer P, Frischknecht M, Niggli UA, & Pfander H (1998) Different carotenoids and potential information content of red coloration of male three-spined stickleback. J Chem Ecol 24: 787–801 [Chap 16]

Wedell N & Karlsson B (2003) Paternal investment directly affects female reproductive effort in an insect. Proc R Soc Lond B 270: 2065–2071 [Chap 26]

Wedell N, Gage MJG, & Parker GA (2002) Sperm competition, male prudence and sperm-limited females. Trends Ecol Evol 17: 313–320 [Chap 20, 22]

Wedell N, Kvarnemo C, Lessells CM, & Tregenza T (2006) Sexual conflict and life histories. Anim Behav 71: 999–1011 [Chap 26]

Welch AM (2003) Genetic benefits of a female mating preference in gray tree frogs are context-dependent. Evolution 57: 883–893 [Chap 27]

Welch AM, Semlitsch RD, & Gerhardt HC (1998) Call duration as an indicator of genetic quality in male gray tree frogs. Science 280: 1928–1930 [Chap 24]

Welham CVJ & Ydenberg RC (1988) Net energy versus efficiency maximizing by foraging ring-billed gulls. Behav Ecol Sociobiol 23: 75–82 [Chap 8]

Wells KD (2001) The energetics of calling in frogs. Pp 45–60 in Ryan MJ (ed) Anuran Communication. Smithsonian Institution, Washington, DC [Chap 16]

Wells MM & Henry CS (1998) Songs, reproductive isolation, and speciation in cryptic species of insects: a case study using green lacewings. Pp 217–233 in Howard DJ & Berlocher SH (eds) Endless Forms: Species and Speciation. Oxford Univ Press, New York [Chap 27]

Wenseleers T & Ratnieks FLW (2004) Tragedy of the commons in *Melipona* bees. Proc R Soc Lond B 271: S310-S312 [Chap 19]

Wenseleers T, Helantera H, Hart A, & Ratnieks FLW (2004) Worker reproduction and policing in insect societies: an ESS analysis. J Evol Biol 17: 1035–1047 [Chap 18]

Wenseleers T & Ratnieks FLW (2006a) Comparative analysis of worker reproduction and policing in eusocial hymenoptera supports relatedness theory. Am Nat 168: E163–E179 [Chaps 18, 19]

Wenseleers T & Ratnieks FLW (2006b) Enforced altruism in insect societies. Nature 444: 50 [Chap 19]

West SA, Griffin AS, & Gardner A (2007a) Social semantics: altruism, cooperation, mutualism, strong reciprocity and group selection. J Evol Biol 20: 415–432 [Chaps 4, 14, 18]

West SA, Diggle SP, Buckling A, Gardner A, & Griffin AS (2007b) The social lives of microbes. Annu Rev Ecol Evol Syst 38: 53–77 [Chap 18]

West SA, Lively CM, & Read AF (1999) A pluralist approach to sex and recombination. J Evol Biol 12: 1003–1012 [Chap 20]

West SA, Pen I, & Griffin AS (2002) Cooperation and competition between relatives. Science 296: 72–75 [Chap 18]

West SA & Sheldon BC (2002) Constraints in the evolution of sex ratio adjustment. Science 295: 1685–1688 [Chap 26]

West-Eberhard MJ (1975) The evolution of social behavior by kin selection. Q Rev Biol 50: 1–33 [Chap 19]

West-Eberhard MJ (1979) Sexual selection, social competition and evolution. Proc Am Phil Soc 123: 222–234 [Chap 14]
West-Eberhard MJ (1983) Sexual selection, social competition, and speciation. Q Rev Biol 58: 155–183 [Chaps 14, 20, 27]
West-Eberhard MJ (1984) Sexual selection, competitive communication and species-specific signals in insects. Pp 283–324 in Lewis T (ed) Insect Communication. Academic Press, New York [Chap 24]
West-Eberhard MJ (1992) Adaptation: current usages. Pp 13–18 in Keller EF & Lloyd EA (eds) Keywords in Evolutionary Biology. Harvard Univ Press, Cambridge [Chap 2]
West-Eberhard MJ (1996) Wasp societies as microcosms for the study of development and evolution. Pp 290–317 in Turillazzi S & West-Eberhard MJ (eds) Natural History and Evolution of Paper Wasps. Oxford Univ Press, New York [Chaps 19, 28]
West-Eberhard MJ (2003) Developmental Plasticity and Evolution. Oxford Univ Press, New York [Chaps 3, 6, 14, 30]
West-Eberhard MJ (2005) Phenotypic accommodation: adaptive innovation due to developmental plasticity. J Exp Zoolog B Mol Dev Evol 304: 610–618 [Chap 28]
Westermarck EA (1891/1921) The History of Human Marriage. Macmillan, London [Chap 31]
Westneat DF (2000) Toward a balanced view of the sexes: a retrospective and prospective view of genetics and mating patterns. Pp 253–306 in M Appolonio, M Festa-Bianchet, & D Mainardi (eds) Vertebrate Mating Systems. World Scientific, Singapore [Chap 23]
Westneat DF (2006) No evidence of current sexual selection on sexually-dimorphic traits in a bird with high variance in mating success. Am Nat 167: E171–E189 [Chaps 2, 7]
Westneat DF & Sargent RC (1996) Sex and parenting: the effects of sexual conflict and parentage on parental strategies. Trends Ecol Evol 11: 87–91 [Chap 26]
Westneat DF & Sherman PW (1993) Parentage and the evolution of parental behavior. Behav Ecol 4: 66–77 [Chaps 20, 26]
Westneat DF & Stewart IRK (2003) Extra-pair paternity in birds: causes, correlates, and conflict. Annu Rev Ecol Evol Syst 34: 365–396 [Chaps 23, 24]
Wheatcroft DJ & Price TD (2008) Reciprocal cooperation in avian mobbing: playing nice pays. Trends Ecol Evol 23: 416–419 [Chap 17]
Wheeler DA, Kyriacou CP, Greenacre ML, Yu AQ, Rutila JE, Rosbash M, & Hall JC (1991) Molecular transfer of a species-specific behavior from *Drosophila simulans* to *Drosophila melanogaster*. Science 251: 1082–1085 [Chap 5]
Wheeler DE (1986) Developmental and physiological determinants of caste in social hymenoptera: evolutionary implications. Am Nat 128: 13–34 [Chaps 19, 28]
Wheeler WM (1911) The ant colony as organism. J Morphol 22: 307–325
White DW, Dill LM, & Crawford CB (2007) A common, conceptual framework for behavioral ecology and evolutionary psychology. Evol Psychol 5: 275–288 [Chap 8]
White G (1789) The Natural History of Selborne, 3rd (1813) ed. Cassell & Company London [Chap 1]
White SA, Nguyen T, & Fernald RD (2002) Social regulation of gonadotropin-releasing hormone. J Exp Biol 205: 2567–2581 [Chap 28]
Whitfield CW, Cziko AM, & Robinson GE (2003) Gene expression profiles in the brain predict behavior in individual honey bees. Science 302: 296–299 [Chap 28]
Whitfield DP (1987) Plumage variability, status signalling and individual recognition in avian flocks. Trends Ecol Evol 2: 13–18 [Chap 15]
Whitfield J (2004) Everything you always wanted to know about sexes. PLoS Biol 2: 718–721 [Chap 20]
Whittingham LA & Lifjeld JT (1995) High paternal investment in unrelated young: extra-pair paternity and male parental care in house martins. Behav Ecol Sociobiol 37: 103–108 [Chap 26]
Widemo F & Owens IPF (1995) Lek size, male mating skew and the evolution of lekking. Nature 373: 148–151 [Chap 24]
Wiens JJ & Morris MR (1996) Character definitions, sexual selection, and the evolution of swordtails. Am Nat 147: 886–869 [Chap 7]
Wigby S & Chapman T (2004) Female resistance to male harm evolves in response to manipulation of sexual conflict. Evolution 58: 1028–1037 [Chap 23]
Wigby S & Chapman T (2005) Sex peptide causes mating costs in female *Drosophila melanogaster*. Curr Biol 15: 316–321 [Chap 22]
Wigby S & Chapman T (2006) No evidence that experimental manipulation of sexual conflict drives premating reproductive isolation in *Drosophila melanogaster*. J Evol Biol 19: 1033–1039 [Chaps 23, 27]
Wilcox AJ, Weinberg CR, & Baird DD (1995) Timing of sexual intercourse in relation to ovulation: effects on the probability of conception, survival of the pregnancy, and sex of the baby. New Engl J Med 333: 1517–1521 [Chap 31]
Wild G (2006) Sex ratios when helpers stay at the nest. Evolution 60: 2012–2022 [Chap 26]
Wild G & West SA (2007) A sex allocation theory for vertebrates: combining local resource competition and condition-dependent allocation. Am Nat 170: E112–E128 [Chap 26]

Wiley RH (1994) Errors, exageration, and deception in animal communication. Pp 157–192 in Real LA (ed) Behavioural Mechanisms in Evolutionary Ecology. Univ Chicago Press, Chicago [Chap 16]

Wiley RH & Poston J (1996) Indirect mate choice, competition for mates, and coevolution of the sexes. Evolution 50: 1371–1381 [Chap 20]

Wilkinson GS (1984) Reciprocal food sharing in the vampire bat. Nature 308: 181–184 [Chap 18]

Williams CL & Meck WH (1991) The organizational effects of gonadal steroids on sexually dimorphic spatial ability. Psychoneuroendocrinology 16: 155–176 [Chap 10]

Williams EE (1983) Ecomorphs, faunas, island size, and diverse endpoints in island radiations of *Anolis*. Pp 326–370 in Huey RB, Pianka ER, & Schoener TW (eds) Lizard Ecology: Studies of a Model Organism. Harvard Univ Press, Cambridge, MA [Chap 3]

Williams GC (1966) Adaptation and Natural Selection: A Critique of Some Current Evolutionary Thought. Princeton Univ Press, Princeton, NJ [Chap 1, 2, 12, 24, 26]

Williams GC & Nesse RM (1991) The dawn of Darwinian medicine. Q Rev Biol 66: 1–22 [Chap 2]

Willis CKR & Brigham RM (2007) Social thermoregulation exerts more influence than microclimate on forest roost preferences by a cavity-dwelling bat. Behav Ecol Sociobiol 62: 97–108 [Chap 17]

Wilson DS (1975) A theory of group selection. Proc Natl Acad Sci USA 72: 143–146 [Chap 18]

Wilson DS (1977) Structured demes and the evolution of group-advantageous traits. Am Nat 111: 157–185 [Chap 4]

Wilson DS (1998) Adaptive individual differences within single populations. Phil Trans R Soc Lond B 353: 199–205 [Chap 30]

Wilson DS & Dugatkin LA (1997) Group selection and assortative interactions. Am Nat 149: 336–351 [Chap 12]

Wilson DS, Clark AB, Coleman K, & Dearstyne T (1994) Shyness and boldness in humans and other animals. Trends Ecol Evol 11: 442–446 [Chap 30]

Wilson DS & Yoshimura J (1994) On the coexistence of specialists and generalists. Am Nat 144: 692–707 [Chap 12]

Wilson EO (1975) Sociobiology: The New Synthesis. Harvard Univ Press, Cambridge, MA [Chaps 1, 2, 14]

Wilson EO & Hölldobler B (2005) Eusociality: origin and consequences. Proc Natl Acad Sci USA 102: 13367–13371 [Chaps 12, 19]

Wilson K & Lessells CM (1994) Evolution of clutch size in insects. I. A review of static optimality models. J Evol Biol 7: 339–363 [Chap 11]

Wimberger PH & de Queiroz A (1996) Comparing behavioral and morphological characters as indicators of phylogeny. Pp 206–233 in Martins EP (ed) Phylogenies and the Comparative Method in Animal Behavior. Oxford Univ Press, New York [Chap 7]

Wingfield JC, Hegner RE, Dufty AM Jr & Ball GF (1990) The "challenge hypothesis": theoretical implications for patterns of testosterone secretion, mating systems, and breeding strategies. Am Nat 136: 829–846 [Chap 15]

Winston ML (1987) The Biology of the Honey Bee. Harvard Univ Press, Cambridge [Chap 28]

Wisenden BD (1999) Alloparental care in fishes. Rev Fish Biol Fish 9: 45–70 [Chap 26]

Witherington BE & Bjorndal KA (1991) Influences of artificial lighting on the seaward orientation of hatchling Loggerhead turtles *Caretta caretta*. Biol Conserv 55: 139–149 [Chap 29]

Witter MS & Cuthill IC (1993) The ecological costs of avian fat storage. Phil Trans R Soc Lond B 340: 73–92 [Chaps 12, 15]

Wolf AP (1995) Sexual attraction and childhood association: a Chinese brief for Edward Westermarck. Stanford Univ Press, Palo Alto, CA [Chap 31]

Wolf JB (2000) Indirect genetic effects and gene interactions. Pp 158–176 in Wolf JB, Brodie ED III & Wade MJ (eds) Epistasis and the Evolutionary Process. Oxford Univ Press, New York [Chap 14]

Wolf JB (2000) Gene interactions from maternal effects. Evolution 54: 1882–1898 [Chap 14]

Wolf JB (2001) Integrating biotechnology and the behavioral sciences. Trends Ecol Evol 16: 117–119 [Chap 14]

Wolf JB (2003) Genetic architecture and evolutionary constraint when the environment contains genes. Proc Natl Acad Sci USA 100: 4655–4660 [Chap 14]

Wolf JB, Brodie ED III, & Moore AJ (1999) Interacting phenotypes and the evolutionary process II. Selection resulting from social interactions. Am Nat 153: 254–266 [Chaps 14, 23]

Wolf JB, Brodie ED III, Cheverud JM, Moore AJ, & Wade MJ (1998) Evolutionary consequences of indirect genetic effects. Trends Ecol Evol 13: 64–69 [Chaps 14, 26]

Wolf JB, Moore AJ, & Brodie ED III (1997) The evolution of indicator traits for parental quality: the role of maternal and paternal effects. Am Nat 150: 639–649 [Chap 26]

Wolf JB & Wade MJ (2001) On the assignment of fitness to parents and offspring: whose fitness is it and when does it matter? J Evol Biol 14: 347–356 [Chap 4]

Wolf JB, Wade, MJ, & Brodie ED III (2004) The genotype-environment interaction and evolution when the environment contains genes. Pp 173–190 in DeWitt TJ & Scheiner SM (eds) Phenotypic Plasticity: Functional and Conceptual Approaches. Oxford Univ Press, New York [Chap 14]

Wolf M, van Doorn GS, & Weissing FJ (2008) Evolutionary emergence of responsive and unresponsive personalities. Proc Natl Acad Sci USA 105: 15825–15830 [Chap 30]

Wolf M, van Doorn GS, Leimar O, & Weissing FJ (2007) Life-history trade-offs favour the evolution of animal personalities. Nature 447: 581–584 [Chaps 12, 30]

Wolf MC & Moore PA (2002) Effects of the herbicide metolachlor on the perception of chemical stimuli by *Orconectes rusticus*. J N Am Benth Soc 21: 457–467 [Chap 9]

Wolff JO (1992) Parents suppress reproduction and stimulate dispersal in opposite-sex juvenile white-footed mice. Nature 359: 409–410 [Chap 31]

Wood SR, Sanderson KJ, & Evans CS (2000) Perception of terrestrial and aerial alarm calls by honeyeaters and falcons. Aust J Zool 48: 127–134 [Chap 13]

Woodley SK & Moore MC (1999) Ovarian hormones influence territorial aggression in free living female mountain spiny lizards. Horm Behav 35: 205–214 [Chap 15]

Woodward J & Goodstein D (1996) Conduct, misconduct and the structure of science. Am Sci 84: 479–490 [Chap 1]

Wooninck LM, Warner RR, & Fleischer RC (2000) Relative fitness components measured with competitive PCR. Mol Ecol 9: 1409–1414 [Chap 21]

Wootton JC, Feng X, Ferdig MT, Cooper RA, Mu J, Baruch DI, Magill AJ, & Su X-Z (2002) Genetic diversity and chloroquine selective sweeps in *Plasmodium falciparum*. Nature 418: 320–323 [Chap 25]

Wray GA (1995) Evolution of larvae and developmental mode. Pp 413–447 in McEdward LR (ed) Ecology of Marine Invertebrate Larvae. CRC Press, Boca Raton, FL [Chap 21]

Wright AA, Santiago HC, Sands SF, Kendrick DF, & Cook RG (1985) Memory processing of serial lists by pigeons, monkeys and people. Science 229: 287–289 [Chap 10]

Wynne-Edwards VC (1962) Animal Dispersion in Relation to Social Behaviour. Oliver & Boyd, Edinburgh [Chap 1]

Wyszecki G & Stiles WS (1982) Color Science: Concepts and Methods, Quantitative Data and Formulae. Wiley, New York [Chap 9]

Yabuta S (2008) Evolution of cross-contextual displays: the role of risk of inappropriate attacks on nonopponents, such as partners. Anim Behav 76: 865–870 [Chap 16]

Yachi S (1995) How can honest signaling evolve? The role of the handicap principle. Proc R Soc Lond B 262: 283–288 [Chap 16]

Yapici N, Kim YJ, Ribeiro C, & Dickson BJ (2008) A receptor that mediates the post-mating switch in *Drosophila* reproductive behaviour. Nature 451: 33–37 [Chap 5]

Yasui Y (1997) A "good sperm" model can explain the evolution of costly multiple mating by females. Am Nat 149: 573–584 [Chap 22]

Ydenberg RC (1998) Behavioral decisions about foraging and predator avoidance. Pp 343–378 in Dukas R (ed) Cognitive Ecology. Univ Chicago Press, Chicago [Chap 8]

Ydenberg RC & Dill LM (1986) The economics of fleeing from predators. Adv Study Behav 16: 229–249 [Chap 13]

Ydenberg RC, Brown JS, & Stephens DW (2007) Foraging: an overview. In Stephens DW, Brown JS & Ydenberg RC (eds) Foraging: Behavior and Ecology, chapter 1. Univ Chicago Press, Chicago [Chap 12]

Ydenberg RC, Butler RW, Lank DB, Smith BD, & Ireland J (2004) Western sandpipers have altered migration tactics as peregrine falcon populations have recovered. Proc R Soc Lond B 271: 1263–1269 [Chap 8]

Ydenberg RC, Stephens DW, & Brown J (2007) Foraging: an overview. Pp 1–28 in Stephens DW, Brown J & Ydenberg RC (eds) Foraging. Univ Chicago Press [Chap 8]

Yeh PJ & Price TD (2004). Adaptive phenotypic plasticity and the successful colonization of a novel environment. Amer Nat 164: 531–542 [Chap 6]

Yip EC, Powers KS, & Avilés L (2008) Cooperative capture of large prey solves scaling challenge faced by spider societies. Proc Natl Acad Sci USA 105: 11818–11822 [Chap 17]

Young AJ, Carlson AA, Monfort SL, Russell AF, Bennett NC, & Clutton-Brock TH (2006) Stress and the suppression of subordinate reproduction in cooperatively breeding meerkats. Proc Natl Acad Sci USA 103: 12005–12010 [Chaps 17, 18]

Young CM (1990) Larval ecology of marine invertebrates: a sesquicentennial history. Ophelia 32: 1–48 [Chap 21]

Young LJ & Hammock EA (2007) On switches and knobs, microsatellites and monogamy. Trends Genet 23: 209–212 [Chaps 5, 28]

Yund PO (2000) How severe is sperm limitation in natural populations of marine free-spawners? Trends Ecol Evol 15: 10–13 [Chap 21]

Yund PO & Meidel SK (2003) Sea urchin spawning in benthic boundary layers: are eggs fertilized before advecting away from females? Limnol Oceanog 48: 795–801 [Chap 21]

Zach R (1979) Shell dropping: decision making and optimal foraging in Northwestern crows. Behavior 68: 106–117 [Chap 1]

Zach R & Smith JNM (1981) Optimal foraging in wild birds? Pp 95–109 in Kamil AC & Sargent TD (eds) Foraging Behavior: Ecological, Ethological and Psychological Approaches. Garland STPM Press, New York [Chap 8]

Zahavi A (1975) Mate selection: a selection for a handicap. J Theor Biol 53: 205–214 [Chaps 4, 16, 23, 24]

Zaklan SD & Ydenberg RC (1997) The body size burial depth relationship in the infaunal clam *Mya arenaria*. J Exp Mar Biol Ecol 215: 1–17 [Chap 8]

Zigler KS, MA McCartney, DR Levitan, & HA Lessios (2005) Sea urchin bindin divergence predicts gamete compatibility. Evolution 59: 2399–2404 [Chap 21]

Zuk M, Rotenberry JT, & Tinghitella RM (2006) Silent night: adaptive disappearance of a sexual signal in a parasitized population of field crickets. Biol Lett 2: 521–524 [Chap 5]

Index

Boldface type indicates text box.

p. 146 Ecological rationality hypothesis

CPSIA information can be obtained at www.ICGtesting.com
Printed in the USA
BVOW031037120912

300133BV00004B/9/P

9 780195 331929